THE IMMUNE SYSTEM

SECOND EDITION

THE IMMUNE SYSTEM
Mental Health and Neurological Conditions

SECOND EDITION

HYMIE ANISMAN
Canada Research Chair in Behavioral Neuroscience, Department of Neuroscience, Carleton University, Ottawa, ON, Canada

SHAWN HAYLEY
Health Sciences Building, Carleton University, Ottawa, ON, Canada

ALEXANDER KUSNECOV
Behavioral and Systems Neuroscience Program, Department of Psychology, Rutgers University, Piscataway, NJ, United States

Academic Press is an imprint of Elsevier
125 London Wall, London EC2Y 5AS, United Kingdom
525 B Street, Suite 1650, San Diego, CA 92101, United States
50 Hampshire Street, 5th Floor, Cambridge, MA 02139, United States

Copyright © 2025 Elsevier Inc. All rights are reserved, including those for text and data mining, AI training, and similar technologies.

Publisher's note: Elsevier takes a neutral position with respect to territorial disputes or jurisdictional claims in its published content, including in maps and institutional affiliations.

For accessibility purposes, images in this book are accompanied by alt text descriptions provided by Elsevier.

No part of this publication may be reproduced or transmitted in any form or by any means, electronic or mechanical, including photocopying, recording, or any information storage and retrieval system, without permission in writing from the publisher. Details on how to seek permission, further information about the Publisher's permissions policies and our arrangements with organizations such as the Copyright Clearance Center and the Copyright Licensing Agency, can be found at our website: www.elsevier.com/permissions.

This book and the individual contributions contained in it are protected under copyright by the Publisher (other than as may be noted herein).

Notices
Knowledge and best practice in this field are constantly changing. As new research and experience broaden our understanding, changes in research methods, professional practices, or medical treatment may become necessary.

Practitioners and researchers must always rely on their own experience and knowledge in evaluating and using any information, methods, compounds, or experiments described herein. In using such information or methods they should be mindful of their own safety and the safety of others, including parties for whom they have a professional responsibility.

To the fullest extent of the law, neither the Publisher nor the authors, contributors, or editors, assume any liability for any injury and/or damage to persons or property as a matter of products liability, negligence or otherwise, or from any use or operation of any methods, products, instructions, or ideas contained in the material herein.

ISBN 978-0-443-23565-8

For information on all Academic Press publications
visit our website at https://www.elsevier.com/books-and-journals

Publisher: Stacy Masucci
Acquisitions Editor: Joslyn Chaiprasert-Paguio
Editorial Project Manager: Billie Jean Fernandez
Production Project Manager: Gomathi Sugumar
Cover Designer: Vicky Pearson Esser

Typeset by STRAIVE, India

Contents

Preface ... vii

Acknowledgments ... ix

1. Multiple Pathways Linked to Mental Health and Illness

A General Perspective of Mental Illness ... 2
Multisystem Coordination ... 3
Genetic Contributions to Illness ... 5
Biological Processes ... 10
Neuroimmune Links ... 15
Microbiota and the Enteric Nervous System ... 16
Personalized Treatment Strategies (Precision Medicine) ... 19
Concluding Comments ... 24

2. The Immune System: An Overview

Introducing Immunity ... 28
The Immune System Consists of White Blood Cells: Leukocytes ... 29
Types of Immunity ... 31
What Actually Induces Immunity? ... 32
The Cells and Organs of the Immune System ... 32
The Lymphoid System ... 34
The Acquired Immune Response ... 36
The Immune Response to Antigen Progresses Through a Distinct Set of Phases ... 38
The Antibody Molecule: An Immunoglobulin ... 43
T Lymphocytes ... 45
Cytokines ... 50
Tolerance and Autoimmune Disease ... 54
Concluding Comments ... 55

3. Bacteria, Viruses, and the Microbiome

Bacterial Challenges ... 57
Viral Illnesses ... 61
Bacterial and Viral Challenges Affect Hormonal and CNS Processes ... 66
Concluding Comments ... 79

4. Pathogenic Factors Related to Mental and Neurological Disorders

Pathogens and CNS Functioning ... 82
Viral Invasion of the CNS ... 85

Lessons From COVID-19 ... 89
Long COVID ... 96
Concluding Comments ... 98

5. Lifestyle Factors Affecting Biological Processes and Health

Risky Behaviors and Illness ... 99
Dietary Factors as Determinants of Health ... 101
Exercise ... 117
Sleep ... 122
Climate Change Affects Physical and Mental Health ... 127
Concluding Comments ... 129

6. Stressor Processes and Effects on Neurobiological Functioning

Vulnerability Versus Resilience ... 131
Attributes of the Stressor ... 132
Stressor Appraisals and Reappraisals ... 132
Neurotransmitter Alterations Provoked by Stressors ... 141
Hormonal Changes Stemming From Stressor Experiences ... 149
Impact of Prenatal and Early Postnatal Events ... 153
Stress and Energy Balances ... 159
Stress and Immune Alterations ... 160
Microbiota, Inflammatory Responses, and Stressors ... 161
Concluding Comments ... 162

7. Stress and Immunity

Implications of Stressors for Immune-Related Diseases ... 163
Dynamic Interplay Between the CNS and Immune System ... 165
Stressor Influences on Immunity: Animal Studies ... 166
Immunological Consequences of Stressor Exposure in Humans ... 179
Stress and Cytokine Production ... 185
Autoimmune Disorders ... 191
Concluding Comments ... 195

8. Prenatal and Early Postnatal Influences on Health

Early Development ... 197
Prenatal Challenges ... 198

vi — Contents

Impact of Adverse Early Life Events — 207
Immunity and Microbial Factors — 215
Epigenetics and Intergenerational Actions — 220
Concluding Comments — 226

9. Depressive Disorders

Defining Depression — 229
Theoretical Perspectives Concerning Depressive Illnesses — 231
Neurochemical Perspectives of Depressive Disorders — 233
Gamma-Aminobutyric Acid and Glutamate Processes in Depressive Disorders — 239
Neurotrophins — 245
Inflammatory Processes and Depressive Disorders — 249
Microbiome-Immune Interactions — 263
Concluding Comments — 267

10. Anxiety Disorders

Neurobiological Factors and Treatment of Anxiety Disorders — 269
Anxiety Disorders — 275
Inflammatory Factors in Anxiety and Anxiety Disorders — 280
Immune and Inflammatory Processes in Specific Anxiety Disorders — 283
Linking Microbiota to the Promotion of Anxiety — 286
Conclusion — 290

11. Posttraumatic Stress Disorder

A Perspective on Trauma-Related Disorders — 293
Acute Stress Disorder — 294
Posttraumatic Stress Disorder — 294
PTSD and Structural Brain Alteration — 295
Theoretical Perspectives on PTSD — 298
Biochemical Correlates of PTSD — 301
Inflammatory Processes Associated With PTSD — 308
Epigenetic Contributions — 314
Treating PTSD — 317
Comorbid Illnesses Linked to PTSD — 321
Conclusion — 322

12. Pain Processes

Pain as the Ubiquitous Malady — 323
Psychological Impact of Chronic Pain — 324
Neurophysiological and Psychological Processes That Accompany Pain Perception — 325
Next Steps in Pain Management Methods: Advances and Caveats — 337
Concluding Comments — 344

13. Autism

Features of Autism Spectrum Disorder — 345
What is Autism: Today! — 346
Brain Development and Autism: Behavioral Deficits Linked to Specific Circuits — 347
Etiology of Autism: The Immune Hypothesis — 349
Concluding Comments — 364

14. Schizophrenia

Overview of Schizophrenia — 365
Defining Schizophrenia — 366
The Pathophysiology of Schizophrenia — 367
The Immune System and the Two-Hit Model of Schizophrenia — 372
Perinatal Infection and Schizophrenia — 379
Autoimmune Disease and Schizophrenia — 383
Microglial Activity and Schizophrenia — 386
Microbiota and Schizophrenia — 387
Concluding Comments — 390

15. Inflammatory Roads to Parkinson's Disease

General Introduction to Parkinson's Disease — 393
Lifestyle and Aging — 398
Experimental Animal Models of Parkinson's — 401
Environmental Stresses and Parkinson's Disease — 403
Genetic Vulnerability — 412
Future Immunomodulatory Treatments and Parkinson's Disease — 414
Concluding Comments — 418

16. A Neuroinflammatory View of Alzheimer's Disease

Alzheimer's Disease Background — 419
Lifestyle Factors and AD — 423
Genetics of AD — 426
Environmental Factors and Alzheimer's Disease — 430
Inflammatory Mechanisms of Alzheimer's Disease — 433
Immunomodulatory Treatments for AD — 436
Concluding Comments — 441

17. Illness Comorbidities in Relation to Inflammatory Processes

Something About Comorbidity — 443
Inflammatory Factors in Comorbid Illnesses — 444
Psychiatric Disorders and Cardiovascular Illnesses and Diabetes — 451
Stroke — 461
Concluding Comments — 467

References — **471**
Index — **579**

Preface

It is difficult to dance concurrently at two weddings. And yet, this is what we tried to do in this book. Broadly, it was our intention to provide an overview of the relationship between inflammatory immune processes and several neuropsychiatric and neurological disorders. This entails introducing behavioral scientists and clinicians to fundamental aspects of immune functioning and how these processes are linked to hormonal, neurotransmitter, neurotrophic, and microbial factors. Simultaneously, we aimed to introduce basic neuroscientists to the chief characteristics of diverse mental illnesses and some of the factors that moderate these conditions. In so doing, we explored the processes by which the brain and immune system, separately and conjointly, contribute to various psychiatric and neurological states.

If it seems that we strayed or overlooked some things, it was more by design and less due to ignorance—although the breadth and depth of the field impose its own form of humility at what there is to know and learn. In fact, our goal was to provide sufficient clarification to what can be quite confusing for the newcomer. We omitted coverage of numerous physical illnesses that likely have their roots in inflammatory processes and which account for why physical, mental, and neurological disorders are so frequently comorbid with one another. In calibrating our focus, we were squarely concerned with knowing what we can say about the impact of immune and inflammatory processes on brain and behavioral functions, as well as the destabilization that might result from this impact. We hope we were successful.

Several basic chapters kick things off, dealing with diverse biological processes, including neurons and glial cells, hormones, neurotransmitters, and neurotrophins, as well as immune functioning, immune-related diseases, microbiota, and the impact of viral and bacterial infections. This is followed by a description of the ways by which various lifestyle factors (eating processes, exercise, and sleep), stressor experiences, and prenatal and early postnatal experiences come to influence (and are influenced by) neurobiological systems that have been implicated in mental disorders. Having described these processes, we then relate them to the most frequent mental illnesses, including depressive illnesses, anxiety disorders, PTSD, autism spectrum disorders, schizophrenia, chronic pain conditions, as well as neurodegenerative disorders, primarily Parkinson's and Alzheimer's disease. Throughout, we offer views concerning how immunity and brain processes are interlinked, with an eye toward the identification of biomarkers relevant to the development of illness and predicting treatment responses, as well as the potential for novel treatment targets. Many of these illness conditions are comorbid with one another and with many physical conditions. Thus, we discussed the links between mental illness and several comorbid illnesses, primarily heart disease, stroke, and diabetes, although, throughout the book, many other comorbid illnesses are covered.

It was not our intent to cover any single topic in exceptional depth or provide an exhaustive set of references (although there are many). There are many books that are dedicated for this purpose, but none that provide an overview of how the immune system has fully integrated itself into psychopathology. At present, there is certainly a need for an integrated approach to this complex issue, and the present effort represents a stab at an intriguing topic. Most certainly, we may not have done justice to all fields equally, so we encourage more informed readers to forgive us our trespasses and invite the curious to enjoy, delight, and marvel at how far we have come since the days when evil spirits were released with the tap of a blunt chisel and faith in a world well-imagined—much like the one we describe in the following pages.

Hymie Anisman, Shawn Hayley, and
Alexander Kusnecov
May, 2024

Acknowledgments

There are many people we ought to thank for their help in many facets of this volume. In fact, there are too many to mention them all without exceeding our page limits. However, Hymie would be remiss not to thank Kim Matheson, Alfonso Abizaid, Robyn McQuaid, Amy Bombay, Zul Merali, and Steve Ferguson for their many thoughtful suggestions in relation to precision medicine approaches, the impact of challenges in vulnerable populations, and the complex molecular biological underpinning of disease. During my 50 years at Carleton University, many graduate and undergraduate students, too many to list here, have made the research fun. I am indebted to every one of them. Early in my career, I was also very fortunate to have mentors, specifically Jane Stewart, Norm Braveman, and Doug Wahlsten, who served as the best role models. Finally, Simon, Rebecca, Max, and Maida were immensely supportive. Shep and Oefa were also helpful to "grumpy" in several ways (I think "grumpy" is a variant of "grandpa," but I might be wrong).

Shawn thanks all the wonderful graduate students that actually conducted the research that underpins many of the concepts talked about within this book, particularly the chapters on neurodegeneration. Most notably, Emily Mangano (now Rocha), Ashley McFee, Faranak Vahid-Ansari, Darcy Litteljohn, Zach Dwyer, Chris Rudyk, and Kyle Farmer have all helped fuel the many neuroimmune studies conducted over the years that have helped keep me excited about research. My colleagues have also been tremendously supportive of my work, with a special shout out to David Park and Steve Ferguson, who serve as models of research excellence and are also wonderful friends. Of course, a hearty thank you also goes out to Olivia and Louise for providing a shining light to come home to each day!

Alex would like to thank his friends and colleagues—especially the late Professor George Wagner, who always found time to listen and offer sage advice. He liked the first edition of this book and would surely have approved of the second. I am also especially grateful to the many students who provided important feedback on the first edition of this book, which was offered in my graduate Psychoneuroimmunology course at Rutgers. Their input confirmed that the topics we cover resonate with up-and-coming biobehavioral scientists in all areas of the behavioral sciences, from clinical and social to the cognitive and brain sciences. Our intellectual lives would collapse without the inspiration and encouragement from our present and former graduate students. In particular, I want to thank Dr. Marialaina Nissenbaum, who weathered the COVID pandemic and plowed on in spite of the various restrictions, rules, and regulations that plagued labs across the globe. If not for her strength and dedication, who knows where things would have ended up? And finally, as always, I want to thank my muse and Delphic Oracle, Wendy, and all the "little" Kusnecov's—Saskia, Aidan, and Simone—all soldiering on and glowing brightly.

We are indebted to April Farr, our first editor at Elsevier, who got us into this, and Nikki Levy and Timothy Bennett, who continued in this capacity. This second edition was put together in collaboration with our latest editors, Joslyn Chaiprasert-Paguio and Billie Jean Fernandez. Their patience and helpfulness throughout the various phases of this project are very much appreciated. We are also indebted to Gomathi Sugumar who coordinated the production of all sections of this edition.

Some of our own research, mentioned in various chapters (sometimes, too often), was made possible by the generosity of the Canadian Institutes of Health Research, the Natural Sciences and Engineering Research Council, the Canada Research Chairs Program, the National Institutes of Mental Health, National Institutes of Drug Abuse, and the Dana Foundation. We can only hope that their generosity will be still greater over the coming year to support the dedicated young researchers who continue to search for ways to diminish and prevent illnesses, thereby enhancing the lives of very many.

CHAPTER

1

Multiple Pathways Linked to Mental Health and Illness

The broad impact of mental illness

Despite the best efforts of scientists and clinicians, mental illnesses are often resistant to pharmacological and behavioral/cognitive treatments. Moreover, even when a positive response is obtained, residual features of the illness may persist, and illness recurrence is notably high. Making matters worse, mental illnesses are frequently comorbid with other psychological problems, as well as numerous physical disorders. In some instances, the comorbid conditions may be reciprocally perpetuating, whereas in other cases, they might be promoted by common elements. These include growth factors and inflammatory immune processes that have been implicated in various mental illnesses and neurological disorders, as well as heart disease, diabetes, stroke, cancer, and an array of immune-related conditions. Treatments for many physical illnesses have been advancing, even if a bit slowly, and the potential for still better outcomes has been bolstered by the recognition that a precision (personalized) medicine approach may be instrumental in determining which patients will be most responsive to particular treatments. This perspective opened the door to precision public health that could potentially predict the most efficacious interventions to benefit whole communities or populations. With new perspectives, new approaches, and new technologies, the modes of treatment and their effectiveness are becoming progressively better.

Yet, there is another side to mental illnesses that needs to be addressed. Despite many treatment advances and the understanding that psychiatric disturbances are medical conditions and not personal failings, mental illnesses have considerable stigma attached to them. This is even more remarkable given that mental illness affects more than 20% of people at some time in their lives, which means that

an awful lot of families are affected by these illnesses. It is hard enough for individuals to deal with their mental illness, but when this is compounded by various forms of stigma, as well as insufficient or ineffective social support, maintaining a positive quality of life becomes much more difficult (Serchuk et al., 2021). If nothing else, diminishing stigmatization of individuals with mental health problems will increase the likelihood that those at risk will seek treatment early.

Aside from public stigma, it is not unusual for self-stigma to occur, frequently accompanied by shame and humiliation, social devaluation, internalization of a negative self-concept, a need to maintain secrecy (Corrigan and Rao, 2012), and a decline in help-seeking. If this were not sufficiently diminishing, individuals with mental health issues frequently must deal with "structural stigma" reflected by inordinately extended waits for treatment as well as biases when they do seek help. Diminished care for other physical illnesses is not uncommon, being misattributed to the manifestation of mental illness. Is it any wonder that the life span of patients with mental illness is reduced?

Over the past decade, there has been a decline in overt stigma, possibly reflecting political correctness, but the problem is still extensive. Attempting to "educate" people about mental illness has not been especially effective in eliminating stigma. Even health professionals are not immune from maintaining stigmatizing attitudes concerning mental illness, and they are also likely to self-stigmatize when they experience symptoms of a mental health problem. It is not unusual for health professionals to have serious concerns about exposing themselves to the judgment of their peers, and they may be reluctant to seek treatment (Brower, 2021). Given that health professionals are affected in this manner, it is hardly surprising that negative attitudes occur in so many others.

The Immune System
https://doi.org/10.1016/B978-0-443-23565-8.00014-4

Copyright © 2025 Elsevier Inc. All rights are reserved, including those for text and data mining, AI training, and similar technologies.

A General Perspective of Mental Illness

Considerable efforts have been devoted to the identification of specific brain regions and neuronal processes that are responsible for the provocation or inhibition of specific behavioral phenotypes. There is merit to this approach, as critical neuronal pathways and brain regions have been identified that contribute to cognitive and emotional functioning, as well as to mental and neurological illnesses. At the same time, complex behaviors involve complicated neural circuits and temporal variations linked to emotional changes that would not readily be discerned by simply assessing single brain regions (Northoff et al., 2020). Accordingly, a systems approach is frequently adopted that involves analyses of the functional connectivity among neural circuits that subserve normal behaviors, which is evident in association with psychological illnesses. It has become increasingly apparent that the activity and functioning of brain neuronal circuits are influenced by peripheral and brain hormones, sympathetic nervous system activity, circulating immune factors, and those that act within the brain (e.g., released by microglia) as well as processes associated with gut bacteria (e.g., Cryan et al., 2019). In essence, there has been a push toward analysis of mental illnesses within the context of a broader systems-based approach.

This opening chapter introduces multiple systems that may contribute to the evolution of physical, neurological, and psychological disorders, and why some individuals may be more vulnerable (or resilient) to such illnesses. As a particular illness may come about owing to several factors, it is understandable that a treatment that is effective for one individual may not be equally effective for a second. The use of personalized (precision medicine) treatment strategies is introduced as these have been used for several physical illnesses and are being incorporated into the treatment of mental health disturbances.

In considering mental illnesses, it is essential to assess not only the diverse processes and mechanisms that are operative within the central nervous system (CNS) itself but also those apparent across several other systems. The focus in this book is on brain-immune system interactions as they pertain to health, which is consistent with the literature of the past decade or so, wherein inflammatory factors have moved from being a marginal consideration to one that is a primary player in a range of brain conditions. Indeed, it is now evident that certain immune cells, and the inflammatory messengers they produce (cytokines), can influence neurotransmission, hormonal release, and even neuronal survival. The involvement of these systems (and their interactions) will be considered in relation to mental health processes as well as comorbid physical illnesses. Finally, genetic, epigenetic, prenatal, and early-life experiences, environmental challenges, and lifestyle factors all feed into these illness-related biological processes and thus need to be considered within any systems-based approach.

It is generally accepted that early negative life experiences (e.g., parental neglect, abuse, and living in poverty), or other types of psychologically toxic environments, together with a failure to cope effectively, would be at or near the top of the list of damaging experiences that can lead to mental illness (Hughes et al., 2017). In concert with these psychological (psychogenic) stressors are the many other "physical" (neurogenic), systemic (biological perturbations), and environmental stressors that affect well-being, including chemical toxicants, industrial contaminants, food additives, and bacterial or viral agents. Given this spectrum of differing challenges, it is easy to get lost in all the specifics, which we wanted to avoid. Thus, we focus on some of the mechanistic commonalities of these varied challenges insofar as they pertain to mental and neurological illness.

Inherent in a systems approach is that in addition to pharmacologically based strategies, psychosocial influences ought to be considered in dealing with psychiatric illnesses (Haslam et al., 2019). The development and recovery from mental illness, "systems" not only encompass intrinsic biological processes but also those related to the individual in the context of their social and physical environment, their experiences, and the influence of their ethnic (including cultural) background (Matheson et al., 2018b).

In evaluating the multiple factors that are involved in mental illnesses, it is also important to adopt a constellation of "omics" that might be used to decipher how diverse systems come together to produce different phenotypes. Unlike the research conducted a little more than two decades earlier, data sets are now immense and becoming progressively larger. Aside from the many genes and gene combinations that can affect health, an enormous number of epigenetic changes and gene mutations appear on genes, any of which have the potential to influence health risks. As well, the presence of inflammatory factors and microbiota (bacteria in and on our bodies) contribute to illnesses, including psychiatric disturbances. All these elements only begin to describe the complex interplay between the multiple processes that determine our phenotype. At every level of analysis, including prenatal and perinatal experiences, psychosocial, experiential, and environmental factors have a say concerning the expression of specific outcomes. Piecing together the interactions of multiple biological networks, at different levels of analysis, can provide a fuller picture of what is going on in the body and the brain (see Table 1.1).

Analyses of the processes leading to or underlying particular illnesses necessarily imply that these can be

TABLE 1.1 Omics

Ome	Composition	Function
Genome	The range of messenger RNA molecules expressed	Genes involved in building proteins and other molecules
Transcriptome	Range of messenger RNA molecules expressed that comprise protein building instructions	Determines which genes are turned on and off
Proteome	Complement of proteins expressed by a cell, tissue, or organism	Application of the genetic instruction manual
Metabolome	Small metabolites in a cell, tissue, or organism, involved in metabolic reactions required for normal growth, maintenance, and function	Chemical reactions that occur within an organism
Epigenome	Chemical changes that influence DNA expression so that genes are turned off or on	Environmental and experiential factors that influence gene activity
Phenome	Characteristics reflecting the actions of the whole genome, proteome, and metabolome	Expression of traits and diseases

Adapted from Hamers, L., 2016. Big biological datasets map life's networks: multi-omics offers a new way of doing biology. Retrieved from https://www.sciencenews.org/article/big-biological-datasets-map-lifes-networks.

accurately identified. Concluding that certain symptoms denote the presence of a particular mental illness is complicated by the fact that many different illnesses share common features. Conversely, a given illness assessed across a population may comprise diverse symptoms to the extent that individuals might present with very few common features. Moreover, an individual's symptom profile may vary over time and may even morph from one condition to another, as in the case of unipolar depression transforming into bipolar illness. Moreover, a given illness may be comorbid with other conditions that can confuse the individual profiles expressed. Aside from the difficulty of differentiating between the symptoms of various illnesses, deciding how illnesses should be treated involves yet another level of complexity. This is attested to by the debate concerning the usefulness of the Diagnostic and Statistical Manual of Mental Disorders (5th ed.; American Psychiatric Association, 2013) and the approach promoted by the National Institutes of Mental Health (NIMH) concerning the use of the Research Domain Criteria (RDoC) in relation to mental illness (Insel, 2014). These issues have considerable bearing on the well-being of affected individuals, but as we will see, many factors contribute to the difficulties associated with diagnosis and treatment. Although we have considerable knowledge concerning the workings of the brain, as well as how and why brain functioning is sometimes disrupted, hence leading to mental illness and cognitive deterioration, this has not been sufficient to consistently cure a malfunctioning brain.

Multisystem Coordination

Although each of our organs has unique functions, they must communicate with one another and with the brain. This is not only essential so that the left leg does not trip over the right but also because the brain may respond to, and conversely, influence the functioning of various hormonal systems, immune activity, and gut-related processes that can affect other organs, such as the liver or kidney and the brain itself.

Just as maintaining general well-being is dependent upon the smooth operations of multiple intersecting systems, the development and emergence of various pathological conditions may reflect a disturbance in any of several nodes within and between these systems. What this implies is that in adopting prevention or treatment strategies for illnesses, a systems approach might be ideal. As we will see in several ensuing chapters, this includes attention to psychosocial processes in the emergence and treatment of mental, neurological, and physical illnesses. This by no means implies that a serious or chronic illness (e.g., heart disease) can be cured by altering the functioning of other systems or focusing solely on brain changes. However, it does suggest that amelioration of some illnesses or symptoms can be facilitated by adjunctive (auxiliary) treatments that deal with systems that affect psychological functioning. In later chapters, we will be dealing with specific illnesses, including those comorbid with psychological disturbances, and this perspective will be considered in greater depth. For the moment, suffice it that a sizable portion of the population experiences multiple illnesses that may or may not be related to a primary mental disturbance, making it that much more difficult to treat disease conditions. Without a diagnosis or target for treatment, effective ways of dealing with the illness, or even masking symptoms adequately, may not be attained.

Sequential or Concurrent Influences on Mental Illnesses

The development of illnesses often reflects an insidious process that progresses over months or years, even though individuals mistakenly come to believe that it occurred suddenly ("One day I was fine, and then bam, I wasn't"). This may not occur in the case of viral illnesses, but the development of most chronic illnesses occurs in progressive phases. For instance, type 2 diabetes may develop owing to a combination of eating the wrong foods, not exercising, being overweight, poor sleep quality, and genetic factors, eventually culminating in diabetes. Some of these very same factors contribute to the buildup of plaque that produces heart disease, and diabetes itself is a good predictor of later heart disease. In some instances, the presence of specific factors may be especially pertinent among individuals who are "at risk" owing to genetic influences, certain experiences, or the presence of other illnesses.

Researchers in several fields have adopted the view that the development of some illnesses involves "two-hits" or "multiple hits." For instance, genetic constituency may be a first hit that placed people *at risk*, but actual illness only occurred with a second hit. This may comprise the actions of a second gene, or it may stem from unhealthy behaviors. The second hit can also include the actions of *toxicants* (foreign materials from man-made sources, such as industrial chemicals released into the environment, or herbicides, insecticides, fungicides, rodenticides, and even food additives) or *toxins* (substances that come from a biological source, such as living cells or organisms that could result in illnesses).

First, second, and multiple hits may comprise current stressful events, repeated exposure to negative stimuli, or may reflect events that occurred prenatally or during early life. Such experiences can profoundly influence the developmental trajectory related to neurobiological, behavioral, and emotional development so that illnesses can materialize many years later. As we will see, negative experiences can promote the suppression of particular gene actions (i.e., through epigenetic changes) and hence affect the phenotypes that would otherwise be apparent. As it happens, some of these epigenetic changes can be passed from one generation to the next so that individuals can be affected by their parent's experiences, and they, in turn, can pass these features on to their children (the sins of the father).

Winning the wrong lottery

Some illnesses occur because of specific genes or gene mutations, or the presence of certain genes coupled with a second hit in the form of an environmental trigger or a stressor. Other illnesses simply occur owing to the accumulation of mutations that appear with aging. There is little question that some forms of cancer have a high heritability rate, and thus, the development of cancer may be linked to inherited genes. Being BRCA1 or BRCA2 positive are well-known genetic risk factor in relation to breast cancer, but many other genes, amounting to several hundred, also contribute in this regard.

Mutations may occur on a random basis, although numerous agents (cigarettes and sun rays) can encourage the development of mutations (mutagenesis). Cancers can certainly develop owing to specific lifestyles or environmental factors, but not everyone who smokes develops lung (or other) cancers, and not every sun-worshipper develops a malignant melanoma. With each cell division, the chance of a mutation occurring increases, and thus, cancerous cells should be more frequent in rapidly reproducing cells than in those that are slower. An analysis of 31 tumor types, in fact, revealed that the lifetime risk for cancers was linked to the rate of total stem cell division (Tomasetti and Vogelstein, 2015). The more rapidly reproducing cells might also be more likely to be affected by environmental triggers, and thus more readily turned into cancer cells. To a significant extent, individuals who develop cancer are the unfortunate winners of the wrong lottery.

Various gene mutations have been identified that predict the subsequent occurrence of some forms of cancer, irrespective of family history of this illness (Natrajan et al., 2016), possibly suggesting that the mutations occurred on a random basis or were engendered by environmental factors. However, if some cancer types occur owing to random mutations, then why is it that many patients may end up being affected by more than a single type of cancer? What are the odds of winning a lottery twice? As it happens, individuals who are affected by several types of cancer are more likely to carry a particular genetic marker, one of the KRAS-variants. Indeed, about 25% of those with cancer carry this mutation, and more than 50% of those carrying this marker develop more than one cancer, which is unfortunately difficult to treat (Moore et al., 2020).

A common approach to determine processes that are linked to illnesses comprises epidemiological analyses that ask what lifestyle, experiential, and genetic factors predict disease occurrence. When systematic differences are detected, they sometimes provide fundamental clues regarding the processes leading to illness and point to interventions to preclude the development of illness or means by which to treat an ongoing disease. In other instances, the question can be turned on its head. Why do some people who ought to be at high risk seem not to develop illnesses? What are the implications of the risk for Parkinson's disease being reduced in association with smoking, and why is the occurrence of cancer among people with schizophrenia or dementia less than that expected based on simple probabilities? Are these inverse relations due to the presence of certain genes or epigenetic effects, or is it a matter that the focus on one illness resulted in symptoms of the other condition being undetected? For instance, those with dementia receive fewer cancer screening tests, making it less likely that both illnesses will be detected. Understanding the psychosocial dynamics associated with diverse illnesses may permit a greater appreciation of the processes that govern their *apparent* occurrence and ways to treat them.

Methods of Evaluating Biological Substrates of Mental and Neurological Illnesses

Potential links between specific neurochemical processes and illnesses have been addressed using several approaches. Each has considerable merit, and the triangulation of methods has been especially useful. These studies have pointed to the complexity of trying to identify the mechanisms responsible for depression, especially as depressive illnesses are biochemically and behaviorally heterogeneous. The data have also pointed to the contribution of several organismic variables (genetic, age, and sex) and experiential factors (ongoing stressors, early-life trauma or neglect, and previous stressful experiences) in the provocation and maintenance of depression, as well as its recurrence following successful treatment. The research approaches adopted to determine the processes associated with mental illnesses have included studies comprising the following:

1. Genetic analyses, including whole genome analyses, assessment of polymorphisms, and epigenetic changes concerning the appearance of illness;
2. Evaluation of hormone and neurochemical factors in blood and cerebrospinal fluid of patients before and after therapy;
3. Imaging studies (e.g., PET and fMRI) that assessed functional changes in specific brain regions or networks in patients versus healthy controls (either by assessing blood flow indicative of cellular activity in certain brain regions or assessing activity or density of particular neurotransmitter receptors), as well as studies that imaged the size of particular brain sites;
4. Determination of brain chemicals and receptors in postmortem tissues of individuals who had died of suicide in comparison to those that had not been depressed and died through causes other than suicide (e.g., sudden cardiac arrest and accidents);
5. Studies in humans that assessed the effectiveness of drug treatments, as well as studies that compared the relative efficacy of several treatments (alone or in conjunction with other agents);
6. Analyses involving animal models in which biochemical effects of treatments were assessed and analyses performed regarding the effectiveness of pharmacological treatments in attenuating behaviors reminiscent of mental illness.

Genetic Contributions to Illness

In a diploid organism, such as humans, the contributions of dominant alleles largely determine the phenotype. However, the *expressivity* (the extent to which a gene influences the phenotype) varies appreciably between individuals even though they might carry the same genotype. A related concept, *penetrance*, refers to the proportion of individuals carrying a particular gene who show the expected phenotype. In essence, carrying a particular gene does not guarantee a particular phenotype, but instead is influenced by interactions with other factors.

Several illness conditions can be determined by single genes, but most complex illnesses involve polygenic actions operating additively or interactively in relation to phenotypic outcomes. Environmental factors, experiences, and the presence of other genes can moderate the phenotypic actions of a particular gene (epistatic interaction). It also seems that a particular gene or set of genes can have more than a single phenotypic outcome (pleiotropy), thereby contributing to several illness comorbidities. Increased gene diversity enhances the ability of organisms to deal with many environmental influences that can have negative consequences. However, the price for this is that gene variants (mutations) also appear that favor the development of disease states. One would think that with so many generations of selection for ideal traits, many of the genes that favor disease would have been selected against, and hence, genetically related illnesses would be far less frequent than they are. No doubt, there has been some selection of this nature, but at the same time, over the course of evolution in which genes for desirable traits were selected, some of these genes may have had downstream consequences that were maladaptive.[1] The classic example of this is that being heterozygous for a particular gene may prevent malaria, but if individuals inherited a copy of the sickle cell gene from both parents they will develop sickle cell anemia. In effect, the sickle cell gene may be evolutionarily conserved because of the advantages it offers, but this also means that some individuals will suffer the consequences of being homozygous for this trait. Likewise, selection for a particular phenotype may have beneficial functions at one time in an individual's life, only to produce negative consequences later. As well, certain phenotypes (e.g., particular biological changes) may be advantageous so long as these occur within a prescribed range. When, for whatever, these features become too high or too low, adverse effects may occur (hormesis).

---

[1] It is intriguing that what make us human and differentiates us from other mammals might not simply reflect the evolution of new segments of DNA but also specific components of the genetic code being deleted, perhaps providing benefits by eliminating elements that would otherwise turn genes off (Xue et al., 2023).

The common view for many decades had been that, for better or worse, the genes carried by an individual were immutable, and thus, what they were born with was what they were stuck with. Indeed, some of the debates in the 1960s and 1970s seemed to be about how much of the phenotypic variance (P_v) stemmed from our genes (G_v) and that produced by the environment (E_v) and the variance attributable to the gene × environment interaction (the latter was often ignored or considered to be negligible). Efforts directed at uncovering the contribution of these elements to a phenotype were not conducted merely to obtain an understanding concerning how these processes came together, but in some cases had serious implications for government policy.

Twin Research

A common method to evaluate the extent to which a trait is genetically inherited (heritability) is through twin studies. Given that *monozygotic* (identical) twins largely have identical genotypes, whereas dizygotic (nonidentical; fraternal) twins share only half their genes, it would be expected that for phenotypes primarily determined by genes, the correlation between monozygotic twin pairs would be greater than that evident among dizygotic twins. In theory, that is correct, but random mutations can affect gene expression and thus reduce the monozygotic similarities. As well, relative to monozygotic twins, dizygotic twins could have more and greater differences in their experiences. Indeed, the prenatal (intrauterine) environment of identical twins may be more similar than that of fraternal twins. Furthermore, monozygotic twins are sometimes treated more alike than dizygotic twins, in a sense being interchangeable, perforce creating greater similarities between monozygotic pairs. The solution to dissociating environmental from genetic contributions was to assess the phenotypes of identical twins who had either been raised together versus those twins that had been raised apart, thereby limiting environmental similarities. This said, the environment of twins reared apart was not entirely different, as most adopted twins ended in relatively good middle- or upper-class environments. It has been argued that many such studies were contaminated (confounded) as twins were not separated at birth and may have lived together for several years, and even when separated, they often communicated with one another in later years (Joseph, 2015).

Twin studies have been used to assess heritability associated with many phenotypes, including physical pathologies, such as heart disease, various types of cancer, diabetes, substance use disorders, and many other physical and psychological disorders. Many studies also explored the heritability of various social attitudes (e.g., socialism, abortion, gay rights, and racial segregation) and dispositional (personality) features (e.g., neuroticism and agreeableness). For the most part, there had not been appreciable agreement concerning the contribution of genetic factors regarding these personality and ideological features. However, with the advent of genome-wide association studies (GWAS), in which the genome was used to identify genetic variations associated with specific phenotypes, multiple genes associated with some conditions could be identified. Combined with sophisticated statistical methods, subtle genetic features have been associated with distinct features of human behavior and characteristics related to personality and the lifestyles endorsed. Personality factors, such as introversion, were associated with genetic influences, as were impulsivity and creativity, and even food preferences (veggies versus high-calorie, fatty foods) seemed to be linked to genetic influences. Likewise, the type of exercise people prefer may involve a genetic component that either additively or interactively operates with environmental influences, and genetic factors are related to whether individuals prefer exercising alone or with others. This is not altogether surprising, considering that genetic factors influencing hormone functioning are tied to social bonding (e.g., oxytocin). As much as the use of GWAS has been useful in linking specific genes or multiple (polygenic) genes to behavioral and pathological conditions, this approach still has limitations. Among other things, this method focuses on the additive effects of multiple genes in relation to specific phenotypes, even though pathologies may stem from the interactive effects of genes, as well as the interactions between genes and environmental contributions.

Molecular Genetic Approaches

Humans have 23 pairs of chromosomes, each comprising about 6 m of DNA coiled into a small packet that can fit into the cell's nucleus. The DNA is made up of several million nucleotides that come in four flavors: guanine (G), adenine (A), cytosine (C), and thymine (T), which in sets of three form amino acids. The double-stranded DNA, and the many genes present, are the template for "transcription" or RNA formation, which in most respects is identical to the DNA (thymine is replaced by uracil in RNA). The strand of RNA transcribed from DNA, referred to as messenger RNA (mRNA) travels through small openings of the nucleus to enter the cell's cytoplasm. Thereafter, another form of RNA, transfer RNA (tRNA), operating with ribosomes, serves in the "translation" process wherein lengthy chains of amino acids form a protein. Depending on the specific sequence of amino acids, a range of different hormones and enzymes, and other complex neurobiological factors are formed, which determine behavioral and psychological outputs.

Only a small fraction of the amino acids of a DNA strand form genes containing the information that contributes to the formation of our phenotypes. These genes, serve like a set of drawings that provide the instructions for a building being constructed. Another sequence of amino acids forms promoter or regulatory regions that provide instructions to the primary gene, essentially telling it when to turn on or off, as well as how and when to interact with other genes. What makes these regulatory processes especially interesting is that psychosocial, experiential, and environmental triggers affect a genes' influence on neurobiological processes and behavioral phenotypes. Contrary to views that had been held only a couple of decades ago, it seems that although the functions of genes are more or less fixed, social and environmental influences may affect whether these genes will reach their potential or be subverted.

Social and Environmental Moderation of Gene Actions

Curiously, although the debates concerning the contribution of genes (nature) versus environment (nurture) to phenotypes were tempestuous, scant attention was given to the role of gene × environmental interactions. As we now know, the environment in which an organism is raised can markedly influence the expression of a gene. Inbred strains of mice, which had been developed through successive brother × sister matings so that every mouse of a given strain is genetically identical to every other mouse, share many behavioral phenotypes. However, even small environmental differences (e.g., where animals were bred) could result in differences in behavioral phenotypes (Crabbe et al., 1999). Likewise, a pregnant mom's experience could affect the expression of genes among her pups, and the early life of rodent pups can affect the later phenotypic expression of their genes (e.g., Szyf, 2019). Similarly, among pregnant women, strong stressors may affect the expression of genes within their offspring, including those that affect inflammatory processes (Cao-Lei et al., 2017).

Mutations in hostile environments

Given the frequency of cell multiplication and genes being duplicated, it can reasonably be expected that some errors will occur; a nucleotide (or string of nucleotides) may be deleted, replaced by another nucleotide, or a small string is duplicated (amplification), or translocations may occur in which a gene fragment from one location appears elsewhere. Some mutations that occur might have negative consequences, and indeed, the term mutation may conjure up expectancies of creatures from the dark lagoon (or Ninja Turtles). Some mutations, such as those caused by

environmental toxicants could result in illnesses, such as some forms of cancer, but others may reside on a portion of the DNA strand that does not seem to have much bearing on phenotypic outcomes.

Importantly, the occurrence of mutations could be a fundamental component of evolution. When environmental changes pose a challenge for organisms, those with mutations that help deal with these conditions might be more likely to survive and pass on their genes, whereas those without the mutation would be less likely to survive and commensurately less likely to pass on their genes. Studies in bacteria indicated that there is still more to it than this. Upon being challenged by conditions that comprised potential starvation or the presence of an antibacterial agent (antibiotic) that could kill them, the rate of *E. coli* bacteria exhibiting gene mutations (mutagenesis) increased appreciably. Evidently, in the context of severe challenges, mutagenesis increases, possibly in an effort for some of the bacteria to survive (Revitt-Mills and Robinson, 2020). To be sure, most bacteria will die in response to an antibiotic assault, but given sufficient bacterial diversity, a very small number may have developed just the right mutation(s), which allowed them to survive and form the core of a bacterial colony, thereby contributing to the formation of more resistant bacteria should the antibiotic again be present.

Although it is usually thought that bacterial mutations within DNA occur on a random basis, it seems that a mutation can also occur within an RNA strand, particularly when the cell is stressed. The mutated RNA gives rise to proteins with enhanced functions and could potentially influence the development of DNA with this mutation in subsequent bacterial generations. This retromutagenesis has been considered a way for bacteria to escape destruction by antibiotics (Morreall et al., 2015). Mutations could similarly increase in humans, especially in relatively hostile environments, and might be preserved within the gene pool, and passed down across generations. This could be very advantageous for humans, but when it happens in bacteria, it may contribute to bacteria becoming resistant to antibiotics, which is very bad news for us.

Gene Polymorphisms

It has been estimated that approximately 10 trillion cells are formed over the average life span, and the sequence of nucleotides ought to be the same in each of these. When DNA is transcribed to mRNA, the sequence of bases should be faithfully reproduced, and to help ensure that this occurs, a built-in proofreading process exists to limit errors. Understandably, however, with so much complexity and activity, some errors will occur.

Acquired mutations may be present owing to environmental influences, such as environmental toxicants, and should the mutations occur within sperm or egg, they can be transmitted across generations, and thus are inherited mutations. A mutation that appears in a significant portion of the population (more than 1%) is referred to as a polymorphism, and if this mutation comprises a change of only a single nucleotide, it is called a single nucleotide polymorphism (SNP; pronounced snip). Just as changing one letter of a word or one word of a sentence can entirely alter its meaning, a change of a single nucleotide at an important portion of a gene can have profound effects on a phenotype. For example, some perturbations may result in certain enzymes, hormones, or hormone receptors not operating as they should. This has provided a means of assessing whether the presence of an SNP (or several SNPs) on a given gene (e.g., coding for a specific hormone receptor) is predictive of the appearance of specific pathological conditions or behavioral styles. As very many mutations occur within every individual, and several SNPs may even be present within a single gene, linking specific polymorphisms to specific phenotypes is often difficult, especially considering that complex behaviors or pathological conditions typically involve multiple genetic contributions. There had been reports indicating relations between certain SNPs and behaviors, but the findings of many early studies, especially those with small sample sizes, were not readily replicated, although greater success was achieved in subsequent studies that included large cohorts.

The difficulty linking specific SNPs to phenotypic outcomes was made more complicated with the realization that the occurrence of a given SNP may vary dramatically across cultures. For example, among Euro-Caucasians, a particular SNP on the gene coding for the oxytocin receptor appears at a rate of 15%–20%, but this same SNP occurs in about 80% of East Asian people (Kim et al., 2011). Furthermore, the phenotypic correlates of a particular SNP might vary in the presence of other genetic factors, as well as in the context of particular earlier experiences. Essentially, certain psychological or physical illnesses might only occur when multiple hits are experienced, and while genetic constitution may represent one or several hits, a psychologically toxic environment, and negative experiences may reflect additional hits that culminate in pathology. As already mentioned, this multihit notion has been accepted in relation to certain types of cancer, Parkinson's disease, and stroke recovery, and likely contributes to the emergence of some psychological disorders, including depression and schizophrenia.

Epigenetics

Although it was eventually accepted that both genes and environment, as well as their interactive effects, contributed to the expression of numerous phenotypes, it had hardly been considered that experiential and environmental factors could actually influence how genes were expressed. Following the lead of cancer researchers, it was demonstrated that in response to particular toxicants, or even some foods, as well as stressors, including those that occurred prenatally (Cao-Lei et al., 2017; Szyf, 2015), epigenetic changes could be engendered which might influence psychological well-being. As depicted in Fig. 1.1, epigenetic alterations can be provoked by a wide constellation of experiential and environmental factors. These epigenetic modifications constitute alterations of gene expression in which the actions of some genes or their promoters (regulatory elements) may be suppressed (or activated) but without the DNA sequence being altered. In essence, a change of phenotype can occur without a corresponding change of genotype. Epigenetic modifications are evident in cell differentiation (e.g., in the formation of different types of cells, such as skin cells versus brain cells throughout prenatal development) and may be responsible for the development of altered immunity (Hoeksema and de Winther, 2016). The emergence of diseases, such as cancer, immune-related illnesses, neuropsychiatric disorders, pediatric disorders, as well as several developmental disorders (e.g., Fragile X, Rett syndrome, Prader–Willi and Angelman syndromes) may likewise occur owing to epigenetic actions. Epigenetic alterations may occur on genes coding for hormones and hormone receptors, growth factors, and immune processes, which have been implicated in psychological conditions, such as depression, posttraumatic stress disorder (PTSD), and substance use disorders (Montalvo-Ortiz et al., 2022; Torres-Berrío et al., 2019).

Epigenetic modifications may come about in several ways, with the most common being DNA methylation, histone remodeling, and noncoding RNA-associated gene silencing, each of which influences how genes are expressed. Methylation refers to the addition of methyl groups to the DNA sequence in which cytosine and guanine bases are connected by a phosphate group (denoted as a CpG site), and as a result, the gene is less transcriptionally active. In addition to methylation, gene expression can be affected by chromatin remodeling as well as histone modifications. Chromatin comprises the complex of proteins (histones) around which DNA is tightly wound to fit into the nucleus. The complex itself is modifiable in numerous ways so that when the chromatin structure is altered, gene expression may be affected. When chromatin is tightly wound, it tends to result in gene expression shutting down, whereas open chromatin tends to be more functional or expressed. Over the past years, the number of epigenetic marks discovered has increased exponentially, and the National Institutes of Health Roadmap Epigenomics provided global maps of regulatory elements, their presumed activators and repressors, which were linked to cell types that might

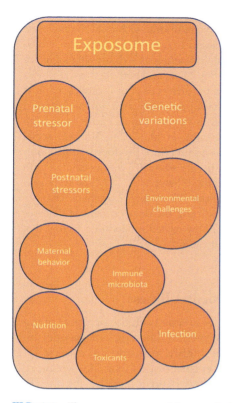 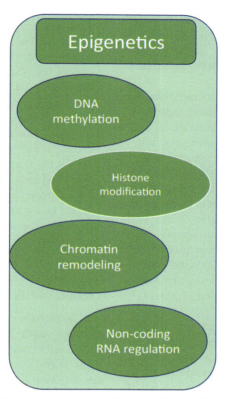 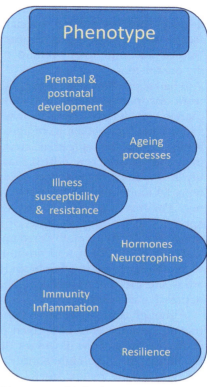

FIG. 1.1 The exposome comprising genetic influences and prenatal and postnatal factors (e.g., nutrition, maternal behaviors, toxicants encountered) can promote epigenetic changes that may involve several processes. In this way, gene expression can be altered without changing the genome itself. Epigenetic changes can stem from methylation of particular DNA sites or through histone modification or chromatin remodeling, as well as by noncoding RNA regulation (sRNA). These epigenetic actions have notable effects on multiple biological processes that favor the occurrence (or limit) numerous physical and psychological phenotypes, shown in the far right panel. If epigenetic changes occur within germline cells, they may persist through cell divisions for the duration of the cell's life and may also occur over multiple generations.

be relevant for the analysis of the molecular basis of many disease conditions (Satterlee et al., 2019). While epigenetic changes had long been thought to be restricted to DNA, these have now been seen in RNA and not only occur at adenine bases within DNA but can also appear at cytosine bases. Specifically, microRNAs (miRNAs) comprise small noncoding RNAs that serve to regulate the expression of target genes. The expression of miRNAs can be regulated by epigenetic processes (methylation) and may contribute to several diseases, particularly cancer. The importance of epigenetic changes in determining diverse phenotypes is well-established, and the trick now is to determine the links between epigenetic alterations and specific pathological conditions, the factors that moderate the occurrence of these epigenetic actions, and whether and how these epigenetic can be erased or preserved (Willyard, 2017).

It is particularly important that if the epigenetic changes occur within germline cells (sperm or egg), these changes can be recapitulated across multiple generations. To assume that a transgenerational effect is occurring and is not due to common factors that lead to epigenetic changes uniquely within each generation, the epigenetic effects need to be demonstrated across at least three (preferably more) generations. There is, in fact, compelling evidence that the experiences of one generation, through epigenetic alterations, can affect phenotypic outcomes in ensuing generations. Indeed, transgenerational epigenetic actions may be fundamental for the fine-tuning of gene activation (and inactivation) processes during embryogenesis (Zenk et al., 2017). Still, the occurrence of transgenerational effects in humans has not frequently been demonstrated, and the phenotypes evaluated were often confounded by genetic, ecological, and cultural factors (Horsthemke, 2018).

The difficulty of linking epigenetic effects and specific phenotypes is still more difficult than first thought. Epigenetic effects are not only exceptionally common, but they can be transient, changing through the lifespan in response to environmental events, such as smoking, exercise, and what we eat, or conversely in response to fasting. Throughout development, epigenetic changes may play a critical role in determining the different types of cells being formed from the original zygote. As about 200 different cell types will eventually be developed, machinery needs to be present to erase genetic marks so that a cell can return to its "naïve" state and then develop into yet another specialized cell. Most genes escape the reprograming process, but about 1% are "imprinted" and thus become permanent features of the organism.

It had at one time been thought that epigenetic changes were uncommon events, and thus when an epigenetic change was detected congruent with a behavioral alteration, these were taken to be linked in some fashion. It is now certain that epigenetic changes occur exceedingly frequently (Koziol et al., 2016), making it difficult to identify a one-to-correspondence with a particular phenotype, let alone assume a causal connection between the two. For instance, 366 separate epigenetic marks were apparent in the hippocampus of individuals who died of suicide (273 hypermethylated and 93 hypomethylated) (Labonté et al., 2013), and no doubt epigenetic changes would be apparent in other brain regions (e.g., hypothalamus and prefrontal cortex) that have been aligned with depressive illnesses.

Although some epigenetic changes may play a causal role in the emergence of a given pathology, others may evolve because of the illness (or distress associated with the illness), or they might simply be bystanders that are neither causally related nor biomarkers that predict illness onset. Aside from the sheer number of epigenetic marks that exist, any given illness likely involves a large number of genes, and thus, it is questionable how significant a single epigenetic mark would be. To complicate analyses further, epigenetic patterns are dynamic and changeable over time (as is their permanence), and thus links to particular pathological conditions are still more difficult to discern, and the probability of false positives increases commensurately (Birney et al., 2016). Longitudinal designs may diminish some of the problems in linking epigenetic marks to the subsequent appearance of certain pathologies, but even these types of studies would not necessarily reveal whether a link is causal or simply correlational. Despite these limitations, identifying epigenetic markers could be used to determine who might be vulnerable to illness and perhaps what treatments might be most efficacious for treatment in any given individual (Szyf et al., 2016).

Biological aging

With multiple experiences that accompany aging, the number of epigenetic marks that occur ought to increase. It was suggested that methylation at sets of CpG sites could provide an index of biological age (referred to as clock CpGs) that is highly correlated with chronological age.

However, under certain conditions, such as in the presence of certain illnesses, biological age acceleration occurs, which is associated with further illnesses as well as with cognitive decline (Horvath and Raj, 2018), and both depressive illness and PTSD. Aside from the effects of poor lifestyles, various stressful experiences have been associated with accelerated biological aging, which may revert to its earlier state with stressor cessation (Poganik et al., 2023). Various social stressors, including discrimination, produced similar outcomes. Predictably, early-life adversity was associated with increased biological aging (Shenk et al., 2022), as were prenatal stressor experiences. Ultimately, epigenetic changes might be useful in predicting the development of illnesses and early mortality and the varied moderating factors that contribute to these outcomes.[2]

Genetic Links to Mental Illness

The picture might seem rosy with the prospect of identifying genetic links to mental illnesses, but skepticism in this context would be highly appropriate, especially in light of the long track record of less than staggering successes. Years ago, we had been fooled by the fact that there are illnesses that can be ascribed to a single major gene (e.g., cystic fibrosis, sickle cell disease, Tay Sachs, Fragile X syndrome, muscular dystrophy, Huntington disease, and Thalassemia), but these tend to be relatively uncommon. Indeed, most complex disorders are regulated by many core genes. But, if one considers the vast array of gene networks in which the actions of one gene interact with others, it becomes clear that most illnesses are likely influenced by dozens or hundreds of genes. The very same thing can be said of epigenetic effects that occur in remarkably large numbers, and their contribution could interact with other genetic actions. As we discuss various psychiatric illnesses in ensuing chapters, we will revisit the genetic and epigenetic factors that are believed to be involved in their occurrence.

Biological Processes

Autonomic Nervous System

The autonomic nervous system (ANS) is involved in the regulation of body organs and processes over which we do not have voluntary control (e.g., heart, gut, and stomach). Regulation of the ANS occurs through the medulla oblongata, and other brain regions involved in this capacity include the hypothalamus, which influences

---

[2] Chapter 6 provides a discussion concerning the use of telomeres (repetitive DNA sequences situated at the ends of a chromosome) as an index of aging, while Chapter 17 discusses changes of inflammation in relation to aging (inflammaging).

energy regulation (e.g., eating), and the amygdala, which contributes to basic emotions, particularly fear and anxiety, that directly or indirectly affect ANS functioning.

The sympathetic component of the ANS, involving the release of epinephrine (adrenaline), and the parasympathetic nervous system (in which acetylcholine is the primary neurotransmitter) act together in the functioning of various organs. These systems are ordinarily in balance with one another, but occasionally environmental triggers will instigate changes so that sympathetic activity predominates, as observed in response to emotionally arousing events that produce an increase in blood pressure and heart rate. In other instances, the compensatory antagonistic system may be overly active, and thus blood pressure and heart rate may become inordinately low. These systems are readily manipulated through pharmacological means, although the focus has been more on modifying epinephrine and norepinephrine activity than on altering acetylcholine.

Sympathetic activation, together with parasympathetic responses, also contribute to the shunting of blood away from those organs that are not necessary to deal with threats, whereas blood flow increases to organs involved in survival (e.g., those required for elevated physical actions). As we will see in greater detail later, the release of epinephrine also increases immune reactivity, setting in motion cells from primary immune organs, such as the spleen, so that they are out doing the work they are meant to do.

The Central Nervous System and Neurotransmission

Experts in cardiovascular, endocrine, or immune system functioning might say that their system is particularly important for well-being. Although they are all special and necessary for survival, most people likely see the CNS, which comprises the brain and spinal cord, as the "first among equals." Through billions of neurons that can communicate with one another, the brain is responsible for sensory and motor processes, judgments and decision-making, sleep and wakefulness, memory, as well as fundamental drives, such as eating, sleeping, defense, and sexual behaviors.

Our 80 billion or so neurons have numerous dendritic branches, with each having many synapses. At birth, each neuron has approximately 2,500 connections, increasing to about 15,000 by 2 or 3 years of age, so that more than 1,000 trillion synaptic connections are present within the brain. Specific experiences can instigate the formation of synapses, which are strengthened as these experiences are repeated and then serve to maintain memories of the events. Those synapses that are fired often are strengthened, whereas those that are hardly activated will be pruned (literally, use it or lose it). In the absence of synaptic pruning, the excessive interconnections could potentially result in illnesses, such as schizophrenia.

Various brain regions serve specific functions, but as we have already said, complex behaviors or pathological conditions are typically determined by neuronal circuits (or systems) comprising several brain regions operating sequentially or in parallel. For example, a syndrome such as PTSD may involve brain regions governing fear (aspects of the amygdala), memory of the trauma (hippocampus and amygdala), and appraisals and judgment (aspects of the prefrontal cortex). Likewise, addictions may involve areas associated with anxiety (aspects of the amygdala), cognitive processes and impulsivity (prefrontal cortex), and reward processes (nucleus accumbens). Indeed, even seemingly basic functions are influenced by systems involving interconnections between several brain regions and multiple neurotransmitters (and hormones), although disruption of any single neurochemical can undermine these behaviors.

More than 100 different neurotransmitters have been identified, including gaseous substances, which stimulate several different receptors that can lead to diverse consequences. Ordinarily, each neuron can release only one type of transmitter, essentially meaning that it sends messages in only one language. However, a given neuron can be stimulated by a variety of different neurochemicals (i.e., it understands many languages). The message sent is not only affected by the specific neurotransmitter released and the specific receptors triggered in specific brain regions, as the rate of neuronal firing can have different meanings for an adjacent receiving neuron. Of the many neurotransmitters present within the mammalian brain, the most abundant is glutamate, which serves to activate neurons that they impinge upon. With so many excitatory neurons abounding and being at play concurrently, considerable organization is needed to ensure that messages are passed along in a coordinated fashion as well as to diminish noise. To this end, other neurotransmitters have an inhibitory effect on adjacent neurons, with gamma-aminobutyric acid (GABA) neurons being the most abundant.

It is often stated that this or that neurotransmitter has this or that action on specific behaviors or illnesses. However, most neurotransmitters and their receptors are present across multiple brain regions and may serve in diverse capacities. When triggering neuronal activity within the prefrontal cortex, a given neurotransmitter, such as norepinephrine, may contribute to appraisals and decision-making, whereas the same neurotransmitter in the hippocampus might be involved in the circuitry linked to memory processes, and in still other areas, such as the amygdala, it may dispose individuals toward vigilance and anxiety. Likewise, dopamine in the nucleus accumbens may contribute to reward processes, but in nuclei of the hypothalamus, it may promote or interact with several hormones to influence energy regulation, feeding, and sexual behavior. For that matter, when a drug is administered to increase a neurotransmitter in one brain area, as in the case of L-DOPA that is administered to increase dopamine in the substantia nigra to reduce Parkinsonian symptoms, neuronal functioning is also affected in other brain regions, such as the nucleus accumbens, which had been operating appropriately all along. The excessive dopamine activity in the latter region may increase the value of rewarding stimuli and may thus encourage the emergence of reward-related disturbances, such as gambling problems. As it happens, dopamine also plays a fundamental role in decision-making, and manipulating nigrostriatal dopamine levels as animals were engaged in specific tasks could trigger a change in their choice behaviors (Howard et al., 2017). As such, dopamine could play a role in both good and bad decisions that are made (e.g., in the case of addictions, including gambling). In considering the actions of various neurotransmitters, it is obviously necessary to distinguish between the multiple brain regions in which they are acting, despite the earlier caution regarding analyses of systems rather than specific brain sites.

Receptor Functions

Upon being released from axon terminals, the neurotransmitter can activate postsynaptic receptors on the adjacent neuron. For any given transmitter, several receptor subtypes can be triggered, each of which can produce different actions and outcomes. Receptors may also be present at the presynaptic end of axons (autoreceptors) that release the neurotransmitter. Upon being stimulated, these autoreceptors serve to tell the neurons to slow their neurotransmitter production. As the amount of neurotransmitter released into the synapse increases, the likelihood of autoreceptors being triggered is increased, and hence, neurotransmitter production is self-regulated.

Once a postsynaptic receptor has been triggered, it sets in motion a cascade of intracellular changes. The first messenger, the neurotransmitter, typically does not enter the cell (although steroid hormones do), so a second messenger system is essential for the message initiated by the neurotransmitter to be propagated within the neuron. When a receptor is bound by a ligand (e.g., a neurotransmitter), conformational changes occur (e.g., G-protein activation), giving rise to second messenger production, leading to activation of intracellular processes. Several different second messenger systems have been identified, including a cyclic adenosine monophosphate (cAMP), a phosphoinositol, and the arachidonic acid system. The prime function of each second messenger is basically the same, primarily activation of intracellular proteins through the process of phosphorylation (addition of a phosphate group, which favors the protein adopting a structural state conducive to signaling). Yet, the presence

of certain biochemical features can distinguish one pathway from one another. As well, the actual impact of a particular pathway being activated depends upon the tissue and system in which it is embedded. Thus, they can potentially serve as therapeutic targets for a wide range of conditions. Indeed, most drugs used to treat depression, schizophrenia, and Parkinson's disease target elements of the key monoamine pathways. Hence, they involve many similarities at the intracellular signaling level.

Turnover and Reuptake

Once a neurotransmitter has been released into the synaptic cleft, it ought to stimulate nearby receptors, after which it is normally eliminated to preclude excessive receptor stimulation. Enzymes present in the synaptic cleft can degrade some of the neurotransmitters that are present, but many transmitters are eliminated by being transported back into a neuron (to be recycled) through a specialized transporter mechanism. This ecologically friendly process, referred to as reuptake, allows for the transmitter to be available for later reuse. The longer the neurotransmitter remains viable in the synaptic cleft, the greater the opportunity to stimulate a receptor. Pharmacological means of altering the efficiency of the neurotransmitter have been developed wherein the time spent in the synaptic cleft is altered. Degrading enzymes can be inhibited so that the transmitter would not be destroyed, or the reuptake of the transmitter back into the neuron can be diminished [e.g., selective serotonin reuptake inhibitors (SSRIs) and selective serotonin and norepinephrine reuptake inhibitors (SNRIs) that are commonly used in the treatment of depression]. Such manipulations are among the obvious ways by which treatments could have their positive effects, but the beneficial effects of the drug treatments may arise owing to downstream actions that occur. In fact, the long latency between drug administration and the onset of clinical benefits of antidepressants resulted in the proposition that some time-dependent cellular effects, independent of the levels of monoamines, must be involved. In this regard, neuroplastic changes downstream of the initial actions of SSRIs might underlie symptom remediation in cases of depression or similar stress disorders, and it was thus proposed that drug treatments ought to target growth factors responsible for the neuroplastic alterations (Castrén and Monteggia, 2021).

Glial Cells

In addition to neurons, the brain is replete with glial cells, largely comprising astrocytes (astroglia), microglia, oligodendrocytes, and ependymal cells. For the longest time, research attention focused almost exclusively on neurons, whereas glial cells were considered secondary players that acted primarily as support cells for neurons, providing nutrients and taking away debris (the help). However, this view turned out to be far too narrow. Astrocytes, the most abundant type of glial cell, are involved in maintaining ion balances within fluid outside of brain cells and are fundamental in the repair of brain and spinal cord neurons. They are also able to communicate with neurons through the release of particular neurotransmitters (e.g., GABA and glutamate). As well, these cells may be important for the repair of brain damage caused by stroke and may even contribute to the formation of neurons. Oligodendrocytes are involved in the myelination of neurons (myelin forms the sheath around axons), which is essential for the rapid propagation of electrical signals down the axon. Ependymal cells line the brain ventricles and the central canal of the spinal cord and contribute to the production and regulation of cerebrospinal fluid (CSF). Junctions between ependymal cells allow for the fluid release across the epithelium so that some exchange occurs between CSF and nervous tissue of the brain and spinal cord. Thus, analyses of CSF obtained through a spinal tap can serve as an observation post of the CNS.

Microglia, develop from myeloid progenitors in the yolk sac, entering the CNS during early embryonic development. Aside from just providing nutrients to neurons and maintaining ion balances within the fluid outside of the brain cells, microglia are the brain's resident immune cells that respond to pathogens and injury, much like peripheral macrophages act in this capacity. Beyond these actions, they are involved in surveillance, programmed cell death, clearance of apoptotic neurons, and neuronal plasticity (Salter and Stevens, 2017). They promote the growth and branching of dendrites and are also essential in the repair of neurons within the brain and spinal cord, and through the release of some neurotransmitters (e.g., GABA), they can communicate with neurons. As well, glial cells contribute to the clearance of neurotransmitters from the synaptic cleft, thereby limiting the damaging effects that would otherwise occur owing to a buildup of some transmitters, such as glutamate. Fortunately, unlike neurons that die off with age, microglia renew themselves several times over the course of a lifetime; thus, the number of such cells remains relatively steady (Askew et al., 2017).

As much as glial cells have multiple positive effects, they can also engender neurotoxic actions through excessive inflammatory immune processes. The microglia respond to a signal coming from the immune system, notably the complement pathway, which prompts them to adopt a phenotype that allows them to engage in defensive behaviors, clear pathogens, and cellular debris, and allows them to prune or get rid of unneeded synapses (Stephan et al., 2012). A disturbance of signaling involving microglia, so that too few (or too many) synapses are pruned, could lead to both developmental and degenerative disorders, including schizophrenia and Alzheimer's (Chung et al., 2015).

The discovery that microglia could release inflammatory immune molecules (cytokines) that had traditionally been viewed as being involved in signaling between peripheral immune cells gave rise to new perspectives concerning processes related to depressive disorders (Wang et al., 2022b). These inflammatory factors were also implicated in recovery from stroke and neurodegenerative disorders, such as Alzheimer's and Parkinson's disease, as well as schizophrenia and developmental disorders. The findings regarding microglia have opened new avenues in understanding illnesses and comorbidities that exist between various disorders and hold promise for effective therapies becoming available that target inflammatory processes (Salter and Stevens, 2017).

Protecting the Brain (Sometimes)

Astrocytes can coax neurons to increase their production of a protein C1q, which is fundamental for pruning to occur. The presence of C1q is essential for early-life neural development and may be essential for synaptic plasticity in the aging brain (Stephan et al., 2013). Reducing C1q or hindering the ability of this protein can limit the initiation of the complement cascade, which comprises a series of small proteins that upon stimulation, can activate one another and contribute to the clearance of pathogens by working together with innate immune cells or antibodies. When this occurs, immune processes are affected, as is the age-related decline of cognition and memory in aging mice (Shi et al., 2015). In line with animal studies, the *C4A* gene, which serves to encode a complement protein downstream of C1q, was linked to synaptic loss and schizophrenia (Sekar et al., 2016).

Following brain injury or the presence of disease, astrocytes changed dramatically (reactive astrocytosis), and a variety of genes were upregulated (Zamanian et al., 2012). It has been debated whether this reflected a destructive or protective effect, given that astrocytes, like microglia, adopt various plastic phenotypic states (Heppner et al., 2015). Indeed, a subset of astrocytes were deemed to be destructive as they were associated with the upregulation of genes linked to the inflammatory cascades that were known to disturb synapses. These astrocytes, which are induced by microglia-secreting cytokines, such as interleukin-1β (IL-1β), tumor necrosis factor-α (TNF-α), and following CNS injury C1q can develop neurotoxic functions, destroying neurons as well as mature oligodendrocytes. In contrast, another subset of astrocytes was associated with enhanced neurotrophic processes and was thus taken as being neuroprotective (Liddelow et al., 2017). In effect, it seems that at low levels, inflammatory cytokines may have neuroprotective effects, but at higher concentrations, they may be neurodestructive.

These negative actions could come about through the promotion of excessively strong inflammatory responses or through the production of free oxygen radicals that promote neuro destruction, thereby giving rise to neurodegenerative disorders.

Some exogenous and endogenous factors that could potentially act in a pathogenic capacity might not readily reach the brain and spinal cord owing to the presence of tightly packed endothelial cells that comprise part of the blood–brain barrier (BBB). In this regard, immune messengers (cytokines) in the periphery that are released when specific immune cells are activated might not reach the brain readily owing to their large size. However, cytokines can influence neuronal functioning at specific sites near the outer reaches of the brain, such as the posterior pituitary and the median eminence. Also, access to the brain parenchyma can occur where the barriers to the brain are less efficient, including around the ventricles (i.e., "circumventricular organs," such as the organum vasculosum of the lamina terminalis and the area postrema). Cytokines can also be ferried into the brain through active transport systems (Banks, 2019) so that neuronal functioning may be affected. Importantly, greater access to the brain may occur following head injury and in the presence of high fever, and gut microbes and their metabolites can affect the BBB so that access to the brain is facilitated (Parker et al., 2020).

Neurotrophins

One of the most remarkable attributes of brain neurons is their plasticity (i.e., their ability to modify their connections, essentially amounting to the rewiring of the brain). Dendritic branching (arborization) and the formation of synapses (synaptogenesis) are promoted by experiences, and acquiring new stimulus–stimulus or stimulus–response associations is due to new synapses being formed and then strengthened with use. In addition to being involved in learning and memory processes, neural plasticity is essential in recovery from stroke and head injury and contributes to emotions and disturbed mood states, such as depression. For synaptic plasticity to occur, neurotrophins (growth factors) are necessary. Particular attention in this regard has focused on brain-derived neurotrophic factor (BDNF), although other neurotrophins, such as fibroblast growth factors (FGF-2) and vascular endothelial growth factor (VEGF) may also be involved (Castrén and Monteggia, 2021). It is also well-established that neurotrophins can stimulate neurotransmitter release and interact with cytokines in promoting mental illness. Table 1.2 provides a listing of several neurotrophins that have been implicated in maintaining mental well-being.

TABLE 1.2 Neurotrophins and Their Function

Neurotrophin	Biological Effect	Outcome
Brain-derived neurotrophic factor (BDNF)	Support survival of neurons; encourages growth and differentiation of new neurons; promotes synaptic growth	Influences memory processes, stress responses, mood states
Basic fibroblast growth factor (bFGF or FGF-2)	Involved in neuroplasticity; formation of new blood vessels; and protective actions in relation to heart injury; essential for maintaining stem cell differentiation	Contributes to wound healing; neuroprotective; diminishes tissue death (e.g., following heart attack); related to anxiety and depression
Nerve growth factor (NGF) and family members Neurotrophin-3 (NT-3) and Neurotrophin-4 (NT-4)	Contributes to cell survival; growth and differentiation of new neurons. Fundamental for maintenance and survival of sympathetic and sensory neurons; axonal growth	Survival of neurons; new neuron formation from stem cells; related to neuron regeneration, myelin repair, and neuro-degeneration. Implicated in cognitive functioning, inflammatory diseases, several psychiatric disorders, addiction, dementia as well as physical illness, such as heart disease, and diabetes
Insulin-like growth factor 1 (IGF-1)	Secreted by the liver upon stimulation by growth hormone (GH). Promotes cell proliferation and inhibits cell death (apoptosis)	Secreted by the liver upon stimulation by growth hormone (GH). Promotes cell proliferation and inhibits cell death (apoptosis)
Vascular endothelial growth factors (VEGF)	Signaling protein associated with the formation of the circulatory system (vasculogenesis) and the growth of blood vessels (angiogenesis)	Creates new blood vessels during embryonic development, encourages the development of blood vessels following injury, and creates new blood vessels when some are blocked. Muscles stimulated following exercise. Implicated in various diseases, such as rheumatoid arthritis, and poor prognosis in breast cancer

Hormones

Much like neurotransmitters, hormones can act as signaling molecules within the brain and in the periphery. In the periphery, hormones are manufactured by a variety of glands and transported to distant organs to regulate physiology and behavior. Some hormones are released directly into the bloodstream (endocrine hormones), whereas others are secreted into a duct and flow either into the bloodstream or spread from cell to cell by diffusion (exocrine hormones). As well, neuropeptide hormones, such as β-endorphin and dynorphin, can be secreted by neurons and activate receptors present within the CNS.

Hormones are fundamental in metabolic processes, eating and energy balances, stress reactions, cell growth and cell death (apoptosis), sexual characteristics and behaviors, and they can have profound actions on brain processes, such as neurogenesis (Mahmoud et al., 2016). Beyond these actions, hormones are involved in cognitive functioning, including learning and memory processes, and may contribute to various mood and motivational states, as well as sex differences that have been related to illnesses, such as heart disease and autoimmune disorders.

It is of particular significance that several hormones (e.g., cortisol and estrogen) influence immune functioning, and drugs that affect hormone functioning have been used to treat a variety of immune-related illnesses. In later chapters, the influence of various hormones will be discussed in more detail, including their roles in stressor and immune-related illnesses as well as inflammatory changes tied to psychiatric disorders.

Neuroimmune Links

It had long been considered that the immune system was independent of brain functioning, which might have contributed to mental illness not being recognized as being tied to immunologically related processes. This chasm was bridged by several findings. Immune functioning was found to be subject to classical conditioning (Ader et al., 1990), and immune activity was influenced by neuroendocrine, neurotransmitter, and other aspects of brain functioning and could thereby influence disease progression. Evidence supporting a link between immune factors and mental illness came from several sources, including reports that activation of the inflammatory immune system (e.g., by immunogenic agents or by stressors) could instigate psychological disturbances. At the same time, mental illnesses involve multiple interactions with innate and adaptive immune processes that might be differentially influenced by environmental, experiential, and genetic factors (Mangino et al., 2017). Thus, even if immune factors were related to mental illnesses, this would no doubt involve complex interactive processes.

We are only beginning to understand the full scope of the connections between the immune system and mental illnesses, including the contribution of gut bacteria (microbiota) that affect inflammatory processes (Cryan et al., 2019). The appreciation that multidirectional processes are at work in affecting brain processes and immune functioning and may contribute to gut-related disorders (Eisenstein, 2016) has led to still other novel approaches and targets being developed in the treatment of mental and physical disorders.

Immune Processes

The primary job of the immune system, as discussed in Chapter 2, is to recognize what is part of us and what is not, and to protect us from foreign particles that could create harm. During prenatal development the immune system learns "what us comprises" and by extension, what is not part of us. In addition to this innate immunity, over the course of postnatal development, the organism will also acquire immunity, in which it learns through experiences about foreign particles, including bacteria and viruses. Thus, the immune system ought to protect us from numerous invaders, although the fact that we get sick with many illnesses, including those in which the immune system turns on the self (autoimmune disorders), speaks to the limitations of these systems. Still, with the different types of immune cells present (neutrophils, monocytes, macrophages, natural killer cells, T helper cells, cytotoxic T cells, and B cells) and the ability to destroy foreign particles floating around in the body or those that have gained entry to healthy cells, we ought to be pretty well set to deal with multiple challenges that can be encountered. As we will see, though, bacteria and viruses do not sit around passively; instead, they go about looking for ways to get through our natural defenses. In fact, under certain conditions, bacteria can subdivide into two distinct populations, one of which is especially virulent, and the "memory" related to this hypervirulent state can be maintained for several generations (Ronin et al., 2017).

Contrary to earlier misconceptions, we know that immune signaling factors (cytokines, chemokines) can also affect brain functioning, and by virtue of several hormonal and neurotransmitter alterations, may influence mood disorders, psychosis, and neurodegenerative disorders. Conversely, many hormones and neurotransmitters affect immune functioning. Given the multidirectional exchanges that involve the immune system, it follows that factors that influence immune activity (e.g., lifestyle, obesity, and sleep) may come to affect psychological state. The prototypical stress hormone cortisol (corticosterone in mice and rats) can have profound actions on immune functioning, which can affect physical and mental well-being. At physiological levels (those ordinarily seen in the absence of a pharmacological treatment), corticoids can limit immune activation (e.g., preventing excessive immune activity in response to an acute stressor), and at pharmacological levels (which are markedly elevated through exogenous factors, such as drug treatments) immune functioning can be drastically suppressed. The latter treatments have been used to act against illnesses in which the immune system attacks the self (autoimmune disorders, such as rheumatoid arthritis and lupus erythematosus). In addition to glucocorticoids, through their actions on immune functioning, estrogen and progesterone can likewise affect autoimmune disorders, which contributes to the female bias that exists in these disorders (Moulton, 2018).

As we will see in Chapter 7, numerous psychological factors can affect immune activity and hence might influence physical illnesses. Beyond these peripheral actions of stressors, brain cytokine variations can also be engendered by a variety of stressors and may interact with immunogenic agents in synergistically increasing brain cytokine functioning. These brain cytokine (e.g., IL-1β) changes, such as those that occur in the hippocampus, have been implicated in fear learning processes (Jones et al., 2018), whereas variations of IFNγ may be related to memory processes (Litteljohn et al., 2014). It also appeared that by undermining BBB integrity, particularly among stress-sensitive mice, access of cytokines to the brain parenchyma was increased, hence influencing the development of illnesses, such as depression (Menard et al., 2017).

Microbiota and the Enteric Nervous System

Until very recently, little attention was devoted to the connections between the gut and the brain, even though the "enteric nervous system" (the nervous system associated with the gut) shares several transmitters that are present in brain. The cells and neuronal composition of the gut are highly dynamic, and like brain cells, it is subject to neurogenesis as well as apoptosis (Rao and Gershon, 2017). The enteric system, which courses through tissues that line the colon, small intestine, stomach, and esophagus, can influence brain functioning, just as the brain can influence gut functioning. Communication between the gut and brain can occur through stimulation of the vagus nerve, which extends from the viscera to the brain stem. As well, enterochromaffin cells that compose 1% of the gut epithelium serve as chemosensors that respond to metabolic and homeostatic signals, which through the release of serotonin inform the CNS of changes that are occurring (Bellono et al., 2017). The enterochromaffin cells are exquisitely sensitive to stimulation and might be one of the first-line defenses against threats and injury. These cells may be overly sensitive and may contribute to inflammation present among individuals experiencing inflammatory bowel disease. In this regard, serotonin within the gut can stimulate 5-HT$_4$ and 5-HT$_7$ receptors, which promote anti- and proinflammatory actions, respectively (Spohn and Mawe, 2017). In light of the many common elements between the enteric nervous system and the brain, and the bidirectional communication that occurs between them, the position was adopted that the enteric nervous system might also contribute to several neurological disorders (Rao and Gershon, 2016).

Gut Microbiota and Health

The trillions of bacteria that colonize the body appear to be regulators of immune processes, and various microbial species can have dramatic effects on brain functioning (some direct and some through their interactions with immune constituents). In recent years, microbiota have become the darlings of the neuroscience world, having been implicated in numerous disease states. Studies in rodents indicated that the links between the microbiome and health are not simply correlational, as manipulating gut bacteria can have health benefits, and several studies in humans have revealed similar outcomes.

While the research related to microbiota has focused on bacteria present within the gut, a variety of microorganisms (e.g., bacteria, archaea, fungi, viruses, and eukaryotes) are not only present within the gut but also in nasal passages, mouth, between our teeth, on the skin, and in the vagina. Cooperation occurs across bacterial species (Rakoff-Nahoum et al., 2016), and as these microbes colonize different parts of the body, new species evolve but behave harmoniously with one another (Silverman et al., 2017a). Varied microbiota species are present in diverse organs, as depicted in Fig. 1.2 (Hou et al., 2022), which have ramifications for health. The presence of local bacteria can profoundly influence brain processes that determine psychological well-being (Hyland and Cryan, 2016). Lifestyles endorsed, such as nutrient consumption, influence microbiota that affect metabolic processes, and could thereby affect physical health (Sonnenburg and Bäckhed, 2016).

There is the view that gut microbes have been fundamental in human evolution. The microbial communities that exist within the gut and the rest of the body are largely dependent upon or influenced by the resources available at these sites. When environments change, so do aspects of the microbiome, which can then influence the host. Those aspects of the microbiome that are most useful for us will be maintained, and those least useful

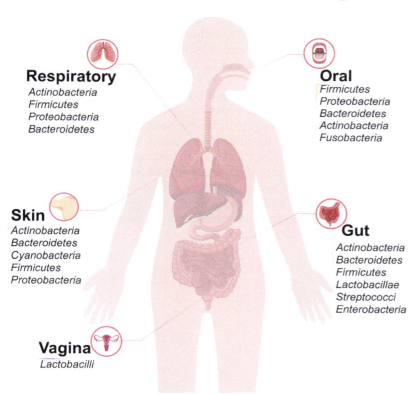

FIG. 1.2 Human microbiota composition in different locations. Predominant bacterial genera in the oral cavity, respiratory tract, skin, gut, and vagina are highlighted. Gut microbiota may communicate with the brain through spinal pathways and activation of the vagal nerve. These microbes can influence several neurotransmitters and growth factors. Short-chain fatty acids (SCFAs) produced in the gut with the consumption of certain foods can affect enterochromaffin cells and may affect brain processes, thereby affecting mood states. *From Hou, K., Wu, Z.X., Chen, X.Y., Wang, J.Q., Zhang, D., et al., 2022. Microbiota in health and diseases. Signal Transduct. Target. Ther. 23, 135.*

ought to fade away. Of the many types of gut bacteria present, some can have positive actions, but others may negatively influence health. In general, microbiota can encourage resistance to colonization by pathogenic species, but when these gut bacteria are eliminated or diminished, such as when mice are bred in a sterile environment or are treated with antibiotics, susceptibility to pathogenic bacteria is elevated (Sassone-Corsi and Raffatellu, 2015). Other forms of bacteria may favor illness occurrence by promoting the expansion and virulence of pathogens (Cameron and Sperandio, 2015).

Co-evolution occurs wherein we influence our microbiome, and our microbiome reciprocally affects what is carried forward across generations of bacteria. It seems we have made a deal with our microbiota; we provide them with a place to live and they help us in multiple ways, ranging from digestion, eliminating toxic xenobiotics (molecules that are foreign to the body), and influencing immune functioning. There is still a great deal that is not known concerning the interactions that occur between microbiota and xenobiotics, but gut microbes are certainly affected by foods, thereby affecting health. When more is understood regarding the functioning of gut microbiota and the genes that control them, it may provide new arsenals to deal with disease vulnerabilities.

Because gut bacteria are fundamental for metabolic processes and efficient immune functioning, alliances form between the many microbial species that inhabit the gut. This balance is necessary so that tolerance can develop to innocuous antigens (e.g., foods that are not contaminated), thereby limiting unnecessary immune activation (Hyland and Cryan, 2016) and playing a key role in regulating aging processes (Popkes and Valenzano, 2020). Gut bacteria are vulnerable to being disturbed yet hardy in the face of some challenges. When bacteria suffer the loss of essential enzymes, others seem to take on the load, even if this means creating a patchwork of multiple components.

The microbiome composition varies considerably across individuals, likely being determined by the host's (us) genetic composition, together with hormonal, metabolic, and immune factors. As well, microbiota are subject to change with environmental influences, including foods consumed, exercise, and other lifestyle factors, and is even affected by changes in our circadian clock. For that matter, some gut bacteria have their own circadian clocks, which are influenced by melatonin variations as well as by changes in body temperature (Paulose et al., 2019). Yet, following microbial changes, this system seems to reset itself, returning to the state it had been in prior to the perturbations encountered.

There has been an explosion of studies suggesting that microbial disturbances may contribute to numerous diseases. It is supposed that various gut bacteria need to be in balance with one another, and when these balances are disturbed (dysbiosis), illnesses may evolve for a variety of reasons, including the disruption of the tight coordination that exists between the microbiome and immune processes (Thaiss et al., 2016b). Not only is dysbiosis connected to greater vulnerability to the impact of pathogens, but it might also contribute to autoimmune disorders, such as multiple sclerosis (MS). Specific bacterial taxa that were elevated in MS patients were able to increase inflammatory responses in peripheral blood mononuclear cells, whereas those that were diminished in MS patients, activated processes that served in an antiinflammatory capacity (Cekanaviciute et al., 2017). Moreover, when certain gut microbiota (e.g., *Akkermansia*) from twins discordant for MS were transferred to mice, signs of MS, including disturbed immune regulatory factors, were apparent only in those that had received the bacteria from the twin with MS (Berer et al., 2017).

Aside from the gut bacteria-immune link, dietary elements are converted by gut bacteria into signals that affect adipose tissue, and when dysbiosis exists, functioning of the intestine, liver, lungs, cardiovascular system, and brain can be affected, and pathology can develop at these sites (Schroeder and Bäckhed, 2016). In view of the links between gut microbiota and foods consumed, dietary interventions could be expected to have health benefits through their actions on gut bacteria (Dahl et al., 2020) (Fig. 1.3).

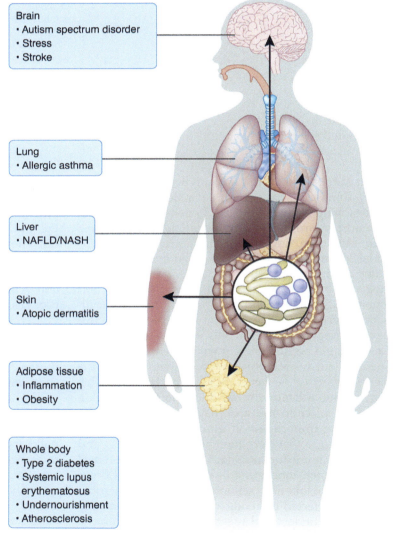

FIG. 1.3 Alterations in composition, diversity, and metabolites derived from the gut microbiota are associated with diseases affecting different organs of the human body. Evidence for a causative role of the gut bacteria is strongest in metabolic disease. Not shown in the figure is that gut dysbiosis has also been linked to inflammatory processes and numerous mental illnesses. *From Schroeder, B.O., Bäckhed, F., 2016. Signals from the gut microbiota to distant organs in physiology and disease. Nat. Med. 22, 1079–1089.*

Although gut bacteria have been implicated in many diseases in humans, the available data have largely been correlational, and there have been only a few syndromes that have been causally tied to microbial factors (e.g., *Clostridium difficile*-related diarrhea; inflammatory bowel disease; some forms of cancer). To a certain extent, the analysis of connections between microbiota and illnesses is complicated by the fact that there is some uncertainty as to what constitutes a healthy microbiome, especially given the pronounced variability that exists across healthy individuals and varies as a function of sex (Holingue et al., 2020). There is also evidence indicating that the diversity of the gut microbiome is appreciable across communities and cultures and differs widely between Euro-Americans and American Indians (Sankaranarayanan et al., 2015), and thus, developing specific remedies for particular illnesses could prove difficult. At the same time, core bacteria need to be present for well-being to be maintained (Valdes et al., 2018), and deep sequencing of gut microbiota revealed some of the "normal" factors that were linked to well-being (Zhernakova et al., 2016). Moreover, genome-guided microbial communities can be established that could act against specific pathogens (Brugiroux et al., 2016). This said, it is unlikely that proper diagnoses will be obtained based *solely* on a bacterial profile, and this variable would be best suited as one of many that are incorporated in the diagnoses and selection of therapeutic strategies. Assessment of microbiota dysbiosis among critically ill patients is another thing entirely given that the changes observed may be secondary to the illness or the treatments (e.g., drugs) administered. That said, microbial disturbances associated with chronic illnesses can instigate the failure of different organs, and recognizing the role of diverse microbes could provide clues in the development of intervention strategies. Furthermore, microbiota and the immune system reciprocally influence one another and are tied to a huge range of pathologies involving brain functioning and physical illnesses. Moreover, like inflammatory processes, microbial factors might contribute to the comorbidity that is so often apparent with many diseases.

Personalized Treatment Strategies (Precision Medicine)

Most individuals encounter events or environments (e.g., chronic stressors) that could promote illnesses, but the nature of the illness experienced varies broadly. Individuals differ with respect to their "weak links" that might have been acquired through experience or through inherited genes, and the nature of the illness that emerges will vary based on these risk factors. Marked individual differences are also seen with respect to the efficacy of treatments. Although some individuals overcome illnesses readily following treatment, there are those who seem to recover slowly, if at all. Once again, numerous factors might contribute to these individual differences, including genetic constitution that affects biological processes, together with psychosocial influences and previous experiences in dealing with illness.

Treatments Based on Endophenotypic Analyses

In the case of some illnesses, a physician can count on a positive response occurring with a particular treatment, but in other instances, the prognosis is not nearly as predictable. Individuals can have similar illness symptoms yet might be differentially responsive to therapies. Some individuals being treated for depressive illness, for instance, may not respond positively to a particular drug but exhibit diminished symptoms in response to an alternative treatment. A second individual, in contrast, might respond to the initial treatment but might not have responded positively to the alternative. A substantial number of depressed individuals ($\sim$30%) are treatment-resistant in that a positive response was not obtained using different classes of antidepressant drugs. Moreover, individuals may respond differently over time following some prescribed treatment. In the case of Parkinson's, some patients respond well for many years to dopamine replacement using l-DOPA, whereas others experience a loss of drug efficacy within a mere 3 or 4 years. Further to this, symptom clusters that co-occur can also influence treatment regimens differently. Indeed, about half of Parkinson's patients are afflicted with comorbid depression, and similarly, many MS patients experience comorbid depression and anxiety, and these symptoms can evolve differently over time. In other instances, people may receive superficially similar diagnoses (e.g., at one time, all breast cancers were viewed as being the same as one another), but the nature of these conditions may actually be very different from one another (e.g., with respect to genetic factors and/or whether they are hormonally responsive). Predictably, the effectiveness of treatments will also differ based on such considerations. Accordingly, if the characteristics of the illness are defined appropriately, more effective treatment strategies can be applied.

A treatment approach was developed some time ago in response to the repeated failures to treat patients effectively. This method, the endophenotypic approach, involved connecting specific genes to disease conditions or specific features of an illness (Gottesman and Gould, 2003). The endophenotypes comprise the "measurable" aspects that link genetic factors and illness symptoms. These genetic contributions may themselves be linked to endocrine, neurotransmitter, immunological,

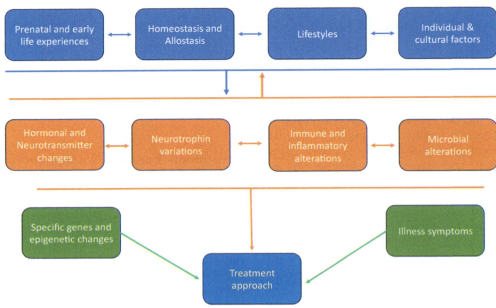

FIG. 1.4 An expanded endophenotypic approach. In addition to focusing on genes and gene mutations in relation to illness symptoms, this approach considers the influence of stressors, early experiences, and cultural factors in moderating neurobiological processes, as well as epigenetic factors, that culminate in disease processes and predict their most effective treatment.

microbial, neuroanatomical, neuropsychological, cognitive, or behavioral factors. Based on these illness features, the selection of effective therapies can be facilitated. Theoretically and practically, genetic and neurobiological factors can serve as biomarkers for illness development, the recurrence of illness, and treatment approaches. There are occasions where an endophenotypic approach may not be possible or practical. The alternative, then, might be one of focusing on dimensions of illness in broader terms, especially as particular symptoms often fall in clusters (e.g., sets or clusters of either neurovegetative or cognitive features). Both approaches could also incorporate several risk factors (premorbid conditions as well as stressor experiences and assorted lifestyle factors) and biomarkers of illness in the endophenotypic analysis (Fig. 1.4).

Linking specific genes to psychological disturbances, which then determines therapeutic strategies, is not an entirely new notion. Precisely, this approach has long been used in treating some forms of cancer and heart disease, but only recently was this proposed for the treatment of mental illnesses. The search for biomarkers of mental illness has recently been going on vigorously, encompassing genetic markers as well as other biological substrates and specific symptoms presented that offer clues concerning the nature of the disturbance. The personalized or precision medicine approach has taken on increasing allure and has posed a significant challenge to traditional methods related to diagnosis and treatment.

The American Psychiatric Association had, over a number of years, released several iterations of the *Diagnostic and Statistical Manual of Mental Disorders* (DSM), which were meant to provide a detailed and systematic description of the symptoms of various psychiatric disorders in the hope of facilitating diagnosis and treatment of a broad set of mental illnesses. Despite the importance of having psychiatrists work from a common playbook, the Fifth Edition of the DSM, published in 2013, was met with considerable criticism in some quarters, pushing for personalized treatment strategies rather than a one-size-fits-all approach.

> If you talk to God, you are praying; If God talks to you, you have schizophrenia. If the dead talk to you, you are a spiritualist; If you talk to the dead, you are a schizophrenic (Szasz, 1973).

The usual approach to diagnosis, including the use of the DSM criteria, has for many years encountered stubborn resistance, and some early theorists found the entire classification system inherently counterproductive. Thomas Szasz, who made a mark in his time, argued against the entire notion of diagnosing "illnesses that people have" as opposed to "symptoms people exhibit." Szasz went so far as to suggest that the labels that are applied to patients, such as "schizophrenia," represent an unprincipled concept, something used for convenience, rather than having any benefits for the affected individual. He even maintained that psychiatry was a pseudoscience with aspirations to become a genuine medically based scientific entity. In his 1961 book, *The Myth of Mental Illness*, Szasz described mental illness as a myth, suggesting that individuals who are labeled, are marginalized by society, and then preyed upon by a coercive system monopolized by drug companies.

While Szasz was a key figure of the antipsychiatry movement of the 1960s and 1970s, his was not an isolated struggle. Others rejected the medical model of mental illness, questioned the use of pharmacological approaches in the treatment of schizophrenia, and were strong in the belief that illnesses were born through experiences (Laing, 1960). During those days, many opponents of psychiatry saw it as controlling and repressive, ascribing certain conditions to the category of mental illness, only to reverse themselves years later. During this period, pharmacotherapy was making the first concerted inroads into pharmacotherapy, and there were many opponents of the psychiatric/medical model, including notables, such as Cooper, Lacan, Basaglia, Lidz, Arieti, and Foucault. These theorists, however, did not hold a candle in relation to the impact of Kesey's book and subsequent movie "One Flew over the Cuckoo's Nest" in promoting public distaste for the then-current medical approaches. The attack on the medical model in psychiatry and drug therapies had abated for some time but was rejuvenated with repeated reports of failed drug treatments, and it was even suggested that this science if it is a science, was entirely unsound.

Dissatisfaction with the DSM and the approach to treatment that it encompassed prompted NIH scientists to offer a multilevel framework for the diagnosis and treatment of mental illnesses. This comprised the RDoC, which followed from the endophenotypic approach (Cuthbert and Insel, 2013). The RDoC was conceived as a dimensional system comprising several layers of analysis ranging from neuronal through to behavioral processes. Instead of beginning by defining an illness and then attempting to treat it, the RDoC approach begins by defining the symptoms presented by patients and then linking these to brain processes in an effort to identify links to pathological conditions, which could inform treatment. The RDoC framework, provided in Table 1.3, comprises five domains:

1. Negative Valence Systems (i.e., response to aversive threats or events).
2. Positive Valence Systems (responses to stimuli or events that are perceived to be rewarding).
3. Cognitive Systems that entail those related to cognitive control, attention, perception, memory, and response selection.
4. Social Processes Systems (attachment, social communication, self-perception, perception, and understanding others).
5. Arousal/Regulatory Systems (arousal and sleep).

On the other axis are the diverse units of analysis that may be tied to the behavioral signs. These include genes, molecules, cells, neural circuits (determined through imaging methods or well-established circuits associated with particular behaviors), physiological processes (e.g., cortisol and heart rate), behaviors, and self-reports (interviews and questionnaires). As in epidemiological analyses, this matrix can provide symptom clusters that are linked to specific units of measurement, such as genes, hormones, or circuits. Should the features of an illness map onto the efficacy of different treatment methods, then this would ultimately prove to be ideal in predicting which of several treatment options would be best for individual patients. In this regard, many illnesses are comorbid with one another (e.g., depression predicting later heart disease); early signs of an illness could also serve as the proverbial canary in the coal mine, informing physicians of further health risks so that interventions are initiated (Anisman and Hayley, 2012b).

The RDoC approach has received considerable support in many quarters, but it has also met resistance. Among other things, psychiatric diagnoses are not as stable as one would hope, and because of the complexity of illnesses, there may be too much room for measurement error, thereby affecting diagnoses. Others have argued that too much emphasis was placed on biological measures and that some behavioral phenotypes were not necessarily a direct result of biological dispositions. Moreover, even if they were, they might vary with many other variables, so identifying all of these might be virtually impossible.

Considerable attention has been devoted to identifying specific genes that might serve as causal factors or markers of many disease conditions, but many problems are likely to be encountered, even with diseases that are seemingly less complicated than those involving the brain. The assumption that genetically based research approaches will take us many steps forward in treatment may be perfectly reasonable, and it could drive precision medicine and the identification of new targets for treatment. The cornerstone of precision medicine has been that large population studies might be able to identify specific associations between genotypes and phenotypes, but with only a few exceptions, these association studies have a limited capacity to predict specific phenotypes in any given individual (Khoury and Galea, 2016). Indeed, as more and more information has come forward regarding the link between genes and disease, it has become eminently clear that the initial hopes might have been a tad optimistic. Aside from complex pathologies involving multiple genes, their functions are not always fixed in stone, and deletion of specific genes may not produce expected outcomes. However, this does not necessarily imply that precision medicine cannot take forward steps. As much as it may be a cliché, humans are not wholly

TABLE 1.3 The RDoC Framework, Listing the Units of Analysis (X-axis) and the Behavioral Domains (Y-axis), Provides a Matrix That Can Potentially Be Used for Predicting Psychological Disorders and, Based on the Matrix of Features, Can Also Serve to Predict the Efficacy of Treatments and Illness Recurrence

The RDoC Matrix

	Unit of Analysis							
	Gene	Molecules	Cells	Circuits	Physiology	Behavior	Self-Report	Paradigms
Domain (Construct)								
Negative valence systems								
Acute threat								
Potential threat								
Sustained threat								
Loss								
Frustrative nonreward								
Positive valence systems								
Approach motivation								
Initial responsiveness to reward								
Sustained responsiveness to reward								
Reward learning								
Habit								
Cognitive systems								
Attention								
Perception								
Declarative memory								
Language behavior								
Cognitive control								
Working memory								
Systems for social processes								
Affiliation and attachment								
Social communication								
Perception and understanding of self								
Perception and understanding of others								
Arousal and regulatory systems								
Arousal								
Circadian rhythms								
Sleep and wakefulness								

Based on Cuthbert, B.N., Insel, T.R., 2013. Toward the future of psychiatric diagnosis: the seven pillars of RDoC. BMC Med. 11, 126.

defined by their genes and instead reflect interactions with multiple experiences, cognitions, wishes, desires, and goals.

This does not necessarily imply that research attempting to link genes and pathology should be abandoned, but it does suggest that we rethink our approach to the processes by which diseases occur and how to treat them. In this regard, a large-scale analysis based on whole exome sequencing (in which sequencing is performed of all the expressed genes in a genome), which was linked to an individual's electronic health records, was able to identify some gene variants that might be tied to illnesses (Dewey et al., 2016). Yet, as most diseases involve multiple gene variants or gene × environment interactions, these approaches can go only so far. Added to this, several distinct gene variants can share common biological functions and thus affect the same disease conditions (Li et al., 2016).

Each of hundreds of genes may contribute to specific diseases, individually accounting for only a small portion of the variance, making it exceptionally difficult to target specific pathways for treatment or even to use genes as biomarkers. Aside from complex illnesses involving multiple genetic contributions, many illnesses may vary in relation to epistatic interaction, referring to one gene having effects on the actions of a second gene or that environmental and social factors may interact with the actions of a gene. We will see that this is precisely the case for many disorders, and understandably, identifying these interactions is unwieldy even in large-scale studies, making a personalized approach orders of magnitude more difficult. For example, using a large cohort of individuals who self-reported whether they had been depressed, numerous genetic variations were identified that might contribute to the illness (Hyde et al., 2016). However, the contribution of some genes to depression may be negligible if interactions with varied experiences, such as current or early-life stressful experiences, are not considered. The very same issues are faced in dealing with other disorders, such as obesity, hypertension, some cancers, and most certainly, other mental illnesses. It ought to be mentioned, as well, that even leaving genetic factors aside, personalized treatments can still have benefits. For instance, features of illness, such as sex and body mass index, and some symptoms expressed, could be used to predict the efficacy of antidepressant treatments.

Yet another issue that warrants consideration is that many genes may have pleiotropic actions. That is, a particular gene can influence two or more phenotypic traits. These pleiotropic actions can occur in parallel or in series. In the latter instance, a given gene can have a particular effect that leads to biological outcomes and several downstream changes that culminate in a pathological condition. Thus, treatments that target one or more steps on this string of biological factors could potentially modify the pathological conditions. However, a parallel pleiotropy could be present, in which a gene comes to affect different neurobiological pathways directly or indirectly and hence provoke more than a single pathology, but these pathologies may be independent of one another. Once more, this makes it difficult to identify which treatments may be best to treat one or another condition.

It is widely accepted, of course, that prevention of illness may often be more effective than treating illnesses once they have occurred. A fundamental benefit of a precision medicine approach is that it might lead to detailed descriptions of risk factors for diseases, and hence programs can be created to prevent illnesses (Khoury and Galea, 2016). Here, however, we face the problem that even when individuals are aware of the problems they could encounter (e.g., being part of a genetically high-risk group), they often do not adopt behavioral changes necessary to preclude illness (Hollands et al., 2016).

This brings us to a critical factor that is incorporated into the RDoC approach that focuses on mental illness but is often overlooked in some aspects of precision medicine involving other disease conditions. For many illnesses, the amount of variance accounted for by genetic factors is considerably smaller than the contribution of behavioral, social, and experiential factors. Too often, however, these are not considered adequately in precision medicine. To what extent do diet, stressful experiences, as well as prenatal or early postnatal factors, and many other psychosocial variables come to affect disease conditions, and does having answers to this have implications for illness prevention and treatment? As we will see in ensuing chapters, it most certainly does, and consequently, having this information may inform best practices.

There is, indeed, a constellation of factors beyond those described in Table 1.3 that are important considerations in the evolution of an illness and the effectiveness of treatments. Kirmayer and Crafa (2014) made the important point that the RDoC approach fails to consider culturally and historically based phenotypes. It was argued that "normality" in relation to mental illness is defined based on what is statistically common or average in a population. Yet, what might be considered "normal" in one culture may not be in a second, and the most effective therapeutic strategies that are best may similarly vary across cultures.[3] They also indicated that the interactions that occur between biological, developmental, and

---

[3] For Indigenous Peoples, traditional medicines may be desired, or better still, a "two-eyed seeing" approach may be more appropriate. This view, first advanced by Mi'kmaq elder Albert Marshall, holds that combining the strengths of Indigenous knowledge and ways of knowing, with the strengths of Western knowledge may be most effective in many contexts, including health delivery.

psychosocial processes that define normal versus pathological conditions are driven by cultural factors and may be dependent on the context in which measurements are made (e.g., concerning Indigenous Peoples vs. Euro-Caucasians). Accordingly, it was argued, as we have earlier (see Fig. 1.4), that a multidimensional approach to understanding health ought to consider a fine-grained analysis of individual experiences, illness narratives, the dynamics associated with interpersonal interactions, and the social contexts in which these occur.

In considering treatment strategies, it is also important to ask whether brain neurobiological processes are reciprocally related to social support and trust and thus will contribute to psychological disturbances and the response to treatment. For instance, do Indigenous People trust non-Indigenous physicians, or do African American patients have greater trust in African American physicians relative to Caucasian health providers? Numerous issues are relevant to how patients fare, but often, these considerations are given short shrift.

Culture and personalized medicine

Culture, which can be influenced by social norms and values, geographical conditions, and environmental exposures, may influence the development and course of disease and may have a profound impact on the efficacy of treatments. It is broadly thought that culture is rooted in ancestry (systems of knowledge), collective history (including trauma), and evolving environmental and social contexts (climate change, colonization, and migration). In essence, culture is a dynamic construct that changes over time (cultural evolution), being modified by a constellation of factors, such as social structures, relationships, beliefs, activities, diet, multiple environmental factors, and the adaptations that these require. The significance of culture varies based on the extent to which individuals see themselves as reflecting their cultural norms. Importantly, as well, is their sense of belonging to the group, their feelings of pride in their group, and the collective esteem it provides (vs. shame, resentment, or anxiety). Together, these aspects of their identity determine whether individuals adopt their culture to deal with adversity (e.g., in seeking social support in the mobilization of collective actions). In essence, cultural factors are instrumental in shaping an individual's core identity, thereby affecting resilience in the face of numerous challenges. Thus, in considering a personalized approach to dealing with pathologies, cultural factors ought to be a key ingredient (Matheson et al., 2018a).

The notion of personalized medicine, despite its presumed limitations, has advanced several steps in recent years. The "high-definition medicine approach" relies on newly developed technological methodologies (e.g., DNA sequencing, advanced imaging, physiological and environmental monitoring), behavioral tracking, and information technologies to promote sophisticated treatment strategies (down the road, these might include genome editing, cellular reprogramming, tissue engineering) (Torkamani et al., 2017). Getting simple precision medicine off the ground has been enormously difficult, and the next step may be a still greater challenge.

Aside from such efforts, the precision medicine approach, which focuses on individual patients, has been extended to apply to whole communities through the application of public health interventions (Khoury et al., 2016). The medical problems stemming from industrialization, overcrowding, various pollutants, new emerging diseases, and many other factors, have grown far too large and too broad to be effectively handled with current approaches. Thus, much greater attention ought to focus on prevention (rather than just treatment), and health efforts need to be addressed at the community level. Related to this, a P4 approach was suggested, which entails four components: predictive (indicating early predictors of illness), preventive (intervening before an illness appears), personalized (identifying markers that are most effective in treatment), and participatory (such that patients have a voice in their treatments, as do physicians and health workers) so that policy and implementation of strategies are enhanced (Hood and Auffray, 2013).

Despite the current limitations, it may be possible at some point to develop appropriate treatment strategies for each individual. Important to this is the understanding that genes likely do not link with one another in any simple fashion. In addition to a regulatory gene influencing the actions of a second gene, networks exist that may be orchestrated by a common element or set of elements, and diverse regulatory elements may interact with one another (Waszak et al., 2015). Mental illnesses and their links to inflammatory processes might represent the interactions among multiple gene networks that control varied symptoms whose presence might provide clues for effective treatments. These gene networks, of course, are likely connected to multiple psychosocial determinants (including context and environmental contributions) that additively or interactively operate with genetic factors in determining the development of various illnesses and influence the best treatment strategies to adopt. Any optimism, however, needs to be tempered given the multiple issues that need to be addressed. It may simply be that at the moment the issues are too complex (and too expensive to adopt) and are still not workable within the primary care context.

Concluding Comments

Mental illnesses are typically viewed as those that involve disturbances regarding how a person feels, thinks, perceives, or acts. These disorders range from disturbances of mood, anxiety disorders, PTSD, schizophrenia, personality, sleep, eating, and sexual disturbances, as well as developmental and conduct disorders. Mental illnesses involve multiple mechanisms; glial and neuronal processes, and there is little question that hormonal, neurotransmitter, neurotrophic, immunological, and microbiome-related processes play into these disorders. Ample data have indicated that mental illnesses are frequently accompanied by organizational changes within the brain and serious mental illnesses may be accompanied by variations of connectivity as well as structural brain changes, including that of the hippocampus, prefrontal cortex, and amygdala. These conditions are most often treated through drugs or various cognitive and behavioral therapies, although other options, such as transcranial magnetic brain stimulation, electroconvulsive shock, and deep brain stimulation (electrical pulses to brain regions in which electrodes have been surgically placed) have been used for some of these conditions.

The diverse features of mental illness are reminiscent of several characteristics that might be evident in "neurological disorders" that involve structural, biochemical, or electrical abnormalities in the brain, spinal cord, or other nerves. The brain changes can be accompanied by symptoms, such as altered levels of consciousness, confusion, paralysis, seizures, poor coordination, muscle weakness, pain, and loss of sensation. Impairments of the brain and mind both comprise neuronal disturbances and are modified still further by psychological and social processes.

Many scientists coming from either a neurological or a psychiatric domain, working with animal models or with patients, have focused on some of the same brain disorders (e.g., epilepsy, Alzheimer's disease, Parkinson's disease, stroke, brain trauma, amyotrophic lateral sclerosis, brain tumors, migraine, multiple sclerosis, lupus erythematosus, autism, and attention deficit hyperactivity disorder). Clearly, there is appreciable overlap regarding the research conducted within neuroscience, neurology, psychology, and psychiatry, making it especially curious that, to a considerable extent, they have existed in independent silos. Moreover, crosstalk and collaborations occur less frequently than one would have expected, although there have been repeated calls for better integration between these fields (Martin, 2002).

Mental disorders are governed to varying degrees by genetic contributions, variations of neurotransmitters, hormones, and growth factors, together with microbial and immune alterations, and they are all modified by experiential and environmental factors. The actions of inflammatory processes and microbial factors on mental illnesses are latecomers to the game, and immune functioning as a player in this regard met considerable resistance, although it seems that the microbial contributions were more readily accepted. Perhaps the ice had been broken with the widespread perspective of inflammation and disease, coupled with the demonstration that microbiota and immune functioning reciprocally innervated one another, and both affect brain processes, making for an easier transition to mental illnesses. Of course, the zeitgeist regarding (w)holistic medicine may have contributed to the acceptance of complex interactions concerning physical and mental illnesses.

CHAPTER

2

The Immune System: An Overview

Protection in a scary world

Immunology as a discipline is a relatively young field. However, given that the principle of immunology is the protection of the host against nefarious biological forces—germs—humans understood well before white blood cells captivated our imagination, the power of contamination, and the incapacitation that arises from spoiled foods, sickly animals, and febrile children. After all, human civilizations came and went not merely because of the expansionist behaviors of emperors and kings but because of plagues that decimated millions of people and altered the geopolitical landscape wherever infection arose. The conquest of Central and South America owed as much to the brutality of the Spanish armies led by Cortes as they did to the foreign microbes they brought from Europe. The lack of immunity among the indigenous peoples led to thousands dying weekly, thereby weakening resistance against the conquistadors. Such historic examples inspired authors like HG Wells to utilize this trick of nature in his classic novel, War of the Worlds, where humanity was rescued by earthly microbes to which the alien invaders had no acquired immunological defense. Of course, the converse did not apply, as it was evident that the Martians did not harbor any extraterrestrial viruses or bacteria that might have vanquished human life on Earth. (Wells made sure to emphasize that the Martians were "clean," having rid their planet of the pesky microbial underworld.) Interestingly, and quite wisely—for all our sakes—NASA has strict decontamination procedures before bringing astronauts back to earth. And, of course, modern travel is strictly regulated to ensure that infectious agents do not cross borders. Such was not the case in the past when the passengers on ships were disease-carrying rats and mice.

But let us focus less on populations and more on individual behavior, as well as on the natural connection between the immune system, behavior, and emotions. One of the strongest human emotions is disgust. It is accompanied by a distinct facial expression in which the nose and mouth are seen to be compressed, and perhaps the tilt of the head

and neck has an avoidant slant as if to distance the person from the source of disgust. The reasons are obvious, as the communicative and personal feelings generated by disgust involve rejection. Much has been written about the biological basis for this emotion, and evolutionary thinking has attributed the fear of contamination as a primary factor. Odious smells and tastes elicit the tell-tale arrangement of facial muscles and feelings associated with disgust as if the offending stimulus is being expunged. It is a rapid learning experience, a basic one-trial learning of place and/or food aversion. The lesson for the individual is that contact and interaction with the inciting stimulus is biologically dangerous. Live snakes incite fear. But a dead snake, decomposing and crawling with maggots, incites disgust. Here, therefore, is the essence of immunological protection, operating in behavioral terms.

Humans and other organisms evolved what seem to be innate behavioral responses to unpleasant odors. When food spoils, the nose is a reliable ally, an expert judge of palatability. Not surprisingly, the olfactory system and connected circuitry in the brain are strongly linked to emotion-generating neural substrates (e.g., piriform cortex and amygdala). Therefore, olfaction, as a sensory system that detects the chemical components of the environment, is a prominent mechanism for avoiding potentially dangerous substances. In performing this function, the olfactory system is now recognized for pattern recognition, very similar to the innate cells of the immune system, which also discriminate between various pathogens based on recognition of specific molecular patterns on "germs." We know this now, but what we understood behaviorally for centuries—that meat that smells "off" can kill you—did not extend to deeper biological scrutiny until the 19th century. Analysis of what it is that is "off" and how those who consume tainted food came to survive engaged a variety of intellectuals who sowed the seeds for the greater discoveries of the 20th century regarding macrophages, lymphocytes, antibodies, and cytokines. So we turn now to this team of players who protect and defend against certain illnesses.

The Immune System
https://doi.org/10.1016/B978-0-443-23565-8.00017-X

27

Copyright © 2025 Elsevier Inc. All rights are reserved, including those for text and data mining, AI training, and similar technologies.

Introducing Immunity

The immune system is a network of blood-derived cellular operations that orchestrates protection against the microbial world.[1] The mechanisms involved in conferring this protection constitute *host defense*, the "host" being the organism in which the immune system resides. Most vertebrate and invertebrate organisms have a system of defense against outside invaders, with the vertebrate immune system providing the complexity and sophistication that emerges in highly evolved mammalian species, such as man. However, rodents, especially mice, have served as the main model for our understanding of immunity, even though recent evidence suggests that translating our knowledge of the murine immune system to humans is not always a simple matter (Beura et al., 2016). Nonetheless, whether we are looking at the immune systems of a mouse or a human, what we see is an intricate network of cells with considerable heterogeneity of function and a strategic anatomical organization. Our overview will focus on the essential elements necessary to appreciate how immunity works. A major goal is to emphasize key principles of immunological functioning, introduce the cast of cellular actors, and offer various details in (hopefully) measured and digestible amounts that will inform later chapters and render them meaningful. To the newcomer, the "stuff" of immunology may seem like a confusing array of different types of cells and molecules that interact in specific locations and for all sorts of different reasons. If, at times, it seems bewildering, rest assured it is! A diffuse cluster of cells circulates through the body at least once each day, in search of microbial trespassers. This part is easy to comprehend. But as with anything, the challenge is in the details.

The immune system is very much a "dynamic" system, a term often used to emphasize a state of recurring functional change. Within this dynamism, there are different levels of functioning. Where a microbial invasion has taken place, the nearest location of available immunological reserves, and over time, the makeup of the cellular and molecular elements that make up an "immune response," are all determinants of a highly coordinated network of signals and cellular interactions. And not just between the cells of the immune system. Movement and migration of cells are also dependent on the local tissue environment and the molecules expressed therein. It is only now that we have come to understand that for this to work normally, the immune system has to rely on communication from neural and hormonal sources. In Chapter 7, Stress and Immunity, in which we discuss the impact of stress on immune function, we explain this interplay of neural-immune interactivity a little further.

This sense of the immune system as a dynamic set of cells working its way around the body in search of danger stands in contrast to the vision we have of the brain. The complexity of this hallowed organ is unquestionable, but it is fixed. The key cells (neurons) trapped in a network of synaptic connections and molecular signals are often compared to a computational system and a switchboard of lights and circuits. To follow the activity of these circuits is no cakewalk, and we are still figuring out how they all operate simultaneously to give rise to ongoing conscious life. Still, information received by the brain is bounced around quickly (in milliseconds or less) and results in rapid decisions and actions. In contrast, the immune system processes information on the run, as cells circulate through the blood, exit into tissue (i.e., extravasate), traverse the tissue, and then return to the blood via specialized vessels that constitute the lymphatic system. A circulating lymphocyte literally has to hang on for its life-extending cellular spindles that latch on to adhesion molecules like a rock climber seeking a foothold—as it rides the turbulent waves and shear stress of pumped blood and seeks to hang back and enter particular tissues to sniff out a vagrant microbe (Abadier et al., 2017).[2] It is all in a day's work, as the immune system tries to contain infection and keep the host healthy.

In making these introductory points, two important issues come to mind. First, the immune system is—like the brain—an *information-processing* system. The second is that by virtue of its inability to sit still, the success of the immune system is based on mobile *surveillance*. As such, it gains access to every nook and cranny of the body,[3] and as with any good information-processing system, it acts on local information in a judicious and effective manner. To this extent, the immune system has to generate action (or, more accurately, reaction), which immunologists refer to as "effector" function. The main

---

[1] In fact, the term "immunity" is derived from the latin word, *immunitas*, which referred to the protection from prosecution (and most likely persecution) that Roman senators enjoyed in ancient times. In legal circles, this term is still used today to signify the pardons and/or light sentences that witnesses may receive for their testimony.

[2] A recent study (Abadier et al., 2017) demonstrated that within blood vessels, mature T cells display a remarkable and highly energetic ability to swing out membraneous protrusions like lassoes or slings, which tether the cell to endothelial adhesion molecules (e.g., selectin). This enables the cells to roll in stable fashion while the blood rushes past within a blood vessel. This operation is similar also to another cell type, the neutrophil, which grapples the "rapids" in similar fashion. The end result is that immune cells can enter nonlymphoid tissue and explore the area for signs of damage or infection.

[3] Historically, there were some exceptions to this rule, in which immune cells were thought to ignore the brain altogether (thereby giving the brain the apparent distinction of being "immune privileged"). But, as we have known for some time—and this book points out—that is simply not true. Leukocyte infiltration of the brain is quite common, both under pathological and normal conditions.

effector functions are cytokine production, cell killing (i.e., cytotoxicity), and antibody production. The cells that mediate these functions are simply called *effector cells*, although each class of effector cells belongs to a particular lineage of immune cells that, on activation, differentiate into a particular effector type. For example, B lymphocytes make antibodies; however, when a B cell has achieved a fully differentiated effector status, it is called a *plasma* cell. In order for immune cells to evolve into an effector state, they require that the information that triggers this transformation fit a particular profile. This information needs to be a specific type of stimulus, one that is cellular and/or molecular in nature and is also identified as being nonself or foreign (i.e., not part of the host). Generally, we think of this stimulus as a microbe—a virus or bacterium. However, in the parlance of immunology, such a stimulus is referred to as an *antigen*, which we discuss in more detail shortly. It is the cognate stimulus for the immune system and elicits the classical type of immune response, in which there is cellular proliferation, development of effector function, and eventually, a denouement and return to a smaller cadre of memory cells.

This is the classical view of how an immune response progresses. It is induced through antigenic stimulation and is part of what is called the *acquired* or adaptive immune system, which culminates in the production of billions of antibody molecules that have highly specific molecular targets on microbial entities (each of these molecular targets is an antigen).

Many of us are now familiar with the spike proteins expressed on the SARS-COV-2 virus, as we describe in Chapter 3. Although viruses consist of many more proteins than this, it turned out that isolating the SARS-2 spike proteins and incorporating them into vaccines, conferred protection against the virus. The spike proteins in this case were antigens, and their easy accessibility (because they are on the surface of the virus) ensured that circulating antibodies would find them. An alternate immune repertoire, known as the *innate* immune response, involves less refined and more global effector arms, and is less concerned with specific antigens than it is with molecular categories (or molecular signatures). One such category includes the so-called pathogen-associated membrane patterns (PAMPs) (Medzhitov, 2009). Recognition of PAMPs in bacteria engages cells of the innate immune system, which automatically generate molecular and cytotoxic responses that can initially decimate the number of invading microbes but, in the process, will recruit cells of the acquired immune system—the T and B lymphocytes. Even if an infection is largely headed off by a robust initial response by the innate immune arm, adding lymphocytes to the fray will result in the making of probably the main effector molecule with which the immune system is associated: Antibody. In the absence of antibodies, infection is likely to be prolonged and unresolved. The antibody attaches to viruses and bacteria that induced them, which tags or flags them for destruction. Without this tagging, pathogens can continue to circulate and multiply, evading the lethal effects of cells in the innate compartment of the immune system. However, with antibodies bound to the microbe, these innate immune cells can attach to the antibody and use this to immobilize the microbe long enough for it to be destroyed. In a scenario like this, the antibody is said to have contributed to "opsonization."

More formal descriptions of acquired and innate immune processes will be provided in the coming sections. As we will see, the multiple types of cellular and molecular interactions that each involves (in a nonmutually exclusive manner) follows a set of commands and instructional guidelines that are determined by genetics, cell-surface molecules, secreted signaling proteins, and various other nonimmune cells (e.g., endothelial and epithelial cells) that can flag down and/or facilitate immune responses. While the key players of the immune system originate from stem cells in the bone marrow (see Fig. 2.1), important functions can be attributed to the stromal and epithelial cells of tissues and the endothelial cells of blood vessels. These represent key sources of information that can guide and recruit immune cells to sites of damage and infection. At these locations, the presence of immune cells forms the basis of tissue-specific inflammatory responses, a major consequence of immunological activity.

The Immune System Consists of White Blood Cells: Leukocytes

The main constituents of the immune system are leukocytes (white cells). In an adult human, a liter of blood typically contains 7–8 billion white blood cells that are in transit, coming from all parts of the body and recirculating back to the same sites as part of a massive and unceasing surveillance operation that persists throughout an individual's lifetime. The word, "leukocyte," is a collective term for a heterogeneous group of cells: Lymphocytes, monocytes, granulocytes (e.g., neutrophils), dendritic cells (DCs), macrophages, and many more (these cells are described further in Fig. 2.1). To say that all leukocytes do the same thing, because they are involved in defending against infection, is to miss the unique contribution that each type of cell brings to the fight against microbial invaders. As a collective, however, they work together in a highly organized and coordinated manner to defend the body against infection. If there are any other purported functions, these are ancillary to this key purpose of protecting the tissues and organs against damage from microorganisms.

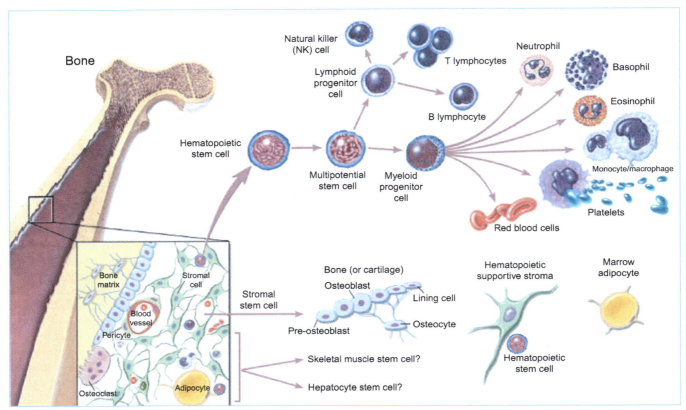

FIG. 2.1 Formation of cells of the immune system from hematopoietic stem cells. As explained in the text, leukocytes originate from a bone marrow-derived multipotent stem cell that itself was derived from a hematopoietic stem cell. The multipotent stem cell differentiates into two sets of progenitor cells that constitute the lymphoid and myeloid cell lineages. The latter gives rise to cells of the innate immune system, while lymphoid cells develop into T and B lymphocytes, as well as Natural Killer cells (large granular lymphocytes), which can kill virus-infected cells and tumor cells. *From https://stemcells.nih.gov/info/Regenerative_Medicine/2006Chapter2.htm. © 2001 Terese Winslow (assisted by Lydia Kibiuk).*

When thinking about the immune system, the metaphor of a military operation often comes to mind. Leukocytes are the sentinels of the body, alert to any form of disruption to the host, and when this happens, they can mobilize quickly and mount an intense campaign through sophisticated forms of molecular signaling. This generates a rich armamentarium of degradative enzymes, highly specific antibody molecules, a plethora of regulatory cytokines, and exponential rates of cellular replication that will ensure that these molecular tools are in abundant supply. These events are not without their collateral impact on the health of the host organism. However, the efficiency with which the immune response is generated and ultimately winds back to a quiescent state ensures recuperation, recovery, and a return to a healthy state.

The evolution and refinement of such a system resulted from pressures that arose due to the manner in which mammalian organisms interact with their environment. There are multiple portals of entry that microorganisms can slip through, and given our need to eat and procreate, oral and genital surfaces are placed at increased risk of microbial encampment and opportunistic entry into the internal milieu for more extensive colonization. Anatomical surfaces that serve sensory functions are also vulnerable, including the eyes and ears, while the skin and its regular cycles of shedding and regeneration limit any major breach, absent a cut or wound that provides a tempting opportunity for airborne and topical (present on the skin) microbes to transcend the epidermal surface and feast on the dermal matrix of cells and fluids. But it is here that classic descriptions of inflammation emerge, describing the redness and swelling indicative of the early stages of an immune response. Tissue damage engages local cells to generate chemokines, signaling molecules that flag down the leukocytes circulating in the blood. The cells extravasate (leave the blood) and roll into the damaged or infected tissue, beginning the process of engaging any extant pathogens while also helping to rebuild the damaged site. These events are highly orchestrated and, in the case of, say, damaged skin ensure that restoration of the epidermis and the underlying dermal and subcutaneous tissue progresses quickly and efficiently so that the risk of infection is eliminated.

The best way to appreciate the nature and function of the immune system is to consider where leukocytes congregate. After all, most biological systems operate within an organ. Gastrointestinal (GI) cells are part of the GI tract, alveolar cells, the lung, and myocardial cells, and the heart, which drives the cardiovascular system and the network of blood vessels that permeate all parts of the body. The latter is critical to our understanding of the immune system and immunity since irrigation of the body with blood not only ensures that all cells will receive molecules necessary for energy (e.g., glucose) but immune cells will be transported to most regions of the body where they are needed to perform defensive and reparative functions. For this reason, the stem cells that originate in the bone marrow, and which are destined to become immune cells, are referred to as *hematopoietic* (blood-forming) cells.

Types of Immunity

As mentioned earlier, immunity from infection is accomplished through twin avenues of cellular reactivity: (1) natural or *innate* immunity and (2) *acquired* (or adaptive) immunity. These are implemented in a staggered fashion, but, in reality, form a cooperative network of interactions that ensure thorough and complete elimination of invading pathogens. The innate (sometimes also called *native*) form of immunity is performed by established defense mechanisms that generally have a low threshold for discriminating between different types of infectious agents and do not show changes in magnitude nor intensity of responsiveness with repeated exposure to the same infectious microorganism. However, these notions of discrimination and unvarying stability of performance should be qualified. For the most part, innate immune responses respond to broad categories of potentially infectious microorganisms. That is, while more refined and surgical forms of attack are the role of cells that are part of the acquired/adaptive immune response, the innate immune system was always perceived as attacking any old microbial organism. However, this notion has been rewritten, as we now know that innate immune cells do possess selective recognition capabilities.

A related concept is memory. Inherent in the term "immunity" is the notion of continued remembrance of events that have passed, for which a form of protection is owed (as befits the original use of the Latin word *immunitas*). The innate immune system does not bother with memory formation. It typically sees things in the same way every time (although see the discussion that follows). Like someone with short-term amnesia, a conversation can be recycled repeatedly, as fresh and alive as it was the first time. So, to some degree, to an innate immune cell, one bacterial cell looks like any other bacterial cell, although in more recent years, we have come to understand that the innate immune system has a categorical manner of recognition, through the recognition of PAMPs. Thus, the responses are consistently the same, catch, kill, and destroy. The same goal is evident for cells of the adaptive immune system, although their role is to be more selective about who gets the chop, and who should have known better about coming back. These cells are bouncers with strong cognitive capacities; they learn and remember (although they are not as beefy as innate immune cells, which tend to be 50% larger than adaptive immune cells). We will have more to say about innate immune cells and their microbe recognition capacity later.

The alternative view of acquired immunity as *adaptive* reinforces the notion of some form of biological improvement over time and experience with particular challenges. It is changing for the better. With acquired immunity, there is specific recognition of a given foreign molecule (antigen) by unique cellular and molecular components of the immune system (viz., receptors), and this recognition (or information) is stored for later retrieval should this foreign molecule ever return. This is a cornerstone of immunity in vertebrate organisms. The chief cells and molecules involved in this specific form of recognition are the *T and B lymphocytes* (also called T and B cells), and the antibody molecules produced by B cells, which we mentioned earlier. It is this form of immunity that reveals the vertebrate immune system to be closely aligned with that of the central nervous system: Both are responsible for information processing, storage, and consolidation of that information, as well as its retrieval. For the adaptive immune system, this information is molecular. When we say that the T and B cells recognize a specific molecule, we must remember that, for example, each bacterial cell will present with a multitude of molecules (or antigens). The hundreds of millions of lymphocytes that are present in the vertebrate organism actually have the ability to differentiate between each of these. It is as if the bacterial cell has become pixelated, and each lymphocyte has the task of recognizing a given unique pixel. In this way, lymphocytes can identify given microbial entities, whether they are viruses or bacteria, and differentiate among them in ways the innate immune cells cannot. This specific form of recognition is what allows lymphocytes to pounce on a returning virus or bacterium much faster and with greater force. We will discuss this notion of specificity shortly and introduce the clonal selection theory, which served to explain the existence of this fine-grained nature of differentiation.

What Actually Induces Immunity?

Generally speaking, immunity is formed against (potentially) infectious microorganisms. But, in practice, as already introduced earlier, the immune response is generated against specific molecules (antigens from the term antibody-generating) that are expressed by microorganisms. In most cases, the chief target molecule needs to be a protein. These molecules possess configurations (or strictly speaking—amino acid sequences or short peptides) that the immune system recognizes as foreign (nonself). A single bacterial cell or virus may express many hundreds or thousands of different antigens, against which the (adaptive) immune system can respond. However, for convenience, when speaking about a given immune response, it is common practice to refer to a bacteria or virus as an "antigen" or antigenic stimulus. For this reason, discussion of antigens can further be broken down into *antigenic determinants* or *epitopes*—specific portions of the conceptual antigen (the typical bacterial cell or virus) that can be bound by an antibody. We will return to this topic later as it addresses a fundamental concern among immunologists—*immune recognition*. Just what do immune cells "see" when they decide to mount a response and enter a battle that potentially may have harmful side effects through inflammation? For example, speaking simply of "antigens" or "pattern recognition receptors" (PRRs)—see section on innate immunity—may not be a sufficient condition generating an immune response. Kagan (2023) has suggested that firing up the immune system may rely first on pathogens making mistakes as they attempt to colonize the host, after which their presence is detected, alarm signals go off, and the full force of the immune armamentarium is called to action.

The Cells and Organs of the Immune System

We noted earlier that the cells of the immune system are a heterogeneous group of cells called leukocytes. As Fig. 2.1 shows, these cells all originate in the bone marrow from a common progenitor cell called the pluripotent hematopoietic stem cell (HSC). These cells differentiate under the influence of locally produced soluble factors into various progenitor cells, two of which give rise to the *lymphoid* and *myeloid* cell lineages. The latter gives rise to cells of the innate immune system, and among these, the most prominent are neutrophils, monocytes, and macrophages. These cells will mature in the bone marrow before entering the blood and localizing in various tissues (e.g., the tissue macrophages).

The lymphoid lineage emerges from the lymphoid progenitor cell and gives rise to T and B lymphocytes, as well as additional cells, such as the large granular lymphocytes (LGLs) that are associated with cytotoxicity functions, and have come to be called Natural Killer (NK) cells (see Fig. 2.1). Of the lymphocytes, only the B cell matures in the bone marrow (and also initially in the fetal liver). In contrast, the T cell initially develops in the bone marrow, but at an early immature state, it exits to enter the circulation, whereby it is sequestered in the thymus.

In the thymus, the cells encounter various influences that drive them to a fully differentiated and mature state, whereby they are able to leave the thymus and populate other lymphoid organs. In their mature state, T lymphocytes are equipped with the molecular tools to allow recognition of the host, which allows them to discriminate between the molecular information of self and nonself. Maturity, therefore, is in part the capacity to respond aggressively only to that which is foreign. Failure to accomplish this results in autoimmune disease. Consequently, events in the thymus are critical to the development of a fully functional T cell repertoire.

The Innate Immune System

We have already provided a sense of what the innate immune system represents. This is a collective term for a system of immediate defenses against the microbial world, and strictly speaking, it can begin with actual physical barriers, such as the skin and mucous secretions (e.g., secretions of the eyes and nose, the phlegm of the lungs, and the mucus lining of the inner intestine). The mucosal surfaces—which secrete a sticky substance called mucous—include the inner epithelial linings of the upper respiratory tract, lungs, intestines, and anogenital regions. In a respiratory infection the phlegm expelled by the lungs of a coughing individual will contain a variety of particulate matter. Not only dust and dead cells, but trapped bacteria and viruses. Therefore, the mucosal line of defense restricts how much foreign matter can enter the internal milieu of the body. Beyond these "walls" invading microbes will need to contend with a second line of defense. These are the cells arising from the myeloid lineage of the hematopoietic system, and which form the innate immune apparatus. The myeloid cells that chiefly comprise the innate system are granulocytes (most commonly neutrophils), monocytes, and macrophages (see Fig. 2.1). Each is typically associated with mediating an inflammatory response by virtue of their ability to contribute to the build-up of regulatory molecules, fluids, and recruitment of other immune cells (e.g., lymphocytes) to a site of infection where they are typically engaged in a cleanup process ("phagocytosis"; literally, "eating up"), and are thus sometimes referred to as phagocytes. When they are engaged in phagocytosis, they are ingesting necrotic tissue and bacterial cells, and throughout this process, liberating humoral mediators of innate immunity, which includes lysozymes and complement, proteins that have cytotoxic capabilities (as we shall learn later in our discussion of antibody).

Additional soluble mediators are secreted which serve to amplify and expand the immune response by alerting the cells of the adaptive immune system. These mediators are cytokines and chemokines, which will both enhance further killing of bacteria and recruit (via the chemokines) other cells (lymphocytes and monocytes) to the site of activity. For example, consider the irritation of a splinter wedged in the skin. If the individual is unable to remove the splinter, the innate immune system will, over time, dissolve the foreign material. There will be redness and swelling, some pain and discomfort, but it will eventually be degraded. This is due to the infiltration of circulating monocytes, as well as activation of any resident macrophages in the dermal tissue, which will create an increased flow of plasma from the blood through the increased gaps between the endothelial cells of the blood vessels. The end result is edema (increased volume of interstitial fluid), which is the reason why an observable area of the body can enlarge dramatically (the swollen finger, foot, or knee) after an injury or infection. Therefore, when we speak of inflammation, we can refer to the clinical description (redness and swelling), but in actuality, the underlying basis of this is an innate immune response that has initiated an immunologic defense and increased leukocyte migration to the site of infection or damage. Moreover, inflammation need not be in external regions of the body but is an ongoing event taking place in the lungs of a smoker, an atherosclerotic cardiovascular system, a cirrhotic liver, an irritable bowel, or an arthritic knee. In all these cases, and many more, the increased presence of innate immune cells and their soluble products is an indication of inflammation.

Specificity and Memory in the Innate Immune System

Two major issues have preoccupied immunologists regarding innate immunity: specificity and memory. The first pertains to whether the innate immune system possesses the same specificity of lymphocytes. The traditional dogma was that innate immunity was nonspecific. However, there was always a nagging question: Just how did neutrophils and other granulocytes recognize a bacterial cell as something worth engaging and destroying? Moreover, when necrotic tissue needs to be removed, how did the innate immune cells manage to recognize that this is self-tissue worth disposing of? The answer to this was arrived at more than 20 years ago when it was proposed—and eventually proven—that innate immune cells expressed pattern recognition receptors (PRRs) for a variety of molecules expressed by microbial agents. These molecules include complex polysaccharides, glycolipids, and lipoproteins that collectively are referred to as PAMPs, although PRRs also recognize nucleotides and nucleic acids (Riera Romo et al., 2016).

The second question that has been posed relates to memory. Do innate immune cells possess memory? In contrast to the adaptive immune compartment, the innate immune system has often been described as having little (if any) memory for previous encounters with microbial agents. Each bacterial cell, whether from *Escherichia coli* or *Staphylococcus aureus*, is dealt with equally, and if infection recurs, the innate immune system will attack these bacteria with the same intensity and speed as it did the first time at least, that had been the long-held view. More recently, a growing body of evidence suggests that innate immune cells, such as neutrophils and NK cells, can respond to PAMPs and generate a short-lived memory for the encounter through epigenetic modification. The form of this memory (which was referred to as "trained memory") is metabolic, to the extent that innate immune cells can respond more strongly and efficiently to later infection, whether it is the same or a related microorganism. The research on innate memory is still on the exploratory ground, but at the very least, there is a sense that neutrophils and NK cells may be trained to be better prepared for future encounters with pathogens (Netea et al., 2015). This may not be the same as the highly specific responses and memory cell formation shown by lymphocytes, but it does suggest that learning is not exclusive to the adaptive immune compartment.

Sterile Inflammation

The induction of inflammation by innate immune cells is part of the normal immune response. While it is often thought to be something to attenuate and eliminate, there is the alternative view that inflammatory cells are also trying to correct areas that induced their activity in the first place. This is a recurring question that keeps popping up in the neural-immune literature, especially in relation to the "inflammatory" events occurring in the brain. How much is an effort to help, and how much help is eventually a burden? Inflammation is essentially a double-edged sword problem and one that has yet to be resolved. However, it was shown that neutrophils were instrumental in rebuilding damaged tissue incurred by thermal hepatic injury, a procedure that produced *sterile inflammation* (Wang et al., 2017b). The neutrophils, in this case, were critical to the reconstruction of blood vessels, which were compromised if these cells were prevented from accessing the damaged tissue. This finding inverted the long-held notion that sterile inflammation, and the presence of neutrophils and macrophages, can delay or interfere with tissue recovery. As such, it reinforces the idea that antiinflammatory treatment may not always be the best option.

Sterile inflammation is a concept introduced to account for cases when there is an accumulation of inflammatory cells and inflammation-related molecules in the absence of actual microbial organisms or infection (Chen and Nunez, 2010). The conditions for induction of sterile inflammation can include a range of factors, including physical injury (e.g., trauma), cancer, autoimmune tissue damage, ischemia, atherosclerosis, and exposure to toxins. Since inflammation is aroused in the absence of PAMPs, it was confirmed that innate immune cells have a recognition mechanism for detecting cells undergoing distress and/or death. The collection of signals generated by damaged tissue is now referred to as danger-associated molecular patterns (DAMPs), and these are recognized by PRRs on neutrophils and other innate immune cells. As can be expected with necrotic cells, liberated intracellular molecules function as DAMPs (e.g., chromatin-associated proteins, heat shock proteins, and purinergic molecules, such as ATP), as well as molecules associated with the extracellular matrix and fragments of enzymatically degraded proteins and certain proteases normally required for tissue repair (Chen and Nunez, 2010). Many of the studies investigating the impact of immune processes in psychiatric patients invoke the concept of inflammation, and, in so doing, may be directly or indirectly alluding to a state of sterile inflammation. However, in light of the findings by Wang et al. (2017b), it may be prudent to exercise caution in just how we interpret studies showing elevations in proinflammatory cytokines and/or circulating neutrophils and monocytes.

In having discussed PAMPs and DAMPs, we should specify precisely the nature of the receptors for these molecules. For some time, macrophages, neutrophils, and monocytes were known to respond to a molecule called lipopolysaccharide (LPS), which is derived from the cell walls of gram-negative bacteria, such as *E. coli*; as such, LPS is also referred to as *endotoxin* (endogenous toxin). This form of innate immune cell activation is a prominent model of immune activation in neural-immune investigations. In recent years, it has been discovered that LPS binds a molecule called Toll-like receptor 3 (TLR3). When this receptor is stimulated, innate immune cells show activation of a transcription factor, NF-kappa B (NF-κB), which results in the further transcription of genes for a variety of proinflammatory cytokines. This explains why LPS is a potent inducer of cytokines such as IL-1β, TNF-α, and IL-6, three prominent proinflammatory cytokines, produced by innate immune cells.[4] The family of TLRs, of which 10 were identified in humans, are now widely recognized as the major recognition receptors for PAMPs and DAMPs.

Further discussion of the innate immune compartment will be incorporated into our description of the adaptive immune compartment. It will be recognized that when innate immune cells engage and phagocytose microbial agents, their immunologic importance does not stop there. For instance, they serve an important and necessary role in ensuring the activation of T cells, and they will be important in capitalizing on the accumulation of antibody molecules generated by B cells. This highlights the fact that segregation of the immune system into innate and adaptive immune components is an arbitrary and temporary measure that helps to explain the workings of the immune system more effectively. However, the immune system, in and of itself, comprises both compartments that work together in a complementary fashion to rid the body of unwanted microbial invaders.

The Lymphoid System

Lymphocytes and certain cells of the innate immune system (e.g., monocytes) circulate throughout the body, performing what are largely surveillance functions. This patrolling behavior is more formally referred to as cell trafficking or migration and is driven by unique anatomical and soluble factors. Although many immune cells are constantly in transit, they nonetheless localize in well-defined and widely distributed organs, the lymph nodes. These nodes drain the local area that surrounds them, and as such, we have nodes located in various strategic locations (see Fig. 2.2). For example, the walls of the gut contain the *peyer's patches*, which are actually small lymphoid nodules that line the walls of the small intestine, and the cells in these regions travel (or drain) to the mesenteric lymph nodes (MLN). These comprise a pearl-like arrangement of lymph nodes that form a daisy-chain of interconnections and are linked to the

---

[4] Interestingly, the name Toll-like receptor is based on the Toll protein that was first identified and named in the fruit fly Drosophila. This protein was initially thought to be critical for the development of *Drosophila*, but eventually it was recognized to be just as important in helping them fight off infection. When the mammalian homolog of the Toll protein was discovered, and similarly identified with host defense, the name Toll-like receptor was adopted, ostensibly to honor the protein which inspired the discovery.

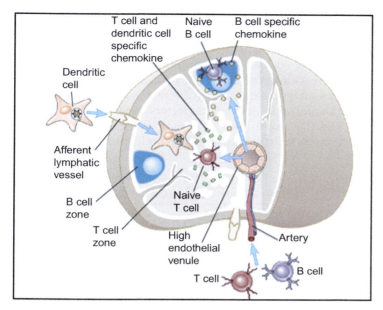

FIG. 2.2 Lymph nodes. The lymphatic system drains via the thoracic duct into the vena cava and into the general blood circulation. Lymphatics drain into and away from lymph nodes, which are distributed in gastrointestinal, abdominal, and cervical regions; additional nodes can be found around the tonsils, under the arms, regions emanating from the small intestines, and behind the knee (so-called popliteal nodes). The figure shows a typical node that displays a point of entry for an *afferent* lymphatic vessel entering the left side and transporting lymphocytes and other cells into the node. Within the lymph node, B and T cells are segregated into distinct locations. Lymphocytes can exit from arteries by reaching a high endothelial venule (HEV), which allows for easy extravasation, and then migrate to different areas of the node as a result of attractive chemicals called chemokines, which are produced in different areas of the node. Note also the migration of dendritic cells (DCs) through the afferent lymph, which captures antigens that enter via the lymphatics. Once the DCs enter the node, they move to the T-cell-rich areas, where they present the antigen and induce T-cell activation. *From Abbas et al., 2007. Cellular and Molecular Immunology, sixth ed. Elsevier Press.*

gut by the mesentery, a fan-like arrangement of connective tissue and capillaries that contain fluid and lymphocytes (this mix of fluid and lymphocytes is called *lymph* and the capillaries from the gut wall to the MLN are the lymphatic vessels). The upper respiratory tract also contains lymphatic portals to a daisy-chain arrangement of lymph nodes, as well as the arm pits, the pelvic and abdominal regions (inguinal lymph nodes), and also the rear of the knee (the popliteal lymph nodes). For example, stepping on a rusty nail that might contain tetanus toxin, will engage local immune cells that can drain to the popliteal lymph nodes. Experimentally, these nodes may swell in size after injection of antigen into the footpad of mice—a measure of delayed type hypersensitivity (DTH) (Smith and White, 2010). This swelling or expansion is due to activation of lymphocytes, resulting in further activation and proliferation (cell division) of the cells in the draining node. The nodes, therefore, represent regions where information gathered by immune cells in the surrounding area can be passed on to cells in the lymph node. This transmission of information (regarding the nature of the antigen) is carried out by a professional antigen-presenting cell (APC) referred to as a dendritic cell (DC). These cells encounter antigens in the tissues and then present this on their surface to lymphocytes. The close confines or packed arrangement of cells in the lymph nodes optimize interactions between lymphocytes and DCs, thereby leading to lymphocyte activation. Both T and B lymphocytes are activated.

In addition to the lymph nodes, the spleen is another major lymphoid organ. The spleen does not represent a drainage site for circulating lymphocytes in the tissues. Rather, it filters the blood and traps antigens and foreign cells that are circulating in the blood. The liver also performs the same function, the difference being that most of the immune cells in the liver belong to the innate component of immunity (viz., macrophages called Kuppfer cells). The spleen, however, contains a full complement of macrophages, DCs, and T and B lymphocytes. Because cells of the spleen are activated in response to an antigen that gains access to the circulation, this lymphoid organ is commonly used in experimental animal studies. This is also due to the fact that it yields a large number of immune cells that can be assayed for multiple parameters. Lymph nodes generally do not yield as many cells, especially when there has been no antigenic stimulation. As noted earlier, if the popliteal lymph node draining the foot of a mouse were collected without any antigen injection into the footpad, the experimenter would be hard-pressed to locate this node, let alone isolate enough cells to conduct useful assays. The node from an immunized mouse, however, would be plump and ripe with millions of lymphocytes. This very much describes the clinical presentation of swollen lymph nodes below and around the jaw in someone with an upper respiratory tract infection. Infection in the throat drains antigens to the local tonsillar lymph nodes.

The *primary* immune organs are the bone marrow and thymus since cells at early stages of development are found there. Once they develop and achieve a fully differentiated state of functional maturity, they will leave the bone marrow and thymus and circulate throughout the blood and spleen, lymphatic vessels, and lymph nodes.

The spleen and lymph nodes are considered to be *secondary* lymphoid organs since they contain mature circulating leukocytes. In this secondary physical terrain, the cells carry out the immune surveillance that ensures the protection of the body.

Circulation of lymphocytes throughout the body is continuous. A given lymphocyte, for example, can circulate and return to its point of origin at least once over a 15- to 48-h period, depending on retention rates in select organs (Ganusov and Auerbach, 2014). The trafficking of cells follows a particular pattern, whereby cells in the blood of major arteries gradually move into smaller blood vessels, such as arterioles, which irrigate the tissue, and then eventually into the vessels of the lymphatic system, a network of capillaries containing lymph[5] (Miyasaka and Tanaka, 2004). When the cells extravasate, they migrate through the tissue, sampling and responding (if necessary) to any signals present on or produced by tissue cells or other immune cells. The cells will then be carried forward into a separate (nonvascular) system of small capillaries that drain tissue fluids and push this unidirectionally through a one-way valve system into larger networks of capillaries (Liao and von der Weid, 2015). These networks of tissue-draining capillaries comprise the lymphatic system since they are filled only with interstitial fluid and leukocytes. Therefore, once in the lymphatic capillaries, the leukocytes have essentially left the tissue parenchyma. In the lymphatics (i.e., lymphatic capillaries), the cells will enter progressively larger lymphatic vessels, some of which will drain directly into lymph nodes. The lymphatic vessel that brings cells to a lymph node is called an *afferent* lymphatic, and the lymph carried forward through this is simply referred to as afferent lymph. To exit the lymph node, the cells are carried forward in efferent lymph and do so through an *efferent* lymphatic. When lymph nodes are daisy-chained (arranged in sequence, as shown in Fig. 2.2), each efferent lymphatic is essentially draining the cells toward the afferent lymphatics of the next lymph node in the chain. In time, the cells will leave the lymph nodes and drain via the efferent lymphatics into the thoracic duct, a large lymph-filled vessel that drains into the superior vena cava and directly into the heart and general circulation.

To summarize, the lymphatic system drains mostly lymph nodes. However, the spleen and bone marrow lack a lymphatic system. Interestingly, it was shown that the brain contains lymphatic vessels in the meninges (Louveau et al., 2017). This, however, appears to drain the contents of brain interstitial fluid or CSF, which also occurs through a *glymphatic* system, it is so-called because it involves glial cells orchestrating efflux of interstitial content out of the brain parenchyma (Plog and Nedergaard, 2017). Historically, this may be the basis for demonstrations of effective immunization against benign proteins, such as ovalbumin, that were delivered into the brain (Gordon et al., 1992). Interestingly, no evidence of afferent lymphatic input to the brain has been documented, although cells and soluble factors can enter the brain via areas that have a weak blood–brain barrier or are actively transported through endothelial cells (Banks, 2015).

The Acquired Immune Response

There Are Two Major Classes of Acquired Immune Responses

Acquired immune responses are mediated by lymphocytes. As we have seen, there are two main types of lymphocytes: The B cells, which produce antibody, and the T cells, which serve regulatory and accessory or helper-like functions, as well as direct cell-killing functions. Traditionally, the two forms of immunity conformed to an adaptive or acquired modality. The first—and the one that enjoyed the most attention during the early, pre-World War II period of immunology's history (Silverstein, 2003)—is *humoral* immunity. The choice of the term "humoral" was used to highlight the notion that protection is mediated by blood-borne soluble factors. In fact, in 1901, the first Nobel Prize in Physiology and Medicine acknowledged the importance of this notion by awarding the prize to Emil Von Behring, who discovered that when serum from animals immunized with tetanus toxin was injected into naïve (nonimmune) recipient animals, these animals (rabbits, actually) survived the challenge with the toxin. Hence, the notion of passive transfer of immunity (or just passive immunity) (Graham and Ambrosino, 2015). The fact that it was produced by serum from immunized animals revealed that the humoral space (i.e., blood compartment) contained protective elements of immunity. This was subsequently found to be a product of B lymphocytes, namely, antibody molecules that are released into the general circulation (lymphatic and blood).

The second type of adaptive immunity, *cellular immunity*, was not recognized until the late 1950s and early 1960s—in spite of several earlier important demonstrations that cells—like antibodies—were capable of providing protection against infection and/or hypersensitivity (Silverstein, 2003). If there was a presiding cellular theory of immunology, it was beholden to Metchnikoff's phagocyte theory of host protection (to which we will come

---

[5] The Latin *lympha* means "clear water," and is derived from interstitial fluid and blood plasma to give it a more translucent appearance. Interestingly, however, lymph collected from the digestive system can have a milky appearance due to high fat content, which is called chyle. The lymph, of course, is not just fluid, but filled with lymphocytes and other leukocytes.

later). However, this essentially took a back seat to the immunochemical tradition pioneered by Behring and Erlich, which revolved around the antibody molecule. Eventually, however, recognition of the importance of the T cell and its special—"helper"—relationship to the B cell (Crotty, 2015) solidified the notion that immunity is very much a cellular affair. The fact that cells mediate protection was demonstrated in a similar manner to that of humoral immunity using passive transfer of lymphocytes from immunized animals to naïve recipients. In this case, no soluble factors or fluids were transferred, only the cells of the immune system. The reason this is attributed to T lymphocytes is because these transferred cells were derived from the thymus and shown to help naïve B lymphocytes (of the recipient) produce antibody to the antigen (against which transferred cells were sensitized or immune). Or, they directly kill bacteria or host cells that contain viral antigens when recipient cells are experimentally incapacitated by irradiation and other procedures. Consequently, cell-mediated immunity is generally synonymous with T-cell-mediated immune responses.[6]

Hallmark Features of the Adaptive Immune Response

The chief distinguishing features of any immune response are (1) specificity, (2) diversity, (3) memory, (4) self-limitation, and (5) self−/nonself discrimination. These hallmark features are shared by T and B lymphocytes (see Text Box and Fig. 2.3), and some of these features were introduced earlier (e.g., specificity and memory). Here, we will elaborate a little further on these distinguishing characteristics of acquired immunity.

Specificity

As already mentioned, antibody molecules are made against unique elements of a foreign protein molecule (antigenic determinants or epitopes). An antibody made against one particular part of a protein, say region X_{ag}, will not bind another part of the protein, say region Y_{ag}; conversely, the latter region, Y_{ag}, will have elicited a response from a subset of B cells that make antibody that will bind this region, but will not bind region X_{ag}. Therefore, within the total pool of B cells in the body, there are cells that are specific to only one antigenic determinant or epitope, and no other. Such a group of B cells is called a clone. And if each epitope—no matter what protein it is derived from—is bound by a unique antibody

molecule by which it is bound, then it follows that there must be as many B cell clones as there are antigenic determinants that the immune system encounters. This reasoning is part of the clonal selection theory proposed by Sir McFarlane Burnet, and for which he received the Nobel Prize (Hodgkin et al., 2007). Further elaboration of this well-established feature of the acquired immune response, but suffice it that not only does this principle of one-antibody/one-epitope matching apply to B cells, but it also extends to T lymphocytes, which possess epitope-specific antigen receptors that are referred to as T cell receptors (TCRs). Similar to B cells, the population of T cells is proposed to consist of a multitude of clones, each of which recognizes with its TCR only one specific antigenic determinant. Therefore, when a B cell clone recognizes epitope X_{ag}, there is a corresponding T-cell clone that also recognizes epitope X_{ag}, and neither clone will recognize any other epitope.

This shared specificity by the X_{ag}-specific B and T-cell clones is mutually beneficial. When T cells that recognize the same antigen as B cells they can serve to help these B cells produce antibody against the antigen. Consider, for example, a mouse that is exposed to a given virus, say herpes simplex 1 (HSV-1). This mouse will generate a T cell and B cell response to the virus. However, if the mouse is exposed, to a second virus, say cytomegalovirus (CMV), there will also be a T- and B-cell response to the virus, but the population of T and B cells will be different from that which responded to HSV-1. Each virus will have a separate group of clones that possess specific recognition of the various antigenic determinants of each unique virus. To be sure, there may be some overlap— or cross-reactivity—to some epitopes shared by each virus, but for the most part each virus will recruit different clonally specific T and B cells.

Diversity

This concept can be readily appreciated in light of the foregoing discussion of specificity and clonal selectivity. In highly specific responses against the many different antigenic determinants encountered by the immune system, it is evident that there is considerable diversity in the potential reactive cells in the B- and T-cell populations. Such diversity allows for a large protein molecule (a toxin, perhaps), as large as it may be, to be responded against by multiple clones, each of which will target different components of the protein. Such diversity of targets is matched by the diverse instructions given to

---

[6] In transfer studies it is important to limit the contribution of the recipient animal's immune cells to an immune response. In a typical set up, lymphocytes (T or B cells) are transferred to the irradiated recipient (mouse A) from another genetically identical animal (mouse B) that has been immunized against an antigen. When mouse A is infected with the antigen (e.g., virus), any response against that antigen could only come from the efforts of the transferred cells of mouse B. It's important that mouse A and B are syngeneic (genetically identical) so that the transferred cells do not respond to the tissue molecules of mouse A, but instead, against the infectious antigen.

each clone. If all targeted the same region, the chances of generating the foreign protein might be minimal if that section does not happen to be a weak point. In essence, there is strength in diversity, and therefore, the reason why the adaptive immune response is so powerful. In addition, any given antigen is bound by antibody from B cells as well as the TCR of cytotoxic T cells (CTLs). In this way, diverse mechanisms of inactivation are being initiated against the same antigen.

Memory and Self-limitation

When lymphocytes respond to their respective antigens, they undergo mitotic activity, undergoing many rounds of cell division or proliferation (or what is sometimes referred to as *transformation*) in order to expand the original population of antigen-specific T and B cell clones. This expansion or proliferation of clones increases the number of cells that are responding against the specific antigen, and once the antigen is eliminated, the expanded lymphocyte population is culled through the generation of signals that promote programmed cell death, or apoptosis. This is an important aspect of self-limitation. This principle recognizes that mechanisms exist to restrain or limit the magnitude and duration of an immune response. For example, during the exponential phase of an immune response, some cells are being restrained and redirected in their behavioral profile, some serving to attack and destroy antigen-laden pathogens, while others are keeping a "leash," so to speak, on the ferocity of such responses, limiting the level of inflammation and potential indirect damage to host tissue. The end result is that the response is sufficiently fine-tuned to allow for necessary and sufficient numbers of cells and effector molecules to be produced. To ensure that this progresses in a manner that does not threaten the organism. As the number of dividing cells accrues, apoptotic mechanisms, typically triggered by cytokines such as tumor necrosis factor (TNF), result in cell death and a dénouement of proliferation. As such, the development of leukemia-like conditions or self-directed immune responses is guarded against, and hopefully, the pathogen that induced the response is eliminated. The regulation of the immune response is a major area of immunological research, and much can be said about it. We will mention some of the regulatory processes that are mediated by cytokines as well as T cells (the so-called T-regs or regulatory T cells).

An important element of the clonal proliferative response to antigen is that once proliferation peaks and apoptosis is initiated, a small population of the reactive clones are retained and remain viable as memory T and B cells. This is a cornerstone of immunity and the bedrock on which vaccination strategies are based. Lymphocytes that have become memory cells are antigen-specific in the same way as their original clones. Although these memory cells are antigen-specific in the same way as the original clones, they are now endowed with additional properties that allow them to be more robust responders should the antigen return. Memory cells respond more rapidly, proliferate in greater numbers, and produce greater amounts of antibodies. In addition, the antibody produced has greater affinity and avidity for the antigen in question.

Self-/Nonself-Discrimination

As we noted earlier, the thymus is a primary lymphoid organ in which T cells from the bone marrow achieve maturity and then enter the general circulation and take up residence in the secondary lymphoid organs. The cells are said to be mature because they now express cell-surface markers that endow them with unique functional properties (e.g., helper or CTLs). More importantly, a T cell remains viable and exits the thymus because it has failed to show reactivity to self-tissue molecules, and recognizes histocompatibility molecules that function to present antigens that are nonself to the TCR (cells that fail to achieve these criteria undergo apoptosis and are removed by phagocytic cells). The mature T cell, therefore, is able to discriminate between molecules that are self and those that are nonself, and in so doing restricts responding against the host. While this intrathymic development and maturation of T cells is one of the most compelling properties of the immune system, it can go wrong, resulting in autoimmune disease.

In summary, these features of the adaptive immune response comprise a high level of plasticity and regulation, and demonstrate highly focused attention to specific molecules (viz., antigens) that are not integral parts of the host. The presence of memory for encounters with antigen ensures accelerated, precise, and expedient elimination of reemerging microbial agents.

The Immune Response to Antigen Progresses Through a Distinct Set of Phases

The adaptive immune response can be subdivided into three distinct phases or steps. These phases progress along a continuum in which the cellular response is first *induced*, and then the intracellular biochemical machinery of the cell is *activated*, and finally, the effector arm of the response is engaged, generating actions designed to eliminate the microbial antigens. The inductive stage is considered as the *cognitive* phase since it involves recognition and processing of antigen, while the closely related activation phase involves signal transduction through cell-surface receptor stimulation with antigen, as well as signaling by other accessory cell-surface molecules expressed on APCs. The activated lymphocyte may be considered "switched on," and will undergo *transformation* and *differentiation*, which includes mitosis and expansion of the original clonal population, changes in individual cell size, and reorganization of intracellular organelles and cytoplasmic space. The cells are now in their differentiated state when they are fully capable of performing effector functions. Not surprisingly, this is referred to as the effector (or elimination) phase, because the lymphocytes now "effect" their influence on antigens and antigen-expressing bodies; in essence, they perform the neutralizing (by antibodies) and cytotoxic

functions for which they evolved. However, effector functions can also include the production of cell-signaling molecules, such as the cytokines, whose actions involve the facilitation of proliferation and differentiation of surrounding lymphocytes that bear the same antigenic-specificity. Consequently, during the effector stage, we can have amplification of the number and range of lymphocyte functions, which is designed to provide a quantitative advantage over the microbial agents that triggered the immune response. One can think of effector function as the final blow (or raining of blows), that a given cell is specialized to perform. Whether this is lysis of a target bacterial cell or virus-infected host cell, or production of soluble factors, such as the cytokines or antibodies, the differentiated cell is imposing an immunologic effect.

Cells of the adaptive immune system

By way of introducing the lymphocytes, we will make a number of declarative points that serve to introduce these cells more fully, but at the same time summarize some points already made. We will then turn directly to the two major classes of lymphocytes: B lymphocytes, which are the precursors of antibody-secreting cells; and T lymphocytes (thymus-dependent), which serve regulatory and cytotoxic functions.

1. Lymphocytes are the preeminent cells in the immune system. Their proper functioning is central to fulfilling its purpose. This is because lymphocytes determine (1) the specificity of immunity and (2) their response orchestrates the effector limbs of the immune system. All other cells (e.g., monocytes/macrophages, dendritic cells, and granulocytes) that interact with lymphocytes serve accessory functions. For example, dendritic cells specialize in the presentation of antigen to lymphocytes, while macrophages, monocytes, and neutrophils produce regulatory cytokines and remove dead and foreign tissue through phagocytosis. The role of these accessory cells in antigen presentation and innate immunity will be addressed shortly.

2. Each individual lymphocyte, as we have seen, exists as part of a set of unique clones precommitted to respond to a restricted range or molecular segment of structurally related antigens. This commitment is due to the presence of cell-surface receptors (antigen receptors), which recognize antigenic determinants (i.e., epitopes). Given that each lymphocyte membrane is peppered with thousands of antigen receptors (Labrecque et al., 2001), all the antigen-specific receptors on a single lymphocyte clone will recognize the same epitope. The eponymously

named molecule, the TCR, is the antigen-binding receptor on T cells, while the antigen receptor on B cells is actually the antigen-specific antibody molecule embedded in the cell membrane. Since each B cell belongs to a clonal group (cells with the same antigen specificity), the antibody that is expressed on the surface of the cell, is specific to only a single antigenic determinant. Once the antigen stimulates the membrane antibody, the B cell undergoes transformation and differentiation and, in the cytoplasm, actively makes and then secretes an antibody that has the same antigen specificity as the membrane-bound antibody.

3. Each lymphocyte clone differs from other lymphocyte clones based on the structure of its receptor binding site. This is the part of the antigen receptor that binds and recognizes the antigenic determinant or epitope. This means that different clones will differ in terms of the epitopes that they will recognize on an antigen. Thus, lymphocytes, as a class of antigen-specific responders, are actually a heterogeneous group of cells. Some estimates put the number of possible different epitopes recognized by lymphocytes at around 1 billion, and commensurately, a comparable number of different sets of clones are present. This also means that only a fraction of the total number of lymphocytes will be engaged in an immune response against any given virus or other microbial agent.

Fig. 2.3 provides a general schematic of the origin, differentiation, maturation, and ultimate effector role of T and B lymphocytes. After reading through the text, a return to this chart will make clear the general roles played by this rich armamentarium of specialized cells.

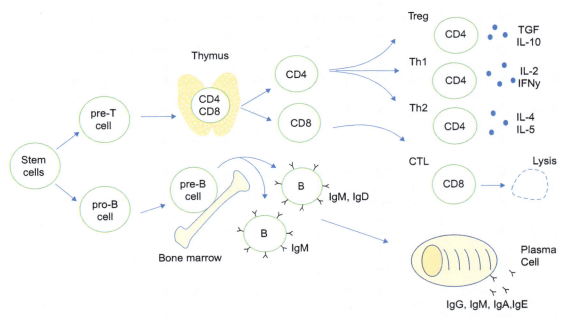

FIG. 2.3 Development of lymphoid cells into effector T and B lymphocytes. From the lymphoid progenitor stem cell, lymphocytes differentiate into pre-T and pre-B cells, which then separately develop in the bone marrow or thymus into mature lymphocytes. In their mature state, they are typically naïve or "virgin" cells until called upon to act if their particular antigen (as per the clonal selection theory) has entered the body. The development of B cells is marked by surface expression of immunoglobulin molecules (especially IgM and IgD), which are essentially antigen receptors. Once B cells have been activated, they produce copious amounts of antibody. In this state, they are known as plasma cells, which will eventually revert into a small subset of antigen-specific memory cells that will typically express most of the antibody isotypes. For T cells, the immature state is marked by the double positive CD4/CD8 expressing state of an immature T cell. In the thymus, such cells will lose one or the other molecule to become either CD4 or CD8 positive, but not both (cells that fail to become either one or the other may die and enter programmed cell death). Those that survive and leave the thymus, will circulate as mature T cells. As can be seen, the CD8 T cells will progress into CTLs, which will mainly attack and lyse virus-infected cells and, along the way, during their fight against infectious antigens, develop into memory cells that can be reactivated more quickly. Similarly, the CD4 T cells differentiate into a variety of T cells that have helper (Th1 and Th2) and regulatory functions (T-reg).

B Lymphocytes and Antibody

In this section, we will take a general look at B cells and their main role in the immune system: Antibody production. This is perhaps the preeminent role that the immune system has in relation to eliminating foreign particles. The cytotoxic removal of bacteria and neutralization of viruses is dramatically facilitated by antibody. Moreover, the activation of B cells propagates a highly energetic process that involves the production of billions of antibody molecules, present in such abundance that antigen would have to be deeply sequestered to escape detection and binding by antibody. Failure to develop antibody responses results in serious pathology. Therefore, we will begin first by discussing B cells and then move on to the antibody molecule and the immunoglobulin class of proteins. Afterward, we will turn to T lymphocytes, the additional line of immunological defense that works to augment B cell responses, as well as promote cytotoxic effector functions that expose viruses and other antigens to the binding properties of antibody.

Development of B Cells

Lymphocytes, whether T or B lymphocytes (we will refer to them interchangeably as lymphocytes or simply "cells") are derived from self-renewing HSCs (see Fig. 2.1). For B cells, most of this development takes place in the bone marrow, although prior to birth, and early during embryonic development, this commences in the fetal yolk sac[7] and after its retardation persists in the fetal liver. However, the bone marrow is the predominant site for B cell development and continues through life. Prior to becoming a fully mature B lymphocyte, a developing B cell passes through various differentiative stages globally called pro-, pre-, and immature B lymphocyte stages. Key events influence the progression of B cells through

---

[7] This provides the initial blood-borne supply of nutrients to the embryo prior to the full development of the placenta; later in gestation, the yolk sac diminishes in size and becomes integrated into the developing gut.

these stages. This includes the presence of growth factors, genetic rearrangement of antigen receptors, and selection of cells with appropriate antigen receptors. This will ensure that the mature population of B lymphocytes that enter the circulation and take up residence in secondary lymphoid organs is fully competent.

B-Cell Induction

B-cell induction occurs through the binding of antigenic determinants to the B-cell receptor. This receptor is an immunoglobulin (Ig) molecule,[8] a heterodimeric protein that possesses antibody activity. Therefore, when we refer to "antibodies," as a functional class of proteins, we should note that these are Ig molecules. Moreover, when integrated into the cell membrane, the Ig molecule serves the initial function of being an antigen receptor. Once Ig molecules are released subsequent to B cell induction (i.e., antigen binding to the membrane-bound Ig molecule that specifically recognizes it) and activation of the cell, the antigen-specific Ig molecules are released into the extracellular environment at extremely high rates, such that a single B cell may release over 10^5 Ig molecules per minute. Since this is equivalent to the release of antigen receptors into the surrounding environment, Ig molecules function as soluble antigen-binding receptors, or as the German scientist Paul Ehrlich coined them in the late 19th century, *antikörper* (literally, antibody). Interestingly, this principle of releasing soluble "receptors" in order to "mop up" circulating ligands for the receptor also exists for cytokines. That is, immune cells will release molecules that act as soluble receptors for cytokines. In this case, the soluble serves in a regulatory capacity to reduce or interfere with the function of cytokines that may be in high abundance or no longer needed.

B-Cell Activation

The receptor-mediated recognition of antigen results in a complex series of intracellular biochemical events (signal transduction) that cause the B cell to undergo clonal expansion and transformation. This type of activation can either be T lymphocyte dependent (involving cognate T cell help), or T-cell independent (cross-linkage dependent).

Cross-linked activation. Cross-linkage forms of B cell activation require that the same epitope is repeated multiple times on some larger molecule or cell body, such that closely associated Ig receptors on the cell surface can bind (or grasp) each of these epitope exposures at the same time (i.e., cross-link the molecule or cell presenting them). The analogy may be that of the fingers (Ig receptors) of a hand all fitting the five spaces (identical epitopes) provided for their entry into a glove.[9] This occurs with viruses, as they expose multiple copies of proteins, thereby repeating identical antigenic determinants at multiple sites on the viral envelope. In addition, common bacteria, such as pneumococci, streptococci, and meningococci, repeatedly express polysaccharides that can bind Ig receptors on B cells. This can serve to anchor a bacterial cell to the B cell and lead to activation of tyrosine kinase activity and signal transduction through the phosphatidyl-inositol pathway and elevated calcium levels. Membrane cross-linkage activation of B cells is considered to be the major protective immune response mounted against infectious agents. It operates in a similar fashion to the PRRs of innate immune cells, such as neutrophils. As such, it serves to eliminate or neutralize antigens in the absence of specific T-cell derived costimulatory signals but does recruit the involvement of other factors, such as complement, a system of circulating proteins with cytotoxic properties that recognize sections of the antibody molecule exposed when it is bound to antigen.

Cognate T cell with the help of B cells. When the antigen binds to the B cell receptor, it is endocytosed and enzymatically digested into peptide fragments. The peptide fragments are loaded onto *MHC class II* molecules (defined later), which migrate to the membrane to be expressed on the cell surface. The MHC class II bound peptide interacts with a specific type of T cell (a CD4[+] helper T cell) via the TCR on T-helper cells, which has the same specificity for the peptide fragment (antigen) as the B cell Ig surface receptor. This results in the T cell synthesizing and secreting a regulatory substance (a *lymphokine*[10] molecule—which is now more commonly referred to as a *cytokine*) that binds to a receptor on the B cell. This promotes the growth and differentiation of the B cell. In fact, it is of interest to note that IL-6 was once known as B cell growth or differentiation factor.

---

[8] The three main proteins in blood are albumin, globulin, and fibrinogen. Globulin—which received its name for its spherical or global tertiary (three-dimensional) shape has three major types, called alpha, beta, and gamma. The gamma globulin fraction is the immune component, and therefore is referred to as immunoglobulin. It is also the basis for the name given to the basic immunoglobulin unit, IgG. As we will learn, other antibodies are multimeric conglomerates of two or more IgG-like molecules, which resulted in the discovery of other Ig molecules: IgA, IgD, IgE, and IgM.

[9] Of course, this analogy might work better in reverse—fingers serving as the epitopes (but you get the picture).

[10] In the early days of immunology, soluble mediators derived from lymphocytes were called "lymphokines." This denoted their status as being messengers produced by lymphocytes (lympho). If a soluble messenger came from a monocyte, it was called a *monokine*. Moreover, while we now have the more collective term "cytokine" to refer to a cell-derived messenger or signaling molecule, the original contextually specific terms, such as interleukin (between leukocytes) have been retained. As a final note, the original Greek meaning from which "kine" was borrowed, relates to movement; therefore, a cytokine is a moving piece of information: a message.

This cytokine is T-cell derived but is also made by cells of the innate immune system, suggesting that these cells can also be recruited to contribute to the activation of B cells. Another cytokine, interleukin-4 (IL-4), is a prominent modulator of B cell antibody production, serving to refine the quality of the antibody produced by B cells as they progress into the effector stage of responsiveness.

Effector B-Cell Response: Plasma Cells

Whether through T cell assistance or independently, activated B cells differentiate into plasma cells, a final state in which they perform the effector function of secreting abundant quantities of antigen-specific antibody molecules. A plasma cell can also be referred to as an *antibody-forming cell (AFC)*. In certain older assays of B cell function, single antibody-producing B cells are called plaque-forming cells (PFC) by virtue of their ability to kill (in the presence of complement) sheep erythrocytes or other cells containing antigen against which the B cell is responding. This assay has been superceded by enzyme-linked immunosorbent methods, such as the ELISPOT, which detects the number of B cells secreting antibody against soluble protein antigens. The ELISPOT assay is extremely versatile and is now used to detect the number of T cells, neutrophils, and monocytes that are releasing a given cytokine. Other more expensive protocols might utilize flow cytometry and cell sorting approaches to quantify the number of cells performing a specific function, such as secretion of a given regulatory molecule.

Memory B Cells

When responding to a specific antigen, during the proliferative or clonal expansion stage (when B cells are differentiating and acquiring the capacity to synthesize and secrete antibodies), a fraction of the responding B cells reverts into a resting, dormant state. These cells are the B cell memory cells. When antigen is encountered for the first time, the response induced (whether B or T-cell mediated) comprises a *primary* response, and the responding cells are said to have been "primed." Since many cells die after the primary response is complete, the reactivation of memory cells by the return of antigen is referred to as a *secondary* response.[11] The memory cells can be referred to as sensitized or primed cells, which

gives them the ability to respond faster and with greater strength to antigen re-exposure. This more accurately reflects the use of these terms since the cells responding are the surviving cells that possess memory for the initial antigen exposure. Finally, it should be noted that the molecular and cellular profile of the secondary response can be quite different from that seen in the primary response. Relative to the primary response, a secondary antibody response is (1) greater in magnitude, (2) more prompt, (3) the antibody has greater affinity and avidity for the antigen, and (4) and is dominated by certain immunoglobulin antibody classes (viz., IgG).

Distribution of Antibody

At the cellular level, antibody molecules are found in the endoplasmic reticulum and Golgi apparatus and are also expressed on the surface of the cell where, as we have already said, they act as receptors for antigen. However, once secreted, antibodies are most abundant in the fluid portion of blood (plasma/serum), although as B cells circulate through lymphatic vessels, as well as traffic to lymph nodes, antibody detection in these areas is expected. However, for evidence of a strong and effective immunization against a given antigen (e.g., vaccination against specific viruses), measurement of serum antibody is the clinical standard.[12] In addition, antibody is present in secretory fluids, such as the mucus of the upper respiratory tract and the GI tract. It is also found in a mother's milk after parturition, which confers passive protection to the neonate against pathogens to which the mother has already developed immunity. The main Ig type in the pool of antibodies found in these areas—which are part of the common mucosal immune system—is IgA.

With respect to effector functions, antibody can bind to the surface of innate immune cells, such as phagocytic cells (e.g., macrophages and polymorphonuclear leukocytes), as well as nonphagocytic cells such as NK cells and mast cells. Antibody is not produced by these cells, but a receptor for antibody is present on their surface which binds a portion of the antibody that is not already dedicated to binding antigen. This capture of antibody by innate immune cells serves to bring them closer to the antigen that antibodies bind, thereby allowing the phagocytic immune cells to engulf and destroy the antigen or microbe that is expressing the antigen. Earlier, we

---

[11] One should note that subsequent responses to further antigenic reexposure are called tertiary or quaternary responses. However, this becomes cumbersome to discuss, suffice it that responses taking place after the primary response are essentially memory responses. When it comes to vaccines, subsequent memory responses are induced by "booster" shots and these are designed to produce the eponymous boost in antibody production.

[12] We should note that when serum is referred to as *antiserum*, this indicates that it contains antibodies against specific antigens. However, using the term "antiserum" is not synonymous with "antibody," which refers to an individual molecule, part of billions present in the antiserum. From a practical standpoint, however, antiserum possesses antibody, and therefore, antigen neutralizing properties. Many techniques in biological research, as well as clinical testing, rely on the application of serum from immunized animals being used to reveal antigens or block certain biological effects. These are considered immunochemical procedures, since they capitalize on actions of antibody in the "raised" serum.

mentioned opsonization. In this scenario, antibody is the opsonin, and the phagocytes ingest the antigen–antibody assembly. This is a highly cooperative function between B cells and innate immune cells, which complete the effector function initiated by antibody binding of antigen. Alternatively, a less positive outcome of antibody binding of surface antigens is that involving mast cells. These cells express a receptor for the IgE form of antibody molecule, which results in the release of mast cell histamine and the development of symptoms associated with allergy (runny nose and teary eyes).

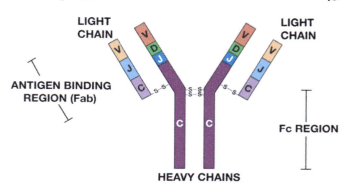

FIG. 2.4 Basic structure of the immunoglobulin molecule. The antigen-binding component (Fab) of the Ig molecule consists of variable amino acid sequences that show binding affinity and avidity to the recognized peptide antigen. This variability in amino acid sequence is dictated by sections of the genome denoted by the letters V, J, D, and C. The amino acid sequence of the Fc region of the molecule remains relatively constant and binds to Fc receptors on macrophages and neutrophils, as described in the text. *From Sompayrac, L., 2016. How the Immune System Works, fifth ed. © 2016 John Wiley & Sons, Ltd, Published 2016 by John Wiley & Sons, Ltd.*

The Antibody Molecule: An Immunoglobulin

Serum contains numerous proteins: albumin, globulins, and other proteins (e.g., acute phase proteins, fibrinogen, and hormones). These proteins can be separated by electrophoresis according to size or charge, with the globulins separating into three distinct zones called alpha, beta, and gamma. The gamma fraction was found to possess immune activity, in that it can bind antigenic molecules. This ultimately revealed that the immune-binding activity of serum was present in gamma globulin proteins, which is where we find antibodies in the molecular form that we call immunoglobulin (Ig).

Structure of Immunoglobulin

The basic unit of an immunoglobulin protein consists of two pairs of identical amino acid chains, which make the basic Ig unit a dimeric protein. The chains of one pair are called *heavy (H) chains*, while the chains of the other pair are called *light (L) chains* (see Fig. 2.4). These distinctions are based on differences in molecular weight between the H and L chains, the former having the greater number of amino acids. Emphasis on referring to a "basic Ig unit" comes from the fact that immunoglobulins can consist of one or more of these basic units.

Various physicochemical properties (amino acid sequences, size, charge, and conformation) of the H chain determine the identity of an immunoglobulin. Thus, there are five main types (actually called *isotypes*)[13] or *classes* of immunoglobulin: IgG, IgM, IgA, IgD, and IgE. The IgG molecule conforms to the basic unit shown in Fig. 2.4. The H chain of the IgG basic unit will have different physicochemical properties from the H chain of the basic Ig units that make up IgM (which has five basic Ig units) or IgA (which has two basic Ig units). That is, what defines an immunoglobulin as IgG, IgA, IgM, and so on is the composition of the H chain.

The IgG and IgM molecules are the most prominent immunoglobulins released during the first 5 days of a primary antibody response, with IgM being most abundant initially but followed, within days, by greater production of IgG. However, during a secondary immune response, IgG is the most prominent antibody produced very early after antigen reexposure, with the exception of mucosal regions, such as the gut, lungs, and respiratory tract, where IgA production is prominent. Still, whether systemically or in mucosal areas, IgG is always a key immunoglobulin, persistently refined by activated B cells to have greater affinity and avidity with prolonged or repeated exposure to antigen. This is why it is the main immunoglobulin produced in secondary immune responses.

Why Do Antibodies Possess Specificity?

Within a given Ig class or isotype, there are regions on each H and L chain that are shared (or *constant*) across all Ig molecules of that class. But there are also regions that are not shared. The regions that differ in amino acid sequence from one immunoglobulin molecule of a given class to another are called *variable* or *V regions*. These variable regions are the binding sites for antigenic determinants. Therefore, if antibody is produced against several different antigenic determinants, within the same class

---

[13] Note that each class of Ig is also referred to as an *isotype*. For example, the statement, "immunization with the protein antigen, ovalbumin, results in the generation of several isotypes of antibody," is merely saying that the immunization has resulted in B cells producing antibody that consists of IgG, IgM, and IgA immunoglobulin forms, with perhaps other forms, such as IgD and IgE, being less abundant in response to the antigen.

of immunoglobulin secreted against these antigens, there will be differences in the amino acid sequences in the variable regions of each antibody molecule. For example, after immunization with a given virus, a sample of antiserum will have a fraction of IgG molecules specific for one epitope (e.g., sequence A of the viral protein envelope), while another fraction of IgG molecules will be specific for a different epitope (e.g., sequence B of the viral protein envelope). However, collectively, all the IgG molecules produced by the B cells responding to the virus will be able to bind to the virus. What will be different among the IgG molecules directed to different epitopes of the virus is the amino acid sequence in the V region of the IgG. However, we should note that some IgG molecules with the same epitope specificity may have additional differences that delegate the IgG to different subsets or isotypes (e.g., IgG1 and IgG2). Interestingly, understanding of antibody structure and function also relates to the types of T cells and cytokines that impact antibody-producing B cells. For example, B cells may be releasing IgG1 against a different epitope, but in the presence of a given cytokine produced by helper T cells, there is a shift to the production of IgG2, a process called isotype switching (although one could just as easily say subtype switching). Similarly, B cells switching from IgM production to mainly IgG is also called isotype class switching.

The different amino acid sequences in the V regions of an antibody Ig molecule determine the *affinity* and *avidity* of the Ig to various antigenic determinants. Affinity and avidity are terms that refer to the degree of specificity (affinity) and strength of binding (avidity) between the antibody molecule and the antigenic determinant. The V region is found in the antigen-binding segment of an immunoglobulin molecule, and this is called the *Fab fragment* (fragment antigen-binding). The constant region is referred to as the *Fc fragment* (since the constant region was shown to crystallize in solution through self-association).

The Function of Antibody Is to Neutralize Antigen

There are three main processes by which antibody serves to eliminate and/or neutralize antigen. These are described as follows.

1. When protein is liberated from a virus or bacteria or simply ingested (e.g., a toxin) or injected (experimentally), it will be known as a soluble antigen and will circulate through the blood and lymph. This soluble antigen will be bound to antibody and form an antigen–antibody pairing that can further be bound by free portions of the Fab region of antibody that is already paired with this same protein antigen. What eventuates is a network of connections called an *immune complex*. The immune complex can be trapped in tissue where local macrophages can bind the complexed antibody via the Fc region (i.e., via Fc receptors on the cell surface). The immune complex will be endocytosed (incorporated into the inner cell compartment), and rendered innocuous by enzymatic breakdown.

2. The degradative fate of immune complexes is aligned with a process called *opsonization*, which was introduced earlier in the section on B cells. The term "opsonin" is derived from the Greek, meaning "prepare to eat." Therefore, antibodies are opsonins, in the sense that they bind antigens or surfaces riddled with antigens, and dress the target body for phagocyte consumption. The target body can be bacteria, viruses, and other insoluble particles. For example, when antibody binds to the surface antigens of a bacterial cell, it exposes the Fc portion for binding to the Fc receptor on phagocytic macrophages. Once the macrophage has "docked" at the antibody-bacteria complex, the bacterial cell is ingested and degraded by intracellular enzymes.

3. An alternative to opsonization is activation of the *complement* system. Complement is a set of approximately 20 proteins that are made primarily by the liver and are available in the blood for a variety of cytotoxic functions. The proteins are modified sequentially to eventually achieve a cytolytic capability. This can be done independently of antibody (in which case, it is part of a humoral innate immune defense) as well as with the aid of antibody. In addition to being synthesized by cells in the liver (hepatocytes), macrophages also generate complement proteins. Overall, complement is an essential and ancient part of immune defense that most likely predated the arrival of antibody-mediated immune responses.

It appears, in summary, that there is a common principle at play when Ag–Ab complexes are formed. Once bound, antigen is summarily degraded and removed by innate cellular (macrophages) and biochemical (complement) mechanisms. In order for this to happen expediently, massive amounts of antibodies are required to ensure that the detection of antigens is inevitable. This is why failure to generate antibody effectively prolongs infection and promotes morbidity.

Immunoglobulin Class Switching

As we noted earlier, the immune response has primary and secondary characteristics. In a primary antibody response, IgM is the most prominent immunoglobulin secreted. However, with time, specific gene rearrangements begin to generate antibody with greater specificity for the various antigenic determinants that induced the B cell response in the first place. The major class (or isotype) of immunoglobulin in this regard is IgG. However, IgA is the predominant isotype at mucosal surfaces (e.g., GI, respiratory, and urogenital tracts).

When antigen is reintroduced to the host, a secondary response is generated (i.e., the memory response), and the predominant class of immunoglobulin will be IgG (if the response is systemic) or IgA (if at the mucosal surfaces). This is not to say that other immunoglobulin classes will not be produced. Rather, the IgG and IgA molecules that are produced in the memory response will be greater in concentration and have greater affinity and avidity, which will allow for faster and more effective binding (immune complexing). Furthermore, the more robust and prolific production of antibodies will ensure that the antigen is bound and subjected to the processes of neutralization and degradation as described earlier.

T Lymphocytes

Development of T Cells

T cells leave the bone marrow in an undifferentiated state and migrate to the thymus gland. Within the thymus, T cells mature into cells that express various distinct molecules (e.g., CD4, CD8, and a TCR) that mediate important T cell functions. Maturation in the thymus is essentially a selection process based on whether T cells possess (1) close affinity for molecules expressed by the major histocompatibility complex (MHC) and (2) the absence of reactivity to molecules that are not encoded by MHC and are expressed on self-tissue. This latter criterion is the basis of *self-tolerance*. Therefore, a mature T cell will not react against non-MHC self-antigens—which are all the proteins and peptides expressed by endogenous host cells—but will strongly recognize MHC-encoded molecules. This optimizes the ability of T cells to recognize foreign antigens (which are not part of the host) and acknowledge that they are being presented to the T cell by host MHC-encoded molecules. This is the molecular basis of antigen presentation and why T cells need to develop a strong affinity to MHC proteins. As we will learn, this MHC requirement allows for selective engagement of different subtypes of T cells.

The CD molecule

A large number of CD molecules (several hundred, in fact) have been identified on lymphocytes and other cells of the immune system (and also applied to nonimmune cells). In many cases, cells are almost always described in terms of the cell-surface markers they express. In some cases, it can get downright confusing since a given CD molecule may be expressed on multiple cell types, for instance, CD2, as a cell adhesion molecule found on T cells and NK

cells. The term "CD" is derived from the expression "cluster of differentiation," to reflect the identification of a unique protein through immunohistochemical procedures (i.e., using antibody recognition). In almost all cases, a particular CD molecule is given a number, in order to simplify nomenclature. This is because many CD molecules initially had different names (e.g., CD2 was once called "T cell surface antigen"), and was eventually found to enhance the binding of APCs to T cells. In many cases, CD molecules are involved in cell–cell interaction and receptor function, but in some cases, the function is unknown. Nonetheless, they do serve as a useful marker of a given cell type, as is the case for T cells, which are all CD3-positive, but as mentioned earlier, not all are CD4- or CD8-positive. Moreover, as described in the text, the CD3 molecule turns out to be more than just a differentiating marker. It forms a complex with the TCR, which ensures antigen recognition. Moreover, most B cells express a marker called CD19, while innate immune cells (macrophages, granulocytes, and dendritic cells) express CD14.

Generating antibodies that specifically recognize these various CD molecules allowed immunologists to differentiate between cell subtypes and to isolate and purify cells that have a particular CD signature. Techniques used to do this are beyond the scope of this chapter, but isolating, for example, $CD3^+/CD4^+/CD8^-$ cells allowed immunologists to determine that when these cells were added to purified CD19+ (B cells) and CD14+ (dendritic cells) cells, antibody production was substantially augmented. From this was born the notion of T cell help, which was confined to CD4 T cells.

On leaving the thymus, all T cells express a TCR, and CD3 molecules are expressed on their cell surface. However, they have further identified characteristics that are functionally important and segregate T cells into two functional classes that express either the CD4 or CD8 molecule. Hence, mature T cells are designated as either $CD3^+/CD4^+/CD8^-$ or $CD3^+/CD4^-/CD8^+$ and can simply be referred to as CD4 and CD8 T cells. These subtypes mediate distinct functions, which we introduced earlier: Helper and cytotoxic functions.

Induction of a T-Cell Response

Generating and inducing a T-cell response first requires a recognition step in which the antigenic determinant is specifically bound to a receptor expressed on the T cell surface, which is referred to as the *T-cell antigen receptor* (or just TCR). Unlike the B cell antigen receptor, the TCR does not disengage from the cell membrane but serves as one end of a coupling event between the T cell and the APC, during which the TCR determines its complementarity to the antigen. This recognition step between antigen and TCR involves a physical interaction between the T cell and another (non-T) host cell that presents the antigen in association with a protein encoded by the MHC. This antigen-presenting non-T cell will also express additional molecules that allow for the induction of a T cell response. These additional molecules are referred to as accessory molecules (e.g., CD28 and B7),[14] and these ensure that T-cell activation will occur.

In most cases, activation of T cells will induce the transcription and translation of the gene for the cytokine, interleukin-2 (IL-2; see Table 2.1 for a description of cytokine functions), along with other cytokines. For instance, with regard to T-helper cells, it was found that these could further be subdivided into what came to be called Th1 and Th2 cells, each with their own particular cytokine profile (e.g., Th1: IL-2 and IFNγ; Th2: IL-4). Subsequently, other subtypes were discovered, and these T cells came to be known as T-reg cells, owing to their powerful regulatory or suppressive functions on other immune cells. They are identified as CD4+/CD25+ and importantly, express the gene for Forkhead box protein 3 (FoxP3) a transcription factor (TF) that suppresses other TFs (e.g., NFAT and NFkB) important in promoting responses by a variety of immune cells, such as CD4 and CD8 T cells, B cells and APCs like dendritic cells, as well as monocytes and macrophages. For this reason, Treg cells are considered important in blocking potential and/or ongoing inflammatory responses whose exuberance or overextended duration may pose a risk of autoimmune disease. Indeed, another regulatory T cell subtype, the Th17 cell, is implicated in the promotion of inflammation, and is subject to Treg-mediated control. This cell was named by virtue of its prominent production of IL-17, and while not entirely threatening, it has been implicated in facilitating pathogenic inflammatory responses that result in disease.

As just alluded to, the presentation of processed antigen in association with a class II MHC molecule is performed by an APC. The major cells that function as APCs are macrophages, B cells, and DCs.[15] Of these cells, the DCs are considered true professional APCs, whose role is exclusively to present antigens. They are prominent in all lymphoid organs, capable of cytokine production, and possess the morphology (i.e., multiple dendritic arborizations) that maximizes the surface expression of MHC class II molecules loaded with antigenic peptides. Special DC cells in the skin, called Langerhans cells, serve as APCs in the epidermal microenvironment, where they can also function as macrophages. Dendritic cells are often considered professional APCs, since early experiments in the 1970s and 1980s showed that they were 10–100 times more powerful than macrophages and B cells in the induction of antigen-specific immune responses.

Overall, *induction* involves a set of cellular and molecular interactions that culminate in an antigenic determinant being specifically recognized by the appropriate T-cell clone (the one T cell in a million that will recognize the particular molecular characteristics of the antigenic determinant that is presented). To understand this more fully, it is necessary to consider more closely the MHC, and the TCR.

The Major Histocompatibility Complex

The MHC is a set (or *complex*) of genes found in all vertebrate mammals and located on distinct chromosomes (viz., chromosome 6 in humans and chromosome 17 in the mouse). The human MHC is also referred to as the HLA region, which is an abbreviation for "human leukocyte antigen." In the mouse, the MHC is known as the H2 region. Skipping between these two species may generate some confusing use of terminology, but essentially, each MHC region is similarly organized into groups of alleles that code for the different MHC molecules that are involved in antigen presentation. To limit the confusion, when not making reference to any particular species, the literature uses MHC,[16] but within a particular species will adopt more specific terms.

The genes (or alleles) of the MHC code for protein molecules that are expressed on the cell membrane play a critical role in the presentation of antigens to T cells. The main protein molecules encoded by the MHC are the class I and class II molecules. There are also MHC class III molecules, but they are more secretory in nature and will not be discussed here.

Class I MHC molecules are expressed on most cells in the body, including immune and nonimmune cells (e.g., brain, liver, heart, lung, and all other major organs). Interestingly, the expression of MHC I molecules in the brain

---

[14] T cells will express CD28, which will bind the B7 molecule (which is a combination of CD80 and CD86) on the antigen-presenting cell (APC). The complete interaction between the TCR on the T cell with the MHC molecule on the APC, along with CD28 on the T cell interacting with B7 on the APC, ensures activation of the T cell. Yes, it becomes a bit dizzying, but see Fig. 2.5 for a schematic illustration of these interactions.

[15] Not only do B cells produce antibody, but they are also capable of expressing MHC II and presenting antigen, this will normally occur in the context of cognate T cell help as was described earlier.

[16] The use of the abbreviation "MHC" sometimes refers to class I and class II molecules encoded by the MHC. For example, "MHC–peptide complex" refers to an MHC molecule bound by an antigenic determinant (usually a peptide fragment of the original, larger antigenic protein). Furthermore, the phrase, "MHC expression," refers to the production and cell surface appearance of the class I and class II proteins encoded by the MHC. Thus, it is important to remember that MHC refers to a unique set of genes, while MHC "molecules," "proteins," or "expression" refers to the products of these genes when they are transcribed and subsequently translated into protein molecules.

is important in shaping neurodevelopment, synaptic density and neurochemical transmission (e.g., glutamate and GABA neurotransmitters). Whether this is related to its immunological function, as discussed below, is unknown, although it highlights the pleiotropic functions of molecules that are associated with the immune system.: Cells of the immune system and cells not of the immune system. For example, MHC class I molecules are found in the brain and are important in shaping the development of the brain. Therefore, in addition to cells of the immune system, we can include neurons, pancreatic cells, kidney cells, epithelial cells, heart cells, and other important tissues, all possessing a functional gene that codes for the production and expression of the MHC class I molecule.

The class I MHC molecules are glycoproteins, which consist of a large alpha chain that is linked to another protein called beta2-microglobulin. The alpha chain is highly variable (i.e., *polymorphic*) at the extracellular aminoterminal end and binds the antigenic peptide and components of the TCR. These regions (whether in the mouse or human) all display very high degrees of polymorphism, which essentially means that the peptide-binding region of the MHC class I molecule is able to bind many different types of amino acid sequences and with different physicochemical properties (e.g., hydrophobic and acidic/basic properties). Finally, class I molecules bind only very small peptides (approximately average size of nine amino acids).

Cells of any type can potentially be infected by a given virus, so long as the cell contains the necessary receptors and/or docking sites that a virus needs to attach and enter the cell. The SARS-Covid-2 virus, for instance, caused havoc because it had a preferential attraction for angiotensin-converting enzyme 2 (ACE2) and other cofactors that allowed it to gain entry into sites expressing ACE2 (e.g., lungs and intestines, although up to 70 different cell types in the body have mRNA for ACE2). This preferential targeting of different cell types explains why viruses are also categorized by their physiological impact, as in for example their designation as respiratory, gastrointestinal, and hepatic viruses. Therefore, the evolution of MHC class I gene expression in most cells of the body allowed for uploading of peptide antigens from any cell type that harbored a replicating virus. In an infected cell, an intracellular protein degradation system driven by proteosomes—enzymes that breakdown proteins—reduces the protein components of viral envelopes into peptides that are transported into the endoplasmic reticulum (ER) of the cell. Within the ER, assembled MHC I molecules bind the viral peptides and migrate to the cell surface where the extracellular alpha chain "presents" the peptides for inspection by CD8+ T cells. Any local or passing CD8+ T cells can interact with the infected cell, and by virtue of their expression of CD8, will recognize the MHC I molecule (a trick it learned in the thymus). However, to induce an antiviral T cell response,

the TCRs distributed across the surface of a given T cell need to come into contact and sample the antigenic peptide on the MHC I molecule.

Antigen presentation via MHC class I occurs after intracellular antigen processing that then loads peptide fragments (of the larger protein antigen) onto the alpha chain of the MHC I molecule. This peptide–MHC assembly is then transported to the surface of the cell for extracellular expression. Thus, when a virus infects a host cell, the virus can be internalized and digested, and it is "bits" (peptide fragments) expressed by the MHC I molecule on the cell surface. Any local or passing CD8+ T cells can interact with the infected cell and, by virtue of their expression of CD8, will recognize the MHC I molecule (a trick it learned in the thymus) and then bring it is TCR around to determine whether it is the clone for the particular peptide being displayed. If there is no recognition of the viral peptide, another CD8 cell—the right one perhaps—will eventually come by and make the correct recognition and swing into action to lyse the infected cell and liberate viral particles for immune-complexing and opsonization.

Since viruses can burrow into practically any type of cell, all cells of the body are susceptible to viral infection. Therefore, the advantage of MHC I expression being ubiquitous is to ensure that the presence of a virus can be advertised to the cells of the immune system, no matter where the virus has taken up lodging. This does require sacrifice, as infected cells are killed in the process of CD8 (and NK cell) cytotoxicity, as well as antibody-dependent cytotoxicity (i.e., opsonization of host cells), and this is why the process needs to progress quickly before viral spreading compromises organ function.

Class II MHC molecules are expressed in a more restricted manner, being found on B cells, macrophages, DCs, epidermal Langerhans cells, thymic epithelial cells, and, in the case of humans, activated T cells. The presence of MHC molecules on epithelial cells of the thymus is essential in determining whether mature T cells are selected for release into the peripheral T cell population. In the brain, MHC II expression is present on microglial cells, but has also been observed on neural progenitor cells. But it is among the immune cells that our understanding of MHC Class II molecules has developed. In fact, it is among the epithelial cells of the thymus that MHC expression is essential in determining whether mature T cells are selected for release into the peripheral T cell population, allowing them to interact with expressed MHC Class I (e.g., if CD8 T cells) and Class II (e.g., if CD4 T cells) molecules on a variety of cell types. In the mouse, class II molecules are encoded by the *I* (for "immune") region of the MHC, containing two subregions, H2-A and H2-E, which code for the various components of the MHC II molecule, said to be I-region-associated (Ia). In humans, there are three sets of genes that code for the class II molecule. These are the DR, DQ, and DP regions of the HLA complex (also referred to as HLA-DR, HLA-DQ, and HLA-DP), each of which codes for the whole MHC class II protein assembly. This means that there can be polymorphism in the MHC II molecule, as well as differential distribution of these polymorphic forms of MHC II in different cells and different body regions (e.g., gut vs. lymph nodes).

The actual MHC II protein assembly consists of two glycoproteins: the alpha and beta chains.[17] Extracellular regions of each chain contain domains that are highly polymorphic (i.e., variable in terms of the consistency of amino acid sequences) since this is where the molecule binds the processed antigenic peptides. In contrast to the class I molecule, peptides of much larger size can be bound by the class II molecule (e.g., 13–25 amino acids in length). Furthermore, MHC II interactions are primarily with CD4$^+$ T cells. The CD4 molecule is the mechanism by which the T cell recognizes an MHC II molecule, and this was the purpose of the education received in the thymus. Graduation from the thymus for CD4$^+$/CD8$^-$ T cells rested on the condition that they acknowledged MHC II molecules and ignored MHC I molecules. As such, CD4 T cells are the chief mechanism by which the immune response to antigen is amplified and regulated.

Summarizing briefly, the major function of MHC-encoded class I and class II molecules is the binding and presentation of peptides to T cells whose receptors are capable of recognizing the MHC–peptide complex. The MHC class I molecules are widespread in most tissue cells, while the MHC II molecule is exclusive to immune cells, which use this in a more "cognitive" manner to direct attention to antigens from foreign entities. As such, professional APCs, such as DCs, will be localized in most areas where interactions with T lymphocytes are most likely, such as in lymph nodes.

The T Cell Receptor

The T-cell antigen receptor (TCR) is a heterodimeric protein (composed of two unique divisions) found on the surface of all T lymphocytes. On most T cells, the TCR consists of the alpha and beta chains and is therefore designated the alpha/beta TCR (A gamma/delta TCR is found on a very small percentage of T cells). The TCR heterodimeric protein is usually complex with several other protein molecules, in particular the CD3 molecule. This TCR-CD3 complex is sometimes referred to as the *functional TCR complex*. Whether a T cell is CD4$^+$ and thereby specialized for recognizing MHC II, or CD8$^+$ and specialized for recognizing MHC I, the peptide antigen being presented on the MHC I or II molecules can be the same and needs to be recognized by the antigen-specific

precommitted T cell clone circulating somewhere in the body. Since CD3 facilitates this TCR-mediated presentation, it makes sense for it to be present in all T cells. As such it is possible to determine the total population of T cells in blood or tissues by using this surface marker.

The TCR$_{\alpha/\beta}$ heterodimer interacts directly with antigenic determinants via variable regions on the alpha and beta chains. The arrangement of these chains is very similar to that of immunoglobulins (an antigen-binding variable region and a nonantigen-binding constant region). The purpose of the TCR—as we have been noting—is to recognize specific antigenic peptides and initiate intracellular signals that will result in cell division (mitosis or proliferation) and differentiation toward an effector state. The CD3 molecule and other surface molecules mediate these intracellular signal transduction mechanisms and activate the T cell.

Antigen Presentation to T Cells

The TCR binds to and recognizes all appropriate antigenic peptides while they are bound to either a class I or class II molecule. However, T cells that are CD4$^+$ will recognize antigen only if it is bound by class II molecules, and T cells that are CD8$^+$ will recognize antigen only if it is bound by class I molecules. Cells that present antigens in the presence of Class II MHC molecules are generally considered professional APCs. Strictly speaking, this applies to DCs, but macrophages and B cells also perform APC functions. For instance, B lymphocytes can bind native antigen (whole and unprocessed) and internalize it by endocytosis. Enzymatically facilitated fragmentation of the antigen then takes place, followed by loading onto a MHC II molecule.

Macrophages, in contrast, are phagocytes that can ingest whole cells, as well as large foreign molecules, subjecting them to enzymatic digestion and fragmentation into smaller elements, such as peptide antigens. These are then loaded onto the MHC Class II molecule and transported to the cell surface. Since macrophages generally do not express MHC class II molecules constitutively, it must be induced. A cytokine critical in doing this is interferon-γ, which is a major inducer of MHC II molecules. This cytokine also renders macrophages more effective in killing tumors. In essence, macrophage APC function combines phagocytosis and molecular signaling to enable the induction of antigen-specific immune responses by T cells.

DCs are so named because of their highly arborized morphology, consisting of an extensive array of thin processes. These serve to both sample their surroundings and make contact with other cells. In many ways, these cells resemble the microglial cells of the brain, also known

---

[17] We should note the alpha chain in the MHC II molecule is not the same one as for the MHC I molecule. Across different proteins with multiple folding subunits, alpha, beta, gamma, etc., are formal designations for distinct sequences (chains of amino acids) within the larger protein.

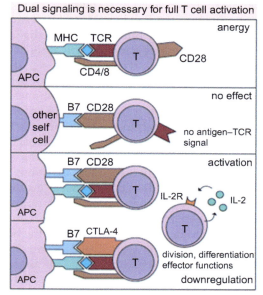

FIG. 2.5 Interlinked molecular interactions required to produce T-cell activation by an antigen-loaded APC. Shown on the left is MHC (whether I or II) presentation of peptide antigen (*blue diamond*) to the TCR, which is complexed to CD3. If the T cell is CD4 positive, and APC expresses MHC II, recognition will take place, and the kinase Lck will be brought closer to the intracytoplasmic tail of the CD3, which will then be phosphorylated, which will facilitate signal transduction. Similarly, if the T cell is CD8 positive, and the presentation is by MHC I, the result will be the same. On the right is an additional costimulatory requirement to ensure that T-cell activation will result in IL-2 production and proliferation of the T-cell. As mentioned in the text, B7 and CD28 are needed to provide additional signals that will fully activate T cells. Without this co-stimulation, cells can become anergic or unresponsive. Finally, when CD28 is replaced by the surface molecule CTLA-4, T-cell activity is reduced or downregulated. *From Male, D. K., Brostoff, J., Roth, D., Roitt, I., 2006. Immunology, seventh ed. Mosby.*

to have APC function, albeit quite limited. The development of DCs occurs in the bone marrow (save for the follicular DCs of lymph nodes) and progresses along the myeloid cell lineage from a precursor cell that can also redirect differentiation toward the development of monocytes. The alternative switch to becoming DCs requires a molecule, Flt3, without which the precursor cell will favor progression toward a monocyte phenotype. The DCs are found in the skin (where they are called Langerhans cells) and in all the various lymphoid organs. As professional APCs, DCs, wherever they may be located, will encounter and ingest foreign proteins and then seek interaction with lymphocytes to "present" antigenic peptides on MHC II molecules. DCs show considerable initiative in performing this role, as they will actually migrate into the lymph nodes, draining the area of antigen exposure, and in an MHC II-dependent manner, trigger the lymphocyte response (Fig. 2.5).

T-Cell Activation and Effector Function

We have learned that antigen recognition by the TCR is through association with either MHC I or II molecules. This is a necessary condition for activation but not a sufficient condition. Additional accessory molecules are required to initiate the drive into the cell cycle (see earlier discussion regarding T-cell expression of CD28). These include CD3, as well as the CD4 and CD8 molecules. The CD3 molecule activates signal transduction by means of its close association with the TCRα/β heterodimer. The CD4 molecule binds to MHC class II, which contributes to cell signaling. Likewise, the CD8 molecule binds the class I MHC molecule and is believed to transduce the cell signal. The CD28:B7 interaction completes this set of molecular interactions and sets up the conditions to initiate T-cell activation and movement toward an effector phenotype.[18]

---

[18] While interlocking of molecules or "handshakes" between the T cell and APC is the prerequisite for activation, a poorly understood conundrum was how this revved things up in terms of T cell effector function, since simply injecting a benign protein (e.g., egg albumin) into a mouse did not generate much of an immune response. Charles Janeway and colleagues resolved this paradox by showing that adjuvants could kick-start a process that ostensibly was in a stalled state. It was eventually realized that adjuvants induced innate immune responses, and in particular, the production of accessory molecules, which turned out to be cytokines. These accessory molecules stimulated the intracellular machinery of T cells that had recognized their antigen, and onward they progressed into a proliferative and cytokine-producing effector state.

T-cell effector functions include: (1) helping B cells generate an antibody response (mediated by CD4$^+$ lymphocytes); (2) eliminating virally infected cells (mediated by CD8 T-cytotoxic lymphocytes); and (3) activating macrophages (mediated by T-cell derived cytokines, e.g., IFNγ). Recall that MHC I expression is ubiquitous throughout the body. Given that viruses can infect nonimmune cells, the immune system evolved a mechanism to recognize foreign viral antigens through the recognition of MHC I molecules. The CD8 T cell, therefore, has the capacity to attack virally infected cells. This will liberate the virus and expose it to antibodies that may already have been generated by B cells. The drawback of CD8 activity is damage to self-tissue, although if regulated appropriately, the response can limit the spread of the virus and, with the aid of antibodies, can remove it from the body.

We noted earlier that the CD4 T cell, also called a T-helper cell, provides "help" to B cells. The discovery and study of two subtypes of T-helper cells—Th1 and Th2—dominated immunological research for many years, and distinguished these subtypes in terms of their cytokine profiles, as well as the extent to which they promoted inflammation. The Th1 cells, for instance, became recognized as a proinflammatory subtype that is less involved in helping B cell function than Th2 cells, which have an antiinflammatory reputation and can fine-tune the nature and affinity of antibody secreted by B cells. Subsequently, other subtypes of CD4+ T cells emerged. This included the T-reg cells (for regulatory T cells) and follicular helper T cells (Tfh cells). The former serve to suppress or inhibit the responses of CD4 and CD8 T cells, and as such have been implicated in the control of inflammatory immune responses. Interestingly, at one time in the history of immunological thinking, there was talk of a "T-suppressor" cell (or a Ts cell), about which little is mentioned today. However, this function is now attributed to Tregs, as they have come to be known, which exert suppressor activity utilizing specific cytokines, such as IL-10 and TGFβ. The other newer CD4+ subtype are the Tfh cells, which appear to be the quintessential B cell helpers. These cells are uniquely attracted to the B cell-rich germinal centers of lymphoid organs (e.g., lymph nodes) and facilitate their proliferation and differentiation into antibody-secreting plasma cells. This function is due to a uniquely expressed receptor on their surface (CXCR5) that allows them to home to the GC, as well as secretion of cytokines that facilitate high-affinity antibody production. Finally, an additional CD4+ T cell, that almost exclusively produces a cytokine called IL-17, has been heavily investigated. This T cell, called a T17 helper cell, is induced by the cytokine IL-6, and plays a role in activating innate immune components. Therefore, from the earlier dichotomy of Th1 and Th2 cells, other CD4+ T cell subtypes have emerged that represent important elements of regulation in the adaptive immune response.

Cytokines

We have learned that lymphocyte activation with antigen requires cell-to-cell contact with attending membrane molecules, such as the TCR-CD3 complex, MHC I and II, and the B7 and CD28 costimulatory molecules. However, there are millions of lymphocytes and just as many circulating monocytes and resident macrophages that all need to be tuned in to the events of antigen retrieval, presentation, and ultimate lymphocyte activation. Their recruitment and involvement are part of the necessary amplification of an immune response, in which antigen-specific lymphocytes proliferate, and antibodies are liberally dispersed to bind and neutralize the bacteria and viruses that express the antigens. For all this to work, there has to be a secondary layer of communication that involves soluble factors that communicate between the individual leukocytes. This secondary layer is the cytokine network. Like neurons that communicate with each other through neurotransmitters that diffuse across a synapse, so are leukocytes informed and instructed to migrate in a given direction, bind to a particular cell, enter a mitotic phase of cell expansion and differentiation, and perform various effector functions such as cytotoxicity and antibody production. All these functions, and more, are induced by soluble proteins called *cytokines*. The name literally means "cell messenger," and these molecules serve in an autocrine (acting on self—e.g., clonal division) and paracrine (acting on bystanders) manner to mobilize cells toward the goal of finding and eliminating pathogens. Within the immune system, cytokines are the chief effector and regulatory molecules of T cells, monocytes, neutrophils, and macrophages. Similarly, while not commonly recognized as cytokine-producing cells, B lymphocytes also are capable of making and releasing cytokines (Lund, 2008). Furthermore, since their discovery in the immune system, cytokines have been found to be made by other cells in the body, including fibroblasts and endothelial cells, and glial cells (and possibly neurons) in the brain. Therefore, while it is recognized that cytokines are an indispensable element of how the immune system goes about its work, it is also recognized that when immune cells produce cytokines, these may act on cellular targets outside the immune system. This renders cytokines as endocrine signaling molecules.

Within the general family of immune-derived cytokines, there are different categories of signaling and regulatory molecules. Historically, cytokines produced by T cells were called *lymphokines*, whereas those from monocytes and macrophages were simply *monokines*. However, for convenience, and the fact that some molecules were both monokines and lymphokines (produced by monocytes and lymphocytes), the generic term *cytokines* was adopted. However, each individual cytokine has retained a unique name, usually reflective of the original function that was attributed to it, or part of another system of nomenclature. For example, many cytokines are called interleukins (i.e., "between leukocytes") and abbreviated, "IL." Many interleukins are produced by T lymphocytes, but monocytes and macrophages also produce interleukins. As an example, interleukin-1 (IL-1) was one of the first major cytokines to be purified and sequenced at the protein and gene level. However, it was first known under different names, such as endogenous pyrogen and thymocyte-stimulating factor. Both names were based on cell-free solutions from stimulated leukocytes producing pyrogenic effects in animals (i.e., inducing fever) or priming thymocytes for mitotic responses to other factors. Ultimately, whether it was endogenous pyrogen or thymocyte-stimulating factor, it eventually became known as IL-1, perhaps the most potent neuromodulatory molecule ever discovered. In fact, the first clue that leukocytes had neural effects was inherent in the pyrogenic effects (a hypothalamic mechanism) of the nonpurified culture supernatant that contained "endogenous pyrogen." No one seemed to catch on, and it was some years before this was more formally demonstrated by Besedovsky and colleagues who demonstrated that IL-1 acted on the hypothalamic–pituitary–adrenal axis. In any case, once IL-1 was coined, others followed suit. The next was IL-2, which was originally known as T-cell growth factor. This was isolated from cultures of activated T lymphocytes, and was able to promote thymocyte proliferation in the presence of IL-1 (since this allowed for greater expression of IL-2 receptors). Since those early days more than three decades ago, many interleukins have been discovered and their genes cloned. What has emerged is an understanding that cytokines operate within a network of mutual regulation in which some cytokines facilitate the production of others, while at the same time suppressing synthesis of cytokines or regulating (either up or down) cytokine receptor expression, the key mechanism by which cytokines affect the functions of their various cellular targets (see Fig. 2.6). Each cytokine has a distinct

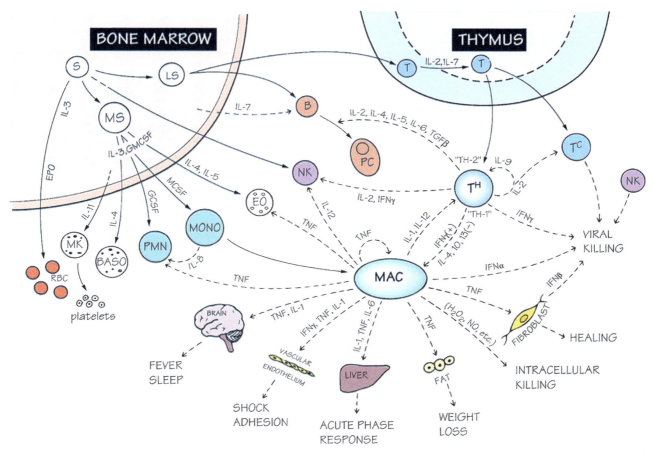

FIG. 2.6 The cytokine network. Cytokines work in all environments where communication between cells is required at some distance, even if cells are present locally. This is why they are hormone- or neurotransmitter-like. In the bone marrow and thymus, they contribute to the differentiation and maturation of cells. Once immune cells are mature and recirculating between blood and lymphatics, cytokines serve as instructional messages. Note the many cytokines released by T-helper cells (T^H in the figure), which target B cells, and the prominence of TNF and IL-1 as targets for both immune and nonimmune targets. These and other cytokines are central players in neural-immune interactions and will be mentioned often. *From Playfair, J.H.L., Chain, B.M., 2013. Immunology at a Glance, 10th ed. Wiley Blackwell, Oxford, UK.*

52   2. The Immune System: An Overview

function, and the compendium of known cytokines has grown considerably. Table 2.1 offers a brief summary of the functions of some of the more prominent cytokines that we will encounter in subsequent chapters.

The major points to be made about cytokines are as follows:

1. Most are not constitutively expressed and in many cases need to be induced;
2. Their actions are mediated through a specific G-protein coupled receptor synthesized and expressed by the target cell;

3. Many cytokines have multiple overlapping functions (i.e., they are *pleiotropic*), which allows for a broad range of effects; while some cytokines have overlapping effects, demonstrating both redundancy and the presence of compensation;
4. There is no one role for cytokines—they can cause cells to proliferate (e.g., IL-2), and they can cause cells to die (e.g., TNF promotes apoptosis); their function is instructional, conferring a particular directive that is needed for a given aspect of the immune response to take place (e.g., signaling macrophages to express

TABLE 2.1 Common Cytokines Encountered in Neural-Immune Investigations

Cytokine	Main Immune Cell Origin	Receptor Types	Most Common Function	Target Cells
Interleukin-1 family				
IL-1 (IL-1α, IL-1β)	Monocytes, macrophages, and other innate immune cells (incl. DCs)	IL-1R1	Inflammation, fever	Endothelium, CNS, hepatic cells, T cells
		IL-1R2		
IL-18	As above	IL-18α	Inflammation	T cells, NK cells, monocytes, neutrophils
		IL-18β	Induces IFNγ by T and NK cells	
Tumor necrosis factor super family				
TNF	Macrophages	TNFRI	Inflammation, cachexia	Endothelium
	NK cells	TNFRII		Neutrophils
	T cells			
Type I cytokine family				
IL-2	T cells	IL-2Rα	Mitotic (proliferation) activity	T, B and NK cells
		IL-2Rβ	T-cell effector and memory formation	
			T-reg generation; NK cytotoxicity	
IL-4	CD4 T cells	IL-4Rα	Isotype switching to IgE, proliferation and differentiation of Th2 cells	B cells
				T cells
	Mast cells			Macrophages
				Mast cells
IL-6	Macrophages	IL-6Rα	Inflammation, acute phase response, B cell differentiation (in plasma cell)	Hepatic cells
		gp130		B cells
				T cells
IL-12 [IL-12A (p35) IL-12B (p40)]	Macrophages	IL-12Rβ1	Th1 differentiation	T cells
				NK cells
				IFNγ induction
			Dendritic cells	Promotes cytotoxicity
				IL-12Rβ2
IL-17 (IL-17A, IL-17F)	CD4 T cells (Th17)	IL-17RA	Increase chemokine and cytokine production by macrophages and endothelial cells	Endothelium
				Macrophages
				Epithelial cells
				IL-17RC
				Group 3 innate lymphoid cells

TABLE 2.1 Common Cytokines Encountered in Neural-Immune Investigation—cont'd

Cytokine	Main Immune Cell Origin	Receptor Types	Most Common Function	Target Cells
Type II cytokine family				
IFNγ	CD4 T cell (Th1)	IFNγR1	Induces MHC II, augments NK cytotoxicity	Macrophages
				NK cells
				IFNγR2
				CD8 T cell
				NK cell
IL-10	Macrophages	IL-10Rα	Suppression (cytokines, MHC II)	Macrophages
				Dendritic cells
		IL-10Rβ		T cells (especially Th1)
				T cells (especially T-reg)
Ungrouped cytokines				
Transforming growth factor-β (TGFβ)	T cells (especially T-regs), macrophages	TGFβR1 TGFβR2 TGFβR3	Suppression (B and T cells; macrophages), Th17 and T-reg differentiation, tolerance promotion	T cells B cells Macrophages

Compilation based on details from Abbas et al. (2015).

more MHC class II molecules, which is one of the directives conveyed by IFNγ, another being to augment the cytotoxic functions of NK cells).

The following table lists some of the major cytokines involved in regulating important immune processes, as well as those that are most commonly studied in psychoneuroimmunology. Immunologists have classified cytokines into superfamilies based on shared structural elements of the receptors they bind. Table 2.1 adopts this grouping, providing descriptions of selected members of (1) the interleukin-1 (IL-1) family, (2) tumor necrosis factor superfamily, (3) type I (e.g., IL-2 and IL-4), and (4) type II (IFN-γ and IL-10) cytokines. Only key functions are listed, and of course, one needs to go to the literature to find many other functions and exceptions for each selected cytokine. Overall, there are close to 100 or more cytokines that have important immunological and non-immunological functions. For example, a recent study focused on mouse immune cells stimulated in vivo by individual injection of each of 86 different cytokines (Cui et al., 2024). Multiple immune cell types (lymphoid and myeloid) showed a similar gene activation profile to the same cytokine (viz., IL-1), while many cytokines activated multiple immune cell types, but which showed differential gene activation profiles to the same cytokine (an example of pleiotropy in which a given cytokine induces different responses in different cell types) (Cui et al., 2024). Consequently, while cytokines can have a broad range of targets, the response of those targets is varied or individualized.

Chemokines

The chemokines are a subset of cytokines. Their categorical name comes from a concatenation of "chemo" from chemotactic and "kine" from cytokine, which indicates the particular role that they play in the recruitment of an immune response. Chemotactic effects have been recognized for a long time, in that the release of certain molecules from a given source acts as attractants for cells that are nearby (Moore and Kunkel, 2019). In principle, the release of a chemokine creates a concentration gradient along which cells will travel or "roll" to reach the area where the chemokine is being released. This can be an effective means to flag down circulating monocytes, for example, and redirect them to a site of infection. The field of chemokine biology has grown dramatically since these molecules play an important role not only in immune function but also in the development of the nervous system. Axonal guidance, for instance, can now be explained in terms of chemokine release.

A special nomenclature exists for the naming of individual chemokines (reviewed in Hughes and Nibbs, 2018). These particular cytokines are polypeptides of 8–10 kDa in size and belong to one of four major families. The family category is based on the location of N-terminal cysteine residues in the polypeptide, which determines the name of the particular chemokine. There is the CC family, containing two adjacent cysteine residues; the CXC family, wherein two cysteine residues flank a different amino acid; the CX3C family, in which three amino acids interdigitate between two cysteine residues; and finally, there is a family of chemokines that contain only a single cysteine residue. Like many other cytokines, the chemokines were identified on the basis of the responses they elicited from various cells. The CC and CXC chemokines are produced by multiple cells of the immune system, as well as fibroblasts, endothelial, and epithelial cells. The latter are important sources of signaling to circulating leukocytes, should an infection be taking place in a given tissue location. There are at least 50 known chemokines, and when they are not referred to by their original functional properties, they are given an "L" suffix, to designate their ligand status (e.g., the leukocyte attractant first named as fractalkine is also referred to as CX3CL1). Interestingly, the number of different chemokine receptors (cited as ranging from 18 to 23) is less than the number of identified chemokines (Hughes and Nibbs, 2018). This implies that multiple chemokines may impact the same receptor, which indeed is the case (Martin et al., 2024).

Chemokine receptors are G-protein coupled and are designated with an R suffix. For example, CX3CR1 is the receptor for fractalkine (CX3CL1), and is found on myeloid cells, as well as microglia in the nervous system. Once chemokine receptors are stimulated, signal transduction changes those elements of a cell that affect motility, including cytoskeleton actin/myosin polymerization. At the same time, cell adhesion and activation are induced—in conjunction with the influence of other cytokines—to facilitate the actions of cells that have migrated to the source of chemokine release. Although chemokine receptors are expressed on all leukocytes, their expression on T cells is particularly important to regulate their homing properties. Given that T lymphocytes are in continual circulation, their movements can be redirected based on chemokines released by sites of inflammation and injury.

As important as T cell recruitment can be, neutrophils and monocytes are often the first cells recruited to the site of an infection. Local macrophages that have been activated by a pathogen can release the chemokine CXCL8, and since this is a ligand for the chemokine receptors CXCR1 and CXCR2, neutrophils that express these receptors will become modified and "brake" as they pass, and redirect their attention to the area of CXCL8 release. This redirection is driven by molecular changes and the expression of adhesion molecules, such as leukocyte adhesion molecules. The neutrophil will then pass between the endothelial cells of a blood vessel (i.e., extravasate), and enter the parenchyma of the infected organ, where it will engage in phagocytic activity and amplification of the immune response through the release of cytokines that act on local and recruited immune cells.

In sum, chemokines are a special category of cytokines that in many ways have come to dominate immunological thinking and development of therapeutic strategies to treat cancer and autoimmune disease. It is difficult to imagine that in a system reliant on constant mobility throughout the body that a soluble mechanism for trafficking and signaling cellular movements would not have evolved. It is yet again testimony to the remarkable complexity of the immune system.

Tolerance and Autoimmune Disease

To this point, we have focused on an overview and description of some of the major elements of the immune system. At the very beginning, we suggested that thinking of the immune system in militaristic terms was useful since it is essentially a defense system. But, in mounting a defense, it is, like any military campaign, subject to mistakes, considerable calamity, and collateral damage. As well, it needs to know who its allies are and who comprises the enemy. The latter is critical; self-annihilation is not part of the "military" objective. Therefore, when we discussed the thymus and the MHC complex, we noted that recognizing the MHC as part of self was important in order to trust self-reports of an external threat (MHC presenting antigen). Furthermore, a peripheral, mature, extra-thymic T cell already has learned not to react (i.e., become activated and develop effector responses) to any other proteins that the body expresses. The consequence of this self-tolerance is that the organism does not develop autoimmune disease, an immunopathological condition in which the immune system treats certain organs, tissues, and/or molecules expressed by these tissues as essentially foreign or nonself.

There are numerous debilitating conditions that are defined as autoimmune diseases. Dysregulation of T cell and macrophage function underlies diabetes, thyroiditis, and rheumatoid arthritis, while B cell auto-antibody responses underlie the immunopathology of myasthenia gravis, systemic lupus erythematosus, and hemolytic anemia. Similarly, multiple sclerosis, a demyelinating disease of the central nervous system, is thought to involve a dysregulated T-cell response and monocyte infiltration of the brain and spinal cord, while other autoimmune disorders, such as inflammatory bowel disease involve both T and B cells (Cheng et al., 2024). Later, we will get to discuss some of the research trying to link autoimmune diseases to conditions like autism and schizophrenia.

A common epidemiological finding is that females are more likely to develop autoimmune disease than men. The greater incidence in women has been estimated to be 75%, with the exception of males showing a greater incidence of type 1 diabetes, psoriasis, and myocarditis. The role of sex hormones has arisen as the primary candidate for why this sex disparity exists. Indeed, it is well established that the cellular composition and responsiveness of the adaptive immune response is greater in females (human and murine), although innate immune components are lower in females than in males (Forsyth et al., 2024). This may bias females in warding off infections, as males are more susceptible to microbial infections. However, this disparity may have introduced a greater risk among females to erroneous targeting of self-tissue antigens, although we still do not know the reasons for why immune cells become auto-aggressive and cause immunopathology. Sex differences with a female bias suggest involvement of sex hormones, such as estrogen, but males still represent a significant proportion of affected individuals and may show a more pronounced and aggressive rate of deterioration, as in multiple sclerosis (Alvarez-Sanchez and Dunn, 2023).

Very little is actually known concerning the etiology of many autoimmune diseases. However, it is generally understood that it may be related to a failure of immune tolerance, a key tenet of effective immune functioning. As discussed earlier, the thymus is critical to the education of T cells into mature phenotypes that do not respond against host (or self) peptides and can only do so in an MHC-restricted manner, having learned to recognize MHC class I and II molecules as legitimate mechanisms for the presentation of antigens. It was established many years ago that removal of the thymus or deficient thymic function resulted in or was associated with autoimmune disease (Shirafkan et al., 2024). We now know that T cells are fundamental in maintaining tolerance against the self. Critical in ensuring this state of affairs are Tregs (discussed earlier), which are now being harvested and engineered in cell therapies designed to maintain tolerance and suppress immune responses in autoimmune and inflammatory conditions (Ho et al., 2024).

Autoimmune disorders also involve the generation of B cell derived autoantibodies, which are antibodies that bind to self-molecules and facilitate cellular damage. Primary examples of this mechanism of pathology are myasthenia gravis and lupus. The development of B cell tolerance takes place in the bone marrow, where any B cells that express self-reactive receptors are removed through apoptosis, or their receptors (the Fab regions of surface immunoglobulins) undergo editing through gene recombination to render them nonself-reactive (Chi et al., 2024). In myasthenia gravis the autoantibodies are directed against the acetylcholine receptor at the neuromuscular junction. This results in impaired control of muscle function and typical symptoms of impaired movement and maintenance of muscle tone, characterized by weakness, droopy eyelids, as well as difficulty swallowing and chewing. In SLE, the autoantibodies are directed against nucleic acids in the nucleus of cells in multiple organs. These are referred to as antinuclear antibodies (or ANAs) and can seriously disrupt organ function.

As mentioned, failure of T cell mediated regulation is central to current theories of autoimmunity. Inflammation can be promoted by buildup of T cells and neutrophils, which release cytokines that create tissue degeneration. As indicated earlier, given that T cells are important in maintaining immunological tolerance, there is general appreciation that T cells, and failed Tregs in particular, are implicated in the etiology and/or perpetuation of MS (Yang et al., 2024b).

The incidence of autoimmune disease has allowed for speculation that to some extent certain behavioral abnormalities (like schizophrenia) are the consequence of an autoimmune process. Indeed, later we will discuss some of the research tying autoimmune disease to conditions like autism and schizophrenia. To be sure, there is no conclusive evidence that mental health disorders are caused by autoimmune mechanisms. However, considered broadly, and expanding the definition of autoimmunity to involve increased and chronic levels of inflammation, it is more than reasonable to suggest that behavioral abnormalities may represent instances of immunopathology. The high incidence of neuropsychological problems that are observed among those who suffer from autoimmune diseases like MS and SLE most certainly speaks to this point.

To conclude, very little is actually known concerning the etiology of many autoimmune diseases. One epidemiological finding is that females are more likely to develop autoimmune diseases than men. The reasons for this are unclear, and an obvious hypothesis is sex hormone regulation. However, for the most part, the reasons why immune cells become auto-aggressive are still uncertain. To some extent, it provides an opportunity to hypothesize that perhaps certain behavioral abnormalities (like schizophrenia) are the consequence of an autoimmune process.

Concluding Comments

It is hoped that sufficient elemental information has been provided to make the remainder of this book easier to read. Where certain information was omitted, it is highlighted in subsequent chapters. These new pieces of information should make sense since the purpose of this chapter was to generate some assumption of knowledge, rendering new terms and processes less confusing. We urge the reader to explore further and deeper the details of immunology, and in addition to the citations that have been used, recommend as starting points the texts by Playfair and Chain (2013), Abbas et al. (2015), and Sompayrac (2016). The latter provides a humorous and gentle overview of the immune system, while Playfair and Chain provide excellent illustrations that offer a pictorial guide of sorts to the immunological landscape. For a more in-depth and clinical approach, Abbas is an excellent introduction, after which readers should be well-versed to venture farther afield.

CHAPTER

3

Bacteria, Viruses, and the Microbiome

Rocking the boat

During the early part of the 19th century, postpartum bacterial infections (puerperal infections) were not uncommon and were fatal in as many as 10%–15% of cases. As odd as it may seem from today's standards of care, at the time, little was understood regarding bacterial infections and proper hygiene within medical settings. Ignaz Semmelweis had the audacity to suggest that postpartum bacterial infections might stem from unhygienic medical staff or medical equipment that had not been cleaned, and he demonstrated that infection frequency could be diminished by having staff wash their hands with chlorinated lime solutions. These suggestions were not well received (were not the hands of doctors always clean?). With Lister's discovery of antiseptics in 1865 (published in 1867), thanks in part to Pasteur's work related to "germ theory," the practice of maintaining cleanliness became paramount in surgical practice. Arguably, antiseptics and anesthetics changed surgery and surgical risks forever. However, poor Semmelweis did not get to see these breakthroughs in medicine, nor was he rewarded for his observations. Instead, he was ostracized from others in his profession. He became progressively more depressed, and his battles with institutionalized medicine eventually landed him in a medical asylum. Upon trying to leave, he was beaten by guards and died 2 weeks later because of internal injuries or gangrene secondary to his injuries.

The situation within hospitals has obviously improved since then, except that hospital-acquired infections have been on the rebound for years, including those that are treatment-resistant. One terrible condition that can afflict patients is sepsis, comprising infection that spreads to the bloodstream, and the resulting inflammatory cascade can damage various organ systems, eventually leading to death. Early treatment with antibiotics and lots of fluids had been the treatment of choice, but it would be far better to prevent the condition as Semmelweis had done regarding postpartum bacterial infections. As it happens, simple procedures, such as increased training for staff and the use of a special observation chart to identify early signs of sepsis, can appreciably reduce its occurrence.

The risk of bacterial infection within hospital settings has gone beyond sepsis. Soon after hospital admission, patients often lose commensal gut bacteria, whereas pathogenic bacteria might rise (McDonald et al., 2016). This could occur because of the stress of illness, the novel hospital environment, a change of sleep pattern, or hospital foods, but whatever the case, this dysbiosis (microbial imbalance) could affect immune functioning, thereby influencing vulnerability to illnesses. With the proviso that resources are available, it may be possible to track individual microbiota that could eventually inform patient vulnerabilities and potentially point to optimal treatment strategies.

Despite the ignominious treatment Semmelweis received while alive, he was treated better posthumously. He became known as the "savior of mothers," and the Semmelweis Klinik, a women's hospital in Vienna, is named after him, as is Semmelweis University in Budapest and the Semmelweis Hospital in Miskolc (Hungary). He might have been pleased by the term the "Semmelweis reflex," which refers to the reflexive rejection of new knowledge that challenges old norms and beliefs.

Bacterial Challenges

Bacteria, which are thought to be among the first life forms that appeared on Earth, are present everywhere, coming in varied sizes and shapes. They typically live in harmony with plants and animals, acting symbiotically, but can also function in a parasitic relationship with other living things. Many of the trillions of bacteria present in animals (and humans) have important beneficial actions, whereas others are less kind, causing a variety of diseases. Under the right environmental conditions, including the temperature and pH, availability of water, oxygen, and a source of energy, bacteria will grow to a particular size and then reproduce asexually through binary fission. Bacteria can double in number every

The Immune System
https://doi.org/10.1016/B978-0-443-23565-8.00020-X

Copyright © 2025 Elsevier Inc. All rights reserved, including those for text and data mining, AI training, and similar technologies.

30 min, and some bacteria can do this within 10 min. Certain bacteria, such as Streptococcus, can stay alive for some time on various external objects (e.g., door handles, toys, and cribs) and thus can represent a relatively persistent threat. Other bacteria, like their viral cousins, are subject to airborne or droplet transmission, or direct physical contact. It seems that different strains within the same bacterial species can engender very different immune responses, which may contribute to the diverse outcomes elicited across individuals (Sela et al., 2018).

Infection stemming from bacteria and viruses can also be transmitted indirectly. For example, carrying an infection on unwashed hands, and depositing these on a surface, which is then touched by another person, can lead to infection being passed along (fecal-oral transmission). As we know from illnesses, such as cholera, dysentery, diphtheria, scarlet fever, tuberculosis, typhoid fever, and viral hepatitis, disease agents can also be transmitted through water, ice, food, serum, plasma, or other biological products. In some instances, a disease (e.g., the bacterial infection syphilis or the parasitic disease toxoplasmosis, as well as viruses, such as HIV and measles) can be passed from a pregnant mother to her fetus. Furthermore, zoonotic diseases, in which infection is transmitted from animals to humans, are a constant threat but can become exceptionally hazardous if they mutate so that they can then be passed between humans.

As we will see in ensuing chapters, viruses and bacteria, by virtue of inflammatory processes being activated, and the downstream actions on many hormones, brain neurotransmitters, and growth factors, may contribute to several psychological disturbances as well as a great number of physical disorders. To a significant extent, common mechanisms account for these varied conditions and may be responsible for the frequent comorbidities that are seen among illnesses. Increasingly, the significance of infection in relation to mental illnesses has been acknowledged. Methods to prevent or control infection are thus essential, but as we will see, the best treatments to ameliorate bacterial infection may destroy commensal gut bacteria, leading to the emergence or exacerbation of other illnesses. The sword clearly cuts both ways.

Antibiotics

The development of penicillin, and other antibiotics to fight bacteria, was undoubtedly among the most important medical discoveries of the first half of the 20th century. Although Alexander Fleming, who identified penicillin obtained from particular molds, is usually given credit for antibiotics, infections had been treated with mold extracts for about 2,000 years. In general, antibiotics either kill bacteria (being bacteriocidal) or inhibit their multiplication (bacteriostatic). They do this by either

preventing bacteria from building cell walls (e.g., by affecting bacterial ribosomes involved in the creation of cell walls) or breaking down the cell walls of bacteria that already exist. Some antibiotics, such as quinolones, disturb DNA and prevent their repair, so that the bacteria are unable to reproduce and thus die off. Based on the response to a gram stain, and the characteristics of the cell walls, bacteria are designated as either gram-positive or gram-negative (the latter being more resistant to antibiotics). When the nature of the bacterial infection is known, a narrow-spectrum antibiotic is used, whereas bacteria that have not been identified are treated with a broad-spectrum antibiotic. The former is preferable as they are less likely to create antibiotic resistance. Some antibiotics can produce uncomfortable side effects, and in some instances, allergic reactions can occur that cause anaphylaxis.

As much as antibiotic treatments can be beneficial, they can also cause health problems by eliminating commensal bacteria and disturbing the microbial balances that exist, thereby increasing susceptibility to assorted diseases. As described in an excellent review (de Nies et al., 2023), although many of these diseases are noncommunicable, such as obesity and diabetes, they can also diminish protection against foreign pathogens, thereby allowing the development of communicable illnesses. The greatest attention regarding microbiota and antibiotic effects has focused on the gut, where the vast majority of bacteria exist, although as we saw earlier, bacteria inhabit multiple body sites, such as the skin, mouth, and vagina, and microbiota disturbances at these locales can provoke adverse outcomes. These actions vary with the nature of the antibiotic agent used since they can affect different microbiota types. In infants and older individuals, lower microbial stability may also exist, so antibiotics may have especially pronounced adverse effects.

Antibiotic Resistance

We grew accustomed to being able to destroy bacteria through treatment with antibiotics, and for some time it had simply been assumed that when one antibiotic failed to do the job effectively, then another could do the trick. Ironically, their very effectiveness contributed to their undoing. As bacteria began to form resistance to antibiotics (reflected by greater difficulty in treating some infectious diseases, lengthier recovery times from infection, and the probability of death increasing) and the first alarms were sounded, a generally cavalier attitude persisted, and most people continued to behave as they had previously. Inevitably, most bacteria followed an effective game plan to get around antibiotics, and they all successively became less effective or entirely ineffective. Antimicrobial resistance has increased exponentially, leading to almost 2 million deaths that could otherwise have been prevented (GBD 2019 Antimicrobial Resistance Collaborators, 2022).

The factors that generated treatment-resistant bacteria comprised the perfect storm. One should never have imagined that bacteria were passive travelers who were simply waiting to be killed by antibiotic agents. Instead, like an opposing army (or groups of terrorists) bent on the host's destruction, some harmful bacteria are clever and vicious, so that with time and experience, they develop resistance to the drugs. It was suggested that in response to stressors, such as nutrient deprivation, microbiota respond in a coordinated manner to deal with the insult. Being a new challenge for bacterial communities, an anti-biotic might result in bacteria rapidly searching for new methods of dealing with the challenge. Ultimately, through a process much like natural selection based on random mutations occurring, bacteria develop resistance to the antibiotic (Jensen et al., 2017b).

The ability of bacteria to become resistant might have been facilitated by the inappropriate use of antibiotics to fight viruses (e.g., strep throat and bronchitis) for which antibiotics are ineffective. In fact, in the face of a serious threat, such as antibiotic treatment, bacterial mutation rates increase appreciably, thereby increasing the proba-bility of a mutation occurring that will protect the bacteria from destruction. Furthermore, when confronted with an antibiotic, especially if the full course of treatment was not adopted (because patients felt better and believed they no longer needed the antibiotic or because they were saving pills in case they were needed at some later time), a few hardy bacteria may survive. This may then give rise to similarly resistant clones, so that over successive gen-erations and increased development of evasion methods, the effectiveness of antibiotic agents diminishes.[1]

The rate of bacterial mutation increases with a person's age as well as with the social environment in which bac-teria find themselves. At the other end of the age spec-trum, babies born very prematurely are at increased risk of illnesses developing. As a matter of course, pree-mies were treated with antibiotics in the mistaken belief that this could not cause harm, but this resulted in a marked decline in the diversity of microbiota and simul-taneously enhanced survival of bacteria that were resis-tant to antibiotics. Thus, should a blood infection subsequently arise, a large proportion of these infants will not fare well, especially as resistance to one type of antibiotic also dials up resistance to other antibiotics (Gibson et al., 2016).

The massive use of antibiotics in farm animals to pre-vent them from developing infections has contributed to resistant bacteria evolving. The antibiotic-infested meats end up on our dinner plates and thus contribute to the development of resistance. Also, there is a good chance that animal waste, laden with antibiotics and antibiotic-resistant bacteria, leeched into waterways which affected humans. Air pollution may also affect commensal bacteria, and influence resistance of some bacteria (*Staphylococcus aureus* and *Streptococcus pneumonia*) to antibiotic treatments (Hussey et al., 2017). There has also been a reasonable concern that some household products, such as the disinfectant triclosan, may contribute to anti-biotic resistance. As a result, it has been banned from hygiene products, such as hand, skin, and body washes, but triclosan and similar agents appear in numerous other products.

In addition to the mutations that are due to the overuse of antibiotics, bacteria have several dirty tricks that they can fall back on. For instance, the genes involved in the development of resistance can be transferred to other cells (conjugation) so that they too will become resistant, although it may be possible to prevent or reverse this action (Lopatkin et al., 2017). Furthermore, in response to an antibiotic, bacteria can go dormant, making them less likely to be attacked (termed persistence). With repeated antibiotic attacks, they essentially "learn" to stay in the dormant state for periods that line up with the antibiotic's actions, but their spores reawaken in response to appropriate signals emerging once it seems safe (Kikuchi et al., 2022). On top of this, bacteria seem to act collectively, coordinating their actions to render maximal toxic effects based on messaging from some external source (quorum sensing), such as the medium in which the bacteria are present (Bridges and Bassler, 2019). Bacterial communities can secrete substances, such as β-lactamase, which can proffer passive resistance to other bacteria that are present in that particular environ-ment. Similarly, bacteria that express the resistance factor chloramphenicol acetyltransferase (CAT) can deactivate antibiotics present in their immediate environment. In essence, the response to an antibiotic could be affected by the present microbial environment (Sorg et al., 2016). It also appears that bacteria, such as Entero-bacteriaceae, that reside in the intestine, may resist

---

[1] The seemingly common-sense perspective concerning overuse of antibiotics was almost universally accepted, even though it has been argued that there was actually limited evidence supporting this contention. Physicians typically prescribe based on precedent, which could actually reflect *overtreatment*, thereby placing patients at increased risk for antibiotic resistance (Llewelyn et al., 2017). Yet, it was demonstrated that 5 days of antibiotic treatment was as effective as 10 days in eliminating community acquired pneumonia (Uranga et al., 2016). The case for limiting the duration of antibiotic treatment has been impressive (Spellberg, 2016). Unfortunately, practioners are frequently reluctant to abandon the dogma that they had maintained for years.

carbapenems used to treat severe bacterial infections. This may not only occur through the selection processes described earlier but also because microbiota reductions may cause diminished metabolites that ordinarily inhibit these bacteria and instead provide nutrients that favor the growth of carbapenem-resistant Enterobacteriaceae (Yip et al., 2023).

Antibiotic resistance is an increasing worldwide threat

The WHO has indicated that antibiotic resistance has become among the most pronounced threats to global health and food security. Several bacterial species can cause illnesses by acting as "opportunistic" pathogens. These common threats comprise species such as *E. coli* and those referred to as the ESKAPE organisms, comprising *Enterococcus faecium, Staphylococcus aureus, Klebsiella pneumoniae, Acinetobacter, Pseudomonas,* and *Enterobacter*. Several species and strains of commensal bacteria will readily be destroyed by antibiotics but may ultimately be replaced by those that are resistant. One of the relatively recent threats has come out of China, where a hypervirulent form of *K. pneumoniae* emerged that was multidrug-resistant as well as highly transmissible (Gu et al., 2018).

Staphylococcus aureus (*S. aureus* or Staph infection) is the best-known and most frequent cause of postsurgical infection, but hospital-acquired infections have also included bacteremia (bacteria in the blood), endocarditis (inflammation of the inner layer of the heart), sepsis, toxic shock syndrome, meningitis, and pneumonia. We had counted on antibiotics, such as methicillin, for the treatment of such conditions, but a bacterial strain evolved that stopped responding to this agent. Hospital-acquired infections, particularly methicillin-resistant *S. aureus* (MRSA) and *Clostridium difficile* have markedly increased over decades throughout Western countries (Turner et al., 2019). These infections occurred in about 2 million patients, leading to between 23,000 and more than 100,000 deaths yearly. Following a hospital stay, one in four older individuals had antibiotic-resistant bugs on their hands, which they could spread elsewhere (Cao et al., 2016). Likewise, antibiotic resistance is relatively more common among diabetic individuals using insulin, people undergoing chemotherapy, or who have burns, cuts, or lesions on the skin, patients undergoing breathing intubation, or who have urinary or dialysis catheters inserted, as well as those with HIV/AIDS or with a weakened immune system owing to still other factors. The immunosuppressive actions of stressors can likewise increase vulnerability to *S. aureus* infection, especially in vulnerable populations, such as older people.

It is worrisome that MRSA has surfaced outside of hospital environs, becoming a community-acquired infection that has become increasingly common within individual homes, infecting meat and poultry. Within the community, about 15% of *S. aureus* cases were MRSA (varying across countries and regions within countries), being attributed to sharing contaminated items, active skin diseases or injuries, poor hygiene, and crowded living conditions. Bacterial coinfection with other illnesses (e.g., with influenza) has created treatment difficulties, as seen with SARS-CoV-2 infection (Elabbadi et al., 2021). Furthermore, antibiotic-resistant bacteria can be "picked up" from other people or foods, as observed among tourists who visit countries where these bacteria are relatively common. The good news, even though it is still a bit limited, is that analyzing the DNA of MSRA can identify those individuals who are at greatest risk of dying because of infection and could potentially facilitate the development of personalized treatment strategies (Recker et al., 2017).

In addition to the threat of bacteria becoming resistant to treatment, the incidence of fungal infection by *Candida auris* has been increasing yearly and resistance has been apparent for several antifungal treatments. The number of cases has still been modest, but the signs of things potentially becoming worse are there to see and better detection and methods to control the spread are needed (Lyman et al., 2023).

Even though it has been some time since new antibiotics have been developed, certainly those that could deal with resistant bacteria, it is exciting that machine learning models seem to be able to determine chemical substructures that can be used in creating antibiotics that are effective in dealing with MRSA and vancomycin-resistant enterococci (Stokes et al., 2020; Wong et al., 2023b).

Some illnesses that we had not thought about becoming resistant to antibiotics are doing just this. Neisseria gonorrhea, which is responsible for gonorrhea, has been showing increased signs of resistance to antibiotics, no longer being responsive to some agents (Unemo et al., 2017). Quinolones, a class of antibiotics that had long been used to treat gonorrhea, have lost their effectiveness, as have other drug classes, such as cephalosporins. The last line of defense, the go-to antibiotic colistin, was found not to be useful in certain cases, possibly owing to a transferable gene mcr-1 (Gogry et al., 2021). This gene can appear in a variety of bacteria, and consequently, they could also develop resistance. The good news on this front is that newly discovered compounds that target ribonucleotide reductase can selectively attenuate gonorrhea (Narasimhan et al., 2022).

It is only a matter of time before other threats emerge for which we have little protection or cure. One of these, Shigella, currently affects upward of 165 million people worldwide, most often being transmitted through the "the fecal-oral" route. Historically, this highly contagious condition was treated successfully with ciprofloxacin, but its efficacy is now questionable (Centre for Disease Control, 2016). An antibiotic-resistant form of typhoid has also evolved, infecting large numbers of people within Asia. As typhoid infections ordinarily occur in as many as 30 million people each year, the spread of a treatment-resistant strain may be devastating to an already illness-ridden population. As much as basic health conditions are required to beat various diseases, the development of treatment-resistant bacteria, together with the lack of funds or global political will, may limit the prevention and treatment of illnesses (WHO, 2016a,b).[2]

Dealing With Antibiotic Resistance

As a first step to combat antibiotic resistance, it might be appropriate to limit the use of these agents for minor bacterial infections. Failing this, alternating doses of antibiotics and changing the specific antibiotics administered with successive infections might be helpful. Combinations involving several antibiotics administered concurrently that can act synergistically have shown promise in dealing with some bacteria, and the use of two compounds, one that shreds the shell of bacteria and the second that potently attacks the exposed bacteria, may be useful in dealing with resistant bacteria (Stokes et al., 2017a,b). As the development of antibiotic resistance has been attributed to the ability of bacteria to limit antibiotic entry into cells, as well as the production of an enzyme, β-lactamase, which can destroy antibiotics, β-lactamase inhibitors have been developed to attenuate resistance (Jiménez-Castellanos et al., 2018).

It is also possible to act on bacterial genes to make them more sensitive to antibiotics. A novel compound, teixobactin, which was isolated from microorganisms present in the soil, was capable of destroying pathogens including *C. difficile*, septicemia, and tuberculosis, without resistance developing. Teixobactin seemed to be effective because it attacks bacteria through multiple methods and resistance to its effects may not be seen for several decades (Shukla et al., 2022). The discovery of Teixobactin's effectiveness was followed by several analogs of this compound, and there is a good chance that this is a first step in the development of new antibiotics. An alternative strategy has been to create new synthetic compounds (e.g., COE2-2hexyl) that attack several aspects of bacteria concurrently (motility, ATP synthesis, respiration, membrane permeability to small molecules), effectively limiting the viability of numerous types of bacteria, without resistance developing (Heithoff et al., 2023).

Several "out-of-the-box" approaches have been used to eliminate bacteria. Efforts have been directed to treat specific conditions by having bacteria turn on one another. Bacteriocins (proteins produced by bacteria to kill their competitors) could be harnessed to kill pathogens while leaving other microbes intact. It similarly appears that certain strains of *C. difficile* are adept at destroying each other (they are competitive strains) by firing a harpoon-like needle through their membrane, which promotes the death of the cell. Thus, the interesting notion was broached that human microbiota could be used as a potential source for the development of novel ways of dealing with bacteria (Kirk et al., 2017). Using a somewhat different approach, resistant bacteria, such as MRSA, could be manipulated by altering features that provide protection from antibiotics. For instance, MRSA is reliant on folate (vitamin B9), and hence, blocking the production of folate can be used as a way of overcoming their resistance to treatments (Reeve et al., 2016).

Viruses have been identified that attack bacteria (termed bacteriophages, or simply phages). These phages appear in mucus, such as in the gums and gut, recognize receptors present in bacteria, and attack them, potentially making them effective in treating multidrug resistant bacteria. Since bacteria will attempt to thwart these attacks, a cocktail of different phages may be an effective therapeutic strategy. The development of viruses that can deal with resistant bacteria, such as MRSA, has been welcomed, and a variety of phages have been identified to deal with several diseases (e.g., Duan et al., 2022a).

In addition to these approaches, nanoparticles have been developed that can produce chemicals effective in destroying otherwise treatment-resistant bacteria (Courtney et al., 2017), and CRISPR-Cas9 could potentially be used to cut out genes from bacteria that show resistance to antibiotic treatments (Tao et al., 2022). The inclusion of AI-based technologies has facilitated the identification of novel antibiotics, and computer algorithms that would inform best treatment approaches may prove fruitful in efforts to overcome treatment resistance.

Viral Illnesses

Several viruses, like their bacterial cousins, contribute to infectious illnesses that have psychological ramifications, which we will consider in greater detail in Chapter 4. Viruses are often said not to be a life form since they are not able to reproduce (as they lack cytoplasm and enzymes required to do so) unless they have the

---

[2] Fortunately, Wellcome and the Bill & Melinda Gates Foundation are providing funding to assess the efficacy of a new TB vaccine candidate, M72/AS01E (M72). If effective, this would be the first new vaccine to prevent pulmonary TB in many decades.

opportunity to use the machinery of a cell to this end. Upon penetrating the host cell's genome, the virus uses it to replicate. Once sufficient replication has occurred, the viruses can force themselves through the host cell's membrane and then infect nearby cells. As the virus has its own complement of genes, it can mutate so that new variants of the virus can appear.

Viral illnesses come in numerous forms that may cause a variety of illnesses, as shown in Fig. 3.1. Viruses can spread from one person to another through various routes (e.g., through the air or body fluids), and they can linger for various amounts of time within external environments. In some instances, a virus can lie dormant within the body for extended periods before re-emerging to induce an illness. Viruses and bacteria can also be transmitted to humans through a vector, such as mosquitoes or ticks, leading to illnesses such as malaria, Zika, West Nile virus, dengue fever, and yellow fever, and severe illnesses have spread to humans through birds, pigs, cattle, and rodents. Vector-borne viruses typically do not make the leap to being transmitted between humans. However, these viruses can mutate, and could potentially be transmitted between humans, leading to diseases such as swine flu, HIV/AIDS, Ebola, and SARS-CoV-2.

The virulence of a microbe varies so that some create mild symptoms, whereas others can have rapid and powerful consequences. How quickly and broadly a virus can spread within a human population is dependent on several factors: (1) how readily it can be passed from one person to another, (2) the route by which it is transmitted (e.g., aerobic transmission obviously occurs more readily than transmission that involves exchange of fluids), (3) the ability of the virus to penetrate the host's tissues and enter cells, (4) the capacity of the virus to inhibit the host's immune defenses, and (5) how well equipped it is in obtaining nutrition from the host. Although it is often thought that transmission ceases after the virus kills the host, passage from one person to the next, as in the case of Ebola, may come about even after death.

In some cases, viruses have nefarious ways of getting around the host's immune defenses. Using particular proteins, they can mask themselves so that they are not readily recognized by immune cells (Holm et al., 2016), and with the assistance of other proteins (neuraminidase), as in the case of influenza virus, for instance, they can counter the attack of NK cells that would otherwise destroy the virus (Bar-On et al., 2014). Fortunately, inhibitors of neuraminidase have been developed to enhance

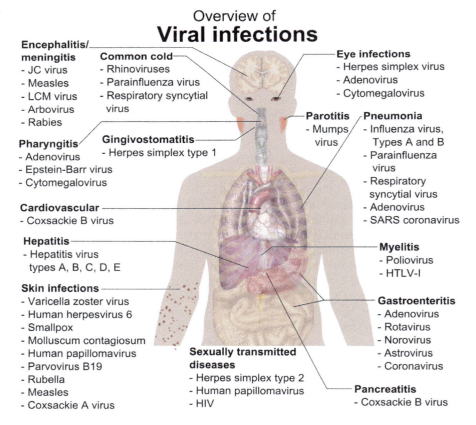

FIG. 3.1 Overview of viruses and their primary targets. *Source: Wikimedia Commons*

the effectiveness of NK cells, and antibodies have been created that act against proteins that limit NK cell activity.

People not to hang with

People react differently to viruses and vaccines. Women, in general, seem to be more reactive to vaccinations, possibly because of hormonal factors increasing immune activity. As well, immune cells encoded by genes present on the X chromosome may contribute to the sex differences and several other genes have been identified that diminish vulnerability to infection in women relative to men (Lipoldová and Demant, 2021). While the greater immune responses among women might seem advantageous, it could also contribute to the greater female disposition toward autoimmune disorders.

Some individuals, often referred to as "superspreaders," seem to be particularly adept in passing on viruses and bacteria. Features of their immune response might be responsible for this facility, or they may have occupations that lead to more social contact either directly or indirectly, or they may simply be especially social, thus encountering a particularly large number of people. Mary Mallon, a cook in the early 1900s, seemed to have been a virtuoso in spreading typhoid, despite not presenting with any symptoms herself. She is now best known as "Typhoid Mary" for having infected 51 people, several of whom died. Today's version of Mary Mallon would be far more dangerous owing to larger populations, crowded conditions, and more efficient travel. Indeed, the Middle East Respiratory Syndrome (MERS) that affected South Korea from May to July of 2015, infected 186 individuals, of which 36 died. It turns out that an individual who contracted the illness transmitted it to another person, who then reported to the hospital with respiratory difficulties. But, as MERS was not yet on the radar, he was not isolated, and over the next few days, he infected 82 others, accounting for 45% of the cases during the outbreak. We have seen similar scenarios play out during the COVID-19 pandemic, especially if individuals were asymptomatic or had mild symptoms. This was abetted by individuals who chose not to be vaccinated, or wear masks in crowded venues.

The Pareto principle, also known as the 80–20 rule, seems to be pertinent to the spread of infection in that 80% of cases transmitted occur through 20% of the people. We can only hope that the potential spreaders choose to be vaccinated, but failing this, we might get lucky, and they will find friends other than us.

Vaccines

For centuries, viral illnesses (as well as bacterial infections) decimated human populations, but the discovery of vaccines to prevent illnesses was an obvious game-changer. Using inactivated or dead virus, the immune system is primed to respond to similar viruses when they are encountered subsequently, thereby preventing the illness from occurring. Despite the effectiveness of many vaccines, others have been less than perfect, varying across individuals, and in some instances, their effectiveness diminishes with age. Genetic mutations may occur within a virus (so that its antigens "drift") and as these mutations accumulate, the virus may be sufficiently altered so that the vaccine no longer recognizes the pathogen. This is what had been occurring with SARS-CoV-2, thus requiring new versions of the vaccine to be made. With each new variant of the virus, its infectivity and its strength may vary, as might its ability to suppress immune functioning.

Influenza viruses come in four forms; influenza A and B are associated with seasonal flu epidemics. Influenza A viruses comprise a large variety of subtypes based on two proteins on the surface of the virus— hemagglutinin (H) and neuraminidase (N). While 18 different hemagglutinin subtypes exist, there are 11 different neuraminidase subtypes. These are numbered sequentially, and the combination of the two is the designated term for influenza viruses (e.g., H1N1 or H3N2). The type of influenza that affects humans varies from year to year. Vaccine makers are often able to anticipate next year's threats, but the accuracy of these predictions is variable. In some years, the vaccine has been very good, but in other years, it has had almost no effect. Even if a yearly vaccine is effective, individuals vary in the extent to which they are "vaccine responders," possibly owing to whether they produce sufficient antibodies to fight future infection.

Some vaccines can be developed readily, as in the case of many seasonal flu vaccines; although effective immunization runs around 50%–60%, developing others is more difficult owing to rapid mutations that occur, as in the case of the H7N9 bird flu virus. This virus spreads from birds to humans and hopefully would not mutate so that the virus spreads between people. However, the CDC has ranked H7N9 and H5N1 at the top of the list of flu strains that could produce a human pandemic, making it essential that new vaccines become available.

It is thanks to mass vaccinations that diseases such as polio have almost been eradicated, as have measles, mumps, and rubella, which also caused many deaths. As most people are likely aware, mRNA vaccines were developed that created proteins, or portions of proteins, which caused the production of antibodies to act against the spike protein of the SARS-CoV-2 virus. The rapid development seemed miraculous, and it was anticipated that they would tame the pandemic, provided that people would actually be inoculated. Fig. 3.2 depicts the immune changes produced by SARS-CoV-2 mRNA vaccines.

Given the moderate efficacy of current vaccines in limiting influenza infection, it might be fruitful to develop new approaches to enhance their effectiveness. It has

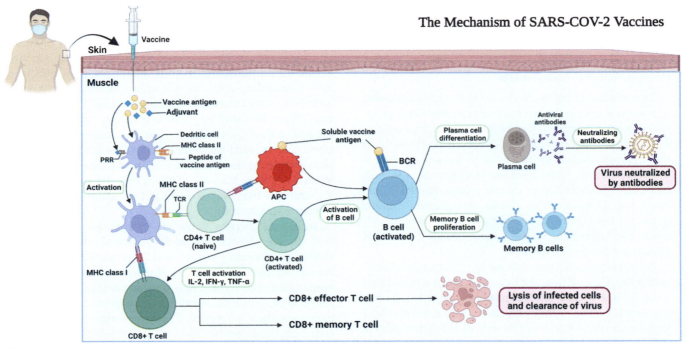

FIG. 3.2 The immune system responds to vaccination. The vaccine is injected intramuscularly, and subsequently, the protein antigen is taken up by dendritic cells activated by pattern recognition receptors (PRRs) to danger signals in the adjuvant. Here, MHC molecules on the surface of dendritic cells can present peptides of the vaccine protein antigen, which in turn leads to activation of T cells via their T-cell receptor (TCR). Following this, B cells can be developed in the lymph node by T cells. This development is a result of the combination of soluble antigen with the B-cell receptor (BCR). The production of plasma cells produces a rapid rise in serum antibody levels over the next 2 weeks. Memory B cells are also generated, thus mediating immune memory. Of note, CD8$^+$ memory T cells proliferate rapidly when they confront a pathogen, and CD8$^+$ effector T cells also have a pivotal role to play in the obliteration of infected cells. *Source: Alagheband Bahrami, A., Azargoonjahromi, A., Sadraei, S., Aarabi, A., Payandeh, Z., Rajabibazl, M., 2022. An overview of current drugs and prophylactic vaccines for coronavirus disease 2019 (COVID-19). Cell. Mol. Biol. Lett. 27, 38.*

been maintained that gut microbiota may have regulatory actions concerning influenza and thus could potentially be harnessed to limit the consequences of influenza infection (Chen et al., 2017a). There have also been efforts to develop peptide molecules that are able to inhibit a variety of influenza strains by grasping onto common features of a set of influenza A viruses (Kadam et al., 2017). The holy grail in this field is the development of a universal vaccine that acts across a broad range of influenza viruses (Wang et al., 2022e). Indeed, we may be part way there with the development of a vaccine that protects against influenza A and B (Feranmi, 2022), and there is hope that a universal vaccine will be developed to deal with SARS-CoV-2 and its subvariants (Appelberg et al., 2022). Still better news has come with the report of the development of an RNA-based vaccine that can act against any form of influenza by targeting the portion of a viral genome that is common in all viral strains. Unlike other vaccines that are based on an immune response being mounted by exposure to a dead or weakened virus (or components of the viral RNA), a different strategy was used. Ordinarily, in response to a virus, the production of small interfering RNAs (RNAi) disturbs viral functioning. However, viruses can cause disease by producing proteins that interfere with the RNAi response. By limiting the viruses' ability to suppress the host RNAi response, in this case, by administering a weakened form of a mutated virus that acts as a vaccine, it was possible to disturb infection produced by the invading virus. This approach may also be effective as a vaccine to prevent infection with other coronaviruses, including SARS-CoV-2 (Chen et al., 2024), but its efficacy in humans remains to be established.

Developing still more effective vaccines may not overcome the widespread hesitancy in being vaccinated. It is unfortunate that despite the availability of vaccines for various infectious viruses, vaccine hesitancy has become increasingly common so individuals leave themselves (and their children) open to illnesses. This is hardly a new phenomenon. Vaccine hesitancy was encountered when smallpox vaccines first became available, and it appeared in response to polio vaccination. There are many reasons (or rationalizations) for individuals choosing not to be vaccinated for common diseases. Frequently, there is mistrust of media and government agencies regarding recommendations to be vaccinated, as observed during the 2009 H1N1 pandemic (Taha et al., 2014). Based on earlier false alarms concerning health threats, including the potential hazards created by MERS and SARS-CoV-1, the government and media may have used up their share of "scary utterances" so that distrust had set in. At the same time, individuals are influenced by social media messages indicating the dangers of vaccination (a pandemic of misinformation), or they may be listening to the unwise advice coming from politicians, a cadre of Hollywood types, or friends with strong, albeit fallacious opinions. A detailed analysis pointed to an extensive set of factors that were linked to individuals choosing whether to be vaccinated (Nowak et al., 2015). Those who opted not to be vaccinated may have based their decisions on earlier experiences, such as having had a negative response to a vaccine or beliefs that the illness (e.g., flu) is manageable. Resistance to vaccination is also attributable to the belief that vaccination recommendations might be correct for others but do not apply to them. Some individuals believe that vaccines are often ineffective or maintain the incorrect and misguided notion that one could get the flu (or another illness) from a vaccine. Those railing against vaccinating their children might also not have had the experience of growing up at a time when illnesses like measles were damaging or killing children, and diseases such as polio were a horrible threat that kept reappearing.[3]

Predictably, those amenable to receiving (flu) vaccination tended to believe that they were flu susceptible and that vaccines were effective. The propensity to be vaccinated was also elevated among older people or those having an existent chronic health condition that might be complicated by becoming ill. Having previously experienced a bad flu or a similar illness was associated with individuals choosing to be vaccinated, as did easy access to vaccination. Also, the intention to vaccinate was particularly high if the recommendation to do so came from a physician or other trusted source.[4]

Some viruses, such as measles, are remarkably effective in spreading so that one person might infect about 90% of people close to them. Other viruses spread less readily so that one person may infect very few others (say, 0.5 people), and thus, the disease will disappear. Fortunately for individuals who choose not to be vaccinated, when enough people in a population are vaccinated, the source for transmission will be diminished, and thus, even potent viruses might not spread (herd immunity). This herd immunity also protects the children whose parents refused to have them vaccinated. However, should antivaxxers be successful in their campaign, so that enough people within a population decide not to be vaccinated (or not have their children vaccinated), a "tipping point" will be reached so that herd immunity no longer protects people who are not vaccinated, or those in whom the inoculation was not particularly effective (i.e., vaccine nonresponders). Because measles is so infectious, approximately 95% of people need to be vaccinated to obtain herd immunity. Owing to increasing hesitation regarding measles, mumps, and rubella (MMR) vaccination, the tipping point has been reached in some places. This has been observed within the United Kingdom and has been alarming within some parts of the EU, where cases increased from 941 cases in all of 2022 to 30,000 from January to October of 2023 (Wong, 2024).

Contrary to the incorrect beliefs of some people, measles can be very dangerous. Should an individual become infected, they may have a tough illness to deal with, and they may also experience serious downstream effects. Specifically, following measles infection, the immune system may be altered, possibly for as long as 2–3 years, so that the risk for other illnesses may be elevated (Mina et al., 2015). Furthermore, if the immune system is not fully developed, as in the case of young children, infection with measles may result in the virus hiding in the body, only to emerge years later to infect the brain.

Many of the factors that influence vaccine hesitancy toward other diseases were apparent with vaccination against SARS-CoV-2. What had not been counted on was the extent to which political influences and ideologies would affect vaccine hesitancy. Aside from this, the collision between safety measures, lockdowns, and economic factors, all contributed to the attitudes concerning vaccination. As well, younger people were often cavalier about being infected since the risk for hospitalization or death was highest among individuals with preexisting health conditions and older people. This contrasts with the 1918 H1N1 Flu pandemic, which affected people of all ages and resulted in many deaths among those 20–40 years of age, although recent evidence obtained through bone analyses suggested that deaths related to that pandemic were more likely to occur among those people who had been frail or already ill (Wissler and DeWitte, 2023). This said one cannot help but wonder to what extent vaccine hesitancy would have prevailed during the COVID-19 pandemic if hospitalization and death were more inclusive.

---

[3] The work of Kahneman and Tversky indicated that individuals are apt to make some seemingly puzzling decisions in certain situations and their theorizing may have much to say about the irrational decision-making processes that are common in relation to whether people choose to be vaccinated.

[4] In some studies, participants are asked about their "intent" to be vaccinated. While this is reasonable, intent doesn't necessarily translate into action (i.e., actually being vaccinated), and so the data must be interpreted cautiously. In fact, vaccination intentions can vary over time. In May 2020, 72% of respondents indicated that when a SARS-CoV-2 vaccine became available they would be inoculated. Curiously, by September, only 51% had intended to be vaccinated (Lewis, 2020). It is essential to identify what factors determine whether intentions become actions.

Bacterial and Viral Challenges Affect Hormonal and CNS Processes

Pathogenic stimuli, such as bacteria and viruses, cause marked effects on glucocorticoids and central neurotransmitters. Many of these changes are comparable to those usually elicited by both psychological and physical stressors, and thus, it was suggested that these systemic challenges were interpreted by the brain as if they were stressors (Anisman and Merali, 2002). In addition to affecting brain neurotransmitters, such as norepinephrine and serotonin (Hayley et al., 2014), immune-activating agents may influence the presence of growth factors (e.g., BDNF) as well as proinflammatory cytokines released by microglia (Audet and Anisman, 2013). As expected, these outcomes vary with sex and age, and at least some of the effects of immune challenge are subject to a sensitization-like effect in that exaggerated responses are evident upon reexposure to a challenge. Moreover, in animal models, bacterial agents and stressors may act cooperatively in producing brain neurochemical changes that favor the development of psychological disorders, such as depression (Anisman, 2009).

As described in Chapter 2, multidirectional communication occurs between immune, autonomic, microbial, hormonal, neurotransmitter, neurotrophin, and other brain-related processes. These systems are so intimately intertwined that actions in one may influence the functioning of others. For example, when mature lymphocytes are not present, ordinary behavioral stress responses might be absent, even in mice that are very stress-sensitive (Clark et al., 2014).

Microbiota

Although it had been suspected for well more than a century, as indicated in Chapter 1, it has only recently been established that bacteria and other microorganisms that exist throughout the human body serve to maintain psychological and physical well-being. What we eat influences our gut microbiota, and gut microbiota can influence what we eat (Danneskiold-Samsøe et al., 2019). When dysbiosis occurs (an imbalance between "good" and "bad" microorganisms) or in the absence of specific types of bacteria being present, an immense range of physical and mental illnesses may follow. The human gut is not equipped to digest many macronutrients that are consumed, such as plant polysaccharides. Thus, commensal bacteria, such as *Lactobacillus* and *Bifidobacterium*, are involved in doing the job, ultimately producing short-chain fatty acids (SCFAs), butyrate, acetate, and propionate, which can enhance immune functioning. Gut bacteria help to break down foods and contribute to the absorption of nutrients. Thus, their presence may help

individuals stay lean, and many useful bacteria are themselves strengthened by fiber-rich foods (Danneskiold-Samsøe et al., 2019). Just as the brain influences gut functioning, brain neuronal activity can be influenced through signaling processes that line the esophagus, stomach, small intestine, and colon. Messages from the gut to the brain may occur through stimulation of the vagus nerve (Cryan et al., 2019), which extends from the viscera to the brain stem, ultimately influencing hormones, such as ghrelin, that affect eating processes. As depicted in Fig. 3.3, gut microbiota functioning may influence immune activity (Fung et al., 2017), and by effects on multiple hormones and brain neurotransmission, may influence mood and reward processes (Lach et al., 2018). In this respect, SCFAs can attenuate stressor-provoked serotonin and BDNF alterations and may thereby influence the function of microglia (Erny et al., 2015), and they may also have neuroprotective effects (Ahmed et al., 2019).

When fiber is not available, microbes may die off, or they may feed on the mucus lining, which ordinarily keeps the gut wall healthy so that "leaky gut" is promoted. Disturbances of gut bacteria and their metabolites can affect immune, neurotransmitter, and hormone systems and may thus influence inflammatory diseases, neurodevelopmental disorders (e.g., ADHD, autism spectrum disorder), and several mental illnesses (Hyland and Cryan, 2016; Schroeder and Bäckhed, 2016). Gut dysbiosis may even affect sensitivity to cocaine reward and thus increase the risk for repeated use (Kiraly et al., 2016). It also appears that some foods, such as emulsifiers (food additives that serve to stabilize processed foods), can have negative health consequences, possibly by altering gut microbiota and the induction of inflammation (Chassaing et al., 2015).

Throughout evolution, various microbes adapted and colonized different parts of the body. Diverse microbiota ordinarily live harmoniously with one another (i.e., commensal bacteria), although some are parasitic, consuming other bacteria. Over the short run, animals with a compromised microbiome can survive, but their ability to do so will be curtailed owing to disturbed immunological functioning. Immune development may be hindered in rodents born germ-free, rendering them more vulnerable to pathologies. Even the response to vaccines may be dependent on the gut microbiome, which contributes to the shaping of immune responses (Lynn and Pulendran, 2017).

Gut Bacteria and Obesity

Given the potent effects of GLP-1 agonists (e.g., Ozempic) in promoting weight loss, it is significant that certain enteroendocrine cells within the gut, notably L cells, secrete GLP-1 and GLP-2, and may be involved in sensing

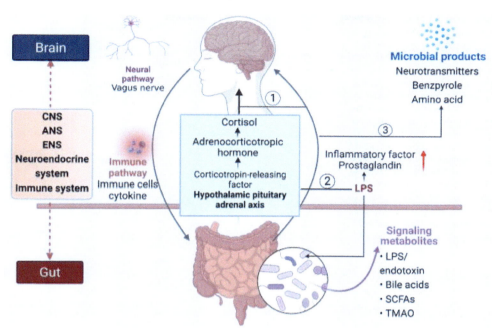

FIG. 3.3 Schematic representation of the mechanisms involved in the crosstalk between the gut microbiota and brain. Gut microbiota can influence the brain directly through stimulation of the vagus nerve, or indirectly through diverse neurobiological processes (e.g., short-chain fatty acids, immune pathways, cytokine activation, and key dietary amino acids, such as tryptophan, tyrosine, and histidine, and can affect numerous neurotransmitters and hormones). In this respect, microbes can synthesize several neurotransmitters (e.g., GABA, norepinephrine, and dopamine) within the gut, which may indirectly affect brain processes. Bacterial neuroactive metabolites and those obtained through diet can affect gut-barrier functioning and enteroendocrine cells of the gastrointestinal tract and pancreas, which cause the release of hormones, immune factors, and microglial functioning that have profound psychological consequences. *CNS*, central nervous system; *ANS*, autonomic nervous system; *ENS*, enteric nervous system; *LPS*, lipopolysaccharide; *TMAO*, trimethylamine N-oxide; *SCFAs*, short-chain fatty acids. *Source: Zou, B., Li, J., Ma, R.X., Cheng, X.Y., Ma, R. Y., et al., 2023. Gut microbiota is an impact factor based on the brain-gut axis to alzheimer's disease: a systematic review. Aging Dis. 14, 964–1678.*

the presence of nutrients, affecting appetite, glucose metabolism, and caloric intake. Microbiota metabolites, particularly within the ileum, influence L cell functioning (Arora et al., 2018), and interactions between L cells and microbiota may be important in maintaining gut-barrier integrity, metabolic homeostasis, and inflammation (Abdalqadir and Adeli, 2022), and may thereby be integral in promoting or limiting metabolic disorders (Zeng et al., 2024). Aside from these actions, GLP-1, together with bile acids and neuropeptide YY, can instigate a host of outcomes associated with the functioning of body and brain processes (see Fig. 3.4).

Aside from these processes, eating and weight changes are influenced by several gut microorganisms and have become targets to diminish obesity. It is not yet fully understood how certain bacteria may promote obesity, although certain bacteria, such as *Fusimonas intestini*, which was found to be especially high in both humans and mice with obesity, produce long-chain fatty acids that favor diet-induced obesity (Takeuchi et al., 2023a). Conversely, diets spiked with specifically modified bacteria (N-acylphosphatidylethanolamines) diminished eating and altered metabolism, thereby lowering adiposity and insulin resistance (Chen et al., 2014). As well, some bacterial families, such as Christensenellaceae, appear in greater numbers among thin individuals than in heavy people, and may causally contribute to this difference. When the bacteria associated with slimness were transferred to other mice, weight gain was diminished relative to mice that had not received this transplantation. Christensenellaceae could be useful in reducing weight (Mazier et al., 2021), but this has yet to be assessed in humans. In addition to affecting obesity, it has been demonstrated that genetically modified bacteria can diminish multiple immune-related disorders (Basarkar et al., 2022).

The link between the microbiome and obesity involves multiple steps, which could be subject to genetic differences across individuals (Duranti et al., 2017), and differ between males and females and are differentially affected by diet. Also, the relationship between microbiota and both brain and immune functioning is moderated by sex, and it seems that the effects of probiotics on behavior may be differentially influenced in males and females (Holingue et al., 2020), possibly being related to hormonal factors (estrogen, in particular) interacting with gut microbes. It is conceivable that diet in men and women will also have different effects on illnesses related to bacteria, and thus, diets meant to treat specific disturbances

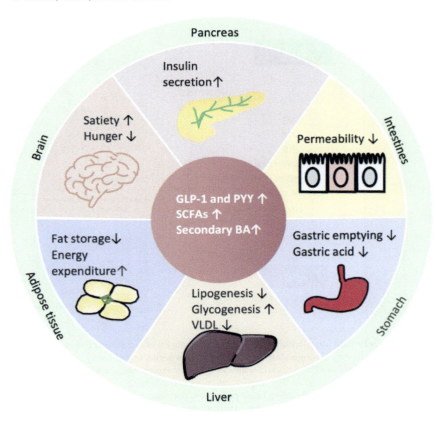

FIG. 3.4 Physiological benefits of gut peptides and gut microbiota metabolites. GLP-1 has multiple physiological functions. It promotes insulin synthesis and secretion in pancreatic β cells and then improves glucose homeostasis, delays gastric emptying and reduces gastric acid secretion, reduces intestinal permeability and bacterial translocation, promotes lipolysis and energy expenditure, increases liver glycogen storage and decreases liver sugar output, and suppresses appetite in the hypothalamus of brain. SCFAs and secondary bile acids also help increase insulin secretion in the pancreas, energy expenditure in adipose tissue, and decrease liver lipogenesis and VLDL (very low-density lipoprotein) output. *Source: Zeng, Y., Wu, Y., Zhang, Q., Xiao, X. 2024. Crosstalk between glucagon-like peptide 1 and gut microbiota in metabolic diseases. mBio 15, e0203223.*

need to be tailored based on sex as well as other individual difference factors that span several domains. In retrospect, it is not surprising that gut bacteria play a prominent role in feeding and energy processes and may contribute to eating- and gut-related disorders as well as psychological disturbances (Dinan and Cryan, 2017).

To assess the contribution of the microbiota to a variety of phenotypes, analyses were undertaken to assess mice born and raised in germ-free environments. These mice seemed not to develop in the usual fashion, in that their immune system was deficient, and the gut of these mice had a smaller surface area and hence could not absorb nutrients as readily as in mice raised in a standard germy environment. The germ-free mice also had leaky intestinal walls, and blood vessels that ordinarily supplied food to the gut wall were relatively limited. Upon receiving microbiota harvested from the intestines of conventionally raised mice, marked changes occurred in the germ-free mice within 2 weeks. Among other things, their body fat content increased, and insulin resistance became apparent even though their food intake was reduced (Bäckhed et al., 2004). Evidently, more food was converted into fat, and hence, these mice gained weight. More than this, the microbiota was integral in the absorption of monosaccharides from the gut lumen (interior of the gastrointestinal tract), resulting in a process by which fatty acids are produced (lipogenesis) and then stored. Moreover, a type of protein, "Fasting-induced adipocyte factor" (Fiaf) was suppressed in the intestinal epithelium, which is essential for triglycerides to be stored within adipocytes. In essence, these studies were among those that led the way in suggesting that gut microbiota are fundamental in moderating energy being obtained from foods and in subsequent energy storage.

Not long after these initial findings were reported, it was demonstrated that in genetically obese (ob/ob) mice, microbial communities could be distinguished from those that appeared in lean animals (Ley et al., 2005). Most prominently, *Firmicutes* were elevated by 50% and the *Bacteroidetes* were diminished to a comparable extent among those who were obese. When microbes harvested from fat and lean mice were fed to germ-free mice, those who received microbes from obese donors exhibited a much greater increase in fat than did those who received microbes from the lean donors (Turnbaugh et al., 2006), pointing to the causal connection between microbial factors and fat storage. It was later demonstrated that predictable phenotypic changes were provoked by the transplantation of fecal microbiota from adult female human twin pairs discordant for obesity into germ-free mice that were maintained on low-fat chow. Specifically, body and fat mass, together with obesity-associated metabolic phenotypes, varied with the fecal bacteria cultures received (i.e., from the heavy or lean twin). Tellingly, when mice that had received an obese twin's microbiota (Ob) were housed with mice containing the lean co-twins microbiota (Ln), the increased body mass and obesity-associated metabolic phenotypes in Ob mice was prevented, which was likely because of lateral transmission of microbiota (Ridaura et al., 2013).[5]

Gut bacteria exist that can favor weight loss, rather than weight gain. For instance, *Akkermansia muciniphila* is far more common among lean than in heavy mice that are prone to diabetes. When these bacteria were fed to the obese mice, they lost weight and the warning signs of type 2 diabetes diminished (Plovier et al., 2017). In humans, prebiotics fed to overweight children, reduce the weight gains that would otherwise appear in growing children, which has important long-term implications given that childhood obesity is often carried into adulthood (Nicolucci et al., 2017). These and similar findings suggest that the gut microbiota might provide a target to reduce obesity and type 2 diabetes (Remely et al., 2016). At the same time, gut bacteria comprise many different subtypes and the specific combinations that are present will influence different phenotypes. Additional research is needed to determine the combination of microbial species that are optimal in predicting the development of prediabetes and the transition to type 2 diabetes (Letchumanan et al., 2022). In this respect, in an analysis of 100 bacterial strains, individually and in combination, differential protection was obtained concerning harmful bacterial pathogens, notably *Klebsiella pneumoniae* and *Salmonella enterica*. Thus, as much as microbial diversity may be essential for well-being, communities of specific bacteria may have especially pronounced protective effects since they consume the nutrients that pathogens need, thereby limiting their growth (Spragge et al., 2023).

Microbiota among super-agers

The microbiome may contribute to both the physical deterioration that accompanies aging, as well as to healthy aging and extreme longevity (Boehme et al., 2023). As individuals age, core microbes decline, as does microbiota diversity (e.g., Ghosh et al., 2022; Wilmanski et al., 2021), varying with lifestyles adopted, as well as cultural and geographical factors (Boehme et al., 2023). Furthermore, the variations of microbiota during aging have been associated with neurodegenerative disorders, leading to the suggestion that manipulations of the aging microbiome might act in a protective capacity in limiting these conditions (Boehme et al., 2023).

Studies in older animals revealed the occurrence of gut dysbiosis, which led to intestines becoming leaky and consequently, released bacterial products promote inflammation and immune dysfunction (Jain et al., 2023). In older humans, especially in response to stressors, gut permeability increases, accompanied by elevated circulating proinflammatory cytokine levels. As well, changes occurred in the levels of a particular microbial family, Porphyromonadaceae, which has been linked to cognitive decline and affective disorders (Scott et al., 2017). It seems that aging may be accompanied by a shift of the microbial community toward a profile reminiscent of that apparent in inflammatory diseases and may contribute to the development of behavioral and cognitive disturbances. In fact, in young rodents that received gut bacteria from old mice, chronic inflammation could be accompanied by elevated leakage of inflammatory bacterial factors into circulation (Fransen et al., 2017).

It is interesting that with normal aging, certain bacterial species disappear, and others become more common, which could contribute to healthy aging. Specifically, dominant species are replaced with subordinate species, and particular bacterial groups (e.g., *Akkermansia*, *Bifidobacterium*, and Christensenellaceae) are more prevalent or enriched. Healthy individuals who were 100 years or older exhibited unique microbial signatures, again varying across cultures and geography (e.g., Sato et al., 2021; Tuikhar et al., 2019). Indeed, their microbiota constituency resembled that of healthy young people (Bian et al., 2017). Functional analyses likewise indicated that the microbiota in centenarians was accompanied by a high capacity of glycolysis and the formation of highly beneficial SCFAs (Wu et al., 2019b).

Among turquoise killifish, which have a relatively short lifespan (4–6 months), several genes located on sex chromosomes were linked to longevity (Valenzano et al., 2015), which might speak to the greater longevity of females. It was particularly interesting that when older fish of this species consumed the poop of younger fish, they lived longer, raising the possibility that some bacterial factors present in young poop produced benefits for the older fish. In studies using a small worm, *Caenorhabditis elegans*, the elimination

---

[5] Gut bacteria produce spores that can survive in open air, and can be transmitted from one person to another, causing dysbiosis in the second individual (Browne et al., 2016). In mice, the disturbance of the skin microbial community can be transferred to cage-mates (Gimblet et al., 2017). It is conceivable that people living within the same home may share a similar microbiome, and thus may share illness vulnerabilities, as well as the propensity for weight gain.

of 29 bacterial genes increased longevity and limited age-related diseases. These effects seemed to have been related to a substance, colonic acid, which affects the worm's mitochondria, thereby altering energy regulation. These data raise the possibility that the link between bacteria and longevity is a causal one, at least in worms, although it is admittedly some distance from worms to humans.

Most studies, including those related to longevity, have focused on the influence of bacteria. However, as we have already seen, the gut harbors a large array of viruses, some of which infect bacterial cells (bacteriophages). They have been implicated in the promotion of gut disorders (Spencer et al., 2022) as well as numerous other diseases through their actions on the release of bacterial endotoxins, regulation of bacteria-related metabolism, and by affecting immune activity (Shuwen and Kefeng, 2022). Of relevance to longevity, in a cohort of Japanese centenarians, viruses were detected that influenced bacteria and modulated metabolic processes, which might have contributed to their long lives (Johansen et al., 2023b).

Factors That Affect Microbiota and Their Implications for Well-Being

The microbial community is negatively affected by poor lifestyles, as we saw in relation to food consumption and obesity. Especially harmful effects are elicited by antibiotics that kill useful bacteria along with those that are not our friends. It was estimated that one in five hospitalized patients experience adverse effects related to antibiotic treatments, including gastrointestinal, renal, or hematologic disturbances. Some common antibiotic treatments (e.g., amoxicillin and azithromycin) taken over just 7 days can have pronounced and long-lasting negative effects on gut microbiota diversity, and when administered early in life, gut hypersensitivity may persist into adulthood (O'Mahony et al., 2014). Yet, in some instances, antibiotic-elicited elimination of bacterial species may produce positive actions. It is likely that the individual's genetic background, along with exposures to environmental stresses, play a role in shaping the microbiome and, hence determining what long-term repercussions result from its disturbance.

The impression might be gained from Fig. 3.5 that each of the main contributing factors independently influences microbial factors and well-being. Ultimately, however, microbial functioning and intestinal immunity are shaped and maintained by multiple interactive processes. Each of the factors shown in the figure affects others, and additively or interactively, gut bacteria will be affected. Furthermore, commensal microflora can affect and interact with immune processes, which can influence nutrition, including the presence of SCFAs and particular vitamins, which then feedback and affect the microbiota (Spencer and Belkaid, 2012). For instance, the presence of microbiota contributes to mast cell functioning

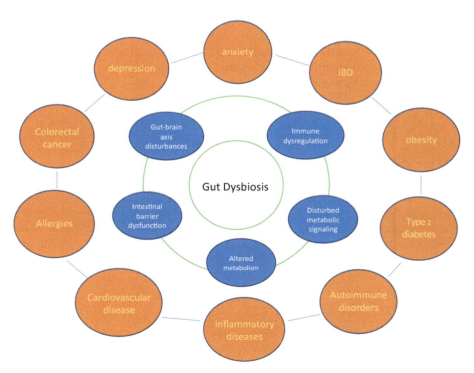

FIG. 3.5 Gut microbiota dysbiosis can produce several biological changes that favor the occurrence of a wide range of metabolic and immune-related disorders.

following the consumption of fat, which can then influence immune activity and instigate allergic and inflammatory responses (Sato et al., 2016). In fact, gut microbiota and immune system functioning in infants mature in tandem, and gut microbiota disturbances during the first year of life can result in several allergic conditions (food allergies, asthma, atopic dermatitis, and allergic rhinitis) that were elevated when children were 5 years of age (Hoskinson et al., 2023).

Gut microbiota and the genes involved in regulating them are exquisitely sensitive to psychological stressors, and their actions can be reversed by oral prebiotics (Bharwani et al., 2017) and by treatments that increase the presence of SCFAs (van de Wouw et al., 2018). Likewise, microbiota alterations can be provoked by prenatal stressors (Yeramilli et al., 2023) and can markedly affect the development of immune processes. Conversely, intake of specific gut microbes can diminish the impact of stressors, such as strenuous exercise, and can limit abdominal dysfunction and discomfort associated with academic stressors. It also appears that in the absence of an inflammatory inhibitor NLRP12, beneficial bacteria were reduced, and disruptive bacteria were elevated, leading to still more inflammation. As expected, increasing the presence of good bacteria could terminate this negative cycle.

Getting around antibiotic resistance to *C. difficile* through fecal transplants

The hospital-acquired antibiotic-resistant bacterium *C. difficile* has proven to be particularly able to be transmitted from one individual to the next. Even using a hospital bed that had recently been occupied by a *C. difficile* patient increased the risk of contracting this condition (Freedberg et al., 2016). Not only was the occurrence of *C. difficile* infection more common among hospital patients treated with antibiotics, but variants of this bacteria emerged that were increasingly destructive owing to their ability to produce a toxin to destroy gut eosinophils that ordinarily act in a protective capacity (Cowardin et al., 2016). Currently, about 3 million cases of *C. difficile* occur yearly, leading to 600,000 deaths., often in association with a variety of illnesses for which patients had been treated. The incidence of *C. difficile* infection has increased over the past 20 years and has become more frequent outside of hospital environments (Marra et al., 2020). Making matters much worse, more toxin-producing strains have been appearing as have cases that involve more than a single strain (Dayananda and Wilcox, 2019).

First-line treatment of *C. difficile* has comprised metronidazole and vancomycin, but when these fail, or *C. difficile* is a recurrent problem, a transplant of fecal microbiota can be used to treat this condition (Sandhu and Chopra, 2021). Specifically, fecal microbiota is obtained from a healthy donor and then transplanted in a purified form (most often by colonoscopy or through the nasogastric route, or more recently through acid-resistant capsules, dubbed "crapsules") to patients with resistant *C. difficile*. This results in the reestablishment of a beneficial bacterial colony and abatement of illness. This approach is not especially novel having been used as early as the 4th century in China to treat food poisoning and diarrhea and was used elsewhere over the centuries to treat gastrointestinal problems.

The 19th-century Russian zoologist Metchnikoff proposed that balances of microbes within the colon, particularly elevated lactic acid bacteria, could lead to gut problems, and Nissle later extended this to include *E. coli* as a protective agent against Shigella and gastroenteritis. Somewhat later, fecal enemas were found to attenuate a form of colitis, and after a few years, fecal microbiota transplants were used in inflammatory bowel disease (Pigneur and Sokol, 2016).

The effectiveness of the procedure seemed to vary based on the microbiota composition of the donor, that of the recipient, and the immune factors of the host. No doubt, other lifestyle factors of the donor and the recipient could influence the efficacy of the procedure so these would need to be considered to obtain the best outcomes, supporting the need for a precision medicine approach (Benech and Sokol, 2020).

With the success of fecal transplants for *C. difficile*, attention increased concerning the influence of fecal transfer in other conditions. Several randomized clinical trials indicated that fecal transplants could alleviate various gut disorders, and one study suggested positive effects in metabolic disorders. Because of the many illnesses associated with microbiome disturbances, assessments have been undertaken to determine the efficacy of this procedure in a variety of other conditions, such as chronic fatigue syndrome/fibromyalgia, multiple sclerosis, obesity, insulin resistance, forms of autism, Parkinson's disease, and depression (Choi and Cho, 2016). Although fecal microbiota transfer has been considered safe, caution needs to be exercised to ensure that donor samples are not contaminated.

Many factors present in feces, including colonocytes, archaea, viruses, fungi, and protists, may be fundamental to the effectiveness of the treatment and could potentially be enhanced by particular probiotics (Spinler et al., 2016). Complete fecal transplants might not be necessary to deal with *C. difficile* as it could be dealt with by transplantation of a cocktail made up of several bacteria (Buffie et al., 2015). Once the precise mix of bacterial and nonbacterial components that lead to positive effects are identified, it will be possible to generate still better treatments more efficiently.

Psychological Functioning Associated With Microbial Changes

The potential involvement of gut microbiota with multiple disease states instigated a great number of studies that traced the links between the microbiome, immune functioning, and brain neuronal changes, which might contribute to psychiatric and neurodegenerative disorders (Bastiaanssen et al., 2019; Kennedy et al., 2017). A meta-analysis that included 59 case–control studies indicated that several psychiatric conditions, notably depression, bipolar disorder, schizophrenia, and anxiety, were associated with a reduction of microbiota that promoted antiinflammatory actions and an increase of bacteria that promoted proinflammatory effects (Nikolova et al., 2021). The development of these illnesses appears to be linked at least to some extent, to epigenetic alterations that may be related to early life experiences associated with maternal diet as well as nutrition and environmental factors. The epigenetic changes related to microbial processes were associated with inflammatory processes and oxidative stress, which could drive the emergence of psychiatric disorders (Nohesara et al., 2023). As we will see in the ensuing chapters, just as certain microbial alterations may contribute to illness vulnerability, others may contribute to resilience in the face of stressors, including those that can lead to PTSD.

Some psychological disturbances may appear owing to variations of serotonin formed in the digestive tract or neurotoxicity brought about owing to increased metabolites of bacterial enzymes, such as D-lactic acid and ammonia (Galland, 2014). There is ample evidence pointing to the microbiota having an impact on depressive-like features in rodent models. For instance, as adults, mice that had been born germ-free exhibited altered dendritic morphology in the amygdala and hippocampus that were linked to depressive-like features (Luczynski et al., 2016a). Likewise, among nonobese diabetic mice, gut microbiota could drive depressive-like symptoms (e.g., social avoidance), which could be attenuated by antibiotic treatment and then resurrected through reconstitution of the microbiota from donor mice (Gacias et al., 2016).

Multiple routes have been identified (as shown in Fig. 3.3) by which microbiota affect the brain and thus could promote psychopathology. Being a major pathway between the gut and brain, the vagal nerve was implicated as a player in accounting for anxiety and depressive-like behaviors (Fülling et al., 2019), possibly acting through inflammatory processes. Moreover, enterochromaffin cells of the gut epithelium can release serotonin and activate CNS functioning (Bellono et al., 2017). Manipulations of the gut microbiota can affect specific serotonin receptors, as well as norepinephrine, dopamine, and GABA activity in limbic brain regions, thereby influencing mood states (e.g., Clarke et al., 2013; Stilling et al., 2015). In addition, gut bacteria can influence neuroendocrine factors (e.g., CRH) and neurotrophins (e.g., BDNF) within stressor-sensitive brain regions that have been tied to the development and maintenance of depressive disorders. Further to this, strong immunogenic agents engender pronounced corticoid responses, which can diminish the effectiveness of the gut barrier (Rutsch et al., 2020). The migration of bacteria out of the gut will thereby be facilitated, which may promote the production of immune signaling molecules (cytokines) that instigate psychological and neurological disturbances.

Consistent with microbiota involvement in depression, among rats highly vulnerable to depressive-like states, probiotic treatment may counter behavioral disturbances that may have been provoked by proinflammatory changes instigated by a high-fat diet (Abildgaard et al., 2017). Moreover, in humans, specific probiotic supplementation may facilitate the alleviation of depressive symptoms (Mörkl et al., 2020). Conversely, antibiotics can affect mood states by altering microbiota and downstream actions on mitochondrial functioning, microglia reactivity, and factors important for neuroplasticity.

Although antibiotics have been a primary concern in relation to microbiota changes and dysbiosis, such treatments are hardly alone in affecting microbiota within the gut as well as elsewhere. Indeed, antipsychotics, antidepressants, nonsteroidal antiinflammatory drugs (NSAIDs), opioids, statins, and proton pump inhibitors (PPIs) influence a wide range of microbiota classes. The secondary effects of some of these may influence the frequent weight gain and visceral fat provoked by antipsychotic medication (Le Bastard et al., 2018).

Microbiota and Immunity

The gut is a dirty place, being the recipient of various foods, some of which may be contaminated, and a good portion of our immune cells operate within the gut. Intestinal immunity comprises collaboration between specific types of immune cells and many different cytokines, various nutritional factors, and commensal bacteria. Moreover, reciprocal communication occurs between gut microbiota and CNS processes, and consequently, gut-level disturbances can instigate psychological, metabolic, and immune-related disorders (O'Mahony et al., 2017a).

Microbial factors are fundamentally involved in the development of immunosuppressive responses generated by T_{reg} cells to dietary antigens. This, of course, is critical for the prevention of excessive immune responses being generated in the face of normal dietary intake. It is thought that gut dysbiosis may promote disturbances of

T_{reg} cells and imbalances of Th1, Th2, and Th17 lymphocytes, which can promote autoimmunity to antigens derived from the diet and may contribute to autoimmune diseases. Even the regulation of neuron myelination within the prefrontal cortex is affected by the microbiome and thus might dispose individuals to susceptibility to multiple sclerosis (Mangalam et al., 2017).

Beyond these effects, gut dysbiosis can affect brain microglia and, through the release of cytokines, can influence psychopathological conditions. This is especially the case with aging, as gut microbiota metabolites, such as choline and trimethylamine, have been implicated in the development of age-related diseases (Hasavci and Blank, 2022). Furthermore, pre- and probiotic manipulations can alter glycemic dysregulation, such as glucose tolerance and insulin resistance, which was accompanied by elevated plasma levels of the antiinflammatory cytokine IL-10. It was thus suggested that microbial manipulations could potentially influence illnesses that involve ongoing inflammatory or metabolic disturbances (de Cossío et al., 2017).[6]

An interesting meta-analysis that considered 10 diseases indicated that the number of microbiota genera varied across illnesses. Whereas some illnesses were associated with the presence of numerous genera, others appeared to be associated with a lack of certain microbiota, and many microbiota species were linked to multiple illnesses. In effect, alterations are not disease-specific and are general vulnerability factors, so other elements contribute to the specific illness that emerges. Such findings might be a step in providing information that could be relevant to the use of probiotics and prebiotics (whereas probiotics comprise living microorganisms, prebiotics are food ingredients that induce the growth or activity of beneficial microorganisms) in specific illnesses.

As much as the findings using germ-free mice are interesting, their relevance for neuroimmune processes may not be as clear as one might like, especially as the blood–brain-barrier (BBB) may be disturbed in these mice (Braniste et al., 2014). Thus, endogenous circulating immune cells and microbial species that might escape the confines of the gut could potentially access the brain. As well, this could result in vulnerability to a range of potentially toxic insults that are present in the periphery following environmental exposure. Remarkably, exposing germ-free mice to the normal microbial constituents of the gut obtained from normally housed mice, reversed BBB deficits. In fact, treating them with the SFCA metabolites from normal microbiota appeared to repair the BBB deficits, such that tight junction integrity was restored.

The fact that constituents of the microbiota send signals that affect BBB functioning and brain homeostasis, has wide-ranging implications for virtually all neuronal pathologies. This is illustrated by the finding that microbiota influence the development of infection following stroke (Stanley et al., 2018). When fecal samples from a mouse that experienced ischemic stroke were transferred to other mice, the size of a stroke-induced infarct was increased in the microbiota recipients (Singh et al., 2016). Conversely, antibiotic treatment reduced stroke damage and diminished the infiltration of inflammatory T cells. Once again, these findings suggest bi-directional communication between the microbiota and brain and are consistent with the view that the translocation of microbial elements might infiltrate the brain or other organs to influence their functioning. It was demonstrated that bacteria (e.g., *Streptococcus pneumoniae* and *Streptococcus agalactiae*) affect nerve cells in the meninges so that the immune response is disturbed, thereby facilitating the development of bacterial meningitis. As it turns out, interfering with calcitonin gene-related peptide (CGRP) or the RAMP1 receptor on macrophages to which it binds, can act against meningitis (Pinho-Ribeiro et al., 2023).

Physical Illness, Immunity, and Gut Bacteria

The link between microbiota, immune functioning, and disease conditions has been broadly supported. Germ-free mice raised in a sterile environment lived longer following a skin graft, possibly because immune functioning was diminished, and hence foreign tissue was not attacked. If these mice received microbes from untreated mice, they rejected the skin graft more readily. In essence, these data point to the importance of the microbiota in determining immune functioning and tissue rejection (Alhabbab et al., 2015). Commensurate with these findings, tissue transplants involving lungs, skin, and intestines, which had been exposed to external influences, were less successful than transplants of tissues that were less directly affected by external microbial factors (Lei et al., 2016).

The complex interactions between the differing microbial species might contribute to autoimmune disorders. Germ-free mice with impaired BBB integrity were more vulnerable to the development of autoimmune pathology in an experimental autoimmune encephalitis (EAE) model of MS. Furthermore, having the normal gut commensal bacteria was essential for mounting CD4+ T-cell and B-cell antibody responses in a model of relapsing–remitting MS (Berer et al., 2011), and thus

---

[6] As we'll emphasize repeatedly, caution should always be exercised in embracing any emerging therapies. This is particularly evident with regards to "hot" areas of research, where there might be the inclination to "jump on the bandwagon." Microbiota have been implicated in a very large number of illnesses, and there may be the concern, as often stated, that when something explains everything, it explains nothing.

germ-free mice might be at greater MS vulnerability. Moreover, among identical twins, discordant for MS, differences were detected in *Akkermansia*, and when microbiota from the affected twin was transferred to mice, an autoimmune-like phenotype was produced together with lower levels of the antiinflammatory cytokine IL-10 (Berer et al., 2017). Paralleling such studies, marked gut dysbiosis was present in MS patients, particularly reductions of *Clostridium* and *Bacteroidetes* species (Miyake et al., 2015). A further interesting aspect related to autoimmune disorder is that *Fusobacteria* increased relapse rate (Tremlett et al., 2016), but successful treatment of MS was associated with elevations of *Prevotella* and *Sutterella* species (Jangi et al., 2016). In effect, various microbial species likely have differing effects on brain processes and their collective impact depends upon a delicate balance between them as well as the metabolites they excrete.

The gastrointestinal tract shares reciprocal connections with the immune system. As we saw in Chapter 2, immune cells have membrane pattern recognition receptors (PRRs) that mediate recognition of damage and pathogen-associated molecules (PAMPs) and damage-associated molecular patterns (DAMPs). These PRRs are rudimentary aspects of the immune system that evolved, in part, as a way of detecting pathogens or other microbial threats and may operate to enable microbiota to communicate with the immune system (Chu and Mazmanian, 2013).

It will be recalled that PAMPs are initial sensors that allow immune cells to recognize microbial presence and determine their pathogenic valence. The toll-like receptors (TLRs) and NODs (nucleotide-binding oligomerization domain-like receptors) are among the most prominent PAMPs found throughout the brain and immune system. They have evolved to recognize specific motifs that characterize bacterial, viral, or fungal invaders. Upon their recognition, very robust intracellular pathways are engaged that give rise to the mobilization of defensive inflammatory (e.g., cytokines), enzymatic, and oxidative (e.g., superoxide) factors, depending upon the nature of the threat. DAMPs act in a similar fashion becoming active following the detection of specific factors that are released in response to cellular distress. Among these distress signals, adenosine triphosphate (ATP), and other purines, along with mitochondrial and intracellular factors are released into the extracellular space in the presence of a damaged cell, creating a "sterile" inflammatory reaction.

A distinction has been made between PAMPs and the largely interchangeable term, microbe-associated molecular patterns (MAMPs), which respond to various microorganisms and act as a bridge between the enteric nervous system and the innate immune system. The MAMPs, by virtue of their effect on immune cells, modulate inflammation (Chu and Mazmanian, 2013) and

might even influence microbial interactions with TLRs. Some of these interactions may not always lead to pathological or inflammatory conditions, which would be in keeping with the symbiotic relationship between the gut microbiota and the host organism.

However, problems may arise when microbial dysfunctions result in improper sensing of microbiome signals. If such protective processes are not doing the job effectively, excessive immune activation and chronic inflammation may evolve (Chu and Mazmanian, 2013). Fortunately, we are blessed with a gene (SIGIRR) that operates to stimulate immune responses that interfere with bacteria forming colonies that would ordinarily have negative health effects. Disruptions of SIGIRR, owing to antibiotic treatments, can cause dysbiosis, wherein the battle for supremacy moves toward the side of the harmful bacteria (Sham et al., 2013).

As we have seen, microbial dysbiosis has been implicated in several diseases that involve gastrointestinal processes and eating disorders, such as anorexia nervosa and bulimia. Beyond these conditions, gut bacteria have also been linked to metabolic dysfunctions (e.g., insulin resistance), ultimately promoting the development of type 2 diabetes, and it has been proposed that the disturbed balance of gut bacteria might serve as a target in the treatment of this illness. Altered bacterial levels have also been associated with increased proinflammatory activity (e.g., IL-1β, IL-17, and IL-18) that exacerbates autoimmune conditions as well as neurodegenerative disorders (Rutsch et al., 2020). Other illnesses might still come about because the wound-healing capacity associated with the microbiome might not be operating properly. A wide range of other illnesses, which will be discussed in ensuing chapters, have also been linked to immune disturbances that might have their roots in microbial dysbiosis and inflammatory processes. These comprise cardiovascular illnesses, periodontal disease, rheumatoid arthritis, and allergies, as well as seemingly unrelated illness conditions, such as chronic kidney disease, uremic toxicity, multiorgan failure, several forms of cancer, and have been associated with the accumulation of amyloid proteins that were linked to neurodegenerative disorders. In this respect, the NLRP3 inflammasome has been suggested as being a key player in the link between microbiota and inflammatory processes that contribute to psychiatric and neurodegenerative disorders (Pellegrini et al., 2020).

Sickness features associated with systemic inflammation stem from brain changes

Sickness behaviors in rodents (diminished social interaction, ruffled fur, hunched posture, inactivity, and sleepiness) are typically associated with the administration of immune-activating agents, such as LPS, and are provoked by IL-1β. These outcomes have been taken to model some of the symptoms associated with depressive disorders as sickness behaviors can be induced by the administration of these treatments directly into the brain, pointing to the involvement of CNS processes in promoting sickness behaviors.

Characteristics of sickness behaviors are frequently apparent among patients experiencing autoimmune disorders, likely reflecting elevated inflammatory immune activation. It seems that in the context of organ inflammation, increased TNF-α levels give rise to monocytes being recruited to the brain, thereby increasing microglial activation and the production of sickness behaviors. The sickness profile associated with liver inflammation in mice can be diminished by a probiotic treatment without affecting the severity of the illness, the actual gut microbiota composition, or permeability of the gut, but were tied to diminished microglial activation and cerebral monocyte infiltration (D'Mello et al., 2015). The sickness behavior and its resolution by probiotics may thus involve brain processes, including microglial activation and limiting recruitment of monocyte-secreting TNF-α within the brain.

Diet, Microbiota, and Cancer

Among the many illnesses related to diet and microbiota, considerable attention has been devoted to the possibility that these factors can influence the development and progression of some forms of cancer and may influence the response to varied cancer therapies (Anisman and Kusnecov, 2022). Before delving into this, we would be remiss if we did not mention that the reliability and validity of some of the data in this area of research were questionable (see discussions in Ioannidis, 2013: Schoenfeld and Ioannidis, 2013). Among other things, it was maintained that the risks and benefits had been claimed much too often, and, in some instances, the single studies conducted reported effects that were too impressive to be credible. Indeed, self-reports of caloric intake were sometimes so extreme as to be "incompatible with life." Aside from this, studies that followed daily food intake were plagued by difficulties in tracking consumption and frequently did not consider the ingredients of meals. As a result, the findings reported have been inconclusive and frequently contradictory. Several foods (tea, milk, coffee, eggs, corn, butter, and beef) have been associated with either increased or decreased cancer incidence, whereas other foods were more reliably associated with elevated or reduced cancer occurrence.

There is little question that genetic factors play a fundamental role in the development of numerous diseases, accounting for 50%–60% of the variance associated with the occurrence of many cancers, but this still leaves a prominent role for many other factors in disease occurrence and progression. This includes epigenetic changes, variations of eating-related hormones, metabolic processes, and microbiota, which speaks to the importance of personalized nutrition to limit disease processes (e.g., Kolodziejczyk et al., 2019) and the view that to an appreciable extent, some forms of cancer may be preventable diseases.

Considerable data have amassed indicating that alcohol consumption was associated with increased occurrence of at least seven forms of cancer (i.e., cancers of the breast cancer, oral cavity, pharyngeal, laryngeal, esophageal, colorectal, and liver, and perhaps even pancreas) and may contribute to early-onset colorectal cancer. Predictably, cancer development was related to the amount of alcohol consumed (Bagnardi et al., 2015), but even modest consumption can produce adverse outcomes. Among individuals who had decided to cease drinking, the risk for oral and esophageal cancer declined. Regrettably, this was not evident in other forms of cancer associated with alcohol consumption (Gapstur et al., 2023).

The impact of moderate and heavy alcohol consumption on cancer occurrence may come about owing to a reduced abundance of gut microorganisms, which can affect immune functioning. Alcohol can also damage gut epithelial cells, thus allowing intestinal microorganisms to enter circulation, promoting inflammation within diverse organs. Besides these actions, even acute alcohol intake can alter cytokine levels among otherwise heavy drinkers (Lee et al., 2021b). Other than microbial and immune actions, alcohol consumption can potentially affect cancer through oxidative stress, which can damage cells and interfere with energy metabolism. Furthermore, alcohol breaks down into acetaldehyde, which can act as a carcinogen, and among postmenopausal women, actions on estrogen activity can foster cancer recurrence.[7]

The case has frequently been made that saturated fats, which are obtained from fat cuts of meat (beef, pork, lamb), processed meats, some dairy products (notably, whole milk, cream, butter, and some cheeses), and many baked and processed foods may undermine health. These negative outcomes have been attributed to the increased low-density lipoproteins (LDL) cholesterol levels produced by these foods, although full agreement in this regard has not been reached. Several prospective cohort studies revealed that increased consumption of red meat and processed meat was accompanied by earlier all-cause mortality (Schwingshackl et al., 2017). There has been increasing evidence supporting the view that gut microbiota can be adversely affected by high saturated fat diets, which can

---

[7] While cigarette packages have routinely carried the warning that smoking causes cancer and heart disease, no such warning appear on bottles of alcohol. While we are reluctant to seem like conspiracy theorists, one cannot help but wonder why alcohol has escaped this notification.

produce an unhealthy metabolic state (Wolters et al., 2019), and through changes in microbiota, a high-fat diet favors gastrointestinal tumor development (Yang et al., 2022a). Considerable debate has continued concerning the impact of meat consumption on the provocation and progression of several forms of cancer. We will not delve into this here, but these issues and related political intrigues are detailed in Anisman and Kusnecov (2022).

There has not been any debate concerning the adverse effects of trans fat, which is created by adding hydrogen to liquid vegetable oils, thereby making them solid. Trans fats have been widely used as an alternative to lard and butter in bread, cookies, and pies, as well as in many fast foods (e.g., french fries and fried chicken). As yummy as these foods might be, trans fats promote increased LDL and reduced "good" high-density lipoproteins (HDL), and increased production of reactive oxygen species (ROS) by mitochondria, which may cause DNA damage. Together, these actions may cause an increased risk of type 2 diabetes, heart disease, stroke, and some forms of cancer.

Unsaturated fats that are found in vegetable oils (e.g., olive and sunflower oil) and are plentiful in a variety of seeds and nuts, are thought to offer health benefits by reducing low-density lipoproteins (LDL) cholesterol. Likewise, some polyunsaturated fats (PUFAs) obtained from fatty fish (e.g., wild salmon, tuna, sardines, herring, trout, and mackerel), soybeans, and some nuts and seeds may provide health benefits. Importantly, different forms of PUFAs may produce diverse effects. Whereas omega-3 PUFAs obtained from fish may reduce disease-causing inflammation, intake of the omega-6 PUFAs might have proinflammatory actions.

High dietary fiber or whole-grain consumption was associated with reduced occurrence of several forms of cancer, including breast, colorectal, and liver cancer (e.g., Reynolds et al., 2019). Further, a diet low in fat but high in fiber was associated with a better prognosis following treatment of some forms of cancer (Rinninella et al., 2020). The benefits derived from high-fiber foods can come about owing to diverse processes. For instance, certain foods, such as cruciferous vegetables (e.g., broccoli, cabbage, cauliflower, collard greens, kale), contain a compound (indole-3-carbinol) that can moderate the actions of a gene that increases the presence of the tumor suppressor phosphatase and tensin homolog (PTEN), which is ordinarily diminished in many cancers. More commonly, it is thought that SCFAs, the primary metabolites of microbiota promoted by high-fiber foods, limit

intestinal inflammation, which might otherwise favor cancer occurrence (e.g., Carretta et al., 2021).

Most Mediterranean diets have been associated with a lower incidence of heart disease and diminished frequency of some forms of cancer (e.g., colorectal and breast cancer).[8] These actions may stem from effects on the microbiota constituency that influence immune and inflammatory processes (e.g., Newman et al., 2019). Indeed, adherence to a Mediterranean diet was associated with elevated diversity of gut microbiota, elevations of SCFAs, diminished levels of proinflammatory cytokines, and reduced oxidative stress (Ghosh et al., 2020). These actions are not restricted to the gut, as a Mediterranean diet is accompanied by distinct changes in the mammary gland microbiome (Shively et al., 2018).

While some people are ardent vegetarians who maintain their lifestyles for decades, most people consume a variety of foods, some of which may be healthy, but others are clearly unhealthy. Prospective analyses that tracked individuals over many years have made it clear that overall diet rather than any single food was most closely aligned to cancer-related and all-cause mortality. Food consumption with high inflammatory potential was most closely aligned with elevated cancer risk, and diets that limited the abundance and diversity of microbiota can favor elevated inflammation that could promote cancer. Specifically, among individuals tracked for 26 years, proinflammatory diets were associated with elevated colorectal cancer, and a prospective study among nurses conducted over 22 years revealed that diets that favored inflammation during adolescence and early adulthood were associated with elevated premenopausal breast cancer (Harris et al., 2017a). Likewise, the occurrence of estrogen and progesterone-negative and HER2-positive breast cancer subtypes was greater in association with a high "diet inflammatory index" that was based on a food questionnaire (Tabung et al., 2016).[9]

Consistent with the actions of certain foods, manipulations that affect microbiota can influence cancer occurrence and the effects of cancer therapies (Sepich-Poore et al., 2021). Disturbing the gut microbiome with antibiotics can exacerbate cancer cell proliferation. Use of an antibiotic for more than 6 months was associated with more frequent colon cancer within 5–10 years, and a modest increase of cancer risk was even evident after a single course of antibiotic treatment (Lu et al., 2022b). Among patients with lymphocytic leukemia, antibiotic treatments reduced progression-free survival stemming from

---

[8] Diets in countries surrounding the Mediterranean may differ in several ways, but typically comprise high intake of whole grains, beans, nuts, fruits, vegetables, herbs, spices, and healthy fats, particularly in the form of extra virgin olive oil, coupled with limited consumption of red meats, sugar, and saturated fat.

[9] In addition to playing a role in cancer occurrence, certain foods (or food groups) may also influence cancer progression and the response to treatments. Moreover, owing to the prodigious growth of cancer cells, they need adequate supplies of nutrients, and thus selectively depriving them of glutamine, methionine, serine, and glycine, could have beneficial effects. Likewise, manipulations that influence essential amino acids may influence the effectiveness of cancer treatments (Kanarek et al., 2020).

cyclophosphamide as well as cisplatin for relapsed lymphoma (Pflug et al., 2016).

It is important to underscore that as effective microbiota may be in cancer prevention, alterations of the micro-ecology of the gut by chemotherapy may cause some bacteria to take on pathogenic characteristics (pathobionts), furthering illness. Several specific bacteria have been identified, notably *Mycoplasma hyorhinis*, *Fusobacterium nucleatum*, and *Gammaproteobacteria*, which can undermine the efficacy of cancer treatments and may favor resistance to therapies that otherwise are effective in cancer treatment (Geller et al., 2017). This outcome can be exacerbated because local microbial factors, as well as those situated within the tumor microenvironment, may have the capacity to alter the chemical structure of therapeutic agents, consequently diminishing their effectiveness (Lehouritis et al., 2015; Zimmermann et al., 2019). Microbial variations can also moderate the influence of radiation and chemotherapies by affecting immune effectiveness (Alexander et al., 2017; Kovács et al., 2020). Microbiota can have these actions by affecting chemotherapy-elicited immune and inflammatory responses, such as monocytes, macrophages, and NK cells that can recognize PAMPs and MAMPs.

Several newer therapies have evolved that focus on ways of either enhancing immune functioning or diminishing the ability of cancer cells to evade attacks by immune factors. One form of immunotherapy is based on the modification of hardwired inhibitory mechanisms that are needed for self-tolerance and for regulating immune responses to avoid immune-provoked damage to healthy tissue. Checkpoint molecules are present on immune cells, which can be turned down by healthy cells, thereby preventing them from being attacked. Cancer cells may similarly be cloaked with these molecules so that they, too, are not assaulted by immune cells. Thus, cancer therapies focused on inhibiting checkpoints in cancer cells, so that the actions of immune cells are unleashed. Several checkpoints have been identified, such as cytotoxic T-lymphocyte-associated antigen-4 (CTLA-4) located on T lymphocytes and the inhibitory checkpoint molecule programmed death-1 (PD-1) present on T cells and B cells. In the latter instance, the programmed death-ligand 1 (PD-L1) on normal healthy cells can bind to the PD-1 checkpoint, thereby inhibiting the signal to suppress immune functioning. Predictably, a combination of PD-1 and PD-L1 inhibitors can produce superior outcomes than inhibiting only one of these (e.g., Han et al., 2020).

Immunotherapy comprising checkpoint inhibitors has been remarkably effective in the treatment of many types of solid tumors that were otherwise untreatable. This said, only a modest percentage of patients (20%–30%) are helped, and only a few more when immunotherapy was administered in conjunction with other treatments, such as chemotherapy, radiotherapy, other immune checkpoint inhibitors, and many other therapies (Yi et al., 2022a). As such, it is especially meaningful that microbiota could enhance the influence of immune checkpoint inhibitors in the treatment of gastrointestinal cancers, renal cell carcinoma, nonsmall cell lung cancer, epithelial tumors, and malignant melanoma. The case has even been made that gut bacteria are fundamental for immunotherapy to have its effects on cancer elimination. Specifically, the therapy causes the translocation of specific gut bacteria into secondary lymphoid organs as well as into certain types of tumors. In doing so, the therapy promotes the activation of dendritic cells and antitumor T-cell responses in the primary tumor and tumor-draining lymph nodes. When gut bacteria are eliminated, say by antibiotic treatments, gut microbiota translocation into lymph nodes does not occur, the related immune changes are absent, and hence, the response to immunotherapy is prevented (Choi et al., 2023a). Overall, it seems that through their actions on immune- and inflammatory-related processes within the tumor microenvironment, microbiota plays a pivotal role in determining anticancer treatments (Bashiardes et al., 2017; Iida et al., 2013).

Among germ-free mice, bacterial reconstitution with fecal material from patients who responded to treatment enhanced T-cell responses, better tumor control, and improved anti-PD-L1 therapy (Matson et al., 2018). A positive outcome, in contrast, was not apparent when fecal material was transferred from patients who had not responded to PD-1 antagonism, although this could be achieved by oral supplementation with *A. muciniphila* (Routy et al., 2018b). Paralleling such findings, transferring fecal microbiota from patients who responded positively to immunotherapy to melanoma patients who had not responded to immunotherapy, tumor reduction was apparent in 6 of 15 patients. In these patients, the abundance of microbes that enhanced T-cell functioning was apparent, as was increased CD8[+] T-cell activation (Davar et al., 2021). A phase I clinical trial similarly indicated positive effects of fecal microbiota transplantation and re-induction of anti-PD-1 immunotherapy among a portion of patients who had previously been refractory to anti-PD-1 therapy (Baruch et al., 2021).

Although a variety of microbiota species could be responsible for the positive effects of microbiota in enhancing immunotherapy, the abundance of *A. muciniphila*, *B. fragilis*, *Bifidobacterium*, and *Faecalibacterium* was associated with improved responses to CTLA-4 or PD-1/PD-L1 therapy (Routy et al., 2018a). Consistent with this, immunotherapy was ineffective among germ-free mice or those that had been treated with antibiotics, but when given *Bacteroides fragilis*, the therapeutic response to CTLA-4 inhibitors was instigated (Vétizou et al., 2015). Moreover, in mice, a diet containing several *Bifidobacterium* species or *A. muciniphila* augmented the efficacy of a PD-L1 treatment (Longhi et al., 2020; Routy et al., 2018b).

In humans, the abundance of microbiota and the presence of specific bacteria predicted longer progression-free survival following immunotherapy in melanoma patients as well as in the treatment of nonsmall cell lung cancer and renal cell carcinoma. The levels of *Bifidobacterium longum*, *Collinsella aerofaciens*, and *Enterococcus faecium* were appreciably elevated in patients who exhibited a positive response to immunotherapy (Matson et al., 2018). Likewise, positive responses to anti-PD-1 therapy among melanoma patients were apparent in those who exhibited greater bacterial diversity and a greater abundance of Clostridiales, whereas a greater abundance of Bacteroidales was apparent in poor treatment responders. The abundance and diversity of microbiota were similarly related to a better response to PD-1 immunotherapy in the treatment of nonsmall cell lung cancer (Nagasaka et al., 2020). The effectiveness of PD-1 immunotherapy in hepatocellular carcinoma was similarly noted in the presence of greater taxa richness. Interestingly, the diversity of microbiota became progressively more pronounced over the course of therapy, and these changes were predictive of treatment efficacy (Zheng et al., 2019b).

One of the greatest problems in cancer therapy is that resistance develops to the treatments administered. The resistance could occur for a variety of reasons (e.g., the ability of cells to pump out therapeutic agents), but it is significant that gut microbiota among treatment responders and nonresponders could be distinguished from one another (Gopalakrishnan et al., 2018; Yi et al., 2019).

Unfortunately, given the marked interindividual differences concerning microbial communities present, it can be difficult to determine what to provide patients with before therapy, especially as the effects of treatments may vary with different types of cancers. Since the clinical response to immunotherapy may also vary with an individual's genetic signature, complex interactions would need to be considered when prescribing microbial treatments. These difficulties become exponentially greater given that microorganisms other than bacteria, such as specific fungi and viruses, could potentially influence the response to immunotherapy. Moreover, it would also be necessary to consider the influence of epigenetic alterations that come about with lifestyles, as well as age and obesity (Baiden-Amissah and Tuyaerts, 2019). Moreover, the presence of common microbial species can vary with dietary practices across cultural and ethnic communities, which can have implications for microbiota involvement in disease occurrence and the response to therapies.

Aside from effects on the therapeutic benefits of microbiota, specific bacteria could influence the activity, efficacy, and toxicity of several therapies. The beneficial outcomes might develop owing to a constellation of changes that are elicited by microbiota. These include the production of bacterial metabolites and changes of oncogenic toxins (Zitvogel et al., 2017), the maturation of T_{17} cells within the intestine and effector lymph nodes, or alterations within the tumor microenvironment (Iida et al., 2013). As well, microbiota can act against graft versus host disease and may thereby affect allogeneic hematopoietic stem cell transplantation in the treatment of hematologic cancers (Shono and van den Brink, 2018).

Caveats Concerning the Potential for Using Microbiota for Health Benefits

With the increased understanding regarding the contribution of microbiota to illness occurrence, one might be seduced into thinking that we may be on the cusp of being able to target microbiota to diminish or prevent illness. Although there have been reports consistent with this perspective, the main positive effects observed were modest, and altering microbiota through diets or probiotics did not have sufficiently powerful effects to moderate systemic inflammation. Furthermore, it has proven difficult to identify specific bacteria that cause the appearance of illnesses, mainly because so many processes are linked to different pathological conditions. This difficulty is compounded by possible interactions with numerous hormonal and gut-related processes, which can be influenced by experiential factors. Accordingly, potential treatments will no doubt have to comprise multiple bacterial changes rather than any one or two bacteria, and the effectiveness of these manipulations may well vary over illness progression (Lynch and Pedersen, 2016).

Even when the mechanisms leading to illness have been identified, manipulating these processes would not necessarily attenuate the characteristics of the disease. Once an illness is sufficiently advanced, or particular factors are well entrenched, simply altering the microbiome can help limit further illness progression but might be insufficient to reverse the already existing damage. Aside from the other actions that have been ascribed to microbiota, they are also involved in the production of metabolites that enter circulation, which can affect various conditions outside of those involving the gut itself. For the moment, firm conclusions regarding the effectiveness of probiotics in the treatment of many illnesses ought to be held in abeyance.

These caveats notwithstanding, there are excellent possibilities of being able to capitalize on individual differences concerning health risks. Perhaps individuals can be stratified based on a few core bacteria, and thus, the broad variability of microbiota may be somewhat more limited. It may be possible to use these broad classes of bacteria in designing ways (e.g., through prebiotics, probiotics, or synbiotics—the latter contains both pre- and probiotics) to enhance gut bacterial functioning (Cani and Everard, 2016). People who consume the same diet, may nevertheless present with glycemic responses that differ appreciably, possibly owing to individual differences in gut microbial composition. Thus, finding appropriate diets for any given individual might benefit from a personalized approach, which could include glycemic responses, microbial factors, and genetic contributions (Zeevi et al., 2015). For instance, among young women, socioeconomic factors, specific food choices, such as fat intake, and the presence of a gene variant (e.g., DRD4 VNTR), together could predict susceptibility to obesity (Silveira et al., 2016). Using a personalized approach to deal with diets and obesity would be enormously difficult (and financially constraining), but given the obesity crisis that seems to be escalating, such an approach could have both short- and long-term benefits.

Optimism has been expressed regarding the potential benefits of microbial supplements in limiting illness occurrence and progression as well as in enhancing treatments. Most studies were conducted in rodents, and the findings might not be translatable to the human condition. In fact, the benefits of probiotic supplements do not necessarily reduce stay in ICU or mortality in hospitalized patients. On the contrary, among some patients in ICU, probiotics increase the risk for *Lactobacillus bacteremia*, which may diminish host defenses and increase illness vulnerability (Yelin et al., 2019). Furthermore, unlike the effects of fecal microbiota transplants that could reconstitute the microbiota following antibiotic administration, probiotics did not consistently produce this outcome, and could even limit the appearance of important microorganisms (Suez et al., 2018). Concerns have also arisen regarding the safety of probiotic supplements in vulnerable populations, such as older individuals, newborns, pregnant women, and those who are immunocompromised, especially as these compounds often do not undergo safety testing. For the moment, it does not appear that probiotic supplementation is beneficial among seriously ill patients, but it remains possible that positive actions might be realized based on a personalized approach to the probiotics consumed (Zmora et al., 2019).

In light of the therapeutic benefits that could potentially be obtained by microbiota manipulation, like so many other over-the-counter supplements with questionable value, probiotic formulations are being offered to treat diverse illnesses without systematic experimental validation of their efficacy or safety. Having been led to believe that probiotics may enhance health, many people consume them despite their limited value unless a person is deficient in these. When microbiota abundance and diversity are not impaired, probiotic supplements are ineffective in colonizing the gut and are eliminated with other undigested waste. Rather than using off-the-shelf probiotics, it might be more profitable to identify specific microbiota or metabolite deficiencies and then consume probiotics accordingly (Zmora et al., 2018). Failing this, proper diets that favor antiinflammatory effects are the best option.

Moderating Variables Concerning Gut Bacteria and Health Outcomes

As we have seen, the positive (or negative) influences of microbial factors may be dependent on a constellation of genetic, experiential, and psychosocial factors (life experiences, trauma, and social learning). Their cumulative effects can also influence the effectiveness of intervention and treatment strategies. Just as stressful events can affect microbiota and/or immune functioning, and hence the provocation of disease, exposure to bacteria can affect the subsequent response to a stressor, thereby affecting physical ailments and psychological disturbances.

Well-being is enhanced when foods eaten fuel microbes are consistent with the needs of the host; however, when these needs are at odds with one another (as occurs in response to sugars and fats), illnesses may ensue. In this instance, microbes may begin to use nutrients that the body requires (e.g., iron) and promote immune and metabolic dysregulation, thereby resulting in inflammatory processes being activated. As depicted in Fig. 3.5, these biological actions may instigate a broad range of physical and psychological disturbances.

Beyond the many linkages between microbiota and physical disorders, gut bacteria may also influence neurogenesis. Specifically, eliminating gut bacteria through antibiotics can diminish the formation of new hippocampal neurons and disrupt performance in memory tasks. Interestingly, mice that had exhibited memory disturbances displayed lower white blood cells, primarily monocytes that carried Ly6Chi as a marker, implicating a link between gut bacteria, aspects of immune functioning, and brain neurogenesis. This connection was confirmed by showing that these outcomes could be reversed by reconstitution with normal gut flora, provided that mice were also able to exercise (using a running wheel) or given probiotic treatments (Baruch and Schwartz, 2016). In addition, disrupting the microbiome in mice through antibiotics affects hippocampal glial reorganization, thereby favoring the development of depressive-like features (Guida et al., 2017).

Concluding Comments

Bacteria and viruses have been constant concerns, but in several ways, the threats they pose have increased appreciably. Many vaccines are not as effective as they ought to be (e.g., in the case of influenza vaccines), and the evolution of antibiotic-resistant bacteria has become more apparent in affecting several existing diseases. Besides the obvious consequences of infection, it has become apparent that activation of inflammatory processes may promote multiple diseases, including physical and psychological disorders.

Because of the health benefits (and risks) associated with gut bacteria, there has been an effort to enlist the microbiome to enhance well-being. In this regard, the lifestyles that are often adopted could produce microbial dysbiosis, thereby promoting psychological disturbances, which can be attenuated, at least to some extent, by pre- and probiotic consumption. The prebiotics that have been used to modify gut-related disorders may have their effects owing to multiple changes that evolve. These include the production of antimicrobial compounds, growth substrates, such as vitamins and polysaccharides released into the internal environment, reduction of the luminal pH, prevention of particular microbes from adhering to epithelial cells, augmented barrier function-

ing, and modulation of immune responses. Prebiotics, such as certain oligosaccharides in human milk, can also inhibit monocytes, lymphocytes, and neutrophils from binding to endothelial cells and might thereby contribute to the relatively low frequency of inflammatory diseases in milk-fed human infants.

As we have discussed, owing to the individual variability that exists regarding microbiota, coupled with the many factors that affect microbiota balances, it is difficult to discern what reflects a harmful versus a beneficial complement of bacteria. But, it is still uncertain which good bacteria to call upon, how much of it is needed, and how to battle with bad bacteria. While the actions of some bacteria are known, many others are hardly understood. By determining individuals and communities of microbes, it may become possible to identify which core microorganisms may have their best prophylactic and therapeutic effects under specified conditions. With the treatment of illness, the intestinal microbial population might need to be individually tailored by diet and other manipulations.

CHAPTER

4

Pathogenic Factors Related to Mental and Neurological Disorders

So many ways to die

Life events are unpredictable, and the view was even expressed that we may lack free will to determine our fate (Sapolsky, 2023). What is predictable, is that none of us will get out of this alive. Fans of the poet and song writer Leonard Cohen will recall that some die by fire, some by water … some by accident … and some by barbiturate … and some by slow decay. Although illnesses, such as heart disease and cancer, which are often related to ageing take an enormous toll, historically, these don't hold a candle to viral and bacterial epidemics and pandemics. Communicable diseases frequently have the greatest adverse effects in the weakest (the very young and relatively old) and those with preexisting health conditions, as well as those living in socially or financially impoverished conditions. However, these illnesses can affect anyone—viruses and bacteria generally don't discriminate. They have one job—to infect us—and they often do this exceedingly well.

There is a lengthy list of pandemics that have ravaged societies worldwide. It isn't our intent to list every pandemic that has been experienced, but to make our point, we will mention the most serious. The Black Death (Bubonic plague) resulted in somewhere between 75 and 200 million people dying from 1346 to 1353 AD, and the earlier Plague of Justinian (a form Bubonic plague) during the period of 541–549, caused the death of 15–100 million people. A third Bubonic plague (1855–1960) killed about 13 million people world-wide, and several other plague outbreaks in subsequent years killed several million more. Several smallpox out breaks (165–180 AD; 735–737; 1519–1520) killed 12–20 million people across different countries, and waves of cholera pandemics between 1865 and 1947 resulted in the death of about 23 million people.

At its peak in 2004, death owing to AIDS exceeded 2 million people yearly but has since declined to about 500,000–800,000. Before the introduction of a vaccine, measles epidemics were fairly common, occurring every 2 or 3 years, resulting in over 2 million deaths.

Various forms of influenza affect more than a billion people annually and somewhere between 300,000 and 600,000 die as a result, although some forms of influenza are far more deadly. The Spanish Flu (1918–1920) killed between 50–100 million people; in the mid-1900s the Russian flu, the Hong Kong Flu, and the Asian Flu each resulted in about 2 million deaths, and the swine flu in 2009 killed about a million people.

Many of us had heard it repeatedly—it's only a matter of time before another pandemic hits again, but we didn't know how severe it would be. The ingredients for this were all in place, such as climate change, altered ecosystems that favored transmission between animal species, and between animals and humans (zoonotic transmission). Whereas SARS-CoV-1 killed fewer than 800 people, more than 7 million died during COVID-19 (SARS-CoV-2) and many more "excess deaths" are attributable to this pandemic. Many of the factors that could limit spread of infection could have been in place (wearing masks, limiting social contact, vaccination becoming available sooner), making it inexcusable that governments were unprepared to deal with COVID-19. Of course, it's simple to be an after-the-fact critic and governments have various needs that need to be accommodated. It also became apparent during the COVID-19 pandemic that many people lacked a basic understanding of viruses and bacteria (or were easily influenced by conspiracy theorists), which could have disastrous consequences, making it that much more important for strategies to be established to enhance public perceptions that could influence individual and population well-being.

The Immune System
https://doi.org/10.1016/B978-0-443-23565-8.00004-1

Copyright © 2025 Elsevier Inc. All rights are reserved, including those for text and data mining, AI training, and similar technologies.

Pathogens and CNS Functioning

As we discussed in earlier chapters, bacterial and viral threats can affect the central nervous system (CNS) both through direct invasion or indirectly by peripheral processes. This chapter will address the mechanisms through which microbial agents are detected and gain entry to the CNS and how they can then interact with neurons, microglia, and astroglia to promote neuropsychiatric and neurological illnesses. In the current pandemic age, there has been a growing awareness of the wide-ranging impact of microbial pathogens and, not surprisingly, an explosion of research focused on how the coronavirus SARS-CoV-2 causes pathology. As humans face further pandemics, it is important to understand the mechanistic underpinnings of pathogens, including how they affect CNS functioning.

Although the blood–brain barrier (BBB) and blood-cerebrospinal fluid barrier provide the most obvious "front-line" protection from pathogenic invasion, penetration into the CNS can still occur. As an example, three very serious neurological illnesses, meningitis, syphilis, and HIV dementia, result from direct migration of pathogens into the brain. Many other microbes have similarly evolved mechanisms that facilitate their transport into the CNS, and viral and bacterial infections have been implicated in many neurological disorders (Ravichandran and Heneka, 2024).

There are three primary means through which pathogens traverse the BBB to enter the CNS; these include transcellular and paracellular routes, as well as adopting a "Trojan horse" strategy. The transcellular route involves direct pathogen interaction with endothelial BBB cells by ligand-receptor binding or through electrical charge interactions. Certain strains of the bacteria, *Escherichia coli*, are believed to use a transcellular route to enter the brain and cause bacterial meningitis. The paracellular route occurs when the pathogen essentially slips between cells lining the BBB, or where microbial organisms disrupt the tight junctions and enter between adjacent endothelial cells. The bacterium that causes syphilis, *Treponema pallidum*, enters the brain through the paracellular route. The third, the trojan horse route, involves microbial transfer by the migration of an infected immune cell, almost always a phagocytic cell, such as a macrophage. The BBB is permeable to phagocytotic white blood cells, and HIV-1 can influence the CNS through infected mononuclear phagocytes that eventually give rise to symptoms of dementia.

In the section that follows, we will discuss several viral and bacterial illnesses, as well as parasitic infections. The number of illnesses described will necessarily be limited, focusing on some of the most frequent and those that clearly have effects on brain functioning and that are influenced by inflammatory processes. Of course, many other conditions could have been discussed, but those included are enough to make our point.

Influence of Viral Infections

As indicated in other chapters in this book, neurodegenerative diseases, such as Parkinson's disease (PD), Alzheimer's disease (AD), and neuropsychiatric illnesses (e.g., depression and anxiety), have been linked to infection. Viral pandemics have been associated with encephalitis, mood disturbances, motor dysfunction, and psychosis, and there is reason to suppose that SARS-CoV-2 might have long-term consequences on such processes (Hayley and Sun, 2021; Troyer et al., 2021). In many cases of neuropsychiatric illness, the pathogen might not directly invade the CNS. Instead, the inflammatory cytokines that are induced by a peripheral infection can result in neurochemical changes that are aligned with emotional and affective disturbances. Likewise, chronic low-level inflammation, possibly owing to subclinical infection, might predispose neuronal networks to dysfunction and possibly eventual degeneration.

National and international organizations that track viruses have been concerned about several very deadly viruses (Nipah virus, Rift Valley fever, Zika virus, Marburg virus, Ebola, and MERS), but other less frightening viral illnesses can cause serious illnesses and death, and many can affect brain processes. Despite the enormous number of viruses that are presumed to exist, a limited number negatively affect humans. Indeed, the human gut is teeming with viruses, which, like their bacterial cousins, are frequently beneficial (Cao et al., 2022c). Viruses that are harmful to humans number about 270, some of which have only recently emerged, whereas others have affected humans for decades or centuries (Forni et al., 2022).

Influenza infections are among the most common experienced by humans. Influenza viruses come in four types, three of which affect humans. Only influenza A viruses are classified based on proteins on their surface [i.e., hemagglutinin (H) and neuraminidase (N)], hence designations, such as H1N1 and H3N2. Influenza A viruses typically cause respiratory infection, resulting in bronchial distress and the production of a constellation of cytokines by immune cells. In particular, the release of IL-1β, IL-6, tumor necrosis factor (TNF), and type I interferons (IFN) contribute to viral clearance but can also have many adverse neuronal and behavioral effects (Düsedau et al., 2021). These cytokines give rise to fatigue, aches and pains, fever, and lack of motivation, referred to as "sickness behaviors." Eventually, adaptive cellular immunity eliminates the virus owing to the maturation of virus-specific cytotoxic and helper T cells.

After replication in the lungs, Influenza A may affect brain processes. In fact, during the H1N1 influenza pandemic of 1918 (the "Spanish Flu"), the incidence of an inflammatory CNS disease then known as encephalitis lethargica occurred relatively frequently, characterized by catatonia, parkinsonism, and psychosis. Since then, it has been learned that several forms of influenza can profoundly influence brain processes. A subtype of the H5N1 virus (i.e., HK483) can spread to the CNS through the olfactory, vagus, trigeminal, and sympathetic nerves (Park et al., 2002), leading to neuroinflammation and neurodegeneration (Jang et al., 2009). Similarly, hippocampal spine loss was provoked by infection with a neurotropic virus (i.e., that can infect nerve tissue), such as the H7N7 virus, and by one that was not neurotropic (H3N2), and these effects persisted for months following recovery from infection (Hosseini et al., 2018). Another nonneurotropic form of influenza infection (A/PR/8/34) likewise provoked behavioral disturbances and concurrently increased microglial activity and that of several proinflammatory cytokines, particularly IL-1β, IL-6, TNF-α, and IFN-α (Jurgens et al., 2012).

T-lymphocyte cells are not only critical for dealing with peripheral infectious agents but also those that occur within the brain. Such brain infections can cause neuroinflammation and swelling that has dire consequences, necessitating T-cell-driven adaptive immunity for long-term microbial clearance. Indeed, mice that lack the crucial T-cell receptor (TCR) that is required for T-cell engagement drastically impaired bacterial clearance in the brain, but this deficit was reversed by the adoptive transfer of either Th1 or Th17 cells (Holley and Kielian, 2012).

Worldwide, rabies and Japanese encephalitis virus (JEV) are responsible for an estimated annual mortality of 13,700 and 25,000 people, respectively (Cheng et al., 2022; Gan et al., 2023). In the absence of treatment, rabies is almost always fatal, whereas cases of JEV are less often fatal, depending on the infected person's age. However, long-term neurocognitive disturbances may occur owing to the production of an uncontrolled inflammatory response and neuronal cell death (Ashraf et al., 2021). As mentioned at the outset of his chapter, HIV/AIDs has caused millions of deaths and may cause a neurodegenerative condition characterized by cognitive (difficulties with attention, concentration, and memory), affective (irritability, depression, and reduced motivation), and motor disturbances, which collectively are known as HIV-associated neurocognitive disorder (HAND). Treatment with a combination antiretroviral therapy (CART) has been effective in markedly reducing deaths, and people with HIV have life expectancies that are roughly comparable to those noninfected. Yet, early in HIV infection, the virus enters the brain, resulting in inflammation that likely contributes to HAND. In fact, the brain can "serve as a sanctuary" for HIV, even after systemic viral replication has been suppressed (Saylor et al., 2016).

Bacterial Infections

Aside from the impact of viruses, as described in Chapter 3, studies in both animals and humans have supported a link between bacterial infection and the presence of numerous cytokines and chemokines with the CSF (Coutinho et al., 2013), which could disturb the BBB, thereby promoting diverse behavioral disturbances (Erickson and Banks, 2018). Bacterial meningitis may result in long-term disturbances of mood, sleep, and cognition, as well as dementia and motor disability (Farmen et al., 2021). By blocking the effects of TNF-α, the effects on the BBB can be prevented, thereby limiting pathogen entry into the brain.

In animal models, administration of lipopolysaccharide (LPS), a component of the outer membrane of gram-negative bacteria, can elicit emotional responses (e.g., anxiety) and cognitive disturbances, likely owing to elevated levels of proinflammatory cytokines in the brain (Mittli, 2023; Zhao et al., 2019a). The gram-positive bacterium, *Staphylococcus aureus*, which is responsible for many hospital-related infections, can similarly elicit anxiety-like behaviors. This effect was associated with microglial activation and release of proinflammatory cytokines in the medial PFC, which is fundamental for cognitive appraisal and processing of stressors (Zou et al., 2022).

Of the many bacterial illnesses in humans, tuberculosis (TB) stands out as being particularly pernicious. This disease occurs through infection by *Mycobacterium tuberculosis* from those infected. The disease is easily spread since only a very small number of bacteria are sufficient to cause infection. It has been estimated that almost a quarter of the world population has been infected, but most often remains latent. In recent years, about 10 million people developed the disease (e.g., 10.6 million in 2022), resulting in about 1.5 million deaths annually. Children are at greatest risk of the disease, as are people with some preexisting conditions (e.g., diabetes, a weakened immune system, such as that in people with HIV/AIDS, as well as in malnourished individuals, and in those who smoke) (WHO, 2023b). TB is second to SARS-CoV-2 as the most prodigious killer worldwide despite being readily prevented and treatable using certain antibiotics, such as isoniazid or rifampicin. This said, multidrug-resistant tuberculosis (MDR-TB) can occur, which can be treated by second-line drugs, although they are not only expensive but can have toxic effects. While the disease occurs worldwide, two-thirds of cases come from eight countries (e.g., India, Indonesia, China,

Philippines, Pakistan, Nigeria, Bangladesh, and the Democratic Republic of the Congo). This should not be taken to imply the absence of the disease in Western countries. Thousands of cases occur in the US, although the frequency has been declining (e.g., Cowger et al., 2019). The favorite target of TB is the lungs, but it is also found in the bones, kidneys, and lymph nodes, as well as within the spine and brain. Owing to CNS infection, cognitive disturbances (notably impaired attention, executive functioning, and working memory) were apparent in 93% of people within 2 weeks of infection, with relatively pronounced impairments occurring in a third of patients (Ganaraja et al., 2021). The severity of symptoms, reflected by latent, active, and treatment-resistant TB, were related to specific immune markers and especially elevated levels of proinflammatory cytokines (e.g., Sampath et al., 2023). Predictably, depressive illness was associated with the occurrence of TB, varying with the severity of this condition, and was most pronounced in MDR-TB patients (Duko et al., 2020; Ruiz-Grosso et al., 2020).

In addition to the impact of specific bacteria, sepsis that is brought about by bacterial infection markedly affects children and adults. Sepsis reflects an extreme response to infection that leads to death in almost 50% of affected adults. This condition is accompanied by elevated levels of proinflammatory cytokines, which may contribute to some of the most severe neurocognitive changes that can be provoked, and their increased levels might make them useful as illness biomarkers (Frimpong et al., 2022). For that matter, the persistent cognitive disturbances evident following recovery from the condition may be related to elevated cytokine levels (Li et al., 2022e).

The development of sepsis following bacterial infection is estimated to affect about 49 million people annually and results in death of 11 million people (e.g., Rudd et al., 2020), leading to the 2017 designation of this as global health priority by the WHO (Reinhart et al., 2017). Sepsis and meningitis, which frequently stem from bacterial infections in neonates, are associated with short- and long-term neurological and cognitive impairments (Cortese et al., 2016). Even modest infection can promote diverse cognitive disturbances. The cognitive impairments may be related to inflammation associated with infection and with hypoxia/ischemia that may occur, which can promote proinflammatory microglial activatiotogether with elevated reactive oxygen and nitrogen species (Sewell et al., 2021).

Both neonatal and maternal factors (e.g., maternal infection) have been linked to sepsis (Chan et al., 2013). Conjugate vaccines against bacterial meningitis due to *Streptococcus pneumoniae* and some forms of influenza have been associated with a reduction of sepsis in high-income countries, and steroids provided as an adjunctive therapy have likewise reduced neurological disturbances. Regrettably, in low- and middle-income countries, the development of neonatal sepsis and high mortality continues (Milton et al., 2022).

Neonatal sepsis induced in animal models revealed long-term behavioral and cognitive disturbances that are apparent in adulthood (Comim et al., 2016). Altering the course of these effects has been difficult. However, treatment with indole-3-guanylhydrazone hydrochloride (LQM01), which has antiinflammatory and antioxidant actions, reduced these effects and reduced microglia activation in the hippocampus of neonatal mice. When treated adult mice were assessed, they displayed reduced anxiety-like behaviors and cognitive impairments (Heimfarth et al., 2020). An alternative strategy focuses on ways of altering microglial functioning. Since microglia are fundamental to illness-related cognitive disturbances (e.g., Yan et al., 2022b), targeting microglia, either by blocking their actions or by altering their M1 (proinflammatory) state to one that is more aligned with an antiinflammatory M2 state, may be effective in attenuating the sepsis-associated encephalopathy (Hu et al., 2023).

Parasitic Infection

Numerous parasites can affect humans, some of which can promote cognitive and emotional alterations. Parasitic infection occurs most frequently in tropical regions of the world and where sanitation systems are poorly developed. However, several parasitic diseases (toxoplasmosis, schistosomiasis, chistosomiasis chistosomiasis tricomoniasis, Chagas disease, and cyclosporiasis) are also present in Western countries. For instance, toxoplasmosis is a widespread parasitic disease caused by *Toxoplasma gondii*, which is most often transmitted from animals to humans or through undercooked foods. One of the most common routes of transfer is through cats who have eaten infected birds or small mammals, so the oocysts in their feces can be transmitted to humans. Toxoplasmosis can affect the brain, and its presence in various brain sites has been associated with anxiety and fear memory (Ihara et al., 2016) as well as cognitive disturbances and the occurrence of schizophrenia (Guimarães et al., 2022). It appeared that toxoplasmosis was associated with elevated indices of inflammation in humans (Egorov et al., 2021), and in animal models, infection was related to elevated mRNA levels of proinflammatory cytokines and the presence of anxiety (Mahmoudvand et al., 2015).

Schistosomiasis caused by blood flukes affects about 250 million people, has been associated with learning and memory disturbances in children, and varied cognitive deficits also occur in adults. In animal studies, such effects were observed even with forms of schistosomiasis that do not enter the brain (Gasparotto et al., 2021). Still more severe disturbances occur after migration of adult worms to the brain where they lay their eggs, which promotes neuroschistosomiasis, a severe form of the disease. Neurological problems arise both in the acute and chronic phases of illness, including the development of encephalopathy, delirium, visual impairment, seizures, motor disturbances, and ataxia. Other severe symptoms occur if eggs are laid within the spinal cord.

Helminths are parasitic worms that can be transmitted to humans through different routes and affect millions of people, especially children. They can be ingested accidentally, or in some instances, worm larvae can penetrate the skin, as in the case of hookworms, schistosomes, and Strongyloides. They can also be transmitted through contact with the soil, frequently affecting children and possibly being responsible for cognitive impairments (Pabalan et al., 2018). These effects may occur owing to perturbations of the gut–brain axis, including effects on brain microglia (Giacomin et al., 2018). It is understandably difficult to identify the contribution of soil-transmitted helminths (STH) to such cognitive impairments owing to the presence of many other conditions that are aligned with cognitive functioning. In fact, a systematic review indicated that when children living in developing countries were treated with antiparasitic agents (deworming), cognitive functioning was not enhanced (Taylor-Robinson et al., 2015).

Malaria is caused by a parasite that infects certain types of mosquitos, which transfer the illness to humans when they feed on them. In 2022, approximately 249 million cases of malaria occurred, which led to more than 600,000 deaths, with 80% being children under 5 years of age (WHO, 2023a). Of several forms of malaria that affect humans, the most common is *Plasmodium falciparum*. Individuals who survived cerebral malaria, which is considered the most severe form of the disease, may develop neurological and cognitive disturbances, as well as mood disorders. Even less severe cases of malaria have been associated with cognitive impairments, especially among children (Rosa-Gonçalves et al., 2022). Fortunately, two vaccines are now available that will hopefully reduce the occurrence of malaria.

Parasitic infection

Intestinal parasites come in three primary forms, protozoa (single celled organisms, such as giardia, cryptosporidium, microsporidia, and isospora), helminths (e.g., tapeworms, hookworms, pinworms, roundworms, whipworms, blood flukes), and ectoparasites (the latter live on skin or outgrowth of the skin; e.g., fleas, lice, ticks, mites). Some parasites consume food that is ingested, so that individuals feel perpetually hungry, whereas other parasites feed off red blood cells, causing anemia.

The presence of parasites within the gut, can cause a wide range of signs and symptoms, such as abdominal pain or tenderness, bloating, nausea or vomiting, itching in the vicinity of the rectum or vulva, and fatigue. As well, these parasites can influence brain functioning and thus affect cognitive and affective processes. Following the administration of the noninvasive parasite Trichuris muris to produce gut inflammation, anxiety was elicited in rodents, which could be attenuated by antiinflammatory treatments, such as a TNF-α antagonist. A probiotic treatment also attenuated the anxiety brought about by Trichuris muris but did so without affecting cytokine or kynurenine levels (Bercik et al., 2010), suggesting that brain functioning might be affected through some other route (e.g., through the vagal nerve).

Parasites, no matter how gross they appear, are among our "old friends" that have evolutionary benefits. Helminths can influence immune functioning to avoid being attacked, and humans have likewise adapted to their presence, even taking advantage of them. Indeed, we have become dependent on their presence, and glycoproteins secreted by some parasites (e.g., filarial nematode parasite) can act against inflammaging (Zhang and Gems, 2021) and helminths can diminish allergies (Cruz et al., 2017), the development of asthma, and perhaps limit autoimmune conditions. Thus, efforts have been made to identify the processes (and molecules) by which they can produce beneficial actions, thereby allowing for the development of effective biopharmaceuticals (Bohnacker et al., 2020).

Viral Invasion of the CNS

Viral infections typically originate in the periphery, usually at mucosal membranes or cell surfaces, resulting in tissue-specific antiviral responses that involve the innate branch of immunity. This promotes a rapid inflammatory response characterized by infiltrating mononuclear immune cells and soluble signaling factors, including inflammatory cytokines. Thereafter, secondary adaptive immunity is engaged, resulting in a more specific and focused attack on the invading pathogen. T and B lymphocytes that have been specifically primed (via antigen presentation) against the viral agent can provide a degree of long-term immunological memory.

The ability of pathogens to directly infiltrate the CNS and infect neural or glial cells is relatively poor for most viral agents. Besides the protection afforded by the BBB, evolutionary pressures have tended to select a CNS environment that is generally not favorable for pathogen survival. Yet, when CNS infection does occur, very substantial neuropathology may ensue. Such CNS infections are often of a zoonotic nature, often resulting in movement from a natural host into a nonnatural or "dead-end host" (i.e., no further human-to-human transmission). Such zoonotic viral infections can be highly virulent in nonnatural hosts, as in the case of rabies and West Nile virus (WNV). Although these zoonotic viruses directly infect neurons, actually "hiding" there, making WNV difficult to detect and its removal by peripheral immune cells and CNS microglia is especially challenging.

Viral entry into the CNS or into cells of the peripheral nervous system (PNS), as mentioned earlier, often involves the pathogen directly entering neural fibers or by way of infected immune cells that essentially act as a type of trojan horse carrying the infection into neural tissue. Essentially, the pathogen is hidden in the immune cell, which interacts with BBB proteins that allow it to traverse this protective hurdle. As an example, the human immunodeficiency virus (HIV) prefers to infect CD4+ T-lymphocytes but can also target macrophages and glial cells, giving it ready access to the CNS. The gp120 and gp41 glycoproteins found on the virus attach to several different leukocyte receptors. This attachment and subsequent viral replication are aided by proinflammatory cytokines and chemokines produced during infection. When HIV-infected macrophages enter the brain, they eventually become perivascular microglia, which can then release toxic factors damaging local neurons. In about 10%–20% of HIV patients, this eventually results in HIV-associated dementia (e.g., Borrajo López et al., 2021). Additionally, some viruses can directly infect cells lining the BBB, including brain microvascular endothelial cells (BMVECs). The RNA viruses, West Nile, hepatitis C virus (HCV), and Epstein-Barr virus can infect BMVECs, resulting in BBB "leakiness" and the infiltration of these pathogens into the brain parenchyma.

Regardless of the site of initial infection, evolutionarily conserved glycoproteins found on viruses are generally first detected by cells of the innate immune system that possess pattern recognition receptors (PRRs). The PRRs are abundant in neutrophils, macrophages, and dendritic cells, and CNS and PNS glial cells also express these microbial detection molecules. Indeed, microglia and, to a certain extent, astroglia express Toll-like receptors (TLRs), which are an important category of PRRs (Garaschuk and Verkhratsky, 2019). Intracellular TLR3 receptors are primary detectors of intracellular viral RNA that is produced during replication in an infected cell. And in the case of SARS-CoV-2, TLR1, 4, and 6 also aid in detection by recognizing viral membrane proteins, such as the spike proteins found on coronaviruses (Choudhury and Mukherjee, 2020).

Many viruses can attach to sensory neuron axonal terminals to enter the PNS. Herpes viruses, including herpes simplex type 1 (HSV-1) and HSV-2, can use this route to enter the PNS and cause long-term infections. Such viruses utilize membrane fusion to penetrate neuronal fiber branches and then retrogradely move to the cell soma by means of axonal transport molecules. In the case of DNA viruses like HSV, the viral genetic material is incorporated within the host nucleus and can remain quiescent, establishing long-term latent infection. Later, cellular stressors can reactivate the virus, resulting in a rapid and robust production of viral progeny and serious infectious events. This is what occurs in the case of shingles, wherein infection with varicella-zoster virus (which causes chickenpox) results in the virus remaining latent in the dorsal root ganglia of the PNS. Subsequent exposure to sufficient stress or immunosuppressive drugs can reactive the virus, causing the clinical rash and extreme pain of shingles (e.g., Schmidt et al., 2021).

The neuromuscular junctions are especially susceptible to certain viral infections, including rabies and poliovirus. These infections are typically acquired from an infected animal or consumption of viral particles. In addition to neurons, the rabies virus infects human and murine astrocytes and microglia. Poliomyelitis, which is characterized by severe motor disability and paralysis, sometimes leading to death, occurs following poliovirus infection of motor neurons and is associated with marked signs of neuroinflammation deep within the CNS. Indeed, paralytic poliomyelitis is characterized by CNS infiltration of polymorphonuclear and mononuclear leukocytes and highly activated microglia. Although the microglia were believed to have come from infiltrating peripheral monocytes rather than local CNS cells, it was subsequently observed that the virus spreads to the CNS through axonal retrograde transport (Ohka et al., 2022).

An interesting case report indicated that elevated CD68- and CD163-immunoreactive microglia were present around plaque-like lesions in a poliomyelitis patient who died approximately 25 years after first experiencing muscle weakness and atrophy. This patient displayed progressive motor disability before eventual death at the age of 80, which led to the suggestion that gradual neurodegeneration may have been driving microglial reactivity over time (rather than the other way around). This last point is interesting considering the ongoing debates around whether neuroinflammatory processes are the primary drivers of neurodegenerative disease (e.g., AD and PD) rather than the inflammatory processes being secondarily recruited in the face of existing neuronal pathology. Moreover, it may have implications for features secondary to polio that may appear long after infection and may be relevant to other diseases that follow this pattern. For instance, postpolio syndrome (PPS) frequently occurs many years after initial infection. The mechanisms responsible for this neurological condition are uncertain. This syndrome not only comprises motor dysfunction but is also associated with neuropsychological disturbances, generalized fatigue, sleep disturbance, decreased endurance, sensory symptoms, and chronic pain (Li Hi Shing et al., 2019). These symptoms are reminiscent of other illnesses (e.g., chronic fatigue syndrome/fibromyalgia) that have been associated with earlier viral infections, such as certain forms of influenza or herpes zoster infection (e.g., Tsai et al., 2014).

Such postacute infection syndromes (PAISs) occur fairly frequently, having been documented in association with about a dozen different viral infections. These conditions share a core set of features, especially extreme fatigue, exertion intolerance, neurocognitive and sensory impairment, flu-like symptoms, unrefreshing sleep, myalgia/arthralgia, as well as several less specific symptoms and more variable in their occurrence. It was suggested that these overlapping symptoms might reflect a common unifying pathophysiology (Choutka et al., 2022).

Viral-Induced Cell Fusion

The fusion of two or more cells can occur during infectious states, wherein the cellular plasma membranes merge to create a multinucleated common cytoplasm. This mass of fused cells, referred to as a syncytium, causes immunosuppression, which can enhance viral entry into host cells, as well as the spread between connected cells. In the absence of infection, syncytia can also occur in very specific cases, such as during wound healing or when gametes fuse during fertilization (Aguilar et al., 2013), as well as when immune cells interact with host immune factors (Efstathiou et al., 2020). The messenger proteins that regulate cell fusion, which is aptly named fusogens, are embedded in human DNA and evolved to deal with viral infections (Martínez-Mármol et al., 2023). Various fusogens regulate normal activities, such as syncytin-2, which causes the immunosuppression that is critical for the prevention of autoimmunity against a developing fetus (Roberts et al., 2021). Unfortunately, viruses can hijack fusogens and use them for their own propagation by disabling host immunity and causing the creation of multinucleated syncytia that allow viruses to flourish. In effect, viral-induced fusogens create a friendly environment by increasing intracellular calcium and iron levels, while, at the same time, reducing interferon-dependent antiviral activity (Frisch and MacFawn, 2020). Like several other viruses, SARS-CoV-2 and viral fusogens have been implicated in fusion between neurons and glia, thereby disturbing neuronal functioning (Martínez-Mármol et al., 2023). While aiding in viral spread, cell–cell fusion and the creation of syncytium also indues cell senescence and may contribute to pathological aging and neurodegeneration (Osorio et al., 2022) that has been associated with viral infections (Filgueira et al., 2021).

Glial Cells and Infection

Microglia express TLRs 1-9, which greatly increase in response to an infectious insult. Considerable heterogeneity exists with respect to microglia expressed throughout the brain, and their TLR expression appears to be highest in brain regions in close proximity to the peripheral vasculature, including in meningeal and circumventricular sites (Chakravarty and Herkenham, 2005). Viral recognition by TLR3 on such microglia precipitates the production of numerous proinflammatory cytokines and chemoattractant molecules (Schilling et al., 2021). Another member of this family, TLR4, serves as a sensing receptor for gram-negative bacteria, as well as binding to molecules produced following tissue injury (e.g., Molteni et al., 2016a). Activated microglia can release protective trophic factors, such as BDNF, GNDF, and TGF-β, which aid in tissue recovery following infectious damage (reviewed in Prowse and Hayley, 2021).

Besides its well-known TLR4 receptor, microglial also possesses CD14, which acts as a co-receptor to maximize the detection of bacterial agents, such as LPS. Interestingly, CD14 gives an almost 100-fold increased ability to detect the gram-negative bacteria LPS, while responding to the endotoxin within physiological limits (to avoid possible autoimmunity) by engaging the intracellular immunomodulatory cytokine, interferon β (Janova et al., 2016).

Astrocytes also express TLR2, TLR3, TLR4, and TLR9 receptors involved in provoking immune responses to infectious insults. Viral infection provokes elevated TLR3 expression on astrocytes, which causes proinflammatory cytokine and chemokine release (Jorgačevski and Potokar, 2023). Also, TLR3 engagement on astrocytes can, like microglia, eventually promote neurorecovery processes by inducing neurotrophic factor expression (Bsibsi et al., 2006). Interestingly, TLR3 and TLR7 are also able to detect "self" RNA and trigger its translocation within the cytosol in response to cellular stress of injury or infection (Fairhurst et al., 2008). Another intriguing aspect of astrocyte signaling is that the TLR9 receptor can influence astrocyte-to-macrophage and astrocyte-to-neuron communication. In fact, inhibiting TLR9 reduced the pro-survival effects of astrocytes on infected neurons and altered the activation state of macrophages (Acioglu et al., 2016; Li et al., 2020a).

Stimulation of TLRs can induce glial migration and proliferation. For instance, LPS promotes astrocyte and microglia proliferation by activation of the TLR2 or TLR4 receptors (Rodgers et al., 2020), whereas TLR3 activation by viral agents can produce the opposite effect and impede proliferation. Regardless of the specific TLR subtype on astrocytes, it is believed that input from microglia is critical for the manifestation of an inflammatory response in astrocytes. In this respect, microglial-produced transforming growth factor-β (TGF-β) was reported to augment astroglia TLR responses (Welser-Alves and Milner, 2013).

Direct CNS Infection

Several viruses can directly infect the brain, and many more can induce neuroinflammatory responses through indirect routes. As depicted in Fig. 4.1, within the brain, microglia interact with other cells, acting in a protective or destructive capacity, in this instance, during viral encephalitis. These actions come about through the functioning of various cytokines released by microglia and astrocytes. Even in the absence of a direct CNS viral invasion, these pathogens can impact neuronal processes by the induction of oxidative stress and inflammatory molecules within the brain. Several viruses from the flavivirus genus, including JEV and WNV, can infect the human CNS. These viral pathogens are recognized by TLR3 and can cause encephalitis and BBB permeability impairment if not properly eliminated. Mice that were genetically TLR3 deficient developed severe inflammatory brain damage that resulted from infiltration of inflammatory immune cells, excessive viral replication in neurons, and devastating levels of proinflammatory cytokines within the brain (Han et al., 2014). In the case of WNV, TLR3 was critical for viral entry across the BBB and into the brain parenchyma. The influence of this single-stranded RNA (ssRNA) flavivirus is often asymptomatic in humans but can be serious in immunocompromised or elderly individuals. TLR3-deficient mice were indeed resistant to lethal encephalitis infection but had elevated viral burden and levels of inflammatory cytokines and other factors in the periphery (Wang et al., 2004). However, viral load and levels of inflammatory factors were reduced in the brains of TLR3-deficient mice. In effect, the presence of TLR3 is beneficial for limiting peripheral viral replication and pathology but has deleterious consequences for the CNS since it appears necessary for allowing the virus entry into the brain. However, the issue is somewhat controversial since it was reported that genetic TLR3 deficiency increased the lethality of the WNV and elevated the viral burden in the brain (Daffis et al., 2008). A subsequent study found that the TLR3 pathway had little role in CNS infiltration of the virus, but TLR3 deficiency did result in less severe immune cell infiltration. There is also evidence that the JEV can engage with TLR2 and TLR7 receptors (Han et al., 2014).

As already indicated, the simplex virus type 1 (HSV-1) is a common neurotropic virus in which exposure typically occurs early in life, with more than 50% of people under 50 years of age being HSV-1 positive (Ayoub et al., 2019). This remarkably common virus first infects epithelial cells, then can travel to PNS sensory neurons, where it can become latent. Stressor or other infections can trigger reactivation of the virus at intermittent points throughout life. The virus can travel along various neuronal axons, eventually reaching the CNS, where viral replication can lead to dangerous encephalitis. TLR3 receptors on microglial cells detect HSV-1 genetic material and induce signaling cascades that promote type I IFN antiviral immunity. The TLR2 receptor also binds glycoproteins components of the virus, resulting in

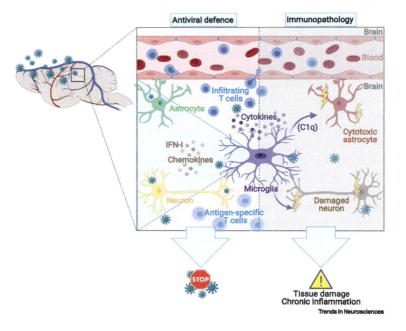

FIG. 4.1 Microglial activation and cellular crosstalk during *viral encephalitis* (VE). During viral infection of the brain, *microglia* fulfill miscellaneous tasks. This schematic overview summarizes the diverse cell–cell interactions and functions of microglia during VE. Activated microglia *phagocytose* cell debris and release cytokines, which activate other immune cells (purple dots). Furthermore, they can confer direct antiviral effects by barrier formation and virus trapping. Microglia also closely interact with other brain-resident cell types, as well as with infiltrating cells. Type I *interferon* (IFN-I) signaling by neurons and astrocytes (brown dots), as well as *chemokine* production by neurons, leads to microglial activation and antiviral defense. *T cells* are recruited to the central nervous system (CNS) and interact with microglia, presumably via MHC, which is a key mechanism to maintain T-cell function within the CNS. However, microglial activation and cellular crosstalk during VE can also exhibit *adverse effects*. Upon complement factor secretion by microglia, astrocytes can develop a neurotoxic phenotype and may confer detrimental effects. Microglia can also perform uncontrolled elimination of synapses, thus harming neurons due to overshooting inflammatory reactions. Nevertheless, microglia and the crosstalk with the depicted cells in the CNS play a central role during antiviral defense that is essential to protect the brain from viral infection. *From Waltl, I., Kalinke, U., 2022. Beneficial and detrimental functions of microglia during viral encephalitis. Trends Neurosci. 45 (2), 158–170.*

cytokine production, especially IL-6, along with potentially toxic reactive oxygen species. These responses are obviously necessary and beneficial in eliminating the pathogen, but the possibility exists that continual or unrestrained antiviral responses might damage delicate neurons within the CNS. In fact, the reactivation of latent brain HSV-1 infection has been implicated in Alzheimer's disease and mild cognitive impairment (Albaret et al., 2023; Marcocci et al., 2020).

Lessons From COVID-19

Our recent experiences in dealing with SARS-CoV-2 have been instructive in multiple ways. Thus, we will spend more time dealing with this virus than we have with others. In many ways, COVID-19 followed a pattern consistent with other viral illnesses, but in several other ways, it differed from any other virus that had previously been experienced. Indeed, SARS-CoV-2 has been described as a "Master of immune evasion" (Rubio-Casillas et al., 2022). Just as the immune system has evolved to limit microbial invasion, the viruses that humans encounter have evolved to escape immune detection and destruction. In this respect, SARS-CoV-2 seems exceptionally skilled in doing so. It can achieve this by disturbing antigen presentation by affecting MHC-1 expression, evading antibody neutralization, undermining mitochondrial functioning, and by suppressing the antiviral actions of interferons. Of course, the frequent variants that evolved each seemed to have their own dirty tricks to get around our defensive measures.

SARS-CoV-2 and influenza spread in similar ways, although SARS-CoV-2 is more infectious than influenza as well as other coronaviruses. They share similar risk factors (e.g., age and preexisting medical conditions), and the viruses both mutate to escape prevention by vaccines. While both viruses share many symptoms, including markedly affecting the respiratory system, SARS-CoV-2 affects multiple organs not affected by influenza and leads to more severe and longer-lasting illness, as well as engendering features not typically apparent with influenza. Both illnesses can lead to some similar complications, but these were more common following SARS-CoV-2 infection (e.g., blood clots, multisystem inflammatory syndrome, and secondary bacterial infection). Further, influenza symptoms appear within 1–4 days following infection but were much longer after SARS-CoV-2, although this has become shorter with subsequent variants. SARS-CoV-2 also appears to be contagious for a longer period.[1] There are people who encounter SARS-CoV-2 infection on multiple occasions. It seems a heritable factor exists that influences the chance of being infected, as well as the severity of the illness (Brown et al., 2024). As SARS-CoV-2 infection also disturbs immune functioning, when a second infection occurs, the symptoms experienced are more severe.

It was discovered early in the COVID-19 pandemic that SARS-CoV-2 enters cells by attaching to the angiotensin-converting enzyme 2 (ACE2) transmembrane protein, doing so through its spike protein. The virus prefers peripheral targets, especially the lungs, as well as the heart and kidneys (Beyerstedt et al., 2021), and as we will see, it can affect the brain. The spike (S) proteins found on the viral envelope of SARS-CoV-2 interact with ACE2 receptors throughout the body (Xu and Lazartigues, 2022). The S protein is made up of two subunits; whereas the S1 subunit has a receptor binding domain responsible for recognizing and binding ACE2, the S2 subunit is responsible for viral cell membrane fusion. Enzymes, particularly transmembrane protease, serine (TMPRSS)2, appear to be involved in the initial processing of SARS-CoV-2, and the S2 portion of the virus facilitates fusion and its entry into the cell (Sallenave and Guillot, 2020). Furthermore, SARS-CoV-2 binds with very high affinity to ACE2 (more so than other coronaviruses), possibly owing to a particular (furin) cleavage site between the S1 and S2 subunits, thereby increasing its infectivity (Sallenave and Guillot, 2020). Many seasonal CoVs (e.g., CoV-229E and CoV-OC43) fail to bind to the ACE2 receptor and are far less pathogenic than SARS-CoV-2, causing relatively mild "cold" symptoms. These seasonal CoVs also prefer to infect the upper respiratory tract rather than the lungs, gut, or brain. In contrast, SARS-CoV-2 causes pneumonia and respiratory failure by invading and damaging

---

[1] The 1957 H2N2 pandemic (sometimes referred to as the "Asian flu") had numerous features that were similar to the SARS-CoV-2 epidemic, including the slow response of the Chinese government to alert other countries of the new viral outbreak. In some ways, that pandemic was worse than COVID-19. The symptoms generally comprised high fever, head, chest, and backache, dry cough, and extreme fatigue. Those who died, most often within 48h of hospitalization, exhibited marked breathing difficulties, low blood oxygenation, and accompanying deathly pallor, coughing blood, and bacterial infections that overwhelmed an already run-down respiratory system. Unlike COVID-19, and more like the 1918 swine flu pandemic, young adults were especially vulnerable to infection, whereas those over 65 years of age were less likely to become infected (possibly owing to earlier infections that provided some immunity); however, if they became infected, it was especially deadly. Despite the fear created among virologists and public health agencies, the public was relatively blasé, and protective measures were rarely adopted, although a vaccine was created that was slightly more than 50% effective. How many people today are aware of the H2N2 pandemic? We have short memories, and as we'll certainly encounter further pandemics, it is essential that policymakers and the general population will have learned some lessons from the COVID-19 experiences (see Honigsbaum, 2020).

multiple respiratory cell types. As already mentioned, ACE2-bearing cells in the gastrointestinal system and brain may also be affected by SARS-CoV-2, causing numerous symptoms ranging from nausea and headache to severe disturbances in mood and cognition. As well, the virus can affect the coronary vasculature, causing plaque formation, thereby increasing the risk of ischemic cardiac complications, and has been associated with increased incidence of several autoimmune disorders.

Lung damage that can occur with SARS-CoV-2 occurs because of the high levels of viral infiltration of this organ owing to substantial ACE2 tissue expression (Han et al., 2021). The ACE2 receptor is found at particularly high levels in alveolar epithelial type II cells in the lung, and damage to these cells causes reductions in surfactant levels and collapse of alveoli (Ziegler et al., 2020). In conjunction with direct damage caused by the viral particles, the TLR-linked inflammatory attempt to contain the microbial invader might also contribute to cellular damage and death. The ensuing infiltration of various leukocytes, such as alveolar macrophages, can place a burden on local cells and may be particularly important in mediating viral spread throughout the lungs. The inflammatory and oxidative factors released by invading cells can cause bystander injury in their attempt to clear the infectious particles. Also, superoxide and other reactive radicals that normally kill invaders can kill otherwise healthy tissues.

The provocation of a "cytokine" storm has been implicated in producing the severe illness associated with COVID-19. The term "cytokine storm" generally refers to a collection of proinflammatory cytokines, notably IL-1, IL-6, TNF-α, and IFNγ, that are released in abnormally high concentrations. The signal pathways associated with infection of several organs and the production of cytokine storm are depicted in Fig. 4.2. Some COVID-19 patients displayed evidence of such a cytokine storm, with elevations of several proinflammatory cytokines and chemokines being apparent (Song et al., 2020). The cytokine contribution might only come into play secondarily, being induced by severe respiratory distress rather than the SARS-CoV-2. Among the proinflammatory cytokines that are involved, IL-6 stands out as a key factor. In the context of viral infection, soluble IL-6 can bind to its transmembrane receptor, IL-6R, creating a complex that can stimulate coagulation factors and foster increased vascular permeability, thereby aiding viral spread (Song et al., 2020). Ultimately, the pathophysiology may be related to aberrant innate immune signaling fueled by proinflammatory cytokines, although it was maintained that cytokine storm is likely not a major factor in widespread pathology in SARS-CoV-2 (Sinha et al., 2020).

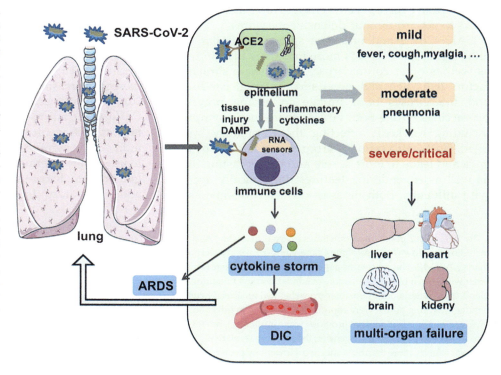

FIG. 4.2 A systemic clinical manifestation of COVID-19. SARS-CoV-2 infects airway epithelial cells or immune cells via binding to ACE2 receptors, causing tissue damage and release of DAMPs, as well as the production of inflammatory cytokines by epithelial cells and immune cells. Then, the crosstalk between epithelial cells and immune cells leads to a wide range of clinical manifestations, from mild forms (e.g., fever, cough, and myalgia) to moderate forms requiring hospitalization (pneumonia and localized inflammation) to severe/critical forms with a fatal outcome that is manifested as pneumonia, ARDS, DIC, CS, and multiorgan failure. DAMP danger-associated molecular pattern, ARDS acute respiratory distress syndrome, DIC disseminated intravascular coagulation. *Source: Yang, L., Xie, X., Tu, Z., Fu, J., Xu D.; Zhou Y., 2021. The signal pathways and treatment of cytokine storm in COVID-19. Signal Transduct. Target Ther. 6, 255.*

Excessive cytokine responses might have particularly dramatic consequences in elderly individuals in whom age-related inflammatory pathology is present. This may be one reason why elderly individuals are much more prone to cytokine-mediated severe organ pathology. Excessive inflammaging is associated with elevated systemic proinflammatory cytokine levels, as well as a reduction in antiinflammatory mechanisms that normally keep such processes in check (Meftahi et al., 2020). By the same token, inflammatory aging is also characterized by excessive reactive oxygen factors, reductions in protective autophagic processes, and alterations in ACE2 expression that impact viral spread (Meftahi et al., 2020). Finally, immune cell senescence is associated with aging and could contribute to a greater ability of viral replication, hence providing a proinflammatory stimulus. As an example, with advanced age, macrophage ability to home in and phagocytize viral antigens is impaired, as is their ability to present antigens to T-lymphocytes and promote adaptive immune responses.

Another peril related to age

How age will affect brain functioning associated with COVID-19 in the long-term is still uncertain. However, microglia in the aged brain are more likely to produce abnormal neuroinflammatory cascades in response to subsequent challenges. The excessive inflammatory aging that is often present in many elderly individuals could result in a subgroup of those infected with SARS-CoV-2 developing further CNS complications and possibly earlier death.

Even in the absence of actual viral penetration into the brain, peripheral immune cells and soluble factors might enter the CNS. A comprehensive postmortem study has provided compelling evidence of infiltrating CD8+ T cells and microglial activation following severe COVID-19 (Schwabenland et al., 2021). Specifically, COVID-19 patients displayed significant CD8+ T cell infiltration, together with signs of activated CD68 and TMEM119 expressing microglia, compared to controls and those who died from respiratory illness unrelated to COVID-19. Overall, 80% of COVID-19 patients expressed increased levels of either diffuse or dense clusters (microglial nodes) of microglia, 68% displayed detectable CD8+ T cell infiltration into the parenchyma, and 36%–44% had the most severe phenotype, which was associated with some degree of axonal damage in the medulla (Schwabenland et al., 2021).

In older individuals, immune T cell infiltration and microglial activation appears to vary between anatomical regions, being particularly evident within the brainstem and cerebellum, and fresh ischemic lesions were observed in a subset of patients (Matschke et al., 2020). A further postmortem study detected prominent CD68 positive microglia that were most apparent in the brainstem of COVID-19 cases. This contrasted with the higher microglia levels in the hippocampus and frontal cortex of a comparison group of Alzheimer's patients (Poloni et al., 2021). Sparse pockets of T-lymphocytes also appeared to infiltrate the CNS, where they tended to cluster around microglial rich regions. Yet, as there was little presence of the actual SARS-CoV-2 virus within the brain, it seems that the virus did not have to breach CNS barriers to influence delicate neural circuitry. Together, the emerging evidence indicates that post COVID-19 brains are characterized by a prominent innate microglial response, with some signs of T lymphocyte driven adaptive immunity, but more subtle changes in neuronal integrity and no obvious signs of encephalitis or frank neurodegeneration. The possibility exists that the virus might exacerbate or sensitize neurons to the acute effects of cerebral stroke, given the increase in ischemic lesions in COVID-19 patients (Matschke et al., 2020). According to a "multihit" framework, with the passage of time and increasing age (and further environmental hits), neuronal damage/degeneration may emerge in post COVID-19 patients.

In parallel with cytokine elevations, a marked reduction of circulating adaptive immune lymphocytes (lymphocytopenia) was observed in severe COVID-19 patients. Moreover, the severity of the lymphocytopenia was negatively related to the elevation of circulating cytokines, IL-6, IL-10, and TNF-α, and was positively correlated with disease severity and progression (Hanna et al., 2020). Hence, as lymphocyte numbers drop, proinflammatory cytokines rise, and disease progresses. The upregulation of T cell exhaustion markers was taken to indicate that the elevated cytokines might be responsible for the lymphocytopenia (Cao, 2020). Also of note was that SARS-CoV-2 antigen was detected in macrophages in the spleen and lymph nodes and that these macrophages promoted high levels of proinflammatory cytokines. It was maintained that these cytokines facilitate viral spread and the promotion of an inflammatory milieu that depletes potentially protective adaptive lymphocyte responses (Coperchini et al., 2021; Jøntvedt Jørgensen et al., 2020). In essence, augmentation of innate macrophage and possibly neutrophil inflammatory cytokine cascades, coupled with impaired activity of protective adaptive immunity (i.e., T and B cells, and possibly NK cells), may drive viral disease and spread.

Viral movement and infectivity may be related to the modification of actin cytoskeletal apparatus via the actin nucleator, Arp 2/3, a large multiprotein complex that acts as a primary nucleator catalyzing the polymerization of actin monomers into the actin filaments required for cellular migration. Importantly, the Arp2/3 complex can also regulate many pathogen-associated functions (Pizarro-Cerdá et al., 2017). In fact, many viruses and bacterial agents can affect actin-based cytoskeletal processes as a manner of propagation. Such pathogens can target the actin cytoskeleton as a means of cellular entry and modulation of the host cell. In some instances, the invader pathogens secrete factors that either disable the cytoskeleton or hijack it to facilitate their entry and propagation (Welch and Way, 2013). Pathogens such as *Listeria monocytogenes* use actin to create protrusions, whereas vaccinia virus utilizes actin polymerization, but in both cases, this facilitates their spread. Thus, targeting the Arp 2/3 complex and other mediators of cytoskeletal regulation might hold clinical utility for antipathogen treatments.

Another unpleasant surprise

The relief experienced when a SARS-CoV-2 vaccine was introduced, even if most people had to await their turn to receive the jab. The hoped-for efficacy of SARS-CoV-2 vaccines was remarkable, but it hadn't been expected that the antibodies to SARS-CoV-2 would disappear within 4 months (after a booster shot) so that individuals were not protected for extended times. Likewise, even though the level of antibody production following natural infection predicted the subsequent illness severity, immunity that developed was impermanent, even though protection was longer lasting among individuals who had been infected and also received the vaccination (Hall et al., 2022). Unfortunately, the broad spread of SARS-CoV-2 resulted in progressively more variants of the virus appearing for which the existing vaccines were largely ineffective (Cao et al., 2022c; Hachmann et al., 2022a), although they seemed to diminish severe illness and death associated with the Omicron variant (Andrews et al., 2022).

Historically, vaccine hesitancy by some groups has been the norm, rather than the exception. Vaccination against influenza hovers around 40%, although there was a small uptick during the COVID-19 pandemic, and a similarly small number were vaccinated against the 2009 H1N1 influenza pandemic (Taha et al., 2013). Still, in the face of a pandemic as severe as COVID-19, it was surprising that vaccine hesitancy was as pronounced as it was, and that many individuals not only advocated broadly against vaccination but also made efforts to prevent others from being vaccinated. There has been debate about the legality and ethics of requiring vaccine passports for people to access some locales, even if this means that allowing the unvaccinated the right to infect the rest of us.

Parenthetically, it was long considered that when a critical number of individuals were vaccinated for some illnesses (e.g., measles, smallpox) the remainder of the population would be protected (herd immunity). Early in the COVID-19 pandemic it was apparent that herd immunity would be ineffective in diminishing viral spread, and it was likely that this could never be achieved (Morens et al., 2022). Nevertheless, this didn't stop some countries from counting on this as a way of controlling the pandemic. When those responsible for public health don't know what they don't know concerning scientific concepts, it's a good bet that negative consequences will follow.

ACE2 and CD209 in Neuroimmunity

In relatively severe cases, SARS-CoV-2 could invade the brain either by a hematogenous-trojan horse route (by infecting infiltrating peripheral immune cells) or through a neural route by binding to olfactory or other neural fibers (Meinhardt et al., 2021). The virus could also raise the risk of stroke or cerebral hemorrhaging (Sánchez and Rosenberg, 2022), thereby causing massive entry of circulating factors into the brain. Within the CNS, the SARS-CoV-2 virus can disrupt the functioning of neurons and glial cells. This said, it had been reported that significant neuroinvasion was not evident in the brain tissue of individuals who died from COVID-19, although robust expression of inflammatory genes was observed, consistent with peripheral inflammatory signals reaching the CNS (Yang et al., 2021c).

Subsequent analyses indicated that the virus could disrupt the BBB integrity to favor CNS invasion (Hernández-Parra et al., 2023) and might damage capillary endothelial cells, possibly leading to increased brain vulnerability to systemic inflammatory or other peripheral factors. Consistent with such findings, analyses of brainstem tissue of individuals who had died at various stages of COVID-19 indicated two different routes by which the virus affects the brain, one of which likely reflected entry to the brain owing to a leaky BBB, thereby promoting neurological disturbances (Radke et al., 2024). Fig. 4.3 shows various routes by which SARS-CoV-2, present in the blood, can affect the brain.

ACE2 is expressed within several brain regions, including the amygdala, cerebral cortex, and brainstem (Lukiw et al., 2022). It appears that SARS-CoV-2 can infiltrate the brain and that CNS neurons and glial cells express the ACE2 receptor, albeit at lower levels than in peripheral organs, such as the lungs (Baig and Sanders, 2020). There is still limited data concerning whether SARS-CoV-2 penetrates the brain at appreciable levels, although SARS-CoV-2 was detected in cortical neurons of humans who died from COVID-19 (Song et al., 2021).

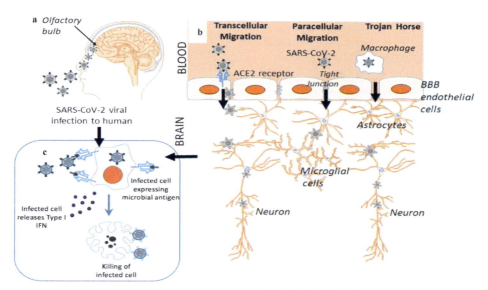

FIG. 4.3 Potential routes of severe acute respiratory syndrome coronavirus 2 (SARS-CoV-2) to the central nervous system (CNS) and preliminary activation of the immune system. (A) Once SARS-CoV-2 is inhaled into the nasal cavity, the virus may travel to the CNS by retrograde axonal transport along sensory and olfactory nerves via the cribriform plate, a bone structure located near the olfactory bulb. In this pathway, SARS-CoV-2 would bypass the blood–brain barrier (BBB). (B) Following a respiratory tract infection characteristic of the virus, SARS-CoV-2 may disseminate into the systemic circulatory system. On reaching the BBB, SARS-CoV-2 may invade host endothelial cells by interaction with the angiotensin-converting enzyme 2 (ACE2) receptor, altering tight junction proteins formed by BBB endothelial cells or phagocytosis by immune cells. These three mechanisms are termed transcellular migration, paracellular migration, and the Trojan horse strategy, respectively. (C) In both pathways, cells infected with SARS-CoV-2 release type I interferons, which alert neighboring and immune cells to the presence of the pathogen. Under normal conditions, infected cells are eliminated by host immune cells to prevent further replication and the spread of SARS-CoV-2. *Source: Achar, A., Ghosh, C., 2020. COVID-19-Associated neurological disorders: the potential route of CNS invasion and blood-brain relevance. Cells 9, 360.*

SARS-CoV-2 in mouse models

Understanding of the mechanisms responsible for various diseases has often relied on animal models. Research of this sort isn't always translatable to humans, and treatments for severe illnesses (e.g., cancer) that are effective in mice may not be apparent in humans. Moreover, some human illnesses are difficult to recapitulate in animal models, making their validity questionable (Anisman and Matheson, 2005).

The ability to evaluate the processes related to SARS-CoV-2 in mice are that much more difficult since the virus is ineffective in producing pathology in rodents owing to differences in their ACE2 receptors compared to humans. Intriguingly, although ACE2 immunoreactivity was absent in mouse lungs under basal conditions, its expression was induced in bronchial cells in response to hypoxia or microbial challenge (Soni et al., 2021). Nevertheless, to optimally model SARS-CoV-2 in the mouse has required genetically designed transgenic expression of the human ACE2 protein (Bao et al., 2020a). The alternative has been to use other coronaviruses in typical mouse strains. One such coronavirus is the mouse hepatitis virus (MHV) which is an enveloped positive-strand RNA virus that is part of the Coronaviridae family. There are several different MHV strains, and much like SARS-CoV2, each has structural envelope proteins, most notably the spike (S) protein (Körner et al., 2020). In at least two prominent MHV strains, specifically JHM and A59, the S protein can be processed just as in SARS-CoV-2, yielding S1 and S2 subunits, that influence infectivity (de Haan et al., 2006). MHV has been studied as a model for the study of CNS coronaviral infection and the various strains can cause a range of disease such as hepatitis, enteritis, as well as demyelination and neuroinflammation.

MHV can infect astrocytes and microglia and can spread throughout the neurovascular system and into the brain parenchyma. Importantly, microglia were critical for restraining brain viral replication in the context of in vivo infection with the neurotropic JHM strain of MHV (Mangale et al., 2020). As well, microglial crosstalk with peripheral immune elements can result in activation of circulating macrophages and other antigen presenting cells, which then interact with viral-specific T cells to orchestrate specific immunity (Wheeler et al., 2018).

In addition to binding to ACE2, SARS-CoV-2 might be able to bind to cluster of differentiation 209 (CD209), also known as, Dendritic Cell-Specific Intercellular adhesion molecule-3-Grabbing Nonintegrin (DC-SIGN) (Brufsky and Lotze, 2020). This is a transmembrane lectin receptor found prominently on macrophages and dendritic cells. However, it is also found in high levels on numerous lung cells, including alveolar, pulmonary capillaries, and vascular endothelial cells. CD209 robustly binds to high-mannose-containing glycoproteins found on viral or bacterial envelopes, which essentially act as PAMPs. High levels of CD209 on various epithelial and endothelial cells can act as alternative receptors permissive to SARS-COV2 infection (Amraei et al., 2021).

CD209 has also been described on several different types of macrophage-like subpopulations, such as microglia, "M2" macrophages, and myeloid-derived suppressor cells. Besides SARS-CoV-2, the CD209 receptor can also bind carbohydrate residues that are found on a range of pathogens, including HIV, Ebola virus, M. tuberculosis, and H. pylori (Garcia-Vallejo and van Kooyk, 2015). Hence, CD209, like ACE2, might be a useful receptor to clinically target to restrict viral spread and possibly to utilize it to follow viral behavior.

Brain Viral Infection

Once in the brain, SARS-CoV-2 infection may have neuroimmune effects either through (a) direct entry into the intracellular compartment of neurons or glia, or by (b) inducing secondary damage from systemically or locally derived inflammatory cells or soluble factors. Early COVID-19 pandemic studies indicated that more than a third of severe cases were accompanied by neurological symptoms. In these patients, particularly high levels of proinflammatory cytokines were evident, as were levels of C-reactive protein (CRP) (Mao et al., 2020; Lagunas-Rangel, 2020). Many of the more serious and hospitalized patients were of advanced age (along with preexisting diseases), which itself could contribute to elevated inflammatory tone. The potential preexisting age-dependent peripheral and central inflammatory state could interact with the viral hit, thereby producing exaggerated collateral CNS damage. In this regard, SARS-CoV-2 is especially effective in provoking senescence of dopamine neurons within the substantia nigra and may thus influence the emergence of Parkinson's disease (Yang et al., 2024a).

When SARS-CoV-2 impacts the brain, numerous CNS symptoms may emerge, such as confusion, seizures, headache, dizziness, impaired consciousness, gait deficits, cerebrovascular pathology, and encephalitis (Najjar et al., 2020). Pathological symptomatology also includes deficits in cranial nerve functioning that can give rise to compromised smell and taste, or the manifestation of elements of autoimmune disease, such as Guillain–Barré syndrome. There may also be other forms of collateral damage, including ischemic stroke stemming from cerebrovascular abnormalities produced by the viral spread and subsequent inflammatory stress upon the cardiovascular system (Helms et al., 2020).

Encephalopathy has been reported in SARS-CoV-2 infected cases and is often associated with delirium, agitation, or motor disturbances. These clinical symptoms are generally found in the context of increased peripheral inflammatory factors, but surprisingly, normal levels typically occur within the CSF. Similarly, there was a notable absence of SARS-CoV-2 viral fragments detected in CSF from severe neurological cases, even in those showing EEG abnormalities (Destras et al., 2020; Helms et al., 2020). Yet, other findings indicated that SARS-CoV-2 has some affinity for the brain, particularly in severe cases. For instance, about 36% of patients hospitalized with severe COVID-19 infections in Wuhan had neurological symptoms (Mao et al., 2020) and this number was 90% in cases in a hospital in France (Helms et al., 2020). A similar study in Germany found SARS-CoV-2 RNA in 53% of the brains of fatal COVID-19 cases (Matschke et al., 2020).

It has been estimated that almost 30% of individuals with COVID-19 develop some signs of neurological or psychiatric illness (Frank et al., 2022), including depression, anxiety, sleep disturbances, and fatigue (Schou et al., 2021), which might stem from ongoing neuroinflammatory activation (Edén et al., 2022). Although viral invasion of the CNS can certainly produce neuropsychiatric symptomatology, it is also possible that a host of viral-induced inflammatory factors are responsible. As we discussed earlier, SARS-CoV-2 proteins might induce inflammatory responses through TLRs, which produce a cascade that results in the stimulation of inflammatory transcription factors, such as NF-κB. Intriguingly, the S1 spike subunit of the virus alone appears to be able to induce such inflammatory signaling and production of cytokines by directly interacting with TLR4 on isolated microglial cells, as well as murine and human macrophages (Olajide et al., 2022). In this respect, exposing microglia to S1 increased the expression of genes associated with microglia/brain macrophage activation markers (Iba1, Cd11b, and MhcIIα), astrocyte activation markers (Gfap), PRRs (Tlr4), inflammasomes (Nlrp3), as well as the protein levels of several proinflammatory cytokines and chemokines (Frank et al., 2022; Nuovo et al., 2021). These actions varied across several brain regions (hypothalamus, hippocampus, frontal cortex), leading to the suggestion that SARS-CoV-2-related proteins could operate like DAMPs in eliciting neuroinflammation (Frank et al., 2022).

While not dismissing this source for cognitive disturbances, it was maintained that the impact of SARS-CoV-2 infection on neurocognitive and some neuropsychiatric disturbances, symptoms may stem from damage of neurons within the orbitofrontal cortex (Crunfli et al., 2022). While these actions may stem from peripheral inflammatory changes, these investigators observed that in a subset of individuals who died of COVID-19, the genetic material of the virus was present in brain astrocytes. It was suggested that astrocyte infection might disturb energy metabolism, thereby limiting fuel for neurons and thereby reducing neuronal viability.

Profiling studies reported that ACE2 was likely expressed in excitatory and inhibitory neurons, along with glial cells (Chen et al., 2021c). While ACE2 was detected in the hippocampus, much higher levels were observed in the olfactory bulb and in pericytes associated with the BBB. Trace levels of ACE2 were likewise found in the substantia nigra, middle temporal gyrus, and posterior cingulate cortex (Chen et al., 2021c). In an intriguing report, SARS-CoV-2 could reliably infect neurospheres and brain organoids, which resulted in reduced synapse formation, further supporting the vulnerability of the brain to SARS-CoV-2 infection (Zhang et al., 2020a). Finally, postmortem analysis of tissue from individuals who had died owing to COVID-19 indicated that ACE2 expression was present throughout the brain, with relatively high levels in the amygdala, cerebral cortex, and brainstem. The fact that the pons and medulla brain regions that control respiratory processes had the highest overall levels raises the possibility that the virus might impact breathing, at least in part, through a central mechanism. Moreover, postmortem analyses revealed ischemia, neuronal loss, necrosis, glial hyperplasia, and demyelination (Paliwal et al., 2020; Sepehrinezhad et al., 2020). Overall, the available data indicate that SARS-CoV-2 has the capacity to infiltrate the brain, provoking an inflammatory milieu that may wreak havoc upon neuronal functioning, including unregulated glutamatergic excitotoxicity and disturbances of monoaminergic activity, which can promote neurological and neuropsychiatric disturbances. At the same time, neurological disturbances associated with COVID-19 can persist even in the presence of an attenuated autoantibody and cytokine response (Michael et al., 2023).

Olfactory and Vascular/Lymphatic Crossroads of Neuroimmunity

Deficits in olfaction and taste are commonly reported during the early stages of COVID-19 disease prior to other symptom emergence and, in a subset of patients, may persist for some time afterward (Heneka et al., 2020). The loss of smell that many COVID-19 patients exhibit is consistent with the possibility that the virus can infiltrate the olfactory nerves and epithelium. It was suggested that the loss of taste and smell induced by SARS-CoV-2 might stem from direct neural infection and that recovery of olfactory function might not fully recover to normal since neural progenitor cells are damaged by the virus (Zhang et al., 2020a). In fact, animal studies supported the possibility of viral entry into the brain via olfactory nerves. Transgenic mice bearing the human form of ACE2 that were infected intra-nasally with SARS-CoV-2 displayed CNS entry via the olfactory bulb (Kumari et al., 2021). Moreover, ACE2 and its protease, TMPRSS2, were expressed on the olfactory epithelium (Bilinska et al., 2020). Thus, SARS-CoV-2 utilizes aspects of the olfactory area to gain CNS entry and infect related brain regions.

Viral invasion of the CNS might occur by way of retrograde synaptic transport (Taylor and Enquist, 2015), which could be a means through which peripheral infections associated with SARS-CoV-2 infection could affect brain functioning (Reza-Zaldívar et al., 2021). SARS-CoV-2 can also gain access to the CNS through the circulatory or lymphatic systems. In this case, the BBB or blood-cerebrospinal fluid barrier can be breached through cellular junctions or via the upregulation or highjacking of transporter molecules/pumps. There is also reason to believe that SARS-CoV-2 can enter peripheral lymphatic vessels and then move into the brain's own specialized glymphatic system (Bostancıklıoğlu, 2020). Once in the brain parenchyma, the virus could spread through perivascular areas, where it may interact with local macrophages or other localized immune cells that have migrated from the periphery. In addition to the direct entry of free viral particles, SARS-CoV-2 might gain entry into the CNS by way of infected immune cells that migrate into the brain parenchyma or perivascular spaces. It is possible that the virus might be released into the extracellular space and interact with microglia and neurons, or alternatively, transmembrane leukocyteglia/neuron interactions could also occur (see discussion in Uversky et al., 2021).

Knowing that people with preconditions were especially vulnerable to the adverse effects of SARS-CoV-2 allowed some people to take measures to protect themselves. But, what was it that resulted in some infected people exhibiting only mild symptoms and about 20% of individuals who tested positive for SARS-CoV-2 to be entirely asymptomatic? As described in Chapter 2, in response to viral infection, HLA proteins situated on the cell's surface alert the immune system so that T cells will recognize and remember them, and then upon later encounters with this virus, they would be destroyed before illness can develop. It turns out that one of the many variants of the HLA gene (HLA-B*15:01) may play a significant role in protecting people from COVID-19 symptoms. If people had previously been infected with a common coronavirus that causes colds, carrying this particular HLA variant provides individuals with some protection against the effects of SARS-CoV-2 infection. The primed T-cell responses among these individuals occur sufficiently quickly and are strong enough to prevent symptoms from appearing (Augusto et al., 2023).

Long COVID

People who only developed mild or moderate symptoms considered themselves lucky. Yet, in some cases, COVID-19 had further surprises in store in the form of long COVID in which a variety of persistent symptoms manifested following infection. About 15% of infected individuals developed long COVID, which in some cases occurred months after initial infection or re-emerged after several months (Montoy et al., 2023). Symptoms of long COVID most often comprised difficulty thinking or concentrating (often referred to as "brain fog"), sleep disturbances, headache, extreme fatigue, muscle aches, shortness of breath, and loss of smell that persisted for at least 3 months. Even though those with preexisting conditions were most likely to develop the most severe illness and long COVID, even mild cases of SARS-CoV-2 infection can have profound long-term effects. Most often, symptoms diminished appreciably after 12 months, but the number of infected individuals who contracted COVID-19 was higher than the 15% estimate, and after 6 months, 22.95% of individuals exhibited symptoms, and even after 2 years, 17.2% were still affected (Ballouz et al., 2023). A systematic review and meta-analysis indicated that long COVID most frequently occurred in females, older individuals, and those with comorbid conditions and was less common among individuals who had been vaccinated (Tsampasian et al., 2023). As well, long COVID was greatest among individuals who exhibited elevated anxiety, worry, perceived stress, and loneliness prior to infection. The lasting effects of COVID-19 on cognitive functioning were pronounced to the extent that the performance of affected individuals resembled that of a person who had aged 10–20 years. This was reflected through accelerated epigenetic aging and through altered telomere length (Cao et al., 2022c). These symptoms often resembled myalgic encephalomyelitis/chronic fatigue syndrome (ME/CFS) and autonomic nervous system dysfunction (dysautonomia), especially postural orthostatic tachycardia syndrome, which could persist for years.

Numerous underlying factors have been implicated in long COVID (see the review in Davis et al., 2023). Imaging analyses of individuals before and at various times after SAR-CoV-2 infection revealed a reduction of gray matter thickness in the orbitofrontal cortex and parahippocampal gyrus and global brain size (Douaud et al., 2022). Overall, it also appeared that persistent COVID may be a neurological disorder associated with disturbed cortical gray matter structures (Rothstein, 2023), together with white matter abnormalities and microvascular changes. Further to this, the persistence of symptoms may have been related to microvascular clot formation and hypoxic neuronal injury. The notion that long COVID is a neurologic disorder (or a set of such conditions) is consistent with the array of neurological symptoms that have been reported, such as increased ischemic and hemorrhagic stroke, encephalitis, episodic disorders (e.g., migraine and seizures), extrapyramidal and movement disorders, musculoskeletal disorders, sensory disorders. The encephalopathy would result in cognitive and memory problems and mental health disorders (Xu et al., 2022).

Aside from the continued neurological and inflammatory alterations that have been reported, long COVID may stem from the persistent presence of viral components within the CNS, which could affect presynaptic morphology and hence synaptic functioning (Partiot et al., 2024). In animal models, the SARS-CoV-2 spike protein was present in the skull marrow, brain meninges, and brain parenchyma, and this was apparent in human postmortem tissues. Moreover, in mice, direct injection of the spike protein into the brain provoked cell death. These findings raise the possibility that the persistence of the spike protein in the brain could contribute to the long-lasting neurological consequences of SARS-CoV-2 infection (Rong et al., 2023). In humans, protein and genetic material from SARS-CoV-2 persisted in the blood and tissue in a subset of individuals ("viral reservoirs") with long COVID, which may drive the symptoms expressed (Proal et al., 2023). For instance, persistent alterations of the complement system involved in the clearance of infection were observed among patients with long COVID, which may have produced tissue damage and influenced the formation of small blood clots that might have contributed to some of the symptoms expressed (Cervia-Hasler et al., 2024). Similarly, long after infection, activated T cells appeared in many tissues (e.g., within the gut wall, lung tissue, certain lymph nodes, and the bone marrow), including sites where they should not be, such as the spinal cord and the brainstem. Among individuals with long COVID, viral remnants may promote persistent elevations of interferons, and the resulting inflammation causes a reduction of 5-HT levels within the gut. In turn, this may impair vagal nerve functioning, thereby disturbing hippocampal functioning and related behaviors and cognitive changes (Wong et al., 2023a).

Beyond these neurological disturbances, long COVID was associated with numerous immune alterations. These have included the presence of autoantibodies, persistent viral presence, unresolved inflammation, and reactivation of viruses (e.g., EBV). Likewise, long COVID was accompanied by altered T cell subsets, including increased CD4$^+$ T cells and apparently exhausted SARS-CoV-2-specific CD8$^+$ T cells. As well, levels of SARS-CoV-2 antibodies were elevated, and it seemed that mis-coordination existed between T and B cell responses related to SARS-CoV-2 (Mina et al., 2023; Yin et al., 2024). Through machine learning, it was possible to distinguish between individuals with and without long COVID based on circulating myeloid and lymphocyte populations and exaggerated humoral responses directed against SARS-CoV-2 (Klein et al., 2023).

The presence of depressive and anxiety symptoms associated with long COVID was accompanied by elevated levels of several proinflammatory cytokines, such as IL-1β, IL-2, IL-12, IL-17, TNF-a, and IFNγ (Bellan et al., 2022). While proinflammatory factors were most reliably associated with long COVID, protracted symptoms were also associated with altered metabolites of the kynurenine pathway and diminished presence of the neurotrophin BDNF (Lorkiewicz and Waszkiewicz, 2021). As well, long COVID was frequently accompanied by a cluster of biological factors that comprised low peripheral oxygen saturation, disturbed body temperature regulation, diminished antioxidant defenses, and elevated oxidative toxicity (Al-Hakeim et al., 2023). In addition to the cytokine elevations, long COVID was accompanied by low levels of cortisol, possibly reflecting the failure of HPA functioning to recover from the illness (Yavropoulou et al., 2022). Long COVID was also associated with autonomic nervous system dysfunction, which may contribute to numerous cardiovascular disturbances over the 1-year period following infection, including cardiac arrhythmias, coronary artery disease, heart failure, pulmonary embolism, chronic obstructive pulmonary disease ischemic stroke, and increased mortality (DeVries et al., 2023).

Aside from psychological and neurological correlates of long COVID, during the year following infection, individuals were at elevated risk of developing several cardiovascular illnesses, such as cerebrovascular disorders, dysrhythmias, ischemic and nonischemic heart disease, pericarditis, myocarditis, heart failure, and thromboembolic disease (Xie et al., 2022).[2] Furthermore, long COVID was associated with autoimmune disorders (e.g., psoriasis and type 1 diabetes) (Chang et al., 2023) as well as conditions that involve the lungs, liver, spleen, kidneys, and gastrointestinal processes (Al-Aly et al., 2023). Aside from the pandemic affecting cancer care, there have been indications that SARS-CoV-2, like several other viruses, might contribute to the development of some forms of cancer (Jahankhani et al., 2023). An analysis of 80 potential health problems among 140,000 people who had been infected with SARS-CoV-2 and 6 million people who had not been infected, indicated that during the 3-month postinfection period, numerous health problems emerged, and death rates were relatively high.

Some of the symptoms of long COVID were associated with immune dysregulation (Mina et al., 2023), and the depressive features were associated with multiple markers of inflammation (Lorkiewicz and Waszkiewicz, 2021). It is of particular interest that following SARS-CoV-2 infection, inflammatory factors were elevated, and immune dysregulation persisted over a 1-year period (Cheong et al., 2023). Moreover, relative to matched controls, individuals who presented with long COVID exhibited pronounced differences in circulating myeloid and lymphocyte populations together with exaggerated humoral responses directed against SARS-CoV-2. In fact, analyses of plasma samples pointed to several inflammatory markers that could predict long COVID with 78% accuracy.

In view of the sensitivity of microbiota to diverse challenges, it comes as no surprise that COVID-19 infection can affect the gut microbiome. It is particularly significant that a symbiotic preparation (SIM01) effectively reduced many features of long COVI (Lau et al., 2023). Aside from bacteria, several fungal species, particularly *C. albicans*, have been associated with COVID-19, which triggers inflammatory immune responses, and antifungal treatments could diminish the symptoms of long COVID (Kusakabe et al., 2023). As the effects of viruses can have effects long after viral clearance, it might reflect the presence of viral RNA, thereby exerting cumulative long-term effects. As well, chronic inflammation could potentially interact with other microbial threats or environmental stressors, yielding additive or synergistic outcomes on a variety of neural or glial systems.

Since it has only been a few years since the beginning of the COVID-19 pandemic, the full measure of the long-term effects on neurological functioning has yet to be determined. It is known from earlier pandemics, such as the 1918 influenza pandemic, that many individuals who developed acute encephalitis later exhibited neurological syndromes comprising motor or cognitive deficits (Foley, 2009). It is certainly possible that individuals who developed COVID-19 may experience sustained low-grade inflammation or other neurobiological alterations that favor neuronal damage. Also, the distress related to uncertainty early in the pandemic, together with the distress created by social isolation, may have impacted neuronal health and could potentially act synergistically with the inflammation associated with infection, as described in Chapter 7.

---

[2] Contrary to the misinformation that has appeared too often, a meta-analysis that included more than 460,000 individuals, COVID-19 vaccination was not associated with increased frisk of depressive illness (Lee et al., 2023).

Concluding Comments

Throughout recorded history, various microorganisms have been the scourge of humankind, decimating large swaths of people across countries, and in many instances, 30%–80% of the population died. Aside from the most notorious pandemics, relatively common bacterial and viral infections, as well as numerous parasitic infections, have been responsible for numerous deaths, disproportionately affecting individuals in poor neighborhoods and in countries where poverty and poor health care are endemic. Although research and clinical analyses associated with these infections have focused on their impact on physical discomfort, these pathogens can directly or indirectly affect brain processes, thereby influencing psychological illnesses, and may contribute to the development of neurodegenerative disorders.

While COVID-19 took an enormous toll on health and mortality, from a historical perspective, its effects were relatively modest. Nonetheless, it is certain that if allowed to progress unchecked, the consequences of a pandemic on morbidity and mortality can be disastrous, and yet resistance to simple preventive measures has too often been neglected. Aside from the severe respiratory distur-bances associated with COVID-19, other illnesses emerged, such as bacterial infections that, in some instances, caused death (Fazel et al., 2023). Likewise, owing to the exaggerated inflammatory response that could accompany COVID-19, systemic complications related to lung and heart functioning could arise, as could the occurrence of ischemic stroke, as well as kidney and spleen dysfunction (Belfiore et al., 2022). Furthermore, autoimmune disorders could be provoked, which could be attenuated by vaccination (Peng et al., 2023). Viral and bacterial illness, through its effects on diverse processes, can also affect the CNS, resulting in neurological and neuropsychiatric illnesses. These effects occur owing to excessive peripheral and brain inflammatory processes being activated, and it also appeared that SARS-CoV-2 could gain entry to the brain, resulting in neuronal damage. Although numerous viral infections can have persistent consequences well after the virus seems to have been eliminated, in the case of COVID-19, these actions have undermined the well-being of a significant proportion of the population. With a greater understanding of the processes that lead to long COVID, new treatments might eventually (hopefully soon) be in the offing (Davis et al., 2023).

CHAPTER

5

Lifestyle Factors Affecting Biological Processes and Health

Once doubt sets in, certainty (and trust) rarely returns

There is little debate concerning the value of moderate exercise in relation to heart disease, type 2 diabetes, and depression, as well as other illnesses. Thus, it is fairly remarkable how little attention many people devote to exercise and instead maintain risky behaviors. Unless it becomes a habit that is undertaken on a systematic schedule, perhaps facilitated by exercising with others, it is difficult to engage in this behavior on a sustained basis. The sad fact is that "doing nothing is simpler than doing something."

There is also certainty regarding the importance of sleep. We have a basic biological need for sleep, even if it gets in the way of work or play. Failure to get enough sleep will likely undermine physical and psychological health, and some negative physical health consequences may be a result of sleep disturbances that accompany stress and depression (Irwin, 2019). There is such a thing as getting too much sleep, and in some instances, sleeping too much might reflect a biological disturbance or an illness being present (Irwin et al., 2016).

As much as eating well is fundamental to good health, it is not overly surprising that individuals engage in poor eating, especially in the face of so many temptations. What could be more reinforcing than some yummy comfort food to deal with life stressors (for some people, alcohol serves as an alternative lousy coping strategy)? At first blush, it might seem surprising that many people barely have an inkling regarding which foods are good for them and which are very bad. Even experts in the field are at odds with one another, frequently providing confusing and contradictory information. In fact, what is healthy and what is not seems to change on a frequent but irregular basis. Butter used to be a no-no, whereas margarine was good. Then, it was argued that margarine was bad, just a step or two removed from plastic, whereas butter became acceptable. Carbs and fatty foods were the scourge that increased cholesterol levels that led to heart disease, then it seemed that they were not that bad, unless individuals were diabetic or at high risk for heart disease. The business with carbs versus fats, or for that matter whether certain types of fat are worse than others is confusing, although, there is general agreement that we should avoid trans fats. Based on data from large-scale prospective studies, every 2% increase in trans fat intake was associated with a 16% greater chance of premature death, and every 5% increase in saturated fat intake was associated with an 8% increase in early death (Wang et al., 2016a).

Adopting a poor lifestyle can have multiple adverse consequences that culminate in diminished well-being through actions on multiple biological processes. Changes to the inflammatory immune system are no doubt among the key agents responsible for multiple illnesses, including those of a psychological nature and those that affect our lifespan and our healthspan. Among individuals who live very long lives (>95 years of age), the difference from others was not so much that these individuals did not become ill, but rather that their illnesses were compressed into a relatively brief period near the end of life.

Risky Behaviors and Illness

Go figure! Actions actually have consequences.

Try telling that to your teenage kid, or to the person carrying a food tray with a double whopper and supersized fries, faithfully offset by a diet Coke. Many illnesses develop because of risky behaviors, particularly poor lifestyle choices, engaging in dangerous behaviors, or those that are outrageously stupid (thereby making the genes for these traits less likely to reappear in the gene pool). These are all common sins of commission, but some individuals will suffer the consequences that come with sins of omission. These include not receiving recommended screening for colon, breast, or prostate cancer (for individuals more than 50 years of age), failing to use sunscreen,

The Immune System
https://doi.org/10.1016/B978-0-443-23565-8.00009-0

99

Copyright © 2025 Elsevier Inc. All rights are reserved, including those for text and data mining, AI training, and similar technologies.

choosing not to be vaccinated against an imminent flu epidemic, not having children inoculated for measles, mumps, and rubella (MMR), or failing to wear a helmet when skiing, skateboarding, or bicycling.

It is not particularly unusual to find that policies and procedures that had been adopted with the most positive intentions turn out to have unintended negative consequences. National, community, and individual-level attempts have been made to promote behaviors to diminish health risks (campaigns to reduce smoking, encourage weight loss, and have children vaccinated against a variety of viral illnesses). These have frequently been successful, but there are instances when unintended consequences act against favorable outcomes. In large measure, this occurs because people have a puzzling tendency to adjust their behaviors based upon their appraised risk of negative outcomes being reduced. According to "risk compensation" theory, when people perceive risks to be high, they behave carefully, but should they perceive personal risk to be low, such as when safety measures have been instituted, they might be less cautious (Peltzman effect). Following the introduction of penicillin to combat sexually transmitted diseases (STDs), such as syphilis, the rates of infection declined precipitously, as those treated were no longer passing on the illness. However, after some time, rates of infection increased once again; because a cure was possible, freewheeling sexual behaviors may have increased and condom use might have declined, increasing the risk of STDs. Likewise, with the introduction of mandatory use of seat belts, some people drove more recklessly, and the introduction of helmets for skiers resulted in the engagement of riskier behaviors on the hills. In the same fashion, when type 2 diabetic patients find that their blood sugars normalize with medications, they may be less attentive to their diet and revert to previous bad behaviors that led to their condition. We saw features of the Peltzman effect early in the COVID-19 pandemic. Individuals who had been infected developed a false sense of security that they would be immune to infection and thus failed to be vaccinated or adopt other protective measures, thereby serving as a reservoir of viral spread (Kaim and Saban, 2022).

Unintended consequences may come about for various reasons. For starters, ignorance of the issue results in it being unlikely that individuals (or groups) can predict all potential outcomes that could occur. Further, problems might not be appropriately analyzed, and the solutions adopted may have been more germane to earlier problems, which were unsuitable for the issue currently being faced. Individuals and governments may also find themselves adopting policies in which basic values interfere with particular behaviors (e.g., the attitude that governments do not have the authority to demand children be vaccinated). On other occasions, government agencies may be reluctant to adopt certain policies or procedures since it could produce fear or panic (e.g., the presence of an imminent threat), and so steps are taken to find alternative strategies, which could end up having negative consequences. Finally, like governments and other organizations, there is a tendency for individuals to favor short-term gains without appropriate consideration of long-term negative consequences.

In its worst form, proposed solutions can give rise to perverse results in which certain actions can lead to solutions that can make situations worse. When antidepressant agents (SSRIs) were linked to elevated suicide risk among adolescents, restrictions were imposed so that these drugs were not offered to young people. As a result, however, physician visits by young people declined, leading to untreated depression and a commensurate increase in suicides that is still being witnessed (Lu et al., 2020).

Blaming the victim: The consequences of smoking

It is unfortunate that there is often a tendency to blame the victims for bad things that happen to them, even if their misfortune was not of their own making. When it is believed that negative outcomes stem from a person's risky behaviors, irrespective of whether they comprise actions or inactions (sins of commission or omission), the sinners are made to suffer doubly. It had been proposed (Lerner and Montada, 1998) that people have a cognitive bias in which they naively see the world as being just (morally fair) so that positive and noble acts will ultimately be rewarded, whereas evil acts will ultimately be punished. In this context, when people suffer from a negative event, it must somehow have been their fault.

It is often assumed that individuals have free will and might choose to smoke, and hence, their illness is self-inflicted. But is this entirely correct (Sapolsky, 2023)? They may have worked in an environment where everyone smoked, and hence, they too took up smoking. It is similarly possible that they inherited genes that were linked to impulsivity and high reward satisfaction, both of which favored addiction, making them particularly likely to engage in risky behavior. In this respect, an analysis that included 3.4 million individuals revealed that many genetic variants were related to smoking (Saunders et al., 2022). It should be underscored here that not every person who develops lung cancer had been a smoker. They may have been unlucky to be born with particular genes or gene mutations or epigenetic modifications that affected their proneness to the disease. Alternatively, they might have been unfortunate to live downwind from a plant spewing pollutants or microparticles that led to cancer or worked or lived in an environment in which second-hand smoke sickened them. Regrettably, the stigma related to smoking-related diseases can be pronounced, so much so that some nonsmokers will not confide that they have this form of cancer.

Smoking is well known to cause heart and lung diseases, but it has been less often considered that smoking may contribute to severe mental illnesses. It turned out that smoking, which often begins during adolescence, is associated with a 250% increase in hospitalization for mental illnesses years later (Balbuena et al., 2023). It is conceivable that genes related to smoking and mental illness share common genes, or smoking may promote neurobiological consequences that favor the development of psychological disorders. For instance, nicotine can reduce serotonin functioning, and can promote inflammation and oxidative stress that influences mental illnesses, and similar outcomes have been associated with e-cigarette (vaping) use (Farrell et al., 2021).

In her reviews related to addictions, Volkow et al. (2010) indicated that drugs, such as alcohol or cocaine, can "hijack" cognitive control circuits, thereby impairing rational thinking, thus allowing for addiction to persist. Perhaps this is true for nicotine use as well. Considering the brain changes that might have occurred in response to these agents, is it reasonable to continue to frown upon patients as if they were sinners?

Dietary Factors as Determinants of Health

Many lifestyle factors are related to various noncommunicable diseases (e.g., type 2 diabetes and heart disease), and it seems that these may also contribute to depressive disorders. An analysis of 290,000 people obtained from the UK Biobank indicated that obtaining 7–9 h of sleep a night and social connectivity was associated with a 22% and 18% reduction in risk of depression, respectively. As well, never smoking, maintaining a healthy diet, regular physical activity, and limited sedentary behaviors were accompanied by lower levels of depression. In a subset of 33,000 participants, healthy lifestyles were related to greater volume of the amygdala and hippocampus, and poor lifestyles were closely aligned with immune and metabolic changes, which were linked to depressive disorders. For instance, elevated levels of markers of circulating inflammation (i.e., C-reactive protein) and triglycerides that are influenced by stressors, as well as loneliness and lack of social support, predicted the occurrence of depression. As much as genetic factors may contribute to affective disorders and numerous other health risks, lifestyles clearly represent a modifiable risk factor that influences mood disorders (Zhao et al., 2023,b).

Diet Adaptations

Over centuries, as humans moved from one type of environment to another, diets changed, and gene adaptations and microbial variations followed. This might have been particularly notable when humans shifted from the hunter-gatherer mode to becoming more reliant on farmed foods. Certain foods may have had positive effects in some cultures, but this same food might have been less beneficial, and even harmful, in other groups. Inuit in Northern Canada and Northern Alaska obtain much more of their calories from fats than do people living much further south but seem to do well with these diets. In fact, the frequency of death by heart disease is about half that of their southern neighbors (this has been termed the Inuit Paradox). To be sure, the fats obtained from wild animals differ from those of grain-fed farm animals, having different cholesterol content, and are free of antibiotics and a variety of other farm-related chemicals, which could account for some aspects of the Inuit Paradox. More than this, selection pressures in far Northern groups may have resulted in genetic changes that facilitated dealing with a diet rich in protein and fats. Indeed, Greenland Inuit carry genetic mutations (that are relatively rare in Europeans), which modulate fatty acid composition, thereby contributing to diminished low-density lipoprotein (LDL) and fasting insulin levels, as well as serving in the regulation of growth hormones (Fumagalli et al., 2015).

The adaptation to certain foods based on selection for specific gene mutations occurs over many generations. Often, the sudden introduction of new foods into a region or culture may engender adverse outcomes. In this regard, Western diets can be hazardous to the health of some Northern groups (type 2 diabetes is almost endemic in some Northern First Nations communities). Likewise, those living in rural parts of Asia and Africa rely on diets that are relatively unique to their groups, but upon migrating to urban settings and encountering unhealthy diets, obesity and various related disorders may follow. Urbanization in some parts of China similarly resulted in the loss of beneficial microbiota, which was associated with increased occurrence of *Escherichia* and *Shigella* (Winglee et al., 2017). In essence, the impact of specific diets may vary depending on the presence of certain genetic influences that favor or act against the development of certain pathological conditions.

The specific foods that ought to be consumed and those that should be avoided seem to change regularly and even nutritional experts have disagreed with one another. Studies that examined the adverse health effects of specific foods have frequently been poorly designed and conducted, which has resulted in findings frequently being questioned (e.g., Ioannidis, 2013; Schoenfeld and

Ioannidis, 2013). Studies of diet concerning health are often unreliable simply because individuals do not accurately recall what they have actually consumed (even using daily eating diaries). These criticisms were fully justified but ample evidence exists that certain diets, notably those that reduce inflammation and favor appropriate gut microbiota balances can have positive effects. Nonetheless, the confusing and contradictory information that has frequently been proffered regarding the foods that should or should not be eaten has left many individuals uncertain about what foods they ought to be consuming to foster good health. It has often been recommended that we consume foods that contain dietary fiber, heart-healthy oils, and low-fat proteins (the latter obtained from fish, poultry, legumes, nuts, and seeds), and avoid foods replete with preservatives, sugar, and sodium. Consumption of foods high in fiber, for instance, may limit a wide range of age-related health disturbances, including cardiovascular illness, cancer, and a reduction of all-cause mortality (Aune et al., 2016). Conversely, diets high in fats may promote obesity, metabolic disorders, and inflammatory-related diseases owing to the release of cytokines from belly adipocytes (Blaszczak et al., 2021). Ultra-processed foods that typically contain additives, such as preservatives, emulsifiers, sweeteners, and artificial colors and flavors, have likewise been associated with a great number of diseases, including heart disease, type 2 diabetes, and several mental disorders (Lane et al., 2024).

The view was frequently advanced that we ought to favor polyunsaturated fats and avoid foods with saturated fats, but it was argued that the available data in this regard have not been all that convincing (Chowdhury et al., 2014), although polyunsaturated fat consumption was associated with a modestly lower incidence of chronic heart disease (Abdelhamid et al., 2020). Indeed, contrary to the standing dogma, it was reported that saturated fats were not linked to stroke, heart disease, or diabetes. Similarly, a prospective epidemiological analysis across 18 countries, revealed that consumption of total fat and individual types of fat were not related to lower mortality owing to heart disease and stroke, whereas high carb consumption was tied to a greater risk of mortality (Dehghan et al., 2017). In contrast, consumption of polyunsaturated fats (e.g., linoleic acid) over 15 years was inversely related to cardiovascular disease and all-cause mortality (Iggman et al., 2016). To an extent, a report from the American Heart Association made sense of some of the inconsistent findings that have appeared (Sacks et al., 2017). It was suggested that research addressing the question needs to look more deeply at the effects of getting rid of saturated fat from the diet, particularly concerning what replaced these fats. In some instances, saturated fats were replaced by refined carbohydrates and sugars, and even trans fats that were worse for heart

health than saturated fats. However, when saturated fats were replaced by polyunsaturated fats, monounsaturated fats, and whole-grain carbohydrates, the risk of heart disease declined.

Individuals wishing to maintain a proper lifestyle, including eating well, may become frustrated when diet gurus provide divergent perspectives, and as a result, trust in them may be lost. Most diets, including those that focus on reducing carbs (Atkins, South Beach, Zone), fat consumption (Ornish, Rosemary Conley), or moderation of macronutrients (Biggest Loser, Jenny Craig, Nutrisystem, Volumetrics, Weight Watchers), are effective, provided that the individual sticks to it, more so if dieting is accompanied by regular exercise (Johnston et al., 2014). An analysis of 14 diets maintained over 6 months revealed that most diets resulted in weight loss, with some producing somewhat better effects than others, and differences existed concerning the presence of high-density cholesterol. It is of significance that except for the Mediterranean diet, the benefits to cardiovascular risk factors (e.g., blood pressure reductions) had disappeared within 12 months (Ge et al., 2020).

This raises the obvious question of whether an ideal diet exists to maintain good health. In fact, analyses of 181,990 individuals obtained from the UK Biobank indicated that a well-balanced diet (rather than those that comprise reduced starch, vegetarian, or high protein/low fiber) seemed to be best for optimal brain (cognitive) functioning. The balanced diet was associated with greater gray matter volume in several brain regions, altered blood-based markers, together with several distinct gene differences. Importantly, dietary modifications to obtain best outcomes required a gradual switch from unhealthy to balanced diets, thereby allowing for effective adjustments (Zhang et al., 2024b).

Just about everyone agrees that sugar from desserts is a definite no-no, and sugar-sweetened beverages are just as bad, and perhaps worse. This gave rise to the manufacture of diet drinks spiked with artificial sweeteners, but we should not be misled about all those diet drinks being healthy substitutes for those containing sugar. Among other things, their continued consumption was associated with an increase in belly size, and by affecting gut microbes they can encourage metabolic syndrome and obesity and might thus contribute to the development of type 2 diabetes and risk for cardiovascular diseases (e.g., Gomez-Delgado et al., 2023). Sugary drinks and artificial sweeteners have also been linked to diminished hippocampal volume, poorer cognitive functioning, dementia, and increased risk of stroke (Pase et al., 2017). Artificial sweeteners have been implicated in the emergence of depression (Lane et al., 2023; Samuthpongtorn et al., 2023), possibly being related to elevated levels of inflammatory factors (Lane et al., 2022). As well, erythritol, a commonly used sugar substitute, was associated with clot formation and adverse cardiac events (Witkowski et al., 2023). Moreover, sucralose (Splenda) can diminish T cell-mediated responses (Zani et al., 2023) and in rodents it was genotoxic, increasing the risk of leaky gut syndrome (Schiffman et al., 2023).

Dietary Fructose

While most people are likely aware that sucrose is not ideal for their diets,[1] they might believe that fructose is somewhat better since it comes from fruits and vegetables, including berries and root vegetables. In the mid-1970s high-fructose corn syrup was introduced as a substitute for sucrose, appearing in soft drinks, juices, and baked goods. Regrettably, the increased use of high-fructose corn syrup corresponded with the elevated appearance of obesity, especially among young people, and consumption of fructose-rich beverages has been implicated in the development of type 2 diabetes, hypertension, coronary artery disease, kidney disease, intestinal inflammatory diseases, and several forms of cancer (Jung et al., 2022).

Fructose is primarily broken down in the liver and is also degraded in the small intestine. The liver uses fructose to create fat but when used excessively, tiny fat droplets may develop within liver cells, culminating in nonalcoholic fatty liver disease, and high-fructose corn syrup could also cause deterioration of the intestinal epithelial barrier (Febbraio and Karin, 2021). As well, the overuse of fructose causes elevation of triglyceride and cholesterol levels, increased proinflammatory cytokine levels, oxidative stress, and insulin resistance (Taskinen et al., 2019). Aside from these actions, by affecting mitochondrial oxidative stress, fructose may reduce adenosine triphosphate associated with energy production, while concurrently blocking energy obtained from fat. The net result is the promotion of obesity and several related diseases, including those associated with aging (Johnson et al., 2023). Ironically, beyond these actions, fructose may promote increased food consumption through the metabolic and brain changes provoked. Among other things, fructose can instigate orexigenic effects, increase the incentive salience of foods, and increase impulsivity (Payant and Chee, 2021).

The gut-liver axis is thought to contribute to illnesses that occur with the use of high-fructose corn syrup, and microbiota may be an essential component in this regard. Fructose may lead to a reduction of microbes that produce needed butyrate, which is associated with elevated inflammation and the development of type 2 diabetes and gut disorders (Jung et al., 2022). Additionally, fructose promoted certain bacteria that fostered inflammation and favored the occurrence of colitis, which could be diminished by the transplantation of good bacteria.

Fructose affects brain mechanisms associated with feeding (Payant and Chee, 2021), and influencing neurotransmitters (dopamine, glutamate) and inflammatory changes may promote anxiety and depression (Chakraborti et al., 2021). In rodents, a high-fructose diet during adolescence produced neuroinflammation and depressive-like behaviors in later adulthood (Harrell et al., 2018). Moreover, a diet high in fructose led to diminished synaptic plasticity and hippocampal neuroinflammation (Li et al., 2019b), and feeding mice a high-fructose diet during gestation and lactation resulted in altered hippocampal long noncoding RNA together with elevated anxiety in offspring (Zou et al., 2023b). It is of practical significance that SCFAs could attenuate the decline of neurogenesis and the microglial activation and neuroinflammation provoked by fructose that had been coupled with a chronic stressor regimen (Tang et al., 2022a).

I can resist anything except temptation
Oscar Wilde

In mice, the equivalent of a "Western diet" together with a sedentary lifestyle, increased levels of proinflammatory monocytes, and activity of brain microglia can impair neuronal survival. This diet was also accompanied by elevated β-amyloid plaque and microglia that express TREM2, which is observed in Alzheimer's disease (Graham et al., 2016). Similar findings in humans should be sufficient to promote good lifestyle practices. However, as much as people might have earnest intentions to eat properly, when faced with a chocolate éclair, or perhaps some nice gelato, resisting may be enormously difficult. Even a person with diabetes might say "Well, just this once, but only a very small bit." Beating temptation is exceedingly difficult, and the response to yummy foods is in many respects reminiscent of an addiction. Indeed, over-eating and drug addiction may share some underlying mechanisms, including activation of dopamine neurons in brain regions that contribute to "liking" and "wanting" that reflect hedonic and homeostatic processes (Morales and Berridge, 2020). Saturated fats can likewise instigate effects on specific brain dopamine receptors within the nucleus accumbens, thereby affecting feelings of reward (Hryhorczuk et al., 2015), hence promoting further eating.

The hormones orexin, ghrelin, and leptin, which play a significant role in food intake, may affect addiction processes by provoking dopamine release at the nucleus accumbens and prefrontal cortex (Abizaid et al., 2014).

---

[1] Excessive use of sucrose, comprising more than six teaspoons of sugar, has been associated with more than 45 adverse health outcomes.

Fibroblast growth factor (FGF)-21 released from the liver likewise communicates with the brain's reward system to affect food preferences and might have clinical applications in the management of diabetes. Beyond such changes, obesity has been associated with dysregulation between the dorsolateral hippocampus and the lateral hypothalamus, which may result in individuals being unable to regulate their emotional responses in anticipation of rewarding meals (Barbosa et al., 2023).

White Fat and Brown Fat in Obesity

Not all fat is the same, nor is all fat bad. Fat located on the belly (abdominal fat) is particularly unhealthy as it contains proinflammatory cytokines, which in excess can favor the development of inflammatory diseases, including diabetes and heart disease. Aside from the location in which fat is stored, the nature of the fat itself is important. Brown fat tends to be readily burned energy, fueled by glucose and lactose, and consequently is easily diminished, whereas white fat is an energy storage tissue that is relatively resistant to being modified by exercise or dieting and is usually considered "bad fat." As infants, humans have appreciable brown fat, which is very much reduced in adulthood, but it seems that adults have much more brown fat than had initially been thought. Fig. 5.1 is a schematic showing the ties between white and brown fat and its effects on microbiota (Schroeder and Bäckhed, 2016).

Some people seem to have an extraordinary ability to activate brown fat, but the reasons for this are uncertain, although clearance of creatinine (a product from muscles) might contribute to this (Gerngroß et al., 2017). It seems that thin mice (and perhaps thin people) can convert white-to-brown fat more readily than obese mice. Among

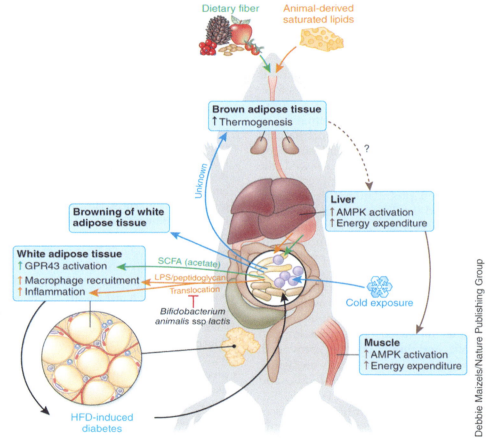

FIG. 5.1 Pathways through which communication occurs between gut microbiota and host adipose tissue. Saturated lipids (*orange arrows*) promote the translocation of several bacteria (Gram-negative bacteria and peptidoglycan) into the circulation, which, in turn, favors CD14- and NOD1-dependent inflammation within white adipose tissue. This may facilitate immune-related disturbances, such as type 2 diabetes and heart disease. Treatment with *Bifidobacterium animalis* can prevent these negative outcomes. Ordinarily, gut bacteria serve to ferment dietary fibers (*green arrows*), thereby producing short-chain fatty acids (e.g., acetate), which then activate the G-protein coupled receptor 43 (GPR43). The microbial composition can be altered by cold exposure (*blue arrows*) so that the browning of white adipose tissue ensues, as is the activation of brown adipose tissue. This promotes increased thermogenesis, thereby affecting liver and muscle function through AMP kinase activation, resulting in elevated energy expenditure. *From Schroeder, B.O., Bäckhed, F., 2016. Signals from the gut microbiota to distant organs in physiology and disease. Nat. Med. 22, 1079–1089.*

people with obesity, the elevated inflammatory response produced by fat works against the conversion of white fat to brown. In essence, being overweight has downstream effects that keep people that way.

There has been the recognition that beige fat, an intermingling of white and brown, also burns quickly and like brown fat serves to maintain body temperature. If it were possible to convert white fat to beige or brown fat, then it might be possible to diminish obesity and the occurrence of metabolic disorders and type 2 diabetes. Several proteins have been identified (e.g., PexRAPO) that can serve in this capacity (Lodhi et al., 2017), and circulating serotonin has been implicated in adult obesity, perhaps because it contributes to brown fat activity. Consistent with this perspective, mice engineered to be deficient of tryptophan hydroxylase 1, an enzyme necessary for the formation of serotonin, did not become obese when fed a high-fat diet, nor did they become insulin resistant, possibly owing to greater energy expenditure and increased burning of sugar present in brown fat (Crane et al., 2015).

Hypothalamic factors may contribute to the conversion of white fat to brown fat. By activation of orexigenic AgRP neurons within the hypothalamus, the conversion of white fat to brown fat can be increased, thereby preventing the negative consequences associated with a high-fat diet (Ruan et al., 2014). Similarly, following a meal, hypothalamic processes are activated by circulating insulin that promotes the browning of fat, which can then be used as a rapid source of energy. Among some people with obesity, the switch that causes white fat to turn brown may be ineffective and hence energy expenditure is not promoted (Dodd et al., 2017). It was also reported that in genetically engineered mice that do not produce folliculin, a substance involved in regulating mitochondria in fat cells, accumulation of white fat was diminished, and the typical weight gain associated with a high-fat diet did not occur (Yan et al., 2016b). Further, the administration of specific drugs that encourage angiogenesis (formation of new blood vessels) can cause white fat-storing cells to turn brown, leading to weight loss. As a bonus, mice treated in this fashion also exhibited a decline in cholesterol and triglycerides and improved sensitivity to insulin (Xue et al., 2016).

Appreciable evidence has indicated that gut bacteria play a pivotal role in obesity. To an extent, metabolic changes that accompany food ingestion may be moderated by specific genes and by epigenetic changes, such as those related to microbiota (Li et al., 2022a). Depleting microbiota (e.g., by antibiotics) promoted fat-burning beige fat within white fat packages, possibly through an increase of certain types of macrophages (Suárez-Zamorano et al., 2015). Moreover, following a meal, gut microbes may produce proteins that signal the release of the hormone glucagon-like peptide-1 (GLP-1), which inhibits eating (Gabanyi et al., 2022). It is more than

passing interest that antibiotic treatment before 2 years of age increases the risk for obesity and disposing individuals to type 2 diabetes (Cox and Blaser, 2015). This could occur because of persistent microbial changes or because gut bacteria may be more vulnerable to later challenges that would otherwise have only modest effects.

Variations in microbial diversity and composition have been tied to obesity, likely owing, in part, to certain gut bacteria influencing host energy harvesting, regulation of metabolic processes, insulin resistance, fat deposition, and the consequent inflammation, as well as central appetite and food reward signaling. In the latter regard, certain bacterial strains and their metabolites can affect brain functioning by stimulating vagal afferents or through immune-neuroendocrine processes, thus favoring the development of obesity (Torres-Fuentes et al., 2017).

Through the actions of SCFAs, gut microbiota can influence energy balances. Whereas butyrate seemed more aligned with burning brown fat, propionate and acetate influenced the browning of white adipose tissue. The actions of the microbiome on energy metabolism may be shaped by early-life factors (and even prenatal events), thereby influencing subsequent hunger, satiety, and metabolic processes related to energy harvest, including the allocation of energy resources to storage versus thermogenesis. Owing to these actions, gut microbiota may contribute to the effectiveness of diverse weight management strategies and is an important component in determining inflammatory tone (Carmody and Bisanz, 2023). In the latter regard, the intake of foods that promoted SCFAs not only reduced peripheral inflammation in aging mice but also decreased expression of genes in microglia related to inflammation (Matt et al., 2018).

Like the involvement of hormones and neurotransmitters, levels of P75 neurotrophin receptors that contribute to neuronal survival and growth might also be involved in energy-regulating processes. Among mice that lacked P75 neurotrophin receptors, weight gain did not occur in mice kept on a high-fat diet (Baeza-Raja et al., 2016). Finally, one of the strongest biological links to obesity is a gene region referred to as FTO, which contributes to how adipocytes (fat cells) function. Two genes within this region (IRX3 and IRX5) control whether adipocytes burn versus store fat, and by manipulating features of these genes, fat storage (or burning) could be altered (Claussnitzer et al., 2016). It is also possible that biological pathways could engender a white-to-brown metabolic conversion that might act against obesity (Moisan et al., 2015). Several other factors have been identified that may contribute to metabolic functioning and novel genetic and epigenetic analyses have been used to identify the aggregation of genetic factors that contribute to diseases that stem from fat depositions (Claussnitzer and Susztak, 2021).

Hormonal Regulation of Eating Processes

The discovery of two hormones, leptin and ghrelin, caused a sea change in how eating and energy regulation was considered. Leptin and ghrelin, released by adipocytes (fat cells) and by cells within the stomach, respectively, enter circulation and through their actions on both the brain and peripheral organs affect food intake, energy expenditure, and adiposity (fat) (Abizaid et al., 2014). As illustrated in Fig. 5.2, when levels of ghrelin increase, eating is promoted, whereas satiety is signaled by elevations of leptin that occur during and following a meal. As both these hormones influence several brain processes, such as HPA axis activity, as well as dopamine and serotonin functioning within frontal cortical limbic regions (Abizaid, 2009; Fulton et al., 2006), it is reasonable to expect that food consumption and the type of food consumed may influence mood states. Given that ghrelin can influence dopamine within the ventral tegmentum and nucleus accumbens, which have been linked to reward processes, craving or anticipatory responses to food might be encouraged (Abizaid et al., 2014).

Ghrelin stimulation of orexin cells in the lateral hypothalamus may likewise contribute to food craving, including the desire for comfort foods following stressor experiences (e.g., Thomas et al., 2022a,b,c). In this regard, craving and the actual consumption of food (i.e., hedonic versus homeostatic drives) might involve distinct processes; ghrelin-provoked dopamine activation within the nucleus accumbens may be associated with craving, whereas other factors (e.g., serotonin), may be aligned with consumption itself (Morales, 2022). Indeed, the nutritional and hedonic value of foods may differentially influence dopamine release from the ventral and dorsal striatum (Tellez et al., 2016). Imaging studies have also indicated that microstructural differences in the nucleus accumbens and caudate were apparent among obese versus average-weight individuals (Samara et al., 2021). Moreover, the links between dopamine receptors and preference for sweet foods evident in lean people were absent in obese individuals, suggesting a disturbance of these connections (Eisenstein et al., 2015).

Given their roles in eating processes, it is hardly surprising that ghrelin and leptin could influence

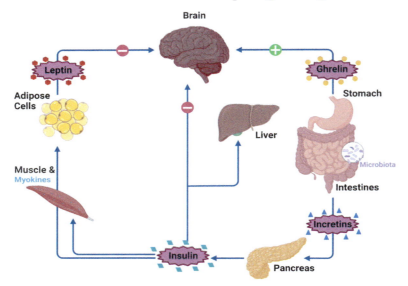

FIG. 5.2 Leptin and ghrelin are key hormones that influence appetite, hunger, and satiety. Elevated levels of ghrelin (right side of the figure) produced by specialized cells that line the stomach and the pancreas stimulate eating and then decline after food has been consumed. As this hormone affects brain reward processes, certain foods will eventually be favored. Leptin secreted by white adipose (fat) cells acts to suppress appetite and regulate energy balances (shown on the left side). When leptin levels decline, appetite and food cravings increase. Moreover, microbiota within the gut that are influenced by diet and exercise (and other psychosocial influences) can affect energy regulation and may influence weight changes. In addition to these hormones, incretin peptides, such as glucagon-like peptide-1 (GLP-1) and glucose-dependent insulinotropic polypeptide (GIP) secreted from enteroendocrine cells in the intestinal epithelium, facilitate the coordination of metabolic responses associated with food ingestion, and thus can be used to treat diabetes and can be used to promote weight loss. Insulin produced and secreted from the pancreas allows glucose to be taken up into cells and affects brain regulatory processes associated with reward processes. Insulin also influences the liver, which is necessary to process the nutrients absorbed from the small intestine. The bile secreted from the liver into the small intestine also serves in the digestion of fat and certain vitamins, as well as brain mechanisms related to reward processes. Through its action on protein synthesis and the availability of glucose, insulin affects the availability of energy to muscles. The muscle cells influence adipose (fat) cells, and in response to exercise, skeletal muscles release myokines (e.g., irisin and IL-6) that contribute to energy regulation and the presence of inflammation. The adipose cells release leptin, which causes feelings of satiety. Several other hormones (not shown in the figure), such as cortisol, insulin, neuropeptide Y, and Neuromedin B, affect eating processes, as do serotonin and dopamine. Aside from these processes, hormones (e.g., estrogen and testosterone) and growth factors also contribute to energy-related processes and fat accumulation and distribution. *From Anisman, H., Kusnecov, A.W., 2022. Cancer: How Lifestyles May Impact Disease Development, Progression, and Treatment. Academic Press, London, UK.*

psychologically based eating disorders. Their actions may be moderated by ovarian hormones (Smith et al., 2022), which might account for sex differences in eating-related behaviors and perhaps eating disorders. Elevated ghrelin levels were detected in anorexia and bulimia nervosa patients, whereas decreased ghrelin was associated with binge eating (Méquinion et al., 2020). Consistent with a causal role for ghrelin in eating or the desire for food, ghrelin administration increased food-related imagery and stimulated feelings of reward, influenced responses to visual and olfactory cues related to food, and enhanced responses to incentive cues (i.e., secondary stimuli that have been associated with reward), thereby affecting craving. It may be significant in this regard that among emotional eaters, the typical ghrelin decline was not apparent following food consumption, suggesting that ghrelin functioning was not operating as it should (Raspopow et al., 2014). Paralleling such findings, MRI images taken before and after eating among women with obesity, indicated that the neo- and limbic cortices and midbrain regions that were activated when hunger was present, persisted after a meal was eaten. Again, these data point to poor regulation of satiety processes being associated with obesity.

Set point and famine preparedness: The impact of intermittent fasting

Each individual has a "set point" so that when weight falls sufficiently, metabolic processes adjust to limit further weight loss, and activation of related hormones concurrently prompts increased eating. When dieters begin to lose weight, their cells go into "famine mode" such that their metabolic rate slows to sustain energy (the body's cells might not have understood that the person was intentionally dieting and thus responded inappropriately). Neurons within the hypothalamus that respond to AgRP together with serotonergic functioning act as a switch that reprograms mitochondrial functioning within brown and beige fat and does so without affecting food intake (Han et al., 2021). Once regular eating is reinstated the "famine mode" may continue so that calories are not readily burned off, and consequently weight gain may occur despite individuals not over-eating. With repeated efforts to lose weight (yo-yo dieting), metabolic disturbances become accentuated and weight loss becomes progressively more difficult to maintain. Thus, it is questionable whether dieting is, in fact, the best solution to losing weight and keeping it off. Indeed, metabolic changes elicited, even after a missed meal, may produce inflammatory responses that encourage fat storage. Aside from this, eating a high-fat diet or being obese can result in mitochondria within adipose tissue being destroyed so that further fat burning is impaired, thereby perpetuating weight gain (Xia et al., 2024).

Although the harms created by yo-yo dieting are well known, this should not be interpreted as meaning that occasional fast days are counterproductive. In fact, in mice and humans, periodic cycles of fasting or a diet that mimics fasting for two or more days can have positive actions, varying with age and sex. Among other things, intermittent fasting or periodic fasting that entails a more prolonged period of not eating could act against diabetes, heart disease, neurodegenerative disorders, autoimmune disorders, and age-related immune-based disorders, including some forms of cancers (Longo et al., 2021), varying with the specific characteristics of the diet. Among people who were overweight or obese, intermittent fasting routines were effective in reducing weight, with alternate day fasting being superior to either a 5:2 diet (normal food consumption for 5 days, and then 2 days of eating just 25% of the usual calorie total), or time-restricted eating (Ma et al., 2024).

Fasting diets can enhance the effects of chemotherapy in the treatment of triple-negative breast cancer (Pateras et al., 2023) and may limit some adverse secondary effects associated with chemotherapy. It seems that caloric restriction or fasting, enables changes in inflammatory processes, thereby limiting diseases that might otherwise be provoked (Youm et al., 2015). This could come about by the promotion of the growth factors that contribute to the formation of blood vessels and activation of a subset of immune cells, specifically antiinflammatory macrophages, which stimulate fat cells so that they burn more fats (Kim et al., 2017b).

Aside from effects on physical illnesses, intermittent fasting, particularly time-restricted feeding, may be effective in alleviating sleep disorders, cognitive disturbances, and some mental health problems, notably depressive illness (Fernández-Rodríguez et al., 2022). These findings are in keeping with the increased BDNF and neurotrophin-3 in the hippocampus of rats that had been maintained on an intermittent fasting regimen (16 h/day) over 3 months (Elesawy et al., 2021). The effectiveness of fasting diets may come about through the activation of cellular processes that govern mitochondrial health, DNA repair, and autophagy (Mattson et al., 2017). This said, it was maintained that restricted eating was associated with an elevated occurrence of death related to heart disease.

Increasing attention has focused on the involvement of the mechanistic target of rapamycin (mTOR) in aging processes and the effects provoked by intermittent fasting. The mTOR network is a signaling hub involved in sensing and integrating intracellular nutrient-related and growth factor signals involved in cell growth, proliferation, apoptosis, inflammation, and, ultimately, the aging process (Weichhart, 2018). It was thus suggested that the mTOR pathway might be fundamentally involved in the antiaging effects of intermittent fasting regimens (Lee et al., 2021a). In response to intermittent days of diminished eating, reparatory processes may be instigated, including changes in gene expression that have been associated with longevity (Wegman et al., 2015). As much as various diets that involve food restriction (e.g., intermittent fasting, fasting-mimicking diets, time-restricted feeding, ketogenic diets, protein restriction, and restriction of specific amino acids) may have beneficial effects that resemble an antiaging process, each may have limitations. Moreover, the evidence supporting antiaging effects of intermittent fasting diets has primarily come from individuals who are overweight or have obesity, and there are insufficient data supporting antiaging actions in nonobese humans.

Although leptin and ghrelin have been the primary players in research regarding energy regulation and food intake, several other peptide hormones contribute to these processes, as indicated in Table 5.1. For instance, the mammalian analogues of the amphibian hormone bombesin (BB), specifically gastrin-releasing peptide (GRP), and neuromedin B (NMB), contribute to the regulation of eating, acting as a satiety signal, and are also

involved in anxiety processes (Merali et al., 2013). As well, insulin plays a pivotal role in energy processes and may contribute to stress responses, and by causing the release of dopamine within the striatal region of the brain, it may contribute to feelings of reward and pleasure (Stouffer et al., 2015). Beyond this, insulin also acts with or on other hormones that affect energy processes. This includes glucocorticoids, leptin, neuropeptide Y (NPY), and regulatory hormones implicated in the development of obesity and metabolic disturbances associated with chronic stressors.

In addition to these factors, a variant of the gene coding for the neurotrophin BDNF may contribute to the development of obesity, so that boosting BDNF protein levels may diminish appetite and obesity, and in collaboration with mitochondrial dysfunction, the gut microbiota and tryptophan-kynurenine metabolism may be a link between obesity and depression (Marx et al., 2021). Further, among women with obesity, μ-opioid receptors in brain regions associated with reward processing were lower than among normal-weight women (Karlsson

TABLE 5.1 Hormones Related to Energy Regulation and Eating

Secreted Hormone	Biological Effect	Behavioral Outcome
Leptin	Produced by fat cells. Influences neurons in hypothalamic regions	Reduces food intake and appetite, and increases energy expenditure
Ghrelin	Synthesized in gut. Affects many of the same brain regions as leptin, but in an opposite manner. Influences brain regions associated with reward	Stimulates food intake and appetite, while reducing energy expenditure. Enhances reward-seeking behaviors; modulates stress responses
Orexin (hypocretin)	Made by cells in the lateral hypothalamus	Involved in the regulation of appetite, arousal, and wakefulness
Insulin	Produced by pancreatic beta cells: Regulates fat and carbohydrate metabolism. Involved in getting glucose from the blood into various body cells and storing it as glycogen	In brain, insulin stimulates hormones that reduce food intake
Bombesin (appears in humans as neuromedin B [NMB] and gastrin-releasing peptide [GRP])	Synthesized in gut and in several hypothalamic regions, and is also found in limbic regions	Acts as a satiety peptide (signals when an individual is full) and is released in response to stress, thereby promoting anxiety
Neuropeptide Y (NPY)	Produced by the gut and in several brain regions, including the hypothalamic arcuate nucleus Increases vasoconstrictor actions of norepinephrine	Increases food intake and reduces physical activity; increases energy stored in the form of fat; blocks nociceptive (noxious) signals to the brain; acts as an anxiolytic agent
Orexin (hypocretin)	Created within the lateral hypothalamus, but orexin receptors are found throughout the brain	Involved in appetite, as well as stress and reward processes, arousal, and wakefulness
α-Melanocyte stimulating hormone (α-MSH)	Produced in the arcuate nucleus of the hypothalamus. Acts as an agonist of melanocortin (MC-3 and MC-4) receptors in brain, including stress-related regions	Reduces appetite and increases energy expenditure as modulated by leptin
Agouti-related peptide (AGRP)	Produced in the arcuate nucleus (by the same cells that produce NPY), and serves as a natural antagonist of MC-3 and MC-4 receptors	Increases appetite and reduces energy expenditure. Modulated by leptin and ghrelin
Glucagon-like peptide-1 (GLP-1)	Manufactured in the gut. Important in stimulating insulin release	Acts in the brain to reduce appetite and to diminish the functioning of reward processes

et al., 2015). Ordinarily, activation of these receptors might subserve reward gained from foods, but when the receptor number is diminished women may eat more as a compensatory response to regain rewarding feelings from food.

Because of the multiple processes associated with eating, a key role of genetic factors in this regard is hardly unexpected, and it might be anticipated that obesity would be linked to genetic processes. Whole genome analyses have identified numerous genes associated with obesity, including several that affect brain functioning (Loos and Yeo, 2022), and about 90 genes present within subcutaneous fat could contribute to obesity and hence to diabetes and heart disease (Civelek et al., 2017). A particular gene, RELM-α (resistin-like molecules) and its proteins, which are highly expressed in infectious and inflammatory diseases, may protect females from obesity. The levels of RELM-α are higher in females than in males, but when RELM-α was deleted, their levels of eosinophils and macrophages declined and they were no longer protected from obesity (Li et al., 2023c).

In addition, epigenetic modifications within germline cells may contribute to obesity and could influence the transgenerational appearance of obesity (King and Skinner, 2020) and diabetes associated with a high-fat diet (Huypens et al., 2016). In this regard, in young animals, an epigenetic change related to a gene associated with impaired glucose metabolism (*Igfbp2*) had ramifications for adult diabetes, and this gene's action was similarly modified in obese humans with diabetes. Thus, the status of this gene might turn out to be a useful biomarker in predicting the development of diabetes (Kammel et al., 2016). It was similarly reported that epigenetic modification of the proopiomelanocortin gene, which is involved in satiety and energy regulation, was associated with an increased risk of obesity, especially in women. Moreover, in a small sample of individuals, a melanocortin 4 receptor agonist (setmelanotide) promoted weight gain in the presence of POMC hypermethylation (Lechner et al., 2023). Genetic polymorphisms also exist that have been associated with increased eating. Some of these may directly influence the development of obesity, such as through the FTO gene (Loos and Yeo, 2022), whereas others may have their effects indirectly.

Stress and Eating

While severe stressors typically cause reduced eating, and moderate-intensity stressors may have such an effect in some people, in others it may increase their propensity to eat, possibly reflecting self-medication through "quick fixes" to diminish distress. Processes related to stress and eating are intricately entwined, so treatments that diminish anxiety, such as benzodiazepines, typically increase food consumption. As well, several stressor-elicited neurochemical changes, notably variations of dopamine and CRH, have been related to eating processes, and along with cortisol, may contribute to stress-related food cravings (Chao et al., 2017). Predictably, hormones fundamental in eating processes (leptin and ghrelin) have marked effects on stress responses, and chronic social stressors could influence leptin levels (Patterson et al., 2013).

From an adaptive perspective, it makes sense that eating and stress experiences would engage antagonistic processes. It would be counterproductive, after all, for an animal under threat to stop for a meal. Yet, in a subset of people, particularly those who use avoidant coping strategies to deal with stressors, negative events can lead to disordered eating, possibly acting as a coping response, much as alcohol might serve in this capacity. Elevated eating is particularly notable among emotional eaters who encounter stressors but receive unsupportive responses from their friends (Raspopow et al., 2013). Among those who adopt eating as a way of coping, the foods selected are typically those that taste good, and individuals may experience cravings for "comfort food" that is high in carbs. The words "I could really use a celery stick right about now" will rarely escape the lips of an emotional eater. It was suggested that hypothalamic proenkephalin changes elicited by stressors are responsible for excessive eating (You et al., 2023), although other brain neurochemical changes that accompany stressful experiences may also contribute to increased eating.

Other brain neurochemical changes that accompany stressful experiences contribute to increased eating. Stressor-provoked activation of dopamine neuronal functioning, which can stimulate reward processes, could cause particular stimuli or responses to be more salient (Berridge and Robinson, 2016). Thus, the positive effects of food in diminishing moderate stress responses may become especially significant, thereby leading to increased consumption of tasty foods upon further stressor encounters. By affecting dopamine functioning, hedonic appraisal of foods high in sugars and fat might favor the development of an "addictive-like" condition so that withdrawal from the palatable diet could lead to a stress-like response (Morris et al., 2016b). Further to this, as ghrelin stimulates dopamine functioning in brain regions associated with reward processes when emotional eaters indulge in comfort foods following a stressor experience, the rewarding feelings might have been potentiated, thus favoring eating when stressors were subsequently encountered. Stressful events may also sabotage an individual's ability to make proper decisions and for those already dieting, good intentions may go by the wayside in the face of stressful events. Functioning of the lateral habenula, and specifically NPY activity in this region, may diminish the reward response associated with activation of the ventral tegmental dopamine response so that animals on a high-fat diet did not overeat. Among chronically stressed mice the lateral habenula did not switch on so that reward signals persisted, thereby allowing for excessive eating (Ip et al., 2023).

There has been the view that cortisol elevations associated with stressors contribute to eating and might be an important element in the provocation of obesity. In this regard, cortisol measured in hair samples (hair grows at about 1.25 cm a month, and thus a sample of about 3.75 cm reflects cortisol accumulation over the preceding 3 months) was directly related to body weight, and it was argued that chronic stressors contributed to obesity (Jackson et al., 2017). In her work linking stressful experiences and eating, Dallman (2010) considered how chronic stressors could have such consequences. In response to stressors, the fast caloric fix provided by sugars and carbohydrates, together with brain neurochemical changes provoked by these comfort foods, provide sufficient energy to facilitate the brain's attempt to limit the distress and anxiety that would otherwise occur. Central to Dallman's reasoning was that with continued distress, persistently high concentrations of corticosterone may increase the salience of pleasurable or compulsive activities (ingesting sucrose and fat) and might thus contribute to obesity. It is of particular significance that glucocorticoids act on the redistribution of fat so that it appears as abdominal fat depots that serve as a storehouse for inflammatory factors, thereby influencing several pathological conditions. As such, dietary interventions may be useful in limiting the course of some of these conditions, including the cognitive decline associated with aging.

Early-Life Experiences

Stressful events experienced early in life may be associated with elevated adult stress reactivity, anxiety, and a preference for comfort foods, thus contributing to adult obesity. In this respect, adult obesity was 30%–50% more frequent among individuals who experienced negative early-life events, particularly abuse (Hemmingsson et al., 2014). The stressful early-life experiences may disturb the functioning of neural circuits associated with emotions and reward processes, including changes in neuronal activity within the nucleus accumbens, amygdala, and medial prefrontal cortex (Levis et al., 2022), which might contribute to the regulation of emotional eating.

Aside from proactive effects on appraisal and coping processes, negative early-life stressors give rise to disturbed neuroendocrine functioning that feeds into appetite regulation and metabolism, negative thinking, poor

emotional regulation, as well as disturbed sleep and cognitive functioning, which also encourage the development of obesity (Hughes et al., 2017). Like early-life stressors, adolescence is also a time that can entrain emotional eating. Consumption of a high-fat diet during adolescence resulted in elevated sensitivity of dopamine reward pathways, and this outcome was subject to a sensitization-like effect so that later stressors promoted reward-related behaviors that may favor obesity.

Obesity and Illness

Considerable research has pointed to the health risks attributable to obesity. However, several studies indicated that being somewhat overweight is not as bad as it had once been thought. A large-scale longitudinal study revealed that a good number of overweight individuals did not present with high blood pressure, insulin resistance, diabetes, low good cholesterol, high bad cholesterol, or high triglycerides, and they were 38% less likely to die early relative to obese individuals who had two or more of these markers (McAuley et al., 2012). Later studies similarly indicated that the relation between weight and premature death was not linear, and has been changing in recent decades (Afzal et al., 2016). These findings were taken to suggest that it might be inappropriate to consider obesity from a singular perspective. In fact, overweight individuals who were cardiorespiratory or aerobically fit had a lower risk of death than less heavy individuals who were not fit (Farrell et al., 2020). A somewhat similar conclusion was reached based on a meta-analysis of 97 studies comprising 2.88 million people, indicating that poor health was related to the extent of the person's obesity. Being somewhat overweight or moderately obese based on a body mass index (BMI) of 25–30 and 30–35, respectively, was not linked to earlier mortality; only frank obesity was associated with earlier death (Flegal et al., 2013). A subsequent meta-analysis and systematic review indicated that studies typically did not differentiate between obesity, hyperinsulinemia, and hyperinflammation which were associated with earlier mortality. As hyperinsulinemia and hyperinflammation may have contributed to health risks, the contribution of obesity alone was uncertain in relation to mortality. A closer examination is needed to dissect the contribution of these factors in accounting for the links between obesity and early mortality (Wiebe et al., 2023).

These findings were unexpected, especially as decades of research had indicated that being moderately overweight was a risk factor for many diseases as well as the effectiveness of treatments for a variety of illnesses. A broad meta-analysis across countries that did not include smokers or people with preexisting diseases indicated that all-cause mortality in those who had greatest obesity was pronounced, more so in men than in women and marked ties to mortality were even apparent in younger people (The Global BMI Mortality Collaboration et al., 2016). Simply being overweight during middle age diminished life span by 3.1 and 3.3 years among males and females, respectively, and in individuals with obesity life span was reduced by 5.8 and 7.1 years. Even gaining a moderate amount of weight before the age of 55 was associated with an increase in chronic diseases and premature death. For a 5 kg weight gain, the risk of type 2 diabetes was increased by 30%, hypertension increased by 14%, cardiovascular disease increased by 8%, obesity-related cancer was elevated by 6%, dying prematurely increased by 5%, and the odds of experiencing healthy aging was reduced by 17% (Zheng et al., 2017). Likewise, a report on 3.5 million people who had been followed for 5.4 years, including many without metabolic problems, indicated that individuals with obesity were at greater risk for coronary heart disease and several other cerebrovascular diseases (Caleyachetty et al., 2017). Furthermore, a 9-year longitudinal analysis that included more than 2.6 million people indicated that the risk of 18 types of cancer was elevated in association with being overweight, and risk was elevated with longer duration and greater degree of being overweight and younger age of onset of being overweight and obesity (Recalde et al., 2023).

It has been maintained that in evaluating the relationship between obesity and health, it is essential to consider how long individuals have been overweight rather than simply focusing on the extent of the person's obesity. Some light was shed on this with reports that individuals who were overweight beginning at an early age were most likely to suffer premature death, whereas individuals who were lean throughout life or who gained weight at midlife were less likely to suffer this fate (Song et al., 2016). An analysis of obesity and mortality that included weight history concluded that sampling weight at one time in a person's life may not provide an accurate reflection of the burden carried, and a better index of how obesity affects health can best be gleaned from the individual's history of being overweight (Stokes and Preston, 2016). A 20-year prospective study conducted with 2,500 men and women in the United Kingdom indicated that about a third of obese people were deemed to be in good condition, showing acceptable blood pressure, fasting blood sugar, cholesterol, and insulin resistance. But, among these seemingly healthy obese individuals, 40% developed risk factors after 10 years, and after 20 years more than 50% fell into the unhealthy obese category (Bell et al., 2015). A 30-year follow-up study of more than 90,000 women likewise indicated that among metabolically healthy women, a shift to an unhealthy phenotype was accompanied by an increased risk of cardiovascular disease (Eckel et al., 2018).

Considerable data have pointed to obesity being tied to the promotion or exacerbation of gastrointestinal (GI), esophageal, breast, renal, and reproductive cancers. The American Society of Clinical Oncology (ASCO) indicated that it would not be long before obesity would become the leading preventable cause of cancer, even exceeding that of smoking. In addition to these disastrous outcomes, several reports have shown that obesity may be associated with resistance to chemotherapy in the treatment of some cancers and has been linked to an increased risk of complications related to surgery and anesthesia, as well as the incidence of cancer recurrence (Pati et al., 2023).

Poor dietary habits and obesity in children predicted adult obesity and contributed to the premature development of adult diseases, including heart disease, hypertension, fatty liver disease, osteoporosis, and immune-related disorders, such as multiple sclerosis. It likewise appeared that maternal obesity during pregnancy could have profound effects on offspring. Maternal obesity in rodents was accompanied by changes in neuronal functioning within brain reward circuits and disturbed social interactions that were reminiscent of human autism. These effects seemed to be related to a microbial imbalance that could be reversed through gut microbial reconstitution (Buffington et al., 2016). It is significant as well that among obese female mice, the presence of metabolic problems can be transmitted through mitochondrial DNA to offspring over three generations, pointing to epigenetic actions (Saben et al., 2016). At the other end of the age spectrum, the combination of obesity and aging represents a double whammy for heart disease, as small blood vessels that feed the heart may become damaged under these conditions, likely owing to inflammatory influences (TNF-α and IL-6), ultimately leading to diabetes and heart failure (Dou et al., 2017).

To this point, we have largely focused on the link between obesity and physical illnesses. However, the chronic inflammatory state produced by cytokine release from abdominal white fat can promote the development of depressive illnesses as well as several comorbid conditions. There is appreciable evidence suggesting that obesity is associated with increased recruitment of processes that affect CNS functioning, promoting activation of microglia and hence the provocation of neurologic disorders. Indeed, data from the English Longitudinal Study of Aging (ELSA) indicated that disturbed executive functioning and memory were linked to body mass and systemic inflammation reflected by C-reactive protein (CRP) levels (Bourassa and Sbarra, 2016). And, elevated BMI was associated with accelerated brain age, diminished thickness and connectivity in default mode- and reward-related areas, as well as gray matter atrophy, likely stemming from systemic low-grade inflammation (Medawar and Witte, 2022). Despite the contentions of several naysayers, there is simply too much information pointing to the health risks associated with being overweight.

Together, the data suggest that "fat but fit" is likely a misnomer, as these individuals may well be at greater risk for pathology. Rationalizations that obesity does not necessarily increase vulnerability to illness in a subset of people might have had the unfortunate consequence of allowing some overweight individuals to take the view that "I may be overweight, but I'm perfectly healthy" giving them license to continue their poor lifestyles.

Obesity and viral illnesses

Having obesity has been associated with elevated risk of clinical complications associated with influenza as well as a poorer prognosis (Andrade et al., 2021), in part owing to dysregulation of immune cell functioning involved in the resolution of inflammation and infection (Shaikh et al., 2022). To the point, the effectiveness of influenza vaccination may be diminished among individuals with obesity. On average, influenza vaccination is only effective 50%–60% of the time. Following influenza vaccination, individuals who are overweight or have obesity are twice as likely to develop influenza relative to normal-weight people. This occurs even though they had built up as many antibodies against the influenza virus as average-weight people. It seems likely that this is related to immune dysfunction secondary to obesity, as well as elevated cortisol levels that may undermine immune efficacy.

As in the case of influenza, individuals with obesity were at elevated risk for COVID-19 infection, the development of relatively severe symptoms, hospitalization, and death (Andrade et al., 2021). These outcomes may have been related to disturbed immune responses and microbiota dysbiosis, the presence of chronic inflammation and oxidative stress, as well as excessive activation of the renin-angiotensin-aldosterone system and that of the angiotensin-converting enzyme 2 receptors in adipose tissue (Dalamaga et al., 2021). Aside from the risk of hospitalization and death after SARS-CoV-2 infection, the waning of humoral immunity following COVID-19 vaccination was accelerated among individuals with obesity, thus leaving them more likely to be infected (van der Klaauw et al., 2023).

Diet and Immunity

Various balances exist within neurobiological and metabolic systems so that one process might be fundamental in the promotion of responses in a second system, whereas another mechanism may operate in an inhibitory capacity to regulate the excitatory effects of the first. For instance, reactive oxygen species (ROS) and other free radicals, which are by-products of cellular metabolism, appear at sites of inflammation where they facilitate the death of unhealthy cells. Antioxidant processes are then engaged to limit the actions of ROS once they have done their job. It appears likely that certain foods (e.g., red beans, blueberries, raspberries, cranberries, and artichokes) increase the production of antioxidants, thereby limiting cellular damage. In contrast, a diet rich in animal fats encourages gut bacteria that are not as beneficial as those that come from a diet rich in plant fibers.

Aside from these actions, dietary factors can also influence microglia, the brain's resident macrophage-like cells. Among mice maintained on a diet rich in fat for 4 weeks, a treatment that increased the number of microglia within the mediobasal hypothalamus, inflammation was promoted along with the tendency to burn fewer calories. The microglia also appeared to recruit immune cells to the mediobasal hypothalamus where they begin to behave like microglia, modulating inflammation. However, if microglia were reduced by experimental drug treatments, mice consumed less and gained less weight, whereas increased hypothalamic glial cells and elevated weight engendered by a high-fat diet could be attenuated by genetically influencing glial cell production. Simply increasing inflammation within the hypothalamus, even in the absence of a high-fat diet, altered eating and weight gain, pointing to the glial involvement in these processes (Valdearcos et al., 2017). Thus, interventions that focus on microglia and inflammatory processes might contribute to the development of treatments to diminish obesity.

It is obviously not for everybody all of the time

"I do not like broccoli. And I haven't liked it since I was a little kid and my mother made me eat it. And I'm President of the United States and I'm not going to eat any more broccoli." So said President H.W. Bush in 1990, having become quite the rebel. The problem with broccoli is that it has no flavor, and getting people to eat it might require changing its taste. However, broccoli has powerful antioxidant capabilities owing to their phenolic composition. There are now efforts to enhance the antioxidant capacity of broccoli and other vegetables (e.g., kale and cabbage) by manipulating their genes.

Virtually every wine drinker is likely able to provide a lecture on the health benefits of wine, perhaps pointing to it increasing gut microbiota diversity (other alcoholic beverages do not do this), or to the presence of the strong antioxidant resveratrol. The antioxidant actions may be beneficial and at low doses, resveratrol caused immune cells to be more responsive to an antigen and can attenuate adverse effects elicited by pathological stimuli, including reduction of oxidative stress in endothelial and smooth muscle cells, vascular remodeling, and arterial stiffness, and can moderate the actions of immune cells (Li et al., 2019a). It has also been maintained that resveratrol may limit the release of inflammatory cytokines, enhance energetic functions of mitochondria, and enhance cognitive functioning by increasing the clearance of β-amyloid peptide. Yet, chronic activation of biological processes by resveratrol may disturb synaptic functioning and dendritic growth, which could disturb cognitive functioning (Yang et al., 2021b). Given that alcohol can have adverse effects, such as being associated with several forms of cancer (Rumgay et al., 2021), it might be better to obtain resveratrol from other sources, such as red grapes (skins), peanuts, cocoa, and dark chocolate, as well as blueberries, bilberries, and cranberries.

Considerable data have supported the view that dietary factors and digestive processes contribute to immune system functioning and immune-related illnesses. The GI tract acts as a barrier that limits potential adverse effects stemming from the foods eaten. Epithelial cells that line the GI tract, along with protective mucous, minimize the passage of damaging molecules into the body, but at the same time permit the passage of beneficial substances. About 70% of our immune cells are located within the gut so that we are protected from potentially dangerous microbes present in food. A molecule present in the gut (gp96) helps in the regulation of the immune system so that inflammatory overreactions that are associated with disease conditions, such as colitis, do not occur (Hua et al., 2017). It seems that among obese women, caloric restriction comprising a diet of 800 kcal/day not only produces substantial weight loss, reduction of plasma glucose, insulin, and leptin in adipose tissue, but also enhances the integrity of the gut barrier, and reduces circulating CRP levels, reflecting diminished inflammation (Ott et al., 2017).

Just as microbiota can affect immune functioning, immunological factors can affect gut bacteria, so the immune and commensal bacteria consistently undergo a degree of rebalancing. Too great an immune response can induce harmful inflammation, favoring the development of colitis and inflammatory bowel disease. However, as the processes that provoke these conditions vary considerably across individuals, treatments of gut-related illnesses, like so many other disease conditions, might require a personalized medicine approach. Related to this, in order for immune functioning to progress efficiently, essential vitamins and minerals obtained from foods need to be present, including zinc, selenium, iron, copper, and folic acid, as well as essential fatty acids and monounsaturated fats. Accordingly, predicting optimal strategies to treat gut disorders ought to consider whether these substances are present in adequate concentrations.

Of course, diets that enhance gut diversity and abundance may foster well-being. The omission of foods, such as fibers, can markedly affect the microbial diversity of the gut, and in mice deprived of fiber, many of the constituents of the gut that had been present earlier were not readily reconstituted or may require weeks or months for this to occur. It is fascinating that the disappearance of particular gut microbiota develops over generations, speaking to the selection pressures that might contribute to cultural differences that are seen in links between diet and health (Sonnenburg et al., 2016).

As described in Chapter 3, changes in the composition of gut bacteria have been associated with numerous illnesses beyond gut disorders. It is noteworthy that medications to treat existent diseases may do so, in part, by altering the gut microbiota. For instance, one of the most effective treatments for type 2 diabetes, metformin, alters the composition of gut microbiota, which may contribute to the beneficial effects of the treatment (Forslund et al., 2015). Indeed, transplanting fecal samples from metformin-treated donors to germ-free mice, enhanced glucose tolerance (Wu et al., 2017a).

Immune factors within the gut walk a tightrope in dealing with bacteria. Immune processes need to be able to destroy anything that is not part of us, yet not destroy beneficial gut bacteria that technically are also not part of us (even if they have been within us for a very long time). To a considerable extent, this sensitive job falls to dendritic cells, which are necessary for an immune response to be mounted and are needed for immune tolerance to develop by activation of T_{reg} cells. Apparently, under some conditions, the dendritic cells that act to suppress immune functioning undergo apoptosis (programmed cell death), and thus will no longer be available to limit immune functioning (tolerance), hence allowing continuous immune activation, inflammation, and pathology (Barthels et al., 2017).

The scoop on vitamins

As essential as vitamins and minerals are for well-being, these micronutrients are not made in the body, and instead are primarily obtained from foods eaten. We had been led to believe that multivitamin supplementation was a good idea, especially in children. For individuals with vitamin deficiencies there is little question that certain supplements are beneficial, and for growing children who were not consuming a varied diet, obtaining sufficient vitamins A, C, and D through dietary supplements might be useful. But, under other conditions, vitamins had few beneficial effects, although several reports indicated that multivitamins could limit cognitive decline among aged people (Vyas et al., 2024).

There has been considerable attention to the involvement of vitamin D disturbances in numerous illnesses. By increasing the absorption of calcium, high doses of vitamin D can be beneficial in reducing bone loss associated with aging (osteoporosis) and might diminish the occurrence of fractures, but too much vitamin D may have the opposite effect. A systematic review also indicated that lack of vitamin D was associated with more frequent upper respiratory infection (Martineau et al., 2017), and a large-scale analysis suggested that vitamin D use in older adults may limit the development of dementia (Ghahremani et al., 2023). Yet, it was argued that vitamin D "is a promiscuous biological candidate, likely to be attracted to any passing half-baked hypothesis (it is a risk factor in search of an adverse outcome)" (McGrath, 2017). As harsh as this criticism might sound, contrary to occasional reports suggesting otherwise, the evidence is weak concerning the usefulness of vitamin D supplementation in the prevention of cancer, respiratory infections, rheumatoid arthritis, and multiple sclerosis, and it is unlikely to diminish susceptibility to asthma, atopic dermatitis, or allergies, as too often proclaimed (Manousaki et al., 2017). Despite these criticisms, there has been a small swing toward the use of vitamin D in treating depressive disorders, but epidemiological and randomized control trials did not support these claims. Reports continue to appear attesting to the importance of vitamin D to improve memory among older individuals (Yeung et al., 2023), and may have small effects on heart health (Thompson et al., 2023). However, in the women's health initiative, a large randomized control trial conducted over 20 years, the incidence of heart problems was moderately elevated.

It was maintained that vitamin D affects inflammatory processes reflected by diminished C-reactive protein and regulation of the antiinflammatory cytokine IL-10 (Krajewska et al., 2022). A detailed review provided by the National Institutes of Health (National Institutes of Health, 2023) indicated that vitamin D supplements can reduce the risk of SARS-CoV-2 infection and could diminish symptom severity. As well, low levels of vitamin D have also been observed among individuals who experienced long COVID, and particularly pronounced symptom severity occurred among patients with low vitamin D at the time of hospital admission (di Filippo et al., 2023).

The limited benefits of daily multivitamin use (as opposed to intake of specific vitamins among individuals with relevant deficiencies), even after this was broadly reported in the media, did not markedly influence the sales of vitamins and supplements. It may be that individuals feel some uncertainty over dropping a long-held habit regarding their vitamin intake. Individuals who have doubts might still say "Well, these vitamins and supplements may or may not improve my health, but taking them surely couldn't hurt," but this assumption might not be entirely accurate. In fact, some supplements seem to have negative health effects with several thousand people ending up in hospital emergency rooms each year.

The risk of death was moderately increased among older women taking daily supplements of iron, copper, magnesium, zinc, or multivitamins, and excessive vitamin intake can cause immune system changes so that infections occur in response to agents to which immunity had previously developed. Calcium supplements have been linked to increased risk of heart attack, and excessive use of vitamin E predicted a 17% greater risk of prostate cancer. Likewise, while antioxidants obtained from foods are sufficient to produce health benefits, antioxidant supplements may increase tumor vascularity (angiogenesis), just as vitamins C and E can have such effects.

Given the many claims and counterclaims concerning the value of vitamins on health, it is difficult to determine which are valid and which are questionable. An interesting piece by Hall (2013) discusses the issue of "uncertainty" in medicine, and in doing so, quotes from Voltaire that "Uncertainty is an uncomfortable position. But certainty is an absurd one." It is remarkable how some naturopaths are as certain as they are that their treatments will definitely heal the patient. In contrast, in making diagnoses and treating patients, uncertainty is not infrequent even among physicians who have been in practice for years.

Omega-3 Polyunsaturated Fatty Acids

Amid the cacophony of voices pointing to one or another food that will prevent illness are those promoting omega-3 polyunsaturated fatty acids (PUFAs). Omega-3 PUFAs have received an unusual amount of attention and have reached fad status, and in the form of eicosapentaenoic acid (EPA) and docosahexaenoic acid (DHA) have been assessed in virtually every known psychiatric illness. They have been offered as adjunctive treatments for major depressive disorders, bipolar disorder, schizophrenia, neurodevelopmental disorders, as well as age-related cognitive decline. It was recommended that omega-3 PUFAs, as well as exercise, S-adenosyl-L-methionine, and yoga, be considered as first or second-line treatments for mild to moderate depression (Ravindran et al., 2016). It is thought that omega-3 contributes to the degradation of cell components that are unnecessary or dysfunctional, which then results in suppression of proinflammatory cytokines, such as IFN-α as well as CXCL-10 secreted by macrophages (Mildenberger et al., 2017). Further, omega-3 fatty acids are converted into cannabinoid epoxides that may promote antiinflammatory effects (McDougle et al., 2017).

Concerns had been raised that high doses of marine omega-3 supplements can increase the risk of atrial fibrillation, but this was not the case when lower doses were consumed (Qian et al., 2023). It should be added that while omega-3 obtained from fish can act against heart disease, it seems that fish oil obtained from supplements may not have similarly beneficial effects. This has not caused purveyors of fish oil supplements to diminish their claims of health benefits that can be derived from their products (Assadourian et al., 2023).

In addition to omega-3 PUFAs, a diet high in polyphenols (e.g., citrus fruits, cocoa, red wine, tea, and coffee) could affect brain functioning, possibly owing to antiinflammatory or antioxidant effects. This outcome could also occur through microbial alterations, as most dietary polyphenols accumulate in the large intestine and are then converted by the gut microbiota to less complex metabolites. Several other supplements and natural products could also have health benefits owing to their microbial actions. Black and green tea influence the growth of *Bifidobacterium* species, which can have positive effects. But they have also been implicated in the growth of nasty bacteria, such as *Helicobacter pylori*, *Staphylococcus aureus*, *Salmonella typhimurium*, and *Listeria monocytogenes*, pointing to the need for proper balances between bacterial communities within the gut (Duda-Chodak et al., 2015). Several plants and mushrooms can similarly influence microbiota and have the benefit of containing fiber and phytochemicals that diminish obesity and type 2 diabetes (Martel et al., 2016).

Meds to Curb Eating

As much as dieting and exercise can be used to reduce weight, many people want or need something much simpler to attain significant weight loss. The pharmaceutical industry has worked diligently to develop profitable weight-reducing compounds. The serotonin-releasing agent fenfluramine had short-lasting effects on weight and was quickly discarded, only to reappear as a composite with another weight-reducing agent, phentermine. The combination of these treatments, fenfluramine/phentermine, commonly referred to by the cutesy name fen-phen, caused heart valve dysfunction and pulmonary hypertension and so was discarded. Another promising agent, Lorcaserin (Belviq), which acts as a 5-HT$_{2C}$ receptor agonist, seemed to be effective in reducing weight by 5%–10% and had received approval as an antiobesity agent. Consistent with the emerging view that eating processes and addiction may involve some overlapping neural circuits, this agent also has potent actions in diminishing some forms of substance use disorders (Higgins et al., 2013). However, the drug was withdrawn because it was associated with an increase in the risk of some forms of cancer occurring.

Despite the setbacks, several drug companies persisted in their efforts to develop compounds that had the potential to reduce weight by effects on microbiota. Inulin propionate, or a type of fiber, inulin, influences gut bacteria and diminishes the activity of neurons within the nucleus accumbens and caudate, which have been linked to reward processes. Paralleling the brain changes, the treatment reduced the appeal of certain foods (Byrne et al., 2016). Still another compound was developed based on the finding that a diet containing fermentable carbohydrates prevented obesity in mice. This protection was lost when mice lacked free fatty acid receptor 2 (FFAR2), but when present, peptide YY which signals satiety, was elevated (Greenhill, 2017). A different strategy for weight reduction was to fool the body into mistakenly thinking that it had already eaten. The drug fexaramine causes bile acid release into the intestine, which ordinarily occurs after people have eaten to promote digestion. It also has the effect of reducing blood sugar and cholesterol and increasing brown fat (Fang et al., 2015).

The involvement of gut bacteria in eating processes and obesity has become increasingly evident, and in mice, manipulations of the microbiome have been useful in moderating obesity and inflammation. Once more, speaking to the value of personalized treatments, the ratio between specific bacteria in the gut (e.g., *Prevotella* vs *Bacteroides*) was predictive of the efficacy of certain diets (Hjorth et al., 2018). It can reasonably be expected that it will not be long before further strategies involving gut bacteria will be developed to produce the most efficacious diets.

Of the various weight-loss medications, particularly pronounced effects are achieved using a glucagon-like peptide-1 (GLP-1) agonist. Agents, such as semaglutide (Ozempic), came to prominence because of its effects in the treatment of type 2 diabetes by stimulating insulin and inhibiting glucagon secretion from pancreatic islet cells. It turned out that GLP-1 agonists have the added benefit of diminishing cravings for high-fat foods (Lin et al., 2020). The effects of semaglutide were sufficiently profound to have been used off-label to reduce weight, and at a higher dose, under the brand name Wegovy, it was approved for the treatment of obesity. Likewise, tirzepatide, the compound in Zepbound, and the diabetes drug Mounjaro produce pronounced weight reductions. Several new drugs, such as retatrutide, which operates through multiple hormonal effects (i.e., acting as an agonist of the glucose-dependent insulinotropic polypeptide, glucagon-like peptide 1, and glucagon receptors) had weight-reducing effects in a Phase 2 trial (Jastreboff et al., 2023) beyond that ordinarily produced by Ozempic. Unfortunately, once individuals stop using agents containing semaglutide and tirzepatide, rebound weight gain may occur (Aronne et al., 2023; Wilding et al., 2022), particularly if lifestyle changes comprising healthy eating and exercise, have not been adopted. The appeal of these quick-fix agents has been remarkable. But, given the need to continue taking these drugs, alternative strategies focusing on lifestyle change may be preferable.

For years, it has been considered that eating processes and drug addictions share several biological processes, and the disposition to sugar consumption can predict the potential for addiction. By affecting dopamine functioning, GLP-1 agonists diminish the rewarding value of alcohol intake and may be effective in reducing reward-related behaviors associated with other addictive substances (Eren-Yazicioglu et al., 2021). In rodents, such treatments reduced alcohol consumption and diminished relapse drinking, likely operating on reward processes through effects on dopamine functioning (Aranäs et al., 2023). Similarly, among individuals with obesity who were using such agents, alcohol intake was reduced (Quddos et al., 2023).

Natures Little Miracles

There seems to be the belief in some circles that if it grows on trees or bushes, or comes directly from the ground, it must be good for us. Some natural products can have very strong effects on the body and the brain and can have potent medicinal properties. The Pacific Yew Tree was instrumental in giving us potent cancer medications. Aloe has been used against burns, and marijuana, deadly nightshade, magic mushrooms, and jimson weed have all been implicated as having some medicinal value. Strychnine and digitalis in low doses can have positive effects for some conditions, but these plant-derived agents can be lethal at somewhat higher doses.

Most natural remedies are not marketed as drugs and thus have not been subject to rigorous, expensive, and time-consuming procedures that are necessary for drugs to come onto the market. Not only have most of the natural products available not been assessed in a rigorous scientific fashion, but information is typically not provided concerning the purity, quality, chemical stability, and active constituents of the products. Similarly, the packaging typically provides little information concerning contraindications, such as side effects or potential risks for some people, such as those taking certain medications. The lax policies regarding herbal remedies, and the lack of comparative studies evaluating natural versus standard medical treatments, have led to the market being saturated by products without genuine value. Ordinarily, after drugs receive FDA approval, a substantial number are later found to have safety concerns. How much more dangerous are products that have absolutely no oversight? There have been notable exceptions to this. Consistent with the long-held belief that extracts of the plant *Hypericum perforatum* (St John's wort) have

antidepressant effects, several studies indicated that it produced positive effects in diminishing mild or moderate-grade depression (Ng et al., 2017) but is less effective in attenuating severe depression. Most studies that have been conducted involved a relatively small sample size, and the observed outcomes tended to be more limited in relatively large trials.

Plants often have nasty chemical means of protecting themselves from herbivorous animals, and they can also affect humans. Some mushrooms are poisonous, and poison hemlock, poison ivy, poison oak, and poison sumac have not been named as they have for no reason. Herbals containing the plant *Aristolochia*, which is widely consumed in some places, can cause kidney disease and cancer in genetically susceptible individuals. As only 5% of the millions of people taking the compound fall ill, and because of the time lapse between consumption and illness occurring, it was difficult for the linkages to be identified (Grollman and Marcus, 2016). In addition to these potential hazards, the risk of bleeding exists with Ginko, and individuals with diabetes should not be taking Asian Ginseng, and many substances should not be used by people undergoing chemotherapy or those with cardiac problems. Some people who swear by supplements are also adherents of other seemingly "natural ways" of obtaining good health, but fortunately, some of these fads, such as colon cleansing, a dangerous practice that causes bloating, nausea, electrolyte imbalances, and possible kidney failure, is just about at its end.[2]

Many individuals, as we will see in the text box, prefer alternative medicines instead of standard treatments, even if these alternatives are unsound and potentially dangerous. Perhaps to increase the appeal of alternative medicine, purveyors of the treatments, particularly herbalists and naturopaths, were pushed to work side by side with hospital doctors. Complementary and alternative medicines (CAM) involve add-on treatments (e.g., relaxation training, acupuncture, or herbal products) to standard medications. While this might provide these treatments with undue credibility, it has been argued that having both approaches available allows patients to reap the benefits that each might offer, and to decrease the possibility of patients going to untrained herbalists. There are several advantages to this approach, particularly as some nutraceuticals can serve as effective adjunctive treatments with standard antidepressant agents, as in the case of S-adenosylmethionine (SAMe) and to some extent methyl folate and omega-3 (Sarris et al., 2016). Further, having standard and alternative options concurrently available may help build confidence among people who are entrenched in traditional ways of healing.

---

[2] The market for supplements has increased remarkably over the past few decades. In 1994 there were about 4,000 supplements but by 2016 there were 80,000. In 2022 the global market for supplements was estimated to be about 164 billion (USD) and was projected to increase by 9% yearly through to 2030. With increases in the number of these substances, so will the risks associated with their unneeded and inappropriate use.

In most instances, there is scant evidence attesting to the effectiveness of CAM treatments. However, there are cases where they might have positive effects owing to their biochemical actions or because they might be powerful placebos, which can, to be sure, be useful. As Deng Xiaoping said in the context of China's economic development, but applicable here as well, "It doesn't matter if a cat is black or white, as long as it catches mice it's a good cat."

The home for natural products

Pseudoscience abounds just about everywhere, assaulting us from magazines and being passed around through multiple internet sites. Testimonials are offered about this or that natural compound that has amazing healing powers, and several physicians have appeared on television peddling various super remedies. Too often, actors are able to influence large numbers of people with the flakiest products ever imagined (Caldwell, 2015). Health food stores peddle books dealing with homeopathy, alternative healing, integrative medicine, acupuncture, the benefits of weekly chiropractic joint adjustments, cupping, organic farming, and reiki therapy (the latter is based on the belief that by touch, the therapist can channel energy into the patient, thereby activating natural healing processes within the patient's body). The use of Traditional Chinese Medicines (TCMs) had, for a time, gained momentum, possibly because it had an aura that made them right for some people. Acupuncture has been in vogue for several years, even though needles stuck in random places yield the same "positive" effects.

So, why do some people so often resort to alternative medicines and are willing to take the advice of actors or politicians concerning what is healthy and what is not, and disparage the work of physicians and scientists on these same issues? To some extent, it may reflect a loss of faith (trust) in physicians and scientists. The sad fact is that in many instances, standard medicines have had a poor track record for some illnesses. Antidepressant agents are claimed to be only a shade better than a placebo, and opioid-based painkillers obviously have risks attached to them. Many cancer therapies are ineffective and come with distressing side effects. The benefits of some standard heart medications (e.g., beta-blockers) have been questioned, and as we have seen, antibiotics have increasingly encountered resistance by bacteria. Furthermore, iatrogenic illnesses (medical errors that comprise omission or commission) are not uncommon and can have severe consequences. If all of this is not sufficiently discouraging, in some countries, patients may encounter long wait lists before they can see a specialist, further encouraging them to seek alternatives. Given the various roadblocks to effective treatment, the movement toward alternative medicines is understandable, but this should not be misconstrued as being the smart thing to do. Aside from these treatments being ineffective, their use may be accompanied by delays in seeking and initiating potentially effective treatments.

Although alternative treatments must be assessed in scientifically sound clinical trials, at times, the issues concerning alternative treatments simply become silly. How many more trials are necessary to show that particular complementary and alternative medicines do not work (Gorski and Novella, 2014)? Homeopathy, for instance, has been among the most impressive scams going for well more than a century and has repeatedly been shown to be ineffective in treating anything. Nonetheless, trials continue to be undertaken to confirm that the treatment does not work. Most of us know the adage that insanity comprises doing the same thing over and over again and expecting different results.

Although several medications are currently available that are remarkably effective in reducing weight, we should not fool ourselves into thinking that these are an alternative for exercise to stay healthy. Given the many benefits of exercise to heart health and immune functioning, using shortcuts to weight loss may not be overly productive. For individuals who have been sedentary for years, and who have only seen a gym from a distance, the identification of processes by which exercise endurance and fat burning can be increased may facilitate weight loss. This would still require that individuals adopt appropriate lifestyle changes, which to some extent could be influenced by inherited genetic characteristics (Wang et al., 2022f). Despite the importance of diet and exercise in maintaining well-being, several research groups have been evaluating calorie restriction mimetics (CR mimetics) and exercise mimetics that, in pill form or specific foods, can be a substitute for dieting and exercise to promote loss of body fat, increase muscle power, limit elevated blood sugars, and enhance heart functioning. These effects could occur by reducing blood sugars or by enhancing mitochondria so that fatty acids are burned more readily. Perhaps something will come from the use of mimetics, but it will be a while before this is ready for human use.

Exercise

Different forms of exercise can have beneficial health effects. Aerobic exercises (distance running, bicycling, and jogging), which increase free oxygen use to meet energy demands, strengthen the heart and lungs, whereas anaerobic exercise (e.g., weightlifting, jumping rope, and high-intensity interval training) to build muscle strength increases bone strength and enhances metabolism. Failing to exercise can have adverse effects that are still greater if individuals engage in sedentary behaviors (Zhao et al., 2020a). Even if exercise is undertaken religiously, lengthy periods of sedentary behaviors need to be avoided as exercise only partially attenuates the adverse effects imparted (Ekelund et al., 2016). The hazards of sedentary behaviors cannot be overstated, as inactivity has been linked to the 10 primary causes of death, as well as 25 other health conditions (Booth et al., 2017). In a follow-up study of about 12.85 years among 481,688 individuals, predominantly sitting at work without adequate physical activity was associated with elevated cardiovascular disease (34%), and all-cause mortality was increased by 16% (Gao et al., 2024).

The combination of diet and exercise can have potent effects on intestinal integrity as well as microbial diversity, varying in thin versus overweight mice (Campbell et al., 2016). It will hardly be surprising that prospective studies reliably revealed that exercise and healthy eating, together with other lifestyle changes, were predictive of reduced incidents of type 2 diabetes, vascular disease, dementia, and all-cause mortality. Adopting a proper diet coupled with maintaining a reasonable exercise regimen acted against cellular senescence, essentially protecting individuals against diseases associated with aging (Schafer et al., 2016). Beyond these actions, exercise and healthy eating can have epigenetic effects that are manifested as reduced inflammation and enhanced health (Ramos-Lopez et al., 2021).

Dancing for health

Some people enjoy exercise, whereas others find it a chore. For them, a reasonable alternative is to engage in dancing. A meta-analysis that comprised 10 studies indicated that dance undertaken by people who were overweight was effective in reducing fat and body mass (Zhang et al., 2024c). A systematic review indicated that among older people, a variety of dancing styles could enhance balance, muscular strength, endurance, and improved cognitive ability (Hwang and Braun, 2015). Like other forms of exercise, among individuals with type 2 diabetes, a dancing regimen increased the viability of neutrophils and reduced the levels of several proinflammatory cytokines (Borges et al., 2019). As little as 60–150 min of dancing a week could reduce chronic pain perception (Hickman et al., 2022).

The benefits of dance may come about for reasons in addition to the exercise it provides. The social component may foster friendships and connections with other people, which can have physical and psychological benefits. As well, the music to which people dance can have multiple positive effects. Among other things, music can enhance various aspects of immune functioning (Rebecchini, 2021), reduce heart rate and blood pressure, diminish neuroendocrine responses ordinarily elicited by chronic stressors, as well as symptoms of mood disorders (Erkkilä et al., 2011). So, if exercise is a drag, get your dancing shoes on.

Exercise and Immunity

The positive effects of exercise have been realized in relation to many physical illnesses, but many issues still need to be addressed. Like most other treatments, pronounced individual differences exist concerning the benefits obtained through exercise, possibly being related to earlier exercise experiences, diet, and specific genes activated. Thus, tailoring exercise for specific individuals based on genetic signatures could inform the ideal exercise routines that ought to be adopted (Chung et al., 2021).

Animal studies generally indicated that moderate-intensity exercise enhanced immune functioning, and wheel running reversed the diminishing production of new neurons in the hippocampus of aging mice (Littlefield et al., 2015) and limited the functional decline of immune system functioning with advancing age (Simpson et al., 2015). It also appeared that exercise comprising voluntary wheel running in mice provoked the slowing of cancer growth. Simply transferring T cells from exercised to naïve mice enhanced their survival (Rundqvist et al., 2020). These actions might have occurred owing to epinephrine activation of NK cells, CD-8 T cells, and monocyte subtypes, as well as their intratumoral infiltration (Pedersen et al., 2016). Additionally, owing to changes in adipose tissue, moderate exercise regimens in humans diminished excessive levels of inflammatory cytokines, thereby limiting the development of illnesses (see Fig. 5.3).

Exercise can promote temporary damage to muscles and a rapid rise of inflammation within muscles. These actions are countered by the provocation of elevated levels of T_{reg} cells that protect muscle mitochondria from damage stemming from IFNγ (Langston et al., 2023). Engaging in exercise can diminish systemic proinflammatory cytokines and might have positive actions in relation to other conditions that are promoted by myokines (small cytokines or peptides released by muscle cells; myocytes) and may affect brain processes, depending on the specific exercise regimens adopted (Wang et al., 2023a). The myokines released from muscles, essentially acting like a hormone, can influence adipose tissue, bone, liver, gut, pancreas, vascular bed, and the brain, thereby influencing metabolic processes, browning of white fat, endothelial cell function, bone formation, and cognitive functioning (Severinsen and Pedersen, 2020).

Aside from these actions, exercise can reduce circulating cytokine levels, which have positive effects among healthy individuals and in some patients with persistent systemic inflammation. This may arise owing to altered metabolic signals, browning of white fat, regulation of innate immune functioning, and increased antioxidant activity. Supporting the contention that exercise may be useful in diminishing features of autoimmune disorders, moderate exercise was accompanied by reduced CRP and IL-6 and diminished symptoms of rheumatic illnesses

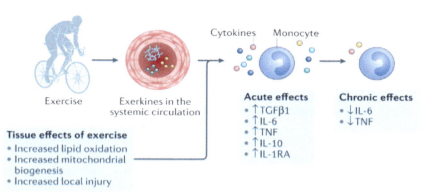

FIG. 5.3 Exercise induces lipid oxidation, mitochondrial biogenesis, and local injury, which stimulates exerkine release into the circulation to influence the immune system. These include proteins (*blue lines*), metabolites (*yellow circles*), and extracellular vesicles (*green circles*), which have a multitude of effects on the immune system (generically represented by a monocyte). Acutely, exercise increases cytokines such as circulating levels of transforming growth factor β1 (TGFβ1) and IL-6 relative to the resting state. This change results in acute inflammation, characterized by increase in tumor necrosis factor (TNF) and IL-6. Once the acute exercise-induced effects have diminished, an increase in antiinflammatory cytokines (such as IL-10 and IL-1 receptor antagonist (IL-1RA)) occurs in response to the acute inflammatory response. Chronic training is associated with a reduction in systemic and tissue inflammation, as characterized by lower circulating levels of TNF and IL-6 in the resting state relative to sedentary individuals. Reduced insulin resistance and tumor growth have been attributed to the effects of chronic training on decreasing systemic and/or tissue inflammation. *From Chow, L.S., Gerszten, R.E., Taylor, J.M., Pedersen, B.K., van Praag, H., et al., 2022. Exerkines in health, resilience and disease. Nat. Rev. Endocrinol. 18, 273–289.*

(Modarresi et al., 2022). Having patients with relapsing–remitting multiple sclerosis engage in high-intensity resistance training likewise reduced proinflammatory cytokine levels and positively influenced fatigue and health-related quality of life (Kierkegaard et al., 2016). A review of randomized controlled trials indicated that some symptoms could be diminished, depending on the form of exercise adopted. Specifically, balance was enhanced by yoga, virtual reality training, and aerobic training, whereas walking, aquatic exercise, virtual reality training, and aerobic training (Hao et al., 2022).

Cognitive and Affective Benefits

Through effects on the microbiome, exercise can engender a persistent enhancement of metabolic and brain functioning. A report that comprised more than 10,100 individuals revealed that moderate and vigorous activity was associated with larger hippocampal volumes as well as greater frontal, parietal, and occipital cortices (Raji et al., 2023). The benefits of exercise may have effects that are evident in children, and some of these actions may be apparent throughout life. Among 10-year-old children who engaged in relatively high levels of physical activity, assessment at the age of 14 revealed greater amygdala volume as well as somewhat greater hippocampal volume relative to less active children (Estévez-López et al., 2023).

Aerobic exercise training and obtaining good sleep may promote multiple cognitive enhancements, including modestly improved attention and information processing speed, executive functioning, and memory (Bloomberg et al., 2023). It is thought that the beneficial effects of exercise on improved cognitive abilities and psychological well-being may stem from enhanced synaptic plasticity, growth of new neurons within the hippocampus, and increased neuronal connectivity (Voss et al., 2015). As well, the altered cognitive functioning associated with exercise may be related to increased platelet activity, notably through the secretion of the chemokine CXCL4/platelet factor 4 (PF4), which may enhance neural plasticity and neural regeneration (Leiter et al., 2023). Furthermore, intense exercise causes the release of lactate from skeletal muscles, which is important for various functions, including neuronal signaling, neuroplasticity, and memory functioning (Xu et al., 2023b).

The reduction of mild or moderate anxiety and depression is among the most pronounced benefits of exercise, and there is evidence that suicide attempts may be reduced. The positive psychological and cognitive effects of exercise may develop through several processes. Most obviously, exercise may diminish distress that otherwise causes impaired mood and cognitive functioning. A program that comprised running with a group (45 min a day over 16 weeks) was as effective as sertraline or escitalopram in diminishing depressive symptoms and had the added benefit of improved weight, blood pressure, and heart rate variability (Verhoeven et al., 2023). Not to diminish the importance of these findings, but it is important to bear in mind that the social support gained from exercising with others could have enhanced the effects of running itself (Haslam et al., 2020).

The impact of exercise on mood states may come about owing to increased growth factors, including

BDNF, IGF-1, VEGF, and GDNF, thereby engendering positive effects on neuronal processes, particularly concerning neurogenesis (Lippi et al., 2020). The gains associated with exercise may also occur by influencing stressor-related neural circuits so that responses to stressors are diminished, possibly by facilitating the activation of the inhibitory neurotransmitter GABA (Schoenfeld et al., 2013).

While most studies concurred with the view that exercise was likely causally linked to brain health, this view has not been unanimous. The possibility has been considered that healthy cognitive functioning can increase engagement in physical activity, rather than exercise, increasing cognitive abilities. However, based on a large-scale GWAS that focused on identifying the directionality of effects, it appeared that regular moderate and vigorous physical activity favored enhanced cognitive functioning, whereas cognitive functioning did not promote altered physical activity (Cheval et al., 2023). Notably, a 30-year prospective study indicated that physical activity during adulthood was accompanied by enhanced cognitive abilities among individuals at 69 years of age (James et al., 2023). To an extent, these effects were linked to early-life experiences, and the possibility cannot be excluded that individuals who exercised regularly also engage in other healthy behaviors that acted against cognitive decline.

Is It the Case of the More, the Better?

The link between exercise and positive health is not linear, and instead comprises an inverted U-shaped (or J-shaped) function, so that the beneficial effects of moderate exercise may be lost in response to excessive exercise. Moreover, adverse effects may develop with intense exercise, including the possibility of ischemic stroke, especially among individuals who have maintained a sedentary lifestyle (Smyth et al., 2022). It was repeatedly demonstrated that as little as 20–30 min of exercise a day could promote multiple health benefits that were accompanied by reduced levels of proinflammatory cytokines, along with the elevation of neutrophils, monocytes, T and NK cell, and their redistribution from storage sites into circulation. The immune cell translocation to the gut, lung, and lymph nodes brought them closer to sites where it was likely that pathogens might be encountered (Simpson et al., 2015). The benefits of moderate exercise were especially important for older individuals, given that immune functioning ordinarily declines with age. Early mortality risk could be reduced by a third through moderate exercise (Matthews et al., 2020). The question has frequently arisen as to how much exercise is ideal to diminish illnesses. There is no question that the amount of exercise varies as a function of numerous

individual differences, even though it has frequently been touted that 7,000–10,000 steps a day is best. A meta-analysis of cardiovascular and all-cause mortality indicated that even 4,000 steps a day could have beneficial effects, and beyond this, every additional 1,000 steps reduced all-cause mortality by 15% (Banach et al., 2023).

Unlike the positive actions of moderate exercise, single bouts of prolonged exercise can transiently disturb multiple aspects of immunity, including disruption of T cell, NK cell, and neutrophil functioning, the balance between pro- and antiinflammatory cytokines, and blunting of the immune responses to specific antigens (Simpson et al., 2015). Furthermore, extreme exercise, often engaged by high-performance athletes, can impair NK cell functioning, and may reduce immunoglobulin within mucosal secretory glands (nose and salivary), thus allowing infectious molecules the opportunity to invade.

These findings promoted the "Open window hypothesis," which asserted that extreme exercise provoked immune disturbances that rendered individuals more susceptible to viral infection. This position was vigorously challenged (Campbell and Turner, 2018) since the immune disturbances that occurred were relatively brief, making it unlikely that individuals would be infected by opportunistic viruses. Indeed, extreme exercise among high-performing athletes resulted in fewer respiratory infections than it did in less well-trained athletes (Walsh and Oliver, 2016). Moreover, although high-intensity exercise in noncompetitive settings was tied to diminished immune cell proliferation, in a competitive situation, cell proliferation was primed, leading to enhanced immune functioning (Siedlik et al., 2016). This difference may have to do with psychological and hormonal processes that accompany the arousal in competitive situations, thereby overriding the effects of exhaustive exercise.

As interesting as the data on elite athletes might be, the findings might not be particularly relevant to less fit people. Moreover, for most people, it would be more instructive to consider the impact of sustained exercise programs. Protracted exercise regimens maintained for 12–15 months by young, healthy individuals were accompanied by diminished inflammation, and a yearlong combined endurance and resistance training regimen, similarly reduced inflammation among obese postmenopausal women (Campbell et al., 2009). Exercise regimens were likewise associated with cytokine reductions among elderly individuals (Zhao et al., 2022).

As expected, when a 15-week exercise regimen was accompanied by a calorie-restricted diet the reductions of inflammatory cytokines IL-6, IL-8, and TNF-α levels in adipose tissue were especially pronounced. As well, combined exercise and healthy dietary practices were accompanied by altered expression of genes coding for adipokines (cytokines released by fat tissue) together with a decline of inflammatory markers within subcutaneous adipose tissue (e.g., Campbell et al., 2017). When statistically controlling for dietary fiber intake change, the reduced CRP associated with exercise was attenuated, indicating that the effects of exercise were determined by other lifestyles individuals had adopted.

Old age ain't no place for sissies
Bette Davis

It is unfortunate that age is ordinarily accompanied by the deterioration of various biological systems, hence favoring the development of multiple diseases. Among other things, genetic stability is under threat by exogenous and endogenous factors, cumulative epigenetic changes can undermine well-being, protein homeostasis (proteostasis) may be undermined, thereby affecting disease vulnerability, mitochondrial functioning deteriorates, and the protective capacity of the immune system declines (immunosenescence) (López-Otín et al., 2023). These actions may be exacerbated by stressors that are frequently encountered among older individuals.

Aging is associated with low numbers of naïve T cells, disturbed proliferative responses to antigen challenges, and a poor ratio between CD4 and CD8 cells. With age, inflammatory factors may be more damaging, and poor vaccine responses are more common. It is of particular significance, certainly to older people, that exercise can enhance immune functioning, and among those who had exercised regularly, morbidity and mortality associated with inflammatory or immunologically related illnesses were diminished (e.g., Turner, 2016).

While common diseases associated with aging (e.g., heart disease, cancers) were studied extensively, limited attention focused on other frequent conditions. Sarcopenia, an age-related loss of skeletal muscle mass and strength, has been recognized as a serious condition in older individuals, especially in the presence of obesity. This condition arises owing to the breakdown of muscle protein, possibly owing to chronic low-grade inflammation (Zhao et al., 2023a). It is worrisome that in recent decades the frequency of sarcopenia has been increasing, affecting about 10%–16% of individuals who are 60 or older. This condition was associated with stressful experiences, the occurrence of depression, and a proinflammatory diet (Xie et al., 2023a), whereas exercise together with diet can offer a degree of symptom alleviation (Colleluori and Villareal, 2021).

Beyond physical illnesses, aging comes with numerous psychological challenges. Social activities and contacts tend to decline owing to mobility problems and the presence of illnesses. Social support systems deteriorate owing to the loss of friends who predeceased them or moved away, and their family members may have dispersed to take opportunities elsewhere. With their social network

disintegrating, loneliness may set in, promoting depression and reducing health and life expectancy (Malhotra et al., 2021). To make things still more distressing, older individuals may experience a loss of control and independence and must rely on others. They may also experience stigmatization and unsupportive social interactions, frequently being patronized, dismissed, and made to feel invisible or even a burden. Worse still, elder abuse in diverse forms is experienced by some older individuals with chronic illnesses and is prominent among those with neurological disorders (Wong et al., 2022). It is hardly a wonder that the frequency of depression and drug treatments to deal with depression and sleep disturbances in older people have been skyrocketing in recent years.

Exercise and Microbiota

Studies in both animals and humans repeatedly indicated that exercise can profoundly influence gut microbiota abundance and diversity. The influence of exercise on gut microbiota varies with age. Among juvenile rats, exercise led to changes in several gut phyla, including increased *Bacteroidetes* and decreased *Firmicutes*, and produced microbial profiles associated with adaptive metabolic changes (Cerdá et al., 2016). When exercise was initiated during early life, microbial processes were especially enhanced, leading to metabolic and brain changes that persisted throughout life (Mika and Fleshner, 2016). It also appeared that in pregnant rodents, exercise could have long-term repercussions on gut microbial diversity and the abundance of numerous microbial taxa in her offspring, although this was less notable if the dam had been maintained on an obesogenic diet (Bhagavata Srinivasan et al., 2018).

Gut immune cell homeostasis and microbiota-immune interactions engendered through exercise can limit illnesses otherwise related to microbiota dysbiosis. Indeed, moderate exercise may produce microbial changes that protect against mental health and cognitive disturbances, obesity, metabolic syndrome, several gut-related diseases (e.g., inflammatory bowel disease), as well as the occurrence of hepatocellular and colorectal cancer (Mailing et al., 2019). Among germ-free mice, the responses were observed when recolonization was undertaken using microbiota from donor mice that had exercised. This treatment also limited the development of chemically induced colitis, possibly through actions on cytokines (Allen et al., 2017).

Like the immune changes associated with exercise, an inverted J-shaped relationship exists between exercise intensity and gut microbiota alterations. Modest exercise can attenuate gut dysbiosis among individuals with obesity (Aragón-Vela et al., 2021) and may limit some illnesses associated with gut microbiota disturbances (e.g., irritable bowel syndrome and inflammatory bowel disorder). In contrast, excessive exercise may promote gut microbial disturbances (Bonomini-Gnutzmann et al., 2022), which may lead to intestinal cell damage and leaky gut syndrome (Costa et al., 2017). Cross-talk occurs between microbiota and mitochondria, so the production of ROS by extreme exercise may elicit changes in microbiota, which may be moderated by genetic influences (Clark and Mach, 2017).

Predictably, the impact of exercise may interact with diet, causing a shift of microbial species otherwise promoted by a high-fat diet in rodents. In obese rats, exercise that led to weight loss comparable to that associated with calorie restriction, modified microbiota, reduced inflammation as well as insulin resistance, fat oxidation, and altered brown adipose tissue (Welly et al., 2016). In humans, moderate exercise was associated with elevated levels of several butyrate-producing microbes. Moreover, among individuals with a sedentary lifestyle, a program of exercise that became progressively more intense over weeks, microbiota and SCFA composition were altered, particularly among lean individuals, and then returned to their earlier state when they again adopted their sedentary lifestyles (Allen et al., 2018). Overall, it appeared that exercise reduced that abundance of potentially harmful bacteria, while increasing beneficial bacteria and their metabolites, although it is still necessary to determine which taxa are most important for the health benefits produced by exercise (Cullen et al., 2023).

Exercise enthusiasts frequently report a "runners high" while engaging in their sport. The rewarding effects of running keep individuals engaged in this activity, but what is it that makes exercise so rewarding? It is believed that activation of dopamine functioning within limbic brain regions may contribute to this. Studies in mice have revealed that exercise may engender microbiota-related production of gut endocannabinoid metabolites, which increase ventral striatum dopamine levels, thereby enhancing the rewarding effects of exercise. Supporting this contention, manipulations that produce microbiota depletion or inhibition of endocannabinoid receptors diminished the motivation to exercise (Dohnalová et al., 2022).

The data have strongly indicated that exercise produces many health benefits, some of which occur by actions on microbiota, immune functioning, and reduced inflammation. The problem is that people too often are not interested in participating in exercise even though they know it has benefits. They find all sorts of reasons (excuses) for not exercising, none of which seem impressive or convincing. Altering an individual's lifestyle can be exceedingly difficult, and there is likely no approach that is suitable for everyone. Still, exercise engagement can be facilitated when self-efficacy can be enhanced, individuals receive appropriate social support that reinforces exercise, and when they identify with an exercise culture (Haslam et al., 2020).

Sleep

Sleep is a basic need that is essential for numerous vital functions. It is necessary for development, serves in a reparative capacity and for energy conservation, effective immune functioning, and removal of brain waste that accumulates over the day. These processes are disturbed with sleep loss, so that physical, psychological, and neurological functioning may be impaired.

Neurobiological Factors Associated With Sleep

In considering the benefits of sleep and the adverse effects of sleep loss, it is important to consider that sleep and circadian cycles are tightly entwined, yet to some degree are independent of one another. The timing of sleep is largely controlled by internal clocks regulated by the suprachiasmatic nucleus located within the hypothalamus, but internal clocks are also present in other brain regions (Hood and Amir, 2017). Disturbances of circadian clock functioning have been associated with impaired hormonal activity and antioxidant production and have been tied to numerous illnesses, including neurodegenerative conditions, such as Alzheimer's and Parkinson's disease. Also, hormones associated with eating processes, including insulin and ghrelin, are tied to circadian clock variations, and their dysregulation was associated with metabolic diseases (Stenvers et al., 2019).

Several clock genes, notably periodic genes Per1 and Per2 and cryptochrome genes Cry1 and Cry2, have been linked to circadian rhythmicity, as have cell cycle genes c-Myc, Wee1, cyclin D, and p21 (Rijo-Ferreira and Takahashi, 2019). Activation of some clock genes, which vary over the circadian cycle, has been linked to changes in hormonal levels and synaptic functioning. Disturbances of sleep and circadian cycle can affect the functioning of these genes, thereby affecting hormonal regulation (Noya et al., 2019). Clock genes have most often been related to neuronal processes, although astrocytes and brain cytokine variations contribute to clock regulation (Brancaccio et al., 2019). Disturbances of the network of clock genes that regulate hormonal processes can be influenced by extraneous factors (e.g., changes in work schedule) and epigenetic processes, thereby affecting age-related diseases.

In addition to varying with circadian cycles, sleep itself is accompanied by neuroendocrine changes, including variations of cortisol and epinephrine, as well as melatonin, prolactin, and growth hormone (Besedovsky et al., 2016). Stressful experiences may upset the operation of clock proteins, and when individuals are sleep-deprived for an extended time, neuronal functioning in cortical brain regions diminishes, essentially overriding the brain's master clock (Muto et al., 2016). The functioning of cortical brain regions was markedly affected by sleep loss, whereas subcortical activity was primarily linked to the circadian cycle (Muto et al., 2016). Thus, while circadian disruptions may contribute to disturbed well-being, sleep loss can also be part of a symptom cluster linked to mood disorders and immune-related disturbances.

Brain activity declines during sleep, and the number of synapses present may diminish, making recently strengthened synapses relatively more prominent, thereby enhancing memory and performance on the ensuing day. Conversely, lack of sleep may give rise to hippocampal atrophy, reduced dendritic length and spine density within the hippocampus, and reduced connectivity (Havekes et al., 2016). As a result, individuals may experience cognitive impairments, including disturbances of attention, learning, memory consolidation, executive functioning, decision-making, stressor appraisals, and emitting appropriate coping responses (e.g., Krause et al., 2017). Neuronal disturbances within the prefrontal cortex that stem from disturbed sleep may also affect appraisals so that rewards are overvalued, whereas losses are undervalued. This, in turn, can influence risky decision-making and impulsivity (e.g., in the context of excessive gambling and relapse in other addictions).

It had long been thought that sleep was regulated by serotonin produced in the anterior raphe nucleus, which sends projections to the preoptic portion of the anterior hypothalamus. In addition, dopaminergic neurons of the ventral tegmental region, which have been linked to heightened arousal and reward processes, appear to be a strong modulator of sleep (Eban-Rothschild et al., 2016), as are GABA and glutamate (Yu et al., 2019). Additionally, sleep homeostasis is influenced by forebrain cholinergic functioning, GABA influences ventral tegmental neuronal activity, galanin neuronal activity within the hypothalamus, and astroglial functioning (Franks and Wisden, 2021). Critically, sleep may be fundamental for neural and behavioral plasticity, which can have cognitive ramifications (e.g., Weiss and Donlea, 2022). The production of adenosine, which contributes to energy regulation, is also a major player in sleep processes (Serin and Acar Tek, 2019). Produced from adenosine triphosphate (ATP), adenosine accumulates as neuronal and glial activity increases. With adenosine receptor stimulation, sleep is provoked, and glycogen energy stores recover. In essence, multiple components distributed throughout the brain operate to regulate sleep and promote diverse restorative functions.

The switch between sleep and nonsleep states, and the sleep period itself, are governed by about 20 neuropeptides and neurotransmitters, many of which are controlled by and correspondingly influence circadian cycles (Richter et al., 2014). As well, movements of gut bacteria over a day may contribute to daily rhythms (Thaiss et al., 2016a), and gut microbes themselves are subject to diurnal rhythms, which can affect sleep and, together with eating patterns, can have health-related consequences.

Sleep, recuperation, and illness

Over the course of the day, essential biological resources may be used up, and sleep facilitates their recovery. During sleep most biological systems are in an anabolic state, which essentially means that skeletal and muscular systems are being rejuvenated, and both hormones and neurotransmitters are being replenished. Also, with the many events experienced on any given day, the number and strength of synaptic connections increases, which entails the use of considerable energy, and restoration of these processes is necessary.

Sleep is essential for the repair of damaged tissues and to diminish accumulating DNA damage within neurons (Zada et al., 2019), as well as for the clearance of waste that had accumulated over the day. Specifically, toxins within the brain are drained into lymph nodes through channels that make up the glymphatic system (Hablitz et al., 2020). Also, owing to shrinkage of glial cells during sleep, openings between cells become larger, allowing more rapid flow of cerebrospinal fluid within the brain and spinal cord, so that toxins can be removed more readily (Cuddapah et al., 2019). These sleep-related processes are particularly important for individuals who have experienced traumatic brain injuries in which the accumulation of waste ordinarily increases.

Lack of sleep could have important ramifications in the provocation of neurodegenerative disorders related to β-amyloid accumulation (Braun and Iliff, 2020). A 25-year follow-up study also revealed that dementia was preceded by earlier short sleep durations (Sabia et al., 2021). Sleep disorders were associated with the cognitive decline in Alzheimer's disease and correlated with cortical β-amyloid presence, as well as phosphorylated tau measured in cerebrospinal fluid (Liguori et al., 2017), and sleep disruption may result in impaired hippocampus-dependent memory consolidation. As well, neurological disorders have been associated with altered neuronal connectivity, which has been linked to sleep disturbances (Kaufmann et al., 2016).

Aside from disturbed cognitive processes, an analysis that included 15 prospective cohort studies indicated that insomnia and the presence of nonrestorative sleep were linked to increased risk for heart attack and stroke, being somewhat more pronounced among women relative to men (He et al., 2017). A meta-analysis of 153 studies, which evaluated more than 5 million participants, similarly indicated that short sleep was associated with increased occurrence of type 2 diabetes, hypertension, coronary heart disease, stroke, obesity, and earlier death (Itani et al., 2017).

Disturbed sleep was associated with poor lifestyle choices, having been linked to increased food intake, possibly owing to the microbiota changes provoked (Hanson et al., 2020) or because of disturbed biological clocks that regulate the release of eating-related hormones (leptin and ghrelin). Increased eating may also occur because food odors are more enticing following sleep loss, or because the ability to deal with temptation is diminished. The link between sleep and eating was reinforced by reports that individuals who slept poorly were more likely to make poor food choices, perhaps stemming from altered neuronal activity within the frontal and insular cortex (Greer et al., 2013). Importantly, the elevated occurrence of cardiovascular disease attributable to either abbreviated or excessive sleep was attenuated among individuals who engaged in about 150 min of moderate or intense exercise a week (Liang et al., 2023), possibly owing to the antiinflammatory effects of a consistent exercise regimen.

Sleep, Emotional Regulation, and Mental Health

It is not just babies that are fussy and grumpy when they do not get enough sleep. Among adults, lack of sleep is associated with poor emotional regulation and emotional intelligence, elevated behavioral reactivity in response to a stressor, amplification of negative appraisals and emotions, elevated impulsivity, as well as increased cortisol levels and amygdala neuronal activity (Thompson et al., 2022). Poor sleep quality among depressed individuals has also been linked to elevated neuronal activity within the dorsal anterior cingulate cortex, accompanied by altered emotion regulatory brain functioning (Klumpp et al., 2017). In essence, with a lack of sleep the anterior cingulate cortex needs to work that much harder to make sense of situations.

Lack of sleep, particularly rapid eye movement (REM) sleep, may have psychosocial and mental health consequences or may be a marker for later mental illness. The impact of strong stressors on mood states may be mediated by sleep alterations and REM sleep disturbances, and sleep loss that follows from stressors may act on hormonal and neurotransmitter processes, as well as synaptic plasticity, thereby provoking psychological dysfunctions (McEwen and Karatsoreos, 2015). Furthermore, among individuals vulnerable to insomnia, the locus coeruleus, the site of norepinephrine cell bodies, is more sensitive to input from the salience network, which can instigate hyperarousal that undermines restful sleep (Van Someren, 2021).

Genetic analyses suggested that sleep and insomnia may comprise multiple complex mechanisms. A genome-wide analysis based on a cohort of more than 113,000 participants identified seven key genes linked to insomnia, as well as sleep disorders, such as Restless Legs Syndrome. The contribution of these genes to insomnia differed between males and females, each having its own predictors, and the genetic correlates of insomnia were also associated with metabolic characteristics and psychiatric disturbances (Hammerschlag et al., 2017).

Impact of Sleep and Sleep Loss on Immune Processes

Normal sleep is accompanied by a decline in T cell functioning as if these cells need to rest and rejuvenate (Besedovsky et al., 2019). In the absence of restorative sleep, the inflammatory marker CRP was elevated, and when individuals were forced to stay awake, their T cell functioning remained high (Besedovsky et al., 2016), and the capacity of the immune system to deal with antigenic challenges may be affected. These actions may be moderated by peripheral norepinephrine and circulating cortisol, which are also subject to circadian cycles.

As observed with several hormonal variations, some immunological changes that occur with sleep are related to circadian factors, whereas others are primarily a reflection of sleep's restorative powers (Besedovsky et al., 2019). For instance, the numbers of leukocytes, granulocytes, and monocytes, as well as the major lymphocyte subsets, including T-helper cells, cytotoxic T cells, activated T cells, and B cells, ordinarily attain their peak in the evening or early portion of the night, and then decline progressively toward the morning, essentially showing the inverse pattern of cortisol. These immune variations appear to be tied to a circadian rhythm, rather than sleep itself. As in the case of adaptive immune responses, innate immunity is altered by nocturnal sleep. Ordinarily, NK cell activity increases over the course of a night's sleep but is diminished among individuals with sleep disturbances (Irwin and Opp, 2017).

Nocturnal sleep ordinarily favors effective inflammatory immune signaling, whereas the absence of sleep is accompanied by elevated levels of inflammatory cytokines (Piber et al., 2022). The production of IL-2, IFNγ, and augmented IL-12 production by dendritic cells and monocytes are affected by nocturnal sleep, rather than circadian factors, whereas the proinflammatory cytokine TNF-α was promoted by circadian influences, and the production of IL-6 was affected by both circadian factors and sleep itself (Irwin, 2019). Just as sleep loss could affect immune functioning, cytokine alterations, particularly variations of IFN-α, can affect sleep. In rodents, administration of proinflammatory cytokines increased NREM sleep, whereas inflammatory cytokine antagonists and antiinflammatory cytokines elicited the opposite effect (Irwin, 2019).

Unlike a night or two of fragmented sleep, which may have limited effects on circulating inflammatory factors, when sleep is entirely prevented, more substantial cytokine variations are instigated owing to alterations of NF-κB, a key transcription control pathway in the inflammatory signaling cascade (Irwin, 2015). The elevated levels of CRP and the proinflammatory cytokines IL-1β, IL-6, and IL-17 associated with partial sleep deprivation over successive nights may normalize after a night of proper sleep (van Leeuwen et al., 2009), and cognitive disturbances may diminish. It was also reported that when sleep is disturbed, responses to stressors may be inefficient, and allostatic overload is more apt to develop. By virtue of the endocrine and immune changes imparted, sleep disorders (which are distressing in themselves) have been implicated in a range of illnesses involving inflammatory processes (see Fig. 5.4).

During some illnesses, particularly those that involve infection, the release of IL-1β and TNF-α promotes or modulates sleep, likely reflecting an adaptive response to facilitate recuperation (Irwin, 2015). In this regard, as proinflammatory cytokines favor the development of fatigue, if seemingly healthy individuals become prone to fatigue and display extended sleep periods, it may well portend something being physically amiss and may predict the occurrence of heart problems (Zhang and Qin, 2023). In this sense, elevations of inflammatory factors might be an illness biomarker, and like stressors, inflammation is linked to a wide variety of illnesses as described in Fig. 5.5.

As vaccines comprise weakened or dead viral particles, they generate an immune response, thus it is possible to determine whether factors, such as stressors or sleep loss (or the presence of a psychological illness), affect the response to a particular challenge. For instance, short sleep duration on two nights prior to administration of the trivalent flu vaccine resulted in reduced antibody titers measured up to 4 months later (Prather et al., 2021). A meta-analysis revealed that insufficient sleep (<6h) was associated with lower antibody titers to vaccination, more so in men than in women (Spiegel et al., 2023). As in the case of influenza vaccination, a full night of sleep loss provoked a persistent reduction in the response to hepatitis B vaccination (Irwin, 2019). There have been a slew of studies showing that sleep loss frequently occurred during the COVID-19 pandemic (Jahrami et al., 2022), but it is uncertain whether sleep loss influenced the effectiveness of SARS-CoV-2 vaccination.

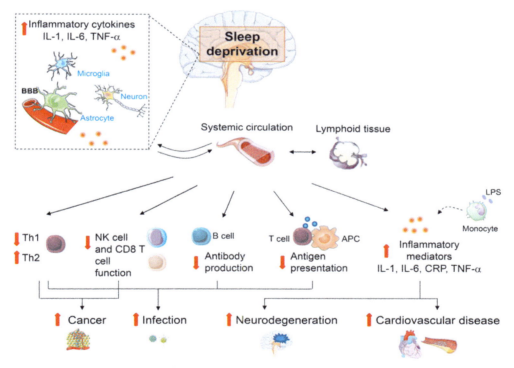

FIG. 5.4 Sleep. deprivation, as induced experimentally or in the context of habitual short sleep, has been found to be associated with alterations in the circulating numbers and/or activity of total leukocytes and specific cell subsets, elevation of systemic and tissue (e.g., brain) proinflammatory markers including cytokines (e.g., interleukins [IL], tumor necrosis factor [TNF]-α), chemokines and acute phase proteins (such as C-reactive protein [CRP]), altered antigen presentation (reduced dendritic cells, altered pattern of activating cytokines, etc.), lowered Th1 response, higher Th2 response, and reduced antibody production. Furthermore, altered monocyte responsiveness to immunological challenges such as lipopolysaccharide (LPS) may contribute to sleep deprivation-associated immune modulation. Hypothesized links between immune dysregulation by sleep deprivation and the risk for immune-related diseases, such as infectious, cardiovascular, metabolic, neurodegenerative, and neoplastic diseases, are shown. Sleep deprivation also favors the development of anxiety and depression (not shown in the figure). The illustrations were modified from Servier Medical Art (http://smart.servier.com/), licensed under a Creative Common Attribution 3.0 Generic License. *APC*, antigen-presenting cells. *From Garbarino, S., Lanteri, P., Bragazzi, N.L., Magnavita, N., Scoditti, E., 2021. Role of sleep deprivation in immune-related disease risk and outcomes. Commun. Biol. 4, 1304.*

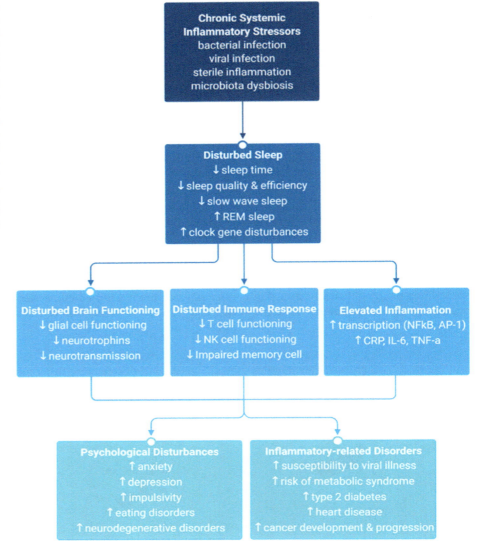

FIG. 5.5 Chronic psychological and systemic challenges promote varied disturbances related to sleep. Sleep disturbances directly increase circulation inflammatory factors, disturb immune functioning, and can exacerbate the actions induced by chronic stressor experiences. These actions may disrupt brain functioning, thereby promoting psychological disorders. The multiple immune- and inflammatory variations stemming from disturbed sleep-related disorders likewise favor the development of several physical illnesses. *From Anisman, H., Kusnecov, A.W., 2022. Cancer: How Lifestyles May Impact Disease Development, Progression, and Treatment. Academic Press, London, UK, based on Irwin, M.R., 2019. Sleep and inflammation: partners in sickness and in health. Nat. Rev. Immunol. 19, 702–715.*

Implications of circadian rhythm for timing of therapies

The fact that immune functioning varies with the circadian cycle has interesting implications for both vulnerability to illnesses and their treatment (Ruben et al., 2019). Specifically, the impact of viral, bacterial, and parasitic infections may vary with the time of day at which these infections first occurred. For instance, when mice were infected with the herpes virus during their resting period, viral replication was an order of magnitude greater than when the viral inoculation occurred just before their active phase. However, in mice in which a gene controlling circadian cycle (Mmal1) was knocked out, viral replication was elevated irrespective of the time of day (Edgar et al., 2016).

It might be possible to exploit what is known about clock genes and daily variations of immune functioning in the promotion of more efficacious treatment strategies for illnesses. Just as illness severity can vary with timing of viral infection, the time of treatments can affect illness outcomes (Ruben et al., 2019). Among older individuals, morning administration of influenza vaccination produced a greater antibody response than did afternoon vaccination (Long et al., 2016), and BCG vaccination against tuberculosis was more effective when administered in the morning (de Bree et al., 2020). Limited evidence is currently available concerning the best time at which SARS-CoV-2 vaccines should be administered, although it is likely that morning vaccination might provide optimal effects.

Circadian rhythm is dramatically altered in critically ill patients, such as those with cancer, which may be related to dysfunction of immune or inflammatory responses. It has been suggested that the clock gene, Per2, acts as a tumor suppressor, so disturbances of this gene's functioning allow uncontrolled cell proliferation, genomic instability, and tumor-promoting inflammation. The idea has also been advanced that epigenetic processes related to clock genes may contribute to cancer development and progression (Masri et al., 2015). Administering treatments in synchrony with an individual's rhythms (chronotherapy) could promote optimal outcomes. This could be achieved by having cancer treatments delivered using programmable-in-time pumps adjusted to an individual's circadian rhythms (Innominato et al., 2014). Certain clock genes may also serve as biomarkers in predicting the benefits of some procedures (e.g., neoadjuvant chemoradiation therapy) in the treatment of some forms of cancer (Zhou et al., 2021a).

Sleep, Diurnal Variations in Relation to Microbiota and Immunity

Given the connections between sleep, microbiota, and immunity, it might be expected that sleep disorders would influence gut microbiota, and conversely, gut dysbiosis might influence sleep processes (Nobs et al., 2019). Jobs that entail shift work, and hence changes in circadian cycles, were associated with disturbed gut microbiota and illnesses, such as gastrointestinal disorders, obesity, disturbed glucose metabolism and metabolic syndrome, and cardiovascular diseases.

Through their derivatives (e.g., butyrate, polyphenols, and vitamins), microbiota can influence host metabolic homeostasis and may be involved in sleep processes. While the diversity and richness of *Bacteroidetes* and *Firmicutes* were directly related to IL-6 as well as to sleep efficiency and sleep duration, other taxa, such as *Lachnospiraceae*, *Corynebacterium*, and *Blautia*, were linked to sleep disturbances. Also, NREM sleep can be appreciably increased by the administration of tributyrin, a form of butyrate (Szentirmai et al., 2019).

Like body cells, gut microbiota have daily rhythms, and multiple environmental factors that affect these diurnal variations can affect physical and psychological well-being (Murakami and Tognini, 2019). When the migration of immune cells out of the gut is disturbed, which is controlled by brain circadian clock processes, the ability to eliminate gut pathogens is impaired (Godinho-Silva et al., 2019). Moreover, manipulations that disrupt circadian rhythms of microbiota can instigate

epigenetic changes that promote a proinflammatory state that has downstream health consequences (Kuang et al., 2019).

Climate Change Affects Physical and Mental Health

Climate change has been attributed to human behaviors, and lifestyles to which we have become accustomed may contribute to this. The "One Health" perspective asserts that intersections between humans, animals, plants, and their shared environment are fundamental to human well-being. The misuse of natural habitats, degradation of ecosystems, the presence of environmental contamination, inappropriate use of certain drugs (e.g., antibiotics), and threats to food and water safety ultimately favor bacterial and viral processes, the appearance of vector-borne infections, and the development of noncommunicable diseases (Mackenzie and Jeggo, 2019). Evaluation of the risks of climate change from a One Health vantage necessarily involves the concurrent or sequential adoption of multiple approaches taken from diverse scientific fields. Limiting climate change will require international government and corporate cooperation, and each individual needs to do their bit. Unfortunately, a 2019 report from the American Psychological Association indicated that although 56% of individuals indicated that climate change was the most important issue facing society, just 40% of individuals adopted behaviors to limit climate change (e.g., often feeling that their actions would have an insignificant impact), and about 30% stated that they had no intention to alter their behaviors.

We have repeatedly seen an increase in the occurrence of hurricanes, forest fires, drought, flooding, and other natural disasters. Severe events considered to be "once in a hundred-year events" are occurring regularly and often. Countless helpless animals have been victimized by these events, and, of course, the ensuing poverty and food insecurity resulted in forced migration among large swaths of people. With climate change, insects have been moving northward, creating greater risks of vector-borne diseases (Lyme disease, Dengue, Zika virus, and Chikungunya virus) as well as zoonotic diseases (e.g., avian influenza, emerging coronaviruses). The warmer, wetter climate will also facilitate the occurrence of various fungal diseases, and with rising sea levels, altered precipitation patterns, and altered temperature of coastal waters, it is likely that the occurrence of water-borne diseases will increase (Phillips et al., 2024). Additionally, agricultural use of pesticides has been fouling aquatic systems, and many foods that are deemed to be healthy may be contaminated, which could affect human health (Rempelos et al., 2021). With further climate disasters, it can be expected that migration of people will increase from rural farming areas to cities, and climate migration will likely appear between countries, which could have psychological and political repercussions.

The WHO has estimated that on a yearly basis, approximately 7 million premature deaths are attributable to indoor and outdoor pollution, although the number may be appreciably greater (Vohra et al., 2021). Aside from visible pollutants, fine particulate matter (smaller than 2.5 µm in width: $PM_{2.5}$) and nitrogen dioxide have been implicated in numerous diseases and diminished lifespan and healthspan (Stafoggia et al., 2022).

Every cell and every organ in the human body can be affected by fine particulate matter, and air pollutants promote oxidative stress, impaired metabolism, pulmonary and systemic inflammation, and impairment of the autonomic nervous system, thereby fostering cardiovascular illnesses, stroke, and chronic obstructive pulmonary disease. Longitudinal studies indicated that fine particulate matter was linked to lung cancer, colorectal, and prostate cancer (Wei et al., 2023), and increased air pollution appears to contribute to the development of antibiotic resistance to bacterial diseases (Zhou et al., 2023d). Moreover, pollutants and excessive heat may contribute to the occurrence of stillbirths, as well as preterm births and low birth weights, which may have further health consequences (Bekkar et al., 2020). Ambient particulate matter experienced during early postnatal development was likewise associated with mitochondrial dysfunction and inflammation stemming from epigenetic changes that were introduced (Isaevska et al., 2021). It is of particular significance that environmental toxicants and social adversity may synergistically affect inflammatory processes and the production of chronic diseases (Olvera Alvarez et al., 2018).

Aside from the physical illnesses provoked by pollutants, climate change has been associated with poor mental health (anxiety, depression) and neurological disturbances (American Psychological Association, 2017). Based on a detailed analysis that included nearly 2 million US residents, it was suggested that a 1°C increase of warming over 5 years was accompanied by a 2% rise in the prevalence of mental health problems (Obradovich et al., 2018). Numerous studies have similarly indicated that depression secondary to pollutants is particularly notable among older people (Wang et al., 2020), and chronic exposure to pollutants can lead to depression among pregnant women, which can result in further complications (Kanner et al., 2021) as well as postpartum depression (Niedzwiecki et al., 2020). It is alarming that in utero exposure to fine particulate matter may affect the later development of psychological disorders in offspring owing to alterations of neuroimmune processes interacting with genetic influences (Li et al., 2021c).

On the other side of the equation, the availability of greenspace has been associated with decreased early mortality risk, even after considering the contribution of socioeconomic status, access to health care, and other social services (Crouse et al., 2017). Also, increased green space was associated with diminished depression among middle-aged and older individuals, and having adequate green space while growing up was associated with diminished occurrence of a wide assortment of psychiatric disturbances, even after adjusting for parental age, history of mental illness, and socioeconomic status (Engemann et al., 2019). Epigenetic aging was associated with the availability of green space, which was related to marked disparities related to socioeconomic class and racial inequalities (Kim et al., 2023c).

While considerable data have pointed to anxiety and depression related to pollutants being associated with neurobiological changes, there is more to the emergence of these mood changes. Young people (16–25 years of age) have frequently reported feeling sad, anxious, and angry owing to the expected consequences of climate change. These feelings were exacerbated by individuals' sense of being powerless, helpless, and guilty of being unable to do anything about climate change. People frequently expressed feeling betrayed by government inactions to thwart the impending disaster (Marks et al., 2021). The "eco-anxiety" (chronic fear of environmental doom) has resulted in some young people reconsidering their plans concerning whether they want to have children. For some groups, such as Indigenous People in Canada, who have been witnessing harm to the land to which they are intimately tied, it has resulted in "ecological grief" (Cunsolo and Ellis, 2018).

The Lancet Countdown commissioned by the Lancet, has provided detailed recommendations to prevent the health consequences of climate change (Watts et al., 2021). A similar Lancet report in 2009 had raised many of the same issues, but the problems have persisted. Perhaps it will be different this time. If not, we will be facing very scary consequences (see Wallace-Wells, 2019). Fig. 5.6 depicts some of the many health consequences that may be elicited by various aspects of climate change, some of which are secondary to immune and inflammatory changes that are provoked (Corvalan et al., 2022).

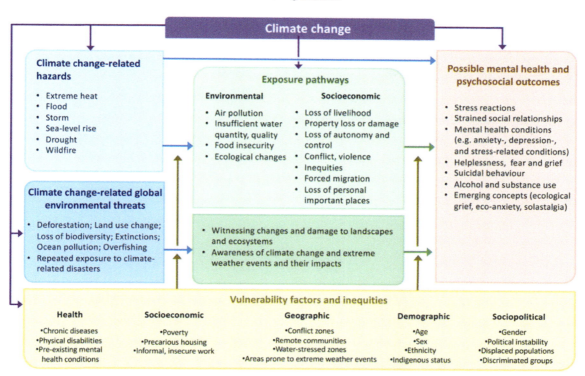

FIG. 5.6 Main interlinkages between climate change and mental health. *From Corvalan, C., Gray, B., Villalobos Prats, E., Sena, A., Hanna, F., Campbell-Lendrum, D., 2022. Mental health and the global climate crisis. Epidemiol. Psychiatr. Sci. 31, e86.*

Concluding Comments

To a significant extent, physical and mental health is dictated by genetic factors, together with stressful experiences and all manner of early-life challenges. Further, what we eat, whether we engage in exercise, and whether we obtain recuperative sleep can interact with stressors and genetic influences in determining well-being. Further, environmental factors can have epigenetic actions that influence well-being and can directly or indirectly affect the microbiome, which can act on immune processes as well as central nervous system functioning, thereby influencing both physical and psychological well-being and the emergence of illnesses. Considering such findings, it is certainly appropriate to focus on the development of treatment strategies that incorporate multiple dimensions, including genetic, biological, and lifestyle factors.

An implicit message that we intended in this chapter is that although there may be some diseases that develop owing to factors outside of our control, many are "preventable illnesses." In the world of retail sales, a common mantra for success has been "location, location, location," and the mantra for health and well-being ought to be "prevention, prevention, prevention." It has been a century and a half since Metchnikoff offered the view that gut-related processes could affect diseases, and it has been more than two centuries since Pott reported that cancer could come about as a result of environmental factors, having linked soot exposure to scrotal cancer among chimney sweeps, which could be prevented by wearing appropriate attire. These and subsequent demonstrations encouraged the development of fields such as epidemiology and preventive medicine. The fundamental importance of these fields is that they contribute to identifying factors linked to disease provocation, which may influence the development of strategies to prevent these illnesses from occurring. With multiple demonstrations pointing to the importance of disease prevention (e.g., through vaccinations or by having people engage in healthier lifestyles), it became certain that health approaches that simply focused on cures for illness were insufficient, and thus, preventive programs evolved still further. In hindsight, it is remarkable that these efforts were slow to be accepted, and, in some respects, they are still in the process of emerging.

CHAPTER

6

Stressor Processes and Effects on Neurobiological Functioning

Happy families are all alike; every unhappy family is unhappy in its own way

This statement from Tolstoy's Anna Karenina gave rise to the "Anna Karenina principle," which is suitable for our exposition regarding the impact of stressors. It can be taken to mean that there is considerable interindividual variability concerning bad experiences and how people react to them. There are any number of ways through which unhappiness can be created, and multiple processes need to be fully operational for well-being to be maintained. To come out of stressful situations relatively unscathed, individuals must be able to assess challenges accurately, consider every possible eventuality, determine the best ways of dealing with these challenges, and then take appropriate actions.

Diverse stressors may be associated with numerous antecedent events and have many consequences. Each person experiencing a chronic unremitting stressor might believe that it's the worst possible condition that anybody could experience. To be sure, each of the tragedies that are encountered is horrible, and comparisons between events are difficult to evaluate. Is a severely debilitating illness equivalent in its impact to the loss of a loved one? Is a business failure more distressing than the shame created by public humiliation? Is posttraumatic stress disorder (PTSD) that stems from a car accident and that which occurs owing to an abusive partner equally distressing? All stressful experiences are disturbing, and each distressing experience is distressing in its own way. As the expression goes, "At night, all black cows look alike."

Still, several fundamental principles may be considered in predicting the types of events that are perceived as most distressing and which are associated with the most negative outcomes. Ordinarily, when a stressor is encountered, an impressive set of behavioral, cognitive, and neurobiological processes can facilitate an individual's ability to contend with the challenge. These include altered activity of an array of hormones, neurotransmitters, neurotrophins (growth factors), inflammatory, immune, and microbial factors, all of which may be regulated by gene-related processes. Experiences with traumatic events, especially those encountered early in life, may also cause the sensitization of neuronal mechanisms so that later stressor responses, for better or worse, are exaggerated. Likewise, such experiences may generate epigenetic changes wherein changes in gene expression foster effective or ineffective physiological and behavioral methods of contending with stressors. Conversely, a constellation of processes can render individuals more resilient.

Understanding stressor actions and stress outcomes is complicated because of how many biological systems are concurrently or sequentially affected by stressors and the very great number of variables that can moderate these actions. As well-equipped as an individual might be to contend with challenges, there is only so much distress that an individual can handle before stress systems become overloaded, culminating in the emergence of psychological or physical pathology.

Vulnerability Versus Resilience

Vulnerability, in the context of illnesses, typically refers to the propensity for an individual or a group (or even a whole society) to be at increased risk for physical or psychological problems emerging in response to environmental or social challenges. Resilience is not quite the opposite of vulnerability, although it is often considered in this manner. Instead, resilience refers to the ability to recover from illness, but it is also used to describe factors that limit or prevent particular stimuli or events from having negative effects. However, the absence of factors that favor increased vulnerability might not translate as elevated resilience. For example, an individual might

---

The Immune System
https://doi.org/10.1016/B978-0-443-23565-8.00018-1

Copyright © 2025 Elsevier Inc. All rights are reserved, including those for text and data mining, AI training, and similar technologies.

carry a constellation of genes that ought to make them relatively resilient to developing illness, but it only takes one severe event (e.g., an aneurysm) to turn that on its head. A person can likewise carry a heavy stress load and have few adaptive biological resources, and hence ought to be at increased risk of becoming ill or for an illness to progress. However, the individual may be blessed with a strong social network or family support system that helps them get through the worst times, and illness may not evolve or symptoms of an existing illness may not progress. There are also instances in which individuals carry genetic mutations that ought to render them vulnerable to illness, and they seem not to have been affected, possibly because they carry yet another genetic attribute that somehow protects them from developing the illness (Chen et al., 2016c).

It's fairly simple for things to break but may be very difficult to repair (the Humpty Dumpty principle). Likewise, it's not overly difficult to identify the constellation of variables that exacerbate the illness-promoting effects of stressors, but it's more difficult to identify the variables that encourage resilience concerning specific illnesses. Perhaps because of this, most studies that assessed the relationship between stressful events and specific pathologies have focused on identifying the characteristics of individuals or the features of the stressors that are associated with illness, whereas fewer have considered the specific factors that promote resilience.

Different views have been offered concerning the elements that lend themselves to resilience, even though it is generally acknowledged that factors or interventions that are effective in limiting distress or illness in one situation may not be equally effective or useful in other situations. Resilience has been related to neural processes underlying reward and motivation (hedonia, optimism), limited responsiveness to negative (fear) situations, and the availability of social resources (altruism, bonding, and teamwork) that minimize the impact of stressors. Resilience has also been described as stemming from the ability to adapt and to be flexible in the context of changes and the ability to solve problems effectively and possessing an optimistic outlook on life. Other views of resilience have included acceptance of change, control, and spirituality (Southwick and Charney, 2018).

Considerable attention has been devoted to individual difference factors (personality) that operate with resilience. The response to illness, for instance, may vary as a function of self-efficacy, self-esteem, self-empowerment, tolerance for uncertainty, optimism, mastery, hardiness, hope, internal locus of control, and acceptance of illness. Being personality traits, these features aren't readily altered (although they can be to some extent), whereas the manner of appraising and coping with illness can be modified (e.g., through cognitive behavior therapy or mindfulness training) so that some of the cognitive consequences of severe illness are affected. Fig. 6.1 lists some of the many factors that additively or synergistically contribute to resilience, but these are only a few of the very many that can act in this capacity.

Attributes of the Stressor

Most people encounter an astounding number of different stressful experiences (stressors), each leading to a variety of possible outcomes (stress or stress response). Stressors may comprise isolated events, or they may be experienced in bunches; often, one stressor experience generates others (financial concerns can create family difficulties, which can instigate health problems). Moreover, stressors can morph from acute events to those that are chronic and intractable, which are more likely to promote illness.

The stressors encountered may comprise those of a processive nature, meaning that they involve cognitive (information) processing, and may entail psychological or physical insults. Another type of stressor that can be encountered is composed of bacterial or viral presence, as well as metabolic changes, collectively referred to as systemic stressors. These challenges can influence brain neurochemical changes, much like those engendered by processive stressors, even though individuals may be entirely unaware of their presence. The brain interprets systemic insults much like it does with other stressors, at least at a neurobiological level, and might thus surreptitiously and insidiously contribute to the provocation or maintenance of mood disorders or physical illnesses that are sensitive to stressor actions (Anisman et al., 2008a, b).

Stressor Appraisals and Reappraisals

Perhaps the most predictable aspects of stressors are their unpredictability and the individual differences that exist in response to various challenges. One individual may interpret a particular event as stressful, whereas a second might not perceive the stressor in the same way, and even if appraised similarly, how they cope with challenges can differ appreciably. Furthermore, when comparable behavioral and cognitive coping strategies are endorsed, biological responses to contend with the challenge may vary across individuals, and consequently, they may be differentially vulnerable to negative health outcomes.

As described in Fig. 6.2, upon encountering a potentially challenging situation, individuals make appraisals as to whether it represents a threat, and whether they have the ability, opportunity, and skills necessary to contend with the challenge. These appraisals can take different forms across people. For example, some individuals

FIG. 6.1 Stress-related resilience can be influenced by a variety of factors. Many of these are outside of the individual's control, but others (e.g., appraisal, coping) are skills and strategies that can be acquired, and thus can be used to act against some of the processes that increase vulnerability to negative stressor effects. The effectiveness of many factors that favor or act against resilience may be context-dependent, varying with different types of challenges, and previous experiences, and can also vary over time as the stressor situation plays out.

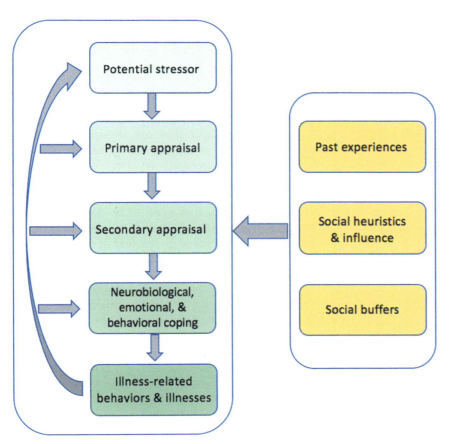

FIG. 6.2 Upon encountering a potential stressor, individuals may ask themselves whether this represents a threat or risk to their well-being. As described by Lazarus and Folkman (1984), following this primary appraisal, individuals ordinarily make a secondary appraisal, assessing whether they have adequate coping resources to meet the demands placed on them. On the basis of these appraisals, specific coping responses to deal with the challenge will be adopted. Social heuristics, referring to ways of interpreting events based on similar or dissimilar experiences, can alter stressor appraisals, and social influences can likewise affect secondary appraisals. The appraisals that are made and coping strategies used may also vary with numerous dispositional (personality) characteristics, such as intolerance for uncertainty, perceived self-efficacy, optimism, hardiness, and locus of control, to name a few. Social buffers (social coping) can also affect other coping methods used. Given accurate appraisals and the adoption of effective coping methods, stressor-related problems ought to be limited. In contrast, appraisal errors or the adoption of poor coping methods may favor the development of pathological conditions.

may appraise (interpret) a stimulus or a condition as a chore that must be endured. Alternatively, they may see it as a challenge and opportunity that will have positive long-term gains or prevent some potential negative outcomes from occurring. Still, other people may interpret such situations as a threat of harm/loss (or bring back traumatic memories), giving rise to negative emotions, such as anxiety or anger.

Appraisals and Misappraisals: Role of Heuristics

In their analyses of decision-making processes, Kahneman and Tversky (1996) offered a theoretical model that has implications for stressor appraisal and coping. Inherent in their view is that biases exist concerning decision-making processes, and thus in some instances, individuals might not behave like rational actors. Decisions made in various situations, and this would certainly apply to stressful circumstances, are often made based on easily accessible information that is tied to earlier experiences or previously established *heuristics* (i.e., rules or shortcuts). These strategies are influenced by several fundamental factors. Decisions can be based on *primed* intuitions that could have developed owing to previous experiences (or second-hand information). Appraisals and decision-making are likewise influenced by how readily decisions come to mind (availability heuristic), and their similarity to other situations that had been encountered (representativeness).

As a situation becomes more complex, making it difficult to appraise all the options available, individuals tend to consider the problem in a simpler way, even if this is inappropriate, or they might just go along with the opinion of their in-group, despite the hazards of doing so. In addition, individuals are affected by "associative coherence" in which a stimulus or event is appraised in a particular way simply because it is consistent with preconceived notions. Having formed a perception of one feature of a situation, related attributions will be applied to this situation ("attribute substitution"). For instance, in the absence of further information, believing that a person is charitable may also result in the attribution of other positive qualities to that person (e.g., kind, warm, and easy to get along with). Many other factors influence decision-making, such as base rate fallacy (in which individuals rely on particular information while ignoring more general information even if it is more pertinent). These are a few of the great number of biases that affect decision-making, only some of which were described by Kahneman and Tversky.

Kahneman brought considerable attention to their views with the lay book "Thinking Fast and Slow" (Kahneman, 2011), in which he discusses dual systems that operate in making decisions. The first comprises an automatic or Fast Thinking system (System 1), whereas the second is a more cognitively based, Slow Thinking system (System 2). Behaviors in decision-making situations are often determined by the automatic, fast-thinking system that has been primed to react in a particular manner.[1] There are many occasions, however, where these rapid responses are apt to be inappropriate, but the cognitively oriented slow-thinking System 2 can limit blunders associated with System 1.

Individuals may find themselves in situations where decisions need to be made, but their knowledge, experiences, and abilities aren't adequate in relation to the pertinent issue. For instance, how certain were we in our decisions about how to react to the threat of an emerging pandemic, or for that matter, the safety of new medications? Many of us know little about certain topics and there are more than a few of us who don't have the first clue about what to do in some threatening situations. Often, we might look for "anchors" that can guide us (e.g., how others have responded to a new medication). Or we might turn to our ingroup members to determine what they think about potential viral threats or vaccine safety, and then behave accordingly (even if the group's choices aren't necessarily the best course of action; in fact, group think tends to favor relatively risky decisions). It's hardly necessary to say that decision-making abilities in stressful situations may be compromised (or augmented) by the mood state engendered by the stressor.

Impact of Stressor Characteristics

Stressors can vary over many dimensions, each of which can influence how they are appraised, and ultimately how coping strategies are affected. Distress increases with the perceived severity of the stressor, and stressors can have cumulative effects so that even day-to-day minor hassles can ultimately have damaging effects. Some stressors that need to be dealt with have already passed (e.g., loss of a loved one), whereas others are anticipatory stressors (e.g., the fear of imminent surgery, or meeting the bully in the schoolyard or workplace). The anxiety created by anticipatory stressors can be exaggerated if the challenge is ambiguous (e.g., the pilot of a plane indicating the need to return to the airport without further explanation) and might be accompanied by disorganized cognitions while the situation plays out.

---

[1] In recent years, the importance of priming has been downgraded as an important component of decision-making, largely in response to overstatements or unreliable findings reported by other researchers, particularly in relation to the influence of stimuli or events that weren't consciously perceived. In contrast, conscious experiences may have considerable sway in some stressful situations (irrespective of whether this is defined as priming), particularly when appraisals and decisions on how to respond must necessarily occur very quickly.

The stressor characteristic that has received particular attention concerns its controllability, as seen in the work concerning learned helplessness (Maier and Seligman, 2016), which is discussed in more detail in Chapter 9 in the context of factors that favor the development of depressive illness. Uncontrollable stressors promote a behavioral profile in which animals cease making attempts to escape from a noxious stimulus, seeming to passively accept the stressor. It was proposed that these animals had essentially learned that "nothing I do matters," leading to a cognitive schema of *"helplessness."* As we'll see, as an alternative to learned helplessness, it was argued that uncontrollable stressors provoke several neurochemical changes that might be responsible for the behavioral disturbances seen in this paradigm.

The predictability of a stressor's occurrence markedly influences stress reactions. Most individuals find the occurrence of unpredictable stressors to be more aversive than those that are predictable, and in animals, unpredictable stressors lead to more pronounced neurobiological changes. Moreover, the effects of uncontrollability and unpredictability can interact to produce greater effects in eliciting anxiety. There are also occasions in which it's uncertain whether a threatening event will actually occur, and this uncertainty can create considerable anxiety. This said, in some instances, uncertainty may also have beneficial effects. In the context of a severe illness that is likely to be fatal, even a small amount of uncertainty may allow a person to maintain a degree of hope so that they can function effectively.

Why is this stressor different from every other stressor?

For animals in the wild, the search for food or approaching the water hole can be fraught with dangers (e.g., from predators awaiting them), requiring some dicey cost-benefit analyses. Thus, interplay exists between the prefrontal cortex responsible for executive functions and the amygdala which is involved in fear and anxiety. Aspects of the basolateral amygdala interact with the prelimbic cortex for fast decisions to be made among rodents placed in an approach-avoidance conflict situation (Burgos-Robles et al., 2017). These regions may likewise be relevant to the bad decision-making that sometimes accompanies mental illness (e.g., whether to continue taking a beneficial drug or not).

It is not unusual to encounter risky situations that involve ambiguity or unknowability, which can increase anxiety. In these instances, neuronal activity increases markedly in brain regions involved in decision-making, possibly to make sense of the situation (Blankenstein et al., 2017). Within a laboratory context, knowing that there is a chance, even a small one, of receiving a painful stimulus (as part of a computer game) greater anxiety was elicited

than in situations in which the painful stimulus would definitely be experienced (de Berker et al., 2016). Humans frequently encounter uncertain events, including waiting for a diagnosis concerning specific illness symptoms and the prognosis if an illness is present. During the early phase of the COVID-19 pandemic, uncertainty concerning infection risk was prominent, and as in other situations, individuals who were unable to deal with uncertainty and anxiety symptoms were often prominent, accompanied by increased information seeking to diminish uncertainty.

In some cases, the information about a threat's possible occurrence can be ambiguous (e.g., are these symptoms of a heart attack or is it indigestion?). Symptoms of this sort are worrisome, but they might not seem sufficiently coherent to prompt emergency medical help. Too often individuals do nothing in the hope that the vague symptoms will simply disappear, which in this instance is obviously not a reasonable strategy. Under these conditions, being high in intolerance of uncertainty may be advantageous. As uncomfortable as ambiguity, uncertainty, and unpredictability may be, the fundamental question arises as to where the cut-off lies between feeling "safe" and feeling sufficiently "at risk" to take action to overcome threats. How unambiguous must symptoms be before an individual decides to take ameliorative action? Knowing an individual's sensitivities about these dimensions can be a useful predictor of how they will respond to certain stressors, and perhaps predict the most effective treatment strategies to reduce their anxieties. However, here again, numerous factors (e.g., previous experiences) moderate the influence of these variables on appraisals and defensive responses.

Although worry and anxiety are troublesome and could promote illness, within limits these cognitive and emotional responses may have adaptive value in that individuals will remain vigilant and ready to respond should a stressful situation escalate. Unfortunately, when these emotions are persistently present owing to inappropriate appraisals and beliefs that a stressor is imminent a good deal of the time, the risk for illness is elevated.

How individuals respond to uncertain and unpredictable events may be tied to when these events are apt to occur. For instance, if some horrific event may or may not occur within the next few seconds or minutes, our intolerance of uncertainty is largely irrelevant as things will be moving too quickly for the uncertainty to have significant consequences. In contrast, uncertainties about somewhat more remote events usually have profound and worrisome effects (will that biopsy show malignancy?), and those who are intolerant of uncertainty suffer the most. Finally, the uncertainty of an event in the distant future might simply fall off the radar as being a potential threat, even though its consequences can be devastating (e.g., predictions of global warming and elevated sea levels within the next 30 or 100 years, or an earthquake hitting in the next 10–20 years). For far-off events, it's just too easy to adopt an avoidance coping strategy. We saw this in relation to the COVID-19 pandemic. Scientists had repeatedly warned that a pandemic of some sort would occur, but since it wasn't certain when this would happen, individuals and governments seemed not to be listening. By making the threat seem more certain (without using scare tactics that often backfire), individuals and health agencies might be convinced that even if they had little control over a pandemic's emergence, they have some control concerning the consequences, thereby encouraging preparedness. Better responses might be obtained if an individual was portrayed as part of an in-group, and that their behavior reflects a moral imperative to protect and assist other group members. Likewise, "nudge" approaches can be used so that through indirect suggestions and positive reinforcement, people are moved so that they behave in a particular fashion (Thaler and Sunstein, 2008).

A less studied feature of some stressors or stressor situations concerns their volatility. It is not unusual for stressors to increase in intensity, depending on several factors. Often changes occur in predictable, graded steps, allowing for preparatory responses to be adopted. However, there are instances in which stressors show volatility, do not follow a predictable trajectory, and expand in just an instant from a few embers to a full-blown conflagration, which is not unusual in the experiences of first responders or soldiers at war.

Coping With Stressors

Numerous methods can be used to cope with stressors, which can be condensed to 15 or so general methods that fall into three broad classes comprising problem-, emotion-, and avoidant-focused coping. Table 6.1 provides a description of several coping strategies that fall into these three categories. Although most coping methods were assigned to one category or another, some of these coping methods aren't comfortable within any specific category, varying with specific stressor contexts, as we'll see later when we discuss social support.

Problem-solving coping methods are often thought to comprise the ideal way to diminish distress to preclude pathology, whereas emotion-focused coping methods are frequently considered to be maladaptive. This view, however, is a bit simplistic as the effectiveness of a particular strategy is situation-dependent. Specifically, emotional coping strategies may have beneficial effects if for no other reason than messaging (through verbal and nonverbal ways) the need for help. Furthermore, this coping method might facilitate an individual's coming to terms with their feelings, and consequently diminishing distress.

Problem-solving efforts are certainly important in helping individuals diminish or eliminate stressors that are controllable. However, in situations that are beyond the individual's control (e.g., dealing with a terminal illness), these efforts might not be productive, although they may offer the illusion that the individual maintains some control over their destiny. At the same time, if a stressor is severe and uncontrollable, people might encounter difficulty planning and initiating actions, and thus problem-coping efforts might be in vain. In fact, rather than adopting straightforward and simple problem-focused approaches, in some instances, exceptionally stressed individuals might engage in excessively complex and unfeasible strategies. The alternative approach, as much as it might be ineffective in other situations, is the adoption of avoidant coping methods (e.g., distraction), allowing the individual to function effectively.

In some instances, nothing can be done to change a situation, and victims need to dig deeply to cope with their challenges. This might comprise a form of cognitive restructuring in which individuals find something positive coming from the trauma. This can emerge in the form of posttraumatic growth (or by finding meaning concerning the situation). In response to severe stressors (e.g., loss of a loved one through accident, terrorism, and suicide), survivors might try to make sense of the event, and at times (e.g., in the case of cancer survivors or those who survived a genocide) some benefit may be derived through these experiences (Park, 2016). Having battled through a severe illness or having engaged in a lengthy caregiving routine (e.g., for a parent with a form of dementia), a person might recognize positive implications of their experience (smelling the roses) or that in some ways they were strengthened or feel that they have grown from the experience. In other cases, efforts are made so that others might gain from their own horrid experiences (e.g., the creation of organizations to raise funds for specific illnesses, supporting gun control, or campaigns against drunk driving). No doubt, these coping methods are very effective for some individuals, but this isn't always the outcome. Engaging in a search for meaning ("meaning-making efforts") doesn't guarantee that the person will actually discover meaning from an event ("meaning made"). Sometimes, simply there isn't anything positive that comes from a tragedy, or it isn't in the person's make-up to find meaning in such a situation. Persistent, but unsuccessful efforts to find meaning might be indicative of a person having an unhealthy preoccupation with a tragedy and is unable to let go as they probably should.

The fact is that in many situations there isn't a single best strategy to deal with challenges, and it's unlikely that a particular strategy will be used in isolation of other methods. The specific cocktail of strategies adopted

TABLE 6.1 Coping Methods.

Problem-Focused Strategies

Problem solving: Finding methods that might deter the impact or presence of a stressor

Cognitive restructuring (positive reframing): Reassessing or placing a new spin on a situation so that it may take on positive attributes. This can entail finding a silver lining to a black cloud

Finding meaning (benefit finding): A form of cognitive restructuring that entails individuals finding some benefit or making sense of a traumatic experience. This might involve emotional or cognitive changes, or active efforts so that others will gain from the experience

Avoidant or Disengagement Strategies

Active distraction: Using active behaviors (working out, going to movies) as a distraction from ongoing problems

Cognitive distraction: Thinking about issues unrelated to the stressor, such as immersing ourselves in our work, or engaging in hobbies

Denial/emotional containment: Not thinking about an issue or simply convincing oneself that it's not particularly serious

Humor: Using humor to diminish the distress of a given situation

Drug use: Using certain drugs in an effort to diminish the impact of stressors, including the physiological (physical discomfort) and emotional responses (anxiety) elicited by a stressor. Some individuals may engage in eating in an effort to cope with challenges

Emotion-Focused Strategies

Emotional expression: Using emotions such as crying, anger, and even aggressive behaviors to deal with stressors

Blame (other): Comprises blaming others for adverse events. This is used to avoid being blamed, or as a way to make sense of some situations

Self-blame: Blaming ourselves for events that occurred. Often associated with feelings of guilt or anger directed at the self

Rumination: Continued, sometimes unremitting thoughts about an issue or event, or replaying the events and the strategies that could have been used to deal with events

Wishful thinking: Thinking what it would be like if the stressor were gone, or what it was like in happier times before the stressor had surfaced

Passive resignation: Acceptance of a situation as it is, possibly reflecting feelings of helplessness, or simply accepting the situation without regret or malice ("it is what it is")

Religion

Religiosity (internal): A belief in god to deal with adverse events. This may entail the simple belief in a better hereafter, a belief that a merciful god will help diminish a negative situation

Religiosity (external): A social component of religion in which similar-minded people come together (congregate) and serve as supports or buffers for one another to facilitate coping

Social Support

Social support seeking: Finding people or groups who may be beneficial in coping with stressors. This common coping method is especially useful as it may buffer the impact of stressors as well as serve multiple other functions in relation to stressors, including providing information as to how to best deal with a particular situation. Social support from ingroup members is generally superior to support from outgroup members

Modified from Anisman, H., 2016. Health Psychology. Sage Publications, London and Matheson, K., Anisman, H., 2003. Systems of coping associated with dysphoria, anxiety and depressive illness: a multivariate profile perspective. Stress 6, 223–234.

might determine the "adaptiveness" of the approach in each situation. High rumination (an emotion-focused strategy) predicts the subsequent development of depression (Nolen-Hoeksema, 1998), depending on the nature of the ruminative content. Depressive symptoms are most apt to occur when rumination is accompanied by other emotion-focused methods, such as self-blame or emotional expression (negative rumination). In contrast, if rumination is accompanied by problem-focused strategies, it might be instrumental in diminishing distress, and negative outcomes might be less likely to develop. Furthermore, having a broad range of coping strategies available, applying these appropriately, and being flexible in both the choice of coping methods and the ability to shift between strategies as the situation demands, may be the ideal approach to deal with stressors. In this regard, the characteristics of stressors may change over time (e.g., as an illness progresses) and individuals must be able to shift coping methods accordingly. Having a narrow range of coping methods and being rigid in maintaining them when they ought to be abandoned will likely be counterproductive (Cheng et al., 2014). Consistent with this view, poor appraisal and coping flexibility have been linked to elevated depressive symptoms (Gabrys et al., 2018), and in the context of COVID-19, social support was linked to coping flexibility, which predicted diminished distress (Tindle et al., 2022).

Poor coping at a cost

In desperate situations, people may adopt desperate solutions to get through their travails. In the face of an incurable illness, there may be a temptation to use complementary alternative medicines, even if there isn't scientific evidence to support their use. For that matter, there have been efforts to adopt a "right to try" strategy when nothing else seems to be effective, essentially allowing terminally ill patients to receive treatments (e.g., biologics or drugs) that are at an early phase of testing and haven't received approval from relevant agencies, such as the Federal Drug Administration. For the patient there is the sense that "there's nothing to lose," but some treatments can promote serious side effects that can make things worse.

In the face of severe stressors, there may also be an increase in the propensity of some individuals to turn to religion, even though in some quarters, this is viewed as being primitive. Yet, religion may be a core component in some individuals' identities and may serve as an effective coping strategy (Ysseldyk et al., 2010). Provided that belief and prayer aren't used as a substitute for potentially effective treatments, even if they serve primarily in a palliative capacity, religion can be effective in diminishing distress. Additionally, through religion, individuals might have gained a social support network that serves to provide solace, peace of mind, distraction, and cognitive reinterpretation (Ysseldyk et al., 2010). Indeed, spirituality has been associated with lower depression and greater neuronal density within the prefrontal cortex and hippocampus (Miller et al., 2014).

Personality and Sex as Moderators of the Stress Response

Personality traits, as we saw in our discussion of resilience, can influence behavioral appraisal, coping, and neurobiological functioning and might consequently affect stressor-related disease conditions. Some individual traits (neuroticism, depressive personality) may favor the development of stress-related pathology, whereas others (optimism, self-efficacy, and hardiness) were linked to enhanced resilience (e.g., Scheier and Carver, 2018).

Several illnesses are more common in women than in men (e.g., depression, anxiety), which might be attributable to women carrying a greater stress load than do men, frequently taking on the responsibility of child-rearing, day-to-day tasks within the home, and they disproportionately serve in a caregiving capacity for family members who are ill (e.g., aging parents). According to the APA's Stress in America survey (2023), self-reported stress was greater in women than in men and they tended to take longer to get over stressor experiences. It also appears that the coping methods used by women differ from those endorsed by men (Matheson and Anisman, 2003) and women may be more likely to internalize stress. These differences, of course, do not diminish the likelihood that the sex differences may be tied to numerous brain neurochemical factors, as well as between the gut microbiota – brain axis that is associated with mental health (Holingue et al., 2020), as we'll see when we discuss specific biological stress responses related to illnesses.

Despite these disadvantages, women in industrialized and nonindustrialized countries continue to outlive men. Aside from estrogen providing women protection against some illnesses (e.g., heart disease), the dimorphism might be due to men being more likely to engage in risky lifestyles, smoke and drink more, and eat more cholesterol-producing foods, which could limit their lifespan. This, however, doesn't explain the longer lifespan observed among females across most mammalian species, which may be tied to distinct sex differences in the hormonal and immunological effects of stressors.

Social Support

There is little doubt that social support can serve as one of the most effective ways of contending with negative events. The usefulness of support can take multiple forms and can change over time, which is obvious among individuals dealing with a serious illness. It can serve in a problem-solving capacity (e.g., obtaining advice; finding alternative treatment strategies for a severe illness) and can also be combined with emotional coping that can facilitate venting or a shoulder to cry on. Having support might also diminish some of the hormonal changes (e.g., cortisol) associated with stressors, as well as limiting stressor-elicited neuronal activation that was otherwise apparent within the dorsal anterior cingulate cortex, the anterior insula, and the amygdala (Eisenberger, 2013). These findings attest to the effectiveness of social support as a stress buffer, which can act against the adverse psychological effects that might otherwise occur.

The source of social support can be important in determining its usefulness. Typically, support is most likely to come from friends, family, and other members of our in-group, and their support is typically more effective than that coming from others (out-group members). Those who share our pain also understand what we might be going through, and hence support groups that include others with similar problems (e.g., support groups for families of those who died of suicide or for the parents of children with cancer) are particularly effective in diminishing distress.

Positivity and well-being

There has been a movement that focuses on "positive psychology" in the promotion of wellness. From this perspective, the absence of distress doesn't necessarily imply the presence of a positive state. The main goal of positive psychology isn't necessarily to cure mental illness or physical disorders, but instead to use positive experiences and thoughts to enhance well-being (Seligman and Csikszentmihalyi, 2000). Part of this entails the promotion of positive emotions, which can enhance individuals' thought-action repertoires so that psychological and social resources are established and strengthened.

Positive states can act prophylactically to prevent the development of stress-related pathology, such as depression, and may act to enhance recovery from illness. Likewise, individuals with ongoing negative experiences exhibit poorer immune responses than those who display positive emotions, although this doesn't necessarily imply that happiness enhances immunity (Barak, 2006). This said, activation of the reward system can enhance both innate and adaptive immune functioning (Ben-Shaanan et al., 2016). As expected, optimism was associated with effective regulation of stress hormones and was accompanied by advantages during cancer progression (Carver et al., 2005) and a meta-analysis indicated that in both ill and healthy populations, positive psychological attitudes were associated with reduced mortality (Chida and Steptoe, 2008). Even though positive psychology can favor increased survival in the face of chronic illnesses, possibly through effects on lifestyles and treatment adherence, it is less likely that it can turn back the clock and produce a cure for chronic, progressive diseases, where damage is already extensive.

Not all researchers believe that positivity is as effective as these reports suggest, and at best might only be useful as an auxiliary treatment for dealing with characteristics related to illness, but not the illness itself. One might also question the meaningfulness of some of the findings reported concerning positive psychology. Although positive life experiences are accompanied by modest elevations of immune functioning, it is questionable whether these limited immune changes have meaningful effects on the ability to fight infection. This said, there have been reports that positive events can diminish the risk of developing upper respiratory infections and colds. There is also the view that the beneficial effects of positive affect come from acting as a stress buffer (Pressman et al., 2019), and as expected, the relationship between positive mood, well-being, and inflammation is most apparent when perceived stress levels are high (Blevins et al., 2017).

Although positive psychology has taken on considerable allure, it has its detractors. Based on a meta-analysis, it was concluded that the magnitude of the effects associated with positive attitudes were, in fact, considerably smaller than initially reported. The benefits attributable to positive psychology were afflicted by statistical problems, conceptual flaws, and inherent biases among researchers in the field (White et al., 2019; Wong and Roy, 2017). In some circles, there seems to be the belief that we're all supposed to be happy, and negativity isn't tolerated for very long. While not denying that positivity can have some very good effects, it needs to be acknowledged that there are limits to its benefits, and forcing positive attitudes might not have the intended consequences. Indeed, positive psychology may foment what Held (2002) referred to as the "tyranny of the positive attitude."

Turning Support on Its Head: The Case of Unsupportive Relations

As useful as social support might be, this is not universally observed, and in some instances, it can be counterproductive. Some people prefer to maintain their independence and privacy and consequently might be reluctant to seek or accept support. For others, the act of seeking or obtaining support might reflect weakness (more so in some cultures), or they resist support since it might make them feel indebted. Even if an individual is hesitant to ask for social support, they may find themselves in a situation where their options are limited and so they reach out to a friend or relative, fully expecting that the support will be forthcoming. It is likely that their expectation will be met, but it's possible that the support will be conditional. Worse still, the request might be rebuffed, or they may encounter an entirely unhelpful response (e.g., blaming the victim, forced optimism or minimizing the victim's distress, disconnecting or distancing in order to avoid hearing about the problems, and bumbling feebly as a result of not knowing what to say). Not having support can be dispiriting, but encountering *unsupportive* interactions can be far more damaging (Ingram et al., 2001). This has been seen among individuals with HIV/AIDS who were abandoned by family members, and among women in abusive dating relationships who were diminished because they stayed in their situation (Matheson et al., 2007), as well as among immigrants from war-torn regions who encountered unwelcoming responses in their new country (Jorden et al., 2009).

It's especially unfortunate that occasions of unsupportive actions not only occur at a personal level but also appear in relation to group behaviors. This is particularly notable when an ongoing injustice is met by silence from witnesses, who might simply "not want to get involved." In the political arena, we've seen this playing out with disadvantaged groups within countries (e.g., Indigenous Peoples within Canada, Australia, the United States, and elsewhere). Diminishing the suffering of the dispossessed and marginalized groups has been a keystone within many philosophies for millennia, but in recent times, this was passionately expressed by Nobel laureates Elie Wiesel ("...to remain silent and indifferent is the greatest sin of all") and Reverend Martin Luther King ("In the end we will remember not the words of our enemies, but the silence of our friends").

Social Rejection and Ostracism

Social rejection and targeted ostracism may be especially painful forms of unsupportive relationships. Although members of a group might consider themselves as a unified social entity (entitativity), some group members may be viewed as not representing them as they would like, and as a result, these individuals might be rejected and denigrated ("black sheep effect") in order to preserve the group's stature. Social rejection is sufficiently strong that even within a laboratory context, a seemingly benign rejection manipulation can elicit negative ruminative thoughts, alter mood, promote hostility, and elicit elevated cortisol levels (McQuaid et al., 2015). Such manipulations also provoked elevated neuronal activity within the dorsal anterior cingulate cortex, much as strong stressors have such effects. It is particularly interesting that receiving or giving support to others can positively influence the same neural circuits (Eisenberger, 2013). The bottom line is that targeted rejection, irrespective of how or when it occurs, can undermine an individual's self-esteem and can promote or exacerbate depressive mood.

Neurobiological Changes as an Adaptive Response to Stressors

Concurrent with behavioral methods of dealing with stressors, an assortment of biological changes occur that serve in an adaptive capacity. Some of these neurobiological changes operate to enhance vigilance and augment preparedness to deal with impending challenges. Others serve to blunt the psychological and the physical impact of stressors, facilitate appraisal, enhance effective coping, favor adaptation to chronic challenges, and activate neuronal and endocrine processes necessary for survival while limiting the actions of still other biological processes that could engender negative outcomes (Sapolsky et al., 2000). Psychological factors related to the stressor situation (e.g., stressor controllability, predictability, and ambiguity) may influence neurochemical functioning, thereby affecting emotional and cognitive processes. The behavioral and biological methods of dealing with stressors operate in tandem, and the effectiveness of behavioral methods diminishes the load placed on biological systems. This may be effective in precluding pathological outcomes over the short term, but with chronic stressor experiences, biological systems may become overwhelmed so that individuals may become more susceptible to pathological outcomes.

Telomeres: An index of getting old or cumulative stressor experiences

The impact of chronic strain might be detectable through neurobiological changes that occur, such as variations of daily cortisol rhythms or cortisol levels, but for the most part, simple biological markers that reflect the cumulative effects of chronic stressors are limited. A possible exception to this involves the length of telomeres that comprise a region at the end of each chromosome, made up of repetitive nucleotide sequences. Like the aglet at the end of a shoelace, these telomeres prevent chromosomes from unraveling.

With each cellular replication that accompanies aging, the telomeres shorten, and thus their length might serve as an index of cellular aging. Shorter telomere length is related to age-related illnesses (Armanios and Blackburn, 2012), although the link between telomere length may be more related to illness than aging itself (Sharifi-Sanjani et al., 2017). Childhood adversities, including lower socioeconomic status or separation from parents, which predicted poorer health, were associated with shortened telomere length (Chen et al., 2019a), and frequent early life infections were associated with shorter telomere length in adulthood (Eisenberg et al., 2017). Cumulative stressful experiences in children, including maternal violence, verbal or physical assault, or even witnessing domestic disputes, resulted in the shortening of telomere length, and prenatal stressor experiences were likewise linked to shorter telomere length in later adulthood (Entringer et al., 2013). Much like so many other stressor-related disturbances, changes in telomere length in adulthood might be subject to epigenetic changes stemming from adverse experiences (Nwanaji-Enwerem et al., 2021).

As in the case of aging, stressors and poor lifestyles (e.g., smoking and high alcohol consumption) were predictive of telomere shortening. The actions of stressors in this regard were especially notable when accompanied by reduced social support, lower optimism, and elevated physiological stress reactivity (Zalli et al., 2014). Although telomere length can be used as an index of biological aging, it seems that the clock can be turned back (somewhat) as the length of telomeres can be rejuvenated through the enzyme telomerase reverse transcriptase. Exercise, for instance, may diminish the impact of aging on telomere length and survival of T cells (Silva et al., 2016).

There were indications that telomere length may be linked to psychological illnesses, such as depression and PTSD (Zhang et al., 2014). Conversely, stress reduction procedures, such as meditation, were associated with increased telomere length (Alda et al., 2016). The processes by which stressors and mood disorders influence telomere length are beginning to be understood, and it seems that changes in glucocorticoids, reactive oxygen species, mitochondria, and inflammation may be pivotal in producing this outcome (Lin and Epel, 2022). Despite the impressive findings, the validity of telomere length as an index of biological aging and vulnerability to illness has been questioned, and alternatives (e.g., accumulation of epigenetic marks) have been offered as viable alternatives. This does not exclude the possible use of telomere length as a component of a broader composite that could predict wellness and longevity.

Chronic Stressors and Allostatic Overload

Many biological changes that occur soon after a stressor encounter serve to maintain equilibrium within and between biological systems and are essential for proper appraisal and behavioral stress responses. As well, they contribute to the distribution of energy resources to ensure the proper functioning of various organs, including the brain. This process, termed allostasis, is reminiscent of the balancing involved in homeostasis, but because of the urgency related to stressors, it involves a much more rapid mobilization of biological resources. Some of the brain processes that are involved in stress responses may have developed shortcuts so that rapid responses can be elicited without having to go through the multistep process of appraisal and reappraisal. When an organism's life is on the line, the luxury of spinning its wheels as to what actions to take simply isn't on the agenda, and innate actions are taken that have developed through evolutionary pressures.

Neurobiological systems are remarkably adept in dealing with moderate, predictable stressors that are repeatedly encountered. However, adequate ways to deal with repeated challenges that are particularly severe, uncontrollable, and unpredictable are exceptionally difficult (e.g., chronic illness or chronic pain). Under such conditions, brain regions responsible for dealing with stressors might not have an opportunity to recuperate, as neurobiological processes may be continuously engaged. As a result, neuronal damage can be accrued (allostatic overload), as seen in the case of excessive cortisol release causing damage to hippocampal neurons (McEwen, 2000). As well, when adaptive neurobiological systems become overly taxed, hormones or neurotransmitters become less available, or receptor sensitivity is altered, which can lead to a variety of pathological conditions. Stressors can also have cumulative additive or interactive effects that might contribute to the development and exacerbation of illnesses, especially when these stressors begin during childhood (Lupien et al., 2016).

In humans experiencing chronic, unremitting stressors (stigmatization, war experiences, and chronic illness), the impact of these challenges may be doubly disastrous, not only because of the continuous nature of the stressor but also because affected individuals lack safe places that would allow for healing to occur. In most instances, the stressors considered are those that are encountered on an individual basis, but it was proposed that insidious challenges at a broader level (e.g., social disturbances, social conflict, and poverty) can provoke "Type 2" allostatic overload, which requires changes of social structures to prevent the development of pathology (McEwen and Wingfield, 2003). This is especially pertinent as social structures and familial challenges (e.g., poverty) may have actions that appear over successive generations (McEwen and McEwen, 2017). The relationship between stressor levels and pathology has been described as an inverted U-shaped function, being dependent on the severity and frequency of the stressor (McEwen et al., 2015). As shown in Fig. 6.3, acute intense stressors may have adverse effects, but as the stressor continues, especially if it is predictable and controllable, an adaptation occurs. At some point, varying across individuals, chronic stressors may undermine resilience and increase the likelihood of systems malfunctioning so that pathology is more likely to occur.

Neurotransmitter Alterations Provoked by Stressors

Before describing the neurobiological effects of stressors, several caveats should be introduced. There is much to be said about assessing the impact of stressors on both human populations and laboratory animals (as well as animals in their natural habitats). Human studies provide a richness of information (e.g., concerning the contribution of psychosocial factors, personality, and appraisal-coping processes) that isn't easily obtained through studies using animals. Likewise, important features in humans, such as rumination or the complex cognitive changes that evolve, can't be recapitulated in animals. Although imaging studies in humans can point to the brain region and circuit involvement in response to or anticipation of stressors, animal studies provide the opportunity to examine brain changes more intensively. Importantly, studies in animals are more amenable to mechanistic analyses concerning causal connections

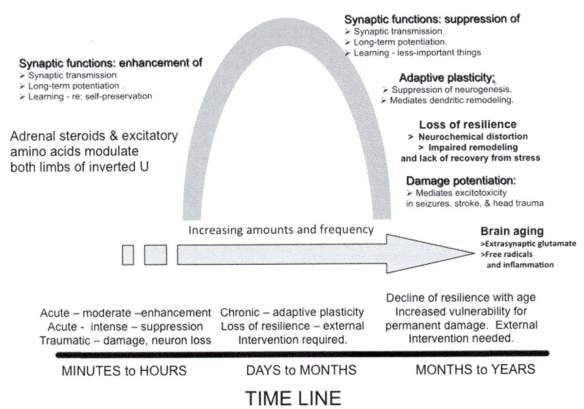

FIG. 6.3 Acute and chronic stressors influence the risk for illnesses, varying with the chronicity of the stressor. The actions of stressors are mediated by glucocorticoids, glutamate, BDNF, tissue plasminogen activator, CRH, endocannabinoids, as well as immune changes and microbiota. With brief stressor exposure, preparatory mechanisms are engaged that might enhance well-being (the left part of the inverted U), although if sufficiently stressful or traumatic, these acute events can be damaging. With continued stressor experiences, a series of further adaptive changes occur to maintain well-being, although under these circumstances social support or other external coping processes are needed to maintain well-being. Eventually, with excessive stressor experiences, these adaptive systems may fail or multiple neurobiological alterations may be instigated that increase the risk for illness. Brain plasticity plays a prominent role in determining how effectively individuals are able to deal with stressors, and as early life experiences are fundamental in shaping many neural connections, they may have a marked influence on the impact of stressors that are subsequently encountered. At the same time, neuronal malleability or plasticity also allows for successful interventions that can limit the destructive effects of earlier negative experiences. *Source: McEwen, B.S., Bowles, N.P., Gray, J.D., Hill, M.N., Hunter, R.G., et al., 2015. Mechanisms of stress in the brain. Nat. Neurosci. 18, 1353–1363.*

between stressors, neurobiological changes, and the emergence of pathology, and allow for the assessment of the moderating effects of numerous variables (e.g., early life experiences) on the impact of stressors on brain neurochemical functioning.

The impact of assorted stressors in animals generally engages several common neurobiological systems (e.g., increasing hypothalamic-pituitary-adrenal (HPA) activity). Yet, certain neurobiological changes are more readily elicited by some types of stressors than others. For instance, psychological challenges that involve higher-order processing (e.g., complex conditioned stressors) might not engage the same neural circuits as those elicited by an innate threat. These challenges may also involve different brain networks relative to those activated by neurogenic or systemic insults (Merali et al., 2004). As we'll see when discussing the effects of stressors on hormonal changes, some stressors (e.g., those that involve public performances and public scrutiny, and hence more likely to elicit emotions, such as shame or anger) are more likely to cause cortisol changes than are other stressors (Dickerson and Kemeny, 2004; Matheson and Anisman, 2012).

Like the behavioral effects of uncontrollable stressors, biological changes may vary over time. Initially, stressors elicit response activation that might reflect anxiety and arousal but are then replaced by processes that lead to a depressive-like condition. In some instances, the effects of the stressor grow with the passage of time, as does the potential for several neurochemical changes. Aside from pointing to the (dynamic) evolution of stress reactions, these findings raise the possibility that the efficacy of treatments to attenuate stress-related illnesses may also differ over the course of illnesses playing out.

Animal models of psychopathology

Given the importance of animal models for both physical and psychological disorders, criteria were established to maximize their validity. Yet, it is important to acknowledge that even under the best of conditions, what happens in rodent models doesn't necessarily translate well to humans. To an extent, the validity of an animal model is reminiscent of a jury trial. A jury might find a defendant to be not guilty, but this doesn't mean that the defendant is actually innocent.

The requirements for an animal model described here are based on the description provided by Anisman and Matheson (2005):

1 Face validity: The animal model should comprise symptoms that are similar to those present in the human condition.
2 Predictive validity: Those treatments effective in ameliorating (or preventing) symptoms in humans should be similarly effective in animal models. Conversely, those treatments that are ineffective in attenuating the human disorder should similarly not be effective in an animal model.
3 Etiological validity: Conditions that provoke the pathology in humans should also do so in the animal model.
4 Construct validity: The presumed biological processes underlying pathology should be similar in the human and animal models.

Meeting all these criteria would be ideal, but often this isn't possible. We might not know, for instance, which treatments are effective in the human condition, and thus the predictive validity can't be met in the animal model. Likewise, the purpose of a study might be to determine the neurobiological underpinnings of a disorder, and thus the construct validity of the model is uncertain. It should also be added that for ethical reasons it isn't possible to assess the effects of traumatic events in animals, like those that humans may unfortunately endure. One of the greatest problems with animal models, as we'll see in considering pathologies such as diabetes, heart disease, and cancer, some laboratory manipulations that were effective in mice are frequently ineffective in humans. Thus, once more, there are limits to what can be gained from animal research. It's the best we have for the moment, and with the evolution of technologies and the use of artificial intelligence, better models will no doubt emerge.

Researchers prefer to believe that within limits, the mouse and rat may be effective in modeling anxiety and depression, particularly given the range of gene manipulations that are possible. Yet, even among genetically identical mice, subtle laboratory perturbations, or the source from which they had been obtained, can dramatically alter the behavioral phenotypes expressed (Crabbe et al., 1999). Added to this, the diet of mice or rats, which can vary across laboratories, can affect the microbiome and various hormonal processes. There is also the issue of selecting the strains that are most appropriate for whatever phenotype is being examined. Some strains, for instance, might be considered appropriate to assess anxiety or depression, whereas others, for several reasons (including their marked hardiness), are less useful.

There have been reports showing genetic similarities between mice and humans (see the ENCyclopedia Of DNA Elements; ENCODE), and several published reports have focused on this issue (see the NIH report Comparing the Mouse and Human Genomes), but there are also distinct differences across species. At the end of the day, mice aren't humans (they aren't even pint-sized rats), and after multiple generations of inbreeding, it's uncertain to what extent they are valid representations of wild mice. Added to this is that laboratory mice are raised in fairly clean environments and hence they hardly represent the dirty world of field mice (or that of humans). They certainly don't have the experiences of wild mice in scavenging for food or adopting strategies to evade predators. In this regard, wild mice that were similar to laboratory mice in many ways exhibited microbiota that differed markedly from their laboratory counterparts. When the laboratory mice received microbiota from wild mice, inflammation was reduced, and survival was enhanced in response to an influenza virus (Rosshart et al., 2017). With these caveats in mind, we can now turn to a description of the neurobiological changes that are introduced by stressful experiences.

Norepinephrine, Dopamine, and Serotonin

Upon first perceiving a potential stressor, an animal (or a human) may be uncertain as to whether this represents a threat, whether it can be dealt with, or how long the potential stressor will persist. Thus, in response to threats, rapid neurochemical responses ought to occur, and then, as characteristics of the stressor are understood (e.g., controllability, chronicity), these neurochemical responses can be adjusted as needed.

In response to stressors, the synthesis and utilization of serotonin, norepinephrine, and dopamine rapidly increase across several brain regions, including the hypothalamus, amygdala, hippocampus, and prefrontal cortex. As the intensity of the stressor increases so does the rate of neurotransmitter release (utilization), which is met by an increase in synthesis, and thus the overall availability of these neurotransmitters does not vary appreciably. If the stressor is relatively intense, and coping methods are unavailable (e.g., in the face of an uncontrollable stressor), the coping burden rests more heavily on biological systems. Thus, monoamine utilization increases further, eventually outstripping the rate of synthesis, and the absolute level of the neurotransmitter declines (varying across brain regions), rendering the organism less able to contend with immediate challenges (e.g., Amat et al., 2014). While controllable and uncontrollable stressors have frequently been found to differentially influence brain monoamine functioning in rodents, at least some of these effects (e.g., dopamine functioning within the prefrontal cortex) are less notable in female rats (McNulty et al., 2023).

Supporting the importance of stressor controllability in determining neurochemical changes, if rats had learned that a stressor was controllable, then the behavioral disturbances and the serotonin variations elicited by a later uncontrollable stressor were prevented. It turns out that exercise in the form of running on a wheel also protects animals against the depressive-like effects of a stressor and precludes the serotonin responses at the dorsal raphe (Clark et al., 2015). It is uncertain whether this reflected exercise enhancing resilience or a consequence of animals having control over their environment.

The neurotransmitter changes elicited by stressors are typically short-lived, normalizing within less than an hour. These actions can be longer lasting in older animals, and stressors that comprise social challenges, such as changes in the social environment or defeat by a conspecific, may have somewhat more persistent effects. It has been suggested that the circuitry associated with certain stressful experiences, particularly social threats, or other innately driven challenges, may be hard-wired and might be accompanied by particularly profound and persistent neurochemical alterations (Krishnan and Nestler, 2011).

In addition to the altered neurotransmitter turnover (i.e., the relative rates of utilization and synthesis), several receptor changes occur in response to intense stressors, including downregulated postsynaptic norepinephrine receptor functioning, possibly reflecting a further adaptation to limit excessive activation of neuronal processes that can be detrimental to well-being. Likewise, an uncontrollable stressor regimen that elicited disturbed behavioral performance was paralleled by elevated firing of serotonergic neurons (Hashimoto et al., 2021) and by altered expression of several serotonin receptors (5-HT_{1A}, 5-HT_{1B}, 5-HT_{2A}, and 5-HT_{2C}) within the dorsal raphe nucleus, the site of serotonin cell bodies. In the case of 5-HT_{1A} receptors (autoreceptors) of the dorsal raphe nucleus, as well as basolateral amygdala 5-HT_{2C} receptors, the changes were more readily induced by an uncontrollable than a controllable stressor (Rozeske et al., 2011).

Given that these neurotransmitter systems affect one another in significant ways, the functioning of norepinephrine, dopamine, and serotonin may need to be integrated for proper behavioral outputs to occur. For example, norepinephrine neurons of the locus coeruleus, through their actions on ventral tegmental dopamine functioning, may influence reward processes and hence may promote the anhedonia associated with depressive disorders (Isingrini et al., 2016). Likewise, infusion of a 5-HT_{1A} receptor agonist directly into the prefrontal cortex resulted in stressor-elicited dopamine and norepinephrine changes being attenuated. Several studies have also demonstrated that serotonin activity or activation of particular serotonin receptors could interact with the neurotrophin brain-derived neurotrophic factor (BDNF) in promoting stress resilience (Leschik et al., 2022). Furthermore, Gamma-aminobutyric acid (GABA) and glutamate alterations may interact with serotonin in determining the behavioral response to stressors (Faye et al., 2018). Assessing the effects of stressors on specific neurotransmitters is important, but as indicated earlier, understanding psychopathology necessitates analyses of system changes.

Chronic Stressor Challenges

It is remarkable how many different adaptive responses develop concurrently or sequentially in response to stressors, accompanied by varied checks and counterchecks aimed at maintaining allostasis. As a stressful experience continues over an extended period or is encountered repeatedly, the initial neurochemical changes are followed by a sequence of further adaptive changes that may facilitate coping and limit the development of stress-related illness. To meet the demands imposed, a compensatory increase in monoamine synthesis occurs, likely owing to the downregulation of autoreceptors, which signals that more transmitters should be produced. Thus, the level of available neurotransmitters normalizes, which may be an effective adaptive response, at least in the medium term, as it may facilitate the organism's ability to deal with ongoing challenges. Yet, when a neurochemical system is overly active, negative secondary effects may develop as indicated in our discussion of allostatic overload. Perhaps as a protective response, further postsynaptic receptor changes are engendered, thereby modulating the actions of the high levels of neurotransmitter released. These adaptive neurochemical changes vary across strains of mice but seem to be less robust in highly stressor-reactive mouse strains, and in older mice.

It is important to consider that not all neurochemical systems operate identically, so allostasis (adaptation) may develop in one system, but this might not occur as readily within a second system or a second brain region. As these adaptations may be driven by genetic, epigenetic, or experiential factors, individual differences can be expected regarding the development of allostatic overload and the nature of the pathology that emerges.

Allostasis may not be maintained readily in response to some stressor experiences. For instance, chronic stressors that are unpredictable and vary from day to day are less likely to result in adaptive changes that preclude excessive neurochemical changes and pathology. Although these outcomes are most likely to occur with relatively intense stressors, they have also been witnessed in response to a chronic mild unpredictable stressor regimen. In response to this challenge, depressive-like symptoms can be provoked, as can cognitive and learning disturbances measured 1-month later. These outcomes were accompanied by cell loss within the dorsal raphe nucleus (the site of serotonin cell bodies) and projections to the prefrontal cortex, possibly being mediated by glutamate processes (Natarajan et al., 2017). The actions of a chronic mild stressor regimen aren't as reliable as one would like, but the source for the different outcomes isn't certain (Willner, 2016).

The scenario outlined here is meant to provide a general schema as to what occurs in response to stressors, and each of the monoamines does not behave identically under all conditions. Nonetheless, such changes have been observed with norepinephrine in brain regions involved in appraisals and executive functioning (prefrontal cortex), dopamine changes in regions associated with reward processes (nucleus accumbens), as well as vegetative or basic life processes (e.g., specific hypothalamic nuclei). Reiterating the obvious, but worth repeating, the stress burden carried by individuals is moderated by age, sex, genetic influences, earlier experiences, and psychosocial factors. Moreover, the response to stressors may also be moderated by the individual's lifestyle and health behaviors, including diet, physical activity, sleep, and substance use.

Previous Stressor Experiences: Stress Sensitization

Even if a stressor experience lasts for only a short time, activation of multiple neuronal systems will be detected within minutes; some of these may persist for less than an hour, whereas others will still be apparent 24h later (Musazzi et al., 2014). It is tempting to assume that once neuronal activity has normalized, the stressor's actions will have similarly ended. However, as discussed in relation to behavioral processes, having been exposed to a stressor, the potential for altered neuronal functioning persists so that neurotransmitter release or receptor processes are more readily induced by later stressor encounters. This sensitization effect occurs even if the second stressor experience is different from the one initially experienced (Anisman et al., 2003). Indeed,

"cross-sensitization" effects occur so that a stressor experience augmented responses to later exposure to drugs, such as amphetamine and cocaine that ordinarily affect monoamine functioning and similar actions have been reported with immune challenges (Anisman et al., 2003). A frequent characteristic of these sensitized responses is that they grow with the passage of time, at least within a 3-week period. The sensitization associated with a stressor has been implicated in the provocation and reemergence of anxiety and addictions.[2]

Several stress-related processes are subject to sensitization. For instance, variations of certain corticotropin-releasing hormone receptors (CRH_2) may contribute to the sensitization of serotonin activity within the raphe nucleus, whereas the cross-sensitization effects that have been reported between stressors and amphetamine might be mediated by glutamate functioning. A reasonable possibility is that the sensitization stems from a time-dependent increase of certain types of receptors within the amygdala (notably glutamate receptors) so that blocking these receptors at the time of the initial stressor experience diminished the amygdala changes (Yasmin et al., 2016). Brain cytokine changes likewise appear to be influenced by sensitization processes, which likely involve still other mechanisms, possibly being related to microglial functioning. The essential point is that the impact of stressors may continue long after the initial challenge has passed.

The impact of powerful stressors in humans can have protracted effects for reasons other than sensitized neuronal changes. The rumination that follows a stressor can have significant and lasting consequences, and stressors may also promote secondary effects (e.g., changes in lifestyle, employment) that can have downstream consequences, particularly in response to reminders of the initial trauma. Indeed, some events, such as being publicly shamed or humiliated, could jeopardize self-esteem and self-efficacy, thereby affecting how individuals deal with subsequent stressors. For that matter, the brain's exceptional neuroplasticity, which serves us so well in acquiring and maintaining information, can work against us when the stressor is one that undermines our self-perceptions, so it is incorporated into one's self-schema.

γ-Aminobutyric Acid

With so much information coming in at any given moment, especially under distressing circumstances, the brain needs to distinguish between relevant versus irrelevant information, and even different forms of relevant information. The inhibitory neurotransmitter GABA, which will be recalled, serves as a brake to

---

[2] In recent years, stressor-related cues that elicit negative reactions in humans have most often been referred to as "triggers."

regulate neuronal activity within other systems, and thus plays an essential function in information gating (Yang et al., 2016). Numerous reports have shown that stressors influence GABA activity within several brain regions. In this respect, GABA neuronal activity within the prefrontal cortex, which may be modulated by dopamine receptor changes, may contribute to the processing of stressful stimuli (Lupinsky et al., 2017). It also appeared that the alterations of GABA neuronal activity within the prefrontal cortex elicited by a chronic mild stressor appeared sooner in females than in males, once again providing a possible mechanism for the greater vulnerability to anxiety and depression among women (Woodward et al., 2023). When GABA functioning is impaired, which may occur owing to epigenetic changes, anxiety may ensue more readily (Persaud and Cates, 2022).

Within the hippocampus and amygdala, GABA levels and the conversion of glutamate to GABA (reflected by elevated functioning of the enzyme GAD65) were typically altered in response to stressors. Within the basolateral amygdala, such effects occurred irrespective of whether the stressor was controllable or uncontrollable, but in a portion of the hippocampus, namely the dentate gyrus, GAD65 was differentially affected by stressor controllability (Hadad-Ophir et al., 2017).

Following a chronic mild stressor regimen that engendered depressive-like behavior, GABA functioning was reduced in the prefrontal cortex and was particularly notable following chronic stressor exposure (Fogaça and Duman, 2019). This outcome was accompanied by variations of GAD67, the vesicular GABA transporter, and the GABA transporter-3 (Ma et al., 2016), as well as a reduction in the number of GABA neurons within the orbitofrontal cortex (Varga et al., 2017). Altered connectivity between GABA and glutamate neurons brought about by chronic stressful experiences has been implicated in psychiatric disorders and, hence, positive therapeutic effects might be obtained through drugs that affect these systems (Duman et al., 2019). Beyond these important effects on mood disorders, GABA appears to be associated with other stressor-related conditions, including inflammatory-related illnesses, such as gastrointestinal disorders and visceral pain (Jembrek et al., 2017).

The influence of GABA on behavioral outputs varies with the type of GABA receptor that is stimulated (i.e., $GABA_A$ versus $GABA_B$). In the case of stressors, particular attention has focused on changes in $GABA_A$, which are influenced by chronic challenges (Poulter et al., 2010). The characteristics of $GABA_A$ receptors differ from that of many other receptors. This receptor comprises a pentameric (5 subunit) protein complex from a set (cassette) of 21 different proteins/genes. Thus, the specific subunits that can make up this complex are enormous, and they can often be distinguished from each other based on their conformation and the resulting docking sites for binding proteins. The specific subunits that comprise a given receptor will determine what stimuli will excite it (e.g., benzodiazepines will primarily affect receptors with particular conformations, specifically those having the α1, α2, α3, α5, or γ subunits). Likewise, stressful events may affect subunits that make up the $GABA_A$ receptor in areas of the brain associated with anxiety and depression (Engin et al., 2018). For instance, chronic social defeat produced an increase of prefrontal cortical and hippocampal $GABA_A$ receptors that contained the α5 subunit (Xiong et al., 2018) and restraint over successive days similarly increased the expression of α1 subunit expression within the prefrontal cortex (Gilabert-Juan et al., 2013). Likewise, $GABA_A$ receptor subunit mRNA expression can be either up- or downregulated in several brain regions of depressed individuals, likely operating through microglial actions (Choudary et al., 2005). Predictably, treatments that affect specific subunits of the $GABA_A$ receptor can attenuate the behavioral and emotional effects of stressors and do so in a sex-dependent manner (Piantadosi et al., 2016).

The characteristics of GABA receptors are also subject to effects stemming from epigenetic and negative early experiences. In this regard, early life stressors, or those encountered during adolescence, affected GABA functioning measured in subsequent adulthood, thereby influencing the course of anxiety (Englund et al., 2021). At the same time, it is important to recognize that $GABA_A$ functioning generates a state-dependent fear/anxiety response in that these responses vary with the animal's internal state, which can be influenced by contextual cues (Radulovic et al., 2017). More to the point, stressful experiences, especially in a relatively young organism, can instigate an inflammatory state related to microglia functioning, which interacts with GABA neuronal activity to generate psychiatric disturbances (Andersen, 2022).

Given the influence of stressors on GABA functioning in animals, and GABA manipulations can influence both anxiety- and depressive-like behaviors, it has been assumed that the link between GABA alterations and mood states is a causal one. This does not belie the common view that serotonin, CRH, and other transmitters play a role in these conditions, as GABA may serve in a regulatory capacity in affecting these factors. Furthermore, $GABA_A$ receptor expression, and responses to stressors, can be affected by ovarian hormones and thus might be key in accounting for sex differences in mood disorders.

Glutamate

Excitatory glutamate functions are moderated by the inhibition produced by GABA, and the two operate together to maintain behavioral functioning. When an imbalance occurred between glutamate and GABA inputs within the prefrontal cortex, neuronal atrophy and loss of synaptic connections were apparent, thus engendering cognitive disturbances and depression (Ghosal et al., 2017). Stressor experiences readily provoke glutamate release, thereby contributing to the establishment and retention of fear responses as well as stressor-related psychopathologies.

Chronic stressors produced especially marked variations of glutamate and GABA that varied across brain regions. Among rodents that were exposed to chronic social defeat, a depressive-like behavioral profile emerged, together with alterations of glutamate and the glutamate-glutamine cycle (glutamine is catalyzed to glutamate by glutaminase), which was particularly marked in females (Rappeneau et al., 2016). Based on electrophysiological findings, it was concluded that a chronic stressor regimen produced inhibition of glutamatergic output neurons within the prefrontal cortex, thereby diminishing the contribution of this brain region to stress reactivity, essentially allowing neuronal activity within other brain regions to predominate (McKlveen et al., 2016).

Consistent with the role of glutamate on stressor-elicited anxiety, a restraint stressor influenced glutamate and cytokine expression within the hippocampus of mice, but the observed effects differed between stressor-sensitive and relatively resilient strains (Sathyanesan et al., 2017). Like these strain-specific effects, a subchronic stressor that selectively provoked depressive-like effects was also linked to variations in the frequency of particular glutamatergic inputs to the nucleus accumbens, which could thus influence dopamine functioning within this region (Brancato et al., 2017).

Glutamate receptors comprise two broad classes, ionotropic and metabotropic, each of which has several receptor subtypes. Within the ionotropic family, N-methyl-D-aspartate receptor (NMDA) and α-amino-3-hydroxy-5-methyl-4-isoxazolepropionic acid receptors are activated by stressors (Marrocco et al., 2014), and early life insults persistently alter NMDA-receptors, thereby influencing fear memory formation or retrieval (Lesuis et al., 2019). The activation of these receptors can promote opposite effects on neurogenesis and neuronal survival, varying with the synaptic or extrasynaptic concentration of glutamate. Upon exposure to a chronic stressor regimen, glutamate levels rise markedly, and it is conceivable that the depressogenic effects of this treatment stem from excessive stimulation of NMDA receptors (Rubio-Casillas and Fernández-Guasti, 2016). Indeed, at sufficiently high levels, glutamate may be neurotoxic and the resulting cell loss may contribute to psychological disturbances.

These effects of stressors, together with indications that antidepressant medications affect glutamate release, encouraged further efforts to target glutamatergic processes in the development of treatments for depression. As we'll discuss in Chapter 9, the glutamate antagonist ketamine can produce rapid antidepressant effects, even among treatment-resistant patients. These positive actions were attributed to the downstream effects of the glutamate changes on the neurotrophin BDNF that enhances neuroplasticity. If nothing else, these findings indicated that perspectives of depression that strictly focused on monoamines were outdated and that it would be more productive to consider the impact of stressors on neuroplasticity, and perhaps the effects on coping processes and cognitive flexibility (Zhang et al., 2017a, b). Indeed, hippocampal neurogenesis influences cognitive flexibility so that individuals are better equipped to deal with changing situations and threats (Anacker and Hen, 2017). Likewise, the lateral habenula has frequently been implicated as being fundamental for flexible coping in response to changing environments (Hones and Mizumori, 2022) and the effects of chronic stressors in promoting depression and anxiety may be related to the activation of a pathway linking the dorsal medial prefrontal cortex and the lateral habenula (Tong et al., 2024).

Neurotrophins

As we've seen, neurotrophins (growth factors) are fundamental for neuronal plasticity, including the arborization and synaptogenesis that contribute to memory processes and have been connected to stress-related mood disturbances. The actions of several growth factors (neurotrophins) are provided in Table 6.2. Each of these has multiple functions and most are affected by stressors and influence the response to diverse challenges.

Acute stressors or even reminders of strong stressors provoked a reduction of BDNF gene expression and protein levels within the hippocampus (Duman and Monteggia, 2006), and were still more prominent in response to a chronic stressor regimen that comprised a series of different challenges (Molteni et al., 2016a, b). Stressors do not uniformly instigate growth factor reductions across stress-sensitive brain regions, including those that might underlie depression. In fact, stressful events may increase rather than reduce BDNF mRNA expression within portions of the prefrontal cortex (e.g., prelimbic, infralimbic, and anterior cingulate) and the ventral tegmentum, the latter being associated with reward processes. Moreover, unlike several neurotransmitter changes, the increased BDNF expression within the anterior cingulate cortex was more pronounced after controllable than uncontrollable stressors (Bland et al., 2007). These BDNF changes might have reflected the adoption of active coping efforts or new learning related to the controllability of the stressful situation, rather than directly stemming from the distress created.

6. Stressor Processes and Effects on Neurobiological Functioning

TABLE 6.2 Growth Factors and Their Functions.

Growth Factor	Biological Effect	Outcome
Brain-derived neurotrophic factor (BDNF)	Supports neuronal survival and growth of neurons; encourages growth and differentiation of new neurons; promotes synaptic growth.	Increased neuroplasticity; influences memory processes, stress responses, and mood states.
Basic fibroblast growth factor (bFGF or FGF-2)	Involved in the formation of new blood vessels (angiogenesis); protective actions in relation to injury and repair; essential for maintaining stem cell differentiation.	Increased neuroplasticity; contributes to wound healing; neuroprotective; diminishes tissue death (e.g., following heart attack); related to anxiety and depression.
Glial cell line-derived neurotrophic factor (GDNF)	Widely distributed in the brain and periphery. Secreted by skeletal muscles and motor neurons, and by astrocytes, oligodendrocytes, and Schwann cells. Encourages the survival of several types of neurons.	Acts strongly on dopaminergic neurons associated with motor functioning and diseases such as Parkinson's. Also influences the proliferation and maturation of cells and may play a role in cancer progression (e.g., neuroblastoma and glioblastoma), as well as major depression and bipolar disorder.
Nerve growth factor (NGF) and family members Neurotrophin-3 (NT-3) and Neurotrophin-4 (NT-4)	Contributes to cell survival, growth, and differentiation of new neurons. Fundamental for maintenance and survival of sympathetic and sensory neurons; axonal growth.	Survival of neurons; new neuron formation from stem cells; related to neuron regeneration, myelin repair, and neurodegeneration. Implicated in cognitive functioning, inflammatory diseases, in several psychiatric disorders, addiction, dementia as well as in physical illness, such as heart disease, and diabetes.
Insulin-like growth factor 1 (IGF-1)	Secreted by the liver upon stimulation by growth hormone (GH). Promotes cell proliferation and inhibits cell death (apoptosis).	Primary mediator of insulin and growth hormone; stimulates cell growth and proliferation; inhibits apoptosis; linked to signaling pathways (e.g., PI3K-AKT-mTOR) associated with some types of cancer.
Vascular endothelial growth factors (VEGF)	Signaling protein is associated with the formation of the circulatory system (vasculogenesis) and the growth of blood vessels (angiogenesis).	Creates new blood vessels during embryonic development, encourages the development of blood vessels following injury, and creates new blood vessels when some are blocked. Muscles stimulated following exercise. Implicated in various diseases, such as rheumatoid arthritis, and poor prognosis of breast cancer.
Epidermal growth factor (EGF)	Found in numerous tissues and is apparent in saliva, plasma, urine, milk, and tears. After binding to its receptor, EGF promotes epidermal and endothelial cell proliferation, differentiation, and survival.	Pivotal in the healing of oral and gastric ulcers, is protective concerning injury that could be provoked by gastric acid, bile acids, trypsin, and pepsin, and could act against damage stemming from chemical and bacterial challenges.
Transforming growth factor (TGF)-β	Part of the TGF superfamily. A cytokine is produced by macrophages and lymphocytes. Induces the transcription of several target genes that influence differentiation, proliferation, and regulation of several types of immune cells.	Regulates inflammatory responses. Plays a key role in gut inflammation, and has been implicated in cancer, autoimmune disorders, infectious diseases, and depressive disorders.
Colony-stimulating factors (CSF)	Binds to specific receptors on hemopoietic stem cells, stimulates differentiation and proliferation of varied types of leukocytes. Comes in several forms, such as macrophage colony-stimulating factor (M-CSF), granulocyte colony-stimulating factor (G-CSF), and granulocyte macrophage colony-stimulating factor (GM-CSF). Promotes proliferation, differentiation, and the survival of macrophages and monocytes.	G-CSF stimulates bone marrow so that stem cells and granulocyte production is elevated. It fosters the proliferation of neutrophils and promotes neurogenesis (growth of neurons) and neuroplasticity. GM-CSF released by diverse immune cells acts as a cytokine that increases macrophages and dendritic cells. Associated with joint damage in rheumatoid arthritis and may enhance immune activity when immune functioning is compromised. M-CSF is implicated in atherosclerosis and kidney failure.
Erythropoietin (EPO)	Produced in the kidney and liver. Promotes the production of red blood cells in bone marrow, and hence increases oxygen supply.	Used to increase blood oxygen carrying capacity in patients with anemia stemming from chronic kidney disease and in patients treated with chemotherapy. However, it may also promote tumor growth. It is probably best known as an illicit means of increasing performance in endurance sports (e.g., long-distance cycling).

The fact that BDNF varies across brain regions is important for understanding its relationship to pathological conditions, and it is possible that stressor-related BDNF changes in different brain sites may be related to diverse emotional and cognitive changes. For instance, hippocampal BDNF might contribute to the memory of negative events, whereas prefrontal cortex BDNF might be related to cognitive changes introduced by negative events. It needs to be considered, as well, that other neurotrophins, such as FGF-2 (fibroblast growth factor), are also influenced by stressful experiences and inflammatory activation, and have been associated with depression (Tang et al., 2018a, b). There is reason to suppose that neural plasticity may be a key factor associated with sensitization effects, much like it influences learning and memory, and might thus contribute to the development of psychopathology. In particular, an initial stressor experience could engender an increase of FGF-2, which could influence the neuronal responses elicited upon later stressor encounters (or in response to amphetamine or cocaine) and might thereby encourage psychological disorders (Flores and Stewart, 2000).

There is yet another important consideration that can't go unmentioned. Among mice that had experienced chronic social stress over a 2-week period, neurons born during the last 5 days of the stressor regimen were elevated relative to that which had occurred in nonstressed controls. In fact, the increased neurogenesis was comparable to that seen when mice were raised in an enriched environment. These findings were surprising as chronic stressors have typically been viewed as resulting in diminished plasticity and neurogenesis. However, in assessing the impact of stressors on neurogenesis it is important to consider the behavioral coping strategies adopted in response to the challenge. In this instance, the analysis comprised cells born during the last 5 days of the social stressor, at which time mice had no longer been attacked, having adopted a submissive position, essentially reflecting adaptation to the otherwise distressing conditions. It was of particular significance that these new neurons were more likely to be incorporated into the dentate gyrus of the hippocampus during the ensuing 10-week poststress recovery period. When these mice were chronically stressed again, these neurons were more likely to be affected. Among other things, spine density and branching nodes were reduced relative to neurons that had not been born during the stress period, and performance in a spatial learning task was impaired. It was suggested that neurons born during a stress period when coping methods were being established were uniquely adapted to deal with future challenges in the form of social stressors. However, it wasn't certain whether the neurobiological changes rendered animals more vulnerable to psychopathology, or conversely, made animals more resilient (and more able to recover) from further challenges (De Miguel et al., 2018).

Hormonal Changes Stemming From Stressor Experiences

Numerous hormones are responsive to stressors and to immune challenges (see Table 6.3). As in the case of neurotransmitters, several receptor subtypes can be activated for many of these hormones, and consequently, different behavioral outcomes can develop. As lifestyle factors, such as exercise, can moderate hormonal stress responses, possibly through actions of GABA neuronal activity, it can reasonably be expected that behavioral and physical pathologies will be similarly affected.

As briefly mentioned earlier, numerous hormones influence inflammatory immune processes, and conversely, immune and cytokine changes markedly affect hormonal processes. For instance, at physiological levels, cortisol can limit immune and cytokine elevations that could otherwise be instigated (e.g., by stressors), and at pharmacological levels (i.e., at doses greater than those produced naturally) immune functioning is suppressed. Conversely, treatments that increased circulating IL-1β levels engender marked cortisol elevations. Thus, in considering stressor-provoked pathology, particularly those involving inflammatory processes, it is necessary to consider that multidirectional links exist between diverse processes. The link between glucocorticoids and immune functioning could occur through various processes, such as altering expression of genes related to inflammation, or by affecting cell mitochondria so that energy needed by immune cells is diminished (Auger et al., 2024).

Within a matter of seconds following the appearance of a potential threat, several brain regions, such as the prefrontal cortex, may be activated for primary appraisals of the situation to be made, and presumably, secondary appraisals involved in the selection of coping strategies follow soon afterward. Autonomic nervous system functioning is similarly set in play so that various organs are altered in preparation for the challenge. Peripheral epinephrine release promotes increased heart rate and blood pressure, thereby providing elevated oxygen to the brain and peripheral organs, and it stimulates the release of immune cells from lymphoid organs so that they're in circulation, where they might be needed.

Corticoid Responses

Stressors ordinarily promote the activation of several brain regions, many of which are involved in the regulation of HPA functioning (Herman et al., 2016). In response to a stressor, the prefrontal and infralimbic

TABLE 6.3 Hormones Related to Stress Responses.

Secreted Hormone	Biological Effect	Behavioral Outcome
Corticotropin-releasing hormone (CRH)	Formed in the paraventricular nucleus of the hypothalamus, as well as several limbic and cortical regions. Stimulates ACTH release from the pituitary gland.	Involved in stress responses, promoting fear and anxiety, and diminishes food intake and increases metabolic rate.
Adrenocorticotropic hormone (ACTH)	Formed in the anterior pituitary gland. Stimulates corticosteroid (glucocorticoid and mineralocorticoid) release from adrenocortical cells.	Stress responses elicited are primarily due to actions on adrenal corticoids.
Arginine vasopressin (AVP)	Released by both the paraventricular and supraoptic nucleus: promotes water reabsorption and increased blood ACTH.	Together with CRH, may synergistically increase stress responses. Influences social behaviors.
Cortisol (corticosterone in rodents)	Released from the adrenal gland. Has antiinflammatory effect, promotes the release and utilization of glucose stores from liver and muscle, and increases fat storage.	Prototypical stress hormone; influences defensive behaviors, memory processes, caloric intake, and may promote preference for high-calorie foods under stressful circumstances (stimulates consumption of comfort foods).
Mineralocorticoids (e.g., aldosterone)	Released from the adrenal gland. Stimulates active sodium reabsorption and passive water reabsorption, thus increasing blood volume and blood pressure.	Aldosterone influences salt and water balance. Excessive sodium and water retention leads to hypertension. Low levels of aldosterone lead to a salt-wasting condition evident in Addison's disease.
Epinephrine (adrenaline) and norepinephrine (noradrenaline)	Produced in the adrenal gland (medulla) and within sympathetic neurons; increases oxygen and glucose to the brain and muscles; promotes vasodilation, increases catalysis of glycogen in liver and the breakdown of lipids in fat cells; increases respiration and blood pressure; suppresses bodily processes (e.g., digestion) during emergency responses; influences immune system activity.	Elicits fight or flight response; in the brain, EPI and NE have multiple behavioral actions related to defensive behaviors (e.g., vigilance, attention).
Beta-endorphin	Secreted from several sites, such as the arcuate nucleus.	Inhibits perception of pain.
Oxytocin	Formed in magnocellular neurosecretory cells of the supraoptic and paraventricular nuclei of the hypothalamus.	Associated with parental bonding, ingroup bonding, maternal behaviors, and a wide range of prosocial behaviors, as well autism, depression, fear, and anxiety.

From Anisman, H., 2016. Health Psychology. Sage Publications, London.

cortex, which are involved in the appraisal of stressful events, may promote inhibition or excitation of other brain regions involved in the stress response. When an event is judged to be a threat, the paraventricular nucleus (PVN) of the hypothalamus is stimulated, resulting in CRH release from cells at the median eminence (situated at the ventral portion of the hypothalamus). The CRH stimulates the anterior portion of the pituitary gland, provoking the release of adrenocorticotropic hormone (ACTH), which then causes cortisol (corticosterone in rodents) release from the adrenal gland. Corticosterone enters the bloodstream and comes to affect several target organs, producing multiple adaptive effects essential for survival. Among other things, cortisol functions to increase blood sugar that is derived from lactate and glycogen present in the liver and in muscle and facilitates the metabolism of fats, carbohydrates, and proteins. The cortisol also reaches the brain where neurons in the hippocampus and hypothalamus are stimulated, which inhibits further CRH release (Fig. 6.4).

In considering the impact of hormonal changes on the development of pathology, it is not only important to consider how readily these are elicited but also how quickly they resolve having served their immediate function. When circulating cortisol reaches the brain, it influences hypothalamic and hippocampal neurons so that further HPA functioning is inhibited. If this self-regulation does not operate appropriately, persistent release of cortisol can occur, which may have detrimental consequences on immune functioning as well as on brain processes. Specifically, the sustained stimulation of glucocorticoid receptors present in hippocampal cells may lead to cell damage. As the hippocampus is a fundamental component of the negative feedback loop, HPA activity may persist, leading to still further cell damage, and eventually the development of cognitive disturbances. Given that hippocampal cell loss and elevated cortisol levels ordinarily increase with age, potentially being exacerbated by chronic stressors, the risk for cognitive impairments is much more pronounced (McEwen and Gianaros, 2011).

pathology (e.g., PTSD). More than these effects, corticosterone may have preparatory actions, facilitating coping with impending stressors, and may promote neuroplasticity, thereby affecting processes (e.g., sensitization) that are important in dealing with subsequently encountered challenges (Sapolsky et al., 2000). Finally, corticosterone has permissive actions that contribute to the provocation or amplification of other stressor-provoked hormonal changes, including those that can limit the immune and inflammatory activity that might otherwise be detrimental to health.

Impact of Chronic Stressors

With repeated stressor exposure, particularly if the stressor is a homogenous one (i.e., the same stressor is administered on successive days), the extent of cortisol released declines. This seeming "adaptation" should not be misconstrued as reflecting the system being unable to respond to a stressor, but instead reflects an adaptive response to diminish excessive activation in response to specific stimuli. Indeed, after repeated exposure to a given stressor, subsequent exposure to a novel challenge promotes an exaggerated corticosterone response (Uschold-Schmidt et al., 2012). In effect, the HPA system remains prepared to deal with novel insults, despite the decline in reactivity to a particular stressor.

Aside from the diminished levels of corticosterone that may accompany chronic stressor experiences, it seems that glucocorticoid receptors may become progressively less sensitive (glucocorticoid resistance) (Walsh et al., 2021). Since glucocorticoid functioning is fundamental in controlling (limiting) autoimmune, infectious, and inflammatory disorders, as well as some psychiatric conditions (e.g., depression), the downregulation of corticoid sensitivity may contribute to aggravation of illness symptoms. Findings such as these make it clear that a more detailed understanding is necessary regarding the links between cytokines and glucocorticoid signaling in the hope of developing novel strategies to deal with the impacts of glucocorticoid resistance.

Stressor-Provoked Cortisol Changes in Humans

There is a disconnect between the impact of stressors in rodents relative to those apparent in humans. Simply taking a rodent out of its cage and placing it in a novel chamber can elicit a 100% increase in corticosterone, and stronger stressors can produce an increase of 400%–800%. These marked effects of stressors may have given rise to the erroneous expectation that this would also occur in humans, but the changes observed are typically relatively modest. Academic examinations, for instance, hardly affect cortisol levels, and even some fairly

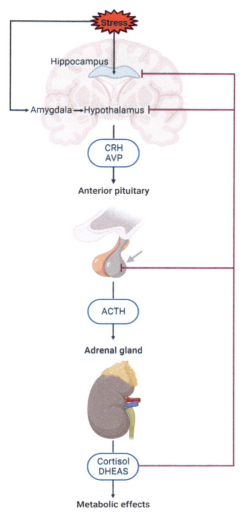

FIG. 6.4 Stressful events cause amygdala activation that provokes activation of the paraventricular nucleus of the hypothalamus. This promotes the release of CRH and arginine vasopressin from the median eminence situated at the base of the hypothalamus. These peptides promote the release of ACTH from the anterior pituitary, which reaches the adrenal gland through the bloodstream, causing the release of mineralocorticoid and glucocorticoids, such as cortisol. The cortisol enters the bloodstream, ultimately stimulating glucocorticoid receptors at the hypothalamus and mineralocorticoid and glucocorticoid receptors at the hippocampus, which serve to limit further HPA activation. Black arrows denote excitatory processes; red lines with blunted ends signify inhibitor processes. *Source: Created using BioRender.com.*

In addition to its actions on hypothalamic neuronal functioning, corticosterone influences neuronal activity at the amygdala, which affects visceral and somatic pain sensitivity among female rodents, and these effects varied with the estrous cycle, likely owing to interactions with ovarian hormones. Glucocorticoid stimulation of the prefrontal cortex can also enhance memory consolidation (Shields et al., 2016), which can have obvious advantageous actions, but as we'll see later, this can also favor the development and maintenance of stress-induced powerful stressors (e.g., in blood samples taken just prior to heart surgery) engendered relatively modest (30%–50%) cortisol elevations (Michaud et al., 2008). In fact, the diurnal cortisol variations that ordinarily occur are more pronounced than those associated with many challenges.

In contrast to these anticipatory stressors, pronounced cortisol increases are evident in the "Trier Social Stress Test" (TSST), which comprises the distress created by public speaking followed by an arithmetic challenge (e.g., sequentially subtract 13 beginning with 1,022). This outcome might be generated because the stressor involves a "social evaluative" threat that promotes shame (Dickerson and Kemeny, 2004) or anger (Matheson and Anisman, 2012). As expected, having social support available may limit cortisol changes ordinarily elicited by stressors, including the impact of the TSST. As expected, having social support available may limit cortisol changes ordinarily elicited by stressors, including the impact of the TSST, although this outcome was not consistently observed. Indeed, in some instances (e.g., when a male partner is present to support a woman), cortisol reactions were more pronounced than in the absence of this support, pointing to the source of support also acting as an evaluative threat.[3]

The Morning Cortisol Response

Like many hormones, levels of circulating cortisol vary over the day, being high in the morning, after which a decline occurs over the course of the afternoon and evening, eventually reaching the lowest levels at about 2300 h. Significant salivary cortisol increases occur (about 40%), within the first 30 min after awakening, and then decline rapidly (Schmidt-Reinwald et al., 1999). The initial cortisol rise is more pronounced among individuals who are experiencing ongoing psychological challenges, and the subsequent cortisol decline may occur relatively slowly. In contrast, morning cortisol levels are diminished among individuals who have been undergoing a chronic stressor or those who have encountered a traumatic stressor that provoked PTSD (Michaud et al., 2008). Once again, the flattening of the cortisol curve might reflect an adaptive response to preclude the adverse effects that could otherwise emerge.

The initial enthusiasm in assessing the morning cortisol response to life stressors seems to have waned, as it is procedurally problematic, because participants might not be reliable in taking saliva samples immediately after awakening and again precisely 30 min afterward. Despite the problems inherent in determining the morning cortisol response, interesting data have been obtained through this procedure. For example, the serotonin transporter gene was associated with variations in the morning cortisol response, and the cortisol changes were marked in stressed elderly individuals (Ancelin et al., 2017). It is possible that the cortisol variations, and related serotonin processes, may contribute to age-related depression.

Cortisol in Relation to Traumatic Experiences

Like the cortisol response that accompanies a chronic stressor, traumatic events that lead to PTSD were associated with cortisol levels that were diminished relative to that of nonstressed individuals or those who had encountered trauma but had not developed PTSD (Yehuda, 2002). The diminished cortisol levels may reflect a protective response to limit the potentially damaging effects that could be elicited by persistent rumination and memories that could readily be rekindled by a variety of triggering cues (see Chapter 11). As we saw in animal studies, this seemingly protective action should not be misinterpreted as suggesting that HPA action is downregulated in all situations. Instead, the diminished HPA response is selective, so that effective activation occurs when necessary.

Ordinarily, injection of CRH promotes ACTH release from the pituitary, and the resulting increase of circulating levels of this hormone promotes cortisol release from the adrenal gland. However, among depressed women who had experienced abuse when they were young, the ACTH release elicited by CRH administration was diminished, indicating that HPA functioning was downregulated. However, when these women experienced a stressor in the form of a social evaluative threat (public speaking), the ACTH release response was exaggerated (Heim et al., 2008). It was similarly reported that cortisol levels were low among women who had experienced PTSD symptoms related to psychological and/or physical abuse in a dating relationship. However, upon encountering reminders of their abuse, cortisol levels markedly increased (Matheson and Anisman, 2012). The downregulated HPA functioning among individuals with PTSD may be protective so that hyper-reactivity to environmental cues would be less likely to result in adverse effects (e.g., on hippocampal cells). At the same time, indiscriminate downregulated HPA functioning might be maladaptive, as activation of this system might be necessary to deal with challenges. Thus, in response to potentially meaningful stressors, activation of brain regions (e.g., prefrontal cortex or amygdala) involved

---

[3] Having a dog present (even one that was not the participant's pet) also reduced the cortisol response in the TSST (Polheber and Matchock, 2014). This speaks to the value of animals as stress buffers (and perhaps the benefit of trading a romantic partner for a loyal pet).

in stressor-appraisal processes may override the HPA blunting, and an exaggerated ACTH or cortisol response will be apparent.

Impact of Prenatal and Early Postnatal Events

There is little question that adverse childhood events (ACEs) can have profound neurobiological and psychological repercussions that persist throughout life. Among rodents that experienced early life stressors, such as repeated brief periods during which they were separated from the dam, corticosterone levels were altered. The nature of the changes varied with the time at which maternal separation was introduced, whether the daily separation was brief or long (e.g., 15 min vs. 3 h), and varied with sex. It was frequently reported that relatively long periods of separation (3 h) resulted in elevated corticosterone levels, although inconsistent findings have been reported (Nishi, 2020), which is not overly surprising given that the effects of the separation procedure vary with sex and the mouse strain examined. Moreover, the effects of the stressor differ as a function of how dams respond to their pups upon being reunited. Specifically, stress responses were less pronounced if moms attended more closely to their pups, reflected by high levels of licking and grooming them. Like maternal separation, a stressor in the form of providing limited nesting/bedding material during early life resulted in elevated corticosterone levels, as well as gut microbiota dysbiosis, more so in females than in males (Moussaoui et al., 2017).

In addition to the corticosterone variations, early life stress could affect neuronal activity in several brain regions, once again varying with the nature of the procedures used (Nishi, 2020). Similarly, the glucocorticoid changes engendered by maternal separation were accompanied by altered microglial functioning that ordinarily contributes to synaptogenesis, synaptic pruning, and axonal growth (Johnson and Kaffman, 2018) and brain circuitry may not develop as it should, leading to the development of psychological and cognitive disturbances (e.g., Milbocker et al., 2021). Microglia responses may become sensitized so that later stressor experiences result in exaggerated cytokine release and HPA functioning. In addition, early life stressors influence immune functioning and increase proinflammatory cytokine functioning that affects the brain, thereby impacting the development of pathology (Brenhouse et al., 2019).

Aside from frank changes in basal corticosterone and the hypercortisolemia that occurs upon further stressor encounters, the behavioral effects of early life experiences are determined by glucocorticoid receptor functioning. These receptor changes are subject to modification by epigenetic processes, thereby affecting the adult responses to stressors (e.g., Szyf, 2019; Szyf and Bick, 2013) as well as

brain processes that involve neurotrophic factors, synaptic structure, and neuroimmune responses (Martins de Carvalho et al., 2021). In this respect, individuals who had experienced ACEs subsequently exhibited a blunted cortisol response to psychosocial stressors together with elevated circulating C-reactive protein (CRP) that reflected inflammation. These changes were accompanied by increased amygdala neuronal activity in response to negative information (Hakamata et al., 2022). Ultimately, these diverse epigenetic actions lend themselves to the development of depressive illness, cognitive disturbances, and biological aging.

Like the actions of early life stressful events, unfavorable conditions experienced by a pregnant female can have lasting consequences on the mental health of offspring, stemming in part from exposure to the fetus being exposed to elevated corticosterone, which affects brain development. Once more, these actions can be influenced by epigenetic programming of glucocorticoid receptors and those that occur with other processes (Saavedra and Salazar, 2021). The multiple hormonal and neuroinflammatory changes produced by prenatal stressors have been linked to cognitive and behavioral disturbances, such as attention and learning deficits. Additionally, these experiences may result in a disposition toward anxiety, depression, and schizophrenia, and as we'll see in later chapters, these effects are moderated by genetic factors and the postnatal care that pups receive from their mother.

Discrimination as a chronic stressor

It won't be surprising to anyone that racial discrimination is a powerful stressor that undermines psychological and physical health in adults and children. Moreover, structural inequities (lower income, living in disadvantaged neighborhoods) experienced by disadvantaged racial groups may become structurally embedded within the brain, which can affect responses to stressors and may favor the occurrence of mental disorders.

Among Black individuals, sensitivity to racism may be elevated, and amygdala reactivity is altered, likely owing to structural inequities, and when PTSD occurred, it was accompanied by decreased connectivity between the amygdala and other brain regions (Webb and Harnett, 2024). The effects of structural inequalities were even seen in children as young as 9 years of age, who displayed lower amygdala and prefrontal cortical volume in direct relation to adversities experienced (Dumornay et al., 2023).

Predictably, cortisol increase can be elicited by threats to an individual's social identity, as observed in response to acute racial discrimination (Matheson and Anisman, 2012) and were elevated in response to a film depicting frank racism (Matheson et al., 2021). As with other chronic challenges, chronic racism tracked over a 20-year period, beginning when Black participants were in grade school, was accompanied by flattening of the diurnal cortisol profile, much as it is with chronic or severe stressors that lead to PTSD (Adam et al., 2015). Consistent with these findings, cortisol levels were elevated the morning following encounters with racial discrimination, and when measured in real time the daily cortisol rhythm was flattened with same-day microaggression (Nam et al., 2022). It has been suggested that the acute awareness of racism and stigmatization that occurs in daily life might result in individuals developing protective mechanisms that act against the damaging neurobiological changes that might otherwise develop.

Aside from cortisol variations, experiences of racism were accompanied by elevated circulating proinflammatory cytokine levels, although this was less apparent if individuals had positive racial identities (Brody et al., 2015). Moreover, among African American women, perceived racial discrimination measured over eight years was associated with elevated levels of proinflammatory cytokines, which predicted the occurrence of chronic inflammatory illnesses (Simons et al., 2021). Perceived discrimination was also accompanied by emotional dysregulation in the form of venting and denial, which were linked to the presence of chronic inflammation reflected by elevated IL-6 and CRP (Doyle and Molix, 2014). Effects such as these also occurred within the brain, as perceived discrimination among ethnic minority members was associated with activation of the anterior cingulate cortex and ventral striatum, which are associated with decision-making and reward processes, respectively (Akdeniz et al., 2014).

The pervasive effects of chronic racism, not unexpectedly, are also manifested in the form of hypertension, blood pressure, cardiovascular reactivity, shortened telomere length (Chae et al., 2014), and increased health risk, particularly mental health problems (Wallace et al., 2016). The effects of discrimination are exceptionally powerful and have effects that are "more than skin deep" (Berger and Sarnyai, 2015).

Corticotropin-Releasing Hormone

In considering base emotions, such as fear and anxiety, attention has frequently focused on subcortical circuits. However, the position has been taken that higher cortical structures interpret these basic (unconscious) feelings into conscious emotional experiences as well as the memories of these experiences (LeDoux and Brown, 2017). In effect, innate fear and defensive survival circuits governed by the amygdala are readily activated by threatening stimuli, and at the same time, cortical circuits are involved in processing the threat. These may comprise sensory cortical regions (e.g., visual cortex), processes related to memory, and those that subserve cognitive appraisals of the stressor (e.g., anterior cingulate cortex

and aspects of the orbital frontal cortex and prefrontal cortex). Through the incorporation of unconscious and conscious experiences of fear, perspectives of how individuals make sense of their situation are broadened and provide a way of approaching complex attributes that are linked to threats (e.g., rumination, existential threats).

As described earlier, the release of CRH from the PVN of the hypothalamus is a pivotal component of HPA functioning, promoting the secretion of ACTH from the anterior pituitary gland. With chronic stressor exposure, arginine vasopressin (AVP) is increased within CRH nerve terminals, and the corelease of CRH and AVP synergistically increases ACTH release. In addition, under these conditions, several other processes are affected (e.g., the density of norepinephrine and glutamate receptors is elevated on CRH cell bodies), which promotes greater downstream effects that favor pathology emerging (Herman and Tasker, 2016).

Aside from the involvement of CRH in HPA functioning, this peptide is present in several other brain regions that are fundamental for stressor-related behavioral changes. Specifically, stressor-related CRH activity can affect locus coeruleus neuronal activity, potentially influencing vigilance in threatening situations. Also, CRH activity within prefrontal cortical regions may contribute to stressor-related appraisals and decision-making, while amygdala and hippocampal CRH changes may be involved in fear responses and fear memories. Amygdala variations of CRH are exceptionally well documented, and changes of this peptide in different portions of the amygdala contribute to the acquisition, expression, and extinction of fear responses (LeDoux, 2012).

The terms fear and anxiety are occasionally used interchangeably despite being different from one another and may involve distinguishable neural circuits. Fear is created by specific stimuli and involves CRH neuronal activity within portions of the amygdala. Anxiety, in contrast, is a mood condition promoted by diffuse stimuli and may be more closely aligned with CRH variations that occur within the bed nucleus of the stria terminalis (BNST), which is a major output pathway that runs from the amygdala to the hypothalamus (Partridge et al., 2016). Later, when we discuss anxiety disorders, the relevance of these regions in dealing with immediate stressors and those that are further off will become clear.

Supporting the involvement of CRH in anxiety, administration of a CRH receptor antagonist could attenuate anxiety- and depression-like behavior provoked by stressors. Likewise, changes in these mood states can be achieved in mice that had been genetically engineered so that the expression of CRH receptors was altered. However, this depended on which of the primary CRH receptors, CRH_1 or CRH_2, were altered. Diminished anxiety was reliably produced by pharmacological antagonism of CRH_1 receptors or by engineering mice so that the CRH_1 receptors were deleted (Dedic et al., 2018).

The data concerning the involvement of CRH_2 receptors in anxiety is less convincing, although it was reported when CRH_2 receptors were knocked out, anxiety levels were altered. In double mutants (where both the CRH_1 and CRH_2 were knocked out), these effects were somewhat more pronounced. It seemed likely that early experiences related to maternal care moderated the actions of these receptors, varying in a sex-dependent fashion. It hasn't received enough attention, but the possibility exists that the two CRH receptor subtypes influence different aspects of anxiety (Reul and Holsboer, 2022). For example, CRH_1 receptors may be involved in the regulation of attention, executive functions, and sleep disturbances, and may contribute to the conscious experience of emotions promoted by stressors. Activation of CRH_2, perhaps in other brain regions, might be more aligned with processes necessary for survival, such as eating, reproduction, and defense. This raises the possibility that patients presenting with anxiety and depression might differentially benefit from agents that act as CRH_1 or CRH_2 receptor antagonists depending on the features of the anxiety condition.

Cannabinoids

There has been growing interest in the use of cannabis (marijuana) to diminish pain perception, symptoms of physical illness, and nausea resulting from cancer treatments, as well as stress-related psychological disorders, such as PTSD and anxiety disorders (e.g., Sarris et al., 2020). However, cannabis can also provoke adverse actions, such as disorientation, drowsiness, confusion, and loss of balance, as well as more serious problems, ranging from dependence, respiratory problems, and psychosis in vulnerable individuals (Gobbi et al., 2019).

The psychoactive component of cannabis, Δ^9-tetrahydrocannabinol (THC), and naturally occurring endocannabinoids (eCBs) bind to specific cannabinoid receptors (e.g., CB_1 and CB_2), which are thought to contribute to the actions of THC. Considerable evidence has pointed to eCB signaling processes being linked to mechanisms underlying neuropsychiatric disorders, such as anxiety, substance use disorders, as well as obesity (Scheyer et al., 2023). There have also been indications that CB_1 receptors within the lateral habenula (which receives input from the lateral septum, hippocampus, nucleus accumbens, thalamus, lateral hypothalamus, and the raphe nucleus), through actions on acetylcholine and serotonin, might also be involved in the expression of memories involving aversive events (Dolzani et al., 2016). The stimulation of CB_1 within the amygdala, through effects on glutamate functioning and by affecting activation of the HPA axis, may be fundamental in mediating emotional responses associated with a stressor and may influence the consolidation of emotional memories (Hill et al., 2018). In fact, administration of a CB_1 receptor agonist soon after a stressor experience enhanced the retention of avoidant response. These and related findings gave rise to the view that in some people an endocannabinoid deficiency may increase vulnerability to illnesses promoted by intense stressors (e.g., PTSD), and they may resort to self-medication with cannabis to deal with trauma-related disorders. It is noteworthy that although intake of cannabis at low doses can diminish anxiety, in some individuals anxiety can be provoked at high doses (Sharpe et al., 2020).

While the focus concerning the effects of cannabis has been on the actions of THC, increasing attention has been devoted to the actions of the second phytocannabinoid, cannabidiol (CBD), which has characteristics like THC but does not contain the psychoactive component needed to produce a high. CBD activates CB_1 and CB_2 receptors as well as $5\text{-}HT_{1A}$ receptors, may affect neurogenesis, and consequently may diminish anxiety and symptoms of depression (García-Gutiérrez et al., 2020). Furthermore, activation of CB_2 receptors contributes to the inhibition of inflammatory responses and might thus be important in dealing with physical disturbances, as well as mood states.

The naturally occurring brain eCBs, anandamide (AEA) and 2-arachidonoylglycerol (2-AG), have been implicated as key players in basal and threat-elicited anxiety. It seems that these eCBs bind to both CB_1 and CB_2 receptors, although AEA only weakly binds to CB_2 receptors. While AEA is believed to mediate basal feelings of calmness as well as anxiety associated with stressor experiences, the function of 2-AG, which is activated somewhat later, is to turn off stress responses. This will be described in greater detail when we discuss anxiety in Chapter 10.

Estrogen

The fact that women are more vulnerable to anxiety, depression, and autoimmune disorders than men raises the obvious possibility that sex hormones contribute to their disposition to these conditions. Estradiol, the predominant estrogen produced in the ovaries has broad physiological actions. Aside from its role in female reproductive processes and sexual development, estradiol influences protein synthesis, fluid balances, gastrointestinal functioning, cholesterol levels, and fat depositions, as well as affecting bone density, arterial blood flow, immune functioning, and multiple brain processes. The impact of stressors varies over the estrous cycle. For instance, ventral tegmental DA neuronal functioning is altered during different phases of the estrous cycle, which affects stressor-provoked reward-related behaviors (Shanley et al., 2023). As expected, cognitive impairments elicited by stressors in female rats were abrogated by estrogen applied to the prefrontal cortex (Yuen et al., 2016).

While not dismissing the importance of sex hormones, other biological processes ought to be considered in accounting for sex dimorphism. Several behavioral and neuroendocrine changes (e.g., cortisol) elicited by stressors are greater in females than in males, and multiple stressor-elicited brain changes (e.g., 5-HT_{1A} receptors) may cause the female bias toward anxiety and depression (Albert and Newhouse, 2019). Consistent with these changes, neuronal functioning in brain regions that govern behavioral and cognitive responses to stressors (e.g., various cortical sites) is decidedly more pronounced in female than in male rodents, and hormonal changes that can influence neuronal growth factors (e.g., BDNF) vary over the estrous cycle. Stressors also have more pronounced effects within the locus coeruleus-norepinephrine system among females, which is accompanied by hyperarousal that may be relevant to the elevated occurrence of PTSD in women relative to men (Bangasser et al., 2016). In addition, connections between the hypothalamus and lateral habenula, likely involving glutamate inputs, have been associated with aversion and mood regulation (Lazaridis et al., 2019). This circuit is subject to modification by estrogen receptors that are present, which might contribute to the greater depression propensity in females (Calvigioni et al., 2023).

To a considerable degree, the greater effects of stressors in female rodents are recapitulated in humans. These stressor actions are dependent on the estrous phase in that neuronal activity in stress-relevant brain regions (amygdala, orbitofrontal cortex) was particularly marked in response to stress-arousing stimuli, and these actions were more pronounced during the early follicular phase (before ovulation during which estradiol levels climb) compared with mid-cycle. Furthermore, stressor-provoked cortisol responses are especially notable during the latter part of the menstrual cycle (luteal phase), a time during which progesterone is particularly high. In line with such findings, among women using oral contraceptives, the influence of a social-evaluative stressor on cortisol release was diminished (Kudielka and Kirschbaum, 2005). Overall, the greater female HPA response to stressors occurs in conjunction with elevated circulating estradiol, whereas progesterone diminishes negative glucocorticoid feedback so that cortisol release is more apt to persist (Oyola and Handa, 2017).

Testosterone

The primary sex hormone in males, testosterone, is formed in the testes, and like estrogen, it is also produced in the adrenal glands to a limited extent. The release of testosterone is influenced by hypothalamic and pituitary processes through a negative feedback mechanism, much like that described concerning cortisol. Aside from the production of male reproductive tissues, and the development of secondary male features, including muscle growth and bone density, testosterone is involved in dominance behaviors (and intraspecies aggression). Levels of this hormone are especially elevated in animals that had won in competitions or were higher in a dominance hierarchy (Ziegler and Crockford, 2017).

By virtue of testosterone's contribution to muscle development, it may be instrumental in helping male animals protect the pack. In humans, this protective tendency (termed parochial altruism) may have adaptive value for the species. At the same time, testosterone may disturb collaborative behaviors and has been related to males being egocentric in that they give greater weight to their own judgments relative to those of others (egocentric choices) during joint decision-making (Wright et al., 2012). At the other extreme, in the presence of low testosterone levels, like that which may occur in some older men, mood changes may occur, such as depressive symptoms as well as anxiety (Zitzmann, 2020).

High doses of androgenic steroids (anabolic steroids) have been used to build muscles and hence improve strength and athletic performance. In addition to increasing muscles in the arms and legs, these steroids also affect the heart muscles and may introduce problems in other organs, particularly the liver, and may adversely affect immune functioning, engender coronary artery disease, and favor the development of skin problems. Through their actions on other hormones and brain neurotransmitters, anxiety can be instigated, and cognitive functioning can be affected. As adolescence is a period that is exceptionally sensitive to the effects of gonadal steroid hormones on brain development (Schulz and Sisk, 2016), the use of anabolic steroids at this time may be especially dangerous.

Stress and Reproduction

In addition to the neurobiological differences elicited by stressors in males and females, proceptive behaviors (e.g., courting) and receptivity (less responsive to overtures) are more affected by stressors in females. Stressors may also influence reproduction because of disrupted ovulation coupled with uterine alterations that impede the implantation of a fertilized egg (Wingfield and Sapolsky, 2003). By promoting the secretion of the opioid peptide β-endorphin, stressors increase the availability of gonadotropin inhibitory hormone (GnIH), which diminishes luteinizing hormone release, consequently promoting infertility. As expected, knocking out the gene for GnIH eliminated the reproductive failure otherwise

TABLE 6.4 Sex Hormones.

Secreted Hormone	Biological Effect	Behavioral Outcome
Testosterone	Male steroid hormone produced in the testis in males and ovaries in females. To a lesser extent it is produced in adrenal glands. Involved in the development and sexual differentiation of brain and reproductive organs; fundamental in secondary sexual features, including body hair, muscle, and bone mass.	Associated with sexual behavior and libido. Linked to aggressive and dominant behaviors.
Dehydroepiandrosterone (DHEA)	In males, produced in adrenals, gonads, and brain. Acts as an anabolic steroid to affect muscle development.	Acts like testosterone. Has been implicated in maintaining youth.
Estrogens (estrone, estradiol, and estriol)	Estradiol is the predominant form of the three estrogens produced in the ovaries. It is the principal steroid regulating hypothalamic-pituitary ovarian axis functioning. It is involved in protein synthesis, fluid balances, gastrointestinal functioning and coagulation, cholesterol levels, and fat depositions. It affects bone density, liver, arterial blood flow, and has multiple functions in brain.	Influences female reproductive processes, and sexual development; important for maternal behavior, maintaining cognition, as well as anxiety and stress responses.
Progesterone	Formed in the ovary; precursor for several hormones; involved in triggering menstruation, and for maintaining pregnancy (inhibits immune response directed at embryo); reduces uterine smooth muscle contraction; influences resilience of various tissues (bones, joints, tendons, ligaments, and skin).	Influences female reproductive processes, and sexual development; affects maternal behavior; disturbs cognitive processes. Has antianxiety actions.
Luteinizing hormone (LH)	Produced in the anterior pituitary gland. In females, an "LH surge" triggers ovulation and development of the corpus luteum, an endocrine structure that develops from an ovarian follicle during the luteal phase of the estrous cycle.	Behavioral changes associated with estrogen or testosterone are elicited indirectly through actions on other steroids.
Follicle-stimulating hormone (FSH)	Secreted from the anterior pituitary gland; regulates development, growth, pubertal maturation, and reproductive processes. Together with LH, it acts synergistically in reproduction and ovulation.	Behavioral changes associated with estrogen or testosterone are elicited indirectly through actions on other steroids.
Prolactin	Secreted from the anterior pituitary; regulated by the arcuate nucleus of the hypothalamus.	Involved in lactation in mammals, and involved in sexual behavior and gratification, influences levels of estrogen and progesterone. Regulates immune functioning, and acts like a growth-, differentiating-, and antiapoptotic factor.

produced by a stressor (Geraghty et al., 2015). In addition to these actions, corticoids released in response to stressors reduce the responsiveness to luteinizing hormone.

Stressors experienced before maturity may influence later adult sexual behavior. Among female juveniles exposed to a stressor, subsequent receptivity is reduced as is behavioral responsiveness to estradiol and progesterone, possibly stemming from an enduring reduction of estrogen receptors in several brain regions (Blaustein et al., 2016). As we'll see when we discuss stress-related immune alterations, infection in a pregnant dam may give rise to a variety of pathological outcomes in offspring, including their maternal behaviors being disturbed and male pups later displaying disrupted sexual behavior. While the lasting behavioral repercussions associated with early life stressors may be a result of estrogen and testosterone variations, as described in Table 6.4, several other sex-related hormones are affected by stressors, which have considerable bearing on mental and physical health.

Prolactin

Prolactin, which is released from the anterior pituitary owing to stimulation by thyrotropin-releasing hormone, is best known for its role in lactation, although it is also related to maternal behaviors, sexual behavior, and sexual pleasure. As well, prolactin has been linked to resilience to stressors, neuroprotection, and neurogenesis, and is important for emotional regulation during pregnancy (Faron-Górecka et al., 2023). Moreover, receptors for prolactin are present within the central amygdala and the nucleus accumbens, supporting the possibility that this hormone contributes to emotional responses, including anxiety and depression. These actions of prolactin may be related to its role in moderating inflammation, which could also influence heart disease and potentially influence the response to viral infection (Faron-Górecka et al., 2023).

Oxytocin

Oxytocin has received considerable attention given its proposed contribution to a variety of prosocial behaviors. Among other things, oxytocin has been linked to social and pair bonding, generosity, trust, altruism, attention to positive cues, and mood (Insel and Hulihan, 1995). It was also suggested that oxytocin might play a significant role in coping with stressors, especially those that involve social processes (McQuaid et al., 2016), and its administration limited stressor-provoked responses, such as cortisol release (Cardoso et al., 2014). Having social support has been associated with stress-buffering actions, which may have been tied to oxytocin functioning (McQuaid et al., 2016). Social defeat increased oxytocin within the BNST (the extended amygdala) of female mice but not in males. However, the administration of oxytocin attenuated the behavioral changes induced by a stressor, but this only occurred in male mice, supporting the view that this hormone contributes to sex-dependent behavioral outcomes (Steinman et al., 2016).

Tend and Befriend Versus Tend and Defend

Oxytocin might play different roles in males and females given that oxytocin was germane to maternal bonding experiences. Specifically, acting together with opioid peptides and gonadal hormones, oxytocin was seen as promoting "tend-and-befriend" characteristics in females, including nurturing behaviors and the development and maintenance of social connections, which serves to augment self-protection (Taylor et al., 2000). The prosocial behaviors attributed to oxytocin are not only apparent in females but also occur in males, although there were several distinguishing features evident between the sexes. Males were more likely to engage in an evolutionarily advantageous "tend-and-defend" characteristic. In addition to altruism in which individuals engage in behaviors to facilitate and encourage the well-being of group members, males also display parochial altruism that involves support of ingroup members, coupled with defense in relation to outgroup members (De Dreu and Kret, 2016). Oxytocin in males thus elicits cooperation and has also been tied to defense-motivated noncooperative attitudes and behaviors to support members of their own group.

A Social Salience Perspective

A related view concerning the function of oxytocin is that in addition to serving as a prosocial hormone, it acts to enhance sensitivity to social stimuli. Thus, the influence of oxytocin would depend upon the context in which oxytocin changes occurred. When oxytocin levels were elevated, positive social interactions would be seen as especially positive and significant, whereas negative social interactions would be viewed as being more negative (Ellenbogen et al., 2013). Accordingly, in the context of negative events, such as early life mistreatment or neglect, the presence of adequate oxytocin functioning might favor damaging outcomes. In contrast, among individuals carrying a polymorphism of the oxytocin receptor, which would be accompanied by diminished social sensitivity, the negative effects of early life abusive events were not as pronounced as they might otherwise have been (McQuaid et al., 2014). In line with this suggestion, in the presence of genes linked to high oxytocin levels, interpersonal stressors predicted elevated depression, whereas a similar outcome was not observed among individuals carrying a polymorphism associated with low levels of this hormone (Tabak et al., 2015). Further to this, among women who had received oxytocin through a nasal spray, forgiving attitudes toward a breach of trust were less prominent, likely because they viewed betrayal as being more profound. Thus, among individuals who had normal levels of oxytocin, administration of this hormone might cause individuals to be overly sensitive to negative social events. It might be expected that if oxytocin were not disturbed among depressed individuals, increasing its levels through exogenous administration might aggravate illness features through elevated sensitivity so that even neutral social stimuli might be more likely to be negatively misinterpreted. From this perspective, although oxytocin could potentially be useful in treating depressive illness, this would depend on the basal levels of this hormone, as well as the functioning of other neurotransmitters (e.g., dopamine) associated with reward and social processes (McQuaid et al., 2014).

Neuropeptide Y (NPY)

Like many other hormones, neuropeptide Y has multiple functions, although it is best known for its role as a vasoconstrictor, in diminishing pain perception, affecting circadian rhythms, and affecting food intake and storage of energy as fat (Abizaid and Horvath, 2008). As well, it serves to diminish anxiety and stress and is thought to be an important contributor to resilience (Russo et al.,

2012b). In Chapter 11, we'll see that NPY has been implicated as a resilience factor relevant to PTSD.

Stress and Energy Balances

Stressors can either increase or reduce eating, depending on their severity and pattern of appearance. As we described earlier, strong stressors typically reduce eating as it would be dangerous for animals to continue their search for food if a stressor or threat were present. It similarly appears that in response to severe stressors, diminished food consumption occurs among most humans. Mild and moderate stressors, in contrast, may be accompanied by increased food consumption in some animal models and among some people (Patterson and Abizaid, 2013). The linkage between stress and eating is also illustrated by the consistent reports that drugs that reduce anxiety (e.g., benzodiazepines) tend to increase eating, whereas treatments that reduce eating are apt to engender anxiety. In fact, it has been exceptionally difficult to create antianxiety treatments without altering food intake and promoting weight gain. Several hormones involved in eating processes and in the regulation of energy balances are described in Table 6.5. It will be recognized that several of these hormones are closely tied to stress processes.

Leptin and Ghrelin

Initiation and cessation of eating, as we saw earlier, are largely determined by ghrelin and leptin, together with other hormones (e.g., neuromedin B and gastrin-releasing peptide) that also contribute to feeding. These hormones are affected by stressors and might contribute to symptoms that accompany stress-related disorders, such as anxiety and depression (Andrews and Abizaid, 2014). In this regard, ghrelin activation may diminish anxiety, thereby allowing animals to maintain "appropriate" food-seeking behaviors, thus making energy available to contend with challenges (Spencer et al., 2015). As expected, leptin increases in response to acute stressors, but when this occurs on a sustained basis, "leptin resistance" could develop, so that leptin will fail to signal "stop eating." As cortisol and ghrelin were linked to self-reported anxiety, it was suggested that the interplay between these hormones was responsible for stress-related eating (Patterson and Abizaid, 2013).

Ghrelin or interactions between leptin and ghrelin may also enhance coping and thus act against the development of depressive-like states (Abizaid et al., 2014). Indeed, in rodents, the depressive-like behavioral disturbances and corticosterone elevations provoked by stressors were largely eliminated by leptin treatment. This said, the connection between both leptin and ghrelin

TABLE 6.5 Hormones Related to Energy Regulation and Eating.

Secreted Hormone	Biological Effect
Corticotropin-releasing hormone (CRH)	Formed in the paraventricular nucleus of the hypothalamus. In addition to being fundamental to stress responses, it diminishes food intake and increases metabolic rate.
Cortisol (Corticosterone)	Released from the adrenal gland it is the prototypical stress hormone; also stimulates caloric intake and may promote a preference for high-calorie foods under stressful circumstances (stimulates the consumption of comfort foods).
Leptin	Produced by fat cells: reduces food intake, and is involved in changes in brain cytokines.
Ghrelin	Produced Produces in gut: serves to stimulate increased eating; may affect reward processes; modulates stress responses.
Insulin	Produced in the pancreas: involves in getting glucose from the blood into various body cells and storing it as glycogen; regulates fat and carbohydrate metabolism
Neuromedin B (NMB) and gastrin-releasing peptide (GRP): bombesin in rodents	Produced in gut and in several brain regions: acts as a satiety peptide (signals when an individual is full) and promotes anxiety.
Melanocyte-stimulating hormone (MSH)	Produced in the lateral hypothalamus: associated with feeding, motivation, and the production of pigmentation that protects skin from UV radiation
Neuropeptide Y (NPY)	Hypothalamic hormone: increases food intake and reduces physical activity; increases energy stored in the form of fat; blocks nociceptive (noxious) signals to the brain; increases vasoconstrictor actions of norepinephrine. Involved in stress resilience.
Incretins: glucose-dependent insulinotropic polypeptide and glucagon-like peptide-1 (GLP-1)	Gut-derived peptide hormones: secretes in response to a meal. They stimulate pancreatic β-cells postprandially, to secrete insulin. Diminishes glucose levels and reduces desire for foods.
Growth hormone (somatotropin)	Stimulates growth, cell reproduction, and regeneration. GH also promotes IGF-1 production and increases circulating glucose and free fatty acids.

and depressive states in humans is less clear, particularly as several additional stressor-sensitive hormones, including CRH, cortisol, neuromedin B, and gastrin-releasing peptide, may act together in affecting mood states. Moreover, the primary function of ghrelin is its capacity to serve as a "survival hormone" that operates in response to extreme nutritional or stressor challenges in which it serves to maintain appropriate blood glucose levels, limit weight loss, and diminish depressive affect (Mani and Zigman, 2017).

CRH and Cortisol

While glucocorticoids and CRH are best known for their capacity to deal with stressful experiences, they are also intimately related to eating and energy processes. The limbic CRH release associated with threats may promote nucleus accumbens dopamine activation, and hence positive stimuli, such as food, may appear more salient or more rewarding. In this regard, stressors may promote a dopamine-mediated craving for comfort foods (a fast caloric fix in the form of sugars and carbs) that might temporarily make individuals feel better. The salience of pleasurable or compulsive activities (ingesting sucrose, fat) is increased in the presence of insulin, and consequently, comfort foods might be still more rewarding. These foods may also be instrumental in providing needed energy to facilitate coping with stressors, and the biological changes provoked by comfort foods might diminish anxiety that could otherwise occur. If, in contrast to a mild challenge, a stressor experience is intense, then dopamine levels will decline and the considerable CRH release will not have comparable effects on eating. Instead, a shift will be provoked away from appetitive responses and toward defensive and vigilant behaviors (Lemos et al., 2012).

It was suggested that linkages between stress, cortisol, altered metabolic processes, eating, and the preference for comfort foods might contribute to obesity as well as the redistribution of stored fat, preferentially appearing as abdominal fat depots (Dallman, 2010). As we'll see repeatedly, the cytokine release from abdominal fat may contribute to inflammatory-related disease conditions, such as diabetes, autoimmune disorders, heart disease, and depression.

Stress and Immune Alterations

Stressful events profoundly influence various aspects of immune functioning. These changes are described in detail in Chapter 7; thus, only a few key points are introduced here that are relevant to interactions with hormonal or neurotransmitter processes. The impact of stressors on immune functioning is influenced by hormonal changes that are provoked and vary with characteristics of the stressor, several genetic and experiential factors, and the immune compartment examined (e.g., in blood versus spleen). In general, acute stressors that are moderately intense result in nonessential functions, such as digestion or reproduction, being suppressed, while immune functioning is enhanced in order for the organism to be able to contend with potential pathogenic risks. In contrast, in response to severe or chronic stressors, it is more likely for both primary and secondary immune functioning to be impaired (Dhabhar, 2009). Such effects are apparent following intense physical injury, and they emerge following psychological challenges (e.g., loneliness, disturbed social stability), indicating that immune disturbances aren't simply a result of tissue damage.

In considering the impact of stressors on hormonal processes, continued stressor experiences resulted in glucocorticoid levels being reduced and/or receptor sensitivity diminished, and consequently the immunosuppressive effects otherwise engendered might be absent. Because of the many challenges that animals experience in their natural habitat, a delicate bit of juggling is needed so that glucocorticoid functioning will operate efficiently, and yet not undermine immune functioning.

In response to acute (or subchronic) stressors (e.g., social disruption), the production of circulating proinflammatory cytokines, such as interleukin-1β (IL-1β) and tumor necrosis factor-α (TNF-α), is increased, and several days of this treatment promoted greater cytokine amounts within lymphoid organs (spleen, lung). The initial rise of proinflammatory cytokines is followed soon afterward by elevated antiinflammatory cytokines, allowing for the balance between different cytokine subtypes to be re-established. Moreover, under conditions of chronic social instability, cytokine presence in the periphery and brain is altered appreciably (Audet et al., 2014).

Stressors can influence concentrations of brain cytokines, although the changes that occur may be distinct from those seen peripherally. Acute stressors provoke an increase of inflammatory cytokine gene expression within the prefrontal cortex, especially if the stressor occurs on the backdrop of an immune challenge (Gibb et al., 2013). Furthermore, when animals are re-exposed to a stressor sometime after an initial challenge, the cytokine response may be greatly exaggerated, even when this involves a very different stressor (Johnson et al., 2002). It seems that encounters with a stressor may prime immune or brain microglia processes to respond more vigorously to later challenges, possibly through the influence of norepinephrine stimulation causing excessive cytokine release that could potentially lead to pathological outcomes (Johnson et al., 2019).

Once cytokine changes occur within the brain, further downstream effects can be engendered, including the provocation of brain neurochemical changes that can promote pathophysiological outcomes. As indicated earlier, cytokine challenges elicit brain neurochemical changes that are reminiscent of those engendered by strong stressors. This includes, among other things, increased monoamine utilization in the prefrontal cortex, central amygdala, and hippocampus, altered GABA and glutamate within limbic and hypothalamic regions, and growth factors in the hippocampus (Audet and Anisman, 2013). Some of these actions could be diminished if animals were pretreated with a nonsteroidal anti-inflammatory drug and could thereby diminish depressive mood (Köhler et al., 2016a). It was surmised that some of the effects of stressors on central neurotransmitter functioning might be governed by an inflammatory response. In this regard, the monoamine changes otherwise provoked by stressors were prevented by inhibiting the actions of IL-1β. Likewise, the behavioral and neuroendocrine effects stemming from a chronic mild stressor, as well as the reduced hippocampal neuroplasticity, were precluded in mice in which IL-1β receptor activity was diminished (DiSabato et al., 2021). The important point here is that stressful experiences can promote immune and inflammatory changes, which operate together with several hormonal and brain neurochemical alterations to produce both psychological and physical pathologies.

Microbiota, Inflammatory Responses, and Stressors

Bacterial communities that inhabit the gastrointestinal tract are not only sensitive to diet, antibiotics, and immune-activating agents but also profoundly affected by physical and psychological stressors (Bharwani et al., 2016). Considerable diversity exists within the microbiome, which is usually taken to be advantageous, but in response to stressors, microbial dysbiosis occurs, so that an unstable and potentially unhealthy community may develop (Cryan et al., 2019). As in so many other contexts, the influence of stressors in rodents varies with sex, being particularly notable in females, which may have implications for the more frequent anxiety and depression seen in human females (Bridgewater et al., 2017).

Multiple factors come together to influence the effects of stressors on microbial processes, which then affect brain functioning. Intestinal bacteria can modulate HPA activity and might thus contribute to stress-related conditions. Among germ-free mice in which basal levels of corticosterone were elevated, stressors provoked an exaggerated glucocorticoid response (Crumeyrolle-Arias et al., 2014), which could be attenuated by colonization with a specific Bifidobacteria species. Furthermore, the transplantation of microbiota from stressed mice to germ-free mice provoked a marked increase in inflammatory responses following infection (Willing et al., 2011). Likewise, stressors encountered during early life influenced the composition of gut bacteria and even affected the later development of stress responses (De Palma et al., 2015). Microbial variations during early life are readily provoked by stressors, and microbiota disturbances during early life can have pronounced ramifications on the functioning of brain regions linked to stress-related behaviors (Stilling et al., 2015). As we'll see in the ensuing chapters, prenatal experiences can also have significant actions in this regard. The influence of stressors on microbial factors is not only pronounced during early life, but with older age the gut becomes more permeable, hence allowing microbiota to enter the bloodstream more readily, leading to increased inflammatory factors (e.g., TNF-α) being present, thereby encouraging poor health (Thevaranjan et al., 2017).

Just as stressor-provoked microbial changes can affect central neurochemical processes, variations of brain functioning can affect the microbiome. Fig. 6.5 describes some of the processes by which multidirectional communication occurs between these processes. Gut microbiota can have positive or negative effects on neurodevelopmental processes and the balance between good and bad bacteria may be accompanied by psychological disturbances, including elevated risk for depressive illnesses or a PTSD-like condition in animal models (Leclercq et al., 2016). Of practical significance, pretreating mice with particular gut bacteria (*Mycobacterium vaccae*) may diminish stressor effects and inflammation and might, thus, attenuate trauma actions, such as the development of PTSD in a mouse model (Reber et al., 2016). Likewise, treatment with prebiotics effectively attenuated anxiety and microbiota changes introduced by stressors (Tarr et al., 2015), and dietary prebiotics similarly improved sleep and generally limited the adverse effects of stressors (Thompson et al., 2017). As predicted, when combined with exercise, the antistress effects of prebiotics were still more pronounced, likely owing to the combined effects on gut microbial processes (Mika et al., 2016). These changes lend themselves to antidepressant actions, including depression secondary to chronic illness.

Like the effects of prebiotics, it was reported that probiotics diminished anxiety- and depressive-like behaviors in rodents and reduced the hormonal and brain neurochemical alterations ordinarily elicited by stressors (Sarkar et al., 2016), and probiotic treatments also increased BDNF changes and enhanced neurogenesis (Sarubbo et al., 2022). There's good reason to suppose that manipulating microbial factors could have significant effects on mood states, but the data are not entirely uniform. Furthermore, although gut microbiota have pronounced behavioral effects in rodents, the clinical value of these manipulations has not been fully worked out, especially in relation to which microbial manipulations are best for any given individual (Kelly et al., 2016a, b).

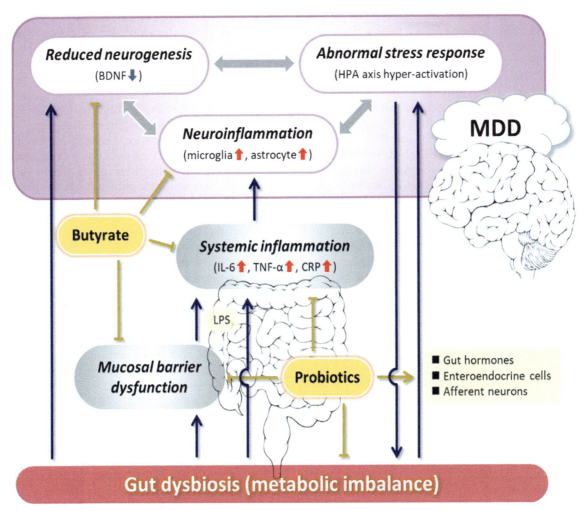

FIG. 6.5 Underlying mechanisms for MDD via the gut–brain axis, and the modulating role of probiotics. Gut microbiota can modulate central nervous system activity through direct and indirect pathways, including endocrine (e.g., cortisol), immune (cytokines), and neural (vagus and enteric nervous system) pathways. Neuroactive bacterial metabolites of dietary fibers, specifically, short-chain fatty acids (SCFAs) affect peripheral and brain inflammation, which affect several brain neurotransmitters and hormonal processes. Hormone variations, such as elevated cortisol, affect immune cells and cytokine secretion, in both the gut and systemically. Significantly, cortisol can affect gut permeability and barrier function, thereby influencing the composition of gut microbiota and exacerbating processes that further influence brain functioning. *Source: Suda, K., Matsuda, K., 2022. How microbes affect depression: underlying mechanisms via the gut-brain axis and the modulating role of probiotics. Int. J. Mol. Sci. 23, 1172.*

Concluding Comments

Stressful experiences engender a range of neurobiological effects that have been linked to numerous pathological conditions. Many of the effects of stressors act in an adaptive capacity but can nevertheless aggravate existing pathologies. As the load becomes progressively heavier by experiences with severe or chronic challenges, the weakest link in the proverbial chain will break, leading to a disease condition. Once this occurs, which can give rise to further physiological alterations, comorbid conditions are likely to emerge. The weak link in this context may be related to genetic dispositions or the impact of earlier stressor experiences, particularly those encountered during early life, essentially reflecting an initial step of a two-hit or multihit phenomenon.

Several different systems (microbial, inflammatory immune, hormonal, neuronal, and neurotrophic) may additively or interactively govern the impact of stressors, thereby determining the development of pathology. Further to this, the nature and extent of the biological responses that occur may be influenced by a constellation of factors related to the characteristics of the stressor, individual appraisal and coping methods, personality factors, and numerous psychosocial determinants. Clearly, this speaks to the individualistic nature of stressor experiences and outcomes, and a precision medicine approach ought to be considered in determining the optimal treatment strategies for stress-related pathology.

CHAPTER

7

Stress and Immunity

The breadth of stressor-related immune disorders

An enormous literature has accrued over the past few decades that addressed the impact of stressors on immunological processes and immunologically related diseases. Of the latter, the most prominent are the various auto-immune diseases and cancer, but stressor-provoked inflammatory changes can also exacerbate asthma, athero-sclerosis, and stroke. In addition, chronic inflammatory immune activation may influence neurodegenerative actions in the brain, and stressor-induced alterations of inflammatory processes could exacerbate the progression of Alzheimer's disease (Piirainen et al., 2017). Findings such as these point to the importance of assessing the mechanisms by which stressors impact the immune system and inflammatory processes, and whether immune perturbations interact with other factors in promoting either the onset or progression of diseases.

It was, at one time, thought that immune-related diseases were confined to infection, autoimmune disease, and cancer, but the range of pathologies attributable to immune alterations has expanded greatly. This intrusion of the immune system into unconventional domains of biomedical investigation reinforced long-held views from psychosomatic research that recognized the strong comorbidity between emotional states, such as anxiety and depression, and conditions such as asthma and heart disease. Even though many diseases do not have a clear etiology, the imputation of genetic, environmental, and experiential factors linked to inflammatory processes have been raised as explanatory models. Given the preponderance of evidence for a variety of immune alterations attributable to stressful experiences, a deeper understanding of this linkage might contribute to preventative and/or therapeutic efforts to treat a wide range of human ailments.

Implications of Stressors for Immune-Related Diseases

Our immune system has adapted to the environment over thousands of years, presumably to protect us against pathogens. About 4500 years ago, selection pressures that promoted antiviral mutations accelerated, likely owing to the increased size and density of populations. While we gained protection against diverse pathogens that were also evolving, the occurrence of autoimmune disorders increased accordingly (Kerner et al., 2023). While our immune system has adapted magnificently to deal with threats, it's far from perfect. We do experience diverse illnesses and are subject to negative effects that arise owing to the activation of inflammatory processes, and the immune system is negatively affected by stressors despite our stress systems have evolved to deal with environmental and psychosocial challenges. The trade-offs that exist between adaptive evolutionary changes and those that favor vulnerability to disease occurrence understandably vary across geographic locations. On the one side, this can make it difficult to conduct some studies, but on the other, it provides an efficient tool to determine which genes are linked to specific pathological outcomes or those that limit disease susceptibility (Tang et al., 2022c).

Immune surveillance occurs broadly, patrolling the entire body, every nook and cranny, but, at one time, it was thought that this did not include the brain, which was said to be "immunologically privileged." However, it is now recognized that multidirectional communication occurs between immune, gut microbiota, hormonal, autonomic, and brain processes. As we've already discussed, microglia serve as a component of the brain's immune system, operating to facilitate the repair of damage, the glymphatic system is fundamental for waste removal from the brain, and the choroid plexus that comprises blood vessels within the ventricles contribute to the immune cells homing to the parenchyma. It also appears that microscopic channels present within the skull permit cerebrospinal fluid to contact the bone marrow, which influences its cellular maturation and migration to the brain. Together, these processes comprise a communication network that facilitates interaction between multiple protective processes that ensure brain health (Castellani et al., 2023).

Immune system signaling molecules, cytokines, have limited access to brain neurons owing to the tight junctions between capillary endothelial cells that line blood vessels, which make up the blood–brain-barrier (BBB). However, as described in Chapter 1, cytokines can gain access to the brain parenchyma so that neuronal functioning may be affected. Under certain conditions (e.g., high fever, head injury) the BBB is affected so that access to the brain is facilitated (Parker et al., 2020). Under stressful conditions, a sterile immune response can be promoted that can have effects on the brain through different routes. In addition, it was demonstrated that monocytes that can reach the brain contain matrix metalloproteinase-8 (MMP-8). Ordinarily, the extracellular space that separates neurons and non-neuronal cells can be affected by MMP-8 under stressful conditions among stress-sensitive mice and was elevated among depressed humans. Importantly, under stressful conditions circulating altered MMP-8 can facilitate infiltration into specific brain tissue, such as the nucleus accumbens, thereby favoring behavioral disturbances, such as those associated with depressive illness. Conversely, by reducing MMP-8 the stressor-provoked behavioral disturbances in mice could be attenuated in parallel with neurophysiological alterations of the nucleus accumbens (Cathomas et al., 2024). Evidently, there are several routes by which stressful experiences can influence access to specific brain regions that might contribute to the emergence of psychological disturbances.

It has become apparent that stressors can influence brain processes that promote growth factors and inflammatory processes. Specifically, stressors can influence specialized extracellular matrix structures (perineuronal nets) that are important for synaptic plasticity (Laham and Gould, 2022), and the impact of chronic stressors on perineuronal nets may come about through the actions of astroglial cells and neurotrophic processes (Coppola et al., 2019). Furthermore, as we observed earlier, resident microglia serve in a macrophage-like capacity in the brain, where they release cytokines, thereby creating a powerful inflammatory response. Activation of microglia occurs in response to stressors, such as bacterial or viral challenges, and in response to physical and chemical insults (e.g., brain injury, concussive injury, seizure, and cerebral ischemia). The elevated brain cytokines might reflect their involvement in a reparatory capacity, but if the cytokine concentrations are too high, they might become neurodestructive, fostering psychopathology and neurodegenerative disorders. This inflammatory state introduces morphological changes characterized by hyper-ramification of microglia, which could influence the occurrence of depression and anxiety (Zhang et al., 2022c). Consistent with this view, in mice a strong stressor that elicited post-traumatic stress disorder (PTSD)-like behaviors was associated with microglial hyper-ramification and neuronal dendritic spine loss in the medial prefrontal cortex and dorsal hippocampus (Smith et al., 2019).

As part of the brain's defensive repertoire, T cells can infiltrate the nervous system, reside in the meninges, as well as extravasate from the blood, and enter the parenchymal spine and brain tissue. The functional consequences of this infiltration have been vigorously debated, with some evidence that this serves in a neuroprotective capacity (Filiano et al., 2017), although the presence of T cells may undermine neural repair and recovery. Irrespective of the specific conditions required to optimize the protective role of T cells in neural injury, it is likely that stressors and autonomic nervous system (ANS) activation influence the migration of T cells into the central nervous system (CNS). Stressors can affect the homing of T cells in the brain and alter their sensitivity to monoamines and neuropeptides. Similarly, monocytes can infiltrate the brain, especially in certain diseases, such as multiple sclerosis (MS) (Ashhurst et al., 2014), in which cycles of relapse and remission can be affected by psychologically stressful daily hassles.

Given that the immune system has the capability of altering CNS function through the influence of proinflammatory cytokines. Identification of the relevant stressor-related immune and inflammatory factors that affect the brain is important in determining the links between stressors and mental disorders, as well as the conditions in which stressors have either positive or negative effects. Depending on the nature of the stressor experienced, immune functioning can either be disturbed or enhanced. The augmentation of immune responses in response to moderate stressors is highly adaptive in protecting the organism from damage that might otherwise occur. As stressors can have both inhibitory and enhancing effects on immune functioning, a closer assessment is warranted concerning which neural-derived factors impact immune cells and how they influence the tenor of immune responsiveness to antigens.

Viral and/or bacterial infections (e.g., rabies and herpes simplex virus) can assault the nervous system, disrupting stable brain functioning and promoting mental health disturbances. Immunological clearance of these infectious agents is paramount but may be compromised if individuals experience stressors that are immunomodulatory. As we will discuss in several ensuing chapters, immune factors have been related to several mental disorders. Depressive illnesses and anxiety may arise owing to the actions of inflammatory processes, and the view has been advanced that schizophrenia is associated with some of the same mechanisms responsible for autoimmune disorders (Jeppesen and Benros, 2019).

Direct and indirect pathways also exist through which gut microbiota modulate peripheral and CNS processes. Through effects on endocrine (e.g., cortisol), immune (cytokines), and neural (vagus and enteric nervous system) pathways, brain processes are affected, leading to behavioral variations. The vagus nerve and functioning of circulating tryptophan levels may contribute to gut-related information being relayed to the brain. In addition, neuroactive bacterial metabolites of dietary fibers, specifically, short-chain fatty acids (SCFAs) also affect the brain and behavior. In response to stressors, several brain neurotransmitters are affected, as are hormonal processes. Several of the hormone variations, such as corticosterone increases, affect immune cells and cytokine secretion, both in the gut and systemically. Significantly, cortisol can affect gut permeability and barrier function, thereby influencing the composition of gut microbiota (e.g., Cryan et al., 2019).

Dynamic Interplay Between the CNS and Immune System

Stress is a physiological state that draws on many different biological resources to allow the organism to deal with the demands of the environment. To a significant extent, reactions to stressors are based on cognitive and sensory information. In cases, where the information generates alarm, the brain is equipped to initiate a cascade of changes that affect the peripheral nervous system and neuroendocrine systems. Stressors promote a rapid increase of norepinephrine and acetylcholine release from the terminal endings of the sympathetic and parasympathetic components of the ANS. Although the branches of the ANS terminate in virtually all organs of the body, the discovery that sympathetic nerve endings terminate in lymphoid organs (see Bellinger et al., 2013) was evidence that the immune system was "hard-wired" so that ANS functioning could affect immune activity. Indeed, it is now well established that various types of leukocytes express noradrenergic and cholinergic receptors so that stressors affect immune functioning through this route. Similarly, when the brain is stressed so that synapses are flooded with neurotransmitters, the connections between the sympathetic nerve terminal and lymphocytes may be affected. This highlights an anatomically

based continuity of function, extended not just to cardiac muscle or liver function but also points to a change in the immune system resulting from the brain's efforts to engage a psychological or physical stressor.

As with receptors for products of the ANS, considerable evidence exists for the presence of hormone receptors on immune cells. Adrenal-derived glucocorticoids (e.g., corticosterone in rats and mice; cortisol in human and nonhuman primates) primarily serve in an anti-inflammatory capacity (Cain and Cidlowski, 2017). During the early years of research on how stressors impacted immune function, this served as the major explanatory mechanism for stressor-induced immunosuppression. Since then, it was found that in addition to the adrenal, the pituitary represents a source of immunomodulatory hormones (e.g., endorphin, prolactin), which can bind to immune cells via specific receptors. To add further to this *potpourri*, peripheral neuropeptides, such as substance P and vasoactive intestinal peptide (VIP), can exercise strong functional effects on immune cells, operating through specific receptors (Ganea et al., 2015). Indeed, an extensive number of endocrine hormones, peptides, and neurotransmitters serve as ligands for selective receptors on lymphocytes, monocytes, polymorphonuclear leukocytes, and dendritic cells (DCs). This serves as testimony to the close functional interplay between the brain and the immune system and reinforces the notion that the immune system evolved to play host to a wide range of products generated by the CNS.

Although this conclusion is certain, it is more difficult to answer how immune-related processes impact CNS function in the provocation of mental illnesses. To answer this, we must consider whether the immune vector that affects the CNS is always the same. Are there behavioral influences that modify the quality and quantity of the immune processes that modify the CNS? For example, if a given cytokine (e.g., interleukin-1 [IL-1]) impacts the brain to induce sickness behavior, does the same amount and duration of IL-1 release exert this effect in all people or just some? And if the answer is the latter, what accounts for the different parameters of concentration and temporal persistence?

The CNS and immune system engage in a dynamic dance in which molecules are exchanged and cellular modifications are made. This functional relationship is as relevant to physiological actions as the relationship between enzymes and substrates is to the biological functions of a cell. The view came about that because of the communication between the immune system and the brain, an immunological challenge resulted in the formation of a neural representation of the organism's current immunological state (referred to as "immunoception"). These neural processes can subsequently be activated, thereby influencing both innate and acquired immune responses. While innate immune responses are largely shaped by evolutionary pressures so that responses are determined by the experiences of the species, the acquired responses are driven by individual experiences. It was proposed that experience-dependent immune regulation promoted a memory trace or "immunogram" that can influence peripheral immune responses (Koren and Rolls, 2022).

Although numerous brain regions can influence immune functioning, the insular cortex may be fundamental in storing immune-related information so that upon subsequent activation, information concerning immune challenges is retrieved and then acted upon (Koren and Rolls, 2022). When animals encounter substances to which they have developed an allergic reaction and hence avoid these stimuli, their behavior can be modified by immune alterations (e.g., by blocking IgG antibodies), which speaks to the perspective that the immune system developed, in part, to facilitate avoidance of hazardous ecological niches (Florsheim et al., 2023). Although the interactions between the immune system and the CNS might have some advantages in dealing with previously encountered immune challenges, activation of the insular cortex can foster inflammatory immune responses that contribute to diverse psychiatric conditions (Gogolla, 2021).

Given the interactions between the immune system and aspects of the brain, the question needs to be addressed concerning what happens to this "dance" when the organism is exposed to psychogenic or neurogenic stressors. If the CNS and immune system are "holding hands," is the grip tightened or loosened? At the molecular level, the study of stress has revealed a loosening of molecular relationships between enzymes and substrates, or between transcription factors and their promoter regions. Researchers can look more closely at these types of relationships when focusing on specific immune cells. However, the higher-order questions concern how this ultimately affects the ability of the immune system to fight off infections, maintain antigenic accuracy, avoid autoimmune pathology, and regulate aberrant neoplastic cells that threaten the development of cancer. This is the traditional rationale for investigating stressor effects on immune function, but newer ones have emerged in light of suggestions that immune processes may impact depression, anxiety, PTSD, autism, and schizophrenia, as well as several neurodegenerative disorders. Furthermore, how do stressor-related changes in immune function affect neurodevelopment? As well is an immune system that has been pummeled by life's stressors able to alter the aging brain and promote neurodegenerative disease? These are all big questions, for which the answers are steadily emerging.

Considering the value of immune alterations provoked by stressors

A caveat in any discussion of stress and immunity is that the literature in this area is far from straightforward, and at times it can be downright perplexing. Straight out of the gate, it ought to be acknowledged that irrespective of whether the measurement of immune function is in vitro or in vivo, it has universally been found that the immune system is perturbed or altered by stressors. There is no argument here. The confusion lies in predicting the magnitude, and even the direction of the change, and determining the clinical significance of some of the changes. Aside from effects on immune functioning, it should be considered whether stressors influence clinical conditions, such as infectious and autoimmune diseases. In this respect, hypothalamic–pituitary–adrenal (HPA) activation can serve in a protective and life-saving capacity during infection, and it may also be necessary to inhibit autoreactive immune responses, thereby limiting the progression of autoimmune diseases. But what else serves in this capacity, and what factors moderate these actions?

Stressor Influences on Immunity: Animal Studies

Recognition that stressors impact the immune system predated any major understanding that the immune system was hard-wired to stress-related processes. However, the relationship of the adrenal to certain immune parameters was recognized many decades ago, although more specific analyses of glucocorticoid receptors present in leukocytes occurred much later. Well before the neural-immune connections were recognized, it was known that the immune system of the laboratory animal reacted strongly to stressful stimuli. The relationship between stress and increased susceptibility to infectious diseases was suspected for some time before being formally demonstrated. Based on a very large number of studies on immune functioning in stressed human and infra-human subjects, we are comfortable with the notion that stressors influence immunity, and there is ample reason to believe that these immune changes may in some fashion contribute to the exacerbation (and provocation) of some illnesses.

Pursuing some measure or indication of immune function, however, is not simple given that there are many different facets to immunological activity. In psychology, the same thing is said of "abnormal behavior," wherein by asking what exactly we mean by "abnormal," the tale of the blind men and the elephant is resurrected as a means to emphasize that perspective matters. How one approaches the question of immune function can similarly be a matter of perspective, but it remains necessary to focus on the inherent differences among immune cells. Few would argue that an immune system is functioning well if an infectious virus is expediently trounced and removed from the body. Conversely, we would conclude poor functioning if auto-aggressive antibodies develop against self-tissue and organ pathology develops, or if cancer cells multiply because immune surveillance mechanisms failed to restrict their growth. These examples are clinical indications of functional and dysfunctional immunological activity, representing critical endpoints of stress research on the immune system. Indeed, it is of fundamental interest to know how a biological system operates under varying environmental conditions (e.g., hot, cold, high altitude, low altitude, and so on), but when we invoke the specter of "stress," the expectation is to consider potential clinical sequelae. So, what we need to ask, in addition to whether stressors modify immune functioning, is the likelihood that such modifications are clinically relevant. As we have already discussed, the connection between the brain and the immune system is natural. If stressors shake up the organism, the way bumps in the road may test a car's suspension, then any clinical repercussions are likely to be due to faulty engineering. Identifying where that fault may lie is the challenge. However, before we look at this more closely, we ought to attend to how we would assess the consequences of stressor exposure on immune function.

Immune Assessment and Stress

Although it would make research far simpler if stressors affected all aspects of the immune system in the same way, this is hardly the case. The diversity of responses across the various aspects of the immune system is likely beneficial in the sense that as one system is diminished by a stressor, a second system may be unaffected, and even influenced in an opposite manner. Because of the heterogeneity of stress responses that can occur, basic research requires the selection of particular parameters of immune function and a detailed assessment of the individual cellular and/or molecular players.

It will be recalled that the immune system is divided into acquired and innate components, both of which need to be considered in relation to stressor effects. The chief cells of the innate immune system are monocytes, neutrophils, and macrophages, and assays for these parameters include phagocytosis, chemotaxis, and cytokine production. For the acquired or adaptive immune response, we rely on measures of lymphocyte function, and here we further divide our attention between T and B lymphocytes. The most prominent measure of the latter is antibody production or simply immunoglobulin release if

the B cell stimulus is not a specific antigen. In so far as T cell function is concerned, mitogen-induced proliferation has been a prominent measure, although antigen-specific T cell responses have also been investigated; both forms of T cell assessment lend themselves to cytokine measurement.[1]

Many assessments of immune capacity have involved in vitro assays. In human studies, the response to infectious agents and immunization have also been assessed. For the most part, however, animal studies lend themselves better to in vivo immune measurements, as well as to the study of infectious disease models and exposure to bacterial toxins. Several animal studies that examined stressor-induced modulation of influenza and herpes simplex virus infections provided an opportunity to consider the immunological mechanisms that are affected by stressors. For instance, T cytotoxic and NK cells are critical in eliminating viral infections, as well as providing insight into surveillance against cancer. Accordingly, in the many studies that examined NK cell function, the clinical implication has often focused on cancer susceptibility.

Stressor Characteristics

Various conditions have served as stressors, which can differentially influence biological and behavioral responses. For example, social disruption through changes in the composition of group-housed animals has proven to be a strong stressor, while the opposite, social isolation, can also serve as a powerful stressor. Diverse stressors can also instigate some significant differences in the neurochemical pathways activated across brain regions. Different psychogenic stressors (e.g., exposure to a fear-provoking stimulus, restraint, and social defeat) can have diverse effects, which may differ yet again from the actions of physical stressors (e.g., electric shock, exposure to cold water). It is of particular importance in these studies to determine the broad physiological perturbations produced by the stressor (e.g., analyses of hormones or neurotransmitters), and how these may be linked to immune system changes. Moreover, it is necessary to identify the factors that act as moderators (e.g., stressor controllability, sex, early life experiences) that are fundamental in determining the nature of the immune responses provoked by stressors.

Aside from the type of stressor, it is important to consider the issue of *acute* versus *chronic* (or repeated) stress. In the current era of stress research, increased attention is being directed to chronic stressors. For the most part, it is understood that the physiological response to a sudden, temporally distinct stressor is adaptive, and unlikely to

---

[1] Mitogens are typically plant lectins that nonspecifically stimulate T cells and B cells and cause them to divide and generate cytokines and antibody; in essence, mitogens are polyclonal stimuli, with no affinity or selective attraction to a particular antigen receptor.

be of much use in revealing pathological changes, although severe stressors, such as those that promote PTSD, may have profound effects. In fact, from a lay perspective, and the popular media in general, a bad day at work or some form of disappointing news is perceived as a biological threat. However, this alarmist view of things is a gross misunderstanding of the importance of biological plasticity and the role it plays in conferring resilience to everyday stressors. Most creatures, humans included, are made of sterner stuff, although minor day-to-day annoyances can under some circumstances have appreciable negative actions, especially if these are superimposed on a chronic stressor experience. Unlike the effects of occasional insults, chronic or prolonged stressor experiences are more problematic, often representing a challenge to behavioral and biological adaptation, and are more likely to promote pathology.

Stressors are a ubiquitous aspect of life, and the greatest danger they pose is often linked to their persistence. It is when they sap adaptive resources that their biological impact, and immunopathological potential, are most likely to be realized. To this extent, it has become fashionable to think of chronic stressors as the true problem that we should be addressing, with acute stressors being part and parcel of manageable physiological fluctuations. Of course, given a sufficient number of acute stressor episodes, even if they are distinguishable and unrelated to one another, their cumulative action can be significant. This is indeed the general story concerning chronic stressor exposure, whereby immune responses that are normally unchanged or modified (e.g., enhanced) by acute stressors eventually come to be suppressed or impaired in some manner. Nonetheless, our understanding of any biological system requires that we know how it reacts when first exposed to an unexpected and distressing stimulus—that is when the organism is acutely stressed. Then it becomes possible to evaluate whether the effects of chronic stressors reflect further adaptive changes or, conversely, the breakdown of these processes.

We considered earlier why immune cells likely have receptors for glucocorticoids, norepinephrine, and other neurally derived neurochemicals, steroids, and peptides. Several compelling studies offered the view that they serve in protecting against autoimmunity as well as regulating and improving immune responses to deal with other challenges (Strehl et al., 2019). It is plausible that in response to stressor-provoked release of neuroendocrine and ANS monoamines, for which immune cells express receptors, the impact on the immune system is less likely to be designed to compromise immune function. More likely, these changes serve to facilitate or regulate immune responsiveness.

The brain, as the ultimate information processor, is equipped to appraise and respond to a multitude of stimuli, ultimately producing a stress response. As a result, strategies are implemented that negotiate the stressor and minimize its potential as a biological or psychological threat. This is not always possible, of course, and in the course of "negotiation," tissue damage may be incurred, as might happen when the twin options of fight or flight result in the former. This is evident in the social confrontation stress model, and good evidence exists that animals subjected to various forms of social disruption display pronounced immune changes (Biltz et al., 2022). In any case, responses to stressors require behaviors that generate steps to distance the organism from the stressor either through withdrawal, elimination of the stressor, or in the case of social contexts, acts of diplomacy, which may or may not involve a struggle.

With the elimination or attenuation of a stressor, the organism can remain at ease until it is compelled to attend to the next threat or challenge. These challenges come in numerous forms, often comprising events that aren't frequently considered to be stressors from a cognitive perspective, but systemic challenges (e.g., sterile inflammation) have effects much like those elicited by psychogenic and neurogenic stressors. In this context, immune functioning may be affected by seasonal changes, temperature fluctuations, and reduced energy intake (e.g., hunger). When chronic stressors involve compound challenges, some of which occur on an unpredictable basis, the immune system's capacity to ward off infectious agents, and self-regulatory functions may be impaired, thereby resulting in autoimmune disease. In humans, stressors can also modify in vivo immune responses to influenza vaccines, and differences in certain cell-mediated immune parameters may exist according to whether a stressor is acute or chronic.

Any instance of chronic stressor exposure bears an inherent acute component (the initial response to challenges of limited duration) and a chronic (or protracted) element, in which that stimulation continues to persist. The acute–chronic dichotomy can be viewed as the equivalent of being thrown an object of heavy weight, which requires the necessary postural and muscular adjustments that absorb the force of the object's impact. However, when this initial acute event is repeated, over and over, it can transform into a persistent state of stimulation that is chronic, and to which the organism needs to keep mounting the same response it generated upon the first encounter. The question is how long such physiological or psychological responses can be sustained at an optimal and effective level before the initial immune change in response to an acute stressor suffers a change that is no longer a reflection of adaptation that is of benefit to the organism.

In an interesting study, among mice susceptible to UV light-induced skin cancer that were stressed for 2 weeks (6 h/day restraint), earlier onset and increased incidence of squamous cell carcinoma were found. This was associated with reduced cutaneous entry of CD4+ T cells that normally restrict cancer growth in this model (Saul et al., 2005). In a follow-up study, imposition of a shorter, seemingly more "acute" stress regimen (2.5 h/day for 9 days), which can be seen as a sub-chronic stressor, the T cell-mediated immune response was augmented, and the incidence and time to skin cancer onset was reduced (Dhabhar et al., 2010). Given that short-term stressor exposure enhanced antitumor immunity, it emphasizes the notion that the stress response can be viewed as an additional defense mechanism that boosts immunoprotection. Indeed, it was argued that physiological changes induced by exposure to a stressor may operate like a natural adjuvant (Viswanathan et al., 2005).[2]

In ecologically relevant situations, successful negotiation of a stressor is dependent on the assessment of the situation according to the natural implementation of cognitive processes. We cannot presume to know what cognitive processes are operating when an animal is exposed to a psychogenic stressor, such as restraint. It certainly appears, however, that whatever cognitive and attendant physiological changes (peripherally and in the CNS) are provoked in an acutely restrained animal, they appear to translate into less threatening immunobiological effects following chronic restraint. Part of the explanation may lie in how a stressor comes to be perceived. Restraint procedures that utilize a constant duration of immobilization have somewhat predictable outcomes to which animals can habituate and learn that release and return to the home cage lies at the end of the experimenter-initiated imposition. It has long been known that measurement of neuronal excitability (as measured by immediate early gene activation) in stress-related brain circuits shows a decline with repeated exposures to restraint. Natural stressors, in contrast, can elicit very different outcomes. Specifically, rodents exposed to repeated social stressors (daily agonistic encounters that result in social defeat) fail to show a significant decline in elevated patterns of neuronal activation in the brain relative to that induced after the first encounter, depending on the defensive strategies mice adopted (Okamura et al., 2022). One way to interpret this is that relaxation of neuronal resources during conflict situations was not possible since the outcome of the conflict involves possible defeat and death. Similar findings are emerging for studies of depression-like behavior such that prolonged social stress (social defeat) imposes a considerable burden on neural adaptation (Laine et al., 2017).

Acute exposure to stressors in the form of social defeat results in activation of the HPA axis and alterations of various aspects of immune functioning (e.g., Adamo et al., 2017; Biltz et al., 2022), presumably by triggering genetically predetermined defense strategies. Assuming the changes are initially adaptive, what might transpire if the stressor occurs again at some later time? Such stressors are potentially lethal, militating against the luxury of reduced vigilance. Instead, the processes that affect immune functioning become sensitized so that exaggerated responses may be elicited upon further encounters with the stressor, and may similarly be introduced when other challenges are encountered. The impact of a naturalistic stressor is not only apparent with changes in peripheral immune processes, having been observed with brain microglial functioning, neurotransmitter, and loss of the anti-inflammatory properties of glucocorticoids, that is, glucocorticoid resistance (Anisman, 2009; Biltz et al., 2022).

Stressor-provoked glucocorticoid, noradrenergic, cholinergic, and neuropeptide changes influence immune function, although the immune system responses to these modulating factors may fluctuate with different types of stressors and may vary with respect to the adaptation that could occur with repeated exposure. We do not know enough at present to determine whether slight alterations in the concentration of hormones and neurotransmitters, reflective of different neural stress states, are processed by immune cells as packets of information that require different immunological adjustments at the cellular and cytokine levels. As we discussed earlier, immunocytes are undoubtedly immersed in an internal milieu that allows them to bind via relevant receptors to various hormones and neuropeptides. These ligands ordinarily circulate and/or are released, likely being receptive to their presence in a manner like the balance of salts and various other ions that govern normal cell biology. However, rapid and/or persistent elevations in the concentration of these ligands represent new information to which cells respond. This may result in several different functional changes (e.g., altered cytokine production), many of which have been documented. If there is a lawful relationship between the pattern and concentration of different neurohormonal factors and the threshold sensitivity of immune cells to these stimuli, it has yet to be fully described.

Assessing Stressor Effects on Immune Function

Evaluating the status of the immune system involves several in vitro and in vivo procedures designed to elicit

---

[2] The implications of this notion are quite interesting in the context of other illnesses. There are usually yearly variations in the success of the flu vaccine. Were the conditions to be identified by which acute stressors enhance immune responses, perhaps this knowledge could be harnessed and applied as a physiological modulator superimposed on flu shots. If this were possible, natural adjuvant mechanisms might boost initial protection and efficacy against influenza exposure. Subjecting people to stressors is obviously not in the cards—but perhaps an acute bout of exercise—a natural challenge that has multiple benefits—prior to immunization would do the trick.

canonical immunological responses, such as lymphocyte proliferation, antibody production, phagocytosis, antigen presentation, and cytokine production. However, the best evaluation of immunity is the retention of stable health in the face of host invasion by viruses and bacteria. Infection with the human immunodeficiency virus offered important lessons and confirmation about the importance of CD4+ T helper lymphocyte function in the fight against opportunistic infections. Since HIV infects and replicates in T cells—most commonly CD4+ T cells—there is a dramatic depletion and destruction of T helper cells. This impairs antigen-specific B cell antibody production, allowing deadly opportunistic infections. As such, the consequences of HIV infection demonstrate the importance of a fully functional immune apparatus.

In evaluating the relevance of stressor effects on immunity, we need to keep in mind whether the immune alterations measured represent a potential breakdown in immunity against infection. Many of the immune parameters assessed under noninfectious conditions comprise mitogenic stimulation of lymphocyte transformation, NK cell activity, and antibody responses to antigens. The clinical relevance of these measures has been established by way of association with additional approaches, such as exposure to viral and bacterial antigens, induction of experimental autoimmune disease, and tumorigenesis.

Analyses of stressor effects on immune functioning have involved a variety of different species and often very different stressors. Typically, rats and mice have been the species of choice, but nonhuman primates have been used on occasion. Infrequently, when agricultural applications are needed, pigs, cattle, and fowl have been evaluated. In the main, the effects of stressors on immune functioning have been consistent across species. Yet, as we will see, in some instances, predicting the outcomes, and even the direction of stressor effects on immunity (i.e., suppression or enhancement), is not always straightforward. In this regard, neurophysiological changes provoked by stressors operate against a highly intricate set of intercellular interactions within the immune system. Furthermore, as described earlier, the immune system is heavily compartmentalized, having unique regulatory requirements operating in several anatomical regions of vulnerability (e.g., immune cells in circulation, those present in local lymph nodes, as well as those present within common mucosal systems comprising the gut, lung, and urogenital tracts). As a result, unique sets of interactions can be expected between the brain and regionally specific immune processes (Powell et al., 2017). This is made still more complex as specific experimental conditions (e.g., species, stressor) and the type of antigen to which animals are exposed can instigate very different immunological outcomes.

Cell-Mediated Immune Responses

Measures of cell-mediated immune responses typically reflect assessment of T lymphocyte operations, although other indices of immune functioning have also been evaluated, such as the presence of neutrophils as well as cytokines released by these cells. An acute stressor that comprised protracted (6 h) restraint was associated with an inflammatory state, reflected by an increase in gene expression related to inflammatory responses, proinflammatory cytokine production, and wound healing. The acute stressor was accompanied by the activation of neutrophils, whereas the number of circulating T and B cells was reduced (Tang et al., 2022b).

More often the effects of stressors on cell-mediated immunity have been examined in vitro, using mitogen-stimulated lymphocyte proliferation assays, since antigen-specific stimulation does not yield a sufficiently robust stimulation index due to a low number of antigen-specific T cell clones. However, the use of certain toxins and transplantation-based antigen-stimulation systems (e.g., mixed lymphocyte reaction) has provided useful information beyond the more artificial conditions of using mitogenic plant lectins. The use of mitogens often revealed suppression of proliferative activity following acute stressor exposure in rats. Yet, the opposite outcome was elicited by an acute stressor in the analysis of the delayed hypersensitivity (DTH) response, an in vivo measure of T cell antigen-specific sensitization and proliferation. This paradigm involves initial sensitization with antigen, followed days to weeks later by challenge with the sensitizing antigen, which results in an inflammatory response characterized by increased redness and swelling of the challenged part of the body (typically the footpad or pinnae of the ear). A comparison of the effects of acute and chronic stressor exposure showed that acute stressor exposure enhanced the DTH response, whereas repeated stressor exposure produced a suppressive effect (Dhabhar, 2014). These findings are consistent with the notion that an acute stressor enhances immune functioning to preserve the organism's well-being, whereas these adaptive effects are overwhelmed with chronic stressor encounters. It is notable that the DTH response could be influenced by stressors that were applied to animals before being sensitized (i.e., given a primary exposure) with the chemical 2,4-dinitrofluorobenzene (DNFB) (Dhabhar, 2014).

Several older studies have been instructive in establishing potential principles by which stressors modify the direction of antigen-specific immune responses. For example, antigen-specific spleen cell memory proliferation to cholera toxin was enhanced by exposure of rats to acute foot-shock (Kusnecov and Rabin, 1993), which was similar to memory DTH responses being augmented by acute restraint (Flint et al., 2001). Acute stressors applied at the time of memory induction (i.e., when cells are naïve and are being stimulated with antigen for the first time) may exert a suppressive effect, but not once memory cells have formed. Indeed, acute restraint can interfere with antigen sensitization of naïve T cells, resulting in reduced memory DTH or in vitro proliferative responses to a recall antigen (i.e., one previously encountered) (Kusnecov and Rabin, 1993; Tournier et al., 2001). Additionally, acute stressor exposure may promote the generation of more effective memory cells from naïve lymphocyte precursors. In this regard, rats exposed to a single stressor session and immunized with the protein antigen keyhole limpet hemocyanin (KLH) around the time of stressor exposure subsequently showed enhanced proliferative responses by memory spleen cells (Dhabhar and Viswanathan, 2005). Since mitogen-induced proliferation was unaffected, it seems that a stressor at the time of immunization promoted better sensitization and development of a greater pool of memory cells, which was reflected by greater leukocyte infiltration of antigen injection sites. In essence, acute stressor exposure proximal to the time of antigen immunization can augment future reactions to the same antigen. These findings differed from those indicating that exposure to an antigen at the time of stressor exposure had the effect of disturbing antigen-specific T cell memory proliferative responses (Kusnecov and Rabin, 1993; Tournier et al., 2001). These different effects could be related to the type and severity of the stressor to which animals had been exposed, or it may be that diverse antigens yield different outcomes in response to stressor challenges. As such, for theoretical and practical purposes, it may be useful to concurrently evaluate the impact of different antigens (benign proteins versus bacterial toxins/viral determinants) in response to stressors. Whatever the case, the essential point is that stressful events may affect immune memory so that later responses are enhanced.

The fact that the response to a prolonged stressor resulted in immune effects moving in an alternative direction relative to those observed after an acute stressor may have implications pertinent to well-being. Such a reversal may be in the organism's interests given that the persistent maintenance of a heightened state of immune reactivity is energetically costly, and the risk for immunological dysregulation may be elevated. As a case in point, the reactivity of the immune system to negative feedback regulation by the HPA axis can be affected by persistent social disruption, thereby allowing for increased inflammation and a greater likelihood of the development and/or exacerbation of autoimmune disease. Glucocorticoids are known to be immunosuppressive, and their ability to exert inhibitory restraint on the magnitude of most immune responses is considered important to ensure that inflammation does not result in disease. Consequently, attenuation of immune responses after chronic stressors may represent a mechanism to protect against dysregulation both within the immune system, as well as between the immune system and regulatory neuroendocrine mechanisms. There are limits to the regulatory actions of glucocorticoids in suppressing immune activity as chronic stressor experiences result in resistance to the glucocorticoid response, thus implicating the involvement of other systems in mediating the impact of chronic stressors (Weber et al., 2017).

The differential impacts of acute and chronic stressors on brain functioning may stem from peripheral inflammatory cytokines accessing the brain and affecting the functioning of astrocytes and microglia. While acute stressors have modest effects in this respect and may serve in an adaptive capacity, chronic stressors cause much more profound effects in promoting neurodestructive cytokine release from microglial and astrocytes. These actions are abetted by increased glutamate release that undermines brain neuronal plasticity (e.g., Rhie et al., 2020).

The foregoing discussion was concerned with in vivo T cell-mediated inflammatory responses. However, as noted earlier, many studies have focused on the impact of stressors on the sterile inflammatory response. Specifically, stressors may cause the peripheral release of norepinephrine, which may promote the production of monocytes by bone marrow, which can interact with DAMPs and MAMPs. This, in turn, promotes activation of inflammatory pathways, notably NF-κB and the NLRP3 inflammasome. As seen in Fig. 7.1, this results in elevated production of inflammatory cytokines and diminished glucocorticoid functioning that otherwise could diminish the actions of the inflammatory cytokines. In addition, stressors may cause microglia to shift to an M1 proinflammatory phenotype, which releases CC-chemokine ligand 2 (CCL2), which can attract activated myeloid cells to the brain. Together, factors favor the occurrence of psychological disturbances, such as depressive disorders (Miller and Raison, 2016).

Several studies demonstrated that spleen and blood lymphocytes from rats exposed to acute conditioned and unconditioned stressors displayed immunoprotective effects, but the increased immune activation can elicit adverse actions (e.g., increasing features of autoimmune disorders). As described in Fig. 7.2, prolonged exposure to stressors (e.g., foot-shock, immobilization, and isolation) shifts these responses in the opposite direction so that health risks become more prominent (Dhabhar, 2014). By increasing proinflammatory factors and reducing aspects of the immune system, chronic stressors favor the occurrence of several pathological conditions.

As much as this formulation is intuitively appealing, data from some studies have not uniformly been consistent with this perspective, as the imposition of restraint or electric shock can still exert a depressive effect on

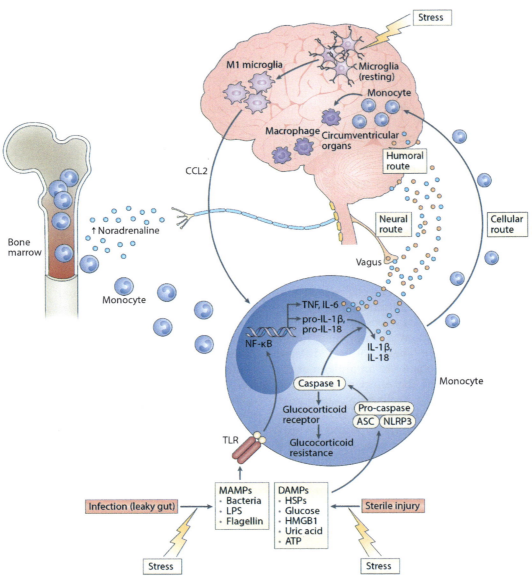

FIG. 7.1 Transmitting stress-induced inflammatory signals to the brain. In the context of psychosocial stress, catecholamines (such as noradrenaline) released by activated sympathetic nervous system fibers stimulate bone marrow production and the release of myeloid cells (e.g., monocytes) that enter the periphery where they encounter stress-induced damage-associated molecular patterns (DAMPs), bacteria, and bacterial products such as microbial-associated molecular patterns (MAMPs) leaked from the gut. These DAMPs and MAMPs subsequently activate inflammatory signaling pathways such as nuclear factor-κB (NF-κB) and the NOD-, LRR- and pyrin domain-containing protein 3 (NLRP3) inflammasome. Stimulation of NLRP3 in turn activates caspase 1, which leads to the production of mature interleukin-1β (IL-1β) and IL-18 while also cleaving the glucocorticoid receptor contributing to glucocorticoid resistance. Activation of NF-κB stimulates the release of other proinflammatory cytokines including tumor necrosis factor (TNF) and IL-6, which together with IL-1β and IL-18 can access the brain through humoral and neural routes. Psychosocial stress can also lead to the activation of microglia to an M1 proinflammatory phenotype, which releases CC-chemokine ligand 2 (CCL2), which in turn attracts activated myeloid cells to the brain via a cellular route. Once in the brain, activated macrophages can perpetuate central inflammatory responses. ASC, apoptosis-associated speck-like protein containing a CARD; HMGB1, high mobility group box 1; HSP, heat shock protein; LPS, lipopolysaccharide; TLR, Toll-like receptor. *Source: Miller, A.H., Raison, C.L., 2016. The role of inflammation in depression: from evolutionary imperative to modern treatment target. Nat. Rev. Immunol. 16, 22–34.*

adaptive functions of lymphocyte proliferation (Kusnecov and Rossi-George, 2002; Maslanik et al., 2012). Accordingly, the interpretation of stressor effects on immune function might need to be reevaluated. This said, studies have differed in the species used, the stressor reactivity between strains of mice, the nature and severity of the stressor, and the timing between stressor exposure and immune assessment. For that matter, not all studies used a true chronic exposure regimen (which theoretically should extend over several weeks, as

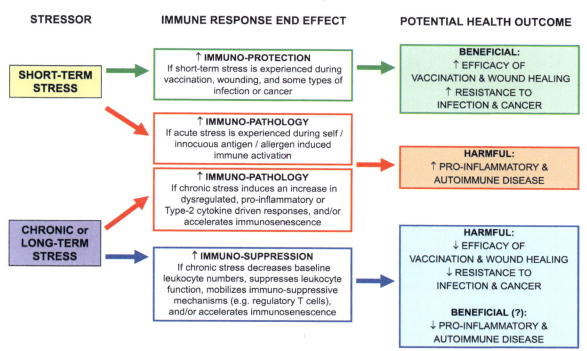

FIG. 7.2 The relationship among stress, immune function, and health outcomes. Acute stressors experienced during vaccination, wounding, or infection may enhance immunoprotective responses. Acute stressors experienced during immune activation in response to self/innocuous antigens or allergens may exacerbate proinflammatory and autoimmune disorders. Chronic stressor-induced increases in proinflammatory or type-2 cytokine-mediated immune responses may also exacerbate inflammatory and autoimmune diseases. Chronic stressor-induced suppression of immune responses may decrease the efficacy of vaccination and wound healing and decrease resistance to infection and cancer. *From Dhabhar, F.S., 2014. Effects of stress on immune function: the good, the bad, and the beautiful. Immunol. Res. 58, 193–210.*

opposed to several days), and in cases where repeated exposure was only for a few days, insufficient time was allowed to elapse to glimpse some form of immune adaptation.

Most studies that examined the impact of acute and chronic stressors focused on T cell functioning, but as we'll see shortly, stressors may profoundly influence innate immune functioning and it seems that pathogenic stimuli can induce memory in monocytes and macrophages (Tercan et al., 2021) or prime these cells so that a more pronounced response can be elicited from them. This may be beneficial in dealing with some threats, but the persistent elevation of inflammatory processes may have detrimental effects, as in the promotion of atherosclerosis. In this regard, the view was offered that chronic psychological stressor experiences can cause the reprogramming of monocytes so that they become hyperresponsive to Toll-like receptor ligands, resulting in a persistent hyperinflammatory phenotype that increases the risk for pathology (Barrett et al., 2021).

One of the most important changes that has come about in the analysis of stressor effects on behavioral and biological processes has concerned the use of varied social defeat paradigms that on the surface appear to be more relevant to psychological distress that occurs in humans. Indeed, repeated (chronic) social defeat in rodents leads to numerous hormonal, inflammatory immune, amygdala neuronal, and microglial changes that are reminiscent of anxiety and depression in humans (Munshi et al., 2020). Moreover, repeated social defeat resulted in reduced T cell functioning and expression of genes associated with T helper cells. This was apparent in mice that were susceptible or resilient to depression, perhaps indicating that aggressive encounters, irrespective of dominance, take a toll on immune functioning. Yet, among susceptible mice, the chronic stressor provoked a greater number of splenic $CD4^+$ and $CD8^+$ T cell numbers that produced IL-17 coupled with a diminished number of regulatory T cells (Ambrée et al., 2019). As well, the levels of inflammatory Ly6C monocytes and spleen-derived $CD11b^+$ cells were elevated in susceptible mice and migrated to the brain more readily in defeated mice (Ambrée et al., 2018). These findings are in line with reports that chronic mild stressors that lead to a depressive-like condition in mice were associated with an imbalance between T_h17 and T_{reg} cell subsets (Niu et al., 2013).

Much like the effects on neurochemical changes induced by stressors, an initial stressor experience may result in sensitization wherein splenic monocyte responses and peripheral cytokines are elevated in response to later stressor encounters. Similarly, as shown

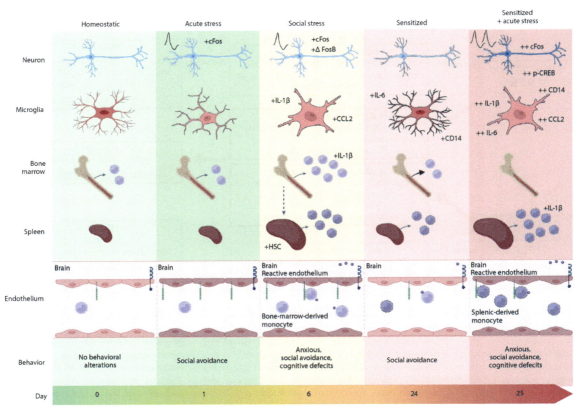

FIG. 7.3 The first box shows a "homeostatic" state in the absence of stress. The second box shows acute stress with one cycle of social defeat (2 h). This induced neuronal (cFos) activation in fear- and threat-appraisal regions and social avoidant behavior. In addition, microglia morphology in the amygdala was increased compared to controls. The third box shows RSD with six cycles of defeat. RSD promotes a reactive (+) endothelium, myelopoiesis in the BM, splenomegaly (HSCs), and neuronal (cFos, ΔFosB) and microglial (IL-1β and CCL2) activation. RSD also leads to behavioral deficits, including anxiety (open field, light dark), social avoidance, and cognitive deficits (Barnes maze, Morris water maze). After 24 days (fourth box), microglia remain in a primed proinflammatory state (IL-6 and CD14), and social avoidance behavior persists. The fifth box shows exacerbation (++) of immune responses, and neuronal activation occurs after an acute stress 24 days after the last cycle of RSD. The spleen becomes a reservoir for immune cells and releases monocytes into circulation following the acute defeat. These spleen-derived monocytes traffic to the reactive endothelium and can signal through IL-1R. Neurons show increased reactivity (increased p-CREB) and activation (cFos) in the fear- and threat-appraisal regions. Primed microglia are activated and release proinflammatory cytokines and chemokines (IL-1β, IL-6, CCL2). Behavioral deficits, including anxiety (open field), social avoidance, and cognitive deficits (Y-maze), are evident after stress sensitization. Created with BioRender.com. *Source: Biltz, R.G., Sawicki, C.M., Sheridan, J.F., Godbout, J.P. 2022. The neuroimmunology of social-stress-induced sensitization. Nat. Immunol. 23, 1527–1535.*

in Fig. 7.3, sensitization effects occur in microglia so that brain cytokine release is elevated upon re-exposure to a challenge weeks later (Biltz et al., 2022). This is accompanied by elevated IL-1b-producing monocytes that may underlie the anxiety provoked (McKim et al., 2018). Significantly, repeated social defeat caused an increase of neutrophils and monocytes in blood, spleen, and bone marrow, whereas T, B, NK, and DCs were reduced. The neutrophil changes were present for at least 6 days after the stressor and were more prominent in stress-reactive BALB/c mice compared to hardier C57BL/6N mice (Ishikawa et al., 2021). Once more, these findings point to the fundamental role of strain differences and the accompanying genetic influences related to stressor reactivity in accounting for the effects of social defeat in provoking immune alterations.

Humoral Immune Responses: B Cell Function and Antibody Production

Assessing the effects of stressors on humoral immunity involves the analysis of B cell function, particularly the ability of B cells to generate antigen-specific antibodies. An important issue in this regard is the timing of stressor application relative to the introduction of the antigen to the host. Prior to antigen exposure in vivo, naïve B cells are largely in a "resting" or quiescent state. This changes quickly once a novel antigen is engaged by the immune system, initiating a dynamic process of interactions between antigen-presenting cells (APCs), T cells, and B cells. Over the next few days, a build-up of antibodies is produced by antigen-specific B cells, which proliferate and differentiate into plasma cells. The immunoglobulin isotype of the antibody will be primarily IgM and IgG, although, at mucosal surfaces, IgA will be dominant. The end result of this antibody accumulation is the binding and removal of antigens. This is a critical aspect of the immune response, involving many cell types, and a temporally precise sequence of events that are potentially susceptible to the neurobiological consequences of stressor exposure. Accordingly, determining the impact of a stressor on the humoral immune response needs to consider the timing of the stressor relative to immunization or the impact of a stressor on the resting state of immune cells before immunization. Indeed, the view emerged that a "neuroimmune window" may be present concerning immune challenges during which stressors may have especially pronounced actions (Dantzer, 2019).

Once an immune response is initiated, the timing of stressors becomes important in determining specific outcomes. In nonimmunized animals, mitogenic responses are reduced by acute stressors, indicating that in this altered context, the arrival of antigen might influence the process, presentation, and induction of antibody-mediated immune responsiveness. In early studies, sheep erythrocytes [i.e., sheep red blood cells (SRBC)] were commonly used as the antigen, mainly because this was the common approach in immunology. When animals received SRBC for the first time and were exposed to an acute stressor (even several days later), the number of antigen-specific antibody-forming cells (AFC) and the concentration of circulating antibodies was reduced. This suggested that well after the initial antigen processing by APCs, stressors can still impact the initial stages of antibody production and antibody-producing B cell proliferation. In contrast, administering the stressor before or at the time of immunization can produce variable effects. In some cases, a one-time application of the stressor at or a day prior to the time of antigen administration failed to alter the number of antibody-producing B cells (Zalcman and Anisman, 1993). In other cases, a single stressor session immediately before or following primary immunization with various antigens (e.g., SRBC, KLH, or tetanus toxoid) either augmented or attenuated antibody production (Kusnecov and Rossi-George, 2002).

As mentioned in the context of acute versus chronic stressor effects, methodological differences across studies need to be resolved with this type of research. Different laboratories did not use the same species (mouse versus rat), stressors, or antigens. Thus, although the B cell antibody response is malleable by stressor perturbation, the outcome is not always predictable. This is no less problematic in the human literature, which we will discuss shortly. Still, one is urged to look for consistencies that could prove useful. For example, in experiments that used KLH as the antigen, enhanced antibody responses were observed after a single, brief (1 h or less) stressor exposure (Shanks and Kusnecov, 1998). However,

inhibitory effects were observed in response to more intense stressors (restraint combined with tail shock), which can also suppress the antigen-specific T and B lymphocyte proliferative response (Gazda et al., 2003). Thus, it seems that primary humoral immune responses may be refractory to acute stressors, especially when these coincide with antigen exposure. More severe stressors may overcome this resistance, especially if the glucocorticoid response is pronounced (Gazda et al., 2003). Overall, however, the evidence is more aligned with the conclusion that cognate interactions between APCs, B cells, and T cells resist acute stressor effects such that the antibody response either remains unchanged or exceeds normal levels. This again supports the notion that in the short term, stressors could serve as an immunological adjuvant (Viswanathan et al., 2005). Importantly, this window of stressor-promoted "adjuvanticity" may be narrow, since application of an acute stressor several days after antigen presentation may suppress AFC numbers. This reduction in the number of AFCs is a deviation from the normal course of the B cell response, suggesting that once B cells are activated and begin generating antibodies, the elevated concentrations of stressor-induced neurotransmitters and hormones play a more important role in modifying the immune response.

The foregoing comments apply to short-term or acute stressor exposures. When mice were exposed to a chronic mild stressor regimen an increase of immature B cells was provoked among mice housed in an enriched environment, whereas this did not occur in mice housed in a standard impoverished condition (Gurfein et al., 2017). The effects of a chronic stressor on B cells varied with the site at which immune responses were examined. Specifically, repeated social defeat led to elevated splenic B cell activation together with an increase of the anti-inflammatory cytokine IL-10. In contrast, within the meninges, which serve as a barrier to protect the brain, B cells were reduced, whereas elevations were observed in certain monocytes (Ly6C^{hi}) that contribute to inflammation (Lynall et al., 2021).

It was similarly observed that in a prolonged stressor, such as mice being isolated or exposed to foot-shock (1 h/day for 2 weeks), and then immunized with SRBC, the antibody response remained unaffected (Shanks et al., 1994). While this suggests some form of physiological adaptation that maintains the stability of the humoral immune compartment, additional manipulations after the chronic stressor experience revealed that this was not necessarily the case. If chronically stressed mice were rested for a few days and then given one more exposure to the foot-shock stressor 3 days after immunization with SRBC antigen, AFC formation was enhanced (Zalcman and Anisman, 1993). As discussed earlier, if naïve (nonstressed) mice were first immunized, and then subjected to a stressor a few days later, the antibody response was suppressed (Zalcman and Anisman, 1993). In effect, a stretch of time during which the organism is working harder than normal to maintain physiological homeostasis might prime the immune system to mount a more pronounced immune response if an additional stressor is experienced after a period of rest.

In the social isolation study mentioned earlier, the initial week of isolation was marked by a suboptimal antibody response, which subsequently normalized (or showed a rebound) after more prolonged isolation. Adaptation of the humoral immune response occurs in the face of persistent changes in an organism's environmental conditions. This might explain why most organisms survive life's interminable cycles of stress and rest, in that preexposure to stress modifies (and may even reinforce) the effects of subsequent stressor experiences on antigen-specific humoral immune responsiveness. In the same way, as the brain learns and adapts to the myriad stimuli to which an organism responds (i.e., displays neuroplasticity), this may similarly occur with respect to how the immune system deals with constant fluctuations or "waves" of neurohormonal levels that are by-products of stressors. But the possibility must also be considered that such adaptations may not occur when stressors are severe and are naturally meaningful.

Stressor Effects on Macrophage Function

Mononuclear phagocytic cells (macrophages) are key elements of the innate component of the immune response. The study of neural–immune interactions is heavily focused on these cells—or the myeloid lineage in general—as their activation is considered proinflammatory and elaborates cytokines (e.g., IL-1) that have potent CNS effects. However, these cells also orchestrate antigen-specific induction and effector mechanisms related to T and B lymphocytes. As the first line of defense against microbial agents, it is important to know how these cells are affected by stressors. This is especially the case given that monocytes and macrophage variations and the secretion of cytokines from these cells provoked by stressors may contribute to psychopathology. In this regard, stressful events promote trafficking of monocytes to the brain, which may promote anxiety (Wohleb et al., 2015).

An important aspect of innate immunity is the ability to display chemotactic responses to sites of infection or inflammation. Mice infected with *Listeria monocytogenes* at the beginning of a 7-day period of daily restraint showed reduced migration of macrophages to the peritoneum where the bacteria were introduced (Zhang et al., 1998). The reductions in migratory behavior may reflect altered responses to chemotactic factors (e.g., chemokines), which are independent of altered phagocytic function. Indeed, stressors can enhance phagocytic and suppressor functions of peritoneal macrophages (Bailey et al., 2007), which is consistent with the notion that the expression of genes linked to the innate immune system is augmented following stressor exposure (Maslanik et al., 2012). In addition, elimination of macrophages, either in vitro or in vivo, removes stressor-induced suppression of spleen cell mitogenic function, as well as restraint-induced enhancement of antigen-specific AFC numbers (Shanks and Kusnecov, 1998). It is possible that by harnessing and enhancing macrophage functioning therapeutic benefits can be realized for diseases, such as some forms of cancer.

Stressor actions on macrophages can influence how T and B cell functions are affected, possibly through changes in macrophage cytokine production. For example, periodic restraint or social disruption increased LPS-induced IL-1β, IL-6, and TNF-α levels in mouse brain, spleen, liver, lung, and peritoneal cells (Engler et al., 2008), and increased TNF-α and IL-12 production in mice infected with toxoplasma gondii and exposed to a cold stressor (Aviles and Monroy, 2001). Likewise, in rats exposed to neurogenic or psychogenic stressors, IL-1β production by alveolar macrophages was increased in response to LPS, and similar actions can be provoked by environmental contaminants (Morimoto et al., 2014).

It also appeared that HPA activation by inflammatory factors may be mediated by prostaglandin synthesis by perivascular macrophages (Serrats et al., 2017). These observations indicate that stressors can elevate proinflammatory cytokines, including those that have the capacity to modify neural and behavioral functions. In addition, these proinflammatory cytokine changes can be instrumental in supporting humoral immune responses to various antigens. In effect, a dynamic interplay between innate and adaptive immune cells may operate during stressor exposure and dictate the direction of a given antigen-specific immune response. In humans who reported elevated perceived stress, accelerated immune aging reflected by monocyte/macrophage inflammation was observed, which was linked to cognitive functioning even among healthy individuals (Casaletto et al., 2018).

Macrophage migration inhibitory factor (MIF), one of the earliest cytokines identified, participates in inflammatory and immune responses and has been implicated in the promotion of inflammatory-related diseases. This cytokine participates in diverse macrophage functions and influences the migration of DCs (Ives et al., 2021), and upon stimulation by glucocorticoids, MIF is released from monocytes/macrophages. In this respect, MIF can attenuate the actions of glucocorticoids in the inhibition of T cell production. In contrast to the many studies that have assessed the effects of stressors on other cytokines, little attention has focused on MIF despite it potentially having effects on multiple illnesses. Nevertheless, MIF has been implicated in depressive illness (Petralia et al., 2020), although studies in animal models have been inconsistent. Of potential significance, basal MIF levels have been effective in predicting the effectiveness of psychotherapeutic interventions, such as mindfulness, in alleviating mild and moderate depressive symptoms (Sundquist et al., 2020).

Stressor Effects on Dendritic Cells

Examining macrophages can provide insight into how stressors might influence antigen presentation. Since DCs represent the chief cells that process and present antigens to T and B cells making it important to determine the impact of stressors on this immune component. A brief review of stressor effects on DC function was provided by Kohman and Kusnecov (2009), and accumulating data since then have indicated that DCs can affect central processes, including microglia (Herz et al., 2017), and may be a player in depressive illnesses through the release of proinflammatory cytokines (Leite Dantas et al., 2021).

As we noted earlier, the application of stressors at the time of immunization with antigen provides the opportunity to gain an indirect sense of how APC function might be affected. An early study indicated that among mice subjected to 8 h of restraint and then sensitized with fluorescein isothiocyanate (FITC), contact sensitivity to FITC—which primarily recruits a specific CD8+ T cell response—was reduced upon testing 5 days later (Kawaguchi et al., 1997). It was also observed that dermal Ia + Langerhans cells (LCs), sampled immediately after stressor exposure, occupied less cutaneous terrain and displayed a more compact, rounded appearance, with less dendritic branching. The LCs are the primary APCs in the skin, and although direct evidence is lacking, it is possible that stressor-induced interference in epidermal DC antigen uptake and processing impaired the antigen-specific T cell response to FITC. In a related study, brief restraint that preceded sensitization with DNFB enhanced the delayed hypersensitivity response to the sensitizing antigen (Saint-Mezard et al., 2003). This effect was dependent on altered DC function, which was demonstrated using two different approaches. In the first, bone-marrow-derived DCs were antigen-pulsed ex vivo (i.e., in which purified DCs are incubated in vitro with antigen for a fixed time) and then passively transferred to mice that were stressed. When these mice were challenged with antigen 5 days later, the delayed hypersensitivity response in the challenged ear pinna was enhanced. In a second approach, fluorescently labeled DCs that were injected into the footpads of mice just prior to restraint were more likely to migrate into the draining popliteal lymph nodes (those closest to the footpads). It was thus concluded that acute stressor exposure augments T cell-mediated contact sensitivity through increased DC antigen presentation.

Further to such findings, mice exposed to acute restraint, and then injected with the sensitizing agent, DNFB, later displayed greater leukocyte infiltration and higher mRNA levels of chemokines and proinflammatory cytokines (e.g., IL-1β, TNF-α, IL-6, and IFNγ). This was associated with a nonselective increase in cell numbers in the dermal tissue and draining cervical lymph nodes, and included mature DCs, macrophages, as well as mature and naïve T cells (Viswanathan et al., 2005). Evidently, in response to an acute stressor, multiple immune factors, including APCs (DCs and macrophages) and T lymphocytes behave in a coordinated manner, thereby facilitating augmented antigen-specific immunoreactivity.

In humans, functional assessments of DC-mediated antigen presentation are uncommon. Nonetheless, it was reported that human circulating DCs, that exhibit a $CD11c^+/CD14^-/CD19^-$ phenotype, are increased during laparoscopic cholecystectomy surgery or physical exercise, followed by a decline to baseline upon termination of the stressor (Ho et al., 2001). The initial rise in DC numbers may reflect a stress-related demand that extrudes sequestered cells from locations with ready access to the vasculature (e.g., the spleen). Afterward, they undergo redistribution to various tissues to maximize their chances of encountering and processing antigens. Unlike these findings, however, a stressful public speaking procedure conducted 24 h earlier, resulted in a reduction in the number of epidermal LCs obtained from cutaneous tissue (Kleyn et al., 2008). In this case, one interpretation is that stress promoted the migration of LCs to the lymph nodes that drained the area where the skin was biopsied.

There appears to be somewhat of a disconnect between the animal and human literature concerning the concordance between in vitro and in vivo measures. Earlier in our discussion of the animal literature, we saw that spleen cells from rodents exposed to a stressor also displayed suppressed mitogen-induced proliferative response. However, injection of antigen at such a time (when mitogen capacity is inhibited) can actually augment the antibody response (Kusnecov and Rossi-George, 2002). This is not always the case, but we do need to be careful in how we use the outcomes of one immunological assessment to predict those of another. Most certainly, in vitro assays are useful as an indication that the behavior of cells has changed but may not shed new light on how the handling of antigens and the response to them transpires in vivo before, during, or after some period of stress. Nonetheless, in healthy human participants, the initial wave of research established that T lymphocyte proliferative capacity and NK cell cytotoxicity, as well as virus-specific T cell cytotoxicity, were particularly affected by stressors (Segerstrom and Miller, 2004). Focusing on just these standard immune measures in healthy subjects, however, does not provide immediate insight into the health relevance of these observations. Focusing more on in vivo dynamics provides a better perspective on this question. As discussed earlier, the studies on viral reactivation and impaired cell-mediated control over viral clearance are particularly relevant, as are studies involving more direct investigation of infectious disease and stress in humans.

Stress and Infection

Analyses of stressor actions on the progression of infectious disease processes have generally revealed elevated susceptibility to replicating viral or bacterial antigens. Stressor-provoked suppression of host defense against influenza and herpes virus infections has been well-established in rodents (Luo et al., 2020; O'Connor et al., 2021), and by limiting leukocyte trafficking, possibly through norepinephrine actions, sick animals are slower to recover (Devi et al., 2021). Similarly, in humans, stressful events influence susceptibility to influenza infection through corticosterone effects on IFN-b (Luo et al., 2020) and a prospective analysis indicated that susceptibility to SARS-CoV-2 infection was tied to stressful experiences (Ayling et al., 2022).

A fundamental issue that ought to be addressed is whether subclinical doses of infectious agents are affected by stressful experiences so that they fulminate into frank infectious disease. It has been reported that exposure to the influenza virus fails to produce symptomatic disease in a large segment of the population, likely owing to the effectiveness of pre-infection T cell immunity (Hayward et al., 2014). Clearly, the emergence of illness is not guaranteed to occur simply because of a pathogen's presence, and symptoms might be more likely to occur with a second hit. Early studies in rodents had revealed precisely this, in that subclinical infection can be encouraged by the presence of stressors to convert into an infectious illness, presumably owing to reduced immune functioning. Later studies confirmed these findings in human volunteers who were challenged with moderate doses of rhinoviruses, the source of the common cold. These studies revealed that those individuals who reported high levels of perceived distress were likely to become symptomatic and exhibited typical signs of infection, whereas those with low-stressor experiences tended to remain asymptomatic (Cohen et al., 1998). Conversely, obtaining social support could buffer against the effects of stressors in limiting upper respiratory infection (Cohen et al., 2015a) and these actions were most pronounced among individuals with dispositional positive affect (Janicki Deverts et al., 2017).

Stressed individuals in these studies showed a greater propensity for somatic sensations to immune-mediated processes (Cohen et al., 2015a). This may be related to augmented immune-mediated effects on afferent or central neuronal systems since individuals, who were exposed to the rhinovirus, exhibited greater intranasal cytokine responses. Specifically, higher levels of intranasal IL-6 and TNF-α were associated with greater levels of stressor-related threat, as well as greater susceptibility to the cold virus (Cohen et al., 2012a). Furthermore, this was directly correlated with increased glucocorticoid resistance, a notion that is increasingly gaining traction as an explanatory mechanism for enhanced inflammation and autoimmunity.

Studies in rodents and humans have assessed the impact of repeated or chronic stressors on susceptibility to viral infection. As observed with other viral infections, stressors have been implicated in the emergence of herpes zoster (shingles), although inconsistent results were obtained in studies that assessed general stressors experienced and those that evaluated relation to specific stressors (e.g., bereavement). This said, using validated measures of perceived stress revealed a marked increase in herpes zoster occurrence (Takao et al., 2018), and a nationwide Danish study revealed that its emergence was directly related to perceived stress (Schmidt et al., 2021). Given the involvement of NK cells in viral clearance, it had been suspected that stressors could influence infection by disturbing NK cell functioning. However, it appeared that although stressful events were related to the emergence of the virus, this was independent of NK cell activity (Kim et al., 2018).

Stressful events can also diminish immunological restraint on the normally dormant Epstein–Barr Virus (EBV) (Glaser and Kiecolt-Glaser, 2005). Ordinarily, cell-mediated immune processes, including T cell cytotoxic activity, suppress latent herpes viruses (e.g., herpes simplex virus and EBV). In the case of EBV, large segments of the population may be infected, but the virus remains dormant within B lymphocytes, likely stemming from ideal immunosurveillance and restricted viral proliferation within the memory B cells where it resides. EBV may also hide within neurons and ultimately lead to neurological conditions, such as Alzheimer's disease and MS (Jha et al., 2015). The mechanism for the reactivation of the virus might have been due to impaired cytolytic destruction of proliferating EBV-infected B cells (Glaser and Kiecolt-Glaser, 2005). It is also possible that MS may be promoted by disturbed immune functioning owing to weakened T cell control, reflected by an increase in circulating anti-EBV antibody titers. Remarkably, in a very large 20-year longitudinal study among US military members, the presence of MS was 32 times greater among those with EBV infection relative to those who had not been infected (Bjornevik et al., 2022).

Like EBV, many people are infected by cytomegalovirus (CMV), which is retained within the body for life.

Symptoms may not be apparent at all, although some individuals present with fatigue, sore throat, swollen glands, and fatigue. This usually dormant virus may precipitate mononucleosis and has been associated with an elevated risk of heart attack and stroke. Among older adults, elevated CMV titers were related to several immune changes, including elevated proportions of aged T and NK cells, and aged T cells were further increased with elevated stress perception (Reed et al., 2019). In addition to stressful experiences being associated with more senescent immune cells, it seemed that when controlling for CMV presence, the tie between stress and immune aging was eliminated (Klopack et al., 2022).

Although many viruses (colds and influenza) are cleared quickly and might not have obvious long-term repercussions, chronic infection and the associated chronic inflammation, may have damaging effects even if people are asymptomatic, as in the case of CMV. Most people may unknowingly carry multiple latent viruses, some of which may be associated with reduced telomere length, indicative of cellular aging (Polansky and Javaherian, 2015), raising the possibility that these viruses may cumulatively shorten lifespan.

In addition to affecting the development of viral illnesses, stressors increase susceptibility to bacterial infection. In part, this may stem from stress hormones, notably norepinephrine, enhancing bacterial motility and facilitating their ability to colonize diverse niches of the host organism (Weigert Muñoz et al., 2022). When mice were stressed through social defeat and then infected with *Mycobacterium tuberculosis*, old mice were initially better at controlling lung inflammation than young mice. However, 60 days after infection *M. tuberculosis* growth was more pronounced in older mice, together with elevated anti-inflammatory IL-10 mRNA and decreased IFNγ mRNA (Lafuse et al., 2022).

Together, these studies indicated that stressful events undermine specific components or operations of the immune system. These are important findings, but it needs to be kept in mind that at the end of the day, the important question is whether stressors increase vulnerability to pathological outcomes, including vulnerability to infection. This said, we'll now turn to the impact of stressors on immune processes that might affect disease occurrence and progression.

Immunological Consequences of Stressor Exposure in Humans

Substantial data in humans amassed concerning the impact of stressors on basic immune functions, antiviral immunity, infectious disease, and the general healing effects of the immune system (Fagundes et al., 2013; Glaser and Kiecolt-Glaser, 2005). Our first line of defense against foreign intruders consists of barriers provided by the skin and the mucosae. Like other barriers against infection, through hormonal influences, stressors can undermine epidermal barrier functioning so that microbes can enter the body (Maarouf et al., 2019). Once microbes have breached mucosal and epidermal barriers, much of the protection available is provided by the immune system and related hormonal processes. The complexity of the various findings that speak to the impact of stressors is in some instances confusing, which is understandable given the heterogeneity of genetic, epigenetic, and experiences that human participants bring to these investigations.

Evaluation of stressor effects in humans comprised determinations of circulating immune substrates, or analyses of the actions of in vitro challenges (antigens or mitogens) on NK cytotoxicity or lymphocyte proliferation. Functional outcomes have also been assessed through analyses of susceptibility to viral infection as we've already discussed, as well as the time for wounds to heal, and analyses of the impact of stressors on responses to vaccines, which in essence are like (inactivated) viral threats.

An early review and meta-analysis of close to 300 studies that assessed healthy humans focused on basic immune parameters derived from peripheral blood leukocytes since more invasive analyses are typically not possible (Segerstrom and Miller, 2004). In this analysis, the various stressors used were classified as being acute, chronic, naturalistic, and what was referred to as "event sequence" and "distant." Bereavement was an example of an "event sequence" stressor that occurred within a year, and potentially possessed persisting residual effects. Chronic stressors were persistent challenges, such as those faced by caregivers, while distant stressors included trauma experienced 5–10 years prior to immune assessment. The outcome of this analysis led to several conclusions that were very similar to some of the reports from the animal literature. The acute stressors, which considered social, cognitive, and emotional domains of experience, generally revealed some consistency in the immune changes that occurred. The strongest effects comprised an increase in NK cell numbers or the related large granular lymphocytes (LGLs),[3] which are associated with NK cell function. The functional significance of the increased NK cell numbers was related to increased cytotoxicity, but this was not related to a fundamental enhancement of cytotoxicity at the individual cell level. Peripheral blood lymphocyte proliferation to mitogens

---

[3] LGLs are morphologically distinct cells of the lymphoid lineage (see Chapter 2). These cells were originally thought to be primarily NK cells, which target tumor cells and virally infected cells. However, a separate subset of LGLs exist which are thymus-related (hence, T-LGLs), and which are seen in certain forms of leukemia. In the studies reviewed by Segerstrom and Miller, as well as other studies on stress and human immune function, LGL numbers and function are generally interpreted in terms of their cytotoxic functions (i.e., their role as so-called NK cells).

that globally stimulate T cells and/or B cells was decreased by acute stressors, and in some studies that examined cytokine production in these mitogen assays, the most consistent effect was increased IL-6 and IFNγ production. In addition, among healthy older individuals, a laboratory stressor was associated with elevated T_{reg} cells in both males and females, which were correlated with depressive mood and physical health status (Ronaldson et al., 2016).

These findings essentially revealed that the actions of acute stressors on two indices of lymphocyte function—proliferation and cytokine production—generated opposing effects. Moreover, humoral immune functioning, reflected by altered immunoglobulins, was not dramatically affected. This is neither the first nor last time that different parameters of immune function would reveal differential changes in the face of a stressor. Unfortunately, this adds to the confusion regarding how to interpret immune changes.

Despite the limitations that exist, it seems that acute psychological stressors of moderate severity usually increase immune functioning, whereas intense or chronic stressors typically disrupt immune activity (Slavich and Irwin, 2014), possibly secondary to hormonal changes that were provoked. It was proposed that the chronic stressors that humans may experience can promote inflammatory changes that foster anxiety and depression and a diverse set of physical illnesses [e.g., rheumatoid arthritis (RA), chronic pain, obesity, metabolic syndrome, type 2 diabetes, cardiovascular disease, and neurodegeneration], which may be comorbid with depression (Slavich and Irwin, 2014). Assessing these relationships necessarily requires analyses of the consequences of chronic stressors in a natural context, preferably in prospective studies.

The immune and health consequences among caregivers have received considerable attention, due to the constant strain of working with people with dementia and other chronic conditions, such as children with developmental disabilities. Caregivers were found to experience elevated allostatic overload relative to noncaregivers, which predicted an increase in future illnesses and disabilities (Gallagher and Kate, 2021). The strain of caregiving for a child with autism spectrum disorder was associated with elevated inflammatory gene expression and shortened telomere length (Lin et al., 2018). Furthermore, among highly stressed mothers of children with autism spectrum disorder effector memory CD8+ and CD4+ T cell percentages were elevated, whereas percentages of naïve CD8+ T cells and central memory CD8+ and CD4+ T cells were diminished relative to control mothers. These findings are consistent with reports that caregiving of aged adults imposed suppressive effects on T cell proliferation and antibody responses to influenza vaccines (Glaser and Kiecolt-Glaser, 2005).

In addition to altered immune cell functioning, chronic distress associated with caregiving was accompanied by dysregulation in the balance between pro and anti-inflammatory cytokines, which could favor the emergence of immune-related disorders (Kim et al., 2022). In this respect, caregiving has often been left to older individuals to attend to an ailing partner. While caregiving has greater effects on old than on younger individuals, it can take a toll on both (Lovell and Wetherell, 2011), even promoting immune aging in younger individuals (Whittaker and Gallagher, 2019). Moreover, this stressor was associated with epigenetic changes associated with HPA regulation and genes that influence inflammation (Palma-Gudiel et al., 2021). Much like other neurobiological responses, immune and cytokine functioning was altered by early life negative experiences and could influence the immune response throughout life (Brenhouse, 2023) and could even affect the immune response among caregivers.

A longitudinal analysis in which caregivers were assessed before and after the death of the care recipient revealed that mitogen-stimulated lymphocyte proliferation declined in the period before the loss and continued during bereavement, albeit at a diminished pace. Regrettably, among socially isolated caregivers, T cell proliferation continued to decline during bereavement, although self-reported health which was poor prior to the loss, recovered afterward (Wilson et al., 2020). Not unexpectedly, among caregivers who engaged in greater negative rumination, which is associated with depression, the diminished antibody response was most pronounced (Segerstrom et al., 2008). Furthermore, like the stress sensitization effects observed concerning brain neurochemical processes, the immune changes associated with caregiving were especially marked if individuals had experienced early life adversities.

Overall, it seemed that caregivers frequently exhibited diverse indices of their immune response being compromised and slower recovery of immune functioning (e.g., Kiecolt-Glaser et al., 2003, 2015). Although caregiving can be distressing and may negatively affect the health of the caregiver, a subsequent meta-analysis that included 30 studies indicated that while immune system functioning was disturbed and inflammation elevated in caregivers, the observed effects were generally weak, and the clinical significance of the immune changes was questionable (Roth et al., 2019). This was in keeping with an earlier analysis which had indicated that CRP and IL-6 were not elevated in most studies that assessed immune markers among caregivers, although therapy to reduce distress was associated with lower IL-6 levels (Potier et al., 2018). At first blush, these reports might appear surprising, and it is important to dig a little deeper to understand why inflammatory changes were often not evident.

It's been known for decades that the impact of caregiving is linked to the personality characteristics of the caregiver (e.g., optimism, neuroticism) as well as to their personal resources, such as self-efficacy, sense of coherence, and perceived social support (Sołtys et al., 2021). For instance, levels of IL-6 associated with caregiving were greatest among individuals with low self-esteem (Sherwood et al., 2016) or who exhibited poor coping self-efficacy (Mausbach et al., 2011). More than this, individuals may differ in the extent to which caregiving is perceived to be a burden, and how they are able to find meaning in their actions.[4] Thus, analyses that do not consider individual difference factors might not provide an accurate reflection of the potential immune and inflammatory actions of the caregiving experience.

As we indicated in our discussion of racial discrimination on hormonal processes, these experiences were accompanied by a lower percentage of naive CD4+ cells and elevated levels of (mature) terminally differentiated CD4+ cells, possibly reflecting an aging immune system that is less capable of fighting invaders. Not unexpectedly, racial discrimination was associated with lower antibody titers following the administration of an influenza vaccine (Stetler et al., 2006). Likewise, in a sample of older adults (>50 years of age), chronic stressor experiences and discrimination were tied to more terminally differentiated CD4$^+$ cells, and lifetime discrimination was associated with a lower percentage of CD8+ cells (Klopack et al., 2022).

The impact of chronic stressors, including discrimination, on physical illnesses may promote psychological disorders, such as depression or PTSD. In this respect, stress-related psychological disorders (e.g., PTSD) were associated with increased occurrence of life-threatening infectious illnesses (Song et al., 2019a), and the nature of the infections to which individuals were most vulnerable varied with sex; men were at elevated risk of skin infections, whereas women were more likely to develop urinary tract infection (Jiang et al., 2019). Of course, stressful experiences may act indirectly on immune aging processes, since the effects observed, such as the limited production of CD4$^+$ cells may be secondary to stressor-related behaviors, such as smoking, drinking alcohol, or failing to engage in exercise. Thus, while the distress of discrimination can act on immune functioning, the adverse effects observed may be abetted by diverse lifestyle alterations that are provoked by these experiences.

While our focus has been on strong chronic natural stressors, such as caregiving, less chronic or severe stressors may affect immune and inflammatory processes. Social stressors, such as public speaking, can promote an arousing and potentially intimidating influence (given the presence of public scrutiny), causing pronounced physiological effects. For students taking exams, the stakes are less about social confrontation and shame, but more about academic success and the opportunities this provides. Even though this type of stressor hardly affects cortisol levels, it was associated with a reduction of NK cells and monocytes in the blood, coupled with a shift towards more immature cells within NK and T cell populations, while mitogen-stimulated levels of IL-6 and TNF-α production were increased. As the duration of the anticipatory stress period lengthened and the academic examination drew closer, the presence of memory cells and monocytes increased, as did levels of IL-17, whereas naïve CD8 T cells declined (Maydych et al., 2017). Other studies similarly revealed reduced in vitro IFNγ but increased IL-6 and IL-10 production. Interestingly, the changes in IFNγ production may be independent of changes in T cell proliferation. Recall that T cell proliferation tends to be reduced by acute laboratory stressors, and using naturalistic stressors, such as social engagement, the same outcome was observed. This differentiation raises the obvious possibility that some parameters of lymphocyte function may be unrelated to one another, again pointing to the need to assess multiple immune-related processes concurrently.

Numerous studies have indicated that the diverse immune and inflammatory changes associated with psychological stressors vary with individual stress perceptions and coping methods adopted (Seiler et al., 2020). As we mentioned earlier concerning the stress of caregiving, immune system functioning was linked to several personality and emotional factors that were often tied to stressful experiences. Specifically, diminished NK cell activity was reported among individuals who were hostile, especially negative, engaged in high levels of rumination, or expressed a depressive mood (Zoccola et al., 2014). Likewise, trait characteristics, such as hostility (e.g., Girard et al., 2016) and disturbed cytokine responses ordinarily elicited by stressors, were limited among individuals with higher self-esteem but were more pronounced among those who felt low in social status. Being optimistic, which has been linked to enhanced health and well-being, was also accompanied by altered stressor-provoked cytokine variations (Brydon et al., 2009).

Summarizing, there is little question that stressful events and characteristics of the stressor can promote immune changes that are linked to the development of illnesses. As depicted in Fig. 7.4, the impact of stressors on

---

[4] There is little question that caregiving can be enormously stressful, and that more attention needs to be devoted to the caring of the caregiver. Some individuals may feel that caregiving (e.g., for an ailing parent) is a chore that was foisted upon them, whereas others may gain meaning and comfort from this experience (Autio and Rissanen, 2018). Not every person is suitable for acting as a caregiver and research examining this question needs to consider these individual differences in relation to immune functioning and for the caregiver's health.

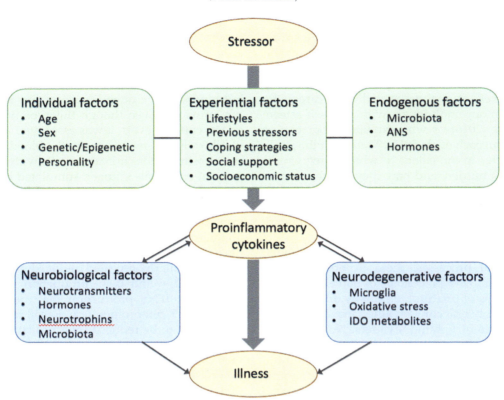

FIG. 7.4 A schematic representation of the relations between stressors, proinflammatory cytokines, and illness, with a particular focus on the individual, experiential, and endogenous factors that may moderate these effects. The capacity of social stressors to promote inflammatory variations that might lead to depression (or other illnesses) may be influenced by the presence of genetic and personality factors, sex, and age. Individuals carrying specific gene combinations or polymorphisms (e.g., variants of IL-6, IL-1β, and TNF-α) may be more vulnerable to the depressive effects of inflammatory activation provided that they also encounter social stressors. Earlier stressor encounters both prenatally and during early life, coping processes (including social support), and gut microbiota variations may influence inflammatory processes and sensitize immune responses to subsequent stressors. The activation of proinflammatory processes may directly or indirectly influence depressive states. Elevations of cytokines may influence monoamine (e.g., 5-HT, NE), hormone (e.g., CRH), and growth factor (e.g., BDNF) activity, which might favor the evolution of depression (conversely, stressor-elicited hormonal and neurochemical functioning may impact cytokine processes). Cytokine elevations may also stimulate the enzyme indoleamine 2,3-dioxygenase (IDO) and promote the release of neurotoxic metabolites, including kynurenic acid, quinolinic acid, or 3-hydroxykynurenine, and cause oxidative stress, culminating in depression. *Source: From Audet, M.C., McQuaid, R.J., Merali, Z., Anisman, H., 2014. Cytokine variations and mood disorders: influence of social stressors and social support. Front. Neurosci. 28, 416.*

immune and cytokine functioning is moderated by several characteristics of the individual, experiential factors, and basal endogenous factors. These interactive effects influence proinflammatory processes that affect a constellation of other neurobiological processes that can promote illness occurrence.

Response to Vaccination

As we've already mentioned, the response to vaccination has been taken as an index of immune functioning and quite obviously has implications for illness prevention. In healthy volunteers who received an influenza vaccination, increased reports of daily stressors were associated with lower antibody titers (Miller et al., 2004). In this study, stress monitoring occurred over 13 days, before, during, and after inoculation. Levels of distress for the 2 days before antigen delivery did not account for the reduction in the antibody response, but the 10 days during which the immune response was being mounted appeared to be critical in influencing the antibody response. Reports such as these are compelling since they imply that the efficacy of an annual vaccine is potentially modified by the psychological state of the individual, and the hardships they may have endured.

These data, in part, extend what has been found in animal studies regarding stressor-induced modulation of humoral immune responses to systemically delivered protein antigens. However, the effects in humans are not as dramatic, nor as predictable. Human studies involving immunizations are constrained by methodological compromises. In these studies, it can be difficult to determine the relationship between onset, duration, and frequency of stressful episodes and measures of antibody. As well, the temporal relationship between stressor experiences and antigen delivery (the vaccine) may be an important influence on the ultimate nature of the antibody response. As such, it can be difficult to explore with any great precision how stressors impact the immune response to vaccinations.

A meta-analysis of the influenza vaccination data (Pedersen et al., 2009) concluded that perceived or chronic stressors (e.g., caregiving) were associated with a reduced antibody titer to vaccination, being evident in both older (mean age 70 years) and younger (mean age 40 years) participants. In most of the studies reviewed, the inoculation involved a cocktail of different viral strains, and the relationship to stressors that existed was mainly apparent in response to the AH1/N1 strain. The reasons for this selectivity are unknown and may relate to preexisting higher antibody levels for other strains. Nonetheless, it appears that experiencing higher stress levels may result in less-than-optimal concentrations of circulating antigen-specific antibodies. Speaking to this, several chronic psychological stressors were associated with immune alterations and altered responses to vaccination. Loneliness and having a relatively small social network were associated with a reduced anti-influenza antibody response (Pressman et al., 2005), and the social rejection experienced during the preceding 6 months was associated with a decline in the antibody response to influenza vaccination (Corallo et al., 2022). Consistent with such findings, lower social cohesion, which is often accompanied by elevated loneliness, was associated with a poorer antibody response to vaccination against COVID-19 (Gallagher et al., 2022).

To conclude this section, the value of many of these studies, and the vaccination studies in particular, is that they extend and provide useful corroboration for the in vitro findings of stressor effects on immune function. However, the relationship between in vitro measures and in vivo immune effects can be complicated. It has always been a pressing question whether a given immunophenotype (e.g., poor in vitro lymphoproliferation) is predictive of impaired responsiveness to in vivo antigen exposure, which remains to be fully determined.

Stress and the COVID-19 Pandemic

The COVID-19 pandemic, as dreadful as it was for many people, taught us numerous lessons that may be important in understanding immune processes related to viral threats, the factors that influence vulnerability to infection, and the appearance of severe symptoms (and death). As described in Chapter 4, what we've learned has implications regarding how to deal with future pandemics that will certainly be encountered. Here, we'll largely restrict our discussion to the psychological and neurological ramifications associated with the pandemic, particularly those related to stressor experiences.

Once the pandemic hit, it was apparent that most governments were woefully unprepared to deal with the challenge. Despite encounters with SARS and MERS in the preceding two decades, and repeated warnings that we would inevitably meet another pandemic, proper steps hadn't been taken in preparation for what could (and did) occur. All the warnings that had been issued had disappeared into the ether. Avoidant strategies to deal with impending stressors typically aren't effective for individuals and they certainly aren't for governments. So, this may have been a teaching moment (a very lengthy one), but we won't know whether governments will receive a passing grade until the next pandemic occurs.

Early in the pandemic, numerous studies determined the predictors of COVID-19 susceptibility and the consequences of the illness or threats of the illness on physical and psychological health. Several early reviews attempted to predict the psychosocial and neuropsychological consequences that might be expected based on earlier knowledge concerning psychosocial dynamics and stress/coping processes adopted to deal with collective challenges (van Bavel et al., 2020). It quickly became clear, however, that dealing with the pandemic was more difficult than anticipated since we initially had no idea what tricks SARS-CoV-2 had in its repertoire that could make prevention difficult. It wasn't fully understood that the virus would be effective in shielding itself from the body's immune system, allowing this invader to replicate prodigiously. Nor was it expected that so many variants would emerge that would allow for the pandemic to stay with us. Back then, the general mantra was "flatten the curve" with the hope that the second wave would not be as catastrophic as it had been during the 1918–19 H1N1 flu pandemic. We hadn't reckoned on encounters with multiple waves.

Like other risk factors, such as being older and having heart disease, diabetes, or immune disorders, vulnerability to SARS-CoV-2 infection and negative outcomes was linked to psychiatric conditions. As well, psychological distress experienced prior to infection was associated with elevated risk for long COVID (Wang et al., 2022d). Having a depressive illness was associated with diminished COVID-19 vaccine immunogenicity and increased risk for adverse outcomes (Ford and Savitz, 2022). Moreover, depressed individuals were more likely to be diagnosed with COVID-19 infection, although to some extent this was also tied to these individuals being more likely to be tested (Meinlschmidt et al., 2022). A large study that included 263,697 fully vaccinated individuals also indicated that SARS-CoV-2 breakthrough was elevated among individuals with psychiatric disorders (Nishimi et al., 2022), and reinfection with the SARS-CoV-2 Omicron variant was elevated among older individuals and those who had exhibited depressive illness

(Li et al., 2022c). Interestingly, a meta-analysis revealed that treatment with SSRIs (especially fluvoxamine) reduced the odds of severe symptoms, although this was not apparent in all studies assessed (Nakhaee et al., 2022). It is not certain why the disparate outcomes occurred or why fluvoxamine was superior to other SSRIs.

The distress created by COVID-19, which included the fear of infection, the impact created by social isolation, and the financial repercussions experienced were related to the increased occurrence of depressive illness and anxiety disorders. For those researchers versed in the ties between stressful events and depression, this was hardly unexpected. Yet, a systematic review revealed that overall depression and anxiety during the pandemic did not increase appreciably relative to that present prior to the pandemic, and the small increases that were apparent occurred primarily in women (Sun et al., 2023e). Importantly, however, many studies did not consider the marked individual differences that were present, and it has become apparent that the psychological effects were most pronounced among people who appraised the situation as most threatening and individuals who tended to be intolerant of uncertainty (McCarty et al., 2023).

Several studies assessed whether the occurrence of depressive disorders was associated with inflammatory immune changes brought about because of the continued distress that had been created by the infection itself. Self-reported depression among infected individuals was accompanied by elevated immune indices, such as white blood cell and neutrophil counts (Yuan et al., 2020). Moreover, among COVID-19 patients, several indices of systemic inflammation were apparent, including elevated peripheral lymphocyte, neutrophil, and platelet counts (Zhang et al., 2023b), which were associated with depressed mood and PTSD. These symptoms were accompanied by lower grey matter in the anterior cingulate cortex and insula and disturbed connectivity within resting state networks (Benedetti et al., 2021a).

Not all the actions of the pandemic, however, could be ascribed to infection. Even in the absence of infection, the distress created by the pandemic was associated with elevated levels of inflammatory factors in the brains of healthy individuals, possibly being related to sterile inflammation. Specifically, through imaging procedures, the levels of translocator protein (TSPO) and myoinositol, which are markers of glial activation, were elevated. The greatest elevations of TSPO in the hippocampus were present in those individuals who reported the greatest physical and mental impact of the pandemic (Brusaferri et al., 2022). As we'll see when we discuss depressive disorders (Chapter 8), these brain changes have been associated with depressive disorders unrelated to SARS-CoV-2 infection. It has frequently been maintained that

inflammatory processes are linked to the development of depression, and blocking the actions of cytokines could reduce the severity of the illness (Benedetti et al., 2021b).

In addition to the immune changes associated with SARS-CoV-2 infection, numerous pathogenic gut microbiota were detected, which were still more pronounced among patients treated with antibiotics. The presence of the microbes was related to COVID-19 severity, and the dysbiosis persisted in patients with long COVID-19 (Yeoh et al., 2021). Moreover, the abundance of certain microbiota that were associated with anxiety and depression in the aftermath of COVID-19 was linked to earlier trauma experiences, including childhood adverse experiences (Malan-Müller et al., 2023).

As we saw earlier, depression associated with long COVID was related to diverse brain changes, such as gliosis within the ventral striatum that was related to inflammation (Braga et al., 2023), and the persistence of depression and anxiety was accompanied by atrophy of limbic brain regions associated with emotional processing, as well as immune cell senescence (Torres-Ruiz et al., 2023). Indeed, COVID-19 patients with self-reported depression showed immune-related disturbances, particularly altered CD4+ lymphocytes, relative to infected individuals who were not depressed (Wu et al., 2021). Moreover, in patients reporting depression or anxiety, alterations were observed in the tryptophan (TRP)-kynurenine (KYN) pathway (Kucukkarapinar et al., 2022), which can promote neuronal toxicity as we'll see when we discuss depression in greater detail (see Chapter 9). It also appeared that blood biomarkers, specifically elevated fibrinogen relative to CRP, as well as elevated levels of D-dimer relative to CRP, were associated with post-COVID cognitive disturbances (Taquet et al., 2023).

In several respects, long COVID was reminiscent of myalgic encephalomyelitis/chronic fatigue syndrome (ME/CFS), which sometimes appears following viral or bacterial infection. The symptoms of long COVID frequently overlap with those of ME/CFS and many of the biological characteristics of long COVID have similarly been observed in ME/CFS, such as reduced cortisol levels, elevated inflammatory factors, and the presence of neuroinflammation. As well, both conditions may involve vascular and endothelial abnormalities, mitochondrial dysfunction, disturbed metabolic profiles, reduced cerebral blood flow, and brainstem abnormalities (Davis et al., 2023). It also appeared that long COVID was accompanied by a fatigued immune system that continued battling with remnants of SARS-CoV-2 or reactivated viruses, such as EBV (Iwasaki and Putrino, 2023; Klein et al., 2023).

With so many symptoms that make up long COVID and the numerous organs affected, efforts to manage this condition may require multidisciplinary teams, especially as there is currently no single effective treatment (Davis et al., 2023). It makes intuitive sense that antivirals would limit COVID-19 symptoms and perhaps act against long COVID if the treatments were initiated soon after infection. However, once long COVID was well entrenched, such treatments were less effective. We've seen it again and again—treatments for various diseases come to us in unexpected ways. Metformin used in the management of type 2 diabetes limited the severity of COVID-19 symptoms and acted against the development of long-term COVID. The effects were most pronounced when taken soon after infection, pointing to the effect occurring through inhibition of a viral mechanism although its effects could come about through the drug's anti-cytokine actions (Bramante et al., 2023). Whether metformin can diminish long COVID symptoms that are already present is an open question.

The consequences of COVID-19 described here are only a few of the many that have been documented. In other sections of this volume, we describe many others as well as the risk inherent to offspring born to infected moms. Attitudes toward SARS-CoV-2 infection have become cavalier and for most people life has returned to normal. It seems that many people have experienced psychological exhaustion in relation to COVID-19 and the attitude has been "I'm done with COVID." However, as Dr. Anthony Fauci had said, "You may be done with COVID, but COVID isn't done with us," and he encouraged the adoption of several measures to limit relatively severe symptoms. Hopefully, the counterproductive attitudes currently seen among many people won't carry over to new pandemics that will certainly be encountered.

Stress and Cytokine Production

The ability of lymphocytes and macrophages to alter synthesis and rates of cytokine production in response to glucocorticoids, norepinephrine, acetylcholine, and various neuropeptides has been well documented. These endogenous factors have the potential to alter the series of changes that culminate in the effector phase of the immune response, be it antibody production, cytotoxicity, or cytokine production. Regulation of T-helper cell cytokines may influence antibody production, as well as the antigen specificity of antibody subtypes. Moreover, the expansion and lytic potential of cytotoxic T cells and NK cells is affected by a variety of T cell and macrophage-derived cytokines, including IFNγ and TNF-α. In addition to this local immunoregulatory impact, stressor

effects on cytokines may influence how they interact with elements necessary to affect neural and behavioral functions. In this particular domain, we have to ask whether stressors will increase cytokine levels to a degree that has greater neural influence, or conversely, precipitate cellular changes that increase the migration of monocytes and T cells into the brain. The details of these dynamics are not altogether clear but are worth considering.

As much as cytokines might be involved in the provocation of psychological or physical disturbances, their actions vary with multiple individual factors, experiences during early and later life, and a constellation of other environmental factors. A schematic representation of such a process is depicted in Fig. 7.4, but as we will see in the ensuing section, the many different pro- and anti-inflammatory cytokines may be differentially influenced by stressors and have diverse consequences.[5]

Stress Effects on Th1 and Th2 Cytokine Responses

The adaptive immune response relies heavily on the balance between Th1 and Th2 cell function. Since a Th2 bias is associated with a drive toward improved support and initiation of humoral immune responses, such as antibody production, it is thought that the predominance of Th1 responses would be counterproductive to the promotion of the antibody response. Therefore, a bias toward proinflammatory processes might operate through the activation of macrophages (if shifted toward Th1 cytokine responses), whereas greater humoral immune responsiveness would operate through a greater shift toward Th2 cytokine output. This can be relevant to immune-related pathology, since exaggerated and protracted skewing in either direction—Th1 or Th2—can influence infectious and autoimmune diseases (Raphael et al., 2015). For instance, during the initial stages of an immune response to a pathogen, Th1 cell cytokine responses are required to drive increased phagocytic functions through activation of macrophages and facilitating antibody-dependent opsonization of the pathogen. Over time, this is down-regulated by Th2 cell cytokines, such as IL-10, which can attenuate the damage that a needlessly protracted immune response can impose on local tissue. Thus, if stressors modify the production of Th1 and Th2 cytokines, this may induce critical imbalances in their mutual counter-regulatory functional relationship, which may result in pathology and inflammatory disease. The key cytokines that have received attention in the context of stress studies include the Th1

---

[5] Levels of cytokines in humans have typically been measured in blood samples. Although cytokine levels in saliva are low, they have been used in studies that examined various illnesses where cytokine levels had been expected to be elevated. It has been demonstrated that acute stressors were similarly associated with elevated IL-6, IL-10, TNF-α, and IFNγ (Szabo et al., 2020).

cytokines, IL-2, IFNγ, and TNF-α, whereas the impact of Th2 is generally attributed to IL-4 and IL-10 production. However, it should be acknowledged that cytokines from either Th cell subtype are synthesized and released by other immune cells. Thus, in assessing circulating cytokines, the cellular source is unknown, and may not originate from T cells. Nonetheless, analysis of the effects of an acute laboratory psychosocial stressor indicated a rapid increase of the transcription factor for NF-kB, whereas that of IL-6 and mRNA of IL-1ß, IL-6, and that of the NF-kB inhibitor, IκBα, was somewhat delayed (Kuebler et al., 2015).

Many of the effects of stressors on cytokine production have been conducted using in vitro approaches, in which harvested leukocytes, whether from blood or lymphoid organs, are stimulated with mitogens or antigens to which animals or humans had previously been primed or sensitized. There have been studies in animals and humans that assessed in vivo elevations in cytokines after stressor exposure (Marsland et al., 2017), and it likewise appears that the cytokine variations are more pronounced among animals that had previously been exposed to a stressor or cytokine treatment (Anisman et al., 2003). In addition to sensitization related to epigenetic changes, it was suggested that this neuroinflammatory priming involves a signal cascade within the brain that involves DAMPs and the NLRP3 inflammasome. Whatever the case, it appears that earlier experiences can promote exaggerated neuroinflammatory responses that can promote behavioral disturbances, and might contribute to psychiatric illnesses (Fleshner, 2013; Frank et al., 2020).

Among rodents that had been exposed to an uncontrollable stressor, later challenge with an immune-activating agent elicited anxiety and depressive-like behaviors and produced elevated peripheral inflammatory responses to a greater extent in females than in males. As well, microglial phagocytic activity was altered by the stressor procedures but the microglial alterations within the hippocampus did not uniformly map onto the sex-dependent peripheral immune changes, perhaps pointing to other processes mediating the greater stressor reactivity evident in females (Fonken et al., 2018).

Interferon-γ

Exposure of animals to acute stressors enhanced IFNγ production in the DTH immune model of antigen-specific responding (Dhabhar, 2014). This has been shown to apply to in vivo measures, but when in vitro assessments were conducted, inhibition of IFNγ production was provoked by acute stressors (Kohman and Kusnecov, 2009). Similarly, when stressors were prolonged or repeated daily, IFNγ production was attenuated in response to a variety of antigens, including tetanus toxin, herpes simplex, and influenza viruses, and less pathogen-associated

stimuli, such as ovalbumin and CD3 crosslinking with a monoclonal antibody. Importantly, in vivo regional lymph nodes displayed an inhibited ability to produce IFNγ after exposure of the animal to a stressor (Sloan et al., 2008).

It was noted earlier that in humans, acute stressors are more likely to increase IFNγ production assessed ex vivo (Marsland et al., 2017). However, the suppression is eliminated after prolonged or more persistent stressor exposure. Under these conditions, stressors either failed to alter the percentage of IFNγ+ CD4 and CD8 T cells or exerted a modest reduction in T cell IFNγ production (He et al., 2014). Interestingly, the persistent stress of academic examinations enhanced IFNγ production by stimulating blood leukocytes (Maes et al., 1998). This contrasts with the findings from animal studies, which showed chronic stressors to inhibit IFNγ expression. Of course, animal studies access cells from lymphoid organs, whereas human studies rely on circulating cells, which might account for the differences observed. Also, the impact of academic examinations has effects that differ from many other stressors, typically failing to produce cortisol elevations.

Interleukin-2

Stressor-induced suppression of IL-2 production was established some time ago in rat and mouse studies. Similarly, psoriasis patients or healthy controls that were exposed to the Trier Social Stress Test, both showed a significant decline in mitogen-induced IL-2 production by peripheral blood leukocytes (Buske-Kirschbaum et al., 2007). This is consistent with research showing that caregivers, who reported high levels of perceived distress, also displayed reduced capacity to generate IL-2 (Bauer et al., 2000). Indeed, adults who had experienced intense trauma, such as emotional neglect, and emotional, physical, and sexual abuse before the age of 16, displayed elevated levels of mitogen-stimulated levels of IL-2, IL-6, and TNF-α (de Koning et al., 2022). However, a meta-analysis revealed that stimulated IL-2 production was unaffected by a social threat (Marsland et al., 2017). Thus, while IL-2 gene activation and cytokine output in humans is susceptible to stressor effects, this is not a guaranteed outcome. In having considered stressor effects on IL-2 and IFNγ, we are in essence gaining a perspective on the response capacity of Th1 cells and affords the opportunity to determine factors that moderate stressor-provoked cytokine responses.

For the most part, commonly employed experimental stressors in animal studies exerted a suppressive influence on Th1 cytokine production, although largely when this was assessed by in vitro restimulation methods. Less research is available on measures of Th1 cytokine production in vivo, which may help to determine whether the implications of the in vitro studies are relevant to Th1-

dependent diseases. As noted earlier, the argument has been made that augmented in vivo Th1 responses occur after acute, but not after chronic stressor exposure (Dhabhar, 2014). This implies a requirement for cognate stimulation of T cells (i.e., with antigen) since mice exposed to prolonged restraint (from 1 week to several weeks) failed to show plasma IFNγ elevations (Voorhees et al., 2013). Given that Th1 cytokines generally need to be induced by a cognate stimulus, this finding is not surprising and suggests that spontaneous elevations of circulating Th1 cell cytokines are less susceptible to the effects of stressors, which was also the case in human studies (Marsland et al., 2017).

Interleukin-4

In considering Th2-type responses, we can begin with IL-4. This cytokine is important in regulating B cell activation and differentiation but is most prominent in promoting IgE antibody responses, which are involved in allergic reactions. In humans, mitogen-induced production of IL-4 was reduced in response to psychosocial and academic stressors (Buske-Kirschbaum et al., 2002), although another social stressor (public speaking task) did not affect mitogen-stimulated IL-4 production. In young children, with or without asthma, the presence of depression or high levels of perceived stress, the levels of IL-4 measured 6 months later were appreciably elevated (Wolf et al., 2008). In other studies, in children with asthma who had experienced chronic family stress, exposure to an acute stressor resulted in elevated production of IL-4, IL-5, and IFNγ, an effect that was not apparent in the absence of family discord (Marin et al., 2009). It was similarly observed that life stress reported by children with atopic dermatitis was directly related to circulating IL-4 levels (Wardhana, 2016). These findings are interesting, but it might be more informative to concurrently assess the IFNγ/IL-4 ratio, which is frequently adopted as an index of the Th1/Th2 ratio. In doing so, this ratio was reduced among healthy young adults who experienced chronic strain (Takemori et al., 2021). Further research is needed to address important interactions between stressors and immunological mechanisms underlying allergic reactions, which in some cases represent a significant life risk.

In some animal studies, restraint failed to affect IL-4 production induced by T cell mitogens, although, in other studies, IL-4 and IL-10 were elevated in response to acute and chronic stressors (Himmerich et al., 2013). Likewise, in mice with induced airway inflammation, IL-4 was elevated following a chronic stressor (Okuyama et al., 2007). However, restraint reduced spleen cell IL-4 responses to herpes simplex virus, and murine exposure to a social stressor imposed a long-term memory deficiency in IL-4 production to a porcine pseudorabies virus (de Groot et al., 2002). The latter was selective for mice that exhibited wounds, with nonwounded mice failing to show any effect on IL-4. These data suggest that infection and injury interact with psychogenic stressors to inhibit Th2 cell cytokine production. The consequences may be an extension of Th1-mediated immunity, which is necessary to drive inflammatory and phagocytic processes that are designed to eliminate pathogens. However, this notion is not in keeping with the known suppressive effects of stressors on IFNγ that were mentioned earlier, relegating the significance of changes in IL-4 production as something yet to be confirmed.

Interleukin-10

If there is one cytokine that could be considered the "glucocorticoid" of the immune system, IL-10 would certainly fit this role. It has been referred to as the master anti-inflammatory regulator of the immune system, and few elements of immune activity are not inhibited by this cytokine (Rojas et al., 2017). As with other cytokines, the production of IL-10 is modifiable by stressors. For example, social disruption among female mice reduced the expression of IL-10 as well as the IL-6/IL-10 and TNF-α/IL-10 ratios and increased the expression of the chemokine CX3CR1 (Díez-Solinska et al., 2022). Likewise, in vitro IL-10 production in response to tetanus toxin, influenza virus, and mitogens was suppressed following prolonged restraint or social disruption (Merlot et al., 2004; Tournier et al., 2001). However, when IL-10 was induced in vivo by an LPS injection, acute swim or restraint increased IL-10 production by spleen cells (Curtin et al., 2009b). In essence, two different scenarios are revealed by observations that focus either on the induction of IL-10 in vitro or in vivo. The in vivo findings are in keeping with the notion that acute stressor exposure can enhance key components of an immune response. In this particular case, the source of IL-10 is most likely macrophages, since LPS does not directly stimulate T cells, but does induce IL-10 in cells of the innate immune system. Additional models are required to focus more closely on T-cell-derived IL-10. Indeed, the development of IL-10-secreting memory T cells responsive to pseudorabies antigen was inhibited by exposure of mice to social stressors (de Groot et al., 2002).

In addition to the peripheral actions observed, a chronic mild stressor regimen provoked multiple brain cytokine changes, including elevated expression of proinflammatory cytokines IL-1β, TNF-α, and IL-6, together with diminished expression of anti-inflammatory cytokines TGF-β and IL-10, so that the inflammatory status was elevated (You et al., 2011). Chronic social instability in female mice similarly reduced IL-10 in the hippocampus (Labaka et al., 2017), and the effects of stressors, as expected, differed between high and low-stress sensitive strains of mice (McWhirt et al., 2019). Chronic social isolation of rats beginning immediately after weaning gave rise to reduced IL-10 protein and mRNA in blood, as well as lower hippocampal IL-10 (Corsi-Zuelli et al., 2019). As we'll see in greater detail in this chapter early life stress in the form of repeated maternal separation provoked marked cytokine changes, varying with the tissues examined. Elevations of blood and nonblood proinflammatory cytokines were produced by maternal separation, which was still more pronounced upon exposure to a further stressor. Maternal separation was associated with increased IL-10, but the effects of further stressors had inconsistent effects. These procedures were associated with reduced IL-10 expression in brain regions coupled with increased levels of proinflammatory cytokines (Dutcher et al., 2020).

An increase in perceived stress among elderly individuals who received an influenza vaccine was associated with an augmented in vitro IL-10 response to influenza antigen restimulation (Kohut et al., 2002). This was in keeping with a subsequent study that examined hepatitis B patients who were partitioned into high and low daily life stress groups. In these patients, the high stress group showed the greatest stimulated IL-10 production (He et al., 2014). In contrast, a social stressor applied in a laboratory setting suppressed in vitro IL-10 production by isolated blood leukocytes (Buske-Kirschbaum et al., 2007). This points out that the particular conditions of stressor exposure and/or induction of IL-10 differentially affect cytokine output.

In some cases, the influence of stressors on IL-10 production was studied in relation to the Th1 cytokines IL-2 and IFNγ, the production of which is reduced by elevations of IL-10. Indeed, high levels of stress among younger caregivers were associated with elevated numbers of IL-10 T regulatory cells, while the number of IL-2 or IFNγ Th1 cells remained unchanged (Glaser et al., 2001). These caregivers had previously exhibited attenuated antibody responses to the influenza vaccine, as well as reduced output of IL-2, which was in keeping with the increase in IL-10-producing T cells. This inverse relationship was also evident among students with high exam pressure, showing elevated production of IL-10 (Maes et al., 2000).

It is evident that stressors can modify IL-10 concentrations, which might exert greater inhibitory control over Th1 cytokine production. If such regulation is inappropriate or untimely in the face of a need to mount an effective immune response, stressor-induced elevations of IL-10 could be considered a risk factor. However, we do not know enough about the role of IL-10 at the interface between stress and disease. To be sure, since IL-2 and IFNγ are not entirely suppressed by stressors, it is possible that some refractoriness to the inhibitory effects of IL-10 may also develop during stressor experiences. At the very least, it should be considered that the observed increases of IL-10 are compensatory changes, serving as a physiological check against excessive inflammation in scenarios that would benefit from augmentation of Th1 cytokine responses. Most certainly we need to know more about this relationship to gain a better understanding of how stress may benefit or compromise immunologically mediated protection against disease.

Interleukin-5

The Th2-derived cytokine IL-5 has received attention in the development, maturation, and recruitment of eosinophils. These cells are granulocytic leukocytes with antimicrobial cytotoxic effector capability and might contribute to a range of immunopathologic situations, including respiratory allergies, asthma, and gastrointestinal disease. In addition, they are involved in the coordination of antigen-specific T and B cell responses. As observed in the case of other Th2 cytokines, academic examinations in students with mild forms of asthma induced pulmonary eosinophilia and increased the production of IL-5 in cells harvested from sputum samples (Liu et al., 2002). Peripheral blood eosinophilia was noted in stressed individuals who presented with atopic dermatitis, suggesting that in allergic situations, stressors can promote the migration of potentially pathogenic eosinophils (Buske-Kirschbaum et al., 2002).

Interleukin-6

The functions of IL-6 have for the most part been recognized as a B cell growth and differentiation factor. It is also involved in fibroblast and neuronal growth and has been viewed as a neuroprotective factor (Sun et al., 2017a), although it is more commonly discussed in the context of various neuronal and systemic pathologies (Rothaug et al., 2016). Alterations in the production and/or release of IL-6 in response to stressors have long been recognized (Marsland et al., 2017; Steptoe et al., 2007) and it has been related to depression in humans.

Animal studies demonstrated that exposing animals to a series of different stressors increased circulating plasma IL-6, with more severe stressors producing greater IL-6 elevations. When animals were exposed to a series of stressors that included handling, exposure to novelty, social disruption, restraint, and then electric shock—a sequence that involves increased intensity of threat—the IL-6 concentrations became increasingly higher (Hale et al., 2003). In earlier animal research, spleen cell IL-6 production induced by a T cell mitogen was increased in lactating (but not nonlactating) female rats exposed to a conditioned psychological stressor (Shanks et al., 1997). This drew attention to IL-6 elevations being subject to unique physiological states in animals. Similarly, in mice exposed to social disruption in vitro IL-6 production was increased in response to LPS, suggesting heightened macrophage reactivity (Stark et al., 2002).

In several studies, the presence of inflammation elicited by LPS, an acute stressor enhanced circulating IL-6 levels measured 1.5h after treatment in the hardy C57BL/6By strain of mouse but had lesser effects in the anxious BALB/cBy strain. In contrast, TNF-α was only modestly affected, whereas IL-10 levels rose dramatically in both strains, although a bit less so in the more anxious strain (Gibb et al., 2011). These actions varied appreciably with a chronic stressor, and with time following the endotoxin treatment. For instance, in the noninbred CD-1 strain, an acute stressor increased plasma IL-6 measured 1.5h after LPS treatment, but by 3h the IL-6 levels were reduced. In contrast, both acute and chronic stressors reduced circulating TNF-α levels at both times, whereas IL-10 was markedly increased 1.5h after LPS treatment in both acutely and chronically stressed mice (Gibb et al., 2013).

In humans, psychological stressors, such as public speaking or exercise, were accompanied by elevated IL-6 and TNF-α together with increased presence of the inflammatory marker C-reactive protein (Zoccola et al., 2014). As well, the chronic strain experienced by parents of young cancer patients showed a stress-related suppression of IL-6 production (Miller, Cohen, and Ritchey, 2022), whereas the short-term distress of exams, public speaking, or exercise enhanced in vitro mitogen-induced IL-6 production (Goebel et al., 2000). A meta-analysis of 29 studies indicated that stressors typically provoked plasma/serum IL-6 elevations (Marsland et al., 2017). When measuring cytokines in plasma, it is assumed that the cytokine may have emerged from activated leukocytes. However, an important measure that needs to be obtained simultaneously with circulating cytokines is the activation state of peripheral blood leukocytes. In the absence of confirmation in the same individuals that circulating leukocytes are actively producing and secreting IL-6, there is the very real possibility that stress-induced plasma IL-6 elevations are emerging from an endocrine-like organ responsive to stressor intensity or from brown fat, which may contribute to inflammatory and neuropsychiatric illnesses (Qing et al., 2020). Other studies with mice showed that an immobilization stressor increased circulating IL-6, and was not associated with the spleen, but was instead linked to increased IL-6 mRNA expression in the liver. This raises the possibility that stressor-induced plasma IL-6 elevations are due to the activation of hepatocytes, a long-recognized source of the acute phase inflammatory response. In addition, cytokines (myokines), most notably IL-6, are released from muscles during exercise and may serve in an anti-inflammatory capacity (Nara and Watanabe, 2021). It was maintained that IL-6 released by muscles may be an important stress sensor that influences endocrine functioning and thus may affect other cytokines (Welc and Clanton, 2013).

Stressful events have also been associated with an increased presence of cytokines, including IL-6, within the brain. While IL-6 and other inflammatory changes were provoked by stressors within blood and brain, the profile of cytokine changes produced by stressors in these compartments was different from one another and also varied across stress-sensitive brain regions, such as the prefrontal cortex and hippocampus (Gibb et al., 2011, 2013). As we've already described, the increased presence of cytokines within the brain may come about owing to activation of microglia, or from the periphery (see Fig. 7.5). It was maintained that the entry of peripheral IL-6 into the brain in mice that had experienced chronic social stress may contribute to the promotion of depressive-like behavior (Menard et al., 2017). As well, the IL-6 changes introduced by repeated social defeat provoked the recruitment of proinflammatory monocytes to the brain, which may have been responsible for the induction of anxiety-like behavior (Niraula et al., 2019). It was similarly maintained that IL-6 was linked to cognitive impairments and metabolic alterations associated with Alzheimer's disease, and thus targeting this cytokine might be an effective strategy to alleviate symptoms (Lyra e Silva et al., 2021). In this respect, stressful events can influence gut microbiota that affect circulating cytokine production, which influence brain functioning, and ultimately mental state (See Figs. 7.6 and 7.7).

Interleukin-1 and Tumor Necrosis Factor

Cytokines, such as IL-1β and TNF-α, may be prominent in the regulation of stress responses (Goshen and Yirmiya, 2009) and may thus contribute to depressive disorders, including the instances of depression that had frequently been observed during the COVID-19 pandemic (de Mello et al., 2022). Stressors affect circulating IL-1b levels and can modify the actions of other treatments that increase the levels of this cytokine. Plasma elevations of IL-1β and TNF-α are readily induced by injection of LPS, which targets TLR4 innate immune cells, such as macrophages. In mice challenged with LPS, the IL-1b response was significantly reduced by an acute restraint session, whereas repeated social disruption augmented both spleen and lung IL-1b and TNF-α responses. Similarly, rats exposed to a single session of inescapable tail shock generated greater amounts of IL-1β and TNF-α in response to an in vivo LPS challenge (Johnson et al., 2002). Moreover, in mice exposed to a relatively severe stressor, circulating levels of IL-1β, TNF-α, and IL-6 were elevated in the absence of an LPS challenge (Cheng et al., 2015). Evidently, exposure to stressors varying in intensity and/or stressor duration results in opposing inflammatory cytokine responses. Whether the source of these cytokines was immunological is not clear. However, stressor-induced elevations in circulating IL-1β and IL-6

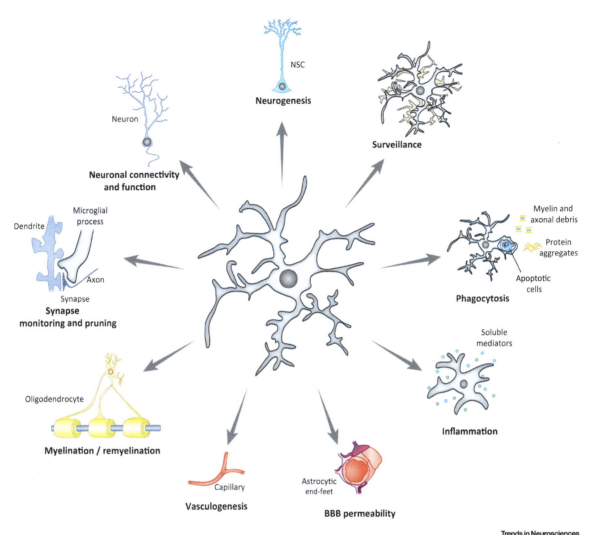

FIG. 7.5 In pathological conditions, microglia perform classical immune functions, such as release of inflammatory mediators and phagocytosis of cellular debris (apoptotic cells, axonal, and myelin waste). In addition, microglia perform several physiological chores that are to some extent altered in pathological conditions, such as synapse monitoring and pruning. Microglia are capable of interacting with other brain cells, impacting their function: neurons and their connectivity; neural stem cells (NSCs) and neurogenesis; oligodendrocytes and myelination/remyelination; endothelial cells and vasculogenesis/revascularization; and astrocytes and blood–brain barrier (BBB) permeability. *Source: Sierra, A., Paolicelli, R.C., Kettenmann, H., 2019. Cien Años de Microglía: milestones in a century of microglial research. Trends Neurosci. 42, 778–792.*

can be blocked by immunological activation with LPS 1 week prior to stressor exposure (Merlot et al., 2004), implying a form of desensitization or tolerance at the cellular level. As LPS targets macrophages, it points to the possibility that such cells are involved in the spontaneous production of proinflammatory cytokines following stressor exposure.

Many more studies have assessed the effects of stressors on immune function and cytokine production. In this section, we highlighted some of the effects produced when individuals are exposed to different types of stressors. Cytokines, such as IL-1β, IL-6, and TNF-α are major activators of the neuroendocrine system and have become prominent points of discussion in theories of depression. It is uncertain whether stressor-induced elevations in cytokines represent a potential neurotrophic influence that affects normal brain function. Most certainly, the demonstration of increased IL-6 entry into the brains of chronically stressed mice establishes the plausibility of this notion. Finally, there are numerous studies regarding the effects of stressors on cytokine induction in the brain. Our focus here, however, has been to address how stressors affect peripheral cytokine components. The overall consensus—always with exceptions—is that stressors can increase the concentration of circulating cytokines. However, in what way this might influence mental health is part of an ongoing investigation in this exciting area.

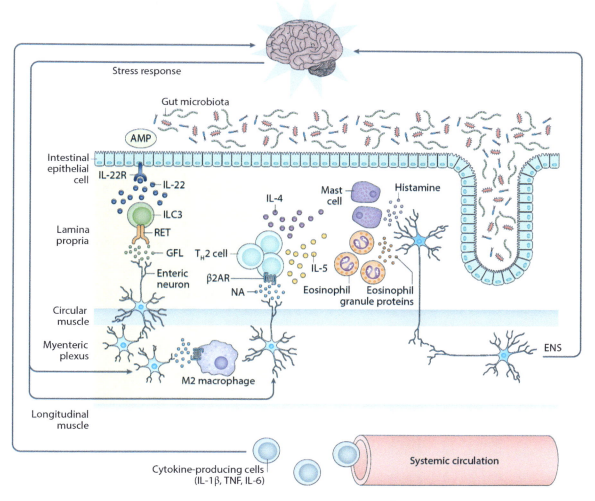

FIG. 7.6 There is complex crosstalk between the immune system, the brain, and the gut. Noradrenergic neurons in the gut release noradrenaline (NA), which ligates the β_2 adrenergic receptor (β_2AR) on macrophages in the myenteric plexus (supporting differentiation toward the M2 phenotype) and T cells, which limits T helper (T_H)1 differentiation (indirectly favoring T_H2 and T_H17 differentiation). Mast cells and eosinophils are found degranulating next to enteric neurons, providing a mechanism for sensory excitation (which can be perceived by the enteric nervous system [ENS] and central nervous system). Gut immune cells (e.g., $\gamma\delta$ T cells) differentiating in the gut can traffic to the brain under some circumstances (e.g., after brain injury). Blood-borne cytokines generated in the gut can also signal in the brain. Enteric glial cells also produce glial-cell-derived neurotrophic factor family ligands (GFL), which stimulate IL-22 production by innate lymphoid cell (ILC)3 (acting on the specific receptor RET). IL-22 acts on an epithelial-restricted receptor (IL-22R) to stimulate epithelial proliferation and antimicrobial peptide production. IL-4 and IL-5 (potentially generated locally by T_H2 T cells) support activation of mast cells and eosinophils, respectively. AMP, antimicrobial peptides. *Source: Powell, N., Walker, M.M., Talley, N.J., 2017. The mucosal immune system: master regulator of bidirectional gut-brain communications. Nat. Rev. Gastroenterol. Hepatol. 14, 143–159.*

Autoimmune Disorders

As we've seen throughout this chapter, stressful events have been associated with immune alterations, that can increase vulnerability to infectious diseases. Some of these illnesses might arise because stressors weaken immune functioning, thereby allowing viruses to do their unsavory business. However, stressors can also exacerbate disorders in which aspects of the immune system attack the self (i.e., autoimmune disorders). The increase in the occurrence of autoimmune disorders has prompted the notion that stressful experiences may promote the onset of some autoimmune disorders, and epidemiological studies have indicated that stressful experiences, including childhood adverse events, may precede the occurrence of autoimmune disorders (Choe et al., 2019). In a cohort of almost 1.5 million individuals with a stress-related disorder, about 50% were more likely to

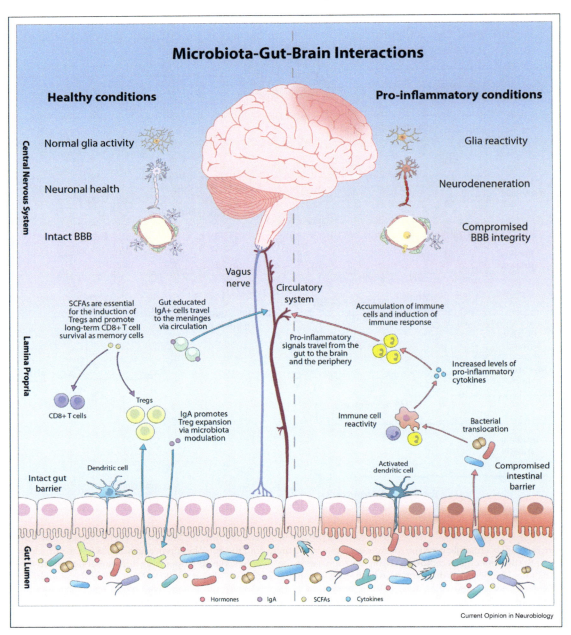

FIG. 7.7 Microbiota–gut–immune–brain axis in healthy and proinflammatory conditions. Communication between the gut and brain may occur through several pathways, such as the vagus nerve and the circulatory system. Microbiota can influence the peripheral inflammatory process, which may ultimately be key in determining pathology as are the actions of brain glial cells. *Source: Ratsika, A., Cruz Pereira, J.S., Lynch, C.M.K., Clarke, G., Cryan, J.F., 2023. Microbiota-immune-brain interactions: a lifespan perspective. Curr. Opin. Neurobiol. 78, 102652.*

experience an autoimmune disorder than were those who did not have a stress-related disorder. The hazard ratio for single and multiple autoimmune disorders was especially elevated among individuals with PTSD but reduced if they had been treated with an SSRI (Song et al., 2018). The impact of stressors could come about through the immune activation elicited, alterations of microbiota functioning (Ilchmann-Diounou and Menard, 2020), and their interaction with genetic factors or environmental toxicants (e.g., Wang et al., 2021a).

The influence of stressors has been examined in relation to many autoimmune disorders, with the most common being MS, systemic lupus erythematosus (SLE), RA, inflammatory bowel disease (IBD), and Type 1 diabetes. Other autoimmune disorders, such as celiac disease, psoriasis, and Sjogren's syndrome, are also affected by stressors. The 65 or 70 other autoimmune disorders identified or suspected have not been examined as extensively in relation to stressor actions, although many can be exacerbated by the same factors that affect MS, SLE, and RA.

Here we will focus primarily on the latter three. In Chapter 17, we discuss stressors, inflammatory factors, and microbiota concerning diabetes and IBD.

It is generally thought that although the processes associated with autoimmune disorders are not identical to one another, they may share common elements, including that they preferentially occur in females (except for Type 1 diabetes and psoriasis), and thus might be affected by many of the same factors. In this regard, autoimmune disorders may be determined by a gene variant that influences Th cells so that ordinary body cells are attacked more readily. Relatedly, cells that are responsible for immune suppression may not be operating as they should, thus allowing for autoimmune disorders to emerge. It was suggested, as well, that still other genetic variants might determine which aspects of the body will be attacked. For instance, inheriting mutations of the "autoimmune regulator" gene (AIRE) may change immune cell competency, thus allowing them to turn on the body's organs (Oftedal et al., 2015). As we'll see, however, each of the autoimmune disorders also has its own unique contributing agents.

Although the sex dimorphism associated with autoimmune disorders has most often been attributed to hormonal influences, other mechanisms exist that may contribute to the differences. The female X chromosome (but not that of males) has a coating that leads to X-chromosome inactivation that can provoke counterproductive immune responses. In this instance, lengthy RNA strands (termed XIST) coil around the chromosome resulting in the attraction of proteins that cause the silencing of some genes on the X chromosome that may be relevant to autoimmunity. More than this, the XIST molecule could provoke inflammatory immune responses, and it also seemed that autoantibodies targeted the proteins that were attracted to XIST. Ultimately, chronic inflation is provoked which causes damage related to autoimmune disorders.

Multiple Sclerosis

MS comprises a disorder in which immune responses are directed toward the myelin surrounding axons of the spinal cord and brain, leading to plaques or lesions. The disorder may present in a form wherein new symptoms appear as discrete attacks, but can be followed by lengthy periods, even years before further incidents are experienced (relapsing–remitting form). Despite symptoms seeming to be in abeyance between episodes, the neurological disturbances may persist and even progress. In a form of the disorder, that of primary progressive MS, remission does not occur following the initial symptoms. In secondary progressive MS ("galloping MS"), in which patients had initially been diagnosed with relapsing–remitting MS experience, a progressive neurologic decline occurs between episodes.

In animal models, such as Theiler's murine encephalomyelitis virus (TMEV) infection, stressors exacerbated several aspects of MS. However, as we saw concerning stressor effects on immune changes promoted by antigenic challenges, the MS variations were tied to the timing of the stressor relative to infection. For instance, when social defeat coincided with infection, the course of the disease was diminished and the extent of inflammation was limited, whereas a similar outcome did not occur when the stressor preceded infection, possibly owing to the actions of IL-6 changes elicited by the stressor (Meagher et al., 2007). Studies using other animal models also indicated that acute stressors may limit MS symptoms, perhaps owing to the immunosuppressive actions of corticosterone. In contrast, chronic stressors seemed to promote illness exacerbation. As indicated earlier, it is unclear how MS evolves, but there are indications from studies in mice that elevated activity of immune-related genes in the inflamed meninges may seep into grey matter, thereby contributing to progressive MS (Gadani et al., 2023). As stressful events influence inflammatory processes, it is possible that these experiences may come to affect this process, thereby furthering cognitive and psychiatric correlates of MS.

The occurrence of MS flares has been attributed to numerous factors, including medication usage, cardiovascular reactivity, baseline heart rate, and stressful life events, accounting for as much as 30% of the variance of symptoms. Of particular interest in the present context is that distressing events were especially common during the period prior to relapse, and once symptoms were present, stressors were accompanied by new brain lesions coupled with the exacerbation of MS symptoms (Burns et al., 2014). These actions may be related to the promotion of several neuropeptides (e.g., CRH, neurotensin), microglial activation, as well as alterations of T_{17} cells, and disturbances of the BBB so that T cell entry into the brain occurs more readily (Karagkouni et al., 2013). Moreover, the course of MS and the occurrence of depression were moderated by the coping strategies that individuals endorsed (e.g., Valentine et al., 2023).

It is particularly noteworthy that viral infection was also associated with MS relapse, possibly stemming from the immune activation that accompanies infection. While not every virus necessarily produces such effects, a study of more than 10 million young adults in the military indicated that the risk of MS was markedly elevated after EBV infection (Bjornevik et al., 2022). Among MS patients, about 25% have antibodies that bind to a protein of the EBV, EBNA1, as well as a protein formed in the spinal cord and brain, specifically that of glial cell adhesion molecule (Lanz et al., 2022).

As viral illnesses may be influenced by stressor experiences, the combination of psychological and viral insults may contribute to MS flares. At the same time, it would be essential to identify the nature of the stressor encountered and its chronicity, as well as the coping resources available, such as social support (Briones-Buixassa et al., 2015). Interestingly, the frequency of stressful experiences, rather than the severity of acute stressors, was most closely aligned with the worsening of symptoms. In fact, although acute stressors of moderate severity could aggravate symptoms, relapse was diminished if individuals experienced a strong acute stressor, such as major surgery or fractures (Mohr et al., 2004). As we've seen, the neurobiological processes associated with moderate versus severe or chronic stressors differentially influence immune and corticoid functioning, and it is reasonable to hypothesize that the effects of stressors concerning MS may involve these mechanisms (e.g., a strong stressor might suppress immune functioning and thus transiently inhibit symptoms).

Systemic Lupus Erythematosus (SLE)

Like the other autoimmune disorders, SLE is considerably more prevalent in women than in men, usually commencing anywhere from 15 to 35 years of age. In addition to genetic influences, environmental factors have been implicated in the development and aggravation of SLE. The environmental triggers ranged broadly, having been attributed to factors such as cigarette smoke and those that affect estrogens. There is reason to suspect that air pollutants, pesticides, and phthalates might also be contributing factors, and it was suggested that infection with EBV may be related to SLE (Harley et al., 2018). The environmental influences on SLE risk may be tied to systemic inflammation, oxidative stress, and the accumulation of epigenetic changes that affect these processes (Barbhaiya and Costenbader, 2016).

The illness can involve an attack on just about any portion of the body, including the nervous system. Thus, comorbidity with multiple illnesses is common, often associated with RA as well as osteoporosis, cardiovascular disease, and varied skin conditions. Characteristics of the illness also include elevated fatigue, pain, rash, fever, abdominal discomfort, headache, and dizziness, which are often antecedents of flares, and neurological and psychiatric disturbances may ultimately appear. At this stage, white matter hyper-intensities are detectable within the brain as are lesions, hemorrhage, cellular atrophy, and microstructural abnormalities. Thus, it is hardly surprising that patients present with impaired executive functioning, speed of information processing, and poor memory.

The development and progression of SLE have been attributed to a variety of biological mechanisms. One view has it that the disease arises owing to impaired clearance of debris comprising dead cells, and the development of antinuclear antibodies that attack portions of healthy cells (Lisnevskaia et al., 2014). Given that blood barrier permeability may be compromised in SLE, autoantibodies could also access brain sites in which damage is present (Dema and Charles, 2016), thereby affecting glutamate (NMDA) receptors, and hence promoting cell death, and behavioral disturbances (Arinuma, 2018). As well, microglia activation may be accompanied by altered inflammation, which can lead to cognitive disturbances and depressive symptoms (Seet et al., 2021).

An enormous number of gene loci have been identified that were linked to SLE as have numerous epigenetic changes. These have involved autoantibodies, IgA deficiencies, loss of mechanisms associated with central and peripheral immunological tolerance, as well as cytokine processes (Kwon et al., 2019). Likewise, polymorphisms related to glucocorticoid functioning (e.g., FKBP4 and FKBP5) that were tied to anxiety and diminished quality of life were also associated with features of SLE (Lou et al., 2021). Determining the genetic and epigenetic factors associated with SLE is enormously difficult given the wide array of features that accompany the disorder. For that matter, the possibility has been raised that SLE is not a single disorder but instead, its manifestation reflects the confluence of several conditions (Rivas-Larrauri and Yamazaki-Nakashimada, 2016), although several core genes may be involved in its appearance. This makes it all the more important to identify multiple markers in addition to those of a genetic and epigenetic profile, in developing approaches to treatment (Fasano et al., 2023) as well as in accounting for the diverse comorbid illnesses associated with SLE.

Having a chronic illness, such as SLE, is often terribly distressing, and encountering day-to-day irritations, particularly those associated with social relationships, could exacerbate symptoms. In this regard, increased stressor perception was predictive of increased disease activity (Patterson et al., 2022) as well as disturbed cognitive functioning. It may be significant that SLE patients seemed to be particularly sensitive to stressors, at least concerning changes in immune and cytokine responses, which is in keeping with the symptom exaggeration elicited by stressors, and the way individuals cope with the distress also influenced health-related quality of life among SLE patients. A prospective analysis revealed that the presence of depression was associated with poorer physical, mental, and social well-being in those with SLE (Dietz et al., 2021), and cognitive behavioral therapy and mindfulness training diminished depressive symptoms and enhanced psychological quality of life. Regrettably, this might not be paralleled by enhanced physical quality of life (Solati et al., 2017).

Rheumatoid Arthritis (RA)

This condition most commonly appears in middle age, but can also appear in children (juvenile arthritis). Even when the disorder initially manifests in adulthood, systemic inflammation and autoimmunity often begin several years earlier (Demoruelle et al., 2014). The features of RA comprise inflammation of the synovial joints, the membrane lining joints, tendon sheaths, and cartilage, and in some cases, ANS disturbances and heart problems may occur. The distress created by the disease is made worse by the chronic pain experienced, as well as general malaise or feelings of fatigue, weight loss, and poor sleep. As we've seen in other autoimmune disorders, symptoms may wax and wane and patients may even experience lengthy periods during which symptoms diminish, before again recurring. Numerous factors, including environmental and genetic processes, likely contribute to the development of RA, and as with other illnesses, there is the view that multiple hits contribute to its occurrence. It has been known for some time that RA is associated with specific antibodies, rheumatoid factors, and citrullinated peptides, often being detectable in advance of clinical signs of the illness. This does not necessarily imply that these are causative agents of RA, but they might be useful biomarkers for disease occurrence (Myasoedova et al., 2020).

As in the case of other autoimmune disorders, day-to-day stressful events may promote symptom exacerbation, as can negative ruminative coping (Lu et al., 2022c), possibly owing to increased inflammatory cytokine functioning. Chronic interpersonal stressors were likewise accompanied by worsening symptoms, together with dysregulation of inflammatory processes, which could undermine treatment efficacy. Analysis of the everyday changes in pain symptoms among workers indicated that on days when they encountered stressful work-related events or high levels of job strain, their pain levels were relatively elevated. It appeared that under stress conditions, changes in cortisol activity and α-norepinephrine receptor functioning played a role in producing arthritic flares (Straub, 2014).

Attempts to understand RA have focused mainly on immune dysregulation, but there are indications that microbial factors contribute to the illness. Specifically, the disorder was associated with increased bacterial antigens within the gut and lung, and even in the tissues surrounding the teeth (Brusca et al., 2014). The specific microbiota that might contribute to RA isn't known, although bacteria from the *Lachnospiraceae* and *Ruminococcaceae* genus *Subdoligranulum* could promote Th17 cell expansion, the production of IgG autoantibodies, and joint inflammation (Chriswell et al., 2022). Given that stressors profoundly affect microbiota, which has been linked to immune alterations, it ought to be considered that this route may be provocative in RA exacerbation (e.g., Gur and Bailey, 2016).

Because of the contribution of stressful experiences and other lifestyle factors in the exacerbation of autoimmune disorders, it might be of appreciable benefit for the treatment of these conditions to include psychological approaches to diminish distress (especially as these illnesses themselves may create considerable stress) and education to enhance illness management and treatment compliance (Lisnevskaia et al., 2014).

Concluding Comments

The immune system, through its exquisitely designed components and interactions with endocrine processes, typically protects from an assortment of challenges. As such, it is predictable that stressors would set in motion a sequence of adaptive immune changes to deal with imminent or ongoing threats. However, there are limits to the capacity to deal with such challenges, and with chronic stressful experiences, the immune system's capabilities may be undermined, thereby influencing vulnerability to bacterial and viral infection. As well, immune dysregulation may exacerbate the symptoms of autoimmune disorders and perhaps cancer as well.

On the surface, assessing the impact of stressors on immune functioning should be easily achievable. However, there is no a priori reason to assume that all facets of the immune system (e.g., innate versus acquired immunity; different immune compartments; varied immune cells that come into play at different times following antigen presentation; components that stimulate or suppress immune processes; the actions of different cytokines) would operate in precisely the same way. Like so many aspects of neurobiological systems, effective immune functioning is dependent on intricate balances between various components of this system as well as the countervailing forces that keep functioning within an effective range. As well, the actions of the immune system are dictated by multiple hormones, and it is subject to psychological influences that affect these neuroendocrine processes.

In describing the actions of chronic stressors on various immune mechanisms, it seemed that inconsistent findings were common, and the effects on some elements of the immune system behaved in strikingly different ways from that evident on other aspects of immunity. At one level, this might be interpreted as reflecting a smorgasbord of outcomes without any apparent overarching schema to assure effective functioning. Alternatively, there may be a master conductor that operates to ensure that balances are maintained within and between components of the immune system, but we have yet to discover how this principal modulator operates.

It is equally possible that the seemingly disparate outcomes provoked by stressors may be more a reflection of our naivety about how the immune system operates, and what subtle aspects of the immune system dictate the outcomes that emerge. As a case in point, it had long been thought that while the adaptive immune system was capable of memory, whereas the innate immune system did not have this capability, this view has been undergoing some modifications. To an extent, cells of the innate immune system may be capable of functional memory in response to particular inflammatory immune signals, reflected by elevated cytokine changes upon reexposure to these stimuli. This "trained innate immunity" may be related to sensitized neuronal processes in the brain (which speaks to the immune system) or epigenetic changes that alter sensitivity to this same challenge at a later time (Salam et al., 2018).

There is yet another essential element that needs to be addressed. As we have seen repeatedly, pronounced interindividual differences exist in every aspect of human physiology and behaviors, as well as how stressors are managed. Some individuals are more apt to exhibit either excessive immune activity, or alternatively diminished responses, leading to illness. Stressors may produce especially pronounced adverse consequences in older individuals (Fali et al., 2018), and it is clear that women and men respond differently to such challenges. It is also likely that the constellation of variables that affect neuroendocrine and brain processes (e.g., early life stressor experiences, personality factors) may come to indirectly affect immune functioning, and hence vulnerability to psychological and physical diseases. Despite our best efforts, many of these illnesses have been refractory to treatment. It is possible that treatments to diminish distress (along with many other lifestyle changes) may be useful in disease prevention and treatment. However, even if this turns out not to be effective in the treatment of illnesses, simply diminishing distress may have immeasurable advantages for the affected individual.

CHAPTER

8

Prenatal and Early Postnatal Influences on Health

A simple twist of fate

An editorial published in the Lancet (2017) made the point that being born into poverty has an enormous impact on health and well-being, and that income and life expectancy are directly related to one another (Chetty et al., 2016). Where an individual sits on an income gradient, as indicated some years ago (Marmot et al., 1978), determines their life trajectory and lifespan. Numerous similar reports indicated that the top 20% of income earners lived a decade longer than those at the bottom of the income ladder. Indeed, children born in poverty generally died 10–15 years before those of the wealthiest 20%, and the interval between birth and death was accompanied by multiple health disparities, including obesity, elevated emergency department visits, mental health issues, and self-harm.

Multiple physiological processes contribute to increased illness and premature death among individuals who had been raised in an unfavorable socioeconomic climate. Poor diet, together with other early life factors, may have influenced the microbiome in a way that lent itself to disease occurrence (Snijders et al., 2016). Also, the chronic strain associated with low socioeconomic status may have produced a defensive response style that promoted greater resistance to glucocorticoid signaling, which then promoted elevated inflammatory immune functioning. The HPA and immune activation are beneficial in response to acute threats, but early-life social stressors may leave a biological "residue" that renders individuals at elevated risk for poor well-being. Fortunately, the effects of such experiences may be modifiable by later social support from family and friends, as well as overall social connectedness (e.g., Cruwys et al., 2019). Likewise, maternal support during early years can attenuate behavior and brain alterations associated with adverse early life experiences (Luby et al., 2016). It is notable that events experienced in childhood, such as those associated with nutritional factors, microbial

state, and psychosocial stressors, are associated with multiple epigenetic actions detected during adulthood, including those related to inflammation, and thus might be linked to illnesses, such as depression, diabetes, and heart disease (McDade et al., 2017).

Early Development

The prenatal period and early postnatal life are precarious times. Cell duplication and the formation of a diverse number of different cells progress rapidly. This process is dynamic in the sense that various endogenous substrates are added as part of the instruction for the developing fetus, and then removed at appropriate times. These modifications necessarily require that particular genes be turned on and off in a well-orchestrated sequence.

Gene transcription that occurs early in embryonic development can instill a stable state that has implications for later physiological functioning (Greenberg et al., 2017). Yet, with so many changes occurring during perinatal development, especially those stemming from multiple rounds of cell multiplication, the risk for chance mutations and those that are instigated by toxicants is considerable, and should such an outcome occur, it will be amplified with cell multiplication. At the same time, dynamic epigenetic processes are thought to be important in the regulation of developmental processes, and here again, there is ample room for problems to occur owing to environmental challenges and stressors.

During prenatal development, immune system functioning is blunted so that the fetus is not swarmed by the maternal immune system, and to some extent, components of the immune system are functional by the second trimester (McGovern et al., 2017). Indeed, during pregnancy a mother's immune functioning may be downregulated, which may account for the alleviation of autoimmune symptoms women may experience during this period, but this may also leave her at risk for infection, which could impact the developing fetus. In addition to variations of peripheral immune functioning, cytokine changes are diminished in the maternal brain (Sherer et al., 2017), which can have profound behavioral consequences that are apparent during the postpartum period. However, among women who were stressed by ongoing depression or anxiety, profound immune disturbances were observed in placental samples at birth, which could affect their well-being. These immune changes included downregulation of transcripts associated with T-cell regulation, innate immune responses, and cytokine functioning (Martinez et al., 2022).

Prenatal Challenges

Particular drugs taken during pregnancy, especially during the first trimester, place the fetus at risk for a variety of adverse outcomes (teratogenic effects) ranging from gross physical abnormalities to profound or relatively subtle neuropsychiatric conditions and developmental disturbances. In many instances, these challenges might not directly cause pathology, but instead increase the susceptibility to such outcomes. The agents that increase the propensity for negative outcomes include pollutants that are frequently encountered, such as second-hand smoke, alcohol, recreational drugs, polychlorinated biphenyls (PCBs), estrogen-related products that have been showing up in water supplies, as well as common household chemicals. Women who had been exposed to organic pollutants were at four times greater risk of developing gestational diabetes, which could have affected their offspring. Also, being exposed to air pollution and high temperatures during the perinatal period (i.e., shortly before and after birth) was accompanied by increased asthma risk (Lu et al., 2022a), as well as childhood emotional problems (Jorcano et al., 2019).

There was a time when the focus of teratogenic effects involved drugs and environmental chemicals, and to some extent immunologically related illnesses. However, it has become clear that a wide range of factors, including moderate stressors, may affect the developing brain, and prenatal stressors may interact with genetic factors in determining the occurrence of developmental disorders as well as pathologies that emerge in adulthood.

Greedy genes influence the fetus

The interactions between a mother and her fetus ought to be harmonious. At the same time, a degree of competitiveness exists between the two. The fetus requires nutrients for its growth and development, while metabolic demands

on the pregnant mother are elevated (Napso et al., 2018). Thus, genomic factors may drive the interactions that exist so that both receive the nutrients needed (Cassidy and Charalambous, 2018), which is an evolutionary advantageous characteristic. While maternal genetic factors are key to this relationship, it seems that paternal genetic influences may be fundamental in ensuring that the fetus receives needed nutrients. These genes, essentially reflecting "greedy genes" cause the release of placental insulin-like growth factor 2 (IGF2), which is involved in reducing insulin sensitivity in the pregnant mother (Lopez-Tello et al., 2023). As a result, the mother's glucose tissue absorbance is diminished, allowing greater nutrient availability for the fetus (e.g., Stern et al., 2021). If, for whatever reason, IGF2 is dysregulated, possibly through disturbances of the estradiol/progesterone ratio, altered prolactin, and TNF-α, as well as reductions of circulating lipids, the fetus might be nutritionally deprived and prenatal development will be slowed.

The Need for Prenatal Resources

Ideally, prenatal development proceeds smoothly with few hazards appearing along the way. However, the fetus can encounter exposure to infection (e.g., influenza in the mother), nutritional challenges, obstetric complications, and stress reactions in the pregnant woman. Each of these events can have lasting effects on the offspring's cognitive, emotional, and social behavior, as well as on processes related to neuroplasticity. Additional risks to the fetus can be encountered if the mom is not knowledgeable about prenatal care.

Experiences tied to poverty within Western countries (e.g., living in distressed neighborhoods, high rate of unemployment, uninsured status, and low education level), and an array of maternal factors (young maternal age, nonmarital status, planned pattern of prenatal care, and late recognition of pregnancy) have been associated with negative outcomes for the fetus so that postnatal life may begin with numerous physiological or psychological disadvantages. There is no question that mothers and offspring in wealthier countries have it better than those in poorer nations. However, obesity and gestational diabetes, which are frequent in wealthier nations, are associated with neurodevelopmental conditions, such as autism spectrum disorder and health problems related to immune dysfunction (Macpherson et al., 2017).

Studies of Prenatal Stress

A wide range of stressors mom experiences can affect the developing fetus, such as extreme challenges, wartime experiences, or natural disasters, as well as more common challenges (e.g., workplace stressors, discrimination, and bereavement). These experiences can have profound effects that become apparent soon after birth and may also affect the infant's developmental trajectory so that pathological outcomes appear in adulthood. The negative consequences of these psychological challenges are greater with more intense challenges but are particularly marked in the offspring of women who encountered multiple adverse experiences. Interpersonal violence experienced during pregnancy, which often appears in a chronic form, was accompanied by sexually dimorphic differences in brain volumes assessed in young offspring. Prenatal abuse was associated with a larger caudate nucleus among males and a smaller amygdala among females (Hiscox et al., 2023), which could have downstream effects on emotional and cognitive functioning. Like interpersonal stressors, the offspring of women who had experienced group trauma (natural disasters) exhibit profound behavioral and cognitive developmental disturbances, some of which may be related to epigenetic actions (Lafortune et al., 2021). An impressive series of studies that examined offspring following a severe ice storm that affected portions of Quebec and Eastern Ontario for weeks (e.g., Lafortune et al., 2021; Paxman et al., 2018) revealed that offspring of mothers who had been pregnant during that time experienced numerous neuroendocrine and metabolic disturbances often accompanied by developmental problems. Likewise, when assessed at 13 years of age, offspring exhibited epigenetic actions that were tied to immune alterations (Cao-Lei et al., 2016).

The risk of premature delivery and low birth weight is increased among women with high anxiety or depression, as well as women who encountered strong stressors during pregnancy (e.g., Langham et al., 2023). The lower birth weights of offspring were likewise observed among women with PTSD relative to offspring of nonstressed mothers, and those of mothers who experienced trauma, but did not develop PTSD (Seng et al., 2011). In effect, the outcomes may not simply be a reaction to the stressor, but instead evolved owing to the mother's psychological or physical responses to these challenges.

Offspring of prenatally stressed mothers exhibited elevated circulating CRH, possibly of placental origin, along with variations of other hormones that pass through the placenta, such as cortisol and the endogenous opioid met-enkephalin. These factors may contribute to fetal development and could precipitate preterm labor and reduced birth weight (Ghaemmaghami et al., 2014). As shown in Fig. 8.1, elevated levels of glucocorticoids during pregnancy may result in several changes within the placenta. Among other things, neuronal and glial changes occur within several brain regions of offspring and may affect the programming of HPA functioning. The placental alterations also include multiple hormonal alterations and epigenetic changes that can influence postnatal development (Krontira et al., 2020).

Unfortunately, preterm infants are at elevated risk of endocrine or metabolic disorders (e.g., diabetes) and illnesses that develop because of diabetes (Paz Levy et al., 2017). As well, several hormonal alterations that were associated with prenatal stressor experiences and the accompanying premature delivery predicted delayed fetal neuromuscular and nervous system maturation, and diminished gray matter volume (Wu et al., 2020). Not surprisingly, these conditions were also accompanied by neurodevelopmental disorders, cognitive disturbances (e.g., delayed language development; attention deficit hyperactivity disorder), and emotional problems that often carried into adulthood (Jeličić et al., 2022).

Prenatal stressor experiences have been associated with diminished adult levels of cortisol and ACTH (Entringer et al., 2015), consistent with that seen in adults diagnosed with PTSD. Moreover, treatments that simulate some stressor actions, notably the administration of the synthetic glucocorticoid betamethasone (used to promote lung maturation in fetuses at risk of preterm delivery), provoked persistent effects on subsequent infant temperament and behavioral reactivity in response to stressors (Savoy et al., 2016). Likewise, among rodents, prenatal administration of the synthetic corticoid dexamethasone, rather than a psychological or physical insult, promoted hypertension and increased basal plasma corticosterone levels in offspring, as well as altered hippocampal glucocorticoid receptors. Significantly, prenatal stressors as well as intrauterine dexamethasone administration resulted in long-lasting changes in hippocampal and nucleus accumbens microglia that favored later anxiety and depression in females (Gaspar et al., 2021).

Epidemiological studies demonstrated that prenatal stressor experiences were associated with numerous physical illnesses, such as disturbed cardiovascular regulation, type 1 diabetes, and metabolic syndrome that might foreshadow adult type 2 diabetes. Moreover, male offspring were at elevated risk of infection-related hospitalization, more so if the mom experienced repeated stressful events, but this outcome was not apparent in females (Robinson et al., 2021). This dimorphism may be due to the maternal stressor altering the sex steroid pathways during fetal development. Also, a prospective analysis revealed that stressors encountered early in pregnancy were associated with greater weight gain in male offspring (but not females) which was mediated by both breastfeeding practices and smoking among moms (Bräuner et al., 2021). The accelerated weight gain can have downstream effects on health, including the development of type 2 diabetes, and could potentially influence other inflammatory-related disorders.

Aside from these effects, prenatal stressors have been linked to psychological disturbances, such as depression, schizophrenia, drug addiction, and eating-related disorders in offspring. It is tempting to attribute these actions

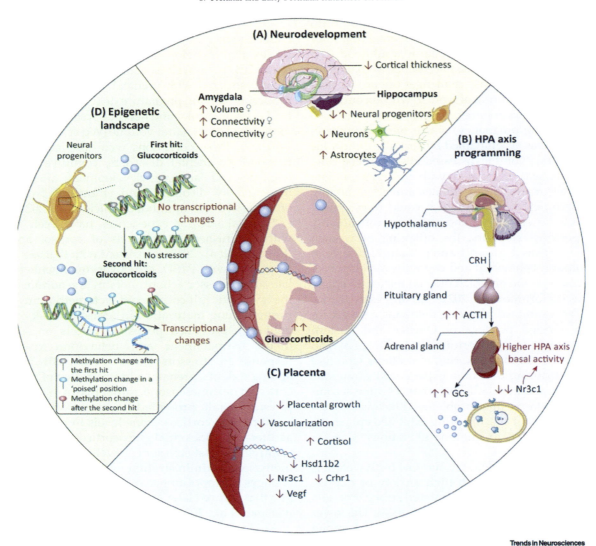

FIG. 8.1 Elevated prenatal glucocorticoid signaling affects fetal health. Prenatal excess of glucocorticoids affects many aspects of fetal biology, from placental biology to HPA-axis programming, neurodevelopment, and the epigenetic landscape. (A) Glucocorticoids impact neurodevelopment by affecting brain regions of the limbic system such as the amygdala and the hippocampus, and by inhibiting the neuronal differentiation process. (B) Glucocorticoids alter the HPA axis and result in increased basal activity of the fetal axis. (C) Elevated glucocorticoids affect the placenta by inhibiting its vascularization and consequently its growth, but also by reducing the expression of Hsd11b2, which forms a protective barrier against maternal glucocorticoids. (D) Elevated glucocorticoids alter the fetal epigenetic landscape, which results in increased transcriptional responses to future stressors. Abbreviations: *ACTH*, adrenocorticotropic hormone; *CRH*, corticotropin-releasing hormone; *Crhr1*, corticotropin-releasing hormone receptor 1; *GCs*, glucocorticoids; *Hsd11b2*, hydroxysteroid 11-β dehydrogenase 2; *Nr3c1*, nuclear receptor subfamily 3 group C member 1; *VEGF*, vascular endothelial growth factor. *Source: Parts of the figure have been modified based on images from SMART (Servier Medical Art, http:// smart.servier.com/). Krontira, A.C., Cruceanu, C., Binder, E.B., 2020. Glucocorticoids as mediators of adverse outcomes of prenatal stress. Trends Neurosci. 43, 394–405.*

to neurobiological disturbances promulgated by the prenatal stressor, but the stressor experiences can also disturb maternal behaviors (e.g., lifestyles adopted) that could have long-term effects on the offspring. Furthermore, some of the actions of the prenatal stressor were not apparent during the normal course of development but were only manifested upon further stressor encounters (e.g., in the case of eating disorders), again supporting the double-hit perspective regarding pathological outcomes.

Depression frequently occurs during pregnancy, estimated to develop in 8%–20% of women, and postpartum depression as we will see in Chapter 9 is not uncommon and has been a serious problem for women and their offspring. Among women who had experienced childhood abuse and subsequently exhibited depressive illness, higher levels of inflammation were observed during pregnancy (Kleih et al., 2022), which could potentially influence the fetus. The processes related to depressive illness, or the stress associated with this condition were accompanied by gut microbiota and immune alterations in offspring (Rodriguez et al., 2021). Elevated anxiety among pregnant women similarly influenced several aspects of immune functioning in offspring. Among other things, this included changes in the ratio between cytotoxic and helper cells, as well as between Th1 and Th2 cells, and alteration of Th17 and T_{reg} cells (Sherer et al., 2022). As anxiety and depression during pregnancy often carry over to the postpartum period, the mood changes after delivery may have contributed to the immune alterations in offspring.

The breadth of the psychological and physical illnesses, depicted in Fig. 8.2, can accompany prenatal stressors and may reflect the diversity of biological changes introduced by these challenges. Alternatively, it is possible that a core set of processes, such as particular hormones and the increase of inflammatory immune factors, play a key role in these disorders.

The impact of mom's prenatal and perinatal diet

Commensal bacteria and their numerous metabolites that originate within the mom influence the fetus. Thus, what the mom eats, the drugs consumed, the toxicants encountered, and the stressors experienced, which can affect her microbiota, may indirectly affect the fetus (Macpherson et al., 2017). Among women who had encountered multiple adverse childhood experiences, gut microbiota disturbances occurred during pregnancy, which may have influenced inflammatory and glucocorticoid responses to stressors. Also, antibiotic treatments can affect the transmission of microbial factors from a pregnant mom to her fetus (vertical transmission) and may thereby disturb protective qualities associated with bacteria (Bäckhed et al., 2015). Hence, offspring can bear the benefits or risks associated not only with the genes passed on to them but also with the microbiota they inherit.

Rats that consumed the equivalent of a high fat/high sugar Western diet during pregnancy had offspring who were inclined to gain excessive weight during the suckling period and were likely to develop an epigenetic profile associated with diabetes. However, this could be reversed if mice were maintained on a low-fat diet following weaning (Moody et al., 2017). The maternal diet during fetal development has also been associated with obesity-related epigenetic changes that can even appear in later generations (Reichetzeder, 2021).

A Western diet in pregnant rodents influenced the subsequent brain reward circuitry of their offspring so that they were more attuned to rewards, which could have contributed to their elevated eating and weight gain (Paradis et al., 2017). This diet was also associated with anxiety in adolescence, possibly by affecting glucocorticoid and inflammatory signaling (Wijenayake et al., 2020). Pregnant women who partake in a high-fat, high-sugar diets were more likely to have children with conduct problems and attention-deficit/hyperactivity disorder (ADHD) symptoms, which were linked to epigenetic changes on the gene coding for IGF-2 within the cerebellum and hippocampus (Rijlaarsdam et al., 2017). Likewise, maternal diets that were rich in omega-6, but low in omega-3 (creating a proinflammatory bias), resulted in offspring that had a smaller brain and displayed disturbed emotional responses in adulthood (Sakayori et al., 2016), and were at increased risk of developing schizophrenia-like symptoms (Maekawa et al., 2017). Conversely, the influence of adverse childhood experiences was diminished among women who reported a high dietary intake of omega-3 fatty acids (Hantsoo et al., 2019a,b).

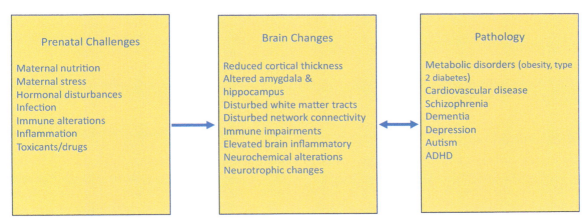

FIG. 8.2 Diverse prenatal challenges can profoundly influence structural brain changes, connectivity, and neurobiological processes, which can affect multiple neurodevelopment, psychological, and metabolic disorders.

Timing of Prenatal Stressors

The teratogenic effects of most compounds are greatest during the first trimester of pregnancy, and stressors can produce marked effects at this gestational stage. However, stressors experienced as late as the third trimester were still accompanied by psychopathology as well as disruption of some aspects of immune functioning. Stressors in humans promote dynamic behavioral and neurobiological consequences that vary over time, and rumination regarding stressors that have been encountered may have pronounced proactive effects. For that matter, a stressor encountered just before pregnancy was linked to shorter gestation and reduced birth weights (Mahrer et al., 2021), speaking to the possible ongoing effects of stressors (e.g., rumination) on fetal well-being.

Although it seems clear that stressful prenatal experiences proactively affect the well-being of the offspring that may continue throughout life, the impact of the prenatal trauma may be confounded with postnatal rearing conditions or other maternal factors. It is similarly possible that prenatal stressors interact with genetic contributions in determining outcomes within the offspring (Abbott et al., 2018). Presumably, genetic, prenatal, and postnatal factors can all influence the offspring, but their specific contributions depend on the phenotype of interest.

Biological Correlates of Prenatal Stress in Humans

A general representation of some of the biological variations associated with prenatal stressors is provided in Fig. 8.3. Aside from the glucocorticoid changes described earlier, psychosocial stressors experienced by a pregnant woman (or rodent) lead to changes in placental and fetal cytokines, reactive oxygen species (ROS), monoamines, neurotrophins, and gut bacteria, which could affect neurodevelopmental processes (Gur et al., 2017), and may promote anxiety in the offspring (Quagliato et al., 2021). While stressors experienced by a pregnant female can be transmitted to the fetus, which can then have consequences that persist throughout life, identifying the individual and complex interactive linkages between biological processes and offspring well-being can be daunting. For that matter, some of the changes observed may reflect adaptive changes that could potentially enhance offspring well-being.

Glucocorticoid Variations

Exposing a pregnant female to a stressor gives rise to elevated glucocorticoid levels within both the mom and the fetus, which can influence the programming of fetal neurons. Among other things, this may be accompanied by structural changes within stress-sensitive brain regions, which can then affect behavioral and cognitive processes (e.g., attention and learning deficits), and favor the development of anxiety and depression. These actions, like so many others, may be moderated by genetic factors, and by the postnatal environment, including the maternal care received. Yet, not every neurobiological change stemming from prenatal stressors should be taken as promoting adverse outcomes. For instance, the adult offspring of rats that had been stressed during the second half of pregnancy exhibited elevated corticosterone reactivity to stressors, increased firing of norepinephrine and dopamine neurons, but reduced firing of serotonin neurons. However, these neurobiological changes were not accompanied by the presence of anxiety or depression, raising the possibility that the neurobiological variations reflected adaptive responses that conferred resilience (Oosterhof et al., 2016).

Sex-Dependent Effects of Prenatal Stressors

Prenatal challenges may result in sex-dependent behavioral and cognitive alterations. The greater anxiety and depression seen among female offspring that had been stressed prenatally could stem from estrogen changes and elevated HPA responsiveness, whereas male offspring are at elevated risk of developing disorders involving attention and cognitive disturbances (Hodes and Epperson, 2019). Paralleling the immune alterations, prenatal stressor experiences were associated with increased occurrence of infectious illnesses (Bush et al., 2021). Essentially, prenatal stressors can have effects on both sexes, but the phenotype detected may be dependent on the specific hormonal processes that were altered and when these occurred during prenatal development.

When encountered early in prenatal development, stressors can instigate subsequent changes in male offspring (Morgan and Bale, 2017). A sensitive period seems to exist in early gestation during which epigenetic programming of the male germ line occurs most readily, thereby affecting specific sex-related phenotypes. Stressors experienced during this period of prenatal development can promote a surge of gonadal hormones, potentially resulting in brain functioning being biased in a sexually dimorphic manner. During postnatal development, these gonadal hormones might promote neuronal changes that affect the expression of sex-specific phenotypes.

In rodents, a chronic stressor experienced before pregnancy was accompanied by reduced GABA and glutamate within the right hippocampus of female offspring, but upon further exposure to a stressor during adolescence, males reacted more strongly (Huang et al., 2016b). Why these sensitization-like effects were apparent in a sexually dimorphic fashion is uncertain, but points to the different stressor vulnerabilities in the two

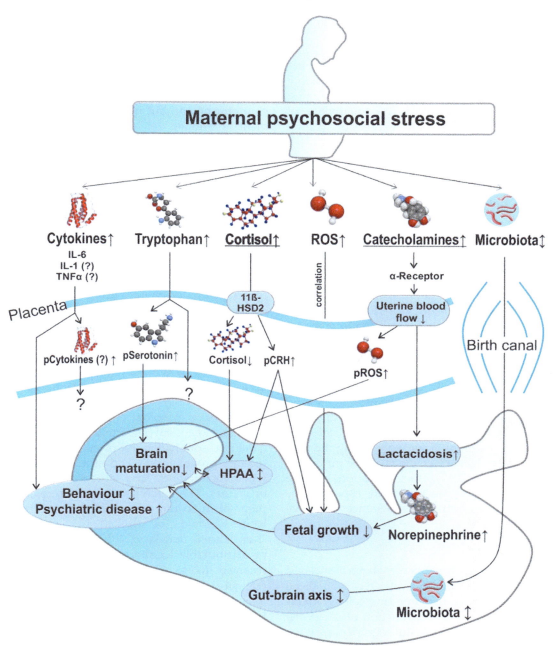

FIG. 8.3 Multiple neurobiological processes are affected by prenatal stressors, which may affect the fetus and thereby influence postnatal development, even affecting adult behavior and physical processes. The involvement of HPA functioning and placental CRH have received particular attention in this regard, although other processes, notably inflammatory mechanisms, monoamine functioning, placental reactive oxygen species (pROS), and microbial factors, all have pronounced consequences on fetal development and postnatal functioning. These consequences of prenatal stressors are fairly broad but represent only a portion of the changes that occur. *From Rakers, F., Rupprecht, S., Dreiling, M., Bergmeier, C., Witte, O.W., Schwab, M., 2020. Transfer of maternal psychosocial stress to the fetus. Neurosci. Biobehav. Rev. 117, 185–197.*

sexes. In this context, the presence of high levels of the inflammatory cytokine TNF-α coupled with low levels of antiinflammatories during pregnancy were associated with increased adult depression in males, but not in females (Gilman et al., 2016). Once more, the source for this dimorphism is not certain but suggests that the greater vulnerability to depression typically seen in females is related to factors other than or in addition to TNF-α, such as other cytokines and the hormonal changes that occur.

Neurotrophins

Just as adult stressors can affect neurotrophins, prenatal stressors have been associated with BDNF variations accompanied by reduced brain cell proliferation, and these actions could stem from epigenetic actions. The effects of acute prenatal stressors can influence BDNF in specific brain regions in a sex-dependent fashion. Specifically, acute prenatal stressors upregulated BDNF expression in the prefrontal cortex more so in females, which appeared to be tied to DNA demethylase of Gadd45β, which influences BDNF expression (Luoni et al., 2016).

Like the BDNF changes, when a pregnant dam encounters a stressor, astroglial FGF-2 gene expression may be affected in her unborn pups and might thereby influence subsequent stressor responses (Choi et al., 2022). Conversely, FGF-2 administered systemically could attenuate the depressive-like effects of a chronic mild stressor (Wang et al., 2018c) and treatments to alleviate depressive symptoms, such as repeated antidepressant or electroconvulsive shock treatment, increased FGF-2 in cortical and hippocampal sites. It seems that prenatal stressors could set the groundwork for later depressive disorders, and other pathological conditions that involve growth factors.

Studies in humans linking prenatal stress, neurotrophins, and later pathology are difficult to conduct, but data collected in "unnatural" conditions, have been instructive. For instance, among pregnant women who experienced war-related trauma, BDNF methylation was elevated in maternal venous and umbilical cord blood, as well as in placental tissue, supporting the possibility that prenatal epigenetic changes can be promoted by stressors (Kertes et al., 2017). The implications of these effects for later well-being are uncertain, especially as many other effects of these stressor experiences can have downstream consequences.

Inflammatory Factors

Prenatally encountered stressors have been associated with immune system disturbances reflected by NK cell activity and mitogen-stimulated immune cell proliferation, and these effects varied as a function of sex (Veru et al., 2014). Also, prenatal stressful experiences were associated with elevated levels of circulating proinflammatory and antiinflammatory cytokines (Entringer et al., 2015), microglial activation markers, and elevated inflammatory cytokines within the prefrontal cortex and hippocampus of offspring (Posillico and Schwarz, 2016). Although IL-6 blockade attenuated the microglial changes, the behavioral alterations persisted, again pointing to the possible involvement of other (or multiple) processes in determining the impact of prenatal stressor challenges (Gumusoglu et al., 2017).

Despite the pronounced adverse actions of prenatal stressors, this should not be misconstrued as indicating that once the dye has been cast, it is not modifiable. The potential negative consequences engendered by having a depressed mother could be diminished if the infant received appreciable contact comfort during the initial weeks of life. Similarly, although the elevated prenatal cortisol levels detected in amniotic fluid (at 17 weeks of gestation) predicted subsequent cognitive impairment, this was primarily apparent among children expressing insecure attachment, but not in children with secure attachment (Bergman et al., 2010). Moreover, prenatal adversities may limit the occurrence of certain epigenetic effects, such as for the oxytocin receptor, perhaps reflecting an adaptation to prenatal stress that would favor postnatal parent-offspring attachment (Unternaehrer et al., 2016).

Impact of Prenatal Infection

There is little question that some forms of infection during pregnancy can have widespread and long-lasting repercussions on the offspring. This can occur through direct transplacental transmission, placental damage, or fetal-maternal hemorrhage, as well as by pathogen transmission via the genital tract (e.g., Megli and Coyne, 2022). Most studies in rodents indicated that prenatal infection (e.g., by LPS) was accompanied by reduced cortical volume, and by anxiety- and depression-like behaviors, and perinatal infection in humans has been associated with these same psychological disorders, as well as schizophrenia and autism (Depino, 2017). Like the effects of prenatal stressor challenges, prenatal immune activation in rodents provoked sex and strain-dependent actions on HPA functioning, and anxiety- and depressive-like behaviors. The behavioral disturbances associated with prenatal infection, as well as the actions of other challenges, could be linked to epigenetic changes (Kundakovic and Jaric, 2017), although they could also be attributable to altered maternal care that the pups received following infection (Penteado et al., 2014). Fig. 8.4 depicts some of the common effects of prenatal infection with influenza A, Zika, and cytomegalovirus infection, and similar effects have also been observed with *Toxoplasma gondii*, rubella virus, and herpes simplex virus. This does not necessarily imply that all viruses have comparable prenatal effects, and as we have seen with SARS-CoV-2 infection, even variants of this virus may have somewhat different effects (Male, 2022). This said, in-utero exposure to SARS-CoV-2 infection was associated with an increased risk for neurodevelopmental disturbances in male offspring at 12 months of age, whereas a similar effect was not evident in females (Edlow et al., 2023). It remains to be established whether prenatal SARS-CoV-2 infection, like that of other viruses, such as influenza, would have adverse psychological effects that persist into adulthood.

There is no shortage of factors that might contribute to the impact of infection during pregnancy. Prenatal

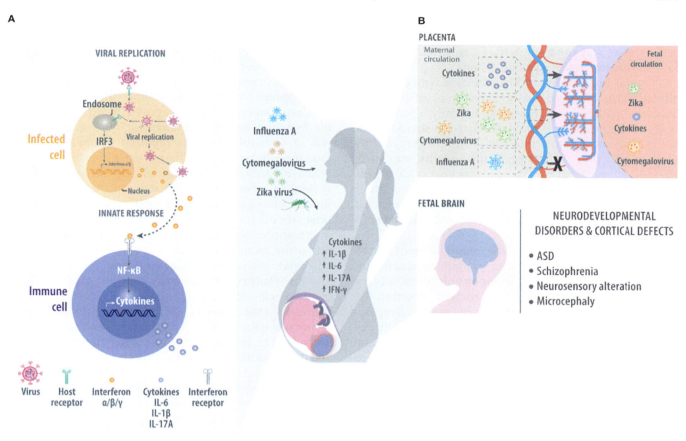

FIG. 8.4 Maternal viral infection. As a result of maternal viral infection, immune responses generate proinflammatory cytokines such as IL-1β, IL-6, and IL-17A. These cross the placenta and generate neuroinflammation in the fetal brain. (A) During viral infection, molecules on the viral surface are recognized and trigger a cascade of events in the infected cell. Host receptors recognize viral molecules or nucleic acids and initiate a signaling cascade that ends with the nuclear translocation of transcription factors such as interferon regulatory factor 3 (IRF3), which induce the synthesis of type I interferons α and β, which are ultimately released into the extracellular environment. There the interferons are detected by cells of the innate immune system. Innate immune system cells, such as macrophages or dendritic cells, recognize pathogen-associated molecular patterns (PAMPs) and damage-associated molecular patterns (DAMPs). They also respond to cytokines such as interferons to activate signaling pathways *via* NF-κB to generate proinflammatory cytokines such as IL-6 and IL-1β. These factors trigger inflammation and recruit more cells to inhibit viral replication or kill infected cells. (B) Maternal infection by the influenza A, cytomegalovirus, or Zika virus generates an increase in maternal proinflammatory cytokines that can reach the fetal circulation through the placenta. In the same way, viruses such as cytomegalovirus or Zika also cross the placenta and directly affect the fetal brain. In the fetal brain, the presence of proinflammatory cytokines or viruses can result in neurodevelopmental disorders and defects in the formation of the cerebral cortex, manifesting effects such as ASD, schizophrenia, neurosensory alteration, or microcephaly. *From Elgueta, D., Murgas, P., Riquelme, E., Yang, G., Cancino, G.I., 2022. Consequences of viral infection and cytokine production during pregnancy on brain development in offspring. Front. Immunol. 13, 816619.*

immune activation led to altered DA and serotonin within the prefrontal cortex and hippocampus (Winter et al., 2009), which might prime neuronal functioning so that the impact of stressors encountered in adulthood led to altered corticoid responses and activation of NMDA receptors, which then affect behavioral stability (Burt et al., 2013). Alternatively, elevated maternal inflammation during pregnancy can influence the fetal brain and thus might reflect a mechanistic link by which stressors and prenatal infection lead to later psychological disturbances in offspring (Graham et al., 2022). Consistent with this, the effects of prenatal infection on the behavior of offspring were accompanied by cytokine variations and altered functioning of Th17 cells (Bergdolt and Dunaevsky, 2019). It seems likely that prenatal immune activation may have behavioral consequences by compromising neural maturation and survival owing to deficiencies of neurotrophins, such as BDNF and vascular endothelial growth factor (VEGF) (Kundakovic and Jaric, 2017).

A meta-analysis of gene expression changes associated with maternal immune activation identified epigenetic changes related to inflammation and GABA functioning (Woods et al., 2021). Also, prenatal LPS treatment increased anxiety in offspring and these actions were carried over to the F2 generation (Penteado et al., 2014). The increased anxiety and elevated stress reactivity in adulthood was accompanied by reduced hippocampal fatty acid-binding protein 7 (FABP7), which is important for neurogenesis (Chen et al., 2022a). Once more, these actions were still apparent in the F2 generation, pointing to intergenerational epigenetic actions.

Prenatal stressors were also associated with epigenetic alterations that favored the transmission of inflammatory immune pathologies transgenerationally (Schepanski et al., 2018). Such effects could be related to altered maternal care stemming from infection. Administration of the viral mimic poly I:C to pregnant rodents influenced subsequent maternal care, which was also evident in F2 offspring. These actions were paralleled by variations of hippocampal gene expression of mineralocorticoid, oxytocin, and G-protein coupled estrogen receptor (Ronovsky et al., 2017).

These are only a few of the many neurobiological changes that are introduced by prenatal stressors, and it remains uncertain which are responsible for the adverse effects evident in offspring. The view was advanced that owing to the extent and breadth of biological changes, a general susceptibility to pathology can be created by prenatal stressors, and the specific pathologies that emerge vary with genetic influences as well as postnatal experiences (Entringer et al., 2015).

Schizophrenia Stemming From Prenatal Viral Infection

For obvious reasons, controlled studies cannot be conducted in humans to assess the impact of viral infection on the development of psychopathology in offspring, but data have been available related to naturally occurring (or illness-related) immune disturbances. It has been known for decades that fetal exposure to syphilis, rubella, herpes simplex, and other viruses can engender varied morphological disturbances, as well as psychological problems, such as schizophrenia (Brown and Derkits, 2010). Given how often and how readily pregnant women can contract viral or bacterially related illness, it is obviously of considerable importance that the effects observed, and how these evolved, be better understood. This was highlighted with the recent Zika outbreak, but it has been an issue for some time in relation to other infections, such as influenza, and now, obviously, with SARS-CoV-2 infection. Understandably, analyses in humans focused on linking prenatal viral infection to the later development of pathology in the offspring can be difficult, particularly if the disturbances, as in the case of schizophrenia, might not emerge (or at least be detected) until mid or late adolescence.

It was long suspected that prenatal infection during influenza epidemics was associated with an increase in the later occurrence of psychosis (Kępińska et al., 2020). It has since become clear that viral epidemics in humans (Rubella, influenza, and herpes simplex virus type 2) were associated with a 500%–700% increase in the birth of children who subsequently developed schizophrenia. More than one-fifth of individuals who had been exposed to rubella prenatally later received a diagnosis of either schizophrenia or a schizophrenia spectrum disorder. Similarly, among women exposed to influenza during the first trimester of pregnancy, a 700% increase in the risk of schizophrenia was reported in offspring, while the intracellular parasite Toxoplasma gondii was associated with a 200% increase of schizophrenia as well as a variety of other neuropsychiatric difficulties. Although several viral and bacterial infections contracted during pregnancy have similarly been linked to schizophrenia in offspring, these are correlational findings and numerous other factors could have been responsible for the observed outcomes. While viral and bacterial insults and the resulting inflammatory changes, including brain cytokine elevations, could potentially affect neuronal processes that provoke schizophrenia, febrile (fever) responses associated with viral insults, and fetal hypoxia secondary to infection, could have lasting effects (Boksa, 2008). In addition to schizophrenia, prenatal infection was also associated with autism, cerebral palsy, and epilepsy, and it was even maintained that these events could affect late-life neurodegenerative disorders. Once more, a prenatal infection might create a general vulnerability for neurodevelopmental disorders, but still other factors dictate which specific pathologies might arise (Harvey and Boksa, 2012).

Given the difficulties of forming one-to-one connections between prenatal infection and later pathological conditions in humans, there have been many attempts made to do this through animal models. Of course, there is no certainty that animal models of schizophrenia reflect the human disorder, particularly as some symptoms cannot be adequately assessed (e.g., hallucinations). Nonetheless, prenatal infection in rodents produced disturbed social interaction in pups, together with attentional disturbances and altered exploratory behaviors. These behavioral phenotypes were accompanied by brain alterations that have been linked to schizophrenia, such as the diminished size of the cortical and hippocampal regions, coupled with an increase in cortical pyramidal cell density (Meyer and Feldon, 2009).

Viral challenges during prenatal development in animals consistently revealed negative effects that were manifested throughout the postnatal period and into adulthood. For instance, an endotoxin administered to pregnant rodents induced a schizophrenia-like profile in offspring, accompanied by decreased myelination of neurons within cortical and limbic brain regions, more so in males than in females (Wischhof et al., 2015). Likewise, the administration of an immune-activating agent to pregnant rhesus monkeys resulted in offspring showing gray matter abnormalities and reduced volume of brain regions linked to psychological disorders, together with subtle cognitive changes (Vlasova et al., 2021).

These actions may come about owing to interactions of inflammatory factors and genetic influences, or they may stem from the induction of epigenetic changes (Bergdolt and Dunaevsky, 2019). The offspring of pregnant rats that had received poly I:C, which mimics a viral insult, exhibited several behavioral disturbances accompanied by elevations of the P2X7 receptor and inflammatory signaling involving NF-κB, NLRP3, and IL-1β. This treatment also reduced the expression of specific $GABA_A$ receptor subunits in the prefrontal cortex and dopamine D2 receptors in the hippocampus, while increasing specific NMDA receptor subunits in the prefrontal cortex, D2 receptor in the nucleus accumbens, and $5\text{-}HT_{2A}$ in the hippocampus (Su et al., 2022b).

As dopamine and glutamate neuronal functioning may be key ingredients in the emergence and maintenance of schizophrenia, it is significant that prenatal infection in rodents was accompanied by dopamine receptor variations and activity (Meyer and Feldon, 2009), as well as changes in glutamate signaling in offspring (Meyer, 2019). The dopamine changes occurred regardless of whether pups were raised by their biological mother (who had been infected during pregnancy) or a surrogate mother who cared for them following birth, indicating that the prenatal infection (as opposed to postnatal maternal care) was fundamental for the neurochemical alterations. Paralleling what occurs in humans, the behavioral disturbances among pups that experienced the prenatal infection only emerged during the adolescent period (Meyer, 2019), and could be diminished by antipsychotic medications that attenuate dopamine functioning.

Not every prenatal infection leads to adverse outcomes. Ordinarily, a small number of maternal cells get through the placenta and embed themselves in the organs of the developing fetus (microchimerism). This may be important for the fetus' and mom's immune systems to become compatible and hence not attack one another. In countries where malaria is endemic, the offspring of mothers who had malaria during pregnancy exhibited signs of having been infected (showing a positive blood smear), but they were less likely to exhibit malaria symptoms (Harrington et al., 2017). How this resistance comes about may be related to the mom's immune system teaching the offspring how to deal with an infection.

During the MERS and SARS-CoV-1 pandemic, women who had contracted these infections were at increased risk of encountering obstetric complications and this was also apparent among women who contracted SARS-CoV-2 infection while pregnant (Male, 2022). Not only was the risk of stillbirth elevated but unvaccinated pregnant women were at increased risk of severe symptoms and death (Stock et al., 2022). Among these women, an inflammatory immune response was elicited in the fetus (Garcia-Flores et al., 2022), but it is too early to know what long-term effects will appear in offspring.

Impact of Adverse Early Life Events

Besides the effects of prenatal stressors, adverse childhood experiences (ACEs) may promote pronounced long-term ramifications on well-being. Children who had experienced physical assault subsequently exhibited a three-fold increase in mood and anxiety disorders (Archambault et al., 2023). Likewise, other forms of abuse and neglect, parental substance use, incarceration, and domestic violence, were associated with long-term mental health disturbances. Among other things, such negative early life experiences can alter the trajectory of stress-relevant neurobiological processes so that stressor reactivity is exaggerated in later adulthood, and the occurrence of psychological disturbances is elevated. There have been indications that ACEs can increase the aging process (reflected by telomere length) and could be aligned with developmental problems (Esteves et al., 2020). Likewise, prolonged separation from parents has been associated with shortened telomere length, which predicted elevated vulnerability to psychopathology (Chen et al., 2019a). Appreciable heterogeneity exists across studies that assessed the ties between childhood adversities and telomere length, possibly reflecting the features of participants included in these studies. Specifically, these associations were most often observed in nonclinical and younger individuals relative to clinically diagnosed older participants (Bürgin et al., 2019). The sequence of events by which ACEs may come to promote disease occurrence and early death is shown in Fig. 8.5.

Beyond abuse and neglect, growing up in poverty has marked efforts on psychological and physical health, which can appear across generations. A longitudinal study conducted over 10 years similarly revealed that the risk of obesity and low cognitive functioning was tied to familial and neighborhood inequality, possibly being related to limited sensory and cognitive stimulation. The behavioral and cognitive delays were linked to structural changes in white matter tracts within the brain (Li et al., 2023e).

Aside from the elevated anxiety and depressive features associated with early life stressor experiences, chronic physical illnesses were frequently evident in adulthood (Secinti et al., 2017). In addition, ACEs have been associated with immune disturbances, increased occurrence of immune-related disorders, such as systemic lupus erythematosus (Feldman et al., 2019) and have also been linked to chronic pain in adulthood

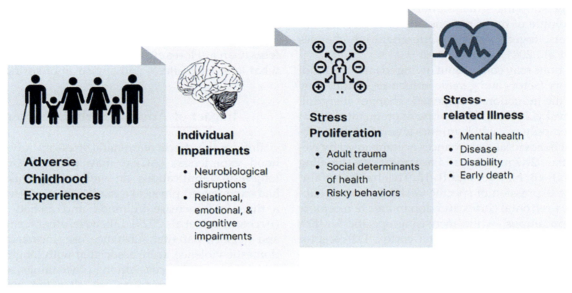

FIG. 8.5 Early life adverse events may influence numerous biological processes as well as emotions, cognitive functioning, and relationships with others. These may form the basis for subsequent poor health-related behaviors and may affect the individual's encounters with further stressors (stress proliferation). Together, these diverse factors may favor the development of stress-related illnesses and premature death. *From Anisman, H., Matheson, K., 2023. An Introduction to Stress and Health. Sage Publications, London.*

(Burke et al., 2017), possibly owing to elevated inflammation. In general, the multiple neurobiological changes associated with adverse childhood experiences may contribute to allostatic overload that manifests as increased illness vulnerability, including the development of age-related disorders (Danese and McEwen, 2012). Significantly, the effects of earlier stressors on mood disturbances are modifiable by antidepressant treatments and by omega-3 polyunsaturated fatty acids (Pusceddu et al., 2015), and as mentioned earlier, enhancing the child's environment can limit the damaging effects that would otherwise emerge.

Stressors that children encounter may comprise "toxic challenges," such as psychological, sexual, or physical abuse, neglect, withholding affection, family disturbances, parental problems (e.g., mental illness, addiction), and poverty. These childhood experiences, which often appear as multiple stressors, profoundly shape behavioral, neurobiological, immunological, and microbial trajectories, and thus may have considerable sway on the appearance of later behavioral and physical pathologies (e.g., depression, anxiety, PTSD), and may even be linked to earlier death (Hughes et al., 2017). Aside from traditional ACEs, a large study of more than 45,000 Swedish individuals indicated that even children who had lost a parent when young were at elevated risk of asthma (in females) and autoimmune disorders (in males), which were mediated, in part, by the development of mood disorders and socioeconomic status (Brew et al., 2022).

Toxic early life experiences can program later behavioral and emotional responses so that responses to new stressor encounters are exaggerated. These experiences can also undermine behavioral styles so that individuals have difficulties forming social relations, become less trusting, experience disturbed self-regulation, and the adoption of poor lifestyle choices, any of which could aggravate already disturbed adaptive neurobiological processes. In addition, strong early life stressors have been associated with impaired decision-making, increased stress reactivity, and resistance to fear responses being extinguished. Given the multiple negative consequences associated with early life abuse, it probably comes as no surprise that the depression that occurs in adulthood among these individuals tends to be especially resistant to positive actions of antidepressant medications (Williams et al., 2016a). Still, among children brought up in low socioeconomic conditions and are more prone to develop chronic illnesses, this outcome is less likely to occur if children receive adequate maternal nurturance, making it essential to develop platforms to protect children from adverse experiences and establish strategies to attenuate the effects of these experiences (Hughes et al., 2017).

Not all early life adversities have negative long-term repercussions. While some stressors can be seen as "toxic" challenges (extreme poverty, psychological or physical abuse, neglect, maternal depression, parental substance abuse, and family violence), modest (tolerable) stressors can have positive actions by facilitating coping and active responses (McEwen and McEwen, 2017). Also, modest stressors experienced at an early age may influence the child's learning of how to deal with stressors, which can have long-term benefits. Similarly, tolerable stressors might prime stress-relevant biological systems so that a later stressor experience will give rise to moderate neurochemical changes that operate to enhance coping efforts (Santarelli et al., 2017). Moderate challenges during early life and adolescence might also prime growth factor functioning, and the resulting enhanced synaptic plasticity could potentially facilitate the ability to contend with subsequently encountered challenges. This raises the question as to whether prenatal stressors, like those encountered early in life, can in some way enhance resilience in response to postnatal adversities that might be encountered.[1] At the same time, differences between children's responses can be shaped by genetic factors, as well as family and community influences. Critically, during especially sensitive developmental periods, steps can be initiated that can affect positive or negative outcomes (Boyce et al., 2021).

Passage of Poor Appraisal and Coping Methods

Traumatic experiences encountered during early life could affect the way individuals perceive the world around them, including how they appraise and interpret later stressful encounters. Such experiences may also be associated with warped internal attributions (i.e., self-blaming and self-criticizing) and cognitive distortions linked to the appraisal of safety, a preoccupation with danger, as well as excessive focus on being able to have control over their own lives. These characteristics, particularly the exaggerated perceptions of future harm, and maintaining this sense of threat and unpredictability about the future, may contribute to the emergence of anxiety, and the generally negative worldview might favor later depressive illnesses (Kalmakis and Chandler, 2015). The seemingly warped (negative) perspective about future harms is not entirely without merit, as individuals who experience trauma are, in fact, at elevated risk of encountering further stressors. As adults, children who experienced childhood abuse were at elevated risk of revictimization experiences (e.g., domestic violence, partner abuse, and rape). This concept of "stress generation" (Harkness and Washburn, 2016) has been known for some time, even appearing in the work of the philosopher Rashi (1040–1105), who asserted that "one misfortune invites another."

Inappropriate appraisals may give rise to poor coping abilities, and children who experience traumatic events, often employ ineffective coping strategies to deal with ongoing challenges. Children and adolescents who experienced community violence, sexual abuse, or maltreatment tended to use emotion-focused and avoidant coping strategies (as well as risk-taking, confrontation, and the release of frustration), rather than problem-oriented coping that might be more effective in eliminating the stressor. Ordinarily, the resources and experiences to appraise stressors accurately are not well-formed among children, and the coping strategies available to them are limited. Early life stressors may further diminish cognitive flexibility and may stunt coping development, and the poor ability to adopt new and more effective strategies is carried into adulthood (e.g., Sheffler et al., 2019).

Chronic or repeated adversities may also give rise to the child attempting to understand why abusive or neglectful experiences are happening to them. These children might attribute the stressor to aspects of themselves and may internalize the belief that the adverse events are due to their shortcomings (Pilkington et al., 2021). Thus, these experiences may influence individual characteristics (e.g., self-perceptions, self-blame, mastery, and self-esteem), which may foster the development of negative cognitive styles, further undermining the use of effective coping strategies to deal with later stressors.

How a child behaves might not always provide information concerning the actual impact of very adverse experiences. Even if children had experienced family discord and violence but displayed little general anxiety, did not imply that the potential for disturbances was not present. Among children who had experienced strong early life stressors, even if they did not show outward signs of distress, depictions of angry faces provoked increased neuronal activity within the amygdala and the anterior cingulate cortex, much like that associated with strong threat responses in other individuals. These brain responses were not a reflection of general reactivity, as the depiction of sad faces did not elicit similar neuronal changes (McCrory et al., 2011). As discussed earlier, stressful experiences may promote neuronal sensitization so that later responses to stressors are exaggerated, and this may be especially notable in children. Indeed, resting frontal EEG asymmetry, which was linked to the presence of inflammation, was most apparent among individuals who had experienced relatively extensive levels of childhood maltreatment (Hostinar et al., 2017).

It is unfortunate that toxic stressor encounters during early life frequently are not isolated incidents but may

---

[1] This type of change is not unique to stressors. "Hormesis" is said to occur when positive outcomes emerge following exposure to low levels of a toxicant, which at high doses has negative effects. Similar effects are seen in other contexts. Low-level brain stimulation by external means, for instance, can protect against the damaging effects that might otherwise be provoked by a subsequent large seizure.

comprise a set of distressing events within a constellation of adverse experiences. It was thus suggested that victimization should not simply be viewed as "an event," but should instead be considered as "a condition" that entails ongoing distress (e.g., Pratchett and Yehuda, 2011). Predictably, perhaps, these stressor experiences all too often extend throughout the lifespan and even across generations (Bombay et al., 2011). As such, when evaluating the impact of childhood stressors on the subsequent emergence of pathology, it might not be productive to limit analyses to a particularly severe stressor, and instead, it might be more appropriate to consider the cumulative effects of multiple lifetime stressors that might have been endured.

Impact of Parenting

Poor parental behaviors come in several forms, such as neglect, disengagement and disorganization, hostility, coercion, and low positive parent-child interactions. Regardless of the form, these behaviors may have profound repercussions on children, including the development of depressive symptoms, suicidality, PTSD, and interpersonal difficulties, often mediated by negative perceptions of the self and the future. It has been known for at least 50 years that poor parenting styles in rhesus monkeys may be recapitulated when these offspring have children. In rodents, as in monkeys, having an inattentive and distant mom leads to negative outcomes in her pups. These long-term actions may stem from persistent stressor-related neuroendocrine alterations, such as CRH or glucocorticoids, or those associated with attachment, such as oxytocin. In fact, oxytocin neuronal functioning may be involved in the transmission of parenting style to offspring (Carcea et al., 2021).

Although early life stressful events can have long-term ramifications owing to the reprogramming of neuroendocrine or neurochemical stress responses, such outcomes might also evolve owing to mom's behavior toward her pups being disturbed. Using a cross-fostering procedure (i.e., pups were transferred from their biological mother to one that displayed either high or low maternal care) and other related paradigms, it was demonstrated that some behaviors of the offspring can be "inherited" from their nursing mother, rather than solely from their biological mother (Anisman et al., 1998).

In humans, poverty during early life (and the multiple hardships that come with poverty) can engender epigenetic effects that influence neuroendocrine and immune functioning in adulthood, which can also be passed on to their offspring. In this regard, genetic factors and parenting interacted to predict the parenting style that offspring later expressed. In essence, experiences within a given generation, together with genetic influences, comprise the ingredients that govern the recapitulation of negative parenting and stressor experiences that appear in the next generation.

Many illnesses, such as type 2 diabetes and heart disease, take years to develop. While these conditions are usually associated with aging, increasing evidence has suggested that stressors in early life or in utero set the stage for the expression of these illnesses. For example, prenatal psychosocial stressors were accompanied by insulin resistance in young adults, and the risk for diabetes was increased among the children of women who lost a loved one during pregnancy (Li et al., 2012). Childhood adversities were also associated with the doubling of subsequent gestational diabetes and a 500% increase in the occurrence of postpartum depression (Madigan et al., 2017). These experiences, in turn, were linked to negative health outcomes in their children, including anxiety and depression.

Early Life Stressors and Hormonal Changes

Early life stressor experiences, like those encountered prenatally, can alter the developmental trajectory of several neurobiological processes, including diverse epigenetic effects that influence glucocorticoid functioning and inflammatory processes, which may have repercussions that endure throughout life (see Fig. 8.6). Among the most prominent effects of adverse early life stressor experiences is altered glucocorticoid functioning, which affects adult behavioral and biological reactivity in response to further stressful experiences. Increased glucocorticoid responses elicited by stressors in early life may also promote glutamate release at the prefrontal cortex, amygdala, and hippocampus, which could favor anxiety in adulthood if these glutamate variations persist (Averill et al., 2020). These actions could come about owing to the sensitization of neurobiological processes so that further challenges promote exaggerated responses. Also, epigenetic alterations can develop in glucocorticoid receptor functioning, hence affecting adult responses to stressors (Szyf, 2019). It also appeared that through actions on glucocorticoids, early life stress or neglect could promote excessive astrocyte-mediated elimination of excitatory synapses, which could result in the development of behavioral disturbances (Byun et al., 2023). Also, glutamate dysfunction stemming from early life negative experiences may increase microglial activation and increased production of inflammatory factors that promote several psychiatric disturbances (Mondelli et al., 2017). These include anxiety, depression, and substance use disorders, and might be a contributing factor in treatment-resistant depression.

Based on early studies, it was assumed that the infant period in rats, spanning 4–14 days, is exceptionally sensitive to stressors, such as separation from the dam,

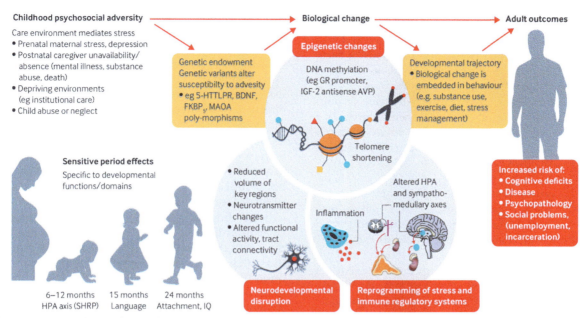

FIG. 8.6 Some of the pathways that mediate exposure to early adversity and adult outcomes. Exposure to adversity early in life interacts with a child's genetic endowment (e.g., variations in genetic polymorphisms), which in turn leads to a host of biological changes across multiple levels. These changes, in turn, influence adult outcomes. *HPA axis (SHRP)*, hypothalamic pituitary adrenal axis (stress hyporesponsive period). *From Nelson, C.A.3rd, Gabard-Durnam, L.J., 2020. Early adversity and critical periods: neurodevelopmental consequences of violating the expectable environment. Trends Neurosci. 43, 133–143.*

especially when this occurs for an extended period, such as 24 h. It was subsequently reported that separation from mom for as little as 3 h a day on 7–10 successive days was sufficient to produce elevated anxiety in pups, even though moms under naturalistic conditions may leave the nest for such periods to forage for food. As much as the notion concerning the effects of maternal separation (for 3 h each day) is intuitively appealing, other investigators found that separation for extended periods did not uniformly have the negative effects initially reported, although several sex-dependent phenotypic effects were observed (White and Kaffman, 2019).

It also turned out that the early life period was not unique for adverse effects of stressors to appear. The window from 10–20 days postnatally was as sensitive to prolonged separation from the dam relative to that seen at earlier periods. Moreover, the stress sensitivity seen at later times seemed to be related to genes linked to reward processes involving the ventral tegmentum, which could confer life-long sensitivity to further stressors (Peña et al., 2017). Using an inflammatory challenge, a window from 14–21 days following birth was also reported in the provocation of anxiety responses, independent of any actions related to maternal influences. Other studies pointed to the juvenile period (adolescence) as having particularly pronounced epigenetic effects on GABA functioning and anxiety (e.g., Ben David et al., 2023). Evidently, multiple windows exist during which stressor experiences can affect different phenotypes, although they might not involve identical processes.

Despite the inconsistencies it was reported that among germ-free mice, separation from the dam for 3 h a day was sufficient to alter HPA functioning and colonic cholinergic neural regulation. Upon bacterial recolonization, anxiety behaviors were introduced along with a microbial profile distinct from that seen in control animals (De Palma et al., 2015). Evidently, separation from the mom can serve as a stressor that renders mice vulnerable to bacterial dysbiosis, which favors anxiety.

Selectivity of the HPA response

The cortisol rise associated with acute stressors has multiple adaptive attributes, but as we discussed earlier, sustained cortisol release may have negative effects, including loss of hippocampal neurons. Thus, there may be benefits for the glucocorticoid response to be blunted under certain conditions, such as in response to chronic challenges. An instance in which HPA functioning is blunted is in the corticoid response seen in pregnant moms and their pups in response to stressors. During the last trimester of pregnancy and during lactation, the HPA response to stressors is diminished in rodent dams. The reduced cortisol response may be attributable to any of several neurobiological processes, including alterations in the function of other hormones, such as oxytocin, prolactin, and opioid peptides, as well as a downregulation in the ability of norepinephrine to promote hypothalamic CRH and arginine vasopressin release (Tu et al., 2005; Walker, 2010). It seems that during lactation, moms are protected from biological changes ordinarily associated with anxiety (although postpartum depression is an obvious deviation from these findings). The suppression in the dam prevents excessive corticosterone from reaching the fetus or pups (the latter through the mother's milk). There is no doubt a fundamental evolutionary need for stress hormone responses to be inhibited during the lactation period, but at the same time, there is also a need for selectivity regarding the effects of different stressors. Thus, in contrast to the cortisol suppression characteristic of threats made toward the dam herself, exaggerated corticosterone release may occur in the mom in response to threats directed toward the pups. While still nurturing pups, animals can "filter" relevant from irrelevant stimuli at least in terms of their offspring's well-being. Accordingly, a threat directed at her offspring, provokes a profound change in cortisol levels, presumably to provide resources essential for her to maintain the safety of her pups. Messing with momma is one thing, targeting her pups is another thing entirely.

The glucocorticoid changes associated with early life stressors are certainly fundamental to well-being, but as we have seen, other processes are also affected by these experiences. Stressors experienced early in life can have enduring effects on sexually related behaviors that are manifested during adulthood, perhaps stemming from persistent changes in estrogen receptors. Mild postnatal infection likewise reduced subsequent receptivity among females, as well as behavioral responsiveness to estradiol and progesterone in adulthood (Blaustein et al., 2016). Early life stressful events may also have marked effects on prolactin which can influence responses to emotional stressors. In the latter regard, prolactin receptors are present within brain regions that control emotional responses (e.g., the central amygdala, the bed nucleus of the stria terminalis, and the nucleus accumbens), and stressor-elicited prolactin functioning may contribute to HPA activation and the development of psychological disorders (Faron-Górecka et al., 2023).

It will be recalled that oxytocin has been linked to social bonding and attention to socially salient environmental triggers, and oxytocin levels and functioning can be influenced by psychosocial stressors. After 6 weeks of isolation, beginning just after weaning, prairie voles displayed high levels of anxiety coupled with enhanced mRNA expression of oxytocin, CRH, and AVP (Pournajafi-Nazarloo et al., 2013). Moreover, the maternal care female rats received during the early postnatal period predicted the subsequent appearance of oxytocin receptors in adulthood (Carcea et al., 2021).

Paralleling the findings in animals, stressful experiences in humans affected oxytocin functioning, and negative early life experiences, including those related to maternal care and emotional abuse, had persistent repercussions on oxytocin throughout adulthood (Sanson and Bosch, 2022). This included lasting consequences on prosocial behaviors, thereby affecting the organism's ability to contend with further psychosocial challenges. There have been indications that such outcomes were due to epigenetic effects, but as already indicated, other actions related to maternal care could ostensibly elicit these effects.

Given oxytocin's presumed role in social interactions, it may be significant that levels of this hormone were diminished among women who had experienced childhood abuse and then subjected to a social stressor, but a different form of distress, specifically, that of childhood cancer, did not produce equally pronounced effects (Pierrehumbert et al., 2010). Thus, it is tempting to surmise that the persistent effects of oxytocin are limited to those that involve social stressors. Furthermore, the negative impact of a prenatal stressor on offspring neural development and behavioral disturbances could be attenuated by a positive postnatal environment, perhaps owing to actions on oxytocin or neuroinflammatory processes. As such, it is likely that the connections between prenatal stressors, oxytocin changes, and social interactions were of a causal nature (Nolvi et al., 2022).

Breastfeeding—more than just an oxytocin hit

Breastfeeding has been recommended as the preferred method of providing infants with nutrition. It has been suggested that breastfeeding be maintained for at least 6 months and then complemented with other forms of nutrition until the infant reaches 2 years of age. Breastfeeding has multiple benefits, including reducing infant mortality by 33% during the first year of life (Ware et al., 2023).

Breastfeeding may facilitate bonding between the offspring and mom owing to the release of hormones, such as oxytocin. Beyond this very important aspect of breastfeeding, breast milk contains nutrients such as fat, protein, vitamins and minerals, and long-chain polyunsaturated fatty acids that contribute to the development of cognitive and motor processes (Jonas and Woodside, 2016). Breast milk also promotes immune responses in the infant which fosters good health. In addition, breast milk contains growth factors and other hormones that affect neurodevelopment, and this form of nutrition also contains immunoglobulins that may be important to preclude some infections. Breastfed infants were less likely to develop gastrointestinal tract problems, respiratory tract infections, obesity, diabetes, and childhood leukemia and lymphoma, relative to their bottle-fed peers (e.g., Ahmadizar et al., 2017), possibly operating through inflammatory processes. Through the presence of particular microRNAs, breast milk can provide preterm infants with a metabolic boost that enhances their ability to thrive (Carney et al., 2017).

Breast milk also contains numerous human milk oligosaccharides (complex sugars) that are an exceptionally rich energy source. These oligosaccharides, which are not easily digested, travel to the gut where they feed good bacteria, which then act to create adhesive proteins that seal the gut so that microbes are kept out of the bloodstream. In addition, there is the belief that breast milk acts against bacteria that promote disease and may act against the development of allergies (van den Elsen et al., 2019). Among infants that were fed breast milk, the appearance of microbiota metabolites increased relative to that measured in formula-fed babies. For instance, increased breastfeeding was accompanied by elevated cholesterol (measured in stool samples), which was directly related to their performance on cognitive tests, perhaps reflecting the influence of fatty acids on the elaboration of brain growth processes. In formula-fed babies, in contrast, the presence of metabolites, such as cadaverine, could have negative effects on cognitive development (Chalifour et al., 2023). Furthermore, among premature infants, some protective intestinal bacteria may not be present and if they are unable to engage in breastfeeding, which is often the case, they might be at increased risk for infection. However, lactoferrin, a protein ordinarily obtained in breast milk, can diminish these adverse effects (Sherman et al., 2016).

Growth Factors and Early-Life Experiences

Early life stressors can disturb hippocampal synaptic plasticity, more so in males than females (Derks et al., 2016), and in line with the presumed involvement of early life experiences in the development of depression, epigenetic modifications related to BDNF may account for the persistent effects observed (Zhou et al., 2023a). Female rhesus macaques who experienced maternal deprivation developed a depressive-like profile that was accompanied by reduced plasma BDNF and nerve growth factor (NGF) levels. These outcomes could be diminished if monkeys were reared by peers, an outcome that was most apparent in females (Cirulli et al., 2009).

Similar studies are difficult to accomplish in humans, but it was reported that features of the mother, notably her attachment insecurity, maltreatment experienced during childhood, and antenatal depressive symptoms, were tied to epigenetic change in offspring. The methylated genes included those involved in hormonal regulation, immune functioning, inflammatory responses, and neurotransmission (Robakis et al., 2022). It was similarly observed that depressive symptoms among pregnant women predicted reduced BDNF methylation in male and female infants (Braithwaite et al., 2015), which might have implications for later mental health.

Humans who experienced negative early life events and who also carried a BDNF mutation (Val66Met polymorphism) were likely to exhibit behavioral biases linked to the development of depression (Aguilera et al., 2009), which were accompanied by diminished gray matter within the subgenual anterior cingulate cortex and in other brain regions that have been implicated in depression (hippocampus, prefrontal cortex). It is thought that the anterior cingulate cortex comprises a mechanistic tie between stressful early life experiences, BDNF polymorphism, and adult depression. It seems, as well, that early life abuse that was associated with later depression and suicide, was accompanied by altered myelination of axons that linked the cingulate cortex and both the amygdala and nucleus accumbens and might thereby affect emotions and reward processes (Lutz et al., 2017). Also, among individuals who had experienced early-life emotional abuse, the amygdala neurons responded very strongly to threats and affected neuronal processes within a greater expanse of brain regions (Klumpers et al., 2017).

The neurotrophic changes elicited by stressors and their implications for depression have received considerable support, but findings in both rodents and humans have suggested that the stress-BDNF-depression linkages are not as straightforward as first believed. The BDNF interaction with life stressors is only one among many others that have been reported, including polymorphisms of the FKBP5 gene that influence glucocorticoid receptor sensitivity as well as polymorphisms related to the 5-HT transporter (Grabe et al., 2016). Just as a more nuanced role for stressor-elicited neurotrophic factors should be considered concerning learning and memory processes, the link between stressor-elicited neurotrophic factors and mood states needs to be evaluated in conjunction with other neurobiological changes that occur.

Morphological changes in the human brain occur with chronic stressor experiences and the presence of developmental and psychiatric disorders. The volume of the amygdala and cortical regions was reduced among individuals who had endured early life abuse or neglect, varying with the nature of the adverse events (Buimer et al., 2022; Dahmen et al., 2018). Negative early life challenges that were linked to variations of brain structure and volume were also associated with internalizing symptoms, such as depressive affect and anxiety, as well as impulse control difficulties. A genome-wide association analysis of 15,640 individuals revealed several genes that were common in depression, schizophrenia, cognitive functioning, insomnia, and smoking, and it appeared that early development was fundamental in determining brain changes (Brouwer et al., 2022). Brain morphological changes that have been tied to previous distressing events similarly appeared to be moderated by genetic influences, although it has been reported, at least with respect to PTSD and depression, that preexisting hippocampal disturbances promoted increased risk for illness. Thus, small hippocampal size may be a core element for an assortment of mental illnesses, but, other factors determine whether and which pathologies will emerge. Specifically, some of the effects of stressors on brain morphology were primarily apparent in the presence of particular gene mutations, such as those related to BDNF, serotonin reuptake, or an enzyme involved in degrading norepinephrine (COMT) (Rabl et al., 2014), each of which has been implicated in the development of depressive illnesses or high stressor reactivity. Hippocampal and other brain changes were also linked to diminished social support and disturbed coping, possibly being related to epigenetic effects. In this respect, epigenetic changes may contribute to brain morphological and functional outcomes, and while these may vary with the nature of the adverse early life events experienced, several were common across diverse negative experiences (Løkhammer et al., 2022).

Being empathetic, some humans are affected simply by witnessing others experiencing a distressing event (even if they might not take action to diminish this). It seems that this also occurs in rodents as well as in other species so that when an animal is stressed, nearby conspecifics display stress responses, although in rodents this might reflect a stress reaction brought about by pheromones, rather than empathy for a buddy. These reactions are fairly powerful such that the offspring of pregnant rats who witnessed another rat being stressed exhibited marked depressive-like behaviors, accompanied by more frequent epigenetic changes and brain morphological variations (Mychasiuk et al., 2011). The brain changes were characterized by diminished dendritic arborization, neuronal and glial cells being reduced within the prefrontal cortex and hippocampus, and limited maturation of hippocampal granule cells, which persisted into adulthood. These stressor conditions influenced cortical brain regions that subserve attention, executive functioning, as well as information processing and memory so that the behavior of the offspring was greatly disturbed.

Early Life Stressor and Brain Neurochemical Variations

As we have seen, adverse experiences early in life elicit pronounced brain neurochemical changes and may cause exaggerated neurochemical responses to subsequently encountered stressors. Indeed, such actions have been reported with numerous neurotransmitters. Throughout life, neuronal ensembles within the nucleus accumbens and prefrontal cortex were hypersensitive to stressors among rodents that had experienced early life adversity, so stressor reintroduction readily promoted the changes

in neuronal activity. However, by inhibiting the activity of nucleus accumbens neurons during early life stress, the effects of later challenges on behavioral alterations were minimized (Balouek et al., 2023).

Early life stressful events can influence the functioning of a specific stress-sensitive pathway that involves connections between the basolateral amygdala and the nucleus accumbens through GABA and CRH activity, and by disturbing dopamine reward processes (anhedonia), lasting behavioral disturbances can be engendered (Birnie et al., 2023). Even a seemingly mild challenge, such as early life social isolation downregulated dopamine activity within the nucleus accumbens and increased kappa opioid receptor reactivity, which could have implications for later pathology (Karkhanis et al., 2016). Given the fundamental role of reward processes in the development of depression, these findings have implications concerning the route by which early adverse experiences may increase adult vulnerability to this disorder. Some of the effects observed seemed to differ between males and females, which may be relevant to sex differences in psychological disorders, including opioid-seeking behaviors (Levis et al., 2022). In addition to these processes, social anxiety elicited by separating pups from their mom was accompanied by reduced 5-HT_{1A} receptor expression in the dorsal raphe nucleus, the site of 5-HT cell bodies (Franklin et al., 2010). When previously stressed mice encountered a stressor during adulthood, tryptophan hydroxylase mRNA expression and that of the 5-HT transporter were elevated (Soga et al., 2021).

Early life stressors produced especially marked and lasting alterations of $GABA_A$ functioning (Perez-Rando et al., 2022), and rats stressed during adolescence subsequently exhibited more pronounced GABA variations within the lateral amygdala (Zhang and Rosenkranz, 2016). Predictably, based on the $GABA_A$ subunit conformational variations stemming from the stressor treatment, previously stressed animals were more sensitive to the impact of later stressors and these actions could be prevented by being raised in an enriched environment (Ardi et al., 2019), which otherwise might have contributed to elevated anxiety that developed subsequently (Salari and Amani, 2017). It also seemed that diminishing inflammatory responses (e.g., through the administration of a cyclooxygenase inhibitor), reduced the behavioral signs of depression, and altered the GABA and dopamine D2 receptor changes ordinarily provoked by an early life stressor (Lukkes et al., 2017).

Adolescence and early adulthood

Transitions from one phase of life to another often require the ability to adapt to new people and circumstances, the development of new social networks and new social identities, and for college-aged individuals, this includes the formation of an adult-like identity, which entails a need for social, economic, and emotional independence (which may clash with the continued need for parental support). This transition often occurs seamlessly, but among some individuals it is destabilizing, distressing, and lonely, leading to high levels of depression and anxiety. Aside from the actions of the transitions themselves, some individuals may experience bullying at various stages of adolescence, which can favor the later development of inflammatory-related diseases (depression, heart disease), as well as elevated drug use (Earnshaw et al., 2017).

The neurobiological changes associated with stressors experienced during adolescence occur in the context of socialization processes, and it might be expected that social instability (and other stressors) at this time could engender pronounced and lasting effects, manifesting as increased adult anxiety and depression. Fear and anxiety responses established during adolescence are difficult to overcome, and adult fear- and anxiety-related disorders often have their roots in moods that developed at these and earlier ages.

Adolescence is accompanied by the reorganization of neuronal and hormonal systems, which may be especially sensitive to stressors. During these times, stressors are apt to produce disturbance of nerve cell growth within the hippocampus, altered CRH_1 and CRH_2 receptor functioning, changes of certain brain neurotransmitters, such as GABA and its receptors, and persistent alterations of glutamate NMDA-receptors (Yohn and Blendy, 2017). Additionally, a chronic stressor administered during the adolescent period in rodents provoked exaggerated expression of NFκB-related genes in the hippocampus of both males and females, but the provocation of hippocampal IL-1β primarily occurred in female rats, which is consistent with the greater female vulnerability to mood disorders (Bekhbat et al., 2019).

To a considerable extent, the response to stressors during adolescence is tied to earlier stressor experiences that influenced inflammatory processes. Among children who encountered adverse events beginning at 2 months of age, inflammatory factors were elevated when they were adolescents (O'Connor et al., 2020). While numerous factors might have mediated this relationship, in some instances the link between early experiences and elevated adolescent inflammation and depression was mediated by elevated BMI that may have been secondary to stressor experiences (Reid et al., 2020).

Immunity and Microbial Factors

Early Life Stress and Immune Functioning

Early life stressful experiences can influence the regulation of proteins within cells, instigate elevated inflammation, and foster progressively greater epigenetic changes, and impaired DNA repair processes, which together drive cellular senescence and ultimately increase vulnerability to age-related disorders (Chaudhari et al., 2022). Numerous reports indicated that various forms of early life stressors in rodents have pronounced effects on later immune and inflammatory processes thereby affecting well-being throughout the lifespan (Cattaneo et al., 2015). The negative effects of early life stressors were observed in response to psychological as well as systemic insults (e.g., among rodents challenged with a bacterial endotoxin or to an influenza virus), and were moderated by adult stressor experiences, social support systems, and coping processes (Fagundes et al., 2013). As we have seen in the context of prenatal stressor effects, some of the consequences of early adversity, including that produced by the administration of an inflammatory agent, could be diminished by environmental enrichment (Landolfo et al., 2023).

Much like the effects of separating pups from the dam on hormonal and neurochemical functioning, a systematic review that included 46 studies concluded that repeated maternal separation resulted in elevated levels of proinflammatory cytokines and greater microglial activation. These effects were particularly pronounced when animals encountered a further stressor in adulthood, which is consistent with early life challenges priming systems to respond to further challenges (Dutcher et al., 2020). It has been suggested that early life stressors push immune cells, especially NK cells, into senescence and are less able to eliminate pathogens (Fernandes et al., 2021). Early life stressful experiences similarly influenced the functioning of microglia, including their phagocytic activities, and both prenatal and postnatal stressors could promote the priming of microglia so that later responses to challenges were augmented (Catale et al., 2020). In this respect, repeated social defeat in juvenile rats increased the subsequent response to an endotoxin challenge reflected by disturbed social interactions and anhedonia, which were related to elevated microglia density and glial reactivity to the challenge (Guerrin et al., 2023).

The effects of early life stress on microglia could occur through several processes (e.g., sensitization of activation processes). For instance, epigenetic reprogramming of innate immune memory of microglia may account for the long-term effects of perinatal stressors and could thereby influence vulnerability to neurodevelopmental disorders (Carloni et al., 2021). Also, the persistent effects of adverse experiences can occur because the disrupted synaptic pruning by microglia can affect excitatory synapses on stress-sensitive hypothalamic CRH-expressing neurons, which can affect the response to later stressors (Bolton et al., 2022). The effects appear to differ between male and female offspring, which once again may be relevant to the sex biases regarding subsequent neuropsychiatric and neurodevelopmental disorders (Garvin and Bolton, 2022).

In addition to microglia effects, a stressor in the form of maternal separation and early weaning resulted in an increase of genes that code for TNF-α and mast cell protease CMA1 in dura mater, one of the layers that make up the meninges that are involved in trafficking of immune cells between the periphery and the brain. When previously stressed mice were exposed to a mild stressor during adulthood, female mice exhibited still greater mast cell alterations and anhedonia reflected by a diminished preference for sucrose. Both these effects were prevented among mice that had received the mast cell stabilizer Ketotifen before the adult stressor was administered (Duque-Wilckens et al., 2022). Given the critical nature of the meninges in protecting the brain from intrusions, these findings provide a viable explanation for how early adversity can increase vulnerability to later challenges.

An alternative to maternal separation to induce early life stress was providing nursing dams with limited bedding and nesting material. Doing so prevents the construction of a proper nest and alters the quality of care provided to pups. An extensive review of this literature indicated that this procedure promoted numerous long-term behavioral changes (anxiety, depression, cognitive disturbances) that resembled those elicited by proximal stressors and induced biological changes that reflected stress and emotional responses. These comprised neuro-endocrine and metabolic changes, delayed maturation of gut microbiota, changes in neurogenesis, and prefrontal cortical-limbic-HPA functioning. Many of these outcomes are sex-dependent and vary with the age of the pups at which the limited bedding was introduced (Walker et al., 2017a), which may account for some of the inconsistent findings that have been reported using this procedure.

Elevated inflammation associated with limited bedding and nesting material has been suggested as a factor responsible for the behavioral and pathological effects observed. To some extent, this conclusion was based on indirect evidence showing that the procedure induces effects that have been tied to inflammatory elevations. Also, the limited bedding and nesting material protocol promoted increased inflammatory gene expression in adipose tissue and elevated microglia density and more amoeboid microglia, which produce inflammatory cytokines and neurotrophic factors (Ruigrok et al., 2021). This early life stressor resulted in mice subsequently displaying morphological changes of certain subtypes of microglia, as well as transcriptomic changes associated with TNF-α. Subsequently, when mice were more than 6 months of age, synaptosomes of mice that experienced early stress were phagocytosed less frequently than in nonstressed mice (Reemst et al., 2022b), indicating less synaptic pruning that is needed for adequate cognitive functioning. Early life stressors often comprise compound challenges that can have still more pronounced adverse effects than single events. For instance, among mice that had been raised with limited bedding and had been exposed to an environmental toxicant (diesel exhaust particles), later social behavior was especially impaired. This was accompanied by the inhibition of microglia phagocytosis within the anterior cingulate cortex (Block et al., 2022), which as we have seen is intimately related to the development of depressive-like behaviors.

Other stressors, such as the early isolation of rat pups, which led to a depressive behavioral profile, were also accompanied by microglial activation and elevated proinflammatory cytokine expression within the hippocampus, and these outcomes could be prevented by the antibiotic minocycline (Wang et al., 2017a). Supporting the importance of cytokines, antidepressant treatment during adolescence reduced the anxiety and depressive features provoked by maternal separation, and concurrently enhanced immune regulation, diminished levels of proinflammatory cytokines (IL-1β), and increased levels of the antiinflammatory IL-10 (Wang et al., 2017d).

Studies in humans have supported the contention that ACEs can elicit inflammatory elevations that favor the later development of pathological outcomes. A meta-analysis indicated that ACEs were directly related to the presence of immune-related markers (CRP, IL-6, TNF-α, IL-1β, and IL-10), and methylation of genes coding for specific hormones or immune factors was frequent (Soares et al., 2021). A systematic review similarly indicated that a variety of childhood stressors could increase circulating proinflammatory cytokines and acute phase proteins (primarily IL-6, TNF-α, and CRP), and this chronic inflammatory state might lend itself to a variety of disease conditions (Kerr et al., 2021). The ex vivo stimulation of lymphocytes indicated that cytokine changes were especially notable among individuals who encountered the most severe ACEs. In these individuals, the IL-2, IL-6, IL-8, and TNF-α were elevated, as were monocyte chemoattractant protein-1 and macrophage inflammatory protein-1α and β, indicating that childhood adversity was associated with dysregulated cytokine function (de Koning et al., 2022). An analysis that included a large number of participants indicated that SNPs related to the P2X7 gene were linked to anxiety. While certain SNPs interacted with early adversity in predicting anxiety, other SNPs were more closely tied to recent stressors (Kristof et al., 2023).

A prospective analysis revealed that early life social adversity predicted elevated inflammation in adulthood (Slopen et al., 2015), and in a unique sample of participants who were followed for an extended period, microbial and psychosocial experiences in infancy and childhood predicted epigenetic changes linked to inflammation. These epigenetic marks were predictive of later heart disease and other illnesses tied to inflammation (McDade et al., 2017), even affecting cellular functioning associated with aging (Ambeskovic et al., 2017). Once again, the actions of early adversity may come about by factors that indirectly affect inflammatory processes. For instance, early life trauma can affect lifestyles that promote both inflammation and depression, so that when controlling for lifestyle, the link between abuse and later depression disappears (e.g., Jonker et al., 2017). This does not mean that early life trauma is not tied to depression but instead suggests that lifestyle factors act as an intermediary in this regard.

Just as stressors affected immune functioning, among rat pups that were treated with an endotoxin, subsequent adult stressor responses were altered, including greater suppression of lymphocyte proliferation and NK cell functioning (e.g., Shanks et al., 2000). Even a single encounter with a pathogen can affect T_{reg} functioning, thereby influencing later susceptibility to immunological tolerance and the development of autoimmune disorders (Yang et al., 2022b). Inflammation provoked during early life similarly increased amygdala neuronal functioning and resulted in impaired auditory fear extinction in adulthood, a feature seen in PTSD (Doenni et al., 2017). These actions were not limited to immune challenges during the very early postnatal period, as immune-activating agents administered 14 days following birth also engendered lasting effects on emotional reactivity in rodents (Dinel et al., 2014). Likewise, early-life LPS administration could promote anxiety during adolescence, accompanied by elevated expression of NLRP3 inflammasome proteins within the CNS (Lei et al., 2017). The adverse effects of pathogenic encounters early in life, including elevated HPA functioning, hippocampal inflammation, and the anxiety and depressive-like behaviors evident subsequently, could be dampened by the tetracycline minocycline (Majidi et al., 2016).

To summarize, persistent cytokine variations related to early life stressors can undermine physical health, promote psychological disturbances, and could potentially render individuals at increased jeopardy for disorders promoted by inflammatory processes. To the point, childhood adversity may increase reactivity to adult stressors, accompanied by increased perceived stress, fatigue, poor quality of life, and depressive symptoms, together with diminished immune cell activity and elevated levels of proinflammatory cytokines. Given these linkages, it was suggested that new strategies ought to be explored to limit the consequences of childhood trauma before clinical symptoms are present, including antiinflammatory interventions and alterations of adaptive immunity (Danese and Lewis, 2017).

Gut and Immunity

Early life stressor experiences may be particularly adept in affecting gut microbiota. Simply disturbing a rodent's nest promotes erratic maternal behaviors and instigates excessive corticosterone levels in pups, coupled with diminished microbial diversity (Moussaoui et al., 2017). Likewise, separating pups from their mom can provoke gut bacterial alterations that may lead to increasing susceptibility to later psychopathological outcomes (Dinan and Cryan, 2017) as well as immune-related disorders, such as inflammatory bowel syndrome (O'Mahony et al., 2017b). Given the sensitivity of the gut at this time, it can be expected that the diet consumed during early life can affect an organism's microbiome and might thus affect CNS processes. Conversely, beneficial effects can be accrued through supplementation with certain bacteria. For instance, pseudocatenulatum CECT7765 was able to reverse the immune and neuroendocrine disturbances, as well as the anxiety associated with early-life maternal separation (Moya-Pérez et al., 2017). In line with such reports, a probiotic (*Lactobacillus fermentum*) could diminish intestinal barrier dysfunction stemming from early life maternal separation, possibly through immune alterations that increased IFNγ and reduced the antiinflammatory IL-4 (Vanhaecke et al., 2017).

Consistent with these reports, supplementation with the prebiotic BGOS during the neonatal period influenced BDNF protein levels during young adulthood (Williams et al., 2016b), and micronutrient supplements provided early in life limited later stressor-provoked corticosterone changes and cognitive disturbances that would otherwise be apparent (Naninck et al., 2017). Likewise, early-life diets that included prebiotics and bioactive milk fractions affected stress-elicited mRNA expression of dorsal raphe nucleus 5-HT_{1A} receptors and largely precluded behavioral disturbances ordinarily provoked by uncontrollable stressor exposure (Mika et al., 2017). Among germ-free mice, hippocampal levels of serotonin and its primary metabolite 5-hydroxyindoleacetic acid were appreciably elevated, primarily in males, suggesting a stress-like profile. Although later colonization with microbiota did not restore serotonin functioning, the elevated anxiety that was otherwise present was diminished, suggesting that these behaviors were independent of microbially related serotonin changes (Clarke et al., 2013).

The presence of bacteria during or soon after birth has important ramifications for the offspring's well-being. As we discussed in earlier chapters, the postnatal development of immune processes is undermined among germ-free mice. Specifically, lymphoid disturbances may be present, and the disturbance of balances between various aspects of the immune system can result in elevated vulnerability to immune-related disorders.

Likewise, among pregnant women, good gut bacteria can positively affect the fetus, and the use of antibiotics by a pregnant woman negatively affects both the mom and her fetus. An analysis conducted among mother-infant dyads that assessed 449 bifidobacterial strains revealed vertical transmission in almost 50% of dyads. To a significant degree, this transmission was influenced by vaginal birth, spontaneous rupture of amniotic membranes, and avoidance of intrapartum antibiotics (Feehily et al., 2023). It has been known for some time that the vertical transmission of microbial protection also comes with vaginal birth when the emerging infant meets a variety of bacteria within the mom's birth canal, whereas the presence of opportunistic bacteria was elevated in C-section babies and for months the gut bacteria that ordinarily enhance immune functioning were diminished (Shao et al., 2019). Cesarean delivery may render offspring more vulnerable to metabolic and inflammatory diseases, including allergies, asthma, and other chronic childhood immune disorders (Bäckhed et al., 2015; Gensollen et al., 2016).

To accommodate the loss of bacteria associated with Cesarean birth, efforts were made to provide newborns with bacteria by applying vaginal swabs (vaginal seeding) from their birth mother (Dominguez-Bello et al., 2016). Although this mode of delivery was associated with altered abundance and diversity of gut bacteria at 3 months following birth, these effects were no longer apparent at 6 months (Rutayisire et al., 2016). It had been reported that vaginal seeding resulted in the normalization of the infant's microbiome (Song et al., 2021), but a randomized controlled trial indicated that the procedure did not provide lasting benefits related to gut microbiota or allergy risks during the first 2 years of life (Liu et al., 2023d). However, there were indications that vaginal swabbing of C-section babies resulted in improved brain development (Zhou et al., 2023b) and infants generally reached developmental milestones without adverse effects being evident (Korpela et al., 2020; Zhou et al., 2023b). Concerns had been raised that the procedure could have the undesirable effect of transmitting pathogens to the infant, coupled with the continued debate regarding the benefits of vaginal seeding, The American College of Obstetricians and Gynecologists has not recommended or encouraged the adoption of this procedure.

There have been reports in which dietary factors were manipulated to alter the microbiota of children born by Cesarean section. As expected, synbiotic treatment normalized gut microbiota (Chua et al., 2017), but the effectiveness of this in relation to other phenotypes is less clear. A meta-analysis confirmed that among the offspring delivered by C-section, probiotic supplementation served to restore bacterial colonies so that the offspring microbiota was aligned with that of vaginally born infants (Carpay et al., 2022).

Vulnerability to Asthma, Allergies, and Other Developmental Conditions

Among neonates, the frequency of infections is relatively high, which has typically been attributed to immune functioning not being sufficiently mature to ward off bugs that happen to be around. This may be correct, but there is more to it than that. Microbial and immune processes are functionally related, and they mature in concert with one another, so that perturbations of one system may affect the other. Even small changes in the intestinal environment can affect microbiota, which will affect immune processes. For instance, the presence of specific microbial organisms has been associated with inflammatory-related conditions. Byproducts of gut microbiota, together with two types of fungi, were associated with increased inflammation, leading to asthma and allergies (Fujimura et al., 2016), and children deficient of several gut bacteria at 3 months of age were at increased risk for developing asthma by age 5 (Arrieta et al., 2015).

Although it is often thought that the greater the bacterial diversity the better, well-being may also depend on the presence of specific bacteria. Children with less diversity but high levels of the genus *Bacteroides* performed better relative to that apparent in the presence of greater diversity and lower *Bacteroide* levels. Thus, there may be an ideal microbial community that includes bacterial diversity and the presence of specific bacteria in predicting later cognitive outcomes (Carlson et al., 2017).

Ordinarily, organisms such as *Lactobacilli* and *Bifidobacteria* appear in the gut within a few hours of birth, typically being obtained from the mother's milk. Also, low numbers of other nonpathogenic bacteria like *Streptococcus*, *Micrococcus*, *Staphylococcus*, and *Corynebacterium* soon colonize the gut. To improve health and limit allergies and hypersensitivities, prebiotics and probiotics, as well as a combination of the two (synbiotics), have been included in infant formulas, but a systematic review concluded that their usefulness is questionable despite animal studies suggesting that the procedure could be helpful (Meirlaen et al., 2021). However, this may be dependent on the specific phenotypes assessed. A systematic review and meta-analysis of randomized controlled trials indicated that symbiotic interventions reduced the occurrence of respiratory tract infections (Chan et al., 2020).

There is also reason to believe that the timing of microbial alterations during prenatal and early postnatal periods can influence the lasting effects on illness vulnerability (Amenyogbe et al., 2017). It seems that the early postnatal period (postnatal days 2–9), the preweaning period (postnatal days 12–18), and the postweaning period (days 21–27) are differentially sensitive to the effects of an antibiotic cocktail in producing long-lasting emotional and neurobiological effects that varied with sex. Aside from affecting the formation of allergies, early-life antibiotic treatments effectively altered circulating immune cells, modified myelin-related gene expression in the prefrontal cortex, and altered microglial morphology in the basolateral amygdala. Likewise, the effects of antibiotic manipulation provoked sex-dependent effects on microbiota depletion, which influenced anxiety in adolescence and adulthood (Lynch et al., 2023). Thus, targeting these phases of development may be a way of preventing later illness if microbial disturbances exist.

Modifying gut bacteria can have marked ramifications on some illness conditions even in the presence of genetic factors that favor illness occurrence. For example, upon being treated with normal gut bacteria from an adult mouse, female mice that were genetically at risk for type 1 diabetes exhibited a marked decline in the occurrence of this illness. Numerous studies have also implicated specific microbiota in the development of atopic dermatitis (eczema) in young children. A longitudinal analysis in children up to 1 year of age indicated that the appearance of this condition was associated with altered diversity or abundance of bacteria within the Firmicutes and Bacteroidetes phyla (Sasaki et al., 2022). Moreover, a meta-analysis of randomized controlled trials indicated that atopic dermatitis in children could be reduced by 4–8 weeks of probiotic treatment (Huang et al., 2017), depending on the specific microbial mix administered. A detailed review, however, indicated that although single and multiprobiotic mixtures could reduce eczema, the magnitude of the effects was not clinically meaningful (Makrgeorgou et al., 2018).

Even if "cleanliness is next to godliness," it does not necessarily limit disease

According to the "hygiene hypothesis," protecting children from every bug can render their immune system less practiced, and hence they may be more vulnerable to infection. Likewise, early experiences with environments that are too sterile can limit the development of immune tolerance to foreign substances, thus allowing for exaggerated responses to some harmless elements. The occurrence of allergies (and asthma) has been attributed to excessive cleanliness (Strachan, 2000), as have other immune-related illnesses, such as multiple sclerosis, inflammatory bowel disease, and depressive disorders.

While subsequent views have not entirely dismissed this notion, it was suggested that the processes presumed to be involved in creating barriers to infection ought to be reconsidered (Bloomfield et al., 2016). Specifically, the position was advanced that disturbances of immunoregulatory processes may be attributable, at least in part, to the failure of encountering microorganisms from our evolutionary past.

Not meeting with these "old friends" has made humans less able to develop optimal immune functioning and hence greater vulnerability to disease states. To be properly trained, the immune system needs to encounter a diversity of bugs, not just infectious pathogens, but also those that are friendly. Encounters with a diverse set of bacteria serve to train T_{reg} cells that are important for immune regulation and prevent excessive immune activity that favors the development of allergies and asthma, as well as autoimmune conditions.

Tight regulation of immune functioning during the early postnatal period is necessary to prevent excessive inflammation that can have adverse actions. For this to occur, young children need early, regular, and frequent exposure to harmless microorganisms (old friends) that have been present throughout evolution and thus are readily recognized by the immune system (Lowry et al., 2016; Rook et al., 2015). These experiences may begin in the uterus and are marked during the perinatal period. Among other experiences, Cesarean delivery, antibiotic abuse, and migration from poor to richer environments in which a change of immunoregulatory organisms is encountered can disturb the microbiome, thereby influencing later health. In contrast, being raised in a rich microbial environment (e.g., living on farms) was associated with reduced occurrence of inflammatory bowel disorder during adulthood, although the greater use of pesticides within such environments may have increased vulnerability to neurodegenerative disorders (Mostafalou and Abdollahi, 2017). It seems that not adapting adequately to microbes may result in exaggerated inflammatory immune responses, thereby influencing vulnerability to physical and psychiatric illnesses. As some of our old friends, including helminths and bacteria, have been diminished in the urban environment, our dependence on microbial factors obtained from our mothers, other people, animals, and specific features of the environment is elevated (Rook, 2023).

The increasing incidence of allergies has been attributed to new external factors (possibly related to changing diets, increased use of antibiotics, or increased presence of pollutants) that cause disturbances related to the balancing of immune system responses (Belkaid and Hand, 2014). Foods consumed are, in a sense, an antigen that should cause the immune system to react against them. However, the mucosal membrane within the digestive system, along with particular immune factors, limit these immune reactions, and permit the development of tolerance to the food antigens. Yet, there are occasions in which the barrier system is disturbed, and hence problems may arise. Certain foods eaten during early life can influence gut bacteria that favor the development of allergies and inflammatory illnesses, and the microbiome of individuals with allergies is distinguishable from those without allergies. For that matter, gut bacteria present during the first few postnatal months could predict food sensitization that appeared at 1 year of age. Aside from the severe allergic responses that can lead to anaphylactic shock (e.g., in response to peanuts and related products), even minor allergies can have profound effects. Possibly owing to elevated inflammatory functioning, these allergies were associated with earlier onset and doubling of the risk of adult heart disease (Silverberg, 2016).

Following exposure to a new pathogen in young children (before age 3), memory T-cells develop, first in the lungs and intestines, and later in blood and lymph tissue. However, as the immune system is insufficiently mature until several years later (4–6 years of age), young children are vulnerable to respiratory infection. It is possible that early exposure to a variety of foods during infancy may diminish the likelihood of children subsequently developing severe food allergies. Be this as it may, in infants, antibody-producing B-cells surrounded by T-cells are present in the lungs, which offers some protection against respiratory infection. This so-called bronchus-associated lymphoid tissue ultimately disappears at about 3 years of age. However, if their presence persists, overreaction may occur to specific antigens, resulting in allergies and asthma (Matsumoto et al., 2023).

During the COVID lockdown, many young children were prevented from obtaining diverse bacteria and microbiota diversity and the presence of specific bacteria was altered among infants during this time (Querdasi et al., 2023). It did not appear that neurodevelopment impairments occurred among infants during this period, although communication patterns were slowed (Hessami et al., 2022). The media has been speculating obsessively as to whether the failure to meet "old friends" during the pandemic may affect children's ability to contend with infection later in life. This is not an unreasonable concern, especially as the full impact of isolation on the well-being of children is still not known.

Epigenetics and Intergenerational Actions

Stressor-Related Epigenetic Effects

As we have seen throughout this chapter, early life experiences, whether good or bad, may affect the developmental trajectory of neurobiological, behavioral, and emotional processes, rendering people vulnerable or resilient to pathology, even decades later. Furthermore, individuals can be affected by the experiences of their parents as well as the toxicants to which they have been exposed, which, in turn, might influence the characteristics of the next generation.

Environmental and experiential factors that promote epigenetic changes that cause the silencing (or activation) of particular genes can occur at any time of life but maybe especially pertinent if they occur prenatally or early in

development (Szyf, 2019). The most notable influence within a rat pup's life is their mom, and thus a dam's behavior during early development (exhibiting good attention to pups vs being neglectful) may influence the expression of particular genes, including those that regulate HPA functioning (Champagne, 2010). Indeed, poor maternal care during early postnatal development was accompanied by elevated methylation of the promoter for the gene regulating hippocampal glucocorticoid receptors, and persistently elevated stressor reactivity. It was similarly observed in mice that repeated prenatal stressor experiences encountered during the first trimester of pregnancy reduced DNA methylation of the gene promoter for CRH and sex dependently increased the methylation of the promoter region of the gene for glucocorticoid receptors (Grundwald and Brunton, 2015). Of potential practical relevance, the impact of early life stressors can be altered by treating rats with either a methyl donor (methionine) or a histone deacetylase inhibitor (trichostatin A) (Weaver et al., 2005).

While not dismissing the potential importance of epigenetic changes associated with stressors, toxicants, and other external influences, there have been problems in identifying which specific epigenetic changes contribute to the effects of adverse early life or prenatal experiences. Many epigenetic marks are present within the genome, any of which could be linked to a given behavioral phenotype. For instance, childhood abuse was associated with 997 differentially methylated gene promoters relative to that evident among individuals who had not experienced early life abuse (Suderman et al., 2014). Aside from the epigenetic modifications related to glucocorticoid receptors, early life events in monkeys were linked to DNA methylation of the gene coding for the serotonin transporter, which might contribute to the emergence of depression and other illnesses (Kinnally et al., 2010). Similarly, in female rodents, poor maternal care was accompanied by methylation of the gene promoter of estrogen receptor alpha (ERα) in the hypothalamus (Champagne, 2010). As this receptor is fundamental for the functioning of estrogen and oxytocin, which contribute to maternal behaviors, it is possible that the silencing of the gene for ERα might contribute to the poor maternal behaviors stemming from impoverished early life care.

As described earlier, linking any given epigenetic effect to specific phenotypes is difficult, certainly in studies with a small number of participants, as in the case of most research involving animals. Furthermore, most of the research that assessed the link between epigenetic changes and specific behavioral outcomes has comprised association studies, precluding conclusions regarding causal connections. Research is obviously needed to assess the effects of reversing epigenetic changes to determine the causal connections between epigenetic effects and particular phenotypes.

Epigenetics in Relation to Environmental Toxicants

The marked cellular proliferation and differentiation that occurs during fetal development makes this a particularly sensitive period for environmental toxicants and stressors to turn genes on or off. Commensurate with this perspective, numerous environmental toxicants (e.g., pesticides and fungicides, dioxin, jet fuel, and plastics) may produce epigenetic effects that can be transmitted across generations, just as trauma and maternal stressors can have such effects. Likewise, numerous prenatal challenges (e.g., methyl mercury, ultrafine particulate matter, diesel fumes, the androgenic fungicide vinclozolin, the estrogenic peptide methoxychlor, and endocrine disrupting chemicals) were accompanied by epigenetic changes within genes coding for BDNF, as well as several immune factors that have been linked to psychological and physical disturbances.

It may be especially significant from an intervention perspective that the impact of toxicants, like that of stressors, can be attenuated by increasing the presence of folate in the dam's diet (McGowan and Szyf, 2010), which has its positive effects by increasing the availability of methyl donors to limit methylation that had occurred earlier. It is interesting in this context that the use of folic acid (and multivitamin) supplements before and during pregnancy predicted a marked reduction in the occurrence of autism (Levine et al., 2018).

The scoop on bisphenol A

Considerable debate has surrounded the hazards created by bisphenol A (BPA), which is present in numerous consumer products and is even present in drinking water. The presence of BPA in humans may occur because the substance migrates from plastics used to make food storage containers and reusable beverage bottles and appears in epoxy resins that serve as protective coatings and linings for food and beverage cans.

Even at low doses, endocrine disruptors, such as BPA, were reported to increase the risk of breast cancer as well as other forms of cancer (Wazir and Mokbel, 2019). Also, developmental BPA exposure was associated with reduced T_{reg} cells and elevated levels of proinflammatory and anti-inflammatory cytokines and chemokines (Xu et al., 2016). As BPA and its analogs affect both innate and acquired immune processes, various immune-mediated conditions can be affected, including hypersensitivity reactions and allergies, and can disturb the microbiome (Kodila et al., 2023). Elevated BPA was also linked to the development of heart disease (Moon et al., 2021), possibly owing to inflammatory changes that were reflected by elevated CRP and increased proinflammatory gene expression

(Tsen et al., 2021). A meta-analysis of epidemiological studies implicated BPA in greater occurrence of type 2 diabetes, general and abdominal obesity, hypertension, and earlier all-cause mortality (Bao et al., 2020b).

In addition to physical illnesses, BPA may influence the development of psychological disorders that begin during early life. BPA in infants may be elevated owing to its presence in plastic baby bottles and may be transmitted to the infant through breastfeeding, leading to slowed postnatal growth (Jin et al., 2020). It seems that prenatal BPA treatment can produce epigenetic modifications in the expression of the genes encoding estrogen receptors within the cortex and hypothalamus of juvenile mice, which was coupled with sex-dependent effects on social anxiety (Kundakovic and Jaric, 2013). Even very low doses administered to pregnant rats caused marked hormonal variations in offspring, as was amygdala expression of receptors coding for estrogen, androgen, oxytocin, and vasopressin, which could affect later stressor responses (Arambula et al., 2017). Consistent with these findings, the pups of rodents that had been treated with BPA during both pregnancy and lactation, exhibited an anxiety-like profile that may have stemmed from glutamate receptor alterations (Zhou et al., 2015a). Likewise, lasting epigenetic changes were observed in the BDNF gene within the hippocampus and blood of mouse pups (Kundakovic et al., 2015), which might be relevant to the later emergence of depressive illness.

Numerous studies in humans have indicated that chemicals that influence sex hormones can have pronounced effects on depressive behaviors in women. In this context, bisphenols and phthalates present in urine early and in mid-pregnancy were associated with lower levels of progesterone, which could predict postpartum depression (Jacobson et al., 2021). For the most part, epigenetic links have been studied among women and their offspring, but there is evidence that epigenetic changes in dad's sperm brought about by phthalates, can also affect offspring (Pilsner et al., 2017).

Stressors encountered during the prenatal period or early life can have profound downstream actions that can carry on throughout life. The alterations of the developmental trajectory may be related to a constellation of individual factors and the nature of the stressors encountered (e.g., psychological challenges versus systemic insults, such as infection). As described in Fig. 8.7, the altered trajectory includes changes in the reactivity of biological processes (sensitization) so that subsequent challenges may have exaggerated neurobiological changes that can promote appropriate allostasis on the one side, or allostatic overload on the other and the development of diverse pathological outcomes. A multihit view such as this has been adopted to account for the emergence of numerous pathologies, and the organism's early experiences and genetic vulnerabilities may be fundamental in determining the emergence of pathology or resilience. As we will now see, these experiences may have intergenerational and transgenerational effects.

Intergenerational and Transgenerational Effects of Early Life Stressors

Environmental and experiential challenges can elicit biological and behavioral effects that may be passed on across generations (transgenerational actions) provided that the epigenetic changes occurred in a germline cell. The stressors encountered, like exposure to pesticides, can have profound effects on hormonal systems, engendering adverse effects that are recapitulated over several generations. Studies in rodents similarly indicated that infection (or maternal immune activation through the viral mimic poly I:C) can promote behavioral and biological effects, including altered prefrontal cortical gene expression, that spanned several generations (Kleeman et al., 2022; Raymann et al., 2023). Identification of the mechanisms by which transgenerational effects develop in response to maternal immune activation may provide a fundamental understanding in the creation of novel preventive strategies to deal with potential neurodevelopmental disorders.

Although there has been considerable attention that focused on the role of glucocorticoid receptors in the intergenerational effects of stressors and toxicants, it seems that estrogen receptor changes, as well as epigenetic effects on genes coding for BDNF also play a role in this regard. When assessed as adults, in rats that had been raised by a stressed caretaker that displayed abusive behaviors, elevated methylation of BDNF-related genes were expressed in the prefrontal cortex. When these rats had litters, the very same BDNF epigenetic profile was apparent in the offspring, pointing to the transgenerational epigenetic effects of early life stressors (Roth et al., 2009). Although epigenetic effects can last throughout life, and be passed across generations, raising rats in an enriched environment, a manipulation that may increase BDNF, reversed the adverse transgenerational actions of a stressor (Gapp et al., 2016). Likewise, exposure to enrichment during the juvenile period could reverse the negative behavioral consequences ordinarily provoked by poor early-life maternal care (Champagne and Meaney, 2007).

Transmission of stressor effects from a mother to her offspring could occur either through epigenetic effects, various actions related to the uterine environment, or actions directly attributable to the dam's behavior (e.g., high anxiety in the dam may be transmitted to pups through her behaviors). There is also the question of whether transgenerational epigenetic effects are transmitted exclusively through the mother, or can the father's experiences also have such effects. It appeared that in male pups that experienced chronic maternal separation, depressive-like behaviors and altered behavioral responses to aversive stimuli were evident upon being stressed as an adult. In the offspring of these males, behavioral and neuroendocrine disturbances were also apparent even though these animals had not been subjected to any particular stressor (Franklin et al., 2010). Likewise, the anxiety- and depressive-like responses elicited by social defeat experienced by male mice were transmitted to their offspring despite these males not being present after the dam was impregnated and hence had no direct effect on their postnatal environment (Dietz et al., 2011).

To determine whether stressed male mice might have caused females to become distressed, thereby affecting their behavior toward the pups, females were impregnated through IVF using sperm from a stressed mouse. Under these circumstances, however, the transmission of behavioral effects from parent to offspring was not present (Dietz and Nestler, 2012). These data suggest that paternal transmission of the stressor effects was, in fact, not linked to epigenetic changes. Yet, stressors experienced by the male could have influenced sperm quality, including sperm motility, and thus might have been least likely to fertilize an egg, hence precluding transgenerational effects that potentially might have occurred. Given these varied findings, there is still some question about the epigenetic involvement of transgenerational transmission of stressor effects stemming from male forebears.

Finally, it is important to underscore that even when epigenetic effects appear, they often peter out over several generations. It had typically been surmised that either this is not a genuine transgenerational effect or one that decays or dilutes with successive passages across generations. It has, however, been shown that genes exist (termed Modified Transgenerational Epigenetic Kinetics) that can switch epigenetic events on and off. Moreover, small RNAs are present that can regulate these genes. What causes the RNAs to behave as they do is uncertain, and it is not known to what extent these actions are modifiable.

Transmission of Psychological and Physical Sequelae of Trauma in Humans

It is well documented that cultural differences exist regarding the occurrence of many diseases. As we have seen, these differences may stem from living conditions (poverty, pollution, smoking, and diet), but ethnic identity was also associated with hundreds of epigenetic differences, about 76% being inherited, whereas the

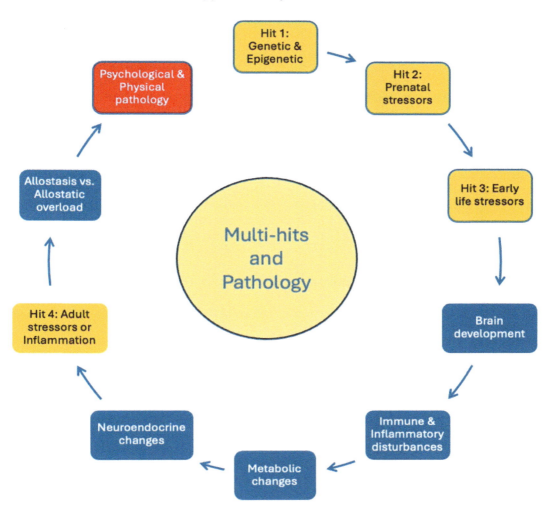

FIG. 8.7 Multihits comprising genetic/epigenetic factors, prenatal and early postnatal factors (timing of stressors, intensity, chronicity) are linked to numerous neurobiological disturbances that can be exacerbated with a hit experienced in adulthood (together with poor appraisals and ineffective coping, as well as aging), culminating in the allostatic overload that favors the emergence of pathology.

remainder came from social and experiential events (Galanter et al., 2017). Only a few studies examined transgenerational (or intergenerational) epigenetic effects in humans.

A meta-analysis that included 23 independent studies indicated that the mental health of the offspring of holocaust survivors was linked to the mental health of their parents, as well as parental gender, perceived parenting, and attachment quality (Dashorst et al., 2019). It was reported that among the children of Holocaust survivors the risk for PTSD (or subthreshold symptoms of this disorder) and abnormal HPA functioning was elevated (Yehuda and Bierer, 2009), and associated epigenetic marks could be detected (Yehuda et al., 2016). For instance, Holocaust experiences were tied to epigenetic changes associated with glucocorticoid receptor functioning (i.e., *FKBP5* gene methylation), an outcome that was also apparent in the children of survivors, which might render them at increased risk for pathology (Yehuda et al., 2016). This outcome was more pronounced if both parents had experienced PTSD (Yehuda et al., 2014). Parenthetically, although second-generation Holocaust survivors did not exhibit an increased disposition toward schizophrenia when it did occur, the severity was more intense among individuals whose parents had been survivors of the Holocaust (Levine et al., 2016).

Although the view was frequently adopted that, as a group, Holocaust survivors and their children were at elevated risk of developing psychological disturbances, especially in response to further stressors, this perspective may be too narrow. In the children of Holocaust survivors, methylation of glucocorticoid genes at some sites occurred less frequently than among the children of individuals who had not been in the holocaust and varied yet again with the offspring's own early adverse experiences (Yehuda et al., 2016). Importantly, an analysis that included more than 38,500 Holocaust survivors and about 35,000 control participants revealed a somewhat paradoxical finding. Specifically, the survivors experienced more serious illnesses than controls, they lived longer (Fund et al., 2019). It is premature to form firm conclusions based on these findings, but the possibility exists that survivors fall into two broad classes—those in whom the Holocaust was aligned with future adverse outcomes, whereas a second subgroup may have comprised uniquely resilient individuals. In essence, it is not productive to paint survivors with a broad-brush concerning illness vulnerability. Some individuals certainly might be at elevated risk for pathology, whereas others may have been (or became) more resilient. In this respect, epigenetic changes associated with trauma could potentially enhance resilience, although these actions could be subverted by other epigenetic changes or experiential factors. The data supporting this have not been recorded in humans, but in mice that had experienced chronic stressors, the expression of two genes (dihydrocaffeic acid (DHCA) and malvidin-3'-O-glucoside (Mal-gluc)) that influenced inflammation and synaptic plasticity were associated with diminished depressive phenotypes (Wang et al., 2018a).

The intergenerational consequences attributable to traumatic experiences are not unique to the children of Holocaust survivors. During the Kosovo war, pregnant women who experienced sexual violence or torture, frequently (72%) developed PTSD symptoms, and among their subsequently born children, elevated methylation was apparent in the *NR3C1* gene and those that coded for BNDF and the 5-HT$_{3A}$ receptor (Hjort et al., 2021). Similar effects were observed at specific alleles of the FKBP5 gene among the offspring of mothers who experienced early-life trauma. This effect was most pronounced in women who had encountered threat-based adverse childhood events but less so if adverse events entailed deprivation (Grasso et al., 2020).

In children of Tutsi mothers who had been pregnant during the genocide that occurred in Rwanda, epigenetic changes were detected in genes coding for leukocytes (Musanabaganwa et al., 2022). Likewise, cortisol levels and glucocorticoid receptors were reduced among Tutsi women who had developed PTSD, as well as in their children. They also expressed greater epigenetic changes in the *NR3C1* and *NR3C2* promoter genes that are involved in glucocorticoid regulation (Perroud et al., 2014).

Like these conflict-related events, intergenerational outcomes have been reported in survivors of the Dutch Hunger Winter, which occurred at the end of World War II, when the Nazi regime prevented people in the western parts of the Netherlands from obtaining food or fuel. Offspring born during this period displayed epigenetic changes within genes that influence birth weight as well as later LDL cholesterol levels (Tobi et al., 2014).

Moreover, individuals who had been prenatally exposed to this condition subsequently expressed epigenetic changes related to insulin-like growth factor 2 (IGF2) (Heijmans et al., 2008), which were linked to the development of type 2 diabetes. Furthermore, offspring of the Dutch Hunger Winter were at elevated risk of poor cognitive functioning as well as the development of depression and schizophrenia (Roseboom et al., 2011).

Somewhat similar findings were observed in the offspring of parents who had lived through the 1959–1961 famine in the Suihua area of China. Among the adult offspring whose mothers were pregnant with them during this period, 31% were hyperglycemic and 11% developed type 2 diabetes, twice that of children born after the famine (Li et al., 2017a). As in the Dutch Hunger Winter studies, the high levels of type 2 diabetes have been attributed to transgenerational epigenetic effects that originated during the famine (Zimmet et al., 2018). Epigenetic changes occurred in the gene coding for IGF2 (Shen et al., 2019) and epigenetic changes that spanned two generations were associated with the capacity of the kidneys to filter excess waste (glomerular filtration rate) (Jiang et al., 2020).

There have, of course, been many occasions in which groups were targeted for annihilation but for various reasons (often political) we have little information regarding the downstream effects of these collective traumas. Attempting to form conclusions based on different collective traumas may not be productive or even appropriate given that they differed in multiple ways. These collective traumas not only varied in the magnitude, ferocity, and duration of the events but also the hatred and motivations that gave rise to them. Importantly, the impacts of trauma could vary with the conditions that prevailed after the trauma, including the availability of "safe spaces" that allowed individuals to heal (Matheson et al., 2020).

The data available have nevertheless been instructive in pointing to potential intergenerational effects that could occur in response to trauma. The notion that trauma experiences may have intergenerational and transgenerational consequences has important social ramifications. Still, there is the question concerning why some epigenetic changes are maintained across generations, whereas others are not, and there is no certainty concerning why transgenerational effects occur in response to some experiences, but not others. There is no certainty that an epigenetic change that occurred is necessarily carried through from childhood to adulthood given that epigenetic effects are modifiable. Furthermore, while some of the transgenerational effects that occur could be related to epigenetic actions, it might be that being born in traumatic times could have affected children for any number of other reasons, such as poverty, food shortages, safety concerns, or the ability of parents to care for their children.

Early life trauma could set in motion epigenetic actions that provoke downstream neurobiological effects and pathological conditions. These experiences could similarly create a behavioral or emotional milieu (e.g., detachment, social rejection sensitivity, or changes in family dynamics), independent of epigenetic factors, which would favor the later development of pathology (Bombay et al., 2014). In fact, in some groups (e.g., Indigenous Peoples in North America, Australia, and elsewhere), negative early experiences were later coupled with multiple toxic situations (poverty, abuse, and drug use), any of which might have had epigenetic consequences that favor pathology. In this regard, society as a whole might serve as an environment, which through epigenetic actions could influence cognitive, emotional, and physical health (Branscombe and Reynolds, 2015). Selecting one event and then attempting to link this to epigenetic changes may not be all that meaningful given that a traumatic event can have multiple sequelae that could cumulatively affect health-related processes.

Collective, Historical Trauma

Unfortunately, far too often, groups of individuals are not only affected by single or multiple traumatic events (collective trauma) but this cumulative emotional and psychological wounding is experienced by groups over generations. According to social identity theory, a sense of self, or identity, is derived from group memberships, and having a particular identity and affiliation with a group (irrespective of whether it comprises race, religion, gender, or one that entails occupation, or being a survivor of chronic illnesses) serves multiple adaptive functions. Social identities take on considerable importance, particularly when the group is challenged, and a variety of adaptive or counterproductive emotions can be elicited (e.g., collective shame or guilt vs collective anger). These threats can also come to provoke individual or group behaviors that can be either constructive (i.e., prosocial) or destructive. In this respect, among individuals living in the Blackfeet reserve (situated in Browning, Montana), the adverse behavioral disturbances and elevated levels of IL-6 and CRP associated with experiencing ACEs were diminished among those who had strong ties to their tribal community (John-Henderson et al., 2020).

Members of racial, religious, or cultural groups may accept the impermanence of their own lives but believe that aspects of their group (e.g., values, morals, beliefs, traditions, and symbols representing the group) will be passed on from one generation to the next. The more they identify with the group, the more threatening they will perceive external threats, leading to efforts to push against the challenge. Thus, a history of collective abuse and threats of extermination may be associated with group members being especially vigilant, sensitive, and reactive to the perceived evil or malign intents of others. Outsiders might view the heightened vigilance of ingroup members as a counterproductive obsession with past trauma (perhaps even resenting the constant referrals to the past collective abuses attributed to the outgroup). Yet, if history is any teacher, then the elevated vigilance might be perfectly reasonable and adaptive.

Carrying the load of collective, historical trauma

The psychological and physical consequences of "collective, historical trauma" experienced by a group can be passed down across generations through multiple processes. Epigenetic changes and parental influences might act in this capacity, but negative outcomes can also occur through other processes. In the years following collective trauma, dislocation creates multiple hardships, living conditions may be exceptionally disturbed, and poverty may be pervasive so that the psychologically toxic environment promotes a wide range of physical and mental illnesses.

As with any stressor, wide individual differences exist regarding the effects of collective trauma. Even within a circumscribed group, such as Holocaust survivors, considerable diversity exists regarding their experiences, and marked physical and behavioral variations occur among survivors and their children. While some survivors might not have seemed to carry the trauma of the Holocaust with them (or, might not have openly expressed their feelings), others frequently displayed one or more of several emotional disturbances, ranging from survivor's guilt, denial, agitation, mistrust, intrusive thoughts, nightmares, disorganized reasoning, difficulty expressing emotions, anxiety, and depression, although, in most instances, these were at subsyndromal levels (Bar-On and Rottgardt, 1998). Whereas some survivors of collective trauma could not stop speaking of their trauma, others rarely spoke of their experiences (a "conspiracy of silence" that was frequently the norm; Danieli, 1998). Even if the trauma was not spoken about, it was nonetheless "silently present in the home." Despite not directly encountering the trauma, children of those who did might incorporate their parent's or grandparent's narratives and images into their own schema (Hirsch, 2001). These narratives might have been passed on through both verbal and nonverbal communication, and when blanks existed in historical accounting, these may have been filled in by the fertile imaginations of children. As it turned out, the children who experienced this silence generally seemed to be more vulnerable to intergenerational transmission of trauma (Wiseman et al., 2002).

The consequences of collective, historical assaults among Indigenous People within the United States and Canada (as well as many other countries) aimed at eliminating a group's identity (taking the Indian out of the Indian) can reflect a "soul wound" that is not readily eliminated (Duran et al., 1998). These experiences can produce social and psychological consequences on families and communities, fostering disturbed social functioning, and an erosion of leadership, basic trust, social norms, morals, and values, often persisting for generations (Bombay et al., 2014). The impact of collective trauma may diminish over successive generations, largely being recalled through rituals and symbols (e.g., holidays and remembrance days) or storytelling. At the same time, "recovery" might never be complete as feelings concerning these traumatic events sit only slightly beneath the surface, reemerging with further reminders (e.g., discriminatory behaviors or other threats to the groups' well-being) of the indignities committed against their group. Once again, the transmission of trauma across generations of Indigenous people can also be influenced by the communications (either silence or continued recapitulation of experiences) between survivors and their children and grandchildren (Matheson et al., 2018a).

As we have already intimated, it is difficult (and inappropriate) to make comparisons between the experiences and consequences associated with collective trauma across different groups. Each collectively traumatized group is different, each has unique experiences, and each has its own consequences. Further, the long-term impacts of collective, historical trauma are not "just" a result of the trauma itself but might also be subject to the healing opportunities (or the lack of such opportunities) that came afterward, particularly if these groups continued to live in substandard conditions and experienced ongoing discrimination. While ingroup and outgroup support and nurturance could potentially act against adverse effects emerging, in the end, traumatized groups may need to find ways of healing through the adoption of a strength-based approach to overcome current problems (Matheson et al., 2020).

Concluding Comments

It might seem curious that stressful experiences in infancy or early childhood, of which we typically have no memory, can have profound and damaging long-term physical and psychological consequences. Early theorists struggled to understand this paradox, postulating that such memories were buried as unconscious thoughts that fought to emerge in some manner, thereby affecting behavior and cognition. They had not considered the impact of prenatal stressors and likely would have had a hard time incorporating the action of these events into a formulation involving unconscious thoughts (Jung, perhaps being an exception, given his views on the collective unconscious). Recent explanations that are more parsimonious and testable include the possibility that the effects of early experiences, acting through neuroplasticity or epigenetic processes, influence the reactivity of neurons, thereby affecting developmental trajectories as well as reactivity in response to later stressors. In essence, the effects of early life stressor experiences may become "embedded" during childhood and then exacerbated by further stressor encounters. These early life challenges can provoke a wide range of pathological conditions, paralleled by changes of reactivity within stress-relevant brain regions, which also predict the efficacy of drug treatments in producing illness remission.

Early life epigenetic programming can play a significant role in determining the appearance of abnormal behaviors, although adaptations in several forms can affect the developmental trajectory that might otherwise occur. The brain's exquisite neuroplasticity is apparent throughout life but is most notable during early life periods. Thus, both positive and negative experiences and environmental influences will have their greatest effects at this time. Because of the brain's plasticity, genetic influences and early experiences are, in a sense, the seeds laid down, whose full blossom might not be realized for some time.

Besides stressors, we have seen that numerous other events experienced prenatally and postnatally, ranging from bacteria and viruses to nutritional factors, can have lasting effects. It ought to come as no surprise that maternal malnutrition and the loss of macronutrients and micronutrients disturb immune functioning in moms during fetal immune development, leaving them both vulnerable to opportunistic infections. In addition, stress responses will be elicited, including HPA activation, which will affect microbial and immune functioning, and thus may further undermine later health (Macpherson et al., 2017). For that matter, the development of adiposity, adult diabetes, and heart disease may have their roots in the uterine environment, possibly through epigenetic changes that occurred at that time. Not only are the repercussions apparent with the development of inflammatory-related physical illnesses, but also with respect to mental health. A strong case has been made that our conceptions of mental health need to be considered more broadly so that in addition to neurotransmitter and growth factors, greater consideration needs to be given to the gut and its commensal microbiota (Jacka, 2017), including actions derived from prenatal and early postnatal experiences.

Just as early postnatal life may be a critical period during which numerous experiences can readily induce epigenetic changes, the prenatal environment may also lend itself to such outcomes (Cao-Lei et al., 2016). Although this view, at least from the perspective of epigenetics, is fairly recent, the "fetal programming hypothesis" has been around for some time, and has been broadened and renamed as the Developmental Origins of Health and Disease (DOHaD) theory (Gluckman et al., 2010). The basic view expressed was that stressful experiences, through a variety of neurobiological processes, such as changes in fetal glucocorticoid levels, can program fundamental components of growth and metabolism, thereby leaving a lasting imprint that is manifested at later times. Increasing research has pointed to the social environment as potentially having effects on epigenetic programming, especially if these occur during prenatal and early postnatal times (Turecki and Meaney, 2016). Moreover, through epigenetic processes, the sins of one generation can be imposed on the next. This said, epigenetic processes are dynamic, and providing nurturing environments can alter previously programmed epigenetic actions as well as psychosocial factors that affect well-being. It has been suggested (Anisman and Matheson, 2023) that in this case, it may be possible that the "bell can be unrung." This might well be a bit of exaggeration, but surely, its sound can be muffled.

CHAPTER

9

Depressive Disorders

The burden of depression

Of the many mental disorders that can affect people, depressive illnesses are the most common, affecting 10%–15% of individuals, but are much higher in some populations. Discouragingly, since 2005, depression has increased by 18% (before COVID-19), rather than declining as one would have hoped, and projections are that the incidence of depression will continue to increase over the next decade. After being successfully treated, illness reoccurs in 50%–70% of cases within 5 years, and it has been maintained that depression ought to be considered a life-long disorder.

Depressive disorders are pervasive among older individuals, with almost half of seniors in residential care homes being affected, and antidepressants have become the most widely used prescription medication in this population. Likewise, depressive illness occurs frequently in young people, with 25%–30% of college-aged individuals experiencing clinically significant depression, which increased during the COVID-19 pandemic. Frighteningly, after automobile accidents, suicide related to depression is the leading cause of death among youth, although gun-related deaths may soon exceed other causes.

The financial burdens of medical care have become an enormous strain across countries, and the cost of care related to mental illnesses is at the front of the pack. This is not only because depression reoccurs frequently but also because depressed individuals utilize medical facilities (for issues unrelated to mental health) more frequently than nondepressed individuals. In some countries, more than 50% of new disability benefit claims are due to mental health issues, far exceeding that of lung, colorectal, breast, and prostate cancer combined, and the amalgam of all infectious diseases doesn't come close to the cost of treating depressive illnesses. It is disturbing that the vast majority (more than 70%) of individuals who are depressed do not receive medical attention.

Given the burden of depression on healthcare systems, institutional support for prevention and treatment has been exceptionally short-sighted, especially as every dollar spent on treatment for depression and anxiety yields a 4-dollar return on investment (WHO, 2016a). It has been said that resource allocation for mental illness is the orphan of the healthcare system, and research concerning mental illness is the orphan of the orphan (Merali and Anisman, 2016).

Defining Depression

From a precision medicine perspective, mental illness and its treatment should not be considered as a syndrome, but instead ought to be broken down into specific symptoms, genes, biochemical processes, and experiential factors (e.g., adverse childhood experiences), and on this basis determine treatment methods. This is especially germane for depressive disorders given the variability of symptoms that are common across individuals and the differential effectiveness of treatment strategies. Even though our inclination is to proselytize for a precision medicine approach, much like the Research Domain Criteria strategy, the Diagnostic and Statistical Manual of Mental Disorders (DSM-5) (American Psychiatric Association, 2013) is still a widely adopted framework and thus we will primarily consider depressive disorders from this vantage.

Table 9.1 describes depression "subtypes" based on DSM-5 features. It is readily apparent that individuals can present with widely different symptoms, as in the case of major typical depression and atypical depression (the latter is now referred to as major depression with atypical features) in which many symptoms may even be opposites of one another. Research in depression has all too often failed to distinguish between subtypes of depression, and having all patients in the same bucket may have contributed to the inconsistent outcomes reported in response to pharmacologically based treatments.

Most mental health disorders, as depicted in Fig. 9.1, begin fairly early in life, corresponding to the maturation of brain regions associated with threat systems,

The Immune System
https://doi.org/10.1016/B978-0-443-23565-8.00006-5

TABLE 9.1 Subtypes of Depression.

Typical depression: Characterized by sadness and/or anhedonia, coupled with symptoms such as reduced eating, weight loss, and sleep disturbances (e.g., early morning awakening), psychomotor agitation, fatigue, feelings of guilt or worthlessness, diminished cognitive functioning, including memory disturbances, and suicidal ideation

Atypical depression: This form of depression is reminiscent of typical depression, being characterized by poor mood or anhedonia, as well as many of the cognitive features of the illness. However, the atypical form of depression is associated with *reversed* neurovegetative features, which include increased eating, weight gain, and increased sleep, as well as a tendency toward persistent rejection sensitivity (often causing impairment of social functioning), feelings of heaviness in the limbs ("leaden paralysis"), and mood reactivity even during well periods

Melancholic depression: Reflected by features of typical depression, but at a severe level. The intense depressed mood is accompanied by marked anhedonia, psychomotor agitation, early morning awakening, excessive weight loss, or excessive guilt accompanying grief or loss

Dysthymia: Symptoms are appreciably less intense than major depressive illness, and a personality disorder is often present. Dysthymia is diagnosed when the low-grade depressive symptoms persist for at least 2 years, although symptoms typically wax and wane. If not treated, major depression may evolve, so that it is superimposed on a dysthymic background. This "double depression" is more difficult to treat

Seasonal affective disorder (SAD): This form of depression is tied to seasons (or duration of light over the course of a day), emerging in the autumn or winter, and then resolving in the spring. This diagnosis is made if episodes have occurred in colder months over a period of at least 2 years, without episodes occurring at other times

Treatment-resistant depression: Typically diagnosed when antidepressant treatments fail to diminish symptoms appreciably, even after repeated (usually three) efforts were made using different compounds

Recurrent brief depression: Characterized by intermittent depressive episodes that occur, on average, about once a month over at least 1 year. The appearance of an episode, which lasts for 2–4 days, does not appear to be tied to any particular cycle (e.g., menstrual cycle). The diagnostic criteria for recurrent brief depression are much like those of major depressive disorder, but can be particularly severe, and may be accompanied by suicidal ideation and suicide attempts

Postpartum depression: Occurs after childbirth and affects about 5%–10% of women. Symptoms are much like those of major depression. The appearance of the illness may be linked to hormonal changes that accompany pregnancy or those that occur in association with pregnancy or childbirth; however, hormonal therapy has not been an effective treatment strategy

Minor depression: Characterized by a mood disorder that does not fully meet the criteria for major depressive disorder, but persists for at least 2 weeks

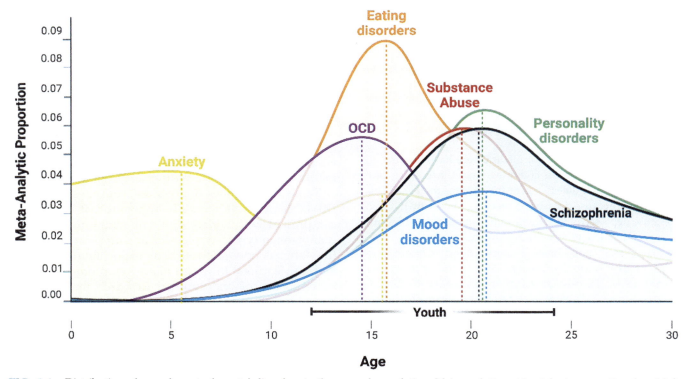

FIG. 9.1 Distribution of age of onset of mental disorders in the general population. Meta-analytic epidemiologic proportion (y-axis) for anxiety disorders (5.5/15.5 years), substance use disorders (19.5 years), schizophrenia/psychotic disorders (20.5 years), eating disorders (15.5 years), personality disorders (20.5 years), obsessive-compulsive (14.5), and mood disorders (20.5 years). The dotted horizontal lines represent the peak age of onset for each diagnostic category. *Source: Uhlhaas, P.J., Davey, C.G., Mehta, U.M., Shah, J., Torous, J., et al., 2023. Towards a youth mental health paradigm: a perspective and roadmap. Mol. Psychiatry 28, 3171–3181.*

hypothalamic-pituitary-adrenal (HPA) axis functioning, forebrain cortical connections, reward systems, and processes associated with social cognition (McGrath et al., 2023; Uhlhaas et al., 2023). Anxiety disorders and depressive illnesses often begin before the age of 16, and at one time, early-onset depression was considered to be a genetic disorder rather than one brought on by psychosocial stressors. However, it is likely that environmental, experiential, and social factors (e.g., relationship problems, abuse, bullying, and cyber-bullying), in conjunction with genetic and biochemical disturbances are responsible for the development of the illness in adolescents.

A depressive disorder may begin with one or two disturbed thoughts and evolve insidiously so that a range of symptoms develop that involve multiple cognitive, behavioral, and neurobiological processes. As in the case of many other illnesses, depressive disorders are more amenable to treatment if caught relatively early, which speaks to the importance of defining biomarkers for depression and focusing to a greater extent on the appearance of depression in young people to thwart the lifelong impact of this illness. No doubt, more will be needed since individuals might not recognize their depression as being abnormal and thus fail to seek treatment. Regrettably, the stigma associated with mental illness undermines help-seeking, and so a good number of people suffer in silence.

Typical Versus Atypical Depression

The distinguishing features of depressive disorders not only vary across illness subtypes but also with respect to their neurobiological correlates. For instance, elevated cortisol levels are more likely to be apparent in melancholic than in atypical depression (Lamers et al., 2020). In contrast, relative to melancholic depression, the atypical form of the disorder has been associated with reduced ghrelin and higher levels of leptin and insulin, as well as elevated circulating levels of several proinflammatory cytokines and C-reactive protein (CRP), and lower levels of brain-derived neurotrophic factor (BDNF) (e.g., Simmons et al., 2020). Differential gene expression analysis likewise revealed a greater enrichment of genes related to inflammatory processes among patients with hyperphagia relative to those with hypophagia or in healthy controls (de Kluiver et al., 2019). Features of depression were also tied to gut-derived metabolites, comprising both short- and long-chain free fatty acids and several acylcarnitines (these are involved in the generation of cellular energy), amino acids, and bile acids (Brydges et al., 2022).

Beyond these features, the response to specific cues differentially influenced depression associated with increased versus decreased eating. Among depressed individuals who exhibited increased eating, food cues promoted elevated neuronal activity within the mesocorticolimbic and parahippocampal regions, whereas patients who presented with reduced appetite exhibited decreased reactivity in portions of the insula (Simmons et al., 2016). The differences between these regions in patients with typical versus atypical depressive features also varied with specific hormonal and neurotransmitter alterations. Specifically, ventral striatal neuronal activity, which is associated with reward processes, was relatively low in those with diminished appetite and highest cortisol levels (Simmons et al., 2020). As we'll see shortly, depressive disorders have been associated with structural brain changes, which can vary with the features of the illness. The severity of depression was inversely related to the thickness of the rostral anterior cingulate cortex, and the thickness of the caudal aspect of the anterior cingulate cortex was diminished in melancholic depression (Toenders et al., 2022). In view of the biological and behavioral differences between depressive subtypes, it follows that they might also be differentially responsive to antidepressant pharmacotherapies.

Theoretical Perspectives Concerning Depressive Illnesses

Both cognitive and neurobiological views of depression have been offered, and within each framework, several theoretical positions have been adopted. These perspectives aren't necessarily independent of one another, and both have implications regarding the treatment strategies that ought to be adopted.

Cognitive Models

Hopelessness

The "hopelessness" hypothesis advanced by Beck (1967) suggests that individuals develop schemas (or perspectives) based on their experiences, which influence the way events are interpreted (Beck, 2008). Negative events experienced early in life might contribute to the development of dysfunctional self-referential schemas ("I'm unworthy" or "This is all my fault, and always will be"). Thus, when individuals encounter challenges at some later time, these schemas promote negative reactions, which become progressively more entrenched. Biased information processing evolves so that memories of past events are viewed more negatively, and individuals may develop pervasive pessimistic expectations of the future. This bias may become increasingly pronounced to the extent that individuals will selectively attend to stimuli that are consistent with their negative perspectives, whereas positive events and evidence inconsistent with their negative appraisals are filtered or ignored. Eventually, streams of negative "automatic thoughts" emerge spontaneously, which reinforces their poor self-esteem and self-worth, and encourages feelings of hopelessness and negative rumination.

Framing Hopelessness in a Neurobiological Context

A neurobiological accounting was provided that mapped onto the cognitive aspects of a hopelessness view (Disner et al., 2011). It was reasoned that multiple brain regions underlie depressive disorders, concurrently operating in both a bottom-up and a top-down fashion. Limbic brain regions were seen as regulating the emotional aspects of depressive disorders (bottom-up components), whereas cognitive disturbances were attributed to the top-down cortical process. According to this model, depression evolves when inhibitory control (e.g., coming from the prefrontal cortex) is limited so that unrestrained neuronal activation occurs at brain sites (e.g., aspects of the amygdala) that allow for the predominance of disturbed emotional processing. In essence, this disorder comprises the abdication of cortical control mechanisms that ordinarily limit negative automatic responses, together with a bias toward negative appraisals of experiences.

While disturbed medial prefrontal cortex functioning was seen as being fundamental for the regulation and control of self-referential schemas, the dorsolateral prefrontal cortex was taken to be responsible for rumination and impaired processing, and biased attention was determined through the ventrolateral prefrontal cortex. Disturbed medial orbitofrontal cortex functioning in depressed individuals was implicated in the appraisal of rewarding events, and the functioning of the lateral orbitofrontal cortex may be fundamental for translating nonreward and punishing events, thereby contributing to the development of a negative sense of self and poor self-esteem (Stalnaker et al., 2015). Also, disturbed functional connectivity involving a frontoparietal network contributes to cognitive inflexibility and ineffective coping.

Aspects of the amygdala were taken to contribute to emotional memory and disturbed hippocampal functioning may contribute to biased episodic memory (i.e., those pertaining to autobiographical events, including times, places, and contextual knowledge) and may foster negative biases and negative information processing. Furthermore, the anhedonia associated with depression was attributed to ventral tegmental and nucleus accumbens neuronal responses being blunted, so that the appraisal and recognition of positive stimuli would be impaired. Thus, depression could be seen as a combination of diminished reactivity to rewarding stimuli and increased responsivity upon failing to receive a reward or to reach desired goals, together with the retention and focus on memories of low reward (Rolls et al., 2020). While newer frameworks have been advanced since the inception of this model, it continues to be among the most detailed and inclusive.

Cognitive Behavioral Therapy

The development of cognitive behavioral therapy (CBT) in the treatment of depression evolved from the hopelessness perspective, with the goal of using behavioral and cognitive methods to diminish dysfunctional behaviors, cognitions, and emotions. Through this therapy, individuals learn to challenge inappropriate and counterproductive patterns and beliefs and to replace cognitive errors, such as overgeneralization, magnification of negatives, catastrophization, and minimization of positives, with thoughts that are realistic and functionally effective. Through a series of steps, individuals learn to challenge their way of thinking and to alter their reactions to entrenched habits and behaviors. Ultimately, patients come to appraise stressors appropriately, contextualizing stressors and their depression, and coping effectively. It was consistently reported that CBT was as effective as pharmacotherapy in treating depression, and when used as an adjunct to standard pharmacotherapy, it had lasting beneficial effects (Wiles et al., 2016). While originally developed to diminish depression, CBT or variants of the procedure that are tailored for specific conditions have found their way into the treatment of several other health challenges, including anxiety disorders, posttraumatic stress disorder (PTSD), eating disorders, obsessive-compulsive disorder (OCD), substance use disorders, and psychotic illnesses.

CBT has contributed to the rise of other therapies, such as mindfulness meditation, which may diminish depression and depressive relapse (Kabat-Zinn, 2019; Segal and Walsh, 2016). Alternative strategies, such as "Behavioral Activation" therapy, encourage patients to focus on activities that are personally meaningful to them based on their values. Although it hasn't been as studied as extensively, Behavioral Activation was reported to be as effective as CBT (Richards et al., 2016). A related approach, Acceptance and Commitment Therapy (ACT) has likewise been effective in treating depressive disorders by having individuals become more aware and focused on their therapeutic goals and more engaged in reaching these by facing problems head-on. Through a combination of mindfulness skills and self-acceptance, individuals become more accepting of their thoughts and feelings and acquire the ability to be psychologically flexible in dealing with negative issues. Although ACT may not be for everybody, several systematic reviews and meta-analyses supported its effectiveness in the treatment of depression (e.g., Bai et al., 2020). This procedure has also been effective in dealing with chronic pain, the distress among parents of children with developmental disabilities, and facilitated outcomes in weight loss programs.

Mindfulness: Flavor of the decade

Think in the moment (as opposed to ruminating over the past or worrying about the future) and don't be judgmental are the mantras of mindfulness meditation. Mindfulness can reflect a dispositional (trait) characteristic wherein some individuals ordinarily engage in mindful behaviors. These individuals exhibit lower emotional reactivity in response to negative experiences and display lower neuroendocrine and inflammatory responses, which are associated with diminished distress among patients with severe illnesses (Marinovic and Hunter, 2022).

A meta-analysis confirmed that positive outcomes were produced by mindfulness in most studies, and when administered to groups, it was as effective as individually administered CBT in diminishing psychiatric symptoms (Sundquist et al., 2017). As encouraging as this seems, many of the published reports were underpowered, and negative findings were less likely to be published (just as this is ordinarily the case with other therapeutic approaches), thereby biasing the conclusions coming from meta-analyses.

Mindfulness training isn't ideal for everybody. Some individuals may be drawn to this type of therapy, whereas for others it might be a turn-off. This shouldn't be viewed as castigation of this procedure. Instead, it simply means that patients need to seek out the treatment with which they are comfortable, just as clinicians need to determine whether their patients are suitably inclined to this therapeutic approach.

Helplessness

The "learned helplessness" hypothesis has been a popular cognitive view of depression (Maier and Seligman, 2016). This position, initially developed based on studies in animals, proposed that when uncontrollable experiences are encountered, people learn that their responses and outcomes are independent of one another and consequently adopt a cognitive perspective that they are helpless in controlling events. If uncontrollable negative experiences are encountered in important situations, which include failure to reach important goals, the feelings of helplessness could eventually lead to depression.

The helplessness hypothesis was not uniformly accepted, especially as little attention was paid to the fact that only a small proportion of individuals who experience uncontrollable stressors become depressed. An elaboration of this hypothesis to deal with the individual differences focused on the cognitive attributions that are made in response to failure (Abramson et al., 1978). Individuals were described as differing in three fundamental appraisal dimensions that were linked to "locus of control" (the extent to which individuals perceive themselves as having control over events that affect them).

Upon encountering a failure experience, individuals make appraisals and attributions as to why this occurred, which in specific combinations would be more or less likely to favor the emergence of helplessness. If individuals form certain internal, stable, and global attributions regarding their inabilities ("It's my fault that I failed; that won't ever change, and I'll continue to fail; I'm no good at anything"), they may develop broad feelings of inadequacy and poor self-esteem. This may culminate in cognitions and feelings related to helplessness, which then encourage the emergence of depressive disorders.

In clinically significant cases of depression, patients often describe themselves as feeling helpless and their situation as hopeless. However, are these feelings symptomatic of the illness rather than being a causal agent? It is conceivable that feelings of helplessness in response to uncontrollable experiences represent one of the early symptoms associated with depression, or that feelings of helplessness might be most readily induced by stressors among individuals with subclinical levels of depression. Despite these caveats, the learned helplessness model was widely adopted as being fundamental to the emergence of depression.

Neurochemical Perspectives of Depressive Disorders

Not long after the introduction of the learned helplessness position, several alternative views were advanced based on the neurochemical changes elicited by uncontrollable events (e.g., Anisman and Zacharko, 1982). Eventually, even the originators of the learned helplessness view came around to propose that the behavioral changes observed stemmed from neurobiological disturbances (Maier and Seligman, 2016). It was suggested that the resting state of the brain favors vigilance/caution, which might reflect an evolutionarily prescribed state to deal with potential hazards, and the passive response to uncontrollable events is not learned but instead reflects adaptive neurochemical processes being excessively strained, which favors the emergence of depression.

As sophisticated as these neurobiological views have been, analyses in humans are typically limited in the extent to which they can provide a sufficiently detailed understanding of the neurochemical changes that govern psychological disorders. As a result, much of what we know about these processes has come from studies using animal models of the disorder. A common critique of this is that depression may be a uniquely human disorder (although other species exhibit depressive-like behaviors with the loss of other animals with which they have bonded). To be sure, it is questionable whether rats and mice have the same feelings that humans do, and whether they are governed by attributions like those associated with depression in humans. While the symptoms of depression in humans can't be fully recapitulated in rodents, several models arguably comprise valid proxies of some aspects of the human disorder, and when used in parallel with human research, a better understanding of depressive illnesses may be possible.

The effects of uncontrollable stressors on later escape performance in rodents had for some time been a favorite model of depression (i.e., the learned helplessness paradigm). Although drug treatments that were effective in treating human depression, such as selective serotonin reuptake inhibitors (SSRIs) attenuated the behavioral disturbances elicited by an uncontrollable stressor, several drugs that were ineffective as antidepressants also had positive effects in this paradigm (Anisman et al., 2008b). Thus, the escape deficits introduced by uncontrollable stressors might not be a valid model of depression and would be more comfortably used as a screen for antidepressant agents.[1] As an alternative to assessing escape deficits engendered by uncontrollable stressors, increasing use of positively motivated behaviors has been adopted to evaluate the impact of diverse stressors. These have included responding to rewarding brain stimulation or intake of preferred snacks (e.g., pieces of cookie or sweetened solutions), which might be useful in assessing the anhedonia associated with a stressor experience. Likewise, disturbances of decision-making or memory were used to mimic cognitive disturbances related to the depressive profile.

Recent efforts have focused on using natural stressors to induce a depressive-like profile in rodents. A social challenge that has proven useful in this regard comprised changes in housing conditions to create social stress (e.g., transferring mice from grouped to isolated housing or vice versa) or exposing rodents to a predator.[2] An increasingly popular approach to model depression and anxiety has been to expose a dominant, aggressive mouse to a subordinate on single or repeated occasions, and then assess the neurobiological and behavioral changes that emerged. In general, paradigms that include naturalistic stressors and involve evaluation of behaviors that align with human depression (e.g., anhedonia) have been instrumental in determining neurobiological effects of stressors, as well as the genetic and epigenetic interactions that operate in mediating or moderating stressor actions.

Monoamine Processes Associated With Depressive Disorders

Identifying the specific processes governing psychological illnesses, as we've repeatedly indicated, is exceptionally difficult given that they involve complex actions of numerous neurobiological processes within several brain regions. Moreover, the processes involved in one form of depression may be distinct from that of a different form of the illness (e.g., dysthymia vs. major depression; typical vs. atypical depression), and even within a particular subtype, vastly different symptoms might be present. As a result, the efficacy of therapeutic strategies also varies between individuals. One gets the sense that we're dealing with several disorders that share common features comprising sadness or the feeling that life is not as pleasurable as it once was, but they differ in many other respects. As a result, the efficacy of therapeutic strategies ought to vary based on the symptoms expressed.

Several biological changes linked to depressive disorders may involve the sensitization of neurochemical processes stemming from earlier stressful experiences, particularly those that were uncontrollable and occurred chronically. The effects of recent stressors may also be primed by negative early life experiences and may vary with several genetic and epigenetic factors. As in the case of many other illnesses, a multihit hypothesis has been advanced wherein early life trauma or specific genetic factors serve as a first hit (or two hits), and a stressor encountered subsequently represents the trigger that takes the individual over the top (or perhaps, more appropriately, it takes the individual to the bottom) into a state of depression.

Norepinephrine

The involvement of norepinephrine in depression was congruent with the finding that stressful events markedly affected this neurotransmitter's activity, which was linked to anxiety that is so often comorbid with depressive disorders. Despite being displaced by serotonin as a primary player in depressive disorders, there has been a bit of a resurgence concerning the involvement of norepinephrine and specific receptors in this illness (Maletic et al., 2017). There is also evidence that norepinephrine alterations provoked by stressful events could steer

---

[1] Behavioral tests, such as passivity that occurs in a forced swim test (Porsolt forced swim test), are useful as screens for antidepressants. Specifically, those treatments that ameliorate the behavioral disturbances *might* be effective as antidepressants, whereas treatments that do not modify the behavioral disturbances likely won't be effective. It is not unusual for this test to be inappropriately used to model depression rather than being used as a screen. There has been a movement to limit the use of forced swim because it is viewed as an unnecessarily intense stressor.

[2] Exposing mice or rats to a predator (or predator scents) has been reported to be effective in some studies, but in others, this manipulation did not produce appreciable effects. It may be that after generations of inbreeding and being raised in a laboratory, these critters lost their innate tendency to react to certain threats.

variations of other neurotransmitters (e.g., dopamine) and hence influence depressive-like behaviors.

Dopamine

The focus on dopamine in depressive illness was based on its role in subserving reward processes coupled with the findings that stressors affect dopamine functioning within key brain regions (ventral tegmentum, nucleus accumbens). In rodents, manipulations that elevated dopamine (DA) functioning effectively reduced the impact of stressors on depressive-like behaviors, and alterations of specific dopamine receptors (e.g., D_1 within cortical regions) could instigate a depressive-like profile (Delva and Stanwood, 2021). Adverse early life experiences were also found to alter the gene expression profile within the nucleus accumbens, and it was surmised that these changes might account for the frequent failures of antidepressants to alleviate depression (Parel et al., 2023).

While stressors initially promote excitation of neuronal activity within the ventral tegmental-nucleus accumbens pathway, with chronic stressors dopamine circuits are overwhelmed, so that coping and reward processing falter, thereby favoring the occurrence of depressive disorders (Douma and de Kloet, 2020). In addition to being involved in reward processes, DA neuronal activity within this circuit might make negative events more salient and thus take on greater significance (Olney et al., 2018). Moreover, while anhedonia may stem from dysfunction of the ventral tegmental-nucleus accumbens pathway, aspects of coping may be related to disturbances involving the ventral tegmental-prefrontal cortex pathway.

Consistent with other sources of evidence, genetic links involving dopamine and depression have been reported. Genes related to the dopamine transporter were tied to depression, and depression in both animal models and in humans was associated with reduced expression of the gene *Slc6a15* on particular dopamine (D_2) receptors in brain regions that regulate reward processes and may be particularly sensitive to stressors (Chandra et al., 2017). Dysregulation of dopamine D_1 and D_2 receptors was also associated with suicide, which may be related to early life adversity (Fitzgerald et al., 2016).

As dopamine-mediated anhedonia is a key symptom of depressive disorders, it had been expected that increasing DA levels through L-DOPA treatment would diminish depressive symptoms. However, this agent was ineffective in diminishing depression, which undercut interest in this transmitter's role in depression (a scientific version of "tossing the baby out with the bathwater"). It has since been demonstrated that several drugs, such as bupropion (Wellbutrin), that increase DA functioning were moderately effective antidepressants, but without side effects related to disturbed sexually related behaviors or weight gain (Patel et al., 2016a).

Considerable data has pointed to an intimate interplay between DA and immune functioning in promoting depressive symptoms. Dopamine may affect various types of immune cells through D_1 and D_2 receptors present on their surface. Whereas D_1 receptors on T cells promote inflammatory cytokine release, D_2 receptor activation promotes the release of antiinflammatory cytokines. As well, DA has immunomodulatory effects through actions on chemokines and oxidative processes, thereby affecting central nervous system (CNS) neuronal activity (e.g., Channer et al., 2023). In fact, upon being activated, DA receptors present on brain microglia and astrocytes cause inflammation, which may contribute to depression as well as to neurodegenerative and neurodevelopmental disorders such as attention deficit hyperactivity disorder (ADHD), as well as schizophrenia. These relationships are bidirectional in that elevated cytokine levels can disturb DA neuronal connectivity within the corticostriatal reward circuitry (Felger et al., 2016).

Beyond the involvement of D_1 and D_2 receptors in inflammatory processes, there is a reason to believe that D_3 receptors may contribute as well. Specifically, depressive-like effects provoked by an endotoxin (lipopolysaccharide; LPS) could be attenuated by administration of a D_3 agonist, which prevented proinflammatory alterations in the ventral tegmentum and nucleus accumbens. As expected, a D_3 antagonist could induce depressive-like effects along with elevated inflammation in brain regions associated with a reward (Wang et al., 2018a, b). Commensurate with the animal findings, in treatment-resistant patients with elevated levels of interleukin-17 (IL-17), administration of the D_3 agonist pramipexole was therapeutically beneficial (Fawcett et al., 2016).

Serotonin

Depressive-like behavioral impairments in rodents produced by stressors are often accompanied by altered serotonin turnover and levels, and SSRIs administered repeatedly attenuated the behavioral deficits elicited by uncontrollable stressors. It had initially been assumed that the beneficial effects of SSRIs came about because of the inhibition of serotonin reuptake, resulting in increased availability of serotonin at the synapse. It is now understood that SSRIs have effects beyond that of inhibiting serotonin reuptake, which might account for their antidepressant actions. For example, in mice genetically engineered so that the functioning of the serotonin transporter (5-HTT) was altered, $5-HT_{1A}$ receptors were also affected. It is possible that 5-HTT dysfunction might be responsible for some features of depression, whereas $5-HT_{1A}$ receptor variations may contribute to other symptoms. Supporting this view, manipulation of postsynaptic

5-HT$_{1A}$ receptors across multiple brain regions influenced a depressive-like state in animal models (Yohn et al., 2017). Imaging studies indicated that 5-HT$_{1A}$ receptors were greater in the hippocampus of depressed individuals who experienced adverse early life events (Bartlett et al., 2023). Importantly, 5-HT$_{1A}$ receptor binding potential could predict the effectiveness of pharmacotherapy in diminishing depressive symptoms.

These findings were consistent with reports indicating that stressful experiences influenced prefrontal cortical 5-HT$_{1A}$ expression, which could be altered by repeated antidepressant treatment. Moreover, the elevated 5-HT$_{1A}$ was tied to epigenetic changes that occurred and stressful experiences provoke methylation of receptor expression, which can limit the efficacy of serotonin-acting antidepressants (Albert et al., 2019). Using specific biomarkers, it may be possible to predict strategies by which manipulations of the inhibitory actions of 5-HT$_{1A}$ autoreceptors (which ordinarily stimulate 5-HT release) and activating 5-HT$_{1A}$ heteroreceptors (that have a stimulating action) could diminish resistance to SSRI acting agents (Vahid-Ansari et al., 2019).

In addition to the involvement of 5-HT$_{1A}$ receptors, it was proposed that 5-HT$_{1B}$ receptors present on cholecystokinin inhibitory interneurons of the dentate gyrus are fundamental for the initiation of a cascade of changes that are responsible for the therapeutic benefits stemming from antidepressant treatments (Medrihan et al., 2017). A case had also been made for 5-HT$_{2A}$ receptor involvement in depression (Carhart-Harris and Nutt, 2017), which was supported by the finding that 5-HT$_{2A}$ receptor abundance was elevated within the anterior prefrontal cortex in postmortem tissue of depressed individuals (Shelton et al., 2009). However, DNA microarray analyses failed to detect 5-HT$_{2A}$ gene differences within the prefrontal cortex of individuals who died of suicide and controls (Sibille et al., 2004). It may be significant that among individuals with particularly elevated feelings of pessimism and hopelessness, 5-HT$_{2A}$ binding was elevated within the dorsolateral prefrontal cortex (Meyer et al., 2003). Despite some inconsistent findings, the case has been made that 5-HT$_{2A}$ alterations, through effects on multiple neurochemical systems, can affect depressive symptoms and might serve as an attractive target in the development of new therapeutic agents (Zięba et al., 2021).

Genetic Links to Depression

It has been known for decades that children of depressed parents are at increased risk of early-onset depression and a high incidence of recurrent depression (Weissman et al., 2016). It had been hoped that once the specific genes that operate in depression were identified, better treatment methods would follow. However, this has proven to be difficult given the biological heterogeneity of depressive disorders. To make the complexity still greater, research in animals and humans indicated that males and females may differ considerably in the biological substrates associated with depression. Among chronically stressed mice, the expression of numerous genes was altered relative to nonstressed mice (e.g., those linked to executive and reward processes), and these occurred in a sex-specific fashion (Labonté et al., 2017). Regulation of proinflammatory cytokines within the hippocampus and amygdala in response to a chronic stressor similarly differed between males and females (Barnard et al., 2019).

Analyses of gene expression in several brain regions in human postmortem tissues revealed marked differences between depressed males and females, although many of the downstream pathways that were linked to depression were comparable in the two sexes. A detailed analysis in humans indicated sex differences in 13.2% of brain proteins, and across multiple psychiatric and neurologic conditions approximately 25% of expressed genes could potentially be responsible for differences in protein abundance that might have been causally linked to the sex differences in illness occurrence (Wingo et al., 2023).

Early genetic studies of depression often comprised relatively small sample sizes, and thus had limited power, possibly contributing to the frequent failures to replicate the findings. Of course, the inconsistent reports may have been due to symptom and biochemical heterogeneity that wasn't considered in many studies. It is equally possible that some nondepressed individuals may well have carried genes that were linked to illness, but their impact was limited by their social and developmental experiences.

These caveats notwithstanding, a polymorphism was identified that linked depression to a gene controlling the enzyme tryptophan hydroxylase-2, which is fundamental in 5-HT synthesis. This polymorphism also predicted diminished SSRI responsiveness, supporting the involvement of this enzyme in affective disorders. A polymorphism was similarly detected on the promoter region of the gene regulating the 5-HT$_{1A}$ receptor that was associated with depression (Albert et al., 2014), and a polymorphism related to the norepinephrine uptake process and that of 5-HT$_{1A}$ was linked to diminished hippocampal volume in treatment-resistant depressed patients (Phillips et al., 2015). It should be underscored that these genetic features were not unique to depression because a polymorphism within the promoter region of the 5-HT$_{1A}$ gene was linked to a range of psychiatric disturbances, including suicidality and substance use. Given that most psychiatric illnesses involve multiple factors, each accounting for a modest portion of the variance, the 5-HT$_{1A}$ polymorphism may be one among many neurobiological factors that contribute to diverse illnesses.

For a time, attention had been devoted to the finding that depression was associated with a gene promotor polymorphism for the 5-HT transporter (5-HTTLPR) (Caspi et al., 2003). Specifically, individuals carrying polymorphism on the alleles of the 5-HTT gene might be especially significant in predicting depression. These alleles can be either long (l) or short (s), based on the long allele having a segment that has 14 repeats, whereas the short allele has 12 repeats. Individuals carrying the short form of the 5-HTT promoter on one or both alleles were reported to be at greater risk for the development of depression if they also experienced major life stressors or early life trauma (Caspi et al., 2003). It seems that among individuals carrying the short allele, amygdala and hippocampal functioning was tonically (continuously) activated, as might be expected under conditions of hypervigilance, threat, and rumination, and the neuronal response to negative events was exacerbated. In essence, inheriting specific genes didn't condemn individuals to depression, but instead created conditions wherein individuals were more sensitive or reactive to stressors, thereby favoring the appearance of depression (Caspi et al., 2010).

As exciting as these reports had been, data inconsistent with the initial findings were frequently reported, and overviews of the relevant literature questioned the reliability of the findings given that most studies were underpowered. To add to the confusion, several meta-analyses came to very different conclusions, depending on whether the analyses included chronic and recurrent depression and whether the studies included in the analyses were tainted by reliance on retrospective, self-report measures of life stressors, rather than being based on objective measures of stressor experiences (Karg et al., 2011). Also, depression associated with the short allele may have been influenced by the presence of anxiety, the age of individuals, as well as the type and intensity of life stressors (Juhasz et al., 2015).

A comprehensive analysis that included 31 independent data sets, meant to be the last word on the subject (even though there's rarely a last work on virtually any subject), did not support the contention that carrying the s allele conferred increased vulnerability to depression in the presence of stressors. This was apparent irrespective of whether current or lifetime illness, childhood maltreatment, or broad stressor experiences during adulthood were considered (Culverhouse et al., 2018). A further analysis based on large data sets and case-control studies similarly did not find evidence of any interactions between polymorphisms and either traumatic life events or negative socioeconomic conditions being linked to depression (Border et al., 2019).

It is tempting to argue that the story linking 5-HTTLPR and depression is actually a nonstory in the sense that the original effects observed were spurious. Yet, the 5-HTTLPR × Stress interaction was apparent in the neuronal connectivity of executive control brain regions and activity within the default mode brain network (van der

Meer et al., 2017). Furthermore, the interaction between stressful events and the short allele for the 5-HTT promoter gene was not limited to predicting depression but was also linked to OCD, anxiety, and PTSD. Ongoing distress associated with chronic pain was also more likely to result in depression among individuals carrying the 5-HTT polymorphism (Hooten et al., 2017). Thus, the combination of the 5-HTT polymorphism and stressor experiences might favor a general susceptibility to illness, which favors a range of illnesses. In essence, it could be argued that there have been too many positive reports concerning the importance of 5-HTT polymorphisms to dismiss them, and there is more to the findings that still need to be unearthed.

Another gene, p11, which functions to move certain serotonin receptors from inside the cell to the cell surface, has been implicated as a player in the development of depression, possibly through interactions with the 5-HT$_{1B}$ receptor (Svenningsson et al., 2013). Postmortem analyses indicated that among male and female depressed individuals who had died of suicide, mRNA expression of p11 was diminished in several brain regions (frontopolar cortex, orbital frontal cortex, hippocampus, amygdala) concurrent with 5-HT receptor mRNA variations in some of these regions (Anisman et al., 2008b). It was proposed that reduced p11 favors depression through the actions of BDNF, and there is reason to believe that by affecting IFN-α, p11 could influence 5-HT$_{1B}$ and 5-HT$_4$ receptor functioning (Guo et al., 2016). In rodents with the p11 gene knocked out, HPA functioning and autonomic reactivity were elevated, and anxiety and depression were increased with stressor exposure (Sousa et al., 2021).

In addition to being related to the presence of depression, p11 levels predicted the effectiveness of antidepressant therapies. For instance, elevated levels of p11 in cytotoxic T cells predicted a better response to treatment with ketamine among treatment-resistant patients (Veldman et al., 2021), whereas low levels of p11 in natural killer cells and monocytes were accompanied by a poorer response to citalopram treatment (Svenningsson et al., 2014). Elevated methylation of the p11 promotor measured in blood predicted a better response to electroconvulsive therapy (ECT), possibly reflecting a characteristic related to depression that makes them responsive to this treatment (Neyazi et al., 2018). Together, the data have supported p11 involvement in depression and circulating p11 could be useful as one of several biomarkers to predict treatment efficacy.

In keeping with the complex nature of depressive illnesses, genome-wide association studies (GWAS) revealed that multiple genes contribute to this disorder. A meta-analysis that combined participants from three large GWAS identified 269 genes, 102 independent variants, and 15 gene sets that governed neurotransmitters and synaptic structure, which were associated with major depression (Howard et al., 2019). Genetic analyses have also implicated inflammatory factors in depressive disorders. Whole exome sequencing indicated that treatment-resistant depression was associated with genes that modulated immune responses, cell survival, and neurodegeneration (Fabbri et al., 2020). A compilation of such studies implicated IL-6-related processes to depression, and based on a Mendelian randomized analysis, it was concluded that this relationship was a causal one (Kelly et al., 2021). Furthermore, high-throughput "omics" approaches, which provided a broad view of treatment-resistant depression, indicated that the factors most closely linked to this condition comprised altered inflammatory mechanisms, disturbed neuroplasticity, and altered neuronal signaling processes (Amasi-Hartoonian et al., 2022).

Effectiveness of Antidepressant Medications

The early findings indicating that SSRIs could ameliorate depression had been met with considerable enthusiasm. It was thus disheartening that the initial optimism concerning the effectiveness of SSRIs was largely wishful thinking, fueled by media hype. Human trials indicated that the treatment efficacy of SSRIs was modest with about 50%–60% responding positively to these treatments. Some studies suggested that drug efficacy was much lower, hardly being distinguishable from that associated with placebo treatment (Kirsch, 2014). Generally, antidepressant effectiveness was modest among patients with moderate levels of depression and was still less effective in reducing severe depression. Understandably, these reports created a considerable stir, and it was often suggested that the case against antidepressants was overstated. Indeed, subsequent meta-analyses suggested that at moderate doses, antidepressants could be effective while limiting side effects that would favor patient dropouts (e.g., Furukawa et al., 2019). In adolescents, SSRIs and other antidepressants, especially when combined with behavioral therapy, yielded relatively good outcomes (Dwyer and Bloch, 2019), although it had been maintained that these agents increased the risk of suicide in this population.[3]

It is of practical significance that among individuals experiencing depression secondary to a chronic illness, such as kidney disease or cancer, SSRIs are largely ineffective in modifying mood (e.g., Ostuzzi et al., 2018) and the effects in patients with traumatic brain injury (TBI) were small (Silverberg and Panenka, 2019). It is concerning that heart attack patients who received SSRIs at the time of discharge were at increased risk of mortality over the ensuing year. In contrast, the alleviation of depression through psychotherapy was accompanied by diminished risk of future heart problems (El Baou et al., 2023). It hardly needs to be said that the choice to use antidepressants under these and other health conditions that promote depression must be made judiciously on a case-by-case basis.

The limited effectiveness of SSRIs, in retrospect, shouldn't have been all that surprising given that the biological underpinnings of depressive illnesses involve multiple mechanisms that vary across individuals. This view prompted multitargeted approaches in the treatment of depressive disorders. Pharmaceutical companies that at one time focused on increasingly more "specific" treatments, possibly to find "the magic bullet" to treat the illness, or hoping to limit side effects, reverted to the development of drug combinations that concurrently affected several neurotransmitters (e.g., serotonin and norepinephrine uptake inhibitors, SNRIs, and triple reuptake inhibitors that affect the three key monoamines). Considerable evidence has attested to the benefits derived from concurrently prescribing several drugs to treat the illness if this is done based on patients' symptoms (Blier, 2016a, b).

Aside from the 2–3 weeks needed for positive effects to emerge, further limitations of standard antidepressant therapies exist. Several drugs are often prescribed before one is found to be effective, and even then, depressive symptoms may not be entirely ameliorated as residual symptoms frequently persist, sometimes presaging illness recurrence. Furthermore, the effects of an SSRI in diminishing the adverse effects of a chronic stressor regimen in mice were most apparent if they were maintained in an enriched environment during the antidepressant treatment. When the two treatments (medication plus enriched environment) were combined, the reduced depression-like behavior was accompanied by elevated BDNF, and HPA axis activity normalized (Alboni et al., 2017).

Findings such as these point to the effects of SSRIs not simply resulting from biological alterations, but instead contextual factors moderate the actions of the drug treatment so that the beneficial effects are realized more readily. This doesn't imply that positive environments are essential for the beneficial effects of the treatments, but it does suggest that SSRIs might be most efficacious when accompanied by positive environments or cognitive types of therapy or having the two therapies provided sequentially.

---

[3] As a result of the concerns, in 2003, a black box warning was issued concerning the safety of antidepressants among individuals under 24 years of age. The unintended consequence of this was that younger people were less likely to seek treatment for depression, so that the steady decline of youth suicide that had preceded the black box warning reversed course and ultimately exceeded the rates prior to antidepressant restrictions (Isacsson and Rich, 2014).

Antidepressant use during pregnancy

It has been estimated that 2%–3% of pregnant women use antidepressants, with SSRIs being the most common. There had been concern that SSRIs and other antidepressant agents could affect the developing fetus. In fact, antidepressant use was associated with preterm delivery and low birth weight, as well as reduced Apgar scores in newborns, increased occurrence of seizures, respiratory distress, and persistent pulmonary hypertension (Tak et al., 2017) as well as a moderate increase of heart abnormalities. There were also indications that antidepressant use by pregnant women might be associated with diminished brain volume in offspring (Koc et al., 2023), and several studies suggested that antidepressants may increase the occurrence of autism spectrum disorder (Boukhris et al., 2016), although the data concerning antidepressant linkages to these disorders have been inconsistent.

Prenatal antidepressant use was associated with more frequent depression in offspring. This outcome, of course, might be related to the heritability of depression rather than the effects attributable to the drugs' use during pregnancy. This said, a long-term analysis of more than 900,000 individuals indicated that antidepressant use among pregnant women was associated with elevated incidence of a range of psychiatric conditions (Liu et al., 2017b), leading to the suggestion that focusing simply on depression in offspring may be too conservative. It is tempting to come down on the cautious side and recommend against the use of antidepressants during pregnancy. Yet, depression during this time may be accompanied by increased oxidative and nitrosative stress biomarkers, and altered neuroimmune activity (Roomruangwong et al., 2018). Thus, left untreated, depression in pregnant women may have diverse negative consequences for the fetus, including depression that occurs years later, together with microstructural changes and reduced amygdala volume (Wen et al., 2017).

All things considered, given the potential risks of SSRIs balanced against those of depression among pregnant and lactating women, SSRIs might still be considered the treatment of choice (Fischer Fumeaux et al., 2019). That said, a reasonable approach for women at risk for depression (based on earlier depressive episodes) would be one of having the depression treated through psychotherapy or other behavioral methods, or perhaps transcranial magnetic stimulation (more on this later) before resorting to drug treatments.

Gamma-Aminobutyric Acid and Glutamate Processes in Depressive Disorders

Interest in gamma-aminobutyric acid (GABA) activity as a player in depressive disorders and comorbid anxiety has been an off-and-on affair. As we've seen, GABA is markedly affected by stressors as are changes in the composition of the $GABA_A$ receptor. Reinforcing the link to depression, GABA levels were relatively low in the cortical regions of depressed individuals and were especially notable among severely depressed (melancholic) patients and were tied to treatment resistance (Godfrey et al., 2018). Likewise, levels of the enzyme involved in GABA synthesis, glutamic acid decarboxylase, were altered within the hippocampus and prefrontal cortex of depressed patients. Imaging studies in unmedicated depressed patients similarly revealed that GABA functioning was diminished within the prefrontal, anterior cingulate, and occipital cortices (see Fogaça and Duman, 2019). A genomic analysis confirmed that among individuals who died of suicide, GABAergic-related gene expression was altered in several cortical and subcortical brain regions, including the prefrontal cortex and hippocampus. However, this was apparent irrespective of whether these individuals had been depressed, and hence the GABA changes were likely linked to suicide rather than depression itself (Sequeira et al., 2009).

The $GABA_A$ receptor, it will be recalled, comprises five subunits that come from a much larger set of subunits. Whether a particular drug (e.g., alcohol or benzodiazepines) influences the receptor is determined by the subunit conformation, and it has been considered that stressors and depression might similarly influence specific GABA subunits. Aside from frank changes in $GABA_A$ receptors, within several limbic regions of depressed individuals who died of suicide, the mRNA expression of $GABA_A$ subunits may be upregulated (Merali et al., 2004; Poulter et al., 2010), but more than this, $GABA_A$ subunit expression "patterns" were altered (Merali et al., 2004). Ordinarily, among nondepressed individuals who had died suddenly of causes unrelated to suicide, the mRNA expression of the various $GABA_A$ subunits was highly correlated with one another within several brain regions (e.g., in the prefrontal cortex, hippocampus, and amygdala). In contrast, among the depressed individuals who died of suicide, the $GABA_A$ subunit interrelations were markedly lower. Coordination between these subunits may be needed for the integration of neural networks and for the neuronal rhythms that ordinarily occur (Poulter et al., 2010). The disintegration of the subunit coordination could thus favor depression being present, although the subunit relations may be secondary to depression.

Among depressed individuals and chronically stressed rodents, many antidepressant therapies, including SSRIs, ECT, and transcranial magnetic stimulation, effectively normalized cortical and plasma GABA levels that were otherwise disturbed. Such actions may come about because GABA neurons contain receptors for diverse serotonin receptor subtypes, raising the possibility that the actions of SSRIs come about owing to effects on GABA activity (Fogaça and Duman, 2019). Indeed, in animal studies, GABA manipulations could diminish symptoms of depression. Specifically, GLO1, an enzyme that reduced methylglyoxal, which acts as a competitive partial agonist at $GABA_A$ receptors, was linked to depressive-like behaviors elicited by stressors. By inhibiting GLO1 for a limited time (5 days), antidepressant effects were achieved, which was associated with hippocampal BDNF variations (McMurray et al., 2017).

Preclinical and clinical data have pointed to the promising actions of neuroactive steroid $GABA_A$-receptor-positive modulators (i.e., that increase binding to GABA receptors) in the treatment of major depressive illness and peripartum depression (Luscher et al., 2023). A potential advantage of targeting GABA receptor functioning is that it can have rapid actions in diminishing mood disturbances. In this respect, a new compound, Zuranolone, which acts as a positive allosteric modulator of the $GABA_A$ receptor, diminished postpartum depression within three days after treatment commencement (Clayton et al., 2023). This compound (marketed as Zurzuvae), which is a synthetic form of allopregnanolone, is the first compound taken orally to obtain FDA approval for postpartum depression, but its cost may be prohibitive for those who are uninsured.

Glutamate

Processes that regulate glutamate clearance and metabolism and morphological changes within several brain areas have been identified that are associated with cognitive and emotional behaviors. In this respect, stressor-provoked glutamate changes involving the prefrontal cortex, hippocampus, and nucleus accumbens were implicated in the production of several illness symptoms, presumably occurring most readily among individuals who carry genetic vulnerabilities. The sources for the glutamate changes in depression have not been fully defined, but they have been attributed to the reduction in mitochondrial energy production or a decline of neuronal input or synaptic strength. A systematic review that included both animal and human studies pointed to mitochondrial disturbances brought about by stressors contributing to depressive disorders (Picard and McEwen, 2018).

The interactions between GABA and glutamate activation patterns could lead to disturbed engagement of prefrontal cortical regions that might underlie depression, as well as its treatment. As described earlier, both depression and chronic stressor exposure are accompanied by atrophy of neurons within the prefrontal cortex and hippocampus, as well as altered connectivity and network functioning. These actions may stem from the excitotoxic effects of glutamate together with the actions of GABA, elevated glucocorticoid levels, and excessive levels of inflammatory cytokines (Duman et al., 2019).

The flexible brain

Depressive symptoms may be determined by a shift in awareness related to the balance between external and internal mental focus, particularly increased self-focus and negative ruminations. Disturbances of inhibitory GABA regulation manifested by an imbalance between the default-mode network (DMN; operative when an individual isn't focusing on the external world, and for all intents the brain is in a state of wakeful rest) and executive networks may result in a shift in focus from external to internal mental content, thereby promoting features of depression (Northoff and Sibille, 2014). As well, hyperconnectivity within the DMN and the "salience network" could promote recurrent negative ruminations, leading to feelings of depression (Jacobs et al., 2014). In this respect, resting-state functional connectivity involving the dorsomedial prefrontal cortex and several other regions related to the DMN were tied to depression and other related disorders. Resting state connectivity within the DMN was also associated with the effectiveness of antidepressant treatments (CBT and antidepressant drugs) and could serve as a marker to predict treatment efficacy (Sun et al., 2023d). Functional connectivity within the DMN differed between adolescent males and females, and it seemed that the weak connectivity in females was aligned with altered reward-related brain activation and the emergence of depressive illness (Dorfschmidt et al., 2022).

From the symptoms expressed by depressed patients, it might be assumed that this illness is accompanied by diminished neuronal functioning or diminished connectivity. Contrary to this perspective, however, depressive disorders may be accompanied by hyperconnectivity involving several brain regions, possibly because of dysfunction related to several genes that direct cells through ubiquitin, which has prominent effects on inflammatory processes. As much as having many connections might be advantageous in some situations, at some point connectivity may become excessive, resulting in disturbed selectivity concerning the interconnections that need to be activated, and hence miscommunication may occur within neuronal circuits. Also, having too many associations in memory circuits could lead individuals into negative ruminative loops where they are unable to turn off counterproductive thoughts—there may be many neuronal roads, but they all seem to lead to the same bad place. Resilience calls for cognitive flexibility so that individuals can move from one strategy to another as the situation demands. As GABA is fundamental in inhibiting messages between neurons, dysfunction related to this transmitter may contribute to the inflexibility and selectivity of neuronal exchanges that are evident among depressed individuals.

Ketamine and Related Fast Acting Antidepressant Compounds

A dramatic shift in the focus of antidepressant treatments came about with the unexpected findings regarding ketamine, an N-methyl-D-aspartate (NMDA) receptor antagonist that is frequently used as a general anesthetic and analgesic in veterinary practice, and as a street drug where it is known as Special K. Formal analyses of the effects of this agent confirmed anecdotal reports of ketamine's antidepressant effects, revealing that approximately 60% of patients who had been considered treatment resistant responded positively to ketamine (Blier, 2013; Krystal et al., 2023) and was as effective as ECT in this regard (Anand et al., 2023). In contrast to SSRIs, the actions of ketamine appear within 24 h, and greater effects are obtained with six infusions administered over 2 weeks (Phillips et al., 2020). The rapid actions of ketamine made it an especially valuable tool in treating depression, as ketamine has potent actions in diminishing suicide risk (Murrough et al., 2015).[4]

Conducting a placebo-controlled double-blind study to assess the effects of ketamine is impractical since patients who received ketamine will recognize that they are in the active drug group based on the feelings they experience. To deal with this, ketamine and placebo were administered to depressed individuals while they were anesthetized for routine surgery. Unexpectedly, both groups reported a decline in their depression, and the better they felt, the more likely they were to believe that they had received the ketamine treatment. It was suggested that patient expectancies influenced outcomes, just as this frequently occurs with other treatments (Lii et al., 2023). Yet, it should be considered that the antidepressant effects observed might have been related to the anesthetic used for surgery. Indeed, the anesthetic isofluorane had antidepressant actions in a subset of patients, possibly through actions like those of ketamine, including effects on BDNF-related processes (Antila et al., 2017).

The positive neurobiological effects of ketamine may stem from the downstream actions of glutamate NMDA receptor antagonism, followed by increased action at α-amino-3-hydroxy-5-methyl-4-isoxazolepropionic acid (AMPA) glutamate receptors. The antidepressant actions of ketamine on glutamate functioning could also occur through an increase of BDNF (Duman et al., 2021) or by affecting adenosine functioning, which impacts glutamate release. As well, the treatment promotes synaptogenesis and spine formation in frontal cortical regions through stimulation of the "mammalian target of rapamycin" (mTOR) complex, which regulates cell growth, cell survival and proliferation, protein synthesis, and transcription (e.g., Zanos and Gould, 2018). Aside from these processes, ketamine may produce its antidepressant actions by producing a disconnection from the individual's surrounding environment yet maintaining internal subjective experiences. This altered state of consciousness essentially results in the dissociation between sensory stimuli from those associated with cognitive functioning and the processing of affective information (Cichon et al., 2023).

The view was also advanced that the antidepressant actions of ketamine stemmed from effects on inflammatory cytokines, which may influence glutamate metabolism, thereby affecting mood symptoms (Haroon and Miller, 2016). Consistent with this, levels of serum IL-6 were useful in predicting ketamine's effectiveness and the benefits of ketamine may have stemmed from its effects on Th17 cells and the balance between Th17 and T_{reg} cells (Cui et al., 2021). Parenthetically, rumination, especially when accompanied by anxiety, is accompanied by elevated levels of circulating proinflammatory cytokines (Szabo et al., 2022). While the inflammation associated with rumination may be tied directly to depression, it may reflect the distress stemming from rumination.

Owing to the many failures encountered, numerous pharmaceutical companies have been reluctant to pursue new treatments for depressive disorders. Focusing on glutamate-acting agents may be one of the exceptions to this given the success seen with ketamine. Since the antidepressant actions of ketamine in humans are present for only a few days but can be modestly extended with repeated treatment, the goal of novel compounds includes increasing the duration of the effects, as well as developing methods of delivery other than by intravenous infusion. A compound derived from ketamine, that is esketamine, administered intranasally, diminishes treatment-resistant depression over the short term (Dean et al., 2021), although it might be less effective than ketamine in diminishing suicidal ideation (Jollant et al., 2023). Among treatment-resistant patients who received esketamine together with an SSRI, high rates of remission were observed after 8 weeks of treatment and were still evident after 32 weeks in about half of the treated patients (Reif et al., 2023).

Several related formulations, such as traxoprodil (CP 101606), GLYX-13, and CGP3466B, were developed that had rapid and potent effects without the negative side effects that other treatments might have elicited, although their positive effects dissipated relatively quickly. Another agent, methoxetamine, also had impressive effects in rodents, producing a rapid onset, and the effects lasted longer than those elicited by ketamine. The prospects for

---

[4] In a clever way of assessing the mood changes associated with ketamine, an analysis was undertaken of 41,000 patients prescribed ketamine or other drugs to deal with chronic pain. The incidence of depression, which is frequently comorbid with chronic pain, was 50% lower among ketamine users relative to those taking other drugs (Cohen et al., 2017a).

improved ketamine-like agents seem promising, even if further analyses of safety and efficacy of long-term use will be required, and it will be necessary to determine for whom the treatments will be most effective.

The buzz on other antidepressant treatments

Ketamine is not the only drug with mind-bending actions that has found its way into the clinic. Magic mushrooms, whose active ingredient psilocybin promotes antianxiety and antidepressant actions when administered with supportive therapy (Carhart-Harris et al., 2016; Davis et al., 2021). Even a single large dose of psilocybin diminished treatment-resistant depression that was sustained over 12 weeks (Goodwin et al., 2022). A multisite randomized controlled trial likewise indicated that psilocybin coupled with psychological support promoted a rapid and sustained reduction of depressive symptoms (Raison et al., 2023).

The benefits of psilocybin (and lysergic acid diethylamide, LSD) may come about through actions on $5\text{-}HT_{1A}$ and $5\text{-}HT_{2A}$ receptor signaling (Carhart-Harris and Nutt, 2017; Nutt et al., 2020). The $5\text{-}HT_{1A}$ activation was said to moderate stress responses, whereas $5\text{-}HT_{2A}$ receptor stimulation promotes cognitive flexibility and hence the capacity to change depression-related perspectives. Because of the chemical structure of psychedelic agents, they might be able to cross the plasma membrane so that they activate intracellular $5\text{-}HT_{2A}$ receptors, thereby increasing neuroplasticity (Vargas et al., 2023). Intriguing data have come from imaging studies indicating that the antidepressant actions of psilocybin were associated with an increase in connectivity between several brain networks that involve $5\text{-}HT_{2A}$ receptor functioning (Daws et al., 2022). Relatedly, psilocybin's effects on depression were associated with greater flexibility between brain networks without an increase in any single network. These actions were still evident 6 months later, perhaps reflecting psilocybin's actions in promoting psychological flexibility in which individuals can break away from fixed patterns of thinking (Garakani et al., 2023).

Several distinct neurobiological processes in addition to the 5-HT receptor changes may contribute to the effects of these agents. These include changes in glutamate and neurotrophins, and the resulting increased synaptic rewiring (Shao et al., 2021). As well, the effectiveness of the treatment was related to changes within key cortical components of the default mode network, essentially reflecting a "reset" of connectivity within this network (Carhart-Harris et al., 2017).

LSD and psilocin (the latter is derived from psilocybin) directly bind with BDNF TrkB receptors to a far greater extent than standard antidepressants and can have antidepressant effects independent of actions on $5\text{-}HT_{2A}$ receptors. Thus, compounds could be created with the potential of psychedelic agents, yet lacking the hallucinogenic actions otherwise produced (Moliner et al., 2023). Indeed, a nonhallucinogenic compound, 2-Br-LSD, could diminish the responses to stressors and attenuate depressive features (Lewis et al., 2023b).

It was observed that 3,4-methyl enedioxy methamphetamine (MDMA) was similarly effective in diminishing depressive symptoms, and MDMA-assisted psychotherapy could reduce features of severe PTSD (Mitchell et al., 2021). It is intriguing that MDMA increased social connectivity, and enhanced closeness to others (Molla et al., 2023), which could potentially diminish depressive symptoms. In addition to their antidepressant actions, psilocybin, LSD, and MDMA have been offered as a treatment for existential distress among individuals confronting life-threatening diseases, such as cancer, as well as in controlling substance use disorders (Vargas et al., 2020). Like psilocybin, positive effects were reported using ayahuasca, also known as yagé, which is consumed as a brew that has dimethyltryptamine present, and purportedly acts on monoamine oxidase (MAO) and BDNF, and thus has antidepressant qualities (Dos Santos et al., 2016). Ayahuasca also has antiinflammatory properties that are associated with its antidepressant actions (Galvão-Coelho et al., 2020).

Increasingly more information has pointed to the impact of psychedelics on inflammatory processes (Flanagan and Nichols, 2022). Emerging evidence indicated that psilocybin may promote antiinflammatory actions, including inhibition of tumor necrosis factor-α (TNF-α) binding to its receptor within the brain, normalizing within 7 days, and IL-6 and CRP were diminished over this time (Mason et al., 2023). It is too soon to attribute the antidepressant effects of psychedelics to inflammatory changes, but this possibility warrants further investigation.

In general, most psychedelic agents have tolerable physical side effects, such as headache, dizziness, and nausea, and suicidal ideation was observed in about 10%–15% of patients (Goodwin et al., 2022). The actions of these compounds were clinically assessed at doses lower than those taken by recreational users and were not evaluated for effects introduced by chronic use that may occur on the street. With the media hype concerning the benefits of psychedelics, individuals who are ill may choose to self-medicate, which can be disastrous. Not only is a therapist important to guide patients, but for individuals with a previous psychotic episode or who have a family history of psychosis, the drugs may cause pronounced destabilization.

Corticotropin-Releasing Hormone

With the recognition that corticotropin-releasing hormone (CRH) and its receptors were present at sites other than the hypothalamus, such as the prefrontal cortex, amygdala, and hippocampus, it was considered that depression and anxiety might stem from CRH changes at these sites (Holsboer and Ising, 2008). Consistent with this possibility, CRH was elevated in cerebrospinal fluid as well as in the prefrontal and dorsomedial prefrontal cortex of depressed individuals who had died of suicide, and mRNA expression of CRH_1 receptors was reduced (Merali et al., 2004). As expected, a single nucleotide polymorphism (SNP) on the CRH_1 gene was associated with a diminished therapeutic response to SSRI treatment among anxious depressed patients, and an SNP on the gene regulating CRH binding protein was similarly associated with a diminished therapeutic response to SSRIs (O'Connell et al., 2018).

Although CRH_1 receptors were thought to play a principal role in depression, it is premature to dismiss a role for CRH_2 receptors in mood disorders. As described in Chapter 5, CRH_2 or the interplay between CRH_1 and CRH_2 receptors may contribute to anxiety and could influence depressive states. In this respect, sex-dependent differences of stressors on CRH_2 receptor expression have been observed, possibly accounting for the male–female differences in depression.

The involvement of CRH in depression may occur owing to its indirect effects on inflammatory processes. Stressor-elicited activation of CRH processes could influence ATP-gated ion-channel P2X7 receptors (P2X7R) that are present in immune cells, thereby increasing brain inflammatory responses that promote depression (Silberstein et al., 2021). Supporting this possibility, the increase of brain microglia and diminished ventral hippocampal neurogenesis provoked by a chronic mild stressor were precluded among mice with the P2X7R knocked out (Troubat et al., 2021).

Despite impressive evidence supporting CRH involvement in anxiety and depression in animal models, the development of related pharmacological treatments to deal with these conditions has been slow in coming. The failure to develop therapeutics based on CRH_1 alterations may be attributable to several factors but if a treatment based on CRH variations is developed, it will be best to identify biomarkers that inform the nature of the biochemical disturbances, and then decide on whether this is an appropriate treatment option.

Arginine Vasopressin (AVP)

Limited attention has been devoted to AVP involvement in depressive illnesses, although AVP was elevated in the cerebrospinal fluid of depressed patients as well as within the locus coeruleus and dorsomedial prefrontal cortex of depressed individuals who died of suicide (Merali et al., 2006). In mice, blocking AVP receptors had both antianxiety and antidepressant-like effects,

and in small trials in humans, a positive effect was obtained with such a treatment, being most apparent among individuals with high cortisol levels (Chaki, 2021). Likewise, AVP V1B receptor antagonists at doses that reduced HPA activity resulted in diminished depressive symptoms (Chaki, 2021), provisionally suggesting that AVP might be linked to depression through HPA changes. Furthermore, the adverse effects of early life events in mice, which may engender a behavioral profile like that of depression and anxiety, have been related to epigenetic changes in a gene coding for AVP (Bodden et al., 2017).

Interestingly, AVP may act conjointly with oxytocin in contributing to social interactions that are relevant to diverse mental and developmental conditions (Abramova et al., 2020). The actions of oxytocin, which may be sex-specific, can be influenced by AVP alterations in the brain, which is consistent with the female bias for depression and anxiety that is ordinarily evident (Rigney et al., 2023).

Oxytocin

Owing to the broad actions of oxytocin and its contribution to health, this hormone was described as "Nature's Medicine," and the many attributes of oxytocin were described in an impressively detailed review (Carter et al., 2020). This included its profound effects on immune and inflammatory processes through which this hormone may act against multiple challenges directed at body tissues and the brain. These actions are fundamental for developmental processes needed in dealing with typical stressors encountered as well as in response to injury and traumatic experiences (Kingsbury and Bilbo, 2019), and serve in an antiinflammatory capacity during birth to protect the fetal brain from the hypoxic conditions that are encountered during labor.

Since oxytocin favors the development of social interactions and might contribute to the salience of stimuli becoming more significant, it might be expected that treatments that increase oxytocin functioning would be useful in diminishing depressive disorders and might be effective as an adjunct to standard medication. In line with this view, oxytocin variations can modulate dopamine functioning within brain regions supporting reward processes and limiting anhedonia (Xiao et al., 2017).

Plasma oxytocin levels were inversely related to depressive symptoms (Thul et al., 2020) and together with vasopressin were implicated in postpartum depression. Methylation of the oxytocin receptor gene at a site located near the estrogen receptor binding region was associated with postpartum depression. Furthermore, epigenetic changes in the oxytocin receptor might have played a role in the link between mood and estrogen production and may contribute to the occurrence of postpartum depression (Kimmel et al., 2016). Consistent with the aspects of this study, abuse in early life was associated with epigenetic changes in oxytocin receptor genes that were linked to psychiatric symptoms displayed in adulthood (Smearman et al., 2016).

The ties between oxytocin and depression aren't straightforward. Specifically, as described earlier, oxytocin enhances sensitivity to social stimuli, and may thereby contribute to the adverse consequences associated with negative social cues. In fact, in the presence of an oxytocin polymorphism and hence diminished oxytocin actions, the negative impact of adverse events, including poor early life care, may be diminished (McQuaid et al., 2014; Senese et al., 2022). This said, a prospective study indicated that among young adults who had received poor parenting, the presence of an oxytocin receptor SNP produced greater depressive symptoms (Keijser et al., 2021). Why such different results were obtained could be due to several factors, including the well-documented cross-cultural differences in the prevalence of oxytocin SNPs, and the specific polymorphism assessed (Senese et al., 2022). As well, the impact of epigenetic changes linked to oxytocin receptors could be influenced by varied psychosocial factors that could affect depression. For instance, among returning war veterans, an epigenetic change in the gene coding for oxytocin was negatively related to depressive illness, but this was overshadowed by the impact attributable to a lack of social connectedness (Warrener et al., 2021). In essence, as important as oxytocin functioning might be, in some instances social factors may promote greater actions than biological processes.

The possibility had been entertained that if oxytocin functioning was disturbed in patients, exogenous administration of oxytocin alone or in combination with other therapies might help diminish depressive symptoms. However, as described in Chapter 6, if depression was not related to disturbed oxytocin functioning, then administration of the hormone might aggravate illness features by increasing sensitivity to or salience of negative stimuli. Thus, in the absence of an understanding of the individual's basal oxytocin functioning, the use of this hormone in a therapeutic capacity can be risky.

Oxytocin alterations may also be a fundamental component of loneliness, which can be distressing, agonizing, and formulaic in promoting various illnesses (Cacioppo et al., 2015). Loneliness has frequently been tied to depressive disorders, and it can have adverse effects beyond that of depression itself, including earlier mortality. Levels of this hormone were implicated as being fundamental in social support that buffered against feelings of loneliness in depressed patients (Tsai et al., 2019) and enhanced the effects of CBT in diminishing loneliness. Furthermore, among older individuals, social engagement was accompanied by lower inflammation, whereas factors that promoted inflammation were associated with greater feelings of loneliness (Walker et al., 2019) and it appeared based on genetic analyses that the oxytocin tied to loneliness might be linked to heart disease (Winterton et al., 2022). A study of 3066 twins further indicated that stressor experiences were tied to loneliness, and that in men and women, both genetic and nonshared environmental factors accounted for the relation to feelings of loneliness (Moshtael et al., 2024).

As alluded to earlier, the actions of oxytocin may be related to effects on inflammatory processes. In rodents, oxytocin administered through a nasal spray reduced LPS-induced inflammation in microglial cells (Yuan et al., 2016) and protected against the depressive-like profile stemming from inflammatory actions that were attributable to adverse early life events (Amini-Khoei et al., 2017). Peripheral inflammatory responses provoked by stimulated macrophages could similarly be affected by oxytocin. By inhibiting inflammatory processes, oxytocin seems to facilitate the healing of damaged tissues and can diminish cardiovascular disturbances (Reiss et al., 2019).

Neuropeptide Y

Neuropeptide Y (NPY) has been implicated in promoting resilience in the face of stressors. In addition to possibly protecting against PTSD, it may limit depression and anxiety. The levels of NPY were lower in depressed patients than in controls and increased in response to antidepressant medications (Ozsoy et al., 2016). NPY alterations in cerebrospinal fluid were related to early life adversity, which was related to depression (Soleimani et al., 2014). Furthermore, greater affective responses to challenges were apparent among individuals with low NPY gene expression, and the presence of an NPY promoter polymorphism was linked to the presence of depressive-like features (Treutlein et al., 2017). Supporting these findings, analyses of brain tissue of depressed individuals who had died by suicide revealed a reduction of NPY mRNA expression and protein in the prefrontal cortex, as well as upregulation of NPY receptors within the hippocampus and prefrontal cortex (Sharma et al., 2022).

There hasn't been a great deal of research concerning the links between NPY and immune activity in depression. Nevertheless, immune cells can release NPY (Chen et al., 2020), and NPY influences cells of both the innate and adaptive immune system and affects the release of cytokines (Dimitrijević and Stanojević, 2013). The position was advanced that centrally, NPY operates to maintain appropriate regulation of emotional responses elicited by peripheral immune activation and might limit depressive-like behaviors by attenuating some of the inflammatory effects provoked by stressors.

Neurotrophins

The volume of the prefrontal cortex and hippocampus may be reduced in association with chronic stressor experiences and the occurrence of major depression (e.g., Belleau et al., 2019), especially in nonremitted patients and in those who experienced repeated or lengthy periods of illness (Nolan et al., 2020). These brain changes are aligned with disturbed coping methods related to decision-making abilities, so individuals generally resorted to previously established behavioral strategies (i.e., biases, habits), rather than framing and adopting new approaches. Paralleling these findings, early life stressful experiences, especially sexual abuse, which favors the development of later depressive illness, were accompanied by reduced grey matter volume within cortical brain regions.

The diminished hippocampal volume associated with depression might be related to lower levels of BDNF. Chronic stressor experiences in rodents, as we saw earlier, can influence neurotrophins, such as BDNF, within several brain regions. This can result in hippocampal neuroplastic disturbances, including interference with dendritic remodeling and a reduced number of synapses, which might contribute to the emergence of depression (Duman et al., 2021). Altered BDNF signaling within the pathway between the ventral tegmental area (VTA) and nucleus accumbens (NAc) reward circuit and altered BDNF receptors (tyrosine kinase B; TrkB) were similarly involved in the provocation of depressive-like outcomes elicited by chronic social defeat (Koo et al., 2016). Congruent with this notion, chronic stressor experiences that elicited depressive-like behaviors in rodents promoted BDNF gene methylation in the dorsal hippocampus while reducing methylation in the ventral hippocampus. Moreover, peripheral and central inflammatory factors may influence BDNF, which then affects neuronal processes that ultimately contribute to depressive disorders (see Fig. 9.2).

BDNF expression and protein levels and those of its TrkB receptor were reduced within the hippocampus and prefrontal cortex of depressed individuals who died by suicide (Park et al., 2019). In a small set of depressed individuals who died of suicide, BDNF mRNA expression varied across brain regions in a gender-dependent manner, being reduced within the prefrontal cortex among females who had been depressed, whereas a reduction of BDNF was noted in the hippocampus of males (Hayley et al., 2015). Reduced BDNF levels within the amygdala were also more prominent in females than in males, and the TrkB receptor for BDNF was lower within the anterior cingulate cortex of depressed individuals who died of suicide (Tripp et al., 2012).

Support for BDNF involvement in depression has come from reports that the positive effects of antidepressants were accompanied by elevated hippocampal neurogenesis in rodents and prevented the downregulation of hippocampal cell proliferation ordinarily provoked by stressors (Malberg and Duman, 2003). As expected, the positive behavioral effects of antidepressants were diminished among mice with targeted deletion of genes for BDNF, indicating that functional BDNF activity is necessary for the positive effects of antidepressants to emerge.

There are several ways by which antidepressants could affect BDNF functioning. Antidepressant treatments can directly influence TrkB receptors, thereby promoting the positive actions that emerge (Casarotto et al., 2021), but more often the positive effects of the treatment have been attributed to a series of changes that promoted neuroplasticity. It appeared that repeated antidepressant treatments and the resulting 5-HT$_{1A}$ receptor activation instigated a signaling cascade that promoted BDNF and vascular endothelial growth factor (VEGF) secretion, leading to increased proliferation of neural progenitor cells together with the differentiation and maturation of young granule cells within the dentate gyrus (Samuels et al., 2015). In essence, the beneficial effect of SSRIs might stem from the progressive changes of neuroplasticity associated with BDNF variations, thus accounting for the lag between treatment initiation and beneficial effects emerging (Johansen et al., 2023a).

Polymorphisms of genes associated with BDNF have supported the involvement of this neurotrophin in depressive illness. The severity of illness in drug-free patients was related to the presence of a specific BDNF SNP (rs6265) (Losenkov et al., 2020), and stressful life events and childhood adversity interacted with a BDNF polymorphism (i.e., the Val66Met polymorphism) in predicting the presence of depression (Zhao et al., 2018). While the presence of a BDNF SNP was not associated with the presence of depression in some reports, it may be more appropriate to assess the presence of the polymorphism in specific features of depression. Individuals with a polymorphism coding for BDNF expressed greater rumination following stressor experiences, which is predictive of a depressive propensity (Hosang et al., 2014). Likewise, the incidence of depression was elevated among adolescents carrying this SNP and who also displayed increased morning cortisol levels. These considerations aside, it is important that irrespective of whether the presence of the BDNF SNP was associated with the occurrence of depression, it was predictive of the response to therapy (Zelada et al., 2023).

In addition to BDNF, a case has been made for fibroblast growth factors (FGF)-2 involvement in depressive disorders. Stressors disrupt FGF-2 functioning, and this growth factor and its receptor were reduced by chronic stressors in rodents and were similarly diminished in the prefrontal cortex of depressed individuals

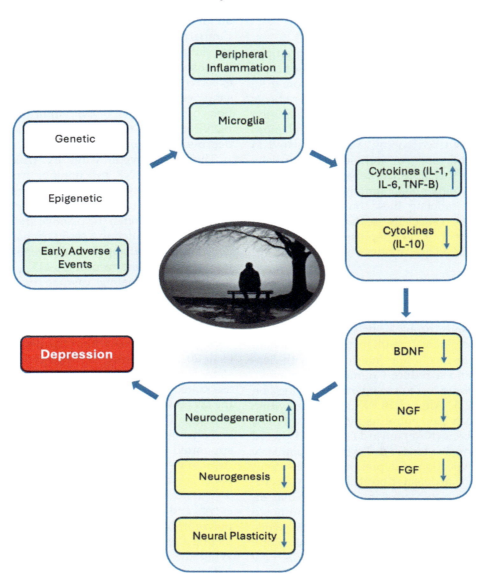

FIG. 9.2 BDNF in the neuroimmune regulation axis of depression. Both peripheral and brain inflammatory factors and the release of proinflammatory cytokines cause a reduction of BDNF, which undermines neural plasticity and neurogenesis, culminating in depression. BDNF, brain-derived neurotrophic factor; IL-1, interleukin-1; IL-6, interleukin-6; TNF-β, tumor necrosis factor-β.

(Evans et al., 2004). Predictably, the administration of FGF-2 directly into the brain of rodents largely eliminated stressor-provoked behavioral disturbances. Conversely, pretreatment with an FGF-2 antagonist prevented the positive effects of antidepressants in attenuating the effects of stressors (Elsayed et al., 2012). FGF-2 alterations modulate microglia activity and inflammation and could thereby induce a depressive-like state (Tang et al., 2018a). As expected, antidepressants increased FGF-2 expression through activation of early growth response factor 1, a transcription factor that regulates FGF-2 (Kajitani et al., 2015).

Although the data are limited, VEGF has been implicated in depressive illness as this neurotrophin was diminished in the hippocampus and prefrontal cortex in both animal models and depressed humans, and a SNP on the VEGF gene was linked to depression (Xie et al., 2017). Chronic social instability in rodents provoked depressive-like behaviors and altered VEGF mRNA in the amygdala and hypothalamus, which were diminished by antidepressant treatments (Nowacka-Chmielewska et al., 2017). It also seems that the presence of VEGF is necessary for BDNF to provide antidepressant actions, and conversely, BDNF is required for VEGF to have its positive effects (Deyama et al., 2019). Not unlike the link between BDNF and inflammatory processes, VEGF alterations may operate with inflammatory cytokines in the elicitation of depressive behaviors as well as in the therapeutic response elicited by repetitive transcranial magnetic stimulation (rTMS) (Valiuliene et al., 2021).

Like other neurotrophins, glial cell line-derived neurotrophic factor (GDNF) may be a player in subserving depression. This neurotrophin is involved in the integration, development, and functioning of adult-born hippocampal neurons (Bonafina et al., 2019). It is widely distributed throughout the brain where it contributes to the regulation of norepinephrine and GABA functioning and affects midbrain dopamine neuronal activity. A postmortem analysis revealed that GDNF was reduced in the amygdala and prefrontal cortex of depressed individuals and antidepressant treatments could increase central GDNF mRNA and protein levels (Maheu et al., 2015).

The involvement of nerve growth factor (NGF) in depressive disorders has not been examined extensively. Nonetheless, NGF may influence stress responses, thereby promoting depression, and exogenous NGF administration diminished signs of depression in animal models (Mezhlumyan et al., 2022) and enhanced the action of antidepressant treatments.

Neurotrophins Measured in Blood

Serum levels of BDNF have often been used as a surrogate index for brain BDNF. However, since BDNF is secreted from immune cells and accumulates in the form of platelet-producing cells (megakaryocytes), it is questionable to what extent peripheral BDNF reflects what occurs in the brain, let alone specific brain regions that may be associated with depressive disorders. This said, a meta-analysis confirmed that depression is accompanied by lowered serum BDNF levels, although it was cautioned that the reported findings were highly variable and studies were often underpowered (Molendijk et al., 2014). The extent of the BDNF reductions was more pronounced among patients with comorbid anxiety, greater suicidal ideation, as well as recurrent depressive episodes. Among depressed patients in whom BDNF was relatively high, the response to antidepressants was more favorable than in patients with low BDNF levels (Wolkowitz et al., 2011), indicating that the effects of SSRIs might require the presence of adequate BDNF levels. In this respect, childhood trauma can influence late-life depression and through actions on neurotrophins, reflected by serum BDNF, can diminish the benefits of SSRIs (Dimitriadis et al., 2019). Furthermore, treatment with antidepressants could increase the presence of serum BDNF, varying with different SSRIs (Zhou et al., 2017a). Importantly, while serum BDNF levels increased with clinical improvement following antidepressant treatment, a similar rise was not evident among patients who did not show clinical improvement, supporting the link between the growth factor and the illness (Kishi et al., 2018). This supposition was reinforced by the finding that serum BDNF and NGF were also

accompanied by reduced amygdala size (e.g., Inal-Emiroglu et al., 2015).

As we saw earlier, brain VEGF changes were observed in animal models of depression; however, serum VEGF in humans was not reliably related to depression, although plasma levels were increased among patients who responded positively to antidepressant treatments (Shi et al., 2020). There were indications that serum NGF levels were increased by antidepressant treatments, but the findings have been mixed, precluding conclusions concerning its role in determining the therapeutic response of these agents (Mondal and Fatima, 2019). There were indications that serum levels of GDNF were also reduced in depressed individuals, although this relation was inconsistently observed. Levels of serum insulin growth factor-1 likewise predicted the appearance of depression in some studies, but once more, this relation was not consistently observed, and the effects of antidepressant medication on levels of this growth factor have been variable (Levada and Troyan, 2017).

Overall, it seems that circulating BDNF (and perhaps VEGF and NGF) is associated with the presence of depression (Shi et al., 2020), whereas the link to other neurotrophins is questionable. Neurotrophins may act in concert with one another in affecting illness and may influence inflammatory immune processes (e.g., Morel et al., 2020), thereby contributing to psychopathological conditions. It should be added that serum BDNF is promiscuous in that it is not only related to depressive disorders but is also evident in patients with schizophrenia, stroke, lupus erythematosus, type 2 diabetes, and age-related cognitive impairments. Each of these conditions is highly comorbid with depression and has been linked to inflammatory processes. Thus, serum BDNF could be a common feature reflecting previous stressors or other factors that contribute to diverse psychopathological conditions.

Epigenetic Processes and Depression

Epigenetic changes associated with multiple neurobiological processes, including those related to the regulation of BDNF expression, have been implicated in the evolution of depression and the response to therapeutic agents. Polymorphisms and epigenetic changes at BDNF regulatory sites were observed in depressed individuals, which were moderated by early-life adversity (Ferrer et al., 2019). BDNF promoter methylation was similarly associated with late-life depression (Kang et al., 2015). Likewise, the gene coding for a microRNA, miR-124-3p, which may be fundamental for brain plasticity, appeared in the postmortem prefrontal cortex of depressed patients, as well as in rodents that received chronic corticosterone to roughly mimic some of the effects of a stressor (Roy et al., 2016). To be sure, the data linking BDNF gene polymorphisms and epigenetic alterations have not been entirely consistent, perhaps owing to the heterogeneous nature of depression and the different methodologies used to assess this relationship (Treble-Barna et al., 2023).

Studies that evaluated the impact of adverse childhood experiences indicated that the occurrence of psychopathology was often associated with methylation of the glucocorticoid receptor gene NR3C1 (Wadji et al., 2021). Among pregnant women who experienced strong stressors, elevated NR3C1 methylation was apparent in male offspring, and reduced BDNF DNA methylation occurred in both male and female offspring (Braithwaite et al., 2015). Moreover, among individuals who died of suicide, and who had a history of early childhood neglect/abuse, ribosomal RNA expression was hypermethylated in the promoter region of the gene for the glucocorticoid receptor within the hippocampus (McGowan et al., 2009).

Aside from these findings, epigenetic changes were reported within the gene coding for the $GABA_A$ receptor within the prefrontal cortex of depressed individuals who died by suicide (Poulter et al., 2008). Furthermore, among school-aged children who experienced severe maltreatment, epigenetic changes were evident in the $5\text{-}HT_{2A}$ receptor (Parade et al., 2017), and early adverse events were accompanied by methylation of the gene coding for 5-HTT (Soga et al., 2021). Although human studies concerning epigenetic changes of genes tied to inflammatory processes in depression have been relatively sparse, data supporting this perspective have been reported (Qi et al., 2022), and CRP-related epigenetic changes are associated with structural brain alterations among depressed individuals (Green et al., 2021). It is germane that blood DNA methylation profiles related to cellular stress responses and signaling mechanisms associated with immune cell migration and inflammation could predict the occurrence of depression six years later (Clark et al., 2020).

Of the various stressors that humans encounter, racial and ethnic discrimination can elicit especially pronounced psychological repercussions, premature aging, and decreased life expectancy. These outcomes were especially prominent among individuals who internalized the anger provoked by discrimination (McKenna et al., 2021). Distress of discrimination has been associated with numerous epigenetic changes that have been related to disease occurrence and may even contribute to the racial disparities reported in chronic pain perception (Aroke et al., 2019). Several studies indicated that discrimination promoted cortisol changes (e.g., Busse et al., 2017), and elevated CRP levels and several proinflammatory cytokines (Lawrence et al., 2022). Likewise, in Latinx women who experienced discrimination while pregnant,

elevated methylation was observed for genes associated with T_{reg} cells and in the promotor region of the gene for TNF-α (Sluiter et al., 2020), as well as in genes for glucocorticoid receptors and BDNF (Santos et al., 2018).

It has been surmised that some epigenetic changes that occur may reflect adaptive changes that favor resilience, whereas others seem to be aligned with pathological outcomes. For instance, epigenetic changes of the NR3C1 gene have reliably been associated with the emergence of pathology, such as depression or PTSD (Zhang and Liu, 2022), whereas chronic stressors provoked methylation of the IL-6 gene and changes related to the blood–brain barrier (BBB), have been associated with enhanced resilience (Dudek et al., 2021). The bottom line is that complex psychological disorders may be subserved by a very large number of genes and their interactive effects, so attempting to find specific genes or epigenetic changes that govern such illnesses is exceedingly difficult.

Neuronal stimulation to attenuate depression

Electroconvulsive treatment has long been known to have positive effects in ameliorating symptoms of depression, and the dangers of the procedure have diminished considerably over several decades. The positive effects of the treatment have been attributed to the activation of restorative and neurotrophic processes, moderation of underactive cortical neuronal activity, and overactive subcortical limbic functioning (Njau et al., 2017).

With the realization that stimulation of brain activity could diminish symptoms of depression, other ways of achieving this were developed. In this regard, rTMS has received increasing attention, not just for depression, but for many other brain disorders (e.g., PTSD, generalized anxiety disorder, pain relief, and substance use disorder). This procedure entails an electrical coil being passed over the surface of the head to create a magnetic field that excites neuronal activity, thereby diminishing depressed mood. A second method, transcranial direct-current stimulation (tDCS), applies a lower current (for a longer duration) to electrodes attached to the skull, which can stimulate sites deeper in the brain.

A meta-analysis indicated that both rTMS and tDCS elicited positive effects comparable to those of other therapies for nonpsychotic depression (Mutz et al., 2019), and was even effective in some treatment-resistant patients. How rTMS has its effects is uncertain, but increasingly more information has been closing in on the identification of the brain processes involved. Ordinarily, the anterior insula, which is involved in the integration of sensory information, sends signals to the anterior cingulate cortex, which is fundamental for emotional regulation. However, in a subset of severely depressed patients, their symptoms were associated with disturbed propagation of brain activity within the salience network, and with alleviation of symptoms following rTMS the typical propagation patterns were restored (Mitra et al., 2023). Alternatively, the antidepressant actions of rTMS might stem from increased glutamate stimulation of the anterior cingulate cortex, and there is evidence that rTMS increases serum BDNF and may reduce proinflammatory cytokines (Zhao et al., 2019b).

Deep brain stimulation (as the name implies, involves direct stimulation of neurons deep in the brain) made its debut with the demonstration that it could be used to attenuate the symptoms of Parkinson's disease (Lang and Lozano, 1998). Subsequent trials demonstrated that this procedure, targeting the subgenual cingulate cortex, could alleviate symptoms in patients with treatment-resistant depression (Mayberg et al., 2005). However, a later randomized, double-blind, crossover trial of subcallosal cingulate stimulation yielded less impressive outcomes (Holtzheimer et al., 2017), and stimulation of the ventral capsule/ventral striatum did not diminish treatment-resistant depression (Dougherty et al., 2015). Somewhat better outcomes were observed in other similarly designed investigations in which DBS was applied for lengthy periods, although positive effects were only observed in a modest number of patients (Merkl et al., 2018; Ramasubbu et al., 2020). In a study in which implanted electrodes were accompanied by a device that could record brain activity at the subcallosal cingulate cortex, reduced depression was associated with unique brain patterns, raising the possibility of being able to predict the efficacy of deep brain stimulation in severely depressed individuals (Alagapan et al., 2023). It was suggested that better clinical outcomes may be realized with improved technological advances, particularly through the concurrent use of imaging or electrophysiological approaches to identify patient-specific networks that would be most amenable to DBS at particular stimulation parameters (Johnson et al., 2024).

Since the inception of deep brain stimulation, other target sites have been proposed for this procedure, and it has been used in the treatment of other conditions, such as neuropathic pain, OCD, bipolar disorder, as well as several neurological conditions. The beneficial effects of deep brain stimulation might come from prefrontal release of glutamate and activation of AMPA glutamate receptors (Jiménez-Sánchez et al., 2016) or by indirectly producing inhibition of neurons that would otherwise favor depressive features. The positive effects of deep brain stimulation on depression were also linked to reduced regional inflammation and altered neurotrophic processes (Dandekar et al., 2019).

The downside of deep brain stimulation is that it involves a surgical procedure, which has multiple drawbacks, even though as far as brain surgery goes, it's one of the less risky procedures. A method has been developed that avoids surgery but can activate specific deep brain regions using a set of individual high-frequency fields that only activate neurons at restricted sites where different electrical fields intersect (Grossman et al., 2017). Several reports in animals and humans have since confirmed the benefits of this procedure for various conditions and improved technologies have made it more precise in limiting off-site actions (Guo et al., 2023c).

Inflammatory Processes and Depressive Disorders

Multidirectional communication occurs between the immune system, hormonal processes, growth factors, and gut bacteria, and could thereby influence a range of physical and psychological illnesses. As we've seen repeatedly, inflammatory processes interact with several hormones in determining some of the features of depressive disorders. In this respect, the metabolic demands stemming from chronic inflammation can also promote motivational disturbances characteristic of depression (Lucido et al., 2021).

It is unlikely that inflammatory processes are involved in all instances of depression but instead would only occur in a subset of individuals. Manipulations of inflammatory processes might be expected to diminish depression in cases where inflammation is elevated but have little effect in cases where noninflammatory factors are responsible for the illness. The corollary is that therapies that do not influence inflammatory processes would be relatively ineffective in attenuating depression that is driven by the presence of inflammation. Furthermore, the biological features of mental illnesses, including depression, may change over time, just as the response to stressors varies from an acute to a chronic condition. Thus, the processes associated with a first episode of depression might be distinguishable from those associated with chronic depression and treatment-resistant depression, or the recurrence of illness following successful treatment. For instance, while first-episode depression may involve innate immunity (e.g., activation of neutrophils and macrophages), adaptive immune processes may play a greater role in chronic depression (Singh et al., 2022), which may have implications for the choice of therapies selected to treat patients.

Impact of Inflammatory Acting Agents on Rodents

With an increased understanding of the multiple functions of inflammatory processes, the view was taken that depression might arise owing to the failure of peripheral and brain regulatory processes that keep inflammation in check (Hu et al., 2022). Just as peripheral immune cells protect against invasive pathogens, glial cells act as guardians of the central nervous system. Indeed, depression has been associated with many proinflammatory cytokines and chemokines measured in blood, together with a reduction of antiinflammatory cytokines. A similar proinflammatory bias was also observed in cerebrospinal fluid as well as in the postmortem brain (Sakamoto et al., 2021).

The administration of immune-activating agents and cytokines elicits a range of behavioral changes in animals. The most broadly studied of these is sickness behaviors, which comprise anorexia, fatigue, reduced motor activity, curled body posture, sleepiness, ruffled fur, and loose bowel functioning. These symptoms are much like those that accompany influenza infection, and in some respects are reminiscent of the atypical form of depression (Dantzer et al., 2008). Low doses of IL-1β or TNF-α administered directly into the brain can elicit the sickness profile, indicating the involvement of central processes in determining such symptoms. If the effects of immune activation were limited to sickness behaviors, the case for depression might not be overly impressive. However, animals treated with LPS or IL-1β also displayed other depressive-like behaviors, including anhedonia and disrupted social interaction (Moieni and Eisenberger, 2018). Importantly, the sickness and depressive behaviors can be distinguished from one another, in that antidepressants preferentially influence the behaviors associated with affective and anxiogenic changes elicited by cytokines over those that involve sickness symptoms (Merali et al., 2003). Moreover, protracted elevations of IL-6 in response to a social stressor accounted for vulnerability to the depressive-like phenotype (Hodes et al., 2014).

Although IL-6 is correlated with depression in humans, in rodents, the administration of IL-6 hardly produces any observable behavioral changes, such as sickness symptoms, but disturbs food-motivated behaviors and reduces dopamine release in the nucleus accumbens (Yohn et al., 2016). Thus, while IL-6 might be responsible for motivational and reward processes associated with depression-related inflammation, other processes account for the neurovegetative aspects characteristic of depression.

Contribution of obesity to depressive disorders

There are laws to thwart negative biases concerning race, gender, and age, but this is less evident in obesity (sizeism), although this may be changing. Weight bias and stigmatization is a powerful stressor that gives rise to physiological changes that frequently accompany chronic stressors, including disturbed immune, hormonal, and metabolic functioning (Daly et al., 2019), which could promote psychopathology. Instead of viewing obesity as an illness brought about by genetic factors and both social and environmental influences, people who have obesity are at times viewed as lacking discipline and willpower. Unfortunately, those who have endured sizeism are more apt to display increased alcohol use, sleep disturbances, poor lifestyle choices, and reduced life expectancy (e.g., Sutin et al., 2015).

Obesity favors insulin and leptin resistance, and inflammation may give rise to depression (Fulton et al., 2022). A meta-analysis of Mendelian randomization studies suggested that obesity is causally linked to depression (Jokela and Laakasuo, 2023). Although inconsistent findings have been reported concerning leptin dysregulation (resistance) in depressive disorders, this may be due to lumping all depressed individuals into a single pool. Analyses conducted in patients with atypical depressive features revealed a tie to leptin dysregulation that was associated with the presence of obesity (Milaneschi et al., 2017), which could promote elevated levels of circulating inflammatory factors.

Abdominal obesity as we saw in earlier chapters has been associated with increased adipokines and the release of cytokines, which favor the development of a constellation of inflammatory-related disorders, such as depression, type 2 diabetes, and heart disease. With the co-occurrence of depression and obesity, circulating inflammatory factors are elevated still further (McLaughlin et al., 2022), potentially favoring greater illness occurrence. When the biological sources of illness are coupled with those related to stigma, the tyranny of obesity is amplified.

Inflammatory Markers Associated With Depression

Early indications of an immune-depression link came from reports that depressive disorder, particularly melancholia, was accompanied by elevated levels of acute phase proteins, such as CRP and alpha-1-acid glycoprotein, as well as proinflammatory cytokines and their soluble receptors (Maes et al., 2011). The assessment of multiple case–control studies revealed that depression was associated with enriched genes for innate immune-related and neutrophil functioning in blood cells (Drevets and Bullmore, 2020). A detailed meta-analysis that included 107 studies similarly concluded that major depressive disorder was related to numerous inflammatory factors, such as elevated CRP, IL-6, TNF-α, IL-12, and IL-18 (Osimo et al., 2020). Such relations were especially notable among depressed patients with suicidal ideation, who displayed a combination of elevated inflammation, reduced neurotrophin concentrations, and increased stressor encounters (Priya et al., 2016). Thus, the identification of relevant inflammatory biomarkers could inform therapeutic strategies to diminish depression, especially among individuals who are otherwise treatment-resistant (Drevets et al., 2022).

The relationship between depression and inflammatory changes is bidirectional. A meta-analysis of prospective studies in community samples indicated that elevated levels of CRP and IL-6 predicted the subsequent occurrence of depression, and the presence of depression was linked to later elevations of inflammatory markers (Mac Giollabhui et al., 2021). Predictably, individuals who were already depressed were more sensitive to stressors, leading to still greater cytokine changes upon encountering further challenges. These relations varied with sex in that higher levels of IL-6 were associated with subsequent depression, particularly among women, and conversely, illness appearance was an antecedent of later cytokine variations (Lamers et al., 2019).

From a precision medicine vantage, it is significant that elevated CRP levels were closely aligned with the vegetative features of depression (e.g., fatigue, restless sleep), which were no longer evident after antidepressant treatment (White et al., 2017). It was subsequently reported that elevated CRP was associated with greater depressed mood, increased appetite and sleep disturbances, fatigue, irritability, and worry. Moreover, the elevated CRP and IL-6 were aligned with the anhedonia characteristic of depression (Li et al., 2022g).

A Mendelian randomization analysis revealed that genetic factors associated with IL-6 were related to both sleep disturbances and fatigue. In this respect, altered sleep, either in the form of early morning awakening or sleeping excessively, is a hallmark of depression, and sleep disturbances may be one of the first signs of depression emerging. As we'll discuss shortly, sleep disturbances can affect cytokine activity (Irwin and Opp, 2017), so these processes may be intimately related to the emergence of depression. Overall, specific inflammatory factors were aligned with multiple features of depression that could, to some extent, be distinguished from one another (Milaneschi et al., 2021).

Although elevated low-grade inflammation was more closely related to late than early-onset depression (Rozing et al., 2019), inflammation associated with depression was apparent among adolescents (Colasanto et al., 2020). Importantly, a longitudinal analysis indicated that serum CRP and IL-6 levels among children at 9 years of age predicted neurovegetative symptoms of depression that were apparent when they were 18 years old (Chu et al., 2019a). Similarly, among individuals who were followed for an average of 11.8 years, elevated CRP and IL-6 predicted later cognitive symptoms of depression (Gimeno et al., 2009). The ties between immune changes and depression may not be a direct one since the immune alterations could be secondary to lifestyles associated with the illness (Slavich and Irwin, 2014).

Further supporting inflammatory involvement in depressive illness, the expression of genes for IL-6 and NK cells was particularly elevated among currently depressed patients relative to nondepressed controls or individuals who had previously been depressed. This was confirmed in a 2-year longitudinal analysis that showed elevated IL-6 gene expression in depressed patients (Jansen et al., 2016). Additionally, genome-wide association analyses and differential expression of genes in transcriptome analyses indicated that genes relevant to inflammatory processes were over-represented among depressed individuals (Sharma, 2016), and numerous gene variants were linked to depression (Wray et al., 2018). Evaluation of the many genes tied to depression indicated that those related to leukocyte regulation were especially prominent as were genes related to neuroinflammation, cytokine functioning, and innate immunity (Tubbs et al., 2020).

Imaging procedures and postmortem analyses supported inflammation involvement in depression. Specifically, among treatment-resistant patients, the connectivity between the ventral striatum and the ventromedial prefrontal cortex was disturbed in association with IL-6 and TNF-α (Rengasamy et al., 2022). Additionally, CRP levels were elevated in depressed individuals, accompanied by reduced grey matter volume in the prefrontal cortex and insula (Opel et al., 2019). As well, the over-representation of inflammatory processes and altered signaling within the hippocampus were observed among depressed individuals (Mahajan et al., 2018).

Aside from the immediate consequences of stressors, the priming of neuroinflammatory processes governed by danger-associated molecular patterns and NLRP3 (inflammasome) can dispose individuals toward psychiatric disturbances. As expected, agents that act as NLRP inflammasome antagonists, such as β-hydroxybutyrate, diminished the behavioral and inflammatory actions of stressors (Yamanashi et al., 2017). In essence, when experiences that give rise to inflammation are followed by stressors, particularly those that are uncontrollable and occur chronically, neurobiological systems are more likely to become overly taxed (allostatic overload), hence favoring the emergence of depression. As described in Fig. 9.3, stress-related inflammatory processes can be activated through different routes, which can affect brain processes linked to depression.

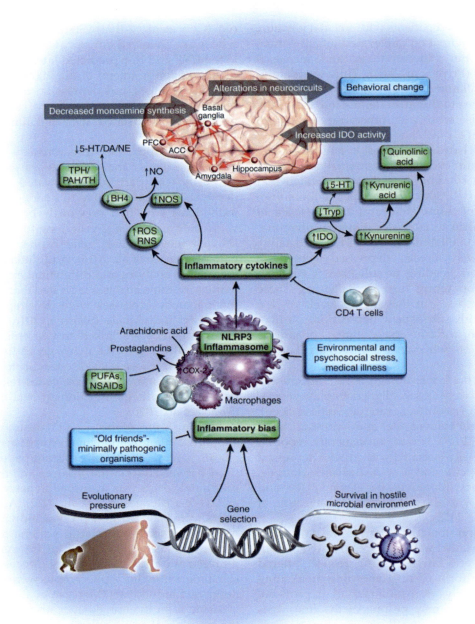

FIG. 9.3 We are the way we are, in part, because our ancient ancestors were able to adapt to their environment. This entailed, among other things, being able to deal effectivity with multiple challenges, including a hostile microbial environment. These microbial processes may have provoked inflammatory changes that were enhanced over successive generations. Like so many other biological processes, inflammatory immune responses are modifiable by numerous psychosocial and experiential influences, thereby maintaining our ability to deal with varied challenges. These exceptional abilities involve a constellation of neurobiological systems, such as innate immune process, that detect pathogenic microorganisms and sterile stressor system receptors. These receptors and sensors moderate caspase-1 (and related factors), which promotes inflammation upon exposure to infectious microbes. Having been activated, this inflammasome triggers the release of several cytokines, which activate enzyme pathways, such as IDO, together with the production of both reactive nitrogen and oxygen species. As much as these changes ought to have been effective in protecting us, they are also able to provoke multiple neurobiological changes that can foster psychological disturbances. Among other things, they promote the release of neurotoxic metabolites of kynurenine, such as quinolinic acid, and disrupt monoamine functioning by limiting the availability of their precursors, including tryptophan (Tryp) and tyrosine, as well as tetrahydrobiopterin (BH4), which is an essential co-factor for enzymes (TPH, PAH, TH) needed for the formation of monoamines. The net result of these changes comprises variations of brain neural circuits, such as those that include the anterior cingulate cortex (ACC) and the prefrontal cortex (PFC), which are needed for survival in a stressful environment, particularly in the integration of immune and behavioral response to both pathogens and predators. When these circuits are overloaded as a result of chronic stressors, these same responses may lend themselves to, as Raison and Miller (2013) indicated, "malaise, melancholy and madness, which are the inflammatory legacy of our evolutionary past". NLRP3: NACHT domain-, leucine-rich repeat-, and pyrin domain-containing protein 3; NO: nitric oxide; NOS: nitric oxide synthase; NSAIDs: nonsteroidal antiinflammatory drugs; PUFAs: polyunsaturated fatty acids. *Source: Raison, C.L., Miller, A.H., 2013. The evolutionary significance of depression in Pathogen Host Defense (PATHOS-D). Mol. Psychiatry 18, 15–37.*

Perinatal depression and inflammation

Estrogen levels have been implicated in depression, having been observed with the use of oral contraceptives and varying with the estrous cycle. Also, the reduced production of ovarian estrogen that occurs during menopause is accompanied by increased risk for various pathologies, such as metabolic disorders, cardiovascular diseases, and depressive illnesses, varying with the age at which menopause occurred and lifestyles that had been adopted (Turek and Gąsior, 2023).

A systematic review of earlier systematic reviews indicated that the most frequent associates of perinatal depression were a history of mental health problems or chronic health conditions, gestational diabetes, preeclampsia, sleep disturbances, stressful life events, prior abusive experiences, marital conflicts, and a lack of social support (Al-Abri et al., 2023). Various hormonal changes that occur at this time may also increase vulnerability to mood changes, and prenatal placental CRH and HPA dysregulation were risk factors for postpartum depression, as were levels of estrogen, progesterone, and oxytocin (Zhu et al., 2022a). Genome-wide analyses similarly revealed multiple gene expression changes during the third trimester of pregnancy, and many genes associated with estrogen functioning were associated with depression assessed 2 months postpartum (Mehta et al., 2021).

Several innate immune changes occur during pregnancy that serve the well-being of the fetus. Pregnancy is accompanied by immune alterations so that the fetus isn't attacked by the maternal immune system. During the second trimester of pregnancy, this is reflected by elevated antiinflammatory cytokines and low levels of proinflammatory cytokines. Toward the end of pregnancy and following childbirth, a further shift occurs so that a proinflammatory state predominates. If inflammatory immune changes are excessive during this period, postpartum depression can be provoked (Sawyer, 2021).

Women with low levels of Th1 and Th$_{17}$ immune cells were particularly likely to develop depression within days of childbirth (Osborne et al., 2020a). This condition may also be accompanied by variations of the NLRP3 inflammasome, and a shift in the Th1 and Th2 balance so that proinflammatory processes predominate and brain microglial activation has been implicated in the promotion of postpartum depression. Postpartum depression was also associated with dysregulation of cytokine and kynurenine pathways (Achtyes et al., 2020), and if mom had previously experienced depression accompanied by elevated levels of IL-6, the risk for this condition was particularly elevated (Worthen and Beurel, 2022). It is significant that a formulation of allopregnanolone, brexanolone, which is used to diminish this condition, was accompanied by a reduction of blood TNF-α and IL-6, and inhibition of inflammatory responses to TLR4 and TLR7 activators (Balan et al., 2023).

Predicting Recurrence of Depression

What makes depressive disorders especially disturbing is the high rate of recurrence, and depressive-like features can be reinstated by specific contextual cues and experiences. Affected individuals might simply have a negative or ruminative personality style that makes them prone to depression. Alternatively, they might have experienced multiple stressors throughout life so that their biological systems are sensitized or primed, so that further stressor encounters; even if these are relatively modest, they promote strong neurochemical responses that push them toward depression. For example, among currently depressed patients and those in remission, a sad mood induction procedure that entailed exposure to negative autobiographical memory scripts resulted in a decline of cerebral blood flow within the medial orbitofrontal and anterior cingulate cortex. And, when depressed patients in remission engaged in autobiographical recall of sad memories, elevated cognitive reactivity and rumination were elicited, coupled with a disturbance in the connectivity of the hippocampal-default network (Figueroa et al., 2017). In line with these findings, when patients with recurring depression were assessed for a long period, viewing a sad film clip was sufficient to provoke changes in medial prefrontal cortex activity, which predicted the rumination propensity and elevated risk for relapse (Foland-Ross et al., 2014).

Essentially, among some patients in remission, certain brain processes remained fragile, so a mood challenge effectively "unmasked" mechanisms that favor depression. The neuronal response to a sad film, especially if it occurs together with rumination, might reflect the need for continued therapy (e.g., CBT) to thwart depressive relapse. Conducting an imaging procedure on each former patient to predict relapse may not be feasible, but other approaches could be, including analyses of hormonal changes elicited by stressor (or hormone) challenges. As well, elevated levels of inflammatory factors can affect brain network activity and might be used as a marker of depression recurrence (Liu et al., 2019).

Neurochemical Effects of Immune Activation in Depressive Illnesses

There is little debate concerning the broad and pernicious effects of excessive inflammatory activation, and various routes were described as to how this came about in the provocation of affective illnesses. Circulating cytokines could stimulate the vagal nerve, which then affects brain processes and might thus contribute to depressive illness. Additionally, proinflammatory cytokines released following immune activation could gain access to the brain, thereby favoring the occurrence of depression. It will be recalled that despite their large size, cytokines can reach the brain parenchyma at sites where the BBB is relatively permeable, notably at sites that surround the brain's ventricular system (circumventricular organs). Also, active transport mechanisms are present that could ferry cytokines from the periphery into the brain, thereby affecting neuronal functioning (Banks, 2016).

It seems that infection, traumatic insults, and strong stressors could undermine the integrity of the BBB, thus permitting greater cytokine passage into the brain, and hence psychological disturbances could emerge (Dudek et al., 2020). For instance, in rats, a chronic stressor increased BBB permeability in the vicinity of the dorsal striatum, which could be exacerbated with aggregation of inflammatory Th$_{17}$ cells, allowing still greater cytokine infiltration and the emergence of depressive-like behaviors (Peng et al., 2022). With greater permeability of the BBB and the intestinal barrier (leaky gut) in response to chronic stressors, factors associated with inflammation are still more likely to affect brain functioning, making it that much more important to establish ways of limiting these actions.

While not dismissing the possibility that peripheral cytokines might be a link to psychological illness, as indicated earlier, a paradigm shift occurred with the finding that cytokines and their receptors were endogenously expressed in the brain, being released by activated microglia (Allen and Barres, 2009). Indeed, brain cytokine concentrations or their mRNA expression was elevated in response to traumatic head injury, stroke, and seizure, thereby promoting depressive disorders. Likewise, depression is not an uncommon postoperative problem, which has been linked to increased pain, delirium, delayed recovery, and recurrences of illnesses, potentially arising because of the activation of inflammatory processes. Generally, stressful events promote microglial activation and elevated cytokine expression (Wang et al., 2022b) together with neuronal processes that elicit depression (Woodburn et al., 2021).

Microglia as inducers of depression

It has been maintained that depression ought to be viewed as a microglial disease (Yirmiya et al., 2015). In response to stressors, circulating monocytes are increased and find their way to the brain where along with cytokines they influence neuroplasticity. Some of the recruited monocytes remain in the brain, adopting characteristics of microglia (Hodes et al., 2015). The cytokines released from brain microglia may have pertinent actions on neuronal processes that govern memory and executive functioning, hence favoring the development of depression.

While modest activation of the microglia NLRP3 inflammasome may have protective effects in relation to illness, stressor-provoked NLRP3 overactivation and the ensuing cytokine release can disturb hippocampal neurogenesis and the incorporation of new neurons into existing neural circuits, thereby promoting the behavioral disturbances observed in inflammatory disorders (Chesnokova et al., 2016). Further to this, in response to stressors, the NLRP3 inflammasome was associated with elevated cytokine release from immune cells, which engenders a neurotoxic response in neighboring cells, leading to their death. NF-κB activation, which influences microglia and the NLRP3 inflammasome, promotes the presence of neurotoxic astrocytes following stressor exposure, and when NLRP3 of microglia was knocked out, the depressive-like behavior in mice was diminished (Li et al., 2022d). Paralleling these findings, among healthy women exposed to a social stressor, alterations of NF-κB were accompanied by disturbed reward processing (Boyle et al., 2023).

Factors that increase the activity of microglia, such as stressors, augment brain cytokine levels, thus promoting depressive disorders (Frank et al., 2019). These actions were marked in aged rats owing to sensitization of hippocampal microglial responses, leading to greater cytokine release (Barrientos et al., 2015). Stressors can also increase TLR2 and TLR4, which contribute to the functioning of pathogen-associated molecules (PAMPs) and damage-associated molecular patterns (DAMPs) that enhance the synthesis and release of inflammatory mediators upon subsequent stressor or inflammatory challenges (Weber et al., 2013). It is of potential practical significance that in rats that received repeated treatment with beneficial bacteria (*Mycobacterium vaccae*), the effects of stressor-provoked microglial priming and the emergence of anxiety were attenuated (Frank et al., 2018).

Microglial disturbances may arise owing to events that occurred during early life. The induction of inflammation at this time may interfere with the subsequent ability of microglia to moderate the actions of glutamate neurons in the anterior cingulate cortex upon later stressor exposure (Cao et al., 2021b). Early life adversity may cause distinct gene expression profiles of microglia, and these experiences were associated with altered expression of genes related to phagocytosis in the dentate gyrus of depressed individuals who had died by suicide (Reemst et al., 2022a). As aspects of the microglial phenotype were preferentially influenced by stressors in females, this might contribute to sex differences in mood disorders and may be relevant to the effectiveness of antidepressant treatments (Martinez-Muniz and Wood, 2020).

At the other end of the aging spectrum, microglial senescence, which occurs with aging and chronic stressor experiences, may instigate neurotrophin alterations that impair neuroplasticity and neurogenesis, which encourages depression and neurodegenerative disorders (Yirmiya et al., 2015). The available data point to microglia-produced inflammatory factors as a potential target for the treatment of stressor-produced cognitive dysfunction. It should be added that although late-life depression has been associated with elevated levels of IL-1β, TNF-α, and IL-6, these relations may have been confounded with the presence of physical illnesses frequent among older people. In fact, in the absence of other health conditions, inflammatory factors were not elevated and auxiliary antiinflammatory treatment did not diminish signs of depression (Luning Prak et al., 2022).

Stress, Inflammation, and Depression

Many of the neurobiological consequences engendered by immune activation are remarkably like those elicited by stressors. Proinflammatory cytokine or bacterial endotoxin (e.g., LPS) administration markedly increased HPA functioning and monoamine activity at hypothalamic and extra-hypothalamic sites, and affected BDNF functioning (Audet and Anisman, 2013). Thus, like psychogenic and neurogenic stressors, inflammatory processes can instigate or exacerbate psychiatric or neurological conditions (Dantzer et al., 2008). Of course, bacterial and viral infections do not elicit appraisals comparable to those provoked by physical or psychological stressors, particularly as individuals might be unaware of inflammatory processes being activated. In this respect, psychogenic stressors stimulate HPA functioning through amygdala activation, whereas systemic stressors may do so through nonlimbic circuits, but in many respects, the end result is the same.

The impact of inflammatory factors varies across brain regions, and the development of a particular pathology is dependent on interactions with other neurochemical processes. Specifically, proinflammatory cytokines interact with CRH, neurotrophins, and enzymatic pathways that govern the production of oxidative species and other neurodegenerative processes, thereby favoring the occurrence of depression (Bakunina et al., 2015). Inflammatory immune activation also increases the utilization of serotonin and norepinephrine in cortical and limbic regions and interferes with dopamine functioning within reward circuits (Felger et al., 2016). With persistent cytokine activation, protracted changes occurred regarding CRH functioning and secretion of adrenocorticotropic hormone (ACTH) and corticosterone, possibly leading to greater strain on physiological systems and hence greater risk for psychological disturbances.

Given that corticoids act to diminish cytokine functioning, it might seem puzzling that depression has been associated with both elevated cortisol and increased cytokine levels. Ordinarily, glucocorticoids limit or prevent the activity of the NF-κB, thereby suppressing the production and secretion of proinflammatory cytokines. However, with continued HPA activation engendered by chronic stressors, glucocorticoid receptor sensitivity may be reduced (Walsh et al., 2021), so that the cortisol inhibition of cytokine functioning is diminished. Furthermore, despite the downregulated effects of chronic

stressors on glucocorticoid functioning, hippocampal microglial functioning can promote elevated inflammatory responses (Picard et al., 2021).

Stressors and inflammatory immune activation synergistically influence brain neurotransmitter activity, thereby eliciting more pronounced behavioral changes. Likewise, the pronounced loss of hippocampal neurons was exacted by the combination of a chronic stressor and an inflammatory challenge. Episodes of depression can similarly exacerbate inflammation, and with each incidence of depression, lower levels of stressor or inflammatory challenges may be sufficient to promote neurotoxic changes that encourage more severe depression (Belleau et al., 2019). Stressful events can likewise result in exaggerated microglial neuroinflammatory and behavioral responses to an immune challenge, although sensitization (priming) may have developed through different processes in male and female rats (Fonken et al., 2018).

Irrespective of whether cytokines are causally linked to depressive illnesses (or suicide), elevated circulating cytokine levels might serve as a marker for these disorders and might also predict treatment efficacy. Indeed, basal TNF-α levels were associated with later depression, and levels of IL-6 were predictive of response to antidepressant treatment (Lombardi et al., 2022). Furthermore, elevated cytokines presence after symptom relief could reflect a biological system that remains perturbed and could signal a disposition for depression recurrence. We underscore once again that being a biochemically heterogeneous disorder, elevated inflammation should neither be expected in all depressed individuals nor in all forms of depression. It seems that elevated inflammation might be more commonly seen in patients exhibiting greater vegetative symptoms (poor appetite, sleep disturbances) and might be more frequent in cases of atypical depression (Colpo et al., 2018). Consistent with this, relative to depressed individuals with low CRP levels, those with high levels of inflammation had greater somatic symptoms, fatigue, greater concentration problems, indecisiveness, and elevated feelings of guilt (Foley et al., 2021).

Like other profound stressors, social rejection can influence cytokines and mood states (Eisenberger, 2013). Provoking rejection in a laboratory context elicited pronounced IL-6 elevations, which was associated with increased neural activity in the dorsal anterior cingulate cortex and anterior insula (brain regions implicated in decision-making and depression) as well as the ventral striatum (associated with reward processes). While these changes were implicated in depression, inflammation could also augment sensitivity to positive social experiences. Evidently, inflammation may both enhance avoidance of danger and the approach to positive social cues, possibly reflecting an adaptive response to diminish distress (Eisenberger et al., 2017).

Inflammation and suicidality

In view of microglial involvement in depression, it might be expected that brain cytokine levels or mRNA expression would differ among depressed individuals who died of suicide relative to that of nondepressed individuals who died through causes other than suicide.[5] While IL-1β, IL-6, and TNF-α mRNA as well as their protein expression were elevated in the prefrontal cortex of teenagers who died by suicide, it seemed that these relations were more readily attributable to suicide than depression itself (Pandey et al., 2014). The mRNA expression of the antiinflammatory cytokines IL-4 and IL-13 was also elevated in the orbitofrontal cortex of women who died by suicide (Tonelli et al., 2008). Congruent with the cytokine variations, TLR3 and TLR4 protein overexpression in the dorsolateral prefrontal cortex was linked to suicide (Pandey et al., 2019) as were NLRP inflammasomes (Pandey et al., 2021). Analyses that involved multiple approaches led to the conclusion that microglial pathophysiology and the elevated expression of inflammatory factors were key components of suicide and suicidality (Suzuki et al., 2019).

A meta-analysis indicated that most of the circulating cytokines that had been assessed (TNF-α, IFNγ, transforming growth factor-β, IL-4, and soluble IL-2 receptors) were not conclusively related to suicidal ideation, whereas IL-6 elevation showed a positive relationship to suicidality (Gananca et al., 2016). It was similarly concluded that IL-1β and IL-6 in blood, cerebrospinal fluid, and postmortem brain samples were most consistently and robustly associated with suicidality (Black and Miller, 2015). In a study of 7.2 million individuals who were followed for up to 32 years, one-quarter of those who died of suicide had previously experienced an infection that was sufficiently severe to require hospitalization, which far exceeded that expected or seen among individuals who had not been similarly infected. Moreover, the risk of suicide increased with more infections experienced (Lund-Sørensen et al., 2016).

Suicidality may be present in conjunction with several mental disorders (depression, anxiety, combined anxiety and depression, nonaffective/psychotic illness), each with its own biomarkers. These included markers linked to neurogenesis, mTOR signaling, the serotonin transporter, and the 5-HT$_{2A}$ receptor, as well as levels of APOE and IL-6. Based on these biomarkers, together with a set of other genes, suicidal intent and future hospitalization could be identified with 90% and 77% accuracy, respectively (Niculescu et al., 2017).

[5]Although recent attention on microglia has focused on their actions related to inflammatory mechanisms, glial cells can influence depression through other processes. For instance, the NG2 form of glia can release FGF-2, which then causes astrocytes to regulate glutamate within the brain. Should NG2 decline owing to genetic factors or stressors, then regulatory processes may be disturbed, culminating in depression (Birey et al., 2015).

Infection and Depression

As described in Chapter 4, diverse viral and bacterial infections, as well as common parasites, can affect the brain, thereby influencing the occurrence of depression and suicidality. In some instances, the risk of depression associated with infection was related to genes linked to immune functioning and nerve development (Ye et al., 2020). Even seasonal allergic rhinitis, which is associated with inflammatory and endocrine responses, was related to depressive symptoms (Trikojat et al., 2017), and treatment-resistant patients were more likely than others to be affected by allergies and autoimmune conditions that reflect elevated inflammatory functioning (Lauden et al., 2021).

Numerous studies showed the elevated prevalence of depression among individuals infected with SARS-CoV-2 (Mazza et al., 2023), and the incidence of anxiety and depression in university students was elevated during the pandemic relative to that evident during the preceding years, especially in females. Using a machine learning approach, several variables were identified (mood score at baseline, family history of depression, state and trait characteristics, sex) that predicted susceptibility and resilience to the development of depressive symptoms in response to SARS-CoV-2 infection (Turner et al., 2023). It is likely that some of the effects of the pandemic might have been attributable to stress related to the inflammatory response associated with infection, the distress created by social isolation, as well as appraisals (including uncertainty of events) and coping strategies that individuals endorsed. The incidence of anxiety and depression was far lower in people who were vaccinated than in those who were not, perhaps reflecting the (temporary) relief from stress enjoyed by those who had been inoculated (Chen et al., 2020).

A causal connection between infection and depression was inferred from studies showing that administration of a low dose of an endotoxin that increased plasma TNF-α and IL-6 levels elicited modestly elevated depressive symptoms and a feeling of "social disconnection" in otherwise healthy people (Reichenberg et al., 2001). Moreover, the intensity of the depressive mood was correlated with the extent of the cytokine rise (Yirmiya, 2000). In a more extensive analysis, mood together with IL-6 and TNF-α was monitored on an hourly basis following endotoxin (*Escherichia coli*) infusion. The endotoxin caused an increase in depressed mood which was moderated by several factors determined at baseline, including trait sensitivity to social disconnection, together with several socio-behavioral factors (Irwin et al., 2019). Transcriptome analyses revealed that perceived stress, sensitivity to social disconnection, and depressive symptoms were related to the endotoxin-elicited activation of proinflammatory transcription control factors (i.e., activator protein-1, NF-κB). It is notable that the mild depressive-like symptoms (e.g., lassitude, social anhedonia) brought on by an endotoxin were attenuated by pretreatment with the SSRI citalopram, but these behavioral changes were not accompanied by peripheral cytokine variations, although this does not rule out a role for altered brain inflammatory changes (Hannestad et al., 2011).

Consistent with the mood disturbances elicited by pathogens, elevated plasma IL-6 levels provoked by vaccination (e.g., typhoid vaccine) were accompanied by mild mood reductions, fatigue, and impaired concentration, which were correlated with altered neuronal activity within the anterior cingulate cortex (Harrison et al., 2009). Vaccination with live attenuated rubella virus similarly provoked a long-lasting (10 weeks) increase in depressed mood among high-risk (low socioeconomic status) teenage girls (Yirmiya et al., 2000). Paralleling these findings, influenza vaccination led to an increase in poor mood, which was considerably greater among individuals with existing anxiety/depression (Harper et al., 2017).

Imaging and Postmortem Analyses

As described earlier, imaging procedures have the potential to identify processes associated with psychiatric and neurological disorders. In a relatively large study of 1,188 participants, of which about 400 were depressed, patients could be distinguished from one another based on symptoms and functional magnetic resonance imaging (fMRI) scans (Drysdale et al., 2017). Depressed patients clustered within several categories based on distinct patterns of dysfunctional connectivity that were apparent within the insula, orbitofrontal cortex, ventromedial prefrontal cortex, and several subcortical areas, which were tied to specific clinical symptoms (e.g., feelings of sadness, hopelessness, helplessness, anhedonia, and fatigue or anergia). Importantly, these profiles could predict the effectiveness of rTMS directed at the dorsomedial prefrontal cortex.

Extending such findings, fMRI analyses in depressed patients with differing levels of inflammation reflected by elevated circulating CRP levels indicated that functional connectivity was diminished in a widely distributed network comprising the ventral striatum, parahippocampal gyrus, amygdala, orbitofrontal and insular cortices, and posterior cingulate cortex (Yin et al., 2019). The relationship between elevated proinflammatory levels and depression was tied to altered functional connectivity and disturbed activation patterns within neural circuits involved in cognitive control, emotion regulation, and reward processing. Imaging studies also revealed that elevated levels of peripheral inflammatory factors were associated with cortical thinning, reduced cortical grey matter, and disturbance of white matter tracts within the circuitry linked to depression, possibly being related to NLRP3 DNA methylation (Han et al., 2022a). Commensurate with these findings, CRP levels among depressed patients were associated with altered neuronal activity within the posterior cingulate cortex and medial prefrontal cortex, as well as functional connectivity between these regions and the hippocampus, which comprises a node of default mode network (Kitzbichler et al., 2021). Elevated blood CRP, together with elevated IL-6 and neutrophil presence, was similarly associated with disturbed connectivity between the insula and the posterior cingulate cortex. The connectivity profile was somewhat different among depressed individuals with high levels of anxiety (or PTSD) who also displayed elevated inflammation, in that dysconnectivity was apparent between the right amygdala and left ventromedial prefrontal cortex (Mehta et al., 2018b). It can reasonably be expected that further alterations would be evident in the presence of other features of depression as we saw in discussing individuals with typical versus atypical symptoms.

The findings described here have supported the view that peripheral inflammation may be a cogent marker of depression, possibly being linked to brain microstructural perturbation and altered functional connectivity. In this respect, the presence of inflammation may influence effective communication relevant to interoceptive states (e.g., sensing emotions and bodily functions) that might be linked to depression (Aruldass et al., 2021).

Traumatic brain injury and depressive illness

Among individuals who experienced TBI, about 50% developed depressive symptoms within 1 year and PTSD frequently occurred (Howlett et al., 2022). A longitudinal analysis of soldiers who had experienced TBI indicated an elevated risk of new-onset mental health conditions, including an elevated risk of suicide, more so in the presence of a substance use disorder (Brenner et al., 2023).

The initial effect of head trauma may comprise necrotic death of brain cells, followed by a second set of damaging actions. These include excitotoxicity, oxidative stress, mitochondrial disturbances, compromised BBB integrity, and elevated inflammation, which together provoke persistent and progressive damage (e.g., Simon et al., 2017). The excessive glutamate activation stemming from head injury could lead to cell loss, and some of the downstream consequences of TBI may arise owing to actions on DAMPs that promote cytokine release from microglia, although peripheral immune changes may also be contributing factors (Simon et al., 2017). During an early phase following a head insult, when various cells are involved in managing recovery from neural damage, the presence of moderate cytokine levels may be advantageous (e.g., Russo and McGavern, 2016). As inflammatory changes subsequently become more pronounced, neurodegenerative and depressive actions can emerge, which can be predicted based on the early appearance of elevated proinflammatory cytokines (Visser et al., 2022). In this respect, although research related to TBI has focused on neurobiological processes that develop following head injury, preinjury brain functionality may influence the chronic effects of TBI, especially among older people (Houle and Kokiko-Cochran, 2022).

The behavioral and cognitive impairments associated with relatively severe TBI appear to involve a frontostriatal network in which the central lateral nucleus of the thalamus is a key component. In a small but very important study it was found that among six patients treated with a form of deep brain stimulation directed at this site, the related neuronal network showed improved processing speed and enhanced functional day-to-day activities (Schiff et al., 2023). The procedure is complex since it entails multiple differences between patients' brain physiology that needed to be identified, and patients received stimulation for 12 h a day for 90 days. This feasibility study opens the door for further studies using improved procedures.

Head injury may influence connectivity between brain regions associated with emotions as well as between neuronal processes involving the thalamus, insula, and subgenual anterior cingulate cortex. As expected, the connectivity profile varied with the nature of the symptoms expressed, notably the cognitive versus the affective symptoms (Han et al., 2015b). As such, analysis of resting-state network mapping might be useful in identifying individuals who would gain from targeting the subgenual anterior cingulate cortex through rTMS (Siddiqi et al., 2023).

In mice that experienced traumatic head injury in which behavioral symptoms dissipated after 1 week, a subset of microglia was still affected for as long as 30 days. When mice were then challenged with an immunological insult, pronounced microglia activation was elicited that corresponded with depressive-like symptoms (Fenn et al., 2014). Consistent with animal studies, the altered levels of inflammatory factors and white matter degeneration in humans can persist for years after injury, and the associated depression and suicidal behavior may persist for just as long (Fisher et al., 2016).

In view of the links between inflammatory immune activation following TBI and neurodegenerative changes, antiinflammatory agents might be expected to have ameliorative effects. However, this was typically not observed and marked negative consequences could emerge (Simon et al., 2017). It is of therapeutic importance that if an SSRI (sertraline) treatment was initiated soon after head trauma,

the incidence of later depression could be diminished (Jorge et al., 2016), possibly through the drug's antiinflammatory actions. However, these benefits were not seen once depression had emerged during the year following TBI (Fann et al., 2017).

Since the effects of head injury vary over time, the actions of treatments might similarly be time-dependent. Being able to identify the time following injury during which inflammatory-promoting cells foster tissue repair versus damage will be important in the development of treatment strategies. It is essential to consider that a primary regulator of inflammation is the presence of T_{reg} cells, but these are present in low numbers in the brain. Using viral vectors, it was possible to deliver astrocyte-targeted IL-2 genes to the brain, thereby increasing the presence of T_{reg} cells and diminishing the impact otherwise produced by TBI in a mouse model (Yshii et al., 2022).

It will be recalled that an essential component of well-being is that waste products be removed from the brain, which is accomplished through the glymphatic system (Hablitz and Nedergaard, 2021). Likewise, effective functioning of the glymphatic system is necessary to overcome the consequences of TBI in which cellular debris and necrotic cells need to be eliminated (Mestre et al., 2020). However, the effectiveness of this process may be compromised by TBI, especially if suboptimal meningeal lymphatic functioning has already been present (Bolte et al., 2020).

Perhaps it isn't intuitively obvious but given that bidirectional communication occurs between the gut and the brain, this linkage may contribute to the impact of TBI. Following TBI, the abundance of several bacterial taxa was altered, including species that have been related to CNS processes (see Taraskina et al., 2022). It appeared that head trauma may instigate gut microbial alterations and peripheral inflammation, which, in turn, exacerbate neuropathology and behavioral disturbances associated with TBI.

Do Treatments That Affect Depression Involve Antiinflammatory Actions?

Ongoing questions have concerned the role of diminished inflammation in accounting for the positive effects of antidepressant medications. In animal models of depression, several antidepressants caused the suppression of humoral and cell-mediated immunity, accompanied by diminished release of proinflammatory cytokines. The most prominent effects of antidepressants comprised a reduction of IL-6 and CRP levels in blood as well as a decline of oxidative stress (Bhatt et al., 2023), and the benefits of SSRIs and SNRIs may come about through their antiinflammatory actions (e.g., Dionisie et al., 2021).

Like SSRIs, the fast-acting NMDA antagonist, ketamine, reduced IL-6 and TNF-α (Li et al., 2017c) and repeated ECT diminished immune and cytokine activity (Zincir et al., 2016). The antiinflammatory action of different antidepressant treatments could stem from their direct effects on macrophages as well as through changes in neurotransmitters sensed by these cells. Moreover, BDNF levels were linked to IL-6 concentrations among melancholic patients (Patas et al., 2014), and BDNF was increased by antidepressants, supporting the presumed involvement of this neurotrophin in the cytokine-depression link.

Nonpharmacological treatments of depression similarly produced the normalization of cytokine levels. Successful psychotherapy was accompanied by reduced levels of serum IL-6 and TNF-α, although levels of these cytokines were not correlated with the severity of depressive symptoms after the intervention (Del Grande da Silva et al., 2016). Mindfulness training was likewise accompanied by downregulated NF-κB gene expression, and among women with a history of interpersonal trauma experiences, mindfulness-based stress reduction was accompanied by a decline of plasma IL-6 (Gallegos et al., 2015). Nondrug interventions to treat depression can produce cytokine effects much like SSRIs, which is consistent with the position that the reduced depression or distress was tied to the inflammatory cytokine changes.

Modifying Depression Through Antiinflammatory Treatments

If depression emerges because of inflammation, then antiinflammatory treatments might be useful in the treatment of this illness, or at least serve as useful adjunctive treatments. Strictly speaking, this isn't necessarily correct since illness progression may involve diverse lifestyle and experiential factors and inflammation simply accompanies but does not contribute to depression. This said, depression associated with several inflammatory disorders was modifiable by immunomodulatory drugs, with efficacious treatments being realized through anti-IL-6 and anti-IL-12/23 antibodies (Wittenberg et al., 2020). Autoimmune disorders have been associated with stress-related disorders, such as depression, and in both socially stressed mice and depressed humans, brain-reactive antibodies were observed, which correlated with the levels of anhedonia. When these antibodies were depleted in mice, a superior response to stressors was evident (Shimo et al., 2023). These findings were in line with reports indicating that treatment of rheumatoid arthritis or systemic lupus erythematosus with drugs that blocked proinflammatory cytokine actions, such as the TNF-α antagonist etanercept, diminished comorbid depressive symptoms (Nerurkar et al., 2019). A meta-analysis that included seven randomized controlled trials indicated that drugs that have been used as antiinflammatory agents in the treatment of autoimmune conditions (e.g., adalimumab, etanercept, infliximab, and tocilizumab) were effective as adjunctive treatments of depression. These effects varied with baseline depressive symptom severity but, importantly, were unrelated to the alleviation of the autoimmune symptoms being treated. These findings are consistent with the view that cytokines are causally related to depression, and that the reduced depression was not simply a reflection of changes in the features of the autoimmune condition (Kappelmann et al., 2016).

Numerous antiinflammatory treatments diminished the symptoms of depression (Sakamoto et al., 2021). The broad-spectrum antibiotic minocycline enhanced the actions of standard medication among treatment-resistant patients who had displayed elevated CRP before therapy, with the effects on depression and the changes of IL-6 being more pronounced in females than in males (Lombardo et al., 2022). Likewise, among patients with low-grade inflammation reflected by elevated CRP levels, the enhancement of antidepressant treatment with add-on minocycline was more pronounced among patients with higher baseline IL-6 concentrations, and with successful treatment, IFNγ was significantly reduced (Nettis et al., 2021). The benefits of adjunctive minocycline were also related to an increase of endogenous IL-1 receptor antagonist (IL-1Ra), complement C3, and intercellular adhesion molecule 1 (ICAM-1), which contribute to inflammatory processes (Walker et al., 2022). In contrast to these findings, however, a large trial showed that minocycline provided no added benefits to antidepressant treatment in alleviating treatment-resistant depression (Hellmann-Regen et al., 2022).

In both rodents and humans, depressive symptoms could be moderately reduced by nonsteroidal antiinflammatory agents (NSAIDs) and through cyclooxygenase-2 (COX-2) inhibitors (Eyre et al., 2015), and the effectiveness of standard antidepressant medication in humans was augmented. Once more, however, several studies that assessed the effects of NSAIDs (either as adjuvant treatments or as monotherapies) revealed that they had negligible effects on depression. The source for the differences between studies is not immediately apparent. Reiterating our earlier comments, if these agents are to have benefits, it would preferentially occur among patients with elevated inflammation prior to treatment. In most studies, considerable heterogeneity existed concerning the patient population assessed and treatment regimens varied, thus limiting firm conclusions from being made (Simon et al., 2023). Without knowing the inflammatory status of patients in experimental paradigms, evaluating the effects of minocycline, NSAIDs or COX-2 inhibitors may be of limited practical value.

Exercise as an Antidepressant: Effects Through Inflammatory Processes

Considerable data have supported the position that physical exercise may diminish depressive characteristics, particularly the affective symptoms of the disorder (Ross et al., 2023). Indeed, exercise was as effective as pharmacological interventions for nonsevere depression (Recchia et al., 2022), although among older people, antidepressants were superior to physical exercise and fewer dropouts were reported. Aside from potentially acting in a therapeutic capacity, moderate regular exercise (even 1 or 2h a week) can have prophylactic effects. Based on a prospective study conducted over 11 years, it was estimated that 12% of depressive occurrences could have been prevented through regular exercise (Harvey et al., 2017).

Engaging in exercise could reduce poor mood simply by acting as a distractor, but there's likely more to it than that. Fig. 9.4 describes several ways by which exercise may produce antidepressant effects. Engaging in exercise influences circulating cytokine levels, particularly those with antiinflammatory actions, and can promote kynurenine clearance (Cervenka et al., 2017), which in relatively high concentrations can ordinarily foster depression. As well, exercise could reverse the diminished BDNF that is associated with depression (Ross et al., 2023) and can attenuate the glucocorticoid effects associated with stressful experiences. Furthermore, exercise can affect brain processes through actions on microbiota, thereby altering cognitive functioning and mood. As exercise and healthy diets may influence different gut bacteria populations, the combination of exercise and consumption of probiotic foods might be more efficacious in enhancing mood.

In instances where depressive disorders occur concurrently with elevated signs of inflammatory activity (e.g., owing to excessive abdominal adiposity), reduced inflammation produced by exercise diminished depression recurrence (Kiecolt-Glaser et al., 2015). Exercise might also promote positive actions by increasing the activity of erythropoietin (EPO), a cytokine that stimulates red blood cell production (made famous as a way of blood doping to enhance performance in endurance sports). EPO increases neuronal functioning, promotes antiinflammatory processes, serves as an antiapoptotic agent and antioxidant, and increases the synthesis and levels of BDNF (Osborn et al., 2013). Although there is evidence that EPO can influence cognitive functioning

and diminish depressive symptoms in an animal model, it is not used as an antidepressant owing to the potential risk of cardiovascular problems and stroke.

Depression Associated With Immunotherapy: The Case of IFN-α

Interferon-α and IL-2 were among the earliest immunotherapies used to treat some forms of cancer (e.g., malignant melanoma), and owing to its antiviral properties, IFN-α had been used in the treatment of hepatitis C. Regrettably, a considerable portion of cancer patients (30%–50%) developed depressive-like symptoms over the course of IFN-α therapy. In some instances, mood changes were sufficiently severe to require treatment discontinuation.

The effects of IFN-α immunotherapy are not limited to the promotion of depressive symptoms and come with a constellation of other neurovegetative symptoms and cognitive disturbances. Patients sometimes report feeling "in a fog," and at higher doses used in treating cancer, patients may experience malaise, especially during the first few days of treatment. Also, concentration and memory may be disturbed, and nonspecific features may occur, such as confusion, disorientation, psychotic-like features, irritability, anxiety, and disturbed alertness. It is possible that IFN-α provokes a nonspecific state (e.g., general toxicity) that favors a depressive-like condition. Yet, the appearance of depression stemming from IFN-α treatment was linked to the later recurrence of depression over a 12-year period, just as depression recurrence is common in depression unrelated to immunotherapy (Chiu et al., 2017).

Not unexpectedly, preexistent depressive symptoms were associated with immunotherapy-elicited depression, as well as elevated levels of IL-6 and salivary cortisol (Machado et al., 2016). Moreover, depression was prominent among individuals with low levels of the 5-HT precursor tryptophan and in patients who displayed relatively marked ACTH and cortisol elevations following IFN-α treatment (Capuron and Miller, 2011; Udina et al., 2013). Vulnerability to depression in response to IFN-α was also elevated among individuals carrying a variant of the $5HT_{1A}$ serotonin receptor (HTR1A-1019G) (Kraus et al., 2007) or a polymorphism on the gene coding for IL-6 (Udina et al., 2013). Tellingly, many of the signs of depression associated with IFN-α treatment were attenuated by antidepressant treatment (Capuron and Miller, 2011) and could be used prophylactically when administered at the start of immunotherapy.

A transcriptome analysis indicated that in patients who became depressed after IFN-α therapy elevated gene expression was related to those linked to inflammation, neuroplasticity, and oxidative stress pathways (Hepgul et al., 2016). Thus, the inclination toward depression with

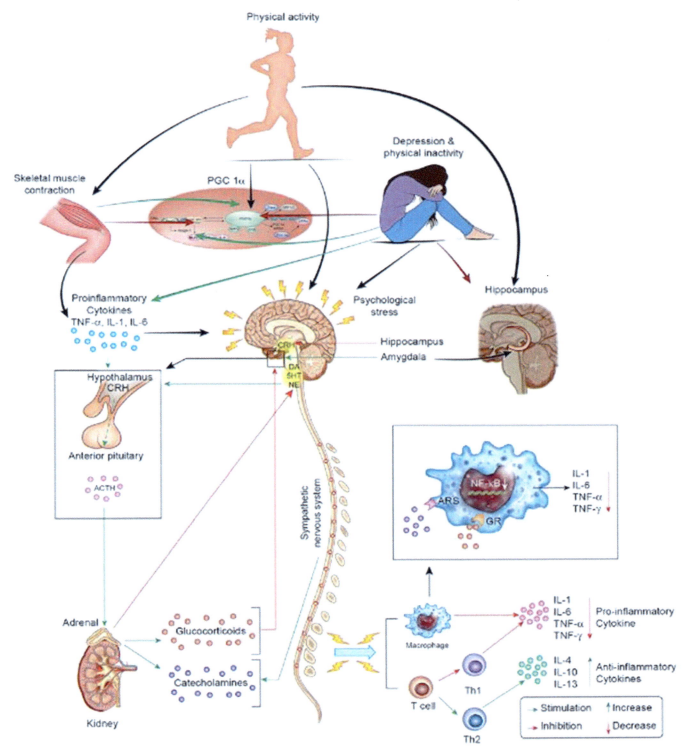

FIG. 9.4 Stress, inflammation, and depression. The HPA and sympathetic nervous systems regulate the response to stressors, i.e., cytokines, psychological stress, and PA. The systemic response to stress is initiated via CRH secretion by the hypothalamus. CRH stimulates the pituitary to secrete ACTH into systemic circulation. In turn, ACTH secretion stimulates the adrenals to release catecholamines and glucocorticoids, factors that collectively induce pro- or antiinflammatory cytokine release. Negative feedback mechanisms limit the process of inflammation in times of health. Conversely, persistent stress leads to dysregulation of the HPA with resultant endocrine disturbances in states of disease, e.g., depression. Stress-related disturbances in neuroendocrine hormones are problematic as they disrupt immune modulation and lead to a proinflammatory state. By acting as an intermittent stressor, PA exerts its central and peripheral neuroprotective effects via several avenues. During PA, muscle contractions induce the release of myokines. These factors increase the expression of PGC-1α and decrease the expression of proinflammatory cytokines at the molecular level. Moreover, PA directly modulates the neurotransmitter level and function (e.g., noradrenergic function), which is important in promoting a pro- or antiinflammatory milieu. Finally, PA increases hippocampal neurotrophic factor levels (e.g., BDNF) to promote hippocampal health and, thereby, promotes stress hormone regulation (e.g., cortisol regulation). *Source: Phillips, C., Fahimi, A., 2018. Immune and neuroprotective effects of physical activity on the brain in depression. Front. Neurosci. 12, 498.*

cytokine therapy may reflect greater sensitivity to IFN-α, which was captured by the broader gene expression changes, and by several processes linked to inflammation and growth factors. At the same time, the actions of the therapy were linked to psychosocial factors. A longitudinal study confirmed that social support was effective in diminishing symptoms and increasing compliance among melanoma patients treated with low-dose IFN-α therapy (Kovács et al., 2015).

Cytokine—Stressor synergies

Sometimes, what happens in a petri dish stays in a petri dish and doesn't generalize to the living mouse. As we know all too well, the effects seen in mice and rats don't always translate easily to humans. In vitro studies indicated that IFN-α provoked CRH release within the amygdala and hypothalamus, but in vivo administration of IFN-α to rodents had limited effects on circulating corticosterone, monoamine levels, and turnover in the brain (Anisman et al., 2007), and elicited moderate GABA and glutamate changes within the hypothalamus and limbic regions. As expected, the modestly reduced serotonin turnover and the depressive effects of the IFN-α treatment were attenuated in animals pretreated with an NSAID (De La Garza et al., 2005).

It might seem curious that IFN-α profoundly affects mood in humans treated for cancer or hepatitis C but has very modest behavioral effects in mice. However, when administered directly into the brain, the effects of IFN-α were marked, promoting increased brain cytokine mRNA expression, elevated hypothalamic neuronal firing, altered serotonin turnover within the prefrontal cortex (De La Garza et al., 2005), and increased 5-HT$_{2C}$ receptor mRNA editing. If IFN-α can access the brain in sufficient concentrations, it affects multiple neurobiological processes that could promote depressive symptoms.

It is of therapeutic relevance that the impact of cytokines on behavioral and neurochemical outcomes in mice is moderated by the background conditions upon which the treatments are administered. When animals were exposed to a psychosocial stressor, and then treated with IFN-α, marked changes in cortisol and central monoamine changes were elicited as were the behavioral signs of depression (Anisman et al., 2007). The very same effects were observed in response to a viral analog (poly I:C), or a bacterial endotoxin (LPS) (Gandhi et al., 2007; Gibb et al., 2008). The fact that synergisms occur between stressors and these immune/cytokine challenges has implications for the development of depressive illness. Within clinical settings, cytokine therapy was administered to patients with severe illnesses and the distress related to these conditions may have interacted with the IFN-α treatment to produce the affective symptoms. In essence, the impact of inflammatory agents in clinical situations might have reflected the conjoint actions of the cytokine and the stress related to illness.

IFN-α Linkages to Neurodestructive Outcomes

A view of the IFN-α depressive actions was based on the reports that cytokines, such as IFN-α, stimulate GTP-cyclohydrolase activity and indoleamine-2,3-dioxygenase (IDO), which influences the rate-limiting step of tryptophan catabolism, thereby reducing serotonin levels. More importantly, by affecting IDO, IFN-α causes kynurenine to form the oxidative metabolites, 3-hydroxykynurenine, and then quinolinic acid (an NMDA agonist), which can have neurotoxic actions, thereby promoting depressive disorders (Maes et al., 2011). According to an elaborated view of the kynurenine-depression hypothesis, depressive illness develops through a combination of inflammatory-immune, oxidative, and nitrosative stress pathways (in which reactive nitrogen and reactive oxygen species cause damage to cells) (Maes et al., 2011). These neurobiological actions stimulate inflammatory pathways that contribute to disturbed synaptogenesis and neurodegeneration, hence promoting depression.

A review of prospective studies in which chronically ill patients were treated with IFN-α indicated that depression levels were elevated between 4 and 24 weeks following the cytokine treatment, which corresponded with variations in kynurenine functioning (Hunt et al., 2020). Furthermore, among hepatitis C patients treated with IFN-α over a 12-week period, cerebrospinal fluid (CSF) kynurenine and quinolinic acid accumulation were associated with elevated depression, which was correlated with increased soluble TNF-α receptor 2 and monocyte chemoattractant protein-1 (Raison et al., 2010). Several reports in animals confirmed this sequence of changes and indicated that a tricyclic antidepressant agent, such as desipramine, reduced the elevated IDO elicited by inflammatory agents (Brooks et al., 2017).

The consequences of kynurenine pathway activation are not restricted to responses related to immunotherapy. Levels of plasma kynurenine were linked to depression (and suicidality) even in the absence of IFN-α therapy and were related to the ratios between glutamine and glutamate (Umehara et al., 2017). Among previously unmedicated depressed patients, plasma TNF-α and the kynurenine/tryptophan ratio were associated with CSF kynurenic acid and quinolinic acid, which were related to the severity of depression and the response to therapies (Haroon et al., 2020). As well, attempted suicide was linked to enzymes in the kynurenine pathway that promoted glutamate receptor excitotoxicity and neuroinflammation (Brundin et al., 2016).

Aside from the actions associated with immunotherapy, stressful events might have their depressive effects by provoking kynurenine elevations. Following a chronic mild stressor that promoted depressive-like symptoms in mice, hippocampal kynurenine levels increased together with activation of the astrocytic NLRP2 inflammasome, which appeared to be dependent on NF-κB activation and IL-1 production. When hippocampal NLRP2 expression was knocked down, the behaviors reflecting depression elicited by kynurenine were abolished (Zhang et al., 2020b).

Beyond effects on depression, oxidative and nitrosative stress pathways were associated with the occurrence of chronic fatigue syndrome (Lucas et al., 2015) and other fatigue-related disorders, diverse autoimmune disorders, cancer, cardiovascular diseases, stroke, obesity, schizophrenia, and neurodegenerative diseases, as well as AIDS dementia (Kanchanatawan et al., 2018; Savitz, 2020). As different as these disorders appear, they share common processes, such as intracellular inflammation, increased production of NF-κB, COX-2, inducible NO synthase, and damage to membrane fatty acids. Moreover, Th_{17} cells and the secretion of IL-17 may be key drivers for these effects (Slyepchenko et al., 2016). These reports have been instrumental in the consolidation of findings concerning responses to stressors, and the diverse comorbid conditions associated with depression (Wohleb et al., 2016).

The data linking IDO activation and depressive symptoms have been impressive, but there have been reports that were inconsistent with this position. Postmortem analysis of the prefrontal cortex of depressed individuals indicated that in the absence of a comorbid medical condition, changes within the kynurenine pathway were less evident. In otherwise healthy individuals, conversion of tryptophan to kynurenine was diminished and mRNA expression of IDO and tryptophan-2,3-dioxygenase was reduced, which correlated with lowered quinolinic acid levels and altered IFNγ and TNF-α (Clark et al., 2016). This isn't overly surprising since multiple factors may cause the appearance of depression, often having little to do with inflammatory processes or alterations within the kynurenine pathway. Besides, the response to IFN-α therapy also varies with the presence or absence of polymorphisms associated with cortisol and BDNF, as well as being moderated by still other factors (Udina et al., 2016). The very fact that IFN-α induces a depressive-like state in only a subset of patients underscores the importance of identifying the numerous factors that predict which patients will do best with certain treatments.

Microbiome-Immune Interactions

It has been more than a decade since depression was first associated with dysfunction of the intestinal mucosa, and the suggestion that some cases of depression might be tied to the leaky gut syndrome. Microbiota dysbiosis was linked to stressor-elicited depression and alterations of specific microbes (some species were elevated, and others were reduced) were greater among currently depressed patients than in those in remission or healthy controls (see Dinan and Cryan, 2016). Many of these microbial alterations were apparent, albeit to a lesser extent, in remitted patients and those in whom mild depression persisted, perhaps being forerunners of illness reoccurrence (Caso et al., 2021). The gut microbial involvement in illness conditions has become part of our broader understanding of the holistic processes that are tied to a broad range of psychiatric and neurodegenerative disorders. It is recognized that microbiota functioning can influence the "social brain" comprising the prefrontal cortex, amygdala, and hippocampus, and through actions on diverse hormones and neurotransmitters, changes in social behaviors can be provoked that can affect depressive mood (Sarkar et al., 2020).

With such a perspective in mind, innovative treatment strategies (fecal microbiota transplantation, dietary interventions, probiotics, prebiotics, synbiotics) have been used to diminish mood disorders and have been used as an adjunct therapy (Liu et al., 2023c). While attempts to treat depression by the concurrent use of antidepressants and microbial manipulations is a realistic goal, it is important to appreciate that microbiota and their metabolites can affect drug absorption, distribution, metabolism, and excretion, thereby affecting the efficacy of pharmacotherapy for depression. Given the pronounced individual differences in the basal microbiota population, their effective (and safe) use might necessitate a precision medicine strategy (Mundula et al., 2023).

Support for the microbiota-depression link came from the finding that symptoms of depression could be elicited by microbiota depletion through antibiotic treatment in rodents (Hoban et al., 2016a) and humans (Dinan and Cryan, 2017). Moreover, transferring gut microbiota from mice that had been exposed to a chronic mild stressor to naïve mice resulted in the appearance of depressive-like symptoms in recipient mice together with disturbed hippocampal neurogenesis and serotonergic functioning, as well as resistance to the effects of an SSRI (Siopi et al., 2020). Remarkably, transferring fecal microbiota from depressed patients to microbiota-depleted rats instigated behavioral and neurobiological features of depression (reviewed in Cryan et al., 2019). Furthermore, fecal transplantation from healthy humans in both clinical and preclinical studies successfully reduced symptoms of depression and anxiety (Chinna Meyyappan et al., 2020), and probiotic treatment diminished the corticosterone, serotonin, GABA, BDNF, and cytokine variations elicited by stressors in rodents (Cai et al., 2022).

Specific microbes have been implicated in the microbiota-depression linkage. The probiotic *Bifidobacterium infantis* attenuated immobility in a forced-swim test (a screen for antidepressant agents) among rats that had experienced early life distress, while concurrently diminishing circulating IL-6 levels. Likewise, chronic *Lactobacillus rhamnosus* elevations altered cortical, hippocampal, and amygdala functioning (Bravo et al., 2011) and diminished stress-elicited corticosterone and behaviors that reflected anxiety and depression. In this regard, early life treatment with *Bifidobacterium pseudocatenulatum* could attenuate the depressive-like effects in mice that experienced early maternal separation, possibly through its antiinflammatory impact (Moya-Pérez et al., 2017). Further to this, the levels of several genera of the *Firmicutes* phylum were reduced in association with depression, as were *Faecalibacterium*, and the low abundance of *Lactobacillus* and *Bifidobacterium* may be particularly relevant in this regard (Aizawa et al., 2016). Consistent with these reports, depression was associated with low bacteria of the Clostridia family, whereas elevated Bacteroides were accompanied by greater anxiety irrespective of the presence of depression (Mason et al., 2020).

Microbiota may contribute to myelination and myelin plasticity within the prefrontal cortex. Axons were hypermyelinated among germ-free mice, which normalized following bacterial colonization, indicating that the microbiome may be fundamental for the regulation of myelin-related processes, which could point to potential targets to deal with psychiatric disturbances (Hoban et al., 2016b). In line with preclinical studies, *Bifidobacterium longum* administered to patients with irritable bowel syndrome altered brain responses to negative emotional stimuli and diminished signs of depression (Pinto-Sanchez et al., 2017). Findings such as these might be instrumental in accounting for the comorbidity often seen between depressive illness and some autoimmune disorders.

A large-scale study identified 124 metabolites related to energy and lipid metabolism that were distinguishable between depressed and nondepressed individuals. These metabolite changes were congruent with gut microbiota alterations in the *Clostridiales* order and the Proteobacteria *Pseudomonadota* and *Bacteroidetes/Bacteroidota* phyla (Amin et al., 2023). As predicted, a probiotic combination comprising *Lactobacillus helveticus* and *B. longum* reduced 24h urinary cortisol output in healthy volunteers, and a 4-week regimen of prebiotic intake was associated with a reduction of the cortisol awakening response (Schmidt et al., 2015). This microbial combination was similarly effective in increasing levels of antiinflammatory cytokines and reducing proinflammatory cytokines (De Oliveira et al., 2023). A symbiotic mix comprising the combination of probiotics and polyphenol-rich prebiotics similarly diminished the inflammatory response in the ileum and the prefrontal cortex that was ordinarily elicited by a chronic stressor regimen, and diminished anxiety and depressive-like behaviors in mice (Westfall et al., 2021). Consistent with these reports, probiotic treatment was accompanied by reduced negative ruminative thinking and self-reported depression and influenced functional brain activity in a cognitive task (Tillisch et al., 2013).

Based on a systematic review, it was concluded that probiotics had positive effects on depressive symptoms and their corresponding neurobiological processes. Also, probiotics could enhance the effects of antidepressants among patients who had shown an incomplete response to the initial therapy (Nikolova et al., 2023). Even if microbiota were found to be instrumental in diminishing clinical levels of depression, the essential questions that remain concern which processes are responsible for these effects, and whether these could be mined to alleviate the mood disorder.

Gut Bacteria and Immune Variations

A theoretical model by which gut-related events could promote psychological disorders is provided in Fig. 9.5. This model offers multiple routes to get from the gut to the brain, including many of those that were discussed in earlier portions of this chapter, such as the actions of stressors, specific microbiota variations, and disruptions of the intestinal barrier. Through their actions on immune functioning and gut barrier integrity, various short-chain fatty acids (SCFAs) promoted by certain microbiota may come to affect depressive disorders (Morris et al., 2016a). As described in the figure, the balance between beneficial and harmful bacteria may be thrown off by stressor experiences or antibiotic intake, which can then affect immune, hormonal, and brain processes. Disturbances in the production of SCFAs by specific microbiota may affect gene expression and inflammation within the CNS and cause increased levels of quinolinic acid and kynurenic acid, thereby influencing brain functioning and behavioral disturbances (Stilling et al., 2016).

Aside from potentially affecting mood states, altered SCFAs together with ketone bodies (e.g., acetoacetate and d-β-hydroxybutyrate) may contribute to obesity, diabetes, inflammatory bowel disease, colorectal cancer, diabetes, and allergies (Kelly et al., 2015). What's more, both gut and brain processes affect autonomic

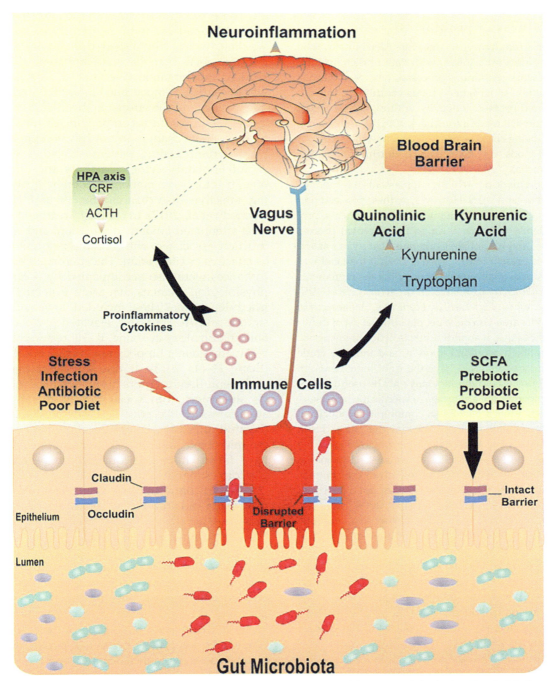

FIG. 9.5 Postulated signaling pathways between the gut microbiota, the intestinal barrier, and the brain. Gut dysbiosis could produce a microbiota-driven proinflammatory state that has implications for inflammatory immune processes that lead to illness. Source: Kelly, J.R., Kennedy, P.J., Cryan, J.F., Dinan, T.G., Clarke, G., et al., 2015. Breaking down the barriers: the gut microbiome, intestinal permeability and stress-related psychiatric disorders. Front. Cell. Neurosci. 9, 392.

functioning, which then influences visceral processes (Taché et al., 2017), thereby promoting the emergence of illness. It was also found that SCFAs in the gut, primarily coming from fruits, vegetables, beans, and legumes, can produce epigenetic effects within cells of the gut lining, and can thus be instrumental in fighting infection (Fellows et al., 2018) and possibly affecting mood states.

The usefulness of probiotics?

Well over a hundred years ago, Élie Metchnikoff, who received the Nobel prize jointly with Paul Ehrlich for their independent work on immune processes, advanced the idea that consumption of fermented foods contributed to well-being and could promote longevity. Although this concept was long ignored, it made a comeback with the understanding that well-being could be promoted by probiotics present in fermented foods (e.g., kefir, yogurt, kimchi, kefir, sauerkraut, kombucha).

Intake of probiotic or prebiotic supplements has been the "in" thing for several years. However, do these pro- and prebiotics really work? Like so many other supplements, probiotics frequently don't go through the approval process that drugs do, and it's not always certain that what's stated on the label is present in the capsule. Even if the label was accurate, how many consumers know what the ingredients signify—*Faecalibacterium*, *Proteobacteria Actinobacteria*, *Alistipes*, *Bifidobacterium*, *Lactobacillus*, *Firmicutes*, *Actinobacteria*, and *Bacteroidetes* aren't on most people's bucket list of things to know. Besides this, if an individual is deficient of specific bacteria, they likely wouldn't know it, and thus the bottle they chose may or may not have the right ingredients. In the end, the purchased probiotics may only be useful if individuals are experiencing a shortage of specific microbial species. If the individual's microbiota community is adequate, probiotic supplements may not colonize the gut and will be disposed of as waste. Eventually, a personalized medicine approach may be possible, focusing on specific microbial deficiencies, but this will not be simple given the complex interactions that exist between host genetics, gut microbiota, and diet (Ratiner et al., 2024; Zmora et al., 2019).[6]

As foods consumed may affect microbiota, immune functioning, hormones, and central neurotransmitters, dietary patterns would be expected to be linked to depressive disorders. Diet-related data are notoriously unreliable, thus studies linking diet with depression ought to be taken with a grain of salt. Several cross-sectional and prospective studies nonetheless pointed to the therapeutic benefits obtained through specific diets as well as diet counseling (Jacka et al., 2017; Zmora et al., 2019).

[6]On its surface, the notion that probiotic supplements can effectively reconstitute gut microbiota following antibiotic use seems a reasonable assumption. However, there has been debate as to whether this actually occurs. Indeed, supplements comprising probiotics may impair microbiota recolonization of the gut (Suez et al., 2018).

Considerable evidence has supported the position that microbiota could have health consequences by affecting immune and inflammatory functioning. Some gut bacteria can produce and deliver neuroactive substances, which may contribute to the emergence of illness. Conversely, the influence of LPS administered during puberty on brain cytokine mRNA expression in rodents could be precluded by probiotic treatments and reduce the enduring depressive and anxiety-like behaviors otherwise provoked (Murray et al., 2019). As expected, following treatment with the tetracycline antibiotic, minocycline, which diminished the response to stressors, gut microbial changes were apparent that promoted reduced inflammation (Wong et al., 2016), although, as we saw earlier, minocycline has antiinflammatory consequences beyond the effects on the microbiome.

Depression or the accompanying distress could potentially promote changes in gut functioning, accompanied by disturbed feeding patterns, together with elevated proinflammatory cytokine activity and altered kynurenine/tryptophan metabolism, thereby further exacerbating affective disturbances (Bercik and Collins, 2014; Kennedy et al., 2017). In contrast, treatment with *L. helveticus* attenuated anxiety, depression, and cognitive dysfunction associated with chronic restraint stress, being as effective as citalopram. The attenuated behavioral disturbances were also accompanied by a constellation of physiological changes that were expected with diminished depression. These included, lower plasma corticosterone and ACTH levels, elevated plasma levels of the antiinflammatory cytokine, IL-10, restored hippocampal 5-HT, and greater hippocampal BDNF mRNA expression (Liang et al., 2015). It also appeared that disturbing peripheral IL-6 receptors normalized microbiota dysbiosis stemming from social defeat, thereby leading to antidepressant-like effects (Zhang et al., 2017a). Given the multidirectional communication that occurs between gut microbiota and other systems, microbial changes can have far-reaching consequences on psychological and gastrointestinal disorders (Bercik and Collins, 2014).

Another connection has been identified between gut microbiota and immune system changes that are relevant to depressive illnesses. Specifically, a subset of T cells, namely intestinal gamma delta T ($\gamma\delta$ T) cells that promote inflammatory responses, are elevated in response to chronic social defeat in mice. The mice that exhibited the greatest stress susceptibility, reflected by social avoidance and diminished consumption of a sweetened solution, had the least diverse microbiota and diminished levels of *Lactobacillus johnsonii*, a form of good bacteria. When mice were treated with *L. johnsonii*, their tendency toward social avoidance declined and the $\gamma\delta$ T cells normalized. As well, when mice were fed the natural antiinflammatory agent pachyman, which is derived from wild mushrooms, the stressor-provoked behavioral disturbance and the elevated $\gamma\delta$ T cell were normalized, perhaps operating through dectin-1, which is involved in innate immune responses (Zhu et al., 2023c).

There has been interest in exploring the therapeutic effects of enhancing microbial communities to enhance health. Although depression has been linked to several microbially derived molecules, human studies must move away from simple association studies and focus on those that are geared toward understanding the causal mechanisms by which microbiota might enhance health.

This includes more detailed analyses of the impact of peripheral manipulations that influence neural, endocrine, and immune mediators, and identification of the epigenetic regulation relevant to the actions of microbiota. Once this is accomplished, microbial manipulations might be possible to treat illnesses through a personalized medicine approach.

Bipolar disorder

Numerous biological processes have been identified that may contribute to bipolar disorders, and several studies have pointed to microbial factors as playing a role in the evolution of the illness (Ortega et al., 2023). Is it a coincidence that many individuals hospitalized with mania had been treated somewhat earlier with antibiotics to treat infection (Yolken et al., 2016)? Acute episodes of mania have been linked to immune activation, whereas the decline of mania was accompanied by reduced levels of inflammatory factors (Dickerson et al., 2013). The severity of bipolar symptoms was related to microbial alterations (Evans et al., 2017), and members of the phylum Actinobacteria and the class Coriobacteria were elevated in bipolar disorder, whereas *Ruminococcaceae* and *Faecalibacterium* were less abundant (Painold et al., 2019). Increasingly, the possibility is being explored that manipulations of gut microbiota may ultimately be effective as an augmentation strategy in the treatment of bipolar disorder (Lucidi et al., 2021).

How immune system changes come to influence bipolar disorder is still vague. Nonetheless, in their impressive review, Ortega et al. (2023) made a strong case for the involvement of inflammatory immune factors in bipolar disorder and offered several reasonable strategies by which symptoms could be managed.

Concluding Comments

The symptoms of depressive illnesses differ markedly across individuals and markedly different neurobiological signatures may accompany these symptom variations. Thus, it is understandable that treatment efficacy will differ across individuals, and some patients will be resistant to treatments. For many patients, the illness may stem from excessive inflammatory immune activation, which might also account for some of the comorbid conditions that are apparent. For these patients, treatments that influence inflammatory processes might be productive.

Even with successful treatment, patients often persist in presenting with residual symptoms, and should they stop the medication, symptoms may re-emerge. The treatments do not necessarily provide a permanent "cure". If the depression stemmed solely from stressor-provoked neurochemical, immune, or microbial disturbances, then drug treatments might have beneficial effects, at least in the short-run, but for sustained benefits, the root of the problem obviously must be identified and treated.

Trial-and-error therapy is not in the patient's best interests and can have harmful effects, possibly encouraging feelings of hopelessness as patients encounter one treatment after another failing to be effective. For precision medicine, so many genetic, biochemical, and behavioral variables would need to be determined for each individual, making the feasibility and sustainability of this approach questionable. On the other side, however, it is essential to ask what the consequences are of not developing an appropriate strategy, especially as depression tied to inflammatory processes has implications for so many other diseases.

In their thoughtful review more than a decade ago, Raison and Miller (2013) tied numerous biological systems to one another, showing how these might be connected to the provocation of mental illness. It was suggested that evolution favored certain inflammatory responses, but today, this inflammatory bias is less well orchestrated and moderated (e.g., owing to lifestyles involving poor eating, lack of exercise, disturbed sleep, as well as exposure to numerous toxicants) and consequently may favor diseases. Being able to identify biomarkers associated with illness may go a long way in facilitating the selection of optimal therapeutic strategies.

CHAPTER

10

Anxiety Disorders

The two sides of anxiety

It's almost certain that all of us will, at some time, experience fear and anxiety. If the anxiety levels are not excessive, then this emotion might have adaptive value, as it keeps individuals prepared and alert for potential dangers and might instigate or contribute to high levels of performance in various milieus (stage presentations, athletic competitions). When anxiety becomes excessive and persistent, behavioral and cognitive responses, as well as social functioning, may be impaired. Moreover, high levels of anxiety are often comorbid with other psychological disorders, such as depression, posttraumatic stress disorder (PTSD), and substance use disorders, and may contribute to physical illnesses, such as heart disease. It is important to distinguish between anxiety and fear. Anxiety comprises worry or uneasiness that is broad and unfocused, whereas fear reflects a response to a real or imagined threat. Anxiety also needs to be distinguished based on it being a trait characteristic (a component of the individual's personality) or a state that is brought about by specific experiences. Several anxiety disorders have been identified [e.g., generalized anxiety disorder (GAD), phobias, social anxiety disorder] that differ in their characteristics. Existential anxiety also exists which may stem from individuals questioning whether their life has meaning or purpose (often termed existential crisis, but it also borders on nihilism). It can also encompass instances in which individuals face uncertainty and worry regarding the survival of their ingroup (e.g., a religion) owing to threats of its annihilation or disappearance through other ways. For highly anxious individuals "the mind never rests," and some patients with comorbid anxiety and depression report that the anxiety is more debilitating than the depression itself.

Neurobiological Factors and Treatment of Anxiety Disorders

Anxiety disorders are among the most common mental illness in Western countries, frequently being linked to a family history of anxiety. In many instances, adult anxiety may have first raised its threatening head in childhood, and if left untreated may persist throughout life. The various anxiety disorders that exist, as described in Table 10.1, can be exceptionally debilitating, often undermining individuals' social and work functioning. Given the frequency of anxiety disorders, it is somewhat remarkable that they haven't gained the notoriety of depression in the public mind. Like depression, the great majority of cases of anxiety disorders are not brought to the attention of physicians, and consequently go untreated.

In assessing anxiety-related processes, it is essential to differentiate between the detection of threats, the response to threats in the form of conscious fear, and the anxiety that arises owing to unconscious processes. Aside from the symptoms related to conscious and unconscious processes differing from one another, the detection, appraisal, and anxiety responses elicited by threats can differ from those that promote fear responses. Furthermore, they may be subject to distinct antecedent factors and predisposing influences, and attenuation of these feelings may require different treatment strategies (LeDoux, 2014; LeDoux and Pine, 2016). Likewise, individual difference factors are associated with anxiety. Much like people who can't handle ambiguity or uncertainty, some individuals generally seem to always see situations negatively (glass half-empty), which is accompanied by worry and rumination, as well as depression. When these worriers were shown a negative image and instructed to put a positive spin on it, they found it difficult to do so, and analyses of their brain activity suggested that their negativity worsened. It seems that negativity is a deep-seated characteristic among some individuals that is difficult to change, although in slow steps, such as through cognitive behavioral therapy (CBT), this becomes possible.

Fear and anxiety are often considered together, although they are distinguishable from one another in several ways, even differing in the brain processes that underlie these emotions. Studies in both animals and

---

The Immune System
https://doi.org/10.1016/B978-0-443-23565-8.00001-6

269

Copyright © 2025 Elsevier Inc. All rights are reserved, including those for text and data mining, AI training, and similar technologies.

TABLE 10.1 Anxiety Disorders.

Generalized anxiety disorder (GAD): Persistent (at least 6 months) anxiety or worry that is not focused on any single subject, object, or situation. It is accompanied by at least three of the following: restlessness (or feeling on edge), easily fatigued, difficulty concentrating, irritability, muscle tension, and sleep disturbance comprising either difficulty falling asleep or staying asleep (or restless sleep). The life-time prevalence of GAD is about 5%, occurring twice as often in females relative to males, frequently first appearing when individuals are in their 20s or 30s, but can initially appear early in life.

Phobias: The presence of significant anxiety (fear) elicited by certain situations (e.g., heights, open spaces, snakes, spiders), activities, things, or people. It becomes a problem when individuals display an excessive and unreasonable desire to avoid or escape from the feared object or situation. Phobias are common, with the estimated prevalence being approximately 8%–9%, and are more frequent in women than in men.

Social phobia and social anxiety disorder: These conditions are characterized by an intense fear of public scrutiny, embarrassment, or humiliation (negative public scrutiny) often being evident across multiple situations, but may be restricted to specific venues. The anxiety is especially pronounced in the presence of elevated scrutiny, such as when an individual is required to speak or perform publicly, or when they must interact with others. Social anxiety is common, occurring in 6.8% of people, equally distributed among females and males. It often begins during childhood, persisting into adolescence and adulthood. In children, it may be incapacitating, to the extent that they are fearful of playing with others or speaking to teachers.

Separation anxiety disorder: Characterized by fear or anxiety in relation to separation from an attachment figure, which is inappropriate for the person's development age. Persistent worries and anxiety about the harm being experienced by the attachment figures could be responsible for the development of the disorder. Children may experience physical symptoms characteristic of distress and may have nightmares. Although symptoms occur most often during childhood, separation anxiety can be diagnosed in adults.

Panic disorder: Characterized by discrete periods (usually 1–20 min) of sudden intense apprehension or terror, with episodes occurring for a period longer than 1 month. Panic episodes are accompanied by additional symptoms, including shortness of breath, chest pain, palpitations, feelings of choking or smothering, nausea/abdominal distress, sweating, trembling, feeling dizzy, unsteady or lightheaded, depersonalization (person feels detached from themselves) or derealization (a feeling of unreality), fear of dying, hot flashes, and a fear of losing control. Panic disorder occurs within 2.7% of the population, being twice as likely to appear in women as in men.

Obsessive-compulsive disorder (OCD): This disorder has been separated from anxiety disorders in the DSM-5, but is nevertheless characterized by the presence of repetitive obsessions that involve distressing, persistent, and intrusive thoughts that provoke anxiety and compulsions. The obsessions may involve a preoccupation with particular thoughts (e.g., with sexual or religious behaviors) that provoke anxiety. The anxiety may be alleviated, for a time, by behaviors being acted out (which eventually become compulsive behaviors). Individuals may display ritualistic behaviors, such as checking and rechecking (e.g., whether the door is locked, or the stove is off) or engaging in specific behavioral sequences. Moreover, obsessions can become more formed so that individuals will interpret certain objects as being significant or "meaningful," or the obsession can take the form of delusional behaviors (e.g., presence of conspiracies). This disorder has a life-time prevalence of 1%–2%, being equally common in females and males, often first appearing in childhood or adolescence.

humans have pointed to the amygdala and the bed nucleus of the stria terminalis (the extended amygdala) being differentially involved in fear and anxiety, respectively (Davis et al., 2010). Recent formulations indicated that the activity of neurons within the bed nucleus of the stria terminalis (BNST) may govern responses to temporally or spatially distant challenges. Thus, a threat that is immediately expected gives rise to increased amygdala neuronal activity, whereas in response to a threat anticipated at a somewhat later time, the BNST is more active (Klumpers et al., 2017). The clinical implications concerning these differences or those related to neuronal connectivity involving these regions remain to be fully determined. However, the differences detected through imaging procedures may prove to be useful in the differential diagnoses of stress-related disorders (e.g., generalized anxiety, social anxiety, PTSD) and could potentially inform the best treatment approaches (Knight and Depue, 2019). Even though the amygdala and the BNST might serve different functions, it was argued that these regions should be viewed as an integrated unit, particularly since they respond similarly to an assortment of threats and challenges, even those that are uncertain

(LeDoux and Pine, 2016), as well as persistent experiences within nonspecific threatening contexts.

There are clearly many differences between the various anxiety disorders, but given that they share behavioral features, there ought to be overlapping neurobiological features associated with them. Multiple neurochemical processes have been linked to anxiety disorders, and many brain regions may contribute to this, including those tied to appraisal processes and memory of aversive or threatening events. For instance, the diminished volume of the inferior frontal cortex was accompanied by relatively high levels of anxiety together with a negativity bias (Hu and Dolcos, 2017), which could potentially graduate to a clinical level of pathology. Moreover, dispositional negativity, hypervigilance, and attentional biases serve to promote the development and maintenance of anxiety, likely involving amygdala, prefrontal cortical, and locus coeruleus functioning (Shackman et al., 2016). Imaging studies revealed that although different anxiety conditions can be distinguished from one another, they also share common features. In this regard, GAD, social anxiety disorder, obsessive-compulsive disorder (OCD), and panic disorder may be accompanied by altered network activity, although default mode network connectivity and the salience network may be comparable in only some of these conditions (Peterson et al., 2014).

Determining the neurobiological mechanisms associated with anxiety disorders has been difficult for several reasons, including that anxiety can be elicited by many stressful stimuli, which as we've seen, can have different neurobiological consequences. This is made still more complicated by the distress that individuals may experience because of their psychological illness, and consequently, it is uncertain whether brain neurochemical alterations are linked to the anxiety disorder or to the distress that occurs. Aside from this, some forms of anxiety may be linked to conscious experiences, whereas others may be tied to nonconscious events (LeDoux and Brown, 2017).

In view of the array of brain regions that contribute to anxiety disorders, an understanding is necessary concerning the connections that tie cortical and subcortical neuronal processes to the effects of aversive experiences. In this regard, it was demonstrated that the failure to control emotions might be mediated by the lateral frontopolar cortex failing to coordinate and control messages coming from the amygdala and sensorimotor cortex. Instead, a shift occurs so that greater activity occurs in the dorsolateral and medial prefrontal cortical regions. As a result, neuronal activity linking the lateral frontopolar cortex and amygdala persists so that even mild emotional challenges can elicit neuronal hyperexcitability, thereby diminishing control of emotional actions (Bramson et al., 2023). Furthermore, the activity of specific neurons within the prefrontal cortex (i.e., parvalbumin interneurons) is altered with a chronic mild stressor regimen, but these actions occurred more readily in female mice, which mapped onto the development of anxiety symptoms (Woodward et al., 2023). Similarly, stressors affect neurons within the claustrum, which comprises a thin sheet of neurons and glial cells that connects the cortex to several subcortical regions (e.g., amygdala, hippocampus), serving as a hub for information from various stress-responsive regions that determine emotional responses (Niu et al., 2022).

In diagnosing and treating anxiety disorders, as in any other psychiatric illness, it would be ideal to have biomarkers or behavioral indices that could direct clinicians to specific treatments. Efforts have been made using neuroimaging and genetic markers to this end, as well as the identification of various hormonal substrates, neurotrophic factors, inflammatory factors (cytokines), and neurophysiological measures (EEG, heart rate variability) that were reliably associated with anxiety. Several markers proved to be tempting candidates in this regard, but a consensus statement from a broad group of researchers suggested that none were sufficient and specific to the extent that they could be used as diagnostic tools (Bandelow et al., 2016).

Neurobiology of Anxiety

To a significant extent, data relevant to human anxiety have come from animal models with a particular focus on the impact of stressors. Unfortunately, the anxiety elicited in many of these models should not be misconstrued as necessarily reflecting anxiety disorders, especially as the manipulations used also elicit depressive-like behaviors and PTSD. For that matter, studies of the behavioral effects of stressors often lump anxiety and depression together given that they may be difficult to disentangle in some animal models, and many of the treatments used to treat anxiety disorders are precisely those that are used to diminish depressive illnesses and PTSD (e.g., selective serotonin reuptake inhibitors; SSRIs).

Anxiety disorders are accompanied by variations in brain morphology and connectivity. Some of the structural variations seen in adults were also evident in pediatric anxiety disorders in which ventromedial prefrontal cortex and left precentral gyrus thickness was increased, possibly being related to disturbed emotional processing. The enlarged size of the striatum seen in many anxiety patients is linked to an inability to deal with uncertainty and ambiguity concerning future threats (Kim et al., 2017c), again speaking to specific symptoms being predicted by particular brain changes. It also seems that specialized cells within the hippocampus may only fire in response to anxiety-provoking stimuli or situations and do so without input from cortical brain regions (Jimenez et al., 2018). Once activated, these cells trigger hypothalamic neurons involved in controlling several stress hormones and affect cardiovascular responses.

Corticotrophin-Releasing Hormone (CRH)

As described in earlier chapters, stressors readily provoke the secretion of CRH from the paraventricular nucleus of the hypothalamus, which instigates hypothalamic pituitary adrenal (HPA) activation and is essential for allostasis. The involvement of CRH in stress responsivity goes beyond HPA functioning, as this hormone's role in fear and anxiety is attributed to changes that occur within specific aspects of the amygdala. It will be recalled that a threat that is expected immediately is accompanied by amygdala neuronal activity, whereas neurons within the BNST (the extended amygdala) are more closely aligned with temporally or spatially distant threats (Klumpers et al., 2017). Stressors also influence CRH activity within the prefrontal cortex and the hippocampus, contributing to appraisals, decision-making, and memory, all of which are involved in the development of anxiety. Moreover, stressor-elicited CRH stimulation of locus coeruleus activity may promote elevated vigilance.

Consistent with the involvement of CRH in anxiety, the provocation of CRH overexpression in the dorsal amygdala of monkeys promoted anxious temperament, and connectivity within components of anxiety circuits was altered (Kalin et al., 2016). Predictably, treatments that increase CRH in rodents typically promoted anxiety, whereas CRH receptor antagonists had the opposite effect (Slater et al., 2016). These actions were dependent on the type of CRH receptor that was activated. Anxiety stemming from the overproduction of CRH was attenuated by pharmacologically antagonizing the CRH_1 receptor subtype and was likewise diminished by genetic deletion of CRH_1 receptors (Reul and Holsboer, 2022). Paralleling such findings, the antagonism of CRH_1 receptors attenuated anxiety associated with inflammation provoked by lipopolysaccharide (LPS) (Sun et al., 2023b).

The functions of CRH_1 and CRH_2 receptors in anxiety have yet to be fully deduced. It seems that CRH_1 receptors primarily contribute to emotional responses, whereas the role of CRH_2 receptors is less clear (Reul and Holsboer, 2022). Still, another view is that CRH_1 receptors mediate emotional as well as executive functions, attention, and learning about emotions. The activation of CRH_2 receptors, in contrast, contributes to stress-related changes in basic functions necessary for survival, such as feeding, reproduction, and defense. From an applied perspective, patients presenting with anxiety and depression might benefit most from treatments that modify CRH_1 receptors, whereas patients with eating disorders would benefit more from treatments that affect CRH_2 receptors. Regrettably, for a variety of reasons, limited headway has been realized regarding the development of CRH antagonists for clinical purposes.

Norepinephrine

Peripheral norepinephrine produces signs of anxiety (e.g., elevated heart rate), which may serve as a signal to the individual (feedback) that they are anxious, thus giving rise to emotional responses. By virtue of effects on HPA functioning as well as neuronal activity within other brain regions, norepinephrine may affect anxiety responses. Specifically, stressor-provoked activation of the locus coeruleus, from which norepinephrine neurons originate, and the ensuing prefrontal cortex activation resulted in heightened vigilance and anxiety (e.g., Borodovitsyna et al., 2018). The locus coeruleus norepinephrine neurons also affect amygdala CRH functioning through which anxiety can be produced (Daviu et al., 2019). The intersection between norepinephrine and

CRH has also been posited to occur through other routes. With repeated stressor exposure, norepinephrine neurons within aspects of the amygdala (basolateral amygdala—BLA) may be sensitized through CRH_1 processes, which then favors the development of hyper-reactivity and anxiety (Rajbhandari et al., 2015). Irrespective of the mechanism, the view that norepinephrine played a role in anxiety was reinforced by the antianxiety effects provoked by β-norepinephrine antagonists (e.g., propranolol).

Serotonin

There has long been the view that serotonin functioning may contribute to the development of anxiety. SSRIs are effective as anxiolytics (e.g., Curtiss et al., 2017), and the functioning of the serotonin transporter gene may be predictive of the efficacy of drug treatments (Lueken et al., 2016). Based on studies in the marmoset monkey, it seems that trait anxiety may be mediated, in part, by amygdala serotonin functioning. The elevated anxiety elicited in these critters by a threat (a human staring at them) could be attenuated by the administration of an SSRI, citalopram, directly into the amygdala (Quah et al., 2020). Studies in rodents have also indicated that specific serotonin receptors (e.g., $5\text{-}HT_{1A}$) are involved in anxiety (Albert et al., 2014; Borroto-Escuela et al., 2021) and there is reason to suppose that $5\text{-}HT_{2A}$ and $5\text{-}HT_{2C}$ receptors also contribute to anxiety and depression.

Serotonin certainly doesn't act alone in generating anxiety, and potent antianxiety effects could be achieved by a combination treatment that affects both norepinephrine and serotonin (Gosmann et al., 2021). Although serotonin and norepinephrine reuptake inhibitors (SNRIs) are effective in diminishing anxiety, at higher doses SSRIs are more effective (Jakubovski et al., 2019). Like CBT, SSRIs and SNRIs are effective in attenuating pediatric anxiety and produce only minor side effects. As anxiety in adults often has its roots in childhood and adolescence, continued treatment may be necessary to thwart anxiety from emerging in adulthood (Patel et al., 2018a).

GABA and Glutamate

One of the more prominent consequences of stressors is the change of gamma-aminobutyric acid (GABA) activity and that of the subunits that make up $GABA_A$ receptors (Poulter et al., 2010). Moreover, threats that elicited anxiety were associated with altered GABA activity relative to that evident during a safe period. Among other actions, GABA interacts with serotonin and CRH, which might also contribute to anxiety symptoms. It also appears that $GABA_A$ receptor expression may be affected by ovarian hormones, such as progesterone (and vice versa), which could account for sex differences in anxiety-related conditions (Gilfarb and Leuner, 2022). These interactions may also account for the heightened anxiety evident at points within the menstrual cycle, during pregnancy, and during the postpartum period. The effectiveness of benzodiazepines in reducing anxiety has been attributed to their actions on $GABA_A$ functioning and agents that augment GABAergic tone, such as valproate, vigabatrin, and tiagabine, diminished anxiety. Although most studies focused on the role of $GABA_A$ in anxiety, elements of $GABA_B$ functioning may also contribute to aspects of anxiety. Behavioral studies indicated that $GABA_B$ modulators influence anxiety and may contribute to the long-term consolidation of contextual memories that influence generalized fear (Lynch et al., 2017). Different aspects of the amygdala and the BNST influence multiple brain regions, thereby influencing anxiety and related behaviors.

As discussed in the context of depressive disorders, GABA and glutamate act in a coordinated fashion, and glutamate can be expected to play a prominent role in anxiety (Nuss, 2015). Diminished anxiety has been associated with glutamate activation and kappa opioid receptor functioning. Upon being activated, kappa opioid receptors cause glutamate release from the BLA inputs to the BNST, and hence anxiety levels increase (Limoges et al., 2022), whereas deletion of kappa receptors within the amygdala produces an anxiolytic phenotype (Crowley et al., 2016). Similarly, glucocorticoid functioning within several glutamate circuits, notably in the forebrain and BLA, is fundamental to the regulation of fear and anxiety (Hartmann et al., 2017). In line with this, the synthetic corticoid dexamethasone administered chronically can undermine the coordination of GABA and glutamate functioning, thereby promoting stress-related anxiety. Glutamatergic hyperactivation in the amygdala, prefrontal cortex (PFC), and hypothalamus have been implicated in the development and maintenance of several anxiety disorders, including GAD, social anxiety disorder, and panic disorder (Boff et al., 2022).

Cannabinoids

Cannabinoid activity has been tied to reduced anxiety and depression, although in a subset of individuals, it can elicit excitation and elevated reactivity. Aside from generalized anxiety, cannabinoids have been implicated as potential treatments for panic disorder, social anxiety disorder, OCD as well as PTSD. The psychoactive component of cannabis, Δ9-tetrahydrocannabinol (THC), acts through endogenous endocannabinoids (eCBs) that bind to particular CB_1 and CB_2 receptors that are located within numerous brain regions, many of which have been tied to anxiety (Hill et al., 2018). As described in Chapter 6, basal and threat-elicited anxiety may be governed, at least in part, by the functioning of two eCBs, anandamide (AEA) and 2-arachodonoylglycerol (2-AG), which serve as the gatekeepers of the stress response (Morena et al., 2016). A calm disposition can be sustained because AEA tonically affects the CB_1 receptor; however, in response to stressors, CRH is released at the amygdala, which ultimately reduces the signaling by AEA, thereby contributing to anxiety. In essence, the role of AEA is to maintain individuals in a relaxed state during nonstress periods and contribute to the HPA stress response and anxiety that stems from stressor experiences. Activation 2-AG is likewise provoked by stressors, largely acting to keep the stress response in check, and may be involved in the adaptation associated with chronic stressors. It was proposed that disturbances in this system may contribute to the development of stressor-induced depression and PTSD (Morena et al., 2016). This view was supported by the finding that experiences with control over a stressor may affect eCB-provoked neuronal activity in forebrain circuits so that resilience in response to later challenges is enhanced (Worley et al., 2018). Thus, 2-AG may be a potential target in the treatment of anxiety-related conditions (Bedse et al., 2020).

Activation of eCBs and CB_1 receptors regulate the release of several neurotransmitters and hormones that influence stress reactions, and the application of a CB_1 agonist into the BLA diminishes stressor-induced HPA activation and might contribute to the consolidation of stressor-provoked emotional memories (Hill et al., 2018). In addition to these effects, eCBs affect hippocampal cannabinoid receptors and modulate synaptic plasticity, thereby altering stress responses (Scarante et al., 2017). As expected, in mice, a stressor in the form of social defeat results in eCB promotion of synaptic plasticity within the nucleus accumbens that might contribute to anxiety and depressive symptoms (Bosch-Bouju et al., 2016).

As described in Fig. 10.1, cannabinoids can inhibit microglial activation, thereby diminishing inflammatory responses within the brain (Lisboa et al., 2016). Similarly, cannabinoids can inhibit cyclooxygenase-2 (COX-2) and thus might act like other COX-2 inhibitors in diminishing inflammation (Patel et al., 2017). It also appears that CB_2 has immunomodulatory actions and enhanced the actions of coadministered bacitracin in fighting against gram-positive bacteria (*Staphylococcus species*, *Listeria monocytogenes*, and *Enterococcus faecalis*), allowing for the use of lower doses of antibiotics (Wassmann et al., 2020).

For the most part, the actions of cannabis had been assessed acutely, and much less information had been available concerning the influence of its long-term use

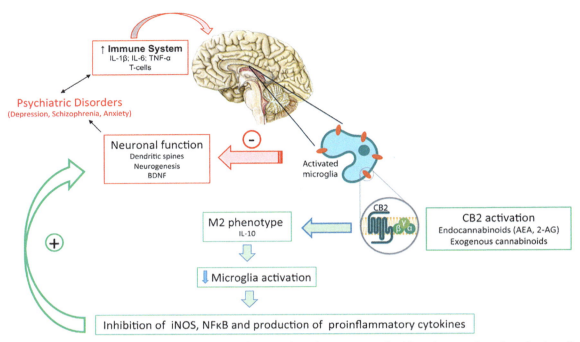

FIG. 10.1 Disorders such as depression, schizophrenia, and anxiety have been associated with an increase in activated microglia as well as neuronal and synaptic damage. Activation of cannabinoids receptors (primarily CB2 on activated microglia) may drive M2 microglia, creating a state in which the proinflammatory factors would be diminished, whereas antiinflammatories increased. As a result, neurons would be protected from damage and psychopathology would thus be reversed or prevented. *Source: Lisboa, S.F., Gomes, F.V., Guimaraes, F.S., Campos, A.C., 2016. Microglial cells as a link between cannabinoids and the immune hypothesis of psychiatric disorders. Front. Neurol. 7, 5.*

in humans. This had largely been because of government prohibitions related to cannabis use, including its clinical testing. With these prohibitions having been reduced in the United States through the Medical Marijuana and Cannabidiol Research Expansion Act, greater understanding of the positive and adverse effects of cannabis will no doubt become better understood. The eCB system is involved in brain development, so THC can affect the maturation of brain regions that govern decision-making and self-control in young people, which may also become apparent upon THC reexposure in adulthood (Ferland et al., 2023). As well, cannabis use during pregnancy can affect offspring, increasing the occurrence of anxiety, aggression, and hyperactivity, together with elevated cortisol and reduced expression of genes associated with immune functioning (Rompala et al., 2021).

Despite the belief that cannabis can act against some mental disorders, a meta-analysis indicated that it is ineffective in the treatment of most psychological illnesses other than slightly reducing anxiety associated with other medical conditions, such as noncancer pain and multiple sclerosis (Black et al., 2019). On the contrary, cannabis can induce persistent functional brain changes, and impaired neuronal plasticity and organization, most prominently in adolescents. It also appeared that even though cannabis produces munchies, its use during adolescence may disturb energy balances, fat cells, and fat storage homeostasis, which can ultimately disturb brain processes that are reliant on adequate nutrient supply (Lin et al., 2023).

In a subset of individuals, powerful cannabis strains could favor the development of schizophrenia (Renard et al., 2018) and early chronic cannabis use may stunt the development of brain white matter. A report that involved more than 69,000 teenagers indicated that moderate cannabis users were 2–4 times more likely to experience major depression or suicidal ideation than nonusers, and they also exhibited cognitive disturbances (difficulty concentrating, slower thoughts) and diverse social problems, especially if cannabis was used chronically (Sultan et al., 2023).

Numerous reports pointed to cannabis being used as self-medication to quell anxiety, depression, insomnia, and pain. In part, individuals may turn to cannabis because other treatments are ineffective. However, cannabis use may create harm among some individuals, which outweighs the benefits obtained. Instances of cannabis overdose reported to poison centers in the US have been increasing yearly for more than a decade, and it is suspected that in some cases these reflected suicide attempts (Graves et al., 2023). Aside from these psychological consequences, a report that included 430,000 individuals indicated that cannabis use, irrespective of whether it was smoked, eaten, or vaporized, was associated with the elevated occurrence of coronary artery disease, myocardial infarction, and stroke, with the odds increasing with greater cannabis use (Jeffers et al., 2024).

Anxiety Disorders

Generalized Anxiety Disorder (GAD)

It is generally accepted that inappropriate threat appraisals or a failure to inhibit responses to nonexistent threats might contribute to the emergence of GAD. Relatedly, this disorder may reflect overgeneralization so that individuals see nonthreatening events as being a threat, which promotes persistent worrying. As communication between the anterior cingulate cortex and the BLA ordinarily serves in the appraisal of threat and safety, dysfunction of these connections could be responsible for persistent threat appraisals. Ordinarily, the anterior cingulate cortex influences amygdala activity and can limit anxiety, so that dysfunction of the white matter tract connecting the anterior cingulate cortex and amygdala (the uncinate fasciculus) would result in the inhibition of the amygdala being lost, hence leading to persistent anxiety (Hur et al., 2019). However, based on a systematic review, it was concluded that generalized anxiety was more complex, being accompanied by deficient prefrontal cortex and anterior cingulate cortex functioning together with disturbed top-down control operations during emotion regulation tasks (Mochcovitch et al., 2014). This pathway was viewed as important for the extinction of fear responses, as well as fear reinstatement by specific environmental triggers, and thus a disturbance in this regard would result in persistent fear and anxiety (Likhtik and Paz, 2015). As well, in GAD patients who were instructed to think about a recent stressful event, and to keep thinking about this ("perseverative induction procedure"), the increased anxiety and worry expressed was associated with lower connectivity between the ventromedial prefrontal cortex and the right amygdala, as well as between the amygdala and thalamus (Makovac et al., 2016). Importantly, connectivity between these regions predicted the later development of GAD.

There is good reason to suppose that cortical GABA, together with midbrain serotonin, $5\text{-}HT_{1A}$ receptors, and 5-HT reuptake processes, influences dopamine functioning, which then affects several anxiety disorders, including GAD. Further, GAD is associated with elevated levels of inflammatory factors, just as this was reported in other anxiety-related disorders, as well as PTSD (Michopoulos et al., 2017). Although most studies showed similar outcomes, a sizable number of studies indicated that markers of inflammation were not elevated in GAD (Costello et al., 2019). Of course, even where cytokine and CRP levels were found to be higher among GAD patients, these studies do not speak to the causal connections between them.

Treatments of GAD

Benzodiazepines can be effective for acute anxiety, as well as for several anxiety-related disorders, including GAD, panic disorder, and social anxiety disorder, and when given together with SSRIs or psychological treatments, still better outcomes may be realized (Starcevic, 2014). However, benzodiazepines are not recommended for long-term use owing to the development of tolerance and physical dependence. There has also been concern that the positive effects of benzodiazepines stemmed from their sedative/hypnotic properties rather than their effects on anxiety. Also, many individuals asked about the impact of benzodiazepines reported numerous persistent symptoms (low energy, distractedness, memory loss) that were distinct from the symptoms for which the benzodiazepines had originally been prescribed (Ritvo et al., 2023). Not all individuals exhibit benzodiazepine-induced neurological dysfunction during benzodiazepine use or in tapering use to withdraw from addiction, but what accounts for the individual differences is uncertain.

As an alternative to benzodiazepines, SSRIs and SNRIs are more commonly used for GAD, especially when depressive symptoms are also present. One of the problems that can be encountered in a subset of patients is that increased feelings of anxiety may be experienced early in treatment with SSRIs, possibly owing to CRH_1 receptor activation within the BNST (Marcinkiewcz et al., 2016). The effectiveness of SSRIs (escitalopram or sertraline) was associated with reductions in plasma cytokine levels, and baseline levels of CRP and IL-6 were predictive of the response to the SSRI (Hou et al., 2019).

Several other 5-HT-acting drugs have also been used in the treatment of GAD, such as buspirone (BuSpar), a $5\text{-}HT_{1A}$ receptor partial agonist that also acts as a dopamine D_2 and α-adrenergic antagonist. The GABA-acting agent, pregabalin (Lyrica), which is known for its effects on neuropathic pain and seizure control, has also been used in treating GAD (Generoso et al., 2017). Although it had been maintained that the risk for dependence was low this is questionable. A related compound gabapentin may have similar effects on GAD, although with fewer side effects, and has been used less often.

In addition to pharmacological treatments, CBT has been used to diminish GAD symptoms (Carpenter et al., 2018) as has mindfulness training (Ghahari et al., 2020). GAD, like other anxiety disorders, is associated with elevated inflammation within several brain sites, although it is uncertain whether the effects of therapies are reliably associated with diminished inflammation (Michopoulos et al., 2017). Indeed, in some studies, effective CBT was not accompanied by reductions of proinflammatory cytokines, IL-6, IL-8, and TNF-α, although other cytokines might have been altered (Santoft et al., 2020).

While the cognitive/behavioral approaches may have relatively sustained effects, the pharmacological route appears to promote a more immediate fix, although even in the case of pregabalin, about 1 week of treatment is needed for ideal effects to become evident. The selection of treatments needs to be done judiciously since none of these treatments are effective for everybody, only producing appreciable remission in 50%–60% of individuals.

Panic Disorder

The symptoms of panic disorder (see Table 10.1), particularly its sudden appearance, have been both puzzling and intriguing. Several cognitive explanations were advanced to account for the development and maintenance of panic disorder (Schmidt and Keough, 2010). An emotion-based perspective suggests that owing to genetic factors or life experiences, some individuals are disposed to overreacting to stressors. These reactions are persistent, possibly owing to Pavlovian conditioning, so that internal sensations may be "catastrophically" misinterpreted as being especially threatening, perhaps owing to disturbances within the prefrontal cortex, which then promotes psychological and physical arousal and elevated perceived threat.

Several potential neurochemical mechanisms were offered to account for panic disorder. These included alterations of neuropeptide factors, such as CRH, arginine vasopressin (AVP), and cholecystokinin (the latter is a gut peptide better known for its role in digestion and satiety). Genetic analyses have also pointed to several variants of the glycine receptor B gene playing a role in panic disorder by affecting a fear network (Deckert et al., 2017). This disorder was also attributed to lower concentrations of GABA in the anterior cingulate cortex and basal ganglia, and serotonin was similarly offered as being a primary contributor to panic disorder. Among individuals with panic disorder, 5-HT$_{1A}$ receptor binding was altered, and indirect support for serotonin involvement in this condition has come from reports that SSRIs can be used to diminish panic disorder (Chawla et al., 2022).

Panic disorder has been associated with impaired activation of portions of the prefrontal cortex in response to passively viewed negatively valenced pictures, suggesting involvement of top-down regulation (Wang et al., 2021b). Similarly, in panic disorder patients, selective dysfunction may occur within the cortico-limbic network in response to emotional stimuli (Oliva et al., 2021). Earlier studies had indicated that upon exposure to emotionally salient cues that comprised anxiety-provoking visual stimuli or threatening words, patients exhibited markedly elevated neuronal activity in the anterior cingulate cortex, posterior cingulate cortex, orbital frontal cortex, and hippocampus, which are involved in appraisal and executive processes (Beutel et al., 2010). It seemed that these altered brain processes normalized following successful psychotherapy. Importantly, however, when successfully treated patients were assessed in an emotional conflict paradigm (patients were exposed to emotional faces and words that were either congruent or incongruent with one another), exaggerated neuronal activity was still evident in the anterior cingulate cortex, dorsal medial prefrontal cortex, and amygdala (Chechko et al., 2009). These findings suggest that either these brain regions do not underlie panic disorder, or that persistent dysfunction of this network is responsible for a high probability of relapse.

Although panic disorder is considered to be a form of anxiety, in several respects this disorder is distinct from forms of anxiety that involve the amygdala. Thus, efforts were made to determine pathways independent of this brain region that might be responsible for the production of panic attacks. The parabrachial nucleus (PBL) situated in the pons, which is a component of the "alarm center" of the brain, may play an important role in promoting some of the features of panic disorder. In mice in which a panic attack was induced, the lateral PBL was activated. This nucleus produces a neuropeptide, pituitary adenylate cyclase-activating polypeptide (PACAP) that stimulates specific receptors within the raphe nucleus, thereby eliciting symptoms of panic attack. As expected, inhibiting PACAP signaling diminished these symptoms (Kang et al., 2024). Based on these findings, it was suggested that manipulations of PACAP might be an effective target to diminish the symptoms of the disorder.

Treatment of Panic Disorder

Treatment with SSRIs and benzodiazepines has been the most common method of dealing with panic disorder. Studies that compared the efficacy of these treatments were typically underpowered and firm conclusions couldn't be made based on the available information. Still, a review of the literature indicated that SSRIs were effective treatments in 5 of 6 studies reported (Chawla et al., 2022). A review of 24 studies of benzodiazepine effectiveness similarly indicated positive effects of the treatment but cautioned that the quality of the evidence was low to moderate (Breilmann et al., 2019). As already mentioned, given the addiction risk of benzodiazepines, its use should be limited to short-term treatment.

Several psychotherapeutic approaches were adopted to treat panic disorder, including psychoeducation, supportive psychotherapy, and psychodynamic therapy, with CBT being somewhat superior to other methods

(Pompoli et al., 2018). This method was particularly effective (60% success rate) when it explicitly dealt with the perceived likelihood of panic occurring, the expected consequences of panic, and an individual's ability to cope with panic. However, CBT was not especially effective when panic disorder was comorbid with PTSD, which is not all that uncommon. Panic disorder has been treated with combined psychotherapy and drug treatments, which often (but not always) yielded better outcomes than either treatment alone (Kyriakoulis and Kyrios, 2023).

Evidence has emerged showing that in drug-free patients assessed during an early phase of panic disorder, serum IL-1β levels were elevated, whereas levels of the anti-inflammatory IL-10 were low (Quagliato and Nardi, 2022). Furthermore, immune alterations that were linked to panic disorder varied over the course of CBT therapy, apparently involving epigenetic changes (Moser et al., 2022).

Obsessive-Compulsive Disorder

As described in Table 10.1, OCD is an anxiety disorder that involves repetitive obsessions (distressing, persistent, and intrusive thoughts or images) together with compulsions to perform specific acts or rituals to diminish the anxiety elicited by the obsessive features of the illness. As social opprobrium might occur, overt compulsive behaviors may be suppressed, but may still play out mentally.

Behaviors that comprise OCD generally fall into several classes or groups, comprising (a) contamination symptoms (obsession over dirt and germs) that promote excessive washing or cleaning; (b) harm-related features wherein individuals focus on potential threats, leading to repeated checking; (c) unacceptable symptoms (including "forbidden thoughts") in which obsessions take the form of aggressive, sexual, or religious thoughts and consequently individuals might engage in compulsive praying or mental rituals; (d) symmetry symptoms in which objects in the individual's surrounding need to appear in a precise and orderly fashion, and thus individuals engage in ordering, straightening, repeating, or counting objects or acts; and (e) hoarding, wherein individuals obsessively collect items despite disturbances to their social functioning. Since some of these behaviors may appear to be abnormal to others, social alienation may occur, which may aggravate an already bad situation.

Given that the specific obsessions and compulsions come in so many varieties, the underpinnings of OCD have been difficult to define. Still, the nature of the disorder, which involves repetitive behaviors, has encouraged the position that OCD reflects a disturbance within a complex loop involving cortical brain regions associated with executive functioning and decision-making (i.e., the anterior cingulate cortex, ventromedial, dorsolateral, and lateral-orbital cortex) together with processes related to reward (nucleus accumbens, caudate). In turn, trajectories from these regions activate the thalamus and the basal ganglia, which then transmit messages back to cortical regions. Since its initial inception, this view was reformulated so that the lateral and medial orbitofrontal cortex have been given key roles in OCD, being responsible for processing information with a negative or a positive valence, respectively (Milad and Rauch, 2011). These regions act together to determine appraisals of threatening and nonthreatening situations and control (inhibiting) the functioning of other cortical regions that contribute to the emergence of anxiety. In addition, OCD was associated with greater connectivity between the left caudate and dorsolateral prefrontal cortex, which was especially pronounced in the presence of depression. Importantly, poor cognitive flexibility in those with OCD was associated with elevated functional connectivity involving the dorsal caudate, dorsal anterior cingulate cortex, and anterior insula (Tomiyama et al., 2019). Across studies, inconsistent findings were reported concerning brain morphological alterations associated with OCD. It turned out that this might be because distinct subtypes of the disorder exist that involve different brain regions. In one subtype, increased grey matter volume was present in cortical and subcortical regions (e.g., within the orbitofrontal gyrus, right anterior insula, bilateral hippocampus, and bilateral parahippocampus and cerebellum), whereas in the other, decreased grey matter volume was noted in some of these regions, including the orbitofrontal gyrus, right anterior insula, and the precuneus (Han et al., 2022b). How these subtypes map onto specific symptoms remains to be determined, although it has been demonstrated that discrete analyses of brain characteristics (e.g., measuring neurite density) can be used to predict features of OCD (Zhang et al., 2023c).

The anterior cingulate and orbitofrontal cortex have frequently been discussed in the context of depressive disorders, but some of its presumed functions, such as identifying cognitive conflict, error monitoring, and decision-making, may also contribute to the development of OCD and could be instrumental in determining best treatment strategies. Ordinarily, when dual and inconsistent messages are received, or when noise comes from external sources, the anterior cingulate cortex is necessary for decision-making. Among individuals with OCD, this brain region is hyperactive in decision-making situations, possibly reflecting the difficulty in making appropriate appraisals and decisions, or it may reflect a disturbance involving improper feedback, thus leading to repeated behavioral responses. In fact, in circumstances in which the situation is ambiguous or relatively unpredictable,

OCD symptoms can be exacerbated among individuals who are intolerant of uncertainty (Pinciotti et al., 2021).

It has been shown that among OCD patients, excess glutamate and reduced GABA functioning were present within the anterior cingulate cortex, and elevated glutamate was observed in the supplementary motor region, even being apparent among healthy individuals with mild compulsive tendencies (Biria et al., 2023). These findings were consistent with earlier reports, supporting the position that glutamate dysregulation may contribute to OCD (Pittenger, 2021) and that GABA concentrations in the orbital frontal cortex (and within the anterior cingulate cortex) of OCD patients were reduced (Zhang et al., 2016).

While not discounting the involvement of such processes in OCD, the view had been taken that it might be profitable to view this disorder from the perspective of an inability to stop certain behaviors rather than one of evaluating what starts these behaviors. An interesting view of some subtypes of OCD is that it reflects the failure of a security motivation system in which individuals need a "feeling of knowing" (e.g., Did I turn off the stove?) before they can move on to other tasks. Should feedback systems not operate as they should, then OCD symptoms, such as repeated checking, would persist (Woody et al., 2019).

Several animal models of OCD have been developed, although these were complicated owing to the diversity of OCD typologies that exist (Does hoarding involve the same processes as persistent checking and do these involve the same processes as obsessive hair pulling or handwashing?). The current animal representations of OCD probably should not be viewed as models in the traditional sense, but instead might reflect behaviors (compulsivity, stereotypy, or perseverance) seen in several psychopathological conditions.

Despite their limitations, animal studies have provided clues concerning the primary neurobiological processes that govern OCD. For instance, knocking out kainate receptor subunits within the dorsal striatum (iontophoretic receptors that are stimulated by glutamate) resulted in mice showing signs of OCD (Xu et al., 2017a). As well, several genes were identified that were tied to OCD. Deletion of the *Sapap3* gene, which is important in communication between neurons, was accompanied by prodigious self-grooming, which could be attenuated by treatment with an MGluR5 antagonist. Moreover, these effects occurred very quickly, rather than the many days needed for SSRIs to exert a positive effect (Ade et al., 2016). Aside from this, dysfunction of microglia, particularly those linked to the transcription factor Hoxb8, have been implicated in OCD, and these actions could be exacerbated by female sex hormones (Tränkner et al., 2019). These actions may involve several brain regions (e.g., amygdala, dorsal hippocampus), and

diverse microglia populations may differentially influence the appearance of some OCD-like symptoms (Nagarajan and Capecchi, 2023).

A detailed and intriguing review identified the presumed processes related to OCD and compulsive behaviors, many of which are observed across diagnostic categories and even intrude on everyday functioning (Robbins et al., 2024). This perspective combines the involvement of learning processes and specific neural circuits (some of which comprise the same brain regions) in accounting for these behaviors. It was suggested that several neural networks mediate components of the compulsive behaviors. These networks largely comprise circuits that involve the orbitofrontal, prefrontal, anterior cingulate, and insular cortices and their connections with the basal ganglia, together with sensorimotor and parietal cortices and cerebellum. These systems include an executive control system that exerts top-down inhibitory control over the other systems. The others are responsible for reward-related processes, a negative affective system that influences coping responses, as well as circuits that influence goal-directed (instrumental) behaviors, habit formation, action monitoring, and the interoceptive system, which involves the accumulation of sensory information. Because of the interconnectivity between these diverse systems, disturbances in any component of these systems may contribute to OCD and compulsive behaviors.

Treatment of OCD

It is common for SSRIs to be used in the treatment of OCD, usually at high doses, which may be especially valuable as OCD and depression are frequently comorbid conditions (Del Casale et al., 2019). Although neuroleptics are generally ineffective in the treatment of OCD, their use in conjunction with serotonin-acting agents can enhance effects in patients who are resistant to SSRIs alone (Zhou et al., 2019). Since GABA and glutamate dysfunction may contribute to OCD features, manipulations of glutamate functioning could have therapeutic effects (Marinova et al., 2017). Although SSRIs continue to be a first-line treatment of OCD, adjunctive treatments that influence glutamatergic or inflammatory processes (e.g., riluzole, ketamine, memantine, N-acetylcysteine, lamotrigine, celecoxib, and ondansetron) may enhance the effects observed even among some treatment-resistant patients (van Roessel et al., 2023). Only limited data are available regarding the effects of antiinflammatory agents on OCD. By themselves, such agents are generally ineffective; however, the NSAID celecoxib and the antibiotic minocycline enhanced the effects of SSRIs (Grassi et al., 2021).

As an alternative to pharmacotherapies, especially when these fail to be effective, repetitive transcranial magnetic stimulation (rTMS) and deep transcranial stimulation (dTMS) could be effective add-on therapies to SSRI treatment. Deep brain stimulation was also effective in ameliorating symptoms in cases of treatment-resistant OCD, depending on the brain region targeted (Rapinesi et al., 2019). As well, appreciable success was achieved through CBT, often accompanied by changes in the neuronal functioning in brain regions that had been implicated in OCD (Poli et al., 2022). Positive outcomes were also obtained with "exposure and response prevention" (ERP) in which patients learn in graded steps to tolerate the anxiety that comes about when they are unable to engage in the compulsive behavior (e.g., Wheaton et al., 2016), and somewhat better outcomes were realized with the combination of CBT and ERP (Reid et al., 2021). There were indications that the combination of an SSRI and CBT could produce enhanced effects, but this outcome was not sustained beyond 16 weeks (Fineberg et al., 2018). An alternative method that has yielded some success has been a learning-based approach in which an attempt is made to alter the associations normally present in response to obsessive thoughts. For instance, an individual obsessed with not touching anything "contaminated by germs" might be "taught" to appraise external objects less negatively by having them pair these objects with neutral thoughts or emotions.[1]

Phobias

The typical definition of a phobia is that it comprises an uncontrollable, irrational, and persistent fear of certain objects, situations, or activities. To an extent, certain phobic-like reactions, especially "biophobias," may be predicated on evolutionary pressures that favor withdrawal from or avoidance of potentially harmful stimuli. Considered from this vantage, there is a basis for the fear of spiders and snakes, as well as microbes (mysophobia), parasites (parasitophobia), and germs. In an extreme form, "disgust" that reflects an emotion that is regulated by the insula serves as an important avoidance mechanism that is used so that particular stimuli are not contacted (e.g., feces, vomit, blood, rotting, or diseased meats). At the same time, biases created by disgust may contribute to the interpretation, expectancies, attention, and memory that contribute to anxiety-related disorders, particularly phobias (Knowles et al., 2019). Parenthetically, disgust, which involves the functional coupling of the insula, anterior cingulate cortex, and amygdala, may also contribute to racial discrimination (Liu et al., 2015b).

Phobias may also develop through classical conditioning processes wherein events or stimuli associated with a negative feeling or emotion may come to elicit fear or anxiety upon later encounters with these stimuli. The processes involved in different types of phobias may not be the same, and certain behaviors may be present that were not necessarily conditioned (e.g., Garcia, 2017). It wouldn't be unreasonable to assume that many of the processes that are associated with other anxiety-related disorders also contribute to phobic responses and involve brain regions that are associated with appraisals, decision-making, fear responses, and their extinction. A phobic reaction is accompanied by neurophysiological responses comprising elevated amygdala and anterior cingulate cortex activity, which are normalized with the alleviation of a phobia through behavioral therapy (Fredrikson and Faria, 2014).

Exposure therapy promotes the extinction of the fear/anxiety response, and gradual desensitization treatment (achieved by diminishing distress in small steps or by having the feared situation or object come progressively closer or more real) has a high success rate. Therapies for phobia frequently involve behavioral/emotional methods to extinguish this conditioned response. To this end, imagery and virtual reality treatments have been used to desensitize individuals by promoting emotional and behavioral change in incremental steps. Likewise, CBT has been offered so that individuals will begin to understand their negative thought patterns, and then to take steps to modify their behavioral and emotional responses.

Social Anxiety

As in the case of other anxiety conditions, social anxiety is accompanied by exaggerated neuronal activity in the amygdala and cortical regions linked to attention and processing of social threats (Hiser and Koenigs, 2018). Moreover, among social anxiety patients without comorbid depression, cortical thickness was increased (e.g., within the salience network that includes the left insula and right anterior cingulate cortex), which was consistent with greater attentional and executive control functions. In addition, self-referential processing comprising self-focused attention is frequently disturbed among individuals with social anxiety disorder, which is associated with hyperactivation of the medial prefrontal cortex and posterior cingulate cortex, components of the default mode network (Yoon et al., 2019). The amygdala may also be related to social anxiety in that the increased left amygdala volume associated with social anxiety disorder was diminished 1 year after successful CBT treatment (Månsson et al., 2017).

---

[1] It can be imagined that individuals with OCD that entailed not touching anything with germs would have suffered miserably during the COVID-19 pandemic, particularly early on when it was thought that infection could be transmitted by handling contaminated objects. The distress created might have exacerbated OCD symptoms, perhaps by increasing inflammatory processes (Nezgovorova et al., 2022).

As in the case of other anxiety disorders, serotonin, norepinephrine, glutamate, and GABA have been implicated in social anxiety disorder. Genetic analyses have frequently pointed to the link between social anxiety disorder and polymorphisms related to the oxytocin receptor and brain-derived neurotrophic factor (BDNF), although the data have not been entirely consistent (Baba et al., 2022). The ties to a single nuclear polymorphism (SNP) related to the 5-HT transporter (SLC6A4) were identified in patients with social anxiety (Forstner et al., 2017), and in response to fearful faces, activation of the insula was greater among individuals carrying the ss allele for the 5-HT transporter than among individuals carrying the ll genotype (Klumpp et al., 2014). However, since social anxiety is often comorbid with depressive illness, it may be difficult to uncouple the link between 5-HT transporter and social anxiety versus that of depression (assuming that this polymorphism is involved in depression; see our earlier discussion in Chapter 9). Moreover, as subtypes (or dimensional differences) exist concerning social anxiety disorders, heterogeneity may exist regarding the neurobiological substrates for this condition.

Increasing evidence has amassed consistent with the proposition that oxytocin may be relevant to social anxiety. In humans, social anxiety was associated with a less secure attachment style in the presence of an oxytocin receptor (OXTR) polymorphism (Notzon et al., 2016). As well, oxytocin levels in adolescents were elevated in association with social anxiety disorder but this was not unique to this syndrome, being elevated in other anxiety disorders (Uzun et al., 2022). The influence of altered oxytocin functioning may be especially prominent under conditions that involve some sort of social challenge. A prospective study of 400 adolescents revealed that the combination of chronic interpersonal stressors and the presence of the CD38 polymorphism (which is associated with lower oxytocin levels rather than changes of the oxytocin receptor) predicted trait social anxiety and symptoms of depression over 6 years (Tabak et al., 2016b). Further, a polymorphism for the oxytocin receptor was related to greater amygdala volume and exaggerated amygdala activity in response to socially relevant face stimuli (Marusak et al., 2015). Similarly, social anxiety was associated with epigenetic changes related to the oxytocin receptor accompanied by increased amygdala responsiveness in a test that involved social phobia-related word processing (Ziegler et al., 2015).

Treatment of Social Anxiety

Supporting the involvement of oxytocin in generalized social anxiety patients, the hyperactive amygdala response ordinarily observed in response to emotional faces was eliminated by oxytocin administration, and the functional connectivity between the amygdala and insula, and the anterior cingulate gyrus was elevated (Gorka et al., 2015). Thus, manipulations of oxytocin (and arginine vasopressin) may serve as targets for social anxiety disorder (Neumann and Slattery, 2016). Yet, as we discussed earlier, oxytocin treatments can have deleterious effects by increasing the salience of negative social cues. Indeed, among individuals with elevated social anxiety, social working memory was disturbed by oxytocin, suggesting that exogenous oxytocin administration may impair social cognitive functioning (Tabak et al., 2016a).

Several reports attested to the effectiveness of SSRIs (e.g., escitalopram) in the treatment of social anxiety disorder, just as these agents have been effective in treating other anxiety conditions. A meta-analysis indicated that SSRIs and SNRIs were the treatment of choice (Curtiss et al., 2017) and yielded positive effects in children and adolescents, as have psychotherapeutic interventions (Weisz et al., 2017). In cases in which SSRIs and SNRIs were ineffective or not well tolerated, the GABA-like agent pregabalin reduced features of social anxiety disorder.

A systematic review confirmed that behavioral treatment strategies may be effective in the treatment of social anxiety, although drug treatments frequently promoted greater effect sizes than behavioral strategies (Bandelow et al., 2015). Of the behavioral approaches, CBT was deemed to be the most effective treatment (Mayo-Wilson et al., 2014), and when administered as group therapy, CBT and mindfulness training were equally effective (Goldin et al., 2021). In line with its treatment efficacy, successful treatment with CBT was accompanied by structural changes within cortical regions linked to emotional regulation (Steiger et al., 2017). Likewise, the amygdala-frontal cortical coactivation profiles could distinguish between effective and ineffective treatments (Faria et al., 2014).

Inflammatory Factors in Anxiety and Anxiety Disorders

Inflammation and Anxiety Responses in Animals

The data linking inflammatory processes and anxiety have largely come from studies in rodents. Individual differences in anxiety in rodents (or differences across strains of mice) have been linked to inflammatory factors, which can vary still further in the presence of stressors. Brain inflammatory responses tended to be more pronounced among rats bred for high anxiety, and in response to neonatal LPS treatment, hippocampal neuroinflammation and microglia morphology were more pronounced in anxious rats (Claypoole et al., 2017). Consistent with these findings, imaging studies in humans indicated that anxiety was accompanied by inflammation within the amygdala, insula, and anterior cingulate cortex (Felger, 2018).

By eliciting peripheral inflammatory responses, stressors can influence brain functions that contribute to anxiety. Through peripheral norepinephrine activation, stressors may promote the release of T cells from secondary immune organs (e.g., spleen), which could trigger brain neurochemical changes that elicit emotional responses. Stressor-elicited anxiety may, alternatively, have stemmed from the microglial recruitment of monocytes to the brain endothelium and the subsequent IL-1β release (McKim et al., 2017). Brain microglial cell changes varied systematically with levels of anxiety across strains of mice, supporting the view that these inflammatory factors account for individual differences in anxiety (Li et al., 2014).

Consistent with a role for specific cytokines in accounting for anxiety, stressors activated brain microglia so that sustained TNF-α release was provoked within the ventral hippocampus. If the gene for TNF-α was knocked out, then the behavioral and synaptic change otherwise produced by the stressor was prevented (Kemp et al., 2022) and the presence of a polymorphism of the TNF-α-238G/A gene was associated with a decreased risk of OCD (Jiang et al., 2018a). Supporting the cytokine-anxiety relation, repeated treatment with the antiinflammatory agent ibuprofen reduced anxiety in several behavioral tests while concurrently reducing the expression of TNF-α and IL-1β, as well as hippocampal BDNF (Lee et al., 2016b).

As in the case of depression, IL-6 may play a prominent role in anxiety. A social stressor that reliably induces an anxious/depressed profile was most prominent in mice that exhibited the greatest IL-6 changes, and the anxiety could be modified by genetic manipulations that altered IL-6 production (Hodes et al., 2014). As expected, the stressor-provoked anxiety/depressive profile and elevated IL-6 levels were normalized with repeated antidepressant (imipramine) treatment (Ramirez and Sheridan, 2016).

Treating mice with an agent that activated the immune system (e.g., LPS), so that cytokines were released into circulation or were released by microglia within the brain, dose-dependently elicited signs of anxiety in several behavioral paradigms. Treatment with IL-1β likewise elicited anxiety, reinforcing the possibility that the actions of LPS might largely be due to this cytokine's release, although it could also involve the synergistic action of IL-1β and other cytokine variations that are provoked by the endotoxin. On the surface, it did not appear that LPS provoked anxiety by causing sickness, as a dose that did not produce sickness behaviors engendered anxiety while concurrently increasing inflammation within the BLA (Loh et al., 2023). Predictably, the LPS-elicited anxiety was attenuated by treatment with an antioxidant

(esculetin) that also had antiinflammatory and neuroprotective actions (Sulakhiya et al., 2016). Together, these findings are consistent with the view that systemic insults can engender anxiety through amygdala stimulation, but other brain regions are also affected by inflammatory factors. For instance, social defeat in mice may influence the locus coeruleus norepinephrine pathway, thereby priming an inflammatory response so that a later challenge increases central amygdala activity and the production of anxiety (Finnell et al., 2019).

Not every bacterial challenge introduces the same effects on anxiety. Although LPS and staphylococcal enterotoxin B (SEB) are both bacterial products that elicit distinct immune responses and provoke elevated amygdala neuronal activity and anxiety responses, only LPS elicited proinflammatory changes within the amygdala. These data suggest that the brain may be sensitive to varying immune challenges, and perhaps the anxiety provoked involves factors other than or in addition to cytokines. There have been indications that superantigens, such as SEB and SEA, which elicit nonspecific T-cell activation and marked cytokine release, might only elicit anxiety if the test situation creates a degree of distress, pointing to the importance of contextual factors in permitting the anxiolytic actions of neuroinflammation to appear (Rossi-George et al., 2004). Consistent with such findings, among mice with the gene for the antiinflammatory cytokine IL-4 knocked out so that proinflammatory processes predominated, anxiety was generally not provoked unless the test situation involved a mild challenge (Moon et al., 2015).

As in the case of bacterial endotoxin, anxiety induced among virally infected mice (e.g., by murine cytomegalovirus) was accompanied by IL-6 and TNF-α elevations, which could be restrained by the presence of corticosterone. Anxiety could similarly be provoked in mice infected with a protozoan parasite or following viral infection, which was accompanied by the elevation of several proinflammatory cytokines within the brain (de Miranda et al., 2011). Congruent with such findings, the administration of the viral mimic poly I:C elicited an anxious-depressed behavioral profile, coupled with altered BDNF expression and activation of the kynurenine pathway (Gibney et al., 2013). In essence, anxiety could be provoked through several routes that elicit an inflammatory immune response.

Further support for inflammatory factors in anxiety has come from the demonstration that diminished short-term anxiety by benzodiazepines also inhibited stressor-provoked IL-6 elevations, the accumulation of macrophages within the CNS, and inhibited trafficking of monocytes and granulocytes in circulation (Ramirez et al., 2016). As we saw earlier, antidepressant agents, such as SSRIs, may also diminish anxiety by affecting immune functioning. Of course, these agents may have

Inflammation and Anxiety in Humans

Anxiety introduced by a stressor in a laboratory context increased circulating interferon gamma (IFNγ) and IL-1β levels (Moons and Shields, 2015). As well, greater circulating CRP in men (but not women) was associated with increased threat-related amygdala activation (Swartz et al., 2017). Paralleling the effects of psychological stressors, endotoxin administration in humans elicited feelings of anxiety (and depressive-like mood) that were accompanied by elevated cortisol levels as well as TNF-α and its soluble receptor (Reichenberg et al., 2001). Congruent with this report, low doses of LPS that elicited subclinical anxiety symptoms also produced elevated IL-6 and TNF-α as well as the antiinflammatory IL-10 (Lasselin et al., 2016). Of particular significance, imaging studies consistently showed that inflammatory-inducing stimuli elicited increased neuronal activity in anxiety-related brain regions, such as the amygdala, insula, and anterior cingulate cortex, possibly reflecting the actions of inflammatory factors on glutamate and monoamines (Felger, 2018).

Anxiety Responses Associated With Chronic Illness

The links between anxiety and inflammatory processes have been inferred from reports that anxiety was associated with a substantial number of illnesses, notably those related to immune disturbances (e.g., autoimmune disorders). Even food allergies were associated with childhood anxiety, particularly social anxiety. In this regard, the evaluation of multiple illnesses revealed biomarkers that were linked to anxiety and depression, and these transdiagnostic markers were linked to therapeutic efficacy (Goldsmith et al., 2023). The analysis of more than 144,000 participants obtained from the UK Biobank indicated that the inflammatory marker CRP and IL-6 were associated with depression and anxiety, and Mendelian randomization analyses suggested that the ties between inflammation and both depression and general anxiety disorder were causal (Ye et al., 2021). This said, inflammatory factors were much more closely aligned with depression than with anxiety (Milaneschi et al., 2021).

Using population-based data, immune-related inflammatory diseases (intestinal bowel disorder, MS, and rheumatoid arthritis) were associated with elevated anxiety. The anxiety across illnesses might reflect the effects of being ill or effects related to depression, but might also reflect common neurobiological changes (Marrie et al.,

2017). Both anxiety and depression were seen among patients with multiple sclerosis, which were tied to increased cerebrospinal fluid (CSF) levels of inflammatory factors. These outcomes were diminished among patients whose symptoms were remitting, and while IL-2 was aligned with anxiety, the levels of IL-1β and TNF-α were more closely linked to depressive symptoms (Rossi et al., 2017). Variations of IFNγ were also closely aligned with anxiety symptoms in systemic lupus erythematosus patients (Figueiredo-Braga et al., 2018). Increased inflammation and the presence of psychiatric symptoms were similarly observed in patients with rheumatoid arthritis, accompanied by elevated levels of IL-17, perhaps being mediated by altered BDNF that prompted proinflammatory elevations (Lai et al., 2021).

Research in animals that allowed for better experimental control than in human studies suggested that the inflammation itself was a prime mediator of the mood changes. Specifically, in an animal model of multiple sclerosis, emotional changes were detected early in the development of the disease and related to elevations of inflammatory mediators and might have involved microglial activation (Gentile et al., 2015). Aside from the effects associated with these illnesses, anxiety elicited by inflammatory pain in mice was accompanied by amygdala TNF-α elevations, which could be attenuated by a TNF-α neutralizing antibody, suggesting causal involvement of this cytokine in producing the anxiety response (Chen et al., 2013). Sepsis in mice was likewise accompanied by anxiety together with elevated plasma TNF-α and that of brain TNF-α, IFNγ, IL-1β, and IL-6 (Calsavara et al., 2013). Tellingly, the anxiety-related behaviors elicited by sepsis were attenuated by inhibiting the connectivity between the central amygdala neurons and those of the bed nucleus of the stria terminalis (Bourhy et al., 2022). In a review largely based on animal studies (given how few studies were available in humans), it was suggested that chronic illnesses alter resident brain microglia and disturb hippocampal neurogenesis. This, in turn, may instigate cognitive and mood disturbances, such as anxiety and depression (Chesnokova et al., 2016).

Despite the sensitivity of both IL-6 and TNF-α in response to stressors, it might be unrealistic to believe that anxiety would exclusively be mediated by these cytokines. Many of these relations disappeared when controlling for lifestyle factors, pointing to the indirect relationship that might exist between anxiety and immune factors. However, other cytokines, such as IL-8 (also referred to as CXCL8, a chemokine produced by macrophages, epithelial, and endothelial cells), were elevated in both current and remitted patients, even after controlling for lifestyles (Vogelzangs et al., 2016).

While not discounting the diverse routes by which anxiety may be linked to inflammatory processes and various illnesses, psychological interventions may be beneficial for patients dealing with disorders, such as inflammatory bowel disease, in which interrelations exist between neural, hormonal, and inflammatory factors. For that matter, poor coping was associated with worse outcomes among some individuals with inflammatory bowel disease, and a combination of CBT and mindfulness training was associated with diminished psychological stress and disease severity (Nemirovsky et al., 2021).

Genetic Correlates of Inflammation Related to Anxiety

Genetic analyses tying anxiety disorders to inflammatory processes have come from linkage studies, genome-wide association studies (GWAS), and candidate gene studies. These reports have implicated the usual suspects in this regard (Gottschalk and Domschke, 2017). Specifically, generalized anxiety was related to several candidate genes associated with monoamine-related factors (5-HTT, 5-HT1A, MAOA) and BDNF, and their interactions with stressful experiences (e.g., early adversity). Together with psychosocial determinants of anxiety, these factors could potentially serve as markers to predict treatment responses to therapeutic drugs.

Cytokine links to anxiety disorders were also related to genetic influences. For example, anxiety was related to a SNP within the gene associated with IL-8 (Janelidze et al., 2015), and in the gene coding for IL-1β and the NF-κB p100 subunit gene. As observed in other contexts, a polymorphism related to the gene coding for IL-1β was associated with anxiety and depression, provided that individuals had encountered a significant life stressor (Kovacs et al., 2016). The presence of a haplotype (several genes that are inherited as a group) related to the proinflammatory cytokine IL-18 was associated with increased central and medial amygdala responsivity. Moreover, in women, the IL-18 haplotype was also linked to symptoms of anxiety and depression that emerged following a stressor encounter (Swartz et al., 2016). As tempting as it might be to assume causal links between the genetic control of immune functioning and anxiety changes, the directionality of any such relations can't be deduced from these data.

Consistent with the epigenetic effects provoked by stressful experiences, global DNA methylation levels were elevated among anxious individuals in comparison to those who were nonanxious. Moreover, among anxious patients, symptom severity was associated with epigenetic processes related to IL-6 gene expression (Murphy et al., 2015), as well as to several other immune-relevant genes, such as the TLR-2 promoter and the inducible nitric oxide synthase promoter (Kim

et al., 2016a). There are likely many epigenetic changes that might similarly be associated with anxiety, including other genes associated with inflammatory processes, but at this time the relevant data are limited.

Immune and Inflammatory Processes in Specific Anxiety Disorders

Several anxiety disorders share common features related to immune functioning, just as common neurotransmitter processes have been identified, but in several respects, they are also distinct from one another. It has been observed that various infectious illnesses (e.g., whooping cough, scarlet fever, diphtheria) were associated with adult anxiety disorders (Witthauer et al., 2014). Indeed, even the distress related to COVID-19 may have contributed to the emergence of anxiety through inflammatory processes (Mazza et al., 2020).

Studies that assessed specific anxiety disorders and inflammatory factors typically involved only a modest number of participants, and different inflammatory immune factors were assessed across studies, thus limiting the conclusions that could be drawn. We emphasize once again that many anxiety disorders are comorbid with depression, and they are accompanied by eating and sleep disturbances, as well as poor lifestyles (e.g., elevated smoking), any of which can engender elevated levels of inflammatory factors, such as CRP and IL-6 (Slavich and Irwin, 2014). As such, the inflammatory alterations observed may be secondary to these symptoms rather than being directly linked to anxiety-related illnesses. This caveat notwithstanding, a large analysis that comprised 353,136 participants of which 12,759 had a history of an anxiety disorder indicated that CRP was related to GAD and panic disorder, although this relation was diminished when accounting for depressive symptoms, multimorbidity, and BMI. In contrast, neither OCD nor phobic anxiety disorders were linked to CRP (Kennedy and Niedzwiedz, 2021). The factors responsible for these differences weren't immediately apparent, but suggest that the disorders may not share all the same processes.

Generalized Anxiety Disorder

Consistent with a link between GAD and inflammation, a detailed review indicated that elevated levels of CRP accompanied this disorder (Orsolini et al., 2023), and elevated CRP levels were present in both current and remitted anxiety patients. A meta-analysis revealed that GAD was most often accompanied by elevated IFNγ and TNF-α, and in some studies, other inflammatory cytokines were elevated as well (Costello et al., 2019).

Analysis of inflammatory factors that were stimulated *ex vivo* indicated that the production of IL-8 was directly related to the severity of anxiety and depressive symptoms (Vogelzangs et al., 2016). As well, the concentrations of IL-2 and IFNγ, as well as the antiinflammatory cytokine IL-4, were lower in cell cultures of GAD patients than in control participants, whereas TNF-α and IL-17 were elevated (Vieira et al., 2010).

Several reports indicated that a constellation of inflammatory cytokines was elevated in anxiety, and the ratio between pro- and antiinflammatory cytokines favored a proinflammatory profile in GAD patients (Hou et al., 2017). The links to cytokine changes in GAD, as well as other anxiety-related conditions, weren't uniformly related to proinflammatory cytokine elevations, suggesting that additional factors are responsible for the occurrence of these disorders (Michopoulos et al., 2017). The source for the differences between studies is uncertain given that an insufficient number of studies evaluated GAD duration in relation to cytokines, or the ties between cytokines and the course of the illness from onset to transitioning to a chronic condition.

Of the various psychosocial interventions to treat anxiety, CBT appeared to be best, frequently being accompanied by reduced inflammation (Shields et al., 2020). It was similarly observed that the elevated cytokine levels and adrenocorticotropic hormone (ACTH) among patients with GAD were limited among patients who had received Mindfulness-Based Stress Reduction (Hoge et al., 2017) However, a meta-analysis of inflammatory biomarkers before and after Mindfulness-Based Stress Reduction revealed only small differences in varied psychiatric conditions (Grasmann et al., 2023).

Panic Disorder

Panic disorder has been associated with elevated levels of CRP and the proinflammatory cytokines IL-6, IL-2, and TNF-α (Moser et al., 2022), some of which were linked to epigenetic alterations (Petersen et al., 2020). Other reviews similarly indicated that panic disorder was associated with elevated serum IL-6, IL-1β, and IL-5, whereas the ties to IL-2, IL-12, and INFγ were less certain (Quagliato and Nardi, 2018). Panic disorder was also related to elevated expression of the immunomodulatory gene T-cell death-associated gene (TDAG8) (Strawn et al., 2018). It has been maintained that dysregulated inflammatory processes, initially stemming from peripheral processes, may affect brain signaling mechanisms that disturb body-to-brain communication and favor conditioned fear responses that come to produce repeated panic attacks (McMurray and Sah, 2022).

Not all reports showed these cytokine alterations in panic disorder patients, perhaps reflecting the frequent analyses conducted in drug-treated patients. Of the few studies that were conducted among drug-naïve patients, elevated serum levels of IL-1β and IL-2 receptors, as well as reduced IL-10 were reported (Quagliato and Nardi, 2022). As well, IL-6 levels were elevated in patients who experienced panic attacks, persisting even after standard pharmacotherapy (SSRIs, SNRIs, or *tricyclic antidepressants*) that had diminished panic symptoms (Kim et al., 2019). Whether panic symptoms returned with treatment discontinuation and whether this varied as a function of IL-6 levels isn't certain.

As we've already seen, to some extent the cytokine profile associated with panic disorder could be distinguished from that associated with GAD (Zou et al., 2020). Panic disorder differs from other anxiety conditions insofar as panic attacks are tied to body sensations that signal the presence of threats to physiological homeostasis, whereas other conditions, such as GAD, are more persistent and appear on most days. It is difficult to gauge cytokine levels during or shortly after panic attacks, which can be achieved in animal studies. This can be done by subjecting animals to CO_2 inhalation, which is known to produce panic attacks in humans. This treatment is associated with behavioral and cardiovascular responses relevant to panic, operating through IL-1β and changes of the subfornical organ, through which cytokines can gain entry to the brain (McMurray and Sah, 2022).

In sum, inflammatory processes have been implicated in panic disorder, and both peripheral and central cytokines may contribute to the occurrence of this condition. Which specific cytokines are core to this relationship is uncertain. As well, the mechanisms by which individuals are primed to develop panic attacks (i.e., having panic disorder versus having a panic attack) may not be identical and the contribution of cytokines in these states has not been fully assessed.

Obsessive-Compulsive Disorder

Much like other anxiety disorders, inflammatory processes have been implicated in the development of OCD (Meyer, 2021). Among some children, OCD or tics may appear suddenly following infection (e.g., after streptococcal infection). This was termed Pediatric Autoimmune Neuropsychiatric Disorders Associated with Streptococcal Infections (PANDAS), and as the name implies, it was considered a form of autoimmune disorder. Several studies revealed that streptococcal infection was associated with an elevated risk of developing OCD, although it was maintained that this was not necessarily exclusive to streptococcal infection (Marazziti et al., 2023). In a subset of children with OCD, proinflammatory cytokines were elevated and autoantibodies were identified that targeted the basal ganglia (Endres et al., 2022). Among children and youth with childhood-onset OCD, levels of IL-1β, IL-6, and TNF-α were elevated in most studies (Fabricius et al., 2022). Studies in unmedicated children with OCD indicated that serum levels of IL-2, IL-6, TNF-α as well as IL-4 and IL-10-α were elevated, although IL-12 which stimulates pro-inflammatory cytokine release was reduced (Çolak Sivri et al., 2018). As well, in children and adolescents with OCD, the levels of Th$_{17}$ cells were elevated, whereas the percentage of T$_{reg}$ cells was reduced, and these differences became still more pronounced with a longer illness duration (Rodríguez et al., 2019). Moreover, elevated levels of several circulating chemokines (e.g., CCKL3 and CXCL8) and soluble TNF receptors were elevated in OCD, and these relations were dependent on the symptom dimensions expressed (e.g., washing, hoarding) (Fontenelle et al., 2012). Consistent with a causal role for inflammatory factors in OCD, the administration of the antiinflammatory agent celecoxib enhanced the therapeutic efficacy of fluvoxamine (Shalbafan et al., 2015).

The occurrence of OCD was associated with several changes in gene expression assessed in monocytes, notably those related to antigen processing and presentation, and leukocyte cell adhesion (Rodríguez et al., 2022). The occurrence of OCD was also associated with a polymorphism for the TNF-α gene, as well as a SNP within the gene for FKBP5 (the glucocorticoid modulator) (Minelli et al., 2013), as well as epigenetic changes of inflammatory-related genes (Cappi et al., 2016). It may be especially important that a familial link appears to exist between the presence of autoimmune disorders and elevated circulating proinflammatory cytokines, which were more often found in mothers of children with OCD (Jones et al., 2021).

A Swedish nationwide survey with more than 30,000 OCD patients revealed that more than 40% of affected individuals were at markedly elevated risk for comorbid autoimmune disorders, even appearing in first-, second-, and third-degree relatives of individuals with OCD (Mataix-Cols et al., 2018). A meta-analysis that comprised more than 31,200 patients with inflammatory or rheumatological diseases indicated the presence of OCD was associated with elevated levels of proinflammatory cytokines (Alsheikh and Alsheikh, 2021). Moreover, IL-1β, IL-6, and TNF-α were associated with OCD and some of the cognitive disturbances that accompany this disorder, such as immediate and delayed recall abilities (Karagüzel et al., 2019).

The presence of adult OCD has frequently been associated with low-grade inflammation and the presence of neural antibodies, as well as comorbid neuroinflammatory and autoimmune disorders (Gerentes et al., 2019). Still, early and late-onset OCD are etiologically distinguishable from one another in several ways in that the early onset type may be more aligned with glutamate transmission and involvement of immune processes. As such, these factors, perhaps in conjunction with microbial alterations may be useful in personalized therapy (Burchi and Pallanti, 2019).

In addition to peripheral factors, the occurrence of OCD was accompanied by inflammation within the orbital frontal cortex (Attwells et al., 2017). As well, imaging analyses of OCD patients indicated elevated microglia activation related to neuroinflammation in the dorsal caudate, orbitofrontal cortex, thalamus, ventral striatum, and dorsal putamen (Attwells et al., 2017). Analysis in an animal model of OCD indicated that neurons and astrocytes within the caudate operate in tandem to produce OCD symptoms, mediated by a protein SAPAP3. Genetic deletion of this protein caused anxiety and compulsive grooming, and delivery of SAPAP3 attenuated these effects (Soto et al., 2023).

In a subset of individuals with OCD, the onset of the disease might have been triggered by bacterial, viral, or parasitic agents, and the possibility was entertained that dysregulation of gut microbial processes could affect brain microglia that promoted OCD (Troyer et al., 2021). It is of both theoretical and clinical significance that treatments that reduce inflammation, such as the antibiotic minocycline, were effective as adjuncts to SSRIs to enhance the therapeutic response in diminishing OCD symptoms (Esalatmanesh et al., 2016).

It warrants mentioning that OCD is characteristically different from GAD and social anxiety disorder, and hence comparable inflammatory responses would not be expected. Even if inflammatory factors are linked to OCD, this relationship is likely moderated by a constellation of factors, and as we mentioned regarding other anxiety disorders, OCD is often comorbid with depressive disorder, making it difficult to dissociate the relative contributions of inflammatory processes to these illnesses.

Social Anxiety Disorder

Although social phobia appears in various immune-related inflammatory diseases (Reinhorn et al., 2020), a systematic review indicated that social anxiety disorder was not reliably related to inflammatory factors, and ties to cortisol, neuropeptides, and neurotrophins were likewise inconsistent (Caldiroli et al., 2023). Still, the links between social anxiety and neuroinflammation may be obscured by characteristics associated with the illness. Social anxiety disorder, which tends to appear relatively early in life, is accompanied by lower levels of CRP and IL-6, especially among women (Vogelzangs et al., 2013), whereas high CRP levels were observed among individuals who were 50 years of age or older. This contrasts with the relationship between childhood maltreatment and social anxiety disorder which was associated with elevated circulating IL-6, especially among individuals with low positive affect and trait mindfulness (Carlton et al., 2021). Supporting the importance of early life events, blood transcriptome analyses pointed to such experiences being linked to adult social anxiety disorder, and were mediated by inflammatory factors (Edelmann et al., 2023). It also appeared that social anxiety disorder was associated with relatively low levels of the antiinflammatory cytokine IL-10, coupled with elevated kynurenic acid (Butler et al., 2022). As indicated by these investigators, longitudinal studies are needed to determine the role of these processes in the emergence of social anxiety disorder, including events that transpired during early life.

Linking Microbiota to the Promotion of Anxiety

Just as inflammatory processes stemming from stressors or tissue damage have been related to depressive illnesses, as described in Fig. 10.2, microbial changes may be related to the development of anxiety (Hu et al., 2022). The focus of research concerning the health implications of microbiota has focused on gut bacteria, but as already indicated, viruses, fungi, archaea, and protists could influence biological processes associated with anxiety, as might parasites that inhabit the human gut. Gut microbiota can influence brain functioning through different routes, which may contribute to a variety of neuropsychiatric disorders, including anxiety and PTSD (Nikolova et al., 2021). Some of the pathways by which this may come about are provided in Fig. 10.3. Additionally, oral microbiota disturbances, such as those associated with periodontal disease, have been implicated in numerous pathological conditions, including anxiety, possibly owing to an increase in circulating cytokines (Martínez et al., 2022).

While diet-provoked alterations of gut bacteria might predict later anxiety (Phillips et al., 2017), the difficulties in having individuals accurately report their food consumption can result in the reliability of the data linking diet and anxiety in humans being questionable. The alternative approach is that of assessing microbiota among patients with specific anxiety disorders. However, even in these instances, problems of interpretation can arise given that the distress created by the illness or altered lifestyles that might have been adopted may be responsible for the microbial changes.

Modifying Anxiety Through Microbial Processes

To assess the influence of microbiota on anxiety in rodents, a common procedure comprised the assessment of behaviors and neurobiological disturbances in germ-free mice. This allows for the introduction of specific microbes at different developmental ages to assess their behavioral actions as well as colonizing these mice with microbiota from conventionally raised mice or those with specific disturbances. However, these animals also display alterations of intestinal epithelial cells, diminished intestinal mobility, increased susceptibility to liver damage, and an underdeveloped immune system, which might make them unsuitable for some types of research. Moreover, the question exists as to whether these mice genuinely reflect an adequate representation of "normal" development. The alternative has been to assess the loss of microbiota (at different ages) stemming from antibiotic treatments, which can provide a rich source of data concerning the microbiota-anxiety link.

Germ-free mice exhibit reduced anxiety (Neufeld et al., 2011) together with lasting brain morphological alterations, including dendritic hypertrophy within pyramidal neurons and spiny interneurons within the BLA, and neurons in the ventral hippocampus were both shorter and less well-branched (Neufeld et al., 2011). Aside from morphological changes, in germ-free mice, as well as those that had gut bacteria diminished by an antibiotic cocktail, marked dysregulation of microRNA occurred within the amygdala and prefrontal cortex, which partially normalized following bacterial recolonization (Hoban et al., 2017a). The influence of microbial manipulations, at least in so far as hippocampal neurogenesis is concerned, varies with age and sex (Scott et al., 2020), which might be relevant to the greater anxiety seen among females.

The diminished anxiety apparent in germ-free mice was accompanied by upregulation in the expression of several glucocorticoid receptor genes and elevated hippocampal serotonin utilization in males (Luo et al., 2018). Anxiety could be reestablished through postweaning bacterial colonization, but the hippocampal serotonin changes remained altered (Clarke et al., 2013), which suggests that other mechanisms are responsible for the anxiogenic effects. Interestingly, the behavioral alterations were not apparent when bacterial reconstitution was conducted in adulthood (Neufeld et al., 2011). Overall, it seems that gut–brain processes may be responsible for the development of adaptive neurochemical stress systems, which may influence anxiety; however, a critical window exists to alter anxiety related to microbial disturbances.

Consistent with these findings, when the presence of gut microbiota was reduced from weaning onward through antibiotic administration, anxiety-like behaviors were diminished. These behavioral changes were accompanied by changes within the tryptophan metabolic pathway, together with altered BDNF, as well as oxytocin and vasopressin expression (Desbonnet et al., 2015). An antibiotic cocktail administered to a highly anxious mouse

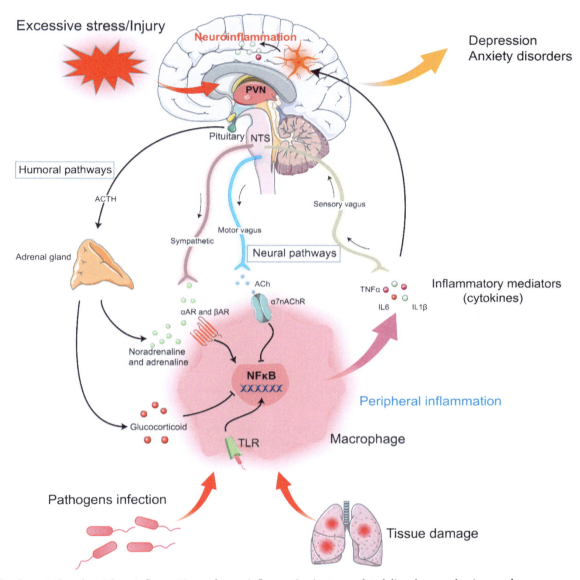

FIG. 10.2 Stress-induced periphery inflammation and neuroinflammation in stress-related disorders: mechanisms and consequences interaction between the immune system, HPA axis, and sympathetic nervous system. Exposure to traumatic and stressful events in individuals may facilitate increased immune activity in both the periphery and the central nervous system (CNS) by activating the HPA axis and the sympathetic nervous system (SNS). HPA axis activation results in the release of glucocorticoids, which modulate the inflammatory response by suppressing the expression of pro-inflammatory cytokines by immune cells. However, overactivity of the SNS increases the release of pro-inflammatory cytokines. These cytokines access the brain via afferent fibers (e.g., vagus nerve) or through the damaged blood–brain barrier to activate microglia, which in turn contribute to neuroinflammation via the secretion of pro-inflammatory cytokines in the brain. *Source: Hu, P., Lu, Y., Pan, B.X., Zhang, W.H., 2022. New insights into the pivotal role of the amygdala in inflammation-related depression and anxiety disorder. Int. J. Mol. Sci. 23, 11076.*

strain (BALB/c) through their drinking water for several days similarly produced an increase in exploratory behaviors, possibly reflecting diminished anxiety. These behavioral changes were accompanied by elevated hippocampal BDNF, whereas little change was apparent concerning autonomic nervous system activity, gastrointestinal neurotransmitter presence, or inflammation. Among BALB/c mice that were colonized with microbiota from mice of the relatively hardy Swiss-Webster strain, exploratory behavior and hippocampal BDNF were elevated. Conversely, when the germ-free Swiss-Webster mice were colonized with microbiota from the BALB/c mice, exploratory behavior was diminished (Bercik et al., 2011). While not definitive, these findings point to the causal relationship between microbiota alteration and the development of anxiety.

Experiences during early life and adolescence may have effects on mood disorders that last throughout life, irrespective of whether this comes about in response to stressors or exogenous administration of agents that

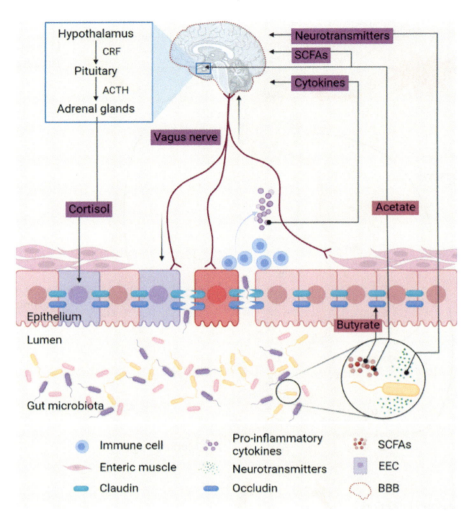

FIG. 10.3 Mechanisms of bidirectional communication between gut microbiota and the brain. A network of entero-epithelial cells (EECs) along the gut wall mediates the bidirectional communication. In response to various stimuli and external cues, the central nervous system (CNS) modulates EECs via vagal efferents and the hypothalamic pituitary adrenal (HPA) axis. Gut microbiota return signals to the brain through different afferent pathways. Microbial metabolites, cytokine induction, and neurotransmitters function via endocrine pathways; vagal afferents form part of the neurocrine pathway. Short-chain fatty acids (SCFAs) produce by bacteria in the gut include acetate, lactate, butyrate, and propionate. SCFAs modulate the integrity of the blood–brain barrier (BBB). Butyrate induces the expression of tight junction proteins, including claudins and occludins, and is therefore important for maintaining gut epithelial barrier integrity. A disrupted barrier encourages the translocation of gut microbiota and their metabolites from the lumen to the circulatory system, resulting in the production of pro-inflammatory cytokines by immune cells, which can lead to changes in cognition and mood. Acetate crosses the BBB and accumulates in the hypothalamus, thereby controlling appetite. The bidirectional flow of information via the gut–brain axis can modify the gut microbiota and modulate behavior, mood, and mental health. *Source: Dicks, L.M.T., Hurn, D., Hermanus, D., 2021. Gut bacteria and neuropsychiatric disorders. Microorganisms 9, 2583.*

produce inflammatory responses. Probiotic consumption during adolescence diminished the inflammatory response to LPS treatment, limited the changes of gut microbiota, and reduced cytokine mRNA expression in the hypothalamus, hippocampus, and prefrontal cortex, so that subsequent anxiety in males and the depressive-like phenotype in females were prevented (Murray et al., 2019). In a follow-up study, LPS administered during puberty (6 weeks of age) reduced microbiota of several microbiota genera, including *Lactobacillus*, and repopulating the gut with *Lactobacillus reuteri* during the pubertal period prevented the emergence of anxiety-like behavior and the elevated stress reactivity.

The importance of microbiota to anxiety has been gleaned from studies that assessed the impact of infection. During the early stages of infection, rodents display anxiety-like behaviors, possibly mediated by inflammatory immune processes that influence CNS functioning via the vagal nerve (Bercik and Collins, 2014). In mice deficient in Caspase-1 [also referred to as interleukin-1 converting enzyme], which ordinarily cleaves proteins, such as precursors of IL-1β and IL-18, anxiety and depression were reduced, as would be expected with reduced levels of proinflammatory cytokines. As predicted, following treatment with a Caspase-1 antagonist, depression and anxiety associated with a chronic stressor were absent. Moreover, in genetically engineered mice that lacked the pro-inflammatory caspase-1 gene, the IFNγ receptor, nitric oxide synthase, and anxiety-like behaviors stemming from chronic mild stress were attenuated in conjunction with altered gut microbiota composition (Inserra et al., 2019). Conversely, overexpression of hippocampal caspase-1 was accompanied by behavioral disturbances, possibly occurring through processes associated with glutamate functioning (Li et al., 2018). Overall, it appears that treatments that reduced the antianxiety actions of caspase-1 inhibition were linked to a rebalancing of several stress–gut bacteria relations (Wong et al., 2016).

Like the effects of other microbiota manipulations, the anxiogenic actions of stressors may stem from the disruption of intestinal microbiota, which may promote the overproduction of inflammatory mediators (Bailey, 2014). This was not apparent among germ-free mice attesting to the role of commensal bacteria in the provocation of stressor-provoked inflammatory and anxiety responses. Commensurate with this finding, gut microbiota dysbiosis produced either by stressors, a high-fat diet, or antibiotics, elicited an anxiety-like profile, which could be attenuated by probiotic treatments (e.g., Park et al., 2021).

Relevant to the microbiota-anxiety linkage, germ-free mice exhibited diminished emotional responses to cues that had been paired with a stressor. These mice also displayed a neural transcription profile that could be differentiated from that of conventional mice, particularly concerning genes thought to be involved in synaptic transmission and neuronal activity, raising the possibility that these factors mediate the link between microbiota and fear/anxiety. In fact, the persistent hippocampal changes and the fear reactions normalized in germ-free mice upon recolonization with microbiota (Hoban et al., 2017b). In line with such findings, reduced anxiety in germ-free mice corresponded with a decrease in central amygdala N-methyl-D-aspartate (NMDA) receptor expression together with decreased $5HT_{1A}$ mRNA and elevated BDNF expression in the dentate granule layer of the hippocampus (Neufeld et al., 2011).

It appears that SCFAs may be pivotal in determining the affective and anxiety-related effects attributable to microbial alterations, doing so through hormonal, neurochemical, and inflammatory mediators (Dalile et al., 2019). As expected, providing mice with probiotics attenuated the anxiety produced by a chronic mild stressor regimen while concurrently reducing hypothalamic expression of IL-1β, NLRP3, NF-κB, and Caspase-1, whereas maintaining mice on a high-fat diet had the opposite effect (Avolio et al., 2019). Moreover, 2 weeks of treatment with a prebiotic, such as 3′Sialyllactose (3′SL) or 6′Sialyllactose (6′SL), limited the microbiota changes otherwise elicited by a stressor and attenuated the anxiety evident in stressed mice (Tarr et al., 2015).

The precise processes that govern the microbiota-behavior linkage remain to be fully worked out, although several candidates have been offered. As 90% of peripheral serotonin is present within the gut, it was a good bet that it might serve in communicating with the brain. It is conceivable that epigenetic changes related to serotonin functioning might contribute to microbial dysbiosis thereby influencing mood states (Stilling et al., 2014). A prebiotic mix administered over 3 weeks attenuated the anxiety provoked by LPS, possibly through modulation of cortical IL-1β and $5-HT_{2A}$ receptor expression within the frontal cortex (Savignac et al., 2016). Similarly, chronic treatment with the *L. rhamnosus* (JB-1) altered GABA mRNA in brain regions associated with both depression and anxiety and reduced stressor-provoked corticosterone and behaviors that reflected anxiety and depression. These actions were not evident in vagotomized mice, implicating the vagus nerve in mediating the behavioral actions associated with microbial changes (Bravo et al., 2011). In addition, effects on the kynurenine pathway may contribute to the development of anxiety (Kennedy et al., 2017) just as it might play a role in the development of depression.

Even though studies in germ-free mice and those treated with antibiotics have shown diminished anxiety, studies in humans have frequently indicated that this treatment may result in increased occurrence of several neuropsychiatric conditions, including anxiety disorders (Dinan and Dinan, 2022; Nikolova et al., 2021). A large population-based analysis indicated that a single course of antibiotic treatment was associated with elevated anxiety and depression (Lurie et al., 2015). Likewise, treatment with the antibiotic fluoroquinolone was associated with increased anxiety disorders (Kaur et al., 2016). These actions may stem from antibiotic treatments influencing diverse hormonal processes and brain neurotransmitter functioning, and the provocation of elevated inflammatory factors within the brain.

Relatively few studies have examined the ties between microbiota and clinical levels of specific anxiety disorders in humans, and while microbiota dysbiosis might play a role in this regard, the limited data available has precluded the identification of specific microbes in promoting anxiety. Still, GAD was related to lower fecal bacterial diversity and the abundance of *Bacteroides* and *Escherichia-Shigella* were elevated with increasing illness severity, whereas the abundance of *Eubacterium* and *Prevotella* were inversely related to GAD (Chen et al., 2019b). Similarly, microbial richness and diversity were disturbed in GAD, as were specific bacteria, such as *Escherichia-Shigella*, *Fusobacterium*, and *Ruminococcus gnavus*) (Jiang et al., 2018b).

Among patients with OCD, the diversity of gut bacteria was reduced, as was the abundance of three butyrate-producing genera (*Oscillospira, Odoribacter*, and *Anaerostipes*). These microbiota alterations were associated with increased CRP levels, but not IL-6 or TNF-α (Turna et al., 2020). Similarly, in stool samples of OCD patients, a modest reduction of bacterial diversity occurred, together with a lower abundance of *Prevotellaceae, Agathobacer*, and *Coprococcus* as well as an elevated abundance of *Rikenellaceae*. The oropharyngeal microbiome of OCD patients likewise differed from that of healthy controls, largely comprising a lower ratio between *Fusobacteria* and *Actinobacteria* (Domènech et al., 2022).

Following studies showing that gut microbiota were linked to social behaviors (Sherwin et al., 2019), microbial differences were observed between patients with social anxiety disorder and age-matched healthy individuals (Gheorghe et al., 2021). Although these data are merely correlational, it was later shown that microbiota may be causally related to the disorder. Specifically, when microbiota were transplanted from patients with social anxiety disorder or healthy controls to mice, these critters subsequently displayed several features of the disorder. Importantly, these mice did not exhibit anxiety unrelated to social fear, suggesting that the fecal transplants uniquely affected this specific anxiety syndrome. Furthermore, in mice showing the social anxiety disorder features, oxytocin expression was affected in the bed nucleus of the stria terminalis (i.e., the extended amygdala) and was immune functioning within the periphery and the brain (Ritz et al., 2024).

While social anxiety disorder in humans was not associated with changes in microbiota diversity, an elevated abundance of *Anaeromassillibacillus* and *Gordonibacter* was present, whereas *Parasuterella* was reduced (Butler et al., 2023). These findings are in keeping with the increasing evidence showing that gut microbiota contributes to brain activity linked to social functioning (Sherwin et al., 2019), which can have implications well beyond anxiety disorders, including disorders related to difficulties associated with social communication and social interactions (e.g., autism spectrum disorder).

In closing, several issues need to be raised (again). First, microbial signatures comprising reduced antiinflammatory butyrate-producing bacteria coupled with enrichment of pro-inflammatory genera have been observed across diagnostic categories, including major depressive disorder, bipolar disorder, schizophrenia, and anxiety, rather than any single disorder (Nikolova et al., 2021). Furthermore, by influencing the NLRP3 inflammasome, microbial influences may be fundamental in promoting neurodegenerative disorders (Pellegrini et al., 2020). Even though these microbial alterations have been associated with the development of diverse pathologies, they might be secondary to these illnesses or associated stress/lifestyle factors. This certainly does not belie the possibility that microbiota disturbances may exacerbate illnesses that are already present or those for which individuals are at elevated risk. Second, insufficient attention has been devoted to the factors that promote illness differences between the sexes. Microbial-hormonal interactions may contribute to sex-related pathology, and with a greater appreciation of the correlates of anxiety conditions, a better handle could be obtained regarding how to treat illnesses through a precision medicine approach (Shobeiri et al., 2022). Finally, we make particular note, as we have in earlier chapters, that while anxiety in humans could be diminished by probiotics supplements, better effects were realized through diets that increased SCFAs (Yang et al., 2019).

Conclusion

There's little question that anxiety can be more than just an uncomfortable feeling and may become exceptionally debilitating. The processes responsible for anxiety and related disorders have yet to be fully defined, although CRH, GABA/glutamate, monoamine, and cannabinoid functioning have all been implicated as possible mediators. Considerable evidence has also supported the role of microbiota and inflammatory immune dysregulation in the provocation of anxiety. These actions could occur owing to the effects of microbial and cytokine effects on brain hormonal, neurotransmitter, and neurotrophin changes. It is unlikely that any one of these processes acts in isolation from others in promoting and maintaining anxiety, and although several contribute to more than a single anxiety disorder, the neurobiology of these disorders is distinguishable from one another.

The risk factors for anxiety disorders comprise genetic components together with psychosocial stressors, and specific hormonal alterations, such as the elevated afternoon cortisol levels common in social anxiety disorder. These disorders have been treated by a variety of pharmacological agents, the most common being SSRIs, benzodiazepines, β-norepinephrine blockers, as well as psychotherapeutic treatments (e.g., CBT, exposure therapy, stress management). But, here again, the symptoms presented vary remarkably across the different anxiety disorders, and even within any disorder, marked individual differences exist regarding the symptoms presented. This is particularly notable in the case of OCD in which some patients may exhibit very different behaviors but share the common feature of anxiety declining when the obsessive behavior is expressed.

Even if inflammatory processes contribute to anxiety, it is uncertain why inflammatory processes would give rise to one or another form of anxiety disorder.

Inflammatory factors likely act to either aggravate an already existing condition or serve as a general challenge to mental health, much as psychogenic stressors act this way, with the specific condition that emerges varying with other factors that occur concurrently. Despite these uncertainties, there is the view that microbial changes or antiinflammatory manipulations could be used as adjunctive treatments for anxiety-related conditions. As more is learned about the specific immune or microbial factors that accompany different forms and symptoms of anxiety, the prospects for adjunctive treatments will be improved.

CHAPTER

11

Posttraumatic Stress Disorder

It ain't over even after it's over

It isn't unusual for people to mistakenly believe that once a traumatic event is over, so is the distress, and consequently individuals should recover. We even have expressions that speak to this, such as "time heals all wounds," and there are occasions when trauma survivors are told, "Just get over it." Time might allow some wounds to heal in some individuals. For others, time moves ever so slowly, and the trauma memories linger and so do the psychological and physical consequences of the trauma. Parents who have lost a child, individuals who experienced the horrors of warfare, or a survivor of genocidal efforts don't simply forget. Some might never speak about their experiences (a conspiracy of silence), whereas others can't stop speaking about it. Although most individuals go on with life, the traumatic memories are often just below the surface, and their effects can re-emerge.

Although some of Freud's views have been abandoned, many of his thoughts on ego defense mechanisms and the impact of early life experiences are still considered to be pivotal. He had postulated, among many other things, that early experiences might mark individuals for life, although he didn't understand how this came about from a biological perspective. After all, he didn't have the luxury of knowing about the workings of neurochemical systems, and the involvement of neuroplasticity in mental illnesses was far off. He nonetheless understood that traumatic memories, especially those from childhood, represented a causative agent that continued to undermine an individual's mental health, often throughout life. This could occur through the unconscious but could also have damaging effects through persistent negative rumination and incidents being replayed.

It is now well established that the occurrence of posttraumatic stress disorder (PTSD) is significant within the general population and is especially frequent among individuals who experienced traumatic experiences as well as chronic stressor encounters. PTSD is often considered a disorder that involves dysregulation of fear-related processes, as well as memories of trauma, which involve alterations of various aspects of the amygdala, hippocampus, and portions of prefrontal cortical circuits. The development of the disorder is linked to previous stressor encounters, especially those experienced during early life, and was tied to the expression of a variety of genes involved in stress processes (Ressler et al., 2022). Likewise, a range of neurobiological factors have been identified that are associated with PTSD. Despite the significant advances in defining the mechanisms associated with PTSD, effective therapeutic methods to attenuate the symptoms of this disorder have been relatively modest, although advances have been occurring, even if these have progressed in baby steps.

A Perspective on Trauma-Related Disorders

For many years, PTSD has been a top-of-mind issue among those working in trauma-related mental health fields, and thanks to media attention, public awareness increased, and the stigma of PTSD has fallen away. Still, some affected individuals continue to suffer in silence or attempt to diminish their distress through self-medication (e.g., use of alcohol). With a greater understanding of PTSD came the realization that the disorder was common among people who encountered a wide range of traumatic events, and it has been estimated that the global lifetime prevalence of this disorder is about 5.6%. Of course, the frequency of trauma experiences varies greatly across countries as does the occurrence of PTSD, and seeking treatment was considerably greater in high-income countries than in low- and middle-income countries (Koenen et al., 2017).

At one time, PTSD was considered to be an anxiety disorder, but this is no longer the prevailing view. Within

The Immune System
https://doi.org/10.1016/B978-0-443-23565-8.00010-7

293

Copyright © 2025 Elsevier Inc. All rights are reserved, including those for text and data mining, AI training, and similar technologies.

the DSM-5, two separate, but related syndromes are described that have been linked to severe or chronic stressor experiences. These comprise acute stress disorder (ASD) and PTSD, which differ from one another in several respects but still share several features.

Acute Stress Disorder

Soon after a catastrophic event, intense emotional reactions may appear, typically diminishing over time, although the emotional upheaval can persist or even become more pronounced. A diagnosis of ASD is applied if symptoms appear within 1 month of a trauma experience and persist for at least 3 days. Should ASD persist for more than 30 days, a diagnosis of PTSD may be applied.

In addition to intense anxiety, a diagnosis of ASD requires the presence of three or more peritraumatic (at the time of the trauma) "dissociative" symptoms. These may comprise (1) a sense of numbing, detachment, or absence of emotional responses, (2) diminished awareness of surroundings (feeling as if in a daze), (3) derealization, in which perceptions or experiences of the external world seem unreal, (4) depersonalization, wherein an individual has the feeling of watching themselves act, but feel that they have no control, and (5) dissociative amnesia, which is characterized by memory gaps, such that individuals are unable to recall information concerning events of a traumatic or stressful nature. Although dissociative features weren't necessarily predictive of later PTSD emerging, these characteristics have been linked to further mental health difficulties and physical illnesses (Boyer et al., 2022). Aside from these symptoms, ASD may also be accompanied by "re-experiencing" the traumatic event by way of recurrent images, thoughts, dreams, illusions, and flashbacks. Individuals may also feel as if they're reliving the traumatic experience. As these features can be promoted by reminder stimuli, individuals often exhibit strong avoidance responses. Features of anxiety and arousal are typically present, including hypervigilance, hyper-reactivity, irritability, impaired concentration, disturbed sleep, and motor restlessness.

Posttraumatic Stress Disorder

A wide range of stressful events not only promote PTSD, including unique or unusual experiences (being held hostage), but also traumatic experiences that are relatively common (car accident, medical complications, being told about a severe medical condition, assault, rape, and repeated exposure to images of traumatic events). In addition, chronic stressors, such as bullying and racial discrimination, can instill PTSD symptoms, as can distal stressors (i.e., when individuals were not directly confronted with the trauma) as seen among US residents following the 9/11 terrorist attacks. PTSD stemming from interpersonal violence (IPV) occurs frequently and may promote complex PTSD (CPTSD) that comprises the symptoms of PTSD coupled with disturbances of self-organization, and adoption of a maladaptive emotion regulation strategy (e.g., Fernández-Fillol et al., 2021). While studies of IPV most often assessed the impact of physical and sexual abuse, psychological abuse can have equally pernicious effects. This has been notable among victims of "narcissistic abuse" in which the abuser uses a series of techniques to achieve their goals of being adored and venerated, with little care for their victim (Arabi, 2023).[1]

A diagnosis of PTSD, based on the DSM-5, is considered within the context of five broad categories. (1) Exposure to actual or threatened death, serious injury, sexual violation, or witnessing others experience these traumatic events; (2) presence of one or more intrusive symptoms (e.g., involuntary distressing memories of the traumatic event(s), or intense psychological distress or physiological reactions triggered by reminders of the trauma); (3) persistent avoidance of trauma-associated stimuli; (4) disturbed cognitions and mood (e.g., impaired recall of an important aspect of the event, persistent and distorted self- or other blame regarding the cause or consequences of the traumatic event(s) (feelings of detachment or estrangement from others); and (5) marked arousal and reactivity associated with the traumatic event(s), reflected as hypervigilance, hyperarousal, or exaggerated startle response.

PTSD Vulnerability and Resilience

At some time in their life, most individuals (>60%) encounter stressors that are sufficiently chronic or traumatic to produce PTSD. Yet only a modest proportion of these individuals develop this disorder. Numerous risk factors have been identified with the development of PTSD including being female, having encountered childhood trauma, experiencing multiple previous traumatic events, preexisting mental disorders, and a history of IPV. A detailed accounting of the demographic and epidemiological variables linked to PTSD, together with the

---

[1] Narcissistic relationships, which can have either females or males as victims, have the perpetrator adopt manipulative behaviors to control their partner. This often begins with "love bombing" (the abuser showers them with flattery and excessive attention) and jealousy that may be misattributed to the perpetrator's adoration. When the behaviors of the perpetrator are questioned, conversations are often shut down and stonewalling is common. Gaslighting is used to the extent that victims frequently question their perceptions, memories, and even their sanity, and perpetrators engage in coercive control.

incidence of varied types of trauma, the biological factors associated with PTSD, as well as its treatments, has been well documented (Shalev et al., 2017).

Fundamental to the impact of stressors is how they are appraised (i.e., perceived level of threat) and what sorts of responses emerge following a trauma experience. As in so many other stress-related conditions, the emergence and continuation of PTSD symptoms were particularly notable among those with low social support, and the use of maladaptive cognitive coping methods (i.e., worry, self-punishment). In the latter regard, in some instances avoidant strategies can be useful, as this might allow for compartmentalization of the stressor experience, thereby limiting memories of the trauma, as well as the cues that encourage these memories. Thus treatments aimed at diminishing maladaptive coping strategies and encouraging support and understanding from a person's social network might limit PTSD symptoms. As well, it may be important to separate emotional responses from the cognitive representation of the event (cognitive-emotional distinctiveness).

Pretrauma Experiences

Stressful experiences, as we've seen, may result in the sensitization of neurobiological systems so that individuals are more responsive to later challenges, essentially creating a disposition toward PTSD. Childhood trauma, IPV, and severe symptoms of an anxiety/affective disorder predicted the development of PTSD and poor remission from the disorder (Gould et al., 2021). In this respect, earlier trauma experiences, including abuse or neglect in childhood, can have lasting effects on the individual's ability to cope with subsequent stressors, thereby influencing the development of PTSD in response to further challenges. One of the most prominent predictors of PTSD is having a history of multiple traumas and having mental health problems before the trauma (Kessler et al., 2018), and individuals who score highly on a neuroticism scale (characterized by constant worrying, chronic anxiety, and over-reactivity to daily negative experiences) are most likely to develop PTSD.

Physical illness and PTSD

The progression of life-threatening illnesses and the response to treatment can be affected by lifestyle factors and stressor experiences. Severe illnesses, such as cancer and cancer-related experiences, can promote psychological disturbances, including PTSD and depression, and are more prominent among individuals with a preexisting psychological disorder, such as depression (De Padova et al., 2021). As such, it would be propitious for cancer treatment teams to determine an individual's psychiatric history

before the initiation of therapy, particularly as the development of a psychiatric condition can undermine the efficacy of the therapy (Cordova et al., 2017).

There have been many reports concerning the development of mental health problems and PTSD during the COVID-19 pandemic. While most individuals did not experience long-term mental health difficulties owing to the pandemic, PTSD occurred among survivors of severe SARS-CoV-2 infection, varying considerably across studies (Nagarajan et al., 2022). The risk for psychiatric disturbances comprised elevated disease severity, symptom duration, and symptoms that were often associated with inflammatory markers (Schou et al., 2021). Given the marked distress created among healthcare workers, the occurrence of PTSD and other mental health conditions was elevated in this population (Scott et al., 2023). As PTSD is not only associated with traumatic events but also chronic stressors, it isn't altogether surprising that COVID has been associated with elevated mental health challenges, including PTSD, and was predicted by elevated levels of IL-6 (Giannitrapani et al., 2023).

In addition to being comorbid with depression and poor quality of life, PTSD is often associated with other pathological conditions tied to immune, HPA, or autonomic disturbances, including diabetes, heart disease, and autoimmune disorders, and may be linked to premature cellular aging. Inflammatory genes are prime culprits that have been implicated as acting in this capacity.

PTSD and Structural Brain Alteration

Although few studies have been able to assess PTSD-related neurochemical changes in the brain tissues of human participants, imaging studies have pointed to variations of connectivity within neuronal networks and changes in brain architecture. Several reports focused on community participants who encountered some form of trauma that led to PTSD, whereas others assessed PTSD among war veterans, including individuals who had sustained traumatic brain injuries of varying degrees. It is important to assess the mechanisms operating improperly among these individuals, but in the analysis of the mechanisms supporting PTSD, the presence of physical injuries, particularly head injury, may be a confounding factor, just as the presence of comorbid illnesses makes it difficult to identify the mechanisms responsible for PTSD. Accordingly, rather than just assessing the effects of trauma and PTSD relative to no-trauma controls, it is essential to distinguish between the brain changes seen among individuals who experienced trauma associated with PTSD relative to those who did not develop PTSD. As well, it has been maintained that PTSD with dissociative features present may be a subtype of PTSD that may involve mechanisms distinct from (or in addition to) those associated with PTSD without dissociation. Thus analyses of brain processes would do well to distinguish these subtypes of PTSD.

An impressive number of reports indicated that PTSD was accompanied by diminished size of the hippocampus and prefrontal cortex, and symptom severity was closely aligned with reduced left hippocampal volume (e.g., Nelson and Tumpap, 2016). Unlike the fairly consistent hippocampal findings, the data concerning amygdala volume in PTSD have been less uniform. The variability was related to subregional differences in connectivity and grey matter volume within the amygdaloid complex. Indeed, PTSD was accompanied by lower right basolateral amygdala connectivity and volume, but elevated centro-medial amygdala connectivity, the latter being most notable with severe PTSD (Aghajani et al., 2016).

Although bilateral amygdala volume was reduced relative to that seen in healthy controls, a meta-analysis indicated that amygdala volume did not differ between individuals with trauma-related PTSD and trauma-exposed individuals without PTSD. Thus trauma rather than PTSD itself may have been the key factor linked to reduced amygdala volume (O'Doherty et al., 2015). A subsequent study by this group revealed that relative to individuals who experienced trauma without developing PTSD, individuals with PTSD exhibited more pronounced reductions of hippocampal and amygdala volume, as well as diminished grey matter in the anterior cingulate cortex, frontal medial cortex, middle frontal gyrus, superior frontal gyrus, paracingulate gyrus, and precuneus cortex. Moreover, PTSD severity was inversely correlated with compromised white matter integrity in pathways linking the amygdala to frontal brain regions, such as the anterior cingulate cortex (O'Doherty et al., 2018). Disturbances of this pathway permit the emergence (disinhibition) of anxiety and fear, which form part of the PTSD profile. Greater amygdala neuronal responsivity was present among individuals with PTSD, and amygdala reactivity 1 month after the trauma predicted the presence of PTSD 12 months later (Stevens et al., 2017b).

Functionally, PTSD was accompanied by hyperactivity within the mid- and dorsal anterior cingulate cortex and the amygdala, whereas hypoactivity was present within the ventromedial prefrontal cortex, which may have been linked to elevated amygdala activity. Those individuals who exhibited particularly high levels of amygdala activity and a progressive decline of neuronal activity within the anterior cingulate cortex were at greatest risk for the emergence of PTSD symptoms and exhibited a poorer prognosis (Stevens et al., 2017b). As we'll discuss later in greater detail, effective psychotherapy was accompanied by upregulation of medial prefrontal cortical functioning, whereas amygdala changes were less evident (Manthey et al., 2021).

Following severe trauma about 40% of individuals may exhibit symptoms of ASD, but fewer than 10% of individuals report the occurrence of PTSD. Being able to identify those individuals who were most at risk for the eventual development of PTSD might be of prophylactic value. Among car accident victims who performed an emotional appraisal task 1 month following a car accident, variations of neuronal activity were initially observed within the left dorsolateral prefrontal cortex, medial prefrontal cortex, and the bilateral inferior frontal gyrus (IFG). Of particular interest was that upon re-evaluation 6 and 14 months after the accident, those individuals who had initially exhibited the greatest right IFG activation displayed the most pronounced reduction of symptoms (Sheynin et al., 2023). As we'll see, these findings were congruent with reports that focused on changes in PTSD that were related to disturbed regulation of specific brain networks.

Rather than focusing on PTSD as a syndrome, it has been instructive to consider specific features of this condition. Analyses of morphological changes indicated that hippocampal abnormalities were most closely aligned with arousal symptoms, whereas amygdala alterations were linked to re-experiencing symptoms (Akiki et al., 2017). Other symptoms seemed to involve diminished cortical thickness that was apparent among veterans with PTSD, as well as in those who scored highly on the Combat Exposure Scale even in the absence of frank PTSD (Wrocklage et al., 2017). Bilaterally, anterior cingulate cortex volume reductions were linked to attentional disturbances and impairments of emotional regulation among individuals with PTSD (O'Doherty et al., 2015). Moreover, it was maintained that PTSD was linked to the inability of cortical regions to regulate subcortical areas responsible for fear responses. Supporting the disconnect between cortical and subcortical involvement, it was demonstrated that among individuals with PTSD white matter integrity was diminished in the uncinate fasciculus as well as in the corpus callosum and corticospinal tract, whereas enhanced white matter integrity was noted in the inferior fronto-occipital fasciculus and inferior temporal gyrus (see Ju et al., 2020).

The nature of the changes detected varied with specific features that accompanied PTSD. For instance, when PTSD was accompanied by depressive illness, symptom severity was associated with reduced grey matter volume in the left dorsomedial prefrontal cortex, and variations within aspects of the amygdala were related to both emotional reactivity and emotional components of memory (Knight et al., 2017). Postmortem analyses also indicated that gene expression (e.g., for glucocorticoid signaling) within single neurons and astrocytes of the dorsolateral prefrontal cortex distinguished between individuals who had been depressed and those with PTSD (Chatzinakos et al., 2023). In essence, while PTSD and depression often co-occur, they may involve independent brain processes that could be related to distinguishable behavioral phenotypes.

Like studies that considered specific features of PTSD, attempts were made to identify the neural network alterations that were involved in the disorder. An interesting review of animal and human studies outlined the different neural circuits associated with PTSD, indicating that the multiple characteristics associated with PTSD involved discrete, yet overlapping neural processes (Fenster et al., 2018). It was posited that PTSD may stem from disturbed regulation of medial prefrontal cortical regions that are part of the default mode network (Clausen et al., 2017). Also, among individuals with PTSD, resting-state functional connectivity was greater between the salience network and the default mode network, the dorsal attention network, and the ventral attention network, which may have accounted for disturbed disengagement seen in a test of spatial attention. Based on a meta-analysis it was concluded that although PTSD involves a common neural network that includes the cingulate cortex, insula, and the parietal, frontal, and limbic regions, differences existed concerning the type of trauma experienced (Boccia et al., 2016). This is consistent with the finding that diverse trauma experiences (e.g., abuse vs. accident) may have different cognitive effects, and varied networks can be engaged in response to diverse stressors.

While not diminishing the importance of considering the impact of different trauma experiences, as well as differences related to sex, the specific symptoms expressed by PTSD patients were attributed to connectivity within brain networks (reviewed extensively in Hinojosa et al., 2024). Aspects of the salience network, notably the dorsolateral cingulate cortex and amygdala, were related to hyperarousal and hyperreactivity, whereas components of the default mode network, specifically the hippocampus and ventromedial prefrontal cortex, were associated with disturbed emotional regulation together with the failure of extinction processes, and intrusive symptoms. The DMN has also been linked to dissociative and re-experiencing symptoms. Aside from PTSD being associated with disturbances within specific networks, symptoms may develop owing to connectivity between networks. For instance, disturbed connectivity between the SN and DMN was associated with PTSD, and it was similarly reported that impaired SN functioning results in the DMN and central executive network not operating together appropriately in response to cognitive demands.

Using machine learning algorithms, it was possible to identify several neural networks associated with PTSD and to differentiate between PTSD in which dissociative symptoms were or were not present (Nicholson et al., 2020). A more detailed analysis that distinguished subtypes of PTSD based on the specific set of symptoms presented described four PTSD subtypes that also differed in the functional neural connectivity profiles that existed. In a 2-year follow-up, the subtypes showed distinct longitudinal trajectories in symptoms and the organization of functional neural activation patterns (Lee et al., 2023c).

Vulnerability to PTSD and structural brain changes has been linked to genetic influences. For instance, altered amygdala functioning in individuals with PTSD was related to the presence of a specific polymorphism in genes associated with dorsolateral prefrontal cortex functioning (Bharadwaj et al., 2018), and epigenetic changes related to glucocorticoid receptors within the hippocampus were also associated with PTSD (McNerney et al., 2018). Likewise, symptom severity and the most pronounced reductions of left hippocampal volume were particularly evident among individuals who displayed a polymorphism for the gene expressing catechol-O-methyltransferase (COMT) an enzyme involved in norepinephrine degradation (Hayes et al., 2017). These findings mapped onto a report indicating that a polymorphism related to the gene coding for COMT also interacted with the actions of childhood trauma in determining increased risk for PTSD symptoms (van Rooij et al., 2016).

A meta-analysis of GWAS that comprised individuals of European ancestry, African, and Native American ancestry, and mixed ancestry identified 95 genome-wide significant loci that were linked to PTSD of which 43 were potentially causally related to the illness. These genes were linked to various neurobiological processes, including neurotransmitter and ion channel synaptic modulators, synaptic structure, and function developmental, axon guidance, and transcription factors, as well as endocrine or immune regulators. Moreover, identified genes were associated with immune functioning, stress responses, as well as fear and threat-related processes (Nievergelt et al., 2024).

When we discuss specific mechanisms that may underlie PTSD in ensuing sections, other genes will be introduced that are pertinent to the disorder. As interesting as the findings are, these analyses would gain if links could be determined between these processes and specific behavioral features associated with PTSD and in the efficacy of various treatments.

As with so many other mental illnesses, PTSD appears to reflect a conglomerate of multiple neurobiological changes involving a constellation of brain regions and neural circuits, and is often comorbid with other conditions, making it difficult to identify biomarkers that are uniquely predictive of PTSD. Thus it might be advantageous to adopt a transdiagnostic approach to determine biomarkers that map onto symptoms (and subtypes of the illness) that are shared across disorders, and on this basis select therapeutic strategies (McQuaid, 2021).

Imaging the child's brain

With brain connections not yet mature, children may be particularly vulnerable to long-lasting changes in brain structure and functioning in response to stressors, thereby influencing later vulnerability to PTSD. It seems childhood maltreatment was accompanied by reduced amygdala volume, leading to the suggestion that early trauma, such as sexual abuse, could disturb normal amygdala development, hence increasing vulnerability to adult PTSD (Veer et al., 2015). Likewise, PTSD stemming from pediatric trauma was associated with clusters of white matter alterations in adults (He et al., 2023b). Maltreatment might not engender immediate brain changes, such as altered hippocampal volume, but may emerge over the course of development, appearing fully in adulthood (Ahmed-Leitao et al., 2016).

The damaging effects of trauma and PTSD in children went well beyond structural changes. For instance, among children who had experienced an earthquake in China in 2008, the map of neural connections within the brain (connectome) varied between individuals who experienced PTSD and those who experienced the trauma but did not develop PTSD. This included diminished local and global network efficiency stemming from disturbances or disconnections between brain regions, and neuronal networks tended to be relatively localized so that connections between diverse brain areas required more synapses or junctions for messages to traverse between networks (Lei et al., 2015).

The impact of childhood abuse can manifest as hypersensitivity to threats and elevated reactivity to further stressors or triggering cues. In later adolescence or adulthood, the negative early experiences are associated with alterations of functional brain activity in response to threatening cues (emotional processing) and those associated with reward, thereby affecting PTSD on the one side and depressed mood on the other (Cisler et al., 2019). As well, among people with a history of abuse, the functioning of neuronal systems associated with perceptual processing, attention, and executive control was altered, and could thereby contribute to diverse psychopathologies.

Theoretical Perspectives on PTSD

The development of PTSD, as depicted in Fig. 11.1, involves multidimensional changes that include genetic and epigenetic dispositions, varied stressor experiences, and several neurobiological processes. Thus it is predictable that subtypes of the disorder exist that may involve somewhat different processes.

FIG. 11.1 Schematic representation showing that severe trauma or chronic stressor experiences, coupled with predisposing vulnerabilities (e.g., genetic and epigenetic factors, early life stress, sex differences), may promote HPA alterations, activation of inflammatory processes, and production of excitotoxicity actions that undermine synaptic connectivity in brain areas that subserve emotional dysregulation, fear memories, and disturbed fear extinction processes. This cycle ultimately contributes to PTSD symptomatology, which itself constitutes a profound chronic stressor that perpetuates the processes fundamental in maintaining and exacerbating PTSD.

PTSD as a Failure of Recovery Systems

Ideally, when stressors are imminent or present, neuronal activity should be elevated, and then decline when the stressor is no longer present. However, owing to the sensitization of neurochemical systems, altered brain functioning related to anxiety, fear, and cognitive processes may persist. Even cues that were related to a traumatic experience were effective in modifying amygdala functioning (Liberzon and Abelson, 2016). Some of the brain alterations associated with PTSD normalized with treatment, but this was not apparent across all regions. As we know from studies in animals, even when "extinction" has occurred, so that anxiety and fear are not apparent, they can readily re-emerge in the presence of potential threats.

PTSD as the Failure to Distinguish Danger (Threat Detection) From Safety

Ordinarily, the functioning of the hippocampus is modulated by inhibitory projections from the entorhinal cortex, which could play a significant role in being able to distinguish between danger and safety (Basu et al., 2016). A fundamental component that determines stress responses is being able to appraise stressors appropriately, which may depend upon the context in which an event occurs. Contextual processing essentially means being able to distinguish between safe and dangerous places, which involves the medial prefrontal cortex, amygdala, and hippocampus. Some aspects of the prefrontal cortex are involved in threat expression, whereas others are involved in threat inhibition and cognitive regulation. These regions interact with the amygdala and hippocampal processes in determining threat learning and contact-specific threats, but with PTSD, these connections are disturbed so that the responses to threats are altered (Alexandra Kredlow et al., 2022). Similarly, dysregulation of the hippocampal-prefrontal-thalamic circuitry results in impaired fear learning, salience, threat detection, emotional and executive regulation, the application of appropriate meaning to various situations, and an inability to turn off danger (fear) cues (Liberzon and Abelson, 2016; Shalev et al., 2017).

PTSD as a Disturbance of Memory Processes

As interesting as reports of diminished hippocampal size in PTSD might be, they typically don't provide any indication of whether these outcomes are causally linked, the direction of this action, or even if both outcomes are related to some other common factor. It was of particular significance that reduced hippocampal size was not only present in a person with combat-related PTSD but also in their co-twin who had not been traumatized or suffered PTSD (Pitman et al., 2006). In effect, having a relatively small hippocampus might increase vulnerability to PTSD in response to severe stressor experiences, rather than stemming from the trauma or the illness. It is therapeutically significant that reduced hippocampal volume was less common among individuals who had shown symptom remission relative to that evident among those with continued PTSD (Apfel et al., 2011a). At the same time, the hippocampus is only one of several regions that are altered owing to the trauma rather than being present before trauma and acting as a vulnerability factor. Specifically, grey matter volume in the anterior cingulate cortex and insula was smaller in a twin with combat experience who developed PTSD than in their co-twin who had not been in a war situation, or in twin pairs who had been in war environments but had not developed PTSD (Kasai et al., 2008).

Given that neuroplasticity is as pronounced as it is within the hippocampus, it has been a prime suspect in the development of the adverse effects of trauma and points to memory processes being involved in PTSD. Having been traumatized by a particular event, it would be reasonable to expect that strong behavioral and neurobiological reactions would emerge in response to stimuli relevant to the trauma. These reactions would be highly adaptive to maintain safety and readiness to respond to similar threats but would be maladaptive if the hyper-responsivity generalized to other situations that bore limited resemblance to the initial trauma event. Ordinarily, when memory is well entrenched, strong responses are elicited by cues that are reminiscent of the primary aversive stimulus. Among those with PTSD, strong reactions are elicited by cues reminiscent of the original trauma, pointing to a well-entrenched memory. However, stress responses also occur in response to vague cues, which might ordinarily be evident if the memory was not well established or inaccurate. The fact that memory processes may be activated even when cues are distinct from a previously encountered traumatic experience may occur because these cues engage the same neural circuits that had been activated by a previously encountered stressor. From this perspective, PTSD might not stem from disturbed memory but instead reflects sensitized and generalized biological responses to stressors. Alternatively, these findings can also be taken to imply that PTSD is not a disorder of strong memories being instigated but instead reflects a disorder in which the normal extinction of fear memories does not occur readily.

The possibility has been considered that the brain representation of trauma memories differs from other memories. Among individuals with PTSD who were presented with a narrative that comprised sad memories (e.g., death of a loved one) their hippocampus was ordinarily engaged. However, when the same people were presented with narratives of their traumatic memories, the hippocampus seemed not to be engaged, and instead marked activation was observed in the posterior cingulate cortex, which serves in processing and integrating information. Individuals were not exhibiting a profile ordinarily associated with recall of the past and in a sense, were responding as if it were a present experience or one in which fragments of prior memories were overriding the ongoing experiences (Perl et al., 2023).

There is another way of seeing the connection between PTSD and memory-related processes. Memories are maintained through "engrams" that comprise the interconnections between particular neuronal connections, and because of the brains' remarkable synaptic plasticity new information can be incorporated into this ensemble. Reminders (cues) of events can sustain these engrams, whereas the decay or interference of the accessibility of these engrams leads to natural forgetting. Maintaining memories or the occurrence of forgetting can be influenced by GABA and glutamate functioning as well as through microglial processes that eliminate engram synapses tied to specific memories. From this perspective, forgetting entails the diminished presence of natural retrieval cues that would ordinarily permit the accessibility of engram cells. This can be highly adaptive so that the brain would not be overwhelmed by information that is not salient or meaningful, which would allow for the greater cognitive flexibility needed in problem-solving efforts. Thus forgetting and learning both reflect adaptive changes to meet the demands of a constantly changing environment (Ryan and Frankland, 2022). Fundamental to our discussion of PTSD, experiential or environmental factors, including traumatic experiences, can influence engram accessibility so that normal forgetting is prevented. In essence, PTSD may comprise the "hijacking" of ordinary forgetting processes so that fear responses persist.

We've provided only a capsule of this intriguing perspective, primarily as it relates to PTSD. However, it has important implications for memory and forgetting processes relevant to memories of pain and psychological disorders (e.g., depression and substance use disorders), as well as episodic memories that are relevant to early experiences (Ramsaran et al., 2023). Focusing on processes relevant to forgetting has implications for pharmacological and behavioral/cognitive therapeutic approaches to treat disorders involving well-entrenched behavioral styles (e.g., in chronic depressive disorders).

This brings us to the question of whether inflammatory processes have any bearing on the model described by Ryan and Frankland (2022), especially as this might pertain to PTSD. Although immune memory is distinct from cognitive and emotional memory processes, aspects of the brain seem to remember immunological and inflammatory insults. Specifically, body inflammation (e.g., experimentally induced colon inflammation) in rodents gives rise to changes in activity within several brain regions, including the insular cortex that serves as a hub involved in communication between the brain and body, and seems capable of storing information concerning the nature and location of activated immune responses. After animals had fully recovered from inflammation, activating the same insular cortical neuronal ensembles (engrams) provoked another bout of intestinal inflammation without affecting other organs or immune functioning. By inhibiting these infralimbic neurons the development of experimentally provoked inflammation could be prevented (Koren et al., 2021a).

Memory engrams for a specific event may be distributed across multiple brain regions that are functionally interconnected. While the connectivity between the hippocampus and cortex may be particularly notable (Roy et al., 2022), especially in the context of fear memories, the insular cortex's involvement in emotional processing and inflammatory-related processes may play a prominent role in PTSD. As we've seen repeatedly, microglia are intimately related to brain inflammatory processes and are fundamental in synaptic plasticity and neural circuit remodeling, and hence contribute to diverse mental illnesses (Appelbaum et al., 2023). It seems that microglia-related inflammation could interfere with memory processes, thereby favoring PTSD being developed and maintained (Enomoto and Kato, 2021).

Memory Reconsolidation

When memories are first established and are present in short-term storage, they are labile and easily disturbed or altered. Once these memories are consolidated and in long-term storage, they are more resistant to being altered. However, upon the events being recalled, the memories are, in a sense, back in a labile state, and once again they can readily be altered. Accordingly, when animals are provided with reminders of a previous stressor experience, thereby bringing the memories back to short-term storage, treatment with a protein synthesis inhibitor can alter or disturb the memory (Nader, 2015). Likewise, when a memory returns to the labile short-term store, novel events and relevant cues can be incorporated into the original memory, and consequently may appear in a modified form when it is reconsolidated and returns to long-term storage (Lee et al., 2017b). By taking advantage of the memory lability when it is being recalled, it might be possible to modify memories relevant to a PTSD-like state through suggestions or by pharmacological treatments (e.g., β-blockers). Moreover, the memory can be modified or dissociated from the related emotional and cognitive responses that would otherwise be elicited. As we'll see later in this chapter, manipulations applied during reconsolidation can attenuate symptoms of PTSD.

Attention and Salience

Subtle impairments related to response inhibition and regulation of attention mechanisms involving the anterior cingulate cortex were implicated as risk factors for PTSD. This brain region, as we've seen repeatedly, plays a pivotal role in decision-making processes, and neuronal functioning at this site is exceptionally sensitive to stressor experiences. It is especially instructive that neuronal activity within the anterior cingulate cortex and dorsal cingulate predicted PTSD severity over both short- and long-term follow-up analyses (Kennis et al., 2017). Moreover, the diminished severity of PTSD symptoms following prolonged exposure treatment was associated with reduced subgenual anterior cingulate cortex and parahippocampal activation during recall of fear extinction (Helpman et al., 2016).

Activation of the salience network may allow specific cues (or threats) to take on greater significance, which may be particularly pronounced among those with PTSD. Activation of the medial PFC response was altered upon the presentation of stimuli that served as reminders of a trauma experience and could even be provoked by narratives of a negative nature that were unrelated to the trauma experience. In essence, frontal cortical activity that ordinarily inhibits amygdala functioning was altered by diverse threatening stimuli beyond those relevant to the trauma that the individual with PTSD had experienced. Among twin pairs, recalling trauma-unrelated imagined events, medial prefrontal cortex activity was diminished in the twin with PTSD relative to that evident in their co-twin without PTSD, supporting the view that the reduced neuronal activity elicited by aversive thoughts is an acquired feature of this condition (Dahlgren et al., 2018).

Sensitized Responses

Simply because an individual has seemingly gotten by a traumatic experience does not mean PTSD will not appear at some later time. The traumatic experience can influence neuronal responsivity so that subsequent stressor events or reminders of negative experiences more readily trigger these responses. The sensitization of these neurons may grow with time, peaking several weeks after the initial trauma (Anisman et al., 2003), which would account for the delayed emergence of PTSD. With further stressor experiences or reminders of the initial trauma, the neuronal network associated with the emotional memories stemming from trauma could be strengthened, thereby diminishing the dissipation of emotional trauma memories and making them more resistant to therapies. From a psychodynamic perspective, the distress associated with trauma might provoke persistent rumination and the replaying of events, which can ultimately have a negative cumulative impact, just as chronic stressors can do so.

The emergence of PTSD may reflect the cumulative effects of multiple trauma experiences, which can also reflect sensitization. When a rodent has been moderately stressed, re-experiencing another stressor gives rise to enhanced memory of the earlier trauma, enhanced anxiety, threat generalization, and resistance to extinction, all of which are characteristic of PTSD. Memory traces associated with consolidation may facilitate generalization, which can enhance PTSD symptoms.

Biochemical Correlates of PTSD

Defining the specific neurochemical and hormonal mechanisms that underlie PTSD is difficult as individuals might not only differ in the specific symptoms expressed, but the neural circuitry activated may vary as a function of the specific stressor encountered as well as synaptic connections laid down based on earlier experiences. The efficacy of drug treatments in attenuating an illness may provide clues as to the mechanisms supporting the disturbance, although the treatments might simply be masking symptoms rather than getting at the root mechanisms responsible for the illness.[2] Unfortunately, treatments for PTSD have not been impressive, achieving a success rate of about 30%, and the drugs used are so varied that they tell us little concerning the disorder's neurochemical underpinnings.[3]

Consideration of biomarkers could lead to better accuracy in selecting appropriate treatments from the array of drugs that can potentially be used. However, none of the available biomarkers has been sufficiently specific to serve as a diagnostic tool or one that predicted treatment efficacy (Bandelow et al., 2016). As PTSD is a biochemically and genetically heterogeneous disorder with many

---

[2] From the perspective of discovery processes it is important to discern whether treatments actually affect mechanisms responsible for the illness or are simply masking symptoms. However, even if the treatments were only masking symptoms, from the perspective of the affected patient, this might be good enough for the moment.

[3] The lengthy list of drugs assessed have included β2 norepinephrine antagonists, dopamine-β-hydroxylase inhibitors, dopamine antagonists, partial dopamine D2 receptor agonists, glucocorticoid receptor agonists, tropomyosin receptor kinase B (TrkB) agonists, SSRIs, COMT inhibitors, GABA receptor agonists, glutamate receptor inhibitors, MAO B inhibitors, N-methyl-D-aspartate (NMDA) receptor antagonists, and fatty acid amide hydrolase antagonists. There was a good rationale for each of these agents being assessed, but in retrospect, it seems like a fishing expedition that went awry.

comorbidities, defining the best treatment strategies will necessarily require some sort of culling to determine the single best set of methods from so many that are possible.

Once more, to determine causal connections between neuronal processes and PTSD, there has been considerable reliance on animal models of disorders, but it is questionable, as we discussed earlier, whether complex human pathologies can be fully simulated in rodents. Some characteristics of PTSD, such as hyperarousal, can readily be assessed in animal models (e.g., by evaluating startle responses to sudden noise), but it is difficult to determine whether rodents are "re-experiencing" the trauma. Despite these difficulties, several behavioral paradigms in animals have been developed to mimic PTSD. However, comorbid features of PTSD often include depression and general anxiety, and in many studies, the behavioral paradigms used to model PTSD are the same as those used to model the latter disorders. As such, it is unclear whether PTSD, depression, anxiety-related features, or some combination of these syndromes was being measured. As well, the predictive validity of animal models was questionable as the treatments used to attenuate features of PTSD (e.g., SSRIs) are also used to diminish anxiety and depression. If this weren't sufficient to undermine some animal models, they often do not incorporate known risk factors for this disorder. Further to this, among individuals who encounter trauma, only a small subset develops PTSD. In animal studies, however, the goal is typically to induce symptoms in as many subjects as possible, usually without adequate consideration of the individual differences that favor the occurrence of illness. It's unfortunate that in adopting this approach, researchers forego the opportunity of detecting individual differences linked to specific symptoms and the mechanisms that might subserve them. Ultimately valid models of PTSD ought to consider the specific symptoms expressed and determine the risk factors that favor (or act against) the development of psychopathology (e.g., Richter-Levin et al., 2019).

These caveats notwithstanding, several reasonable animal models of PTSD have been developed to simulate human experiences.[4] One approach to model PTSD has entailed exposure to a stressor, followed by stressor re-exposure sessions (or stressor cues) over several weeks (Olson et al., 2011). Mice treated this way exhibited a pronounced startle reaction, diminished social interaction, exaggerated aggressiveness toward an intruder, and elevated resistance to extinction of the fear response. Still, another model involves a single prolonged exposure to a compound stressor that varies over the course of a session, followed by a lengthy quiet period that allows for the sensitization of neuronal responses to take hold. This model meets many of the criteria of an animal model for PTSD in humans and has allowed the identification of factors that moderate the effects observed and may help identify some of the mechanisms associated with the behavioral disturbances and might be instrumental in pointing to treatments for PTSD in humans.

Norepinephrine

Numerous hormones and neurotransmitters have been implicated in subserving PTSD, including norepinephrine, 5-HT, GABA, BDNF, glucocorticoids, and cytokine activity within several brain regions (see Fig. 11.2). This makes it difficult to determine which of these (or their combinations) might be relevant to the identification of harm and the generalization of such appraisals (Sun et al., 2023c), which are relevant to the development and maintenance of PTSD features.

Considerable evidence from both animal and human studies implicated disturbed norepinephrine functioning in the development and treatment of PTSD. Among individuals with PTSD, peripheral norepinephrine accumulation was increased, and elevated norepinephrine receptor sensitivity was evident in some brain regions (Al Jowf et al., 2023). As well, in a prospective analysis conducted among police officers (deemed to be a high-risk population) increased peripheral norepinephrine utilization following a critical incident predicted later development of PTSD symptoms (Apfel et al., 2011b) and the interaction of peripheral norepinephrine and cortisol were predictive of intrusive memories among individuals with PTSD (Nicholson et al., 2014). In line with this, patients with PTSD exhibited greater indices of autonomic reactivity in response to aversive stimuli accompanied by elevated locus coeruleus neuronal activity, which might account for the hypervigilance and exaggerated startle response characteristic of PTSD (Naegeli et al., 2017).

The actions of locus coeruleus norepinephrine can influence PTSD through other neurochemicals involving several brain regions. For instance, in rodents, elevated locus coeruleus norepinephrine activity promoted basolateral amygdala neuronal firing and prevented the extinction of a conditioned fear response (Giustino et al., 2020). In contrast, blocking β-norepinephrine receptor activity within the amygdala attenuated the hyperarousal. As PTSD in humans is more frequent in women than in men, it is significant that estrogen can enhance the actions of locus coeruleus norepinephrine activity

---

[4] Frequent attempts capitalized on natural responses to predators (e.g., rats' responses to a cat or fox odor) to produce a profile that resembled PTSD. As mentioned earlier, the suitability and reliability of such predator models have been questioned, since the normal response of mice (and rats) raised in cages for generations may have been bred out of these critters. At the least, these models appropriate selection of highly reactive strains of mice or rats.

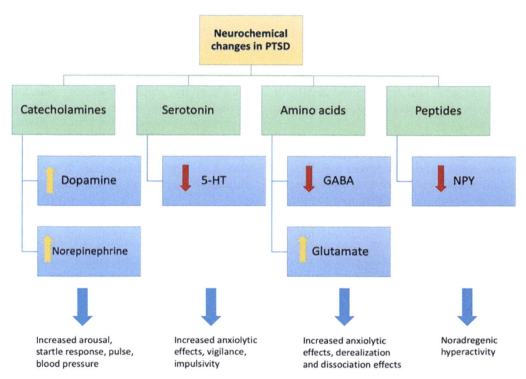

FIG. 11.2 Neurochemical changes associated with PTSD. High levels of dopamine and norepinephrine have been observed in PTSD causing increased blood pressure, anticipation, and astonishment response. A low level of serotonin (5HT) in PTSD increases anxiolytic effects. GABA is an important inhibitory neurotransmitter and an alteration of the GABA receptor system results in a decreased level of GABA in PTSD. On the other hand, high levels of glutamate in PTSD can cause excitotoxic effects leading to dissociation phenomena. Also, reduced levels of neuropeptide Y (NPY) in PTSD contribute to noradrenergic hyperactivity. *Source: Nisar, S., Bhat, A.A., Hashem, S., Syed, N., Yadav, S.K., et al., 2020. Genetic and neuroimaging approaches to understanding post-traumatic stress disorder. Int. J. Mol. Sci. 21, 4503.*

and can influence the CRH response to norepinephrine to a greater extent in females than in males (Bangasser et al., 2019). As expected, repeatedly exposing rats to a predator resulted in the sensitization of basolateral amygdala α_1-norepinephrine receptors through actions on CRH_1-related processes (Rajbhandari et al., 2015).

In addition to the amygdala involvement in PTSD, in predator models, norepinephrine activity was elevated in both the prefrontal cortex and hippocampus, and the extinction of the fear response was disturbed by interfering with norepinephrine functioning in the medial prefrontal cortex (Fitzgerald et al., 2015). The involvement of norepinephrine in PTSD has been supported by numerous pharmacological studies. Manipulations that attenuated norepinephrine functioning in rodents, such as activation of inhibitory autoreceptors or blockade of postsynaptic α_1-NE receptors, acted against the PTSD-like symptoms (Olson et al., 2011). Likewise, the β-blocker propranolol diminished the retention of fear memory, possibly by altering glutamate receptors (particularly GluA1 subunits) within the lateral aspect of the amygdala. The potential involvement of norepinephrine in mediating PTSD was also supported by reports that propranolol could diminish the stressor-provoked disruption of fear extinction if it was administered soon after fear conditioning (Fitzgerald et al., 2015). The extinction of fear memories in an animal model of PTSD was not only linked to elevated norepinephrine but could be achieved through diminished dopamine functioning within the prefrontal cortex and amygdala. Furthermore, while corticosterone has also been implicated in PTSD, its functioning could be separated from that of norepinephrine. Specifically, the release of norepinephrine in the prefrontal cortex predicted arousal symptoms that appeared 1 month later, whereas corticosterone within the prefrontal cortex and hippocampus was linked to the immediate behavioral and emotional responses to stressors (Kao et al., 2015).

Commensurate with the findings in animal studies, in humans, treatment with the α_1-norepinephrine receptor antagonist prazosin among those with PTSD reduced trauma nightmares, avoidance, and hypervigilance (Simon and Rousseau, 2017), and the postsynaptic α_1-adrenergic receptors inhibitor doxazosin similarly diminished PTSD symptoms (Rodgman et al., 2016). Like α_1-antagonists, treatments with an α_2-agonist, such as clonidine or guanfacine, which ultimately diminished postsynaptic NE receptor activation, could reduce symptoms of PTSD. Moreover, a positive correlation was found between CSF norepinephrine and behavioral symptom expression among veterans with PTSD, which was absent in veterans who had not experienced traumatic stress or those who had been treated with prazosin (Hendrickson et al., 2018).

Norepinephrine's role in the development of PTSD may involve the consolidation of fear memories by affecting diverse interconnected processes. Stressor-provoked activation of glucocorticoid receptors and norepinephrine release within the amygdala can promote the mobilization of glutamate (AMPA) receptors, which may be involved in "firming up" fear memories (Aubry et al., 2016). As we'll see shortly, treatments that affect NE receptors can influence the reconsolidation of memories associated with PTSD and when administered soon after trauma or trauma-related cues, PTSD symptoms can be thwarted. This may occur because disruption of specific signaling processes interferes with pathologic emotional memories, but without affecting declarative memory (i.e., those associated with processing of facts and events).

Serotonin

Altered stressor-provoked serotonin functioning through the actions of diverse receptors across several brain regions may contribute to anxiety, depression, and the development of PTSD. In previously stressed animals the appearance of PTSD symptoms was elevated upon later exposure to a strong challenge, as was amygdala 5-HT$_{2C}$ receptor presence. When the functioning of these receptors was blocked, the signs of PTSD were prevented, raising the possibility that targeting these receptors may be a viable strategy to diminish PTSD (Baratta et al., 2016). In addition, stressors increase the expression and functioning of 5-HT$_{2A}$ receptors, possibly reflecting an adaptive response. However, upregulation of these receptors in response to severe trauma can also result in the formation of associations that may promote overreaction to cues that signal danger (Murnane, 2019). Furthermore, PTSD characteristics in rodents were accompanied by elevated expression of the 5-HT$_{1A}$ receptor at the dorsal raphe nucleus, and conversely, knocking down the 5-HTT transporter (SERT) gene within the dorsal raphe nucleus diminished the PTSD symptoms (Wu et al., 2016).

A meta-analysis that included 14 studies indicated that individuals carrying the short allele of the 5-HTT gene were at increased risk of developing PTSD, especially in the context of having experienced childhood trauma (Zhao et al., 2017d). For instance, among survivors of the World Trade Center attack, PTSD symptoms were more common among carriers of the short allele, and neuronal activity within posterior cingulate cortices was inversely related to episodic memories and self-reflection in response to images of the attack (Olsson et al., 2015). As depression and PTSD are often comorbid, it is not surprising that PTSD would be associated with a gene mutation related to serotonin reuptake. It is uncertain, however, whether this gene mutation is present among individuals with PTSD who do not present with comorbid depression.[5]

Selective serotonin reuptake inhibitors (SSRIs) are among the most common treatments used for PTSD (Williams et al., 2022b), and it has often been assumed that the positive effects observed (as limited as they are) stem from actions on serotonin processes. However, as reported with depressive illnesses, the beneficial effects could be due to downstream changes that occur (e.g., on growth factors, such as BDNF). It is also possible that SSRIs affect anxiety and depressive symptoms, making it appear as if PTSD has diminished, even though fear memories are unaltered. Indeed, among mice exposed to a PTSD-eliciting stressor, some of the amygdala disturbances persisted even after SSRI treatment (Han et al., 2015a).

Corticotropin-Releasing Hormone (CRH)

As CRH activity and CRH receptors within aspects of the amygdala and prefrontal cortex are intimately linked to anxiety and fear, it isn't a great leap to suspect that CRH might be related to PTSD. In mice, a strong stressor experienced during adolescence elicited behavioral signs of stress reactivity, accompanied by HPA dysregulation and elevated CRH$_1$ receptor expression (Li et al., 2015a). Likewise, among mice that displayed a symptom profile reminiscent of PTSD, long-lasting upregulation of CRH$_2$ mRNA expression occurred within the bed nucleus of the stria terminalis, whereas knocking down this receptor's expression diminished susceptibility to PTSD-like characteristics (Henckens et al., 2017). Moreover, if CRH$_1$ and CRH$_2$ gene expression was transiently elevated during early life, predator exposure subsequently increased avoidant behaviors in adulthood (Toth et al., 2016).

Blocking CRH activity had effects as strong as those associated with SSRIs in diminishing PTSD symptoms

---

[5] In view of the controversy concerning the involvement of the short allele for 5-HTT interacting with stressful experiences in predicting depressive illnesses (described in our discussion of genetic correlates of depression in Chapter 9), interpretation of findings related to other stressor-provoked disorders, such as PTSD, must be approached cautiously.

(Philbert et al., 2015). In particular, a CRH_1 receptor antagonist administered prior to or 30 min after stressor exposure (during the period when memory was being consolidated) diminished stress-related behavioral disturbances measured days later and was also effective in altering CRH activity within regions of the hippocampus. These outcomes were accompanied by upregulation of BDNF and pERK1/2 protein (the latter contributes to cellular responses to neurotrophins, cytokines, and mitogens), possibly indicating their involvement in the actions of the CRH receptor manipulations (Kozlovsky et al., 2012). Interestingly, CRH was also tied to remote but not recent fear memories, as they could be disrupted by a CRH_1 receptor antagonist, possibly by enhancing AMPA receptor (GluR1) signaling within the dentate gyrus. In this regard, fear conditioning was accompanied by long-lasting GLuR1 and GluR2 expression changes within the anterior cingulate cortex and ventral hippocampus, which could affect amygdala functioning, thereby contributing to fear memories (Shultz et al., 2022).

Data in humans supporting CRH involvement in PTSD have been sparse, although cerebrospinal fluid CRH levels were elevated in some patients with either PTSD or depression (Bangasser and Kawasumi, 2015). Genetic analyses indicated that SNPs related to CRH_2 receptor expression may have limited the occurrence of PTSD in women (Wolf et al., 2013) and several SNPs associated with CRH_1 receptors were aligned with the occurrence of PTSD following trauma associated with exposure to a hurricane (White et al., 2013). It is relevant that in a subset of women, methylation of the CRH_1 receptor could serve as a marker that predicted the treatment response to a CRH_1 receptor antagonist GSK561679 (Pape et al., 2018). A GWAS revealed that CRH_1 was associated with PTSD, and positive relations were observed between PTSD and CRH_1 within the amygdala, hippocampus, and frontal and anterior cingulate cortex (Stein et al., 2021a). Few studies assessed CRH receptor changes associated with specific PTSD symptoms. However, a GWAS among veterans indicated that CRH_1 receptors were tied to posttraumatic re-experiencing symptoms among European Americans, although this association was not apparent in African Americans (Gelernter et al., 2019).

Glucocorticoids

Given the glucocorticoid changes induced by stressors, considerable attention was devoted to assessing cortisol involvement in the provocation and maintenance of PTSD. The interest in cortisol was increased given its possible involvement in fear/trauma memories (consolidation, retrieval, extinction, and reconsolidation) that may be relevant to PTSD. It might have been thought that cortisol, acting in an adaptive capacity, would be particularly elevated in association with PTSD. On the contrary, many reports revealed reduced cortisol levels among individuals with PTSD (e.g., Yehuda, 2002), particularly if early life trauma had been experienced (Yehuda and Seckl, 2011). In this respect, an earlier review had indicated that conditions that might lead to PTSD, such as chronic stressors and early life abuse, were frequently accompanied by diminished cortisol levels (Michaud et al., 2008). This outcome might be related to the elevated sensitivity of hippocampal glucocorticoid receptors involved in the regulation of HPA functioning (Szeszko et al., 2018). The fact that PTSD is accompanied by reduced cortisol does not necessarily imply that the hormone reduction stemmed from PTSD, and the presence of low cortisol or a blunted cortisol response may contribute to the development of the disorder. Consistent with this possibility, in rats with genetically low cortisol levels, the volume of the hippocampus was reduced, fear extinction was impaired, and rapid-eye movement sleep was disturbed. If rats received behavioral treatments to reduce the learned fear, subsequent corticosterone treatment attenuated the behavioral disturbances (Monari et al., 2023).

In addition to the diminished cortisol levels, PTSD was associated with a change in diurnal cortisol secretion. Ordinarily, the morning spike of cortisol levels (from awakening to about 30 min afterward) that accompany increased life stressors were reduced in patients with PTSD, whereas evening cortisol levels were elevated. The flattened daily cortisol changes were notable among individuals with the most pronounced arousal symptoms and to some extent it was evident in the presence of emotional numbing (Garcia et al., 2020). Although these relationships do not indicate causality, it was considered that treatments that normalize glucocorticoid receptor sensitivity might have positive therapeutic effects (Somvanshi et al., 2020).

Among soldiers who were about to be deployed (i.e., pretrauma), blunted stressor-elicited levels of cortisol were predictive of the later occurrence of PTSD if testosterone levels were also reduced (Josephs et al., 2017). Evidently, the link to PTSD involved more than a single hormone, which might speak to some of the inconsistent results that have been reported, especially as testosterone ordinarily acts to suppress cortisol. Furthermore, although basal levels of cortisol might be important in predicting who will be most vulnerable to illness, hormone functioning might be best assessed under a modest challenge to understand how the system will operate if more severe stressors are encountered.

Simply because HPA functioning is downregulated among individuals with PTSD doesn't necessarily imply that this system is incapable of responding. As described earlier, among previously abused women who displayed diminished cortisol levels, a novel challenge or reminders of their abuse, generated particularly elevated ACTH or cortisol levels (Heim and Nemeroff, 2002; Matheson and Anisman, 2012). It was surmised that the downregulated HPA functioning among women who had previously been traumatized might have been instrumental in preventing hippocampal cell loss that would otherwise occur. Yet, reduced corticoid functioning could be counterproductive when individuals have to deal with further challenges. Thus in the face of meaningful stressors, relevant brain regions would be engaged, overriding the processes that produce the downregulated HPA system (Matheson and Anisman, 2012).

Although reduced cortisol levels have been widely observed, this relationship was not uniformly apparent. The source for the diverse findings between studies is not immediately apparent, and it is important to know whether differences were evident in women using oral contraceptives that could affect the cortisol response. Moreover, it was frequently not considered that subgroups of PTSD exist (e.g., presence vs. absence of dissociative symptoms) in which HPA responses might differ. Further to this, neurobiological processes that govern PTSD symptoms may vary with the nature of the trauma experienced (e.g., those primed by early adverse events vs. a sudden catastrophic event experienced in adulthood). The takeaway from these studies is that it may be counterproductive to assume that the neurobiological correlates among all individuals with PTSD would be identical, much as the symptom profile differed among them.

Among its many other functions, glucocorticoid stimulation of receptors in the amygdala and hippocampus may be involved in emotional memory formation and might be instrumental in the reconsolidation of such memories. Studies in animals supported the view that corticoid levels soon after trauma may predict the later development of PTSD, and that cortisol treatment could augment memory of negative or arousing events. When corticosterone was administered to rodents shortly after initial fear conditioning, cue-specific memory consolidation and hippocampal long-term potentiation were augmented varying with the strain of mouse tested, possibly reflecting differences in trait anxiety and dendritic arborization in the basolateral amygdala (Brinks et al., 2009).

Norepinephrine and glucocorticoids both peripherally and within the brain may act collaboratively in supporting the development and maintenance of PTSD symptoms by promoting glutamate functioning, thereby affecting memory consolidation. As well, repeated exposure to a predator provoked persistent sensitization of α_1-norepinephrine receptors within the basolateral amygdala through a mechanism that involved CRH_1 receptor activity (Rajbhandari et al., 2015). The variations of HPA functioning and hypothalamic-sympathetic processes may feedback to promote several brain changes, yielding the cognitive and emotional features that are part of the PTSD syndrome.

Cortisol administration may reduce symptoms of PTSD, although considerable differences exist between individuals, possibly reflecting genetic or epigenetic influences on HPA functioning (Meir Drexler et al., 2016). Whole genome analysis in postmortem tissue revealed that PTSD was accompanied by low expression of a particular gene, SGK1, within the prefrontal cortex. This gene encodes serum and glucocorticoid-regulated kinase 1 and seems to contribute to marked behavioral disturbances indicative of a depressive-like state coupled with high levels of fear (Licznerski et al., 2015). It also influences the regulation of multiple stress-related processes, including neuronal excitability, hormone release and functioning, regulation of monocyte/macrophage migration, inflammatory processes, cell proliferation, and apoptosis.

Overall, the available data are consistent with the perspective that PTSD is associated with cortisol actions being amplified owing to enhanced glucocorticoid receptor sensitivity in several brain regions even though circulating cortisol levels may be diminished. Most often, cortisol alterations related to PTSD have been attributed to HPA and hippocampal functioning; however, this condition is also accompanied by alterations of neuronal activity within the amygdala, anterior cingulate cortex, and insula. These sites are components of the brain network associated with the salience of environmental stimuli and may underly several PTSD characteristics, notably hypervigilance, elevated attention toward potentially threatening stimuli, and failure to integrate sensory information and emotional processes.

Gamma-Aminobutyric Acid (GABA)

When levels of GABA in specific brain regions are low, diminished inhibition of neuronal activity may favor a fear or PTSD-like state (Huang et al., 2023a). This supposition was supported by reports that PTSD symptoms were associated with reduced GABA in the prefrontal cortex and insula (Rosso et al., 2021), and lower GABA and glutamine in the anterior cingulate cortex among trauma-exposed veterans (Sheth et al., 2019). Consistent with this, SNPs on the gene (GAT-1) that codes for the transporter that removes GABA from the synaptic cleft were associated with PTSD irrespective of the presence of depression or substance use disorder (Bountress et al., 2017).

A prospective analysis over 1 month following deployment among military personnel revealed that the development of PTSD was accompanied by altered plasma GABA levels (Schür et al., 2016), and among combat veterans presenting with PTSD, the presence of benzodiazepine-responsive GABA receptors was reduced. This is in keeping with the finding that the binding of benzodiazepines to GABA receptors was altered in association with PTSD (Reuveni et al., 2018) as well as reports that the sensitivity to substances that stimulate GABA$_A$ receptors were diminished, and benzodiazepines became less effective among PTSD patients.

Although benzodiazepines administered after the trauma did not deter the pathology from evolving, the GABA-analog pregabalin reduced the severity of PTSD symptoms and augmented the effects of antidepressants (Baniasadi et al., 2014). The case for GABA involvement in PTSD was strengthened by the finding that GABAergic neuroactive steroids (or agents that enhance their synthesis) facilitated the extinction of fear memories (Rasmusson et al., 2017). In humans, the GABA$_B$ agonist baclofen was also an effective add-on agent in SSRI treatment of PTSD.

Glutamate

As glutamate is involved in anxiety, it might be reasonable to expect that this excitatory transmitter would also contribute to PTSD or changes in anxiety associated with PTSD, and perhaps influence memory and consolidation processes associated with this condition. As we've seen, at low concentrations, glutamate stimulates neural functioning, but at high concentrations that can be provoked by traumatic events or chronic stressors, it can be neurotoxic.

Imaging studies indicated that among recently traumatized individuals, the ratio of glutamate to its precursor glutamine in the dorsal anterior cingulate cortex was predictive of cognitive-affective symptoms of PTSD (Harnett et al., 2017). Similarly, PTSD symptoms, particularly upon reexperiencing the trauma, were accompanied by indications of hippocampal glutamate excess together with compromised neuron integrity (Rosso et al., 2017). Further, in traumatized youth, glutamatergic functioning was diminished so that the connectivity between the prefrontal cortex and the amygdala was impaired (Ousdal et al., 2019). Beyond the changes in glutamate concentrations, PTSD was accompanied by greater availability of specific glutamate receptors (mGluR5), especially when avoidant symptoms were prominent (Holmes et al., 2017) and when patients reported suicidal ideation (Davis et al., 2019).

Fear memories associated with PTSD were tied to an imbalance between GABA and glutamate within the hippocampus, but not in other regions, such as the prefrontal cortex (Gao et al., 2014). At the same time, analyses of fear memories related to glutamate can be complicated because different types of NMDA receptors might be aligned with diverse types of fear memories (e.g., cue-specific vs. contextual memories). Moreover, the influence of glutamate receptor alterations may vary across portions of the cortex (e.g., prelimbic and infralimbic), and could even act in opposition to one another.

The control of stress-related epigenetic and gene transcriptional responses involving hippocampal functioning may stem from distinct glutamatergic and glucocorticoid-driven processes, which are regulated by both GABAergic interneurons and limbic inputs. The position was also advanced that neuronal activity stimulated by glutamate and norepinephrine might cooperate in affecting attention, and memory consolidation, thereby influencing the development of PTSD (Abdallah et al., 2017). Interactions between glutamate and cortisol might likewise contribute to PTSD, and interventions based on alterations of these systems may interfere with stress-related memories. The altered synaptic strength and disturbed synaptic connectivity associated with traumatic events might be a biomarker of PTSD and could potentially be a causal component of the disorder (Averill et al., 2022).

Neuropeptide Y (NPY)

Data from animals and humans have indicated that NPY confers resilience so that PTSD is less likely to develop in reaction to trauma (Kautz et al., 2017). Ordinarily, stressors inhibit NPY functioning within the CA1 region of the hippocampus. In rodents, the anxiolytic actions of NPY are mediated by neuronal changes within this region and hippocampal NPY injection diminishes anxiety elicited by a stressor (Li et al., 2017a). Likewise, manipulations that affect PTSD symptoms in rodents (e.g., environmental enrichment and exposure therapy) influenced NPY-Y1 receptors within the basolateral amygdala, implicating this receptor as a possible target in the treatment of this disorder (Hendriksen et al., 2014).

A chronic variable stressor that ordinarily leads to behavioral disturbances in rodents was accompanied by diminished amygdala NPY levels. However, in animals with seemingly better methods of dealing with a stressor, the levels of NPY in the hippocampus, amygdala, and BNST were elevated, and behavioral resilience in a model of PTSD was associated with elevated NPY levels (Cohen et al., 2012b). Moreover, particularly pronounced reductions of NPY were found in stressed females, which is in keeping with the greater vulnerability to PTSD among females than males (Nahvi and Sabban, 2020).

In an animal model, NPY administration limited the occurrence of later stressor-provoked PTSD-like outcomes and could attenuate the effects of a previously administered PTSD-inducing stressor (Serova et al., 2017). Consistent with these findings, the anxiety and hyper-reactivity elicited by the odor of a predator were accompanied by reduced hippocampal and amygdala NPY expression, whereas NPY administered directly into the brain attenuated the behavioral disturbances that were otherwise apparent (Cohen et al., 2012b). In line with the therapeutic potential of NPY, when administered soon after a traumatic stressor, it prevented the later development of PTSD features in rodents (Sabban et al., 2016). Similarly, when administered intranasally, an NPY agonist prevented the development of depressive-like behaviors and dysregulation of the CRH/HPA system seen in models of PTSD (Serova et al., 2017). Paralleling this finding, early intervention using intranasal NPY attenuated the CRH and behavioral changes that were otherwise apparent and prevented stressor-elicited dysregulation of HPA functioning, possibly by reinstating effective negative feedback inhibition through glucocorticoid receptor alterations (Laukova et al., 2014). Aside from HPA alterations, NPY reduced the actions of stressors in the promotion of CRH and NE release within the amygdala, which could affect PTSD symptoms. Beyond these actions, stressor-produced behavioral changes were promoted by alterations of cannabinoid receptors (CB_1) accompanied by NPY alterations within the amygdala and infralimbic prefrontal cortex, which could be attenuated by altering these processes, suggesting that their co-occurrence subserved PTSD symptoms (Maymon et al., 2020).

Studies in humans supported the contention that NPY serves to enhance resilience and acts against the development of PTSD. Across several situations, individuals with high NPY levels tended not to develop stress-related disturbances as readily as those with low levels of this peptide (Schmeltzer et al., 2016). Moreover, baseline NPY levels and that stimulated by an α_2-norepinephrine autoreceptor antagonist were lower among individuals experiencing PTSD than in healthy volunteers who had not experienced trauma (Rasmusson et al., 2000).

Among soldiers with high levels of NPY, symptoms of PTSD were unlikely to develop following combat (Sah et al., 2014), and NPY levels were directly related to symptom improvement and positive coping in veterans with a history of PTSD (Yehuda et al., 2006). Predictably, NPY levels among Special Forces soldiers were higher than in non-Special Forces soldiers, and increased NPY was elicited during mock interrogations, which was negatively associated with the development of dissociative symptoms (Morgan et al., 2000). Consistent with such reports, lifetime PTSD was marginally elevated in the presence of an NPY polymorphism (Ferić Bojić et al., 2019).

Considering such reports, it has been suggested that NPY expression could be used as a biomarker to identify individuals at risk for this illness, and resilience to PTSD could be augmented by targeting NPY and glutamate (NMDA) functioning (Horn et al., 2016). It isn't certain how NPY comes to have positive effects, although it may be tied to memory extinction, possibly involving the infralimbic cortex in which neuronal activity is diminished among PTSD patients (Vollmer et al., 2016).

Brain-Derived Neurotrophic Factors (BDNF)

Given the importance of BDNF in other stress-related conditions, it isn't at all surprising that alterations of this neurotrophin were implicated in the emergence or maintenance of PTSD. In rats, chronic and intense stressors were accompanied by reduced BDNF levels and altered expression of H3K9me2, an epigenetic marker of the BDNF gene. In response to a strong stressor that increased H3K9me2, behavioral changes reminiscent of PTSD were provoked, accompanied by diminished dendritic branching in the hippocampus and prefrontal cortex (Zhao et al., 2020c).

In combat-exposed individuals, PTSD was related to elevated methylation of the BDNF promoter gene (Kim et al., 2017d) and a polymorphism of the BDNF gene (rs6265) was associated with elevated PTSD vulnerability (Hu et al., 2021b). A BDNF polymorphism was similarly associated with negative memory biases in individuals with PTSD. This polymorphism interacted with childhood trauma so that PTSD was more apt to develop, accompanied by altered cortical thickness (Jin et al., 2019a).

A meta-analysis revealed lower blood levels of BDNF in those with PTSD relative to non-PTSD controls (Mojtabavi et al., 2020). In some reports, the altered BDNF was even apparent years after trauma exposure (Wu et al., 2021). It may be significant, however, that among trauma-exposed individuals without PTSD, BDNF increased over time, possibly acting in a protective capacity. Based on studies in animals, it was proposed that BDNF, operating together with glucocorticoids, might influence the development of PTSD by affecting the processing of salience related to stressor experiences (Chakraborty et al., 2021).

Inflammatory Processes Associated With PTSD

Stressors markedly affect immune functioning, as well as circulating and brain cytokines, raising the possibility that these factors contributed to features of PTSD.

Symptoms that were thought to reflect PTSD in animal models were associated with increased levels of inflammatory factors within the circulation and in the brain. This was apparent with repeated social defeat and in paradigms that involved predator exposure (Deslauriers et al., 2017). Furthermore, the monocyte and IL-6 response to LPS, as well as the IL-6 response to initial social defeat, predicted the development of subsequent PTSD-like features (Hodes et al., 2014). As expected, knocking down IL-1 receptors in the endothelium among mice that were exposed to chronic social defeat diminished anxiety behaviors, which were accompanied by reduction of circulating monocytes, macrophage recruitment to the brain, and microglia activation (Wohleb et al., 2014). Supporting the involvement of inflammatory factors, in rodents, antiinflammatory agents (e.g., chronic NSAID treatment, or administration of minocycline) reduced anxiety-like behaviors, and it was maintained that these treatments administered soon after trauma might preclude the development of PTSD-like features (Deslauriers et al., 2017).

The data in humans have largely aligned with those obtained from animal studies. Based on a broad review that included 65 published reports, it was concluded that immune alterations, elevated proinflammatory processes, and diminished levels of antiinflammatory factors were associated with PTSD (Sun et al., 2021). Moreover, PTSD was associated with a reduction in T_{reg} cells, expansion of activated T cells, and changes in peripheral leukocyte sensitivity in response to glucocorticoids, as well as low-grade inflammation (Dell'Oste et al., 2023; Peruzzolo et al., 2022). In a small trial, a subset of T_{reg} cells was reduced in patients with PTSD (Jergović et al., 2014) and a similar study indicated that the trauma-related reduction of T_{reg} cells might be accompanied by elevated capacity to influence CD4+ cells, likely through inhibition of the antiinflammatory cytokine IL-10 (Sturm et al., 2020). Alterations of T_{reg} cells may be especially significant, potentially accounting for the frequent comorbidities between PTSD and several autoimmune disorders (Neigh and Ali, 2016).

Beyond the changes in immune cell functioning, an array of inflammatory factors was associated with PTSD. Among patients with primary PTSD, a constellation of numerous proinflammatory cytokines was elevated relative to age- and gender-matched healthy controls, and spontaneous production cytokines in isolated peripheral blood mononuclear cells were correlated with symptom severity (Gola et al., 2013). Reviews of the relevant literature pointed to PTSD being closely aligned with elevated levels of IL-1β, IL-6, and TNF-α (Lee et al., 2022a), although considerable heterogeneity existed between studies (Dell'Oste et al., 2023). Significantly, the inflammatory changes were related to PTSD rather than trauma itself. Specifically, proinflammatory cytokine elevations were greater among trauma (earthquake) survivors who developed PTSD relative to those who had experienced the trauma but had not developed the disorder (Wang et al., 2019).

The elevated inflammatory cytokines were detected with different types of stressors that promoted PTSD, including IPV and terrorism experiences, and a combination of multiple indices of inflammation was related to PTSD severity (Fonkoue et al, 2020). While PTSD was accompanied by elevated levels of circulating IL-6 and its soluble receptor (Newton et al., 2014), resilience to trauma stemming from urban violence was associated with elevated levels of the antiinflammatory IL-10 (Teche et al., 2017). It also appeared that while illness duration was related to IL-1β levels, illness severity was more aligned with elevated IL-6 (Passos et al., 2015). Further supporting the link to inflammatory factors, normalization of immune and cytokine activity occurred with the decline of PTSD symptoms following trauma-focused therapy (Morath et al., 2014).

In civilian populations (e.g., among nurses who experienced trauma) the presence of inflammation predicted later PTSD occurrence (e.g., Sumner et al., 2018), and elevated cytokine levels in the immediate aftermath of motor vehicle accidents were associated with the later development of PTSD. Counterintuitively, although PTSD has been associated with elevated circulating levels of proinflammatory cytokines, in blood samples taken at the hospital emergency department about 3 h after the traumatic event, low levels of TNF-α and IFNγ were found in individuals who developed chronic PTSD relative to individuals who did not develop PTSD or those who recovered (Michopoulos et al., 2020). In a subsequent study that followed the same protocol, it was again observed that low cytokine levels predicted nonremitting PTSD. In this report, women were more likely to develop chronic PTSD than men, and proinflammatory cytokines and testosterone levels were reduced, whereas cortisol and progesterone were elevated (Lalonde et al., 2021). Steroid hormones, in conjunction with inflammatory factors, evidently contribute to the risk of PTSD, which may have significant implications for ways of attenuating the actions of trauma.

There had been the question of whether the presence of elevated proinflammatory cytokines disposed individuals to PTSD upon encountering trauma. A detailed meta-analysis revealed that several proinflammatory cytokines were elevated in patients with PTSD, and a subgroup analysis indicated that this was apparent when comorbid depression was excluded. Moreover, when controlling for the presence of depression, higher baseline levels of IL-1β predicted amplified perceived stress in response to later trauma experiences (Schrock et al., 2021), and among military personnel with PTSD, the proinflammatory cytokine elevations were present even after controlling for early life stressors and adult depression (Lindqvist et al., 2017). These findings do not eliminate some sort of linkage between these disorders. Gene expression analyses implicated inflammatory factors in both PTSD and depression and several genes were identified that might be responsible for the common phenotypes (Garrett et al., 2021).

It has frequently been reported that elevated blood CRP levels, which are released by the liver in response to inflammation, were associated with PTSD. Levels of CRP among male marines before deployment were elevated in those who subsequently developed PTSD (Eraly et al., 2014), and in monozygotic twins discordant for PTSD, both CRP and Intercellular Adhesion Molecule-1 (ICAM-1) were present in low concentrations in membranes of endothelial cells and leukocytes in the twin without PTSD but were elevated in the twin with PTSD (Plantinga et al., 2013). Critically, the tie between CRP and perceived stress was diminished in individuals with high levels of social support (Shimanoe et al., 2018). Longitudinal studies among veterans revealed that predeployment elevations of peripheral inflammation predicted postdeployment PTSD. A machine learning approach that included numerous biological and cognitive variables among soldiers who were to be deployed to combat in Afghanistan could predict the occurrence of subsequent PTSD. In addition to behavioral and cognitive indices associated with PTSD (sleep quality, anxiety, depression, sustained attention, and cognitive flexibility), epigenomic, immune, and inflammatory factors were cogent predictors of the disorder (Schultebraucks et al., 2021).

Consistent with other neurobiological changes that we discussed in earlier chapters, childhood adversities were associated with elevated IL-6 and TNF-α, and in clinical samples, the association between childhood trauma and CRP was still greater (Baumeister et al., 2016). The downstream effects that occurred in association with abuse or neglect were similarly observed among economically disadvantaged children (Liu et al., 2017a). Indeed, being raised and living in poverty is tantamount to a chronic illness that favors the occurrence of diverse pathologies, with elevated PTSD being one of these.

PTSD was linked to the presence of several SNPs related to genes that code for CRP. In a large community sample of severely traumatized individuals, elevated PTSD symptoms, primarily hyperarousal, were evident in individuals carrying a gene polymorphism (rs1130864) that was accompanied by high CRP levels (Michopoulos et al., 2015). Similarly, in individuals who developed PTSD related to physical and/or sexual violence, the presence of another CRP polymorphism (rs2794520) was associated with greater symptom severity and poorer cognitive function, and the relationship between PTSD severity and CRP levels was similarly moderated by a SNP (rs3091244) in the CRP gene promoter region (Miller et al., 2018). Further, among individuals who had also experienced early life adverse events, the polymorphism predicted greater avoidance symptoms (Otsuka et al., 2021). In a large community sample of individuals who had been severely traumatized, another CRP gene polymorphism (rs1130864), which was accompanied by elevated CRP levels, predicted elevated PTSD symptoms, primarily hyperarousal (Michopoulos et al., 2015). Finally, PTSD symptom severity among US Army veterans was directly related to CRP levels, which was mediated by an epigenetic change of AIM2, a gene that had previously been linked to CRP levels. Findings such as these gave rise to the suggestion that in addition to simply serving as an illness biomarker, CRP might act causally in promoting the illness through yet unknown processes (Friend et al., 2022).

Much of our discussion has focused on the presence of elevated proinflammatory levels in peripheral blood among individuals with PTSD. Peripheral cytokines can access the brain under some conditions, thereby promoting central inflammation, hence favoring trauma-associated PTSD. Beyond infiltration from the periphery, in animal models of PTSD several proinflammatory cytokines, primarily coming from microglia, were elevated in the prefrontal cortex, amygdala, and hippocampus (Muhie et al., 2017). It will be recalled as the brain's resident macrophage cells, microglia serve to protect the CNS from infectious agents, damaged neurons, as well as plaques, and they serve to eliminate unnecessary synapses. Ordinarily, the neuroinflammatory response stemming from microglial activation may have protective effects, through the recruitment of monocytes into the brain to deal with infection or injury. However, in response to chronic or severe stressors, the excessive production of proinflammatory and neurotoxic factors (e.g., free radicals, nitric oxide, and superoxide) gives rise to neuronal injury and cell death (De Pablos et al., 2014). Fig. 11.3 shows some of the microglial and inflammatory effects of trauma that can undermine neuronal functioning and contribute to behavioral disturbances, including PTSD.

The emergence of PTSD could involve the interplay between microglial inflammation and the actions of glucocorticoids (Li et al., 2023d). These processes operate bidirectionally so that glucocorticoids can inhibit the actions of microglia, and cytokines release from microglia and astrocytes increase HPA activity. Additionally, glucocorticoids and TNF-α cooperatively regulate toll-like receptor 2 gene expression (Li et al., 2023d), and glucocorticoids may contribute to the priming of microglial

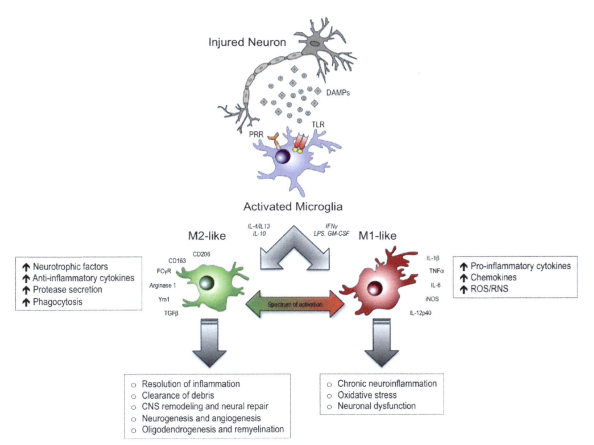

FIG. 11.3 In response to DAMPs and other extracellular signals released by injured neurons, microglia can become polarized toward M1-like and M2-like activation states that can have distinct roles in neurodegeneration and tissue repair. M1-like microglia are characterized by upregulated expression of phenotypic protein markers such as IL-1β, TNFα, IL-6, iNOS, and IL-12p40. They release pro-inflammatory cytokines, chemokines, and free radicals that impair brain repair and contribute to chronic neuroinflammation, oxidative stress, and long-term neurological impairments. In contrast, M2-like microglia upregulate protein markers such as CD206, CD163, FCγR, arginase 1, Ym1, and TGFβ. M2-like microglia release anti-inflammatory cytokines, neurotrophic factors, and proteases, and they have increased phagocytic activity. M2-like microglia promote immunosuppression and resolution of M1-mediated neuroinflammation, and participate in CNS remodeling and repair by modulating neurorestorative processes such as neurogenesis, angiogenesis, oligodendrogenesis, and remyelination. Abbreviations: DAMPs, danger-associated molecular patterns; PRR, pathogen recognition receptors; TLR, toll-like receptors. *Source: Loane, D.J., Kumar, A., 2016. Microglia in the TBI brain: the good, the bad, and the dysregulated. Exp. Neurol. 275, 316–327.*

proinflammatory responses elicited by stressors. Moreover, with chronic elevations of glucocorticoids hippocampal microglial inflammatory responses enhanced the response to an immune challenge (Frank et al., 2014).

The findings in humans concerning the link between PTSD and microglial inflammatory factors are less clear. Although peripheral markers were elevated among patients with PTSD, within the prefrontal cortex–limbic circuit, PTSD was inversely related to TSPO, a marker of microglia. Thus despite the elevated inflammation in the periphery, this may not translate to elevated brain microglia (Bhatt et al., 2020). Whether the general absence of microglial activation in PTSD was unique to the brain regions examined hadn't been explored. Nor was it clear whether the absence of microglial changes was related to the timing of trauma and duration of PTSD. Despite these negative findings, microglia can be primed so that inflammatory factors give rise to exaggerated activation of microglial-related processes so that vulnerability to PTSD is increased (Neher and Cunningham, 2019).

Anxiety and PTSD Related to Head Injury

The prevalence of PTSD associated with traumatic brain injury (TBI) was about 15% in civilian populations with considerable variability being apparent across studies and was comparable among individuals who experienced relatively mild versus moderate/severe TBI. The risk for PTSD following mild TBI was moderated by a constellation of variables, including race/ethnicity, history of mental health problems, education obtained, and how the injury occurred (Stein et al., 2019), as well as polygenic risk factors that had been determined based on a whole genome analysis (Stein et al., 2023). Like other injuries (e.g., wounds), TBI may promote depression, anxiety, and PTSD, which often predict poor overall recovery. Indeed, among military veterans who had suffered TBI the occurrence of substance use disorder was markedly higher than in those who had not experienced head injury and suicide occurred sooner (Brenner et al., 2023).

Disturbed connectivity between regions comprising the default mode network was apparent among individuals who experienced TBI and exhibited PTSD, possibly reflecting disruption of circuits associated with attention and emotional regulation (Nathan et al., 2017). Moreover, among patients with TBI and PTSD, white matter integrity was reduced in the uncinate fasciculus, which connects the orbitofrontal cortex to the anterior temporal lobes, including the amygdala (Santhanam et al., 2019). In contrast to these reports, there have been indications that concussive injury, somewhat counterintuitively, was accompanied by elevated connectivity between particular brain regions, frequently being linked to the nature of the symptoms expressed. Specifically, among individuals who showed the poorest prognosis, connectivity between the thalamus and other brain regions was elevated, almost as if the thalamus was overcompensating for an inability to transmit information (Woodrow et al., 2023). Also, in TBI patients with relatively severe PTSD, connectivity was elevated between the amygdala and hippocampus and reduced in the cingulate cortex (Sydnor et al., 2020). It was said that the smaller volume of several brain regions (composed of the insula, superior frontal cortex, rostral, and caudal cingulate) measured 2 weeks after TBI predicted the occurrence of PTSD 3 months after the injury (Stein et al., 2021b).

Like the brain alterations, plasma IL-2 and IL-6 levels within 24 h of injury were elevated in TBI patients relative to controls who experienced an orthopedic injury. The elevated IL-2 levels were related to more severe symptoms 1 week later, possibly pointing to this cytokine being a useful marker to predict the course of symptoms (Vedantam et al., 2021), although these would need to be assessed for a more protracted time to obtain a full measure of the cytokine's predictive ability.

A systematic review indicated that poor psychological outcomes were associated with elevated levels of CRP, IL-6, TNF-α, and IL-10 (Malik et al., 2022). The elevated cytokine levels could be of peripheral origin or may stem from brain astrocyte or microglia activation. Either way, it might be expected that PTSD would emerge owing to cytokine elevations, as described in Fig. 11.4. In this respect, TBI may prime microglia, so that a subsequent second hit in the form of a stressor, activation of an inflammatory peripheral immune response, or a further head injury (often seen, for instance, in some contact sports) increases the risk of neuropsychiatric and neurodegenerative conditions (Witcher et al., 2015).

It has become clear that TBI and PTSD share numerous neurobiological features. These include white matter tract abnormalities and grey matter changes within the basolateral amygdala, hippocampus, and prefrontal cortex that favor common PTSD symptoms. Also, both conditions are accompanied by neuroinflammation, oxidative damage, and excitotoxicity that promote neuronal death, axonal injury, and dendritic spine dysregulation. It was thus suggested that neurorestorative therapeutics comprising antiinflammatory, antioxidant, and anticonvulsant agents could diminish the behavioral and cognitive disturbances associated with these comorbid conditions (Kaplan et al., 2018). Although TBI is associated with a constellation of neurocognitive disturbances, there has been a paucity of data concerning the effects of SSRIs in treating affected patients, possibly owing to the risk of provoking epilepsy (Christensen et al., 2019).

It had at one time been recommended that cognitive and physical rest was a basic healing strategy for mild TBI and concussions. This view was altered with the accumulating evidence indicating that rest beyond 3 days is of limited value and may actually be counterproductive. Instead, following the precept of "exercise as medicine," it was recommended that concussed individuals resume their usual activities after a brief rest, provided that these activities can be tolerated. This perspective has taken root so that the consensus now is that mild exercise be undertaken within a few days of experiencing a concussive injury (Silverberg et al., 2020).

Genetic Influences

Several immune and cytokine-related genetic factors have been linked to PTSD, as has dysregulation of gene networks related to innate immunity. A study that comprised 30,000 PTSD cases and 170,000 controls suggested that heritability of PTSD ranged from 5% to 20%, and genes linked to dopamine and immune pathways were related to PTSD, differing between sexes. Although PTSD and depression shared common genetic features, it was likely that some of the gene loci identified were specific to PTSD (Nievergelt et al., 2019). By integrating data from blood-based transcriptomic and genomic data related to trauma, it was possible to identify driver genes that were specifically tied to PTSD or depression (Wuchty et al., 2021).

Another large GWAS that included more than 180,000 participants and a second cohort of 133,000 participants revealed several gene loci linked to PTSD and/or lifetime

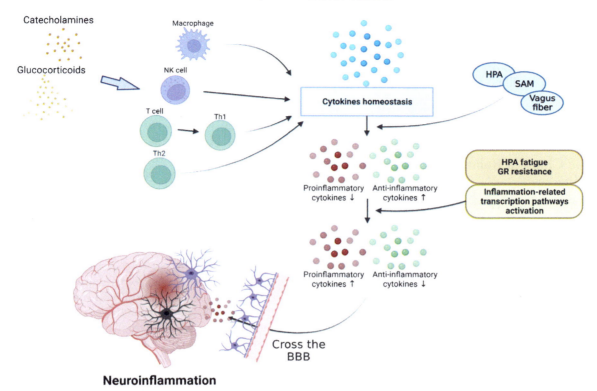

FIG. 11.4 Effect of inflammatory cytokines in PTSD. PTSD interferes with cytokine homeostasis and inhibits the secretion of antiinflammatory cytokines through the activation of the HPA axis, SAM axis, and vagus nerves. Continuous PTSD can induce HPA fatigue and increase glucocorticoid receptors, inducing inflammatory reactions and causing neuroinflammation by crossing the BBB. HPA, hypothalamic–pituitary–adrenal; SAM, sympathetic–adrenal–medullary system. *Source: Lee, D.H., Lee, J.Y., Hong, D.Y., Lee, E.C., Park, S.W., et al., (2022a). Neuroinflammation in post-traumatic stress disorder. Biomedicines 10, 953.*

trauma. Several genetic correlations were also apparent between PTSD and specific traits, such as neuroticism, mood swings, loneliness, depressive symptoms, irritability, feeling fed up, and risk-taking. Not unexpectedly, some gene loci were also related to other psychiatric conditions (Maihofer et al., 2022). These included genes associated with COMT, the serotonin transporter gene (SLC6A4), and NPY, which interacted with childhood adversity (Maul et al., 2020).

Earlier broad assays indicated that the genetic relationships to PTSD appeared across trauma types and that some of these were consistent in males and females (Breen et al., 2018). This analysis revealed that before and following deployment among soldiers, a set of coregulated genes was tied to the overexpression of interferon. In addition, PTSD has been associated with increased NF-κB, as well as cell survival and differentiation. Genes encoding proteins of the NF-κB family (e.g., p65 and c-Rel) were elevated among individuals with PTSD (Gupta and Guleria, 2022) and in women who had experienced childhood abuse and later presented with PTSD characteristics. Not only was NF-κB activity elevated compared to controls, but this transcription factor was positively correlated with the severity of symptoms, possibly being attributable to reduced glucocorticoid sensitivity of immune cells (Pace et al., 2012).

Several transcriptome-wide analyses assessed mRNA expression that was related to PTSD. These studies have implicated a constellation of genes related to stress processes (e.g., BDNF, glucocorticoid signaling), and the aggregation of multiple differentially expressed genes between those with or without PTSD. Transcriptome analyses among veterans also revealed relations between PTSD and cytokine processes and toll-like receptors, and gene expression often differed from the pre- to the postdeployment period (Mehta et al., 2018a).

A polymorphism in the gene coding for TNF-α was associated with later PTSD, increasing the possibility that inflammatory reactivity may be a precondition that increases vulnerability to PTSD (Bruenig et al., 2017). A substantial number of other immune factors and hormones have been linked to PTSD, many of which could potentially serve as biomarkers for this condition (Michopoulos et al., 2016), although none were specific to this disorder. Most have been linked to other mental conditions (e.g., depression, anxiety, and schizophrenia) as well as physical illnesses (chronic heart disease, type 2 diabetes). Still, a constellation of biomarkers and life events (e.g., stressor history) might be useful in predicting which individuals will be most vulnerable to PTSD.

Epigenetic Contributions

Epigenetic Changes Related to Glucocorticoid Functioning

Considerable evidence amassed, primarily from preclinical studies, indicating that stressors can engender epigenetic changes that contribute to both psychological and physical pathologies. Traumatic events were, as predicted, linked to DNA methylation in promoters of glucocorticoid receptors, which were accompanied by hypoactive HPA functioning in individuals with PTSD (Labonte et al., 2014). Because of the influence of glucocorticoids on fear memory consolidation, and its potential role in the development of PTSD, epigenetic changes related to glucocorticoid receptors may be especially relevant (Maddox et al., 2015). At the same time, epigenetic changes have been identified in genes that code for diverse neurobiological changes, such as inflammation, cytotoxicity, and functioning of kynurenine, that were related to PTSD (Smith et al., 2020). Several of the identified epigenetic changes were also linked to other psychiatric conditions and might contribute to the comorbidities that frequently appear with PTSD (Blacker et al., 2019).

Epigenetic changes can occur at any time in life, but those that occur during childhood are particularly adept at creating such outcomes (Szyf, 2019), and might increase vulnerability to PTSD upon subsequent trauma experiences (Torres-Berrío et al., 2019). Individual differences in response to traumatic events have been linked to preexisting risks for PTSD, which may result in differential DNA methylation of genes that govern endocrine functioning.

Low cortisol levels, perhaps owing to earlier adverse experiences that provoked specific epigenetic changes, could predict the emergence of PTSD and might be at the root of later responses to traumatic events. Some of the early reports that assessed this relationship indicated that low cortisol levels observed very soon after trauma predicted the subsequent development of PTSD (Yehuda, 2002). Similarly, in the adult offspring of Holocaust survivors, low glucocorticoid levels were associated with alterations of glucocorticoid-regulated genes and those tied to immune/inflammatory processes (Daskalakis et al., 2021).

More than other processes, epigenetic changes associated with PTSD were those associated with genes relevant to cortisol or cortisol receptors, and in this respect,

both the *NR3C1* gene (which encodes the glucocorticoid receptor) and the *FKBP5* gene (a co-chaperon protein that regulates glucocorticoid receptor sensitivity) have been the focus of greatest attention (Sheerin et al., 2020). The expression of the *NR3C1* gene was proposed as a good biomarker for PTSD with sensitivity and specificity being 62.5% and 89.8%, respectively (González Ramírez et al., 2020).

Adverse early life experiences were associated with adult PTSD among individuals carrying an *FKBP5* polymorphism (Morrison et al., 2019), and a systematic review indicated that most studies that examined early life adverse events revealed elevated methylation of the glucocorticoid receptor gene *NR3C1* promoter (Turecki and Meaney, 2016). Methylation of the *NR3C1* promoter region was accompanied by an increased risk for PTSD and specific symptoms of the disorder (Labonte et al., 2014). Women who had experienced IPV and methylation of the *NR3C1* gene exhibited increased reactivity of prefrontal cortex neuronal activity, which was also linked to their responses upon watching videos of parent–child interactions (Schechter et al., 2015). More than this, a prospective study among women who experienced IPV revealed methylation of the *NR3C1* gene and that this epigenetic action was frequently apparent in their children, which may have rendered them at increased risk of pathology (Cordero et al., 2022). Aside from epigenetic changes associated with childhood trauma, a meta-analysis that included 33 studies concerning prenatal epigenetic risks for PTSD indicated that a polymorphism related to BDNF and of *FKBP5* and methylation related to *NR3C1* mRNA were most commonly tied to the disorder (Pierce and Black, 2022).

As indicated earlier, epigenetic changes in sperm or ova that come about owing to trauma experiences can be passed on intergenerationally and perhaps transgenerationally. Although it is often assumed that these changes are accompanied by psychological disorders, as described in our discussion of depressive disorders, some epigenetic alterations stemming from adverse events may have adaptive consequences (Dudek et al., 2021). It has been argued that neurobiological sequelae of trauma need not be viewed in a negative context, and even embitterment generated by trauma and passed on across generations may have compensatory positive actions (Lehrner and Yehuda, 2018), including ways of making meaning of the trauma experienced (Kidron et al., 2019).

Epigenetic Changes Related to Inflammation

In light of the many epigenetic changes that occur following stressor experiences, it comes as little surprise that among individuals with PTSD, epigenetic changes were present in genes related to immune functioning. Gene methylation was increased in several genes associated with inflammation, and plasma levels of IL-2 and TNF-α, and the antiinflammatory cytokine IL-4, were elevated among individuals who experienced earlier childhood abuse and overall life stress. A relatively large epigenome-wide association study of PTSD in three civilian cohorts identified 4 epigenetic sites within the AHRR gene that were involved in both immune processes and kynurenine functioning (Uddin et al., 2018). DNA methylation related to yet another cytokine, IL-18, was elevated among armed forces personnel who developed PTSD after deployment (Rusiecki et al., 2013). Unfortunately, limited information is available regarding epigenetic changes associated with resilience, although it was suggested that high resilience was accompanied by diminished inflammatory responses (Mehta et al., 2020).

In soldiers assessed before and after combat, PTSD was associated with epigenetic marks on genes (HEXDC, and MAD1L1) associated with the human leukocyte antigen (HLA), a component of the major histocompatibility complex (Snijders et al., 2020). A similar analysis also revealed that the severity of PTSD symptoms was related to epigenetic alterations in genes relevant to oxidative stress and immune activity (Katrinli et al., 2022a). Furthermore, among US Army veterans, recent and lifetime PTSD epigenetic changes were observed, including those related to immune functioning and cell signaling (Montalvo-Ortiz et al., 2022). Also, the Absent in Melanoma 2 (AIM2) gene was related to inflammation and anxiety, and that methylation at a specific site (cg10636246) was connected to the tie between CRP and PTSD. In veterans, the association between PTSD and cytokines (IL-6 and IL-10) was mediated by methylation of AIM2 and might be an important factor that governs comorbidities between PTSD and other disorders (Hawn et al., 2022).

Within peripheral blood mononuclear cells of patients with PTSD, several miRNA variations relevant to proinflammatory cytokines were detected, in that Th1 and Th17 cells were elevated, whereas T_{reg} cells were diminished (Zhou et al., 2016). Similarly, increased methylation occurred in the promoters of several genes related to inflammatory processes (e.g., IFNγ and IL-12), and the expressions of these proinflammatory cytokines were regulated by miRNAs (Bam et al., 2016a). In war veterans with PTSD, relative to controls, expression differences of 326 genes and 190 miRNAs measured in peripheral blood mononuclear cells were detected, and epigenetic changes of genes linked to inflammatory processes were likewise present (Bam et al., 2016b). Furthermore, in individuals with PTSD, epigenetic dysregulation was noted in the TP53 gene (best known as a tumor suppressor gene), which promoted the downregulation of miRNA that resulted in the proinflammatory Th17 phenotype predominating (Busbee et al., 2022). It is meaningful that some epigenetic changes measured in blood mirrored several actions within the prefrontal cortex (Logue et al., 2020)

and postmortem analysis of individuals with PTSD indicated the presence of epigenetic changes within the orbitofrontal cortex that were related to immune and inflammatory processes (Núñez-Rios et al., 2022).

It isn't certain which systems modified through epigenetic processes are responsible for the emergence of PTSD in any given individual. Glucocorticoid receptor changes associated with early-life trauma might be relevant to later pathology, as might BDNF gene methylation within the hippocampus, which has been linked to fear-related memory processes. The case was also made that epigenetic processes relevant to hippocampal and amygdala functioning are responsible for the formation and stabilization of fear-based memories, which may be tied to PTSD (Zovkic and Sweatt, 2013). Despite the individual differences that are present, as a whole, the data point to the presence of a relationship between epigenetic factors and the presence of PTSD. Importantly, studies in animals have made it clear that although epigenetic effects may be persistent and even be transmitted over generations, they are subject to modification with time and experience, thus the potential adverse consequences of trauma, at least with respect to the suppression of gene expression, could potentially be undone.

Epigenetic aging in PTSD

As individuals age, epigenetic marks accumulate, so that these can be used to estimate biological age. Likewise, epigenetic changes within clock genes have been used as an index of biological aging. Several diseases, such as heart disease and some forms of cancer and their treatment, have been linked to epigenetic aging involving changes related to elevated CRP and IL-6. Using a deep learning approach, inflammatory indices (e.g., the chemokine CXCL9) could serve as markers of biological age that predicted the occurrence of heart disease (Sayed et al., 2021). Like the links to physical illnesses, psychiatric disorders, such as PTSD, were associated with epigenetic age and increased expression of inflammatory genes in the brain (Wolf et al., 2021), and the influence of trauma on epigenetic aging could be diminished by having effective social support (Mehta et al., 2022).

Through a variation of the usual epigenetic aging index, it was observed that PTSD was associated with accelerated biological aging, and when measured 3 years later, epigenetic age was correlated with PTSD symptom severity and indices of T cell senescence (Yang et al., 2021d). Other reports, however, have indicated that while epigenetic aging and inflammatory processes could serve as markers to predict mortality, these processes were independent of one another (Cribb et al., 2022). It is especially relevant that among people who presented to an emergency department following trauma, their epigenetic age predicted the subsequent development of PTSD, as well as the diminished volume of subregions of the amygdala. Thus not only can PTSD favor epigenetic aging, but epigenetic aging may be a marker for the development of the disorder (Zannas et al., 2023).

Are Microbiota Related to PTSD?

Microbiota disturbances provoked by stressors could affect neuronal functioning, including gene expression in excitatory neurons, and remodeling of postsynaptic dendritic spines, as well as affecting glial functioning. Prenatal and early life stressors could also influence microbiota, which could affect adult fear extinction, and microbiota recolonization during a discrete neonatal developmental window could preclude the fear extinction profile (Chu et al., 2019b). Mice susceptible to the impact of the single prolonged stressor (to model PTSD) were associated with elevated levels of fecal microbiota with a proinflammatory phenotype together with diminished levels of claudin-5 protein expression that may reflect elevated blood–brain barrier permeability (Tanelian et al., 2022). The elevated inflammation and altered microbiota associated with intense stressors were accompanied by diverse brain changes in mice that had been deemed "susceptible" (Wang et al., 2023b). Specifically, an increased abundance of certain bacteria could influence prefrontal cortex dopamine neurotransmission (elevated dopamine turnover and dopamine D3 receptor expression) by increasing elevated levels of the l-tyrosine-derived metabolite p-cresol in the prefrontal cortex of highly susceptible mice. Conversely, limited p-cresol-induced dopaminergic dysfunctions in the prefrontal cortex were found in mice that did not develop a PTSD-like condition, implicating this process in resilience (Laudani et al., 2023).

The data supporting stressor-provoked microbial change in subserving PTSD have been suggestive, albeit limited. Among PTSD patients, the levels of several microbial phyla were lower than in individuals who experienced trauma but were not diagnosed with PTSD (Malan-Muller et al., 2022). An oral microbiota signature was likewise observed among soldiers who had been to war and experienced PTSD, and these effects varied with other experiential and environmental factors (Shomron et al., 2022). Among frontline care workers in Wuhan who had been treating COVID-19 patients (relative to secondary healthcare workers who did not treat infected patients) the distress experienced was associated with several microbial changes that were related to anxiety, depression, and the occurrence of PTSD. While most microbiota normalized over time, some of the microbiota changes were relatively long-lasting and predicted PTSD persistence (Gao et al., 2022). Although these data do not speak to a causal connection between the two, the possibility cannot be excluded that microbial alterations,

through actions on brain processes, may come to affect the emergence and maintenance of PTSD. Despite the convincing data indicating that PTSD was associated with a proinflammatory microbial milieu, attenuation of symptoms by cognitive therapy was not accompanied by microbiota normalization (Voigt et al., 2022). Once more, these findings might suggest that microbiota are not causally related to the development of PTSD, but persistent microbiota dysbiosis may be a marker for continued vulnerability to the adverse effects of further stressors.

It has been maintained that diet and exercise, which affect both microbiota and inflammation, could serve to limit PTSD symptoms (Pivac et al., 2023) even though the connections between microbiota, inflammatory alterations, and PTSD might be indirect. For instance, among veterans with PTSD, poorer food choices were adopted, perhaps as a coping response to limit emotional distress (Escarfulleri et al., 2021). The adoption of a poor diet in response to stressors encourages obesity and elevated abdominal adipose tissue (belly fat), which is accompanied by elevated levels of circulating proinflammatory cytokines. While highly speculative, this sequence of changes might account for the greater propensity for PTSD in those with obesity (van den Berk-Clark et al., 2018). By adopting appropriate diets and maintaining active behaviors, the adverse effects on microbiota could potentially be thwarted.

If PTSD were related to microbial disturbances, and the specific nature of the microbiota deficiency could be identified, then probiotic or prebiotic treatments, alone or in combination with usual treatments, might have positive effects in alleviating PTSD symptoms. Until a few years ago scant data were available relevant to this. There had been hints concerning the ameliorative effects of such treatments, but in the main, there had been insufficient evidence to arrive at definitive conclusions. Since then, an oral microbial signature was identified that was correlated with specific PTSD symptom severity (e.g., intrusiveness, arousal, reactivity, anxiety, memory difficulties, and idiopathic pain) among Israeli veterans who had been to war (Levert-Levitt et al., 2022). In a small trial, probiotic supplementation with *Lactobacillus reuteri* reduced plasma CRP levels and limited the increased heart rate in veterans with PTSD or persistent postconcussive symptoms in response to a psychosocial stress test, although diminished subjective stress appraisals weren't necessarily affected (Brenner et al., 2020). Likewise, fewer PTSD symptoms were experienced among individuals who maintained a Mediterranean diet, whereas consumption of processed and red meats was associated with greater PTSD symptoms. It appeared that the Mediterranean diet was accompanied by elevated *Eubacterium eligens*, which may have acted in a protective capacity, whereas red and processed meat were associated with diminished levels of this bacteria (Ke et al., 2023).

A meta-analysis of patients who experienced multiple traumas or TBI indicated that probiotic supplements did not influence CRP, IL-6, or stay in a hospital ICU (Noshadi et al., 2022). Despite these discouraging findings, it is probably premature to abandon the notion that probiotics could attenuate PTSD symptoms. As in the case of other illnesses, clinical trials need to distinguish patients based on their microbiota signatures and apply microbial therapies accordingly. Importantly, supplements ordinarily contain a limited range of microbes, and individuals may be better off obtaining diverse microbiota from natural foods.

Treating PTSD

Cognitive and Behavioral Therapies

Several behavioral, cognitive, and psychotherapeutic methods, such as cognitive behavioral therapy (CBT), family or interpersonal therapy, and trauma management therapy, have been used to treat PTSD. It generally seemed that CBT, trauma-focused psychotherapies, and prolonged exposure therapy were most effective when administered soon after the trauma (Lee et al., 2016b). Some of these treatments may also be effective in children and adolescents and are preferable to drug treatments in young people (McGuire et al., 2021a). Only a cursory overview of these approaches is provided here, but more detailed descriptions are available in numerous reports (e.g., Lewis et al., 2020). Also, an overview of interventions for complex traumatic events (as opposed to single-event PTSD), such as childhood sexual abuse, domestic violence, armed conflict, and forcible displacement, is available (Coventry et al., 2020).

Cognitive Behavioral Therapy

Individualized CBT has been among the most effective of the behavioral approaches to treat PTSD (Lewis et al., 2020). Several biomarkers could potentially be used to predict psychotherapy outcomes, including genotypes linked to serotonin and glucocorticoid sensitivity and metabolism, greater baseline heart rate, and heart rate responses that might be indicative of fear habituation (e.g., Colvonen et al., 2017). Also, elevated IL-6 and IL-10 reactivity associated with a laboratory social stressor predicted the subsequent response to therapy (Renner et al., 2022). The case was made that pretreatment activation and density of brain regions involved in appraisal, memory, emotional regulation, and decision-making (amygdala, hippocampus, as well as anterior cingulate cortex and insula) were associated with treatment response (Colvonen et al., 2017), and thus, volumetric measures of specific brain regions could be used to predict treatment efficacy.

Mindfulness-based therapies have also been viewed as holding promise in the treatment of PTSD, possibly in an adjunctive capacity. These actions may be linked to the restoration of connectivity of brain regions associated with the default mode network and both the executive and salience networks (Boyd et al., 2018). Positive effects of mindfulness stress reduction have been observed among combat veterans and among women who experienced relationship abuse. Among responders to mindfulness therapy, an increase in methylation of the *FKBP5* gene was observed, whereas a decrease in methylation was apparent among nonresponders (Bishop et al., 2018).

Exposure Therapy

Prolonged exposure therapy has seen some success in the treatment of PTSD. This approach, which follows from desensitization methods, involves the individual re-experiencing the traumatic event through remembering and engaging with it, rather than avoiding the memories. The goal is to extinguish persistent emotional and cognitive responses to danger cues, which otherwise has allowed individuals to continue in their excessive responses to potential danger signals. In progressive steps, the memory and cues that had elicited the powerful negative cognitive, emotional, and physiological responses might be diminished. Significantly, among individuals who responded positively to prolonged exposure therapy, baseline hippocampal volume was greater than it was among treatment nonresponders (Rubin et al., 2016). Thus based on pretreatment markers, some patients may be better suited for certain treatments. Regardless of whether exposure therapy or CBT is undertaken, the best outcomes are obtained a short time after the trauma (Shalev et al., 2016).

Among patients who were successfully treated through prolonged-exposure therapy, changes were detected regarding the epigenetic changes related to *FKBP5* and *NR3C1*, but these did not necessarily parallel one another. Interestingly, methylation of the *NR3C1* gene in PTSD patients predicted the response to prolonged exposure therapy but was unaltered by the treatment itself. Conversely, methylation of the *FKBP5* gene promoter region was not predictive of the impact of the treatments but nonetheless declined in treatment responders (Yehuda et al., 2013). It was subsequently reported that prolonged exposure therapy that diminished PTSD symptoms was associated with reduced methylation of the *FKBP5* gene, whereas this did not occur among treatment nonresponders (Yang et al., 2021e).

A related approach, that of Narrative Exposure Therapy, comprises emotional exposure to the memories of traumatic events and having patients reorganize their emotional responses into a coherent narrative (Wilker et al., 2023). Interestingly, *NR3C1* methylation at a specific CpG site (cg25525999) increased among individuals who showed a positive treatment response to this therapy.

Eye Movement Desensitization and Reprocessing

Another form of therapy, that of eye movement desensitization and reprocessing (EMDR), consists of a series of sessions during which patients are asked to focus on a vivid image of the traumatic event, while at the same time engaging in eye movements in which they track a finger that moves across their visual field. It is thought that while the memory of the trauma is in short-term storage EMDR allows for an association to be made with non-threatening stimuli or it permits the dissociation of the traumatic memory and the emotions ordinarily elicited by these memories.

At first blush, this procedure may sound a bit flaky and might not readily be accepted as a legitimate method of treatment. Yet, considerable evidence supported its effectiveness (Lewis et al., 2020), including among those who received the treatment soon after a traumatic experience (de Roos et al., 2017).[6] The benefits of EMDR have been reported with a wide variety of trauma experiences (e.g., acute traumatic events, natural disasters, war outcomes in veterans and civilians, women who experienced birth trauma, and complex traumatic events, which comprised encounters with several traumatic events). A meta-analysis suggested that EMDR was just as effective as CBT and was superior in eliminating some symptoms (Khan et al., 2018), and when EMDR was preceded by prolonged exposure therapy, still greater reductions in PTSD symptoms were realized (Van Minnen et al., 2020).

A systematic review of EMDR, prolonged exposure therapy, and cognitive processing therapy (a form of CBT) indicated that each of these therapies enhanced posttraumatic growth, but EMDR had greater effects on brain functioning (Pierce et al., 2023). EMDR-like manipulations were associated with a reduction of neuronal activity in brain regions associated with emotional processing (Thomaes et al., 2016), and the volume of the left amygdala increased following EMDR treatment, an outcome that was not provoked by prolonged exposure therapy. As expected, EMDR was accompanied by improved fear extinction responses accompanied by changes in neuronal connectivity in fear circuits (amygdala and hippocampus) and the insula (Rousseau et al., 2019).

Pharmacological Therapies

Selective Serotonin Reuptake Inhibitors

Pharmacological interventions, including SSRIs and SNRIs, have been a primary treatment for PTSD. Even though the success achieved by most agents has been moderate (Williams et al., 2022b), some antidepressants may be useful for specific PTSD symptoms, such as nightmares and difficulty sleeping (e.g., Zohar et al., 2017). Although both SSRIs and prolonged exposure therapy provide some relief from symptoms, the combination of the two did not engender a better outcome than either treatment alone (Rauch et al., 2019). Likewise, the combination of CBT and SSRI treatment most often did not promote greater effects than either treatment administered alone. However, more sustained benefits could potentially be derived from the combination therapy.

Norepinephrine Receptor Antagonists

As described earlier, treatments that modified norepinephrine activity during the early posttrauma period could prevent the development of PTSD (Giustino et al., 2016), but were less effective in modifying these features once PTSD was well entrenched. To an extent, compounds that diminished norepinephrine release, such as α_1- and β-receptor blockers (e.g., prazosin and propranolol, respectively), reduced symptoms of PTSD (Pitman et al., 2012). However, the specific symptoms abolished varied with different treatments. For example, the α_1 adrenergic antagonist prazosin reduced trauma-related nightmares and insomnia, whereas the β-blocker propranolol was more useful in attenuating the emotional disturbances associated with traumatic memories. Accordingly, it was maintained that PTSD would be especially amenable to a treatment that affected both receptors concurrently.

Once PTSD had been established it was amenable to being altered when individuals recalled the trauma experience (during reconsolidation). In a 6-week, double-blind, placebo-controlled, randomized clinical trial, propranolol administered just before the memory recall procedure, appreciably reduced PTSD symptoms (Brunet et al., 2018). A similar protocol yielded positive effects, but these occurred primarily with more severe PTSD and tended to improve over the weeks following the final therapy session (Roullet et al., 2021). However, the effectiveness of norepinephrine antagonists in attenuating the reconsolidation of memories related to PTSD has been inconsistent. While a meta-analysis indicated

---

[6] Patients waiting in hospital after a motor vehicle accident who were provided with reminders of their experience and asked to play the computer game Tetris, PTSD development diminished (Iyadurai et al., 2016). Engaging in this task, which entails high visuospatial demands, presumably disturbs the consolidation of trauma memories and hence PTSD symptoms could be prevented.

that the data generally didn't support a role for propranolol in disrupting the reconsolidation of memories and diminishing PTSD (Steenen et al., 2016), a subsequent meta-analysis countered this conclusion (Pigeon et al., 2022). Despite impressive findings in several well-controlled studies, debate has continued concerning the effectiveness of the procedure, and firm conclusions regarding the effectiveness of propranolol and similar drugs in attenuating fear reconsolidation may have to await further data becoming available. Nonetheless, the possibility of targeting memory and memory reconsolidation to deal with PTSD is likely a viable option, particularly if propranolol is administered repeatedly.

Dynamic neurobiological changes in PTSD and the timing of treatments

In addition to the diversity of neurochemical processes that could underlie PTSD, the characteristics of the disorder evolve with time following the trauma, and dynamic variations of neurobiological processes might also occur with time. The neurochemical alterations apparent soon after a trauma, presumably acting in an adaptive capacity to protect behavioral and physical integrity, may be replaced (or overwhelmed) by harmful neurobiological changes (e.g., receptor dysregulation or cell loss). It follows that the effectiveness of treatment strategies would also vary over the different phases of the disorder. Accordingly, a given treatment, such as β-blockers and corticoid manipulations, administered during a window of opportunity soon after the traumatic experience (or during reconsolidation), can diminish the signs of PTSD, but when administered in the days and weeks following trauma it is less effective in preventing the appearance of PTSD (Argolo et al., 2015). Also, the nature of the stressor (familiar vs. unfamiliar) and the timing of treatments relative to trauma were linked to the behavioral and glutamate changes that occurred (Nasca et al., 2015). This is consistent with the view that illnesses, such as PTSD, require a multitargeted approach, but more than this, we are suggesting that the treatment or treatment combinations effective at a particular phase of the illness might be less efficacious at a second phase in the evolution of the condition.

Glutamate Manipulations

Support for glutamate as an important player in PTSD initially came from the finding that the partial NMDA glutamate receptor agonist D-cycloserine (Seromycin) facilitated fear extinction in animals and could have similar actions in patients with PTSD (Inslicht et al., 2022). While D-cycloserine did not enhance the effects of CBT, it had promising effects in human PTSD patients when combined with exposure therapy, although the magnitude of the changes was only moderate (Mataix-Cols et al., 2017).

The NMDA antagonist, ketamine, which is so effective in treating recalcitrant depression, was also posited as potentially being effective in attenuating PTSD symptoms, possibly by enhancing fear extinction through effects on mTOR signaling (Girgenti et al., 2017). Ketamine produced a rapid decline of symptoms in patients with chronic PTSD and the effectiveness of ketamine could be enhanced with repeated treatments (Albott et al., 2018). Based on a meta-analysis of the few studies available, it was provisionally suggested that ketamine might be effective in diminishing symptoms in chronic PTSD but did not have this effect when administered relatively early in PTSD development (Du et al., 2022).

There had been the notion, dating back at least four decades, that 3,4-methylenedioxy-methamphetamine (MDMA), known on the street as "molly" or "ecstasy," could be used clinically. A damper was placed on this because MDMA was considered a Schedule I illicit drug, and some researchers believed that the drug simply wasn't ready for prime time. With interest increasing in the use of psychedelic agents to alleviate mental illnesses, a small number of trials assessed the effects of MDMA in diminishing symptoms of PTSD. It appeared that MDMA could enhance the effectiveness of psychotherapy among patients with chronic PTSD that was refractory to treatments (Bahji et al., 2020). Moreover, MDMA-assisted therapy could promote marked and lasting reductions of PTSD symptoms even in the presence of comorbid conditions, such as depression and substance use disorders, as well as in individuals who had encountered childhood abuse (Mitchell et al., 2021). The enhanced efficacy of MDMA when combined with psychotherapy might occur because of the additive effects of the treatment or because MDMA allows for better connections with the clinician.

It is believed that MDMA treatment diminishes the distress elicited when trauma events are recalled, and the drug is effective in limiting the emotional responses related to the trauma event, overriding negative cognitive appraisals (Feduccia and Mithoefer, 2018), possibly reflecting the actions of increased BDNF that may be provoked. In addition to its actions on glutamate, MDMA could also have beneficial actions through effects on serotonin, norepinephrine, and dopamine in brain regions associated with fear- and appraisal-related processes (Amoroso, 2015). As well, reduction in PTSD symptoms following treatment with MDMA was associated with elevated methylation at a site of the *NR3C1* gene (Lewis et al., 2023a). Psilocybin, like MDMA, may have positive effects in the treatment of PTSD (Khan et al., 2022). However, to the consternation of many scientists, in the summer of 2024 the FDA did not approve the use of MDMA (plus psychotherapy) in the treatment of

PTSD, indicating that the risks associated with the drug outweighed the benefits. This was a controversial decision and the last word on the issue has likely not been spoken.

Cannabinoids

Aside from affecting anxiety cannabis might reduce symptoms of PTSD, and the high use of cannabis by individuals with stress-related psychological disturbances may reflect an effort at self-medication (Hill et al., 2018). With the loosening of restrictions concerning cannabis use, it has been possible to conduct a small number of controlled studies, even though these were often underpowered. Some of the purported effects of cannabis have been based on anecdotal reports or observational studies that don't speak to whether causal links exist between cannabis use and changes in PTSD since some of the effects could be related to actions on comorbid conditions (e.g., anxiety disorders). Despite these limitations, the case was made that endogenous endocannabinoid insufficiency may be an ingredient that disposes individuals to pathologies related to traumatic events, and thus appropriate targeting of endocannabinoid systems could potentially be used to treat PTSD.

Supporting the use of cannabis in the treatment of trauma-provoked disorders, a retrospective analysis of veterans with treatment-resistant PTSD suggested that cannabis could diminish symptoms (Nacasch et al., 2023). Likewise, THC used as an add-on treatment was associated with diminished hyperarousal, improved sleep quality, and reduced frequency of nightmares, and the synthetic THC analog, nabilone, produced similar effects (Steardo et al., 2021). A systematic review supported the contention that cannabis and synthetic cannabinoids can reduce some symptoms of PTSD, such as anxiety, memory-related processes, and disturbed sleep (Orsolini et al., 2019), and may reduce aversive memories (Raymundi et al., 2020). Another review of the relevant literature indicated that among PTSD-affected individuals, THC, cannabidiol, and synthetic cannabinoids that targeted CB1 and CB2 receptors enhanced sleep quality and quantity, reduced hyperarousal, and augmented extinction of the fear response (Forsythe and Boileau, 2021).

As much as these reviews have supported the benefit of cannabis for PTSD, most studies were nonrandomized, were generally of low quality, and may have been liable to biased outcomes owing to participant self-selection (Rehman et al., 2021). Making matters worse, pitted against the positive findings reported, a randomized placebo-controlled trial of cannabis use (at different concentrations) did not reduce PTSD any more than a placebo (Bonn-Miller et al., 2021), and it was even reported that cannabis can foster intrusive symptoms

(Metrik et al., 2022). In this regard, it is concerning that while cannabis is generally well tolerated, in a subset of individuals it may promote psychiatric disturbances, such as depression, anxiety, psychosis, and substance misuse, and may promote ineffective coping, and these effects were most pronounced among individuals who initially used during adolescence (Volkow et al., 2014). In fact, during the COVID-19 pandemic, the use of cannabis among individuals with PTSD resulted in the severity of comorbid depression increasing (Murkar et al., 2022). What it boils down to is that evaluation of the potential benefits of cannabis for PTSD needs to be measured against the harms that can be created.

Antiinflammatory Agents

Stressor-elicited circulating and brain inflammatory processes have been implicated as a potent marker of PTSD, particularly if depression was also present. Certain gene expression markers might, in theory, be used to predict the development of PTSD, but once more, the vast array of genes that could act in this capacity has made this difficult. Yet, sets of blood-based gene markers achieved 90% accuracy in predicting PTSD among deployed soldiers (Tylee et al., 2015).

The most obvious antiinflammatory strategy to treat the disorder is the use of NSAIDs, which have benefits, albeit modest, in depressive disorders and diminishing anxiety. As indicated earlier, repeated treatment with antiinflammatory agents (ibuprofen, minocycline) reduced the expression of hippocampal IL-1β, TNF-α, and BDNF, and in an animal model treatment with ibuprofen diminished PTSD symptoms. The actions of these drugs were particularly effective when administered soon after the trauma (Levkovitz et al., 2015). Regrettably, the influence of NSAID treatments in humans has been limited, and in some studies that have been conducted, an NSAID did not reduce PTSD symptoms (Grau et al., 2022).

Consistent with the inflammatory involvement in PTSD, symptomatology could be affected by immune-acting agents. Detailed reviews of inflammatory factors in PTSD provided fundamental information concerning the influence of antiinflammatory treatment in alleviating symptoms (Katrinli et al., 2022a). The use of these agents seems promising, but the available data are still sparse and important gaps in the information remain to be filled.

Glucocorticoid Manipulations

Given the pronounced effects of stressors on cortisol and the findings concerning the changes of cortisol and glucocorticoid receptors in PTSD, attention to this hormone in therapy for PTSD has received considerable attention. For the most part, treatment with hydrocortisone did not prevent the development of PTSD when given in advance of trauma (e.g., among patients who received the drug before cardiac surgery). However, systematic reviews supported the contention that hydrocortisone could diminish symptoms of PTSD (Bisson et al., 2021), especially if administered relatively soon after traumatic injury (see Florido et al., 2023). Even though the effects of hydrocortisone were relatively modest (Bisson et al., 2021), it could augment the actions of exposure therapy (Yehuda et al., 2015). In evaluating the effects of hydrocortisone in the context of reconsolidation of trauma memories, positive actions were observed, but this was primarily evident in patients with severe physical illnesses (Astill Wright et al., 2019).

The data concerning the effectiveness of cortisol manipulations in the treatment of PTSD have generally not been especially encouraging. Most often, patients in these studies involved select populations (e.g., war veterans) or included patients who experienced diverse forms of trauma (universal prevention). Rarely were the effects of the treatment assessed in subtypes of PTSD based on the nature of the stressor encountered, whether patients had experienced early life adverse events, or displayed dissociative responses, which have been linked to hypocortisolism. In this respect, the influence of hydrocortisone may interact with progesterone in modifying specific symptoms of PTSD, such as involuntary aversive memories, which speaks to the sex differences in the occurrence of PTSD. It is not unrealistic to expect that differences might exist regarding the most efficacious therapies based on such characteristics, once again pointing to the need for a precision medicine approach for optimal treatment.

Transcranial Magnetic Stimulation

Transcranial magnetic stimulation, which was effective in treating depression, was promising in mitigating PTSD and generalized anxiety disorder symptoms (Cirillo et al., 2019), and was effective when administered during the consolidation of fear extinction (Van't Wout et al., 2017). Among veterans with comorbid PTSD and depression, rTMS reduced PTSD symptoms to the extent that 46% no longer met the threshold criteria for this condition (Madore et al., 2022), and was effective among veterans with PTSD who had experienced TBI (Philip et al., 2023). A variant of rTMS, that of intermittent theta-burst stimulation (iTBS), which can be administered quickly (about 3 min), similarly reduced PTSD symptoms (Philip et al., 2019).

Whether rTMS has its anti-PTSD effects by altering inflammatory processes isn't certain. In rodents, it can affect microglia and inflammation within the hippocampus and prefrontal cortex of chronically stressed mice and provoke a switch from the dominance of the pro-inflammatory M1 phenotype to the antiinflammatory M2 phenotype (Zuo et al., 2022). Just as some of the benefits of rTMS in the treatment of depression were accompanied by modification of inflammatory processes similar mechanisms might be operative in PTSD.

Comorbid Illnesses Linked to PTSD

Comorbid illnesses, as we will discuss in Chapter 17, are frequent between various psychiatric illnesses, as well as between mental illnesses and a range of physical conditions. This not only complicates analysis of the mechanisms responsible for a specific illness but is a problem for a patient's health and treatment. In the case of substance use disorder, which is a frequent comorbid condition of PTSD, the effectiveness of treatments is less favorable, although promising effects were observed using integrated psychosocial therapies (Flanagan et al., 2016). Depression is likewise exceptionally common among individuals with PTSD, as is the frequency of suicide, although this may vary between civilian and noncivilian populations and may be tied to other extenuating factors (Shalev et al., 2017). It is not uncommon for PTSD to be comorbid with both schizophrenia and bipolar disorder, and several inflammatory genes were upregulated in association with these disorders (Pandey et al., 2015). However, it is uncertain whether these reflect causal connections, pleiotropic actions, or simply that the inflammatory changes emerge with stress or general illness.

PTSD patients with and without comorbid depression could be distinguished from one another based on their ACTH levels following a challenge with the synthetic corticoid dexamethasone. Both illnesses share several common inflammatory disturbances, including elevated proinflammatory cytokines or factors that modulate them (e.g., glucocorticoid functioning), which might be responsible for their comorbidity. Likewise, PTSD was associated with low levels of adiponectin, a hormone linked to inflammatory illnesses, possibly through its actions on proinflammatory cytokines, particularly TNF-α (Zhang et al., 2016).

It is instructive that PTSD is accompanied by increased vulnerability to immune-based illnesses, such as autoimmune disorders, ranging from psoriasis to rheumatoid arthritis (e.g., Gupta et al., 2017). Indirectly related to these findings, whole genome analyses among soldiers returning from war zones revealed that PTSD was linked to a gene, *ANKRD55*, which has been associated with autoimmune and inflammatory disorders, such as multiple sclerosis, rheumatoid arthritis, and celiac disease (Stein et al., 2016). Given these multiple comorbidities, it is no wonder that PTSD is also associated with earlier mortality.

Considerable data indicated that inflammatory factors are a common denominator for a variety of illnesses that have been linked to PTSD, but long-term prospective studies will be needed to determine whether the relationships are bidirectional and what factors operate along with inflammatory factors in predicting these conditions. The changes in immune and cytokine functioning associated with PTSD may be related to norepinephrine or serotonin changes, which promote the release of immune cells from secondary immune organs. Also, as we saw in earlier chapters, serotonin modulates immune activity and SSRIs can be used to affect immune activity.

Pain and PTSD

Aside from the comorbidities that exist between PTSD and other mental disorders, numerous studies revealed that chronic pain may be a companion of PTSD. While chronic pain can promote PTSD, in other cases the presence of PTSD may cause an exaggeration of pain responses, possibly because of excessive release of proinflammatory cytokines. Whatever the case, both conditions are difficult to treat, and their combination is still more challenging. As some neural circuits associated with physical pain and those associated with emotional distress overlap, the possibility exists that therapeutic strategies that target this circuitry might be helpful (Scioli-Salter et al., 2015). In this regard, it has been maintained that TNF-α may contribute to the co-occurrence of PTSD and allodynia, and treatments to diminish the actions of this cytokine may be useful in alleviating both conditions (Dib et al., 2021).

Conclusion

The development of PTSD had initially come to the attention of the public owing to the many wounded soldiers who had returned home from various conflicts with severe psychological disorders. However, PTSD is a much broader problem, frequently appearing in civilian populations. At one time, PTSD was considered a condition that developed owing to trauma that was outside the norm of human experiences. This notion was abandoned with reports that PTSD was widespread in response to a variety of traumas that were clearly within the normal course of human experiences.

Understanding the mechanisms that operate in the development of PTSD has been hampered by numerous factors, including the lack of appropriate animal models for the disorder (although this has changed in recent years), and differentiating the processes underlying PTSD from the many comorbidities that accompany this disorder. It is equally clear that although many treatment options have been offered for PTSD, including behavioral and pharmacological approaches, none have turned out to be particularly impressive, although in most instances considerable variability was apparent. Like so many other conditions, PTSD is a multidimensional disorder, and hence it's unlikely that one treatment suits all, nor that any single treatment is sufficient to deal with this condition.

It seems that PTSD is a problem linked to memory and overgeneralization, difficulty distinguishing safe from unsafe stimuli, and difficulties in extinguishing fear reactions. Dynamic processes operate during the course of an illness, changing over time, and the effectiveness of treatments could potentially vary in this regard. The involvement of inflammatory factors in PTSD is a relative newcomer to this field. To some extent, it's difficult to understand how immune functioning could be germane to this mental condition unless it is considered that stressors, especially traumatic events, can have actions on glial processes and cytokine release that occurs. Many peripheral immune correlates have been identified in PTSD, but it is questionable which, if any, plays a causal role in the illness or which might sustain this condition once it has developed. Given the paucity of data, it is even difficult to identify which of the inflammatory factors could serve as markers of PTSD development and the response to treatments.

Before closing off this chapter, an additional thought should be added at the risk of sounding a bit too preachy. In Chapter 1, we mentioned some of the historical views that guided (or misguided) research and treatment of mental illnesses. One perspective that has been expressed concerning mental illness is that it is a normal response to an abnormal world (situation). The original statement in this regard, provided by the great theorist and Holocaust survivor Viktor Frankl in his 1959 book *Man's Search for Meaning* (initially published under the title *From Death-Camp to Existentialism*), was that "an abnormal reaction to an abnormal situation is normal behavior." Those with PTSD suffer greatly, to be sure, but given the trauma experienced, it's neither unusual nor a reflection of weakness. As the frequency of trauma related to diseases, wars, forced immigration, abusive relationships, natural disasters, those brought on secondary to climate change, and pandemics that are likely to reoccur, we can probably count on many more cases of PTSD and related pathologies occurring. Failing our ability to limit traumatic events, we need to be ready to deal with the normal mental health issues that will evolve.

CHAPTER

12

Pain Processes

Pain as an adaptive response?

Pain is a sensation that affects most people at some time. Humans may have inherited several genes from Neanderthals that could influence pain perception, which could have adaptive value. Unfortunately, the presence of mutations on these genes may increase sensitivity so that otherwise moderate pain can morph into severe pain perception (Faux et al., 2023). An important element of pain is that it provides feedback concerning the occurrence of potential tissue damage so that appropriate defensive actions can be taken to preclude further harm and allow for healing to occur. Yet, one wonders whether the pain that acts as a warning needs to be as intense and chronic as it is all too often. Nature could have been much, much kinder.

Pain responses inform others of an ongoing problem and that their help might be needed. As well, shared pain may serve as a trigger that brings people together so that they form a tighter and more cooperative group (Bastian et al., 2014). When people use social resources in response to painful stimulation, pain-relevant areas of the brain are activated to a lesser extent than they are otherwise. Empathy and social support can act as a pain buffer, possibly reflecting the adaptive nature of pain communication, group membership, and social support.

Studies in mice indicated that they share their pain; when one animal watches another in distress, their own pain sensitivity increases (Smith et al., 2016), and to some extent this also occurs in humans. Watching others endure trauma or pain can affect us emotionally and may affect the same brain circuits that are excited when pain is actually felt. It was suggested that such reactions stem from the release of opioids within the amygdala and periaqueductal gray (Haaker et al., 2017).

Pain as the Ubiquitous Malady

Chronic illness and chronic pain are experiences that nobody wants to endure, yet there's hardly anybody who hasn't experienced some form of acute pain, and in Western countries, about 20%–30% of people experience intense chronic pain. The impact of chronic pain (defined as lasting longer than 6 months, or longer than it ought to take for normal healing to occur) on well-being is pronounced, particularly as it gives rise to other illnesses, and can reduce life span. Chronic pain sometimes recedes over time, but frequently it remains stable even in the absence of signs of tissue or nerve damage. The severity of pain may worsen with time (chronic progressive pain), as in the case of rheumatoid arthritis or some types of cancer, and in some instances, the pain arises from unknown causes (idiopathic pain).

Classification systems have been devised to have a uniform perspective on pain, which has included consideration of the body region affected, as well as the intensity, duration, and pattern of pain occurrence. But there's more to pain perception than just that. Pain sensations come in various flavors that can be clinically meaningful, as they influence treatment and the recovery process. *Neuropathic* pain arises owing to damage to nerve endings, appearing as numbness along the course of an affected nerve, or by a burning or heavy sensation. This form of pain can occur because of infections, toxins, and particular diseases, as well as by poor nutrition and excessive alcohol use. As well, damage to tissues can be detected by specialized sensory nerves that create *nociceptive* pain. This type of pain may be felt as sharp, aching, or throbbing sensations, possibly having developed secondary to some other condition, such as the growth and spread of tumors that crowd other body organs or create blood vessel blockage. It may also appear in response to thermal or chemical stimulation (intense heat or cold, or to noxious substances that stimulate particularly sensitive body regions) or to mechanically elicited damage (cutting, crushing). Pain has also been classified as being somatic (coming from bones, muscles, and other soft tissues) or comprising visceral pain (coming from internal

The Immune System
https://doi.org/10.1016/B978-0-443-23565-8.00002-8

323

Copyright © 2025 Elsevier Inc. All rights are reserved, including those for text and data mining, AI training, and similar technologies.

organs). In both instances, painful stimuli activate nociceptors that have specialized nerve endings, which can detect temperature, pressure, and stretching within and around affected tissue.

A chronic pain condition, complex regional pain syndrome (CRPS), can develop following a limb injury. In this syndrome, a burning sensation or the feelings of pins and needles are often present, reflecting *allodynia* (elevated sensitivity to stimuli that ordinarily do not produce pain) that can spread beyond the site of injury (e.g., from a part of the hand to the entire arm). When no confirmed nerve damage is identified, Type I CRPS is diagnosed, whereas the Type II form is associated with identified nerve injury. The development of CRPS seems to be mediated by brain neuroplasticity, activation of peripheral nociceptors, interactions with inflammatory processes, and autonomic nervous system functioning (Taylor et al., 2021), as well as disturbed functional brain connectivity involving the insula (Kim et al., 2017a).

Psychological Impact of Chronic Pain

Experiencing pain day-after-day takes an enormous toll on psychological health and encourages other comorbid physical conditions, and addiction may develop to opioid medications or alcohol as part of an effort to cope with the pain. Chronic pain forces behavioral changes as individual's activities might revolve around accommodating their physical discomfort. Even sleeping can be difficult, and the resulting changes in alertness may increase pain sensitivity (Alexandre et al., 2017). Sleep disorders have been associated with elevated pain perception, and there is reason to believe that procedures to enhance sleep may diminish pain perception. Predictably, chronic pain can engender various psychological disorders (anxiety, depression, posttraumatic stress disorder; PTSD), which can also morph into catastrophizing that exacerbates chronic pain. The perception of pain can be influenced by multiple organismic and experiential factors, as well as biological processes, some of which are shown in Fig. 12.1.

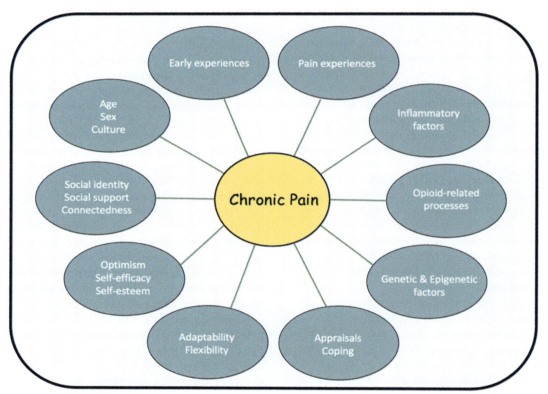

FIG. 12.1 A wide range of risk factors have been identified that might contribute to chronic pain. These comprise innate genetic influences, specific polymorphisms, or epigenetic changes that developed as a result of earlier pain experiences or other environmental influences. Genetic factors could act additively or interactively with acquired processes in affecting pain perception. Differences in pain perception can also be related to the presence of particular neurotrophins. As well, early-life stressors or pain experiences, sensitization of neuronal processes, or memory imprints of pain, which affect multiple neurochemical systems, may alone or in combination influence the response to painful stimuli (Burke et al., 2017). Furthermore, characteristics of the individual (age, sex), including a large constellation of cognitive changes and personality factors (optimism, neuroticism, self-esteem, and self-efficacy) can affect pain processing. Gender, cultural, social, and developmental factors contribute not only to pain perceptions, which could be related to genetic influences, but also to the marked individual differences in the ability to cope with pain. They might also be driven by the engagement of neuronal circuits involved in emotional appraisals, including circuits involving the anterior cingulate cortex. It's also a good bet that genetic and epigenetic factors related to inflammatory processes contribute to differences in pain perception. In this regard, inflammatory pain experienced early in life can also alter hippocampal-dependent memory processes, and this outcome can be worsened by other chronic stressor experiences.

Chronic Pain Leading to Depression and PTSD

In response to chronic pain, some individuals develop a pattern of negative thinking and diminished coping flexibility to deal with their pain. By promoting counter-productive coping methods (e.g., rumination, wishful thinking, social withdrawal, and substance use) and feelings of helplessness, chronic pain can create a milieu that fosters psychological disturbances to the extent that 50%–80% of individuals with chronic pain experience significant depression. In some instances, social interactions necessarily need to be curtailed, so that individuals may find themselves with fewer social supports that might ordinarily help them cope with the pain and diminish depression. With physical health deteriorating, the occurrence of anger, suicidal ideation, or suicide attempts increases. Understandably, chronic pain would instigate depression, but not all individuals react this way, and it seems that genetic factors might also contribute to the evolution of depression associated with chronic pain.

The sensory component of chronic pain can be dissociated from its affective components, with the latter likely being determined by the functioning of the anterior cingulate cortex (Barthas et al., 2015). In addition, variations of neuronal activity within the mesolimbic dopamine circuit, which has been associated with reward salience, motivation, and mood state, may be accompanied by altered pain perception. Moreover, this circuit might contribute to the responsiveness to opioids and to the pain-reducing effects of antidepressants, as limited as these might be.

In addition to affecting mood, chronic pain conditions have been associated with other CNS processes. For instance, pain associated with chronic knee osteoarthritis was associated with brain aging, together with negative affect, passive coping methods, and catastrophizing (Johnson et al., 2022). In a subsequent analysis the accelerated brain aging associated with chronic knee osteoarthritis was accompanied by memory decline and incident dementia. These actions were especially prominent within the hippocampus, and the relationship between chronic pain and brain aging was mediated by a gene (SLC39A8) within microglia.

Psychological and Contextual Factors Influence Pain Perception

It's been known since the seminal work of Melzack and Wall (1965) that psychological factors moderate pain perception, and pain perception may vary with the context in which it is experienced. For instance, individuals who had been in a motor vehicle accident or who experienced rape can develop a chronic pain condition that is exacerbated by further negative psychological factors. Such effects were especially marked among individuals carrying a polymorphism of the FKBP5 gene that is fundamental in regulating HPA-axis functioning, or the gene that codes for corticotropin-releasing hormone-binding protein (Linnstaedt et al., 2016). In effect, an interaction exists between physical and psychological factors in determining long-term pain perception.

Given that multiple mechanisms are responsible for some forms of pain, a *neurological pain signature* was sought based on brain imaging procedures. Developing such a signature could be exceptionally valuable in identifying the source of allodynia, which could then be incorporated into personalized treatment strategies. As worthwhile as this might be, owing to the limits of technology, it might not always be possible to identify subtle characteristics that comprise the neural signature for certain types of pain. Thus false negatives will appear on some occasions, and patients might not receive appropriate and necessary treatments, fail to obtain time off from work to recover, or might even be denied insurance benefits. The issue had become sufficiently serious to have clinicians, brain imaging experts, and those practicing neuroethics and law, to caution that the methodologies are not foolproof, and they have discouraged their use in the diagnosis of chronic pain (Davis et al., 2017).

Neurophysiological and Psychological Processes That Accompany Pain Perception

For several decades, the *gate control theory* was the leading perspective regarding pain processes (Melzack and Wall, 1965). It was proposed that information is transmitted along fibers from the site of injury to the dorsal horn of the spinal cord (where sensory fibers synapse onto spinal neurons), which then send pain signals to projection neurons, through which they are transmitted to the brain. Fibers were hypothesized to exist that could produce either fast or slow signals (the Aβ fibers and C fibers, respectively), and nerve fibers could have either activating or inhibitory effects, as well as actions that comprised the inhibition of inhibitory effects. These varied neuronal inputs would result in the figurative opening or closing of a "gate" within the spinal cord, which would determine pain sensations. Painful stimuli influence nonmyelinated C fibers and respond to strong stimuli, resulting in deep pain. The myelinated Aβ fibers, in contrast, are responsible for rapid, shallow pain, and are stimulated by touch, pressure, or vibration. Importantly, these fibers can dampen painful sensations associated with activation C fibers (e.g., lightly rubbing a wounded area can diminish the pain that is otherwise perceived). Later studies by other investigators identified additional fibers that contributed to pain perception. The Aδ fibers carry cold, pressure, and acute pain signals, and certain fibers and receptors were identified that were activated exclusively by stimuli that could damage the skin. As well, different C-fibers could be activated by nociceptive stimuli that involved diverse stimuli (e.g., thermal, mechanical, chemical), which activated nociceptors. Many other additions to pain processes were subsequently identified, including processes that occurred within distinct portions of the brain, and the involvement of different neurotransmitters in modifying pain perception.

Of particular significance, at least from the perspective of the present discussion, is that a central state was hypothesized so that emotional and cognitive processes moderate processes that determine pain perception. Attributes of the central state that could affect pain perception were subsequently elaborated to include sensory-discriminative (intensity, location, quality, and duration), affective-motivational (unpleasantness), and cognitive-evaluative processes (appraisal, cultural values, context, and cognitive state) (Melzack and Casey, 1968). The incorporation of these processes, together with inhibitory neuronal processes, allowed for an accounting of pain-related phenomena (e.g., placebo effects, phantom limb pain, and the influence of distraction) that wouldn't fit comfortably into an analysis that was limited to spinal processes.

This theorizing occurred well before a good understanding was available concerning diverse neurotransmitter processes, and knowledge of different fiber types was still limited. The initial perspectives have since been updated and reformulated. Nonetheless, the Gate Control theory was pivotal in the understanding that multiple processes, particularly those of a psychological nature, could influence pain sensations. This gave credibility to the development of cognitive treatments in modifying pain perception (Moayedi and Davis, 2013). In this respect, pain is typically considered to involve a multidimensional process involving interactions between sensory inputs and their translation in the brain. Much like other neural systems, those involving pain are subject to plasticity so that the perception or interpretation of pain can be affected by mood, attention, earlier pain experiences, expectancies, and memories of pain.

Several formulations have advanced earlier positions, incorporating several new elements, but maintaining central state as a core feature, as described in Fig. 12.2 (Grace et al., 2014). This formulation described the role of sensory neurons and their functioning and the involvement of glutamate signaling as well as other neurotransmitters in mediating pain processes. The functions of diverse brain circuits and their complexity have been described in numerous overviews of the topic (e.g., Kuner and Kuner, 2021). Through the innervation of spinal cord fibers, the thalamus may play a key role in acute pain detection, whereas its role in chronic pain is less clear, although corticothalamic loops could play some role in chronic pain processes. Neocortical regions, particularly the prefrontal, cingulate, insula, and somatosensory cortices, were viewed as being integral to the processing of sensory, emotional, and cognitive components linked to pain. And, owing to its involvement in memory processes, the hippocampus may affect pain perception, and aspects of the brain associated with anxiety, notably the central amygdala and the basolateral amygdala, through connections with multiple brain regions, including the prefrontal cortex, hippocampus, hypothalamus, and brainstem areas, can modulate pain perception. As well, ventral tegmental–nucleus accumbens functioning, which is involved in the response to aversive stimuli and those related to reward may affect pain perception. Identifying the processes that promote chronic pain becomes an order of magnitude more difficult to determine given the involvement of a very wide range of neurotransmitters, neuropeptides, growth factors, and those associated with inflammatory processes.

Sex differences in pain perception

Females are more sensitive to painful stimuli (Mogil, 2016), are more often affected by chronic pain than men, and tend to be less responsive to analgesic treatments. The sexual dimorphism in chronic pain processes is conserved across species and may be relevant to the development of pain-reducing treatments in both sexes.

Based on analyses in rats, it seemed that the elimination of sex hormones in females by ovariectomy before puberty created the pathological pain phenotype apparent in males. Furthermore, males and females differ in the ways by which the immune system contributes to pain perception. Among males, microglia within the spinal cord are involved in pain perception, whereas T cells are more involved in this capacity among females (Sorge and Totsch, 2017). Similarly, following peripheral nerve injury, a male-specific inflammatory microglia subtype may be generated, and the transcriptional changes in some microglial populations corresponded with upregulation of the gene that codes for apolipoprotein (APOE). In humans, a subpopulation of microglia may be related to pain processes, and APOE polymorphisms are related to the presence of chronic pain (Tansley et al., 2022b).

Aside from promoting cell growth and synaptic plasticity, the neurotrophin brain-derived neurotrophic factor (BDNF) may be associated with neuropathic pain associated with the activation of glial cells within the spinal dorsal horn (Zhou et al., 2021b). Speaking of the sex difference associated with inflammation-provoked allodynia, disturbances of neuronal excitability in rodents occur within the dorsal horn among males but not females. This sexual dimorphism was also observed in the response of glutamate (NMDA) receptor activation by BDNF. Spinal cord tissue from human organ donors likewise revealed that the application of BDNF influenced neural excitation in superficial dorsal horn neurons obtained from males but not females (Dedek et al., 2022).

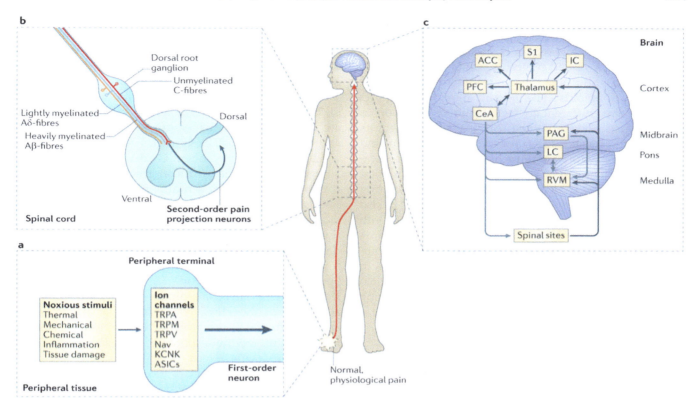

FIG. 12.2 Processes related to pain perception. (A) Peripheral nociceptive stimuli of various sorts will activate sensory neurons (first-order primary afferent neurons). The peripheral terminals of these neurons contain ion channels that include several "transient receptor potential" channel subtypes (TRPA, TRPM, and TRPV), sodium channel isoforms (Nav), potassium channel subtypes (KCNK), and acid-sensing ion channels (ASICs), which can cause membrane depolarization. (B) The resulting activation of nociceptive Aβ- and C-fibers triggers glutamate release at central synapses located within the spinal dorsal horn. The axons of these nociceptive neurons project to the brainstem and thalamic nuclei through the anterolateral system. The firing rate and the specific fibers activated code for the quality, intensity, and duration of the painful stimulation. Nociception may be influenced by several mechanisms, including glutamate and neuropeptides, such as substance P and calcitonin gene-related peptide (CGRP). Glutamate activates AMPA (α-amino-3-hydroxy-5-methyl-4-isoxazole propionic acid), kainate, and NMDA receptors, which differentially contribute to different types of pain. Nociceptive signaling at the level of the spinal cord can be altered by activation of local GABA and glycine inhibitory interneurons, excitatory glutamatergic interneurons, as well as descending serotonergic and noradrenergic efferents from the brain. The second-order spinal projection neurons connect to supra-spinal sites in the brainstem and thalamus, which then project to cortical and subcortical regions, which ultimately encode the perception of the multidimensional pain experience. (C) Several brain regions may contribute to pain perception, with each having somewhat different functions. As depicted in Panel C, neurons from the thalamus project to a variety of cortical and subcortical regions that govern sensory-discriminative [e.g., somatosensory cortex (S1)], emotional responses [anterior cingulate cortex (ACC), amygdala (CeA), and insular cortex (IC)], as well as cognitive processes [prefrontal cortex (PFC)]. As well, pain modulation occurs through brainstem sites (gray arrows), including the periaqueductal gray (PAG), locus coeruleus (LC), and rostral ventromedial medulla (RVM). *Source:* Grace, P.M., Hutchinson, M.R., Maier, S.F., Watkins, L.R., 2014. Pathological pain and the neuroimmune interface. Nat. Rev. Immunol. 14, 217–231.

Opioids and Opioid Receptors

The powerful analgesic effects of opioids come from their action on the endogenous peptides enkephalins, dynorphins, and endorphins, which stimulate mu (μ), delta (δ), and kappa (κ) receptors, as well as the less studied opioid receptor like-1 (ORL1). In addition to these receptors being triggered within the brain, opioid receptors have been identified on immune cells, which could influence pain perception, and inflammation that occurs within the CNS was associated with elevated μ receptor expression (Cuitavi et al., 2023).

As we all know, although opioids are effective in alleviating many types of pain, their addictive potential has had disastrous consequences, including a vast number of overdose deaths. In 2010 about 21,100 deaths in the United States were caused by opioids, but this dramatically increased, so 68,600 deaths were reported in 2020 and has continued to rise since then. While opioid deaths frequently stemmed from street use, a significant number of deaths resulted from prescription use to diminish chronic pain (Seth et al., 2018). Continued attention has been directed toward the development of better treatments for pain reduction, but without the addiction potential ordinarily generated by opioid-acting agents.

Considerable evidence has been amassed, indicating that μ-, δ-, and κ-opioid receptors play a prominent role in pain perception and the development of addictive properties of opioids. The μ- and δ-opioid receptors seemed to have different effects on emotional responses (Nummenmaa and Tuominen, 2018). While μ receptors might be responsible for features related to pain and addiction, δ-receptors are more aligned with the regulation of emotional responses. In addition to modulating pain sensitivity, κ receptors may be useful in buffering the effects of stressors (Jacobson et al., 2020). It also appeared that μ receptor activation was associated with the actions of cannabinoids, nicotine, and alcohol. Although ongoing efforts have attempted to create compounds that affect the different opioid receptors without the hazards that they ordinarily produce (e.g., Ehrlich et al., 2019), alternative pain-reducing strategies are being developed that do not involve opioid receptor actions.

Cannabinoids

The assertion that cannabis and cannabidiol (CBD) can reduce pain or at least reduce its unpleasantness has promoted research to determine its potential as a tool in the antipain arsenal. Pain associated with spinal cord injury was not diminished by cannabis (Thomas et al., 2022c) and the analgesic effects of cannabis on other forms of pain may have been overhyped. Yet, meta-analyses of randomized controlled trials revealed that cannabis could reduce chronic neuropathic pain by about 30% (Sainsbury et al., 2021), although there was considerable variability across studies. Noninhaled medical cannabis generally produced a small reduction of pain, enhanced physical functioning, and disturbed sleep quality was reported among chronic pain patients (Wang et al., 2021c). In fact, among individuals who used cannabis to reduce chronic pain, more than half reported that it was sufficiently effective to allow them to reduce the use of prescription opioid and nonopioid use, and over-the-counter pain medications, as well as physical therapy (Bicket et al., 2023).

The psychoactive component of cannabis, delta-9-tetrahydrocannabinol (THC), binds to CB_1 receptors within the paraventricular nucleus of the hypothalamus, hippocampus, prefrontal cortex, and amygdala. Upon being activated, the eCB system influences the release of Gamma-aminobutyric acid (GABA) and glutamate, thus inducing a variety of behavioral outcomes, including altered stress responses. The effects of THC on emotions could moderate the impact of painful stimuli, and the pain-reducing qualities of cannabinoids may stem from their effectiveness in reducing the unpleasantness rather than actually diminishing the pain sensation (De Vita et al., 2021). As pain perception is dependent on contextual factors and mood, the analgesic effects of cannabinoids may also be related to their actions on anxiety or its stress-buffering effects (Morena et al., 2016). This said, it is important to emphasize that while low doses of THC have an antistress action, the opposite effects are associated with higher doses (Childs et al., 2017), and the optimal dose likely varies across individuals.

Cannabinoids may inhibit GABAergic input from the amygdala onto periaqueductal grey projection neurons, thereby reducing pain sensations (Winters et al., 2022). Endocannabinoids may also interact with nerve growth factor and BDNF, thereby affecting pain sensitivity (Luongo et al., 2014). In addition, the link between amygdala and hypothalamic functioning involved in stress responses is modifiable by variations of eCB functioning. Specifically, the application of a CB_1 receptor agonist to the basolateral amygdala diminished stressor-provoked activation of the HPA axis, whereas pharmacological antagonism of CB_1 receptors increased basal HPA activity and increased the response to psychogenic stressors. The pain-reducing effects of cannabis may also occur through the activation of CB_1 receptors at brain sites that govern fear and pain processing (e.g., amygdala, hippocampus), and may be linked to neuroimmune interactions that stem from inflammatory changes.

Although it was initially thought that its positive actions were limited to reducing neuropathic pain, cannabis might diminish adverse actions brought about by inflammation. The analgesic effects of cannabis might come through the activation of toll-like receptors expressed on dendritic cells and macrophages, and a decline of the proinflammatory IL-17 producing T cells or the chemokine CCL-2 (also known as monocyte chemotactic protein1) (Fitzpatrick and Downer, 2017). There has indeed been a growing movement toward the use of medicinal marijuana to reduce pain associated with illnesses that involve inflammatory factors. These have included pain associated with multiple sclerosis and symptoms of Crohn's disease. Relatedly, cannabinoids influence brain processes that could affect brain-gut axis functioning, thereby modifying gastrointestinal symptoms, including pain (Sharkey and Wiley, 2016).

It had been hoped that the pain stemming from some types of cancer could be diminished by cannabis and anecdotal reports had supported this. In animals, cannabis and CBD seemed to be beneficial as an adjunctive cancer therapy and could diminish sleep problems and gastrointestinal disturbances engendered by chemotherapy, and even diminished cancer cachexia and anorexia that is so difficult to attenuate. Unfortunately, controlled studies in humans indicated that cannabis was largely ineffective in diminishing cancer-related pain. It was nonetheless maintained that if the THC:CBD constituents of cannabis are well balanced, it might be effective as an adjunct therapy to reduce cancer-related pain and to reduce intake of other medications (Aprikian et al., 2023).

A degree of tolerance develops to the pain-altering actions of CB_1 receptor stimulation, and preclinical studies suggested that treatments that target CB_2 receptors could diminish pain, but without the development of tolerance or adverse psychotropic effects since these receptors are primarily expressed on immune cells (Woodhams et al., 2015). In fact, in the presence of inflammation, CB_2 expression within the brain and in activated spinal cord microglia was elevated and CB_2 manipulations promoted antiinflammatory outcomes. In this regard, treatments that activate these receptors can suppress the proliferation and activation of microglia, and it is reasonable to believe that this approach could be adopted to diminish pain resulting from elevated inflammatory microglial activity (Xu et al., 2023a).

Inflammatory Processes

There is no question that inflammatory factors are associated with chronic pain conditions, and could serve as a link to comorbid depression, anxiety, or PTSD that is often apparent. Insults to peripheral tissue give rise to immune processes being activated, but the nature of the activated cells may differ with the type of insult or injury sustained. Preclinical studies indicated that the characteristics of the immune cells activated could predict changes in nociception. While peripheral inflammatory factors could certainly affect pain processes, centrally acting cytokines, including both IL-1β and TNF-α released from microglial cells, could promote hyperalgesia (Grace et al., 2014), which can be exacerbated by stressors. As well, by influencing excitatory glutamate and inhibitory GABA actions, both acute and persistent pain can be provoked (see Fig. 12.3). As expected, intrathecal administration of an IL-1 receptor antagonist (IL-1Ra) effectively diminished pain, and the broad-spectrum tetracycline antibiotic, minocycline, reduced chronic neuropathic pain, including that associated with peripheral injury, chemotherapy, and diabetes, likely owing to its action on inflammation (Zhou et al., 2018).

As one of the immune cells that engage pathogens early in response to pathogens, macrophages may contribute to pain processes by releasing proinflammatory cytokines that stimulate specialized primary sensory neurons that sense pain. This interaction is bidirectional in that nociceptors secrete neuropeptides and chemokines (e.g., CCL2) that activate macrophages, thereby exacerbating pathological pain. At the same time, macrophages may also contribute to pain resolution by releasing GPR37, which diminishes inflammatory pain by altering the antiinflammatory cytokine IL-10. In addition, interactions between macrophages and nociceptors may be mediated by microRNAs (e.g., let-7b and miR-711) that can activate nociceptors (Chen et al., 2020).

As much as opioids can curb some forms of pain, there may be a significant downside to the effects of opioids on inflammatory processes. Specifically, rats that had received a weeklong course of morphine treatment to diminish pain stemming from traumatic injury subsequently exhibited a more sustained chronic pain condition, possibly owing to elevations of TNF-α, IL-1β, TLR4, and NLRP3 at the site of spinal cord injury (Ellis et al., 2016). In essence, the short-term gain attributable to opioids was accompanied, literally, by long-term pain. A somewhat similar situation was found with the actions of antiinflammatory agents to mitigate pain. As increased inflammation is associated with elevated pain, treatments that reduce inflammation have pain-reducing actions over the short term. However, antiinflammatory agents may have the unexpected consequence of favoring persistent pain being present years later. How this occurs isn't entirely certain but a role for neutrophils may contribute to this type of paradoxical effect. Ordinarily, early in the course of inflammation, neutrophils play an important role in pain management, such that the initial upregulation of the inflammatory response had downstream effects (adaptive or compensatory) so that chronic pain did not develop. However, prevention of the neutrophil response (e.g., by antiinflammatory drugs) did not permit the beneficial changes to develop (Parisien et al., 2022).

Glial Functioning

As already mentioned, microglia play a fundamental role in pain processes, operating at the site of tissue damage in the periphery and at the dorsal horn within the spinal cord where microglia are responsive to cytokines and chemokines that are released when primary afferents are damaged (Salter and Stevens, 2017). We'll now dig a little deeper into the function of glial processes in pain perception.

Analyses of intracellular processes have been assessed to determine the processes that are related to pain, but it was uncertain how this came about without other sensations being affected. It seems that certain projection neurons that carry pain signals to the brain are surrounded by extracellular matrix structures (referred to as perineuronal nets), which can be degraded by activated microglia following peripheral nerve damage, leading to selective elevation of pain (Tansley et al., 2022a). As well, visceral pain has been attributed to the actions of spinal microglia and astrocytes (Long et al., 2022). Provoking gut inflammation causes prostaglandin E_2 release from enteric glial cells, which results in the sensitization of local neurons so that elevated responses occur to specific stimuli, thereby promoting abdominal pain. The impact of glial

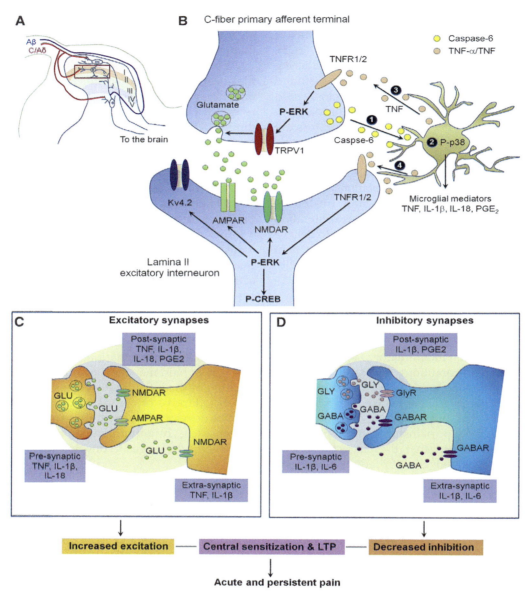

FIG. 12.3 Schematic illustration of microglial regulations of acute and persistent pain in spinal cord pain circuit via multiple mechanisms, including presynaptic mechanisms and postsynaptic mechanisms and regulations of both excitatory and inhibitory synaptic transmission (A) Schematic of pain circuit in the spinal cord dorsal horn (laminae I–V). Nociceptive input is carried by small C- and Ad-afferents (red), whereas tactile input is conducted by large Ab-fibers (blue). (B) Painful insults result in hyperexcitability of primary sensory neurons and increased release of caspase-6 from the central terminals of primary afferents (Step-1), leading to p38 phosphorylation (P-p38), activation (Step-2), and subsequent TNF release from microglia (Step-3). Activation of TNF receptors at presynaptic terminals causes glutamate release via ERK and TRPV1. Activation of TNF receptors at postsynaptic neurons also results in ERK phosphorylation (P-ERK) and activation. P-ERK drives central sensitization via positive regulations of NMDAR and AMPAR and negative regulation of potassium channel Kv4.2, as well as phosphorylation of the transcription factor CREB. In addition to TNF, activation of p38 via various microglial receptors, as shown in Panel B, also increases the secretion of the inflammatory mediators IL-1β, IL-18, and PGE2 to facilitate pain. (C and D) Regulation of excitatory and inhibitory synaptic transmission by microglia mediators at presynaptic, postsynaptic, and extrasynaptic sites. These regulations drive central sensitization, as characterized by decreased inhibitory synaptic transmission and increased excitatory synaptic transmission, as well as long-term potentiation in the pain circuit, leading to acute and persistent pain. In addition, BDNF release from microglia was also implicated in disinhibition and neuropathic pain. *Source: Chen, L., Teng, H., Jia, Z., Battino, M., Miron, A., et al., 2018. Intracellular signaling pathways of inflammation modulated by dietary flavonoids: the most recent evidence. Crit. Rev. Food Sci. Nutr. 58, 2908–2924.*

activation, however, only produces these actions in the presence of inflammation (Morales-Soto et al., 2023).

Once peripheral neuronal damage occurs, reactive gliosis is engendered, which comprises nonspecific reactive changes of astrocytes, microglia, and oligodendrocytes. As a result of cytokine release, immune cell infiltration and astrogliosis occur together with several other changes, including inflammatory mediators being released from reactive microglia, and activation of some toll-like receptors (TLR-2 and TLR-4) that may promote persistent pain (Lacagnina et al., 2018). Aside from actions within the periphery and spinal cord, cytokines released by brain microglia contribute to gliosis, and the gliopathy that occurs in the spinal cord and peripheral nervous system might contribute to chronic pain, including that seen in some disease conditions, such as autoimmune disorders (Mifflin and Kerr, 2017). Relatedly, the view was advanced that chronic pain is much like an autoimmune condition, in which autoantibodies that recognize self-antigens modulate the actions of nociceptive neurons, thereby promoting persistent pain (Lacagnina et al., 2021).

Peripheral damage can induce allodynia, but glial cells within cortical circuits were likely responsible for this action (Kim et al., 2016c) given that pain perception was modifiable by blocking the activity of glial cells. As well, neuropathic pain stemming from spinal cord injury, which is often intractable, has been linked to inflammatory processes, which can be somewhat abated by antiinflammatory agents (Walters, 2014). It also appears that certain types of astrocytes may cause damaging effects and induce chronic pain through the provocation of reactive astrogliosis, which comprises an increase in the number of astrocytes owing to the destruction of nearby neurons (Li et al., 2020b). In this respect, peripheral monocytes and microglia act synergistically in modulating chronic pain, and agents that prevent microglia from proliferating soon after injury could reduce pain (Peng et al., 2016).

It may be particularly significant that following an immune challenge, transcriptional activity, or epigenetic modifications within microglia, may be elevated so that the response to later challenges is augmented. For instance, priming (sensitization) of pain responses, which can be elicited by a variety of factors, such as stressor occurrences, could instigate chronic pain. Paralleling such findings, allodynia elicited by an endotoxin (lipopolysaccharide, LPS), could be augmented by a stressor, such as abdominal surgery, which could be attenuated by minocycline (Hains et al., 2010). Like several neurotransmitters and hormones, cytokine levels and functioning are subject to a sensitization effect (Anisman et al., 2003), making it possible that these inflammatory factors are related to the priming effects that have been reported concerning pain processes. In this regard, early life adversity can influence pain-related processes so that neurobiological functioning is altered, and later nociception is increased (Salberg et al., 2021).

Although inflammatory-related pain ought to diminish with a natural decline of inflammation, its resolution is not always accompanied by reduced pain. It seems that macrophages are actively involved in attenuating inflammatory pain, doing so by transferring mitochondria to sensory neurons. Specifically, M2-like macrophages can infiltrate the dorsal root ganglia where the cell bodies of sensory neurons reside, and together with other processes contribute to pain resolution (van der Vlist et al., 2022). From this perspective, treatments that interfere with M2 macrophage functioning may allow pain to persist.

Beyond these actions, an initial inflammatory bout associated with pain may instigate hyperalgesic priming, which increases the expression of a mitochondrial protein (ATPSc-KMT), resulting in mitochondrial and metabolic disturbances within sensory neurons. These actions may undermine pain resolution given that limiting these processes (e.g., by knocking down the expression of the gene for ATPSc-KMT) can diminish the development of chronic pain. As we've seen in so many other conditions, this may reflect a process through which the sensitization of inflammatory-related processes favors a switch from an acute to a chronic pain condition and facilitates pain reoccurrence at later times (Willemen et al., 2023).

Given the apparent involvement of glia and cytokines in pain processes, it was considered that they could potentially represent an ideal target for pain management. Modifying glial-neurotransmitter interactions could potentially initiate beneficial changes through the altered release of antiinflammatory factors, thereby precluding or limiting neurotoxicity and may diminish sensitized pain responses so that "normal" pain is experienced (Tiwari et al., 2014). As we'll see shortly, manipulating inflammatory processes has become a focus for pain management, with signs of success being apparent in some conditions.

To summarize, in response to tissue damage, a constellation of brain pro- and antiinflammatory cytokine changes occurs. If a persistent imbalance occurs in favor of proinflammatory factors, chronic pain may emerge, whereas treatments that increase antiinflammatory factors (i.e., IL-1Ra, IL-4, and IL-10) may diminish hypersensitivity. The link between glial cells and cytokines in relation to pain processes certainly ought to be considered within a broader context, and there is still information that needs to be uncovered concerning glia and pain processes that are unrelated to inflammation. This is especially the case since immune responses that occur within the nervous system are not only involved in pain provocation but may also contribute to its resolution (Fiore et al., 2023). In particular, T_{reg} cells that are essential in maintaining adequate balances between pro- and antiinflammatory factors can moderate pain processes (Bethea and Fischer, 2021), as might maresins that regulate macrophage phenotypes and thus affect inflammation. As well, resolvins and protectins obtained through various omega-3 fatty acids can diminish the production of interleukin-6, TNF-α, and VEGF, thereby affecting inflammatory pain (Chen et al., 2018). As further data are obtained concerning individual difference factors related to pain perception, it may become possible to direct treatment strategies appropriately. For instance, age is accompanied by longer-lasting inflammation, and might thus contribute to differences in pain perception. Likewise, sex, culture, and experiences may similarly operate in modulating pain-related inflammatory processes. Ultimately, these factors will need to be considered in developing therapeutic strategies that are effective for all affected individuals.

Microbiota and pain

Given the influence of microbiota on neuroendocrine, neurotransmitter, and inflammatory processes, it is hardly unexpected that gut processes may also be related to pain perception, particularly visceral pain. Gastrointestinal pain, such as that associated with irritable bowel syndrome, may persist even after tissue injury is no longer present, possibly reflecting hypersensitivity of serotonergic enterochromaffin (EC) cells, which can be directly contacted by gut microbiota (Bayrer et al., 2023). In addition, gut microbiota metabolites can influence the blood-brain barrier so that the infiltrating immune cells can affect brain microglia processes (Guo et al., 2019). Conversely, antibiotics that severely diminish microbiota can reduce somatic pain that stems from diabetes, chemotherapy, and nerve injury (Ma et al., 2022).

Neuropathic pain in rodents was associated with a distinct set of microbial changes in which specific microbes were elevated and others reduced (Chen et al., 2021e), and the inflammatory changes resulting from gut dysbiosis may instigate pain. Consistent with this view, visceral hypersensitivity occurs in germ-free mice, accompanied by elevated gene expression of cytokines and with Toll-like receptors within the spinal cord, which normalized following recolonization with microbiota from ordinary mice. It is significant that the visceral sensitivity was associated with a greater volume of periaqueductal grey, and diminished volume of the anterior cingulate cortex, both of which have been linked to pain perception (Luczynski et al., 2017).

Gut dysbiosis early in life can have proactive effects that influence visceral pain in adulthood (O'Mahony et al., 2017a), and in female rodents, early-life stressors could have effects on adult visceral pain that varied with the estrus phase (Moloney et al., 2016). It is believed that visceral pain in such instances occurs through the effects of specific gut bacteria on TRPV receptors, together with inflammatory factors, protease release, polyunsaturated fatty acids, and short-chain fatty acid (SCFA) production. As well, lipids, such as N-acylethanolamine (NAE) and SCFAs are fundamental in altering inflammatory processes, thereby influencing pain mechanisms (Russo et al., 2017). It may still be a bit early to be certain, but treatments that act against pain, such as opioid and endocannabinoid variations, could potentially involve gut microbiota changes (O'Mahony et al., 2017b).

Emotional Processing Affects Pain Perception

Pain processing is influenced by cognitive appraisal processes, and it appears that the anterior insula and the anterior cingulate cortex serve as part of the salience network involved in the interpretation of impending pain (Uddin, 2015) and in physiological states associated with future pain (Livneh et al., 2020). As we saw in our discussion of depression and PTSD, the insula serves as a hub where sensory information is processed and is integrated with emotional, cognitive, and motivational processes, and it may be fundamental in the transition from acute pain states to that of chronic pain (Labrakakis, 2023).

Consistent with this, the insula is activated in response to noxious stimuli and direct stimulation of the insula creates pain sensations. Conversely, the insular cortex connectivity among chronic pain patients is diminished with the administration of pain medications (Labrakakis, 2023). The expectation of pain, which is accompanied by increased activity within the anterior cingulate cortex and other frontal regions thought to underlie appraisal and decision-making processes, may underlie the perceived unpleasantness of a stimulus. Among patients who had electrodes implanted in the orbitofrontal cortex (to which the insula projects), chronic refractory neuropathic pain could be predicted by neural activity within this region, supporting its fundamental role in pain perception (Shirvalkar et al., 2023).

The nociceptive somatosensory system comprises two interconnected branches. One of these, the exteroceptive branch, is involved in sensing and appraising threats and is integral in initiating actions to avoid possible injury. When injury is incurred, the interoceptive branch becomes involved to elicit protective behavioral measures that foster healing and diminish pain perception (Ma, 2022). It was posited that the insular cortex functions to integrate the information from these branches. An alternative accounting regarding the functions of the insula concerns the view that both a conscious and nonconscious pain state exists wherein nociception can occur subconsciously without apparent pain perception. The subconscious is involved in injury avoidance, but when pain sensation is sufficiently intense (passing an individual's threshold) conscious pain emerges (Baliki and Apkarian, 2015).

To some extent, the neural circuitry associated with controllable pain can be distinguished from that of uncontrollable pain, much as stressor controllability can differentially influence the functioning of several neurotransmitters. Healthy participants were exposed to a task in which they could adjust the temperature applied to their hand, whereas in the uncontrollable situation, the temperature level was identical but was not under the participant's control. Even though the temperature was identical in both situations, the temperature was rated as being higher in the uncontrollable situation than during the controllable pain trials. This perception was paralleled by elevated activation of pain-processing regions such as the insula, anterior cingulate cortex, and thalamus. Moreover, increased connectivity occurred between the anterior insula and medial prefrontal cortex in the uncontrollable situation, whereas greater negative connectivity was apparent between the dorsolateral prefrontal cortex and insula in the controllable situation. To be sure, these findings might not necessarily reflect the actions under natural conditions involving more intense pain. Nevertheless, the findings speak to the involvement of different circuits associated with subjective pain perception.

Although somewhat controversial, it was suggested that physical and emotional pain shared neural circuitry involving the anterior cingulate cortex (Lieberman and Eisenberger, 2015). When participants received a nasal spray (placebo) that would allegedly diminish their emotional pain, this was associated with a change in their neuronal activation profile (Koban et al., 2017). Further to this, just as physical pain influences neuronal activity within the anterior cingulate cortex, it seems that feelings of social exclusion (rejection) are accompanied by increased neuronal activity within this region (Lieberman and Eisenberger, 2015).

The insular cortex is activated in association with empathy related to pain felt by others (Soyman et al., 2022). Simply witnessing rejection experienced by another person, which presumably elicits empathy in the witness (emotional pain), also activates aspects of the cingulate cortex, as does viewing pictures of a romantic ex-partner and thinking about rejection by this partner (Kross et al., 2011). The psychological aspects of pain perception on others (empathy) may involve glutamatergic projection from the insula to the basolateral amygdala (Zhang et al., 2022a).

Such findings have been consistent with the view that the anterior cingulate cortex is part of a "neural alarm system" that is sensitive to both physical and social threats (Lieberman and Eisenberger, 2015). The finding that transcranial direct stimulation applied to the right ventrolateral prefrontal cortex reduced the pain of social rejection (Riva et al., 2015) was consistent with the belief that emotional pain is modifiable by activation of neuronal processes, even if these are somewhat diffuse.

Consistent with the view that psychological and physical pain are linked, treatment with a mild antiinflammatory acetaminophen that reduces physical pain diminishes the psychological pain of social rejection while concurrently reducing the neural responses within the dorsal anterior cingulate cortex and anterior insula (DeWall et al., 2010). Similarly, acetaminophen could reduce empathy associated with viewing another person's pain (Mischkowski et al., 2016). Interestingly, as much as acetaminophen could reduce social pain, this occurred primarily among individuals who expressed high levels of forgiveness (Slavich et al., 2019). It turned out that acetaminophen could also diminish the affective response to positive experiences that others enjoyed (Mischkowski et al., 2019), which raises the possibility that acetaminophen influences the affective component tied to social behavior irrespective of its valence.

Although physical and psychological pain elicited similar brain activation profiles, detailed analyses revealed that they were not identical, just as differences in connectivity were apparent in other brain regions. Even though the dorsal anterior cingulate cortex may be relevant to "survival-relevant goals," activation of this brain region is also apparent in response to nonstressful conditions, including attention, emotion, reward expectancy, skeletomotor, and visceromotor activity (Wager et al., 2016). Thus activity within the dorsal anterior cingulate cortex should not be considered to uniquely reflect either physical or emotional pain, especially as different forms of pain might engage somewhat different neural circuits. This said, even if different challenges have distinct pain profiles, they might have overlapping features, and emotional pain can modify different physical pain sensations through actions at the anterior cingulate cortex and insula. After all, the insula is involved in numerous functions, has been implicated in mediating depressive disorders, and may be responsible for the frequent comorbidity between depression and pain perception (Doan et al., 2015).

In addition to the anterior and dorsal cingulate, the midcingulate also appears to be involved in nociceptive processes. The position was taken that this region does not mediate pain sensations, but by acting through a broad cortical and subcortical network, particularly serotonergic projections to the spinal cord, the midcingulate may contribute to sensory hypersensitivity (Tan et al., 2017). The progressive changes that can occur in serotonergic functioning within this descending pathway might contribute to the transition from acute pain to a more troubling chronic pain condition.

Placebo Responses

Placebo effects comprise positive responses to treatments that don't have direct organic actions on physiological processes. Typically, placebo effects are discussed in the context of drug responses, but they should also be considered in nonpharmacologic treatments, including mechanical or electrical devices to reduce pain perception, acupuncture needles (e.g., when inserted into inappropriate locations), as well as faith healing. Placebo effects can be exceptionally powerful, to the extent that the placebo response can often match the effectiveness of low morphine doses in dealing with mild or moderate levels of pain. Furthermore, a portion of the remedial effects of "genuine" treatments, such as those of antidepressants, is attributable to placebo effects. Although such outcomes are commonly discussed in relation to mood changes and pain perception, they have also been reported concerning agents that purportedly act as muscle relaxants or drugs that influence blood pressure. Remarkably, symptoms of neurological conditions, such as Parkinson's disease, can be modestly abated, possibly owing to increased dopamine release within the striatum (e.g., Quattrone et al., 2018).

From a research perspective, a placebo effect is problematic as it confounds the actions of a genuine treatment. For that matter, simply recruiting participants for a drug trial, or interactions that occur among participants in such a trial, is sufficient to alter outcomes, which has considerable bearing on how data are interpreted concerning the actual efficacy of therapies. As most researchers and clinicians know, individuals frequently modify their behavior when they are aware that they are being observed (Hawthorne effect), which can influence the effects of drug treatments and placebo responses. Yet, a case can be made that understanding the basis of placebo may facilitate their use in clinical practice. This may be evident in a placebo run-in phase of a clinical trial that may inform the effectiveness of subsequent drug treatment.

As much as placebos may have an important place in treatment, their usefulness shouldn't be overstated as only 25%–50% of patients display analgesic responses, and the placebo response for antidepressants is still lower. Many factors play into the effectiveness of placebos, including optimism and altruism, previous experiences with medications, and the presence of certain genes. The expectancy of positive outcomes, even if these are not explicit (individuals might not consciously be aware of these expectations), is especially important in creating a placebo response. For instance, if patients experienced side effects that were provoked by the treatments, they were more likely to believe they were in the drug arm of a placebo-controlled trial, and thus more likely to show a positive outcome (Berna et al., 2017). Conversely, in some instances, nocebo responses may occur, in which ordinarily effective drugs have no positive effects or when symptoms worsen or harmful effects emerge among patients expecting that negative effects would arise.

The entire social milieu that goes with treatments (the presence of a doctor or a nurse, and the hospital environment) may contribute to a placebo response emerging.

For that matter, having a trusted doctor (or with whom the patient identified) was associated with greater pain relief (Losin et al., 2017). Likewise, when patients were treated by a physician who expressed support and empathy, a degree of pain reduction was achieved that was accompanied by altered neuronal activity within the dorsolateral and ventrolateral prefrontal cortex, as well as somatosensory regions (Ellingsen et al., 2023). Just as contextual and environmental stimuli are potent in affecting the reoccurrence (reinstatement) of drug addictions as well as the reestablishment of depressive and PTSD symptoms, these same features may contribute to placebo effects.

The specific environmental factors that instigate a placebo response can be subtle, so that patients may be unaware of them. For example, pain relief was greater among patients informed when a morphine drip began than in patients who were unaware of when this occurred. Moreover, simply seeing another patient obtain pain relief following treatment resulted in their own response to this agent being enhanced, supporting the suggestion that social comparisons, social interactions, and social learning can promote a placebo response (Benedetti, 2014). Furthermore, pain reduction obtained through a placebo treatment for several days persisted even after patients had been informed that they had been receiving a placebo. Likewise, even when it was made clear to participants that they were in a placebo condition (in a test of analgesia), they responded as if they had received an active pain-relieving treatment (Rosén et al., 2017). Also, patients who were aware that they were taking a placebo in conjunction with a standard treatment reported a sizable reduction of lower back pain, again pointing to the importance of the general environment and the patient-doctor connection in promoting pain relief.

Neurobiological Correlates of Placebo Responses

Given the multimodal attributes of pain relief, it was of interest to determine whether particular brain changes accompanied placebo responses with the view that this could inform the development of novel targets to reduce pain. As it turned out, placebo analgesia responses were accompanied by a rich network of brain changes, including activation of pain and stress-related subcortical neuronal processes, such as the connections between the midbrain periaqueductal gray and limbic brain regions. Marked neuronal changes were also detected at brain sites that govern how pain is construed (e.g., anterior cingulate, prefrontal, orbitofrontal, and posterior insula) and in brain regions associated with the motivation to take action (Zunhammer et al., 2021). As well, placebo responses involved the integration of neuronal

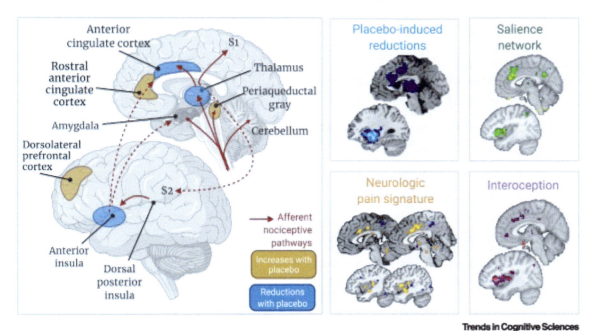

FIG. 12.4 Placebo analgesia is associated with variations in brain responses to noxious stimuli that can be visualized in humans using fMRI or positron emission tomography (PET) imaging. (Left) The regions that are most commonly activated by painful stimuli include the anterior cingulate cortex (ACC), insula, primary, and secondary somatosensory cortex (S1 and S2), and thalamus. Each of these is targeted by afferent nociceptive pathways and contains nociceptive neurons. fMRI studies that compare placebo administration with a control treatment (i.e., no expected pain relief) indicate that pain-related responses in the ACC, thalamus, and anterior insula are reduced with placebo (blue), whereas placebos elicit increases in activation in modulatory regions (gold) including the dorsolateral prefrontal cortex, rostral ACC, and the opioid-rich periaqueductal gray. (Upper middle) Meta-analysis of studies of placebo analgesia indicates reliable reductions in the dorsal ACC, anterior insula, and thalamus (Atlas and Wager, 2014). (Lower middle) Despite placebo-induced reductions in pain-evoked responses in a subset of pain-related regions, placebos do not elicit reliable modulation of the neurologic pain signature (NPS), which is a brain-based pattern that can reliably distinguish between responses to painful and nonpainful stimuli and is sensitive and specific to pain. This suggests that placebos might modulate nonspecific affective and cognitive processes rather than affecting nociception. (Right) Studies of placebo analgesia and other forms of pain modulation must carefully distinguish pain from salience processing (upper right) and interoception (lower right) because pain is highly salient and requires interoception, and the brain networks that process pain overlap substantially with the salience network and interoceptive processing. Images of placebo-induced reductions (upper middle) were adapted with permission from Atlas and Wager (2014). *Source: Atlas, L.Y., 2021. A social affective neuroscience lens on placebo analgesia. Trends Cogn. Sci. 25, 992–1005.*

activity within the prefrontal and anterior cingulate, and the periaqueductal gray (Vachon-Presseau et al., 2018), possibly reflecting emotional and cognitive processes as well as those associated with executive functioning and reward processes that affect pain perception. As described in Fig. 12.4, analgesia elicited by placebo treatment was paralleled by activation of brain regions fundamental for emotional and anxiety responses (e.g., anterior cingulate cortex, anterior insula, amygdala) as well as reward processes (nucleus accumbens) (Atlas and Wager, 2014; Atlas, 2021). It is especially significant that a clinical placebo response could be predicted based on resting-state brain connectivity determined by fMRI, which can be important in clinical applications. For example, in patients with chronic knee osteoarthritis, connectivity involving the right parahippocampal gyrus predicted placebo responders (95% correct) as well as the magnitude of the response (Tétreault et al., 2016).

Pain-related placebo responses were thought to be mediated by endorphins, as the opioid antagonist, naloxone, attenuated the placebo response on postoperative dental pain. However, several types of placebo responses can be provoked by diverse treatments, involving varied neurobiological mechanisms. Some of these might contribute to feelings of relief or reward, whereas others might be more aligned with expectancies, or to the actual nature of the disturbance being treated (e.g., Nestler and Waxman, 2020). As alluded to earlier, the placebo responses may develop owing to classically conditioned effects on varied neurobiological processes, and effects related to inflammatory immune functioning are no exception. As well, the individual's beliefs about a treatment they were receiving not only influenced neuronal functioning but also produced discernible differences related to the dose of the treatment that was believed to have been received. These findings have implications for the effectiveness of other manipulations to alter behaviors, such as those related to antidepressant therapies, and the efficacy of treatments to diminish substance use disorders.

No, you definitely don't feel my pain

Most people are empathetic to others who are experiencing pain, but there are instances in which the pain of others is dismissed and may even be met with unsupportive responses. Even though pain can emerge because of neuronal sensitization so that benign stimulation may elicit pain (allodynia), reports of chronic, unexplainable pain, may be met with skepticism. To an extent, the same holds for psychosomatic (psychogenic) pain which is diagnosed when other causes of pain have "seemingly" been eliminated. Unfortunately, stigmatization is often encountered by individuals experiencing psychogenic pain, which may exacerbate poor mood in pain sufferers.

For some time, chronic fatigue syndrome and myalgic encephalomyelitis (CFS/ME) had fit this framework. This illness is characterized by severe fatigue, widespread pain, depression, and a constellation of other symptoms that weren't attributable to a specific cause. The diagnosis of this condition was hampered because its etiology was uncertain, and an objective diagnostic test didn't exist. Patients with CFS were said to have their stress-response systems continuously on alert, which ultimately led to physical and psychological disturbances. Understandably, many patients became increasingly embittered, especially as they had to weave their way between attitudes (even blunt accusations) that implied that they were either malingering or that the illness reflected psychological instability. Not infrequently, patients were stigmatized by their own GPs and some psychiatrists seemed to be reluctant to deal with patients with "undiagnosable" conditions that might be secondary to personality disorders (e.g., borderline, histrionic), and hence exceptionally difficult to treat.

Skip ahead just a few years. It became clear that the occurrence of CFS/ME symptoms was not *just* due to psychological disturbances. Chronic fatigue syndrome was associated with bilateral white matter atrophy and disturbances within the arcuate fasciculus (fibers that connect the temporal and inferior parietal cortex to the frontal cortex), which could potentially be used as a biomarker of CFS/ME (Zeineh et al., 2014). As well, CFS/ME was accompanied by reduced grey matter in pain-processing brain regions and could be influenced by physical activity. This syndrome was also associated with inflammatory immune activation that stimulated the basal ganglia, which is involved in reward processes, cognitive functioning, and motor acts. Indeed, a constellation of cytokines, including IL-1β, IL-6, TNF-α, IFN-α, and TGF-β, were implicated in the emergence of CFS symptoms (Montoya et al., 2017). As expected, systemic inflammation provoked by the administration of a low dose of LPS increased pain sensitivity relative to that evident among placebo-treated individuals. The elevated pain sensitivity was accompanied by diminished neuronal activity within the rostral anterior cingulate and prefrontal cortex but was elevated within the anterior insular cortex, which contributes to affective and interoceptive pain (Karshikoff et al., 2016). As many of these cytokines were similarly associated with autoimmune disorders (e.g., rheumatoid arthritis, systemic lupus erythematosus, and Sjögren's disease), and other illnesses in which fatigue was a prominent feature, it is conceivable that they were responsible for that aspect of the illnesses. Parenthetically, in many ways, features of long COVID are like those of CFS/ME and it was suggested that chronic inflammation may be a shared underlying mechanism for both these conditions (Tate et al., 2022).

Aside from cytokine variations, CFS/ME was accompanied by a low diversity of gut bacteria, including a deficiency of butyrate-producing bacteria (Guo et al., 2023a), and based on their gut bacteria composition, it was possible to distinguish individuals who were affected with CFS/ME from those who were not. Several studies indicated that CFS/ME was associated with gut dysbiosis and alterations of specific gut and oral microbiota (Lupo et al., 2021), and interactions between immune functioning and microbiota may contribute to CFS/ME (Vogl et al., 2022). The gut bacterial disturbances could also account for the frequent comorbidity that exists between CFS/ME and intestinal bowel disease, although the two conditions could also be differentiated from one another based on gut bacterial composition (Giloteaux et al., 2016).

A recent analysis of CFS/ME that involved the collaboration of 75 researchers and clinicians across numerous disciplines indicated that the disorder involved multiple intersection factors, many of which were consistent with earlier reports (Walitt et al., 2024). Their findings indicated that CFS/ME was associated with immune system functioning such that naïve B cells were elevated, whereas memory B cells were reduced. Moreover, a marker of immune exhaustion (PD-1 functioning) on CD8 T cells was elevated just as this was observed with chronic infection. Aside from immune alterations, microbial diversity was diminished, and dopamine metabolites were reduced in cerebrospinal fluid, which was associated with cognitive disturbances and poor motor performance. As well, autonomic functioning was disturbed, reflected by the resting heart rate being elevated, heart rate variability reduced, and heart rate responses to exercise disrupted. It was suggested that the ME/CFS may have originated with infection, perhaps reflecting the persistent presence of the pathogen in the body. Importantly, it was concluded that psychiatric disorders were not a major component of ME/CFS.

So, we're now at the point where CFS/ME is widely believed to be a genuine illness, even if its etiology is still not well understood. A sure cure for CFS/ME isn't available, but CBT and graded exercise therapy were helpful for some patients (Sharpe et al., 2015). The findings that CFS/ME seemed to be a medically accepted illness didn't initially lead to a decline in the stigma that so often existed. To deal with the stigma and to legitimize it as a physical illness, in 2015 CFS/ME had its name changed to "systemic exertion intolerance disease." In essence, people who had been considered neurotic or had personality disorders could take solace in their physical illness being legitimized.

Like CFS, we have been seeing a similar scenario play out concerning multiple chemical sensitivity (MCS), a condition associated with low-dose chemical exposures. Even though this condition has been associated with altered functioning within diverse brain regions and was comorbid with several illnesses (e.g., fibromyalgia, asthma, migraine, and stress/anxiety), it has frequently been considered to come about because of psychological disturbances. Too often patients were stigmatized and ostracized, even being denied accommodation for their disability, and educating the public and health workers has been an uphill battle (Molot et al., 2023).

Next Steps in Pain Management Methods: Advances and Caveats

The management of chronic pain hasn't been nearly as effective as one would have hoped, and disparities have been reported concerning the under-treatment of pain based on race and gender (Hoffman et al., 2016). Together with the risks associated with opioid treatments, this has encouraged a broader search for pharmacological and nonpharmacological methods to reduce chronic pain.

Drug Treatments to Diminish Pain

The efficacy of many drugs varies with the nature of the pain being experienced. For instance, whereas corticoid injections at the site of musculoskeletal injuries can be used to reduce pain stemming from local inflammation, pain stemming from nerve damage might be best treated by compounds that affect GABA neuronal functioning, such as pregabalin (Lyrica). The anticonvulsant medication gabapentin also turned out to be effective in many cases of chronic neuropathic pain and chronic CFS/ME. Unfortunately, pregabalin has been associated with increased occurrence of deaths, and gabapentin can promote experiences like those of opioids and benzodiazepines and has become increasingly misused.

Alternative methods of dampening pain have included ways of affecting afferent pain fibers and their connection to nerves in the spinal cord dorsal root ganglion. Specifically, manipulations of a receptor ($Na_v1.8$)

that is involved in the transmission of pain signals between nerves have been examined to determine whether this could diminish pain perception. A compound, provisionally dubbed VX-548, which potently blocks the sodium channel associated with $Na_v1.8$ signaling, diminished acute pain owing to surgical procedures for abdominoplasty (tummy tuck) (Jones et al., 2023b), but it remains to be determined whether the treatment is effective for other forms of pain. Like $Na_v1.8$, manipulations of $Na_v1.7$ sodium ion channels, which influence the endings of pain-sensing nerves (i.e., nociceptors), can be regulated by a protein CRMP2 that can dial down sodium entry, thereby diminishing pain. These effects, which have been observed in several species, can diminish mechanical allodynia associated with nerve injury and peripheral neuropathy stemming from cancer chemotherapy (Gomez et al., 2023).

Acute pain experiences

The most widely used pain management treatments comprise nonopioid drugs containing acetaminophen (e.g., Tylenol), NSAIDs, or COX-2 inhibitors (also an NSAID), even though their effectiveness is often limited. While NSAIDs reduce pain associated with inflammation related to some conditions, several detailed reviews of the literature indicated that acetaminophen had only a modest effect on acute lower back pain and tension-type headaches, and only transiently reduced postoperative pain and did so in only a third of patients (e.g., McNicol et al., 2016). To make matters worse, many people inadvertently misuse acetaminophen, sometimes leading to serious health consequences. Some agents may pose risks to cardiovascular health, and their overuse may provoke gastrointestinal damage and bleeding, disturbed kidney health, and may have serious effects on children in whom liver damage and death could occur (Reye's syndrome) when taken while infected with a viral infection (e.g., influenza or chicken pox).

The NSAID Ibuprofen (Advil, Motrin) fairs relatively well in diminishing some cases of acute pain, particularly those linked to inflammation (Varrassi et al., 2020). However, it only helps a minority of people in diminishing tension headaches and is about as (in)effective as acetaminophen in alleviating acute back pain. The combination of acetaminophen and ibuprofen provides superior postoperative (dental) pain relief than either treatment alone (Derry et al., 2015). Acetylsalicylic acid (aspirin) is also used to manage pain, fever, and inflammation, although its effectiveness as a pain reliever is thought to be less than that of ibuprofen. Its ability to control fever, owing to its actions on prostaglandins is well established, and it is used to reduce acute and chronic inflammation, although only short-term use is recommended given its potential for inducing intestinal bleeding.

Opioids

It is widely assumed that severe pain can be relieved by drugs from the opioid family of drugs (e.g., morphine, diamorphine, buprenorphine, oxymorphone, oxycodone, hydromorphone, and fentanyl). While opioids can be effective in diminishing acute pain (e.g., following surgery), they may only be modestly effective in the treatment of chronic noncancer pain (Busse et al., 2018), including lower back pain for which opioids are frequently prescribed (Jones et al., 2023a). As mentioned earlier, just a few days of opioid treatments can put glial cells into a heightened state of activation, provoking inflammatory responses within the spinal cord. The combination of the initial pain signal and the later opioid treatment gives rise to a signaling cascade, including activation of IL-1β release, which promotes increased responsiveness of neural processes, leading to increased pain perception that can persist for months. So, the very treatments meant to diminish pain intensity may have the effect of increasing pain duration (Grace et al., 2016).

Because of the addiction potential, the CDC prepared guidelines for physicians concerning opioids being prescribed for adult patients experiencing chronic pain (apart from that associated with cancer treatment, palliative, and end-of-life care) with an eye toward consideration of when the benefit of opioid use outweighs risks, and whether alternatives to opioids are a reasonable option (Dowell et al., 2016). Unfortunately, in some countries where long delays in treatment are often encountered, patients may be kept on opioid-acting agents for extended periods, thereby promoting the development of addictions.

Not every person who is treated with an opioid will develop an addiction and many variables determine who will or will not be a victim. Personality characteristics and previous experiences with drugs have been tied to substance use disorder, as have the drug dose used, the route of administration, how quickly the compound's rewarding effects appear, what other drugs may be on board, the context in which the drug is taken, and the expectations concerning the effects that will be obtained (Volkow et al., 2017). In the latter regard, the rewarding effects attributable to the drug as well as its addiction potential may be less notable when taken for a medical condition relative to that apparent when it is used to obtain pleasurable feelings.

The opioid epidemic and the large number of deaths (increasingly attributable to fentanyl, which is orders of magnitude more powerful than other agents) is hardly news. Overdoses are the top cause of death among younger individuals, outstripping car accidents. The media has done its work in bringing the problems into public focus, but unfortunately, it may have the unintended consequences of denying opioid-based medications to those who are in great need. It would be ideal to have a set of biomarkers that could inform pain specialists who would benefit most from specific treatments and who would be most at risk for developing an addiction.

Dealing With Opioid-Based Treatments

Efforts have been made to develop new and better treatments to diminish pain yet limit the odds of substance use disorder. Compounds were developed based on the transformation of opioid agents so that they carried reduced tolerance liabilities. As indicated earlier, drugs have been developed to antagonize the μ-opioid receptor so that the analgesic effects of opioids would be maintained, but the euphoria and reward induced by these opioids would not be present (e.g., Sutton et al., 2016). Still, another method has been that of using compounds that act as Nociceptin/orphanin FQ peptide receptor agonists and that of a μ-opioid receptor peptide (MOP). Thus pain-relieving effects would be obtained with lower doses of the MOP, but with diminished risk of addiction (Ziemichod et al., 2022). Other agents have been developed that block calcium channels fundamental for electrical nerve conduction, or act on both calcium channels and endocannabinoid receptors, thereby diminishing inflammatory pain and tactile allodynia. Other approaches that are being evaluated have involved manipulations of TRPV4 and TRPA1 receptors, which are thought to sense particular types of painful stimuli involving inflammation, such as joint pain and abdominal pain, respectively (e.g., Giorgi et al., 2019).

Itch and scratch

Typically, an itch appears transiently, leading to a scratch reflex. The cause for an itch might be due to a mild skin irritation, or a moderate allergic reaction following contact with substances that are released from plants, such as poison ivy or poison oak. Aside from body and head lice, and insect (mosquito) bites, itch can be provoked by viral illnesses (e.g., chicken pox, herpes), skin conditions (eczema), or severe allergic reactions (e.g., in response to bee or hornet stings). Burns and sites of surgical wounds may also be accompanied by a hard-to-relieve itch, and in some chronic conditions, such as kidney or liver failure, diabetes, and cancer, incessant itchiness may appear.

The neural circuitry associated with itch comprises some of those involved in pain perception, so that altered pain sensitivity is present around an itchy area, and individuals who are congenitally insensitive to pain tend not to feel itch. Moreover, stimuli that promote pain and itch activate spinal neurons, and both sensations are transmitted through the spinothalamic tract, and involve the activation of many of the same brain regions (Akiyama and Carstens, 2013).

At the same time, the processes related to itch can be dissociated from those related to pain, as treatments that reduce itch do not necessarily affect pain perception (Pitake et al., 2017).

Receptors exist that are sensitive to pain, itch, or both, and while antihistamines reduce itch stemming from mild allergic reactions, they hardly affect itch stemming from diabetes or kidney failure. Likewise, opioids, which reduce pain, may cause a worsening of itch, again suggesting that the two involve independent processes. Nonetheless, there has been interest in capitalizing on the links between itch and pain in the development of new targets to treat chronic pain.

Itch is elicited by activation of somatosensory neurons that express a particular ion channel TRPV1 (transient receptor potential cation channel subfamily V member 1). It seems that a neurotransmitter, natriuretic polypeptide b (Nppb), is present in a subset of receptors present in TRPV1 or TRPV4 neurons. Upon these receptors being activated, vigorous scratching responses are elicited (Chen et al., 2016f), but even in the absence of Nppb, the hormone gastrin-releasing peptide (GRP) could elicit marked scratch responses, pointing to its involvement in itch/scratch processes (Mishra and Hoon, 2013). A compound, triazole 1.1, a κ-opioid receptor agonist, can attenuate pain and itch, and does so without producing euphoria, hence limiting the addiction potential (Brust et al., 2016). Several other compounds have since been developed that operate as κ agonists to reduce pain and itch without undesirable side effects (Mores et al., 2019). Such compounds would obviously be ideal for pain prevention given the current opioid epidemic.

Sensitization of TRPV1 and TRPA1 has been implicated in MCS, in which nociceptive pain is a symptom, possibly owing to the perpetuation of elevated inflammation (Molot et al., 2023). Whether MCS is related to itch/scratch perception and responses are uncertain, but the broad symptoms associated with this condition and their role in pain processes make this a possibility.

Antidepressants for Pain Relief

Serotonergic mechanisms, both centrally and peripherally, may influence pain processes, and several 5-HT receptors within the CNS play a fundamental role in this regard (Cortes-Altamirano et al., 2018). As well as the involvement of 5-HT_{1A} and 5-HT_{2A} receptors, striatal dopamine D_2/D_3 receptors may contribute to the top-down regulation of neuropathic pain (Martikainen et al., 2018). Likewise, serotonergic involvement in diverse pain syndromes (e.g., irritable bowel syndrome, fibromyalgia, and migraine) may be moderated by estrogen functioning (Paredes et al., 2019). Descending serotonergic neurons

may also be involved in chronic primary pain (CPP) that arises in association with several diseases, even though they are not associated with identifiable structural or specific tissue pathologies (Tao et al., 2019).

Antidepressants may diminish back pain, postoperative pain, and neuropathic pain (Ferreira et al., 2023). An extensive meta-analysis that included 176 studies, most of which comprised randomized controlled trials, assessed the relationship between 25 antidepressant therapies and pain perception associated with fibromyalgia, neuropathic pain, and musculoskeletal pain. Positive effects were largely restricted to duloxetine and to a lesser extent milnacipran (Birkinshaw et al., 2023). In reports in which pain reduction was achieved by antidepressants, this occurred in tandem with diminished depressive symptoms, supporting the importance of mood states in modifying pain perception. As it happens, reductions of chronic pain and depression are regulated by a specific gene (RGS9), and RGS9-2, the protein for which it codes, and antidepressants that regulate chronic pain and depression can influence neuronal functioning within the nucleus accumbens, possibly by affecting RGS9-2 processes (Mitsi et al., 2015).

This isn't the sole process by which psychological factors or antidepressants may affect chronic pain sensitivity. Glutamate likely plays a role in pain perception, and treatments that diminish glutamate receptor functioning (notably mGluR1) could be useful to diminish chronic pain. As the inhibitory actions of GABA have also been implicated in pain perception (Luo et al., 2021), it may also serve as a target for pain management.

Psychedelics in pain management

Because of the actions of psychedelic actions in attenuating depression and PTSD, the question arose concerning how broad the effects of these agents might be, including whether they could act against pain. In this respect, a low nonpsychedelic dose of LSD reduced pain and unpleasantness assessed in a cold pressor test pain that persisted over a 5-h period (Ramaekers et al., 2021). This was in keeping with earlier studies indicating that LSD could diminish phantom limb pain, as well as reports that LSD could diminish migraine and cluster headaches, and cancer pain (Dworkin et al., 2022).

It may be recalled that LSD and psilocybin have their hallucinogenic actions by stimulation of the 5-HT_{2A} receptor, and these agents may affect pain processes by acting on these receptors and by disturbing functional connections in brain regions that subserve chronic pain (Castellanos et al., 2020). While the effects of psilocybin and LSD may stem from direct antinociceptive actions, it is equally possible that they have their effects through psychological processes, such as increasing tolerability, promoting pain acceptance, or engendering a sense of spirituality (Zia et al., 2023). As well, perceived pain reduction may be related to reduced depression that often accompanies chronic pain, or by affecting synaptic plasticity and functional connectivity between brain regions that contribute to pain perception (Elman et al., 2022).

It may be too early to prescribe psychedelic agents in the treatment of chronic pain. However, as pain is often intractable, alternative approaches are necessary. Exploring the analgesic actions of LSD, psilocybin and other hallucinogenic agents warrants further attention, and addressing how these agents operate might contribute to understanding concerning pain processes generally and the mechanisms that contribute to comorbid disorders related to pain.

Manipulating Inflammatory Process to Modify Pain

With the realization that inflammatory factors could affect brain or spinal processes, thereby influencing pain perception, new drug targets became available to reduce pain by diminishing inflammation. These comprised agents that affected local proinflammatory signaling, stimulation of antiinflammatory processes, inhibition of specific immune mediators, and blocking cytokine or chemokine receptors (e.g., using agents to inhibit TNF-α or antagonize IL-1β receptors). These methods aren't without problems, as inhibiting cytokine functioning may leave individuals vulnerable to some illnesses. Still, for the moment, these approaches are among the many techniques being considered in the search for chronic pain treatments.

Inhibition of Proinflammatory Signaling

Several antiinflammatory drugs (minocycline, propentofylline, ibudilast, and methotrexate) diminished allodynia in preclinical models of neuropathic pain (e.g., Fujita et al., 2018). By acting as a toll-like receptor 4 antagonist they may have positive effects on pain related to diabetic neuropathy, CRPS, and neuropathic pain stemming from multiple sclerosis (Patten et al., 2018), as well as in treating migraine headaches (Liu et al., 2022b). It was similarly observed that purinergic factors could potentially affect pain perception. Purinergic receptors influence numerous cellular functions, such as promoting the proliferation and migration of neural stem cells, regulation of immune functioning (Cekic and Linden, 2016), and are effective in exciting microglia and cytokine release (Tsuda, 2017). Following peripheral nerve damage, pain hypersensitivity may develop owing to the activation of purinergic P2X4 receptors located

on microglia. The SNRI duloxetine (the best of the antidepressants to reduce pain) has an inhibitory effect on P2X4 receptors and may thereby serve as an antiallodynic agent (Stokes et al., 2017b).

The purine nucleoside adenosine, which is involved in energy processes, may also affect pain processes. Altering the activity of adenosine A3 receptors, which stimulates GABA, was able to turn off the pain signal (Salvemini and Jacobson, 2017). Likewise, pharmacologically targeting A_{2A} and A_{2B} receptors on immune cells and glia increased the presence of the antiinflammatory IL-10 and decreased the levels of the proinflammatory TNF-α, thereby attenuating peripheral nerve injury-nociceptive hypersensitivity for an extended period and could diminish postoperative and visceral pain (Zhou et al., 2023c).

Repetitive transcranial magnetic stimulation and deep brain stimulation

Better known for its effects in ameliorating depression, repetitive transcranial magnetic stimulation (rTMS) has also been used to alleviate lower back pain, neuropathic pain associated with spinal cord injury, and phantom limb pain (Moisset et al., 2016). Pain related to mild traumatic brain injury could be diminished by rTMS, although studies of this sort usually involved a small number of participants. Several reports also attested to the efficacy of rTMS for both episodic and chronic migraine, but this has been somewhat controversial, and a meta-analysis indicated that little benefit was obtained through magnetic brain stimulation (Shirahige et al., 2016).

There have been limitations concerning the efficacy of rTMS, often stemming from procedural factors. The precision of the coil's placement may be imperfect, and in the early studies, the current penetrated only a small distance into the human brain, and in some instances, portions of the cortex could not be adequately stimulated. However, improved methods have been developed that provide better outcomes.

Deep brain stimulation (DBS), which has been used for various pathological conditions, was similarly effective in ameliorating neuropathic and phantom limb pain, as well as intractable chronic headache and pelvic pain (Alamri and Pereira, 2022). These actions could be achieved by targeting specific brain regions, such as the sensory thalamus, periaqueductal gray, periventricular grey matter, and the anterior cingulate cortex. As well, DBS that targeted the ventral striatum/anterior limb of the internal capsule had positive actions in relation to the affective sphere of pain among poststroke pain syndrome patients (Lempka et al., 2017). Given that depressive mood can exacerbate chronic pain sensitivity and reactivity, it may be profitable to modify the affective component of pain syndromes to diminish severe discomfort.

Behavioral and Cognitive Approaches to Reduction of Chronic Pain

As pain perception is influenced by psychological and experiential factors (e.g., previous pain experiences, mood state), there has been a push for therapies based on psychological and cognitive treatments, particularly methods to modify perspectives and expectancies of pain relief. The effectiveness of these strategies is, to a considerable extent, dependent on patients having an action-oriented attitude, including being ready for change and taking charge of pain management. Treatment adherence is also fundamental in determining whether positive effects will be obtained, and pain-related beliefs, particularly self-efficacy, were particularly important. Some of these approaches have been moderately useful and have been beneficial when combined with pharmacological attempts at pain reduction.

Hypnosis may promote pain relief, and self-hypnosis was also effective as a pain-reducing procedure. The actions of hypnosis could not entirely be attributed to placebo-like effects but were nonetheless related to expectancies created in highly suggestible people (Thompson et al., 2019). Not surprisingly, the positive impact of hypnosis varies with the nature of the pain experienced. Few patients with spinal cord injuries obtain pain relief, somewhat better results were promoted in patients with multiple sclerosis, and relatively good outcomes were seen in patients with phantom limb pain (Jensen et al., 2014). Hypnosis was also effective in pain associated with irritable bowel syndrome as well as pain secondary to breast cancer. Several brain changes accompany the pain reduction associated with hypnosis, including diminished "connectivity" within circuits related to pain and pain perceptions, including the interior insula (Jensen et al., 2017a).

Guided imagery, which entails individuals visualizing and focusing intensely on a place or event that promotes positive thoughts and feelings as well as feeling relaxed and at peace, can transiently reduce pain (Carpenter et al., 2017). The usefulness of this procedure is seen when it is a component of integrative therapies that include music, art, massage, therapeutic play, and distraction. In combination with other approaches, guided imagery has been used with some success in children and adolescents dealing with cancer, and in diminishing symptoms among patients with CFS/ME.

Biofeedback training typically comprises feedback based on autonomic nervous system responses (heart rate, blood pressure, and galvanic skin response) or brain activity (EEG) to gradually promote particular physiological or behavioral outcomes. This procedure has been used to diminish anxiety and to promote calmness, which may be linked to altered heart functioning. Feedback can be used to alter activity within the anterior cingulate cortex, but this was not necessarily accompanied by changes in chronic pain perception (Bucolo et al., 2022).

Consistent with the often-used adage of "exercise as medicine," forms of exercise may reduce neuropathic pain. Exercise engagement in over 10,700 individuals assessed in two waves 7–8 years apart indicated that exercise was associated with increased pain tolerance, with greater exercise having the most pronounced actions (Årnes et al., 2023). Multiple processes were implicated in accounting for such effects, such as the actions at peripheral nerves, the spinal dorsal horn and dorsal root ganglion, brainstem, and higher brain centers (Kami et al., 2017). The pain-reducing effects of exercise might likewise stem from action on brain inhibitory processes, and variations of glutamate, serotonin, and opioids (Lima et al., 2017). In addition, variations within the mesolimbic reward system may contribute to changes in pain perception and modulation as might the effects on immune and inflammatory processes (Rice et al., 2019). However, it was cautioned that it is important to distinguish between voluntary exercise that may be rewarding from the effects of forced exercise that may be aversive (Kami et al., 2017). Thus the analgesic effects of exercise in humans may vary based on the individual's appraisal of exercise being a chore versus a positive experience.

Cognitive Behavior Therapy, Acceptance and Commitment Therapy, and Mindful Meditation

Several brain regions implicated in pain perception involve cognitive and emotional processing, and pain brought about by various physical insults can be exacerbated or diminished by cognitive processes and emotional responses (see Fig. 12.5; Bushnell et al., 2013). Catastrophizing, negative appraisal, depressive affect, and the tendency to inhibit the expression of anger have all been associated with elevated pain perception among individuals with chronic illnesses, such as cancer. Conversely, diminished pain among cancer patients could be promoted by psychosocial interventions and by promoting empowerment, augmenting feelings of self-efficacy, and through the availability of adequate social support. Each of these factors, individually, might have modest effects on pain perception, whereas multimodal treatment strategies that include specific medications, as well as cognitive therapies, can provide still better pain-attenuating actions.

Several studies indicated that CBT and mindfulness were effective in diminishing some forms of pain (Pardos-Gascón et al., 2021). Being faced with chronic pain may promote anxiety and depression, which can aggravate pain perception, and a degree of pain relief could be achieved solely through treatments to reduce anxiety and depression. Because pain perception and psychological disturbances may be intertwined, perceived pain reduction using pharmacological treatments might be enhanced among patients who initially obtained psychological therapies to diminish disturbed emotional states.

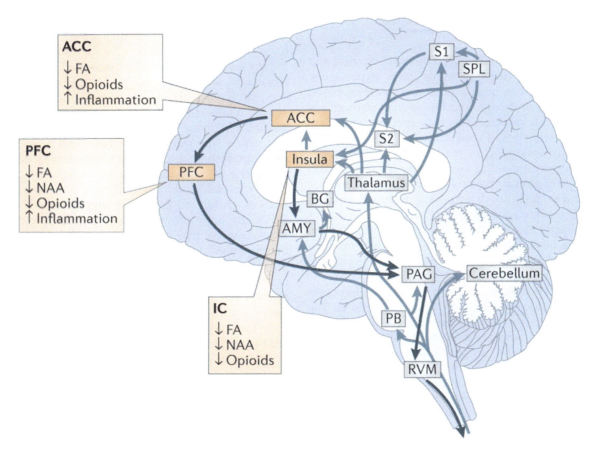

FIG. 12.5 The three cortical regions that consistently show decreases in grey matter are the anterior cingulate cortex (ACC), prefrontal cortex (PFC), and insula (IC). Studies have also identified changes in white matter integrity in these regions; such changes are manifested by decreased fractional anisotropy (FA), which suggests that there is a decrease in white matter health. Molecular imaging studies show decreases in opioid receptor binding in patients with chronic pain in all three regions. Studies using in vivo proton magnetic resonance spectrometry show chronic pain-related decreases of the neuronal marker N-acetyl aspartate (NAA) in the frontal cortex and the insula. Finally, rodent studies show increased neuroinflammation in the ACC and PFC. Black arrows show the descending pathways; gray arrows show afferent pain pathways. *Source: Bushnell, M. C., Ceko, M., Low, L.A., 2013. Cognitive and emotional control of pain and its disruption in chronic pain. Nat. Rev. Neurosci. 14 (7), 502–511.*

Mindful meditation affects pain perception, possibly through changes in neuronal activity within the insula and anterior cingulate cortex, which not only govern attention and emotional responses but also interoceptive awareness (Jensen et al., 2014). As with so many other actions of this method, the outcomes observed varied between novice and expert practitioners; however, even limited mindfulness training could affect hippocampal and parietal cortical functioning, which mapped onto a decline in pain perception (Zeidan et al., 2015).

Despite reports pointing to the benefits of mindful meditation, a systematic review and meta-analysis indicated that many of the published reports were of mixed methodological quality and that there was, in fact, limited support for the position that mindfulness interventions were useful for chronic pain (Bawa et al., 2015). This conclusion notwithstanding, for some individuals, mindfulness meditation could represent a viable alternative to opioid-based therapy for chronic pain (Jacob, 2016), and may help patients taper their use of high doses of opioid-acting agents.

As we saw in discussing placebo effects, the effectiveness of analgesic agents is dependent on patient beliefs about their pain, and their perceived self-efficacy about pain management. Negative attitudes at the beginning of treatment predict relatively unsuccessful outcomes using CBT, whereas pain relief was obtained more readily among individuals who reported a high level of social support and who were actively involved in treatment, and when CBT was offered as part of a chronic pain management program. In essence, maximal positive effects could be attained if patients maintained the belief that their pain could be managed and that they actively engaged in achieving pain relief, rather than being passive bystanders waiting for relief to come. At the same time, patients must avoid the negative cognitive perspectives that might have developed owing to a track record of failed treatments. Moreover, when pain relief is obtained, it is important that patients attribute this to cognitive therapy.

In acceptance and commitment therapy (ACT) to regulate pain, individuals are trained to become more aware and more focused on their therapeutic goals, and to develop proactive behavioral patterns (e.g., thinking or *acting* based on the presence of pain) instead of simply focusing on reduction of their pain symptoms. This approach was useful in attenuating features secondary to chronic pain (e.g., sleep disturbances) or actual pain perception, and produced benefits in diminishing pain sensitization syndromes, possibly being mediated by psychological flexibility, optimism, self-efficacy, or adherence to values (Galvez-Sánchez et al., 2021).

Painful memories and memories of pain

Memories of pain can influence decisions that patients make about later medical procedures, and re-exposure to the context in which pain was experienced may promote elevated sensitivity to pain. In this regard, chronic pain was accompanied by epigenetic changes within the prefrontal cortex that could serve as a "genomic memory" to influence later responses to painful stimuli (Alvarado et al., 2015).

Adverse childhood events, which so often predict later depression, were associated with increased occurrence of chronic pain in adulthood as well as pain-related disability (Bussières et al., 2023). It seemed that these stressful experiences were associated with FKBP5 gene expression changes and altered right hippocampal volume, which were tied to chronic musculoskeletal pain (Lobo et al., 2022). Furthermore, early life infection reflected by elevated CRP levels was associated with increased spinal cord excitability together with increased evoked brain activity in response to tactile and noxious stimulation. The increased sensitivity persisted after the inflammation had diminished, supporting the animal studies that intimated that immune disturbances during early life can have lasting effects on pain sensitivity (Cobo et al., 2022).

Pain memories may be fundamental in acute pain transitioning to a chronic pain state. The question arose as to whether memories of pain perceptions in humans are modifiable much as other types of memory can be altered. Specifically, these memories can be modified just as memories related to PTSD can be altered (i.e., having the memories be recalled and at the same time administering a treatment that dissociates the memory of the pain from the affective component). It was reported that hyperalgesia was labile so that reactivation of spinal pain pathways could promote pain processes in a manner like memory reconsolidation, being modifiable by a protein synthesis inhibitor administered during a narrow time window following a second exposure to pain-provoking stimuli (Bonin and De Koninck, 2014). In the same way, among children with strong memories of earlier pain, having them repeatedly recall positive aspects associated with the experience can foster beneficial effects (Marche et al., 2016).

Pain and fear are catastrophically entwined. Fear memories can cause hyperalgesia in patients with chronic pain. In rodents, cells that are active in the medial prefrontal cortex are thought to form an engram of fear memories that may be involved in chronic pain. Modifying such fear memories can influence pain perception, raising the possibility that targeting the fear engram could potentially diminish chronic pain in humans (Stegemann et al., 2023). Just as individuals may retain memories of their pain, negative or positive experiences in response to an analgesic medication influenced the behavioral responses to subsequent treatment, which were accompanied by changes of neuronal activity within brain regions that encode pain and analgesia (Kessner et al., 2014).

There is another form of memory that is relevant to pain perception. Injury experienced in early life may cause "neonatal nociceptive priming" so that exaggerated pain perception is felt upon injury later in life. It appears that macrophage activation in response to the initial insult may be fundamental in accounting for this process, operating through the p75 neurotrophic factor receptor (nerve growth factor receptor) in affecting inflammatory processes. Remarkably, the "pain memory" could be induced by transferring macrophages from a previously injured animal to a naïve host. The possibility exists that specifically targeting p75 neurotrophic receptors macrophage could diminish the priming effect, thereby limiting the augmented pain induced by subsequent injury (Dourson et al., 2024).

Alternative, Complementary, and Integrative Medicine

For those who experience chronic pain that isn't ameliorated by standard pain treatments, it isn't uncommon for alternative medicines to be sought. In part, this may reflect desperation, but also because individuals want to feel a semblance of control over their destiny. In fact, alternative treatments have been used to assuage pain more than for any other ailment. Despite many reports debunking alternative treatments and the hazards that these approaches entail, for better or worse, alternative medicines and complementary medicine have maintained their popularity in some circles.

Pain relief was sought through acupuncture, spinal manipulation, mobilization, cupping, and massage techniques. The relief these treatments provide is typically limited (possibly reflecting a placebo response). These are only a few of the great number of unproven therapies that have been offered for virtually any ailment that exists. We've reached the ironic point where, in some circles, effective vaccines are considered ineffective and even dangerous, whereas homeopathy and other outrageous procedures are readily accepted even though they have repeatedly been found to be ineffective.

When approaches such as mindfulness are combined with other treatments, often as part of Complementary and Alternative Medicine (CAM) approaches within hospital settings, the likelihood of perceived pain reduction is increased (e.g., Deng, 2019). Should patients choose such approaches, the downside isn't particularly bad so long as they aren't adopted as an alternative to those that could have genuine positive effects. Given how difficult it is to treat chronic pain with the current methods available, it's likely that alternative therapies will continue to be used. One supposes that if patients pay for these treatments out of pocket, then the cognitive dissonance created may keep them as believers.

Concluding Comments

Chronic pain is unquestionably an exceptionally debilitating condition, and patients often have little recourse in efforts to remedy this. The toll taken by chronic pain is considerable, not only affecting the quality of life but also the development of other illness conditions (e.g., depression, addiction). Inflammatory processes have been implicated as being fundamental in the emergence of some forms of pain, and these same processes, as we've seen repeatedly, may contribute to comorbid psychological disorders as well as the emergence of physical illnesses. Although some forms of chronic pain cannot easily be thwarted, in many instances they can become manageable through behavioral and pharmacological treatments. In this respect, lifestyle endorsed may influence chronic pain perception. For instance, as we've seen repeatedly, dietary factors can influence inflammatory processes, and increasingly it was maintained that maintaining an antiinflammatory diet can limit chronic pain.

The most common means of limiting pain are through agents with high addiction potential. It is doubly unfortunate that in some countries, patients with chronic pain that is treatable (e.g., surgery for disc problems) often encounter remarkably long wait lists, all the while consuming opioid-based agents, thereby increasing the risk for addiction. Socialized medicine is the norm within Western countries (the United States being only partway there), which certainly has improved the health condition of many people, although inequities persist as a function of wealth, race, culture, and social settings. When treatment becomes unavailable (or severely delayed) for serious, but treatable pain conditions, often leading to addiction, this signals the existence of problems within medical systems that need to be resolved, sooner rather than later.

Substance use related to pain has been a serious problem that has led to ruined lives and frequent deaths. As a result, restrictions have increased on the use of several medications, but this has led to the unintended consequence of many people with chronic pain being denied access to the medications that they need. This has created an unacceptable state in which individuals may resort to suicide to end their pain. Indeed, suicidal ideation occurred in 18%–50% of people experiencing chronic pain. Although numerous risk factors exist that may increase suicide risk (e.g., family history of suicide, comorbid depression), it was argued that "mental defeat" may be a key ingredient in this respect (Themelis et al., 2023). Although the guidelines that have been offered for the use of opioids to treat chronic pain have been commendable in many respects, in some instances the policies have been inconsistent and frequently exceeded the restrictions on the use of these agents (Kroenke et al., 2019). Finding ways around the existing quagmire won't be easy, but abandoning the many million patients who suffer intractable pain isn't the solution. It's time for scientists, clinicians, and government agencies to rethink the approaches to pain management.

CHAPTER

13

Autism

The changing face of autism over a century

Autism is the most prevalent developmental disorder of behavior in the current biomedical era. This condition has been recognized for many years and only recently became a separate diagnostic entity in major psychiatry classification systems [e.g., International Classification of Diseases (ICD), Diagnostic and Statistical Manual (DSM) of the American Psychiatric Association]. Since its initial use, the term "autism" has experienced a variety of dramatic transformations. It was first used in 1911 by the German psychiatrist, Eugene Bleuler, who applied the term to schizophrenia. In his view, "autism" was a mental strategy or defense, in which unwelcome thoughts and circumstances were supplanted by fantasy and hallucinations. In this conceptualization, "autism" denoted a rich and vibrant inner mental life, if perhaps marked by the turbulence and psychopathology that is often witnessed in schizophrenia. Later, in the 1960s, the term took on new meanings, describing mental life that was bereft of fantasy and imagination. This turnaround in what was meant by "autism," created the notion of poverty or the absence of normal mental life. Indeed, this appeared to be how the term was used when applied to a particular syndrome observed in children by Kanner in 1943 and elaborated upon subsequently (Kanner, 1946). He was struck by a unique constellation of behavioral features in 23 children who presented with marked social withdrawal and an inability to form conventional relations with people, beginning very early in life. Additional features, such as isolation from others, a proclivity for sameness and repetitive behavior (including echolalia), cases of mutism, and a lack of reactivity to surrounding events (which he noted, led parents to suspect deafness), all converged on Kanner coining what he saw as "early infantile autism." Interestingly, it was noted that symptoms overlapped with those observed in

schizophrenia and may have initially eclipsed the emergence of autism as a unique, developmental disorder, which did not receive its own separate entry until the third version of the DSM in 1980. Even today, clinicians may still grapple with defining whether someone has autism or early-onset schizophrenia.

Features of Autism Spectrum Disorder

A diagnosis of "autism" or "autism spectrum disorder" (ASD) now encompasses a far less restrictive conceptualization than it had in earlier years, referring to a range of social and emotional behaviors that are incompletely developed and/or inaccurately or inappropriately expressed.[1] Intelligence, cognition, and motor function are similarly incorporated into this terminology, widening the range of potential problems that might be identified in children eventually diagnosed with ASD. The DSM-5 applies the broad term ASD and incorporates conditions previously considered as distinct entities: Aspergers's disorder, childhood disintegrative disorder, Rett's disorder, and pervasive developmental disorder. What unifies these conditions are the core symptoms of (1) deficient social communication and interaction and (2) repetitive patterns of behavior, interests, and activities.

The prevalence of autism has increased over the past several decades. This was prominent from the 1970s to the 1990s, partially driven by increased awareness of childhood developmental delays, including those involving language and social behavior. Clinicians swept up by this interest, expanded the range of potentially relevant symptoms thought to constitute autism, thus casting a wider diagnostic net, which likely accounted for the dramatic increase in the perceived prevalence of autism.

---

[1] The term is derived from the Greek "auto" for self. The suffix "–ism" implies a state of self-directed actions, and this very much captures the isolated nature of the autistic individual, who fails to engage with others, and in extreme cases, directs harmful attention to themself.

The Immune System
https://doi.org/10.1016/B978-0-443-23565-8.00013-2

Copyright © 2025 Elsevier Inc. All rights are reserved, including those for text and data mining, AI training, and similar technologies.

For example, in 1992, for every 10,000 children examined, 19 cases of autism were recorded in the United States, a number that spiked to 90 (per 10,0000 children) in 2006, and more recently stands at around 100 per 10,000 children, varying considerably across countries and ethnic groups (Zeidan et al., 2022). According to the CDC, the frequency of ASD in 2020 had climbed still higher and occurred four times more often in boys than in girls (Maenner et al., 2023). There has been some debate concerning the frequency of autism, although it is broadly accepted that inaccurate estimates can have serious policy repercussions.

The sharp rise in numbers has given way to speculation that we are in the midst of an "autism epidemic," and explanations for the spike vacillate from overly inclusive diagnostic criteria to the contribution of environmental and genetic risk factors. All are likely to be relevant to varying degrees, and an intensive research agenda has focused on determining the factors that favor the occurrence of ASD. Much of this focuses on identifying its etiology, including the presence of mutations in candidate genes. This approach has a solid foundation, given that autism and autism-like traits have strong familial associations. While individual gene variants provide little risk, combinations of many different gene variants contribute to ASD within a range of 15%–50% (Vorstman et al., 2017). Moreover, frank neurodevelopmental disorders linked to genetic mutations (e.g., fragile X syndrome; tuberous sclerosis, and Rett's disorder) present with behavioral symptoms that overlap with those observed in autism, although they can be distinguished from one another based on their neurophysiological signatures (Neklyudova et al., 2022). While this supports the notion that ASD is driven by a strong genetic contribution, this position must be tempered by the likelihood that nongenetic factors play an important role, since any affected individual can display a unique set of genetic mutations.

Aside from genetic influences, ASD etiology may also encompass multiple environmental factors, including chemical toxicants, environmental pollutants, and infectious disease (Bilbo et al., 2018), and occurs more often with greater maternal and paternal age. As well, specific diets and nutrients have been implicated in the emergence of ASD, possibly owing to their microbial effects (Sorboni et al., 2022), and specific diets could potentially influence core symptoms of the disorder (Yu et al., 2022). Not surprisingly, there is no single factor that has emerged as a game-changing explanation for the development of ASD. This is likely the case because of the heterogeneity of autism. Indeed, a survey of the many reviews on specific aspects of autism can be dizzying in the range of factors that may or may not be important in contributing to this developmental disorder. In this chapter, after briefly reviewing some of the history and major positions maintained regarding the etiology and neuropathology of autism, we will address factors relevant to the immune system.

The immune system is now considered a potential variable that is either dysregulated in autism or contributes in uncertain ways to the etiology of the disorder. We will highlight several different approaches taken to examine the relationship between the immune system and autism. Although autism is a distinctly human disorder, animal research has proven an ally in generating conceivable hypotheses for how neurodevelopment may either be facilitated or impaired by immune alterations that occur around the perinatal period. This research approach will be addressed again in Chapter 14, as behavioral animal models of both disorders are quite similar.

What is Autism: Today!

One feature of autism that distinguishes it from childhood-onset schizophrenia is that it is usually evident by the second or third year of life, and thereafter is accompanied by a range of developmental delays in different compartments of behavior, such as language and motor function. In about 35% of cases, however, developmental progress and acquisition of important communication and motor skills appear to achieve expected and normal milestones, but then, usually between 18 and 36 months, marked and persistent regression to poor functioning appears. A prominent feature of autism is impaired social behavior, characterized by aloofness, disinterest, or odd and unexpected styles of interaction. Children also display poor recognition or perception of the intentions and feelings of others, which together may reflect disturbed social cognition. This represents one of the three core features of autism, or "triad of impairments," identified, which comprise social interaction, communication, and repetitive/stereotypic behavior, which vary somewhat with age and sex (Van Wijngaarden-Cremers et al., 2014). This understanding of autism has endured to the present day.[2]

Social interaction has been a particularly strong research and clinical focus, given its importance in learning, affective experience, and communication. As such, it influences the creation of various representations of the self across different social contexts. It is here that deficits or deviations from normative aspects of behavior can be

---

[2] In the DSM-5, social interaction and communication are combined as one core feature. However, given the heterogeneity of behavioral problems presented by ASD children, it is simply a starting point to an inevitable parsing process.

particularly distressing to parents. Moreover, research on the genetics and neurobiology of social cognition and/or social behavior has considered the identification of factors important in exercising trust (and conversely distrust), promoting affiliative behavior, perceiving faces, and differentiating between biological organisms and inanimate objects. The human brain is essentially wired to initiate and maintain social behavior, and social interactions might be a basic need, so individuals will crave social interaction when deprived of it (Tomova et al., 2021). However, in ASD, the neurocircuitry associated with social interactions, primarily comprising the amygdala, orbital frontal cortex, temporoparietal cortex, and insula, may not be operating adequately (Weston, 2019). This view was based on a substantial number of studies in humans, and many animal studies have modeled autism-like problems by focusing on social behavior. There is more to say about this but suffice for the moment that several studies reported changes in animal social behavior that were linked to immunogenetics, as well as to key cytokines (Filiano et al., 2017).

Brain Development and Autism: Behavioral Deficits Linked to Specific Circuits

As already mentioned, the onset of autism typically occurs before the age of 2 years, when the development of the brain is still dynamic, and the full assembly of circuits and synaptic connections is years away from its final state. Indeed, children can vary in their rate of development, and therefore, appearances can be deceiving. The preternaturally gifted eventually fall back to the norm, while the seemingly delayed or immature may work their way to the head of the pack. Consequently, determining whether a 2-year-old child has autism can be difficult, and requires diagnostic restraint, although experienced clinicians presented with a full display of the core symptoms (delays in language and social interaction) can confidently determine suspected autism by ages 2–3 years. Analyses of brain scans have raised the possibility that diagnosis can be comfortably determined within the first 2 years (Hazlett et al., 2017). However, this study, like several others, involved participants from ASD-prone families, and thus predictions were made on the likelihood of a pending appearance of autism in the scanned infants, making it less likely that false positives would occur. The use of brain scans is based on the long-standing belief that autism is a neurodevelopmental disorder in which there is abnormal brain growth and brain size. This said, a systematic review that comprised 17 neuroimaging and 43 behavioral studies indicated that the prediction of ASD is beset by considerable variability across studies, and consequently the sensitivity and specificity of the measures varied accordingly (Geng et al.,

2020). It was suggested that this might have been attributable to the heterogeneity concerning targeted populations (i.e., general population vs. high-risk groups), and the age of children when the features were collected. As such, it is still premature to rely on these measures for clinical appraisals but may nonetheless form the core of multimodal features that might enhance prediction accuracy. In this respect, the use of artificial intelligence (AI), machine learning (ML), and deep learning (DL) techniques has appreciably enhanced ASD diagnostic abilities based on imaging procedures (Helmy et al., 2023). This could presumably be improved through the incorporation of other neurobiological predictors and behavioral measures.

Early support for neurobiological underpinnings of ASD came from postmortem and electroencephalographic (EEG) brain analyses that revealed alterations in cell numbers, cell morphology, impaired neural circuits, regulatory function, neurochemical stability, and regional and functional interconnectivity (e.g., Sperdin et al., 2018).

The physical substrates for variations in functional connections within and between brain regions are the quantity of synaptic input and dendritic spine numbers (i.e., spine density). Functional connectivity MRI cannot address this level of refinement or resolution, but can measure the frequency and intensity of dynamic changes—or oscillations—that occur across distinct cortical and subcortical neuronal groups. This can generate time-stamped correlations that provide some notion of the functional connections operating within or between various brain sites during a specific behavioral event, be it cognitive, emotional, or sensorimotor. It was thought that in autism the connections between brain regions were overly busy or synaptically dense (i.e., "over-connectivity"), whereas, at another level, there was insufficient or weak connectivity (i.e., "under-connectivity"). Evidence has been offered for both scenarios, with under-connectivity evident for more distant connections (e.g., anterior-posterior connections within the brain), whereas over-connectivity appeared to be more prominent within localized areas. While this understanding has become the norm, confusion and various inconsistencies were noted, since intrinsic functional over-connectivity can vary according to intraindividual characteristics during data collection. For instance, whether participants close or open their eyes during resting state data collection in fcMRI can reveal or dissolve any intrinsic connectivity variations. In this respect, differences in connectivity patterns between ASD and control participants vary across the cingulate gyrus, based on whether their eyes are closed or open (Nair et al., 2018). While these procedural details may have a significant impact on how to interpret functional connections in ASD subjects, there appears to be little dispute that coordinated activity within the brain is different for ASD individuals. The challenge is in knowing why this happens.

The neural eccentricities noted in ASD underlie the behavioral deficits that occur at specific developmental stages. For example, language development progresses along a well-recognized trajectory that involves vocabulary building, sentence construction, grammatical rule development, and conceptual expression and understanding—all initiated during infancy by active listening and extraction of word sounds and their meaning. In children with autism, significant delays in language development and expression frequently occur (Baird and Norbury, 2016), as do delays and abnormalities in motor development, such as alterations in gait, and performance of repetitive or perseverative movements. The motor abnormalities are likely linked to known disturbances in the striatum and the cerebellum. The latter is responsible for fine motor coordination and temporal resolution of movements. However, the cerebellum has important connections to other regions of the forebrain, including the striatum (or basal ganglia) and the frontal cortex. Together with some of the motor deficits seen in autism, cerebellar dysfunction is considered a prominent neuroanatomical feature of the disorder (Okada et al., 2022).

Finally, the core feature of autism, which comprises a deficit in social behavior, has received particular attention and was instrumental in encouraging the development of the relatively young field of social cognition. This line of inquiry has been approached from multiple perspectives. One obvious strategy, when considered in the context of autism, is to understand the neurobiology of social behavior. Deficits of social behavior among ASD children are characterized by a poverty of eye contact and a lack of directed attention and/or awareness related to mutual interactivity. This represents a failure of engagement and cooperative behavior among children with autism, or at the other extreme, exhibiting overt displays of aggression, which can take the form of self-injurious behavior. The affectless state or overt angry outbursts belie problems with emotion regulation, which are consistent with studies showing disturbances in the development of the limbic system, including regions associated with impaired social judgment and social perception. In neurological patients who sustained damage to the amygdala, more trusting interactions and/or perceptions of strangers are common. Thus, it was suggested that in ASD, dysregulation of the amygdala may contribute to deficits in social interaction (Sato et al., 2020).

The amygdala receives multiple inputs from brain areas that process sensory information, as well as those involved in assigning meaning to information (e.g., the hippocampus and the anterior cingulate cortex), and executing decisions based on this meaning (e.g., inputs from the prefrontal cortex). Since the amygdala interacts reciprocally with these brain regions, decisions and actions are dependent on the quality of these interactions. The application of this knowledge to social cognition and autism has minimized regional bias and single locus attribution when reaching for neuroanatomical and neurobiological explanations for autism. Nonetheless, the well-documented cerebellar abnormalities, as well as recent evidence of abnormal cell numbers and synaptic connections in specific cell layers of the frontal cortex, highlight that the behavioral problems in autism very likely arise from impairments of interregional connections or pathways, and from within the nodes that make up these networks (e.g., Kelly et al., 2020). In this respect, based on animal models it was suggested that impaired social behaviors occur owing to dysfunction of a broad cortical network comprising the medial prefrontal cortex, anterior cingulate cortex, and particularly the insular cortex, which is fundamental in the integration of information related to social interaction, social decision-making, and emotional responses. In addition, it was proposed that dysfunction of subcortical regions associated with reward processes contribute to ASD (nucleus accumbens and ventral tegmental area). These circuits may be important in social deficits, possibly reflecting a failure to experience social reward. The disturbed behaviors are presumably related to hormones associated with social connectivity (e.g., oxytocin) and reward processes (dopamine) (Sato et al., 2023).

Neural circuits that subserve motor behavior may also be impaired in autism. The cortico-striatal circuit regulates voluntary motor behaviors and habit formation. Disturbances in this circuit may result in stereotypic and repetitive behaviors and may arise from overstimulation of the nigrostriatal dopaminergic pathway, which is known to promote stereotypic behaviors. Maintenance of these behaviors may require some form of reward or reinforcement, which can be provided by dopamine pathways from the midbrain to the nucleus accumbens (i.e., the mesocorticolimbic dopamine system). What we already understand about this system in driving reward, and its interactions with the nigrostriatal dopamine system (in shaping habits), may explain the persistence of stereotypic behavior. As much as this pathway in ASD and schizophrenia became evident some time ago (Bissonette and Roesch, 2016), it is important to recognize that while this pathway may contribute to social reward, it is likely just one of many contributing factors.

The circuits that are involved in attention and social perception include the parietal-frontal and temporo-frontal systems. The former is particularly relevant in attention to spatially distributed stimulus information and the accurate determination of stimulus location. Visual attention is modulated by collicular nuclei in the midbrain that guide eye movements. This ensures the cognitive capture of visual information so that its meaning and significance can be determined. Children with autism display poor eye contact and perception of gaze, which is a primary aspect of social interaction. Visual scanning of faces by ASD children is seldom directed at the eyes, with attention given more to the lower regions of the face (Chita-Tegmark, 2016). Children with autism not only show aberrant processing of gaze but also fail to recognize emotional expressions and perform poorly in the recall of previously seen faces. Moreover, when presented with different gaze conditions, there is poor attribution of intention (e.g., seeing another gaze at candy does not imply a desire for a sweet, when most children would conclude otherwise). Examples such as these represent instances of theory of mind, and the failure of such mental processes to operate as they should in ASD. These types of problems may reside in poor neural processing in the superior temporal sulcus, as well as amygdala disturbances, ostensibly because gaze processing possesses strong emotional valence (Rutishauser et al., 2015).

The point to take away from all this is that the behavioral problems in ASD align with many of the key circuits driving cognitive-emotional and motor behavior. This is consistent with what is already known about brain development in this condition: physical changes in the development of the brain and related alterations in cell numbers and neuronal connectivity. Since this can range in severity, distribution, and frequency, it has naturally revealed marked heterogeneity in the display of symptoms. However, it does raise problems about etiology, which has also plagued other areas of psychiatry (e.g., schizophrenia and mood disorders). When disorders are syndromal, the symptoms frequently do not coalesce in a sufficiently consistent fashion to make the search for causes simple. This problem is compounded by the symptoms of ASD overlapping with numerous concurrent behavioral problems, including epilepsy, sleep disorder, depression, obsessive-compulsive disorder, aggression, and self-injurious behavior (Khachadourian et al., 2023). As we indicated earlier, a precision medicine approach might prove useful in defining the links between symptoms and particular neurobiological processes, which ultimately may be relevant to treatment methods.

Etiology of Autism: The Immune Hypothesis

The foregoing section characterized the brain of an ASD individual as possessing different levels of connectivity and absolute variations in synaptic and spine density. We will now consider what an examination of the immune system can tell us about why this happens. Several reviews have focused purely on the immunological changes in ASD individuals or have addressed immunological factors that might drive the neurodevelopmental aberrations observed (Ashwood et al., 2006; Careaga et al., 2017; Edmiston et al., 2017; Hughes et al., 2023; Meltzer and Van de Water, 2017). An enormous number of neurochemical alterations have been implicated in ASD (e.g., glutamate, GABA, serotonin, dopamine, oxytocin, arginine-vasopressin, orexin, opioid peptides) (Marotta et al., 2020), too many to discuss here. However, as we'll see in this chapter as well as in Chapter 14, they may interact with inflammatory processes in affecting the symptoms of ASD.

A prominent hypothesis of ASD is that antecedent autoimmune processes interfere with neuronal development and impair behavior. This is distinct from autoimmune effects operating throughout childhood, and which undermine the normal quality of functional activity in the brain. We will consider autoimmune phenomena, and the strength of this evidence, after first addressing immunogenetics and autism, a topic that frequently underlies explanations of autoimmune dysregulation. Much of the research regarding immunological factors that might affect neuronal development is aligned with the same themes present in schizophrenia research. To the extent that it is largely indistinguishable from investigations that examine early life immune events and how they may affect alterations in social, cognitive, and emotional behavior in animal models, this aspect of autism-related research will be considered when we discuss schizophrenia (Chapter 14).

Research on autoimmune antecedents highlights the possibility that immune system functioning may already be compromised among individuals with ASD, although the extent to which this explains the onset or development of ASD is uncertain. Dysregulation in the immune system is likely to affect neurodevelopment prenatally and postnatally, doing so in a highly individualized manner. Critical periods of influence, and what types of aberrant immune processes might be of the greatest impact, might be linked to variations of several cytokines. To be sure, immune dysregulation may derail neurodevelopment in only a subset of people, while many others who suffer from altered immunity remain resilient or perhaps succumb to other problems unrelated to ASD. Confronting such questions, or caveats, is now more prominent, especially in the face of the growing number of genetic risk factors that are being identified. It does not necessarily generate treatment or preventative strategies but does highlight the possibility that the etiology of ASD may be unique and in need of personalized approaches to treatment (Hodson, 2016). The problem of indeterminate etiology due to heterogeneity of gene candidates has arisen in the case of schizophrenia, as well as mood disorders, where confidence in heritable risk factors is moderate or high. It is nonetheless possible that certain inherited immune-related genes might contribute to ASD.

Immunogenetics

There is little doubt that the development of ASD involves the interplay between environmental and genetic factors (at last count, more than 800 genes were implicated in ASD), including epigenetic changes, as well as chromosomal alterations. These genetic factors may influence neuronal growth, dendritic spine profiles, and hormonal and immunological processes, all of which play into the occurrence of ASD (Genovese and Butler, 2023). Considerable research has focused on cytokine involvement in producing a neuroinflammatory milieu, as have immune molecules involved in antigen presentation and inflammatory cellular phenotypes. In this respect, the inflammatory molecular pattern associated with ASD might be promoted by oxidative stress responses and mitochondrial system functioning. Although the specific cytokines that contribute to ASD have not been definitively identified, both preclinical and clinical studies have pointed to elevated IL-6 and IFNγ stemming from maternal immune activation (MIA) as being involved in the disorder, operating in conjunction with extrinsic factors (e.g., Majerczyk et al., 2022). Increasing attention has likewise focused on members of the IL-17 family that are activated in association with MIA. In fact, IL-17 has been causally related to the development of ASD given that antibodies that inhibit IL-17A signaling could diminish ASD symptoms in animal models (e.g., Wong and Hoeffer, 2018).

The main genetic approach in the immunological domain has focused on human major histocompatibility complex (MHC) genes, which have been linked to several diseases, and in particular, autoimmune disorders. Recall from Chapter 2 that, unlike the mouse system, the nomenclature for the MHC is based on the human leukocyte antigen (HLA) system of classification. The HLA region is found on chromosome 6 and encodes over 200 different proteins or antigens that serve important immune system functions, including cytokines, receptors, transcription factors, and features that allow the immune system to recognize host tissue and foreign antigens (e.g., on microbes, tissue transplants). Interest in how the HLA figured in autism was piqued by early studies that noted greater similarity in HLA antigen profiles between parents of children with autism relative to parental pairs with non-ASD children. This was considered important, since less sharing of HLA antigens between parents was thought to provide greater protection for the fetus, resulting in reduced chances of rejection and/or other complications. However, it is uncertain how this disparity in HLA confers such protection.

Moving away from parents and focusing on the child, reports pointed to HLA antigen expression being different in ASD children. In this regard, children with ASD may carry specific haplotypes, i.e., DNA variants long

a chromosome that are inherited together. In this instance, ASD individuals are more likely to carry the 44.1 HLA haplotype (also known as B44-SC30-DR4), which contains alleles for A2 and B44, DR1*04 (DR4), as well as a deletion of complement C4B (Torres et al., 2016). These alleles are found in Class I, II, and III regions of the HLA, each of which corresponds to genes that affect different aspects of immune functioning. Moreover, in determining which Class I alleles were carried more frequently in ASD, it was found that HLA-A2 was more common. The functional significance of these associations is not known, but it should be noted that HLA-A2 is associated with autoimmune and neurodegenerative diseases (Torres et al., 2016). Furthermore, a reduction of naïve CD4 T cells and an increase in CD4 T cells bearing a memory phenotype were reported in ASD children, and this was linked to ASD children being more likely to carry HLA-A2 and HLA-DR11 alleles (Ferrante et al., 2003).

Other immune genes associated with HLA Class I molecules have been linked to ASD (reviewed by Torres et al., 2016). For instance, killer-cell immunoglobulin-like receptors (KIRs) are found on natural killer (NK) cells, and certain HLA products of the Class I region serve as ligands for KIRs. The presence of HLA and KIR genes in the placenta suggests that HLA-KIR interactions may occur prenatally to influence the development of the fetus. However, KIR genes appear at increased frequency in ASD individuals, suggesting that HLA-KIR interactions are more likely to be important in the developing organism. Indeed, individuals with ASD have a higher frequency of activating cB01/tA01 KIR gene haplotype, and similarly have a higher frequency of the cognate ligand HLA-C1$_k$, which activates the KIR haplotype. Thus, carriers of these two haplotypes are more likely to show increased NK cell activation. How this contributes to the search for an etiology to the neurodevelopmental basis of ASD is not known, but it does offer some genetic support for possible immune dysregulation in autism. Finally, complement proteins have received considerable attention concerning neurological and psychiatric disorders, and in particular neurodegenerative conditions (Morgan, 2015), and were linked to autism. We noted earlier that a region of the MHC codes for complement proteins (in the HLA this is the Class III region), which may contribute to ASD. Interestingly, the increased risk for autism has been linked to the complement system, which consists of multiple blood-borne proteins that are activated in sequence to eventually implement cytotoxic effects on pathogenic cells. One of these complement proteins, C4B, is coded by the eponymous gene, *C4B*, the absence of which increases the risk for autoimmune disease and was also found in over a third of ASD individuals, compared to less than 10% of healthy controls (Mostafa and Shehab, 2010). The argument was made that when ASD individuals show a deficit in the C4B allele, and reduced amounts of circulating C4B, it may reflect an increased risk for autoimmune disease (Estes and McAllister, 2015). The challenge is to extend these associations and show direct evidence that an autoimmune process is directed at specific targets that would derail brain development in children.

Considerable evidence indicated that children who develop ASD are more likely to have mothers with circulating antifetal brain autoantibodies (Edmiston et al., 2017). Studies in both animals and humans have pointed to an increase of maternal brain-reactive antibodies in mothers of a child with ASD relative to that of mothers of an unaffected child (Bagnall-Moreau et al., 2023). It is believed that activation of the maternal immune system during gestation can disturb neurodevelopment owing to maternal autoantibodies that recognize proteins in the developing fetal brain (Jones and Van de Water, 2019). Many of these actions may stem from or interact with genetic and epigenetic factors that promote immune system dysregulation (see detailed review in Erbescu et al., 2022). While prenatal stressors and the presence of autoantibodies have both been implicated in the development of ASD, it seems that they do so independently of one another (Costa et al., 2023). Nonetheless, these systems may be relevant points of intervention or prevention in the development of ASD (Beversdorf et al., 2019)

Animal studies indicated that maternal antibrain autoantibodies altered species-typical social interactions with unfamiliar conspecifics (peers), as well as an increase in frontal lobe size and white matter density (Bauman et al., 2013). Similarly, in mice exposed prenatally to antibrain maternal autoantibodies, brain overgrowth and enlargement of cortical neurons occurred, which pointed to an autism-like neuroanatomical phenotype (Martínez-Cerdeño et al., 2016). Among ASD children, there is evidence for the generation of antineuronal autoantibodies, including antibodies that bind cerebellar antigens (Wills et al., 2009). It is thus possible that genetic deficits, such as those in the MHC, which increase the risk for autoimmune disease, may also be the basis of antineuronal antibody formation, and possibly impaired brain development.

Autoimmune Disease and Autism From a Broader Perspective

As we have observed, it has been a long-standing hypothesis that the development of autism is driven by autoimmune processes (Estes and McAllister, 2015), and MIA is associated with the development of autism. Children showing marked social impairments were more frequent among mothers with a history of allergies and asthma (Patel et al., 2018b), and the incidence of autism was associated with other sources of inflammation, such as maternal exposure to small particulate matter ($PM_{2.5}$) (Yu et al., 2023). A systematic review and meta-analysis indicated that prenatal exposure to ambient air pollution, particularly $PM_{2.5}$ and nitrogen dioxide, was associated with an increased risk of autism (Flores-Pajot et al., 2016). Further to this, animal models supported the view that in utero maternal brain-reactive antibodies can cause persistent behavioral or cognitive phenotypes, although it is likely that neurodevelopmental disorders arise owing to the interplay between antibodies, genetic contribution, and environmental factors. As well, microbial alterations, sex chromosomes, and gonadal hormones may contribute to ASD, which may be relevant to the sex biases frequently observed (Gata-Garcia and Diamond, 2019).

Early studies had indicated that the average number of autoimmune diseases was greatest in families that had children with ASD. The odds ratio for having a child with ASD moved from 1.9 to 5.5 as the number of autoimmune diseases multiplied within a family. Moreover, approximately one in five mothers with an ASD child reported having an autoimmune disorder. A review of the relevant literature concluded that ASD risk increased with the presence of familial autoimmune disease (Wu et al., 2015), and was particularly notable in the presence of maternal autoimmune disease (Chen et al., 2016e). In this regard, certain aspects of ASD and autoimmune disorders share aberrant genetic factors that are linked to mTOR signaling (Trifonova et al., 2021).

A meta-analysis that included 10 studies, nine of which were case-control studies (9,775 pregnant autoimmune cases and 952,211 pregnant healthy controls), examined the link between maternal autoimmune disease and ASD diagnoses in the offspring. The studies were conducted globally, with most coming from the United States, Canada, and Europe, with two reports from Asia. The overall conclusion of the review was that the presence of autoimmune disease during pregnancy was associated with a 30% increase in the risk of an ASD diagnosis in the offspring, with the main contribution coming from autoimmune thyroiditis (Chen et al., 2016e). Yet another large study conducted in Finland with 1.2 million people, including 4,600 cases of ASD, revealed that in addition to autoimmune thyroiditis, the likelihood of ASD appearing was related to autoimmune disorders involving the central/peripheral nervous system, the skin/mucous membrane, and/or respiratory disturbances (Spann et al., 2019). Since these diseases were associated with the familial connection to ASD, it seems that alternative mechanisms—such as genetics—may contribute to ASD. This was reinforced by the findings that no significant association existed between maternal autoimmune disease and ASD. Moreover, a connection to ASD did not appear in cases where systemic lupus erythematosus (SLE), inflammatory bowel disease (IBD), psoriasis, and rheumatoid arthritis (RA) occurred during pregnancy.

Registry Studies of Autoimmune Disease

Large epidemiological studies that relied on well-organized nationwide registers have served to determine whether a familial history of autoimmune disease represents a risk factor for the development of ASD. An analysis conducted within three Swedish registries compared 1227 ASD children with 30,295 controls matched for sex, birth date, and birth hospital (Keil et al., 2010). Information regarding whether either or both parents of each child had an autoimmune disorder was obtained a priori through the Swedish Hospital Discharge Register, with diagnoses of up to 19 different autoimmune diseases being identified. An odds ratio analysis revealed that children with ASD had a marginally higher likelihood of having parents with autoimmune disease, and of these, type 1 diabetes, ulcerative colitis, and Crohn's disease were most strongly associated with ASD. However, a different set of autoimmune diseases were associated with ASD in a Danish study. Specifically, of 3,325 ASD children (1,089 with infantile autism diagnoses), and 26 autoimmune disorders examined, children with ASD were most likely to have mothers with RA or celiac disease, and in the case of infantile autism, a family history of type 1 diabetes (Atladottir et al., 2009). It was interesting that the strongest associations were for non-CNS autoimmune diseases, suggesting that a familial history of auto-aggressive immune responses directed at the CNS is not as much of a risk factor as other conditions. In later studies, however, autoimmune disorders during pregnancy were associated with an array of psychiatric and neurodevelopmental disorders, not only including ASD but also schizophrenia, OCD, and attention-deficit hyperactivity disorder (He et al., 2022).

It is somewhat paradoxical that RA is an ASD risk since pregnancy is typically associated with amelioration of RA symptoms in 60%–75% of pregnant women with RA before pregnancy. Given that autoantibodies are known to transfer to the fetus during gestation, failure to inhibit key autoimmune processes in preexisting autoimmune disorders may influence fetal neural development. Not enough is presently known to determine whether mothers who gave birth to children who developed ASD were more symptomatic or had flare-ups during pregnancy and whether this predicts offspring ultimately diagnosed with ASD. At the same time, it is significant that a two-sample Mendelian randomization (MR) study using publicly available data indicated that there was no causal connection between RA and risk for ASD (Lee and Song, 2022).

Regarding other autoimmune diseases, amelioration during pregnancy is not as common as for RA but certainly has been observed. SLE in particular represents a potential source of complication for pregnancy, influencing outcomes like preterm labor or fetal death (Sangah

et al., 2023). A Canadian study examined the frequency of ASD births in 509 women diagnosed with SLE before pregnancy and 5824 normal, healthy control women (Vinet et al., 2015). In the SLE group, 1.4% of the births resulted in a diagnosis of ASD, as opposed to an incidence of 0.6% in the healthy controls. This is a relatively small association and is in keeping with the absence of an association between SLE during pregnancy and ASD outcomes (Chen et al., 2016e). The latter point was reinforced in another study involving pregnant women with antiphospholipid syndrome (APS) or SLE (Abisror et al., 2013). The results were modest and inconclusive, with no ASD diagnoses among the offspring of SLE mothers, while 0.08% of the offspring of the APS mothers were diagnosed with ASD. As in the case of RA, the two-sample MR study mentioned earlier (Lee and Song, 2022) indicated that SLE was not causally related to ASD.

Given the severity of SLE and the relationship to symptoms during pregnancy, these results are somewhat surprising. Specifically, an analysis of the literature (Bundhun et al., 2017) revealed that pregnancy in women who have SLE was strongly associated with an increased risk of requiring cesarean delivery, episodes of preeclampsia, hypertension, as well as increased risk of spontaneous abortion or premature birth, and classification of infants as "small for gestational age." Lupus flares occur during pregnancy and may pose a particular health and pregnancy risk, such as preeclampsia (Lateef and Petri, 2017). Given the multiple effects of SLE, it is predictable that behavioral risks for the offspring are significant, and several prominent neurobehavioral deficits were observed, particularly learning disorders and dyslexia (Yousef Yengej et al., 2017).

As mentioned earlier, other maternal autoimmune diseases linked to ASD include autoimmune thyroiditis (Brown et al., 2015). Serum samples obtained from mothers who had children with ASD and those with healthy offspring were assayed for antibodies against thyroid peroxidase (TPO-Ab), an indicator of autoimmune thyroiditis. For pregnancies that resulted in ASD cases, 6.15% of mothers were positive for TPO-Ab, whereas in those mothers who did not give birth to children diagnosed with ASD, 3.54% had detectable TPO-Ab. Furthermore, relative to TPO-Ab-negative pregnancies, the chance of obtaining an ASD birth was increased by 80% in the presence of TPO-Ab-positive pregnancies. Once again, while these results are statistically encouraging, they show only a modest association between maternal autoantibody and a possible influence on autism. The autoantibody data are not necessarily sufficient to explain the presence of autism in offspring, since TPO-Ab was present in pregnancies that did not result in an ASD diagnosis. Thus, as with other studies, it did not appear that maternal autoimmune disease drove the development of an ASD phenotype in the developing fetus.[3]

Immune Dysregulation in ASD Individuals

Many studies concerning the links between immune function and ASD focused on lymphocyte function, the capacity for cytokine production, and antibody (or immunoglobulin) production. Abnormalities in cytokine production, as well as altered basal levels of circulating cytokines, may have several implications. Cytokines are necessary to regulate the various activities of lymphocytes, dendritic cells, and macrophages/monocytes, and may influence neural and behavioral functions. This can operate in the adult nervous system to produce behavioral disturbances, but in the nascent nervous system of a young child, the impact of cytokines is likely to be more profound.

The organization and patterning of the nervous system continue well into adolescence, and immune dysregulation can disrupt this process. However, to determine the relevance of immune dysregulation to ASD itself, the information gathered needs to demonstrate an immunological effect on relevant brain and behavioral functions. With certain types of immune measures, this is not always clear. For example, suppression of T cell mitogenic function does not provide evidence of a proinflammatory response that can potentially impact CNS function. To be sure, it is possible that reduced cell division may underlie insufficient T cell involvement in a given inflammatory process, which could potentially influence the brain. However, this is pure speculation and is based on an ill-defined or characterized premise. In short, data like these (viz., T cell proliferative function) are important but limited. At best, one can simply identify an aberrant state of immunophysiology in ASD. Much of the information on immune function in ASD is of this nature—associative and lacking in explanatory power regarding ASD etiology. Nonetheless, eruptions of immunological activity may exacerbate certain ASD conditions as well as several medical and psychiatric co-occurrences that might point to common pathophysiologic mechanisms that favor the development of clinically diverse subgroups of ASD. At the same time, certain forms of immune activity might have benefits. Febrile conditions typically arise from inflammatory responses, and children with ASD, who present with fever show reductions in a variety of behavioral symptoms, including communication (Grzadzinski et al.,

2017). In studies of this nature, there is typically no information concerning potential immune mechanisms, or even whether fever was linked to immunological factors. However, it is an intriguing phenomenon, and one that deserves further exploration, especially as it may reveal important clues regarding which brain mechanisms are seemingly corrected (albeit temporarily), thereby opening some avenues for therapeutic development.

Cytokines and Autism

Brain Measures of Cytokine and/or Microglial Cells

Cytokines hold an important place in the pantheon of immunological factors that might disturb or impair CNS function, and we will look closely at some attempts to link cytokines to the autistic brain. In a relatively early study, cerebrospinal fluid (CSF) and serum samples were collected from individuals diagnosed with autism, as well as age-matched control subjects with non-ASD diagnoses (Zimmerman et al., 2005). None of the cytokines pursued (IL-1β, IL-2, IFNγ, and TGFβ) were detected in the CSF of those with autism, although trace amounts of soluble IL-1 receptor antagonist and IL-6 were detected in a few participants in both the autism and control groups, which was also the case for measures of soluble tumor necrosis factor receptors I and II (TNFRI and TNFRII).[4] In contrast, a small increase in soluble TNFRII was present in the serum of the autism group. Overall, this study was notable for its access to proximate material from the brain (i.e., CSF), although the lack of any cytokine elevations, or even detectable amounts of these cytokines failed to support an activated inflammatory cytokine network in the CNS of children with autism.

The same group of researchers (Vargas et al., 2005) assessed levels of glial fibrillary acid protein (GFAP) and HLA-DR (i.e., MHC II expression), which identify astrocytes and microglia, respectively, in postmortem brains from autistic and control brains. The number of GFAP and HLA-DR-expressing cells in the cerebellum was elevated in the autistic brain sections, and this was associated with neuronal cell loss. Moreover, the anterior cingulate gyrus, an area heavily involved in bidirectional communication between the prefrontal cortex and other deeper structures that drive motivational and emotional behaviors, showed elevated IL-6, IL-10, and TGFβ1. A systematic review of postmortem studies indicated that ASD was associated with elevated neuroinflammation, reflected by elevated cytokines and chemokines, likely stemming from astrocytes and microglia (Liao et al.,

---

[3] A report that added 18 more genes to the hundreds being investigated suggested that each ASD individual may have a unique etiology. While this is consistent with the principles of precision medicine, it makes the search for critical elements still more difficult to identify. The involvement of autoimmune factors is just one piece of the puzzle, and although relevant, it does seem to have a relatively modest effect in human studies.

[4] Soluble receptors for different cytokines, such as IL-1 and TNF, represent a regulatory system that "mops" up and inactivates excess amounts of cytokine. An increase in such receptors may reflect an increased level of the cytokine to which they bind.

2020). It was maintained that epigenetic changes stemming from prenatal factors and an assortment of postnatal environmental challenges promote dysregulation of glutamatergic signaling as well as an imbalance between excitatory and inhibitory pathways, which cause microglial release of inflammatory factors that favor ASD appearance (e.g., Bhandari et al., 2020). Speaking to this chain of alterations, chronic inflammation is associated with activation of the kynurenine pathway, which is accompanied by elevated neurotoxic metabolites, as well as altered glutamatergic functioning. Thus, the possibility was proposed that in the presence of genetic vulnerability, processes related to kynurenine functioning together with inflammatory mechanisms may contribute to ASD in a subset of individuals (Savino et al., 2020).

Studies of postmortem brain tissue reinforced the view that ASD was associated with aberrant epigenetic profiles linked to GABAergic, glutamatergic, and glial dysfunction, particularly within the frontal cortex, and transcriptomic analyses indicated that the disorder was associated with aberrant synaptic processes and immune pathway disturbances (Fetit et al., 2021). It likewise appeared in earlier studies that increased concentrations of TNF-α and IL-6, as well as granulocyte-macrophage colony-stimulating factor (GM-CSF), a growth-promoting cytokine with chemokine-like actions, were present in the frontal cortex samples of autistic individuals (Li et al., 2009). In addition, elevations were observed in IL-8 (which functions as a chemokine) and IFNγ. Finally, measures of cytokine mRNA in the frontal gyrus of the autistic brain revealed modest elevations in IL-1β and IFNγ, and a more robust increase for IL-6.

Analyses of this type suggest that some form of inflammation-related activity is present in the postmortem brain tissue of individuals with autism. However, we should note that measures were taken in the brains of individuals who died traumatic deaths (e.g., drowning), and who did so many years after the onset of autism. Consequently, while offering important clues, the etiology of autism is not obvious from these studies. They do, however, point to target molecules to investigate in very young children, although sampling in this age group is highly restrictive.

While not diminishing these caveats, a meta-analysis indicated that most postmortem analyses were consistent with the position that neuroinflammation was associated with ASD reflected by greater astrocyte and microglia activation and elevated levels of cytokines and chemokines (Liao et al., 2020). A review that included studies in humans and animal models indicated that ASD stemmed from dysregulated immune system functioning, together with synaptogenic growth factors. Moreover, ASD was associated with several genes linked to autoimmunity and those related to immune functioning. In this respect, genome-wide mRNA screening identified multiple upregulated sets of genes associated with immune signaling (Ohja et al., 2018), and long noncoding RNAs associated with NF-kB were detected among children with ASD (Honarmand Tamizkar et al., 2021). Furthermore, in children diagnosed with ASD, activation of toll-like receptors (TLR2 and TLR4) provoked a unique profile of immune activation and change of immune regulator genes, as well as failure in the downregulation in the expression of genes associated with a prolonged response of monocytes (Hughes et al., 2022). Consistent with these findings, a bioinformatic analysis of ASD-related data sets revealed the importance of two hub genes, that of fatty acid-binding protein 2 (*FABP2*) and Janus kinase 2 (*JAK2*). These core genes were tied to particular immune alterations, notably FABP2, which is associated with memory B cells and CD8 T cells, and JAK2, which is related to a wide array of immune factors, including monocytes, naïve CD4 T cells, CD4 activated memory T cells, CD8 T cells, activated dendritic cells, $\gamma\delta$T cells, and T_{reg} cells (Wei et al., 2022). Speaking to this, using a ML classification model based on mRNA expression data from the peripheral blood, numerous genes, primarily those associated with immune-related pathways, could be used to predict ASD with 86% accuracy (Tang et al., 2023b).

Together, these findings indicate that the activity of innate brain immune cells may be more vigorous in ASD, driving increased cytokine production. Moreover, elevated cytokine levels (IL-6, TNF-α, and IFNγ) in the periphery and CSF were associated with ASD, whereas the expression of several other cytokines, specifically IL-1, IL-2, transforming growth factor-β (TGF-β), and GM-CSF, was reported to be altered in some studies but not in others. As such, certain cytokines might serve as biomarkers to determine the presence of ASD (Xu et al., 2015).

The finding that cytokines such as TGFβ, IL-6, and IL-10 are elevated in the autistic brain is especially interesting, but the question remains as to what types of neurobiological functions they mediate. However, it is worth noting that TGFβ1 has assumed a neuroprotective, antiinflammatory role in the brain (as it does in the immune system), and IL-10 is the canonical antiinflammatory cytokine that regulates levels of proinflammatory activity in the immune system and may do so in the brain. With regard to IL-6, we should note that this cytokine generally receives a bad rap, and typically gets lumped with the proinflammatory gang of cytokines (like IL-1β and TNF-α), when in fact, it can also function in an antiinflammatory capacity and mediates neuroprotective influences (Kummer et al., 2021). Elevation of these cytokines in postmortem tissue both implies that neuroinflammatory activity may be elevated in the brain of an autistic individual and suggests the generation of antiinflammatory mechanisms.

Cytokine Measures in the Periphery

We have focused in some detail on research using postmortem brain and CSF samples since these studies offer relatively direct indices of intra-CNS inflammatory profiles. Measures of immunological factors, such as cytokines and chemokines in serum or plasma, provide the next level of investigation that may be relevant to cytokine impact on brain function. There are, however, several caveats to be made here. In the absence of infection or other immune activators, typical concentrations of some cytokine in serum or plasma tend to be very low, and in some cases may be undetectable. Statistical tools and data transformation are applied in some studies to enable some form of meaningful interpretation to be taken from the work. This is often done in the psychosocial literature but is rarely of much concern in immunological studies, where frank increases in a given cytokine or chemokine measure are of greater value since minimal or trace levels introduce ambiguity and uncertainty regarding the effect of a particular variable. Still, mechanisms exist for proinflammatory cytokines entry to brain parenchyma (Banks, 2015). Thus, it is conceivable that a slight elevation in a given cytokine may increase its potential interaction with neurons and glial cells once the cytokine leaves the blood. Studies that report on circulating cytokine levels typically allude to this possibility, but this is highly speculative.

Serum levels of IL-1, IL-6, IL-12, IL-23, and TNF-α were reported to be elevated in autistic individuals (Ricci et al., 2013), but this was largely a statistical conclusion, as many ASD subjects had undetectable cytokine levels, or were at the same level as control sera. ASD was also linked to IL-18, a cytokine with a growing reputation (Businaro et al., 2016). Interleukin-18 is part of the IL-1 family of cytokines, and its principal immune sources are monocytes, macrophages, and dendritic cells. It is a prominent IFNγ inducer and has been associated with a variety of pathological conditions. Elevations of IL-18, or its intrinsic production in the CNS (ostensibly by microglial cells), may contribute to the accumulation of amyloid precursor protein (APP), which has been found in the brains of autistic individuals. Moreover, IL-18 reactivity was present in the brains of individuals with viral encephalitis and tuberous sclerosis, in which ASD behavioral symptoms are common (Businaro et al., 2016). This was less directly linked to the ASD samples measured, but it was noted that ASD serum samples showed significantly lower IL-18 concentrations relative to control samples. Those less than 10 years of age tended to show higher values of IL-18, which dropped by 33% in the older ASD samples. Although IL-18 is more highly expressed by activated glial cells, the peripheral levels of IL-18 are inversely related to brain elevations.

Reviews of the various circulating cytokines measured in ASD indicated that levels of IL-1 did not differ from control individuals. The ties to IL-2, IL-6, and IFNγ were inconsistent, being elevated in children with ASD in some studies, but this was not apparent in other reports (e.g., Masi et al., 2017). Discrepancies in the measurement of cytokines limit the ability to form conclusions regarding whether peripheral cytokine elevations occur in ASD. Many laboratories tend to run cytokine arrays in the hope of observing one or more cytokines that might be systematically high in ASD. When reductions in circulating cytokines are present, as has been reported, it can be difficult to conceptualize the formation and/or persistence of autism-like behavior in terms of a cytokine impact on CNS function. This is because the main premise for assessing cytokines is predicated on the notion that they modify neural and behavioral functions. Consequently, when cytokines are reduced or undetectable, this fails to support the hypothesis that a proinflammatory cytokine mechanism contributes to autism-like symptoms. As such, while low levels may be reflective of the general status of blood cytokine levels in ASD individuals, as a mechanistic explanation for behaviors typical of ASD, the lower values hold little explanatory power. Overall, there do not appear to be consistent patterns regarding the relationship between circulating cytokines and ASD.

Based on a systematic review that assessed possible molecular biomarkers of ASD, an equally pessimistic conclusion was reached. The review included analyses of cytokines, growth factors, measures of oxidative stress, neurotransmitters, and hormones, as well as neuroimaging (e.g., fMRI analysis), and neurophysiological analyses (e.g., EEG and eye tracking). The studies examined were highly heterogeneous regarding individual features and the tissues assayed. As well, most studies were underpowered and could only detect biomarkers with large effect sizes. In the end, no biomarker could be identified that informed or justified clinical trials related to ASD (Parellada et al., 2023).

Despite the inconclusive data regarding the contribution of cytokines and other biological indices being reliable biomarkers of ASD, it might be profitable to approach the issue from another perspective, acknowledging that undue reliance on peripheral measures may be counterproductive. In general, MIA conjointly promotes microglial activation, oxidative stress, and mitochondrial dysfunction, which together may favor neuroinflammation and hence neurodevelopmental disturbances in offspring. Specifically, following prenatal infection, the increased production of proinflammatory cytokines may enter the placenta and hence fetal circulation. This, in turn, can activate resident immune cells, which cause elevated proinflammatory cytokine production that can enter the fetal brain, leading to microglial activation. Congruent with this, immune activation provoked by maternal viral infection was associated with the risk of ASD being elevated (Jiang et al., 2016), which was accompanied by elevated presence of proinflammatory cytokines (Estes and McAllister, 2015; Krakowiak et al., 2017a). In addition to the direct actions of cytokines, they can favor elevated oxidative stress and mitochondrial dysfunction, which together can promote disturbed brain development and neuropathologies (Usui et al., 2023; Zawadzka et al., 2021).

While these suppositions are based on the presence of MIA linked to viral or bacterial infection, comparable outcomes can occur in the absence of frank infection. As discussed earlier, environmental or individual factors (e.g., toxin exposures, maternal stress, and maternal obesity) can additively or synergistically promote sterile inflammatory responses that contribute to ASD among genetically high-risk individuals (Bilbo et al., 2018). Even if inflammatory factors are linked to ASD, this might only be a component in a subset of individuals (Mead and Ashwood, 2015). This may be of fundamental therapeutic significance since the identification of subtypes of ASD could direct interventions on an individual basis.

It was maintained that MIA can promote the downregulation of genes related to ASD, including those at cortical sites that are associated with synaptic processes. At the same time, MIA can provoke upregulation of numerous genes (e.g., those involved in translation initiation, cell cycle, DNA damage, and proteolysis processes) that influence developmental processes that affect multiple key neural developmental functions. In effect, ASD risk associated with MIA may stem from dysregulation of genes that are fundamental for neurodevelopment beginning prenatally (Lombardo et al., 2018).

While researchers frequently seek critical pre- and postnatal windows that are especially amenable to disease vulnerability, the development of autism may comprise a multistage, progressive prenatal disorder involving disturbances of cell proliferation and differentiation of neuronal brain development that begins early in pregnancy. Further neuronal processes are subsequently affected, including disruption of neural migration and maturation, as well as neurite outgrowth and synaptic development (e.g., Prem et al., 2020), ultimately resulting in impaired neural network functioning involving multiple brain regions. The timing of these prenatal cascading disturbances may contribute to postnatal phenotypic differences, but ways to identify individual factors that are tied to molecular and cellular differences responsible for ASD remain to be identified (Courchesne et al., 2020). Still, with improved strategies to identify malfunctions that occur during prenatal development together with a greater understanding of postnatal risk, it may eventually be possible to find individualized prophylactic and therapeutic strategies in the treatment of ASD.

Ex Vivo Measures of Immune Function and Autism

The emphasis on measuring cytokines in serum/plasma and CSF is based on the expectation of observing spontaneous production and release of cytokines in ASD individuals. However, the source of the stimulus and cells involved is not known when cytokine elevations are reported. Moreover, it is never clear whether the measures would be consistent over multiple sampling periods, nor whether the actual behavior of the ASD individual accounts for the elevation. As discussed, when we considered the ties between stress and immunity, psychological stressors can elevate circulating cytokine levels, and given that ASD individuals may be emotionally labile, under some circumstances this might account for changes in cytokine concentrations in the blood.

These points will be relevant for our next discussion of the relationship between immunity and ASD. Much of this concerns measurement of lymphocyte response capacity using in vitro stimulation procedures. Consequently, caution should be exercised in how to interpret these studies, particularly since the response of immune cells ex vivo is not necessarily predictive of in vivo functional capacity. In these studies, blood is collected from ASD individuals, and leukocytes are processed for analysis. A common procedure, as described in Chapter 2, is to conduct a mitogen assay to determine the proliferation of immune cells that are challenged with an antigen. It had been reported that blood leukocytes from autistic individuals displayed diminished proliferative responses to the T cell mitogens PHA and Con A, and the mixed T and B cell stimulus, Pokeweed Mitogen (PWM), indicating that T cell proliferative capacity can be depressed in autism (e.g., Ashwood et al., 2006, 2011).

Other approaches to assess the functional status of the immune system include quantitation or relative amounts of lymphocytes and their subtypes. Various surface markers on lymphocytes can provide clues as to their prior experience. Cells can be naïve, not having been activated by antigen, or they can exhibit signs of prior activation. Blood leukocytes from autistic individuals have more pronounced reductions in $CD4^+$ T cells, as well as in the percentage of naïve CD4 T cells (designated $CD4^+/CD45RA^+$). A percent reduction in naïve T cells might reflect the presence of more activated, or experienced T cell populations. However, autistic individuals present with higher surface levels of HLA-DR on T cells, which are expressed once activated by antigen. As such, while T lymphocytes from autistic individuals can appear to be functionally normal (i.e., proliferation is neither augmented nor depressed), their DR+ status might suggest prior antigenic experience. In some cases, this activation status can appear incomplete, and more akin to that seen in autoimmune diseases, since the DR+ T cells of autistic individuals may also lack IL-2 receptor (CD25) expression. Because various autoimmune diseases are associated with a greater number of DR^+ T cells that are negative for IL-2 receptor expression, it is possible that in the autistic individual, T cells exhibit more autoimmune-like features (Głądysz et al., 2018).

As with any evolving hypothesis, data may emerge that punch holes in what would seem to be a nice and neat story. Evidence was gathered in ASD individuals that the percentage of $CD4^+$ and $CD8^+$ memory T cells was similar to that determined in healthy sibling controls (Saresella et al., 2009). While this suggests that more activated T cell phenotypes are not universal in ASD, the study itself does not outright dismiss the notion that propensity toward greater reactivity or sensitivity to activation resides among T cells in ASD individuals. In those with ASD and their healthy siblings, there was an additional interesting observation. The distribution of activated T cell phenotypes (CD4 and CD8), when compared to healthy, age-matched nonfamilial control subjects, was quite different. Specifically, both those affected with ASD and their healthy siblings showed more CD8 naïve T cells, but fewer terminally differentiated (i.e., effector) and memory CD4 and CD8 T cells (Saresella et al., 2009).

How can these data be interpreted? The presence of naïve cells is an indication of possible overcompensation of thymic CD8 T cell output, or simply a lack of antigenic opportunity among this T cell subset population, which is supported by the reduction of effector memory T cells. Moreover, terminally differentiated CD4 T cells are considered effector cells with little proliferative capacity, and reduction in this population further supports reduced levels of antigenic encounters. Whatever the case, these data together with other evidence discussed earlier, point to the idea that T cells in autistic individuals present with varying degrees of experiential history with antigenic encounters. Whether these antigens are linked to the self or prior infectious encounters is not known. Concerning the latter, infectious encounters in the first 2 years of life were not found to differ between those diagnosed with autism and nonautistic populations. Consequently, greater antigenic load due to increased incidence of combatting infections may not be the answer. Yet, based on a large register-based cohort study, childhood infections were associated with an elevated incidence of ASD (varying with age of infection), even after considering heritable and nonheritable factors (Karlsson et al., 2022). Moreover, a review of the literature suggested that certain viral infections (rubella, cytomegalovirus, herpes simplex virus, varicella zoster virus, influenza virus, zika virus, and severe acute respiratory syndrome coronavirus 2) during critical developmental windows were associated with increased ASD risk (Al-Beltagi et al., 2023).

The immunological similarity between ASD individuals and their siblings implies a hereditary link to an ASD diagnosis that may influence the development of a more reactive immune system. The outcome of this reactivity, however, does not have to result in autoantibodies to CNS antigens, since significant concentrations of antibodies directed against a small selection of up to eight different CNS antigens in the plasma was not observed in association with ASD (Saresella et al., 2009). Of course, this does not exclude the presence of other antibodies that might be directed against the CNS. Indeed, older research had already alluded to the possibility of higher autoantibody levels that were directed against brain antigens (Ashwood et al., 2006).

Many immune cells and immune-related signaling molecules are found in the developing nervous system and contribute to healthy neurodevelopment. Several components of the innate immune system, including Toll-like receptors, cytokines, inflammasomes, and phagocytic signals, are critical contributors to healthy brain development. Accordingly, dysfunction in innate immune signaling pathways has been functionally linked to neurodevelopmental disorders, including autism and schizophrenia (Zengeler and Lukens, 2021). Speaking to this, in some individuals with ASD, anomalies were observed in humoral and cellular immunity assessed in blood and CSF. Ultimately, being able to link immune system dysregulation to specific ASD features may go a long way in defining immune biomarkers relevant to this disorder and could even inform therapeutic strategies on an individual basis (Głądysz et al., 2018).

Effector Immune Function and Autism

To this point, we have been considering the distribution of leukocyte subtypes in the blood of ASD individuals. While useful, this reveals little about function. Earlier, we talked about deficiencies of mitogen-induced proliferation in ASD. However, T and B lymphocyte expansion is only one aspect of effector function, which evolves to produce and amplify several other critical effector functions. This includes cytokine production, cytotoxic activity, T cell regulatory function (e.g., helper functions, or suppressor functions), and antibody production. These are specific to the adaptive arm of the immune system, which has been the major focus in immunological assessments in ASD individuals.

Leukocyte Production of Cytokines

Earlier, we discussed studies that measured in vivo cytokine levels in ASD serum, CSF, and postmortem brain tissue samples. However, in vitro cytokine production by T lymphocytes and cells of the innate immune system, such as monocytes and polymorphonuclear leukocytes, has also received some attention. In nonactivated T cells, the level of most cytokines in the cytoplasm is ordinarily very low. However, after activation, these intracellular proinflammatory cytokine concentrations can increase substantially, whereas that of antiinflammatory cytokine production was reduced. Findings such as these were interpreted as evidence that children with autism have a greater Th2 cell functional capacity and lower Th1 cell function. Yet, it has been reported that in children with autism, the balance between pro and antiinflammatory cytokines from CD4 T and CD8 T cells rested on the side of antiinflammatory actions (Krakowiak et al., 2017a; Saresella et al., 2009). However, a review of the immunological literature, which included cytokine analyses, noted that there is no consensus on whether the pattern of cytokine production in ASD is skewed toward an inflammatory or antiinflammatory profile (Meltzer and Van de Water, 2017). Nonetheless, this may be an important aspect of immunity to explore further, as evidence gathered in the mouse has suggested that the presence of IFNγ-secreting T cells in the meninges may influence prosocial behaviors (Filiano et al., 2016, 2017). Consequently, immunologic profiles that suppress IFNγ output may contribute to the development of an autism-like behavioral profile.

In summary, several lines of evidence have linked the immune system to autism, which are presented in Fig. 13.1. At present, much of the data reveals little about the etiology of autism, other than perhaps the influence of antineuronal antibodies, whether of maternal origin or arising from the individual with autism. However, it is obvious that some form of immune dysregulation is operating in ASD, although there is little coherence and uniformity to the data that had been collected. This is a sentiment expressed by others in the field (Careaga et al., 2017), who have pointed to the possibility that there may be an immune endophenotype that is present in a subset of ASD individuals. However, this needs to be isolated from the general ASD population to allow for a more focused investigation of immunologic contributions to the disorder. This view is likely

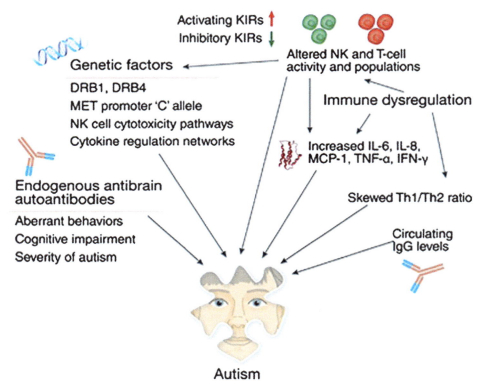

FIG. 13.1 Ongoing immune dysregulation persists in autism. After birth and at least throughout childhood, an individual with ASD may have endogenous antibrain autoantibodies, separate from any maternal IgG, which correlate with aberrant behaviors and impaired development. There also exists a broad picture of ASD-related immune dysregulation, including an increased inflammatory cytokine milieu (e.g., IL-6, IL-8, and MCP-1), thus leading to an increased, proinflammatory Th1/Th2 ratio. T-cell and NK-cell populations may also be skewed, displaying a shift in cell subpopulations. NK cells in particular show an increased baseline activity but a decreased response to activation, rendering the cells unable to properly respond to stimuli. NK cells interact with activating and inhibitory KIRs, many of which are genetically linked to ASD. Other genetic factors include the oncogene MET and members of the diverse family of HLA genes. The broader background of immunogenetic factors of ASD includes multiple networks of the immune system, such as pathways that regulate cytokines and NK cells, which together constitute a broad, endogenous environment of atypical immune regulation and response. *Source: Meltzer, A., Van de Water, J., 2017. The role of the immune system in autism spectrum disorder. Neuropsychopharmacology 42(1), 284–298.*

to gain some favor, given the heterogeneity of symptoms and states of severity that exist within the population of diagnosed autistic individuals.

Gastrointestinal Immunity and Autism

Increased attention has focused on gastrointestinal functions in autism and whether this is related to the development and/or exacerbation of autism-like symptoms. This interest intensified based on findings that the gut microbiome is capable of influencing neural functions (see Chapter 3). Numerous reviews of this literature pointed to mechanistic studies that examined brain and behavioral changes, which suggested a functional link between bacterial entities in the gut and the immunological apparatus in the small intestine. Most certainly, there are many other ways that the microbiome might affect brain function, and some of the data are relevant to autism. For example, germ-free mice exhibit deficits in social interactions, anxiety, and motor function, which could be corrected by replacement with microbial-rich fecal material (Bruce-Keller et al., 2018). The degree to which immune factors are involved in providing recovery of behavior is not clear. As we have observed, T cell-derived IFNγ may facilitate social behavior (Filiano et al., 2016). This study involved a thorough dissection of T cell factors and adhesion molecules to establish that a loss of social exploration and interaction was dependent on the production of IFNγ by T cells that resided in the meninges. Similar approaches are needed to investigate whether intestinal bacteria shape the immunological profile of gut immune processes to maintain normal behavioral functions.

The gut immune system is perhaps the most abundant of the regionally distributed components of the common mucosal immune system (which also incorporates the lungs, upper respiratory tract, and urogenital regions). Altered microbiota may either impact or be a result of immunological activity in the duodenum and small intestine. Research into the possibility that autism is linked to GI processes has not been extensive, but it is known that GI distress is frequently comorbid with autism. This may influence the behavioral symptoms of the autistic individual, and as such may also drive further changes in the gut and associated immune system. To some extent, dietary manipulations have been reported to improve symptoms, but many of the observations are anecdotal and based on parental observations. For example, a dairy or wheat-free diet (referred to as a casein-free and gluten-free [cf/gf] diet) may have benefits (Quan et al., 2022), which might suggest food sensitivity or the presence of food allergies.

Several hypotheses were offered regarding the importance of the gut in autism and the potential causes of GI symptoms. These include the *leaky gut hypothesis*, *dysbiosis*, and *autism colitis*. The leaky gut hypothesis, described in Chapter 3, refers to a possible deficit in intestinal permeability that compromises physical and immunological defenses against possible toxins and pathogens encountered through the diet. A leaky gut could potentially influence ASD by affecting toll-like receptors (e.g., TLR-4) and NF-κB (Li et al., 2023b). Alternatively, macromolecules (e.g., milk proteins) may be able to diffuse through the intestinal wall and sensitize resident immune cells, which then results in chronic allergic-like reactions and physical discomfort. Aside from the intraepithelial and lamina propria lymphocytes, macrophages, and dendritic cells of the gut wall, there are also more organized groups of immune cells present in the peyer's patches, the quasi lymph nodes of the small intestine. These represent the immune defense of the gut and drain into the mesenteric lymph nodes (MLN), where further processing and general immune reactivity take place. Other more nonspecific defense mechanisms operating in the gut include antimicrobial enzymes produced by specialized local cells (e.g., Paneth cells found in the intestinal crypts), which degrade gram-positive and gram-negative bacteria. These types of defenses limit the entry of luminal bacteria through the gut epithelium. Furthermore, the colonic epithelium can maintain gut homeostasis and an antiinflammatory state.

Evidence for a breach of these defenses or at the very least, signs of a "leaky" gut has been pursued using the Lactulose/Mannitol ratio measure. Mannitol is normally absorbed by the gut, but entry of lactulose, a larger molecule, is less permissive. This approach is used clinically for testing gut permeability, and there is some support for individuals with autism having increased gut permeability, but this is not a well-replicated finding (Samsam et al., 2014), and a leaky gut is not present in all ASD subjects.

Metabolite levels of bacterial origin (e.g., lipopolysaccharides and indoles) have been found in the blood and urine of ASD children, and it seems that this may also be accompanied by disturbed gut–blood barrier functioning. The alterations of bacterial metabolites could trigger core features of ASD and could also affect multiple biological processes (e.g., mitochondrial dysfunction, oxidative stress, altered blood–brain barrier integrity, as well as structural changes in several brain regions) that influence ASD symptomatology (Srikantha and Mohajeri, 2019). Such actions could come about through diet, and as depicted in Fig. 13.2 may engender numerous biological changes that could favor the development of ASD. These actions may stem from interactions between gut bacteria and host genetics, including the presence of single nucleotide variations related to immune processes (Liu et al., 2021b).

Dietary consistency is likely to maintain some sort of status quo in the makeup of the intestinal flora but this could be disturbed in ASD. In children with

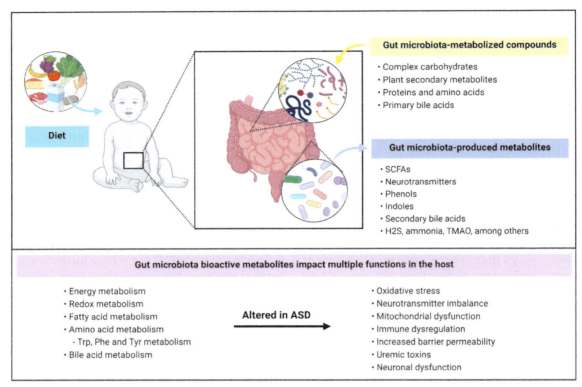

FIG. 13.2 Gut-bacterial dependent metabolic pathways and their influence in ASD. Diet is a major contributing factor that influences gut microbiota composition and function. Intestinal bacteria metabolize compounds that are unabsorbed by the gastrointestinal tract including (i) complex carbohydrates, (ii) plant secondary metabolites, (iii) proteins and amino acids that come from digestive secretions that escape host digestion, and (iv) primary bile acids. Bacterial metabolism of these compounds results in the production of bioactive metabolites such as SCFAs, neurotransmitters, and indolic and phenolic compounds. Several metabolic pathways were reported to be altered in ASD such as redox and aromatic acid metabolism. In turn, metabolic dysfunction mediated by gut microbiota may lead to other complications in individuals affected by ASD such as impaired neuronal function, oxidative stress, and increased intestinal and blood–brain barrier permeability. SCFAs, short-chain fatty acids; H2S, hydrogen sulfide; TMAO, trimethylamine N-oxide; Trp, tryptophan; Phe, phenylalanine; Tyr, Tyrosine. *Source: Peralta-Marzal, L.N., Prince, N., Bajic, D., Roussin, L., Naudon, L., et al., 2021. The impact of gut microbiota-derived metabolites in autism spectrum disorders. Int. J. Mol. Sci. 22, 10052.*

regressive autism, treatment with the antibiotic vancomycin resulted in temporary attenuation of their behavioral symptoms (see Jyonouchi, 2009). Because antibiotics may shift the makeup of commensal bacteria, the implication of this finding is that vancomycin had corrected a dysbiotic state in ASD children. Yet, antibiotics destroy both "good" and "bad" bacteria, making it uncertain what may actually be occurring in children with ASD. In fact, maternal antibiotic use or antibiotics taken during early life was associated with an increased risk for the development of autism (Njotto et al., 2023), although a systematic review revealed considerable variability across studies (Łukasik et al., 2019).

Assessment of colonic content in regressive ASD individuals has revealed newer emergent clostridial bacterial species and a preponderance of nonspore-forming anaerobes and microaerophilic bacteria. Similarly, there is a greater incidence of *Clostridium histolyticum* in the intestinal flora (Ding et al., 2017). This type of bacteria is a recognized producer of illness-inducing toxins and potentially can account for GI distress and other symptoms in ASD individuals. Such distress can impose a burden on the immune system in ASD, resulting in impaired production of innate immune cytokines, including IL-1β, IL-6, IL-12, and IFNγ (Cristofori et al., 2021; Jyonouchi et al., 2011). Some of the cytokines, specifically IFN-γ and IL-6, were associated with the elevated presence of several pathogenic gut microbiota that were detected primarily in individuals with ASD, whereas the presence of several beneficial microbiota was reduced (Cao et al., 2021c). As body compartments like the gastrointestinal system are functionally entwined with innate immune responses, the GI distress experienced in ASD may perpetuate both physiological and behavioral problems.

The presence of potentially toxic bacteria in the gut of ASD children focuses attention on the protective tools present in the gastrointestinal system. Aseptic or sterile conditions need to be maintained outside the enteric system to prevent infectious disease. Immunological surveillance within the intestinal wall ensures that any bacterial or parasitic infiltrate is eliminated, and in recent years, more information has been gained concerning just how the immune system monitors the state of the intestinal flora (Powell et al., 2017). This prepares and ultimately engages the immune system when potentially pathogenic gut dysbiosis takes place. Dendritic cells, the professional antigen-presenting cells of the immune system, are a particularly important source of immune surveillance in the gut. These cells widen the normally tight junctions between intestinal epithelial cells and extend their dendritic branches into the gut lumen, where they can sample for the presence of novel changes in microbial antigenicity and microbial by-products. The small intestine is also rich in another type of antigen-processing cell, namely, M (for *microfold*) cells. These cells are abundant in the follicle-associated epithelium, where they present or transport antigens to lymphocytes in the peyer's patches. This presentation may induce immune responses as well as promote oral tolerance to food antigens.

Oral tolerance is a case of immunological tolerance that is best demonstrated experimentally. Animals that receive foreign proteins for the first time via the gastrointestinal tract (i.e., by feeding) will display reduced immune responsiveness to the antigen when it is subsequently injected systemically. In contrast, animals injected with the antigen without prior exposure to the protein through feeding display enhanced T and B cell immune responses to the antigen. Evidently, initial encounters with foreign proteins through oral ingestion induce tolerogenic mechanisms (i.e., capable of producing immune tolerance) that suppress the adaptive immune response to food antigens. The precise cellular interactions involved in producing tolerance to enteric antigens are currently being determined. Dendritic cells of the lamina propria (nonpeyer's patch region of the small intestine) can be transported from the intestine to the MLNs via draining lymphatics. Retinoic acid, a vitamin A metabolite, is released by dendritic cells in the MLNs and local stromal cells (related connective tissue cells). This induces expression of gut-homing receptors (e.g., alpha4/beta7 integrin and CCR9) on activated T cells that, in conjunction with TGFβ and FoxP3 expression, influence the development of Treg cells. These cells then migrate back to the lamina propria, where IL-10 derived from macrophages coaxes Treg expansion and subsequent egress from the gut to extra-enteric lymphoid compartments (e.g., spleen and lymph nodes) and the induction of systemic antigen-specific tolerance (Pabst and Mowat, 2012). As autoimmune and inflammatory responses involve the generation of Th17 cells, it is pertinent that the induction of bile acids and retinoic acid by gut bacteria may serve to inhibit the generation of Th17 cells (Paik et al., 2022), thereby establishing an antiinflammatory state. Interestingly, excess IL-17 production was found in ASD, along with enhanced expression of the IL-17 receptor on monocytes (Nadeem et al., 2018), which may contribute to a proinflammatory state. Moreover, restoration of vitamin A deficiencies, which reduces IL-17, improved behavioral symptoms in ASD children (Guo et al., 2017). Despite an increased understanding of gut immunoregulatory systems, however, it is not known whether intestinal tolerogenic and antiinflammatory mechanisms are deficient in autism patients. Nonetheless, given the presence of food hypersensitivity in subsets of ASD patients and the high prevalence of gastrointestinal complaints (Kang et al., 2014), this represents an important area of immunological research to pursue.

Scant research has been conducted in which gastrointestinal cells have been examined for immunological properties in ASD patients. Yet, the opportunity for such research has been provided by clinical examinations of patients with colitis and Crohn's disease, which are highly comorbid with autism (Lee and Song, 2022). In fact, up to 50% of ASD patients show inflammation of the gut, which is characterized by lymphoid nodule hyperplasia, increased infiltrating eosinophils (which are cells of the myeloid lineage), and lymphoid aggregates (Kang et al., 2014). For a more specific examination of the properties of gut-related immunopathology, one can turn to a report by Ashwood et al. (2006), who obtained tissue samples from ASD patients with ileocolonic lymphoid nodular hyperplasia (LNH). They were able to stimulate isolated CD3+ T cells to induce cytokine production, which revealed enhanced concentrations of TNF-α and IL-12 production, but lower IL-10 levels. Furthermore, gut biopsies obtained from autistic individuals revealed increased numbers of CD3+ T cells that were positive for TNF-α and IFNγ. Whether these immunological characteristics are etiologically linked to autism is uncertain. However, they may contribute to an exacerbation of symptoms, and in cases of remission, might influence autistic regression.

Microbiota in Relation to ASD

It has become increasingly apparent based on studies in rodents that disturbances of gut microbiota can disturb social behaviors, whereas reconstitution of gut bacteria reestablishes social behaviors, although it is uncertain from an evolutionary perspective how this linkage first developed (Sherwin et al., 2019). Beyond affecting social behaviors, increasingly more evidence is amassed, indicating that gut microbiota contributes to ASD and that the microbiota disturbances may begin during gestation and early life (Taniya et al., 2022). A thorough review suggested that in the majority of studies, ASD was accompanied by microbiota alterations (Kelly et al., 2017). It has nevertheless been cautioned that the available data have been limited and the findings variable. Still, there were reports that this disorder was associated with particular microbiota species, notably *Bacteroides vulgatus* and *Clostridium bolteae* (Bastiaanssen et al., 2019), as well as diminished diversity of the gut microbiota community. Several studies also pointed to gut microbiota variations at the genus level among children with ASD (Strati et al., 2017). It appeared that features of the gut–brain axis were associated with heterogeneity that existed concerning ASD phenotypes, characterized by alterations of amino acid, carbohydrate, and lipid profiles. These features were associated with alterations of several microbiota genera, notably *Prevotella, Bifidobacterium, Desulfovibrio,* and *Bacteroides*, which were aligned with brain gene expression changes and proinflammatory cytokine profiles (Morton et al., 2023).

It is of practical importance that bacterial diversity in saliva and dental samples were similarly lower in children with ASD relative to controls, as well as lower levels of *Prevotella, Selenomonas, Actinomyces, Porphyromonas,* and *Fusobacterium* (Qiao et al., 2018). Although there have been few studies that assessed this relationship, its simple accessibility (relative to gut microbiota) might turn out to be a good method of assessing microbiota associated with illness. Importantly, oral bacteria can affect those present in the gut (and can translocate there) and the dysbiosis engendered may cause a reduction of Th17 cells and an increase in the M1/M2 macrophage ratio, which influence inflammation (Kobayashi et al., 2020).

It ought to be clear at this juncture that microbiota could potentially affect autism symptoms by different routes, including through their well-established effects on immune processes that come to affect brain functioning (e.g., Doenyas, 2018; Needham et al., 2018). As we have observed, most studies have been correlational and comprised a small number of participants. Nevertheless, some reports provided fundamental information concerning the links between microbiota and behaviors related to autism. In a study with a limited number of participants, the transfer of fecal microbiota resulted in the diminution of autism symptoms together with the attenuation of GI symptoms (Kang et al., 2017), which were still apparent 8 weeks after the transfer. In a follow-up study, most of the benefits continued to be present 2 years later (Kang et al., 2019) as were fecal microbiota metabolites (Qureshi et al., 2020). Fecal transplants provide individuals with multiple bacteria and metabolites but are there specific bacteria (or sets of bacteria) that can diminish ASD features? Based on studies in mice, it was demonstrated in a double-blind, randomized, placebo-controlled trial that *Lactobacillus reuteri* can reduce the social deficits present in ASD (Mazzone et al., 2023). As the treatment did not affect other features of ASD (e.g., repetitive behaviors), overall autism severity, general microbiome composition, or the overall

immune profile might suggest that ASD involves multiple processes but *L. reuteri* affects only some of these.

The importance of microbiota to autism was underscored by the previously mentioned finding that the antibiotic vancomycin, which acts only in the gut, provided a degree of symptom alleviation, although these actions were short-lasting. Moreover, the fact that fecal transplants could have positive effects supports the importance of microbiota in ASD. It is still premature to begin broader analyses of the effects of microbiota transfer (even in oral form) on ASD symptoms, and studies have only received approval for a small number of individuals (not in young children). Nonetheless, based on the available data, it was suggested that the gut microbiome might be a viable target in the treatment of ASD (e.g., Yang et al., 2018). However, identifying the specific bacteria and their metabolites that should be manipulated is still uncertain, making fecal transplants, which influence many types of microbiota, a better bet than probiotic treatments that are more restricted in their actions (Taniya et al., 2022). At the same time, excessive levels of the SCFA propionate can promote ASD-like effects. There is also the issue of whether optimal treatments would have to be applied early in life before autism symptoms are present (e.g., Watkins et al., 2017). While not diminishing these caveats, the focus on the transfer of microbiota in the treatment of ASD has been gathering steam, and efforts are being made to find strategies based on microbiota transfer without dependence on obtaining samples from healthy donors.

ASD and Food Allergies

A final word needs to be said regarding food allergies. These have been categorized as IgE-mediated or non-IgE-mediated. Atopic conditions (i.e., those that involve allergic reactions) are typically immune-mediated, with IgE responses serving as the trigger that liberates symptoms such as redness, swelling, itching, sneezing, and runny nasal and lacrimal (i.e., tear ducts) discharge. Mast cells are a major target for IgE, and the release of histamine and other molecules by these cells produces the foregoing allergy symptoms. Other types of allergic reactions, such as asthma, can involve T cells that have become hypersensitized to their respective allergenic molecules. Atopic conditions are quite common in ASD individuals, and although food allergies are frequently reported, ASD is also associated with asthma, eczema, and psoriasis (Theoharides et al., 2016). Interestingly, examination of atopic symptoms in those ASD individuals who also have comorbid GI difficulties has not revealed a uniquely higher incidence of IgE-mediated food allergies (Jyonouchi, 2009).

With regard to non-IgE-mediated food allergies, peripheral blood leukocytes from ASD individuals do react to common dietary proteins, such as whole cow's milk protein (and its major derivatives, including casein and beta-lactoglobulin), and the wheat protein, gliadin. Specifically, leukocytes obtained from individuals with ASD who have pronounced GI symptoms responded with higher TNF-α and IL-12 production when challenged with milk protein antigens and gliadin (Jyonouchi et al., 2011). This was not uniquely due to the presence of GI problems, as those ASD individuals who were negative for GI symptoms also responded with significantly greater TNF-a and IL-12; however, unlike the GI-positive ASD individuals, they were significantly less responsive to gliadin and casein and beta-lactoglobulin. In effect, GI symptoms in ASD individuals are associated with a broader panel of food antigen sensitivity.

Overall, despite a paucity of close immunological assessment of ASD patients with GI disturbances, there is sufficient information to consider seriously the hypothesis that gut food processing, as well as gut-associated immune surveillance, may be involved in autism. Whether these conditions are antecedent to the development of autism or evolve concurrently with or following the onset of the disorder remains to be determined. Such problems can emerge from 18 months to 3 years of life (Bresnahan et al., 2015), but can become less severe as children with ASD grow older (>6 years) and display less food hypersensitivity (Jyonouchi, 2009). This may be due to the maturation of oral tolerance mechanisms, which would be in keeping with other states of developmental lag present in autism.

Perinatal Immune Activation

For many years, there has been a stirring and at times emotional debate regarding whether vaccinations contribute to the development of autism. The pediatric and neurological medical communities have vociferously rejected this notion, citing a lack of convincing scientific evidence. Nonetheless, the debate continues (fueled by several actors and politicians), as do investigations into the impact of vaccinations on psychiatric conditions (Leslie et al., 2017). In addition to childhood vaccinations, there has been intense interest in the immune response of the mother during pregnancy. The autoimmune disease literature, already discussed earlier, is one part of the probe into the impact of a mother's immune system on the developing embryo or fetus. Another approach is to consider the presence of maternal infection and the immune system's response to this on intra-uterine neurodevelopment and subsequent postnatal behavioral development. Epidemiological observations precipitated experimental investigations into this question, and for the most part, much of the current literature comprises animal studies. Because much of the animal literature pertains to neurodevelopmental alterations

and behaviors that not only include those relevant to autism but also schizophrenia and many other psychiatric disorders, discussion of this literature is presented in the chapter on prenatal and postnatal manipulations and their impact on behavior as well as Chapter 14. Suffice it to say that studies that examined influenza vaccines taken by women during pregnancy have not found a significant relationship between vaccination and risk for autism among their offspring (Zerbo et al., 2017). Vaccinations, however, are not equivalent to actual infection, which has a different time course, affects behavior, and can vary in the level of immune cell recruitment and responsiveness. There does not appear to be strong evidence, at present, that this is a neurological risk for the developing fetus. Consequently, the CDC continues to support and urge vaccinations in pregnant women to ensure prevention of infection, which, in being a different animal altogether, could be a greater threat than receiving a vaccine.

Autism and links to immunity

The research concerning autism in relation to immunity, and particularly to vaccination, has had a shady history, thanks to the controversial reports by Wakefield suggesting that vaccination with MMR (mumps, measles, Rubella) led to autism. We needn't go through this sketchy story as it's likely well-known to readers. Suffice it that it has had enormous adverse effects as parents stopped immunizing their children, thus leaving them vulnerable to diseases. It may also have hampered legitimate research evaluating the impact of viral factors on the provocation of autism and other conditions that may be linked to inflammatory processes.

It's likely that a single factor doesn't lead to all instances of autism, although prenatal or early life experiences (e.g., very premature birth and the hypoxia that goes with it) might contribute to this disorder, as might epigenetic changes related to neurobiological processes. A credible view of autism has emerged that immune-related disturbances in pregnant women might be at the root of autism in a substantial number of cases (Estes and McAllister, 2015). These immune-related alterations include MHC Class I molecules, microglia, complement factors, and pathways downstream of cytokines. A large study of more than 2.3 million people in Sweden indicated that maternal hospitalization with infection during pregnancy was accompanied by a 30% increase in the occurrence of autism in offspring. Among women who had experienced a viral infection during the first trimester of pregnancy, a 300% increase in autism was later detected, and when the infection occurred in the second trimester, a 40% increase occurred (Atladóttir et al., 2010). A large meta-analysis that included 36 independent studies indicated that the ties to autism were evident across several forms of infection irrespective of the time during pregnancy at which this occurred (Tioleco et al., 2021).

Much as previously reported for schizophrenia, it was proposed that it is not the virus itself that causes the problem, but the strong immune responses on the part of the mother or the presence of fever that produces the collateral damage in the offspring. Although it is thought that proinflammatory factors, especially IL-6, may be responsible for the link to autism, the relationship between maternal infection and autism may be determined by genetic factors and their interaction with environmental influences.

Concluding Comments

In general, as with many other investigations, the relationship of the immune system to autism has proven to be complex. It is difficult to determine cause and effect when individuals with autism are sampled well after neurodevelopmental delays have already been instigated. Moreover, it is virtually impossible to separate the immune alterations in autism from neurohormonal influences on the immune system as a function of their aberrant behavior and changes in neural tone.

One of the challenges of human studies is to assess the immune status of newly born infants and form predictions regarding the likelihood of whether a particular immune profile is indicative of the future development of autism. However, the immune system is developing prenatally, and at birth contains a full complement of adaptive and innate immune components ready to receive antigen stimulation, which will further shape the immune system. How prenatal immune development is shaping neurodevelopment is not well known, and of course, human studies are not in a position to address this question with as much detail as animal studies. The animal studies, however, are limited concerning this question. To be sure, there is a growing movement that identifies microglial cells as shapers of synaptic density, although this is something that very likely is most important during the postnatal period. At present, the research data are strongly suggestive of immune dysregulation in autism. However, to what extent this is another index of aberrant physiological behavior in the ASD individual is still a matter of speculation. Longitudinal studies of immune function in ASD will help resolve the issue of whether the various immunological deviations are permanent or temporary states.

CHAPTER

14

Schizophrenia

The extra burden of schizophrenia

Understanding schizophrenia and identifying its etiology has long engaged thinkers, philosophers, scientists, and clinicians. Although it is defined by the loss of contact with reality, within the general public it had been misidentified as multiple personality disorder or what many termed "split" personality. Even before this, it was thought of as a malady of the blood or the result of meddling by the devil and other possessive spirits. Among other confusions, schizophrenia was wrongly thought to be an aggression disorder with patients having a predominantly violent predisposition, an impression that led to the perception that schizophrenia causes criminal behavior. Not uncommonly, a large proportion of individuals with mental health challenges who fit the diagnosis of schizophrenia were incarcerated, which is more a problem of poor public health policy than the defining trait of criminals. An additional misunderstanding—often propagated by popular films—is that some people with schizophrenia are intellectually brilliant, bordering on genius, or at least display a high level of intelligence. As with many mental health challenges, there are high-functioning individuals, although for schizophrenia, the harsh statistic is that close to 20% of people diagnosed with the disorder are unemployed, and in some cases with little prospect of gaining employment. This is due to significant deficits in cognitive functions such as impaired attention and working memory, and diagnostically has become a core symptom domain of schizophrenia. For all these reasons and more, schizophrenia remains the most debilitating and disturbing mental health disturbance reflected by the magnitude of functional disruption and its relatively poor prognosis. Of course, it takes an exceptional toll on family members, and of all psychiatric disorders assessed, it has the highest societal cost per patient.

Overview of Schizophrenia

There are presently no entirely adequate explanations for the etiology of schizophrenia, and the disease continues to remain elusive in relation to preemptive intervention and/or eradication. In contrast to the views of clinicians and psychiatrists before the 1950s, who approached etiology from a psychodynamic perspective inspired by Freudian psychoanalytic thinking, the disease overwhelmingly is now considered to have a biological origin. This was inspired by early familial studies of heredity, and in the past decade, large-scale genetic studies. However, the presence of genetic mutations—many of which are inconsistent across different individuals with schizophrenia, and which makes the task of blaming specific genes more difficult—is not a sufficient condition for the appearance of the disease. Most certainly, it may potentially be a necessary requirement, but complete emergence of the disease appears to occur in conjunction with particular environmental conditions. These additional conditions are considered under the general rubric of "stress," and as such, are championed by the diathesis-stress—or the related "two-hit"—hypotheses of schizophrenia. Into this theoretical framework, one can integrate the immune system and the relationship of immune-related cells and molecules (e.g., microglia and certain cytokines) to neurodevelopment and neural modulation.

The legitimacy of an immune approach to schizophrenia has steadily increased over the past few decades. Many studies have attempted to link the immune system to psychosis, but for a variety of reasons, the bulk of these studies lacked well-controlled longitudinal assessments that spanned the first 20–30 years of life. Mainstream views had not fully embraced the role of the immune system in the etiology of schizophrenia. However, this changed with a series of landmark studies that examined a large array of genes and located a significant number of

The Immune System
https://doi.org/10.1016/B978-0-443-23565-8.00011-9

variations within the MHC region (i.e., the HLA genes) of chromosome 6. There is now a concerted effort to learn more about how the major histocompatibility complex (MHC) region figures in the etiology of schizophrenia. One such role appears to involve the gene for the C4 component of complement,[1] which may be involved in synaptic pruning (Sekar et al., 2016).

In this chapter, we will describe some of this evidence and consider new avenues that might be worth exploring. Following a brief overview of the nature of schizophrenia and efforts to determine etiology and optimal therapy, we will turn our attention squarely to how the immune system fits into this complex and seemingly intractable disease. To an extent, we will face questions like those addressed in assessing autism. Are autoimmune elements directed at the central nervous system responsible for schizophrenia? Is immune functioning in schizophrenia different from that of healthy individuals, and do these vary with features of the illness? Do prenatal and early-life postnatal immune perturbations account for the appearance of schizophrenia? And finally, can the immune system be targeted with antiinflammatory drugs to attenuate clinical symptoms of schizophrenia?

Defining Schizophrenia

The term schizophrenia did not exist until early in the 20th century when Bleuler coined the term as a combination of the Greek words for "split" (viz., *schizo-*) and "mind" (viz., *phrene*). Bleuler was urged by the need to differentiate the condition from Kraeplin's earlier formulation of "dementia praecox," which involved early (childhood) loss of mental faculties due to a suspected degenerative process; this placed Kraeplin firmly in the biological camp concerning the etiology of psychological disorders. Bleuler contended that the disorder emerged later than childhood or early adolescence and was not necessarily unremitting and degenerative. There may have been merit to this general understanding of mental illness, given the individual differences in its presentation. The conceptualization that Bleuler formulated appealed more to psychoanalytic forms of thinking since greater emphasis was placed on psychological anomalies, as opposed to organic explanations that sought causality in unique biological or constitutional states of the individual. Nonetheless, throughout the 20th century, there was

a gradual movement to restore Kraeplin's more biocentric conceptualizations, and over time schizophrenia gained prominence as a neurodevelopmental disorder.

Schizophrenia has multiple domains of dysfunction, including aberrant forms of thought and behavior, as well as distorted sensory and perceptual experiences, which converge on a diagnosis. The key elements of schizophrenia comprise hallucinations, delusional and/or disorganized thinking and speech, grossly disorganized (or catatonic) behavior, and paranoid ideation. These are often referred to as *positive* symptoms, not because they are highly valued, but because they stand out as bizarre and unusual in the context of normal cognitive and social experience. A constellation of additional factors may then gravitate about the core presence of psychosis, which is also emblematic of schizophrenia. This typically includes a set of "negative" symptoms, so-called for the general failure to express various normal cognitive and emotional behaviors. This includes absent or reduced verbal communication (i.e., poverty of speech or alogia) and loss of affect. Negative symptoms, commonly comprise blank, featureless facial expressions, as well as anhedonic and avolitional states. This plethora of "a"-based prefixes served as a useful mnemonic, since schizophrenia appeared to subscribe neatly to the four A's rule, comprising altered associations, impaired affect (flat or inappropriate), ambivalence, and autistic isolation (Insel et al., 2010). These features reflect a loosening of *associations* (often in thought) considered critical in a diagnosis of schizophrenia.[2] In addition to these features, the illness may be accompanied by disturbed sleep patterns, dysphoria, hostility, derealization (feeling that surroundings are not real), depersonalization (feeling detached or disconnected from the self), lack of insight regarding one's illness, and deficits of social cognition. As with many psychiatric conditions, these symptoms need to be differentiated from possible organic causes, such as trauma due to injury or stroke. However, once an organic basis is dismissed, the presence of negative and positive symptoms, as well as psychosis, can point the clinician toward a diagnosis of schizophrenia, or some variant of this disorder.

It is not unusual to find the terms "psychosis" and "schizophrenia" used interchangeably, even though these are distinctly different from one another. Psychosis refers to losing touch with reality, reflected by hallucinations, delusions, and agitation, brought about by factors

---

[1] Recall from Chapter 2 that the various genes that line up on chromosome 6 to form the MHC have as their neighbors genes for key components of the complement system. Mutations in these genes may affect how the nervous system is sculpted, and if not done right might facilitate the development of schizophrenia.

[2] Historically it was not uncommon to think of the child with autism as having a schizophreniform disorder (i.e., a seemingly less full-blown form of schizophrenia). However, the differentiating factor is that schizophrenia appears in its frank, clinical form much later than the core features of autism, which are diagnosed after 3 years. Still, schizophrenia, autism, and bipolar disorder share common patterns of gene expression, although distinct differences were also detected (Gandal et al., 2018).

such as substance use, sleep disorders, traumatic brain injury, neurodegenerative disorders, brain tumors, or stroke. Schizophrenia, in contrast, is associated with disturbed thought processes, emotions, and behavior that appear in various phases, beginning with a prodromal phase in which symptoms (e.g., loss of interest in activities, social withdrawal, or difficulty concentrating) develop gradually over months or years. Thereafter an active (acute) phase emerges during which psychotic symptoms may appear. During a residual phase, symptoms may have abated but features of the illness (difficulty focusing, withdrawal) may persist. Whereas episodes of psychosis may never reappear, schizophrenia is a life-long condition. Features of schizophrenia can vary over the course of the illness and are often viewed as a spectrum of conditions in which patients may be classified into broad diagnoses (schizoaffective disorder, schizophreniform disorder, and schizotypical personality disorder) that comprise schizophrenia spectrum disorders.

The Pathophysiology of Schizophrenia

Genetic Contributions

As schizophrenia often runs in families, early theorists assumed a strong genetic component underlying the disorder. Early research that included twin studies indicated that the heritability of the disorder was about 40%. Later studies relying on sophisticated genetic analyses similarly supported a strong genetic component, although it has not been simple to identify specific genetic factors that accounted for the disorder. Difficulties in coming to grips with the processes aligned with schizophrenia stem, in part, from the broad heterogeneity that exists concerning symptoms, course, and outcomes of the disorder. Moreover, the characteristics of the illness and its polygenic features overlap with neurodevelopmental disorders (e.g., autism, ADHD), bipolar disorder, and schizoaffective disorders (Owen et al., 2023; Smeland et al., 2020). Schizophrenia involves multiple risk alleles, including common and rare genetic variants that are associated with neurodevelopment, neuronal excitability, and synaptic functioning, together with inflammatory immune system processes, which ultimately cause dysfunctional information processing.

Genome-wide association studies (GWAS) revealed many genes that were associated with schizophrenia, which together accounted for only a modest portion of the variance associated with this disorder (Trubetskoy et al., 2022). A systematic review similarly revealed that genome-wide variations among patients diagnosed with nonaffective psychosis revealed that genetic/familial risk for illness was linked to early age of onset, illness chronicity, and the presence of functional impairments (Taylor et al., 2023). As schizophrenia has been tied to an individual's age, GWAS assessed the contribution of polygenic risk scores and copy number variant analyses in a large cohort of young people of European ancestry. Moderate SNP-based heritability was found ($\sim$20%) even though a significant locus could not be determined, and the number of variants present was not related to earlier illness onset (Sada-Fuente et al., 2023).

Genetic analyses have pointed to particular neuronal and synaptic processes across multiple brain regions being associated with schizophrenia. An extensive systematic review indicated that schizophrenia was moderately related to epigenetic changes in genes involved in the regulation of neurotransmission, neurodevelopment, and immune processes (Smigielski et al., 2020). This review indicated that methylation of genes involved in other psychiatric disorders (e.g., BDNF, COMT, SLC6A4, and HTR2A), such as depression, was apparent in schizophrenia. A large-scale study also indicated that schizophrenia and autism shared several risk genes (Rees et al., 2021). Moreover, SNPs associated with schizophrenia were evident in autism, although the linkage was smaller than with depression, bipolar disorder, and ADHD (Romero et al., 2022). The very fact that schizophrenia has behavioral, neurobiological, and genetic features that overlap with other disorders may prove to be an asset in defining the processes responsible for the emergence and treatments of the disorder (Owen et al., 2023).

The capacity of GWAS in using large samples has provided important information in defining the processes related to schizophrenia, but the findings have limitations that cannot be ignored. Beyond those already mentioned, these studies did not include a sufficiently diverse population, thereby limiting the generalizability of the findings. The data were often obtained from biobanks that may not represent the population at large. And, too often, this was associated with limited phenotyping concerning disorders, frequently being based on self-reports rather than being clinically determined. Questions have also arisen concerning the causal connection that exists between the presence of pathology and identified genes, which has implications for the development of therapies aimed at targeting gene-related biochemical processes (Derks et al., 2022). Beyond these difficulties, it is generally understood that the contribution of multiple genes associated with complex disorders likely involves complicated interactions between many genes (epistatic interactions), thus enormous sample sizes are needed to determine the relations. Increasingly larger studies have been conducted so that new variants are being discovered. Many of the variants and epigenetic changes might not be directly related to schizophrenia but may instead reflect actions secondary to lifestyles adopted by schizophrenic patients (e.g., smoking diet, obesity, and drug therapies). Furthermore, as the features of schizophrenia can vary over the course of the illness, and with the accumulation of environmental and experiential influences, longitudinal studies will be needed to disentangle the contribution of diverse genetic and other causes of illness (Tam et al., 2019). These are only some of the limitations of GWAS outlined in reviews of the issues (Derks et al., 2022; Tam et al., 2019). Some of these issues will hopefully be overcome, but for the moment they are obstacles in understanding mechanisms associated with schizophrenia and its treatment.

Biochemical Processes

Although our focus is largely on the immunological hypothesis of schizophrenia, immunity is essentially a distal variable since mental functions are driven by neurons. Proximal influences on neuronal and synaptic functions typically represent factors within and between neurons. Much of the effort to try and understand the cause of schizophrenia has focused on neurochemical processes. As we indicated with other disorders, the features and underlying processes related to schizophrenia with either negative or positive symptoms vary with the progression of the illness. Based on functional neuroimaging and genetic analyses, it appeared that illness progression was associated with increasingly greater dysfunction of higher cortical regions relative to subcortical areas. These changes were accompanied by more disturbed external sensory gating (filtering out redundant or irrelevant stimuli) together with the disturbed balance between internal excitation-inhibition (Jiang et al., 2023).

The Dopamine Hypothesis

Of the monoamines, altered functional properties of dopamine and its receptors have figured prominently as an explanatory tool for the symptoms of schizophrenia. According to the "dopamine hypothesis," schizophrenia is due to dopamine over-activity in widespread areas of the brain. It was believed to account largely for the positive symptoms of the disorder. To a considerable extent, this view followed the finding that chlorpromazine could diminish the features of the disorder.

The success of chlorpromazine in attenuating the more bizarre and disruptive symptoms of psychosis-related disorders resulted in the development of other drugs that influenced the symptoms of the disorder. Paralleling these pharmacotherapeutic developments was a better understanding of the monoamine neurotransmitters, which eventually led to the proposal that antipsychotics worked by blocking monoamine actions, particularly antagonism of the dopamine D_2 receptor. Together with evidence that dopamine agonists could produce psychotic-like states, and that dopamine was important in brain pathways that regulated cognitive, emotional, and motor functions (viz., the mesocorticolimbic and nigrostriatal pathways), it became evident that the positive symptoms of schizophrenia could be due to excessive release of dopamine and/or heightened dopamine receptor stimulation.

This version of the dopamine hypothesis was eventually modified to deal with certain nagging problems regarding the precise efficacy of antipsychotics, and the disconnect between cognitive deficiencies (e.g., hypofrontality)[3] and negative symptoms with more active dopamine release and D_2 receptor transmission. As opposed to predominant *hyper*-dopamine activity, which was felt to be exclusively in subcortical regions, such as the striatum (which regulates voluntary motor functions, as well as habitual, repetitive, and stereotypic movements), room was made for evidence that suggested *hypo*-dopamine activity in the frontal cortex. To date, much of the molecular imaging work in patients has tended to point in this direction: too much dopamine below the cortex; and not enough in the cortex (Howes and Shatalina, 2022). How this inverse dopamine involvement in different regions of the brain developed continues to be a driving question and may be linked to neurodevelopmental disruptions. Genetic factors could, no doubt, contribute to these actions, as might epigenetic changes. Indeed, among patients diagnosed with nonaffective psychosis, blood leukocyte D_1 receptor hypermethylation was observed, which was linked to stressful experiences, and D_1 methylation in brain tissue was observed in the dorsolateral prefrontal cortex (Garcia-Ruiz et al., 2020). Whether these epigenetic changes predict the development of schizophrenia or are secondary to the illness is uncertain.

The drug, chlorpromazine, was developed by Paul Charpentier in 1951 and released for clinical investigation as a way to potentiate general anesthesia. Soon afterward, it was used to treat bipolar patients, as it reduced manic symptoms, such as agitation and excitement. It caught the attention of American psychiatrists who administered the drug to psychotic patients to reduce positive symptoms

---

[3] Hypofrontality refers to reduced or deficient function in the frontal lobe of the brain. Although it originally was derived from observations of a physiological deficiency, such as reduced blood flow, it can sometimes be applied to the associated deficits in psychological functions deemed unique to the frontal lobe. For instance, executive functions (attention, decision-making, emotional regulation, and working memory) in the prefrontal areas of the frontal lobe are less well implemented, and therefore, behavioral indicators of *hypofrontality*.

and increase clarity of thought and emotional stability. Chlorpromazine soon achieved celebrity status and ushered in several decades of pharmaceutical development that saw the persistent use of neuroleptic drugs. Because there was a reduction in motivational or volitional deficits in behavior, along with improved control of emotional lability (without obvious sedation), the term "neuroleptic" was used to describe chlorpromazine and other drugs that were soon developed. This was reinforced by animal studies, in which these drugs suppressed spontaneous movements, reduced initiative and interest in the environment, suppressed emotional responses, and lowered aggressive and impulsive behavior. With the advent of chlorpromazine, it was evident that behavior could be controlled, and it also implied that psychotic and mood disorders emerged from unspecified biological dysfunction. The nature of this dysfunction gave birth to the dopamine hypothesis, which, to an extent, has endured to this day. This position was supported by the effectiveness of dopamine-acting agents in ameliorating positive and negative symptoms (more so the positive symptoms), depressive symptoms, and enhancing quality of life. These effects were more notable among first-episode relative to multiepisode patients. Yet, the effectiveness of these agents in treating acute schizophrenia has not been overwhelming. About 23% of patients exhibited a "good" response, while a "minimal" response was attained in 51% (Leucht et al., 2017). Among patients who come off their medication, the risk of relapse may be high, varying with the presumed cause of illness, making it essential for treatment discontinuation to be tailored to individual risk factors (Zipursky et al., 2020). Regrettably, upward of 30% of patients are resistant to the treatments for whom clozapine, as second generation "atypical" antipsychotic, may be more effective than other therapeutics (Haddad and Correll, 2018). However, even here poor responses are often encountered, and finding alternatives has been difficult largely owing to heterogenous underlying mechanisms associated with the disorder (Wada et al., 2022). In view of the treatment difficulties encountered, greater attention focused on other models of the disorder and alternative treatment strategies that might be more effective than agents that acted primarily on dopamine processes. Among other things, greater focus was placed on the involvement of cortical and subcortical systems and the contribution of GABAergic and glutamatergic functioning, especially the balance between cortical excitation and inhibition that emerges during neurodevelopment, including the possible excessive pruning of synapses (Howes and Shatalina, 2022; Wada et al., 2022).

Enter Glutamate …

The dopamine hypothesis had enjoyed a long run, and despite having certain problems, it persisted and expanded into a more general neurochemical theory of schizophrenia that now includes glutamate, and to a lesser degree, gamma-aminobutyric acid (GABA) (Howes et al., 2017). The glutamate hypothesis was designed to remedy problems associated with perspectives that focused primarily on dopamine functioning. The chief premise was that the pathophysiology of schizophrenia involves insufficient glutamate activity, or NMDA receptor *hypofunctioning*. This was supported by studies that showed NMDA antagonists could produce negative, positive, and cognitive symptoms that are aligned with a schizophrenia-like phenotype (e.g., McCutcheon et al., 2020), whereas neuroleptics treatments increased glutamate within the insular cortex, primarily among treatment responders (Sonnenschein et al., 2022). Similarly, among treatment-resistant patients, several genetic variations associated with NMDA receptor function were related to glutamate levels in the anterior cingulate cortex (Griffiths et al., 2023). Rare genetic variants, including those that influence glutamate signaling, as well as DNA transcription and chromatin remodeling, have likewise been associated with schizophrenia (Farsi and Sheng, 2023). The altered glutamate functioning associated with schizophrenia may result from the actions of astrocytes, the primary brain regulatory glia. Specifically, astrocytes may do so by affecting glutamate synthesis, clearance of glutamate by excitatory amino acid transporters (Mei et al., 2018), as well as through the release of inflammatory factors that affect neuronal processes.

Among patients deemed to be ultra-treatment resistant, cortical thickness was reduced bilaterally in frontal, temporal, parietal, and occipital gyri, which were correlated with glutamate and its metabolite glutamine (Shah et al., 2020). The cortical thinning was also associated with patients' poor insight concerning their condition, which may contribute to their frequent nonadherence to treatment protocols, which undermines positive treatment outcomes (Kim et al., 2023b).

Schizophrenia and psychosis have been associated with reduced resting state glutamate alterations within the anterior cingulate cortex, varying when individuals engage in cognitive tasks. Depending on the features of the illness, the relationships between glutamate levels and brain activity may be altered by therapeutic agents (Zahid et al., 2023). Even though imaging studies have consistently shown glutamate and GABA variations in association with cognitive dysfunction in schizophrenia (and depression), additional analyses are still needed to identify their presence at various stages of the illness, and the influence of medications at these times (Reddy-Thootkur et al., 2022). This can similarly be said concerning the ties between glutamate and GABA brain functioning in relation to positive and negative symptoms, as well as in distinguishing between patients who are treatment-responsive or resistant. As well, the question of individual differences that exist among schizophrenic patients needs to be explored in greater depth. In this respect, considerable variability of glutamate and its metabolites occur in several brain regions among patients, and in some brain regions the variability was more pronounced among younger, more symptomatic patients, which were tied to positive and negative symptoms. Being able to identify the causes for this variability may be instrumental in defining patient subgroups and optimal therapeutic strategies (Merritt et al., 2023).

From this perspective, a disturbed excitatory system in the brain creates imbalances in the integration of neuronal firing patterns such that there is a bias in the flow of excitatory and inhibitory influences on key behavioral actions, such as response-inhibition, maintenance of working memory, regulation of emotion, and control over perceptual functions. These actions may have a neurodevelopmental origin, given that during the preadolescent years there is a gradual balancing of inhibitory and excitatory synaptic functions in the prefrontal cortex (Howes and Shatalina, 2022). A thorough analysis of genome-wide studies that identified well over 100 genetic abnormalities pointed to a general loss of balance between excitation and inhibition in the brains of schizophrenic patients (Devor et al., 2017), which included glutamate, GABA, dopamine, serotonin, acetylcholine, and opioids (Sundararajan et al., 2018).

If there is a major pathophysiological profile of schizophrenia, this is dominated by a general appreciation that dopaminergic transmission is dysregulated, and may be tied to altered functions of regulatory neurotransmitters that are essentially responsible for traffic control in the brain (see Fig. 14.1). To this end, glutamate and GABA, our "green light" and "stop sign," respectively, take part in converging simultaneously on most cells in the brain, and by doing so, dictate whether the target cell will generate a response (glutamate wins) or not (GABA wins). This is the general scenario in virtually all areas of the brain, and an important part of the neural circuits driving many of our behaviors. There is no mystery that schizophrenia is a condition characterized by competing neural commands and scrambled bits of neurochemical information. In essence, genetic abnormalities that control many of the slow and fast forms of transmission in the central nervous system offer a good explanation for the origins of the discordant and disorganized nature of the behaviors displayed in schizophrenia (Devor et al., 2017). It is knowing precisely how we get from genetic variability to neurochemical dysregulation that is the over-arching problem—the solution to which remains elusive.

Not unexpectedly, several hormones have been implicated in the development or maintenance of schizophrenia. The sex differences in the occurrence and clinical course of the disease implicated gonadal steroid hormones acting in a protective capacity (Papadea et al., 2023), and diminished oxytocin levels or receptor functioning may contribute to the affective and social symptoms of schizophrenia (Goh and Lu, 2022). The identification of these and other hormones in subserving specific symptoms of schizophrenia has opened the door for the development of new pharmacotherapeutic targets that may be useful in treating the disorder (Doshi et al., 2023).

The Two-Hit Hypothesis

And so we come to the two-hit hypothesis. Two prominent features of schizophrenia that might interact with the immune system have been emphasized: genetics and development. Both are in keeping with what is likely to be the most promising approach to unraveling the cause of schizophrenia (Howes and McCutcheon, 2017). As in the case of other illnesses that are thought to be regulated by "two hits," this perspective is conceptual, since "two" (hits) could be three, four, or more factors that all interact in some way to generate disease. As we have seen in relation to other psychiatric and neurodegenerative disorders, this hypothesis adopts an interactive approach to the origins and emergence of schizophrenia. To the extent that genetic abnormalities may be necessary for schizophrenia to develop, they do not appear to be sufficient. More than one gene is involved in schizophrenia, and the culprit genes may exert their influence at critical stages of development, both prenatally and after birth. Development is a dynamic process, during which nothing is complete until a particular age-related milestone has been reached; any additional changes are layered into a mosaic of biological mechanisms that drive growth toward functional maturity. Throughout this process, periodic fluctuations in specific gene function may fully express a given abnormal phenotype or weaken the ability of specific biological systems to adapt to discrete environmental events, which may lead to pathology.

What we know about schizophrenia fits neatly into the two-hit hypothesis. Genetic risk is one factor, whereas developmental critical periods sensitive to neurobiological perturbation serve as the second—and stressors (or environmental events) serve as the third. In the latter categories, we can include alterations in immune system functioning. No one element seems to be sufficient to explain the etiology and progression of the disorder, but in combination, something might give (akin to the concept of allostatic overload that we discussed concerning stressor effects on the development of psychological disturbances). In what follows, we will attend to some of the features of immunity that have been related to schizophrenia.

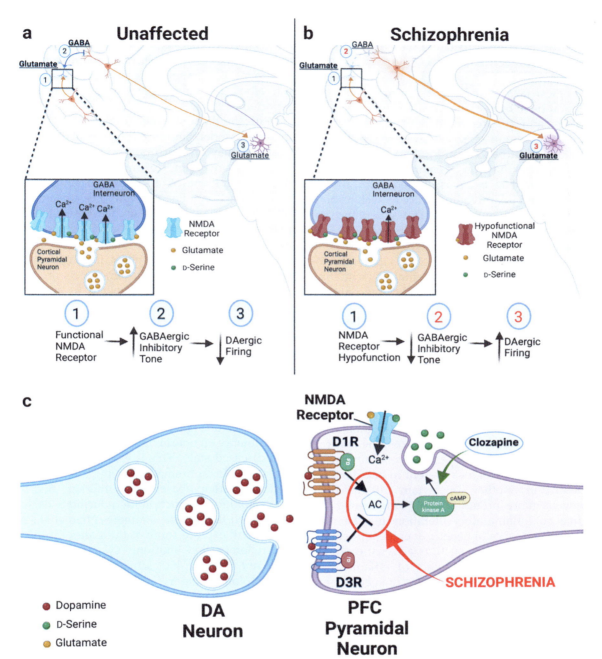

FIG. 14.1 Interplay between dopamine and glutamate systems is disrupted in schizophrenia. (A) In unaffected individuals, striatal dopamine (DA) levels are modulated via a polysynaptic circuit: (1) Cortical pyramidal glutamatergic projections stimulate GABAergic interneurons via stimulatory NMDA receptors. Upon activation by local increases in synaptic glutamate and co-substrate D-serine, NMDA receptors conduct Ca^{2+}, triggering increased GABAergic interneuron firing (inset). (2) GABA released by these interneurons lowers the firing rate of cortical glutamate neurons that project onto DAergic midbrain neurons. (3) The resulting glutamate tone from these stimulatory projections to the midbrain enables homeostatic control over DAergic firing in the mesolimbic pathway to finely modulate striatal DA levels. (B) According to the NMDA receptor hypofunction hypothesis, in schizophrenia, disruptions in DA-glutamate communications within this circuit drive abnormal increases in striatal DA to produce schizophrenia pathology. Hypofunctional NMDA receptors expressed by GABAergic interneurons diminish the amount of Ca^{2+} influx in response to glutamatergic stimulation (inset). This significantly dampens GABAergic inhibitory tone and increases firing by the midbrain-projecting secondary glutamatergic neurons. The resulting increases in midbrain DA neuron activity raise striatal DA levels, contributing to the positive symptoms of schizophrenia. Red numbers indicate pathway steps affected by schizophrenia pathology. (C) In pyramidal neurons of the prefrontal cortex (PFC), the combined actions of dopamine (DA) on stimulatory D1 (D1R) and inhibitory D3 (D3R) receptors modulate activity levels of adenylate cyclase, the enzyme responsible for cAMP synthesis. Tight control over intracellular cAMP enables neurons to control the release of D-serine and finely tune local glutamatergic neurotransmission. In schizophrenia, we propose that disturbances in coordination between D1R- and D3R-mediated cAMP signaling alter dopaminergic regulation of D-serine release (in red). This consequently produces pathologic changes in the downstream glutamatergic circuitry. Conversely, treatment with the antipsychotic drug clozapine raises D-serine release (in green), providing a potential therapeutic mechanism for ameliorating the disturbances in the dopaminergic modulation of D-serine release. *Source: Buck, S.A., Quincy Erickson-Oberg, M., Logan, R.W., Freyberg, Z., 2022. Relevance of interactions between dopamine and glutamate neurotransmission in schizophrenia. Mol. Psychiatry 27, 3583–3591.*

The Immune System and the Two-Hit Model of Schizophrenia

Immunological factors and the development of neuroinflammation have been considered a viable influence in the etiology and/or precipitation of schizophrenia. Of course, not all cases of schizophrenia are attributable to inflammatory disturbances, and not all aspects of the immune system act in the promotion of the disorder. This said, schizophrenia has been associated with disturbances of innate and adaptive immunity as well as both humoral and cellular immunity. The emergence of schizophrenia has been linked to peripheral immune irregularities as well as altered microglia functioning within the brain. The identification of specific immunological disturbances may be instrumental in facilitating precision therapies, despite so many other mechanisms being tied to the disorder.

We will dig more deeply into the ties between inflammatory-immune processes and schizophrenia, but before we do so several caveats need to be introduced. Altered levels of inflammatory factors, as we saw in several earlier chapters, have been associated with anxiety, depression, PTSD, and bipolar disorder. How is it that so many conditions, several of which are distinctively different from one another, all involve inflammation? There are, of course, several explanations that can accommodate these findings. Specific proinflammatory and antiinflammatory cytokines present in one disorder may be absent in another, or certain inflammatory changes may be more pronounced or long-lasting. As well, the impact of inflammatory alterations may be more prominent in the presence of genetic or epigenetic conditions or specific experiences, such as adverse early-life encounters. Furthermore, the impact of inflammation may interact with or produce alterations of specific neurotransmitters (e.g., dopamine and glutamate) or neurotrophins. Indeed, dopamine receptor alterations situated on lymphocytes may be disturbed among schizophrenic patients, varying with illness severity (Penedo et al., 2021). Ultimately, to use these inflammatory changes as markers of illness or to identify optimal treatment strategies, thorough analyses will be necessary that allow for the differentiation of how inflammatory processes are tied to specific aspects of pathologies.

Animal-based immune activation models have often concerned maternal immune system changes, and how these propagate a sequence of physiological events that impact the developing fetus. This work has largely supported a link between inflammatory factors and schizophrenia, although the processes by which this occurs have yet to be entirely uncovered. Nonetheless, new pieces are being introduced regularly; these often comprise better characterization of the problem and a clearer path to prevention and better treatment. Most certainly,

efforts to repurpose certain antiinflammatory drugs (such as acetylsalicylic acid, celecoxib, and minocycline) for the treatment of schizophrenia have begun, and some modest, albeit inconclusive, signs of promise have been reported.

With the data gathered from genetic studies, and an increased understanding of the predictive power of prodromal behavioral factors, schizophrenia seems to arise from the convergence of multiple influences that favor the development of disorganized and/or anomalous neural functions. For example, the current prominence of the glutamate hypothesis has raised the notion that glutamate-induced neurotoxicity together with dopamine activity may involve interactions with inflammatory processes. Among other findings, it seems that dopamine functioning, through effects on cytokine and kynurenine processes, is involved in the regulation of inflammatory operations. Likewise, disturbed glutamate activity has been related to microglial activation and central inflammatory processes that were tied to schizophrenia (de Bartolomeis et al., 2022). In fact, in a subgroup of schizophrenic patients that displayed high levels of inflammation, binding to glutamate (NMDA) receptors was particularly elevated (Rahman et al., 2022). In essence, inappropriate or prolonged release of cytokines in the brain and/or periphery may provide a destabilizing influence on neural functions operating in the context of a diseased state. This is especially true of treatment-resistant schizophrenia and being able to identify these patients, perhaps based on biomarkers tied to immune and cytokine dysfunction, may facilitate early and individualized therapeutic strategies (Jiao et al., 2022).

We will explore some of the issues more closely, but we reiterate that immune system functioning alone may not fully explain schizophrenia. More likely, the immunological theory of schizophrenia helps to bridge the contributions of genetics and the environment. As we indicated earlier, the specific genes that trigger the disorder have not been identified to a degree of confidence that serves to predict accurately whether an individual will develop schizophrenia. Important discoveries have been made linking mutations in the MHC region, as well as other more prominent mutations, such as in the genes for neuregulin and transcription factor 4, which are important in neuronal development and differentiation. In addition, genes for neurotrophins, serotonergic neurotransmission, cell adhesion molecules, sodium channels, and dopamine system functioning have been linked to schizophrenia (Zai et al., 2017).

Early studies had indicated that schizophrenia was associated with elevated plasma levels of IL-6 and complement component 3 (C3C) and C4, and positive acute phase proteins, including haptoglobin. Several other components of the immune-inflammatory system in mediating schizophrenia were subsequently identified, which pointed to the possible role of regulatory compensatory immune responses in subserving the disorder (Roomruangwong et al., 2020). Numerous studies likewise revealed elevated circulating inflammatory factors (e.g., CRP, IL-6, TNF-α) among schizophrenic patients. Moreover, markers of inflammation, including the neutrophil/lymphocyte ratio, monocyte/lymphocyte ratio, and platelet/lymphocyte ratio, were related to the appearance of schizophrenic symptoms (Bhikram and Sandor, 2022), which did not change following a full course of treatment. As well, monocytes have been tied to schizophrenia, varying with the presence (or absence) of a mood disorder (Hughes et al., 2021). Among individuals with early-onset schizophrenia, analyses of monocytes indicated elevated functioning of genes enriched for NF-κB signaling pathways, which influence the expression of various proinflammatory genes, whereas those of glucocorticoid response pathways were downregulated (Kübler et al., 2023). Importantly, in patients with greater inflammation based on a group of seven markers, elevated hippocampal, amygdala, putamen, and thalamus volumes were reported as was grey matter thickening (Lizano et al., 2021).

Specific single-nucleotide variants have been identified that were linked to schizophrenia (Kato et al., 2023) as were epigenetic changes that were related to synaptic functioning, neurotransmission, myelination, energy production, oxidative stress, and inflammatory mechanisms (Bilecki and Maćkowiak, 2023; Smigielski et al., 2020). Specific epigenetic changes were also determined that could predict the early onset of schizophrenia (Srivastava et al., 2022). For instance, epigenetic disturbances associated with glutamate 2/3 receptor changes were observed in the prefrontal cortex of some schizophrenic patients, which could potentially contribute to treatment resistance in response to neuroleptics (Matrisciano, 2023). Aside from these predictors, the occurrence of schizophrenia has been related to small noncoding RNA molecules (microRNAs; miRNAs) that can affect glutamate, GABA, serotonin, and BDNF, as well as inflammatory and oxidative stress processes (Zhang et al., 2023a). Of several miRNAs that were associated with schizophrenia, one in particular (miRNA-26a) exhibited especially good predictive power and could potentially be used as an early biomarker for illness development (Shafiee-Kandjani et al., 2023).

Analysis of recent onset versus established schizophrenia indicated that microglia-related inflammation, determined through binding of translocator protein (TSPO), indicated that in drug-free recent-onset patients, microglial functioning was disturbed, which was not evident in patients that had received antipsychotic therapy. As well, it appeared that in the presence of established schizophrenia, inflammation in the anterior cingulate and orbital frontal cortices was inversely related to

positive symptoms, and directly related to negative symptoms (Conen et al., 2021). In a similar earlier study by this group of researchers, which was conducted in patients 2–6 years following diagnosis, the availability of TSPO in several prefrontal cortical regions was not elevated in medication-free patients but was in medicated patients. Yet, TSPO availability was directly related to negative symptoms (Holmes et al., 2016).

As described in Chapter 2, Toll-like receptors (TLR) are involved in immune and inflammatory processes and are fundamental for damage/danger recognition patterns (DAMPs). It appears that some TLRs, notably TLR3, TLR4, and TLR7, were expressed at particularly high levels among schizophrenic patients, whereas expression of other TLRs (TLR1, TLR2, TLR4, TLR6, and TLR9) were downregulated (e.g., Kozłowska et al., 2019). Finally, viral and bacterial infections, including syphilis, have been linked to psychosis and dementia, while neuroimaging approaches have highlighted increased inflammation in the brains of schizophrenic patients. As previously reviewed (Feigenson et al., 2014), the role of inflammation in schizophrenia is very likely just one element of a range of other potential influences—some from the environment, some from constitutional anomalies—that converge on the individual and skew development toward an abnormal phenotype.

The development of synapses, particularly within the cortex, is highly dynamic during preadolescence. Grey matter volume expands dramatically during these early years, followed by a synaptic pruning phase that creates a seemingly optimal level of functional connections. The frontal cortex does not fully attain a stable state of synaptic connections until the early 20s. Until then, the cortex is sufficiently plastic but vulnerable to toxic damage and/or modifying environmental experiences. In schizophrenia, the early cortical changes appear to be unregulated and hence sustain more pronounced reductions in neurons and synapses. Whether an aberrant immune system contributes to this is an open question, although it seems that lower grey matter volume in regions associated with cognitive functioning in excited schizophrenic patients (i.e., characterized by persistent pacing, agitated behavior, repetitive movements, or speaking) was linked to altered cytokine functioning. As well, proinflammatory cytokines were related to white matter density among drug-free individuals experiencing a first episode of psychosis (Serpa et al., 2023).

In addition, immunological disturbances may impact postnatal neurodevelopmental stages in preadolescence. Neuroimaging data have revealed excessive cortical thinning in very early-onset schizophrenia, and that cortical degeneration can persist into adulthood (Cobia et al., 2012). It is not unusual to observe fluctuating changes in cortical volume during early life and a longitudinal analysis revealed that among high-risk youth aged

10–17, those who transitioned to psychosis exhibited marked reductions of cortical thickness, which was especially notable among younger individuals (Fortea et al., 2023).

Animal studies suggested that microglial cells influence synaptic pruning, which is part of a normal set of operations in the developing brain, and aberrant pruning may contribute to neurodevelopmental disorders (Mordelt and de Witte, 2023). Whatever other mechanisms are involved in tending to the number and quality of synapses that are formed in the brain, it seems that immunological factors—whether through cytokines or other means—represent a potential source of disruption to this process. In this respect, lowered levels of BDNF may interact with inflammatory factors in promoting the development of schizophrenia (Mehterov et al., 2022).

Cytokines, C-reactive Protein, and Schizophrenia

Several studies indicated that the inflammatory marker C-reactive protein (CRP) was highly correlated with the severity of schizophrenia and elevated CRP was accompanied by resistance to treatment (Fond et al., 2019). The elevated levels of circulating CRP were associated with disturbed attention and cognitive functioning and reduced cortical thickness and were related to the presence of negative symptoms (Orsolini et al., 2018).

Patients with elevated CRP who exhibited negative symptoms (sometimes referred to as "deficit" patients) displayed poorer cognitive functioning (memory, attention, and language performance) than healthy controls or individuals with less intense negative symptoms (Wang et al., 2022a). These findings are in keeping with reports concerning the links between CRP and specific features of the illness (Boozalis et al., 2018). However, a detailed systematic review that included 53 studies indicated that elevated CRP was also aligned with positive symptoms of the disorder, as well as microbial disturbances and cognitive impairments (Fond et al., 2018). It seems that elevated CRP is associated with both negative and positive symptoms among adolescents within the general population who had exhibited psychotic features that hadn't yet transitioned to an acute schizophrenic episode (Khandaker et al., 2021). Once schizophrenia becomes more intense and chronic, the profile of CRP with symptom features may change.

Paralleling the findings concerning CRP, schizophrenia was linked to several cytokines, varying with the course of the disorder. Indices of antioxidant status and levels of IL-6 and TNF-α were identified in first-episode patients (Fraguas et al., 2019). As well, relatively severe schizophrenic symptoms corresponded with elevated serum concentrations of IL-6, IL-8, and the soluble form of the IL-2 receptor (sIL-2R) (Dahan et al., 2018). In a comprehensive review designed to determine whether schizophrenia presents with a unique immunophenotype, it was noted that elevated circulating IL-1β, IL-6, IL-12, TNF-α, TGFb, and sIL-2R were frequent in both first episode and chronic forms of schizophrenia, whereas changes in IFNγ were more variable (Miller and Goldsmith, 2017). Likewise, levels of IL-2, IL-6, and IL-4 were elevated in the first episode of schizophrenia, as were levels of IL-1β. Moreover, elevated IL-6, which is involved in initiating and sustaining inflammatory responses, was associated with negative schizophrenia symptoms and the recurrence of symptoms (Cheng et al., 2023).

Among first-episode patients, levels of the chemokine CCL22 were elevated after controlling for the use of antipsychotic medication, BMI, and smoking. In these patients, CCL22 levels were associated with the presence of hallucinations and disorganization but did not appear to be tied to brain glial cell activity (Laurikainen et al., 2020). Not unexpectedly, levels of several other chemokines (eotaxin-1, MIP-1β, and IL-8) were also elevated with repeated schizophrenic episodes (Frydecka et al., 2018). A review of this literature confirmed that several chemokines (CXCL8/IL-8, CCL2/MCP-1, CCL4/MIP-1β, CCL11/eotaxin-1) were elevated in the blood of schizophrenic patients, and CXCL8 was high in cerebrospinal fluid. Similarly, altered chemokine levels were noted in peripheral immune cells, and altered genes coding for chemokines and their receptors were detected in the brain (Ermakov et al., 2023).

It seems that a broad and somewhat confusing array of cytokines was associated with schizophrenia. This is not altogether surprising given the heterogeneous nature of the disorder, often influenced by early-life experiences, diverse features of the gut microbiome, brain microglial factors, and varying with characteristics of the illness (Dawidowski et al., 2021). Cytokine variations were also related to the symptom profile patients exhibited. A review of pertinent studies indicated that the severity of positive symptoms of schizophrenia was directly related to many of these same cytokines (IL-1β, IL-6, IL-8, IL-10, IL-17, and IFNγ) and inversely related to IL-2 and TNF-α. Negative symptoms were also directly related to several of these cytokines (IL-1β, IL-6, TNF-α, IFNγ, and TGF-β), whereas an inverse relationship was associated with IL-17. There have also been reports that certain cytokines may be related to both positive and negative symptoms, although IL-6 and TNF-α elevations were more closely aligned with total negative symptoms (Goldsmith et al., 2018). Not unexpectedly, cytokine levels varied with disease duration and symptom severity, and in the presence of aggressive symptoms, cognitive abilities were correlated with levels of certain cytokines (Dawidowski et al., 2021). While not dismissing a role for other cytokines, it has been suggested that IL-6 may play an especially prominent role in schizophrenia and may be fundamental in the initial emergence of the disorder. A meta-analysis indicated that among individuals at risk for psychosis based on the presence of prodromal features, only high levels of IL-6 were present (Misiak et al., 2021).

The question has often arisen as to whether disturbed blood cytokine levels could predict the later appearance of schizophrenia among individuals who exhibit prodromal features of the illness and are thus at high risk for the development of pathology. A review of 15 studies indicated that plasma levels of IL-1β, IL-7, IL-8, and matrix metalloproteinase (MMP)-8 predicted transition to psychosis, whereas this was not evident with CRP or IL-6 (Khoury and Nasrallah, 2018). A subsequent report indicated that among individuals who transitioned to psychosis, IL-1β and IL-17 levels were significantly higher than in healthy controls or individuals who did not transition to psychosis (Ouyang et al., 2022).

Ambitious efforts have been undertaken to determine whether plasma analytes that reflected inflammation, oxidative stress, various hormones, and metabolism predicted the emergence of schizophrenia. Early reports regarding blood biomarkers were short on details regarding specific cytokines that might represent key therapeutic targets. Since then, it has been demonstrated that assessing blood samples of individual psychotic patients against a database of known inflammatory factors linked to psychosis might prove to be a useful predictive tool for schizophrenia onset (Perkins et al., 2015). Moreover, machine learning models may facilitate precise and generalizable prediction of psychosis in clinical high-risk young patients (Koutsouleris et al., 2021b). By including clinician risk assessments, structural magnetic resonance imaging data, and polygenic risk scores, the development of psychosis could be predicted with high accuracy (Koutsouleris et al., 2021a). In a follow-up undertaken at 9 and 18 months, it was possible to distinguish individuals based on positive and negative symptoms, as well as the presence of depression. A subgroup could also be identified that exhibited brain volume reductions associated with negative symptoms, and polygenic risk scores associated with schizophrenia (Dwyer et al., 2022). While it may be relatively early to form definitive conclusions, this approach could potentially be used to disentangle the contribution of various inflammatory factors in the development of schizophrenia at various phases of this disease and to predict optimal therapeutic strategies on an individual basis.

Unlike the many studies that assessed cytokine levels in plasma or serum, far fewer studies did so in CSF samples. Still, IL-1β and kynurenic acid were elevated in patients with schizophrenia or bipolar disorder relative to healthy controls, and IL-6 and IL-8 were elevated in CSF of patients with schizophrenia or major depression (Wang and Miller, 2018). As well, CSF levels of IFNb were elevated in patients diagnosed with either schizophrenia or bipolar disorder (Hidese et al., 2021), and IL-8 levels were elevated among individuals with schizophrenia spectrum disorders (Runge et al., 2021).

Inflammatory factors have been associated with altered volumes of specific brain regions. For instance, the level of IL-6 was associated with cortical thickness (Wu et al., 2019a), and elevated levels of IL-6 and TNF-α were associated with altered grey matter (Quidé et al., 2021), although contradictory findings have also been reported. Interestingly, postmortem analyses of protein or mRNA levels in the cortex of schizophrenic individuals also showed upregulation of IL-1β, IL-6, and IL-8 (Fillman et al., 2013), and prefrontal cortical protein and mRNA expression of TNF-α and IL-6 were increased, whereas that of the antiinflammatory IL-10 were reduced (Pandey et al., 2018). Similar effects were observed in other brain regions, although variability was evident concerning the cytokines most aligned with schizophrenia (Webster, 2023). Overall, studies across these different tissues have provided confidence regarding which cytokines might be useful targets for therapeutic intervention (Dawidowski et al., 2021).

As we've already mentioned, through their inflammatory actions, alterations of brain microglia and astrocytes may contribute to the development of schizophrenia (Dietz et al., 2020). This was apparent in postmortem analyses (Trépanier et al., 2016) and in imaging studies (Marques et al., 2019). Although there have been several inconsistencies across studies, the overall evidence supported the position that microglia activation is unusually elevated in a subset of patients (Zhou et al., 2023a).

Activation of microglia has been associated with diverse brain changes related to schizophrenia, such as abnormal synaptic pruning, elevated mortality of astrocytes, oligodendrocytes, neurons, and increased presence of proinflammatory cytokines. The elevated microglia in schizophrenic patients are correlated with poorer cognitive performance which may be due to the increased production of proinflammatory cytokines. This may be especially important not only because cognitive disturbances are present in most schizophrenia patients but also because cognitive decline may precede the primary illness symptoms or occur early in the course of the disease (Fatouros-Bergman et al., 2014). Microglial activation and the release of cytokines and free oxygen radicals may be provoked by psychological stressors that cause cognitive impairments associated with schizophrenia. Thus, therapies that target specific receptors on microglia might be effective in a prophylactic or therapeutic capacity to diminish features of schizophrenia (Zhou et al., 2023a).

From a mechanistic perspective, the value of these diverse findings is a bit uncertain, as similar types of changes in circulating cytokines have been noted in depression and bipolar disorders. The cytokine differences from healthy controls may reflect the stress typically experienced by patients or by lifestyles that had been maintained. Attempts to address this, as we've already discussed, have been through prodrome studies that can determine whether circulating cytokines can be used as biomarkers of pending psychotic breakdowns.

It is instructive that the effects of various therapeutic agents, especially neuroleptics, can influence cytokine levels. The concentrations of IL-1β, IL-4, IL-6, and IL-8 were elevated during an acute phase of schizophrenia, and at a 6-month follow-up elevated levels of IL-6 and lower levels of IL-8 were associated with greater improvement of negative symptoms. At this time, greater IFNγ levels were still apparent among patients with excitatory symptoms (He et al., 2020). Given the presumed importance of IL-6 in schizophrenia, attention has focused on changes in this cytokine following pharmacological therapies. A meta-analysis indicated that the high levels of IL-6 in schizophrenic patients were reduced following antipsychotic treatment, although it is unclear whether this was secondary to reduced symptoms or involvement of IL-6 in the beneficial effects achieved (Zhou et al., 2021c). A narrative review had indicated that IL-6 levels were not necessarily reduced following antipsychotic medications, perhaps indicating that this cytokine reflects a trait characteristic of schizophrenia (Dawidowski et al., 2021), whereas IFNγ might be a state marker (Halstead et al., 2023). In this respect, there is evidence that elevated IL-6 levels may be related to treatment-resistant illness, more so in females than in males (He et al., 2023a).

A meta-analysis that included drug-free patients with first-episode schizophrenia or psychosis again indicated elevated levels of several cytokines, but after neuroleptic treatment, most declined appreciably (IL-1β, IL-6, IFNγ, and TNF-α and antiinflammatory IL-4, IL-10), although not always reaching that of healthy controls. Moreover, levels of proinflammatory IL-2 and IL-17 remained persistently high (Marcinowicz et al., 2021). It is conceivable that the persistent elevations of these cytokines, together with IL-6, were associated with the development of a chronic condition or presaged treatment resistance.

An analysis of second-generation antipsychotic treatments revealed that TNF-α was reduced following risperidone treatment, whereas similar effects were not elicited by clozapine. These effects also varied with the characteristics of the illness in that risperidone reduced IL-6 and TNF-α in chronically ill patients but not in first-episode illness (Patlola et al., 2023a, b). These features of illness may ultimately be significant in selecting therapeutic strategies (e.g., inclusion of antiinflammatory agents as an adjunct therapy).

Immunologic Impact of Prenatal and Adverse Childhood Experiences: Preclinical Studies

The question of particular importance is whether prenatal or early postnatal infection or stressor experiences contribute to the emergence of later psychopathology. Earlier, we discussed human investigations and epidemiological analyses that suggested a link between maternal infection and schizophrenia, but these studies do not speak to possible causal relations that may exist. We now consider experimental approaches using animal models of immune activation, which could provide a better indication of possible causal contributions of inflammatory factors to disease occurrence. In general, immunobiological and psychosocial manipulations in the young animal have a significant impact on later-life endocrine, immune, and behavioral functions, and the emergence of schizophrenia-like symptoms during postnatal development could be elicited by prenatal immunologic activation and/or infection (Estes and McAllister, 2015). This research capitalizes on the notion that the immune response influences the developing nervous system of the embryo and/or fetus. A variety of outcomes have been assessed, including dopaminergic, GABAergic, and glutamatergic changes, as well as several tests of altered cognitive and emotional behavior. It is understood that the full spectrum of schizophrenia symptoms is difficult to replicate in animals, and to some extent, the behaviors that are assessed may apply to other conditions, such as autism and depression. For example, changes in social exploration, hedonic capacity, anxiety, learning and memory, and prepulse inhibition (PPI) can be found in a range of psychiatric conditions. Thus, while the effects of maternal infection or immune activation are relevant to schizophrenia, it is probably more helpful to view these findings as indicators of the contribution of prenatal infection to specific endophenotypes that cluster with other symptoms relevant to a diagnosis of schizophrenia.

An important perspective here is that an animal model is useful for understanding schizophrenia if it can reproduce aspects of the condition, even if these are not unique to the disorder. As described in our discussion of animal models, it is unlikely that any single test will reproduce all aspects of schizophrenia, and it is recognized that this is an impossible goal given some of its distinctly human aspects (e.g., formal thought disorder and hallucinations). Furthermore, given that significant postnatal and early adolescent environmental events are likely to be superimposed on intrauterine neurodevelopmental dysregulation due to prenatal infection, additive or synergistic interactions likely trigger truly abnormal behavioral changes that warrant psychiatric intervention.

The form of immune stimulation used in most experiments that assessed the impact of maternal immune activation (MIA) largely relied on molecular agents that activate the innate immune system. Most studies utilized either the endotoxin LPS or poly I:C, a double-stranded RNA that mimics viral infections. Both these immunogenic molecules stimulate macrophages, monocytes, and neutrophils with TLR3 being preferentially stimulated by Poly I:C, whereas LPS tends to stimulate TLR4. As noted earlier, LPS stimulates monocytes and macrophages to produce a range of proinflammatory cytokines that can have a variety of neurobehavioral effects. Poly I:C similarly induces proinflammatory cytokine production and exerts neural and behavioral effects, including increased turnover of monoamine neurotransmitters in the brain (Gibb et al., 2011; Meyer and Feldon, 2012). However, in normal adult mice, differences have been observed in the profile of neurochemical changes after challenge with LPS or Poly I:C. For example, LPS can have pronounced effects on brain monoamine neurotransmitter alterations, although dopaminergic changes are minimal in the prefrontal cortex. Similarly, the effects of Poly I:C appear to be relatively modest or absent in the prefrontal and limbic brain regions (Gandhi et al., 2007). Postnatal behaviors assessed in animals from mothers that have been subjected to MIA have included exploratory behavior, social interaction, cognitive function, and sensorimotor gating, many of which were affected (Meyer and Feldon, 2012). Interestingly, immune challenge with Poly I:C on gestational day 9 (GD9), but not GD17, resulted in impaired sensorimotor gating, as measured by the acoustic PPI procedure, while injection of Poly I:C on GD17 impaired working memory (Meyer and Feldon, 2012). Evidently, immune activation can cause differential phenotypic changes in the nervous system, depending on critical stages of embryonic development.

Of particular relevance to any preclinical animal study purporting to identify antecedents to schizophrenia-like abnormalities is evidence of neuroanatomical and neurotransmitter and receptor alterations. It is noteworthy that in rat offspring from mothers challenged with LPS on gestational day 15/16, dopamine D_2 receptor expression in the medial PFC, and numbers of D_2 receptor-expressing cells were reduced on postnatal days 35 and 60 (Baharnoori et al., 2013). This was consistent with murine postnatal dopaminergic changes in the prefrontal cortex and hippocampus pursuant to maternal Poly I:C challenge (Meyer and Feldon, 2012). Changes in the brain GABAergic system of offspring have also been reported after MIA. This includes a reduction in prefrontal cortical and hippocampal concentrations of GABA and glutamic acid decarboxylase (GAD), the synthetic enzyme needed to make GABA. These changes corresponded with maturation-dependent alterations in the prefrontal GABAergic transcriptome (Richetto et al., 2014), and as such support various neurotransmitter models of schizophrenia that may have their origins during the prenatal period.

Extending these neurochemical analyses are electrophysiological approaches used to determine neurodevelopmental alterations in synaptic activity in offspring from LPS-challenged pregnant rats. A common method for studying synaptic plasticity involves high-frequency electrical stimulation of presynaptic neurons to generate enhanced electrophysiological effects in postsynaptic neurons (i.e., long-term potentiation). This typically involves glutamate signaling and likely reflects the physical basis of memory formation. Conversely, downregulation of the postsynaptic glutamate receptor can attenuate postsynaptic electrophysiological responses after high rates of presynaptic stimulation, an effect referred to as long-term depression (LTD). Interestingly, glutamatergic signaling is altered in offspring from mothers given LPS. Specifically, LTD in the hippocampus was impaired because of disturbed glutamate NMDA receptor signaling (Burt et al., 2013), which agrees with evidence that prenatal immune activation alters glutamate functioning. These findings lend support to the glutamate hypofunctioning hypothesis of schizophrenia that was discussed earlier.

Finally, the normal operation of the brain is obviously dependent on efficient communication between different regions that specialize in cognitive, emotional, and motoric functions. This is mediated through well-defined neuroanatomical circuits that integrate cortical activity with subcortical information processing (e.g., sensory events). Prenatal challenge of pregnant rats with Poly I:C can produce asynchronous EEG activity in the prefrontal cortex and hippocampus in offspring (Dickerson et al., 2010). Given that synchronous activity (i.e., temporally coincidental activity) in these two areas underlies successful working memory performance, the asynchronous activity may reflect altered cognitive capacity. A review of the literature concerning the effects of prenatal Poly I:C treatment supported the contention that this treatment administered early in pregnancy elicited behavioral and molecular phenotypes reminiscent of schizophrenia and ASD (Haddad et al., 2020). As well, epigenetic changes of genes that influence GABA functioning were elicited by prenatal Poly I:C, which might account for the behavioral disturbances evident in offspring (Labouesse et al., 2015).

Impact of Viral Infection

A comprehensive review has implicated numerous viruses [specifically influenza virus, herpes virus 1 and 2 (HSV-1 and HSV-2), cytomegalovirus (CMV), Epstein-Barr virus (EBV), retrovirus, coronavirus, and Borna virus] encountered during the first or second trimesters of pregnancy in the emergence of schizophrenia. These effects have most often been attributed to elevated cytokine levels during gestation since they can cross the placenta and influence the fetal brain. Predictably, viral infection may be linked to environmental influences, such as living in densely populated regions and experiencing income inequality, and may be related to the presence of specific genes or epigenetic actions that affect the response to viruses (Kotsiri et al., 2023).

Epidemiological data has energized the immunological hypothesis for developmental origins of schizophrenia, and there have been reports in rodents indicating that aside from known immunologic changes (e.g., cytokine expression) that can occur in the fetal brain, infectious agents (e.g., influenza virus) can traverse the placenta and gain access to the developing organism. However, this interpretation is not altogether conclusive, as others have found no evidence for cross-placental migration of the influenza virus into the fetal brain. Although this issue is largely unresolved and may depend on a variety of different factors, the consensus is that altered neurobiological functioning following maternal viral infection is the result of maternal immune reactivity.

A meta-analysis indicated that elevated CRP or IL-8 was associated with subsequent schizophrenia in the offspring (Zhang et al., 2018a). How these inflammatory changes come to favor schizophrenia isn't certain, but it may be significant that MIA may result in IL-6 elevations in offspring, which promotes increased dopamine functioning (Aguilar-Valles et al., 2020). Supporting the link between maternal viral infection and behavioral disturbances in offspring, maternal influenza virus infection of rhesus monkeys reduced the amount of white matter in the cerebellum and the grey matter density in the prefrontal cortex, frontal cortex, cingulate, insula, parietal cortex, and the superior region of the temporal cortex (Short et al., 2010). Likewise, in nonhuman primate offspring of Poly I:C treated mothers assessed at 3.5 years of age, dendritic morphology was disturbed in the dorsolateral prefrontal cortex (Weir et al., 2015). Furthermore, this manipulation resulted in a reduction of grey matter in the prefrontal and frontal cortices at 6 months, which was still apparent upon subsequent evaluation at 45 months, and at that time frontal white matter volumes were likewise reduced (Vlasova et al., 2021). These brain changes were accompanied by specific cytokine alterations, such as increased levels of IL-1β, IL-6, and TNF-α measured at 1 year of age, and at 4 years of age, IL-1β was still elevated (Rose et al., 2017).

Moreover, the possibility exists that prenatal immunologic influences are more profound in genetically vulnerable individuals, and there is evidence for altered sensorimotor gating and cognitive deficits in animals that were infected or immunologically challenged as neonates. Taken together, the findings have pointed to the immune response and/or inflammation as

being a significant factor in the multifactorial conceptualization of the two-hit hypothesis and schizophrenia development.

Impact of Prenatal Stressors

Epidemiological and preclinical studies have implicated prenatal stressful experiences as a risk factor for the evolution of schizophrenia, varying with the timing of these experiences and by sex, possibly by affecting brain microglia (Mawson and Morris, 2023). It was surmised that early-life stressful experiences may cause systemic inflammation, resulting in greater brain access to inflammatory factors and consequently elevated brain microglia activation. The combination of these peripheral and brain changes may contribute to the emergence of schizophrenia and may account for comorbid metabolic abnormalities that have been observed (Nettis et al., 2020). It was similarly proposed that prenatal and early postnatal factors could result in a shift from antiinflammatory M2 predominance to that of the proinflammatory M1 form. With persistent inflammation, several brain alterations may be provoked, such as neuronal loss, reduced dendritic spines, and myelin degeneration, and thus the use of antiinflammatory agents might be profitable as an adjunct treatment for schizophrenia (Messina et al., 2023).

Studies in rodents indicated that stressful early-life experiences promoted persistent disturbances of hippocampal microglial functioning, notably reduced synaptosome phagocytic capacity (Reemst et al., 2022a, b). Moreover, downregulated miRNA functioning, specifically, that of miR-125b-1-3p, was observed in the hippocampus of rats that had experienced prenatal stress and in the blood of adult humans who had experienced childhood trauma, which may have been a marker for the development of schizophrenia (Cattane et al., 2019). It has also been reported that early-life stress may increase neutrophil extracellular traps (NETs) that comprise networks of extracellular strings of DNA that originate from neutrophil granules, which can bind pathogenic proteins. The presence of NETs was elevated in schizophrenic patients who had experienced early-life abuse, and in rats that experienced adolescent stressors (Corsi-Zuelli et al., 2022). As NETs have also been associated with autoimmune disorders, the possibility was entertained that they might be a factor involved in the comorbidity frequently observed between these conditions and schizophrenia.

As described in earlier chapters, early-life stressors have been associated with epigenetic changes in genes that code for glucocorticoids, BDNF, and immune-related processes. Such studies in rodents and humans have implicated epigenetic changes (e.g., of NMDA-related genes) in elevating the risk for schizophrenia (Cattane et al., 2020). The development of schizophrenia stemming from adverse childhood experiences has been ascribed to altered glutamate and GABA functioning, as well as inflammatory and oxidative stress processes, and the interplay between gut microbiota and immune functioning, which may be aligned with genetic and epigenetic influences (Cattane et al., 2020).

Perinatal Infection and Schizophrenia

Early-life adverse experiences in humans have been associated with elevated IL-1β and TNF-α during subsequent adolescence, which was associated with the development of schizophrenia. As well, stressors encountered early in life were associated with elevated IL-6 which was associated with altered DMN connectivity in schizophrenic patients (King et al., 2021). Moreover, among individuals who experienced early-life sexual abuse and were subsequently diagnosed with schizophrenia, levels of CRP, IL-6, and TNF-α were elevated, a profile that differed from that evident in bipolar patients (Quidé et al., 2019). A review of similar studies indicated that adverse early-life experiences were associated with subsequent levels of IL-6 and TNF-α being elevated, BDNF levels and hippocampal volume reduced, and the occurrence of schizophrenia elevated (Miljevic et al., 2023). Based on longitudinal cohort studies among individuals from the general population, it was suggested that circulating indices of inflammation and oxidative stress, as well as CSF cytokine levels, could be useful markers in identifying the risk of schizophrenia related to early-life adversity (Upthegrove and Khandaker, 2020). These reports, and the hypotheses that emerged from them, focused attention on the prenatal effects of environmental stressors. This included the role of prenatal infection, and the immune response of the pregnant mother, in provoking neurodevelopmental disorders, such as schizophrenia. There are now hundreds of studies in humans and animal models that argue for the potential influence of the maternal immune response on neural and behavioral development.

Prenatal Infections

In discussing the effects of various prenatal and early postnatal insults on later pathology in Chapter 7, bacterial and viral infections were linked to psychological disturbances in adulthood. Indeed, the link between viral infection and schizophrenia was fairly impressive, and it was maintained that the effects of maternal immune challenge stemmed from elevated levels of proinflammatory cytokines, including IL-1β, IL-6, TNF-α, monocyte chemotactic protein 1, and leukemia-inhibiting factor (Izvolskaia et al., 2020). As well, the effects of prenatal infection have been attributed to persistent changes related to alterations within apoptotic pathways, oxidative stress, and diminished antioxidant processes.

Of the various infections that have been addressed in humans (e.g., rubella, herpes simplex), influenza has received the most scrutiny. Following on from epidemiologic reports indicating increased incidence of schizophrenia among offspring of pregnancies that coincided with major influenza epidemics, analyses were undertaken using archived maternal serological data from births recorded between 1959 and 1966. Antibody to influenza was measured in the serum to confirm exposure to influenza virus during pregnancy. The appearance of antibodies during the first half of gestation suggested that exposure to influenza at this time was associated with a three-fold increase in risk for later development of schizophrenia among the offspring. This risk increased seven-fold among those who showed antibody appearance during the first trimester (Brown et al., 2004). This was the first empirical evidence using a nested case-control design that an immune response to influenza during pregnancy increased the risk for the development of schizophrenia among the offspring. Other infectious agents presenting during pregnancy, such as *Toxoplasma Gondii* (*T. Gondii*) and herpes simplex virus type 2 (HSV-2), were also linked to subsequent diagnoses of schizophrenia in the offspring (Brown and Derkits, 2010). Furthermore, the odds ratio for the development of schizophrenia from pregnancies that involved diverse infections (rubella, mumps, respiratory infection, measles, and polio infection during pregnancy) was variable (ranging from 0.62 to 3.58) (Scharko, 2011). However, a problem with this type of research is that the presence of antibodies, such as those related to influenza, can be detected for lengthy periods following infection, raising the possibility that serological testing may have uncovered viral infection that had occurred before rather than during pregnancy. Accordingly, prospective studies ought to employ indices of infection during pregnancy beyond simple serological antibody analyses.

If maternal infection poses a risk to the offspring, what is the mechanism? Is it the maternal immune response, or direct infection of the fetus? Much of the attention on maternal immune effects on postnatal psychiatric disorders is focused on viral infections, but owing to the physical and immunological blockade set up by the placenta, it is rare for the fetus to be infected during a maternal viral infection. This said, on occasion, certain viruses [e.g., CMV, herpes simplex virus-2 (HSV-2), or rubella] can breach the placenta and infect the fetus, causing impaired development of neurosensory systems, learning deficits, and in the case of the zika virus, microcephaly (Racicot and Mor, 2017). Such events are infrequent, and in the United States, congenital rubella is virtually nonexistent, and the prevalence of fetal exposure to CMV is exceptionally low (0.05%), and 95% of the 0.06% rate of neonatal

HSV infections are developed postnatally. This is likely due to the lack of adequate receptors for different viruses in the placenta, an evolutionary benefit that provides an effective impediment to the transmission of viruses across the placenta. In contrast, maternal susceptibility to other viral infections is higher, depending on the type of virus involved. Specifically, maternal influenza virus infections are around 40% and HSV-2 infections range from 18% to 22% (Racicot and Mor, 2017).

If there were an infection-related insult to the developing fetus, it would more likely come from the maternal side of the placental interface between mother and fetus. Moreover, given the rich armamentarium of innate, as well as adaptive immune cells in the decidual tissue of the uterus, the immune system seems the most likely threat to the developing fetus, although, one should be prepared to question the virtue of allowing an immune apparatus to engage an infectious microbial antigen without also ensuring protection of the fetus. The cytokine and antibody cascade that would ensue as a direct assault on an invading virus should hardly serve to compromise safe and healthy fetal development, and progression toward timely parturition. Intuitively, it makes little sense to accuse the immune system of interfering with intrauterine development. Nonetheless, as argued concerning obstetric complications, given the high degree of genetic abnormalities in schizophrenia, it is likely that special cases of vulnerability to the maternal immune response might affect the developing fetus.

One possibility in this respect concerns the influence of soluble factors, such as cytokines and antibodies. Indeed, elevated levels of inflammatory cytokines, such as TNF-α and IL-8, are significantly elevated in the serum of mothers whose offspring later develop schizophrenia. An extensive review of this literature pointed to multiple processes that may contribute to schizophrenia developing in response to viral infections of various types (Kotsiri et al., 2023). Viruses may influence the expression of genes that promote elevated cytokine levels, or pathogens may interfere with normal neural development. For instance, TNF-α elevations stemming from infection can produce neurotoxic actions that interfere with cell growth. Alternatively (or in addition), increased complement protein, which is involved in innate and adaptive immunity, may engender excessive synaptic pruning (e.g., Allswede and Cannon, 2018). These observations support the notion that during gestation maternal immune factors can be a vulnerability factor for developing schizophrenia. Irrespective of the viral infection considered, or the immune processes examined, as described in Fig. 14.2, it is certain that their impact varies with numerous variables, such as sex of the offspring, genetic influences, prenatal stressors encountered, dietary ingredients that influence microbial factors, and perhaps parental age (Hall et al., 2023). While many women experienced SARS-CoV-2 infection during pregnancy, it is still too soon to know all the postnatal consequences on the developing fetus. However, early indications are that prenatal SARS-CoV-2 infection may be accompanied by developmental delays in about 10% of infants assessed at 12 months of age (Shook et al., 2022).

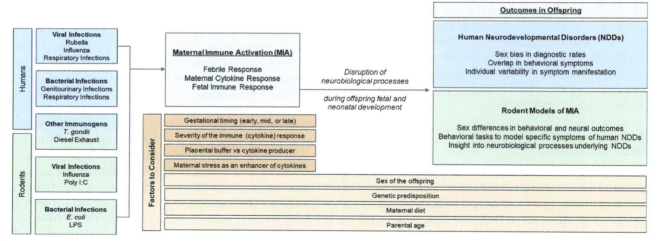

FIG. 14.2 Proposed model of the association between maternal immune activation (MIA) and neurodevelopmental disorders (NDDs). Blue boxes represent human pathogens and outcomes related to MIA. Green boxes represent rodent immunogens and outcomes related to MIA. Orange boxes represent factors that should be considered in rodent models of MIA and their relevance for human NDDs. Dark orange boxes represent factors that are related to the gestational immune response, whereas light orange boxes represent factors that are related to the immune response both during gestation and in postnatal offspring. *Source: Hall, M.B., Willis, D.E., Rodriguez, E.L., Schwarz, J.M., 2023. Maternal immune activation as an epidemiological risk factor for neurodevelopmental disorders: considerations of timing, severity, individual differences, and sex in human and rodent studies. Front. Neurosci. 17, 1135559.*

Postnatal Infections

Newborns are particularly susceptible to all manner of insults, including nutritional imbalances, and of course, infection. This may be superimposed on a genetic constitution that may be present (with significant anomalies or variations), some of which are genomic, and others that are epigenetic (Weber-Stadlbauer, 2017). Whereas prenatal immune events might alter postnatal behavioral propensities, additional immunologic events during the early (and even late) postnatal period may continue to disrupt development and tip things in favor of abnormal behavior. But how does infection after birth affect the developing nervous system and promote schizophrenia-like phenotypes? Surprisingly, little is known about this compared to prenatal immune events. Most studies involve animals, and these often use postnatal nonimmunologic stressors (e.g., maternal separation) that have been linked to epigenetic and behavioral alterations. In contrast, the analysis of the impact of postnatal immune changes on schizophrenia-relevant neurobiological changes, such as reduced hippocampal neurogenesis, is rare relative to maternal immune activation studies. While, several reports suggested that immune activation during early life can enhance the risk of adult psychiatric disorders, such as schizophrenia, the available data are limited, which is certainly an important gap in knowledge that needs to be addressed.

In response to infection, postnatal and childhood immune responses are superimposed on a sensitive and incompletely developed CNS. Normal and pathological MRI data show brain development to be a highly dynamic process right into adulthood, with changes in white and grey matter proceeding through regionally determined phases of cellular expansion followed by pruning and refinement of synaptic connections. Cortical grey matter increases during early childhood, then enters an elimination stage during which the cortex thins and seems to reach maturity sooner in fundamental areas of sensory and motor processing, being delayed in association cortices and in the prefrontal cortex (Sydnor et al., 2021). Concurrent with the maturation of cortical brain regions, neural projections grow along predetermined pathways, various neurotrophins increase synaptic connections and neuronal plasticity, and neurotransmitter maturation follows a well-laid-out progression. Challenges comprising activation of inflammatory processes (including the provocation of sterile inflammation) and activation of Toll-like receptors can undermine neurodevelopment during the fragile neonatal period. Aside from effects on neuronal processes, factors that influence brain microglia can promote several neurodevelopmental disorders, such as autism and schizophrenia (Patlola et al., 2023a, b). Upon moving past the neonatal period and into childhood, excessive exposure to stressors, such as maltreatment and abuse, can promote later mental health problems, including psychosis and schizophrenia, which are related to broad cerebral white matter disturbances (Xie et al., 2023b). Relatedly, stressors can alter cortical development (Howes et al., 2017), thus providing reasonable support for the notion that periods of dynamic neuronal sculpting can be derailed by traumatic and disturbing life events.

It will be recalled that infection can modify cognitive functioning, motor performance, motivational behavior, social investigation, and other sickness-related behaviors, which are among the many premorbid signs for the development of schizophrenia. Delirium and mental confusion are not uncommon in serious infections but are commonly thought to be a by-product of a medical condition. Nonetheless, given certain predisposing mental conditions and potentially malleable dynamic maturation of the brain in young individuals, infection operating as a stressor may introduce enduring changes in brain and behavioral development that side with the appearance of a psychiatric disorder. In the case of schizophrenia, proof of concept has been provided by clinical studies.

A review of the literature revealed a significant link between viral infections of the CNS in childhood and the future development of adult schizophrenia, which were likely moderated by genes (and their variants) that affected the risk of schizophrenia (Wahbeh and Avramopoulos, 2021). Likewise, epigenetic changes secondary to maternal infection or other challenges have been implicated in the emergence of schizophrenia (Khavari and Cairns, 2020). Infections that were most strongly correlated with later development of schizophrenia were CMV and mumps. Interestingly, an epidemiological analysis of almost 2 million children in a Swedish registry revealed that infections during pregnancy increased the likelihood of both psychotic development in the offspring and higher rates of childhood infection (Blomstrom et al., 2016). Notably, this study controlled for confounding variables previously linked to the development of psychosis, such as urban births, socioeconomic status, winter births, and being small for gestational age. It appeared that although maternal infection alone did not account for the development of schizophrenia, interactions with maternal (but not paternal) psychiatric complications predicted later psychosis in offspring. Again, this is in keeping with the two-hit hypothesis, emphasizing that children destined for a diagnosis of schizophrenia may reach this point as a function of converging influences of prenatal and postnatal experiences.

As discussed in Chapter 5, repeated exposure to various stressors can modify immune function and increase susceptibility to infection. Stress and emotional lability are common in schizophrenia, and while we have noted increased parameters of inflammation in schizophrenia, this may actually be secondary to the behavioral disorder. That is, the burden of having schizophrenia alters immune function. Among children and adults with diagnoses of schizophrenia or nonaffective psychosis, relapse during or after remission is often associated with a recent infection. In effect, infection may be viewed as a precipitant of psychiatric symptoms, such as hypermania, as well as being a consequence of stress reactions experienced. For the former condition, we know that infection and associated changes in proinflammatory cytokines can impact stress pathways in the brain. As such, it is not unlikely that infections could precipitate relapse or even initial conversion to psychosis during the prodromal phase. Incurring infection can arise from poor hygiene, especially during a psychotic period, or the presence of certain negative symptoms (e.g., avolition, depressed mood) that compromise self-care. Alternatively, a preexisting immunological dysregulation may be present, emerging during development, or altered by the various prodromal traits observed in people who eventually experience a psychotic episode.

Toxoplasmosis: **Toxoplasma Gondii** *Infection*

Attention has been given to the possibility that toxoplasmosis, an infection due to the coccidian protozoan, *T. Gondii*, causes schizophrenia in a subset of individuals. Particularly relevant to neuropsychiatric concerns is that this parasite has neurotrophic properties and can replicate in the brain (Wohlfert et al., 2017). The preferential host for *T. Gondii* is cats, which become infected by virtue of exposure to mice, birds, and other infected animal species. The parasite replicates in cats and is present in oocysts shed through feces such that household litter or outdoor regions can become areas of *T. Gondii* exposure for humans and other animals. It has been urged that pregnant women, young children, and any immunosuppressed individuals avoid exposure to specific areas (such as cat litter) or undercooked meat that might contain *T. Gondii*-containing oocysts. At least 10% of the US population over the age of 6 years has been exposed to *T. Gondii*, while in parts of the world with a warm, humid climate, over 95% of the population is positive for toxoplasmosis. Most people exposed to *T. Gondii* are asymptomatic due to effective immune surveillance. However, in immunocompromised individuals, infection results in flu-like symptoms (fever, weakness, muscle aches, and pains), and in more severe cases, toxoplasmosis can lead to neurologic problems (e.g., seizures and encephalitis).

Reviews of various studies that examined the relationship between serological levels of *T. Gondii* antibodies and schizophrenia revealed a significant link between infection and schizophrenia (Rantala et al., 2022).

Similarly, register-based studies in Denmark revealed that higher serum levels of antibodies from pregnant mothers and infants were predictive of the future development of schizophrenia (Contopoulos-Ioannidis et al., 2022). *T. Gondii* in pregnant women is a risk factor for the development of schizophrenia in offspring implies a possible sensitivity of the developing nervous system to toxoplasmosis, and it is important to consider that postnatal exposure to *T. Gondii* may similarly alter neural development either directly or via the immune response. Acute *T. Gondii* infection can directly alter dopamine metabolism, kynurenic acid activity, and glutamate signaling (Haroon et al., 2012) and can effectively colonize neuronal cells as well as astrocyte and microglial cells. Additionally, infection can alter cellular migration, cytokine production and release, and regulation of neurotransmitter activity. Indeed, reactive astrocytes and microglia initiate a series of complex signaling cascades in response to *T. Gondii*, generating greater levels of prostaglandins and members of the transforming growth factor, interleukin, and interferon families (Wohlfert et al., 2017).

The effect of *T. Gondii* on the peripheral immune response may also play a role in the development of mental health challenges. Given the prominence of schizophrenia-associated gene variants occurring in the MHC region, it is notable that the response to *T. Gondii* infection appears to be influenced by specific genes in this region (Conneely et al., 2019). Asymptomatic hosts infected with *T. Gondii* typically have elevated immune responses to counter the infection, and this creates a dynamic environment in which parasite and host alter their respective environments to seek mutual advantages. To escape immune surveillance, *T. Gondii* exerts an inhibitory effect on proinflammatory cytokines, while simultaneously augmenting antiinflammatory and/or regulatory cytokines. This is coupled with reduced production of nitric oxide, which can promote cytotoxic effects, thereby allowing the parasite to propagate within the host cells. In the brain, control of parasitic replication and protection of neurons is supported by infiltrating T cells, which can then limit infection (Wohlfert et al., 2017). As an act of self-preservation, *T. Gondii* infection reduces the number of cytotoxic CD8+ T cells, an effect that may underlie the promotion of cognitive disturbances in schizophrenia (Bhadra et al., 2013).

In sum, there is strong evidence that *T. Gondii* infection—as a neurotrophic parasite—can interfere significantly with neural and behavioral functions. In the brain, it can potentially contribute to increased glutamate neurotransmission, contribute to excitotoxic effects, and interfere with dopaminergic functions (Romero Núñez et al., 2022). Given the presence of high antibody titers to *T. Gondii* in schizophrenia, it is not unreasonable to consider that genetically and environmentally produced vulnerabilities interact with toxoplasmosis and precipitate psychosis. At present, we are only at the stage of hypothesizing such possibilities, and while the evidence does seem to be pointing in this direction, alternative views have been expressed.

Endogenous Retroviruses

A potential bridge between environmental stressors and the later development of schizophrenia is the human endogenous retrovirus type-W (HERVs). HERVs comprise a family of heritable retroviruses that are found in the CNS and are associated with elevated proinflammatory cytokine levels, and with disorders, such as bipolar disorder and schizophrenia (Slokar and Hasler, 2015; Tamouza et al., 2021). The expression of HERV-W, however, is variable and typically dormant but can be activated by specific environmental triggers, which can lead to an immune response against HERV-W-associated retroviral envelope proteins. These can induce a proinflammatory response consisting of elevated IL-1β, IL-6, and TNF-α, which as we saw are often associated with schizophrenia (Slokar and Hasler, 2015).

Increased HERV-W element expression in schizophrenia has been obtained through measures of serum, CSF, and tissue analyses, and is prominent in both first episode and chronic forms of psychosis. During embryonic development, the reduction of specific epigenetic modifications may create an environment conducive to the reactivation of HERV-W elements. The potentially elevated expression induced by environmental stressors aligns with the multifactorial model embodied by the two-hit hypothesis. Responses to HERV-W can have both developmental effects, as well as acute inflammatory consequences once development is complete. Influenza infection during pregnancy may also help activate a second proinflammatory cascade in the embryo through induced expression of HERV-W elements, and subsequently alter neuronal development (Perron et al., 2012). This may also sensitize individuals to respond to immune stressors at later times, creating a cyclic reaction, whereby future stressors and infections initiate additional HERV-W expression and subsequent inflammatory events. Indeed, TNF-α, BDNF, and the dopamine receptor D_3 may be elevated during the immune response to HERV-W, which may contribute to altered neuronal communication during the later phases of schizophrenia (e.g., Yan et al., 2022a). Finally, interactions have been noted with toxoplasmosis, in that greater expression of HERV-W can occur in some cases of schizophrenia with evidence of prior *T. Gondii* infection (Slokar and Hasler, 2015). Once again, it seems that the convergence of different infectious and immune elements may be complicit in precipitating and/or maintaining schizophrenia.

Autoimmune Disease and Schizophrenia

Just as autism was frequently associated with autoimmune disease (see Chapter 12), so was schizophrenia. This idea had been circulating for some time and was dusted off, so to speak, in the early 1990s, reviving earlier thinking about the possible importance of an autoimmune process facilitating, or at the very least, maintaining schizophrenia symptoms (Ganguli et al., 1995). Although interest in this relationship waned, with more observations on this front, there has been a resurgence of interest in the autoimmune hypothesis (Mayorova et al., 2021). Large epidemiological studies reported strong relations between varied autoimmune disorders [e.g., multiple sclerosis (MS) and lupus erythematosus] with psychosis and schizophrenia. These actions may be related to shared genetic risk factors, specific inflammatory or microbiota perturbations, as well as the T- and B-cell dysregulation in these conditions (Jeppesen and Benros, 2019). It is similarly possible that disturbed T_{reg} cell activity, which may be involved in autoimmune disorders (Corsi-Zuelli and Deakin, 2021), promotes dysregulation of microglia, thereby affecting dopamine and GABA functioning, as well as synaptic pruning, hence promoting schizophrenia.

Epidemiological research has shown that a high proportion of people with schizophrenia have strong associations with several autoimmune diseases. A 30-year population-based register study revealed that a history of autoimmune disease increased the likelihood of schizophrenia by 29%–45%, and this was increased to 60% with the inclusion of infection-related hospitalizations (Benros et al., 2011). Moreover, the incidence ratio for schizophrenia was virtually tripled (from 1.29 to 3.4) in cases of combined multiple infection and autoimmune disease. These studies also confirmed that the parents of schizophrenic individuals presented with a somewhat higher incidence of autoimmune disease than did the parents of healthy individuals (Nevriana et al., 2022), supporting the contention that associations between autoimmune disease and schizophrenia have a heritable component.

Genome-wide analyses revealed common variants between schizophrenia and several autoimmune disorders, specifically inflammatory bowel disease, Crohn's disease, ulcerative colitis, psoriasis, and systemic lupus erythematosus (SLE) (Pouget et al., 2019). Further to this, allelic variations in the MHC were consistently linked to autoimmune disease (Mokhtari and Lachman, 2016). As well, autoantibodies made against the NMDA glutamate receptor are prominent in schizophrenia patients.

A meta-analysis of cohort studies confirmed that autoimmune disorders involving the nervous system were associated with the occurrence of schizophrenia, varying with the nature of the autoimmune condition (Cao et al., 2023a). Unfortunately, association studies offer few clues as to why schizophrenia is associated with autoimmune disease. However, the data are consistent with the notion of immunological dysregulation and possible impairment of tolerance to self-antigens. Autoimmune diseases, in and of themselves, are notorious for eluding a well-defined etiology. The presence of autoantibodies explains the cause of organ or tissue damage in autoimmune disease, but what prompted B cells to make antibodies directed at self-antigens, and to relinquish the various control mechanisms that delete such B cells, is not known.

Despite the seemingly strong associations between autoimmunity and schizophrenia, several observations should give pause to any convictions regarding the autoimmune hypothesis. In particular, CNS-directed autoimmune diseases, such as MS and SLE, are not preferentially more comorbid in schizophrenia than in the general population. This poses a challenge to the concept of CNS-reactive autoimmune diseases being a necessary precursor for psychosis. Comorbidity of mental illness can indeed be high in MS, with close to one-half of MS patients reporting some level of anxiety, depression, bipolar disorder, or schizophrenia (Marrie et al., 2017). But much of this is accounted for by clinical depression and not schizophrenia. This is surprising, given that a hallmark feature of MS, that of demyelination, is observed in schizophrenia, while whole brain atrophy and ventricular enlargement—classical neuropathological indices of schizophrenia—are also observed in MS (Zivadinov et al., 2016). To be sure, these observations underlie different causal processes and follow different timelines,[4] but it must be considered that if schizophrenia involves antineural autoantibody production, other co-regulatory elements are required to induce psychosis.

Rheumatoid Arthritis and Schizophrenia

It is the general approach of most investigations to seek a positive association between autoimmune disease and schizophrenia. Remarkably, rheumatoid arthritis (RA) is one autoimmune disease that seems to share a strong *negative*, or mutually exclusive, relationship with schizophrenia. In fact, it was noted years ago that among the various comorbidities present in RA, there may be a low proportion that present with schizophrenia, although this may vary across SCZ patient subgroups

that were stratified based on their cytokine profiles (Eaton et al., 2022).

RA, as we discussed in Chapter 7, is an inflammatory disease with a worldwide prevalence rate of 0.5% in the adult population, and a heritability component of about 15%–30%, as determined by monozygotic twin concordance studies, may be as high as 60% based on GWAS (Dedmon, 2020). This is not a disease of adolescence, like schizophrenia, since most cases of RA tend to emerge after the age of 30. A prospective study indicated that patients with RA were less likely (by 25%) to display signs of psychosis (such as paranoid ideation), suggesting that the immunopathology present in RA somehow retards or inhibits the development of cognitive and emotional disruptions inherent in schizophrenia (or vice versa).

Investigations looking for clues among susceptibility genes for RA in the MHC have proposed only tenuous links to schizophrenia. For example, single nucleotide polymorphism (SNP) analyses in TNF-associated and other regions of the MHC showed only weak associations to psychosis, while a more direct selection of five RA-associated genes, including those for the cytokine-regulating transcription factor, NF-κB, did not find significant overlap with schizophrenia (Watanabe et al., 2009). The importance of such findings was that the *absence* of certain allelic variants normally observed in RA confers a risk to the development of schizophrenia, although how this operates is uncertain. An alternative view was that schizophrenia and RA have different forms (alleles) of similar genes, as well as SNPs in up to eight different genes (including two in the HLA region) (Wang et al., 2015). These allelic variations might exert pleiotropic effects (i.e., individual genes give rise to more than one particular phenotype) so that genes related to RA preclude the appearance of other phenotypes, including schizophrenia. A meta-analysis of GWAS revealed an inverted association between RA and schizophrenia that was tied to chromosome 6 of the HLA, whereas no association was apparent at loci outside the HLA region (Zamanpoor et al., 2020).

Autoimmunity, Inflammation, and the Glutamate Hypothesis of Schizophrenia

Psychopharmacological treatment of schizophrenia has fostered several hypotheses based on potential alterations in neurotransmitter signaling. As described earlier in this chapter, the *glutamate hypofunction* hypothesis of schizophrenia has gained increasing prominence. It is

---

[4] The onset of MS usually occurs after age 30, and most prominent in women. Schizophrenia has a much earlier onset and may even present with prodromal symptoms well before the initial psychotic episode. This is typically in late adolescence and early 20s. Interestingly, females tend to have a delayed onset of schizophrenia, most commonly in their mid-20s, which aligns with the commencement of the risk period for MS.

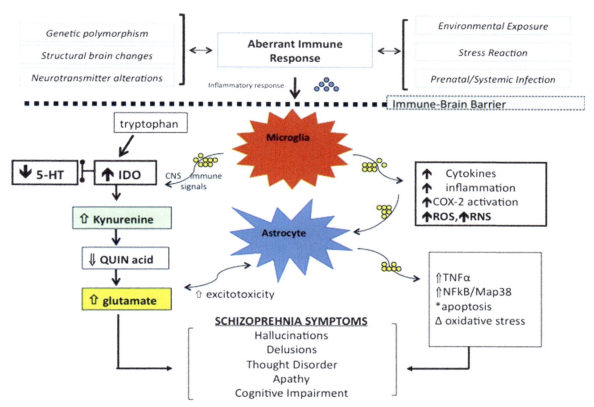

FIG. 14.3 Neuroinflammatory mechanism involved in schizophrenia and linked through the kynurenine pathway (CRP—C-reactive protein; IDO—indoleamine 2,3-dioxygenase; 5-HT—serotonin; RNS—reactive nitrogen species; ROS—reactive oxygen species). *Source: Watkins, C.C., Andrews, S.R., 2016. Clinical studies of neuroinflammatory mechanisms in schizophrenia. Schizophr. Res. 176, 14–22.*

believed that low levels of glutamate signaling can explain the symptomatology associated with schizophrenia, since glutamate antagonists can cause negative symptoms, and developmental abnormalities involve poor glutamate activity. This has been supported by in vitro and in vivo studies that showed dysregulation of the glutamate signaling systems in humans and animal models of schizophrenia, and many genes implicated in schizophrenia are involved in glutamate signaling (Trubetskoy et al., 2022). The dopamine system is still relevant here, but it is a question of how it is controlled by aberrant glutamate signaling. To understand how this scenario may come into play as part of an autoimmune process that affects the glutamate system, we can briefly inspect the known neural circuitry in the dopamine pathway, and how it is influenced by the activities of glutamate and GABA neurons (see Fig. 14.3 and text box).

The inflammation argument for glutamate dysregulation is based on the disruptive influence of cytokines and glial activity. One of the better-studied signaling cascades that regulate this interaction is the kynurenine system, which was briefly described earlier in relation to depressive illnesses but will be covered here in the context of schizophrenia. It will be recalled that in this pathway, kynurenine (KYN) is converted to tryptophan (TRY) with the help of indoleamine 2,3-dioxygenase (IDO), expressed predominantly by CNS astrocytes, or tryptophan 2,3-dioxygenase (TDO), predominantly expressed by microglia. Through a signaling cascade, various metabolites are formed, including kynurenic acid (KYNA), a naturally occurring NMDA antagonist, and quinolinic acid (QUIN), a natural NMDA agonist. Tellingly, the levels of KYNA in the CSF or tissue of schizophrenia patients were higher than in those of healthy controls (Cao et al., 2021a), thereby indirectly supporting the glutamate hypofunction hypothesis. In addition, impaired tryptophan metabolism and the associated generation of kynurenic acid are associated with the deterioration of cognitive symptoms in schizophrenia (Sapienza et al., 2023).

The kynurenine signaling system is tied to the immune system in that its two major catalysts, TDO and IDO, are both produced by glial cells, which become upregulated in response to inflammation and stress. In addition, levels of IDO and KYNA are increased in response to proinflammatory cytokines TNF-α and IFNγ that are upregulated in schizophrenia. An inflammation-induced elevation in KYNA may then alter glutamatergic signaling through noncompetitive antagonism, as well as by downstream signaling through α7 nicotinic acetylcholine receptors and GABAergic interneurons (Feigenson et al., 2014). This can result in extensive neurotransmitter dysregulation, including altered levels of glutamate, dopamine, acetylcholine, and GABA, that could be relevant in causing or exacerbating symptoms in schizophrenia, as described in Fig. 14.3.

Bidirectional communication occurs between inflammatory processes and dopaminergic functioning such that dopamine can regulate inflammatory processes, including kynurenine and cytokine functioning, which may contribute to schizophrenia (de Bartolomeis et al., 2022). Dysregulation of the tryptophan/kynurenine pathway may be introduced by diverse inflammatory processes, which ultimately affect neurotransmitter functioning, thereby promoting distinguishable cognitive and psychotic symptoms (Pedraz-Petrozzi et al., 2020). In this respect, analyses of postmortem tissues revealed alterations of kynurenine within the cortico-cerebellar-thalamic-cortical circuit, which was linked to neurotransmitter changes and schizophrenic symptomatology (Afia et al., 2021). As individuals with schizophrenia spectrum disorders and their siblings both exhibited elevated KYNA levels relative to healthy controls, this points to an inherited component concerning kynurenine functioning, perhaps rendering their unaffected siblings at elevated risk (Noyan et al., 2021).

Although the involvement of kynurenine in the development of pathology has most often been attributed to the neurotoxic effects of its metabolites, it is also involved in cellular energy production, which is needed when an immune response is being mounted. As well, kynurenine influences NMDA receptor signaling and glutamatergic neurotransmission, thereby favoring psychopathology. Furthermore, kynurenine may regulate endocrine, metabolic, and hormonal systems, which may come to affect cognition, metabolic function, and aging processes, hence increasing the risk of psychiatric disorders emerging (Savitz, 2020).

Before closing off this section, it is important to point out that although kynurenine and tryptophan metabolites were altered in the brain, the levels frequently differed from those detected in serum and plasma. Furthermore, while the increased KYN/TRP ratio in serum was related to schizophrenia, this was not apparent in plasma. As such, conclusions drawn concerning kynurenine and psychopathology need to be concerned with the compartment in which it was measured (Almulla et al., 2022).

Anti-NMDA Receptor Antibodies and Schizophrenia

An alternative perspective on glutamatergic hypofunctioning involves a B-cell-mediated autoimmune response that interferes with NMDA receptor signaling. There is no better direct form of immunological inactivation of a given molecular signal than the blocking or antagonistic actions of antibodies. The detection of anti-NMDA-R antibodies in schizophrenia has varied across studies. In a cohort of 121 first-episode patients, 10% showed multiple types of NMDA-R antibodies (Steiner et al., 2013), while only 3% were detected in another large cohort of first-episode patients (Lennox et al., 2017). It appears that detecting antibody for NMDA-R is a fickle business but seems to be more reliable than looking for other antineuronal antibodies, which are equally present in schizophrenic and healthy subjects (Lennox et al., 2017). An instructive piece of information is that toxoplasmosis may be a co-requisite condition for the appearance of anti-NMDA-R antibodies in both animal models and in first-episode patients with schizophrenia (Contopoulos-Ioannidis et al., 2022). This means that better confirmation of anti-NMDA-R antibody status may be enhanced by the consideration of infection history. Overall, however, the ability of anti-NMDA-R antibodies to help buttress an immunological hypothesis for glutamate hypofunctioning is still a work in progress.

Microglial Activity and Schizophrenia

The role of microglia in schizophrenia is beginning to be better understood. Microglia produce, and respond to, many different cytokines, and several antipsychotic drugs limit microglial activity and inflammatory cytokine production, which implicate microglia in the etiology of schizophrenia. A review of postmortem analyses revealed consistent evidence for greater microglial density in individuals who had been diagnosed with schizophrenia (van Kesteren et al., 2017) and a similar review indicated that activated microglia are present in schizophrenia (and in bipolar disorder and depressive illness), whereas markers of oligodendrocytes were reduced in several brain regions (Liu et al., 2022a). As well, among individuals with schizophrenia elevated presence of inflammatory-related genes was found in several brain regions, particularly the hippocampus (Lanz et al., 2019). Yet, the presence of inflammation in postmortem tissues was heterogeneous, precluding firm conclusions from being offered (Ai et al., 2023). Still, it appears that greater microglial proliferation occurs in situ, possibly through increased recruitment of myeloid precursor cells. It is uncertain, however, whether grey matter reductions in schizophrenia are related to increased microglial engagement in phagocytic activity, and removal of synapses and neurons (Sellgren et al., 2019). It is conceivable that the increased microglia in schizophrenia might be inherent to the neurobiology of the disorder or a byproduct of the stress of living with this condition. Since stress can increase microglial cell activation (Howes and McCutcheon, 2017), this is not an unreasonable explanation. Animal studies of both prenatal and early-life infections or stress can prime microglia to be more responsive to later life stressors and infections (van Kesteren et al., 2017), and once activated, the composition of glutamate receptors on microglial cells can result in augmented proinflammatory cytokine release.

Microglia can extensively regulate the glutamatergic signaling pathway. They express and respond to glutamate, have dynamic interactions with NMDA receptors, and alter glutamatergic synaptic function when activated. Furthermore, microglia regulate synapses through a highly regulated pruning process, which can include chemokine signaling and complement (Parellada and Gassó, 2021). Synapses are normally culled in an activity-dependent manner from early life into late adolescence, essentially preserving high-load synaptic connections at the expense of extraneous, low-use synaptic terminals. Individuals with schizophrenia have a disproportionate decrease in dendritic spines beginning around adolescence (Glausier and Lewis, 2012). This may represent aberrant synaptic pruning, which could lead to disruptions in several neurotransmitter systems. The overlap of critical periods of change in the density of spines and active synapses with the first onset of schizophrenia symptoms (or during the prodromal period) implicates pruning as a potentially altered process in schizophrenia (Sellgren et al., 2019). Whether aberrant microglial activity is at the heart of this process remains to be determined. Fig. 14.4 provides an overview of the presumed inflammatory interactions by which microglia may contribute to the emergence of schizophrenia.

Microbiota and Schizophrenia

Although microbiota has received considerable attention in stress-related disorders, such as anxiety and depression, and in gut-related inflammatory disorders, research assessing the link to schizophrenia has been more limited. If it is accepted that microbiota affect immune functioning and that immune processes contribute to schizophrenia (and to autism), then it's essential to at least introduce the potential relationship between these illnesses and microbial factors. We will cover this literature briefly, but excellent reviews on this topic as well as the relationship between microbiota and autism are available (e.g., Kelly et al., 2017). As we indicated in earlier chapters, microbiota alterations have been associated with various brain neurochemical, hormonal, neurotrophin, microglial, and immune changes, any of which could affect psychological state. Indeed, an impressive case for one or more of these processes being mediators

between microbiota and mental health disturbances. As tempting as it might be to experimentally follow up on these possibilities, it needs to be kept in mind that microbiota are sensitive to stressful events, high or low exercise regimens, diet, and smoking, all of which accompany schizophrenia. Likewise, antipsychotic agents that most patients were receiving might influence microbiota. Thus, it is probably premature to make firm conclusions concerning the connection between microbiota and schizophrenia.

Patients diagnosed with schizophrenia exhibit microbial differences from controls at the family, genus, and species level, and specific bacteria vary with the illness characteristics expressed. In medication-free schizophrenic patients numerous microbial changes were reported, including several facultative anaerobes (that can survive in the presence or absence of oxygen) that are rare in a healthy gut. Also, bacteria present in the oral cavity were more abundant in patients with schizophrenia, possibly reflecting the interplay between oral bacteria and those present in the gut. As we've already seen, gut bacteria and associated SCFAs influence the synthesis of neurotransmitters, such as glutamate, GABA, and nitric oxide, as well as the presence of kynurenic acid, which differed between schizophrenic patients and healthy controls. Gut microbiota were altered after 3 months of antipsychotic treatment although the restoration was incomplete. Later studies indicated that gut microbial changes were related to symptom severity and the presence of negative symptoms (Nocera and Nasrallah, 2022). Despite these impressive findings, considerable variability generally existed across studies, which is not surprising given the marked microbiota variability that is ordinarily observed among healthy individuals, and microbial changes would only be expected in a subset of schizophrenic patients. As a result, consensus has not been reached as to the microbiota profile that is characteristic of schizophrenia, although a few species have been identified that tend to be associated with the illness as well as in patients at risk (Novais et al., 2021).

The severity of symptoms was directly related to the abundance of gut *Lactobacillus* bacteria (Kelly et al., 2017), and *Lactobacillus*, *Bifidobacterium*, and *Ascomycota* were elevated in the mouth and pharynx of schizophrenic patients (Castro-Nallar et al., 2015). Likewise, schizophrenia was associated with greater heterogeneity of gut alpha diversity (referring to species diversity at a local scale), together with greater appearance of oral pathogenic taxa, such as *Veillonella* and *Prevotella*. Some bacterial taxa correlated strongly with proinflammatory cytokine levels together with hippocampal gliosis, demyelination, and excitatory neurotransmission. In addition, specific bacteria, notably *Lactobacillus* and *Megasphaera* were elevated in first episode schizophrenia, which

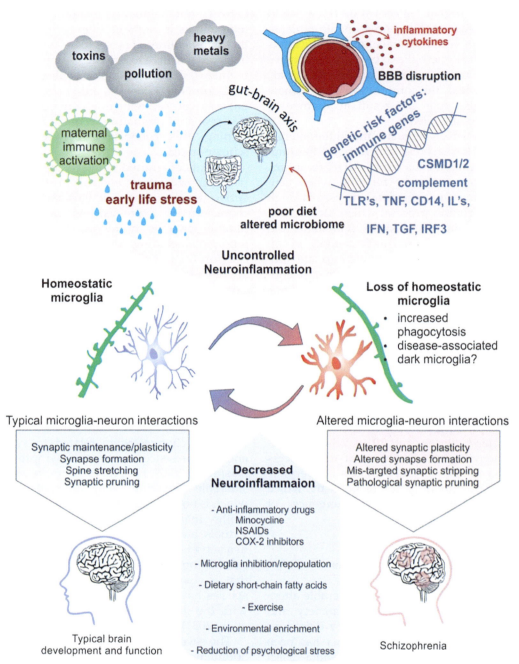

FIG. 14.4 Neuroinflammation-induced changes in microglia that are implicated in SCZ pathogenicity. Risk factors for SCZ that alter microglial function and enhance neuroinflammation include pollution, stress, nutrition-induced gut–brain axis dysbiosis, viral infection, maternal immune activation, genetic predisposition, and cytokine secretion. Homeostatic microglia perform their immune sentinel role by interacting with neurons to guide circuit wiring during development. In an increased inflammatory milieu, loss of microglial homeostasis perturbs microglia–neuron interactions that may cause altered plasticity due to pathogenic synaptic formation, synaptic stripping, and pruning. Therapeutic approaches that promote homeostatic microglia through the reduction of neuroinflammation via antiinflammatory drugs, microglial inhibition and repopulation, improved nutrition, environmental enrichment, and prevention of psychological stress could be potentially exploited to limit exacerbation of SCZ. *Source: Comer, A.L., Carrier, M., Tremblay, M.È., Cruz-Martín, A., 2020. The inflamed brain in schizophrenia: the convergence of genetic and environmental risk factors that lead to uncontrolled neuroinflammation. Front. Cell. Neurosci. 14, 274.*

corresponded with serum tryptophan levels, kynurenic acid, SCFAs, and alterations in the glutamate-glutamine-GABA cycle (Nuncio-Mora et al., 2023). It was likewise indicated that the severity of schizophrenic symptoms was related to the abundance of *Veillonella* and *Lachnospira*, and a microbial panel (that comprised *Aerococcaceae*, *Bifidobacteriaceae*, *Brucellaceae*, *Pasteurellaceae*, and *Rikenellaceae*) distinguished schizophrenic patients from healthy individuals. Interestingly, in male patients, the appearance of *Candida albicans*, a fungal factor, was elevated in schizophrenia (Severance et al., 2016), and the appearance of *C. albicans* antibodies declined in association with probiotic treatment, as did schizophrenia symptoms, although only to a modest extent (Severance et al., 2017). One particular microbe, *Streptococcus vestibularis*, was frequently present in schizophrenic patients, and when it was administered to mice they exhibited numerous behavioral disturbances, which were associated with immune dysregulation (Zhu et al., 2020). Likewise, germ-free mice that received microbiota from schizophrenic patients through fecal transplants exhibited hippocampal glutamate and GABA alterations and schizophrenic-like behaviors (Zheng et al., 2019a).

It has been maintained that gut dysbiosis may diminish protective factors and concurrently increase neuronal and synaptic disturbances aligned with schizophrenia (Yuan et al., 2019). In this respect, a metabolomic analysis comparing schizophrenic patients and healthy controls identified 18 differentially abundant metabolite clusters that were linked to higher proinflammatory cytokine levels, and lower antiinflammatory metabolites (e.g., oleic acid and linolenic acid). In effect, particular bacteria and their metabolites seemed to affect inflammatory cytokines through their actions on amino acids and fatty acids (Fan et al., 2022a).

It seems that microbiota can influence adult hippocampal neurogenesis and various manipulations that affect the gut microbiome (e.g., prebiotics, probiotics, and antibiotics, as well as diets that affect them) could potentially affect schizophrenia-related processes. Likewise, therapeutic interventions to treat the disorder may come to affect the gut microbiota, which might play some role in the effectiveness of the treatments. There is indeed reason to believe that microbiota alterations stemming from therapeutic agents could influence adult hippocampal neurogenesis, possibly acting through microbial metabolites, immune pathways, endocrine signaling, and neuronal functioning (Guzzetta et al., 2022). It follows that microbiota treatments could be used to optimize therapeutic outcomes and limit side effects that might otherwise emerge (Vasileva et al., 2022). It is similarly possible that therapeutic agents to treat schizophrenia may alter gut and oral microbiota, which could contribute to the treatment's efficacy. After 24 weeks of risperidone treatment, the number of fecal *Bifidobacterium* and *Escherichia coli* was elevated, whereas reductions were evident in fecal *Clostridium coccoides* and *Lactobacillus*.

We underscore that microbiota disturbances have been associated with numerous pathological conditions aside from schizophrenia and psychosis, including depression, anxiety disorders, and bipolar disorder, as well as autism. Transdiagnostic analyses have revealed similar microbiota signatures across several conditions.

Generally, this was characterized by beta diversity (i.e., the frequency of different communities within the gut) that comprised a reduction of antiinflammatory butyrate-producing bacteria, together with enrichment of proinflammatory genera (McGuinness et al., 2022; Nikolova et al., 2021). Through a two-sample Mendelian randomization analysis, it seemed likely that the gut microbiome caused the appearance of diverse pathologies (Ni et al., 2022). Particular microbial alterations may be a second hit that disposes individuals to pathologies that are determined by genetic or epigenetic vulnerabilities, although the findings may come about because of changes of lifestyle often resulting from mental health challenges.

As described in Chapter 7, prenatal maternal diet and microbial changes secondary to maternal stressors and immune-related illnesses have been linked to schizophrenia, and indeed, some of the effects of altered microbiota disturbances in rodents may not appear until adolescence, much like schizophrenic symptoms appear at this time. It has been suggested that elevated the presence of genes that modulate the immune system may promote neuropathological responses upon dysbiotic challenges or the presence of specific bacteria. These actions may occur prenatally or during early postnatal periods, culminating in disturbed normal neurodevelopment among genetically predisposed individuals (Severance and Yolken, 2020).

Although microbial and inflammatory processes may contribute to structural brain changes and distinct neurotransmitter disturbances, several inconsistencies have been encountered in this regard, possibly being related to the existence of inflammatory subgroups that determine features of schizophrenia. In fact, several schizophrenia types have been identified that could be distinguished from one another and from healthy controls based on the presence of specific inflammatory factors (low inflammation, elevated CRP, elevated IL-6/IL-8, elevated IFNγ, and elevated IL-10). To a considerable extent, the subgroups also exhibited grey matter volume in distinct brain regions and impairments of cognitive performance (Alexandros Lalousis et al., 2023). It is possible that similar clusters may exist concerning specific neurotransmitter disturbances, microglial functioning, neurotrophins, and other processes that have been linked to schizophrenia. Using data of this sort in a machine learning approach and considering the many features of schizophrenia that were discussed earlier (e.g., first episode vs chronic illness, presence of negative and positive symptoms) as well as microbial alterations, it may become possible to devise best therapeutic strategies. Fig. 14.5 describes some of the pathways by which microbial disturbances may contribute to the development of schizophrenia.

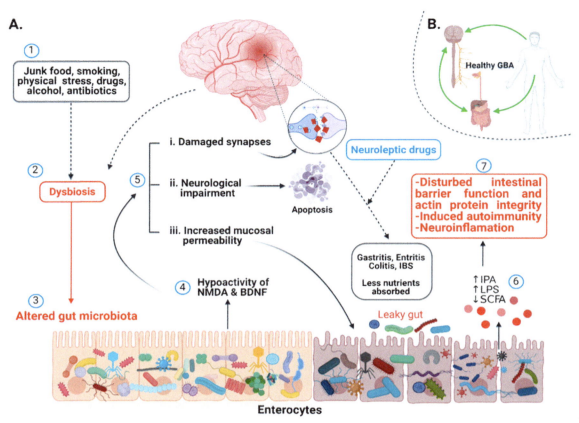

FIG. 14.5 Communication between the gut microbiota and brain in schizophrenia. (A): (1) Junk food, frequent use of drugs, and lack of exercise result in gut microbial dysbiosis. (2) Gut microbial dysbiosis means an alteration in gut microbial species. (3) Decrease in healthy gut microbiota and increase in pathogenic species. (4) Direct influence of altered gut microbes causes the hypoactivity of NMDA and BDNF receptors. (5) Hypoactivity of NMDA and BDNF receptors results in damaged synapsis, neurological impairments, and increased intestinal membrane permeability as indicated by solid arrows. Consequently, abolishment of spinogenesis, gastritis, enteritis, colitis, and irritable bowel syndrome occurs (indicated by dotted arrows). (6) Altered microbial products such as indole propionic acid (IPA), lipopolysaccharides (LPS), and short-chain fatty acids (SCFA). (7) Anomalous expression of microbial products leads to dysfunction of the intestinal barrier as well as induces autoimmunity and neuroinflammation. (B): Normal gut microbiota is crucial to maintain the gut–brain axis. *Source: Munawar, N., Ahsan, K., Muhammad, K., Ahmad, A., Anwar, M.A., et al., 2021. Hidden role of gut microbiome dysbiosis in schizophrenia: antipsychotics or psychobiotics as therapeutics? Int. J. Mol. Sci. 22, 7671.*

Concluding Comments

This chapter has provided an overview of some of the major areas that have sought answers for the etiology of schizophrenia in the inflammatory or immunologic domain. To some extent, the inclusion of the immune system as a pathophysiologic index is reasonable enough. This would place the immune system in a category with diabetes and metabolic disturbances, as well as cardiovascular diseases, all of which may be sequelae of a discordant and stressful lifestyle and/or poor neural regulation of systemic biological processes needed for good health. These are important variables to control and evaluate, and perhaps even investigate as potentially linked to the etiology of schizophrenia.

The immune system may be more than just an accessory biological entity altered and mismanaged in the context of an aberrant set of neural functions. It may actually be an important impetus for change in the CNS that is already weakened and susceptible as a result of inherited gene variations. This is based on research regarding the impact of prenatal immune activation, as well as studies that linked infections to exacerbation and possibly the triggering of psychotic episodes. In the meantime, as researchers try to nail down the precise immunological effects on the development and appearance of psychotic disorders, there are opportunities to try modulating inflammatory processes in the hope of improving symptoms and the quality of life for patients and family members. The use of minocycline as an adjunct to risperidone to augment the elimination of negative symptoms (Zhang et al., 2018b) and the use of nonsteroidal antiinflammatory drugs (NSAIDs) as an adjunct to ongoing antipsychotic treatment are slowly increasing, with preliminary trials suggesting they might be helpful (Fitton et al., 2022). Perhaps the strongest argument came from a small number of double-blind, placebo-controlled studies, which suggested that NSAIDs reduced the severity of positive and negative symptoms. However, other reports suggested that NSAIDs were less effective than targeting serotonin receptors, and in a large study of over 16,000 patients treated with an antipsychotic, the use of NSAIDs increased the risk of relapse (Köhler et al., 2016b). These findings point to the need for greater precision and control over the use of different types of NSAIDs (e.g., celecoxib, acetylsalicylic acid, ibuprofen), as well as an assessment of the inflammatory profile of individual patients, especially as the efficacy of NSAID treatment may be predicated on inflammation being present (Fitton et al., 2022). Moreover, such treatments may be provided earlier, if prodromal research identifies high-risk groups in need of antiinflammatory treatment. A review of randomized controlled trials that assessed the actions of numerous antiinflammatory agents (16 in all) that have been used as adjunct treatments revealed that only aspirin, estrogens, minocycline, and N-acetylcysteine were effective, and these were primarily useful for early-phase schizophrenia or first-episode psychosis (Çakici et al., 2019). As we suggested earlier, the ultimate benefits of specific strategies may hinge on the identification of subgroups of patients and the specific symptoms with which they present.

CHAPTER

15

Inflammatory Roads to Parkinson's Disease

A multihit framework for Parkinson's disease

A growing consensus has revolved around a "multihit" hypothesis for Parkinson's disease (PD). The first "hit" could be genetic in that gene polymorphisms, such as LRRK2, PINK, and DJ-1, have been implicated in the disease. Further hits, in the form of environmental and inflammatory insults, may contribute to the genesis of PD (Fig. 15.1). In the case of idiopathic PD (which accounts for greater than 90% of cases of the disease), heavy metals, organic pollutants, several pesticides, and even viral/bacterial infections can induce pathology in animals and these challenges have been linked to the disorder in humans. Additive or even synergistic interactions between genetic factors and stressors could shape the evolution of pathology.

Several lines of evidence indicated that microglial-mediated neuroinflammatory processes play a prominent role in PD and that environmental stressors can impact neurons through inflammatory mechanisms. In the broadest sense, environmental insults can prime neuroinflammatory cascades, resulting in augmented responses that, over time (possibly in conjunction with genetic vulnerabilities), can damage the susceptible midbrain dopamine neurons that normally are lost in PD. Of course, many processes independent of inflammation are also likely to be involved in this disorder.

General Introduction to Parkinson's Disease

This chapter focuses on the various environmental and genetic elements that contribute to Parkinson's disease (PD), emphasizing the role of the inflammatory immune system in the development, progression, and possibly the treatment of PD. Environmental triggers, such as pesticides and infectious agents, in concert with genetic vulnerabilities, most notably LRRK2 mutations, have well-documented proinflammatory consequences. Animal models have revealed that the administration of chemical or other toxicants that provoke robust neuroinflammatory effects, including activation of the brain immunocompetent microglial cells and the production of cytokines that serve as inflammatory signaling molecules, can engender effects reminiscent of PD (e.g., Tansey et al., 2022).

The brains of many PD patients display similar inflammatory signatures, and appreciable effort is being devoted to devising pharmacological treatments that tame these destructive inflammatory cascades, while still preserving beneficial immune processes. Such treatments could include specific antiinflammatory drugs or immunomodulatory agents that shift microglia and/or other immune cell states toward a protective or recovery mode. Throughout this chapter, we take a broad view of PD, focusing not only on the obvious primary motor pathology but also considering its common comorbid features, such as depression, as well as the many lifestyle factors that could contribute to the diverse facets of the disease.

PD is the second most common age-related neurodegenerative disorder, occurring in up to 0.2% of the population by 60 years of age and sharply rising to 4% in later years (Hirsch et al., 2016). The disease is characterized primarily as a movement disorder, in which patients display the four cardinal motor symptoms: resting tremor, rigidity, slowness of movement (bradykinesia), and postural instability. These motor disturbances emerge because of the progressive degeneration of dopamine (DA) neurons in the substantia nigra compacta (SNc) region of the midbrain.

The SNc DA neurons project to the caudate and putamen regions of the brain, forming the nigrostriatal pathway, which regulates many aspects of motor functioning. In rodents, the caudate and putamen are fused and collectively referred to as the striatum. This striatal (or in humans, caudate-putamen) brain area is considered a master regulator of many aspects of motor functioning and is a fundamental relay and integrative center for many other brain regions, which at a systems level are collectively referred to as the basal ganglia. Other key basal ganglia regions that communicate with the striatum

The Immune System
https://doi.org/10.1016/B978-0-443-23565-8.00015-6

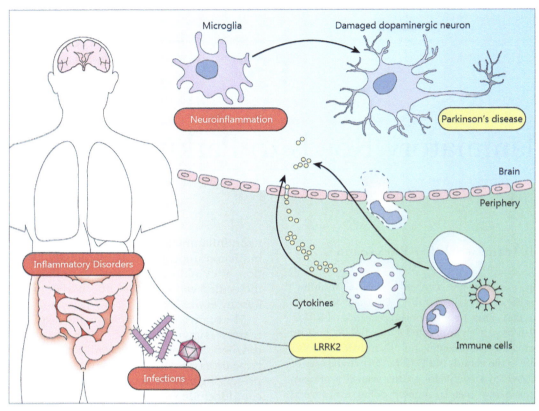

FIG. 15.1 Environmental factors such as inflammatory bowel disease or infections can trigger neuroinflammation and contribute to the pathogenesis of Parkinson's disease. The presence of LRRK2 mutations exacerbates the proinflammatory state of the immune cells from the periphery. Infiltration of monocytes, T cells, or cytokines through the blood–brain barrier can induce the activation of microglia in the brain. The neuroinflammatory environment affects the dopaminergic neurons in the substantia nigra, contributing to the neurodegeneration. *From Cabezudo, D., Baekelandt, V., & Lobbestael, E., 2020. Multiple-hit hypothesis in Parkinson's disease: LRRK2 and inflammation. Front. Neurosci. 14, 376.*

include the internal and external globus pallidus and subthalamic and ventrolateral thalamus. Signals originating from these areas ultimately project to the motor cortex to facilitate voluntary movement.

Another defining characteristic of PD comprises the presence of Lewy body inclusions (Del Tredici and Braak, 2012). Lewy bodies (named after Fritz Heinrich Lewy, who first characterized them over 100 years ago) are abnormal intracellular clumps of protein that are somewhat analogous to the senile plaques that characterize Alzheimer's disease (AD). These inclusions are present throughout the brain, being found in both the soma and dendrites of neurons. Lewy bodies are primarily composed of accumulated misfolded protein aggregates that principally encompass α-synuclein, as well as the limited expression of other proteins including parkin and ubiquitin. The α-synuclein protein is important for sending synaptic signals between neurons, but in PD, this protein becomes abnormally activated (through the addition of a phosphate group; i.e., they become hyperphosphorylated) and tend to clump together.

Although considerable effort has focused on the role of α-synuclein Lewy bodies in PD, it remains unclear whether these actually cause the initial development of the disease or are secondarily produced and are more important for disease progression. There is the view that Lewy bodies might not represent a pathological process but instead reflect compensatory "clean up" mechanisms that are aimed at the repair or disposal of damaged neurons. That said, support for this hypothesis is sparse, and considerable data amassed consistent with the view that Lewy bodies are part of the deleterious degenerative process. In this regard, most studies have suggested that abnormally activated α-synuclein somehow causes the misfolding of proteins, resulting in their pathological clumping, or interferes with the normal breakdown of cellular debris or metabolic products (Kawahata et al., 2022). In either case, the abnormal protein clusters interfere with basic neuronal functions required for signaling and hence motor and cognitive functioning.

One of the most intriguing aspects of α-synuclein associated Lewy body pathology is that these inclusions appear to be transmissible in a manner similar to viral or prion agents (Kordower et al., 2008). Indeed, monomeric α-synuclein species are present in PD plasma and cerebrospinal fluid, and there is evidence that they may spread from transplanted embryonic PD tissue (Brundin et al., 2017). Moreover, animal models of PD have involved "seeding" the brains of rodents with mutant α-synuclein fibrils, which resulted in a spread of α-synuclein pathology akin to what occurs in PD patients (Thakur et al., 2017). The ability for α-synuclein to spread is consistent with the progression of Lewy body inclusions in PD, which are believed to follow an anatomical spread from the posterior to frontal parts of the brain (Rietdijk et al., 2017). It was even proposed that Lewy body pathology in PD might first begin in the gut and then later spread to the brain (Lionnet et al., 2017). Thereafter, Lewy pathology may occur in the brainstem and only later moves to the basal ganglia motor system. In later stages of disease, widespread distribution of the Lewy bodies is evident throughout the cortex, thalamus, and hypothalamus.

The notion that the spread of PD pathology occurs through distinct stages, first occurring outside the basal ganglia and later spreading to motor regions, was first described by Braak et al. (1996). The crux of the "Braak hypothesis" concerning the evolution of PD is that it may result from the formation of Lewy bodies in peripheral tissues, which spreads to the brainstem and then to basal ganglia and forebrain regions (e.g., Borghammer et al., 2021). Most intriguingly and of relevance for this chapter is the view that pathogens entering through nasal membranes and/or the gut might play a role in catalyzing the misfolding of the α-synuclein protein, resulting in the formation and spread of Lewy bodies (Sato et al., 2016).

Parkinson's might begin with peripheral immune processes

The presence of Lewy bodies in the periphery and brainstem before they spread to the midbrain (Fig. 15.2) is consistent with the possibility that peripheral immune processes might be involved in the early stages of disease initiation. It is also consistent with viral or bacterial routes to disease, as these typically are first encountered through mucosal membranes within nasal, intestinal, or respiratory tracts. These microbes could then incubate in peripheral tissue, or in some cases, directly invade the brain. In fact, PD is known to have many comorbidities, such as gastrointestinal and olfactory disturbances, as well as being accompanied by depressive and anxiety symptoms. These clinical manifestations are influenced by inflammatory cytokines and related factors, supporting the contention that multisystem disturbances may be at the core of the disease.

Parkinson's Comorbidity and Stress Connections

Although motor function deficits are the most recognizable clinical features of PD, as already indicated, a large proportion of PD patients also experience comorbid psychological disturbances, along with autonomic, gastrointestinal dysfunction, and olfactory problems (Litteljohn et al., 2017). These comorbid symptoms typically precede the onset of motor decline or essential tremor (Gustafsson et al., 2015), suggesting that pathological processes are simultaneously operating across multiple brain and body sites. As degeneration progresses,

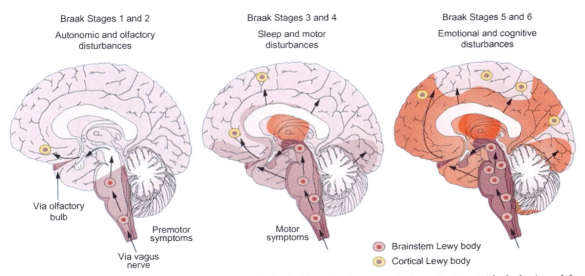

FIG. 15.2 Spread of Parkinson's pathology. Pathology (e.g., pathological Lewy body aggregates) may begin outside the brain and then spread to the brainstem via vagal or olfactory routes. Within the brain, pathology then spreads forward toward the midbrain and, finally cortical layers. As depicted by the different Braak stages, differing symptoms emerge coincident with this pattern of spread of pathology from the brainstem (Stages 1–2) to the motor regulatory basal ganglia brain regions (Stages 2–4) to finally the higher cortical areas that promote cognitive and other symptoms (Stages 5–6). *From Doty, R.L., 2012. Olfactory dysfunction in Parkinson disease. Nat. Rev. Neurol. 8, 329–339.*

symptoms begin to include cognitive deficits, most notably problems with executive functioning, and frank dementia may eventually occur in a subgroup of PD patients. These symptoms are major determinants of poor quality of life among PD patients, and collectively, they might even influence the progression of the disease.

The majority of PD patients do not display dementia or obvious memory loss, at least not until the very end stages of the disease (Hanagasi et al., 2017). What the patients do show is very specific executive deficits involving difficulty planning and organizing for future events, as well as set-shifting problems (Jin Yoon et al., 2021). Set shifting comprises the ability to exercise flexibility in switching between one task and another or shifting strategies in problem-solving situations (e.g., within the Wisconsin Card Sorting Task). Essentially, PD seems to be associated with impaired cognitive flexibility, which has been predominantly ascribed to prefrontal cortex functioning. Animal and human studies have confirmed dysfunctional frontal cortex functioning in PD, which is related to a loss of dopaminergic and noradrenergic input to this region, together with the local accumulation of Lewy bodies and inflammatory factors (Caspell-Garcia et al., 2017). A small subgroup of PD patients that eventually develop dementia are characterized in the separate classification of Parkinson's disease dementia (PDD). These individuals experience the "worst of both worlds," clinically sharing symptoms from both PD and AD.

Depression in Relation to Parkinson's Disease

About 77% of PD patients display psychiatric comorbidity, with depression being the most common (Patel et al., 2023), being evident in approximately 40%–60% of cases. Although depression sometimes occurs with other neurological disturbances, it is notable that depression in PD typically precedes the diagnosis and motor disability. In contrast, in other neurological illnesses, such as stroke, depression often arises after the specific primary disease event and is likely caused by the inflammatory and other neurobiological processes associated with the traumatic event (see Chapter 9; Depressive Disorders). In earlier chapters, the point was made that loneliness is a powerful chronic stressor that may produce adverse effects beyond that of depression. In a study of about 910,600 participants who were assessed at a 15-year follow-up, 2,822 developed PD. The incidence of the disorder was more common with increasing feelings of loneliness, even after controlling for numerous lifestyle factors and the presence of other illnesses (Terracciano et al., 2023).

The two primary comorbid psychiatric symptoms, depression and anxiety, are not unique to PD, being the most common comorbidities across many neurological conditions, including AD, cerebral stroke, and multiple sclerosis. As mentioned in earlier chapters (see Chapters 3 and 5), they are also commonly associated with cardiovascular and metabolic conditions, such as heart disease, diabetes, and general illness. Because of their evolutionary importance for threat detection and social bonding, the emotional brain circuits linked to depressive illness might be especially sensitive to any perturbations in brain functioning. Brain processes that govern emotional states are easily disturbed in many threatening situations because they specifically evolved to be highly plastic and highly responsive to anything that challenges biological homeostasis.

At first blush, depression would be an obvious response to a PD diagnosis and the major changes in quality of life that come with this. Certainly, in some cases, this distress is exactly what precipitates the depression. Yet, considerable evidence points to alternative causes in many cases, especially as depression is often evident long before the first motor symptom presentation (Dallé and Mabandla, 2018). Imaging studies revealed that comorbid depression and anxiety in PD was related to widespread differences in functional connectivity between the cortex, hippocampus, caudate, and thalamus (Dan et al., 2017). Besides the degeneration of midbrain DA neurons, other neurotransmitter systems are affected in PD. The primary norepinephrine (NE) producing neurons in the locus coeruleus also degenerate in PD, perhaps developing early in the disease processes (Buddhala et al., 2015). The loss of these NE neurons might contribute to the subsequent loss of midbrain DA neurons, possibly owing to the reduced availability of essential growth factors that are required for proper neuronal survival. Whatever the case, the loss of brainstem NE has cascading effects throughout the brain, which could adversely influence the functioning of emotional brain circuits. The monoaminergic dysfunctions within multiple interconnected cortical, limbic, and brainstem regions likely play an important role in comorbid depressive pathology. As well, cognitive disturbances associated with PD may be mediated by DA and acetylcholine (Chen and Zhang, 2024). As will be discussed shortly, the damage to the various neurotransmitter systems is intimately related to neuroinflammatory processes.

A Korean longitudinal cohort study followed 98,296 elderly people for over 5 years to assess whether depression was a risk factor for the development of PD (Yoon et al., 2023). Of 839 individuals who developed PD, 230 experienced comorbid depression. There was almost a 50% increased risk of developing PD development in those with depressive symptoms, increasing with the number of depressive symptoms. Another large-scale study followed 1,342,282 patients with depressive disorder over 6 years and assessed the impact of exercise on the risk of developing PD (An et al., 2023). During the study, 8901 PD cases developed, but the risk of PD developing in depressed patients was reduced in those individuals who maintained regular physical activity. This is consistent with depression as a PD risk factor and supports the view that exercise can diminish the risk of illnesses related to inflammatory processes. Further evidence connecting depression and PD came from an epigenome-wide association analysis that revealed gene set enrichment of immune system pathways and interactions between genes that code for cytokine receptors. Numerous epigenetic marks were also identified in blood and brain (substantia nigra, putamen basal ganglia, or frontal cortex), with the IFNγ and T-cell receptor signaling pathway being most closely aligned with depression (Paul et al., 2023).

Stress in Relation to Parkinson's Disease

No discussion of PD would be complete without considering the many aspects of stressor experiences relevant to depression and possibly for some primary motor deficits. It is important to re-emphasize the position that stressors, ranging from psychological to chemical to immune processes, collectively contribute to PD, and as such, it is difficult to parse out the effects of any single type of stressor. That said, some stressors (e.g., chemical pesticides) might have more direct damaging effects on motor brain regions, whereas psychological stressors might be more aligned with neurochemically related emotional symptoms. Importantly, the impact of a given stressor experience may depend upon its temporal proximity to other stressors, its severity, and duration, as well as the actual disease stage and state of the brain's different microenvironments at the time of exposure.

When lag periods are imposed between stressor encounters, such as immune challenges (cytokines, LPS), drugs of abuse (amphetamine), and laboratory stressors (restraint, footshock), sensitized (augmented) or desensitized (diminished) brain and behavioral responses can be realized (Anisman et al., 2008a,b). This is illustrated by ischemic tolerance, in which a very small stroke induced in rats limits the damage induced by a second, larger stroke that follows within a discrete period (usually 1–2 days), whereas such effects are not apparent after a longer delay (Li et al., 2017d). Likewise, in neonatal piglets, LPS pretreatment 4 h prior to an ischemic event increased damage, whereas 24 h following continuous treatment with the endotoxin, the extent of damage was reduced (Martinello et al., 2019). Consistent with these findings, immune priming (with bacterial or viral challenges) enhanced the impact of later pesticide exposure on midbrain dopamine neurons, but this sensitizing

effect was only evident within certain time intervals (Mangano and Hayley, 2009). Pretreatment with LPS or the proinflammatory cytokine, TNF-α, likewise time-dependently sensitized rodents to the effects of later re-exposure to the same challenge, but also a different (restraint) stressor (Hayley et al., 2001).

Various stressors may differentially influence PD symptomatology, varying with the state of the brain at the time of exposure. By the state of the brain, we are referring to the current level of functioning of neurons and the extent of prodeath processes that might be engaged, together with the activation state of microglia and astrocytes. Given the varied microenvironments across different brain regions, it follows that the stressor effects will also vary across regions. A further layer of complexity is that these different brain regions are connected and thus a change in one region could reverberate through these circuits.

Stressors stimulate HPA axis activity, resulting in glucocorticoid release that enters the brain parenchyma and is neurotoxic at high enough levels. Of interest is that the corticoid-inducing hypothalamic peptide, CRH, may alter the activity of brain mast cells and astrocytes, which collectively can influence blood–brain barrier (BBB) permeability and can thus increase neuronal pathology, as observed with experimental autoimmune encephalitis. Moreover, as shown in Fig. 15.3, stressors may activate resident microglia, leading to the enhanced secretion of proinflammatory factors (Delpech et al., 2015). These cytokines can themselves increase BBB permeability (e.g., via upregulating endothelial cell adhesion molecules), which has the dual effect of further enhancing cytokine production at vascular sites (i.e., a positive feedback loop) and augmenting peripheral immune cell trafficking across the BBB.

Since psychological, immunological, and chemical stressors were able to induce BBB disruption, these challenges would place the already vulnerable brain in contact with a host of potentially dangerous substances. As will be discussed in upcoming sections, a variety of environmental factors, including pesticides, heavy metals, and microbial agents, could enter the brain through a compromised BBB to promote local neuroinflammatory consequences. Thus, stressors could directly or indirectly affect brain functioning, which, in turn, could create a microenvironment in the brain that (1) further contributes to prodeath processes operative in PD, thereby increasing motor impairment, and (2) contributes to comorbid symptoms, such as depression and anxiety, that might exacerbate primary motor symptoms.

Chronic psychological distress may also influence symptom presentation owing to the loss of neurotrophic factors, which normally provide neuronal support (Anisman and Hayley, 2012a,b). Certain stressors might

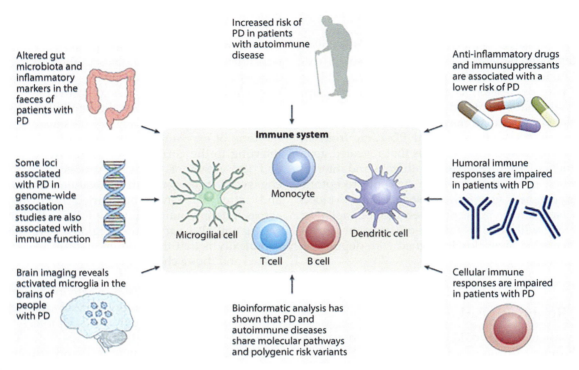

FIG. 15.3 Evidence for involvement of the immune system in PD. From Tan, E.K., Chao, Y.X., West, A., Chan, L.L., Poewe, W., Jankovic, J., 2020. Parkinson disease and the immune system – associations, mechanisms and therapeutics. Nat. Rev. Neurol. 16, 303–318.

also influence PD by promoting proinflammatory effects through the activation of brain microglial or by fostering the mobilization of peripheral immune cells (Kam et al., 2020). Interestingly, extended chronic restraint (8h of restraint 5days a week for 2–16weeks) caused the loss of SNc dopamine neurons, with progressively more neuronal disturbances occurring over time (Sugama et al., 2016). It was similarly demonstrated that DA neurons of the ventral tegmental area (VTA) were vulnerable to cell death with a repeated restraint stressor, leading to the suggestion that damage to the VTA reward pathway might be involved in the comorbid depression observed in PD (Sugama and Kakinuma, 2016).

Psychological stressors can influence the responses provoked by subsequent immune challenges, including augmented microglia reactivity and proinflammatory factors. Indeed, augmented IL-1β, IL-6, and TNF-α responses, along with sickness behaviors, were evident in response to LPS in mice that had been exposed to a strong stressor 4days earlier (Johnson et al., 2003). It appeared that the stressor-induced sensitization of the LPS inflammatory response was related to the modulation of the microglial phenotype. Specifically, the stressor primed microglia so that they upregulated NLRP3 inflammasome expression and the release of endogenous "danger" signals, such as HMGB-1 (Weber et al., 2015). Using in vitro procedures, it was demonstrated that an inflammatory challenge to cells by LPS, primed them through epigenetic changes involving histone H3K27 acetylation so that the response to a later challenge with manganese, which may be a Parkinson's related neurotoxic stressor, was markedly enhanced. Conversely, inhibiting H3K27 deposition prevented the primed inflammatory response elicited by a later challenge (Huang et al., 2023b). These findings suggested that the brain senses psychological stress in a manner akin to the way it responds to pathogenic challenges and that stressful events predispose biological systems to be more vulnerable to future insults.

Lifestyle and Aging

The one unequivocal risk factor for PD is age, in which the deleterious effects of virtually all environmental insults are elevated. However, for now, the focus will be on how age, in combination with other typical lifestyle factors (and not environmental insults per se), can collectively influence the risk of developing PD. In this respect, the incidence of PD has generally increased over the past half-century, which could simply be a manifestation of people now living longer. Yet, the increased PD incidence could be fueled by lifestyle changes, including changes in diet, a more sedentary lifestyle, alcohol, and drug use, greater reliance on chemical additives, and the presence of an increasing number of synthetic pesticides and other pollutants (Chin-Chan et al., 2015), as well as antibiotics increasingly appearing in foods.

As already mentioned, the typical age of onset of PD is about 60–65, and generally, the incidence of new cases drops sharply after 80 years of age. This narrow window regarding the age of onset is consistent with the time-dependent unfolding of genetic-driven processes and/or the slow accumulation of multiple environmental hits throughout the lifetime. However, there are cases of PD that occur as early as the 30s. As we will see shortly, early-onset PD is more often associated with specific genetic mutations (Schormair et al., 2017) than the typical "idiopathic" or "sporadic" cases, in which the origins of the disease are unknown. The absence of a clear genetic link in most PD cases has been one of the driving factors that spurred the search for environmental causes. Of course, thinking that PD is caused by a single environmental factor would be short-sighted. A more realistic view is that genetic factors set the stage for disease, but the "wear and tear" produced by multiple environmental challenges ultimately contribute to the development of PD (Rudyk et al., 2015). This is the crux of the multihit hypothesis, which will be discussed in upcoming sections.

The incidence of PD is higher in men than women, at a ratio of about 1:5. Although this has been taken to suggest a protective effect of estrogen in PD (Vegeto et al., 2020), the alternative is that men adopt poorer lifestyles or are more likely to be exposed to toxicants or physical insults in the workplace (Adamson et al., 2022; Bellou et al., 2016). There is also evidence concerning geographic distribution, with PD occurring more frequently in low- and middle-income countries (Schiess et al., 2022), although it occurs less frequently in parts of Africa and South America compared to Europe or North America. It is significant that PD prevalence varies across African subregions, being lowest in eastern countries, which have a relatively low life expectancy, and higher in the more affluent North African countries (Callixte et al., 2015). Similarly, in the more industrialized regions of South America (e.g., Brazil), PD rates were comparable to those in Europe and North America. Thus, life expectancy might account for differences in PD distribution, or it might be related to lifestyles in more Westernized countries.

Less obvious and indeed, somewhat surprisingly, PD risk factors include lower smoking rate and being engaged in certain occupations, such as farming. Of course, with these types of studies, it is difficult to determine causal connections, which is compounded by the long prodromal state that exists for PD. It is thought that the PD prodeath disease processes might begin decades before the first motor symptoms appear. Indeed, the presence of Lewy body markers may be present in middle-aged people many years before frank symptoms of PD are present (Kok et al., 2024). The long incubation time for PD is quite scary in that certain people could be "ticking time bombs," with a gradual decay in brain tissue

occurring over time. Yet, viewed from another perspective, such a long prodromal phase could potentially allow time for interventions to be undertaken. Of course, this rests on finding the best prophylactic treatment(s) at the best time, which requires a better understanding of the early disease processes. This necessitates the identification of biomarkers to identify who is at risk of developing the disease. It would also be necessary to have biomarkers that speak to the subtype of PD that could develop (e.g., Rodriguez-Sanchez et al., 2021), and knowing the nature of potential comorbidities would also be exceptionally valuable, as the different flavors of primary and comorbid disease might call for different intervention or treatment targets.

Few studies have examined the impact of exercise on PD, and those that have generally revealed modest, albeit significant, links between the extent of physical activity and the likelihood of developing PD. For instance, self-reports of regular high levels of physical activity before age 65 or participation in competitive sports before age 25 cut the risk of developing PD in half (Shih et al., 2016). While early studies had indicated that daily exercise modestly reduced risk in men, but not women, a later meta-analysis indicated that exercise reduced the risk of PD in women when measured over lengthy periods (Portugal et al., 2023). Although not all exercise regimens produce the same effects, many of the reported studies did not control for confounding variables, including smoking and diet.

A clear relationship has not emerged concerning the link between dietary factors and PD, although there were indications that specific foods might have some bearing on PD. An association was not observed between PD and consumption of fruits and vegetables, but a modestly reduced risk was associated with consumption of meat, fish, and polyunsaturated fat, whereas a strong positive relationship was apparent between milk consumption and PD risk (Kyrozis et al., 2013). As we will see shortly, the relationship between specific foods and PD may be mediated by gut microbial changes. Some evidence, however scant, suggested that particular vitamins and minerals could play a role in PD development. Reduced PD occurrence was linked to a high dietary intake of vitamin E, possibly being related to its antioxidant potential. While some studies did not find a link between iron and PD, others reported increased disease risk with high dietary iron levels (Liu et al., 2023b), perhaps being related to its well-known oxidative stress effects (Jiang, Wang, Rogers, & Xie, 2017). Several studies also revealed significant but modest positive associations between dietary zinc and copper levels and the occurrence of PD (Stelmashook et al., 2014).

Perhaps the most surprising relationship with lifestyle factors was the robust inverse association between smoking and PD. More than 20 case–control studies, as well as prospective studies, reported that smoking (particularly current smoking) was associated with a reduced risk of PD. For instance, a 26-year follow-up study of over 8,000 men found a 60% reduced risk of developing PD, and this effect was dose-dependent, being inversely correlated with the amount smoked (Grandinetti et al., 1994). A meta-analysis that examined 48 studies found 40% or 60% reduced PD risk for past or current smokers, respectively, compared to those who never smoked (Hernán et al., 2002). A more recent report confirmed reduced PD risk in smokers, and this effect appeared to be dose-dependent, being greater in heavy smokers (adjusted HR: 0.49) than in light smokers (aHR: 0.80) (Jung et al., 2023). A study that explored the potential interaction between genes and environment found that the synaptic-vesicle glycoprotein 2C (SV2C) locus may mediate the inverse relationship between smoking and PD. It was further discovered that mutant flies that overexpressed the PD α-synuclein gene and received nicotine treatment were protected from normal neuronal loss and deficits in locomotion. Most importantly, the beneficial effects of nicotine were prevented when SV2C signaling was ablated, thereby confirming its biological role in the nicotine effects (Olsen et al., 2023).

Understandably, there has been a push to study the potential protective effects of nicotine or other active compounds found in cigarette smoke. Several animal studies showed that certain nicotine dosing regimens lessened the impact of PD-relevant toxins and that chronic nicotine treatment was neuroprotective in rodent and monkey PD models that involved lesioning with dopaminergic toxicants (Huang et al., 2009). Moreover, chronic oral nicotine administration in monkeys prevented the striatal terminal loss, particularly that of specific cholinergic receptors, following nigrostriatal damage (Quik et al., 2006). These actions may stem from pathways downstream of cholinergic receptors (notably α7-nAChRs), including PARP-1 and caspase-3 (Lu et al., 2017). Thus, nicotine might act on terminals to preserve their functioning and possibly promote compensatory sprouting. This is consistent with findings showing that the cell bodies and terminals of nigrostriatal dopamine neurons degenerate at different rates in PD and that the mechanisms responsible for pathology in these two biological compartments may be autonomous.

In an attempt to translate preclinical findings to patients, clinical trials using nicotine have been undertaken (Villafane et al., 2017). Discouragingly, nicotine did not provide symptomatic relief in 40 PD patients who received either transdermal nicotine therapy or placebo over 39 weeks (Wood, 2017). Likewise, no benefit was obtained by PD patients treated with nicotine for a total of 45 weeks (with dose escalation up to 90 mg/day and then de-escalating before clinical testing following drug washout) (Villafane et al., 2017). A more recent clinical trial likewise failed to find any benefit of a one-year transdermal nicotine treatment program (ClinicalTrials.gov number, NCT01560754). Of course, in these studies, the PD patients may have been past the critical period where nicotine might be beneficial, or alternatively, heterogeneity of disease manifestation and mechanisms of cell death across the patients might dilute the magnitude or ability to detect any clinical effects. In essence, nicotine might be useful in a prophylactic capacity but not in diminishing symptoms in already existent PD. Thus, biomarkers to identify potential PD candidates during the prodromal disease phase might help triage the different subtypes of disease that might be required for efficacious nicotine treatment.

Much like smoking, a meta-analysis of 13 studies revealed that regular coffee consumption reduced the relative risk of PD (Hong et al., 2020). The potential protective effects of these beverages might stem from the well-known antagonistic properties of caffeine upon adenosine, which was limited in the presence of a polymorphism of ADORA2A and CYP1A2 genes, which influence DA functioning and influence caffeine metabolism, respectively (Chuang et al., 2016). Interestingly, caffeine was neuroprotective against 6-hydroxydopamine (6-OHDA) lesions in rodents, which was related to an antiinflammatory effect, characterized by reduced IL-1β and TNF-α levels (Machado-Filho et al., 2014). Furthermore, the direct application of caffeine on cultured macrophages promoted an antiinflammatory phenotype in these cells (Shushtari and Abtahi Froushani, 2017), and caffeine similarly reduced the inflammatory and oxidative stress effects of hypoxia in neonatal rats (Endesfelder et al., 2017).

Several studies, however, failed to find a link between caffeine consumption and PD, possibly owing to the differential dose effects of caffeine. A U-shaped dose relationship was reported regarding caffeine and PD in women such that the most protective effects were observed with intermediate amounts of coffee consumption (one to three cups/day) (Ascherio et al., 2001). Thus, gender effects may be at play, with estrogen possibly serving a mediating role in the biological implications of adenosine receptor blockade by caffeine.

In contrast to smoking and caffeine consumption, alcohol has not consistently been linked to PD incidence. A meta-analysis of 16 high-quality reports revealed a weak protective effect in seven studies, but two showed a negative effect, whereas the remaining seven showed no effect (Bettiol et al., 2015). Such a range of outcomes underscores the complex nature of human research studies, particularly in isolating alcohol consumption from other factors. Subsequent studies similarly indicated that alcohol intake was unlikely to be related to PD (Domenighetti et al., 2022), although

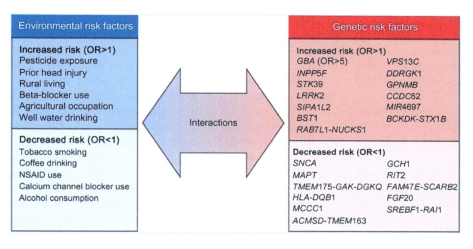

FIG. 15.4 Risk factors and PD. Summarized are the environmental and genetic factors that either increase (Odds Ratio > 1) or decrease (Odds Ratio < 1) in the risk of developing Parkinson's disease. Pesticide exposure and head injury are most strongly linked to increased risk, while smoking and coffee drinking are most robustly related to decreased risk. A number of different genes are correlated with increased or decreased risk. Many of these genes are very rare, but the most common, LRRK2, also is a robust inflammatory immune system regulator. *From Kalia, L.V., Lang, A.E., 2015. Parkinson's disease. Lancet 386, 896–912.*

it was cautioned that such a relationship, if there is one, might vary with the extent and style of alcohol consumption. Specifically, chronic binge drinking may affect processes, such as elevated oxidative stress, glutamate excitotoxicity, and provocation of an inflammatory profile, that could favor PD development (Kamal et al., 2020). A general summary of the environmental and genetic factors positively and negatively related to PD is shown in Fig. 15.4.

Experimental Animal Models of Parkinson's

This section will introduce the most common and best-validated animal models for PD. The majority of studies were conducted in rodents, although several involved nonhuman primates, typically in late-stage preclinical testing of new drug treatments. A growing number of studies used lower sentient organisms, most notably roundworms and fruit flies, but these are generally restricted to experiments utilizing very specific genetic manipulation that require more basic neuronal systems and thus will not be dealt with here.

Classic Toxicant-Based Models: 6-OHDA and MPTP

The most commonly used animal models of PD are those that involve toxicant exposure, notably 6-hydroxydopamine (6-OHDA) and 1-methyl-4-phenyl-1,2,3,6-tetrahydropyridine (MPTP). These well-validated models have provided a wealth of data regarding the degenerative process in PD and are ideal for testing new therapeutic regimens. The former (6-OHDA) is a dopamine analog that is taken up by the dopamine transporter and acts as a potent inducer of oxidative stress. In fact, molecules akin to 6-OHDA are naturally produced as part of the auto-oxidative processes that characterize normal dopamine metabolism.

The dark side of dopamine

The high oxidative nature of dopamine metabolism suggests that high concentrations of dopamine may be neurotoxic, which raises a potential dark side to treatments, such as L-DOPA, that artificially raise dopamine levels. So, efficient metabolic processes are essential for the breakdown of such products and for the detoxification of any oxidative or nitrogen radicals. There is evidence that these processes might become somewhat weaker with age. Similarly, intracellular protein handling and detoxification (e.g., via the proteasome) are also thought to be compromised with advanced age. It also seems that the dopamine neurons in the SNc are appreciably more vulnerable than others, such as those in the ventral tegmental neurons. This is not unique to this neurotransmitter, as glutamate at high concentrations causes excitotoxicity. Thus, as the toxicologist Paracelsus said, "The dose makes the poison."

Studies that showed cellular toxicity of l-DOPA were generally done in vitro, and little in vivo evidence exists to support toxicity. Indeed, a study that profiled the potential toxicity of 20-day continuous in vivo infusion of L-DOPA and carbidopa formulations (the latter allows for greater dopamine entry into the brain rather than being broken down in the periphery) found no adverse effects, except for modest inflammatory changes that resolved after a few days (Ramot et al., 2017).

For about 50 years, 6-OHDA has been used to induce highly reproducible Parkinsonism. The administration of 6-OHDA through infusion into either the SNc, striatum, or medial forebrain bundle induces degeneration of the nigrostriatal system. However, because 6-OHDA does not cross the BBB, it must be infused directly into the brain to produce its toxic consequences. Though there are several variations in the infusion paradigms, generally, the dopaminergic neurons and terminals are lost within 5 days after 6-OHDA infusion. When injected into the SNc, cell death is thought to begin somewhat earlier than when it is delivered to striatal terminals.

Of the animal PD models, 6-OHDA induces the most dramatic motor outcomes, including bradykinesia, disturbed coordination, and gait imbalances (Baldwin et al., 2017). One of its most noticeable and reliable effects is that animals lesioned on only one side of the brain exhibit abnormal unilateral turning, rapidly circling toward the noninjected contralateral side. Unilaterally compromised DA motor regulatory functioning also occurs in human PD patients who typically display unilateral hand tremors, and autopsy confirmed that SNc DA loss tends to be greater on one side, thus giving rise to the asymmetry of pathology (Riederer and Sian-Hülsmann, 2012).

In contrast to 6-OHDA, MPTP readily crosses the BBB and hence is administered systemically (typically intraperitoneally), with mice and monkeys being the species of choice. This synthetic agent is converted into its active toxic form, MPP+, by the enzyme MAO-B within astrocytes. The MPP+ is then extracellularly released and taken up into DA neurons via the dopamine transporter (Richardson et al., 2005). Importantly, MPTP induces nigral DA neuronal loss in humans virtually identical to that observed in the typical idiopathic cases of PD. We know this because of an unfortunate event that occurred in California, wherein a small group of individuals self-administered a drug that was contaminated with MPTP. These people rapidly developed acute Parkinsonian symptoms that were responsive to L-DOPA (just like idiopathic PD) and upon their death years later, it was apparent that they had lost a substantial number of SNc DA neurons (Langston et al., 1999). This is a case of a "naturally occurring experiment" in humans that resulted in the subsequent development of an animal model, essentially reflecting the "bedside to bench" translation of knowledge. One drawback of the MPTP model is that the type and severity of motor pathology varies across species and even across rodent strains (Dauer and Przedborski, 2003). Also, while some PD features are typically observed following MPTP, most notably the reduced basal movement (bradykinesia), other features of PD, particularly the tremors, are not seen. Another weakness of the MPTP model (like that of 6-OHDA) is that it does not cause Lewy body inclusions. No such inclusions were observed in rodents, and monkeys displayed only slight intraneuronal α-synuclein inclusions even after 10 years of treatment (Halliday et al., 2009).

Despite these shortcomings, MPTP is a useful tool for assessing cellular and molecular mechanisms of DA neuronal death. It has highly reproducible neurodegenerative effects on the nigrostriatal system, which have been ascribed to its ability to induce energetic cell failure and oxidative radical generation owing to blockade of electron transport by targeting mitochondrial complex I (Jackson-Lewis and Przedborski, 2007). However, as will be discussed shortly, MPTP, like pretty well all animal models of PD, also provokes substantial neuroinflammatory consequences centered around enhanced microglial proinflammatory and prooxidative activities (Pisanu et al., 2014). Supporting a causal connection to a PD-like condition, enhancing microglial phagocytosis and the consequent reduction of microglial-mediated neuroinflammation can result in the conversion of microglia so that they take on protective functions. Thus, this might ultimately promote a viable strategy to diminish neurodegenerative diseases (Gao et al., 2023a).

While microglia have been the focus of considerable research, using a mouse model of PD, it was shown that border-associated macrophages, rather than microglia, were responsible for the neuroinflammatory response associated with the disorder. It appeared that deletion of MHC2 from brain macrophages (i.e., both microglia and border-associated macrophages) prevented inflammation and the neurodegeneration associated with α-synuclein. Conversely, α-synuclein overexpression promoted microglia and border-associated macrophages that favored a proinflammatory profile. Moreover, paralleling the findings in mice, in human PD samples CD4$^+$ and CD8$^+$ T cells adjacent to border-associated macrophages were altered, perhaps reflecting their penchant for promoting a proinflammatory milieu (Schonhoff et al., 2023).

One final point that bears mention is that besides their individual effects, joint actions of multiple toxicants or stressors and toxicant challenges could influence the development of PD. Animal models of PD using 6-OHDA have shown that chronic stressors, both before and after 6-OHDA lesioning, accelerated the rate and enhanced the magnitude of damage to the nigrostriatal system, and concurrently exacerbated motor impairments (Hemmerle et al., 2014). These findings are similar to reports of enhanced dopamine neuronal loss in MPTP-treated mice that were exposed to a chronic mild stressor. Moreover, anhedonia was only observed in MPTP-treated mice that were also exposed to a stressor (Janakiraman et al., 2016). Importantly, psychosocial stressors experienced by humans resulted in striatal dopamine alterations assessed through positron emission tomography (PET), which corresponded with subjective feelings of threat (Bloomfield et al., 2019). Together, these findings raise the possibility that psychologically relevant stressors experienced before and during the course of PD could affect both neurodegenerative processes and the primary motor symptoms, as well as certain nonmotor or comorbid neuropsychiatric manifestations.

Environmental Stresses and Parkinson's Disease

Mutations in genes, including *SNCA* (PARK1/4), *LRRK2* (PARK8), *parkin* (PARK2), and *PINK1* (PARK6), have been implicated in the manifestation and progression of PD (Blauwendraat et al., 2020). Yet, familial-related PD accounts for only a small percentage of cases, with the vast majority being idiopathic and sporadic in nature (Deleidi and Gasser, 2013). Several factors have been associated with a higher risk of sporadic cases of PD development, including age, genetic polymorphisms, stress, and cumulative exposure to environmental factors (Breckenridge et al., 2016). In addition to heavy metals (lead and manganese), air pollutants, head trauma, viral infections (Breckenridge et al., 2016; Lee et al., 2016c), and exposure to pesticides have been implicated in PD (Ritz et al., 2016), possibly being related to the provocation of inflammatory processes.

Pesticides and Parkinson's Disease

Support for pesticide exposure in disease provocation has come from epidemiological studies primarily conducted in agricultural communities (Ritz et al., 2016). An epidemiological analysis that included 288 pesticides found links between 10 different pesticides and PD based on toxicity evident in dopamine neurons (Paul et al., 2023). Several compelling lines of evidence suggested a role for specific pesticides, such as the organic insecticide rotenone and the nonselective herbicide paraquat, as major risk factors for disease development (Baltazar et al., 2014). These epidemiological findings are supported by animal studies (both primate and rodent) demonstrating that the administration of pesticides, such as paraquat and rotenone, induces many of the neuropathological and behavioral features characteristic of PD (Rudyk et al., 2015). These included motor behavioral disturbances, reduced striatal fiber density, microglia activation, oxidative stress, and aggregated protein inclusions (Bobyn et al., 2012). Moreover, following the priming of inflammatory processes, subsequent paraquat treatment dose-dependently induced the loss of dopamine neurons in the SNc, reminiscent of what occurs in PD (Mangano and Hayley, 2009). Interestingly, paraquat has a chemical structure that is strikingly similar to the active MPTP metabolite, MPP+ (Dauer and Przedborski, 2003), and can gain entry into the CNS via a neutral amino acid transporter present at the tight junction BBB. Upon entry into the brain, the toxicant is distributed throughout the prefrontal cortex, hippocampus, olfactory bulbs, striatum, and SNc (Peng et al., 2007).

In the CNS, paraquat can promote a central inflammatory cascade through microglia activation and can subsequently gain entry to neurons to directly perturb intracellular functions (Mangano et al., 2012; Rappold et al., 2011). Once in the neuron, paraquat can disrupt calcium homeostasis and mitochondrial electron transport chain complexes (i.e., complex I and IV), as well as increase mitochondrial membrane permeability (Huang et al., 2016a). These effects collectively (1) reduce ATP energy production, thereby starving the neuron, (2) increase excitotoxicity, and (3) promote apoptotic factor release. Paraquat can also increase the presence of lipids that favor a proinflammatory milieu (Tong et al., 2022), cause vesicular damage, induce endoplasmic reticulum stress, and even react with α-synuclein, resulting in further neuronal dysfunction and damage (Cochemé and Murphy, 2008). A review of several reviews concluded that despite the frequent association between paraquat and the development of PD, the toxicant did not cause the illness (Weed, 2021). This is to be expected as it is unlikely that a substantial number of PD cases stem simply from paraquat exposure. It is more likely that paraquats simply act as one hit among many that are encountered over time.

Rotenone is an organic insecticide derived from a South American plant root and, like paraquat, has been associated with an increased incidence of PD (Tanner et al., 2011). It is highly lipophilic, can cross the BBB, and acts as a potent inhibitor of the mitochondrial respiratory processes (Greenamyre et al., 2010). This compound can provoke the loss of SNc dopamine neurons, as well as a loss of noradrenergic locus coeruleus neurons (Duty and Jenner, 2011), and induces α-synuclein aggregations reminiscent of Lewy bodies. One major problem of the rotenone model is that there have been difficulties replicating the size of the lesion-induced. Of course, all toxicant models encounter this issue to some degree. As well, the route and timing of administration have a profound impact on the nature of the lesion produced (Fig. 15.5). In this respect, an analysis of the effects of rotenone revealed especially telling findings that speak to the progressive actions of this agent. Among middle-aged rats that received rotenone treatment for 5 consecutive days, motor and postural effects were observed, which abated after about 9 days. However, beginning approximately 3 months later, motor abnormalities reappeared that progressively worsened, which corresponded with microglial activation, nigral dopamine neuronal loss, and the subsequent accumulation of α-synuclein in substantia nigra and frontal cortical neurons (Van Laar et al., 2023). Given the presence of these features and the progressive changes that occurred, it was maintained that this might be an ideal model to assess PD processes and treatments.

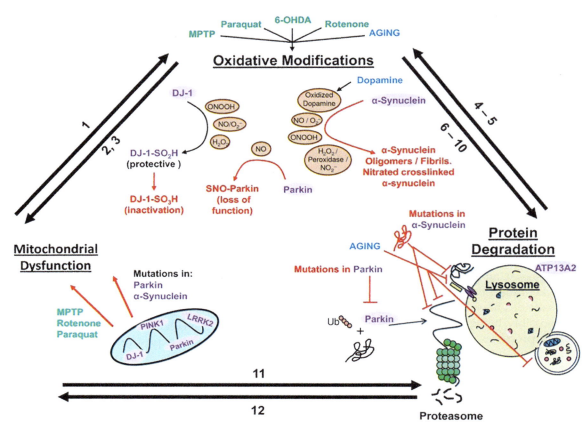

FIG. 15.5 Oxidative stress, mitochondrial pathology, and impaired protein handling together form a "Bermuda triangle" of related processes that can give rise to PD. Multiple toxicants can impact these processes; most notably, MPTP, paraquat, rotenone, and 6-OHDA, along with the general wear and tear of aging. The numbers shown on the arrows depict the following interrelated processes: (1) mitochondrial respiration failure that generates oxidative radicals. (2) Antioxidants like SOD protect against mitochondrial toxins. (3) The protein DJ-1 might also protect against mitochondrial failure. (4–5) Impaired protein degradation elevates sensitivity to α-synuclein and oxidative stress. (6–10) Oxidative stresses and α-synuclein impair protein folding and degradation leading to faulty protein aggregates. (11–12) Autophagy and proteasome inhibition further contributes to protein aggregates and metabolic derangement. *From Malkus, K.A., Tsika, E., Ischiropoulos, H., 2009. Oxidative modifications, mitochondrial dysfunction, and impaired protein degradation in Parkinson's disease: how neurons are lost in the Bermuda triangle. Mol. Neurodegener. 4, 24.*

Infectious Diseases and PD

It has been suggested that several neurodegenerative disorders, including PD, may have their root in the presence of viral pathogens (Blackhurst and Funk, 2023). The view was expressed that diverse neurological conditions may be subserved by inflammatory processes involving the activation of inflammasomes and the consequent activation of IL-1β and IL-18, which in high concentrations promote neuronal damage (Ravichandran and Heneka, 2024). This idea stemmed from early observations of Parkinsonian clusters following various infectious outbreaks. The earliest evidence linking infectious agents to PD dates to the highly virulent influenza epidemic of 1918, after which a sizable number of infected survivors developed PD-like symptoms. These individuals displayed shuffling gait and difficulty with coordination, and later autopsy revealed marked inflammation within the brain, coupled with DA neuronal loss (Ravenholt and Foege, 1982). It appeared that these people had severe encephalitis, which likely caused degeneration of dopaminergic neurons (Casals et al., 1998).

Further studies continued to report cases of PD that were associated with viral infections. It was even suggested that intrauterine influenza infection could lead to PD, but this has never been directly demonstrated. However, viral infection using the Japanese encephalitis virus, Herpes C, and H5N1 influenza virus in rodents provoked the loss of SNc DA neurons (Jang et al., 2009). In addition, there have been cases, albeit rare, of postencephalitic parkinsonism following infections. Studies exploring the infectious origin of PD have been informed by the Braak hypothesis, which, it will be recalled, posits that PD pathology begins in the olfactory bulb and the gut. These locations could act as portals for pathogenic entry. It is, indeed, noteworthy that olfactory and gastrointestinal disturbances are two classic early symptoms in the prodromal stage of PD.

The possibility has been considered that SAR-CoV-2 might come to affect PD, but such a link, if there is one, will not be apparent for years or decades as the infected population ages. This said, it was reported that three relatively young individuals (35–58 years) developed PD following infection (Iravanpour et al., 2023). Moreover, monkeys intranasally infected with SARS-CoV-2 developed α-synuclein accumulation in the midbrain, together with activated microglia and signs of invading T cells (Philippens et al., 2022).

At a mechanistic level, infectious agents are first detected in the body or brain by pattern-recognition receptors (PRRs) that recognize microbe-specific molecular signatures. There are different classes of PRRs, such as Toll-like receptors (TLRs), Nod-like receptors, RIG-I-like receptors, and C-type lectin receptors. TLRs are the most important type of PRR within the brain, being present at the cell surface or intracellularly on the endosome, lysosome, or endoplasmic reticulum. They recognize and bind to microbe molecular lipid, lipoprotein, protein, and nucleic acid sequences found on pathogens. TLR3 is a critical detector of viral infection, and TLR3 activation increases α-synuclein, together with proinflammatory cytokines and complement protein C3 (Thomas et al., 2023). The TLR3-dependent provocation of α-synuclein was downstream of complement C3 activation, indicating an important link between viral infection, the complement pathway, and α-synuclein pathology.

The Gut and PD

A body-first route to PD has been proposed wherein the first pathological changes (primarily involving α-synuclein) might occur outside the brain. Growing evidence suggests that the peripheral enteric nervous system that innervates the intestinal tract is key to the promotion of PD. It is believed that misfolded and pathological oligomeric and fibril forms of α-synuclein migrate from the gut to the brainstem and then forward to the basal ganglia nigrostriatal system. The intestinal enteroendocrine cells respond to bacterial and other pathogenic signals and can then send signals to inform the brain through enteric nerves. Recent findings indicated that exposing enteroendocrine cells to TLR and free fatty acid receptor agonists increased intracellular and extracellular α-synuclein levels (Hurley et al., 2023). This raises the possibility that local gut cells could act as a point of origin for α-synuclein production when faced with immunological insults. Ultimately, increases in gut α-synuclein could result in aggregation and conversion to toxic forms and their eventual spread into the brain.

Increasingly, more studies have provided evidence that gut microbiota can influence the development of PD pathology, including changes in motor functioning and the presence of α-synuclein. For instance, α-synuclein overexpressing transgenic mice raised in a germ-free environment or treated with antibiotics to deplete their microbiota displayed diminished microglial activation and motor deficits (Sampson et al., 2016). This effect could be reversed by the administration of short-chain fatty acids secreted by certain microbiota. Moreover, the transfer of microbiota (through fecal transplants) from healthy mice to those with PD features elicited by MPTP protected them by suppressing inflammatory processes (Sun et al., 2018). A similar study in humans indicated that fecal microbiota transfer from healthy donors could produce modest improvements in PD motor and nonmotor features (Xue et al., 2020). A subsequent report, again involving a small number of participants, only produced transient objective signs of improved motor functioning, although patients expressed improvements and intestinal function was enhanced (DuPont et al., 2023).

The involvement of gut processes in PD is even more intriguing given that gastrointestinal disturbances are common in the disease. The symptoms evident in PD include constipation, abdominal bloating, and nausea, which could be secondary to CNS pathology, although it is also possible that peripheral processes, such as inflammation, are primary driving factors. This is particularly the case considering that the microorganisms that colonize the gut play an important role in shaping immunity, which could then impact CNS processes. Consistent with the gut as a potential early location for PD pathology, intestinal infection with *L. monocytogenes* provoked α-synuclein oligomerization in the ileum. Moreover, this was associated with neuronal mitochondria dysfunction (Magalhães et al., 2023).

Gut feelings and Parkinson's disease

Being the most prominent reservoir of microbes in the body, it is not surprising that the gut has been implicated in numerous brain-related health issues. PD is no different, and there is reason to think that the gut-brain axis is important for this disorder. Besides the obvious gastrointestinal symptoms evident in PD, gut microbes are known to be critical for the proper development of the BBB, as well as for "training" immune cells. In this regard, the composition of the gut microbiome can influence how well the BBB forms, which is presumably a result of evolutionary pressure. As a result of a "leaky" BBB, environmental toxicants would have greater access to the brain and, thus, a greater likelihood of producing neurodegenerative actions. Importantly, endogenous microbes play a key role in modulating immune cell maturity, and environmental stressors that alter the microbiome in favor of pathological microbes can tip the scales in favor of an inflammatory milieu. Further, stressors that limit the complexity of microbial species present in the gut might result in impaired BBB functioning as the normal evolutionary-derived pressures from microbes are diminished.

Inflammatory Mechanisms of Parkinson's Disease

A major shift in thinking about brain-immune interactions and the development of PD has evolved, with an increasing number of studies emerging over the past 20 years linking various inflammatory immune processes to PD, although this view was initially met with considerable skepticism. The dogma regarding the immune-privileged brain dominated much of neuroscience research until only four decades ago. In some ways, the brain certainly is a privileged organ, but this does not mean that it is entirely off-limits to the immune system. As we have seen in earlier chapters, cytokines and immune cells can gain entry to the brain parenchyma, albeit in limited concentration (Filiano et al., 2017). The acceptance that immune processes influence brain functioning has led to the flourishing field of neuroimmunology. Although PD research was slow to catch up, the pendulum has swung sharply in favor of neuroinflammatory mechanisms being a mainstream player in PD pathology, as well as in virtually all neurological and neuropsychiatric disorders. As with so many other diseases, multiple environmental stressors (ranging from chemical, microbial, and psychological challenges) collectively contribute to PD and its comorbidities by affecting neuroimmune processes. Genetic vulnerabilities set the stage for the impact of these insults, but multiple roads may converge at the neuroinflammatory interface to modulate PD pathology. We do not discount other non-immune processes, such as apoptotic or other intracellular prodeath pathways, but these too can be influenced by the inflammatory milieu.

Adaptive Immune Processes in Parkinson's Disease

The innate and adaptative immune systems orchestrate inflammatory processes. Although the innate system has received particular attention in the development of PD, several studies implicated adaptive immune processes in the disease. Aside from the possible involvement of inflammatory factors in the development and progression of PD, the view emerged that the disorder could be viewed from a lens that focuses on similarities to autoimmune disorders. While PD is not an autoimmune disease in the strict sense, it does have certain immune elements in common with such conditions. Most notably, the up-regulation of CD4+ and CD8+ positive T cells in the nigrostriatal system has been found in human PD brains and was evident in animal models of the disease (Alberio et al., 2012; Cebrián et al., 2014). It is thought that peripheral immune T cells can, to some extent, infiltrate the brain and interact with microglia, placing them in a highly active proinflammatory state that may be toxic to DA neurons. In fact, in animal models, treatments that diminished neuronal damage (i.e., were neuroprotective) reduced T-cell infiltration into the midbrain (Gendelman and Appel, 2011).

PD pathology was linked to the levels of the different T-cell subsets in that higher Th1 levels and reduced Th2 levels occurred compared to age-matched controls (Chen et al., 2015a). Further, Th1 lymphocytes were particularly toxic to DA neurons in MPTP-treated mice, whereas the Th2 cells had the opposite effect, being neuroprotective (Olson et al., 2015). This is in keeping with the view of the Th1-Th2 dichotomy, which considers Th1 cells as driving proinflammatory and enzymatic cascades that are highly damaging to DA neurons, whereas Th2 lymphocytes favor the release of antiinflammatory cytokines, such as IL-4 and IL-10, along with beneficial growth factors.

In addition to such T-cell variations, patients with familial inherited PD displayed somewhat higher peripheral serum levels of α-synuclein autoantibodies compared to controls (Papachroni et al., 2007). Curiously, while elevated autoantibody levels were observed in both plasma and CSF of mild and newly diagnosed cases of idiopathic PD, these levels were somewhat lower in the more severe cases of PD (Horvath et al., 2017). Such autoantibodies normally are involved in clearing pathological proteins, but in the case of PD, this process breaks down, resulting in the accumulation of α-synuclein rich Lewy body inclusions. Indeed, genetically induced expression of an A53T mutant bearing α-synuclein in midbrain dopamine neurons promoted the infiltration of T cells (Karikari et al., 2022). The T cells were specifically directed against A53T-α-synuclein peptide fragments found within DA neurons, resulting in the attack and destruction of these neurons. Hence, it is likely that Th-B cell-dependent responses directed at α-synuclein could be protective if the protein is in the extracellular milieu but destructive when it is internalized as Lewy bodies within the intracellular space.

Besides α-synuclein, other autoantibodies may be involved in PD, including those reactive against heat shock proteins and myelin-associated glycoprotein (Papuć et al., 2015). In particular, elevated serum levels of IgG and IgM antibodies that were reactive against the β-crystalline heat shock protein and myelin basic protein were observed in PD patients, although this relationship was not apparent between autoantibody levels and PD (Maetzler et al., 2014). It has been proposed, as already indicated regarding other PD-related biological processes, that some autoantibodies may have evolved as a protective "clean up" mechanism to prevent toxic protein aggregates (Nagele et al., 2013), and in this case, the elevations present in PD might represent a failed attempt to rid the brain of potentially toxic aggregates. The case could be made, however, that although antibody-dependent toxicity could remove faulty or damaged proteins, the associated inflammatory reaction might also produce collateral damage harming otherwise healthy adjacent cells.

Moving beyond CD4+ mediated antibody-dependent humoral immunity, we now turn to the evidence for T lymphocyte-mediated cellular toxicity in PD. Indeed, CD8+ cytotoxic T cells have been implicated in PD, possibly because these cells have access to dopamine neurons for direct cell-to-cell toxicity. Predictably, in animal models of PD involving MPTP or 6-OHDA treatment, α-synuclein overexpression and increased infiltration of CD8+ cells into the brain were observed (Thakur et al., 2017). Although the data concerning the actual CD8+ T-cell levels in the PD brain are scant, alterations were reported within the periphery; curiously, these amounted to both increases and decreases, depending upon the particular study (cf. Baba et al., 2005; Jiang et al., 2017). As well, increased infiltration of CD8+ T cells was accompanied by increased IFNγ, consistent with an inflammatory phenotype (Baba et al., 2005).

Given the importance of CD4+ and CD8+ cells in PD, it follows that interventions aimed at capitalizing on these cells might have therapeutic potential. Consistent with this, an immunization strategy that boosted T-cell infiltration appeared to be neuroprotective in an MPTP model of PD (Benner et al., 2004). In this report, mice were immunized with copolymer 1 (which is a random mix of polypeptides that carry the amino acids found in myelin and is used to treat MS), and the primed T cells were then given to a second cohort of MPTP-treated mice (adoptive transfer), which conferred protection against dopamine neuron loss. Moreover, depleting these mice of T cells reversed the protective effects (Benner et al., 2004).

Studies that isolated the CD4+ CD25+ T cells (i.e., T-reg cells) suggested they might be responsible for neuroprotective consequences. Direct application of T-reg cells conferred protection against the neurotoxic effects of MPTP (Kosloski et al., 2013). Early studies in rodents indicated that T-reg cells that are essential in limiting autoimmune disorders can play a role in the development of a PD-like condition, and manipulations of their activity can attenuate symptoms provoked by nigrostriatal DA neurodegeneration (Reynolds et al., 2010). Subsequent research confirmed that peripheral blood samples from PD patients contained T-reg cells that were less effective in limiting immune functioning, which might contribute to excessive inflammation and exacerbation of PD symptomatology. It was especially significant that in a large population-based case–control study, the risk of PD development was diminished among individuals taking the immunosuppressants azathioprine and mycophenolate (Racette et al., 2018). It was suggested that treatments that act against autoimmune disorders, such as multiple sclerosis, might be effective in limiting the progression of PD (Baird et al., 2019), as

might approaches that employ mesenchymal stem cell therapy.

Intriguingly, transplantation of autologous regulatory T-reg cells improved the survival of grafted dopaminergic neurons (Park et al., 2023b). This is important because less than 10% of such grafted neurons survive, possibly owing to the neuroinflammatory response caused by "needle trauma" associated with the surgical procedure itself. The T-reg cells can reduce this inflammatory trauma, thereby enhancing the effectiveness of the procedure. Further evidence comes from studies showing that adoptive transfer of T-reg cells that were induced by GM-CSF (a potent stimulator of myeloid progenitor cells) downregulated the microglial inflammatory response elicited by MPTP, suggesting an antiinflammatory function in the context of the neurotoxin (Kosloski et al., 2013). Systemic administration of GM-CSF alone was also sufficient to induce neuroprotection in a paraquat-based toxicant model of PD, which was related to the downregulation of the microglial response (Mangano et al., 2011). GM-CSF also has trophic and antiapoptotic effects and induces the promotion of other growth factors, such as BDNF (Mangano et al., 2011), raising the possibility that trophic effects could also be at play in determining the beneficial effects of GM-CSF.

A novel idea that has been advanced regarding the involvement of adaptive immunity in PD is that neoepitopes are created in the brain during the disease state that could fuel the infiltration of neurotoxic peripheral immune cells that could subsequently promote the degradation of DA neurons. For instance, mitochondrial dysfunction and toxic metabolic byproducts might damage DA neurons, creating novel cell fragments that act as neoepitopes immunologically recognized as foreign. This could then result in microglia taking up the epitopes and transmitting a signal for the mobilization of peripheral immune cells, which then invade the brain to encounter the neoepitope being presented by the glial cell (see Fig. 15.6).

Along the same lines, the possibility was considered that specific mitochondrial antigens might trigger some degree of autoimmunity in PD. This concept of mitochondrial antigen presentation (MITAP) essentially posits that mitochondrial damage elicits mitochondrial-derived vesicles (MDVs) carrying antigenic fragments to migrate to the cell surface for presentation in MHC I molecules and subsequent recognition by T lymphocytes (Matheoud et al., 2016). This would result in the damage and eventual death of compromised dopamine neurons that undergo MITAP. Normally, the proteins PINK1 and Parkin act on MITAP by regulating mitophagy and MDV transport (Matheoud et al., 2016). In this regard, PINK1 is a kinase that normally phosphorylates both ubiquitin and Parkin on the surface of damaged mitochondria, thereby regulating mitophagy or the degradation of faulty/damaged mitochondria before they cause

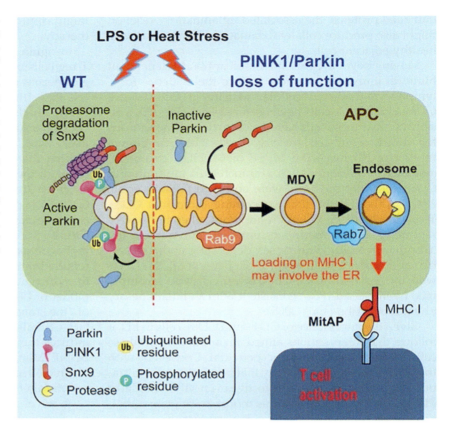

FIG. 15.6 Mitochondrial antigen induced autoimmunity and PD. Environmental stressors (in this case LPS or heat stress) can impact mitochondrial functioning. Normally, the Parkinson's-linked genes *Parkin* and *PINK1* act to regulate mitochondrial functioning. These proteins ensure that any damaged mitochondria are removed through the process of being tagged by ubiquitin (ubiquitinated) for removal by the proteasome. However, when *Parkin* or *PINK1* are altered through genetic mutation, this process breaks down. The end result is that antigens bud off the damaged mitochondria and are placed in an endosome, and then presented on the cell surface via an MHC I molecule. This attracts T cells, which can initiate a destructive autoimmune attack that kills the cell. *From Matheoud, D., Sugiura, A., Bellemare-Pelletier, A., Laplante, A., Rondeau, C. et al., 2016. Parkinson's disease-related proteins PINK1 and parkin repress mitochondrial antigen presentation. Cell 166, 314–327.*

cellular pathology (Lazarou, 2015). Hence, PD linked to PINK1 and Parkin mutations may cause accumulation of damaged mitochondria, which leads to MITAP and subsequent immune-targeted death of dopamine neurons.

Is PD an autoimmune disease?

In considering the involvement of immunity in PD, it was considered that PD might be an autoimmune disease. While the bulk of evidence has favored a role for the innate branch of immunity in PD, several recent convincing studies implicated the adaptive immune response, suggesting that T-cell-dependent autoimmunity might be at play. One account had it that the presentation of mitochondrial antigens might trigger an autoimmune response against SNc dopamine neurons (Matheoud et al., 2016). Having evolved from ancestral endosymbiotic bacteria, mitochondria may be directly targeted by virulence factors of intracellular pathogens, and mitochondrial dysfunction and fragmentation resulting from the activation of the innate immune system at the gut level, trigger innate immune responses in midbrain neurons, which include α-synuclein oligomerization and neuroinflammation.

With mitochondria being thought to have come from ancient bacteria, their antigens may be hidden from the immune system, as any breach could lead to autoimmunity. This is consistent with the considerable data showing mitochondrial dysfunction in PD. Furthermore, it was demonstrated that PINK and PARKIN genes regulate mitochondrial antigen presentation, and mutations in these proteins might confer PD risk by removing the normally protective ability of these proteins to limit antigen processing and presentation. Likewise, the concept of molecular mimicry, wherein certain endogenous neural motifs might be falsely recognized as foreign, has been applied to account for PD pathology. Whatever the case, the microglia are undoubtedly at the center stage in terms of neuroimmune mechanisms.

Microglia and Parkinson's Disease: The Master Innate Neuroinflammatory Orchestrator

Microglia normally provide neuronal support through the release of trophic factors, including BDNF, as well as by clearing cellular debris or metabolic byproducts through phagocytosis (Prowse and Hayley, 2021). During a relatively quiescent resting state, microglia have an immobile ramified morphology characterized by a small cell body with long, thin processes that extend out into the extracellular milieu. These processes possess pathogen-associated molecular pattern (PAMP) receptors or damage-associated molecular pattern (DAMP) receptors, allowing them to detect and deal with microbial or other threats (Venegas and Heneka, 2017). Damaged or stressed cells release ATP into the extracellular space, which microglia can detect and place them in a "clean up" mode for debris removal (Davalos et al., 2005). While removing cellular debris, microglia also extend their processes to shield healthy neurons from dying tissue and release attractant cytokines (chemokines) to recruit neighboring microglia (Gao and Hong, 2008).

Although there is substantial evidence that microglial-driven chronic inflammation is important for the genesis and the progression of PD (Ramirez et al., 2017), it is not clear what precisely drives microglial reactivity. One strong possibility is that it is a reaction to the accumulation of α-synuclein-rich Lewy bodies. Misfolded α-synuclein that comprises a core feature of these pathological inclusions induces strong microglial activation, initiating inflammatory cytokine expression (Zhang et al., 2024a). In essence, microglia likely recognize these pathological features as DAMPs and respond accordingly. Thus, different precipitating challenges (e.g., pesticides, heavy metals, microbial agents, and psychological stressors) might be interpreted as either DAMPs or PAMPs. If such a response becomes chronic or excessive, then pathology may ensue.

Microglia are adept at producing inflammatory factors designed to neutralize threats. Once activated, microglia may be responsible for phagocytic processes and release cytotoxic factors to ingest and eliminate noxious compounds. Microglial release of proinflammatory cytokines (i.e., IFN-γ, TNF-α, IL-6, and IL-1β) and oxidative radicals (superoxide, nitric oxide) comprises their core reaction to PAMPs and DAMPs. Some of these same inflammatory factors were observed in postmortem PD patients or their CSF (Kunze et al., 2023). In fact, in comparison to inflammation observed in arthritic joints, the levels of some of these factors appeared to be even greater in PD (McGeer and McGeer, 2004). While short-term microglial activation is beneficial, problems arise when microglia are in a chronic or permanently active state so that these normally defensive molecules begin to damage otherwise healthy tissue, particularly DA neurons within the SNc. As indicated earlier, these specific DA neurons are far more vulnerable to microglial inflammation than other neuronal subtypes found throughout the brain (Kostuk et al., 2019). LPS infusion into the SNc had far more damaging effects on local DA neurons relative to those in the frontal cortex, hippocampus, thalamus, or hindbrain. It was speculated that this enhanced sensitivity stemmed from the higher density and a greater basal oxidative stress state of microglia in the SNc. Consistent with this perspective, the differences in the sensitivity to MPTP evident between mouse strains were correlated with SNc glial density, and the vulnerabilities to MPTP could be reversed by manipulating the number of glial cells present in vitro (Smeyne et al., 2005). As well, enhanced iron and melanin content is believed to factor into the vulnerability of SNc neurons. Ultimately, the SNc neurons themselves may have an intrinsic vulnerability, and the SNc microglia are phenotypically different or more reactive than microglia from other brain regions.

Dopamine neurons of the SNc have a very high basal oxidative potential and are exceptionally metabolically active. These features make them prone to energetic disturbances and the generation of high levels of potentially destructive oxidative radicals as byproducts of normal cellular reactions. These neurons also contain high levels of iron in PD patients, which may provide a substrate for inflammatory molecules with which they can react (Biondetti et al., 2021). This is consistent with the epidemiological evidence described earlier, which showed that exposure to heavy metals is accompanied by an increased risk of PD. As well, exposing mice to iron or manganese augmented the neurodegenerative effects of paraquat or MPTP (Peng et al., 2007).

The SNc microglia may have more of a "hairpin trigger" than adjacent brain regions. The extent of microgliosis (or essentially the activation and mobilization of microglia) in response to many different challenges is often more robust in the SNc than in other brain regions. Besides the magnitude of the microglial response, the phenotype or characteristics of the factors expressed by these cells also appear to differ in PD. Specifically, microglia are generally polarized toward an M1 state in the disease, wherein they produce more inflammatory cytokines and oxidative factors and may be locked into this state for extended periods (Tang and Le, 2016). Time course studies indicated that microglia activation often occurs before dopaminergic degeneration in the face of neurotoxin administration and that they can display a phagocytic phenotype that could engulf compromised dopamine neurons (Marinova-Mutafchieva et al., 2009).

An interesting route to PD pathology involves the intersection between microglial-driven inflammation and α-synuclein processes. This is supported by the finding that α-synuclein was associated with HLA+ microglia in postmortem PD brain tissue (Orr et al., 2005). Moreover, microglial activation, using inflammatory stimuli, promoted the spread of α-synuclein pathology and neuronal degeneration (Yi et al., 2022b), and it seems that the PAMPs used to detect pathogens might fuel the development of synucleinopathy. In fact, TLR2 was localized to α-synuclein-rich Lewy bodies and promoted α-synuclein accumulation, whereas TLR2 inhibition countered this effect (Dzamko et al., 2017). Also, priming isolated microglial cells with α-synuclein greatly influenced the inflammatory responses to subsequent immune insults, promoting greater cytokine release and a prominent shift in their morphological state (Roodveldt et al., 2013). Not surprisingly, the direct application of α-synuclein into the SNc markedly activated microglia and induced dopaminergic neuronal death. Together, these findings point to a reciprocal relationship between microglial inflammatory and α-synuclein processes such that one likely feeds into the other to fuel the development of pathology (see Fig. 15.7).

The impact of α-synuclein on the microglial inflammatory and phagocytic state differs across brain regions. Adenoviral-driven expression within the midbrain produced brain region-specific microglial activation profiles. Surprisingly, the CD11b+ microglia within the midbrain displayed an antiinflammatory transcriptomic profile, whereas striatal CD11b+ cells showed a proinflammatory profile (Basurco et al., 2023). Infiltration of monocytes/macrophages was also apparent with these cells, and local microglia phagocytize dopaminergic neurons. However, in addition to their neurotoxic contributions, microglia phagocytosis may provide protective effects by facilitating clearance of misfolded and aggregated α-synuclein. This is in keeping with the "double-edged" sword dichotomy characterization of microglia that has frequently been described. Emerging evidence pointed to a failure of the microglial protective mechanisms in PD, which might be instigated by α-synuclein. For instance, α-Syn inhibited the initiation of autophagy

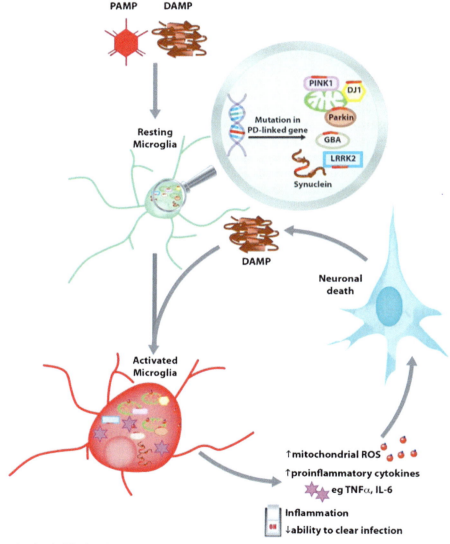

FIG. 15.7 Microglial activation in PD. LRRK2, α-synuclein, Parkin, Pink1, and glucocerebrosidase (GBA) mutations can prime microglia to be more reactive to microbial agents (PAMPs) or even the sterile inflammation provoked by endogenous danger signals (DAMPs). This can lead to enhanced oxidative radicals and proinflammatory cytokines. *From Dzamko, N., Geczy, C.L., Halliday, G.M., 2015. Inflammation is genetically implicated in Parkinson's disease. Neuroscience 302, 89–102.*

(breakdown of cellular waste) in microglia, thereby reducing their ability to clear pathological aggregates (Tu et al., 2021). An autophagy deficiency would also result in many microglia becoming senescent (Choi et al., 2023b), which would render them dysfunctional and prone to a deleterious role.

Cytokines in Parkinson's Disease

Cytokines have been implicated in virtually all neurodegenerative and neuropsychiatric diseases. As we have seen, cytokines and their receptors are expressed endogenously in the CNS, and mounting evidence suggests this is largely due to de novo synthesis by microglia. However, beyond glial cells, it was suggested that neurons might also be capable of synthesizing small amounts of proinflammatory and antiinflammatory cytokines during disease or distress.

The general picture that has emerged is one in which high levels of proinflammatory cytokines are viewed as deleterious within the brain. This is supported by the strong link between chronically or very high levels of TNF-α, IL-1β, and IFN-γ and the incidence of brain pathology (and reactive microgliosis). Conversely, antiinflammatory agents (e.g., minocycline, nonsteroidal, and antiinflammatory drugs) have generally been associated with neuroprotective consequences in a range of preclinical animal models (Subramaniam and Federoff, 2017). However, as we have seen with other pathologies, low physiological levels of cytokines may induce neuroprotection and adaptive neuroplasticity (e.g., via the release of free radical scavengers and trophic factors) (Litteljohn et al., 2014). Complicating the picture, cytokines typically display a high degree of redundancy, pleiotropy, synergy, and even antagonism. The complex nature of cytokine physiology has led to the view that cytokines are best considered as a network of biologically active mediators whose collective output determines physiological and pathological consequences.

Cytokines have been implicated in oxidative and excitotoxic pathways that produce neuronal damage and death. The cytokine-provoked induction of the prooxidant/inflammatory enzymes COX-2 and inducible nitric oxide synthase (iNOS) results in the production of free radicals, such as superoxide and peroxynitrate, which cause oxidative damage to DNA, proteins, and lipids (Mangano et al., 2012). Beyond this, certain proinflammatory cytokines, such as IL-1β and TNF-α, can also trigger the activation of caspase-dependent apoptotic pathways and excitatory glutamatergic signaling (Bakunina et al., 2015). Such inflammatory cytokines are well placed to cause neuronal damage through multiple pathways, and consequently, designing novel treatment options is highly complex.

Cytokines may also contribute to PD by augmenting BBB permeability, which can facilitate the central infiltration of immune factors and pathogens that can have vast and widespread effects on CNS functioning (Coureuil et al., 2017). As described earlier, exposure to the bacterial endotoxin, LPS (which robustly promotes proinflammatory cytokine expression) caused exaggerated dopamine neuron and terminal loss in the SNc and striatum, respectively, upon subsequent exposure to the pesticide paraquat (Mangano and Hayley, 2009). Thus, inflammatory stimuli may prime or sensitize microglia, so that later challenges provoke exaggerated responses that could potentially exacerbate the neurodegenerative process.

IFNγ and PD: A Specific Inflammatory Link

Among the proinflammatory cytokines linked to PD, particularly strong evidence suggests the importance of interferons (IFNs). It will be recalled that these are broadly divided into either type I IFNs, including the IFN-α and IFN-β isoforms, or the type II IFNγ. The latter is biologically active as a homodimer and, upon binding to its receptor, induces various intracellular signaling factors. Although it was initially believed that IFNγ is secreted exclusively from NK cells and Th1 lymphocytes, it appears that antigen-presenting cells, such as dendritic cells and macrophages, also produce IFNγ. Additionally, cytokines are synthesized within the brain by activated microglia and astrocytes, which may contribute to PD.

The IFNs are fundamental mediators of both innate and adaptive immune responses to microbial infection, including antigen processing, activation of macrophages, NK cell effector functions, and stimulation of antigen-specific T cells. In addition to antiviral actions, IFNγ plays an important role in host defense against certain bacterial (e.g., mycobacteria), fungal, and parasitic pathogens, as well as having antitumor properties. Like other cytokines, IFNγ mainly signals through JAK and STAT intracellular proteins, which is a fundamental signaling pathway involved in inflammatory processes and whose dysregulation is associated with autoimmune disorders and several forms of cancer (Hu et al., 2021a).

Ligand binding induces a conformational change in the IFNγ receptor-1/2 chains, which leads to the sequential activation through phosphorylation of JAK1 and JAK2, which then promotes the recruitment, phosphorylation, and homodimerization of STAT1 (Schroder et al., 2004). These STAT1 dimers translocate to the nucleus and bind to IFNγ-activation sites in the promoter region of IFNγ-responsive genes. IFNγ-STAT1 signaling can then initiate or suppress the transcription of a host of immunologically relevant target genes. Additionally, IFNγ signaling can produce STAT1 heterodimers, which can bind to IFN-stimulated response element promoter regions to regulate further rounds of transcription.

Given its potent inflammatory actions, several endogenous mechanisms exist to naturally inhibit IFNγ levels. These include (1) degradation and/or recycling of the IFNγ:IFNγ receptor-1 complex, (2) disruption of JAK phospho-activity and/or targeting of JAKs for proteasomal degradation by suppressors of cytokine signaling-1 (SOCS-1) and SOCS-3, and (3) dephosphorylation of IFNγ receptors and STAT1 by the protein tyrosine phosphatases (Majoros et al., 2017). Similarly, glucocorticoids and the antiinflammatory cytokines IL-4, IL-10, and TGFβ inhibit IFNγ activity/production (Majoros et al., 2017).

In addition to its other actions, IFNγ induces the activation of various inducible inflammatory enzymes, in particular, iNOS, COX-2, and nicotinamide adenine dinucleotide phosphate (NADPH) oxidase (Majoros et al., 2017). Although these inflammatory enzymes confer protection against pathogenic invasion, their chronic activation has been linked to behavioral disturbances and impaired structural and functional neuroplasticity (Litteljohn et al., 2008). Of note, IFNγ signaling may also drive the downregulation of neuroprotective species in glial cells, which could increase progenitor cell and neuronal vulnerability to oxidative and inflammatory damage (Litteljohn et al., 2017). For example, IFNγ might contribute to paraquat-induced neurodegeneration, in part by mediating an early occurring reduction in central BDNF (Mangano et al., 2012) and by promoting astrocytic toxicity (Hashioka et al., 2015). Moreover, IFNγ directly damages cultured neurons through induced glutamatergic neurotoxicity (Mizuno et al., 2008), and IFNγ deficiency protects against the toxic effects of MPTP treatment, which was tied to microglial inflammatory processes (Mount et al., 2007). Most of the deleterious effects of IFNγ were mediated by JAK-STAT1 signaling, and it also appeared that reactive astrocytes could diminish their supportive actions and may promote the secretion of neurotoxic factors, complement components like C3, and certain chemokines that facilitate immune cell recruitment into the brain. Of particular relevance to neurodegenerative disorders, the abundance of proinflammatory reactive astrocytes increases with age (Lawrence et al., 2023).

Unlike IFNγ, the related interferon, IFN-β, is generally believed to have neuroprotective consequences for PD. Analyses based on GWAS have genetically linked dysregulated IFN-β signaling to idiopathic PD, and the relationship was particularly strong for the dementia PDD subtype (Magalhães et al., 2021). In this respect, IFN-β null mice developed a PDD-like syndrome that was accompanied by cognitive disability, dopaminergic neuron death, abnormal protein accumulation, and a deficit in autophagy (Ejlerskov et al., 2015). It was subsequently shown that a deficit in IFN-β activity results in mitochondrial DNA (mtDNA) damage. Remarkably, infusion of such damaged mtDNA to healthy mice induced PDD-like neurodegeneration, along with motor and cognitive pathology that spread to several brain regions (Tresse et al., 2023). The neurotoxicity of the damaged mtDNA was linked to the coactivation of TLR9 and TLR4, resulting in oxidative stress and the spread of mitochondrial dysfunction. These data suggest that mtDNA is being recognized as a DAMP (via TLR pattern-recognition receptors), raising the possibility that the well-known mitochondrial deficits evident in PD and PDD might spread between neurons in an infectious-like manner that is modulated by IFN-β processes.

Genetic Vulnerability

Factors responsible for PD likely exist on a spectrum with familial linked forms (e.g., SNCA, LRRK2, DJ-1, Parkin, and PINK1) that result in early provocation at one end, and purely environmental impact at the other, leaving the bulk to the interactive effects of genetic vulnerability and environmental influence (Ritz et al., 2016; Schlossmacher et al., 2017). As not all genes associated with PD give rise to Parkinsonian symptoms, together with reports that many polymorphisms have been linked to PD implicate the influence of environmental factors in the occurrence of the disease. Evidence from animal research and epidemiological studies over the last decade has supported gene–environment interactions in PD pathophysiology (Lee et al., 2016c).

Several mechanisms have been offered by which genetic vulnerability enhances susceptibility to environmental insults (e.g., toxicant exposure), hence resulting in PD. An epidemiological study revealed an increased risk of developing PD in individuals who possessed either the rs1045642 or rs2032582 polymorphisms in the ABCB1 gene in frequent pesticide sprayers (Narayan et al., 2015). Importantly, the ABCB1 gene encodes for P-glycoprotein, which acts as a cellular efflux transporter of lipophilic compounds across the BBB. Thus, variations in P-glycoprotein would be expected to impact the transport of toxicants to contact nigral dopaminergic neurons. Indeed, these polymorphisms were associated with altered P-glycoprotein expression on epithelial cells lining the BBB, which resulted in elevated concentrations of neurotoxic substances (i.e., xenobiotics, such as pesticides) penetrating the brain (Narayan et al., 2015). Moreover, the use of paraquat sprayers was accompanied by an 11-fold increase in risk for PD, but only in individuals who also lacked the glutathione S transferase T1 (GSST1) gene, which is primarily responsible for detoxifying xenobiotic compounds (Goldman et al., 2010).

Other polymorphisms have similar PD-relevant effects. Individuals possessing polymorphisms in the DAT/SLC6A3 are at increased risk of developing PD when also exposed to pesticides (Kelada et al., 2006).

It was hypothesized that the DAT/SLC6A3 polymorphisms enhance pesticide binding to DAT and entry into dopamine neurons (Kelada et al., 2006). Carriers of the DAT SNPs, rs2652511 and rs2937639, exhibit an increase in DAT expression in striatal regions (van de Giessen et al., 2009), and variations of the transporter can enhance susceptibility to paraquat entry into neurons and subsequent toxicity (Ritz et al., 2016). Thus, again, it seems that these polymorphisms alone do not cause disease but increase susceptibility by modifying the impact of environmental challenges.

LRRK2 in Relation to PD and Other Inflammatory Disorders

LRRK2 is a large multimeric protein (286 kDa) with several distinct domains and is unique in that it has both guanine triphosphate hydrolysis (GTPase) and MAP kinase functions (Cookson, 2015). Substantial evidence suggests that enhanced kinase activity of the gene is a major contributor to the pathogenesis of LRRK2-related PD (e.g., the LRRK2 G2019S mutation increases kinase activity $\sim$ twofold to threefold) (West, 2017). However, not all LRRK2 variants are associated with increased kinase activity (Rudenko et al., 2012). Moreover, other activities, including GTPase functioning and its ability to act as "structural scaffolding" for protein–protein interactions, might be involved in its links to PD.

The LRRK2 mutations are inherited in an autosomal dominant manner but usually display low penetrance; not all carriers of a given LRRK2 mutation develop PD, and among those who do, considerable clinical and neuropathological variability exists (Zimprich et al., 2004). These findings led to the suggestion that environmental events, such as toxicant exposure and chronic stressors, as well as genetic factors (e.g., other PD-associated genes or loci), may be triggers or modifiers of LRRK2-related PD (Karuppagounder et al., 2016; Litteljohn et al., 2017). Consistent with this view, the LRRK2 gene has been implicated in both familial and sporadic forms of PD (Cookson, 2017). Indeed, while LRRK2 mutations are the most frequent known genetic cause of the familial form of PD, several LRRK2 polymorphisms are also associated with a heightened risk of developing the more common sporadic form of the disease, varying with sex and ethnicity (Park et al., 2023a).

The LRRK2 genetic mutation can be a key genetic factor that causes a circumscribed early-onset form of the disease or can act as a vulnerability factor in which some other insult produces the later age-dependent form of the disease. The finding that individuals who possess the G2019S mutation but do not display PD showed alterations in brain metabolism is consistent with the position that the gene might be a vulnerability factor. Imaging

work confirmed that nonmanifesting carriers of the LRRK2 G2019S mutation displayed altered brain regional activity and interregional functional connectivity (van Nuenen et al., 2012). Although the LRRK2-related genetic form of Parkinsonism has been considered to closely resemble sporadic PD, there have been indications that LRRK2 mutations might cause an altered neuropsychiatric phenotype (Hayley and Litteljohn, 2013). Specifically, PD patients and asymptomatic carriers harboring the LRRK2 G2019S point mutation (glycine-to-serine substitution at position 2019) present with higher rates of depressive and disturbed cognitive symptoms. Yet, a study with over 500 G2019S PD patients revealed slightly lower depressive scores compared to idiopathic non-G2019S patients (Marras et al., 2016). These variations might stem from the fundamentally different populations used in the studies that are known to differ with respect to these mutations.

Owing to its inflammatory regulatory functions, LRRK2 sits at the crossroads that influence numerous illnesses in addition to PD, such as Crohn's disease, leprosy, and diabetic neuropathy, as well as HIV-1-associated neurocognitive disorders (Herrick and Tansey, 2021; Lewis and Manzoni, 2012). Paralleling human studies, animal models have provided evidence that LRRK2 increases the risk for inflammatory bowel disorders. Specifically, mice bearing the G2019S LRRK23 mutation exposed to the gut inflammatory agent, dextran sodium sulfate, displayed increased colonic expression of α-synuclein, TNF-α, and TLRs (Lin et al., 2022), along with increased microglial activation (Dwyer et al., 2021). An in-depth discussion of the proposed role of LRRK2 in these conditions exceeds the scope of the present work (but see Greggio et al., 2012). However, it bears mentioning that all these LRRK2-associated conditions have a prominent immuno-inflammatory component in common with PD.

LRRK2: A Novel Regulator of CNS-Immune System Interactions

Despite the clear importance of LRRK2 in PD, the normal role(s) of LRRK2 remains elusive. Nonetheless, LRRK2 has provisionally been implicated in a diverse range of cellular functions, including autophagy, cytoskeletal dynamics, intracellular membrane trafficking, synaptic-vesicle cycling/neurotransmission, and the inflammatory response (Cookson, 2017). The protein might also normally influence microglial morphology since the LRRK2 G2019S mutation can induce changes in microglial branching patterns, and this, in turn, causes excessive pruning of neighboring dopaminergic neurons (Zhang et al., 2022b), which would dramatically impact neuronal transmission. Furthermore, the remodeling of microglial fibers and their pruning capability can greatly impact their inflammatory activity and "immune monitoring" capacity, resulting in certain cerebral areas being highly susceptible to pathogenic stimuli.

As it relates to LRRK2 kinase functions, the growing list of suspected or validated LRRK2 kinase substrates includes LRRK2 itself (auto-phosphorylation), MAP kinases (which are immediately upstream of JNK and p38 MAP kinase), as well as NF-κB (Gloeckner et al., 2009). Similarly, NF-κB transcriptional activity and vascular cell adhesion molecule expression were exaggerated by LRRK2 G2019S overexpression. This mutation also potentiated leukocyte chemotaxis (Moehle et al., 2015) and enhanced the activity of several major intracellular inflammatory signaling pathways, namely NF-κB and the MAP kinases, p38, and JNK (Usmani et al., 2021). Thus, LRRK2 might promote PD pathology by activating these pathways through kinase-dependent phosphorylation.

In addition to being expressed in a variety of peripheral organs (e.g., kidney, lung, liver, heart, and spleen), LRRK2 is readily detectable in neurons over a wide range of brain regions. Included here are the SNc and striatum (surprisingly low in the former but high in the latter), as well as the hippocampus, PFC, locus coeruleus, and various hypothalamic and amygdaloid nuclei. This led to the suggestion that beyond its role in nigrostriatal motor function, LRRK2 may contribute to the regulation of emotional and cognitive processes through altered dopaminergic activity (Litteljohn et al., 2017).

More recently, LRRK2 was found to directly regulate peripheral T and B lymphocyte activity. Replacement of mutant LRRK2 with the wild-type form of the protein in T- and B-lymphocytes reduced the neuroinflammatory and neurodegenerative responses of LPS exposure in mutant LRRK2 mice (Kozina et al., 2022). This indicates that LRRK2 has a clear intracellular function within these lymphocytes. The presence of the G2019S or R1441G LRRK2 mutation in lymphocytes alone was sufficient for LPS-induced neurodegeneration, which was prevented by blocking peripheral IL-6. These data suggest that peripheral immunity might have a primary role in the origins of dopaminergic degeneration in PD.

Subcellular localization studies indicated that LRRK2 associates with membranous and vesicular structures, including mitochondria, lysosomes, endosomes, lipid rafts, and transport/synaptic vesicles (Migheli et al., 2013). Interestingly, the cell types that most highly express LRRK2 are immune cells, mostly monocytes, macrophages, B-lymphocytes, dendritic cells, and microglia (Wallings and Tansey, 2019). Given the prominent role of LRRK2 and its mutations in the inflammatory response (Dzamko et al., 2017), silencing LRRK2 in macrophages and/or microglia attenuated LPS-induced NF-κB transcriptional activity, iNOS, and COX-2 expression,

along with TNF-α, IL-1β, and IL-6 release (Moehle et al., 2012). Conversely, overexpression of the PD-linked LRRK2 R1441G mutation exacerbated many of these endotoxin-induced effects (Gillardon et al., 2012). LRRK2 also plays an active role in the regulation of microglial migration and phagocytosis and may be particularly important in microglial responses to pathological α-synuclein aggregates (Feng et al., 2023).

Among the proinflammatory cytokines, IFNγ stands out as a preferential inducer of LRRK2. This cytokine markedly increases LRRK2 expression in human peripheral blood mononuclear cells, particularly CD14+ CD16+ monocytes (typical of inflammatory conditions), as well as in circulating macrophages and brain-resident microglia (Rui et al., 2018). Thus, inflammatory-driven stimulation of IFN-γ can mobilize LRRK2 expression and may explain how IFNγ knockout was protective against MPTP or paraquat exposure (Mangano et al., 2011; Mount et al., 2007).

Finally, LRRK2 plays an important role in autophagy and, hence, could be important for Lewy bodies or other pathological aggregates that occur in PD. Mutant versions of LRRK2 can bind to the lysosome and interfere with the degradation of proteins, most notably α-synuclein, which can lead to the pathological accumulation of Lewy body inclusions (Orenstein et al., 2013). Additionally, LRRK2 interacts with rab5b and rab7, preventing proper fusion of autophagosomes to lysosomes, resulting in decreased protein degradation (Roosen and Cookson, 2016). Taken together, these studies suggest that LRRK2 acts through multiple intracellular substrates to interfere with protein recycling and processing, which could ultimately produce inclusions such as those seen in PD. Although this could be how LRRK2 directly affects dopamine neurons, the fact that LRRK2 is found at much higher levels in microglia and immune cells and is involved in a variety of immune processes indicates that LRRK2 can also secondarily impact neurons through either resident microglia or infiltrating leukocytes.

Future Immunomodulatory Treatments and Parkinson's Disease

Preclinical work has pointed to different treatments that are neuroprotective in PD. Unfortunately, as we have seen with so many other illnesses, none of these treatments has successfully translated into clinical applications that stop or even appreciably slow the disease. Part of the problem may reside in the differences in the complexity of the human brain compared to experimental animals. Similarly, the degree of epigenetic changes between species is substantial. When comparing transcriptional profiles in 15 different tissues between humans and rodents, there were more similarities between the different tissues within each species than between the species.

These problems are difficult to overcome, yet a major issue that could be more easily addressed experimentally concerns better modeling of different stages of PD by adopting specialized treatments that suit the particular disease state. As we have indicated with other diseases, animal preparations that model early, middle, and late disease states, as well as representative comorbid symptoms, such as depression and anxiety, might be of particular use. In this way, the efficacy of different treatments might be tested, allowing for the determination of whether differential effects appear depending upon disease stage and type of comorbid features present. Furthermore, multiple concomitant treatments that target different disease processes may be required to produce clinically meaningful outcomes. Such combined treatment strategies are now commonplace for psychiatric conditions, such as depression, and such a comprehensive approach may also be warranted for PD.

In the sections that follow, we will describe potential emerging treatments that act upon (at least in part) immunological processes to influence PD pathology. These treatment strategies will be interpreted within the framework of their efficacy at different stages of PD pathology and with different symptom presentations. We will also emphasize how some treatments might be used together or might interact with existing standard therapies, such as l-DOPA, that are used to manage motor symptomatology without actually affecting underlying disease mechanisms. Finally, it might be fruitful to combine immunomodulatory treatments with emerging genetic cell-based strategies.

The three main types of immunomodulatory treatments that we will focus on are antiinflammatory drugs, trophic cytokines, and LRRK2 inhibitors. With each of these strategies, the objectives are to stabilize or prevent further degeneration and promote some degree of functional recovery by facilitating neuronal plasticity and compensation within neural circuitry.

Antiinflammatory PD Treatments

The curious finding that rheumatoid arthritis patients who were on chronic high NSAID doses, displayed a particularly low incidence of Alzheimer's disease (Auriel et al., 2014), kick-started the notion that antiinflammatory drugs might be effective in treating neurodegeneration. This spurred preclinical and clinical trials assessing the utility of NSAIDs as neuroprotective agents, quickly spreading to the PD area, given the substantial neuroinflammatory component implicated in the disorder. Unfortunately, most clinical trials using NSAIDs for PD have provided disappointing results (Samii et al., 2009),

although regular NSAID use was more commonly observed among asymptomatic carriers of *LRRK2* variants than it was among individuals who were symptomatic carriers of this variant (Fyfe, 2020).

A meta-analysis supported a modest reduction of symptoms with ibuprofen but not with other NSAIDs (Gagne and Power, 2010). Despite the generally discouraging findings, the combination of high levels of NSAIDs, together with smoking and high coffee intake, was associated with the greatest reduction of PD risk (Powers et al., 2008). It seems likely that NSAIDs or other antiinflammatories would not be useful for treatment in existing PD patients but might have value as a prophylactic treatment in vulnerable populations. Limiting inflammation during the early incubation stages of the disease might allow for some degree of endogenous plasticity, thereby affecting PD age of onset (Gabbert et al., 2022). However, as with so many other diseases, once the PD pathology has taken hold, the point of no return may have passed, and recovery would be exceedingly difficult to achieve simply through antiinflammatories.

Another caveat concerning the use of broad-spectrum antiinflammatory agents is that they not only limit harmful inflammatory processes but also beneficial inflammatory cascades that might be critical for recovery or reparative processes. Cytokines, for example, while harmful in high doses, are required in low concentrations for normal brain homeostasis (Subramaniam and Federoff, 2017). Instead of simply applying broad-spectrum antiinflammatory treatment to shut down microglia, it might be more efficacious to target the modulation of the microglial phenotype. Thus, an immunomodulatory rather than antiinflammatory approach might be fruitful. This would represent a departure in the conceptualization of the disease, and, instead of demonizing all inflammatory processes, accepting that certain microglial states could promote CNS repair. This is most apparent when considering the M1 and M2 dichotomy and all the intermediate states. As indicated earlier, M2-like states (which are characterized by the release of antiinflammatory cytokines and trophic factors) might release protective trophic factors, as well as help with the clearance of potentially destructive plaques. This is in opposition to the more proinflammatory M1 phenotype. Thus, in a way, this could be viewed as the immunomodulation of microglia toward a neuroprotective phenotype.

It is important to underscore that M1 and M2 are not discrete phenotypes (states), but that microglia exist on a continuum between these two extremes. For that matter, the panoply of different microglial activation states might not even exist on a strictly linear continuum but rather reflect various qualitatively different offshoots. Moreover, different microglia expressing these varied states can co-exist in the same brain region.

Consequently, at any one time, there may be a complex tug-of-war between the different phenotypic states. Clearly, in the later stages of the disease, the M1-like inflammatory microglia are the winners and predominate lesion sites.

When in an M2 state, microglia can engulf and remove debris or damaged cells that display phosphatidylserine, which essentially acts as an "eat me" signal (Xia et al., 2015). Removal of compromised cells and synapses would presumably help strengthen the functioning of healthy cells. At the same time, M2 microglia up-regulate antioxidants, such as glutathione and heme oxygenase-1, which help detoxify potentially harmful reactive oxygen species that typically characterize PD. These microglia further assist neurons by releasing trophic growth factors, including GNDF, NGF, and TGFβ, particularly in the presence of the antiinflammatory cytokines, IL-10 or IL-4 (de Bilbao et al., 2009). Finally, other transmembrane receptors upregulated on M2 microglia, including Ym-1 and Arg-1 (which have been implicated in wound healing), also foster recovery from injury by positively modulating the extracellular compartment (Cherry et al., 2014). Since the M2-like state acts toward processes aligned with cell survival, repair, and removal (via phagocytosis) of damaged tissue, a potential goal of therapy could be to encourage microglia to adopt a more M2-like state, rather than attempting to blunt their overall activity.

The M1–M2 perspective came from in vitro studies involving cultured microglia in which IFNγ and IL-4 promoted these respective states. An analogous situation may not be reflected in vivo since microglia cannot be neatly grouped into these two states (Ransohoff, 2016). Microglia may co-express both M1 and M2 markers simultaneously, but the levels and precise roles of these factors vary with the stimulus/insult and state of the microenvironment (Perego et al., 2013). Irrespective of how the range of microglia states are described, the important and unequivocal point is that microglia sometimes help and sometimes hinder whatever pathological state exists. Thus, even if the M1–M2 dichotomy was fully accepted, the basic premise of immunomodulatory regulation of microglia for neuroprotection is still valid. However, the critical issue is when and how microglia can be modulated to produce positive effects. This brings us to the notion of specific target proteins on or related to microglia that can be pharmacologically or genetically manipulated, and once more, LRRK2 may be an ideal target for this.

LRRK2 Inhibition and Recovery

There are four distinct processes through which LRRK2 can act and hence could serve as targets for pharmacological inhibitors. These comprise (1) kinase-mediated phosphorylation, (2) GTPase-mediated control of G protein cascades, (3) scaffold proteins, and (4) modulation of protein translation. Numerous small-molecule LRRK2 inhibitors have been developed, which primarily focused on interfering with kinase functioning (Hatcher et al., 2017). For instance, in preclinical trials, the LRRK2 kinase inhibitor, PF-06447475, halted α-synuclein induced neurodegeneration and neuroinflammation (Daher et al., 2015). Other early broad-spectrum LRRK2 kinase inhibitors, LRRK2-INI, GNE-7915, and G2019S, were likewise shown to have good BBB penetrably and in vivo potency (Qin et al., 2017; Zhao et al., 2020b).

In addition to kinase inhibitors, there has also been a focus on developing GTPase inhibitors. As an example, the compound, FX2149, acts by inhibiting the LRRK2 GTP binding site (Li et al., 2015c), thereby preventing GTPase activity (which involves signaling by switching on/off a pathway through the shuttling between GTP and GDP). Ordinarily, GTPase activity is critical for facilitating G protein signaling and inhibiting GTP hydrolysis, which would essentially prohibit protein shuttling between a resting and active state, thereby locking normal LRRK2-mediated signal transduction. Certain LRRK2 mutations (specifically those that affect its ROC-COR domain) impair these processes (Gilsbach and Kortholt, 2014) and thus could influence PD symptomatology.

A study that screened 640 LRRK2 inhibitory compounds revealed that only three actually reversed motor disability and dopamine neuron loss in LRRK2-G2019S transgenic flies (that normally have degeneration of their dopamine neurons). Of the three, the lipid-lowering compound lovastatin had the greatest efficacy and was further found to induce antiapoptotic Akt/Nrf signaling, while inhibiting caspase 3 activity (Lin et al., 2016). Lovastatin also had antiinflammatory effects, blocking the release of the proinflammatory cytokines following 6-OHDA administration (Yan et al., 2015). Another statin, simvastatin, promoted an antiinflammatory M2-like microglia state in a model of cerebral stroke and had in vivo and in vitro neuroprotective and antiinflammatory effects in MPTP and high-dose LPS models of PD (Ghosh et al., 2009). Thus, besides their beneficial anticholesterol effects, statins might also convey protective properties for PD, and at least some of these actions could be linked to LRRK2 processes. In essence, statins, such as simvastatin, may limit the loss of dopaminergic neurons in a mouse model of PD (Battis et al., 2023). It is still too early to ascertain whether regular statin use in humans might diminish PD risk.

Trophic Cytokines

Aside from their impact on immune processes, many cytokines are potent inducers of the growth factors GDNF and BDNF, which could potentially be harnessed for therapeutic purposes. Major problems with the administration of GDNF or BDNF stem from their poor ability to cross the BBB, as well as the many deleterious side effects (including pain sensitivity and potential for tumor growth) that have limited their clinical application (Allen et al., 2013). In contrast, CSF cytokines, such as granulocyte macrophage-CSF (GM-CSF), as well as some hematopoietic cytokines, readily cross the BBB and have fewer complicating effects. Importantly, these cytokines are potent trophic factor inducers, and preclinical studies established their neuroprotective potential (Maurer et al., 2008). Interestingly, the correlation between the low incidence of PD in NSAID-treated arthritic patients may not be due to the actual drug treatment but rather that the autoimmune disease itself increases endogenous GM-CSF levels to cope with the injury. This would also help explain why the NSAID clinical trials in PD were unsuccessful.

GM-CSF acts through the JAK–STAT pathway to influence cellular proliferation and inhibit the actions of proinflammatory cytokines and can influence phosphorylation (activation) of STAT3 and STAT5, thereby promoting neuroprotective effects upon local DA neurons. This outcome may occur either through the production of trophic or antiapoptotic factors or by buffering the impact of extracellular excitotoxic and oxidative species. After nuclear translocation, STAT3 and STAT5 can regulate the transcription of antiinflammatory and growth factors, as well as the antiapoptotic factor, bcl-2, together with the antioxidant, manganese superoxide dismutase.

Several studies indicated that GM-CSF might have neuroprotective actions in PD-like pathology (Olson et al., 2021a,b). In particular, systemic GM-CSF administration prevented the loss of DA neurons associated with MPTP (Kim et al., 2009) and acted additively with another cytokine, IL-3, to block the neurodegenerative effects of 6-OHDA (Choudhury et al., 2011). Moreover, GM-CSF had neuroprotective consequences against paraquat exposure alone and in mice that were also pretreated with LPS before receiving the paraquat (Mangano et al., 2011). Strikingly, some of the neuroprotective effects of GM-CSF might be related to the mobilization of peripheral immune cells, as it promoted the protection of SNc dopaminergic neurons by stimulating adaptive immunity through the release of T-reg cells (Kosloski et al., 2013). The T-reg cells normally act as suppressors of inflammatory cells and are thought to be critical for the induction of immunological tolerance, thereby warding off autoimmunity. In fact, the adoptive transfer of GM-CSF primed T-reg cells was sufficient to protect against MPTP-induced cell death (Kosloski et al., 2013).

Besides preventing frank neuronal loss, GM-CSF may have particularly marked effects on neuronal recovery when administered following CNS lesion. Delayed administration of GM-CSF following spinal cord injury promoted axonal regeneration and increased BDNF expression, which could facilitate neuroplastic processes, reflected by dendritic sprouting and neurogenesis. Similarly, GM-CSF induced the differentiation of neural stem cells into mature neurons and could promote the re-innervation of the striatum after 6-OHDA had already caused a lesion, potentially making this a clinically important finding (Farmer et al., 2015). This raises the possibility that trophic cytokines, such as GM-CSF, could promote a degree of neural recovery after the disease has already taken hold by bolstering endogenous neuroplastic processes. Early trials with GM-CSF (sargramostim) in humans showed promising effects in PD, likely owing to increased T-reg numbers and functioning, which functionally protects dopamine neurons (Olson et al., 2021a,b, 2023).

Turning to the hematopoietic cytokine, most notably erythropoietin (EPO), it is clear that this factor might hold promise in the treatment of PD. EPO readily crosses the BBB and is widely used to treat anemia. In addition to EPO itself, several synthetic analogs and receptor agonists that have antioxidant and antiinflammatory effects (Punnonen et al., 2015) may also have clinical benefits for PD. Most of the studies that assessed the brain effects of EPO have focused on its ability to stimulate cognitive processes, as well as its potential use as a protective agent in the context of stroke. Similarly, EPO treatment stimulated hippocampal neurogenesis and improved memory and spatial learning in rodents (Wang et al., 2017c). This growth factor also protected hippocampal neurons from stressor-induced apoptosis, and through such actions might have cognitive-enhancing effects, as indicated by improvements in several neuropsychological indices of cognitive performance (Peng et al., 2014). Preclinical data also suggested its possible use to promote neuronal recovery, been found to diminish the actions of MPTP-induced Parkinson's symptoms by rescuing damage to mitochondria that would otherwise occur (Rey et al., 2021), and could dopamine loss in 6-OHDA and MPTP animal models of PD. Likewise, it reduced 6-OHDA-induced apoptosis and dysregulation of Bcl-2, Bax, and Caspase-3 in striatal neurons. Interestingly, EPO also enhanced the survival of transplanted neural precursor cells into the SNc (McLeod et al., 2006), suggesting its potential utility as an adjunctive agent administered together with stem cells. Some evidence also suggested that EPO might have clinical value for the treatment of comorbid nonmotor PD symptoms, such as autonomic functioning, fatigue, mood, and attention.

The homodimerization of two EPO molecules and their subsequent binding to their receptor, results in the activation (phosphorylation) of several intracellular adaptor proteins (e.g., JAK2 and STAT5). These signaling cascades promote antiapoptotic factors, cell differentiation, cellular growth, and modulation of plasticity and can occur on either neurons or glial cells. In parallel with

STAT5 activation, EPO can signal through the mTOR pathway. Ordinarily, mTOR influences protein translation at ribosomes. Within the brain, mTOR has been implicated in several aspects of synaptic plasticity, such as BDNF production, dendritic remodeling, neurogenesis, and synaptogenesis (Russo, Citraro, Constanti, & De Sarro, 2012). It might also exert protective effects by modulation of autophagy, which is dysregulated in PD and likely contributes to toxic Lewy body formation. Interestngly, EPO reversed the ability of rotenone to down-regulate the autophagy markers, Beclin-1, AMPK, and ULK-1 in cultured SH-SY5Y cells (Jang et al., 2016). In parallel, EPO also reversed the rotenone-induced α-synuclein expression (Jang et al., 2016), suggesting that EPO may prevent toxic protein aggregation.

As much as EPO may have benefits for the treatment of various illnesses, it comes with significant risks. Among other issues, EPO is associated with diverse vascular problems and can increase arterial pressure, as well as favoring cancer development and progression by enhancing angiogenesis (Maiese et al., 2008). Still, EPO is currently being investigated for some conditions where the benefits exceed the risks.

Concluding Comments

There is little question that multiple players contribute to the development of PD, including disturbances of inflammatory processes. As in the case of so many other diseases, heterogeneity exists regarding the processes leading to PD and its symptomatology. This underscores the importance of personalized approaches to treatment that consider the multiple mechanisms that ought to be targeted, which could vary over the course of the illness. In the final section of this chapter, we considered several novel emerging therapies that revolve around the theme of immunomodulation. In the main, these strategies target microglia and peripheral immune cells in an attempt to foster antiinflammatory or trophic phenotypes.

A problem that will require some change in the way we approach PD is that virtually none of the preclinical successes have been translated into meaningful clinical treatments. Some of the reasons for the lack of translation of neuroprotective treatments from animals to humans have already been mentioned. In addition to these, a primary mechanism for neuronal death in human tissue was absent in the rodent. In humans, dopamine contributes to its own demise by accumulating an oxidative form of dopamine that is toxic to the cell's mitochondria and lysosomes; however, this toxic dopamine metabolic pathway was not evident in animal models of PD (Burbulla et al., 2017). Caution needs to be exercised with an eye toward an optimal balance between mechanistic animal experiments and clinically relevant human studies. Hopefully, their combination can inform us about the best possible way forward in understanding and treating PD especially as the basal ganglia circuitry is very well delineated in both the human and animal models.

There is reason for optimism given the emerging technological advances and the many new drugs in development. An exciting advance was the use of deep brain stimulation (DBS), which revolutionized the treatment of tremor and bradykinesia in PD. To be sure, this is an invasive procedure, and its efficacy tends to wane over time but improved methodologies have come about that may facilitate the use of this and related strategies (as discussed in Chapter 8: Depressive Disorders). What is fundamentally needed is a treatment that can stabilize existing neurons, thereby protecting them from further degeneration and some means of enhancing compensatory mechanisms to overcome the dramatic cell loss that has already occurred. We have touched upon the use of trophic cytokines or immunomodulatory agents that could be used in this capacity. However, these factors have pleiotropic actions and could lead to unpredictable outcomes, particularly as it is difficult to regulate their optimal levels to minimize unwanted effects.

Ultimately, it would be desirable to have ways of preventing the disease from taking hold in the first place. This requires excellent biomarkers to aid in the identification of individuals at risk for the disease. This is a major problem as most PD cases are idiopathic and, at best, elevated risk can only be predicted based on exposure patterns (e.g., pesticide applier, welder) and carrying certain polymorphisms (e.g., G2019S LRRK2 mutation). Recent advances in genetic and proteomic multiplex screens might hold the key to finding a profile or collection of markers that together determine disease risk. Likewise, it might also be possible to determine which profiles indicate the severity of disease progression, as well as the type of comorbidities that might appear. It should also be kept in mind that the disease processes involve elements outside the brain. As described earlier, in addition to cytokines, microbiota might fill this bill. Finally, the LRRK2 gene is key for several immune and metabolic processes. So, this should be included in a systems-level view of the disease and the intricate connections between the various nodes within the overall system.

CHAPTER

16

A Neuroinflammatory View of Alzheimer's Disease

Are our brains being inflamed as we age?

The human lifespan is finite essentially because the body eventually wears out but with drastically different patterns of aging depending upon the individual's genetic constitution and the diverse challenges encountered. The brain may be especially vulnerable to the ravages of age, as it is the most energetically expensive organ in the body. A large proportion of energy is shuttled to the brain and the mitochondria present in neurons are especially active, hence producing free radicals and other potentially harsh byproducts. Neurons are exquisitely sensitive to a wide range of environmental stressors, which can also take a toll on cell functioning and survival. At the same time, the brain's glial cells, primarily microglia and astrocytes, respond rapidly to environmental challenges, and over time, they may adopt long-term "active" states, wherein they can chronically release factors that foster an inflammatory environment. These inflammatory processes could prematurely age the brain by placing excessive demands on various neuronal circuits. Pathological protein aggregates, amyloid plaques, may fuel inflammatory pathology still further (see Fig. 16.1). Considerable debate currently exists as to whether these plaques should be the main targets for treating disease. Whatever the case, the possibility exists that antiinflammatory agents, and even vaccines, could be used to combat the neurobiological processes that contribute to Alzheimer's disease (AD).

The past decade has seen an exponential increase in published papers supporting a link between immune factors and neurodegeneration. Just 20 years ago, the role of inflammatory immune factors in AD was considered controversial, whereas now it is pretty much considered mainstream. There has been a growing consensus that immunocompetent microglial cells and the inflammatory and oxidative factors they release are involved in some sort of cell-to-cell combat in the diseased brain. What is not clear is whether such processes are involved in the genesis of the disease, or are secondarily turned on by dying neurons. Understanding these processes and placing them in the context of potential novel treatment strategies is a key focus of research to curtail or prevent the progression of AD.

Alzheimer's Disease Background

Alzheimer's disease is the most common neurodegenerative disease throughout the world. It strikes at the very core of what makes us who we are, namely, our memories and sense of self. The disease ravages the elements of the brain that subserve cognitive functions, particularly the encoding of new memories, and eventually the recall of those that are older. The disease is currently incurable, and at best, symptoms can be slowed among some individuals, but only by mere months! As in the case of PD, a wealth of evidence has accumulated implicating inflammatory immune processes in AD (Blackhurst and Funk, 2023; Ravichandran and Heneka, 2024). In fact, Alois Alzheimer first remarked over a century ago on the large number and distinct morphology of microglial cells present in the AD brain. This chapter will outline the disease characteristics and current theories of AD and then delve into specifics regarding peripheral and central immune processes and their relevance to brain pathology. We will end with a discussion of potential new immunotherapeutic strategies that might hold some clinical utility.

Typically, AD patients first reach the clinic after some aspects of their memory begin to fail. This usually involves forgetfulness related to common daily events, along with problems with autobiographical memory. The individual may forget names and significant life events, or may even become temporarily lost on their way home from a friend's house. Of course, we all show some slight memory lapses from time to time, but in the

---

The Immune System
https://doi.org/10.1016/B978-0-443-23565-8.00005-3

419

Copyright © 2025 Elsevier Inc. All rights are reserved, including those for text and data mining, AI training, and similar technologies.

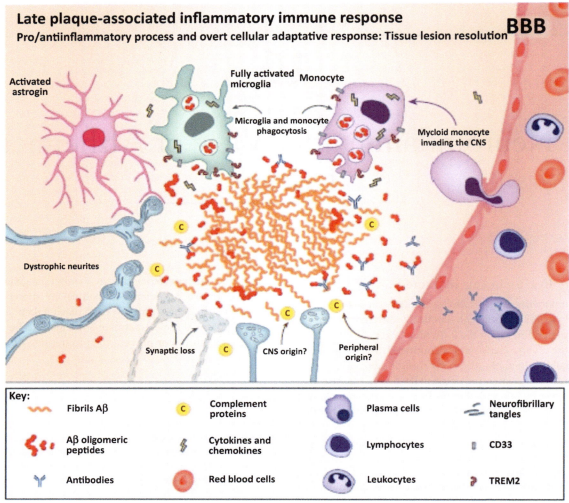

FIG. 16.1 β-amyloid plaques promote inflammation in the Alzheimer's brain. The centrally located β-amyloid plaque (containing pathological fibrils and oligomeric peptides, which are depicted as the orange and red "squiggles" in the middle of the figure) creates a microenvironment that activates microglia and to an extent astroglia (upper left in figure), which not only can help break down plaques but also can damage neurons or their projections (e.g., causing pathological dystrophic neurites; lower portion of figure). Plaques themselves can also damage neurons and cause synaptic loss. Concomitant with these brain changes, monocytes/macrophage immune cells can be recruited from the periphery (shown on the right). These cells can also interact with plaques to help with their disposal but again, like glia, they also can contribute to synaptic loss and the creation of abnormal (dystrophic) neurites and eventually neuronal degeneration. *From Cuello, A.C., 2017. Early and late CNS inflammation in Alzheimer's disease: two extremes of a continuum? Trends Pharmacol. Sci. 38, 956–966.*

AD patient, progressive memory loss eventually extends to all facets of life. Often, within only a few years, the person requires total supervision and essentially has completely lost their memory faculties, sense of self, and the ability to interact meaningfully with the world. As well, more than a quarter of AD patients typically experience a range of psychiatric symptoms, ranging from anxiety and depression to psychosis and agitation. Similarly, several nonpsychiatric features are associated with AD, including sleep and hormonal disturbances, along with motor and gastrointestinal problems. At this stage, other secondary health issues may arise, including infections, blood clots, cardiovascular problems, and pneumonia that eventually result in death.

Not all individuals showing dementia necessarily are afflicted by AD, as dementia can arise from cerebrovascular insults, certain drugs or toxicants, or may temporarily arise from neuropsychiatric states. The main difference is that the insidious progressive nature that characterizes AD may not be present in other forms of dementia. For example, another condition involving dementia, Korsakoff's syndrome, arises from chronic alcohol use (and is related to the vitamin B deficiency this causes), but the pathology is far more circumscribed and its course far less progressive than AD. Indeed, about 25% of individuals may show some degree of recovery.

A separate functional diagnostic category related to AD, that of mild cognitive impairment (MCI), is reserved for cases too "mild" to receive an AD diagnosis and in which obvious dementia is not apparent. Instead, diminished cognitive functioning is present to the extent that it interferes with daily tasks, but is not disabling. Although some individuals remain in this condition without showing appreciable progression of pathology, a subgroup of individuals eventually develop full-blown AD. Thus, there may be a prodromal state reflected by MCI, raising the possibility that early therapeutic interventions might limit MCI progression to AD.

The earliest memory deficits evident in AD are thought to come about because of the loss of neurons and their connections within the parahippocampal gyrus located within the temporal lobe. Within this brain area, the perforant path, a dense fiber bundle comprising axons connecting the tri-synaptic hippocampal circuit with the cortex, is hit hard early in the disease process. This pathway is thought to be critical for the construction (or consolidation) of new memories, as well as acting as a pathway for the retrieval of older memories.

Hippocampal Processes and AD

Both postmortem AD human tissue and that obtained from animal models of the disease show a profound loss of neurons and axons within the hippocampus and the adjacent entorhinal cortex. Although this region may not be the initial source of disease pathology, it is believed to occur early and to represent a tipping point in disease progression, giving rise to the first signs of cognitive impairment. As such, a large proportion of human and animal AD studies have focused on this area, with fewer, but still many studies assessing forebrain cortical regions, which also play an important role in the disease.

The hippocampus proper is comprised of a tri-synaptic circuit, which involves the relay of information between the CA1 (*Cornu Ammonis*), CA3, and dentate gyrus subregions. Briefly, cortical information arrives from the entorhinal cortex via the perforant path to the first synapse on CA1 pyramidal neurons. These cells send dense bushy fiber projections (mossy fibers) to CA3 neurons, which project via their Schaffer collaterals, to the dentate gyrus. The small rounded granule neurons of the dentate gyrus ultimately feedback onto CA1 and CA3, as well as feed-forward to the subiculum and then out through the perforant path to the cortex. The reverberation of signals within this circuit is critical for memory consolidation and the appropriate processing of input from multiple sensory modalities. Disturbances in this circuit could give rise to many of the dementia features that characterize AD (e.g., Igarashi, 2023).

In addition to the temporal lobe, forebrain cortical and subcortical regions are also greatly affected in AD. For instance, forebrain cholinergic neurons of the nucleus basalis of Myenert are susceptible to degeneration in AD (Mieling et al., 2023). These nuclei project diffusely to other cortical regions, which also show substantial neuronal loss with disease progression. Together, these regions are important for learning and memory, as well as emotional regulation and processing, so their loss would be expected to affect many cognitive functions and contribute to the vast AD symptomatology.

The hippocampus and temporal cortical neurons are known to be particularly plastic, such that they readily exhibit long-term potentiation (LTP), which is thought to represent learning at the most basic, electrochemical level. LTP is characterized as the increased neuronal excitability (both electrically and chemically) following high-frequency stimulation (Kandel et al., 2014). Essentially, the increased neuronal firing evident with LTP reflects the "memory" of neurons for a particular stimulus, such that exposure to the same stimulus can now produce more robust neural activity. This is believed to be the most elemental neural basis for memories and LTP processes are impaired in numerous AD animal models. As an example, β-amyloid exposure impaired LTP by disrupting intracellular calcium handling or through the α-amino-3-hydroxy-5-methyl-4-isoxazolepropionic acid (AMPA) glutamate signaling pathways. As well, mice deficient in the AD-linked gene, ApoE, exhibited disturbed LTP (Valastro et al., 2001). Several different insults, including chronic stressors or immune challenges [e.g., lipopolysaccharide (LPS) or IL-1β], also interfere with LTP processes and subsequently cause learning or memory problems (Abareshi et al., 2016). Thus, while being a fundamental cellular correlate for memory processes, LTP is not uniquely important for AD and likely reflects one of many processes that are disturbed in the disease.

Adult hippocampal neurogenesis is a fundamental plastic process that is disrupted in AD. While neurogenesis is critical for the initial development of the brain, there are a couple of neurogenic niches that maintain low levels of neurogenesis throughout adulthood. These are primarily limited to the subventricular zone (around the lateral ventricles) and the dentate gyrus region of the hippocampus. Interestingly, postmortem tissue from early AD patients displayed reduced hippocampal neuronal maturation, but an actual increase in stem cell precursors. This may represent an attempt by the hippocampus to compensate for the loss of neurons or a last-ditch effort to preserve tri-synaptic functioning.

Genetic-based animal models of AD have been associated with disturbances of adult hippocampal neurogenesis (Radad et al., 2017). For instance, transgenic expression of the AD-linked presenilin-1 gene disturbed basal hippocampal neurogenesis, as well as the normal elevation observed with environmental enrichment (Hollands et al., 2016). Similarly, overexpression of a pathological form of the tau protein (which forms a critical component of the AD neurofibrillary tangles) reduced adult neurogenesis. This is important given, as we will see shortly, that a hyperphosphorylated form of the tau protein contributes to the characteristic neurofibrillary tangles that are evident in the AD brain.

Besides genetic factors, chronic inflammation observed in AD might contribute to the suppression of hippocampal neurogenesis (Valero et al., 2014). In this regard, both central and systemic injection of LPS, which promotes inflammation, reduced adult hippocampal neurogenesis. Moreover, this antineurogenic effect was related to elevated hippocampal microglial reactivity and could be attenuated by antiinflammatory nonsteroidal antiinflammatory (NSAID) drugs. The most likely mechanism through which LPS disturbs neurogenesis is through the production of proinflammatory cytokines. Consistent with this suggestion, transgenic overexpression of IL-6 diminished adult hippocampal neurogenesis, as well as reduced neurite length in surviving neurons (Bowen et al., 2011). Similarly, TNF-α administration reduced the number of neurogenic neurons, whereas knockout of the TNF type 1 receptor increased the proliferation of hippocampal cells. Accumulating evidence has also implicated IL-1 as a negative regulator of adult neurogenesis, as central IL-1β administration reduced adult hippocampal neurogenesis, whereas the infusion of IL-1 inhibitors had the opposite effect (Koo and Duman, 2008). It has, indeed, been maintained that IL-1 might underlie the antineurogenic effects of acute stressors, since an IL-1 receptor antagonist (IL-1Ra) administered before the stressor was applied, prevented its impact on neurogenesis (Koo and Duman, 2008). Similarly, chronic mild stressor exposure elevated hippocampal IL-1β, and these effects were attenuated in mice that overexpressed IL-1Ra (Goshen et al., 2008).

Amyloid Plaques and Neurofibrillary Tangles in AD

The two hallmark pathological features of AD comprise the presence of senile plaques and neurofibrillary tangles, which accompany the degeneration of neurons that characterize the disease. The so-called "senile" plaques largely consist of a truncated form of β-amyloid as a core component and are believed to have toxic effects on neurons and their synapses. It has even been reported that β-amyloid induces alterations in neuronal firing, leading to synaptic disturbances (Yavorsky et al., 2023). Of particular significance to inflammation, β-amyloid itself has dramatic microglia-activating effects within both in vitro and in vivo studies, causing the microglial-driven release of inflammatory and oxidative factors.

The amyloid precursor protein (APP) is normally cleaved by the α-, β-, and γ-secretase enzymes resulting in β-amyloid peptides. In the case of AD, cleavage is altered, leading to an abnormal accumulation of the 42-amino acid form of β-amyloid, which has a high rate of fibrillization and insolubility. This form of β-amyloid (and possibly the 40 amino acid form) helps create the formation of diffuse and eventually dense-core plaques. These plaques also contain abnormal neuronal processes (dystrophic neurites), as well as various inflammatory and acute-phase proteins, such as α2-macroglobulin, IL-1β, IL-6, and intercellular adhesion molecules (Jorfi et al., 2023). The amyloid plaques aggregate within multiple cortical areas and the hippocampus, brainstem, and basal ganglia. Whereas diffuse amyloid plaques are commonly present in the brains of cognitively intact elderly people, dense-core plaques, particularly those with neuritic dystrophies, are the type found in patients with AD dementia (Sharoar et al., 2019).

A three-stage model for β-amyloid plaque progression in AD has been proposed (Braak and Braak, 1995). It was suggested that aberrant β-amyloid proteins first accumulate in frontal, temporal, and occipital cortices, which progress to cortical association areas and hippocampus, before finally spreading to all subcortical areas, most notably the cerebellum, striatum, hypothalamus, subthalamic nucleus, and thalamus. The earliest signs of β-amyloid fibril accumulation occurred in the precuneus, medial orbitofrontal, and posterior cingulate cortices (Palmqvist et al., 2017). The affected individuals were in the earliest preclinical AD stages, thus amyloid deposition and the associated mechanisms could serve as biomarkers of the disease. These brain regions comprise much of what has come to be known as the default mode network (DMN), which fuels the "loss of self" and engenders a disturbance of consciousness beyond that attributable to damaged memory circuitry (Aberizk et al., 2023). These specific brain region disturbances are consistent with reports indicating that DMN activity can be used to differentiate AD patients from healthy agers. Also, the DMN disturbances were more pronounced in females than in males, which aligns with the more frequent occurrence of AD in women (Ficek-Tani et al., 2023).

The second hallmark histopathological feature of AD comprises the presence of neurofibrillary tangles, which are primarily composed of a hyperphosphorylated and misfolded form of the microtubule-associated protein, tau, along with a few other microfilament proteins. Also, dendritic and axonal degradation gives rise to neuropil threads that accompany the tangles. The neurofibrillary tangles impair the axonal transport of vital proteins and lead to abnormal accumulation and/or degradation of proteins. Tangle accumulations are thought to begin at the medial temporal lobe (primarily at the entorhinal cortex), spreading to limbic regions (hippocampus, amygdala, and thalamus), and then to sensory cortical areas and possibly the nigrostriatal system (Braak and Braak, 1995; DeTure and Dickson, 2019).

It seems that the most common neurodegenerative states, namely, AD, PD, and multiple sclerosis, share the common histopathological feature of having some aberrant plaque-like brain deposits. As just discussed, the AD brain possesses senile plaques, and as we learned in Chapter 15, the Parkinson's brain is littered with Lewy bodies (which are essentially α-synuclein rich plaques) and multiple sclerotic plaques characterize multiple sclerosis. Different mechanisms give rise to plaque formations and it is unclear whether the plaques play a primary or secondary role in each of these diseases and the possibility exists that the plaques might arise from failed protective processes. It is tempting to speculate that common fundamental biological processes are involved including the inflammatory immune system. As we outlined in Chapter 15, substantial evidence points to innate (and to a certain degree adaptive) immune cells and their soluble factors in the genesis and progression of neurodegeneration. Before we dive into these processes further, we will first turn to the lifestyle and genetic factors that have been implicated in AD. As we discussed in relation to PD, a multihit hypothesis for AD has been maintained, wherein multiple "hits" over the lifetime interact with the genetic constitution to shape the evolution of the disease (Patrick et al., 2019). We posit that the inflammatory immune system is the common thread "weaving" these various "hits" together into what is manifested as the primary disease and its comorbidities.

β-amyloid: To target or not to target

B-amyloid has become the pivotal, and yet controversial target for treatment options. Although amyloid is a key feature of senile plaques and its soluble form can have toxic effects, some investigators believe that it is not a viable target. Indeed, all methods that have been used to remove plaques, ranging from vaccines to drugs, have met with failure in clinical outcomes, despite the many preclinical studies that have shown benefits to using these treatments. Even in vaccination studies that successfully reduced β-amyloid plaque burden, there was no appreciable cognitive benefit. It has been argued that the timing of vaccine administration and the stage of AD disease pathology might be important in these clinical studies. Nonetheless, efforts have been made to develop nonamyloid-directed treatment strategies. Some of these aim to limit microglial activation in the AD brain. These include targeting cannabinoid receptors on microglia or targeting the complement C1q receptor that induces inflammatory effects. However, it would be counterproductive to completely "turn off" microglia, as their normal brain functions are critical for many housekeeping duties. For that matter, microglia might help clear senile plaques.

Lifestyle Factors and AD

The majority of AD cases occur after 65 years of age becoming increasingly prevalent until the late 80s. Although early onset cases can occur among individuals in their 30s or 40s, these are rare. When considering the impact of lifestyles, it is important to bear in mind that these are acting within the context of an aging individual and their effects may vary across the lifespan. In this respect, several studies across different countries have indicated that living in disadvantaged neighborhoods at midlife was associated with an increased risk of later Alzheimer's disease (Dintica et al., 2023), possibly owing to greater stressor encounters and poor diet, which favor diseases associated with Alzheimer's disease. Diverse stressors can have especially negative effects on brain health when they occur early in life (Huang et al., 2023c). This is also true for stressor experiences during the later stages of life (Knezevic et al., 2023) when the degree of plasticity and resiliency is already taxed and when compensatory processes are diminished by the aging process itself. So, it is likely a complex scenario that plays out, much as in PD, in which the multiple hits faced over the entire lifespan collectively act to determine AD risk. Significantly, these actions may be more pronounced in women than in men, which might contribute to the greater incidence of the disorder in females. As we have seen in other health conditions, having social support and connections serves to buffer the effects of stressors, thereby acting against illness occurrences. This was similarly evident with mild cognitive impairments and the development of AD (Mahalingam et al., 2023).

Many aspects of the aging process itself may predispose individuals to the development of AD, including a general reduction of blood flow to the brain or increasing inflammation (Dong, Maniar, Manole, & Sun, 2017). Of course, even healthy aging is associated with the accumulation of β-amyloid plaques, though not to the extent observed in AD. Nevertheless, the brain hypoperfusion that occurs in the elderly first begins during early to middle adulthood and progresses with age. The deterioration of cerebral blood flow limits vital nutrient supplies for the high energetic demands of neurons, and identification of altered blood flow may be an early marker of the development of AD (Swinford et al., 2023).

A strong link exists between cerebrovascular disease and dementia in general and AD in particular. It was reported that dementia may follow a first cerebral stroke and was still higher with repeated stroke occurrences (Pendlebury and Rothwell, 2009). Predictably, hypertension and diabetes (which are major risk factors for stroke) were also associated with AD (Tang et al., 2023a,b). Accordingly, when hypertension is evident in middle age, the risk of dementia is much higher than when it first occurs later in life (Saeed et al., 2023). This could stem from the greater severity and chronicity of high blood pressure or it could be a sign of some other processes at work, such as elevated inflammation. It is known that hypertension is associated with the increased production of proinflammatory cytokines, C-reactive protein (CRP), and prostaglandins. Hypertension occurs owing to blockages or constrictions in blood vessels and elevated platelets and leukocytes contribute in this regard.

Aging itself is associated with a degree of elevated inflammatory tone throughout the brain, characterized by an increase in reactive microglia (Lourbopoulos et al., 2015). Inflammatory factors related to microglia can have long-term central nervous system (CNS) effects and might even confer sensitization so that they augment neuronal and behavioral responses to a variety of subsequently encountered psychogenic, neurogenic, or systemic stressors (Anisman and Hayley, 2012a). In effect, early or midlife inflammatory processes might proactively increase the risk for later pathology.

Type 2 diabetes is associated with a twofold increased risk of AD, possibly owing to elevated inflammation or because the metabolic changes in diabetes disrupt enzymes that are critical for the clearance of extracellular β-amyloid (Maiese, 2023). In diabetic-prone transgenic mice that expressed metabolic pathology showed elevated levels of the soluble β-amyloid species (Infante-Garcia et al., 2016). Intriguingly, insulin treatment improved scores on cognitive tests in early stage, mild AD patients, and pioglitazone (an antihyperglycemic diabetes medication) administration had some (albeit small) cognitive benefits that were associated with reduced microglial activation and β-amyloid oligomer deposition (Gad et al., 2016). These findings are consistent with animal studies showing that intranasal insulin administration reduced hippocampal lesion size following traumatic injury (Brabazon et al., 2017). It should be mentioned, as well, that insulin growth factor-1 (IGF-1) was neuroprotective in several animal models related to AD, possibly stemming from both its trophic and antiinflammatory effects (George et al., 2017). In line with these findings, a prospective study that included 369,711 participants revealed a U-shaped relationship between serum IGF-1 and increased risks of dementia and stroke. While these findings do not speak to possible causal connections, IGF-1 may nevertheless be a biomarker for AD risk (Cao et al., 2023b).

Perhaps not surprisingly, given its obvious ties to diabetes, obesity has been linked to AD, likely stemming from increased inflammation associated with adipocyte presence (Ly et al., 2023). Yet, low body weight has also been linked to elevated AD risk especially in the presence of APOE ε4, although the relationship observed with low body weight may be more a reflection of neuropsychiatric disturbances than the development of AD (Morrow et al., 2023). It could also be that the drastic changes from ideal weight are prompted by some underlying unknown medical condition.

As we have seen, adipocytes are a rich reservoir for proinflammatory cytokines, such as TNF-α and IL-6, which could influence neurodegeneration (Engin, 2017). Conversely, the rapid loss of adipose tissue could remove this reservoir, but could also liberate these inflammatory cytokines into circulation and the spike in circulating cytokines could have negative effects throughout the body and CNS. Another interesting element to consider is that many environmental toxicants, including pesticides, polychlorinated biphenyls (PCBs), and methylmercury, are all highly lipophilic compounds that readily accumulate in adipose tissue (Artacho-Cordón et al., 2016). With aging, there is often a loss of some long-standing fat cells, possibly resulting in high levels of toxicants that were embedded in these cells now entering circulation, thereby heightening inflammatory and oxidative stress on the brain and body.

Diet, along with physical and mental activity levels, is high on the list of lifestyle factors important for AD. A long-term follow-up study indicated that an antiinflammatory diet (e.g., Mediterranean diet) was associated with slower biological aging and may have been responsible for the reduced risk of dementia (Thomas et al., 2024). Lower AD risk was associated with consumption of fish, nuts, tea, and turmeric (Jaroudi et al., 2017). The omega-3 polyunsaturated fatty acids and eicosapentaenoic acid (EPA) found in fish might be responsible for their beneficial effects given their potent antioxidant and antiinflammatory actions. Indeed, EPA may limit excessive microglial and T-cell responses. Another prominent omega-3, docosahexaenoic acid (DHA), is a critical phospholipid in brain membranes. As such, it might be expected that omega-3 intake would have positive effects on AD, but the clinical utility of omega-3s has been discouraging. Placebo-controlled trials, in fact, revealed no significant effect on cognition. This is not altogether surprising, as the beneficial effects of fatty acids would be expected to be modest, and if anything, would be more suited to a preventative or prophylactic role, rather than as a clinical treatment. In this respect, it was maintained that dietary omega-3 fatty acids, by reducing stress responses, may primarily have positive effects among individuals ordinarily experiencing high stressor levels (Hartnett et al., 2023).

Some research, although limited, has supported a neuroprotective role for curcumin, which is the source of the turmeric spice found in many Indian curries. This was supported by the substantially lower rates of AD in India, coupled with epidemiological evidence showing that Asian people who regularly consumed turmeric-rich curry, performed better on the Mini-Mental State Examination (cognitive test), compared to those who did not or very rarely consumed curry. Once again, these are only correlational data and causation cannot be inferred. It is nonetheless significant that even low doses of curcumin modulated microglial proliferation and differentiation and produced neuroprotective properties (Sharma et al., 2017). As it turns out, curcumin readily crosses the blood-brain-barrier (BBB) and reduces the β-amyloid plaque burden in AD transgenic mice (Sundaram et al., 2017). Although it is unclear exactly how curcumin might have such benefits, its antiinflammatory and antioxidant actions are well-documented. Curcumin is, in fact, a potent inhibitor of the prostaglandin-synthesizing enzyme, COX-2 (much like NSAIDs), and reduces the inflammatory transcription factor, NF-κB (Deng et al., 2014). These effects have downstream consequences that result in diminished microglial activation as well as the release of proinflammatory cytokines and reactive oxygen species.

Gastrointestinal problems are frequent in AD patients and increasing evidence has indicated that the gut-brain axis influences processes that could be important for neurodegeneration. Intriguingly, fragments from gram-negative bacteria that are normally found in high concentrations within the gut (*Bacteroides fragilis* and *Escherichia coli*) were evident in the AD brain (Zhao et al., 2017c). Fungal infections and accumulated herpes simplex virus DNA were likewise found in the hippocampus and cortex of AD patients (Sochocka et al., 2017). Thus, some aspects of the disease process might be related to the translocation of gut microbiota into circulation and subsequently, the brain parenchyma. Bacterial product movement would be associated with the stimulation of various elements of the immune system. For instance, within a drosophila fly model, immune hemocytes (phagocytic innate immune cells found in invertebrates) were critical for gut-brain communication and mediated AD-like pathology. Specifically, enterobacteria infection mobilized hemocytes that infiltrated the brain and promoted TNF-α dependent neurodegeneration, and conversely, hemocyte depletion attenuated inflammation and neurodegeneration (Wu et al., 2017b).

As we have repeatedly mentioned, the potential role of probiotics in affecting brain health has gained substantial momentum. In the case of AD, a mixture of lactic acid bacteria and bifidobacteria (SLAB51) had beneficial effects in a transgenic AD mouse model. In particular, SLAB51 increased gut levels of the antiinflammatory bacteria, *Bifidobacterium*, while reducing the more proinflammatory species, *Campylobacterales*. This was associated with behavioral improvements in conjunction with reductions of cerebral β-amyloid levels and hence, diminished plaque pathology (Bonfili et al., 2017). However, 12 weeks of probiotic administration in AD patients revealed that while the supplements reduced serum CRP levels, they had no effect on inflammatory or oxidative stress markers, and resulted in little improvement on several measures of cognition (Akbari et al., 2016).

Several reviews have drawn attention to the gut-brain axis in relation to virtually all neurological conditions, reaching the consensus that bacterial gut species can generally affect CNS functioning (see Chapter 3). Essentially, the available evidence favors a situation in which dysbiosis or shift in the bacterial (and likely viral and fungal) makeup of the gut impacts the brain, thereby contributing to neurological or neuropsychiatric illnesses (e.g., Williams et al., 2024). Perhaps the most dramatic evidence for a causative role of the gut microbiota in brain functioning comes from studies showing that transplantation of the fecal microbiome from one organism to another can also transfer CNS pathology. This has not only been accomplished with fecal transplants between animals but also from humans to animals. For instance, fecal transplantation of human gut microbiota from AD patients into healthy rats induced deficits in neurogenesis and cognitive impairment that were thought to resemble that of AD patients (Grabrucker et al., 2023). The specific microbiota changes responsible for this outcome are unclear and an overall shift is likely important with the various constituents acting collectively. However, several clear differences were noted between the AD patients and healthy age-matched controls. Specifically, in AD patients, abundance of the Firmicutes phyla was reduced, whereas Bacteroidetes phyla were elevated. While the gut microbiome composition was directly related to the presence of β-amyloid (Aβ) and tau pathological biomarkers early in the disease process, it was not tied to indices of neurodegeneration. This said, using machine learning microbiota presence was found to be effective in predicting AD status (Ferreiro et al., 2023). Furthermore, a reduction in butyrate-producing bacteria was noted in AD, which is important given that butyrate has been associated with a reduced β-amyloid burden (Marizzoni et al., 2020). Yet, the evidence of microbiota-modulating neuronal degeneration is still scant and any connection would likely be complex and indirect. Whatever the case, the emerging gut-brain story is fascinating and might foster novel perspectives; although it is likely that these will be relevant to potential prophylactic actions in AD, it is difficult to envisage positive therapeutic outcomes.

For physical and mental activity, the evidence is pretty simple, "use it or lose it." The more physically and mentally active individuals have the lowest risk of developing dementia. There are exceptions and an active lifestyle holds no guarantee of any sort. Nonetheless, systematic reviews of prospective studies indicated that regular moderate exercise was accompanied by about a 28% and a 45% reduction in the occurrence of dementia and Alzheimer's disease, respectively (Meng et al., 2020). Initiating exercise earlier in life may have protracted benefits. For instance, starting regular cardio fitness training in middle age reduced the likelihood of late-life dementia by 50%. As well, modest cognitive benefits were apparent in patients who initiated exercise regimens after a diagnosis of mild AD (Farina et al., 2014). This is in keeping with the animal literature indicating that aerobic exercise in rodents limited β-amyloid plaque development and improved performance on spatial memory tasks (Tapia-Rojas et al., 2016).

Both mental and physical activities promote processes that could be protective and stave off dementia, notably, the increased blood flow to the brain and the upregulation of trophic and antiinflammatory factors (e.g., Wang et al., 2023c), depending on the intensity of the exercise regimen (Lee et al., 2023a). Also, exercise may instigate an increase in the muscle-derived hormone, orisin, which may limit β-amyloid accumulation. Regular exercise, as we saw in Chapter 5, augmented hippocampal-dependent spatial learning in rodents, and such effects were related to increased brain-derived neurotrophic factor (BDNF) levels, along with elevated adult neurogenesis and dendritic branching. Similarly, regular, voluntary aerobic exercise reduced proinflammatory CRP and IL-6 levels (Tyml et al., 2017).

In line with the potential protective effects of increased mental activity, low education was associated with increased AD risk. This has fueled the so-called "cognitive reserve" hypothesis, which holds that the more cognitively advanced and correspondingly more complex synaptic projections present, the greater the protection against dementia symptoms developing (Stern, 2012). It was posited that by the time "intellectually stimulated" individuals first show clinical symptoms, they might already have significantly more pathology than other AD patients (i.e., their extra cognitive reserve managed to keep cognitive processes together despite pathological changes). So, it might be that intellectual activity does not necessarily slow the biological processes operative in AD, but rather the intellectually stimulated brain has more alternative and rich neural connections that can, for a time, keep memory disturbances from being manifested.

A final lifestyle issue that may be especially significant concerns reports that fragmented sleep was associated with poor cognition and increased AD risk. Adequate sleep might play an important role in affecting the cytokine milieu (Irwin, 2019) and in maintaining homeostasis and metabolism of β-amyloid in the AD brain (Kastanenka et al., 2017). Indeed, sleep disturbances were associated with widespread pathology in positron emission tomography (PET) scans of AD patients (Liguori et al., 2017). Interestingly, as well, blood and cerebrospinal fluid (CSF) levels of β-amyloid itself show a circadian pattern of distribution (Cicognola et al., 2015). The duration and quality of sleep may be the two most important characteristics to consider as these were most closely correlated with the buildup of β-amyloid in the prefrontal cortex and even the density of cortical neurofibrillary tangles in AD patients (Fjell et al., 2017).

It is uncertain how sleep mechanistically affects β-amyloid levels, but as already mentioned, disturbed sleep is associated with multiple immune and inflammatory alterations (Irwin, 2019), which might contribute to cell loss. Likewise, sleep deprivation altered microglia and inflammatory processes and increased β-amyloid levels (Ju et al., 2017). Animal studies have also raised the possibility that sleep disruptions might impair the glial ability to metabolize β-amyloid. Among other things, sleep deprivation impaired the ability of astrocytes to interact with ApoE and help clear β-amyloid (Yulug et al., 2017). The impact of sleep deprivation on cognition and β-amyloid is not restricted to AD, having been reported among non-AD individuals. Indeed, poor quality of sleep or sleep deprivation was accompanied by increased brain β-amyloid aggregation and elevated β-amyloid in cognitively healthy adults (Yulug et al., 2017). Thus, the elevations of β-amyloid might be considered as an additional "hit" on the road to AD.

An additional point of caution to bear in mind is that it is tempting to overendorse the impact of lifestyle factors, especially diet and exercise, in virtually all disease states. This is raised in the current context of an extreme focus on the panacea-like effects of probiotics and associated dietary factors. Indeed, dietary companies and consultants spreading nonscientific "snake-oil" claims are easily digested by the population (especially the most vulnerable). We have indicated in several chapters that even nutrition gurus are not fully in agreement concerning the best food to consume and which present risks to health. Similar nonscientific claims are often made by companies that toss in the term *neuroscience* in their pitch to sell supplements as "brain food" recipes that allegedly prevent everything from depression to AD. That said, as we have just discussed, there are credible scientific data showing modest benefits of dietary and other lifestyle factors in limiting the development of AD.

Genetics of AD

Although AD has a higher familial prevalence than PD, the majority of cases are not strictly genetic, but vulnerability genes may still play a role. The early onset form of AD, which accounts for less than 1% of AD cases, has been associated with autosomal dominant inheritance, whereas the more common late form generally has a much more complicated genetic pattern of inheritance appearing to involve many low penetrance genes (Lanoiselée et al., 2017). However, the APOE e4 allele has been linked to both forms of AD and as will be discussed shortly, along with two "inflammatory genes" TREM2 and CD33, which have also been implicated in AD.

Advanced age allows for the accumulation of damage accumulated from multiple hits encountered throughout life, hence leading to the age-related formation of AD. It is equally possible that some genetic program(s) unfold in a time-dependent manner. Essentially, a genetic program may exist that only becomes activated at advanced ages. Another twist on this idea is that such genetic program(s) might only become activated after certain environmental exposures are encountered. It might be that some threshold of basic damage must be incurred before symptomatology arises; but, once certain biological processes are in motion, it may be extremely difficult to limit disease progression.

Analyses of genetic loading often assess genes in terms of early-onset versus late-onset AD. In general, there is much more evidence for multiple genes being operative in the less common early-onset form of the disease (Andrade-Guerrero et al., 2023). Indeed, genetic linkage studies have identified mutations in APP on chromosome 21q, PSEN1 on 14q, and PSEN2 on 1q in early-onset AD (Cruchaga et al., 2017). As well, APP and the PSENs have been implicated in late-onset AD, essentially acting as vulnerability factors for the more commonly observed cases.

It also appears that APP processing can give rise to amyloid pathology that has been linked to AD. As described earlier, the APP gene found on chromosome 21 is a transmembrane protein that is processed by three different enzymes, α, β, and γ secretases. Three forms of the protein result from alternate splicing: APP695, APP751, and APP770. The first isoform is mainly found in neurons, whereas the other two are located throughout the body. It is the neuronal form that is thought to be dysregulated in AD. This mutation can cause faulty enzymatic APP processing by the β-secretase and γ-secretase, producing truncated pathological Aβ fragments that can form the core of amyloid plaques (Siegel et al., 2017; Hampel et al., 2023).

The β-amyloid 42 species cleaved from APP has received particular attention in relation to AD pathology owing to its toxicity stemming from its hydrophobic nature and propensity to form fibrils. The pathological β-amyloid typically accumulates in extracellular senile plaques, but can sometimes accumulate within the neuron to form intraneuronal inclusions in AD. Such inclusions have been found in the AD hippocampus early in the disease, before major senile plaque deposition (Kadokura et al., 2009).

Most of the more than 40 APP mutations that are found in over 100 different families are dominantly inherited. Most of these mutations occur in the vicinity of β- and γ-secretase sites on the gene. The most studied APP mutation, the Swedish APP Mutation (KM670/671NL), results in dramatically increased β-amyloid levels together with widespread cortical atrophy and ventricular enlargement (Balakrishnan et al., 2015). A double APP mutation that was found in a Swedish family increased β-amyloid production by modifying beta-secretase enzymatic activity. Similarly, London and Flemish mutations in APP lead to deficits in γ-secretase activity, culminating in the accumulation of β-amyloid-rich senile plaques (Acx et al., 2017).

It had been believed at one time that APP and its cleaved β-amyloid proteins were simply abnormally created pathological proteins with little "normal" physiological functions. This was subsequently shown to be incorrect, with some studies pointing to the role of APP proteins in the modulation of synaptogenesis, cell adhesion, and transport, as well as intracellular signaling (Antonino et al., 2022). Given its heavy processing, the different intracellular fragments likely have markedly different roles. Such effects are most likely dose-dependent given that high levels of intracellular β-amyloid are extremely toxic, whereas lower more physiologically relevant levels enhance neuroplasticity, facilitate neurochemical signaling, and buffer against neuronal toxicity stemming from excessive heavy metals, such as iron or copper. Hence, pathological plaque evolution might reflect normally protective endogenous processes that have gone awry or protective processes that have been overtaken by some other independent disease mechanisms.

Besides APP, mutations of PSEN1 are associated with familial AD (Kelleher and Shen, 2017). The PSEN1 and PSEN2 genes are vital components of the γ-secretase complex and are localized on the endoplasmic reticulum. PSEN1 acts as the catalytic subunit of gamma-secretase with over 200 pathogenic mutations being reported for the gene (Szaruga et al., 2017). These mutations have been associated with increased levels of pathological β-amyloid 42 that form the core of senile plaques. Indeed, astrocytes derived from AD patients who had a PSEN1 mutation produced increased β-amyloid (Oksanen et al., 2017). Hippocampal pathology and accumulation of β-amyloid peptides were confirmed in rodents bearing the mutant form of the gene. The PSEN2 gene mutations are somewhat rarer, with 13 pathological mutations reported that were associated with altered β-secretase activity (Lanoiselée et al., 2017). Like PSEN1, these PSEN2 mutations were associated with the accumulation of senile plaque formation and AD-like neurodegeneration.

The strongest genetic risk factor for typical nonfamilial late-onset AD is possession of the ε4 allele of the APOE gene. The primary function of APOE is the shuttling of cholesterol around the body and brain, but may also contribute to the regulation of inflammatory processes and synaptic plasticity (Liu et al., 2013; Yamazaki et al., 2019). Strong evidence has linked the ε4 allele to increased AD risk, which was gene dose-dependent, being more than 10 times higher in homozygous individuals that possess two ε4 alleles. Indeed, APOE ε4 homozygotic AD patients displayed particularly excessive hippocampal atrophy and senile plaque accumulation that was associated with microglial reactivity (Schreiber et al., 2017). The ε4 allele has also been linked to less efficient Aβ metabolism and increased inflammation in the brain (Dong et al., 2017a).

APOE ε4 has been implicated in increasing risk for the early onset form of the disease and is the gene most closely aligned with late-onset AD (Sun et al., 2023f). Large-scale genetic linkage studies have implicated the e4 allele of the ApoE gene in both early- and late-onset AD (Di Battista et al., 2016). Among homozygous e4 carriers, an increased risk of AD was evident, whereas heterozygous individuals only displayed enhanced risk in the context of a family history of AD. Thus, the e4 allele might interact with other genetic variables to modulate risk. In contrast, the e2 APOE allele can reduce disease risk (Chung et al., 2016), although the reason behind this is uncertain.

Whatever the case, APOE functioning appears to be important for the formation of β-amyloid plaques and the associated histopathology. There may be several mechanistic reasons for this, but the two that immediately stand out are the cholesterol handling and the inflammatory role of APOE. Regarding the former, accumulating data have indicated that cholesterol metabolism is altered in AD, which might contribute to vascular blockages or other blood vessel problems that could influence the development of AD (Shahbazi et al., 2017). The involvement of inflammatory processes is supported by the finding that genetic overexpression of APOE e4 (in knockin mice) increased the inflammatory response to LPS administration and these mice also exhibited spatial memory deficits, along with a profound loss of synapses and dendritic spines. Even in an unchallenged state, e4 expression can promote deficits in neuronal plasticity and AD-like memory problems. Similarly, in humans, APOE e4 was associated with a greater in vivo inflammatory response, although this varied across brain regions as well as with sex (Yan et al., 2021). These findings have obvious implications for the development of therapeutic strategies to deal with AD based on APOE e4 markers.

In addition to the genes that are secondarily linked to inflammatory processes, genetic studies have begun to identify specific single nucleotide polymorphisms in genes that directly control microglial functioning, including TREM2 and CD33 (Liu et al., 2023a) and KAT8. Interestingly, TREM2 can have inhibitory as well as excitatory effects on macrophage and microglial functioning that might be important for AD. It appears that TREM2 might be important for phagocytic potential and recruitment of peripheral immune cells. Increased expression of TREM2 was found on macrophages that congregated around senile plaques in APP mutant mice (Raha et al., 2017), possibly reflecting TREM2 involvement in immune cell clearance of β-amyloid plaques (Singh et al., 2019). Thus, TREM2 mutations, as shown in Fig. 16.2, might underlie

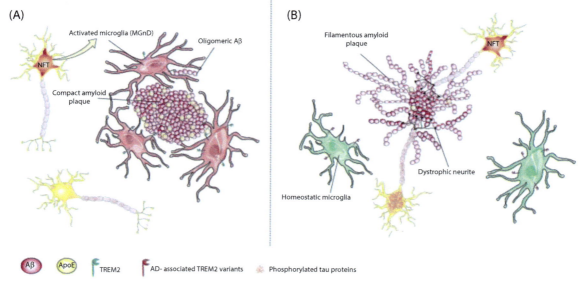

FIG. 16.2 Schematic summary of the role of TREM2 and its variants in AD. (A) Functional TREM2 has been suggested to allow microglia activation (by amyloid and NFTs for example), promote microglia clustering around plaques, amyloid uptake (early stage of the disease), and plaque compaction through binding to plaque-associated ApoE or directly to oligomeric Aβ. (B) AD-associated TREM2 variants resulting in TREM2 partial loss-of-function abolished microglia clustering around plaque and phagocytic activity. These changes could be caused by a blockage of microglia in homeostatic stages because of less plaque-associated ApoE or other reasons. The consequences are filamentous plaques associated with increased dystrophic neurites and a possible increase of tau pathology (in early stages). *From Gratuze, M., Leyns, C.E.G., Holtzman, D.M., 2018. New insights into the role of TREM2 in Alzheimer's disease. Mol. Neurodegener. 13, 66.*

the inability of microglia to properly clear senile plaques (Colonna and Wang, 2016). This makes sense given the well-known role of TREM2 in mediating phagocytosis, but beyond this, it may influence microglial proliferation and the release of cytokines. Indeed, TREM2 is a major regulator of actin remodeling in microglia that is required for their ability to adopt an active "M1-like" proinflammatory state (Sasaki, 2017).

Assessment of microglia in vitro revealed that when these cells were induced to overexpress TREM2, they were far better at phagocytizing damaged neurons and β-amyloid (Krasemann et al., 2017). Similarly, TREM2 expression was critical for the macrophage engulfment of bacteria. Importantly, the enhanced phagocytic potential induced by viral expression of TREM2 benefitted cognitive outcomes in AD mouse models. Conversely, AD mice that carried multiple pathogenic mutations, APP and 5XFAD that normally display marked microglial activation and β-amyloid deposition, failed to do so in TREM2-null mice (Golde et al., 2013). Microglia from TREM2 mutants were also deficient in their ability to bind to phospholipids and express proinflammatory cytokines.

The presence of a second microglial-related gene, cluster of differentiation 33 (CD33), has been linked to AD. The CD33 protein is a transmembrane receptor found on microglia and certain peripheral immune cells, notably macrophages, neutrophils, and mast cells. It is a member of the sialic acid-binding immunoglobulin family and like antibodies, it is critical for the recognition and removal of biological threats. Just as the removal of waste products may be fundamental in limiting damage associated with other conditions (e.g., following head injury), MRI analysis indicated that the suboptimal functioning of this system was accompanied by elevated amyloid depositions (Kamagata et al., 2022). Furthermore, the presence of CD33 was associated with a failure to clear extracellular toxic aggregates and hence, accumulation of senile plaque, and overexpression of CD33 impairs β-amyloid phagocytosis (Griciuc et al., 2013). Several genome-wide association studies (GWAS) implicated the CD33 gene in both early- and late-onset AD (Dos Santos et al., 2017). Importantly, CD33, along with APOE, are the only two genes confirmed to be involved in late-onset AD using GWAS and family-based analyses (Siddiqui et al., 2017; Wang et al., 2017e).

The expression of CD33 was elevated in microglial cells found in the brains of AD patients and the levels were related to the degree of clinical decline (Jiang et al., 2014). Furthermore, individuals with probable AD or MCI and who also possessed CD33 single nucleotide polymorphisms (SNPs), which have been implicated in AD, displayed reduced cortical thickness and diminished hippocampal volume (Wang et al., 2017e). This finding is consistent with the possibility that CD33 might be involved in the initial pathology underlying the development of cognitive impairment and the evolution to full-blown AD. Indeed, in the case of AD, CD33 deficits appeared to be apparent in the recognition and clearance of β-amyloid plaques (Zhao, 2019).

There is evidence that different CD33 gene variants might convey either increased or reduced AD risk. It appears that CD33 processing can result in two splice variants that differentially influence late-onset AD risk, with a full-length form that elevates risk, whereas a shorter version lacking the sialic acid-binding domain has protective effects (Siddiqui et al., 2017). The protective isoform is associated with reduced β-amyloid levels, which was posited to be related to shifts in the cellular location of CD33. Specifically, a shift from being transmembrane-bound to associating with intracellular proteins would result in less inhibitory control over microglia, and hence greater potential for plaque clearance.

Yet another interesting facet of CD33 functioning is that it may interact with other AD vulnerability genes. Specifically, CD33 modulated the impact of TREM2 on AD risk, suggesting a convergence of genetic vulnerability genes (Chan et al., 2015; Griciuc et al., 2019). This is particularly interesting since both of these genes are mediators of innate immunity and microglial functioning. In fact, CD33 influences the relative overall TREM1/TREM2 expression ratio, as well as altering specific TREM2 transmembrane levels. Moreover, the increased TREM2 expression observed in conjunction with the CD33 AD risk allele enhanced β-amyloid plaques and infiltration of peripheral immune cells (Chan et al., 2015). Conversely, inhibition of TREM2 functioning in an AD mouse model diminished microglia pathology and infiltration of inflammatory immune cells (Jay et al., 2015).

Once again, the jury is out on whether TREM2, CD33, or other mutations of immune factors are ultimately beneficial, deleterious, or benign in AD. As we emphasized earlier, it is unwise to simply brand inflammatory processes as being either "good or bad," and it is much more useful to consider the specific aspects of inflammatory responses within the context of the particular disease stage and comorbid features of AD. This approach should emphasize, for example, the important beneficial consequences of microglia in scavenging β-amyloid species and plaques, while at the same time acknowledging the potential deleterious effects of proinflammatory microglia that could be releasing large concentrations of unstable oxidative species. Exactly what aspects of the disease processes that favor certain polarized microglial activation states over another is still being worked out. It is in these phenotypic specificities and their relation to the microenvironment in which they are embedded that may determine pathological and hence, clinical outcomes.

Given that no therapeutic agent had been found to reverse the course of AD, increased emphasis was devoted to identifying biomarkers that predict the later development of the disorder in the hope of finding prophylactic strategies to delay its progress. Longitudinal studies revealed several markers that could predict the development of AD years later. For instance, among individuals with mild cognitive decline, the concentrations of cerebrospinal concentrations of neuronal pentraxin 2 (NPTX2), a protein involved in synaptic function, predicted later AD (Umaña et al., 2021). Likewise, particular proteins within blood samples (GDF15) of middle-aged individuals predicted the development of AD many years afterward (Walker et al., 2023), and among individuals with preclinical AD, the levels of plasma P-tau217 were associated with the subsequent cognitive decline. A study that included more than 500,000 individuals from the UK biobank indicated that elevated levels of inflammatory factors modestly predicted later AD occurrence (Mekli et al., 2023). Similarly, among cognitively healthy individuals, blood markers that comprised both elevated amyloid-beta and abnormal astrocyte activation reflected by elevated glial fibrillary acidic protein (GFAP) subsequently developed AD (Bellaver et al., 2023). Several other such biomarkers have been identified but it remains uncertain what the best strategies might be to significantly diminish the development of this disorder.

Environmental Factors and Alzheimer's Disease

In contrast to PD, there is a relatively weak link between environmental toxicant exposure and AD. That said, long-term exposure to small particulate matter (2.5 µm or less) was associated with a modestly increased risk of AD (Peters, 2023). Similarly, smoking, which results in exposure to hundreds of fine toxic particles, was related to elevated AD risk (Wallin et al., 2017). This is particularly interesting as smoking is strongly linked to *reduced* PD risk. However, the inverse link in PD is much stronger and supported by multiple studies, whereas the evidence for the direct relation to AD is weak and data are sparse (e.g., Zhu et al., 2023d).

Organophosphate pesticides disrupt acetylcholine receptor binding within the brain and may be especially relevant for AD since this results in cognitive problems (Sánchez-Santed et al., 2016). Supporting a link to AD, organophosphates increased oxidative stress and neuroinflammation as well as tau protein phosphorylation (Torres-Sánchez et al., 2023). Although high-dose organophosphate exposure can cause severe disability or even death, of more relevance to AD are the changes that might occur with long-term low-dose exposure. A metaanalysis showed that low-dose pesticide exposure that included organophosphates and other common pesticides was associated with a modest, but significantly increased risk of AD (Yan, Zhang, Liu, & Yan, 2016).

Much like PD, epidemiological data have linked pesticide exposure to AD. Elevated blood levels of pesticide metabolites [e.g., dichlorodiphenyldichloroethylene (DDE)] correlated with a greater likelihood of an AD diagnosis (Bible, 2014), and acute toxic doses of organophosphate pesticides were associated with the appearance of dementia (Sarailoo et al., 2022). Interestingly, perturbations of the PON1 gene for paraoxonase 1 (an enzyme involved in the metabolism of organophosphates) were associated with an increased risk of AD and other neurological disorders (Khalaf et al., 2023). As well, occupational pesticide exposure was correlated with an elevated risk of developing vascular dementia (Agilli et al., 2015). Parenetically, in contrast to PD, limited data are available linking the insecticide, rotenone to AD, although it was directly toxic to cholinergic neurons in brain slices (Ullrich and Humpel, 2009). As well, in mice, this pesticide disturbed BBB functioning, accompanied by deficits in learning and memory (Guo et al., 2022b).

Heavy metals, including iron, zinc, copper, and aluminum have been anecdotally and experimentally linked to AD to some degree (Mateo et al., 2023). A meta-analysis that included eight studies of 901 relevant studies of AD and aluminum exposure found that of eight studies that met the basic inclusion criteria, increased AD risk was found among individuals with chronic exposure to heavy metals (Wang et al., 2016d). In this regard, several studies revealed correlations between the incidence of dementia and aluminum exposure that was presumed to come from drinking water or dietary sources. A 1.5-fold increase in AD was observed in an area where aluminum water levels were high (0.11 mg/L compared to 0.01 mg/L in other areas), and a linear dose-response relationship was found between aluminum exposure concentration and risk of AD, with a peak of 1.46-fold elevated risk being observed in those individuals estimated to have been exposed to >0.2 mg/L. Chronic exposure to aluminum through drinking water led to the accumulation of the metal throughout the brain of rodents, with the highest concentrations occurring in the hippocampus and cortex. Likewise, dietary exposure to aluminum resulted in a buildup of the metal within the brain. Aluminum can enter the cell and accumulate in the nucleus by binding phosphate groups on the DNA, thus potentially influencing genome integrity and apoptotic processes. Furthermore, a postmortem analysis revealed higher levels of aluminum in the brain tissues of individuals with familial Alzheimer's disease (Mold et al., 2020). And, prolonged aluminum exposure can influence inflammatory processes that have been linked to AD (Bondy, 2016).

Despite the considerable evidence implicating aluminum in AD, we would be remiss if we did not indicate that other studies failed to find a relationship between aluminum exposure and AD risk (reviewed in Inan-Eroglu and Ayaz, 2018). A large prospective study did not find a significant relationship between aluminum intake and AD occurrence. However, within a subsample of participants a moderate relationship was observed between aluminum intake and AD among individuals with APOE e4 markers (Van Dyke et al., 2021). The involvement of aluminum in AD has been controversial, but because of the many actions of aluminum on brain processes, it may be premature to exclude the role of this metal in the development of this disorder (Huat et al., 2019).

Like aluminum, several other metals, such as iron, copper, and zinc, were associated with β-amyloid plaques in brain tissue from AD patients, and at a mechanistic level, these metals can interact with β-amyloid to influence their aggregative behavior and potential toxicity. For instance, aged mice showed copper deposits that were associated with elevated β-amyloid levels in neural capillaries (Singh et al., 2013). Furthermore, AD transgenic mice (APPsw/0) also displayed elevated β-amyloid and copper, both in the capillaries and the brain parenchyma. Such metal ions might influence β-sheet formation and assemblage of fibrils in β-amyloid plaques. It has been proposed that an imbalance between brain copper and zinc concentrations may contribute to cognitive decline and the emergence of AD (Sensi et al., 2018). A detailed review indicated that of various heavy metals assessed (aluminum, arsenic, cadmium, mercury, lead, iron, zinc, copper, calcium, manganese, and magnesium), only increased copper was associated with increased AD risk, whereas an inverse relationship was present with zinc levels (Babić Leko et al., 2023).

The fact that environmental toxicants might contribute to AD has informed the development of specific toxicant-based animal models aimed at recapitulating some basic AD symptoms. These have often focused on how lesions of the hippocampus or frontal cortex give rise to deficits in learning and memory. In this respect, intrahippocampal infusion of the potent excitotoxin kainic acid caused the loss of glutamate-producing CA1 and CA3 pyramidal cells, along with cholinergic neurons. The neurodegeneration was associated with marked cognitive deficits, most notably a profound inability to navigate a Morris water maze, which is indicative of a spatial learning disability, much as this occurs in humans. The impact of such excitotoxins appeared to be moderated by genotype since mice with the AD vulnerability gene, APOe e4, were more sensitive to kainic acid neurotoxicity. It is of interest that although overexpression of mutant PSEN-1 augmented kainic acid-induced hippocampal neuronal loss, it had no impact on pathology induced by cerebral stroke (MCAO) (Grilli et al., 2000), indicating specificity regarding the gene–toxicant interaction.

Before closing this section, one additional issue warrants consideration. The vast majority of studies have focused on the effects of heavy metals and pesticides on the individual's propensity to develop AD, but little attention was devoted to the effects of these metals on epigenetic changes and the intergenerational effects that could occur. As described in earlier chapters, it is known that exposure to some environmental agents can affect individual well-being and may favor the occurrence of epigenetic actions that can have intergenerational consequences. It is especially interesting that when pregnant rats were exposed to a toxicant, and this manipulation was repeated over five successive generations, each involving a different environmental agent (e.g., a fungicide in one generation, jet fuel in the next, and DDT in the ensuing generation), novel epigenetic changes occurred in each generation and the negative effects were compounded. By the fifth generation, kidney and prostate problems, as well as obesity, increased by as much as 70% (Nilsson et al., 2023). Whether similar effects occur in other diseases, including AD, has not been determined but may be worth considering.

Pathogens and AD

The idea that infection could lead to AD dates back to the early 1900s when Alois Alzheimer first noticed the similarity between the symptoms exhibited by what came to become known as AD and those that had dementia that arose from syphilis infection. More recent evidence has pointed to the possibility that infectious agents that spread to the brain might be involved in the origins of AD. A meta-analysis reported a 4–10-fold increase in the occurrence of AD in those with spirochetal or *Chlamydophila pneumoniae* infection (Maheshwari and Eslick, 2015). Remnants of the bacteria that cause Lyme disease and pneumonia were also found in the brain tissue of deceased AD patients. Likewise, an increased risk of AD (a fourfold increase in odds ratio) was reported in patients who carried an increased bacterial or viral burden comprising infection with *Borrelia burgdorferi, C. pneumoniae*, and *Helicobacter pylori* (Bu et al., 2015). Similarly, DNA sequencing and polymerase chain reaction (PCR) analyses revealed the presence of a variety of fungal species in the AD brain (Alonso et al., 2014). Further human postmortem studies reported signs of pathogen exposure (e.g., HSV1) in conjunction with amyloid plaques and neurofibrillary tangles (Itzhaki et al., 2016). Likewise, herpes simplex virus (HSV) infection has been implicated in AD-like pathology (Devanand, 2018). Indeed, AD brain tissue can confer pathology when transferred to healthy mice or nonhuman primates

(Clavaguera et al., 2013), indicating that the disease could spread in a manner akin to a typical infectious or prion-like disease.

Animal studies have supported a "prion-like" spread of β-amyloid pathology that occurs with long incubation periods. The brains of marmoset monkeys that had been infused with cerebral tissue from AD patients displayed dense β-amyloid plaques that were not simply a result of aging (Maclean et al., 2000). Similarly, while oral, intravenous, intraocular, and intranasal administration of β-amyloid-containing brain extracts had no effect, the placement of β-amyloid-contaminated steel wires in the brain yielded β-amyloid pathology in mice (Eisele et al., 2009). This finding suggests that direct contact with the brain is required for the spread of plaques and that the process is less effective than typical infections that can be propagated through other routes. Interestingly, central inoculation with brain tissue from transgenic AD mice could transfer β-amyloid pathology to another mouse and this effect was strain-dependent (Meyer-Luehmann et al., 2006). Curiously, there was no tau pathology in these animal studies, indicating that although β-amyloid might be transferable between individuals and spread throughout the brain, this alone was not sufficient to produce full-blown AD pathology, although the possibility exists that with further aging the pathology might emerge.

Any infectious agents that breach the BBB might interact with other factors to collectively augment AD risk. This was the case for individuals with the ApoE4 allele who showed a 12 times higher AD risk when they were also exposed to HSV infection (Bu et al., 2015). Thus, infections might be a vector, which in conjunction with other hits (including genetic vulnerability), increases AD risk. It is also possible that certain genetic polymorphisms might be common vulnerability factors that influence the impact of infectious insults on AD. For instance, the APOE e4 allele is known to modulate susceptibility to infection, and genes for viral receptors have been found in the AD brain (Fujioka et al., 2013). The possibility has also been entertained that a latent virus that is reactivated by environmental stressors could give rise to AD. Moreover, epigenetic changes related to immune functioning have been detected in AD patients, such as the CXCR3 receptors on T-cells that facilitate their entry into the brain. It was considered that these particular epigenetic changes might have been linked to viral infection, environmental pollutants, or other lifestyle factors (Ramakrishnan et al., 2024).

It has not received sufficient attention, but it was reported that infectious illnesses, such as influenza and other viral illnesses, have been associated with the later occurrence of AD as long as 15 years later. This was particularly notable if the viral illness was accompanied by pneumonia and was still greater if encephalitis had developed (Levine et al., 2023). Like the links to influenza, herpes infection (both oral and genital herpes, HSV1 and HSV2, respectively) can promote amyloid aggregation, and it has been considered that antiviral therapies following HSV infection (e.g., using valacyclovir) might act in a protective capacity (Devanand, 2018).

β-amyloid and infection link

A radical idea that is beginning to gain traction is that β-amyloid acts in an antimicrobial capacity. According to the Antimicrobial Protection Hypothesis (and later the Amyloid Cascade Hypothesis), β-amyloid deposition is seen much like an early innate immune response that is involved in entrapping and neutralizing invading pathogens. While β-amyloid drives neuroinflammation to fight the infection (Novoa et al., 2022), when this becomes a chronic condition, neurodegeneration may be provoked (Moir et al., 2018). Consistent with this perspective, β-amyloid aggregation is induced by the same pathogens that were linked to AD, including HSV, HIV, spirochetes, and chlamydia, and genetic knockout of β-amyloid reduced survival rate against these infections (Itzhaki et al., 2016). Furthermore, infecting the brain with pathogens accelerates β-amyloid deposition in AD transgenic mice and in *Caenorhabditis elegans* (Kumar et al., 2016). It is believed that β-amyloid can fight off pathogens by increasing their clumping together (agglutination) and entrapment, much in the same way that immune cells engulf or neutralize invading pathogens. The Aβ oligomers bind to pathogen cell wall carbohydrates and promote agglutination and engulfment of pathogens. In this respect, β-amyloid might act as an innate immune protective antimicrobial peptide, which sharply contrasts with previous notions that β-amyloid was strictly a pathological peptide in the context of AD.

It has been suggested that β-amyloid might be acting in a manner similar to the cathelicidin and defensin families of antimicrobial peptides. These are found in many immune cells, most notably macrophages and neutrophils, and serve as critical mediators of innate immune defense. These antimicrobials can kill pathogens by, among other things, "punching" holes in microbial membranes causing the leakage of essential ions and nutrients. It is known that antimicrobial peptides, such as LL-37 (which is a member of the cathelicidin peptide family), can exhibit both protective and destructive actions, depending upon the particular situation (Kahlenberg and Kaplan, 2013). Thus, while LL-37 normally acts as an important innate immune defender against infection, excessively elevated levels can be toxic. β-amyloid might have effects like LL-37, in that both molecules bind to microbial cell walls and are structurally very similar (Kumar et al., 2016). That said, even if β-amyloid is protective at certain stages of disease (such as early on by fighting off infections), it may be toxic at later disease stages. Perhaps there is a tipping point at which β-amyloid levels become dysregulated so that their antimicrobial functions become harmful to delicate brain tissues.

It has been considered, as already mentioned, that β-amyloid might spread through the brain in a manner analogous to how pathogens spread. In fact, the amyloid precursor protein (APP) may bind to pathogens, as recently shown for SARS-CoV-2, where APP binding enhanced the cellular entry of the virus (Chen et al., 2023).

Inflammatory Mechanisms of Alzheimer's Disease

Role of Peripheral Inflammation

Although inflammation likely plays a role in AD, the question remains whether it plays a primary provocative role or is secondarily involved in modulating the course of illness, or perhaps both. In this context, it is significant that increased circulating inflammatory cytokines, such as IL-6, were reported to occur in AD patients about 5 years before the onset of disease (Engelhart et al., 2004). Likewise, in individuals with modest AD symptoms, centrally mediated sickness behaviors (e.g., malaise, fever, aches, and pains) were accompanied by elevated levels of circulating TNF-α and IL-6 (Holmes et al., 2011). Normal aging is associated with increased levels of inflammatory factors, which could become progressively greater with the development of AD. In this regard, it has been suggested that immune functioning and cytokine presence could predict the transition to a pathological state. In a cohort of AD patients that had been followed for over 20 years, those who displayed elevated signs of inflammation (e.g., increased systemic leukocyte counts) in middle age had a significantly lower hippocampal volume (~5% reduction) later in life (Walker et al., 2017b). Further reports using genetic and proteomic approaches to profile inflammatory markers, including systemic leukocytes, allowed for an impressive (>90%) ability to predict future cognitive pathology and eventual AD diagnosis (Delvaux et al., 2017). Thus, it is likely that peripheral inflammatory processes are in some manner related to disease progression. In this regard, it is telling that the frequent comorbidity that has been observed between AD and depressive disorders was accompanied by numerous shared genes, including those related to inflammatory processes (Guo et al., 2022a).

Accumulating evidence has supported the possibility that peripheral inflammation may come to promote inflammatory processes within the brain, which can give rise to neurodegeneration and cognitive deficits. In animals, repeated systemic LPS injections caused spatial memory deficits, coupled with increased β-amyloid levels in the hippocampus (Kahn et al., 2012). Peripheral poly I:C (a viral mimic) administration essentially had the same effect (Weintraub et al., 2014), indicating that viral and bacterial agents are both capable of producing AD-like pathology.

Specific cytokines induced by viral or bacterial agents might underlie their impact on the brain in AD. Consistent with this suggestion, systemic TNF-α administration produced cognitive dysfunction and when administered in the context of a neurodegenerative model of prion disease, it elicited exaggerated behavioral and inflammatory responses (Hennessy et al., 2017). Similarly, IL-1β provoked cognitive deficits, and at the cellular level could suppress LTP (Lynch, 2015), suggesting that this cytokine could underlie some of its effects on memory. In this regard, analyses of several cytokines in AD patients indicated that IL-1β was more closely aligned with the disorder than other proinflammatory cytokines (Ng et al., 2018). Moreover, IL-1β altered APP processing and promoted tau phosphorylation, which has been linked to senile plaque and fibrillary tangle formation. Conversely, an IL-1 receptor-blocking antibody inhibited tau pathology and diminished cognitive deficits evident in AD mice (Kitazawa et al., 2011).

Role of Microglia

Microglia, which are the actual source of most brain cytokines, are of paramount importance in AD pathology. Damaged cells in AD typically release ATP and when the accumulation is excessive in the extracellular compartment, this can be sensed by the microglial purinergic receptors, which modulate the activation state of these cells. Microglial overactivation could be damaging to neurons owing to excessive inflammatory processes, but as already mentioned their underactivation might be equally destructive (in the context of TREM2 and CD33 mutations) (Leyns et al., 2017; Siddiqui et al., 2017) owing to the failure to properly clear toxic β-amyloid species. Accordingly, in considering inflammatory microglial involvement in AD, it might be productive to be mindful of the remarkable plasticity of microglia, as well as how specific phenotypic states may be beneficial under certain conditions, but deleterious in others.

With advanced age, microglia and astrocytes tend to show impaired or at least altered functioning and exhibit diminished reactivity, reduced phagocytosis, as well as impaired overall motility and migration to sites of damage (Uddin and Lim, 2022). Microglia senescence could occur because these cells are highly active, with proliferation occurring throughout the life cycle, and are consequently prone to DNA damage. Alternatively, the increased myelin fragmentation that can occur with age could give rise to microglia dysfunction by interfering with their intracellular lysosomal functioning (Safaiyan et al., 2016). Besides phagocytic deficits, aged microglia displayed elevated reactions to their usual trigger stimuli, such as ATP (Miao et al., 2023). Whatever the case, their diminished state could contribute to AD pathology owing to a lack of protection from microbial insults or an impaired ability to clear toxic aggregates.

In vitro cell culture preparations revealed that exposing old microglial cells to the culture medium obtained from new microglia resulted in an enhanced clearance of β-amyloid that had been added to the culture (Daria et al., 2017). The addition of the trophic cytokine, GM-CSF, also provoked such effects, raising the possibility that diminished microglial functioning that may occur with age can be reversed by trophic cytokines released by younger glial cells. It is significant that in vivo PET imaging methods using the marker, PK11195, have further allowed for the identification of an increased number of highly activated microglia within the brains of living AD patients (Parbo et al., 2017).

Autopsied brain tissue from AD patients as well as from animal models of the disease, revealed signs of robust neuroinflammation. This invariably involved marked microglial activation (as determined by morphological stains) along with proinflammatory cytokines and/or other inflammatory enzymes (see Kinney et al., 2018). Of note, most activated microglia tended to be especially dense around the β-amyloid senile plaques. This could reflect inflammatory microglia being provoked by the plaques, or conversely, that inflammatory cells fueled the development of the plaques. The combination of plaques and immune cells could represent an aberrant deleterious state or they might represent a protective clean-up response.

Inflammatory insults are often combined within existing genetic animal models to produce a more convincing AD-like state. For example, systemic LPS treatment augmented the pathological consequences observed in a triple transgenic model of AD (Valero et al., 2014). Paradoxically, it was also found that LPS priming increased the clearance of β-amyloid plaques and improved AD-like behavioral pathology (Qin et al., 2016), indicating that inflammation can be a double-edged sword, being either deleterious or beneficial depending on intervening factors. It seems that priming phagocytic cells, notably peripheral macrophages or brain microglia using specific immune agents, can place the cells in a state that has them migrate to and then decorate the surface of β-amyloid senile plaques (Fig. 16.3), subsequently promoting their breakdown through phagocytosis and the release of lytic enzymes (Michaud et al., 2013). Additionally, studies using mice expressing a green fluorescent protein in newly created cells, indicated that LPS administration caused the production of new macrophages derived from the bone marrow, which then migrated to the brain to attack β-amyloid plaques (Michaud et al., 2012). Further, it appeared that infiltrating immune cells and local microglia act as beneficial "clean-up" cells, but become overwhelmed at later stages of disease owing to excessive plaque buildup and a constant onslaught of inflammatory cytokines (El Ali and Rivest, 2016).

These findings are particularly fascinating in light of the revelation that, in contrast to previous dogma, the resident microglia are of distinct embryonic origin, different from peripheral macrophages. Indeed, microglia are thought to arise from the yolk sac earlier in development than the macrophages, which emerge from bone marrow stem cells (Elmore et al., 2014). After microglial colonization of the brain during development, most newly generated microglia during the lifetime of the organism come from resident brain stem cells, with only a small minority coming from bone marrow (Bruttger et al., 2015). Accordingly, microglia and peripheral macrophages, although sharing many functions, may have different roles in sculpting brain processes. In the case of AD plaques, it could be envisioned that resident microglia have a more direct and precise impact on the microenvironment, likely acting to shape rapid synaptic processes and deal with metabolic byproducts. In contrast, peripheral macrophages may only be recruited after some threshold of pathology has been reached or when microglia become overwhelmed, and in this case, the invading macrophages might be tasked with the removal of plaques. Whatever the case, using a mouse model, it was demonstrated that the ADS phenotype can be altered by the transplantation of hematopoietic stem and progenitor cells (Mishra et al., 2023).

While supporting the notion of increased microglial activity in AD, postmortem studies on AD brain samples generally pointed to a complex interplay between genotype, microglial phenotype, and clinical state. Microglia express APOE, and the APOE4 isoform may be fundamental in driving AD pathogenesis (Blumenfeld et al., 2024). Genetic variants of the APOE gene could influence microglia functioning and the protein is, indeed, increased in microglia within genetic AD models (Holtman et al., 2015). Although it is unclear how APOE acts in microglia, it is suspected to be a contributor to innate immune responses, including control of inflammatory cytokines (TNF-α, IL-6) and production of reactive oxygen species (Iannucci et al., 2021). Similarly, as described earlier, the TREM2 gene, which has been implicated in AD, is an important regulator of microglial functioning. TREM2 normally diminishes the microglial release of inflammatory cytokines but enhances their phagocytosis of pathogens or damaged cells (Kleinberger et al., 2014). The R47H TREM2 mutation that increases AD risk also happens to reduce TREM2's phagocytic ability (Wang et al., 2015), which is consistent

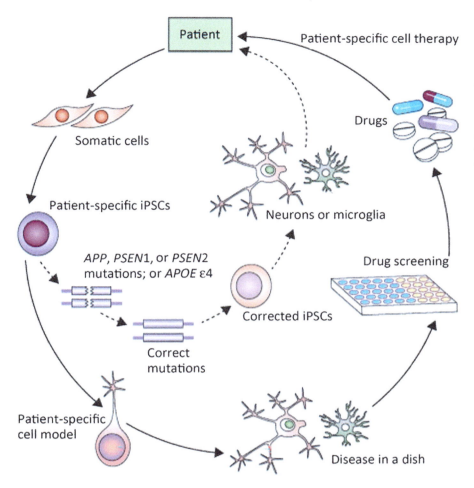

FIG. 16.3 Emerging genetic technology in the personalized treatment of Alzheimer's disease. Stem cell reprogramming techniques allow for Alzheimer patient somatic (e.g., skin) cells to be reprogrammed into neuronal cells, which can then be used to access mechanisms of degeneration and potentially to determine the efficacy of new drugs. This approach also allows for genetic engineering to correct any mutations (e.g., APP, PSEN, or APOE ε4) and then transplant cells back into the same patient. *From Hunsberger, J.G., Rao, M., Kurtzberg, J., Bulte, J.W., Atala, A. et al., 2016. Accelerating stem cell trials for Alzheimer's disease. Lancet Neurol. 15, 219–230.*

with the evidence for a role for microglia in cleaning up β-amyloid aggregates and plaques.

Somewhat troubling are the findings concerning the in vitro and in vivo effects of TREM that appear to be at odds with one another. For instance, TREM2 knockout reduced levels of inflammatory genes in a genetic AD mouse model (Wang et al., 2015). Conversely, LPS-stimulated primary microglia that lacked TREM2 displayed increased proinflammatory cytokine levels. Furthermore, while TREM2 knockout diminished β-amyloid plaques in APPPS1-21 mice (Jay et al., 2015), the opposite was seen in 5xFAD mice (Wang et al., 2015). Once again, these findings point to the complexity that exists concerning the multifaceted role of TREM2 in microglia (Qin et al., 2021). Several TREM2 variants may play different roles in the development of AD, and TREM2 can act in very different ways, depending on the presence of other factors and through crosstalk with several signaling pathways that may ultimately influence microglial proliferation, phagocytosis, and elevated cytokine production.

To complicate matters further, the correlation between microglial and clinical state also varies with the specific microglial marker being assessed. This is not surprising considering that these markers subserve differing microglial functions. For instance, while CD68 is more aligned with microglial phagocytic functions, HLA-DR is typically associated with antigen presentation. Further, the commonly used marker, IBA1, is thought to be required for actin bundling and hence, microglial motility. In line with this, brain samples from a large aging study showed that a reduction of IBA1 and an increase of CD68 and HLA-DR were related to the presence of dementia (Minett et al., 2016). The presence of the APOE e4 risk factor was associated with increased CD68 and HLA-DR levels that were linked to dementia, whereas the e2 allele was tied to high levels of IBA1 and protection against dementia (Kim et al., 2017e). The IBA1 reduction may reflect a failure of microglia to respond and mobilize to fight potentially harmful plaques, whereas the CD68/HLA-DR elevations might reflect an overzealous response to plaques that could damage adjacent neurons.

Substantial evidence going back several decades revealed the presence of complement proteins in AD senile plaques. The complement system, it will be recalled, comprises more than 30 proteins that are important in neurogenesis, synaptic pruning, apoptosis, and neuronal plasticity, and may influence neuroinflammation that favors AD development (Shah et al., 2021). Complement proteins sequentially activate one another (usually following antibody binding), causing the lysis or phagocytosis of some microbial or other foreign invader. Direct in vitro evidence revealed that β-amyloid aggregates can directly trigger the complement cascade inducing the production of numerous anaphylatoxins including, C3a and C3b, along with C5, and the ultimate production of the attack complex (Shah et al., 2021). Furthermore, in vivo studies indicated that C1q was responsible for the toxic effects of soluble β-amyloid oligomers (Hong et al., 2016), and C1q-mediated amyloid-dependent reductions in hippocampal LTP. The C3 complement protein was similarly elevated in the AD brain, whereas C3 deficiency was neuroprotective and reversed functional impairments, although curiously, it was associated with elevated cerebral plaque load (Shi et al., 2017). This again fits with the "double-edged sword" view regarding immune processes. In this case, C3 may be an important element in the clearance of β-amyloid plaques; yet, it has other inflammatory actions that are destructive in the AD brain. It was suggested that the actual plaque load is less important than the microglial reactions to the plaques and that this microglial reaction is fundamental to the critical loss of hippocampal synapses (Shi et al., 2017).

The mechanisms through which β-amyloid plaques interact with complement are not entirely known but may initially involve interactions between C1q and the first few residues of the β-amyloid molecule, and C3 might similarly bind to the aggregated β-amyloid fragments. Given these interactions, efforts were made to assess the impact of inhibiting complement activation in AD, and various protease inhibitors and small molecules aimed at interfering with the β-amyloid binding site on C1 have been evaluated as possible ways of deterring the initial appearance and progression of AD (e.g., Bohlson and Tenner, 2023). It is unlikely that complement-based strategies would reverse well-entrenched AD, but might slow disease progression if identified early. However, it was demonstrated in genetically engineered mice that repopulation of microglia could diminish cognitive and synaptic deficits associated with AD, likely being related to altered BDNF expression (Wang et al., 2023d).

In addition to microglia, there is evidence that astrocytes might contribute to AD through their impact on inflammatory processes, particularly IL-1β. Indeed, triple transgenic AD mice were protected from β-amyloid pathology and cognitive deficits by using an IL-1 receptor neutralizing antibody, which was attributed to astrocytic and not microglial functioning (Lim et al., 2015). As well, astrocytes secrete C3, which binds to the C3a receptor on microglia, resulting in augmented inflammatory drive and β-amyloid pathology in APP-transgenic mice (Lian et al., 2016). Hence, astrocytes might act in parallel with microglia to modulate complement signaling in the context of β-amyloid pathology.

Immunomodulatory Treatments for AD

Antiinflammatory Treatments

The first indication that antiinflammatories might be useful for AD treatment, as described earlier, came from the clinical observation that NSAIDs prescribed for rheumatoid arthritis patients show relatively low rates of developing AD (Miguel-Álvarez et al., 2015). Animal studies further indicated that the COX-2 inhibitory drug, ibuprofen, decreased the activation of the inflammatory kynurenine pathway, which was associated with cognitive improvement in APPSwe-PS1 mice. Administration of the COX-1 inhibitor, SC-560, to transgenic AD ($3 \times$ Tg-AD) mice diminished β-amyloid plaques and tau hyperphosphorylation, along with improving cognition (Choi et al., 2013). The drug also reduced microglial activation but facilitated their phagocytic potential. Similarly, the COX-2 inhibitor, NS-398, improved cognition and reduced microglial activation and β-amyloid pathology. This same COX-2 drug also rescued TNF-α production in brain slices obtained from TgAPPsw mice (Quadros et al., 2003).

Despite the promising actions on brain inflammatory processes, the clinical efficacy of general antiinflammatory drugs has failed to show any benefit in AD patients or elderly individuals with a family history of the disease. For instance, treatment with naproxen or celecoxib for 1–3 years had no significant influence on cognitive functioning in elderly individuals with a family history of AD or in recently diagnosed AD patients. Research trials confirmed the absence of clinical benefit for cognition after 1 year of daily treatment with the NSAIDs, naproxen, indomethacin, or rofecoxib, in mild-to-moderate AD patients (e.g., de Jong et al., 2008). Based on a systematic review and meta-analysis, it was concluded that overall NSAIDs provided no benefit in diminishing AD symptoms (Miguel-Álvarez et al., 2015). The lack of clinical efficacy could be related to the fact that the individuals already had AD, whereas NSAIDs might only be useful in a prophylactic capacity if taken well before symptoms were apparent (Policicchio et al., 2017). In this respect, antiinflammatory agents (e.g., Nimesulide, resveratrol, and citalopram) could influence overexpression of APP, BACE1, COX-2, NCT, and p-Tau, and could be used to diminish symptoms during the early stages of AD (Montero-Cosme et al., 2023). Relatedly, as in other diseases, it might be appropriate to use NSAIDs through a precision medicine approach based on individual characteristics (O'Bryant et al., 2018), although some NSAIDs may be more effective than others.

Speaking to this, despite the frequent failures of NSAIDs having any positive effects, it was reported that, unlike other NSAIDs, diclofenac was alleged to have positive effects in diminishing AD symptoms by acting on brain microglia, as could NSAIDs in the fenamate group (Stopschinski et al., 2023). Fenamates, a form of NSAID, which comes in several forms (e.g., mefenamic acid) have been used to produce pain relief and in several other illnesses in which inflammatory factors are significant and have also been used in an effort to reduce AD progression (Hill and Zawia, 2021). Through their actions on the NLRP3 inflammasome, agents in this class can have beneficial actions in animal models of memory loss associated with β-amyloid as well as in transgenic mouse models of AD (Daniels et al., 2016). Whether these agents have positive effects on AD in humans will not be known in the absence of large-scale longitudinal studies in affected patients or those with mild early symptoms, and as we have frequently indicated, analyses of the benefits of antiinflammatory agents will likely necessitate a precision medicine approach (Hampel et al., 2020).

Cytokines and Trophic Factors

Whatever strategy is adopted, it likely requires a clinical agent that can access large portions of the brain and probably modulate several different pathways relevant to pathology. In the case of AD, this would likely mean targeting inflammatory and other prodeath processes, in conjunction with those aligned with the accumulation of faulty amyloid aggregates. We will briefly cover three relevant cytokines in this section, GM-CSF, IL-1β, and CXCL1, with the rationale being that they reflect the most relevant cytokine categories in AD pathology. This is not meant to suggest that other inflammatory cytokines are not involved in AD, especially as IL-6 and TNF-α have been implicated in the disorder (Souza et al., 2021; Torres-Acosta et al., 2020).

The unfortunate lack of clinical efficacy of NSAIDs in AD has resulted in further analyses concerning why there is such a low prevalence of AD in patients with rheumatoid arthritis. One possible explanation is that elevated levels of endogenous trophic factors in these patients might buffer them against AD. In a mouse model of AD, several studies indicated that the trophic cytokine GM-CSF may play a neuroprotective role (e.g., Kiyota et al., 2018). In addition to elevated levels of several

proinflammatory cytokines in their synovial fluid, arthritic patients also express increased concentrations of GM-CSF, possibly suggesting that it confers protection against neuronal and synaptic loss (Avci et al., 2016). Consistent with this possibility, systemic GM-CSF reduced β-amyloid deposition and cognitive deficits, and inhibiting GM-CSF through specific antibodies also reduced β-amyloid deposition (Volmar et al., 2008). Of course, caution should be exercised given that GM-CSF can increase BBB permeability (by reducing claudin-5 and ZO-1 BBB tight junction proteins) and facilitate peripheral monocyte infiltration (Shang et al., 2016). Using a trophic cytokine, such as GM-CSF, is currently a contentious issue and further data will be required before this can be resolved. It might be that GM-CSF is effective in only a subgroup of patients, but might be deleterious in others.

Targeting the proinflammatory cytokine, IL-1β, to modulate inflammation in AD has received interest, particularly given that IL-1β polymorphisms were associated with AD. Moreover, several studies reported elevated IL-1β or its intermediate signaling factors within the brain and CSF of AD patients. Importantly, pathologically elevated levels of IL-1β levels have been associated with plaque evolution and the rate of cognitive decline among clinically diagnosed MCI patients. The increased brain levels of IL-1β activity are thought to occur early in the AD brain and might represent a novel biomarker for early detection of AD (Dursun et al., 2015). In animal studies, several transgenic AD models demonstrated the age-dependent emergence of increased IL-1β expression that correlated with β-amyloid plaque deposition and spatial memory deficits (Stampanoni Bassi et al., 2017). Mechanistic studies provided further evidence that members of the IL-1β family may fuel the development of senile plaques and neurofibrillary tangles, and conversely, that the endogenous antagonist, IL-1Ra, may limit such processes (Italiani et al., 2018). It appeared that IL-1β promoted faulty APP cleavage, resulting in the accumulation of pathological β-amyloid deposits, whereas IL-1β inhibition using selective antibodies prevented tau pathology and improved cognition in an AD animal model (Kitazawa et al., 2011).

Compounds have been produced that target IL-1β, including rilonacept, a long-acting IL-1β receptor fusion protein, and canakinumab, a fully humanized anti-IL-1β monoclonal antibody. In animal models, IL-1Ra was reported to act within the hippocampus to reverse β-amyloid-induced disruption of LTP and cognitive dysfunction (Prieto et al., 2015), and following hippocampal IL-Ra infusion, the development of AD-like pathology was moderately diminished. It was particularly interesting that the transplantation of neural precursor cells that genetically overexpressed IL-1Ra produced even more impressive results, reversing the cognitive deficits in a Tg2576 genetic AD mouse model

(Ben-Menachem-Zidon et al., 2014). It could be that positive effects occurred with the transplanted cells themselves and the IL-1Ra further augmented such outcomes.

One problem with an IL-1β inhibitory clinical approach is that (as is the case for almost all aspects of immunity) this cytokine can have very different consequences depending upon its tissue levels. Varying with concentration, IL-1β may promote diverse immune processes, some of which are beneficial and some deleterious within the context of an AD-like disease state. For instance, although reducing IL-1β could have positive consequences, genetically driven overexpression of IL-1β enhanced plaque clearance by microglia (Ghosh et al., 2013), indicating a beneficial effect owing to its ability to enhance a microglial phagocytotic state.

The chemokine, fractalkine (CXCL1), which signals through its CX3CR1 receptor on microglia to inhibit phagocytosis, could be another important mechanism involved in plaque clearance. In line with this suggestion, knockout of CX3CR1 exacerbated pathology in transgenic hAPP-J20 AD mice, and genetic fractalkine overexpression reduced tau pathology and prevented neurodegeneration in Tg4510 mice that normally show tau pathology (Merino et al., 2016). Thus, CX3CR1 might be a viable target for treating AD. However, as discussed earlier, cytokines often have complex biphasic actions in regulating CNS processes, and may also have unintended consequences in other pathologies (e.g., promoting the growth rates of many different cell types, including those that could potentially become cancerous), thus caution needs to be exercised in considering cytokine manipulations.

Vaccine Therapy

Owing to the alarming number of people who develop AD, coupled with repeated failures of therapies to alleviate symptoms, increased attention focused on the prevention of the disease or ways to delay its appearance. Modifying lifestyles has been encouraged, particularly the adoption of diets that do not favor inflammation. The intriguing possibility was likewise explored concerning the potential development of vaccination against AD pathology. Given the substantial evidence indicating a deleterious role for β-amyloid in AD, considerable work focused on using either active or passive immunization strategies to fight the disease by targeting β-amyloid to induce its clearance.

Immunotherapeutic approaches were adopted in the hope that the individual's immune system could be utilized to produce lasting elevated concentrations of polyclonal antibodies that can attack β-amyloid plaques. To this end, heavy investments have been made in developing vaccines against β-amyloid fragments that

result in the clearance of amyloid plaques. This has included both active (direct injection of the amyloid or hyperphosphorylated tau antigen) or passive (administration of specific amyloid or tau primed antibodies) immunization strategies. In both cases, the β-amyloid or tau-specific antibodies were shown to infiltrate the brain and bind to monomers or oligomers of the respective antigens, as well as to decorate the surface of plaques and/or neurofibrillary tangles (NFTs). Microglia and macrophages can then interact more efficiently with antibody-coated aggregates to promote their clearance (e.g., Vogt et al., 2023).

It was first reported more than two decades ago that inoculating transgenic AD mice with β-amyloid resulted in a dramatic clearance of plaques, coupled with improved cognitive functioning (Schenk et al., 1999). This was soon followed by animal studies showing that vaccine-provoked plaque clearance was associated with improved cognitive functioning. Of interest from an inflammatory perspective was that the vaccination also modulated microglia, which were thought to be critical for plaque clearance and subsequent cognitive improvement (Bukhbinder et al., 2023).

The early findings prompted clinical trials, the first of which used active immunization against the full toxic β-amyloid 42 peptide fragment (AN1792). Unfortunately, significant cognitive changes were not observed (Bayer et al., 2005). Follow-up analyses revealed a significant reduction of β-amyloid plaques in the treated AD patients, but once again, this was not accompanied by clinical benefits. Even patients with virtually complete plaque removal still displayed severe dementia with no evidence of improved survival (Holmes et al., 2008), thus casting doubt on whether existing plaques contribute to cognitive decline. Finally, the immunized patients show a down-regulation of microglial activity, with variations in the different subpopulations of microglia; CD68, CD64, and CD32 positive microglia were reduced, while IBA-1 positive microglial cells were unaffected.

Of considerable concern was that serious side effects were noted. Meningeal encephalitis and vascular hemorrhages were noted in several inoculated patients, and several patients died during the trials. A substantial number of proinflammatory Th1 lymphocytes had infiltrated the AD brains following vaccination, which may have been responsible for the harmful inflammatory side effects. Hence, more recent efforts sought to develop "cleaner" or more specific vaccines that lacked these side effects. Accordingly, improvements in vaccines have aimed to increase plaque-directed antibodies, while diminishing general off-target T lymphocyte-mediated inflammation. This entailed synthesizing highly specific antigens that represent very confined immunogenic portions of the β-amyloid molecule.

The neuroinflammation caused by clinical vaccination using synthetic β-amyloid (AN1792) was thought to result from recruited proinflammatory Th1 lymphocytes. A more truncated form of β-amyloid, together with an adenovirus vector encoding GM-CSF, provoked an anti-inflammatory Th2 response, along with antibodies that coated the amyloid plaques (Kim et al., 2005). Other vaccines included vanutide cridificar (ACC-001), together with a specific (QS-21) adjuvant, which was assessed for safety and efficacy in Phase 2 clinical trials. While ACC-001 was safe and well tolerated, it did not affect cognitive functioning or volumetric brain measures.

Microglia are likely a major part of the process through which vaccination could clear β-amyloid plaques. The presence of β-amyloid was apparent in microglial cells from immunized mice and the microglia of patients immunized with AN1792 (Nicoll et al., 2003). It was posited that the Fc receptors found on the surface of microglial cells trigger the phagocytosis of β-amyloid. All microglia express multiple classes of Fc receptors, and the nature of the reaction provoked depends upon the ratio of Fc receptors activated, as well as the therapeutic antibody isotype (Vidarsson et al., 2014). Specifically, IgG types found in vaccinations decorate the β-amyloid plaques, which are then recognized by the Fc microglial receptor, resulting in degradation of the β-amyloid species.

In addition to phagocytosis, activating Fc microglial receptors provokes proinflammatory cytokine release and activation of the complement cascade. The proinflammatory cytokines could help in plaque breakdown through the production of lytic enzymes or other chemical inflammatory enzymes. The role of complement in AD pathology is not entirely understood, as described earlier, but has nonetheless been suggested to be a useful therapeutic target. Interestingly, the same antibody isotypes that bind to Fc, also showed a high affinity for C1q, raising the tantalizing possibility of interactions between these systems, especially as C1q could potentially contribute to microglial and astrocytic proteins that are elevated in Tau that increase with age (Dejanovic et al., 2022). As described earlier, C1q triggers the activation of proteins that eventually engage the membrane attack complex, rupturing the membrane and degrading cellular integrity. Thus, complement activation observed in the AD brain could reflect a failed attempt to clean up damage, and thus bolstering this system could aid this process.

Interesting evidence has pointed to a role for complement in the astrocyte-microglia-neuron tripartite synapse in AD. It was posited that β-amyloid can activate NF-κB within astrocytes, resulting in C3 release, which can then activate microglial phagocytosis (Lian et al., 2016). Astrocytes express C5a and C3a, and microglia similarly express complement receptors CR1, CR3, and CR4 that can act as phagocytic receptors for complement-attached complexes. There has even been the suggestion that neurons, under certain conditions, can express complement receptors.

There are many reasons that clinical immunotherapy trials have failed, including problems with side effects, low immunogenicity, and poor pharmacokinetics of the antibody. A novel vaccine has been developed that takes into account these past problems. Specifically, an active particle-based vaccine that targets the β-amyloid epitope, PP-3copy-Aβ1-6-loop123, produced very high antibody titers that were specific for this PP region (Fu et al., 2017). The vaccine cleared β-amyloid plaques in rodents and enhanced cognitive performance in the Morris water maze test of spatial learning and memory, and importantly, no toxicity or proinflammatory T-cell effects were observed. It was also noted that immunization earlier in the course of AD-like pathology had more profound effects than when treatment commenced at a time when pathology was severe. Hence, immunotherapy may be more effective early in the disease or possibly during a prodromal stage.

The mechanisms through which vaccination might clear β-amyloid were proposed to involve either a direct effect of antibodies reaching plaques within the brain, or through the "peripheral sink" pathway (Fu et al., 2017). In the first instance, antibodies bound to their β-amyloid target interact with the Fc receptors found on microglia or infiltrating phagocytes, which would elicit the phagocytosis of the antigen–antibody complex. According to the peripheral sink route, the antiamyloid antibodies generated in the periphery that interact with blood or lymph β-amyloid would result in a concentration gradient effect, favoring the movement of brain β-amyloid out into the periphery. This could then lead to the easier clearance of plaques and diminished burden within the brain.

Although it was long held that β-amyloid senile plaques are primary players in the genesis of AD pathology, as indicated earlier, this view has been challenged, particularly in light of poor correlations between the extent of plaques and the degree of cognitive disability in AD. Further, the soluble β-amyloid oligomers are more closely aligned with AD symptomatology than the β-amyloid fibrils that comprise senile plaques. As senile plaques might reflect a compensatory response, the view was advanced that disequilibrium between β-amyloid fibrils-oligomers-monomers is the key to AD pathology. It should also be considered that β-amyloid normally has a useful role in synaptic functioning (Musardo and Marcello, 2017) and hence, indiscriminately blocking its function could have unwanted effects.

The limited success of amyloid-β-targeting therapies has led to a shift in efforts by targeting the pathological tau proteins found in neurofibrillary tangles, rather than the β-amyloid plaques. Several preclinical studies demonstrated that antibodies can enter neurons and colocalize with pathological tau (e.g., Shamir et al., 2020). It is thought that antibodies binding intracellular tau can prevent the tau protein from disrupting various intracellular processes. In particular, this could prevent endosomal membrane disruption and the formation of toxic aggregates. At the same time, the antibody-tagged tau would be easier for cellular degradation by way of lysosomal enzymes. It may also be possible to design specific antibodies that target different elements of tau, including the monomeric, oligomeric, or aggregated forms, as well as their phosphorylation state.

The first of such Phase 1 clinical trials, involved the active AADvac1 vaccine, which targets N-terminal tau fragments. This synthetic peptide utilizes tau protein amino acids 294–305, which are linked to an adjuvant that comprises keyhole limpet hemocyanin (a metalloprotein from the keyhole limpet sea snail). Of the four clinical trials of AADvac1 that have been completed, three were in mild-to-moderate AD patients and one in more severely affected patients. The vaccine was safely tolerated by most individuals with only 2 of 30 withdrawing because of serious side effects. Importantly, 29 of the 30 patients developed significant antibody titers over the 12-week trial. Moreover, in a 72-week trial, MRI revealed a tendency toward slower atrophy and a modest decline of cognitive disturbances among individuals showing high titers (Novak et al., 2018). This was confirmed in a Phase 2 trial (Novak et al., 2021) and in a subanalysis, it was noted that brain atrophy was slowed in the temporal cortex, whole cortex, and right and left hippocampus (Cullen et al., 2024). Despite the limited cognitive improvement, a prophylactic approach might be more beneficial. Currently, an updated vaccine, ACI-35.030, includes a second adjuvant that is designed to further boost the T helper cell response and consequently augment the tau-directed antibody response.

In addition to the aforementioned active vaccines, passive-acting vaccines (synthetically designed or altered antibody molecules) are also being developed. Bepranemab (UCB0107) is one such monoclonal IgG antibody that binds to tau at amino acids 235–250 found in the microtubule-binding domain. This vaccine blocked tau seeding in rodent models of tauopathy. Phase I clinical trials found no safety issues, and a phase II clinical trial is currently ongoing (expected to finish in 2025). In a systematic review of this literature that considered seven types of vaccines, it was concluded that early data were promising but studies with a larger number of participants will be needed to reliably determine the effectiveness and safety of these procedures (Thakur et al., 2023).

One of the great difficulties associated with antibody treatments is that the agents used have difficulty passing the BBB and entering the brain (Pardridge, 2020). Thus, while anti-Ab antibodies can reduce Ab in animal studies (Van Dyck et al., 2023), entry to the brain may not be sufficient to allow for clinically significant behavioral and cognitive changes. Using focused ultrasound, the BBB could be opened so that levels of an antibody (aducanumab) to targeted brain regions were elevated by five to eight times that observed in the untreated brain regions (e.g., Leinenga et al., 2021). In a human trial comprising three patients, repeated aducanumab infusions following focused ultrasound were markedly effective in reducing Aβ. Unfortunately, during the extensive follow-up analyses conducted at various times after each treatment and over 18 months, cognitive improvements were not observed (Rezai et al., 2024). Larger trials conducted over lengthier periods will be needed to determine whether AD progression is stymied. Another compound, lecanemab, was found to slow the progression of AD by about 27%, although there is debate as to whether this reflects a clinically significant improvement (Chowdhury and Chowdhury, 2023).

Rather than focusing on pathological attributes (i.e., β-amyloid or tau), a very different approach was adopted in which senescent or dying cells were immunotherapeutically targeted. This was done using a vaccine designed to target senescence-associated glycoprotein (SAGP), thereby ridding the brain of compromised cells to stem the flow of pathology, much as would be expected with pathogenic infections. This SAGP vaccine had beneficial effects in other age-related conditions (atherosclerosis, Type 2 diabetes) in mice. The SAGPs are most highly expressed in glial cells and part of their benefit may be to eliminate proinflammatory microglia and astrocytes. Indeed, the SAGP vaccine reduced proinflammatory brain response in conjunction with diminished amyloid deposits in mice (Hsiao et al., 2023). It is to be expected that other such vaccines will emerge once better mechanistic targets for AD are identified.

As a final point, if viral infections contribute to AD, then it follows that vaccinations that target such pathogens might lessen the risk of AD development. Relatively recent systematic reviews and meta-analyses indicated that influenza vaccination was associated with a diminished risk of dementia (Sun et al., 2023f; Veronese et al., 2022) and more annual influenza vaccinations had the strongest relationship to reduced AD occurrence. Interestingly, it was not just influenza vaccination that had such effects. Irrespective of age or gender, a variety of vaccinations (e.g., rabies, tetanus, diphtheria, and pertussis (Tdap), herpes zoster, hepatitis A and B, and typhoid) were associated with reduced AD risk (Wu et al., 2022b).

Combined therapy approach for AD

As in many brain conditions, a combination approach that is tailored for each patient will likely be the future of treatment methods. In the case of AD, this might require vaccines against both tau-tangles and amyloid plaques and possibly antiinflammatory agents that specifically modulate the microglial phenotype. Of course, it will also be necessary to identify and treat the disease at a relatively early stage, when the brain still has sufficient neuroplasticity, making it capable of rebounding. This will require well-validated biomarkers, probably comprising a panel that provides an overall disease signature. The final piece of the puzzle might be the identification of specific combinations of environmental insults that certain people need to avoid, based on the specific polymorphisms they possess. This said, in many individuals, certain genetic alterations may give rise to disease regardless of exposure history. In such cases, future gene editing strategies might be required. This almost "science fiction" approach could be feasible in the future, particularly given the marked advances in gene technology, such as CRISPR-Cas9, that allow deletion or modification of certain genes (especially given the recent advances in diminishing off-target effects). It may even be possible to harvest the patient's somatic cells and reprogram them to healthy neurons that can be transplanted back into the patient's own brain (Fig. 16.3). Thus, one could conceivably have their microglia or neurons express a factor for which they might be deficient, owing to a specific polymorphism. At this point, it is hard to imagine such strategies will not eventually be attempted in humans.

Concluding Comments

We have attempted to cover basic well-established evidence, as well as emerging novel findings, regarding the factors that contribute to the onset and evolution of AD as well as potential therapeutic strategies. Animal models have been important in our understanding of how particular processes map onto the wide range of neurobiological disturbances evident in AD patients. At the same time, human studies help pinpoint potential environmental triggers of disease and possible useful biomarkers, but in the main, these studies are only correlational.

The animal models have primarily involved overexpression of some combination of genes implicated in AD. This has resulted in rodents showing cognitive deficits and the accumulation of β-amyloid deposits, and these effects appeared to be age-dependent, usually emerging after 6 months. There have also been models in which specific toxicants were injected into the hippocampus, causing cognitive deficits, possibly reflecting those that occur early in AD when the parahippocampal

gyrus is first compromised. While these models have recapitulated some aspects of the disease, they have failed to reflect other aspects, such as the disintegration of higher-level cortical processes that are unique to humans. A critical element that profoundly differs between the human and rodent models is the extent of cortical involvement. The rodent cortex is profoundly underdeveloped (being lissencephalic) compared to the highly convoluted, richly developed human cortex. Consequently, a substantial proportion of the higher cognitive functions ascribed to the cortex, which are disturbed in AD, is absent in the rodent. Furthermore, most of the neuronal loss in the rodent is subcortical, compared to the greater cortical involvement in humans. The simple fact is that mice are not humans, and as in other illnesses, positive effects in rodents frequently do not translate into meaningful effects in humans. It has been said that "if mice were humans, cancer would have been cured thousands of times." The same holds for AD.

Determining which mechanisms to focus on in AD research is no easy task and judging by the complexity of processes covered in this chapter, it is likely that multiple processes, which vary over the course of disease progression, should be targeted. Determining which and when is exceedingly difficult and will surely draw upon emerging technological advances made in the fields of genetics, neuroimmune pharmacology, proteomics, and areas related to microbiome functioning. It would be short-sighted to simply focus on one aspect of the disease and neglect the complexity of glial-neuron and immune-neuron interactions. It should be clear that microglial and immune cells may be involved in processes related to AD and targeting these may yield fruitful therapeutic options. Yet, as discussed earlier, this treatment route is complicated by the fact that there are beneficial and deleterious elements to the inflammatory immune response in AD. As we have seen, microglial cells can release toxic factors that are damaging to neurons, but at the same time, these cells also perform valuable phagocytic responses that may help rid the brain of β-amyloid plaques. Similarly, invading peripheral immune cells (and local microglia and astrocytes) can release trophic factors that foster recovery, while at the same time engage in pro-death intracellular signaling pathways. In essence, it may be necessary to target discrete subpopulations of cells, considering that their importance to AD may change over the course of the disease.

The heterogeneity of AD is yet another issue that likely needs to be addressed for large-scale clinical translation to be successful. The complexity of the genetics and environmental stresses involved in the provocation of the disease may contribute to the differing symptom profiles and progression of the disease. Thus, finding biomarkers that allow the streamlining of specialized treatments for patient subgroups could be essential. These biomarkers, however, might be most useful if they could also indicate when treatments ought to be adjusted or changed at different stages of disease progression. Of course, determining what mechanisms to target to prevent the disease from taking hold in the first place will be the greatest challenge.

At this time, pretty much all clinical trials for AD have largely been disappointing. Many variables could account for such discouraging results, including the use of poor animal models of AD, the failure to target useful mechanisms, the heterogeneity of the disease, and problems with drug bioactivity/bioavailability/distribution. What is needed as well are very specific acting pharmacological or genetic tools that allow manipulation of the various neural and glial circuits involved in AD. This will be driven by technological advances that allow us to delve deeper and more specifically into the various brain circuits that are disturbed in AD. Undoubtedly, this will fuel clinical trials with safer and more precise modes of action that can be tailored to suit the particular patient profile. Finding the means to harness the reparative potential of immune cells and their soluble factors is a potentially important route for future therapies. That said, considerable basic research is still needed to better understand not only the mechanisms of disease but also yet uncovered mechanisms of recovery and how basic immune factors interface with both the healthy and diseased CNS.

CHAPTER

17

Illness Comorbidities in Relation to Inflammatory Processes

Illness comorbidity and what this implies

Although everybody becomes sick at some time, there seem to be individuals who develop more illnesses than others. They may be the inheritors of genes that increase their vulnerability to illnesses, or they might have occupations that bring them into contact with viruses (doctors, nurses, school teachers) or various toxicants that can cause multiple illnesses. It is equally possible that once a chronic illness develops, it may promote the development of other illnesses. Alternatively, several illnesses may share certain biological features so that when one illness appears, it is a harbinger for the development of other conditions.

Mood disorders have been associated with a constellation of other chronic illnesses. More than 50% of depressed patients subsequently reported chronic pain, 33% experienced respiratory illnesses, cardiovascular illnesses were elevated by as much as 50%, diabetes and arthritis occurred in more than 10% of depressed individuals, and chronic fatigue-fibromyalgia was not uncommonly observed. The latter effects appeared to be linked to adverse childhood experiences and the cytokine changes that were induced. Also, the risk of Alzheimer's disease doubled among individuals with a history of depression, and the greater the number of cardiac risk factors present, the greater the presence of brain myeloid deposition.

Schizophrenia was associated with increased occurrence of OCD, panic disorder, depression, and substance use, as well as viral illnesses (e.g., hepatitis). In the case of multiple sclerosis (MS), patients frequently suffer hypertension, type 2 diabetes, epilepsy, inflammatory bowel disease (IBD), chronic lung disease, fibromyalgia, and especially high rates of depression and anxiety.

Something About Comorbidity

Most illnesses do not come about owing to a single factor but instead evolve because of many genes and epigenetic changes, various biological processes, cultural influences, early negative experiences, sex, as well as a constellation of other psychosocial and environmental factors. Some of these contributing agents are independent of one another but may come together, additively or synergistically, to promote pathology. Likewise, a given biological change may undermine protective processes (e.g., inhibition of a tumor suppressor), thereby permitting the actions of a second factor (genetic or environmental) to cause the development of illness. Similarly, a specific negative experience or gene polymorphism may serve as a "first hit" that primes biological systems so that a second hit has a greater impact in promoting health disturbances.

As indicated in Fig. 17.1, two illnesses, such as depression and heart disease, may be related to one another in different ways, and each may be related to a common environmental factor, such as stressor experiences. Although a stressor may come to promote depression, its link to a physical illness may occur through either direct or indirect processes. In each of the instances described in the figure, overlapping inflammatory mediators may contribute to both conditions although they may come about through related or unrelated mechanisms. This depiction is certainly simplistic and hardly addresses all the factors that contribute to these comorbidities but may nonetheless be useful in defining the intersections between disease conditions.

Comorbid illnesses might stem from common underlying features, such as particular genes, and they may

---

The Immune System
https://doi.org/10.1016/B978-0-443-23565-8.00003-X

443

Copyright © 2025 Elsevier Inc. All rights are reserved, including those for text and data mining, AI training, and similar technologies.

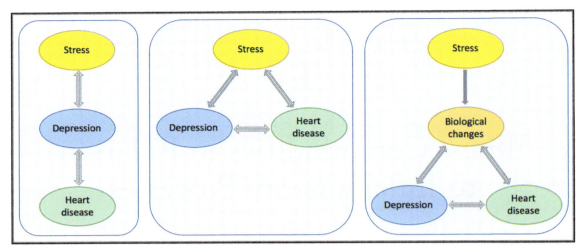

FIG. 17.1 Comorbidity does not necessarily imply that one illness caused another, although it certainly might. For example, an environmental event, such as a stressor, could give rise to depression, which then promotes heart disease. The heart disease may, in turn, exacerbate depression and distress, so that the disease cycle is perpetuated (left panel). It is equally possible that the two illnesses might be provoked by common environmental triggers (e.g., stressors), and then act upon one another (middle panel). Finally, in a third scenario, a particular environmental trigger may instigate biological processes, say activation of inflammatory or microbiotic factors, so that other neurobiological disturbances evolved (e.g., hormones, neurotransmitters, and growth factors) that could favor the development of the two illnesses, but these conditions could affect one another. But, even if they do not, the biological mediating factors could serve as a predictor (biomarker) of subsequent illness. In some cases, they might be useful in predicting specific illnesses emerging, whereas in other instances, they might reflect general risk factors, and still other markers would need to be identified that predict specific disease outcomes. It is of particular significance that if one illness appears, it may itself be a marker that the individual is at increased risk for other types of illnesses, perhaps signaling that preventive measures ought to be taken to diminish further illnesses developing. *From Anisman, H., 2021. Health Psychology: A Biopsychosocial Approach. Sage, London.*

also emerge owing to many lifestyle influences that are shared among individuals with specific diseases. For instance, depression may be accompanied by poor diet, smoking, sedentary lifestyles, and failure to receive medical attention. As a result, other illnesses may emerge that create problems for treatment and make it difficult to evaluate the neurobiological processes that are responsible for the primary illness. Likewise, as described in Fig. 17.2, depressive illnesses may be tied to other medical conditions because they share specific biological features or varied environmental risk factors. The routes by which comorbid illnesses develop may have implications for illness prediction and treatment.

In this chapter, we focus on comorbidities associated with depression and anxiety and several other mental health challenges. The links to heart disease, stroke, and diabetes receive particular attention. Psychological factors and lifestyles endorsed can affect cancer progression, but this is not dealt with here as we have done so in an earlier companion book (Anisman and Kusnecov, 2022).

Inflammatory Factors in Comorbid Illnesses

The Case of Autoimmune Disorders

Stressors, depression, and anxiety have been linked to numerous immune-mediated inflammatory diseases. These include illnesses stemming from immune functioning being compromised, or those in which the immune system turns on the self. The ties between stress-related disorders, including depression and PTSD, and the occurrence of autoimmune disorders, have been associated with genetic factors, and indeed, a genome-wide association study (GWAS) revealed 10 common genes and five shared functional modules between stress-related illnesses and autoimmune disorders (Zeng et al., 2023).

Stressful events and depression may promote flares among patients with MS and can interfere with coping abilities. Relative to healthy individuals, those with rheumatoid arthritis were more likely to experience depression and suicidality. Suicide attempts were three times greater among individuals with arthritis who had also experienced early life stressors (parental domestic violence or sexual abuse) compared to those who only experienced arthritis, pointing to the interactive effects of inflammatory disorders and other stressor experiences (Fuller-Thomson et al., 2016). This relationship is bidirectional; just as rheumatoid arthritis may promote depression, the presence of chronic inflammation associated with protracted stressors or with depression may promote or exacerbate rheumatoid arthritis (Lwin et al., 2020). In this regard, depression or anxiety often occurred years before individuals had developed rheumatoid arthritis, MS, or IBD (Marrie et al., 2019). This relationship might reflect shared underlying processes or the immune

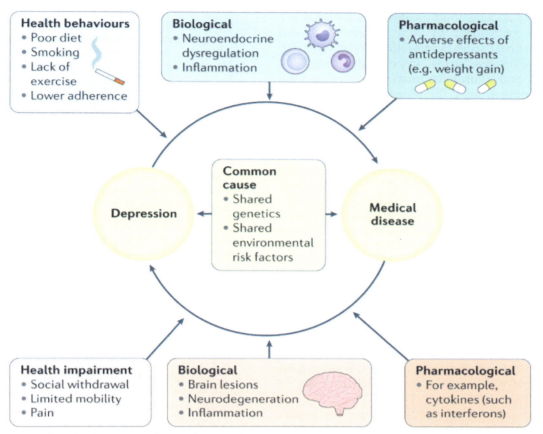

FIG. 17.2 Depressive disorders have been tied to a wide range of physical illnesses (type 2 diabetes, heart disease, autoimmune disorders, cancer). These illnesses may be independent of depression but are affected by common mechanisms (e.g., the presence of inflammation) or they may stem from the same psychosocial and experiential factors, environmental challenges, and genetic and epigenetic alterations. It is equally possible, for instance, that depressive disorders may give rise to specific biological changes that favor disease occurrence. *From Gold, S.M., Köhler-Forsberg, O., Moss-Morris, R., Mehnert, A., Miranda, J.J., et al., 2020. Comorbid depression in medical diseases. Nat. Rev. Dis. Primers 6, 69.*

dysregulation caused by anxiety may have promoted autoimmune conditions.

With the progressively increasing occurrence of autoimmune disorders in recent years, the possibility was entertained that Western diets and microbial disturbances contributed to their development (Christovich and Luo, 2022). By altering immune functioning and creating a persistent inflammatory milieu, gut dysbiosis could promote and maintain rheumatoid arthritis. In some individuals, altered intestinal permeability associated with gut dysbiosis predated the appearance of rheumatoid arthritis (Romero-Figueroa et al., 2023). Conversely, in an animal model, treatment with the bacteria *Prevotella histicola* could prevent or reduce the appearance of arthritis, seemingly operating through regulation of dendritic cells, myeloid suppressors, increased production of gut T_{reg} cells, and elevated transcription of IL-10 (Marietta et al., 2016). In addition to imbalances between proinflammatory and antiinflammatory cytokines, several other factors have been implicated in the ties between microbiota and joint damage associated with rheumatoid arthritis. This has included a lack of microbiota-produced SCFAs, bile acids, and tryptophan metabolites that maintain intestinal barrier integrity, as well as molecular mimicry between bacterial and host epitopes (Reyes-Castillo et al., 2021).

As in the case of rheumatoid arthritis, depression occurs in more than 50% of individuals with MS. No doubt, some cases of depression may arise owing to the distress associated with the illness, including individuals' worry about their future well-being. At the same time, several brain changes among MS patients could predict the subsequent occurrence of depression, such as lesions within the temporal, superior frontal, and superior parietal lobes, and the arcuate fasciculus that connects them (Mustač et al., 2021). Once again, this relationship may be bidirectional given that depression may precede the frank appearance of MS. Even in the absence of brain lesions and demyelination, synaptic disturbances related to inflammation were observed in animal models and the brain tissue of individuals with MS (Bruno et al., 2020).

Although common genetic determinants of MS and depression have not definitively been identified, the two conditions share several features. These include several neuroendocrine abnormalities and the presence of microglial pathology, as well as gut dysbiosis, increased intestinal barrier permeability, and chronic oxidative and nitrosative stress. Importantly, peripheral inflammation and neuroinflammation appear in both conditions (Bruno et al., 2020; Morris et al., 2018), and dysfunction of Th17 cells may be pertinent to their cooccurrence (Melnikov and Lopatina, 2022) as might activity of IL-17-producing γδ T cells and macrophage/microglial functioning.

The obvious question arose concerning the possible benefits that might be obtained for depression and autoimmune disorders through antidepressant drugs. Animal models indicated that the SSRI fluoxetine could reduce signs of MS, but the data from human studies have been less encouraging (Grech et al., 2019). This is not altogether surprising given the heterogeneous nature of MS in which multiple biological processes play into the occurrence and progression of the disorder. This is compounded by depression and its symptoms likewise varying appreciably across individuals. Assessing the impact of antidepressants requires their evaluation on subtypes of MS over the course of their progression, and whether the effects are consistent among patients with varying forms and severities of depressive illness.

In addition to the comorbidity between depression and these autoimmune disorders, among individuals experiencing psoriasis, it was not uncommon for anxiety and depression to be present, typically being attributed to the distress created by individuals experiencing the damaging skin condition. To an extent, this is likely accurate, but it also seems that anxiety and depression may aggravate this autoimmune disorder so that the skin condition worsens. The distress, depression, and cooccurring inflammation created by psoriasis have been linked to the development of psoriatic arthritis (a condition similar to rheumatoid arthritis), possibly acting through IL-17 functioning (Zafiriou et al., 2021). Moreover, psoriasis was accompanied by a substantial increase in the subsequent appearance of type 2 diabetes, which has been linked to inflammatory processes, and it was also associated with increased coronary microvascular dysfunction (Piaserico et al., 2023), which predicts a poorer prognosis of cardiovascular illness.

Inflammatory bowel disease (IBD): The prototypical brain-gut axis disorder

Crohn's disease and ulcerative colitis, the two main forms of IBD, are autoimmune disorders associated with chronic inflammation of the gastrointestinal tract. About 7 million people are affected by IBD worldwide, with prevalence being greatest in high-income countries, such as the United States. Since 1990, it has also been increasing in newly industrialized countries in Africa, Asia, and South America. Numerous factors influence the development of gut disorders, including the impact of acute and chronic stressors (see Fig. 17.3), which affect multiple processes, including actions on microbiota and immune functioning (Leigh et al., 2023).

Among the many puzzling illness comorbidities that have been identified, the fact that migraine sufferers are more likely to develop Crohn's disease and ulcerative colitis is surprising, although it is possible that medications used to ameliorate migraine pain could increase IBD risk. Likewise, the tie between IBD and schizophrenia may seem unexpected. A Mendelian randomization analysis that included 70 single nuclear polymorphisms indicated that schizophrenia was causally related to IBD, whereas IBD was not causally linked to schizophrenia (Qian et al., 2022). Crohn's disease was similarly linked to depressive symptoms, poor sleep quality, and impaired cognitive functioning, and elevated levels of inflammatory factors were prominent in depressed patients and in those with Crohn's. In some individuals, psychological disturbances, such as anxiety and depression, precede the gut disturbances, but in others, the gut symptoms appear first. Aside from indicating interrelations between gut disturbances and psychological factors, the order of illness appearance may be significant in treatment decisions. The link between IBD and depression has been attributed to inflammatory processes, oxidative and nitrosative stress factors, as well as tryptophan catabolites. As expected, the level of depression in patients with Crohn's disease was diminished by reducing inflammation through anti-TNF-α treatment (Bisgaard et al., 2022).

In view of the gut disturbances characteristic of IBD, it was not unexpected that reduced microbiota diversity and diminished stability of the microbiota, as well as their metabolic activity, were linked to depression and anxiety (Liu et al., 2020). Disturbed gut microbiota composition together with elevated levels of inflammatory factors were prominent in patients with Crohn's disease (Tavakoli et al., 2021), and the presence of symptoms, which may come and go, corresponded with changes of inflammation and gut microbiota variations (Halfvarson et al., 2017). In fact, in high-risk individuals, several bacterial taxa were identified that predicted Crohn's that appeared as long as 5 years later (Garay et al., 2023). Significantly, supplementation with butyrate-producing bacteria may have positive effects, and fecal microbiota transplantation may be helpful for many patients (Caldeira et al., 2020).

Stressful experiences have been implicated in the development of IBD, and stressors can trigger flares of the disease, likely acting by the promotion of gut microbiota dysbiosis, intestinal dysmotility, disturbed intestinal barrier functioning, as well as several immune and hormonal alterations (Sun et al., 2019). Also, stressful experiences promote

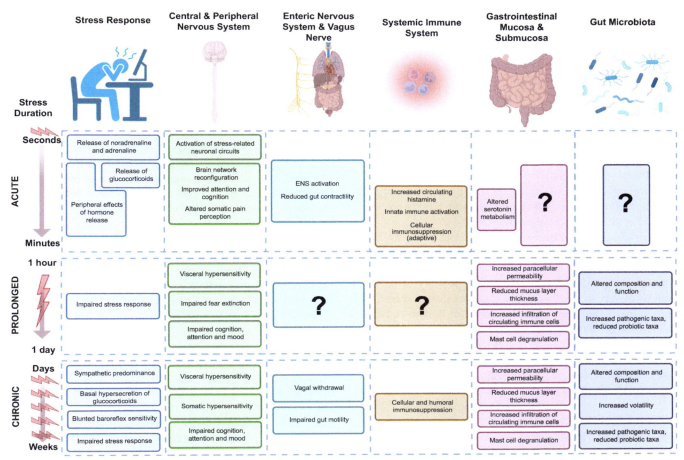

FIG. 17.3 The effect of acute, prolonged, and chronic stress on the pillars of the microbiota–gut–brain axis. Activation of the hypothalamic–pituitary–adrenal and sympatho-adrenomedullary axes following a stressor will trigger a cascade of physiological changes along the microbiota–gut–brain axis that are time-dependent. Acute stress will trigger the release of stress hormones (namely, glucocorticoids and catecholamines) and the activation of the enteric nervous system, reducing gut contractility. These acute changes are associated with innate immune activation, suppression of the adaptive immune system, increased circulating histamine, and an alteration of serotonin metabolism in the mucosa and submucosa. Comparatively, little is known about how acute stress impacts gastrointestinal physiology and gut microbiome. Prolonged stressor exposure of greater than 1h and less than 1 day also impairs the stress response, with mixed effects on glucocorticoid and catecholamine expression. This type of stress induces visceral hypersensitivity and impaired fear extinction. While little is known about the effects of prolonged stress on the enteric nervous system and systemic immune system, the gastrointestinal mucosa and gut microbiota are affected similarly to chronic stress. Chronic stress impairs vagal signaling and enteric function, altering gut motility. Visceral hypersensitivity is induced. Both cellular and humoral immunities are suppressed, and the gut barrier exhibits reduced mucus layer thickness, increased paracellular permeability, and infiltration of circulating immune cells as well as increased mast cell degranulation. The gut microbiome exhibits reduced stability and altered composition and function. Specifically, the relative abundance of pathogenic or contextually detrimental bacteria is increased while beneficial bacteria are relatively reduced. *From Leigh, S.J., Uhlig, F., Wilmes, L., Sanchez-Diaz, P., Gheorghe, C.E., et al., 2023. The impact of acute and chronic stress on gastrointestinal physiology and function: a microbiota-gut-brain axis perspective. J. Physiol. 601, 4491–4538.*

IBD flares through changes in enteric nervous system functioning that promote inflammation, possibly by affecting interactions between glucocorticoids and inflammatory cytokines (Schneider et al., 2023). The IBD connection to brain processes was also derived from the numerous genetic correlations that existed between intestinal and psychiatric disorders, some of which were linked to the microbiome (Gong et al., 2023). In this respect, the expression of genes associated with immune processes was linked to the clinical manifestations of the disorder among patients at different stages of IBD (Peters et al., 2017), indicating that the underlying processes may vary over the course of the illness.

Gut microbiota alterations can influence brain microglia functioning (Cordella et al., 2021), and the induction of colitis in animal models could affect brain microglial activation, which could promote anxiety and depression (Masanetz et al., 2022). Also, the provocation of intestinal inflammation to model IBD in mice led to arrested hippocampal neurogenesis, which might thereby influence behavioral outcomes, such as depression.

Perturbing the gut microbiota of rodents during early life through antibiotic treatment could induce visceral hypersensitivity during adulthood (O'Mahony et al., 2014). These and similar findings led to the perspective that IBD and other gastrointestinal disorders may have their origins in childhood adverse experiences operating through inflammatory processes that influence intestinal permeability, enteric nervous system development, and actions on hormonal systems (Pohl et al., 2015).

Irritable bowel syndrome (IBS) is distinct from IBD. The former may be related to the disorganization of muscles involved in moving food through the digestive tract, disturbances of nerves in the digestive tract, possibly involving gut serotonin, so that overreaction occurs during digestive processes, or because of gut microbial dysbiosis. A large GWAS of IBS identified six genetic susceptibility loci linked to IBS, of which four were also associated with mood and anxiety disorders, likely owing to shared pathogenic pathways rather than the psychological disturbances being responsible for the emergence of IBS (Eijsbouts et al., 2021).

Aside from other processes associated with illness comorbidities, increasingly greater attention has focused on the molecular signaling pathway, cGAS/STING, which comprises GMP-AMP synthase (cGAS) and stimulator of interferon genes (STING) that plays a fundamental role in triggering immune responses that act against pathogenic agents. cGAS/STING has been implicated in cellular senescence, aging processes, and neuroinflammatory and neurodegenerative disorders, possibly by affecting brain microglia (Gulen et al., 2023; Paul et al., 2021). Critically, in rodents, STING inhibitors could enhance some forms of memory, muscle strength, and endurance (Gulen et al., 2023). Aside from this, cGAS/STING has been implicated in inflammatory processes tied to cancer and autoimmune disorders (Chen et al., 2016b) and chronic pain by affecting sensory neurons and glial cells (Wu et al., 2022a), as well as sterile inflammatory-based illnesses, such as heart disease, nonalcoholic fatty liver disease, obesity, and IBD. Based on these reports, the cGAS/STING has become a target for moderating diverse inflammatory diseases through selective small-molecule inhibitors of this pathway (Decout et al., 2021). It also appears that manipulations of STING signaling, through its actions on neuroinflammation, can limit depressive symptoms associated with chronic stressor exposure (Duan et al., 2022b).

Drug Repurposing

Just as illnesses are often comorbid with one another, there are many instances in which treatments that were initially developed for one purpose are effective for a second condition. In some instances, the connection between the two can be predicted because they share underlying processes, whereas in other cases the links between the two are not immediately obvious. These dual drug effects may offer clues regarding the mechanisms responsible for each of the illnesses.

There are numerous such occurrences, and we will only highlight a few of these. Therapeutics that are used for their psychological or neurological impact, have found their way into the treatment of multiple other conditions. It is well known that some SSRIs initially developed to treat depression are especially effective as antianxiety treatments. Surprisingly, several drugs that have been used as either antidepressants (e.g., SSRIs) or antipsychotics, were also helpful in eliminating C. difficile and salmonella. Gabapentin and its cousin pregabalin, which are used as antiepileptic agents, are commonly used to diminish anxiety disorders and neuropathic pain. While those findings may not be overly surprising, the benefits of other agents have been unexpected. Many people are probably aware that sildenafil (Viagra) was initially developed to treat hypertension. Its effects were not particularly impressive and ended up being used effectively for erectile dysfunction. More recent real-world data have indicated that the use of this agent was associated with a 30%–54% reduction in the occurrence of Alzheimer's disease.

One of the most well-known compounds, the GLP-1 receptor agonist semaglutide (Ozempic) developed to treat type 2 diabetes turned out to be especially effective as a weight loss treatment. Because this agent (and related compounds) reduces craving, it is being evaluated to determine whether they can be used to diminish substance use disorders. Also, since these agents enhance glucose utilization and diminish inflammation throughout the body and the brain, they might act against serious heart disease and perhaps slow the progression of Alzheimer's disease, and recent evidence indicated that it might be useful in diminishing symptoms of Parkinson's disease. While the impacts of these agents may be due to reduced inflammation associated with being overweight, the antiinflammatory effects may occur before significant weight loss has been achieved. Through such actions, these drugs may limit the development of multiple illnesses associated with elevated inflammation (see Lenharo, 2024).

Considering the breadth of illnesses that have been linked to elevated inflammation, it can be expected that antiinflammatory drugs created for one immune-related condition might have a positive effect on a second immune-based disorder. Thus, Gleevec used to treat some cancers, may turn out to be useful in treating type 2 diabetes; cancer-acting agents that target the protein BRD4 might act against heart disease; histone deacetylase inhibitors that are used in treating some cancers, could be used to diminish psoriasis; and CART-T therapy used in treating some forms of cancer may be effective in limiting MS. Likewise, the β-blocker propranolol that had been used as an antihypertensive has been offered as a way of limiting soft tissue sarcomas, and pan β-blockers (a specific type of receptor antagonist) markedly increased the effectiveness of immunotherapy in the treatment of melanoma.

It has been said that one of the most effective treatments for type 2 diabetes, metformin, is a wonder drug. This agent primarily acts by reducing glucose production by the liver, diminishing glucose absorption from the intestines, enhancing insulin sensitivity, and reducing inflammation. In addition to its effects on diabetes, metformin may slow the aging process, diminish heart disease, and seem to be useful in the treatment of drug-resistant breast cancer. And, it is currently being investigated to determine whether it can delay the development of Alzheimer's disease and can prevent lung cancer. Moreover, the combination of metformin and the antihypertensive syrosingopine could also prevent cancer cells from receiving the nutrients needed for them to survive.

Repurposing drugs can have enormous benefits in the treatment of "off-site" disorders, and in some instances, their effectiveness may be more important or more effective than the condition for which they were initially developed. Repurposing drugs has the benefit of having previously been tested extensively for side effects. Of course, the cost of bringing a drug to market for treating a second illness is appreciably reduced (although this does not necessarily mean that the patient will see these savings). The repurposing of drugs is hardly new, and many drugs are prescribed as off-label medications because they have been found to have strong alternate effects. In some cases, these agents are prescribed because the usual meds have not been working or because tolerance has developed.

Predicting and Increasing Longevity

Humans have sought every imaginable method of extending life. The search for the fountain of youth dates back at least five decades BCE, but it was made famous by the Spanish explorer Juan Ponce de León in the 16th century and the famous painting of the fountain by Lucas Cranach the Elder. There may well be ways of increasing longevity, but it is unlikely to be related to mythic fountains or superstitions.

Numerous candidates may be germane to biological aging and multiple factors may contribute to this. Specific genes may contribute to accelerated aging, possibly by compromising cell metabolic processes, and aging could be promoted by the cumulative harmful effects of environmental challenges on DNA. It has likewise been suggested that biological aging might be related to the loss of certain cells, particularly those that are long-lived and hence are more likely to be affected by cumulative DNA damage. Likewise, long genes are more apt to be affected by multiple factors than shorter genes, which may contribute to biological aging (Soheili-Nezhad et al., 2024), although at this time there is no evidence for long genes causing age-related disturbances.

As we discussed earlier, the shortening of telomere length has been thought to reflect aging, although this has become somewhat controversial, and epigenetic clocks might be more useful in identifying accelerated aging. Through a multiomic method to integrate genomic, transcriptomic, and metabolomic data, numerous associations were identified that were associated with epigenetic age, including several lymphocyte subpopulations that were tied to longevity (Mavromatis et al., 2023). Based on the notion that biological age is based on aging clocks, studies have been undertaken to assess the impact of various lifestyle changes, the use of specific supplements, and drug treatments on clock functioning.

Ordinarily, cells can only multiply a certain number of times, but when they stop multiplying (senescence) they do not necessarily die. These senescent cells may release harmful proinflammatory molecules that can undermine health and longevity. Thus, longevity may be related to the presence of senescent cells and by determining the multiple mutations that act to suppress the development of senescence, it may be possible to preclude these processes and hence favor longer life. It was demonstrated in mice with genetically engineered T-cells that a procedure used in treating some forms of cancer (i.e., chimeric antigen receptor: CAR T-cells), can be used to target senescent cells, thereby diminishing inflammation and related diseases (Amor et al., 2024). Since T-cells have memory capacity, a single administration of CAR T-cells may have lasting effects, thereby protecting diverse organs against the adverse effects stemming from a build-up of senescent cells. In fact, in a small trial, CAR T-cell therapy was used to diminish severe lupus erythematosus (Mackensen et al., 2022).

Different organs within the body do not necessarily age in parallel with one another. Based on specific proteins measured in blood, it appeared that the age of some organs occurs at an accelerated rate, which is associated with greater disease prevalence and premature death. However, when one organ ages at an accelerated rate, this may influence aging associated with other organs (Oh et al., 2023; Tian et al., 2023). Thus, early identification of accelerated organ aging may allow for approaches to limit morbidity and early death.

From the vantage of cell aging, longevity might be viewed as a reflection of not dying from a wide number of diseases, including those related to disturbed cell metabolism, stem cell exhaustion, inhibition of cell multiplication, disturbances of housekeeping actions (e.g., autophagy), and altered inflammatory processes. Conversely, long life may stem from the protective effects of particular genes (or particular mutations and epigenetic changes) or lifestyles adopted. In this regard, living long may require a particularly effective immune system that does not decay excessively with age. The analysis of immune-specific patterns in peripheral blood mononuclear cells of super-agers (mean age of 106) revealed shifts of the immune complement (e.g., distribution of noncytotoxic to cytotoxic cells, increased B-cells relative to CD4$^+$ T-cells) that reflected highly adaptive changes that had occurred to earlier insults (Karagiannis et al., 2023). In essence, favorable health and longevity may be tied to an individual's capacity to maintain and restore immune functions that diminish inflammation associated with infectious diseases. Maintaining this high level of immunocompetence diminished some virally related illnesses, including influenza, as well as survival in response to sepsis and COVID-19 infection (Ahuja et al., 2023).

Based on data obtained from over 116,000 people, a set of 16 genetic markers was identified that predicted longevity, whereas other gene combinations predicted shortened lifespan (McDaid et al., 2017). A cross-species analysis revealed common genes that might somehow contribute to longevity. One gene in particular, b-cat1, seemed to be especially notable in this regard, since blocking its action increased the lifespan of C. elegans (a type of worm frequently used in research) by 25%, possibly because this increased the availability of important amino acids (Mansfeld et al., 2015). Working on the assumption that a single gene is unlikely to account for considerable variance in longevity, an alternative was developed in which patterns or "signatures" of biomarkers were identified that predicted aging or disease conditions that affect longevity (Sebastiani et al., 2017).

Yet another factor that was implicated in the aging process is the amino acid taurine, which has multiple important functions, especially protection against pathologies related to mitochondrial disturbances. Over the lifespan, taurine levels decline by as much as 80%, and taurine supplementation in rodents can act against diverse pathologies, such as metabolic syndrome, cardiovascular diseases, neurological disorders, and some forms of cancer (Jong et al., 2021). Moreover, taurine supplementation enhanced bone health, metabolic phenotypes, and immune-related profiles, while reducing cellular senescence, mitochondrial and DNA damage, and the presence of chronic inflammation (Singh et al.,

2023). The beneficial effects of taurine could come about by affecting energy expenditure so that a modest decline in weight loss is promoted; it may reduce oxidative DNA damage and can affect gut microbiota (see McGaunn and Baur, 2023). It is not certain whether taurine supplementation has a downside since adequate longitudinal studies have not been conducted. At the same time, given the diverse biological processes affected by taurine, using it as a supplement needs to be approached cautiously, especially as it is not known what doses of taurine might be beneficial on the one side or a health risk on the other. As exercise increases taurine levels, and certain foods (e.g., shellfish, meat) can provide taurine, for the moment these seem to be better alternatives than the use of supplements.

Balances between molecular processes related to growth factors and stress response pathways (e.g., mTOR and the p53 DNA damage response pathway) may play a fundamental role in determining cellular senescence, and manipulating these balances can affect biological aging. As well, members of the sirtuin deacetylase/ADP-ribosyltransferase/deacetylase family (typically referred to simply as sirtuins) have been linked to longevity, presumably through their actions on stem cell regulation or epigenetic processes (Yu and Dang, 2017). Sirtuins may also have positive effects by affecting glucose and fat oxidation, increasing resveratrol, regulating macrophage renewal (Imperatore et al., 2017), or inhibiting inflammation that otherwise takes a toll on various organ systems. Among the benefits of sirtuins is their capacity to diminish lung inflammation and limit the development of chronic obstructive pulmonary disease (COPD), as well as other illnesses, such as cancer (Guarente, 2014).

The suggestion arose that the transfer of "young blood" to older individuals might have beneficial effects (vampire therapy). Young blood may provide ingredients that enhance cognitive performance in old mice, even being able to reverse the cognitive effects attributable to aging (Villeda et al., 2014). These outcomes were accompanied by transcriptional changes tied to enhanced hippocampal synaptic plasticity and increased dendritic spine density. Conversely, transfer of old blood to younger mice reduced hippocampal neurogenesis and impaired performance reflecting neuromuscular ability (Rebo et al., 2016). This procedure also promotes cell senescence in young animals, whereas the clearance of senescent cells that accumulate with aging can rejuvenate old blood and improve tissue health (Jeon et al., 2022).

It has since been demonstrated that systemic treatment with platelet-derived chemokine platelet factor 4 (PF4; also referred to as CXCL4) transferred from the blood of young to aged mice reduced age-related hippocampal neuroinflammation and promoted synaptic plasticity-related molecular changes as well as enhanced cognition

(Schroer et al., 2023). Like these findings, cardiosphere-derived cells (cardiac stem cells) from young rats injected into old rats improved heart functioning and exercise capability, and their cardiac cell telomeres were lengthened (Grigorian-Shamagian et al., 2017). It is possible that such gains can be obtained without actually transferring young blood to an older organism. Ordinarily, within the bone marrow where stem cells reside, inflammation provoked by IL-1β may be overwhelming so that aging features become more prominent. In rodents, treatment with the IL-1 antagonist, anakinra, which is used to treat rheumatoid arthritis, appeared to be able to return blood to a younger state (Mitchell et al., 2023). It remains to be determined whether comparable changes will be evident in humans.

An alternative to reverse aging has focused on ways of restoring DNA methylation patterns (demethylation), transcript profiles, and tissue function that align with that evident in a young organism. By affecting cell growth, proliferation, and differentiation, a series of six cocktails were developed, which could re-establish a young genome-wide transcript profile and essentially reverse transcriptomic age (Yang et al., 2023a). Whether this translates to living organisms rather than cell cultures remains to be established.

One of the difficulties of studying longevity in laboratory-based animal models is that most species do not live very long, and those that do, typically are not amenable or available for scientific research of this sort. The rockfish is an exception. It can live for as little as 10 years or as long as 200 years. A genetic analysis of different rockfish species revealed that their longevity was related to both insulin and flavonoid signaling (Treaster et al., 2023). The findings concerning insulin-related genes speak to earlier research that implicated insulin-related processes and energy metabolism to longevity, while the genes related to flavonoids are relevant given that they appear to influence inflammatory processes (Chen et al., 2018).

Over the centuries, lifespan has progressively increased. During the Bronze Age and Iron Age, lifespan was estimated to have been just 26 years, and through the 12th to 19th centuries, if a person was fortunate enough to survive childhood death, which claimed almost half of children, a person had a reasonable chance of living to about 50. By 1950, lifespan had increased to about 50–60, varying considerably across countries (being linked to the country's GDP), and life expectancy at birth in 2021 was estimated to be about 79 years in the United States, whereas, in other Western countries, it hovers around 82 years. In part, the increase in lifespan over the past five decades has come from improved lifestyles and better medical ways of preventing death, although this has not been matched by a comparable improvement in brain health.

Finally, efforts have often focused on being able to predict a person's lifespan based on a constellation of factors. In the main, these efforts have only been marginally effective given how many variables enter this equation. However, using sophisticated machine learning algorithms, it was possible to predict premature death with 78% accuracy in a Danish registry (e.g., Savcisens et al., 2024). As much as such findings are impressive considering that many events are difficult to predict, in a sense it is also scary, being reminiscent of the notion, based on quantum theory, that the future of the universe is preordained (see Chen, 2023). This is not a new concept and has been tossed around for years, raising the question of free will. If everything is predetermined, then do we mere humans have any say in our futures? In his thoughtful (and entertaining) book, Robert Sapolsky makes an interesting case that free will may well be an illusion (Sapolsky, 2023).

Psychiatric Disorders and Cardiovascular Illnesses and Diabetes

The comorbidities discussed to this point are only a few of the very many that are directly or inversely related to other illnesses. We will now address the links between psychiatric disorders and heart disease, stroke, and diabetes, which are some of the more serious illnesses that affect large segments of the world population. Most premature deaths attributable to heart disease were considered preventable conditions and although heart disease has been declining modestly in most developed countries, it is still stubbornly high, largely owing to poor lifestyle choices (smoking, alcohol consumption, and poor diet) (Vaduganathan et al., 2022). Of the various forms of heart disease that can occur, hypertension and coronary artery disease (CAD) are the most pernicious. More than 70% of all deaths stem from noncommunicable diseases, with heart diseases accounting for more than 32% of global deaths. It is alarming that since 1990 the incidence of acute myocardial infarction has been increasing in younger individuals (<45 years of age) (Arora et al., 2019). Here, we will explore some of the psychological and inflammatory processes that contribute to several forms of heart disease, primarily hypertension and chronic artery disease. We will then examine psychiatric comorbidities linked to stroke and diabetes, for which inflammatory factors are a common denominator.

Hypertension

Chronically elevated blood pressure affects about 20% of people, differing with age, sex, ethnicity, and a constellation of lifestyle factors, such as exercise, diet, weight,

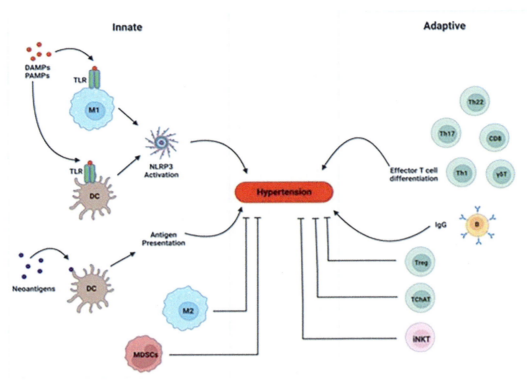

FIG. 17.4 Role of immune cells in the pathogenesis of hypertension. Hypertension-related DAMPs and PAMPs activate TLRs and NLRP3 inflammasomes in M1 macrophages and DCs contributing to inflammation. Neoantigens are processed and presented by DCs to B- and T-cells that lead to differentiation of plasma cells and effector T-cell subsets (CD8+ T cells, Th1, Th17, Th22 cells, and γδ T-cells). MDSCs, M2 macrophages, Tregs, TChAT, and iNKT cells prevent the formation of proinflammatory cytokines, attenuate inflammation, and attenuate hypertension. Abbreviations: *DAMPs*, danger-associated molecular patterns; *DC*, dendritic cells; *iNKT*, invariant natural killer T-cells; *M1*, proinflammatory macrophages; *M2*, antiinflammatory macrophages; *MDSCs*, myeloid-derived suppressor cells; *NLRP3*, NOD-, LRR-, and pyrin domain-containing three inflammasomes; *PAMPs*, pathogen-associated molecular patterns; *TChAT*, choline acetyltransferase-expressing CD4+ T-cells; *Th*, T-helper cells; *Treg*, regulatory T-cells. *From Navaneethabalakrishnan, S., Smith, H.L., Arenaz, C.M., Goodlett, B.L., McDermott, J.G., Mitchell, B.M. (2022). Update on immune mechanisms in hypertension. Am. J. Hypertens. 35, 842–851.*

sleep profile, and stressor experiences. It may also develop owing to ongoing medical conditions, such as thyroid or kidney disease (secondary hypertension). Hypertension may not be accompanied by obvious symptoms, and if left untreated, damage can occur to the arterial walls so that plaque accumulation increases and thus arteries narrow and harden (atherosclerosis).

Heart-related disorders have typically been considered in the context of peripheral changes, primarily autonomic nervous system functioning, but CNS processes also contribute to the development of various forms of heart disease (see Kwon, 2023). Hypertension is influenced by cortical and limbic brain areas, which regulate hypothalamic mechanisms that may influence sympathetic nervous system activity and cardiac functioning. Neuronal activity within the cingulate cortex and insula, which are involved in emotional processing of stressors, as well as the locus coeruleus, which has been associated with vigilance, have been linked to cardiovascular functioning, as has disturbed neuronal connectivity within the hippocampus (e.g., Kraynak et al., 2018).

Beyond these processes, as shown in Fig. 17.4, hypertension has been linked to diverse immunological changes, including activation of the NLRP3 inflammasome and the accumulation of macrophages and T-cells in blood vessel walls, especially in perivascular tissue (Navaneethabalakrishnan et al., 2022). The provocation of innate and adaptive immune system changes and the resulting cytokine elevations (IL-6, IL-17, TNF-α, and IFNγ) may contribute to vascular senescence and renal damage associated with hypertension and heart failure. Thus, it was suggested that circulating T_{reg} cells, CD4 cells, CD8 T-cells, and the ratio of CD4 to CD8 T-cells might serve as biomarkers to predict hypertension and heart failure (Rai et al., 2020).

Many genes contribute to the emergence of hypertension, with each accounting for a small portion of the variance. Indeed, about 100 genes have been linked to hypertension, many of which were also associated with CAD (Warren et al., 2017). A large-scale analysis that included more than one million people of European ancestry identified 535 loci that were associated with blood pressure, some of which were related to lifestyles endorsed (Evangelou et al., 2018). Further analysis indicated stronger genetic associations with hypertension among females than males and that some of the strongest relations involved genes associated with immune functioning (Zucker et al., 2023).

Heart disease, viral illness, and vaccination

Viral illnesses have been associated with cardiovascular problems. Myocarditis, which comprises inflammation of the heart muscle, may arise owing to a viral infection (e.g., influenza). The presence of Epstein–Barr virus, coxsackievirus, parvovirus B19, adenovirus, human herpesvirus 6, and cytomegalovirus were similarly identified in biopsied tissue of patients who had been affected by cardiac inflammation (Badrinath et al., 2022). Influenza can also precipitate acute myocardial infarction, and death may occur owing to elevated blood pressure, especially among individuals with preexisting hypertension. While not dismissing the importance of inflammatory factors in myocarditis, it is possible that a virus could directly cause this condition even before inflammation is evident (Padget et al., 2024).

In addition to protecting individuals from infection, the risk of death by myocardial infarction is diminished by as much as 45% among individuals who have been vaccinated against influenza (MacIntyre et al., 2016). Also, mortality over 1 year following myocardial infarction (or in high-risk CAD patients) was diminished by influenza vaccination (Frøbert et al., 2021). Similarly, a nationwide cohort study conducted in Denmark over nine flu seasons indicated that among individuals with hypertension, vaccination was associated with cardiovascular deaths, death owing to myocardial infarction and stroke, as well as all-cause mortality (Modin et al., 2022). The cardiac benefits of influenza vaccination may stem from a shift from a proinflammatory to an antiinflammatory state (Aïdoud et al., 2022) or the production of antibodies that can interact with bradykinin receptors, leading to vasodilation and greater myocardial oxygen supply.

As we have frequently heard, individuals with certain preexisting conditions fared most poorly following SARS-CoV-2 infection and this was notable among those with hypertension, likely owing to the provocation of a cytokine storm (Su et al., 2022a). As described in Chapter 4, the angiotensin-converting enzyme 2 (ACE2) serves as a functional receptor for coronaviruses, such as SARS-CoV-2, which uses spike glycoprotein to bind to these receptors situated on target cells. Individuals with hypertension (and other cardiovascular problems) have a greater number of ACE2 receptors, making them more vulnerable to SARS-CoV-2 infection and more prone to severe illness and death associated with excessive production of proinflammatory cytokines. Patients with hypertension were also more likely to develop renal damage and further cardiovascular problems following SARS-CoV-2 infection (Su et al., 2022a).

Media attention during the COVID-19 pandemic occasionally focused on the possibility that the mRNA vaccine could induce myocarditis, and for a brief time might have dissuaded some people from being vaccinated, despite myocarditis and pericarditis occurring in fewer than 0.01% of vaccinated people (Gao et al., 2023b). The risk of heart problems was far outweighed by the health benefits obtained by vaccination. More than this, however, COVID-19 vaccination was associated with a marked reduction in the occurrence of various cardiovascular disturbances.

Sympathetic reactivity and chronic stressor experiences have reliably been tied to the occurrence of hypertension, being moderated by a constellation of psychological factors related to coping methods used, personality traits, previous stressor experiences, and lifestyles endorsed. Ethnicity has likewise been linked to hypertension, and while this may be related to genetic influences, the cultural differences in hypertension are also attributable to other variables, such as poverty, lifestyle factors, and persistent encounters with racism (Hill and Thayer, 2019).

Hypertension typically becomes more frequent as individuals age; thus, it is concerning that it has been appearing more frequently in young people. To some extent, this might be attributable to the frequent adoption of sedentary behaviors and excessive consumption of foods containing high levels of fat and sodium coupled with low levels of potassium. In this respect, sugar-sweetened drinks, ultra-processed foods, and those that contain trans fat undermine heart health. Diet-related microbiota alterations may also be associated with the development of hypertension (Katsi et al., 2019) and the hypertensive effects attributable to chronic stressors may stem from the microbial changes provoked by such experiences.

Coronary Artery Disease (CAD)

In response to damage to the endothelium (the thin layer of cells that lines the inner portion of blood vessels), monocytes, macrophages, and T-cells, infiltrate the site of damage, releasing cytokines that encourage inflammation and facilitate plaque formation made up of cholesterol, fat, calcium, and fibrin. If dead and dying cells are present, macrophages will ingest them, but some cells carry a message, a CD47 protein, which essentially tells the macrophage "Don't eat me" and the accumulation of these dead cells also favors plaque formation.

Importantly, CD47 can also promote angiogenesis (new blood vessel sprouting), which can fuel aberrant vascular development. As expected, antibodies against CD47 facilitated the clearance of diseased vascular tissue, and this treatment diminished atherosclerosis in a mouse model (Kojima et al., 2016).

With recurrent endothelial damage, plaque appearance becomes progressively greater, eventually restricting blood flow to the heart. Symptoms may not be noticeable until blood flow is appreciably restricted (more than 75%), whereupon individuals might feel chest pain (angina pectoris) owing to insufficient oxygenated blood reaching the heart. Myocardial ischemia and angina are initially transient, resolving with discontinuation of behaviors that place a load on the heart (stable angina) or through medications that increase blood supply (e.g., nitroglycerine placed under the tongue). With illness progression, symptoms may occur even with minimal energy output or when individuals are at rest, and ischemic periods become more persistent. This unstable angina may predict arrhythmias (disturbed heart rhythm) and myocardial infarction.

Factors That Promote Heart Disease

Numerous risk factors for CAD have been identified, such as high blood pressure or the presence of diabetes wherein elevated blood sugars can foster both endothelial damage and elevated levels of inflammatory factors. Diverse psychosocial influences have likewise been associated with the development of CAD. This included adverse childhood experiences, chronic stressors encountered during adulthood, depressive illness, anxiety, anger, and hostility coupled with low social support, as well as lower socioeconomic class and job strain. Predictably, the risk for the development and progression of CAD increases further when several of these factors are present concurrently. As we will see, many of these influences on CAD operate by favoring inflammation.

Adiposity, cytokines, and heart disease

Obesity may be accompanied by the presence of excess cholesterol and fat (dyslipidemia), disturbed glucose tolerance, metabolic syndrome, and type 2 diabetes, all of which incrementally contribute to CAD. Fat located around the midsection, as we saw earlier, is notoriously rich in inflammatory cytokines, which may determine the link between obesity and heart disease as well as other illnesses mediated by inflammatory factors. In men and women who had obesity in middle life, the age span was reduced by 5.8 and 7.1 years, respectively.

Although bad cholesterol (LDL) was long considered to be responsible for the link to heart disease, only half of the people who had heart attacks displayed elevated cholesterol, leading to the view that some other factor, either alone or in combination with cholesterol, was fundamental in the promotion of heart problems. While it was generally believed that good cholesterol (HDL) did not fully make up for excessive bad cholesterol (LDL), it seems that a variant of the gene coding for the functioning of the HDL receptor (Scavenger receptor BI: SR-BI) increases the risk for heart disease despite high HDL levels (Zanoni et al., 2016).

As described in Chapter 5, despite the seemingly overwhelming evidence indicating that being overweight was associated with heart disease, it had been argued that having moderate obesity did not create cardiovascular risks (Flegal et al., 2013). However, many studies that supported this position failed to consider the cumulative health risks associated with being overweight for years. While maintaining an exercise regimen could reduce health risks, this did not fully reverse the adverse effects of being overweight or obese for sustained periods (Valenzuela et al., 2022). For that matter, children and adolescents who are obese, perhaps being linked to stressful experiences, continue to exhibit biological risk factors relevant to later heart disease. A large Swedish registry study of twins concluded that obesity stemming from psychosocial or experiential influences were more likely to experience heart disease than individuals whose obesity was linked to genetic factors (Ojalehto et al., 2023).

Genetic Factors in CAD

Since heart diseases run in families, it might be assumed that inherited risk factors contribute to this outcome, although these relations may be tied to shared environmental factors (e.g., socioeconomic status, diet). While not excluding these influences, a systematic multiomics integration approach identified 13 core genes that were related to cardiomyopathy, with the TNNI3K gene, which encodes a protein that belongs to the MAP kinase family of protein kinases, playing a particularly prominent role (Rong et al., 2022). A GWAS similarly identified multiple genetic mutations that were tied to heart failure (Rasooly et al., 2023) and a similar analysis indicated that multiple genes accounted for almost 30% of the variance associated with CAD (Khera and Kathiresan, 2017). A mutation in the gene *ANGPTL3* has also been associated with very low triglyceride and LDL levels, and efforts to diminish heart disease have been made to develop treatments that target the actions of this gene (Musunuru and Kathiresan, 2017). Likewise, a variant of *PCSK9* (which was associated with reduced LDL) led to the development of a monoclonal antibody that seemed to have promising effects on heart disease (Natarajan and Kathiresan, 2016). Inflammatory-related processes may similarly contribute to heart functioning, and it might turn out that treatments that target both cholesterol and inflammatory factors will yield still better outcomes.

As important as the effects of genetic factors might be, their presence does not guarantee that individuals are bound to develop heart disease. Genes may interact with lifestyle factors and stressors experienced, thereby influencing CAD. For instance, four genetic variants have been identified that interacted with stressor experience that influenced mortality linked to heart problems (Svensson et al., 2017). As well, considerable evidence has pointed to the contribution of epigenetic changes in the provocation of hypertension, CAD, and heart failure (Shi et al., 2022). Even among individuals at high genetic risk for heart disease, the development of cardiovascular disturbances may be modifiable by adopting appropriate preventive actions. Conversely, among individuals living in disadvantaged neighborhoods, epigenetic changes may develop that are tied to stress reactivity and inflammatory processes, and might thus favor atherosclerosis.

As we saw in relation to hypertension, marked differences in CAD exist across ethnic and cultural groups. Relative to white males of European descent, African American men were more likely to encounter vascular disturbances, as well as type 2 diabetes, and to die from CAD. The health disparities may be related to genetic influences and lifestyles adopted, as well as to lifetime stressor experiences, including racial discrimination (Sims et al., 2020). The risk of heart disease is generally low among people from East Asian countries; however, the risk of cardiac illnesses was elevated among the children of people who migrated to Western countries, likely reflecting the adoption of unhealthy lifestyles.

Stressor Effects

Persistent psychological distress (e.g., chronic caregiving, bereavement, and job strain) was associated with new diagnoses of CAD and was predictive of greater mortality among individuals with stable CAD. A prospective analysis over 8.6–11.9 years that included about 118,700 individuals indicated that psychosocial stress was accompanied by increased CAD, stroke, and earlier mortality. Even though a variety of factors (e.g., age, sex, education, marital status, location, abdominal obesity, hypertension, smoking, diabetes, and familial CAD history) can have adverse health consequences, the association between stress and CAD and earlier mortality was apparent after adjustment for these variables (Santosa et al., 2021).

The impact of chronic stressor experiences associated with CAD may occur through the actions of several brain processes, such as interconnections between amygdala functioning and HPA and sympathetic nervous system activity (Osborne et al., 2020b). Individual differences linked to heart disease were similarly related to changes in functional connectivity between the amygdala and the perigenual anterior cingulate cortex, which were related to elevated circulating IL-6 levels. Also, in response to stressors, circulating inflammatory factors are increased and together with lipids bound to the endothelium favor plaque formation that produces atherosclerosis (Fioranelli et al., 2018). While most research assessed chronic stressors in the years preceding CAD, adverse events experienced early in life may influence the developmental trajectory related to cardiovascular functioning, so adult atherosclerosis was more likely to occur (Kellum et al., 2023).

Of the many chronic stressors that can affect heart disease, social rank within the workplace appeared to be especially significant and was predictive of a constellation of other illnesses, such as type 2 diabetes, certain forms of cancer, gastrointestinal illnesses, chronic lung disease, and not surprisingly, depression and suicide. A particularly germane feature related to the increased occurrence of CAD, and early mortality was the "job strain" experienced (Marmot et al., 1978). This amounts to high job demand together with low decision latitude, and later studies indicated that the effects can be exacerbated by high work efforts and low rewards as well as a perceived lack of justice (unfairness). The impacts of these factors were exacerbated if they occurred on a backdrop of other ongoing stressors that promoted anxiety, irritability, and marital problems, as well as the adoption of poor lifestyles that might have been secondary to job-related distress.

Inflammatory Processes in Heart Disease

Factors that promote inflammation, including total pathogen burden (total number of infections experienced), predict the development of heart disease. As shown in Fig. 17.5, inflammaging, the age-related increase of proinflammatory markers in blood and tissues, was associated with numerous health risks and was especially notable with CAD (Ferrucci and Fabbri, 2018). Aging is also accompanied by varied microbial alterations, increased appearance of proinflammatory biases, elevated oxidative stress, cellular senescence, immune cell dysregulation, and chronic infections, all of which increase vulnerability to pathology. Furthermore, aging and stressor experiences synergistically influence epigenetic changes of the FKBP5 gene that promotes inflammation by affecting NF-κB, culminating in elevated risk for CAD (Zannas et al., 2019), and inflammation related to brain microglia activation may contribute to hypertension and myocardial infarction (Wang et al., 2022c).

The NLRP3 inflammasome was considered fundamental in linking psychological stress to depression and may contribute to heart disease. It will be recalled that toll-like receptors (TLRs) recognize pathogen-associated molecular patterns (PAMPs) as well as

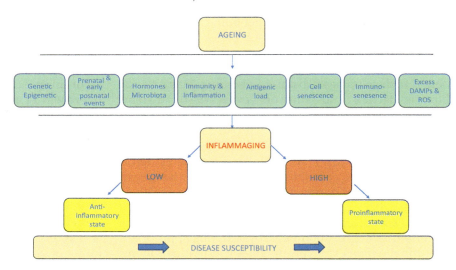

FIG. 17.5 Susceptibility to various inflammatory-related disorders increases with age. Prenatal and early postnatal experiences, hormonal and microbiota alterations, and elevated inflammation may be linked to lifestyle factors (e.g., nutrition, exercise, stressor encounter) that may affect health and well-being. With age and experience, epigenetic changes increase in frequency, and their presence across multiple tissues can be used to predict cellular aging. These numerous factors come together that lead to elevated inflammaging wherein proinflammatory factors exceed the presence of antiinflammatory factors, which may then influence susceptibility to various diseases.

damage-associated molecular patterns (DAMPs), leading to the release of "alarm" cytokines, such as IL-1β, TNF-α, and IL-6. In the presence of heart damage, even if it is relatively modest, DAMPs promote the activation of a sterile immune response, which can produce still greater heart disturbances (see Fig. 17.6). These and other inflammatory cytokines are potentially responsible for the comorbidity between depression and heart disease and may also contribute to a constellation of other illnesses, notably diabetes, stroke, chronic pain, obesity, diabetes, rheumatoid arthritis, multiple sclerosis, asthma, and Alzheimer's disease.

Just as activation of the kynurenine pathway was offered as a mechanism by which cytokines may come to promote depression, this pathway could also affect heart disease, and might thus account for the comorbidity between depression and CAD (Halaris, 2017). Specifically, kynurenine metabolites synergistically generate free radicals that can promote endothelial dysfunction that eventually engenders cardiovascular problems (Halaris, 2017), just as these metabolites can promote cell loss within the brain.

Predicting Heart Disease Through Inflammatory Markers

Considerable interest has focused on identifying biomarkers that are indicative of impending heart problems. Elevated CRP released from the liver in response to inflammation has been used for some time as a marker for later heart disease. A follow-up study over a period of about 13.9 years indicated that CRP, calprotectin (associated with gut inflammation), and the acute phase protein, neopterin, predicted heart disease and early mortality, but after controlling for other common risk factors, CRP was the only marker that independently predicted cardiovascular mortality (Løfblad et al., 2021).

Elevated levels of a protein, troponin, also occurred in approximately 40% of individuals with type 2 diabetes and stable ischemic heart disease, and they were at twice the risk of dying of heart disease or stroke over the ensuing 5 years (Everett et al., 2015). As well, cardiac problems, such as heart failure, were accompanied by the increased appearance of two heart hormones, atrial natriuretic peptide (ANP) and brain natriuretic peptide (BNP), whose release is stimulated by cytokines. These hormones may serve as markers to predict the development of heart failure and could point to therapies to limit its occurrence (Sangaralingham et al., 2023).

Depressive Illness and Heart Disease

Mental illnesses of virtually every sort (schizophrenia, bipolar disorder, anxiety disorder, and major depression) have been genetically tied to immune-related disorders and are associated with a dramatic increase in the risk for CAD, as well as elevated risk of heart attack and stroke. Although severe or chronic depression is most closely aligned with heart disease, even moderate symptoms of depression, possibly reflecting a first hit (or an existing vulnerability), were accompanied by a 75% increase in CRP levels and the development of CAD following subsequent trauma (Murdock et al., 2017). Speaking to the proactive effects of stressors and affective illnesses, childhood and adolescent depression were also predictive of the later development of heart problems.

A large prospective cohort study indicated that the risk of cardiovascular and ischemic heart disease mortality was directly related to graded differences in depressive symptoms, which were frequently linked to lifestyles endorsed (Zhang et al., 2023d). Evaluation of about 72,000 participants indicated that the presence of anxiety and depression increased subsequent cardiovascular risk factors by 38%, which frequently predicted major adverse

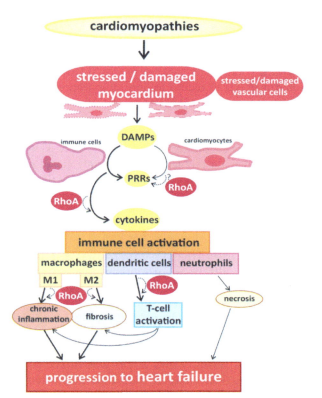

FIG. 17.6 Cardiomyopathies of different causality converge in myocardial damage, inducing immune cell responses and promoting disease progression. Cardiomyocytes and blood vessel cells that are stressed or injured release danger signals with "danger-associated molecular patterns" (DAMPs), which are recognized by "pattern recognition receptors" (PRRs), expressed by immune cells and cardiomyocytes. The activation of PRRs induces the migration and activation of (more) immune cells via cytokine signaling. In damaged myocardium, classically activated M1-type and alternatively activated M2-type macrophages are found, which are associated with chronic inflammation and fibrosis, respectively. Furthermore, dendritic cells and costimulated T-cells contribute to the development of inflammatory and fibrotic remodeling. Infiltrating neutrophils are associated with increased necrosis. These processes accelerate the progression of cardiac dysfunction. RhoA-dependent signaling is essential for effective immune cell activation and might also play a role in the transduction of danger signals and receptor activation in cardiomyocytes. In addition, RhoA signaling is involved in macrophage polarization and signaling pathways mediating the interaction between dendritic cells and T-cells, contributing to proinflammatory and antiinflammatory remodeling and finally, heart failure. *From Kilian, L.S., Frank, D., Rangrez, A.Y., 2021. RhoA signaling in immune cell response and cardiac disease. Cells 10, 1681.*

cardiovascular events (Civieri et al., 2023). A longitudinal analysis similarly indicated that the risk of cardiovascular morbidity and mortality was especially marked among individuals with a history of major depressive disorder, being most pronounced in association with atypical depression (Penninx, 2017). Aside from favoring the occurrence of future cardiac problems, the presence of hopelessness and pessimism was accompanied by slower recovery following cardiac events (Kop and Mommersteeg, 2014).

Depression and CAD share many features, including altered sympathetic and parasympathetic functioning, and a bias in favor of proinflammatory over antiinflammatory processes, which may contribute to the comorbidity between these conditions (Halaris, 2017; Khandaker et al., 2020). The links to depression and heart disease have largely comprised elevated IL-1β, IL-6, TNF-α, and reduced levels of the antiinflammatories IL-4 and IL-10. As well, IL-18 released from cardiac muscle cells (cardiomyocytes), may contribute to the development of heart disease. The cytokine alterations can affect HPA dysregulation and sympathoadrenal hyperactivity, which may lead to increased vasoconstrictive tone, and platelet activation, which ultimately favor the occurrence of heart disease.

While depression exacerbates heart disease and limits recovery, successful treatment of depression does not necessarily eliminate heart problems. Once cardiac problems are present, probably after years in the making, it may be too late to institute a cure simply by diminishing depressive symptoms, although doing so might limit further damage from occurring. Thus, the initiation of prophylactic strategies, such as regular exercise, which acts against inflammation may be ideal in limiting the development of both depression and heart disease (Pope and Wood, 2020).

Although our focus here has been on the link between depressive illnesses and heart disease, similar ties were also reported with other mental health disorders. For instance, PTSD stemming from chronic stress among soldiers who had been in combat, which is often comorbid with depression, was related to an increase in later heart disease (Krantz et al., 2022). Analysis of a Japanese health database that included more than 4.1 million individuals indicated that schizophrenia was associated with elevated occurrence of myocardial infarction, heart failure, atrial fibrillation, and angina pectoris, more so in women than in men (Komuro et al., 2024). As in the case of depression, the ties between schizophrenia and heart disease were mediated by elevated proinflammatory cytokine levels (Goldsmith et al., 2016).

Loneliness and heart disease

Feelings of loneliness can become pervasive and all-encompassing, provoking distress and a sense of being disconnected from others (see Chapter 9, Depressive Disorders). Loneliness has been recognized as being a profound health risk and may be more dangerous than smoking or obesity in heart disease. A systematic review and meta-analysis that included 16 longitudinal studies revealed that loneliness and poor social relationships were linked to a 29% increase in the risk of incident CAD and a 32% increase in the risk of stroke (Valtorta et al., 2016). The pathways that link social isolation and loneliness to heart disease have not been adequately assessed, and limited data are available that focus on populations that were at high risk of illness. As such, little research has been conducted that evaluated interventions to diminish the negative consequences of social isolation and loneliness (Cené et al., 2022).

The link between loneliness and heart disease was especially notable among older people and in people with disabilities, and low levels of social contact were accompanied by diminished volume of several brain regions, which appeared to be partially mediated by the presence of depressive symptoms (Hirabayashi et al., 2023). In older individuals, many of whom had lost their social connections (friends may have died, or moved away) and often suffer multiple indignities, including age discrimination, loneliness was not only identified as being instrumental in promoting heart disease, vascular disease, and stroke, but also a variety of immune-related disorders (Holt-Lunstad et al., 2015). The relationship between loneliness and heart disease was particularly evident among women in whom social isolation and loneliness cooccurred (Golaszewski et al., 2022). Polygenic risk factors associated with depressive illness and loneliness appeared to act pleiotropically (i.e., genes affecting more than a single phenotype) in promoting CAD among women (Dennis et al., 2021).

Loneliness has been linked to elevated inflammation (Vingeliene et al., 2019), which may be related to the physical and psychological disturbances associated with this condition (Cacioppo et al., 2015). It can be expected that in populations at risk for illnesses, the distress and provocation of inflammatory alterations stemming from loneliness might exacerbate their occurrence and disturb recovery. Predictably, during the COVID-19 pandemic, feelings of loneliness and being socially isolated were associated with chronic inflammation (Koyama et al., 2021), which could influence the development of pathology.

As miserable as loneliness might be, these feelings might be evolutionarily conserved, possibly because they might trigger the motivation to connect with group members (Cacioppo et al., 2014). Indeed, social connection reflects a basic need much as food and water are needed to sustain us. After experiencing social isolation, individuals reported craving for companionship, which was accompanied by activation of specific brain regions in response to social cues (Tomova et al., 2020). It was demonstrated that feelings of loneliness were associated with elevated activity within the default mode network, which might have fostered cognitive responses (e.g., reminiscence and imagination) to fill the social void that is present (Spreng et al., 2020).

Considering the powerful effects of loneliness on disease occurrence, the American Heart Association issued a statement emphasizing "the need to develop, implement, and test interventions to improve cardiovascular and brain health for individuals who are socially isolated or lonely."

Likewise, an editorial in the Lancet (2023) explicitly made this point and reinforced the view that health agencies ought to tackle the loneliness epidemic through ways of enhancing connectivity between people. It would be ideal to find methods to reduce feelings of loneliness, perhaps by altering the way individuals process their social environment. Diminishing loneliness can be exceptionally difficult and a single approach to do so is unrealistic, and despite the contribution of the default mode network, there is no uniform brain activity profile that is uniformly characteristic of loneliness, and instead, each person processes the world in idiosyncratic ways (Baek et al., 2023).

Personality Factors

The potential contribution of psychological factors to heart disease came to the fore with reports that heart disease was especially frequent among individuals with a Type A personality, characterized by being highly competitive, impatient, rushed, and hostile (Friedman and Rosenman, 1971). This view was broadly adopted, but as exciting as the notion had been, it seems it was overstated, although a component of this personality style that comprised hostility and anger was linked to coronary problems, especially if individuals directed their anger inwardly (Sahoo et al., 2018). Also, an analysis of about 484,200 people obtained through the UK Biobank indicated that levels of sociability and diligence (generally reflecting personality features of extraversion and conscientiousness) were associated with a lower risk of myocardial infarction, whereas personality features aligned with nervousness were associated with increased risk of myocardial infarction (Dahlén et al., 2022).

As an alternative to the Type A personality, it was proposed that heart disease is more tightly linked to the distressed (Type D) personality (Denollet et al., 2010). This personality type was characterized as being particularly attentive to negative stimuli and tending toward the expression of negative emotions, such as depressed mood, anxiety, anger, hostile feelings, worrying, and high stressor reactivity. Type D individuals are socially immature in the sense that they are uncomfortable with strangers, generally tense and insecure with other people, and they typically inhibit self-expression in social interactions (Kupper and Denollet, 2018). Thus, they may not be equipped with the social coping needed to deal with many life stressors. Individuals with a Type D personality often engage in unhealthy lifestyles (limited physical exercise, smoking tobacco, elevated body mass), and seem to be less able to adopt strategies to enhance their well-being.

Several heart disturbances have been identified among individuals with a Type D personality. In a community sample, ventricular arrhythmias were more common among Type D individuals, and disturbed endothelial functioning was likewise observed (Denollet et al., 2018). Elevated presence of coronary artery plaques was similarly associated with the presence of type D personality, especially among individuals with the social inhibition component of this personality type (Compare et al., 2014). The neurobiological processes associated with the type D individual have not been extensively assessed, although levels of TNF-α and CRP were elevated. While the type D personality predicts the development of heart disease, insufficient data are available concerning the mediating role of inflammatory factors underlying this linkage.

Sex Differences in Heart Disease

Relative to men of the same age, premenopausal women are less apt to develop heart disease, but they catch up after menopause. Estrogen provides protection against heart disease; however, when estrogen levels decline following menopause, or in association with menstrual irregularities, the risk for CAD increases. As in men, CAD was less common among women employed in administrative positions than among those in the lower ranks. While job satisfaction could act as a buffer to diminish life stressors that might otherwise be detrimental to heart health, when women faced the additional load of work at home and taking care of a family, CAD risk was elevated. Women also display greater neuroendocrine and inflammatory changes in response to stressors, which may favor the occurrence of heart disease. In fact, it was suggested that stressors encountered by the fetus during pregnancy may influence the sex-dependent development of both heart disease and depression, likely owing to hormonal, immune, and metabolic processes that are affected throughout life (Goldstein et al., 2019).

Knowing the symptoms of a heart attack is obviously important to survive such an event. Many of these symptoms are the same in men and women, but there are also some very marked differences. Chest pain is common in both sexes, but among women, indigestion, nausea and vomiting, fatigue, dizziness, and back pain are more common than in men. These symptoms are often misidentified, leading to women taking longer to get to a hospital, making for poorer outcomes.

Sleep Disturbances, Inflammation, and Cardiovascular Disease

Cross-sectional and prospective studies indicated that both short and long sleep durations were frequently linked to hypertension, CAD, and stroke, although CAD was more closely tied to short sleep durations and poor sleep quality (Lao et al., 2018). A prospective study over 6 years indicated that very short sleep duration (<5 hours) was associated with CHD (Sadabadi et al., 2023). This is a relatively short time frame, and such effects would likely be more prominent over a longer time frame. Indeed, the negative consequences of sleep disturbances are long-lasting in that sleep problems during adolescence were predictive of heart disease in adulthood.

The effects of poor sleep on heart disease might be secondary to several effects of short sleep duration, such as elevated blood pressure, obesity, metabolic syndrome, diabetes, and depression. As well, immune and cytokine disturbances associated with sleep disruption have been associated with hypertension, CAD, and all-cause mortality (Azevedo Da Silva et al., 2014). Furthermore, as disturbed sleep is common among depressed individuals, frequently accompanied by elevated levels of inflammatory factors, this could contribute to the comorbidity between heart disease and depression. Obtaining restful sleep can enhance immune functioning by facilitating the migration of immune cells to lymph nodes where they meet antigens, allowing the immune cells to be trained so that their potential can subsequently be realized (Martínez-Albert et al., 2024).

Microbiota and Heart Disease

Given the links between microbial factors and inflammatory processes, it has been considered that the microbiome contributes to the emergence of hypertension and CAD, just as this was evident in relation to other illnesses. This perspective was bolstered by the knowledge that the presence of excess body fat and microbial dysbiosis are risk factors for cardiometabolic diseases (Aron-Wisnewsky and Clement, 2016), possibly involving neuroendocrine processes and their effects on immune functioning. Consistent evidence has emerged in line with this view, particularly in the relation between disease conditions and the presence of trimethylamine N-oxide (TMAO), a metabolite derived from gut microbes. Among mice maintained on a Western diet, the body weight increase that occurred was accompanied by dyslipidemia and elevated TMAO levels. In these mice, blood pressure was tied to elevations of several proinflammatory cytokines and reduced levels of antiinflammatory cytokines. Many of these effects were diminished if mice received a TMAO inhibitor, pointing to the causal connection between TMAO and heart problems (Chen et al., 2017b). Furthermore, in mice that lacked a gene for the enzyme, flavin-containing monooxygenase 3 (FMO3), which is responsible for the conversion of TMAO to its active form, obesity was not present, and could thus limit the occurrence of CAD. Among germ-free mice and following antibiotic-induced suppression of gut microbiota, TMAO production was reduced, and atherosclerosis was less likely to occur (Liu and Dai, 2020). Likewise, reducing TMAO levels through diet could diminish the development of plaque in mice susceptible to atherosclerosis.

Among patients with stable CAD, elevated TMAO levels predicted greater mortality risk and could potentially serve as a biomarker for disease progression (Zhang et al., 2021). By generating TMAO, gut microbiota may directly contribute to both platelet hyperreactivity and enhanced potential for thrombosis, as well as the occurrence of heart failure (Anderson et al., 2022). As elevated TMAO is present in association with heart disease and obesity, it was thought that TMAO and related metabolic pathways were responsible for their comorbidity (Tang and Hazen, 2017). Type 2 diabetes was also associated with high TMAO levels (Dambrova et al., 2016), raising the possibility that TMAO might be a link between obesity and diabetes, which can increase the risk of CAD.

Gum disease in relation to heart disease

Microbiota balances in the mouth ought to favor good health, whereas an oral dysbiotic milieu may be associated with pathological outcomes (Tuganbaev et al., 2022). This was prominent in the presence of inflammatory gum (periodontal) disease, which has been associated with later depression, metabolic disturbances, and a 2–3-fold increase in heart disease (Larvin et al., 2021). A meta-analysis that included 26 studies indicated that gum disease was highly comorbid with hypertension and somewhat less with CAD. Stroke and heart failure were likewise associated with gum disease, although these relations were relatively small (Leng et al., 2023). It appeared that the link between gum disease and hypertension was mediated by elevated inflammation reflected by high CRP levels (Muñoz Aguilera et al., 2021). Despite these associations, linking gum disease and heart conditions needs to be considered cautiously given that the disorders share other features, including aging and smoking habits, which might account for the comorbidity.

Preventing and Treating Heart Disease

Given the influence of inflammatory factors in the development of heart disease, antiinflammatory manipulations have been proposed as a preventive treatment. By diminishing systemic inflammation, regular exercise and certain diets can have marked positive effects on heart health. Randomized controlled trials indicated that the antiinflammatory effects of omega-3 reduced the incidence of major adverse cardiac events, and regardless of the stage of CAD, supplementation with omega-3 fatty acids prevented myocardial infarction (Shen et al., 2022a). Pooled data from 19 studies across several countries indicated that seafood and plant-derived omega-3 fatty acids were linked to moderately lower incidence of fatal CAD (Del Gobbo et al., 2016).

Unlike the influence of foods containing omega-3, a meta-analysis indicated that omega-3 fatty acid supplements were unrelated to a decline in coronary disease and other major vascular events (e.g., Aung et al., 2018). Similarly, neither eicosapentaenoic acid (EPA), docosahexaenoic acid (DHA), nor alpha-linolenic acid (ALA) supplements provided benefits for cardiovascular health or mortality (Abdelhamid et al., 2020). The results of these studies are clear; however, the analyses included individuals who had been consuming omega-3 supplements for only 1 or 2 years, which may be insufficient to determine the effects on an illness that develops over many years. Furthermore, in some instances, studies did not exclude individuals with current health problems, including heart disease.

Statins have often been prescribed to reduce LDL cholesterol levels, particularly among older individuals and those with diabetes. Among individuals with elevated LDL levels, but without any apparent heart problems, statins reduced the occurrence of heart disease by about 25% when reassessed in a 20-year follow-up (Vallejo-Vaz et al., 2017). A meta-analysis of 28 randomized trials indicated that statin therapy reduced the likelihood of a stroke, heart attack, and mortality linked to CAD by about 20% (Cholesterol Treatment Trialists' Collaboration, 2019). The JUPITER trials similarly indicated that statin use was accompanied by a marked decrease in surgery for blocked arteries, and diminished risk for heart attack and stroke (Ridker et al., 2017).

Despite the enthusiasm for statin use, controversy exists regarding their usefulness. While statins may have statistically significant benefits, in absolute terms the number of individuals that gain from their use is actually very small (Byrne et al., 2022). Moreover, statin use in limiting heart disease among women was limited, and no benefits were accrued in individuals over 65 years of age. The debate concerning statin use has not been resolved, and in some respects has escalated, which has resulted in many people choosing not to take statins for fear of side effects, even though the benefits of statins outweighed the harms that could be provoked (Cai et al., 2021). It was maintained that rosuvastatin and atorvastatin are superior to other statins in reducing harmful cholesterol but there is still the need to determine who would benefit most from one treatment over another. The American College of Cardiology and the American Heart Association, perhaps to stay on the safe side, continue to recommend the use of statins to limit or prevent cardiovascular disease among individuals with high LDL cholesterol levels.

Several effective nonstatin alternative treatments were developed to diminish cholesterol, while some are still in preclinical analyses. One of the more interesting methods of dealing with this comes from the finding that two proteins, Aster B and Aster C, are fundamental in transporting cholesterol from intestinal membranes that line the gut, eventually affecting its circulation. Through a cholesterol-lowering drug, Ezetimibe, the ability of Aster B and Aster C is blocked and hence might have the capacity to limit atherosclerosis (Ferrari et al., 2023).

Once hypertension has developed, individuals frequently rely on medications to control blood pressure. Years ago, these comprised norepinephrine β-receptor blockers but were since replaced by angiotensin-converting-enzyme (ACE) inhibitors that acted on the hormone angiotensin I, as well as by angiotensin II receptor blockers (ARBs). Unfortunately, a significant portion of patients are resistant to treatment, especially older individuals (Noubiap et al., 2019). Another approach to limit heart disease has involved the delivery of nanoparticles that course through the bloodstream breaking plaques apart. These nanoparticles need a target so that they can act most expeditiously, and as macrophages and other immune cells typically gather at the site of plaque formation, nanoparticles have been created that bind to molecules present on the surface of macrophages, thereby reducing plaque. Related approaches have been developed to repair heart tissue following heart attack and for early detection and treatment of CAD. Irrespective of the illness being considered, it is certain that no treatment is effective for everyone. Advances in imaging and the development of biomarkers may allow patient stratification so that therapies can be administered on a personalized basis.

Stroke

Stroke comes in several forms that vary in severity and the long-term effects provoked. Transient ischemic attack (TIA) reflects a brief interruption of blood to some aspect of the brain, leading to symptoms much like those that accompany more serious stroke. However, a TIA typically does not have permanent functional effects, although it might be a risk factor for more serious stroke. A TIA should be distinguished from a silent cerebral infarct, more commonly referred to as a silent stroke, which may have lasting cognitive and mood-altering effects. Multiple events of this nature may create damage to cells, and among individuals with cerebrovascular problems, lasting cognitive disturbances may occur (Summers et al., 2017).

Hemorrhagic and ischemic are the two most serious forms of major stroke. The hemorrhagic type, which makes up about 10%–15% of cases, comprises a blood vessel rupture so that bleeding occurs directly into the brain or the subarachnoid space surrounding brain tissue. This form of stroke, which has been associated with hypertension, is generally not treatable, but it was observed that IL-27 may have the effect of shifting neutrophils from causing damage to brain functioning to that of promoting recovery. Thus, by harnessing neutrophils

appropriately, it might be possible to treat hemorrhagic stroke patients (Zhao et al., 2017a), although this does not imply anything like a return to normal functioning. As the release of cytokines and chemokines from microglia following hemorrhagic stroke can worsen outcomes, the possibility was considered that treatments to diminish these inflammatory responses may provide therapeutic benefits (Xiao et al., 2020).

Ischemic stroke, which is the most common type of stroke, occurs because of blood vessel blockage owing to a thrombosis or arterial embolism (clot, fat globule, or gas bubble), or because of cerebral hypoperfusion (general reduction of blood supply). If "clot busters," such as tPA (tissue plasminogen activator), are administered within the "golden hours," usually considered to be within 3 h following a stroke, the damage to the brain might be limited (some positive effects can be obtained up to 4.5 h following stroke occurrence). Because of its antiinflammatory properties, aspirin use soon after a minor stroke can reduce the risk of a major stroke later. Predictably, in rodents, agents that limit the inflammatory response, such as an IL-1 receptor antagonist (IL-1Ra) facilitated recovery from stroke and encouraged neurogenesis.

Stroke and Depression

Just as depression predicted later CHD, this condition was also associated with subsequent stroke occurrence, possibly being related to inflammatory factors or midbrain microstructural changes (Sun et al., 2016b). It is also clear, not surprisingly, that depression frequently occurs following a stroke, especially among individuals who had previously experienced a major depressive episode. Likewise, various anxiety disorders have been reported following a stroke. Aside from the psychic drain created by depression in stroke patients, its occurrence signals a poor prognosis for functional recovery and future global functioning, as well as being predictive of further strokes and increased mortality over the ensuing few years (Robinson and Jorge, 2016). Antidepressant treatments reduced the incidence of poststroke depression (Zhou et al., 2020), and when administered early, both physical and cognitive recovery from stroke was somewhat enhanced (Robinson and Jorge, 2016).

The damaging effects that follow stroke can go on for some time owing to the excessive release of the excitatory neurotransmitter glutamate causing cell death (Shen et al., 2022b). Delayed cell death has been the focus of efforts to mitigate the infarct size following stroke. Treatments that modify inflammatory cell infiltration, limit excessive glutamatergic signaling, and buffer mitochondrial energetic functioning, are all potential strategies for achieving this goal. Additionally, as in the case of damage to neurons elicited by other traumatic events, microglial release of cytokines increases following stroke. The cytokines might ordinarily operate in a beneficial capacity, helping in the clearance of debris, diminishing infection, and promoting cell growth. However, the build-up of cytokines within a brief poststroke period may act in a neurodestructive capacity. Specific inflammatory cytokines could be fundamental in accounting for disturbed functional recovery among individuals who experience poststroke depression and could serve as predictors of later stroke and mortality.

Following a stroke, astrocytes and neural stem/progenitor cells (NSPCs) could be fundamental for healing and recovery by modulating inflammation and facilitating the formation of a new blood supply (angiogenesis) (Lindvall and Kokaia, 2015). It seems, however, that following stroke the proinflammatory M1 microglial subtype predominates over the M2 that secretes antiinflammatory cytokines (Zhao et al., 2017b), which may result in incomplete functional recovery. Accordingly, developing antiinflammatory treatments that home-in on injured brain sites could promote improvement after stroke.

This brings us to still other alternative forms of treatment that have been in the cards for some time. In animals with brain damage provoked by stroke, the administration of transplanted human stem cells together with a protein, 3K3A-APC, that spurs the transplanted cells to become functional neurons, had beneficial actions (Wang et al., 2016c). Administration of naturally occurring complement C3a peptide may similarly be useful in stroke recovery as it can increase nerve growth and synapse formation, although to date the research has been restricted to mice (Stokowska et al., 2023).

As gut microbiota alterations can affect inflammatory processes, there is reason to expect that treatments that influence microbial functioning (e.g., diets that increase SCFAs) might have positive effects in attenuating the damage associated with stroke (Peh et al., 2022). This is not to say that gut bacteria are in some fashion related to the primary effects of stroke, but only that the poststroke period involves dynamic inflammatory changes, which could be affected by microbial processes, thereby influencing recovery.

Taking advantage of synaptic plasticity in reversing the effects of stroke

Neuroplasticity is an important element in fostering functional recovery following a stroke. Activity-dependent rewiring and strengthening of synapses can be encouraged by various behavioral manipulations. Soon after a stroke has occurred, various gene expression and protein changes occur, including increased production of growth factors (such as BDNF, insulin-like growth factor, and nerve growth factor, which augment synaptogenesis) and cytokine activity, which could be acting in a reparative capacity. Conversely, low levels of BDNF soon after stroke occurrence predicted the subsequent development of depression (Xu et al., 2018a).

Stroke rehabilitation treatment during the first few months following a stroke might be helpful by encouraging synaptic plasticity. Through the engagement of active behaviors during this time, synapses controlling the relearning of various skills are augmented, thereby facilitating recovery. If partial neuronal functioning is present, appreciable restoration of neural circuit activity can be achieved, likely owing to the establishment and strengthening of compensatory rewiring and remapping of the neural circuitry by activation of growth factors. In this respect, aerobic exercise and environmental enrichment may have significant rehabilitative effects through the neuroplastic changes provoked. While aerobic exercise facilitates motor recovery, acts against depression, and limits degradation that occurs with normal aging, more can be done to encourage recovery from stroke, especially the disturbed cognitive impairments. Multimodal approaches have been adopted to attenuate the consequences of stroke, including transcranial direct current brain stimulation (Allman et al., 2016), exercise, maintaining an antiinflammatory (Mediterranean) diet, and environmental enrichment, the use of video games, and virtual reality-based exercise. Although optimal effects of rehabilitation procedures are realized if these procedures occur within 2–3 months of stroke, it was suggested that these effects can be enhanced using broader therapy that encourages synaptic plasticity, and some recuperation can still occur for some time following stroke (Corbett et al., 2017).

Diabetes

Diabetes has been associated with the development of heart disease and several other illnesses. Diabetes comes in several forms, namely, type 1, type 2, and gestational diabetes. The most obvious symptoms of diabetes comprise elevated thirst and hunger, frequent urination, weight loss, and elevated blood sugar levels. As glucose levels vary over the day, as well as from day to day, diagnosis of diabetes relies more heavily on glycated hemoglobin (A1c) levels, which reflects the average blood sugar level over the preceding 3 months.

Gestational Diabetes

This form of diabetes is typically identified through health screening during pregnancy and thus may go undiagnosed among women who do not regularly visit an obstetrician. Gestational diabetes occurs in about 9% of pregnant women and can ultimately lead to preeclampsia (high blood pressure and high levels of protein in the urine) if the condition is not treated. Women who developed gestational diabetes were also more likely to display symptoms of postpartum depression. The risk of developing this form of diabetes was elevated among women who were depressed during the initial two trimesters of pregnancy. Moreover, having obesity before pregnancy and weight gain during pregnancy was associated with an increased likelihood of gestational diabetes developing, and was associated with type 2 diabetes occurrence years later (Chen et al., 2021b), as well as early mortality (Hinkle et al., 2023). Related to the influence of weight changes during pregnancy, gut microbiome composition during weeks 24–28 of pregnancy could predict gestational diabetes (Guo et al., 2023d).

Type 1 Diabetes

Type 1 diabetes, which typically appears in the young (peaking at about 14 years of age), is an autoimmune disorder, possibly brought about by genetic factors, dietary influences, chemical agents, and viruses. This condition develops owing to the dysfunction of pancreatic beta cells so that insulin is not produced, which is necessary for cells to take up sugars from the blood. In the absence of insulin, cells do not receive adequate nutrition, and elevated sugars in the blood cause damage to cells of various organs. Among its other functions, insulin ordinarily is involved in the elimination of defective mitochondria. However, interruption of insulin signaling may result in mitochondrial recycling being disturbed so that old power generators continue to be used, including those that may be damaged. Ultimately, this could influence aging processes and impaired mitochondrial functioning within neurons, which may favor the development of neurological disorders (Hees et al., 2024).

To preserve the health of normal cells, the immune system increases the presence of immune checkpoints, which either turn immune signals up or down, thereby preventing T-cells from attacking normal tissues. Checkpoints are diminished in patients with type 1 diabetes, and in a mouse model, alterations of the PD-1/PD-L1 (best known in relation to cancer immunotherapy) in pancreatic cells can influence type 1 diabetes (Falcone and Fousteri, 2020). The onset of type 1 diabetes may also be preceded by modification of certain immune cells, namely, MAIT lymphocytes, which can recognize and respond to microbiota. These lymphocytes, which are ordinarily involved in mucosal homeostasis, may be undermined so that bacterial entry into (and out of) the gut is increased, leading to an autoimmune response (Rouxel et al., 2017).

The mainstay for the treatment of type 1 diabetes is insulin administration, which allows for glucose to be taken up into cells and concurrently reduces inflammation by affecting the NLRP inflammasome (Chang et al., 2021). A cure will hopefully be found so that exogenous insulin administration will not be needed. For instance, an individual's stem cells could potentially be modified so that they begin to make insulin. Encapsulated stem cell-derived beta cells can be transplanted, allowing for insulin release, without these cells being attacked by the immune system, which would occur in the absence of the stem cells being encapsulated (Keymeulen et al., 2023).

Type 2 Diabetes

The prevalence of type 2 is about 6%–11% of the global adult population, most often appearing as people age, but may appear in younger individuals, and has been increasing appreciably year over year. Regrettably, individuals with diabetes die approximately 6 years earlier than those without diabetes, and among those who developed diabetes relatively early (e.g., before 50 years of age), lifespan was still more abbreviated (Emerging Risk Factors Collaboration et al., 2023).

Even though metabolic syndrome is a frequent precursor of type 2 diabetes, it is curious that it seems to be taken less seriously than it should. This condition is diagnosed when insulin resistance (i.e., cells have become less responsive to insulin) is accompanied by at least three additional features, such as abdominal obesity, high blood pressure, elevated fasting blood glucose, decreased HDL cholesterol, and high triglycerides. This syndrome may also be accompanied by the upregulation of proinflammatory cytokines stemming from the presence of elevated visceral fat. Prediabetes and early diabetes might be managed by diet, exercise, and maintaining low weight. However, as the illness continues, a variety of medications are necessary to keep blood sugars within a normal range, but diabetes is a hard beast to conquer, especially if patients are resistant to advice.

Diabetes and Its Comorbidities

Diabetes is a wicked illness that engenders a constellation of secondary disturbances that comprise circulatory problems, neuropathy, blindness, and impaired wound healing that can lead to lower limb amputations. And, as we have already seen, it is also highly comorbid with a broad range of other illnesses and conditions, most notably heart disease, stroke, depression, kidney disease, fatty liver disease, immune-related disorders, and some types of cancer. Many of these conditions have also been linked to dyslipidemia (elevated plasma cholesterol and triglycerides), as well as increased inflammatory activity, oxidative stress, endothelial disturbances, and hypercoagulability, which are all correlated with one another. These factors may come together to produce microvascular and macrovascular complications, leading to nephrological and cardiovascular disturbances, which can be predicted by endothelial, inflammatory, and procoagulant biomarkers (Domingueti et al., 2016).

That is hardly the end of it: Individuals with diabetes may also experience reduced volume of temporal, prefrontoparietal, motor, and occipital cortices (Yoon et al., 2017). The age-related decline in the volume of the ventral tegmentum was exacerbated among individuals with type 2 diabetes, which was also tied to cognitive decline, especially among patients who had been diabetic for longer times (Antal et al., 2022). Brain atrophy and the accompanying cognitive decline were identified as a risk factor for the development of dementia, primarily in women. Among older individuals, the cognitive deficits associated with type 2 diabetes may be related to vascular disturbances, as well as to oxidative stress and inflammation stemming from gut microbial dysbiosis (Xu et al., 2017b).

When more than a single comorbid illness is present, they additively or synergistically promote further health disturbances. For instance, when depression accompanies diabetes, the risk of dying prematurely because of cardiovascular illness is appreciably elevated (Farooqi et al., 2019). This may occur owing to the conjoint biological actions of these different diseases, as well as the distress they create. The presence of stress-reactive cytokine growth differentiation factor-15 (GDF-15), which contributes to the regulation of inflammatory responses, is elevated in a wide number of diseases that involve cardiovascular functioning. Thus, its increase among individuals with type 2 diabetes could serve as a useful biomarker in predicting the development of diabetes-related heart disease (Berezin, 2016). Recent evidence has also implicated a metabolic protein, mitochondrial pyruvate carrier (MPC) as being involved in comorbidities associated with type 2 diabetes. This protein is involved in cellular energy through glycolysis, but when metabolic functioning is dysregulated, inflammation and excessive immune activation can be provoked, thereby promoting other inflammatory-related diseases, and may have contributed to the severe responses to SARS-CoV-2 infection (Zhu et al., 2023a).

Given the dangers associated with diabetes, it is unsettling that both type 1 and type 2 diabetes in youth have been increasing progressively since 2002. Type 1 diabetes increased yearly by about 2% and type 2 diabetes that was associated with obesity increased by 5% annually. These rates were still greater among African and Asian Americans and Hispanic people (Bloomgarden and Rapaport, 2023).

Factors That Favor the Development of Type 2 Diabetes

Multiple interlinking factors contribute to the development of type 2 diabetes and the response to treatments

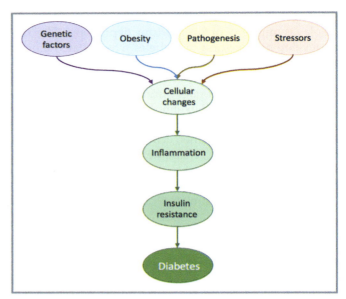

FIG. 17.7 Genetic factors, diet, obesity, pathogenesis (immune and microbial factors) factors, and stressors, come together to favor inflammatory changes that promote insulin resistance and diabetes.

(Fig. 17.7). In addition to the influence of lifestyles, chronic viral infection has been associated with the occurrence of type 2 diabetes (Turk Wensveen et al., 2021). It is well documented that type 2 diabetes was tied to worse outcomes in response to SARS-CoV-2 infection, and it also appeared that this infection was associated with an appreciable increase in the incidence of diabetes (Naveed et al., 2023).

Genetic Contributions

Many genes have been identified that are linked to dysfunction of pancreatic beta cells (Dooley et al., 2016), and numerous genes (possibly exceeding 150) have been associated with type 2 diabetes. Other factors linked to diabetes, such as obesity, may similarly be regulated by genetic factors, thereby indirectly contributing to the development of this disorder. Based on an analysis of 3,000 genes that distinguished healthy controls from those with type 2 diabetes, 168 major hubs (key genes with multiple links) seemed to be particularly germane to this disease, and certain core genes (e.g., SOX5) appeared to be primary contributing factors (Axelsson et al., 2017). A meta-analysis of genome-wide association studies identified 139 common variants, as well as four rare variants, of which 33 genes were thought to be functionally related to type 2 diabetes (Xue et al., 2018).

Although it is often thought that heart disease is a consequence of diabetes, the possibility was explored that these illnesses might share common genetic predictors. In line with this supposition, a GWAS involving more than 260,000 participants revealed 16 previously unreported genetic links between these illnesses. Of these, eight genes were tied to elevated diabetes, which also conferred elevated risk for heart disease, and at least two of these genes were implicated as targets for treating both diseases (Zhao et al., 2017b). It also seems that when certain genes lose function, as in the case of SLC30A8, which is involved in insulin secretion, a protective effect is evident (Flannick et al., 2014). If this gene's expression could be manipulated or the protein that it produces altered, this could potentially serve as a way of managing diabetes.

Numerous tissue-specific epigenetic changes that could influence metabolic syndrome and diabetes have been identified, including within pancreatic islet cells, skeletal muscle, adipose tissue, and the liver (Ling and Rönn, 2019). Such effects were also detected in a gene coding for "G-protein pathway suppressor 2" (GPS2) that influences macrophage functioning within fat cells, which may be linked to the development of type 2 diabetes. Consistent with this suggestion, genetically engineered mice lacking GPS2 in macrophages, displayed inflammation of adipose tissue and insulin resistance developed upon mice being fed a high-fat diet (Fan et al., 2016).

A review of this topic pointed to epigenetic processes being linked to metabolic traits and type 2 diabetes. This was apparent in animal studies and indications of this were apparent in humans so that blood-based markers might be amenable to a precision medicine-based treatment approach as well as to predict the possibility of complications (Ling et al., 2022). Noncoding RNA (ncRNA) and epigenetic regulation may also contribute to obesity and type 2 diabetes, raising the possibility of targeting these processes to achieve therapeutic benefits.

Impact of Stressors

It is unlikely that stressors cause type 1 (autoimmune) diabetes, although serious stressful events in childhood were accompanied by a tripling in the risk for this condition (Nygren et al., 2015). To be sure, the data in this domain are sparse, but it does seem that chronic stressful experiences can exacerbate symptoms once diabetes is present. The case for stressors in the development of metabolic disorders and type 2 diabetes is much stronger. A longitudinal analysis indicated that moderate or high stressor experiences were associated with a 2.3-fold increase in the risk for type 2 diabetes over a 3-year period (Harris et al., 2017b). As expected, diabetes was tied to job-related stressors, such as burnout, occupational class, and chronic distress at work or home (Xu et al., 2018b). As in the case of heart disease, type 2 diabetes was associated with high job strain and the perceived mismatch between efforts expended and rewards received (Li et al., 2021b).

Consistent with the involvement of stressors and depression in the provocation and aggravation of type 2 diabetes, the availability of effective coping strategies, including social support, could act against the development of diabetes and augment treatment adherence and diabetes control. In concert with these findings, the use of effective coping strategies together with high self-efficacy, self-esteem, and optimism, was accompanied by better glucose control. Likewise, mindfulness training modestly reduced A1c levels and reduced diabetes-related distress (Ni et al., 2021). It is of obvious practical significance that among diabetic individuals over 60 years of age who had been diagnosed with depression, treating the mood condition was accompanied by reduced symptoms related to other illnesses.

As type 2 diabetes takes years to develop, it is difficult to determine whether stressful life experiences during certain periods were critical for its appearance. Nonetheless, early life stressors could potentially contribute to the subsequent emergence of this condition, even after adjusting for socioeconomic status and obesity in adulthood. In this respect, early-life emotional abuse was associated with an elevated occurrence of type 2 diabetes, which was mediated by the occurrence of depression that had been linked to the abuse (Atasoy et al., 2022). Psychosocial stressors such as experiencing a natural disaster prenatally were likewise associated with insulin resistance in young adults, and the incidence of diabetes was elevated among the children of women who lost a loved one during pregnancy.

Endocrine changes associated with stressors may directly or indirectly contribute to the development of diabetes. Stressors affect hormones associated with food and energy regulation (e.g., ghrelin, leptin, and cortisol), which influence dopamine processes associated with rewarding perception, thereby further encouraging the consumption of tasty snacks rich in carbs (Abizaid et al., 2014). Ironically, insulin insufficiency or insulin resistance can reduce the rewarding feelings derived from comfort foods, and as in the case of drug addictions, individuals might compensate for this by indulging in these foods to a still greater extent to regain the pleasure they previously obtained from these snacks. The net result of these stressor actions, among other things, is the development of visceral fat, which increases the availability of proinflammatory cytokines and instigation of type 2 diabetes.

Inflammatory Immune Factors and Type 2 Diabetes

Elevated leukocyte levels and a chronic increase of proinflammatory cytokines secondary to stressors and increased abdominal fat, can disturb pancreatic beta cell functioning, leading to insulin resistance and thus metabolic syndrome and diabetes (Alfadul et al., 2022). A meta-analysis indicated that the development of diabetes was linked to elevated circulating cytokines, particularly IL-6 levels (Bowker et al., 2020), and treatments that inhibit IL-1β, TNF-α, and NF-κB, can enhance the secretion of insulin from isolated islet cells (Nordmann et al., 2017). Importantly, a review of 16 randomized controlled trials indicated that NSAIDs reduced the signs of type 2 diabetes (Li et al., 2023a), and inflammatory reductions related to both diet and exercise were linked to the reduced development of type 2 diabetes (Yang et al., 2023b). Moreover, omega-3 treatment was found to be accompanied by a decrease in CRP and may have positive effects in type 2 diabetes, possibly through actions on inflammatory processes (Forouhi et al., 2016).

Diabetes and heart disease, as mentioned earlier, may be linked through the presence of inflammatory factors. Type 2 diabetes is accompanied by elevated cholesterol (and triglycerides) that drive monocytes to form macrophages. The macrophages engulf lipids (cholesterol and triglycerides) that have adhered to artery walls, and the build-up of this complex contributes to atherosclerosis. These actions are helped along the way by a form of protein kinase C, which contributes to both macrovascular and microvascular complications. However, one form of protein kinase C, specifically protein kinase Cδ, may defend against inflammation. Thus, by selectively manipulating this form of PKC, it may be possible to modify atherosclerosis (Li et al., 2017b).

Obesity and Diabetes

Adiposity, as we have seen, is accompanied by elevated inflammatory stores, so when individuals with obesity encounter stressors, exaggerated inflammatory responses occur, potentially leading to type 2 diabetes and depression, as well as contributing to heart disease and other vascular illnesses. In this respect, depression appears to be causally related to type 2 diabetes (the converse direction was not significant), seemingly mediated by increased body mass index (Piaserico et al., 2023). Given these linkages, prevention or treatment of diabetes could be expected by acting on these processes. For example, the generation of healthy blood vessels within adipose tissues could be achieved by vascular endothelial growth factor B (VEGFB), and blocking the VEGF receptor-1 could result in reduced inflammation and enhanced insulin functioning (Robciuc et al., 2016). As well, through actions on hypothalamic neural stem cells and synaptic plasticity, FGF1 might ultimately be developed as a method of controlling glucose levels (Gasser et al., 2017). It also appeared that the continued release of the inflammatory substance leukotriene B_4 (LTB4) from visceral fat can promote metabolic disturbances. Significantly, improved metabolic health could be achieved among obese mice that were treated with an LTB4 receptor antagonist, which was initially developed to treat inflammatory disease (Li et al., 2015b).

Environmental toxicants may contribute to type 2 diabetes

Increasing evidence has revealed a role for environmental toxicants (e.g., arsenic and cadmium) in the development of type 2 diabetes. Pesticides and endocrine-disrupting chemicals (e.g., bisphenol A and phthalates) increase the risk for diabetes by altering beta cell metabolism through actions on estrogen receptor signaling (Bonini and Sargis, 2018). In this respect, fetal exposure to bisphenol A may have long-term repercussions, including the development of type 2 diabetes (Farrugia et al., 2021). It has likewise been suggested that "forever chemicals" comprising per- and polyfluoroalkyl substances (PFAS) that are present in numerous consumer products (nonstick coatings, various fabrics, and food packaging) may contribute to diabetes (Roth and Petriello, 2022), possibly by increasing inflammation and oxidative stress. Furthermore, numerous studies across several countries have indicated that elevated fine particulate matter ($PM_{2.5}$) and other air pollutants were associated with type 2 diabetes (Ren et al., 2023).

Gut Bacteria Influence the Emergence of Diabetes

In genetically susceptible individuals, gut bacteria could trigger an autoimmune response against beta cells, thereby producing type 1 diabetes. As expected, antibiotic treatment in young mice provoked the equivalent of type 1 diabetes, apparently owing to altered gut microbiota (Livanos et al., 2016). Supporting a causal link between these factors, in female mice genetically at risk for type 1 diabetes, treatment with normal gut bacteria obtained from adult male mice, reduced the incidence of diabetes by 85%. These actions were accompanied by elevated testosterone, reduced islet inflammation, and diminished autoantibody production, pointing to the importance of sex hormones in collaboration with microbiota in the production of type 1 diabetes (Markle et al., 2013).

Considerable data have supported a causal link between microbiota and both metabolic syndrome and type 2 diabetes (Zhou et al., 2022) and microbiota modifications have been proposed as a way of diminishing health disturbances that stem from diabetes. In humans, insulin resistance was associated with increased fecal carbohydrates, such as glucose, fructose, galactose, and mannose as well as elevated abundance of gut bacteria from the *Lachnospiraceae* order. Conversely, insulin resistance and monosaccharide levels were lower among individuals with a greater abundance of Bacteroidales-type bacteria. Importantly, type 2 diabetes was linked to elevated intake of antibiotics in the years preceding the onset of this condition (Mikkelsen et al., 2015), and depending on the microbiota present, fecal microbiota transplants could diminish type 2 diabetes (Ding et al., 2022). Based on such findings, it was maintained that specific microbiota might be an effective predictive marker for the development of type 2 diabetes, and modifying them might preclude diabetes among individuals who are still prediabetic (Takeuchi et al., 2023b).

A high-fat diet may promote gut dysbiosis, which causes mucosal inflammation and chronic systemic inflammation, which then favors the occurrence of type 2 diabetes. Consistent with this view, the administration of an antiinflammatory agent, 5-aminosalicylic acid (5-ASA), often used in the treatment of IBD, acted against insulin resistance that developed among mice that had been maintained on a high-fat diet (Luck et al., 2015). Furthermore, a genetically engineered form of the gut bacteria *Lactobacillus* secretes glucagon-like peptide-1 (GLP-1), which reduces glucose levels among individuals with type 2 diabetes (Duan et al., 2015), again pointing to the gut microbiota linkage to this condition.

Especially interesting findings came after reports that gut microbiota were altered among patients with diabetes and that treatment with the diabetic medication metformin was associated with an increase of several bacterial species that were tied to improved metabolism. When microbiota were taken from patients before and after metformin treatment, and then transplanted to mice, the microbiota that had been transformed by metformin led to positive outcomes, suggesting that the benefits of the drug came through microbial processes (Wu et al., 2017a).

Fig. 17.8 depicts some of the pathways linking diet, microbiota, inflammatory processes, endocrine functioning, and brain processes (Calder et al., 2017). These same links may be fundamental in the provocation of type 2 diabetes. Indeed, a meta-analysis that included 68 trials in diabetic patients indicated that supplementation with pro/pre/synbiotics enhanced glucose homeostasis (Paul et al., 2022b). Increasingly, treatments to deal with metabolic syndrome and type 2 diabetes have involved approaches that targeted the microbiome to define specific bacteria that should be altered to treat metabolic disorders.

In addition to beneficial effects that could be gained from prebiotics or probiotics, "postbiotic factors" that are derived from bacteria can also have a positive impact on health, and muramyl dipeptide (MDP), a component of some types of bacteria, lowered adipose inflammation and diminished glucose intolerance in obese mice (Cavallari et al., 2017). As described earlier, dietary supplementation that increases SCFAs may limit weight gain, enhance energy expenditure, and thus augment insulin sensitivity, thereby acting against the development of diabetes. Based on a small clinical trial, it appeared that fecal transplants obtained from a lean donor diminished insulin resistance in about half of the obese men involved in this trial, and the clusters of bacteria that were present in the recipient could predict whether the fecal transplant would be effective (Kootte et al., 2017).

Pharmacological Treatments for Type 2 Diabetes

It would be a reasonable assumption that a great number of prediabetic individuals will say that they can ward off the illness through exercise and a proper diet. That is likely correct in theory, but the reality is another issue entirely. The expression that "The road to hell is paved with good intentions" did not develop for no reason. Ultimately, many individuals have had to rely on medications to manage their diabetes.

Metformin has been an important first-line treatment for type 2 diabetes. If it is not fully effective, additional treatments may be prescribed. Several options are available in this regard, each operating through somewhat different routes. An effective treatment for type 2 diabetes comprises a sodium-glucose cotransporter-2 (SGLT2) inhibitor that increases glucose excretion in urine. As it turns out, an SGLT2 inhibitor (e.g., empagliflozin) was associated with an appreciable reduction of heart failure and death from heart disease (McGuire et al., 2021b), although some agents in this class have been linked to increased risk of toe or foot amputation. Sulfonylureas drugs may be prescribed to increase insulin release from beta cells, or alpha-glucosidase inhibitors can be used to prevent digestion of carbohydrates, and injection of the GLP-1 agonists (e.g., Ozempic and related compounds) markedly diminishes type 2 diabetes and is used to promote weight loss and may have positive effects in limiting heart disease (Richardson et al., 2023).

Concluding Comments

Comorbidities occur between numerous illnesses, being more the rule than the exception. In some instances, one type of illness (or the distress created by this condition) may cause a second illness. Alternatively, illnesses may involve common environmental or experiential factors (e.g., adverse early life events) or they may share underlying biological processes (e.g., the presence of inflammation). Often, research assessing the mechanisms and processes associated with a particular illness makes efforts to limit the contribution of comorbid conditions by statistically controlling for their influence. This is often appropriate, but doing so ignores the possibility that evaluating comorbid illnesses may provide important clues concerning the evolution of the illnesses and their underlying mechanisms.

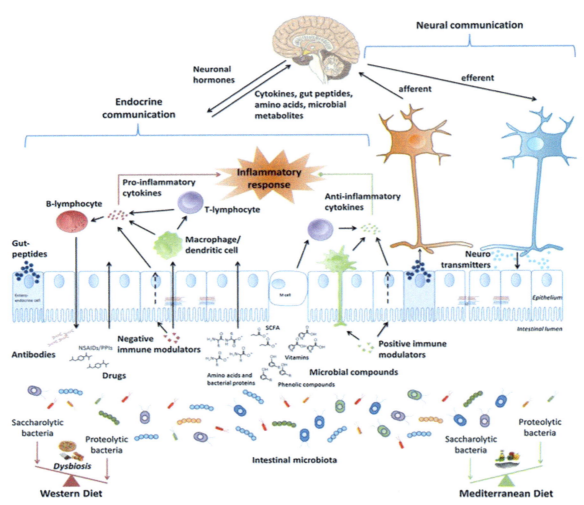

FIG. 17.8 Overview of general mechanisms by which the gut microbiota affects host intestinal epithelium, immune-inflammatory response, and brain functioning. The epithelial layer consists of a single layer of epithelial cells that are sealed by tight junction proteins preventing paracellular passage. The connective tissue close to the epithelial cells (lamina propria) contains a large number of immune cells, both of the innate immune system (e.g., macrophages, dendritic cells, and mast cells) and the adaptive immune system (e.g., T-cells, antibody-producing B-cell-derived plasma cells). In addition, cells of the central and enteric nervous system are innervated in the lamina propria. Factors affecting intestinal barrier function include food-derived allergens, (pathogenic and commensal) bacteria, and microbial compounds [lipopolysaccharides, metabolites such as short chain fatty acids (SCFA), tryptophan-related metabolites, neurotransmitters, and peptides] as well as nonsteroidal antiinflammatory drugs (NSAIDs) and proton pump inhibitors (PPIs). When activated by immune modulators, lymphocytes release antiinflammatory and/or proinflammatory cytokines which trigger or regulate an inflammatory response. In addition, the released cytokines signal the brain to activate immunomodulatory mechanisms like the cholinergic antiinflammatory pathway, the HPA axis as well as the SNS. Furthermore, intestinal neurotransmitters or their precursors can modulate functions of the central nervous system. The neuronal efferent activation may also impact directly the epithelium and the gut microbiota composition. *From Calder, P.C., Bosco, N., Bourdet-Sicard, R., Capuron, L., Delzenne, N., et al., 2017. Health relevance of the modification of low grade inflammation in ageing (inflammageing) and the role of nutrition. Ageing Res. Rev. 40, 95–119.*

Psychological disorders, especially depression and anxiety, are frequently comorbid with multiple illnesses, which is of therapeutic significance and has been important in the prognosis of other conditions (e.g., poststroke depression informing later stroke and early mortality). The comorbid conditions related to depression may involve the activation of specific genes, epigenetic changes, and mechanisms related to inflammation. Irrespective of the processes, type 2 diabetes and heart disease that are linked to depression are among several preventable illnesses that evolve owing to poor lifestyle choices secondary to feelings of depression. The likelihood of type 2 diabetes and heart disease evolving can be diminished by favoring foods low in carbs, exercising regularly, getting enough sleep, and being able to cope effectively with distressing events. However, once these illnesses develop, they are difficult dragons to slay, particularly as many people simply continue with some of the same behaviors that brought them to the ill state. Thus, treatment of diabetes and heart disease ought to involve education, self-management training, social support, and coaching to diminish those factors that interfere with self-control, and a greater focus on mental health.

We focused on the comorbidities involving only a small number of illnesses, but a very great number of comorbidities exist, as we have seen concerning psychiatric and neurodegenerative disorders. Many illnesses share common elements, including specific genetic factors, as well as microbial, inflammatory, and hormonal disturbances, raising the possibility that the shared features may be a common denominator responsible for the emergence of comorbid illnesses. Inflammatory factors may be particularly important in this regard, and it has been suggested that NF-κB, which is involved in cellular responses to a variety of stimuli (e.g., stressors, bacterial and viral challenges, and cytokines) may play a prominent role in the development of disturbed immune functioning, inflammatory illnesses, autoimmune disorders, cancer, responses to viral illnesses, and the evolution of depressive disorders. Many chronic illnesses ought to be subject to modification by treatments that diminish inflammation, and it may be productive to develop a slate of biomarkers to predict the later development of illnesses before they become intractable. The presence of some illnesses may themselves be markers for the development of other illnesses, and it would be propitious for physicians to attend to these to prevent the development of comorbid conditions.

References

Abadier, M., Pramod, A.B., McArdle, S., Marki, A., Fan, Z., et al., 2017. Effector and regulatory T cells roll at high shear stress by inducible tether and sling formation. Cell Rep. 21, 3885–3899.

Abareshi, A., Anaeigoudari, A., Norouzi, F., Shafei, M.N., Boskabady, M.H., et al., 2016. Lipopolysaccharide-induced spatial memory and synaptic plasticity impairment is preventable by Captopril. Adv. Med. 2016, 7676512.

Abbas, A.K., Lichtman, A.H., Pillai, S., 2015. Cellular and Molecular Immunology, eighth ed. Elsevier.

Abbasi, S.H., Hosseini, F., Modabbernia, A., Ashrafi, M., Akhondzadeh, S., 2012. Effect of celecoxib add-on treatment on symptoms and serum IL-6 concentrations in patients with major depressive disorder: randomized double-blind placebo-controlled study. J. Affect. Disord. 141, 308–314.

Abbott, P.W., Gumusoglu, S.B., Bittle, J., Beversdorf, D.Q., Stevens, H.E., 2018. Prenatal stress and genetic risk: how prenatal stress interacts with genetics to alter risk for psychiatric illness. Psychoneuroendocrinology 90, 9–21.

Abdallah, C.G., Sanacora, G., Duman, R.S., Krystal, J.H., 2015. Ketamine and rapid-acting antidepressants: a window into a new neurobiology for mood disorder therapeutics. Annu. Rev. Med. 66, 509–523.

Abdallah, C.G., Averill, L.A., Krystal, J.H., Southwick, S.M., Arnsten, A. F., 2017. Glutamate and norepinephrine interaction: relevance to higher cognitive operations and psychopathology. Behav. Brain Sci. 39, e201.

Abdalqadir, N., Adeli, K., 2022. GLP-1 and GLP-2 orchestrate intestine integrity, gut microbiota, and immune system crosstalk. Microorganisms 10, 2061.

Abdelhamid, A.S., Brown, T.J., Brainard, J.S., Biswas, P., Thorpe, G.C., et al., 2020. Omega-3 fatty acids for the primary and secondary prevention of cardiovascular disease. Cochrane Database Syst. Rev. 3, CD003177.

Aberizk, K., Sefik, E., Addington, J., Anticevic, A., Bearden, C.E., et al., 2023. Hippocampal connectivity with the default mode network is linked to hippocampal volume in the clinical high risk for psychosis syndrome and healthy individuals. Clin. Psychol. Sci. 11, 801–818.

Abildgaard, A., Elfving, B., Hokland, M., Lund, S., Wegener, G., 2017. Probiotic treatment protects against the pro-depressant-like effect of high-fat diet in Flinders Sensitive Line rats. Brain Behav. Immun. 65, 3342.

Abisror, N., Mekinian, A., Lachassinne, E., Nicaise-Roland, P., De Pontual, L., et al., 2013. Autism spectrum disorders in babies born to mothers with antiphospholipid syndrome. Semin. Arthritis Rheum. 43, 348–351.

Abizaid, A., 2009. Ghrelin and dopamine: new insights on the peripheral regulation of appetite. J. Neuroendocrinol. 21, 787–793.

Abizaid, A., Horvath, T.L., 2008. Brain circuits regulating energy homeostasis. Regul. Pept. 149, 3–10.

Abizaid, A., Luheshi, G., Woodside, B.C., 2014. Interaction between immune and energy-balance signals in the regulation of feeding and metabolism. In: Kusnecov, A.V., Anisman, H. (Eds.), The Wiley Blackwell Handbook of Psychoneuroimmunology. John Wiley & Sons Ltd, Chichester, UK, pp. 488–503.

Abramova, O., Zorkina, Y., Ushakova, V., Zubkov, E., Morozova, A., Chekhonin, V., 2020. The role of oxytocin and vasopressin dysfunction in cognitive impairment and mental disorders. Neuropeptides 83, 102079.

Abramson, L.Y., Seligman, M.E., Teasdale, J.D., 1978. Learned helplessness in humans: critique and reformulation. J. Abnorm. Psychol. 87, 49–74.

Achar, A., Ghosh, C., 2020. COVID-19-Associated neurological disorders: the potential route of CNS invasion and blood-brain relevance. Cells 9, 2360.

Achtyes, E., Keaton, S.A., Smart, L., Burmeister, A.R., Heilman, P.L., et al., 2020. Inflammation and kynurenine pathway dysregulation in post-partum women with severe and suicidal depression. Brain Behav. Immun. 83, 239–247.

Acioglu, C., Mirabelli, E., Baykal, A.T., Ni, L., Ratnayake, A., et al., 2016. Toll like receptor 9 antagonism modulates spinal cord neuronal function and survival: direct versus astrocyte-mediated mechanisms. Brain Behav. Immun. 56, 310–324.

Acx, H., Serneels, L., Radaelli, E., Muyldermans, S., Vincke, C., et al., 2017. Inactivation of γ-secretases leads to accumulation of substrates and non-Alzheimer neurodegeneration. EMBO Mol. Med. 9, 1088–1099.

Adam, E.K., Heissel, J.A., Zeiders, K.H., Richeson, J.A., Ross, E.C., et al., 2015. Developmental histories of perceived racial discrimination and diurnal cortisol profiles in adulthood: a 20-year prospective study. Psychoneuroendocrinology 62, 279–291.

Adamo, S.A., Easy, R.H., Kovalko, I., MacDonald, J., McKeen, A., et al., 2017. Predator exposure-induced immunosuppression: trade-off, immune redistribution or immune reconfiguration? J. Exp. Biol. 220, 868–875.

Adams, N.M., Grassmann, S., Sun, J.C., 2020. Clonal expansion of innate and adaptive lymphocytes. Nat. Rev. Immunol. 20 (11), 694–707.

Adamson, A., Buck, S.A., Freyberg, Z., De Miranda, B.R., 2022. Sex differences in dopaminergic vulnerability to environmental toxicants - implications for Parkinson's disease. Curr. Environ. Health Rep. 9, 563–573.

Ade, K.K., Wan, Y., Hamann, H.C., O'Hare, J.K., Guo, W., et al., 2016. Increased mGluR5 signaling underlies OCD-like behavioral and striatal circuit abnormalities in mice. Biol. Psychiatry 80, 522–533.

Ader, R., Felten, D., Cohen, N., 1990. Interactions between the brain and the immune system. Annu. Rev. Pharmacol. Toxicol. 30, 561–602.

Afia, A.B., Vila, È., MacDowell, K.S., Ormazabal, A., Leza, J.C., et al., 2021. Kynurenine pathway in post-mortem prefrontal cortex and cerebellum in schizophrenia: relationship with monoamines and symptomatology. J. Neuroinflammation 18, 198.

Afzal, S., Tybjærg-Hansen, A., Jensen, G.B., Nordestgaard, B.G., 2016. Change in body mass index associated with lowest mortality in Denmark. JAMA 315, 1989–1996.

Aghajani, M., Veer, I.M., van Hoof, M.J., Rombouts, S.A., van der Wee, N.J., et al., 2016. Abnormal functional architecture of amygdala-centered networks in adolescent posttraumatic stress disorder. Hum. Brain Mapp. 37, 1120–1135.

Agilli, M., Aydin, F.N., Kurt, Y.G., Cayci, T., 2015. Assessment of paraoxonase 1 activity in patients with Alzheimer's disease and vascular dementia. Am. J. Alzheimers Dis. Other Dement. 30, 437–438.

Aguilar, P.S., Baylies, M.K., Fleissner, A., Helming, L., Inoue, N., et al., 2013. Genetic basis of cell-cell fusion mechanisms. Trends Genet. 29, 427–437.

Aguilar-Valles, A., Rodrigue, B., Matta-Camacho, E., 2020. Maternal immune activation and the development of dopaminergic neurotransmission of the offspring: relevance for Schizophrenia and other psychoses. Front. Psychiatry 11, 852.

Aguilera, M., Arias, B., Wichers, M., Barrantes-Vidal, N., Moya, J., et al., 2009. Early adversity and 5-HTT/BDNF genes: new evidence of gene environment interactions on depressive symptoms in a general population. Psychol. Med. 39, 1425–1432.

Ahmadizar, F., Vijverberg, S.J.H., Arets, H.G.M., de Boer, A., Garssen, J., et al., 2017. Breastfeeding is associated with a decreased risk of childhood asthma exacerbations later in life. Pediatr. Allergy Immunol. 28, 649–654.

Ahmed, S., Busetti, A., Fotiadou, P., Vincy, J.N., Reid, S., et al., 2019. In vitro characterization of gut microbiota-derived bacterial strains with neuroprotective properties. Front. Cell. Neurosci. 13, 402.

Ahmed-Leitao, F., Spies, G., van den Heuvel, L., Seedat, S., 2016. Hippocampal and amygdala volumes in adults with posttraumatic stress disorder secondary to childhood abuse or maltreatment: a systematic review. Psychiatry Res. Neuroimaging 256, 33–43.

Ahuja, S.K., Manoharan, M.S., Lee, G.C., McKinnon, L.R., Meunier, J.A., et al., 2023. Immune resilience despite inflammatory stress promotes longevity and favorable health outcomes including resistance to infection. Nat. Commun. 14, 3286.

Ai, Y.W., Du, Y., Chen, L., Liu, S.H., Liu, Q.S., Cheng, Y., 2023. Brain inflammatory marker abnormalities in major psychiatric diseases: a systematic review of postmortem brain studies. Mol. Neurobiol. 60, 2116–2134.

Aïdoud, A., Gana, W., Poitau, F., Fougère, B., Angoulvant, D., 2022. Does the influenza A vaccine have a direct atheroprotective effect? Arch. Cardiovasc. Dis. 115, 331–334.

Aizawa, E., Tsuji, H., Asahara, T., Takahashi, T., Teraishi, T., et al., 2016. Possible association of Bifidobacterium and Lactobacillus in the gut microbiota of patients with major depressive disorder. J. Affect. Disord. 202, 254–547.

Akachar, J., Bouricha, E.M., Hakmi, M., Belyamani, L., El Jaoudi, R., Ibrahimi, A., 2020. Identifying epitopes for cluster of differentiation and design of new peptides inhibitors against human SARS-CoV-2 spike RBD by an in-silico approach. Heliyon 6, e05739.

Akbari, E., Asemi, Z., Daneshvar Kakhaki, R., Bahmani, F., Kouchaki, E., et al., 2016. Effect of probiotic supplementation on cognitive function and metabolic status in Alzheimer's disease: a randomized, double-blind and controlled trial. Front. Aging Neurosci. 8, 256.

Akdeniz, C., Tost, H., Streit, F., Haddad, L., Wüst, S., et al., 2014. Neuroimaging evidence for a role of neural social stress processing in ethnic minority–associated environmental risk. JAMA Psychiatry 71, 672–680.

Akiki, T.J., Averill, C.L., Wrocklage, K.M., Schweinsburg, B., Scott, J.C., et al., 2017. The association of PTSD symptom severity with localized hippocampus and amygdala abnormalities. Chronic Stress. https://doi.org/10.1177/2470547017724069.

Akiyama, T., Carstens, E., 2013. Neural processing of itch. Neuroscience 250, 697–714.

Al Jowf, G.I., Ahmed, Z.T., Reijnders, R.A., de Nijs, L., Eijssen, L.M.T., 2023. To predict, prevent, and manage post-traumatic stress disorder (PTSD): a review of pathophysiology, treatment, and biomarkers. Int. J. Mol. Sci. 24, 5238.

Al-Abri, K., Edge, D., Armitage, C.J., 2023. Prevalence and correlates of perinatal depression. Soc. Psychiatry Psychiatr. Epidemiol. 58, 1581–1590.

Alagapan, S., Choi, K.S., Heisig, S., Riva-Posse, P., Crowell, A., et al., 2023. Cingulate dynamics track depression recovery with deep brain stimulation. Nature 622, 130–138.

Alagheband, B.A., Azargoonjahromi, A., Sadraei, S., Aarabi, A., Payandeh, Z., Rajabibazl, M., 2022. An overview of current drugs and prophylactic vaccines for coronavirus disease 2019 (COVID-19). Cell. Mol. Biol. Lett. 27, 38.

Al-Aly, Z., Agarwal, A., Alwan, N., Luyckx, V.A., 2023. Long COVID: long-term health outcomes and implications for policy and research. Nat. Rev. Nephrol. 19, 1–2.

Alamri, A., Pereira, E.A.C., 2022. Deep brain stimulation for chronic pain. Neurosurg. Clin. N. Am. 33, 311–321.

Albaret, M.A., Textoris, J., Dalzon, B., Lambert, J., Linard, M., et al., 2023. HSV-1 cellular model reveals links between aggresome formation and early step of Alzheimer's disease. Transl. Psychiatry 13, 86.

Al-Beltagi, M., Saeed, N.K., Elbeltagi, R., Bediwy, A.S., Aftab, S.A.S., Alhawamdeh, R., 2023. Viruses and autism: a Bi-mutual cause and effect. World J. Virol. 12, 172–192.

Alberio, T., Pippione, A.C., Zibetti, M., Olgiati, S., Cecconi, D., et al., 2012. Discovery and verification of panels of T-lymphocyte proteins as biomarkers of Parkinson's disease. Sci. Rep. 2, 953.

Albert, K.M., Newhouse, P.A., 2019. Estrogen, stress, and depression: cognitive and biological interactions. Annu. Rev. Clin. Psychol. 15, 399–423.

Albert, P.R., Vahid-Ansari, F., Luckhart, C., 2014. Serotonin-prefrontal cortical circuitry in anxiety and depression phenotypes: pivotal role of pre- and post-synaptic 5-HT1A receptor expression. Front. Behav. Neurosci. 8, 199.

Albert, P.R., Le François, B., Vahid-Ansari, F., 2019. Genetic, epigenetic and posttranscriptional mechanisms for treatment of major depression: the 5-HT1A receptor gene as a paradigm. J. Psychiatry Neurosci. 44, 164–176.

Alboni, R.M., van Dijk, S., Poggini, G., Milior, M., Perrotta, T., et al., 2017. Fluoxetine effects on molecular, cellular and behavioral endophenotypes of depression are driven by the living environment. Mol. Psychiatry 22, 552–561.

Albott, C.S., Lim, K.O., Forbes, M.K., Erbes, C., Tye, S.J., et al., 2018. Efficacy, safety, and durability of repeated ketamine infusions for comorbid posttraumatic stress disorder and treatment-resistant depression. J. Clin. Psychiatry 79, 17m11634.

Alda, M., Puebla-Guedea, M., Rodero, B., Demarzo, M., Montero-Marin, J., et al., 2016. Zen meditation, length of telomeres, and the role of experiential avoidance and compassion. Mindfulness 7, 651–659.

Alexander, J.L., Wilson, I.D., Teare, J., Marchesi, J.R., Nicholson, J.K., Kinross, J.M., 2017. Gut microbiota modulation of chemotherapy efficacy and toxicity. Nat. Rev. Gastroenterol. Hepatol. 14, 356–365.

Alexandre, C., Latremoliere, A., Ferreira, A., Miracca, G., Yamamoto, M., et al., 2017. Decreased alertness due to sleep loss increases pain sensitivity in mice. Nat. Med. 23, 768–774.

Alexandros Lalousis, P., Schmaal, L., Wood, S.J., Reniers, R., Cropley, V.L., et al., 2023. Inflammatory subgroups of schizophrenia and their association with brain structure: a semi-supervised machine learning examination of heterogeneity. Brain Behav. Immun. 113, 166–175.

Alfadul, H., Sabico, S., Al-Daghri, N.M., 2022. The role of interleukin-1β in type 2 diabetes mellitus: a systematic review and meta-analysis. Front. Endocrinol. (Lausanne) 13, 901616.

Alhabbab, R., Blair, P., Elgueta, R., Stolarczyk, E., Marks, E., et al., 2015. Diversity of gut microflora is required for the generation of B cell with regulatory properties in a skin graft model. Sci. Rep. 5, 11554.

Al-Hakeim, H.K., Al-Rubaye, H.T., Al-Hadrawi, D.S., Almulla, A.F., Maes, M., 2023. Long-COVID post-viral chronic fatigue and affective symptoms are associated with oxidative damage, lowered antioxidant defenses and inflammation: a proof of concept and mechanism study. Mol. Psychiatry 28, 564–578.

Allen, N.J., Barres, B.A., 2009. Neuroscience: glia—more than just brain glue. Nature 457, 675–677.

Allen, S.J., Watson, J.J., Shoemark, D.K., Barua, N.U., Patel, N.K., 2013. GDNF, NGF and BDNF as therapeutic options for neurodegeneration. Pharmacol. Ther. 138, 155–175.

References

Allen, J.M., Mailing, L.J., Cohrs, J., Salmonson, C., Fryer, J.D., et al., 2017. Exercise training-induced modification of the gut microbiota persists after microbiota colonization and attenuates the response to chemically-induced colitis in gnotobiotic mice. Gut Microbes 1, 1–16.

Allen, J.M., Mailing, L.J., Niemiro, G.M., Moore, R., Cook, M.D., et al., 2018. Exercise alters gut microbiota composition and function in lean and obese humans. Med. Sci. Sports Exerc. 50, 747–757.

Allman, C., Amadi, U., Winkley, A.M., Wilkins, I., Filippini, N., et al., 2016. Ipsilesional anodal tDCS enhances the functional benefits of rehabilitation in patients after stroke. Sci. Transl. Med. 8, 330re1.

Allswede, D.M., Cannon, T.D., 2018. Prenatal inflammation and risk for schizophrenia: a role for immune proteins in neurodevelopment. Dev. Psychopathol. 30, 1157–1178.

Almulla, A.F., Vasupanrajit, A., Tunvirachaisakul, C., Al-Hakeim, H.K., Solmi, M., et al., 2022. The tryptophan catabolite or kynurenine pathway in schizophrenia: meta-analysis reveals dissociations between central, serum, and plasma compartments. Mol. Psychiatry 27, 3679–3691.

Alonso, R., Pisa, D., Marina, A.I., Morato, E., Rábano, A., et al., 2014. Fungal infection in patients with Alzheimer's disease. J. Alzheimers Dis. 41, 301–311.

Alsheikh, A.M., Alsheikh, M.M., 2021. Obsessive-compulsive disorder with rheumatological and inflammatory diseases: a systematic review. Cureus 13, e14791.

Alvarado, S., Tajerian, M., Suderman, M., Machnes, Z., Pierfelice, S., et al., 2015. An epigenetic hypothesis for the genomic memory of pain. Front. Cell. Neurosci. 9, 88.

Alvarez-Sanchez, N., Dunn, S.E., 2023. Potential biological contributers to the sex difference in multiple sclerosis progression. Front. Immunol. 14, 1175874.

Alwani, A., Maziarz, K., Burda, G., Jankowska-Kiełtyka, M., Roman, A., et al., 2023. Investigating the potential effects of α-synuclein aggregation on susceptibility to chronic stress in a mouse Parkinson's disease model. Pharmacol. Rep. 75, 1474–1487.

Amasi-Hartoonian, N., Pariante, C.M., Cattaneo, A., Sforzini, L., 2022. Understanding treatment-resistant depression using "omics" techniques: a systematic review. J. Affect. Disord. 318, 423–455.

Amat, J., Christianson, J.P., Aleksejev, R.M., Kim, J., Richeson, K.R., et al., 2014. Control over a stressor involves the posterior dorsal striatum and the act/outcome circuit. Eur. J. Neurosci. 40, 2352–2358.

Ambeskovic, M., Roseboom, T.J., Metz, G.A.S., 2017. Transgenerational effects of early environmental insults on aging and disease incidence. Neurosci. Biobehav. Rev. pii: S0149-7634(16)30714-X.

Ambrée, O., Ruland, C., Scheu, S., Arolt, V., Alferink, J., 2018. Alterations of the innate immune system in susceptibility and resilience after social defeat stress. Front. Behav. Neurosci. 12, 141.

Ambrée, O., Ruland, C., Zwanzger, P., Klotz, L., Baune, B.T., et al., 2019. Social defeat modulates T helper cell percentages in stress susceptible and resilient mice. Int. J. Mol. Sci. 20, 3512.

Amenyogbe, N., Kollmann, T.R., Ben-Othman, R., 2017. Early-life host–microbiome interphase: the key frontier for immune development. Front. Pediatr. 5, 111.

American Psychiatric Association, 2013. Diagnostic and Statistical Manual of Mental Disorders, fifth ed. American Psychiatric Publishing, Arlington, WA.

American Psychological Association, 2017. Mental Health and Our Changing Climate: Impacts, Implications, and Guidance. [Online]. Available: https://www.apa.org/news/press/releases/2017/03/mental-health-climate.pdf. (Accessed March 2023).

American Psychological Association, 2023. Stress in America 2023. https://www.apa.org/news/press/releases/stress/2023/collective-trauma-recovery. (Accessed January 2024).

Amin, N., Liu, J., Bonnechere, B., MahmoudianDehkordi, S., Arnold, M., et al., 2023. Interplay of metabolome and gut microbiome in individuals with major depressive disorder vs control individuals. JAMA Psychiatry 80, 597–609.

Amini-Khoei, H., Mohammadi-Asl, A., Amiri, S., Hosseini, M.J., Momeny, M., et al., 2017. Oxytocin mitigated the depressive-like behaviors of maternal separation stress through modulating mitochondrial function and neuroinflammation. Prog. Neuro-Psychopharmacol. Biol. Psychiatry 76, 169–178.

Amor, C., Fernández-Maestre, I., Chowdhury, S., Ho, Y.J., Nadella, S., et al., 2024. Prophylactic and long-lasting efficacy of senolytic CAR T cells against age-related metabolic dysfunction. Nat. Aging 4, 336–349.

Amoroso, T., 2015. The psychopharmacology of ±3,4 methylenedioxymethamphetamine and its role in the treatment of posttraumatic stress disorder. J. Psychoactive Drugs 47, 337–344.

Amraei, R., Yin, W., Napoleon, M.A., Suder, E.L., Berrigan, J., et al., 2021. CD209L/L-SIGN and CD209/DC-SIGN act as receptors for SARS-CoV-2. ACS Cent. Sci. 7, 1156–1165.

An, J.H., Jung, J.H., Jeon, H.J., 2023. Association of physical activity with the risk of Parkinson's disease in depressive disorder: a nationwide longitudinal cohort study. J. Psychiatr. Res. 167, 93–99.

Anacker, C., Hen, R., 2017. Adult hippocampal neurogenesis and cognitive flexibility—linking memory and mood. Nat. Rev. Neurosci. 18, 335–346.

Anand, A., Mathew, S.J., Sanacora, G., Murrough, J.W., Goes, F.S., et al., 2023. Ketamine versus ECT for nonpsychotic treatment-resistant major depression. N. Engl. J. Med. 388, 2055–2064.

Ancelin, M.L., Scali, J., Norton, J., Ritchie, K., Dupuy, A.M., et al., 2017. The effect of an adverse psychological environment on salivary cortisol levels in the elderly differs by 5-HTTLPR genotype. Neurobiol. Stress 7, 38–46.

Andersen, S.L., 2022. Neuroinflammation, early-life adversity, and brain development. Harv. Rev. Psychiatry 30, 24–39.

Anderson, K.M., Ferranti, E.P., Alagha, E.C., Mykityshyn, E., French, C.E., Reilly, C.M., 2022. The heart and gut relationship: a systematic review of the evaluation of the microbiome and trimethylamine-N-oxide (TMAO) in heart failure. Heart Fail. Rev. 27, 2223–2249.

Andrade, F.B., Gualberto, A., Rezende, C., Percegoni, N., Gameiro, J., Hottz, E.D., 2021. The weight of obesity in immunity from influenza to COVID-19. Front. Cell. Infect. Microbiol. 11, 638852.

Andrade-Guerrero, J., Santiago-Balmaseda, A., Jeronimo-Aguilar, P., Vargas-Rodríguez, I., Cadena-Suárez, A.R., et al., 2023. Alzheimer's disease: an updated overview of its genetics. Int. J. Mol. Sci. 24, 3754.

Andrews, Z.B., Abizaid, A., 2014. Neuroendocrine mechanisms that connect feeding behavior and stress. Front. Neurosci. 8, 312–322.

Andrews, N., Stowe, J., Kirsebom, F., Toffa, S., Rickeard, T., et al., 2022. Covid-19 vaccine effectiveness against the omicron (B.1.1.529) variant. N. Engl. J. Med. 386, 1532–1546.

Anisman, H., 2009. Cascading effects of stressors and inflammatory immune system activation: implications for major depressive disorder. J. Psychiatry Neurosci. 34, 4–20.

Anisman, H., 2016. Health Psychology. Sage Publications, London.

Anisman, H., Hayley, S., 2012a. Illness comorbidity as a biomarker? J. Psychiatry Neurosci. 37, 221–223.

Anisman, H., Hayley, S., 2012b. Inflammatory factors contribute to depression and its comorbid conditions. Sci. Signal. 5, pe45.

Anisman, H., Kusnecov, A.W., 2022. Cancer: How Lifestyles May Impact Disease Development, Progression, and Treatment. Academic Press, London, UK.

Anisman, H., Matheson, K., 2005. Stress, anhedonia and depression: caveats concerning animal models. Neurosci. Biobehav. Rev. 29, 525–546.

Anisman, H., Matheson, K., 2023. An Introduction to Stress and Health. Sage Publications, London.

Anisman, H., Merali, Z., 1999. Anhedonic and anxiogenic effects of cytokine exposure. Adv. Exp. Med. Biol. 461, 199–233.

Anisman, H., Merali, Z., 2002. Cytokines, stress, and depressive illness. Brain Behav. Immun. 16, 513–524.

Anisman, H., Merali, Z., 2003. Cytokines, stress and depressive illness: brain-immune interactions. Ann. Med. 35, 2–11.

Anisman, H., Zacharko, R.M., 1982. Depression: the predisposing influence of stress. Behav. Brain Sci. 5, 89–137.

Anisman, H., Zaharia, M.D., Meaney, M.J., Merali, Z., 1998. Do early-life events permanently alter behavioral and hormonal responses to stressors? Int. J. Dev. Neurosci. 16, 149–164.

Anisman, H., Ravindran, A., Griffiths, J., Merali, Z., 1999. Endocrine and cytokine correlates of major depression and dysthymia with typical or atypical features. Mol. Psychiatry 4, 182–188.

Anisman, H., Hayley, S., Merali, Z., 2003. Cytokines and stress: sensitization and cross- sensitization. Brain Behav. Immun. 17, 86–93.

Anisman, H., Poulter, M.O., Gandhi, R., Merali, Z., Hayley, S., 2007. Interferon-alpha effects are exaggerated when administered on a psychosocial stressor backdrop: cytokine, corticosterone and brain monoamine variations. J. Neuroimmunol. 186, 45–53.

Anisman, H., Du, L., Palkovits, M., Faludi, G., Kovacs, G.G., et al., 2008a. Serotonin receptor subtype and p11 mRNA expression in stress-relevant brain regions of suicide and control subjects. J. Psychiatry Neurosci. 33, 131–141.

Anisman, H., Merali, Z., Hayley, S., 2008b. Neurotransmitter, peptide and cytokine processes in relation to depressive disorder: comorbidity of depression with neurodegenerative disorders. Prog. Neurobiol. 85, 1–74.

Antal, B., McMahon, L.P., Sultan, S.F., Lithen, A., Wexler, D.J., et al., 2022. Type 2 diabetes mellitus accelerates brain aging and cognitive decline: complementary findings from UK Biobank and meta-analyses. elife 11, e73138.

Antila, H., Ryazantseva, M., Popova, D., Sipilä, P., Guirado, R., et al., 2017. Isoflurane produces antidepressant effects and induces TrkB signaling in rodents. Sci. Rep. 7, 7811.

Antonino, M., Marmo, P., Freites, C.L., Quassollo, G.E., Sánchez, M.F., et al., 2022. Aβ assemblies promote amyloidogenic processing of APP and intracellular accumulation of Aβ42 through Go/Gβγ signaling. Front. Cell Dev. Biol. 10, 852738.

Apfel, B.A., Ross, J., Hlavin, J., Meyerhoff, D.J., Metzler, T.J., et al., 2011a. Hippocampal volume differences in Gulf War veterans with current versus lifetime posttraumatic stress disorder symptoms. Biol. Psychiatry 69, 541–548.

Apfel, B.A., Otte, C., Inslicht, S.S., McCaslin, S.E., Henn-Haase, C., et al., 2011b. Pretraumatic prolonged elevation of salivary MHPG predicts peritraumatic distress and symptoms of post-traumatic stress disorder. J. Psychiatr. Res. 45, 735–741.

Appelbaum, L.G., Shenasa, M.A., Stolz, L., Daskalakis, Z., 2023. Synaptic plasticity and mental health: methods, challenges and opportunities. Neuropsychopharmacology 48, 113–120.

Appelberg, S., Ahlén, G., Yan, J., Nikouyan, N., Weber, S., et al., 2022. A universal SARS-CoV DNA vaccine inducing highly cross-reactive neutralizing antibodies and T cells. EMBO Mol. Med. 14, e15821.

Aprikian, S., Kasvis, P., Vigano, M., Hachem, Y., Canac-Marquis, M., Vigano, A., 2023. Medical cannabis is effective for cancer-related pain: Quebec Cannabis Registry results. BMJ Support. Palliat. Care 13, 1285–1291.

Arabi, S., 2023. Narcissistic and psychopathic traits in romantic partners predict post-traumatic stress disorder symptomology: evidence for unique impact in a large sample. Personal. Individ. Differ. 201, 111942.

Aragón-Vela, J., Solis-Urra, P., Ruiz-Ojeda, F.J., Álvarez-Mercado, A.I., Olivares-Arancibia, J., Plaza-Diaz, J., 2021. Impact of exercise on gut microbiota in obesity. Nutrients 13, 3999.

Arambula, S.E., Jima, D., Patisaul, H.B., 2017. Prenatal bisphenol A (BPA) exposure alters the transcriptome of the neonate rat amygdala in a sex-specific manner: a CLARITY-BPA consortium study. Neurotoxicology. pii: S0161-813X(17)30209-7.

Aranäs, C., Edvardsson, C.E., Shevchouk, O.T., Zhang, Q., Witley, S., Blid Sköldheden, S., Jerlhag, E., 2023. Semaglutide reduces alcohol intake and relapse-like drinking in male and female rats. EBioMedicine 93, 104642.

Archambault, É., Vigod, S.N., Brown, H.K., Lu, H., Fung, K., Shouldice, M., Saunders, N.R., 2023. Mental illness following physical assault among children. JAMA Netw. Open 6, e2329172.

Ardi, Z., Richter-Levin, A., Xu, L., Cao, X., Volkmer, H., Stork, O., Richter-Levin, G., 2019. The role of the GABAA receptor Alpha 1 subunit in the ventral hippocampus in stress resilience. Sci. Rep. 9, 13513.

Argolo, F.C., Cavalcanti-Ribeiro, P., Netto, L.R., Quarantini, L.C., 2015. Prevention of posttraumatic stress disorder with propranolol: a meta-analytic review. J. Psychosom. Res. 79, 89–93.

Arinuma, Y., 2018. Antibodies and the brain: anti-N-methyl-D-aspartate receptor antibody and the clinical effects in patients with systemic lupus erythematosus. Curr. Opin. Neurol. 31, 294–299.

Armanios, M., Blackburn, E.H., 2012. The telomere syndromes. Nat. Rev. Genet. 13, 693–704.

Årnes, A.P., Nielsen, C.S., Stubhaug, A., Fjeld, M.K., Johansen, A., et al., 2023. Longitudinal relationships between habitual physical activity and pain tolerance in the general population. PLoS One 18, e0285041.

Aroke, E.N., Joseph, P.V., Roy, A., Overstreet, D.S., Tollefsbol, T.O., Vance, D.E., Goodin, B.R., 2019. Could epigenetics help explain racial disparities in chronic pain? J. Pain Res., 701–710.

Aronne, L.J., Sattar, N., Horn, D.B., Bays, H.E., Wharton, S., et al., 2023. Continued treatment with tirzepatide for maintenance of weight reduction in adults with obesity: the SURMOUNT-4 randomized clinical trial. JAMA 8.

Aron-Wisnewsky, J., Clement, K., 2016. The gut microbiome, diet, and links to cardiometabolic and chronic disorders. Nat. Rev. Nephrol. 12, 169–181.

Arora, T., Akrami, R., Pais, R., Bergqvist, L., Johansson, B.R., et al., 2018. Microbial regulation of the L cell transcriptome. Sci. Rep. 8, 1207.

Arora, S., Stouffer, G.A., Kucharska-Newton, A.M., Qamar, A., Vaduganathan, M., et al., 2019. Twenty year trends and sex differences in young adults hospitalized with acute myocardial infarction. Circulation 139, 1047–1056.

Arrieta, M.C., Stiemsma, L.T., Dimitriu, P.A., Thorson, L., Russell, S., et al., 2015. Early infancy microbial and metabolic alterations affect risk of childhood asthma. Sci. Transl. Med. 7, 307ra152.

Artacho-Cordón, F., León, J., Sáenz, J.M., Fernández, M.F., Martin-Olmedo, P., et al., 2016. Contribution of persistent organic pollutant exposure to the adipose tissue oxidative. Environ. Sci. Technol. 50, 13529–13538.

Aruldass, A.R., Kitzbichler, M.G., Morgan, S.E., Lim, S., Lynall, M.E., et al., 2021. Dysconnectivity of a brain functional network was associated with blood inflammatory markers in depression. Brain Behav. Immun. 98, 299–309.

Ascherio, A., Schwarzschild, M.A., 2016. The epidemiology of Parkinson's disease: risk factors and prevention. Lancet Neurol. 15, 1257–1272.

Ascherio, A., Zhang, S.M., Hernán, M.A., Kawachi, I., Colditz, G.A., et al., 2001. Prospective study of caffeine consumption and risk of Parkinson's disease in men and women. Ann. Neurol. 50, 56–63.

Aschwanden, C., 2020. The false promise of herd immunity for COVID-19. Nature 587, 26–28.

Ashenafi, S., Loreti, M.G., Bekele, A., Aseffa, G., Amogne, W., et al., 2023. Inflammatory immune profiles associated with disease severity in pulmonary tuberculosis patients with moderate to severe clinical TB or anemia. Front. Immunol. 14, 1296501.

Ashhurst, T.M., van Vreden, C., Niewold, P., King, N.J., 2014. The plasticity of inflammatory monocyte responses to the inflamed central nervous system. Cell. Immunol. 291, 49–57.

Ashraf, U., Ding, Z., Deng, S., Ye, J., Cao, S., Chen, Z., 2021. Pathogenicity and virulence of Japanese encephalitis virus: neuroinflammation and neuronal cell damage. Virulence 12, 968–980.

Ashwood, P., Wills, S., Van de Water, J., 2006. The immune response in autism: a new frontier for autism research. J. Leukoc. Biol. 80, 1–15.

Ashwood, P., Krakowiak, P., Hertz-Picciotto, I., Hansen, R., Pessah, I.N., Van de Water, J., 2011. Altered T cell responses in children with autism. Brain Behav. Immun. 25, 840–849.

Askew, K., Li, K., Olmos-Alonso, A., Garcia-Moreno, F., Liang, Y., et al., 2017. Coupled proliferation and apoptosis maintain the rapid turnover of microglia in the adult brain. Cell Rep. 18, 391–405.

Assadourian, J.N., Peterson, E.D., McDonald, S.A., Gupta, A., Navar, A.M., 2023. Health claims and doses of fish oil supplements in the US. JAMA Cardiol. 8, 984–988.

Astill Wright, L., Sijbrandij, M., Sinnerton, R., Lewis, C., Roberts, N.P., Bisson, J.I., 2019. Pharmacological prevention and early treatment of post-traumatic stress disorder and acute stress disorder: a systematic review and meta-analysis. Transl. Psychiatry 9, 334.

Atasoy, S., Johar, H., Fleischer, T., Beutel, M., Binder, H., et al., 2022. Depression mediates the association between childhood emotional abuse and the onset of Type 2 Diabetes: findings from German multi-cohort prospective studies. Front. Psychiatry 13, 825678.

Atladottir, H.O., Pedersen, M.G., Thorsen, P., Mortensen, P.B., Deleuran, B., et al., 2009. Association of family history of autoimmune diseases and autism spectrum disorders. Pediatrics 124, 687–694.

Atladóttir, H.O., Thorsen, P., Østergaard, L., Schendel, D.E., Lemcke, S., et al., 2010. Maternal infection requiring hospitalization during pregnancy and autism spectrum disorders. J. Autism Dev. Disord. 40, 1423–1430.

Atlas, L.Y., 2021. A social affective neuroscience lens on placebo analgesia. Trends Cogn. Sci. 25, 992–1005.

Atlas, L.Y., Wager, T.D., 2014. A meta-analysis of brain mechanisms of placebo analgesia: consistent findings and unanswered questions. Handb. Exp. Pharmacol. 225, 37–69.

Attwells, S., Setiawan, E., Wilson, A.A., Rusjan, P.M., Mizrahi, R., et al., 2017. Inflammation in the neurocircuitry of obsessive-compulsive disorder. JAMA Psychiatry 74, 833–840.

Aubry, A.V., Serrano, P.A., Burghardt, N.S., 2016. Molecular mechanisms of stress-induced increases in fear memory consolidation within the amygdala. Front. Behav. Neurosci. 10, 191.

Audet, M.C., Anisman, H., 2010. Neuroendocrine and neurochemical impact of aggressive social interactions in submissive and dominant mice: implications for stress-related disorders. Int. J. Neuropsychopharmacol. 13, 361–372.

Audet, M.C., Anisman, H., 2013. Interplay between pro-inflammatory cytokines and growth factors in depressive illnesses. Front. Cell. Neurosci. 7, 68.

Audet, M.C., McQuaid, R.J., Merali, Z., Anisman, H., 2014. Cytokine variations and mood disorders: influence of social stressors and social support. Front. Neurosci. 28, 416.

Auger, J.P., Zimmermann, M., Faas, M., Stifel, U., Chambers, D., et al., 2024. Metabolic rewiring promotes anti-inflammatory effects of glucocorticoids. Nature 629, 184–192.

Augusto, D.G., Murdolo, L.D., Chatzileontiadou, D.S.M., Sabatino Jr., J.J., Yusufali, T., et al., 2023. A common allele of HLA is associated with asymptomatic SARS-CoV-2 infection. Nature 620, 128–136.

Aune, D., Keum, N., Giovannucci, E., Fadnes, L.T., Boffetta, P., et al., 2016. Whole grain consumption and the risk of cardiovascular disease, cancer, and all-cause and cause-specific mortality–a systematic review and dose-response meta-analysis of prospective studies. BMJ 353, i2716.

Aung, T., Halsey, J., Kromhout, D., Gerstein, H.C., Marchioli, R., et al., 2018. Associations of Omega-3 fatty acid supplement use with cardiovascular disease risks: meta-analysis of 10 trials involving 77,917 individuals. JAMA Cardiol. 3, 225–234.

Auriel, E., Regev, K., Korczyn, A.D., 2014. Nonsteroidal anti-inflammatory drugs exposure and the central nervous system. Handb. Clin. Neurol. 119, 577–584.

Autio, T., Rissanen, S., 2018. Positive emotions in caring for a spouse: a literature review. Scand. J. Caring Sci. 32, 45–55.

Avci, A.B., Feist, E., Burmester, G.R., 2016. Targeting GM-CSF in rheumatoid arthritis. Clin. Exp. Rheumatol. 34, 39–44.

Averill, L.A., Purohit, P., Averill, C.L., Boesl, M.A., Krystal, J.H., et al., 2017. Glutamate dysregulation and glutamatergic therapeutics for PTSD: evidence from human studies. Neurosci. Lett. 649, 147–155.

Averill, L.A., Abdallah, C.G., Fenton, L.R., Fasula, M.K., Jiang, L., et al., 2020. Early life stress and glutamate neurotransmission in major depressive disorder. Eur. Neuropsychopharmacol. 35, 71–80.

Averill, L.A., Jiang, L., Purohit, P., Coppoli, A., Averill, C.L., et al., 2022. Prefrontal glutamate neurotransmission in PTSD: a novel approach to estimate synaptic strength in vivo in humans. Chronic Stress (Thousand Oaks) 6. 24705470221092734.

Aviles, H., Monroy, F.P., 2001. Immunomodulatory effects of cold stress on mice infected intraperitoneally with a 50% lethal dose of Toxoplasma gondii. Neuroimmunomodulation 9, 6–12.

Avitsur, R., 2017. Prenatal fluoxetine modifies the behavioral and hormonal responses to stress in male mice: role for glucocorticoid insensitivity. Behav. Pharmacol. 28, 345–355.

Avolio, E., Fazzari, G., Zizza, M., De Lorenzo, A., Di Renzo, L., et al., 2019. Probiotics modify body weight together with anxiety states via pro-inflammatory factors in HFD-treated Syrian golden hamster. Behav. Brain Res. 356, 390–399.

Axelsson, A.S., Mahdi, T., Nenonen, H.A., Singh, T., Hänzelmann, S., et al., 2017. Sox5 regulates beta-cell phenotype and is reduced in type 2 diabetes. Nat. Commun. 8, 15652.

Aydin, Ö., Nieuwdorp, M., Gerdes, V., 2018. The gut microbiome as a target for the treatment of type 2 diabetes. Curr. Diab. Rep. 18, 55.

Ayling, K., Jia, R., Coupland, C., Chalder, T., Massey, A., et al., 2022. Psychological predictors of self-reported COVID-19 outcomes: results from a prospective cohort study. Ann. Behav. Med. 56, 484–497.

Ayoub, H.H., Chemaitelly, H., Abu-Raddad, L.J., 2019. Characterizing the transitioning epidemiology of herpes simplex virus type 1 in the USA: model-based predictions. BMC Med. 17, 57.

Azevedo Da Silva, M., Singh-Manoux, A., Shipley, M.J., Vahtera, J., Brunner, E.J., et al., 2014. Sleep duration and sleep disturbances partly explain the association between depressive symptoms and cardiovascular mortality: the Whitehall II cohort study. J. Sleep Res. 23, 94–97.

Baba, Y., Kuroiwa, A., Uitti, R.J., Wszolek, Z.K., Yamada, T., 2005. Alterations of T-lymphocyte populations in Parkinson disease. Parkinsonism Relat. Disord. 11, 493–498.

Baba, A., Kloiber, S., Zai, G., 2022. Genetics of social anxiety disorder: a systematic review. Psychiatr. Genet. 32, 37–66.

Babić Leko, M., Langer Horvat, L., Španić Popovački, E., Zubčić, K., Hof, P.R., Šimić, G., 2023. Metals in Alzheimer's disease. Biomedicines 11, 1161.

Bäckhed, F., Ding, H., Wang, T., Hooper, L.V., Koh, G.Y., et al., 2004. The gut microbiota as an environmental factor that regulates fat storage. Proc. Natl. Acad. Sci. USA 101, 15718–15723.

Bäckhed, F., Roswall, J., Peng, Y., Feng, Q., Jia, H., et al., 2015. Dynamics and stabilization of the human gut microbiome during the first year of life. Cell Host Microbe 17, 690–703.

Badrinath, A., Bhatta, S., Kloc, A., 2022. Persistent viral infections and their role in heart disease. Front. Microbiol. 13, 1030440.

Baek, E.C., Hyon, R., López, K., Du, M., Porter, M.A., Parkinson, C., 2023. Lonely individuals process the world in idiosyncratic ways. Psychol. Sci. 34, 683–695.

Baeza-Raja, B., Sachs, B.D., Li, P., Christian, F., Vagena, E., et al., 2016. p75 neurotrophin receptor regulates energy balance in obesity. Cell Rep. 14, 255–268.

Bagnall-Moreau, C., Spielman, B., Brimberg, L., 2023. Maternal brain reactive antibodies profile in autism spectrum disorder: an update. Transl. Psychiatry 13, 37.

Bagnardi, V., Rota, M., Botteri, E., Tramacere, I., Islami, F., et al., 2015. Alcohol consumption and site-specific cancer risk: a comprehensive dose-response meta-analysis. Br. J. Cancer 112, 580–593.

Baharnoori, M., Bhardwaj, S.K., Srivastava, L.K., 2013. Effect of maternal lipopolysaccharide administration on the development of dopaminergic receptors and transporter in the rat offspring. PLoS One 8, e54439.

Bahji, A., Forsyth, A., Groll, D., Hawken, E.R., 2020. Efficacy of 3,4-methylenedioxymethamphetamine (MDMA)-assisted psychotherapy for posttraumatic stress disorder: a systematic review and meta-analysis. Prog. Neuro-Psychopharmacol. Biol. Psychiatry 96, 109735.

Bai, Z., Luo, S., Zhang, L., Wu, S., Chi, I., 2020. Acceptance and commitment therapy (ACT) to reduce depression: a systematic review and meta-analysis. J. Affect. Disord. 260, 728–737.

Baiden-Amissah, R.E.M., Tuyaerts, S., 2019. Contribution of aging, obesity, and microbiota on tumor immunotherapy efficacy and toxicity. Int. J. Mol. Sci. 20, 3586.

Baig, A.M., Sanders, E.C., 2020. Potential neuroinvasive pathways of SARS-CoV-2: deciphering the spectrum of neurological deficit seen in coronavirus disease-2019 (COVID-19). J. Med. Virol. 92, 1845–1857.

Bailey, M.T., 2014. Influence of stressor-induced nervous system activation on the intestinal microbiota and the importance for immunomodulation. Adv. Exp. Med. Biol. 817, 255–276.

Bailey, M.T., Engler, H., Powell, N.D., Padgett, D.A., Sheridan, J.F., 2007. Repeated social defeat increases the bactericidal activity of splenic macrophages through a Toll-like receptor-dependent pathway. Am. J. Phys. Regul. Integr. Comp. Phys. 293, R1180–R1190.

Baily, M.T., 2016. Psychological stress, immunity, and the effects on indigenous microflora. Adv. Exp. Med. Biol. 874, 225–246.

Baird, G., Norbury, C.F., 2016. Social (pragmatic) communication disorders and autism spectrum disorder. Arch. Dis. Child. 101, 745–751.

Baird, J.K., Bourdette, D., Meshul, C.K., Quinn, J.F., 2019. The key role of T cells in Parkinson's disease pathogenesis and therapy. Parkinsonism Relat. Disord. 60, 25–31.

Baker, J.F., Cates, M.E., Luthin, D.R., 2018. D-cycloserine in the treatment of posttraumatic stress disorder. Ment. Health Clin. 7, 88–94.

Bakunina, N., Pariante, C.M., Zunszain, P.A., 2015. Immune mechanisms linked to depression via oxidative stress and neuroprogression. Immunology 144, 365–373.

Balakrishnan, K., Rijal Upadhaya, A., Steinmetz, J., Reichwald, J., Abramowski, D., et al., 2015. Impact of amyloid β aggregate maturation on antibody treatment in APP23 mice. Acta Neuropathol. Commun. 3, 41.

Balan, I., Patterson, R., Boero, G., Krohn, H., O'Buckley, T.K., et al., 2023. Brexanolone therapeutics in post-partum depression involves inhibition of systemic inflammatory pathways. EBioMedicine 89, 104473.

Balbuena, L., Peters, E., Speed, D., 2023. Using polygenic risk scores to investigate the evolution of smoking and mental health outcomes in UK biobank participants. Acta Psychiatr. Scand. e13601.

Baldwin, H.A., Koivula, P.P., Necarsulmer, J.C., Whitaker, K.W., Harvey, B.K., 2017. Step sequence is a critical gait parameter of unilateral 6-OHDA Parkinson's rat models. Cell Transplant. 26, 659–667.

Baliki, M.N., Apkarian, A.V., 2015. Nociception, pain, negative moods, and behavior selection. Neuron 87, 474–491.

Ballouz, T., Menges, D., Anagnostopoulos, A., Domenghino, A., Aschmann, H.E., et al., 2023. Recovery and symptom trajectories up to two years after SARS-CoV-2 infection: population based, longitudinal cohort study. BMJ 381, e074425.

Balouek, J.A., Mclain, C.A., Minerva, A.R., Rashford, R.L., Bennett, S.N., et al., 2023. Reactivation of early-life stress-sensitive neuronal ensembles contributes to lifelong stress hypersensitivity. J. Neurosci. 43, 5996–6009.

Baltazar, M.T., Dinis-Oliveira, R.J., de Lourdes Bastos, M., Tsatsakis, A.M., Duarte, J.A., et al., 2014. Pesticides exposure as etiological factors of Parkinson's disease and other neurodegenerative diseases—a mechanistic approach. Toxicol. Lett. 230, 85–103.

Bam, M., Yang, X., Zhou, J., Ginsberg, J.P., Leyden, Q., et al., 2016a. Evidence for epigenetic regulation of proinflammatory cytokines, interleukin-12 and interferon gamma, in peripheral blood mononuclear cells from PTSD patients. J. NeuroImmune Pharmacol. 11, 168–181.

Bam, M., Yang, X., Zumbrun, E.E., Zhong, Y., Zhou, J., et al., 2016b. Dysregulated immune system networks in war veterans with PTSD is an outcome of altered miRNA expression and DNA methylation. Sci. Rep. 6, 31209.

Banach, M., Lewek, J., Surma, S., Penson, P.E., Sahebkar, A., et al., 2023. The association between daily step count and all-cause and cardiovascular mortality: a meta-analysis. Eur. J. Prev. Cardiol. 30, 1975–1985.

Bandelow, B., Reitt, M., Röver, C., Michaelis, S., Görlich, Y., et al., 2015. Efficacy of treatments for anxiety disorders: a meta-analysis. Int. Clin. Psychopharmacol. 30, 183–192.

Bandelow, B., Baldwin, D., Abelli, M., Altamura, C., Dell'Osso, B., et al., 2016. Biological markers for anxiety disorders, OCD and PTSD: a consensus statement. Part I: neuroimaging and genetics. World J. Biol. Psychiatry 17, 321–365.

Bangasser, D.A., Kawasumi, Y., 2015. Cognitive disruptions in stress-related psychiatric disorders: a role for corticotropin releasing factor (CRF). Horm. Behav. 76, 125–135.

Bangasser, D.A., Wiersielis, K.R., Khantsis, S., 2016. Sex differences in the locus coeruleus-norepinephrine system and its regulation by stress. Brain Res. 1641, 177–188.

Bangasser, D.A., Eck, S.R., Ordoñes Sanchez, E., 2019. Sex differences in stress reactivity in arousal and attention systems. Neuropsychopharmacology 44, 129–139.

Baniasadi, M., Hosseini, G., Fayyazi Bordbar, M.R., Rezaei Ardani, A., Mostafavi Toroghi, H., 2014. Effect of pregabalin augmentation in treatment of patients with combat-related chronic posttraumatic stress disorder: a randomized controlled trial. J. Psychiatr. Pract. 20, 419–427.

Banks, W.A., 2015. The blood-brain barrier in neuroimmunology: tales of separation and assimilation. Brain Behav. Immun. 44, 1–8.

Banks, W.A., 2016. From blood-brain barrier to blood-brain interface: new opportunities for CNS drug delivery. Nat. Rev. Drug Discov. 15, 275–292.

Banks, W., 2019. The blood-brain barrier as an endocrine tissue. Nat. Rev. Endocrinol. 15, 444–455.

Bao, H., Asrican, B., Li, W., Gu, B., Wen, Z., et al., 2017. Long-range GABAergic inputs regulate neural stem cell quiescence and control adult hippocampal neurogenesis. Cell Stem Cell 21, 604–617.e5.

Bao, L., Deng, W., Huang, B., Gao, H., Liu, J., et al., 2020a. The pathogenicity of SARS-CoV-2 in hACE2 transgenic mice. Nature 583 (7818), 830–833.

Bao, W., Liu, B., Rong, S., Dai, S.Y., Trasande, L., Lehmler, H.J., 2020b. Association between bisphenol a exposure and risk of all-cause and cause-specific mortality in US adults. JAMA Netw. Open 3, e2011620.

Barak, Y., 2006. The immune system and happiness. Autoimmun. Rev. 5, 523–527.

Baratta, M.V., Kodandaramaiah, S.B., Monahan, P.E., Yao, J., Weber, M.D., et al., 2016. Stress enables reinforcement-elicited serotonergic consolidation of fear memory. Biol. Psychiatry 79, 814–822.

Barbhaiya, M., Costenbader, K.H., 2016. Environmental exposures and the development of systemic lupus erythematosus. Curr. Opin. Rheumatol. 28, 497–505.

Barbosa, D.A., Gattas, S., Salgado, J.S., Kuijper, F.M., Wang, A.R., et al., 2023. An orexigenic subnetwork within the human hippocampus. Nature 621, 381–388.

Barnard, D.F., Gabella, K.M., Kulp, A.C., Parker, A.D., Dugan, P.B., Johnson, J.D., 2019. Sex differences in the regulation of brain IL-1β in response to chronic stress. Psychoneuroendocrinology 103, 203–211.

Bar-On, D., Rottgardt, E., 1998. Reconstructing silenced biographical issues through feeling-facts. Psychiatry 61, 61–83.

Bar-On, Y., Seidel, E., Tsukerman, P., et al., 2014. Influenza virus uses its neuraminidase protein to evade the recognition of two activating NK cell receptors. J. Infect. Dis. 210, 410–418.

Barone, F.C., Arvin, B., White, R.F., Miller, A., Webb, C.L., et al., 1997. Tumor necrosis factor-alpha. A mediator of focal ischemic brain injury. Stroke 28, 1233–1244.

Barrett, T.J., Corr, E.M., van Solingen, C., Schlamp, F., Brown, E.J., et al., 2021. Chronic stress primes innate immune responses in mice and humans. Cell Rep. 36, 109595.

Barrientos, R.M., Thompson, V.M., Kitt, M.M., Amat, J., Hale, M.W., et al., 2015. Greater glucocorticoid receptor activation in hippocampus of aged rats sensitizes microglia. Neurobiol. Aging 36, 1483–1495.

Barthas, F., Sellmeijer, J., Hugel, S., Waltisperger, E., Barrot, M., 2015. The anterior cingulate cortex is a critical hub for pain-induced depression. Biol. Psychiatry 77, 236–245.

Barthels, C., Ogrinc, A., Steyer, V., Meier, S., Simon, F., et al., 2017. CD40-signalling abrogates induction of RORγt+ Treg cells by intestinal CD103+ DCs and causes fatal colitis. Nat. Commun. 8, 14715.

Bartlett, E.A., Yttredahl, A.A., Boldrini, M., Tyrer, A.E., Hill, K.R., et al., 2023. In vivo serotonin 1A receptor hippocampal binding potential in depression and reported childhood adversity. Eur. Psychiatry 66, e17.

Baruch, K., Schwartz, M., 2016. Circulating monocytes in between the gut and the mind. Cell Stem Cell 18, 689–691.

Baruch, E.N., Youngster, I., Ben-Betzalel, G., Ortenberg, R., Lahat, A., 2021. Fecal microbiota transplant promotes response in immunotherapy-refractory melanoma patients. Science 371, 573–574.

Basarkar, V., Govardhane, S., Shende, P., 2022. Multifaceted applications of genetically modified micro-organisms: a biotechnological revolution. Curr. Pharm. Des. 28 (22), 1833–1842.

Bashiardes, S., Tuganbaev, T., Federici, S., Elinav, E., 2017. The microbiome in anti-cancer therapy. Semin. Immunol. 32, 74–81.

Basílio-Queirós, D., Mischak-Weissinger, E., 2023. Natural killer cells-from innate cells to the discovery of adaptability. Front. Immunol. 14, 1172437.

Bastiaanssen, T.F.S., Cowan, C.S.M., Claesson, M.J., Dinan, T.G., Cryan, J.F., 2019. Making sense of the microbiome in psychiatry. Int. J. Neuropsychopharmacol. 22, 37–52.

Bastian, B., Jetten, J., Ferris, L.J., 2014. Pain as social glue: shared pain increases cooperation. Psychol. Sci. 25, 2079–2085.

Basu, J., Zaremba, J.D., Cheung, S.K., Hitti, F.L., Zemelman, B.V., et al., 2016. Gating of hippocampal activity, plasticity, and memory by entorhinal cortex long-range inhibition. Science 351, aaa5694.

Basurco, L., Abellanas, M.A., Ayerra, L., Conde, E., Vinueza-Gavilanes, R., et al., 2023. Microglia and astrocyte activation is region-dependent in the α-synuclein mouse model of Parkinson's disease. Glia 71, 571–587.

Battis, K., Xiang, W., Winkler, J., 2023. The bidirectional interplay of α-Synuclein with lipids in the central nervous system and its implications for the pathogenesis of Parkinson's disease. Int. J. Mol. Sci. 24, 13270.

Bauer, M.E., Vedhara, K., Perks, P., Wilcock, G.K., Lightman, S.L., et al., 2000. Chronic stress in caregivers of dementia patients is associated with reduced lymphocyte sensitivity to glucocorticoids. J. Neuroimmunol. 103, 84–92.

Bauman, M.D., Iosif, A.M., Ashwood, P., Braunschweig, D., Lee, A., Schumann, C.M., Amaral, D.G., 2013. Maternal antibodies from mothers of children with autism alter brain growth and social behavior development in the rhesus monkey. Transl. Psychiatry 3, e278.

Bauman, M.D., Iosif, A.M., Smith, S.E., Bregere, C., Amaral, D.G., Patterson, P.H., 2014. Activation of the maternal immune system during pregnancy alters behavioral development of rhesus monkey offspring. Biol. Psychiatry 75, 332–341.

Baumeister, D., Akhtar, R., Ciufolini, S., Pariante, C.M., Mondelli, V., 2016. Childhood trauma and adulthood inflammation: a meta-analysis of peripheral C-reactive protein, interleukin-6 and tumour necrosis factor-α. Mol. Psychiatry 21, 642–649.

Baune, B.T., Adrian, I., Arolt, V., Berger, K., 2006. Associations between major depression, biplor disorders, dysthymia and cardiovascular diseases in the general adult population. Psychother. Psychosom. 75, 319–326.

Bavel, J.J.V., Baicker, K., Boggio, P.S., Capraro, V., Cichocka, A., et al., 2020. Using social and behavioural science to support COVID-19 pandemic response. Nat. Hum. Behav. 4, 460–471.

Bawa, F.L.M., Mercer, S.W., Atherton, R.J., Clague, F., Keen, A., et al., 2015. Does mindfulness improve outcomes in patients with chronic pain? Systematic review and meta-analysis. Br. J. Gen. Pract. 65, e387–e400.

Bayer, A.J., Bullock, R., Jones, R.W., Wilkinson, D., Paterson, K.R., et al., 2005. Evaluation of the safety and immunogenicity of synthetic Abeta42 (AN1792) in patients with AD. Neurology 64, 94–101.

Bayrer, J.R., Castro, J., Venkataraman, A., Touhara, K.K., Rossen, N.D., et al., 2023. Gut enterochromaffin cells drive visceral pain and anxiety. Nature 616, 137–142.

Beck, A.T., 1967. Depression: Clinical, Experimental, and Theoretical Aspects. Harper & Row, New York.

Beck, A.T., 2008. The evolution of the cognitive model of depression and its neurobiological correlates. Am. J. Psychiatry 165, 969–977.

Bedse, G., Hill, M.N., Patel, S., 2020. 2-Arachidonoylglycerol modulation of anxiety and stress adaptation: from grass roots to novel therapeutics. Biol. Psychiatry 88, 520–530.

Bekhbat, M., Howell, P.A., Rowson, S.A., Kelly, S.D., Tansey, M.G., Neigh, G.N., 2019. Chronic adolescent stress sex-specifically alters central and peripheral neuro-immune reactivity in rats. Brain Behav. Immun. 76, 248–257.

Bekkar, B., Pacheco, S., Basu, R., DeNicola, N., 2020. Association of air pollution and heat exposure with preterm birth, low birth weight, and stillbirth in the US: a systematic review. JAMA Netw. Open 3, e208243.

Belfiore, M.P., Russo, G.M., Gallo, L., Atripaldi, U., Tamburrini, S., et al., 2022. Secondary complications in COVID-19 patients: a case series. Tomography 8, 1836–1850.

Belin, M.F., Didier-Bazes, M., Akaoka, H., Hardin-Pouzet, H., Bernard, A., Giraudon, P., 1997. Changes in astrocytic glutamate catabolism enzymes following neuronal degeneration or viral infection. Glia 21, 154–161.

Belkaid, Y., Hand, T.W., 2014. Role of the microbiota in immunity and inflammation. Cell 157, 121–141.

Bell, J.A., Hamer, M., Sabia, S., Singh-Manoux, A., et al., 2015. The natural course of healthy obesity over 20 years. J. Am. Coll. Cardiol. 65, 101–102.

Bellan, M., Apostolo, D., Albè, A., Crevola, M., Errica, N., et al., 2022. Determinants of long COVID among adults hospitalized for SARS-CoV-2 infection: a prospective cohort study. Front. Immunol. 13, 1038227.

Bellaver, B., Povala, G., Ferreira, P.C., Ferrari-Souza, J.P., Leffa, D.T., et al., 2023. Astrocyte reactivity influences amyloid-β effects on tau pathology in preclinical Alzheimer's disease. Nat. Med. 29, 1775–1781.

Belleau, E.L., Treadway, M.T., Pizzagalli, D.A., 2019. The impact of stress and major depressive disorder on hippocampal and medial prefrontal cortex morphology. Biol. Psychiatry 85, 443–453.

Bellinger, D.L., Nance, D.M., Lorton, D., 2013. Innervation of the immune system. In: Kusnecov, A.V., Anisman, H. (Eds.), The Wiley Blackwell Handbook of Psychoneuroimmunology. Chichester, UK, John Wiley & Sons Ltd.

Bellono, N.W., Bayrer, J.R., Leitch, D.B., Castro, J., Zhang, C., et al., 2017. Enterochromaffin cells are gut chemosensors that couple to sensory neural pathways. Cell 170, 185–198.e16.

Bellou, V., Belbasis, L., Tzoulaki, I., Evangelou, E., Ioannidis, J.P., 2016. Environmental risk factors and Parkinson's disease: an umbrella review of meta-analyses. Parkinsonism Relat. Disord. 23, 1–9.

Ben David, G., Amir, Y., Tripathi, K., Sharvit, L., Benhos, A., Anunu, R., et al., 2023. Exposure to juvenile stress induces epigenetic alterations in the GABAergic system in rats. Genes (Basel) 14, 565.

Ben Nasr, M., Tezza, S., D'Addio, F., Mameli, C., Usuelli, V., 2017. PD-L1 genetic overexpression or pharmacological restoration in hemotopoietic stem and progenitor cells reverses autoimmune diabetes. Sci. Transl. Med. 9, eaam7543.

Benech, N., Sokol, H., 2020. Fecal microbiota transplantation in gastrointestinal disorders: time for precision medicine. Genome Med. 12, 58.

Benedetti, F., 2014. Drugs and placebos: what's the difference?: Understanding the molecular basis of the placebo effect could help clinicians to better use it in clinical practice. EMBO Rep. 15, 329–332.

Benedetti, F., Mazza, M.G., Cavalli, G., Ciceri, F., Dagna, L., Rovere-Querini, P., 2021a. Can cytokine blocking prevent depression in COVID-19 survivors? J. NeuroImmune Pharmacol. 16, 1–3.

Benedetti, F., Palladini, M., Paolini, M., Melloni, E., Vai, B., et al., 2021b. Brain correlates of depression, post-traumatic distress, and inflammatory biomarkers in COVID-19 survivors: a multimodal magnetic resonance imaging study. Brain Behav. Immun. Health 18, 100387.

Ben-Menachem-Zidon, O., Ben Menachem-Zidon, O., Ben-Menahem, Y., Ben-Hur, T., Yirmiya, R., 2014. Intra-hippocampal transplantation of neural precursor cells with transgenic over-expression of IL-1 receptor antagonist rescues memory and neurogenesis impairments in an Alzheimer's disease model. Neuropsychopharmacology 39, 401–414.

Benner, E.J., Mosley, R.L., Destache, C.J., Lewis, T.B., Jackson-Lewis, V., et al., 2004. Therapeutic immunization protects dopaminergic neurons in a mouse model of Parkinson's disease. Proc. Natl. Acad. Sci. USA 101, 9435–9440.

Benros, M.E., Nielsen, P.R., Nordentoft, M., Eaton, W.W., Dalton, S.O., Mortensen, P.B., 2011. Autoimmune diseases and severe infections as risk factors for schizophrenia: a 30-year population-based register study. Am. J. Psychiatry 168 (12), 13031310.

Ben-Shaanan, T.L., Azulay-Debby, H., Dubovik, T., Starosvetsky, E., Korin, B., et al., 2016. Activation of the reward system boosts innate and adaptive immunity. Nat. Med. 22, 940–944.

Bercik, P., Collins, S.M., 2014. The effects of inflammation, infection and antibiotics on the microbiota-gut-brain axis. Adv. Exp. Med. Biol. 817, 279–289.

Bercik, P., Verdu, E.F., Foster, J.A., Macri, J., Potter, M., et al., 2010. Chronic gastrointestinal inflammation induces anxiety-like behavior and alters central nervous system biochemistry in mice. Gastroenterology 139, 2102–2112.

Bercik, P., Denou, E., Collins, J., Jackson, W., Lu, J., et al., 2011. The intestinal microbiota affect central levels of brain-derived neurotropic factor and behavior in mice. Gastroenterology 141, 599–609.

Berer, K., Mues, M., Koutrolos, M., Rasbi, Z.A., Boziki, M., et al., 2011. Commensal microbiota and myelin autoantigen cooperate to trigger autoimmune demyelination. Nature 479, 538–541.

Berer, K., Gerdes, L.A., Cekanaviciute, E., Jia, X., Xiao, L., et al., 2017. Gut microbiota from multiple sclerosis patients enables spontaneous autoimmune encephalomyelitis in mice. Proc. Natl. Acad. Sci. USA 114, 10719–10724.

Berezin, A.E., 2016. Diabetes mellitus related biomarker: the predictive role of growth-differentiation factor-15. Diabetes Metab. Syndr. 10, S154–S157.

Bergdolt, L., Dunaevsky, A., 2019. Brain changes in a maternal immune activation model of neurodevelopmental brain disorders. Prog. Neurobiol. 175, 1–19.

Berger, M., Sarnyai, Z., 2015. "More than skin deep": stress neurobiology and mental health consequences of racial discrimination. Stress 18, 1–10.

Bergman, K., Sarkar, P., Glover, V., O'Connor, T.G., 2010. Maternal prenatal cortisol and infant cognitive development: moderation by infant-mother attachment. Biol. Psychiatry 67, 1026–1032.

Berna, C., Kirsch, I., Zion, S.R., Lee, Y.C., Jensen, K.B., 2017. Side effects can enhance treatment response through expectancy effects: an experimental analgesic randomized controlled trial. Pain 158, 1014–1020.

Berridge, K.C., Robinson, T.E., 2016. Liking, wanting, and the incentive-sensitization theory of addiction. Am. Psychol. 71, 670–679.

Bertolini, F., Robertson, L., Bisson, J.I., Meader, N., Churchill, R., et al., 2022. Early pharmacological interventions for universal prevention of post-traumatic stress disorder (PTSD). Cochrane Database Syst. Rev. 2.

Besedovsky, L., Dimitrov, S., Born, J., Lange, T., 2016. Nocturnal sleep uniformly reduces numbers of different T-cell subsets in the blood of healthy men. Am. J. Phys. Regul. Integr. Comp. Phys. 311, R637–R642.

Besedovsky, L., Lange, T., Haack, M., 2019. The sleep-immune crosstalk in health and disease. Physiol. Rev. 99, 1325–1380.

Bethea, J.R., Fischer, R., 2021. Role of peripheral immune cells for development and recovery of chronic pain. Front. Immunol. 12, 641588.

Bettiol, S.S., Rose, T.C., Hughes, C.J., Smith, L.A., 2015. Alcohol consumption and Parkinson's disease risk: a review of recent findings. J. Parkinsons Dis. 5, 425–442.

Beura, L.K., Hamilton, S.E., Bi, K., Schenkel, J.M., Odumade, O.A., et al., 2016. Normalizing the environment recapitulates adult human immune traits in laboratory mice. Nature 532, 512–516.

Beutel, M.E., Stark, R., Pan, H., Silbersweig, D., Dietrich, S., 2010. Changes of brain activation pre- post short-term psychodynamic inpatient psychotherapy: an fMRI study of panic disorder patients. Psychiatry Res. 184, 96–104.

Beversdorf, D.Q., Stevens, H.E., Margolis, K.G., Van de Water, J., 2019. Prenatal stress and maternal immune dysregulation in autism spectrum disorders: potential points for intervention. Curr. Pharm. Des. 25, 4331–4343.

Beyerstedt, S., Casaro, E.B., Rangel, É.B., 2021. COVID-19: angiotensin-converting enzyme 2 (ACE2) expression and tissue susceptibility to SARS-CoV-2 infection. Eur. J. Clin. Microbiol. Infect. Dis. 40, 905–919.

Bhadra, R., Cobb, D.A., Weiss, L.M., Khan, I.A., 2013. Psychiatric disorders in toxoplasma seropositive patients—the CD8 connection. Schizophr. Bull. 39, 485–489.

Bhagavata Srinivasan, S.P., Raipuria, M., Bahari, H., Kaakoush, N.O., Morris, M.J., 2018. Impacts of diet and exercise on maternal gut microbiota are transferred to offspring. Front. Endocrinol. 9, 716.

Bhandari, R., Paliwal, J.K., Kuhad, A., 2020. Neuropsychopathology of autism spectrum disorder: complex interplay of genetic, epigenetic, and environmental factors. Adv. Neurobiol. 24, 97–141.

Bharadwaj, R.A., Jaffe, A.E., Chen, Q., Deep-Soboslay, A., Goldman, A. L., et al., 2018. Genetic risk mechanisms of posttraumatic stress disorder in the human brain. J. Neurosci. Res. 96, 21–30.

Bharwani, A., Mian, M.F., Foster, J.A.S., M.G., Bienenstock, J., et al., 2016. Structural & functional consequences of chronic psychosocial stress on the microbiome & host. Psychoneuroendocrinology 63, 217–227.

Bharwani, A., Mian, M.F., Surette, M.G., Bienenstock, J., Forsythe, P., 2017. Oral treatment with Lactobacillus rhamnosus attenuates behavioural deficits and immune changes in chronic social stress. BMC Med. 15, 7.

Bhatt, S., Hillmer, A.T., Girgenti, M.J., Rusowicz, A., Kapinos, M., et al., 2020. PTSD is associated with neuroimmune suppression: evidence from PET imaging and postmortem transcriptomic studies. Nat. Commun. 11, 2360.

Bhatt, S., Dhar, A.K., Samanta, M.K., Suttee, A., 2023. Effects of current psychotropic drugs on inflammation and immune system. Adv. Exp. Med. Biol. 1411, 407–434.

Bhikram, T., Sandor, P., 2022. Neutrophil-lymphocyte ratios as inflammatory biomarkers in psychiatric patients. Brain Behav. Immun. 105, 237–246.

Bian, G., Gloor, G.B., Gong, A., Jia, C., Zhang, W., et al., 2017. The gut microbiota of healthy aged chinese is similar to that of the healthy young. mSphere 2. pii: e00327-17.

Bible, E., 2014. Alzheimer disease: high serum levels of the pesticide metabolite DDE—a potential environmental risk factor for Alzheimer disease. Nat. Rev. Neurol. 10, 125.

Bicket, M.C., Stone, E.M., McGinty, E.E., 2023. Use of cannabis and other pain treatments among adults with chronic pain in US states with medical cannabis programs. JAMA Netw. Open 6, e2249797.

Bigley, A.B., Simpson, R.J., 2015. NK cells and exercise: implications for cancer immunotherapy and survivorship. Discov. Med. 19, 433–443.

Bilbo, S.D., Block, C.L., Bolton, J.L., Hanamsagar, R., Tran, P.K., 2018. Beyond infection - maternal immune activation by environmental factors, microglial development, and relevance for autism spectrum disorders. Exp. Neurol. 299 (Pt A), 241–251.

Bilecki, W., Maćkowiak, M., 2023. Gene expression and epigenetic regulation in the prefrontal cortex of schizophrenia. Genes (Basel) 14, 243.

Bilinska, K., Jakubowska, P., Von Bartheld, C.S., Butowt, R., 2020. Expression of the SARS-CoV-2 entry proteins, ACE2 and TMPRSS2, in cells of the olfactory epithelium: identification of cell types and trends with age. ACS Chem. Neurosci. 11, 1555–1562.

Biltz, R.G., Sawicki, C.M., Sheridan, J.F., Godbout, J.P., 2022. The neuroimmunology of social-stress-induced sensitization. Nat. Immunol. 23, 1527–1535.

Binzer, S., Jiang, X., Hillert, J., Manouchehrinia, A., 2021. Depression and multiple sclerosis: a bidirectional Mendelian randomisation study. Mult. Scler. 27, 1799–1802.

Biondetti, E., Santin, M.D., Valabrègue, R., Mangone, G., Gaurav, R., et al., 2021. The spatiotemporal changes in dopamine, neuromelanin and iron characterizing Parkinson's disease. Brain 144, 3114–3125.

Birey, F., Kloc, M., Chavali, M., Hussein, I., Wilson, M., et al., 2015. Genetic and stress-induced loss of NG2 glia triggers emergence of depressive-like behaviors through reduced secretion of FGF2. Neuron 88, 941–956.

Biria, M., Banca, P., Healy, M.P., Keser, E., Sawiak, S.J., et al., 2023. Cortical glutamate and GABA are related to compulsive behaviour in individuals with obsessive compulsive disorder and healthy controls. Nat. Commun. 14, 3324.

Birkinshaw, H., Friedrich, C.M., Cole, P., Eccleston, C., Serfaty, M., et al., 2023. Antidepressants for pain management in adults with chronic pain: a network meta-analysis. Cochrane Database Syst. Rev. 5, CD014682.

Birney, E., Smith, G.D., Greally, J.M., 2016. Epigenome-wide association studies and the interpretation of disease-omics. PLoS Genet. 12, e1006105.

Birnie, M.T., Short, A.K., de Carvalho, G.B., Taniguchi, L., Gunn, B.G., et al., 2023. Stress-induced plasticity of a CRH/GABA projection disrupts reward behaviors in mice. Nat. Commun. 14, 1088.

Bisgaard, T.H., Allin, K.H., Keefer, L., Ananthakrishnan, A.N., Jess, T., 2022. Depression and anxiety in inflammatory bowel disease: epidemiology, mechanisms and treatment. Nat. Rev. Gastroenterol. Hepatol. 19, 717–726.

Bishop, J.R., Lee, A.M., Mills, L.J., Thuras, P.D., Eum, S., et al., 2018. Methylation of FKBP5and SLC6A4 in relation to treatment response to mindfulness based stress reduction for posttraumatic stress disorder. Front. Psychiatry 9, 418.

Bisson, J.I., Wright, L.A., Jones, K.A., Lewis, C., Phelps, A.J., et al., 2021. Preventing the onset of post traumatic stress disorder. Clin. Psychol. Rev. 86, 102004.

Bissonette, G.B., Roesch, M.R., 2016. Development and function of the midbrain dopamine system: what we know and what we need to. Genes Brain Behav. 15, 62–73.

Bjornevik, K., Cortese, M., Healy, B.C., Kuhle, J., Mina, M.J., et al., 2022. Longitudinal analysis reveals high prevalence of Epstein-Barr virus associated with multiple sclerosis. Science 375, 296–301.

Black, C., Miller, B.J., 2015. Meta-analysis of cytokines and chemokines in suicidality: distinguishing suicidal versus nonsuicidal patients. Biol. Psychiatry 78, 28–37.

Black, N., Stockings, E., Campbell, G., Tran, L.T., Zagic, D., et al., 2019. Cannabinoids for the treatment of mental disorders and symptoms of mental disorders: a systematic review and meta-analysis. Lancet Psychiatry 6 (12), 995–1010.

Blacker, C.J., Frye, M.A., Morava, E., Kozicz, T., Veldic, M., 2019. A review of epigenetics of PTSD in comorbid psychiatric conditions. Genes 10, 140.

Blackhurst, B.M., Funk, K.E., 2023. Viral pathogens increase risk of neurodegenerative disease. Nat. Rev. Neurol. 19, 259–260.

Bland, S.T., Tamlyn, J.P., Barrientos, R.M., Greenwood, B.N., Watkins, L.R., et al., 2007. Expression of fibroblast growth factor-2 and brain-derived neurotrophic factor mRNA in the medial prefrontal cortex and hippocampus after uncontrollable or controllable stress. Neuroscience 144, 1219–1228.

Blankenstein, N.E., Peper, J.S., Crone, E.A., van Duijvenvoorde, A.C.K., 2017. Neural mechanisms underlying risk and ambiguity attitudes. J. Cogn. Neurosci. 29, 1845–1859.

Blaszczak, A.M., Jalilvand, A., Hsueh, W.A., 2021. Adipocytes, innate immunity and obesity: a mini-review. Front. Immunol. 12, 650768.

Blaustein, J.D., Ismail, N., Holder, M.K., 2016. Review: puberty as a time of remodeling the adult response to ovarian hormones. J. Steroid Biochem. Mol. Biol. 160, 2–8.

Blauwendraat, C., Nalls, M.A., Singleton, A.B., 2020. The genetic architecture of Parkinson's disease. Lancet Neurol. 19, 170–178.

Bleau, C., Karelis, A.D., St-Pierre, D.H., Lamontagne, L., 2015. Crosstalk between intestinal microbiota, adipose tissue and skeletal muscle as an early event in systemic low-grade inflammation and the development of obesity and diabetes. Diabetes Metab. Res. Rev. 31, 545561.

Blevins, C.L., Sagui, S.J., Bennett, J.M., 2017. Inflammation and positive affect: examining the stress-buffering hypothesis with data from the National Longitudinal Study of Adolescent to Adult Health. Brain Behav. Immun. 61, 21–26.

Blier, P., 2013. Exploiting N-methyl-d-aspartate channel blockade for a rapid antidepressant response in major depressive disorder. Biol. Psychiatry 74, 238–239.

Blier, P., 2016a. Neurobiology of depression and mechanism of action of depression treatments. J. Clin. Psychiatry 77, e319.

Blier, P., 2016b. Neurotransmitter targeting in the treatment of depression. J. Clin. Psychiatry 74, 19–24.

Block, C.L., Eroglu, O., Mague, S.D., Smith, C.J., Ceasrine, A.M., et al., 2022. Prenatal environmental stressors impair postnatal microglia function and adult behavior in males. Cell Rep. 40, 111161.

Blomstrom, A., Karlsson, H., Gardner, R., Jorgensen, L., Magnusson, C., et al., 2016. Associations between maternal infection during pregnancy, childhood infections, and the risk of subsequent psychotic disorder—a Swedish cohort study of nearly 2 million individuals. Schizophr. Bull. 42, 125–133.

Bloomberg, M., Brocklebank, L., Hamer, M., Steptoe, A., 2023. Joint associations of physical activity and sleep duration with cognitive ageing: longitudinal analysis of an English cohort study. Lancet Healthy Longev. 4, e345–e353.

Bloomfield, S.F., Rook, G.A., Scott, E.A., Shanahan, F., Stanwell-Smith, R., et al., 2016. Time to abandon the hygiene hypothesis: new perspectives on allergic disease, the human microbiome, infectious disease prevention and the role of targeted hygiene. Perspect. Public Health 136, 213–224.

Bloomfield, M.A., McCutcheon, R.A., Kempton, M., Freeman, T.P., Howes, O., 2019. The effects of psychosocial stress on dopaminergic function and the acute stress response. elife 8, e46797.

Bloomgarden, Z., Rapaport, R., 2023. Diabetes trends in youth. J. Diabetes 15, 286–288.

Bluett, R.J., Báldi, R., Haymer, A., Gaulden, A.D., Hartley, N.D., et al., 2017. Endocannabinoid signalling modulates susceptibility to traumatic stress exposure. Nat. Commun. 8, 14782.

Blüher, M., 2020. Metabolically healthy obesity. Endocr. Rev. 41, 405–420.

Blumenfeld, J., Yip, O., Kim, M.J., Huang, Y., 2024. Cell type-specific roles of APOE4 in Alzheimer disease. Nat. Rev. Neurosci. 25, 91–110.

Bobyn, J., Mangano, E.N., Gandhi, A., Nelson, E., Moloney, K., et al., 2012. Viral-toxin interactions and Parkinson's disease: poly I:C priming enhanced the neurodegenerative effects of paraquat. J. Neuroinflammation 9, 86.

Boccia, M., D'Amico, S., Bianchini, F., Marano, A., Giannini, A.M., et al., 2016. Different neural modifications underpin PTSD after different traumatic events: an fMRI meta-analytic study. Brain Imaging Behav. 10, 226–237.

Bodden, C., van den Hove, D., Lesch, K.P., Sachser, N., 2017. Impact of varying social experiences during life history on behaviour, gene expression, and vasopressin receptor gene methylation in mice. Sci. Rep. 7, 8719.

Boehme, M., Guzzetta, K.E., Wasén, C., Cox, L.M., 2023. The gut microbiota is an emerging target for improving brain health during ageing. Gut Microbiome (Camb). 4, E2.

Boff, T.C., Soares, S.J.B., Lima, M.D.M., Ignácio, Z.M., 2022. Glutamate function in anxiety disorders and ocd: evidence from clinical and translational studies. In: Glutamate and Neuropsychiatric Disorders: Current and Emerging Treatments. Springer International Publishing, Cham, pp. 539–570.

Bogenschutz, M.P., Ross, S., 2017. Therapeutic applications of classic hallucinogens. Curr. Top. Behav. Neurosci. https://doi.org/10.1007/7854_2016_464.

Boger, H.A., Granholm, A.C., McGinty, J.F., Middaugh, L.D., 2010. A dual-hit animal model for age-related parkinsonism. Prog. Neurobiol. 90, 217–229.

Bohlson, S.S., Tenner, A.J., 2023. Complement in the brain: contributions to neuroprotection, neuronal plasticity, and neuroinflammation. Annu. Rev. Immunol. 41, 431–452.

Bohnacker, S., Troisi, F., de Los Reyes Jiménez, M., Esser-von Bieren, J., 2020. What can parasites tell us about the pathogenesis and treatment of asthma and allergic diseases. Front. Immunol. 11, 2106.

Boksa, P., 2008. Maternal infection during pregnancy and schizophrenia. J. Psychiatry Neurosci. 33, 183–185.

Bolte, A.C., Dutta, A.B., Hurt, M.E., Smirnov, I., Kovacs, M.A., et al., 2020. Meningeal lymphatic dysfunction exacerbates traumatic brain injury pathogenesis. Nat. Commun. 11, 4.

Bolton, J.L., Short, A.K., Othy, S., Kooiker, C.L., Shao, M., et al., 2022. Early stress-induced impaired microglial pruning of excitatory synapses on immature CRH-expressing neurons provokes aberrant adult stress responses. Cell Rep. 38, 110600.

Bombay, A., Matheson, K., Anisman, H., 2011. The impact of stressors on second generation Indian Residential School survivors. Transcult. Psychiatry 48, 367–391.

Bombay, A., Matheson, K., Anisman, H., 2014. The intergenerational effects of Indian Residential Schools: implications for the concept of historical trauma. Transcult. Psychiatry 51, 320–338.

Bonafina, A., Trinchero, M.F., Ríos, A.S., Bekinschtein, P., Schinder, A.F., et al., 2019. GDNF and GFRα1 are required for proper integration of adult-born hippocampal neurons. Cell Rep. 29, 4308–4319.

Bondy, S.C., 2016. Low levels of aluminum can lead to behavioral and morphological changes associated with Alzheimer's disease and age-related neurodegeneration. Neurotoxicology 52, 222–229.

Bonfili, L., Cecarini, V., Berardi, S., Scarpona, S., Suchodolski, J.S., et al., 2017. Microbiota modulation counteracts Alzheimer's disease progression influencing neuronal proteolysis and gut hormones plasma levels. Sci. Rep. 7, 2426.

Bonin, R.P., De Koninck, Y., 2014. A spinal analog of memory reconsolidation enables reversal of hyperalgesia. Nat. Neurosci. 17, 1043–1045.

Bonini, M.G., Sargis, R.M., 2018. Environmental toxicant exposures and type 2 diabetes mellitus: two interrelated public health problems on the rise. Curr. Opin Toxicol. 7, 52–59.

Bonn-Miller, M.O., Sisley, S., Riggs, P., Yazar-Klosinski, B., Wang, J.B., et al., 2021. The short-term impact of 3 smoked cannabis preparations versus placebo on PTSD symptoms: a randomized cross-over clinical trial. PLoS One 16, e0246990.

Bonomini-Gnutzmann, R., Plaza-Díaz, J., Jorquera-Aguilera, C., Rodríguez-Rodríguez, A., Rodríguez-Rodríguez, F., 2022. Effect of intensity and duration of exercise on gut microbiota in humans: a systematic review. Int. J. Environ. Res. Public Health 19, 9518.

Booth, F.W., Roberts, C.K., Thyfault, J.P., Ruegsegger, G.N., Toedebusch, R.G., 2017. Role of inactivity in chronic diseases: evolutionary insight and pathophysiological mechanisms. Physiol. Rev. 97, 1351–1402.

Boozalis, T., Teixeira, A.L., Cho, R.Y., Okusaga, O., 2018. C-reactive protein correlates with negative symptoms in patients with schizophrenia. Front. Public Health 5, 360.

Border, R., Johnson, E.C., Evans, L.M., Smolen, A., Berley, N., et al., 2019. No support for historical candidate gene or candidate gene-by-interaction hypotheses for major depression across multiple large samples. Am. J. Psychiatry 176, 376–387.

Borges, L., Passos, M.E.P., Silva, M.B.B., Santos, V.C., Momesso, C.M., et al., 2019. Dance training improves cytokine secretion and viability of neutrophils in diabetic patients. Mediat. Inflamm. 2019, 2924818.

Borghammer, P., Horsager, J., Andersen, K., Van Den Berge, N., Raunio, A., et al., 2021. Neuropathological evidence of body-first vs. brain-first Lewy body disease. Neurobiol. Dis. 161, 105557.

Borodovitsyna, O., Flamini, M.D., Chandler, D.J., 2018. Acute stress persistently alters locus coeruleus function and anxiety-like behavior in adolescent rats. Neuroscience 373, 7–19.

Boroujeni, M.E., Simani, L., Bluyssen, H.A.R., Samadikhah, H.R., Zamanlui Benisi, S., et al., 2021. Inflammatory response leads to neuronal death in human post-mortem cerebral cortex in patients with COVID-19. ACS Chem. Neurosci. 12, 2143–2150.

Borrajo López, A., Penedo, M.A., Rivera-Baltanas, T., Pérez-Rodríguez, D., Alonso-Crespo, D., et al., 2021. Microglia: the real foe in HIV-1-associated neurocognitive disorders? Biomedicine 9, 925.

Borroto-Escuela, D.O., Ambrogini, P., Chruścicka, B., Lindskog, M., Crespo-Ramirez, M., et al., 2021. The role of central serotonin neurons and 5-ht heteroreceptor complexes in the pathophysiology of depression: a historical perspective and future prospects. Int. J. Mol. Sci. 22, 1927.

Bosch-Bouju, C., Larrieu, T., Linders, L., Manzoni, O.J., Layé, S., 2016. Endocannabinoid-mediated plasticity in nucleus accumbens controls vulnerability to anxiety after social defeat stress. Cell Rep. 16, 1237–1242.

Bostancıklıoğlu, M., 2020. SARS-CoV2 entry and spread in the lymphatic drainage system of the brain. Brain Behav. Immun. 87, 122–123.

Boukhris, T., Sheehy, O., Mottron, L., Bérard, A., 2016. Antidepressant use during pregnancy and the risk of autism spectrum disorder in children. JAMA Pediatr. 170, 117–120.

Bountress, K.E., Wei, W., Sheerin, C., Chung, D., Amstadter, A.B., et al., 2017. Relationships between GAT1 and PTSD, depression, and substance use disorder. Brain Sci. 7.

Bourassa, K., Sbarra, D.A., 2016. Body mass and cognitive decline are indirectly associated via inflammation among ageing adults. Brain Behav. Immun. 60, 63–70.

Bourhy, L., Mazeraud, A., Costa, L.H.A., Levy, J., Rei, D., et al., 2022. Silencing of amygdala circuits during sepsis prevents the development of anxiety-related behaviours. Brain 145, 1391–1409.

Bourmistrova, N.W., Solomon, T., Braude, P., Strawbridge, R., Carter, B., 2022. Long-term effects of COVID-19 on mental health: a systematic review. J. Affect. Disord. 299, 118–125.

Bowen, K.K., Dempsey, R.J., Vemuganti, R., 2011. Adult interleukin-6 knockout mice show compromised neurogenesis. Neuroreport 22, 126–130.

Bowker, N., Shah, R.L., Sharp, S.J., Luan, J., Stewart, I.D., et al., 2020. Meta-analysis investigating the role of interleukin-6 mediated inflammation in type 2 diabetes. EBioMedicine 61, 103062.

Boyce, W.T., Levitt, P., Martinez, F.D., McEwen, B.S., Shonkoff, J.P., 2021. Genes, environments, and time: the biology of adversity and resilience. Pediatrics 147, e20201651.

Boyd, J.E., Lanius, R.A., McKinnon, M.C., 2018. Mindfulness-based treatments for posttraumatic stress disorder: a review of the treatment literature and neurobiological evidence. J. Psychiatry Neurosci. 43, 7–25.

Boyer, S.M., Caplan, J.E., Edwards, L.K., 2022. Trauma-related dissociation and the dissociative disorders:: neglected symptoms with severe public health consequences. Dela J. Public Health 8, 78–84.

Boyle, C.C., Cole, S.W., Irwin, M.R., Eisenberger, N.I., Bower, J.E., 2023. The role of inflammation in acute psychosocial stress-induced modulation of reward processing in healthy female adults. Brain Behav. Immun. Health 28, 100588.

Braak, H., Braak, E., 1995. Staging of Alzheimer's disease-related neurofibrillary changes. Neurobiol. Aging 16, 271–278.

Braak, H., Braak, E., Yilmazer, D., de Vos, R.A., Jansen, E.N., Bohl, J., 1996. Pattern of brain destruction in Parkinson's and Alzheimer's diseases. J. Neural Transm. (Vienna) 103, 455–490.

Brabazon, F., Wilson, C.M., Jaiswal, S., Reed, J., Frey, W.H., et al., 2017. Intranasal insulin treatment of an experimental model of moderate traumatic brain injury. J. Cereb. Blood Flow Metab. 37, 3203–3218.

Braga, J., Lepra, M., Kish, S.J., Rusjan, P.M., Nasser, Z., et al., 2023. Neuroinflammation after COVID-19 with persistent depressive and cognitive symptoms. JAMA Psychiatry, e231321.

Braithwaite, E.C., Kundakovic, M., Ramchandani, P.G., Murphy, S.E., Champagne, F.A., 2015. Maternal prenatal depressive symptoms predict infant NR3C1 1F and BDNF IV DNA methylation. Epigenetics 10, 408–417.

Bramante, C.T., Buse, J.B., Liebovitz, D.M., Nicklas, J.M., Puskarich, M. A., et al., 2023. Outpatient treatment of COVID-19 and incidence of post-COVID-19 condition over 10 months (COVID-OUT): a multicentre, randomised, quadruple-blind, parallel-group, phase 3 trial. Lancet Infect. Dis. S1473-3099(23)00299-2.

Bramson, B., Meijer, S., van Nuland, A., Toni, I., Roelofs, K., 2023. Anxious individuals shift emotion control from lateral frontal pole to dorsolateral prefrontal cortex. Nat. Commun. 14, 4880.

Brancaccio, M., Edwards, M.D., Patton, A.P., Smyllie, N.J., Chesham, J. E., et al., 2019. Cell-autonomous clock of astrocytes drives circadian behavior in mammals. Science 363, 187–192.

Brancato, A., Bregman, D., Ahn, H.F., Pfau, M.L., Menard, C., et al., 2017. Sub-chronic variable stress induces sex-specific effects on

glutamatergic synapses in the nucleus accumbens. Neuroscience 350, 180–189.

Braniste, V., Al-Asmakh, M., Kowal, C., Anuar, F., Abbaspour, A., et al., 2014. The gut microbiota influences blood-brain barrier permeability in mice. Sci. Transl. Med. 6, 263ra158.

Branscombe, N.R., Reynolds, K.J., 2015. Toward person plasticity: individual and collective approaches. In: Reynolds, K.J., Branscombe, N. R. (Eds.), Psychology of Change: Life Contexts, Experiences, and Identities. Psychology Press, New York, pp. 3–22.

Braun, M., Iliff, J.J., 2020. The impact of neurovascular, blood-brain barrier, and glymphatic dysfunction in neurodegenerative and metabolic diseases. Int. Rev. Neurobiol. 154, 413–436.

Bräuner, E.V., Lim, Y.H., Koch, T., Mori, T.A., Beilin, L., et al., 2021. Sexdependent associations between maternal prenatal stressful life events, BMI trajectories and obesity risk in offspring: the Raine Study. Compr. Psychoneuroendocrinol. 7, 100066.

Bravo, J.A., Forsythe, P., Chew, M.V., Escaravage, E., Savignac, H.M., et al., 2011. Ingestion of Lactobacillus strain regulates emotional behavior and central GABA receptor expression in a mouse via the vagus nerve. Proc. Natl. Acad. Sci. USA 108, 16050–16055.

Breckenridge, C.B., Berry, C., Chang, E.T., Sielken, R.L., Mandel, J.S., 2016. Association between Parkinson's disease and cigarette smoking, rural living, well-water consumption, farming and pesticide use: Systematic review and meta-analysis. PLoS One 11, e0151841.

Breen, M.S., Tylee, D.S., Maihofer, A.X., Neylan, T.C., Mehta, D., et al., 2018. PTSD blood transcriptome mega-analysis: shared inflammatory pathways across biological sex and modes of trauma. Neuropsychopharmacology 43, 469–481.

Breilmann, J., Girlanda, F., Guaiana, G., Barbui, C., Cipriani, A., et al., 2019. Benzodiazepines versus placebo for panic disorder in adults. Cochrane Database Syst. Rev. 3, CD010677.

Brenhouse, H.C., 2023. Points of divergence on a bumpy road: early development of brain and immune threat processing systems following postnatal adversity. Mol. Psychiatry 28, 269–283.

Brenhouse, H.C., Danese, A., Grassi-Oliveira, R., 2019. Neuroimmune impacts of early-life stress on development and psychopathology. Curr. Top. Behav. Neurosci. 43, 423–447.

Brenner, L.A., Forster, J.E., Stearns-Yoder, K.A., Stamper, C.E., Hoisington, A.J., et al., 2020. Evaluation of an immunomodulatory probiotic intervention for veterans with co-occurring mild traumatic brain injury and posttraumatic stress disorder: a pilot study. Front. Neurol. 11, 1015.

Brenner, L.A., Forster, J.E., Gradus, J.L., Hostetter, T.A., Hoffmire, C.A., et al., 2023. Associations of military-related traumatic brain injury with new-onset mental health conditions and suicide risk. JAMA Netw. Open 6, e2326296.

Bresnahan, M., Hornig, M., Schultz, A.F., Gunnes, N., Hirtz, D., Lie, K. K., Lipkin, W.I., 2015. Association of maternal report of infant and toddler gastrointestinal symptoms with autism: evidence from a prospective birth cohort. JAMA Psychiatry 72 (5), 466–474.

Brew, B.K., Lundholm, C., Caffrey Osvald, E., Chambers, G., Öberg, S., et al., 2022. Early-life adversity due to bereavement and inflammatory diseases in the next generation: a population study in transgenerational stress exposure. Am. J. Epidemiol. 191, 38–48.

Bridges, A.A., Bassler, B.L., 2019. The intragenus and interspecies quorum-sensing autoinducers exert distinct control over Vibrio cholerae biofilm formation and dispersal. PLoS Biol. 17, e3000429.

Bridgewater, L.C., Zhang, C., Wu, Y., Hu, W., Zhang, Q., et al., 2017. Gender-based differences in host behavior and gut microbiota composition in response to high fat diet and stress in a mouse model. Sci. Rep. 7, 10776.

Brinks, V., De Kloet, E.R., Oitzl, M.S., 2009. Corticosterone facilitates extinction of fear memory in BALB/c mice but strengthens cue related fear in C57BL/6 mice. Exp. Neurol. 216, 375–382.

Briones-Buixassa, L., Milà, R.M., Aragonès, J., Bufill, E., Olaya, B., Arrufat, F.X., 2015. Stress and multiple sclerosis: a systematic review considering potential moderating and mediating factors and methods of assessing stress. Health Psychol. Open 2. 2055102915612271.

Brody, G.H., Yu, T., Miller, G.E., Chen, E., 2015. Discrimination, racial identity, and cytokine levels among African-American adolescents. J. Adolesc. Health 56, 496–501.

Brooks, A.K., Janda, T.M., Lawson, M.A., Rytych, J.L., Smith, R.A., et al., 2017. Desipramine decreases expression of human and murine indoleamine-2,3-dioxygenases. Brain Behav. Immun. 62, 219–229.

Brouwer, R.M., Klein, M., Grasby, K.L., Schnack, H.G., Jahanshad, N., et al., 2022. Genetic variants associated with longitudinal changes in brain structure across the lifespan. Nat. Neurosci. 25, 421–432.

Brower, K.J., 2021. Professional stigma of mental health issues: physicians are both the cause and solution. Acad. Med. 96, 635–640.

Brown, A.S., Derkits, E.J., 2010. Prenatal infection and schizophrenia: a review of epidemiologic and translational studies. Am. J. Psychiatry 167, 261–280.

Brown, A.S., Begg, M.D., Gravenstein, S., Schaefer, C.A., Wyatt, R.J., et al., 2004. Serologic evidence of prenatal influenza in the etiology of schizophrenia. Arch. Gen. Psychiatry 61, 774–780.

Brown, A.S., Surcel, H.M., Hinkka-Yli-Salomaki, S., Cheslack-Postava, K., Bao, Y., et al., 2015. Maternal thyroid autoantibody and elevated risk of autism in a national birth cohort. Prog. Neuro-Psychopharmacol. Biol. Psychiatry 57, 86–92.

Brown, H.K., Ray, J.G., Wilton, A.S., Lunsky, Y., Gomes, T., Vigod, S.N., 2017. Association between serotonergic antidepressant use during pregnancy and autism spectrum disorder in children. JAMA 317, 1544–1552.

Brown, K.L., Ramlall, V., Zietz, M., Gisladottir, U., Tatonetti, N.P., 2024. Estimating the heritability of SARS-CoV-2 susceptibility and COVID-19 severity. Nat. Commun. 15, 367.

Browne, H.P., Forster, S.C., Anonye, B.O., Kumar, N., Neville, B.A., et al., 2016. Culturing of 'unculturable' human microbiota reveals novel taxa and extensive sporulation. Nature 533, 543–546.

Bruce-Keller, A.J., Salbaum, J.M., Berthoud, H.R., 2018. Harnessing gut microbes for mental health: getting from here to there. Biol. Psychiatry 83 (3), 214–223.

Bruenig, D., Mehta, D., Morris, C.P., Harvey, W., Lawford, B., et al., 2017. Genetic and serum biomarker evidence for a relationship between TNFα and PTSD in Vietnam war combat veterans. Compr. Psychiatry 74, 125–133.

Brufsky, A., Lotze, M.T., 2020. DC/L-SIGNs of hope in the COVID-19 pandemic. J. Med. Virol. 92, 1396–1398.

Brugiroux, S., Beutler, M., Pfann, C., Garzetti, D., Ruscheweyh, H.J., et al., 2016. Genome-guided design of a defined mouse microbiota that confers colonization resistance against Salmonella enterica serovar Typhimurium. Nat. Microbiol. 2, 16215.

Brundin, L., Sellgren, C.M., Lim, C.K., Grit, J., Pålsson, E., et al., 2016. An enzyme in the kynurenine pathway that governs vulnerability to suicidal behavior by regulating excitotoxicity and neuroinflammation. Transl. Psychiatry 6, e865.

Brundin, P., Dave, K.D., Kordower, J.H., 2017. Therapeutic approaches to target alpha-synuclein pathology. Exp. Neurol. 298, 225–235.

Brunet, A., Saumier, D., Liu, A., Streiner, D.L., Tremblay, J., Pitman, R. K., 2018. Reduction of PTSD symptoms with pre-reactivation propranolol therapy: a randomized controlled trial. Am. J. Psychiatry 175, 427–433.

Bruno, A., Dolcetti, E., Rizzo, F.R., Fresegna, D., Musella, A., et al., 2020. Inflammation-associated synaptic alterations as shared threads in depression and multiple sclerosis. Front. Cell. Neurosci. 14, 169.

Brusaferri, L., Alshelh, Z., Martins, D., Kim, M., Weerasekera, A., et al., 2022. The pandemic brain: neuroinflammation in non-infected individuals during the COVID-19 pandemic. Brain Behav. Immun. 102, 89–97.

Brusca, S.B., Abramson, S.B., Scher, J.U., 2014. Microbiome and mucosal inflammation as extra-articular triggers for rheumatoid arthritis and autoimmunity. Curr. Opin. Rheumatol. 26, 101–107.

Brust, T.F., Morgenweck, J., Kim, S.A., Rose, J.H., Locke, J.L., et al., 2016. Biased agonists of the kappa opioid receptor suppress pain and itch without causing sedation or dysphoria. Sci. Signal. 9, ra117.

Bruttger, J., Karram, K., Wörtge, S., Regen, T., Marini, F., et al., 2015. Genetic cell ablation reveals clusters of local self-renewing microglia in the mammalian central nervous system. Immunity 43, 92–106.

Brydges, C.R., Bhattacharyya, S., Dehkordi, S.M., Milaneschi, Y., Penninx, B., et al., 2022. Metabolomic and inflammatory signatures of symptom dimensions in major depression. Brain Behav. Immun. 102, 42–52.

Brydon, L., Walker, C., Wawrzyniak, A.J., Chart, H., Steptoe, A., 2009. Dispositional optimism and stress-induced changes in immunity and negative mood. Brain Behav. Immun. 23, 810–816.

Brynge, M., Sjöqvist, H., Gardner, R.M., Lee, B.K., Dalman, C., Karlsson, H., 2022. Maternal infection during pregnancy and likelihood of autism and intellectual disability in children in Sweden: a negative control and sibling comparison cohort study. Lancet Psychiatry 9, 782–791.

Bsibsi, M., Persoon-Deen, C., Verwer, R.W., Meeuwsen, S., Ravid, R., Van Noort, J.M., 2006. Toll-like receptor 3 on adult human astrocytes triggers production of neuroprotective mediators. Glia 53, 688–695.

Bu, X.L., Yao, X.Q., Jiao, S.S., Zeng, F., Liu, Y.H., et al., et al., 2015. A study on the association between infectious burden and Alzheimer's disease. Eur. J. Neurol. 22, 1519–1525.

Buck, S.A., Quincy Erickson-Oberg, M., Logan, R.W., Freyberg, Z., 2022. Relevance of interactions between dopamine and glutamate neurotransmission in schizophrenia. Mol. Psychiatry 27, 3583–3591.

Bucolo, M., Rance, M., Nees, F., Ruttorf, M., Stella, G., et al., 2022. Cortical networks underlying successful control of nociceptive processing using real-time fMRI. Front. Pain Res. (Lausanne) 3, 969867.

Buddhala, C., Loftin, S.K., Kuley, B.M., Cairns, N.J., Campbell, M.C., et al., 2015. Dopaminergic, serotonergic, and noradrenergic deficits in Parkinson disease. Ann. Clin. Transl. Neurol. 2, 949–959.

Buffie, C.G., Bucci, V., Stein, R.R., McKenney, P.T., Ling, L., et al., 2015. Precision microbiome reconstitution restores bile acid mediated resistance to Clostridium difficile. Nature 517, 205–208.

Buffington, S.A., Di Prisco, G.V., Auchtung, T.A., Ajami, N.J., Petrosino, J.F., et al., 2016. Microbial reconstitution reverses maternal diet-induced social and synaptic deficits in offspring. Cell 165, 1762–1775.

Buimer, E.E.L., Brouwer, R.M., Mandl, R.C.W., Pas, P., Schnack, H.G., Hulshoff Pol, H.E., 2022. Adverse childhood experiences and fronto-subcortical structures in the developing brain. Front. Psychiatry 13, 955871.

Bukhbinder, A.S., Ling, Y., Harris, K., Jiang, X., Schulz, P.E., 2023. Do vaccinations influence the development of Alzheimer disease? Hum. Vaccin. Immunother. 19, 2216625.

Bundhun, P.K., Soogund, M.Z., Huang, F., 2017. Impact of systemic lupus erythematosus on maternal and fetal outcomes following pregnancy: a meta-analysis of studies published between years 2001-2016. J. Autoimmun. 79, 1727.

Burbulla, L.F., Song, P., Mazzulli, J.R., Zampese, E., Wong, Y.C., et al., 2017. Dopamine oxidation mediates mitochondrial and lysosomal dysfunction in Parkinson's disease. Science 357, 1255–1261.

Burchi, E., Pallanti, S., 2019. Diagnostic issues in early-onset obsessive-compulsive disorder and their treatment implications. Curr. Neuropharmacol. 17, 672–680.

Bürgin, D., O'Donovan, A., d'Huart, D., di Gallo, A., Eckert, A., et al., 2019. Adverse childhood experiences and telomere length a look into the heterogeneity of findings-a narrative review. Front. Neurosci. 13, 490.

Burgos-Robles, A., Kimchi, E.Y., Izadmehr, E.M., Porzenheim, M.J., Ramos-Guasp, W.A., et al., 2017. Amygdala inputs to prefrontal cortex guide behavior amid conflicting cues of reward and punishment. Nat. Neurosci. 20, 824–835.

Burke, N.N., Finn, D.P., McGuire, B.E., Roche, M., 2017. Psychological stress in early life as a predisposing factor for the development of chronic pain: clinical and preclinical evidence and neurobiological mechanisms. J. Neurosci. Res. 95, 1257–1270.

Burns, M.N., Nawacki, E., Kwasny, M.J., Pelletier, D., Mohr, D.C., 2014. Do positive or negative stressful events predict the development of new brain lesions in people with multiple sclerosis? Psychol. Med. 44, 349–359.

Burt, M.A., Tse, Y.C., Boksa, P., Wong, T.P., 2013. Prenatal immune activation interacts with stress and corticosterone exposure later in life to modulate N-methyl-D-aspartate receptor synaptic function and plasticity. Int. J. Neuropsychopharmacol. 16, 1835–1848.

Busbee, P.B., Bam, M., Yang, X., Abdulla, O.A., Zhou, J., et al., 2022. Dysregulated TP53 among PTSD patients leads to downregulation of miRNA let-7a and promotes an inflammatory Th17 phenotype. Front. Immunol. 12, 815840.

Bush, N.R., Savitz, J., Coccia, M., Jones-Mason, K., Adler, N., et al., 2021. Maternal stress during pregnancy predicts infant infectious and noninfectious illness. J. Pediatr. 228, 117–125.e2.

Bushnell, M.C., Ceko, M., Low, L.A., 2013. Cognitive and emotional control of pain and its disruption in chronic pain. Nat. Rev. Neurosci. 14 (7), 502–511.

Businaro, R., Corsi, M., Azzara, G., Di Raimo, T., Laviola, G., et al., 2016. Interleukin-18 modulation in autism spectrum disorders. J. Neuroinflammation 13, 2.

Buske-Kirschbaum, A., Gierens, A., Hollig, H., Hellhammer, D.H., 2002. Stress-induced immunomodulation is altered in patients with atopic dermatitis. J. Neuroimmunol. 129, 161–167.

Buske-Kirschbaum, A., Kern, S., Ebrecht, M., Hellhammer, D.H., 2007. Altered distribution of leukocyte subsets and cytokine production in response to acute psychosocial stress in patients with psoriasis vulgaris. Brain Behav. Immun. 21, 92–99.

Busse, D., Yim, I.S., Campos, B., Marshburn, C.K., 2017. Discrimination and the HPA axis: current evidence and future directions. J. Behav. Med. 40, 539–552.

Busse, J.W., Wang, L., Kamaleldin, M., Craigie, S., Riva, J.J., et al., 2018. Opioids for chronic noncancer pain: a systematic review and meta-analysis. JAMA 320 (23), 2448–2460.

Bussières, A., Hancock, M.J., Elklit, A., Ferreira, M.L., Ferreira, P.H., et al., 2023. Adverse childhood experience is associated with an increased risk of reporting chronic pain in adulthood: a systematic review and meta-analysis. Eur. J. Psychotraumatol. 14 (2), 2284025.

Butler, M.I., Long-Smith, C., Moloney, G.M., Morkl, S., O'Mahony, S.M., et al., 2022. The immune-kynurenine pathway in social anxiety disorder. Brain Behav. Immun. 99, 317–326.

Butler, M.I., Bastiaanssen, T.F., Long-Smith, C., Morkl, S., Berding, K., et al., 2023. The gut microbiome in social anxiety disorder: evidence of altered composition and function. Transl. Psychiatry 13, 95.

Byrne, C.S., Chambers, E.S., Alhabeeb, H., Chhina, N., Morrison, D.J., et al., 2016. Increased colonic propionate reduces anticipatory reward responses in the human striatum to high energy foods. Am. J. Clin. Nutr. 104, 5–14.

Byrne, P., Demasi, M., Jones, M., Smith, S.M., O'Brien, K.K., DuBroff, R., 2022. Evaluating the association between low-density lipoprotein cholesterol reduction and relative and absolute effects of statin treatment: a systematic review and meta-analysis. JAMA Intern. Med. 182, 474–481.

Byun, Y.G., Kim, N.S., Kim, G., Jeon, Y.S., Choi, J.B., et al., 2023. Stress induces behavioral abnormalities by increasing expression of phagocytic receptor, MERTK, in astrocytes to promote synapse phagocytosis. Immunity. S1074-7613(23)00318-7.

Cabezudo, D., Baekelandt, V., Lobbestael, E., 2020. Multiple-hit hypothesis in Parkinson's disease: LRRK2 and inflammation. Front. Neurosci. 14, 376.

Cacioppo, J.T., Cacioppo, S., Boomsma, D.I., 2014. Evolutionary mechanisms for loneliness. Cognit. Emot. 28, 3–21.

Cacioppo, J.T., Cacioppo, S., Capitanio, J.P., Cole, S.W., 2015. The neuroendocrinology of social isolation. Annu. Rev. Psychol. 66, 733–767.

Cai, T., Abel, L., Langford, O., Monaghan, G., Aronson, J.K., et al., 2021. Associations between statins and adverse events in primary prevention of cardiovascular disease: systematic review with pairwise, network, and dose-response meta-analyses. BMJ 374, n1537.

Cai, T., Zheng, S.P., Shi, X., Yuan, L.Z., Hu, H., et al., 2022. Therapeutic effect of fecal microbiota transplantation on chronic unpredictable mild stress-induced depression. Front. Cell. Infect. Microbiol. 12, 900652.

Cain, D.W., Cidlowski, J.A., 2017. Immune regulation by glucocorticoids. Nat. Rev. Immunol. 17, 233–247.

Çakici, N., van Beveren, N.J.M., Judge-Hundal, G., Koola, M.M., Sommer, I.E.C., 2019. An update on the efficacy of anti-inflammatory agents for patients with schizophrenia: a meta-analysis. Psychol. Med. 49, 2307–2319.

Caldeira, L.F., Borba, H.H., Tonin, F.S., Wiens, A., Fernandez-Llimos, F., Pontarolo, R., 2020. Fecal microbiota transplantation in inflammatory bowel disease patients: a systematic review and meta-analysis. PLoS One 15, e0238910.

Calder, P.C., Bosco, N., Bourdet-Sicard, R., Capuron, L., Delzenne, N., et al., 2017. Health relevance of the modification of low grade inflammation in ageing (inflammageing) and the role of nutrition. Ageing Res. Rev. 40, 95–119.

Caldiroli, A., Capuzzi, E., Affaticati, L.M., Surace, T., Di Forti, C.L., et al., 2023. Candidate biological markers for social anxiety disorder: a systematic review. Int. J. Mol. Sci. 24, 835.

Caldwell, T., 2015. Is Gwyneth Paltrow Wrong About Everything? Penguin, Toronto, ON.

Caleyachetty, R., Thomas, G.N., Toulis, K.A., Mohammed, N., Gokhale, K.M., et al., 2017. Metabolically healthy obese and incident cardiovascular disease events among 3.5 Million men and women. J. Am. Coll. Cardiol. 70, 1429–1437.

Callixte, K.T., Clet, T.B., Jacques, D., Faustin, Y., François, D.J., et al., 2015. The pattern of neurological diseases in elderly people in outpatient consultations in Sub-Saharan Africa. BMC Res. Notes 8, 159.

Calsavara, A.C., Rodrigues, D.H., Miranda, A.S., Costa, P.A., Lima, C.X., et al., 2013. Late anxiety-like behavior and neuroinflammation in mice subjected to sublethal polymicrobial sepsis. Neurotox. Res. 24, 103–108.

Calvigioni, D., Fuzik, J., Le Merre, P., Slashcheva, M., Jung, F., et al., 2023. Esr1+ hypothalamic-habenula neurons shape aversive states. Nat. Neurosci. 26, 1245–2023.

Cameron, E.A., Sperandio, V., 2015. Frenemies: signaling and nutritional integration in pathogen-microbiota-host interactions. Cell Host Microbe 18, 275–284.

Campbell, J.P., Turner, J.E., 2018. Debunking the myth of exercise-induced immune suppression: Redefining the impact of exercise on immunological health across the lifespan. Front. Immunol. 9, 648.

Campbell, P.T., Campbell, K.L., Wener, M.H., Wood, B., Potter, J.D., et al., 2009. A yearlong exercise intervention decreases CRP among obese postmenopausal women. Med. Sci. Sports Exerc. 41.

Campbell, S.C., Wisniewski, P.J., Noji, M., McGuinness, L.R., Häggblom, M.M., et al., 2016. The effect of diet and exercise on intestinal integrity and microbial diversity in mice. PLoS One 11, e0150502.

Campbell, K.L., Landells, C.E., Fan, J., Brenner, D.R., 2017. A systematic review of the effect of lifestyle interventions on adipose tissue gene expression: Implications for carcinogenesis. Obesity (Silver Spring) 25 (Suppl 2), S40–S51.

Cani, P.D., Everard, A., 2016. Talking microbes: when gut bacteria interact with diet and host organs. Mol. Nutr. Food Res. 60, 58–66.

Cao, X., 2020. COVID-19: immunopathology and its implications for therapy. Nat. Rev. Immunol. 20, 269–270.

Cao, J., Min, L., Lansing, B., Foxman, B., Mody, L., 2016. Multidrug-resistant organisms on patients' hands: a missed opportunity. JAMA Intern. Med. 176, 705–706.

Cao, B., Chen, Y., Ren, Z., Pan, Z., McIntyre, R.S., Wang, D., 2021a. Dys-regulation of kynurenine pathway and potential dynamic changes of kynurenine in schizophrenia: a systematic review and meta-analysis. Neurosci. Biobehav. Rev. 123, 203–214.

Cao, P., Chen, C., Liu, A., Shan, Q., Zhu, X., et al., 2021b. Early-life inflammation promotes depressive symptoms in adolescence via microglial engulfment of dendritic spines. Neuron 109, 2573–2589.

Cao, X., Liu, K., Liu, J., Liu, Y.W., Xu, L., et al., 2021c. Dysbiotic gut microbiota and dysregulation of cytokine profile in children and teens with autism spectrum disorder. Front. Neurosci. 15, 635925.

Cao, X., Li, W., Wang, T., Ran, D., Davalos, V., et al., 2022a. Accelerated biological aging in COVID-19 patients. Nat. Commun. 13, 2135.

Cao, Y., Yisimayi, A., Jian, F., Song, W., Xiao, T., et al., 2022b. BA.2.12.1, BA.4 and BA.5 escape antibodies elicited by Omicron infection. Nature 608, 593–602.

Cao, Z., Sugimura, N., Burgermeister, E., Ebert, M.P., Zuo, T., Lan, P., 2022c. The gut virome: a new microbiome component in health and disease. EBioMedicine 81, 104113.

Cao, Y., Ji, S., Chen, Y., Zhang, X., Ding, G., Tang, F., 2023a. Association between autoimmune diseases of the nervous system and schizophrenia: a systematic review and meta-analysis of cohort studies. Compr. Psychiatry 122, 152370.

Cao, Z., Min, J., Tan, Q., Si, K., Yang, H., Xu, C., 2023b. Circulating insulin-like growth factor-1 and brain health: evidence from 369,711 participants in the UK Biobank. Alzheimers Res. Ther. 15, 140.

Cao-Lei, L., Veru, F., Elgbeili, G., Szyf, M., Laplante, D.P., King, S., 2016. DNA methylation mediates the effect of exposure to prenatal maternal stress on cytokine production in children at age 13½ years: Project Ice Storm. Clin. Epigenetics 8, 54.

Cao-Lei, L., De Rooij, S.R., King, S., Matthews, S.G., Metz, G.A., et al., 2017. Prenatal stress and epigenetics. Neurosci. Biobehav. Rev. pii: S0149-7634(16)30726-6.

Cappi, C., Diniz, J.B., Requena, G.L., Lourenço, T., Lisboa, B.C., et al., 2016. Epigenetic evidence for involvement of the oxytocin receptor gene in obsessive-compulsive disorder. BMC Neurosci. 17, 79.

Capuron, L., Miller, A.H., 2011. Immune system to brain signaling: neuropsychopharmacological implications. Pharmacol. Ther. 130, 226–238.

Carcea, I., Caraballo, N.L., Marlin, B.J., Ooyama, R., Riceberg, J.S., et al., 2021. Oxytocin neurons enable social transmission of maternal behaviour. Nature 596, 553–557.

Cardoso, C., Kingdon, D., Ellenbogen, M.A., 2014. A meta-analytic review of the impact of intranasal oxytocin administration on cortisol concentrations during laboratory tasks: moderation by method and mental health. Psychoneuroendocrinology 49, 161–170.

Careaga, M., Rogers, S., Hansen, R.L., Amaral, D.G., Van de Water, J., Ashwood, P., 2017. Immune endophenotypes in children with autism spectrum disorder. Biol. Psychiatry 81 (5), 434–441.

Carhart-Harris, R.L., Nutt, D.J., 2017. Serotonin and brain function: a tale of two receptors. J. Psychopharmacol. 31, 1091–1120.

Carhart-Harris, R.L., Bolstridge, M., Rucker, J., et al., 2016. Psilocybin with psychological support for treatment-resistant depression: an open-label feasibility study. Lancet Psychiatry 3, 619–627.

Carhart-Harris, R.L., Roseman, L., Bolstridge, M., Demetriou, L., Pannekoek, J.N., et al., 2017. Psilocybin for treatment-resistant depression: fMRI-measured brain mechanisms. Sci. Rep. 7, 13187.

Carloni, E., Ramos, A., Hayes, L.N., 2021. Developmental stressors induce innate immune memory in microglia and contribute to disease risk. Int. J. Mol. Sci. 22, 13035.

Carlson, A.L., Xia, K., Azcarate-Peril, M.A., Goldman, B.D., Ahn, M., et al., 2017. Infant gut microbiome associated with cognitive development. Biol. Psychiatry. pii: S0006-3223(17)31720-1.

Carlton, C.N., Garcia, K.M., Sullivan-Toole, H., Stanton, K., McDonnell, C.G., Richey, J.A., 2021. From childhood maltreatment to adult inflammation: evidence for the mediational status of social anxiety and low positive affect. Brain Behav. Immun. Health 18, 100366.

Carmody, R.N., Bisanz, J.E., 2023. Roles of the gut microbiome in weight management. Nat. Rev. Microbiol. 21, 535–550.

Carney, M.C., Tarasiuk, A., DiAngelo, S.L., Silveyra, P., Podany, A., et al., 2017. Metabolism-related microRNAs in maternal breast milk are influenced by premature delivery. Pediatr. Res. 82, 226–236.

Carpay, N.C., Kamphorst, K., de Meij, T.G.J., Daams, J.G., Vlieger, A.M., van Elburg, R.M., 2022. Microbial effects of prebiotics, probiotics and synbiotics after Caesarean section or exposure to antibiotics in the first week of life: a systematic review. PLoS One 17, e0277405.

Carpenter, J.J., Hines, S.H., Lan, V.M., 2017. Guided imagery for pain management in postoperative orthopedic patients: an integrative literature review. J. Holist. Nurs. 35 (4), 342–351.

Carpenter, J.K., Andrews, L.A., Witcraft, S.M., Powers, M.B., Smits, J.A. J., Hofmann, S.G., 2018. Cognitive behavioral therapy for anxiety and related disorders: a meta-analysis of randomized placebo-controlled trials. Depress. Anxiety 35, 502–514.

Carretta, M.D., Quiroga, J., López, R., Hidalgo, M.A., Burgos, R.A., 2021. Participation of short-chain fatty acids and their receptors in gut inflammation and colon cancer. Front. Physiol. 12, 662739.

Carter, C.S., Kenkel, W.M., MacLean, E.L., Wilson, S.R., Perkeybile, A. M., et al., 2020. Is oxytocin "Nature's Medicine"? Pharmacol. Rev. 72, 829–861.

Carver, C.S., Smith, R.G., Antoni, M.H., Petronis, V.M., Weiss, S., et al., 2005. Optimistic personality and psychosocial well-being during treatment predict psychosocial well-being among long-term survivors of breast cancer. Health Psychol. 24, 508–516.

Casaletto, K.B., Staffaroni, A.M., Elahi, F., Fox, E., Crittenden, P.A., et al., 2018. Perceived stress is associated with accelerated monocyte/macrophage aging trajectories in clinically normal adults. Am. J. Geriatr. Psychiatry 26, 952–963.

Casals, J., Elizan, T.S., Yahr, M.D., 1998. Postencephalitic parkinsonism—a review. J. Neural Transm. 105, 645–676.

Casarotto, P.C., Girych, M., Fred, S.M., Kovaleva, V., Moliner, R., et al., 2021. Antidepressant drugs act by directly binding to TRKB neurotrophin receptors. Cell 184, 1299–1313.e19.

Cash, A., Theus, M.H., 2020. Mechanisms of blood-brain barrier dysfunction in traumatic brain injury. Int. J. Mol. Sci. 21, 3344.

Caso, J.R., MacDowell, K.S., González-Pinto, A., García, S., de Diego-Adeliño, J., et al., 2021. Gut microbiota, innate immune pathways, and inflammatory control mechanisms in patients with major depressive disorder. Transl. Psychiatry 11, 645.

Caspell-Garcia, C., Simuni, T., Tosun-Turgut, D., Wu, I.-W., Zhang, Y., et al., 2017. Multiple modality biomarker prediction of cognitive impairment in prospectively followed de novo Parkinson disease. PLoS One 12, e0175674.

Caspi, A., Sugden, K., Moffitt, T.E., Taylor, A., Craig, I.W., et al., 2003. Influence of life stress on depression: moderation by a polymorphism in the 5-HTT gene. Science 301, 386–389.

Caspi, A., Hariri, A.R., Holmes, A., Uher, R., Moffitt, T.E., 2010. Genetic sensitivity to the environment: the case of the serotonin transporter gene and its implications for studying complex diseases and traits. Am. J. Psychiatry 167, 509–527.

Cassidy, F.C., Charalambous, M., 2018. Genomic imprinting, growth and maternal–fetal interactions. J. Exp. Biol. 221, jeb164517.

Castellani, G., Croese, T., Peralta Ramos, J.M., Schwartz, M., 2023. Transforming the understanding of brain immunity. Science 380, eabo7649.

Castellanos, J.P., Woolley, C., Bruno, K.A., Zeidan, F., Halberstadt, A., Furnish, T., 2020. Chronic pain and psychedelics: a review and proposed mechanism of action. Reg. Anesth. Pain Med. 45 (7), 486–494.

Castrén, E., Monteggia, L.M., 2021. Brain-derived neurotrophic factor signaling in depression and antidepressant action. Biol. Psychiatry 90, 128–136.

Castro-Nallar, E., Bendall, M.L., Perez-Losada, M., Sabuncyan, S., Severance, E.G., et al., 2015. Composition, taxonomy and functional diversity of the oropharynx microbiome in individuals with schizophrenia and controls. PeerJ 3, e1140.

Catale, C., Gironda, S., Lo Iacono, L., Carola, V., 2020. Microglial function in the effects of early-life stress on brain and behavioral development. J. Clin. Med. 9, 468.

Cathomas, F., Lin, H.Y., Chan, K.L., Li, L., Parise, L.F., et al., 2024. Circulating myeloid-derived MMP8 in stress susceptibility and depression. Nature 626, 1108–1115.

Cattane, N., Mora, C., Lopizzo, N., Borsini, A., Maj, C., et al., 2019. Identification of a miRNAs signature associated with exposure to stress early in life and enhanced vulnerability for schizophrenia: new insights for the key role of miR-125b-1-3p in neurodevelopmental processes. Schizophr. Res. 205, 63–75.

Cattane, N., Richetto, J., Cattaneo, A., 2020. Prenatal exposure to environmental insults and enhanced risk of developing Schizophrenia and Autism Spectrum Disorder: focus on biological pathways and epigenetic mechanisms. Neurosci. Biobehav. Rev. 117, 253–278.

Cattaneo, A., Macchi, F., Plazzotta, G., Veronica, B., Bocchio-Chiavetto, L., et al., 2015. Inflammation and neuronal plasticity: a link between childhood trauma and depression pathogenesis. Front. Cell. Neurosci. 9, 40.

Cavallari, J.F., Fullerton, M.D., Duggan, B.M., Foley, K.P., Denou, E., 2017. Muramyl diapeptide-based postbiotics mitigate obesity-induced insulin resistance via IRF4. Cell Metab. 25, 1063–1074.

Cebrián, C., Zucca, F.A., Mauri, P., Steinbeck, J.A., Studer, L., et al., 2014. MHC-I expression renders catecholaminergic neurons susceptible to T-cell-mediated degeneration. Nat. Commun. 5, 3633.

Cekanaviciute, E., Yoo, B.B., Runia, T.F., Debelius, J.W., Singh, S., et al., 2017. Gut bacteria from multiple sclerosis patients modulate human T cells and exacerbate symptoms in mouse models. Proc. Natl. Acad. Sci. USA 114, 10713–10718.

Cekic, C., Linden, J., 2016. Purinergic regulation of the immune system. Nat. Rev. Immunol. 16, 177–192.

Cené, C.W., Beckie, T.M., Sims, M., Suglia, S.F., Aggarwal, B., et al., 2022. Effects of objective and perceived social isolation on cardiovascular and brain health: a scientific statement from the American Heart Association. J. Am. Heart Assoc. 11, e026493.

Centre for Disease Control, 2016. Antibiotic Resistance Threats in the United States, 2013. Retrieved from http://www.cdc.gov/drugresistance/pdf/ar-threats-2013-508.pdf.

Cerdá, B., Pérez, M., Pérez-Santiago, J.D., Tornero-Aguilera, J.F., González-Soltero, R., et al., 2016. Gut microbiota modification: another piece in the puzzle of the benefits of physical exercise in health? Front. Physiol. 7, 51.

Cervenka, I., Agudelo, L.Z., Ruas, J.L., 2017. Kynurenines: tryptophan's metabolites in exercise, inflammation, and mental health. Science 357, eaaf9794.

Cervia-Hasler, C., Brüningk, S.C., Hoch, T., Fan, B., Muzio, G., et al., 2024. Persistent complement dysregulation with signs of thromboinflammation in active Long Covid. Science 383, eadg7942.

Chae, D.H., Nuru-Jeter, A.M., Adler, N.E., Brody, G.H., Lin, J., et al., 2014. Discrimination, racial bias, and telomere length in African-American men. Am. J. Prev. Med. 46, 103–111.

Chaki, S., 2021. Vasopressin V1B receptor antagonists as potential antidepressants. Int. J. Neuropsychopharmacol. 24, 450–463.

Chakraborti, A., Graham, C., Chehade, S., Vashi, B., Umfress, A., et al., 2021. High fructose corn syrup-moderate fat diet potentiates anxiodepressive behavior and alters ventral striatal neuronal signaling. Front. Neurosci. 15, 669410.

Chakraborty, P., Chattarji, S., Jeanneteau, F., 2021. A salience hypothesis of stress in PTSD. Eur. J. Neurosci. 54, 8029–8051.

Chakraborty, S., Gonzalez, J.C., Sievers, B.L., Mallajosyula, V., Chakraborty, S., et al., 2022a. Early non-neutralizing, afucosylated antibody responses are associated with COVID-19 severity. Sci. Transl. Med. 14, eabm7853.

Chakraborty, C., Sharma, A.R., Bhattacharya, M., Lee, S.S., 2022b. A detailed overview of immune escape, antibody escape, partial vaccine escape of SARS-CoV-2 and their emerging variants with escape mutations. Front. Immunol. 13, 801522.

Chakravarty, S., Herkenham, M., 2005. Toll-like receptor 4 on nonhematopoietic cells sustains CNS inflammation during endotoxemia, independent of systemic cytokines. J. Neurosci. 25, 1788–1796.

Chalifour, B., Holzhausen, E.A., Lim, J.J., Yeo, E.N., Shen, N., et al., 2023. The potential role of early life feeding patterns in shaping the infant fecal metabolome: implications for neurodevelopmental outcomes. NPJ Metab. Health Dis. 1, 2.

Champagne, F.A., 2010. Early adversity and developmental outcomes: interaction between genetics, epigenetics, and social experiences across the life span. Perspect. Psychol. Sci. 5, 564–574.

Champagne, F.A., Meaney, M.J., 2007. Transgenerational effects of social environment on variations in maternal care and behavioral response to novelty. Behav. Neurosci. 121, 1353–1363.

Chan, G.J., Lee, A.C., Baqui, A.H., Tan, J., Black, R.E., 2013. Risk of early-onset neonatal infection with maternal infection or colonization: a global systematic review and meta-analysis. PLoS Med. 10, e1001502.

Chan, G., White, C.C., Winn, P.A., Cimpean, M., Replogle, J.M., et al., 2015. CD33 modulates TREM2: convergence of Alzheimer loci. Nat. Neurosci. 18, 1556–1558.

Chan, C.K.Y., Tao, J., Chan, O.S., Li, H.B., Pang, H., 2020. Preventing respiratory tract infections by synbiotic interventions: a systematic review and meta-analysis of randomized controlled trials. Adv. Nutr. 11, 979–988.

Chandra, R., Francis, T.C., Nam, H., Riggs, L.M., Engeln, M., et al., 2017. Reduced Slc6a15 in nucleus accumbens D2-neurons underlies stress susceptibility. J. Neurosci. 37, 6257–6538.

Chang, Y.W., Hung, L.C., Chen, Y.C., Wang, W.H., Lin, C.Y., et al., 2021. Insulin reduces inflammation by regulating the activation of the NLRP3 inflammasome. Front. Immunol. 11, 587229.

Chang, R., Yen-Ting Chen, T., Wang, S.I., Hung, Y.M., et al., 2023. Risk of autoimmune diseases in patients with COVID-19: a retrospective cohort study. EClinicalMedicine 56, 101783.

Channer, B., Matt, S.M., Nickoloff-Bybel, E.A., Pappa, V., Agarwal, Y., et al., 2023. Dopamine, immunity, and disease. Pharmacol. Rev. 75, 62–158.

Chao, A.M., Jastreboff, A.M., White, M.A., Grilo, C.M., Sinha, R., 2017. Stress, cortisol, and other appetite-related hormones: prospective prediction of 6-month changes in food cravings and weight. Obesity 25, 713–720.

Chassaing, B., Koren, O., Goodrich, J.K., Poole, A.C., Srinivasan, S., et al., 2015. Dietary emulsifiers impact the mouse gut microbiota promoting colitis and metabolic syndrome. Nature 519, 92–96.

Chatzinakos, C., Pernia, C.D., Morrison, F.G., Iatrou, A., McCullough, K.M., et al., 2023. Single-nucleus transcriptome profiling of dorsolateral prefrontal cortex: mechanistic roles for neuronal gene expression, including the 17q21.31 locus, in PTSD stress response. Am. J. Psychiatry 180, 739–754.

Chaudhari, P.R., Singla, A., Vaidya, V.A., 2022. Early adversity and accelerated brain aging: a mini-review. Front. Mol. Neurosci. 15, 822917.

Chawla, N., Anothaisintawee, T., Charoenrungrueangchai, K., Thaipisuttikul, P., McKay, G.J., et al., 2022. Drug treatment for panic disorder with or without agoraphobia: systematic review and network meta-analysis of randomised controlled trials. BMJ 376, e066084.

Chechko, N., Wehrle, R., Erhardt, A., Holsboer, F., Czisch, M., Sämann, P.G., 2009. Unstable prefrontal response to emotional conflict and activation of lower limbic structures and brainstem in remitted panic disorder. PLoS One 4, e5537.

Cheetham, N.J., Penfold, R., Giunchiglia, V., Bowyer, V., Sudre, C.H., et al., 2023. The effects of COVID-19 on cognitive performance in a community-based cohort: a COVID symptom study biobank prospective cohort study. eClinicalMedicine 62, 102086.

Chen, E.K., 2023. Does quantum theory imply the entire Universe is preordained? Nature 624, 513–515.

Chen, G.Y., Nunez, G., 2010. Sterile inflammation: sensing and reacting to damage. Nat. Rev. Immunol. 10, 826–837.

Chen, X., Zhang, Y., 2024. A review of the neurotransmitter system associated with cognitive function of the cerebellum in Parkinson's disease. Neural Regen. Res. 19, 324–330.

Chen, J.F., Xu, K., Petzer, J.P., Staal, R., Xu, Y.H., et al., 2001. Neuroprotection by caffeine and A(2A) adenosine receptor inactivation in a model of Parkinson's disease. J. Neurosci. 21, RC143.

Chen, J., Song, Y., Yang, J., Zhang, Y., Zhao, P., et al., 2013. The contribution of TNF-α in the amygdala to anxiety in mice with persistent inflammatory pain. Neurosci. Lett. 541, 275–280.

Chen, Z., Guo, L., Zhang, Y., Walzem, R.L., et al., 2014. Incorporation of therapeutically modified bacteria into gut microbiota inhibits obesity. J. Clin. Invest. 124, 3391–3406.

Chen, Y., Qi, B., Xu, W., Ma, B., Li, L., et al., 2015a. Clinical correlation of peripheral CD4+-cell sub-sets, their imbalance and Parkinson's disease. Mol. Med. Rep. 12, 6105–6111.

Chen, Y.W., Lin, P.Y., Tu, K.Y., Cheng, Y.S., Wu, C.K., Tseng, P.T., 2015b. Significantly lower nerve growth factor levels in patients with major depressive disorder than in healthy subjects: a meta-analysis and systematic review. Neuropsychiatr. Dis. Treat. 11, 925–933.

Chen, J., Wright, K., Davis, J.M., Jeraldo, P., Marietta, E.V., et al., 2016a. An expansion of rare lineage intestinal microbes characterizes rheumatoid arthritis. Genome Med. 8, 1.

Chen, Q., Sun, L., Chen, Z., 2016b. Regulation and function of the cGAS–STING pathway of cytosolic DNA sensing. Nat. Immunol. 17, 1142–1149.

Chen, R., Shi, L., Hakenberg, J., Naughton, B., Sklar, P., et al., 2016c. Analysis of 589,306 genomes identifies individuals resilient to severe Mendelian childhood diseases. Nat. Biotechnol. 34, 531–538.

Chen, S.G., Stribinskis, V., Rane, M.J., Demuth, D.R., Gozal, E., et al., 2016d. Exposure to the functional bacterial amyloid protein curli enhances alpha-synuclein aggregation in aged Fischer 344 rats and Caenorhabditis elegans. Sci. Rep. 6, 34477.

Chen, S.W., Zhong, X.S., Jiang, L.N., Zheng, X.Y., Xiong, Y.Q., et al., 2016e. Maternal autoimmune diseases and the risk of autism spectrum disorders in offspring: a systematic review and meta-analysis. Behav. Brain Res. 296, 61–69.

Chen, Y., Fang, Q., Wang, Z., Zhang, J.Y., MacLeod, A.S., et al., 2016f. Transient receptor potential vanilloid 4 ion channel functions as a pruriceptor in epidermal keratinocytes to evoke histaminergic itch. J. Biol. Chem. 291, 10252–10262.

Chen, C.J., Wu, G.H., Kuo, R.L., Shih, S.R., 2017a. Role of the intestinal microbiota in the immunomodulation of influenza virus infection. Microbes Infect. 19, 570–579.

Chen, K., Zheng, X., Feng, M., Li, D., Zhang, H., 2017b. Gut microbiota-dependent metabolite trimethylamine N-oxide contributes to cardiac dysfunction in western diet-induced obese mice. Front. Physiol. 8, 1–9.

Chen, L., Wilson, J.E., Koenigsknecht, M.J., Chou, W.C., Montgomery, S. A., et al., 2017c. NLRP12 attenuates colon inflammation by maintaining colonic microbial diversity and promoting protective commensal bacterial growth. Nat. Immunol. 18, 541–551.

Chen, L., Teng, H., Jia, Z., Battino, M., Miron, A., et al., 2018. Intracellular signaling pathways of inflammation modulated by dietary flavonoids: the most recent evidence. Crit. Rev. Food Sci. Nutr. 58, 2908–2924.

Chen, X., Zeng, C., Gong, C., Zhang, L., Wan, Y., et al., 2019a. Associations between early life parent-child separation and shortened telomere length and psychopathological outcomes during adolescence. Psychoneuroendocrinology 103, 195–202.

Chen, Y.H., Bai, J., Wu, D.I., Yu, S.F., Qiang, X.L., et al., 2019b. Association between fecal microbiota and generalized anxiety disorder: severity and early treatment response. J. Affect. Disord. 259, 56–66.

Chen, W.C., Liu, Y.B., Liu, W.F., Zhou, Y.Y., He, H.F., Lin, S., 2020. Neuropeptide Y is an immunomodulatory factor: direct and indirect. Front. Immunol. 11, 580378.

Chen, L., Zheng, W.H., Du, Y., Li, X.S., Yu, Y., et al., 2021a. Altered peripheral immune profiles in first-episode, drug-free patients with schizophrenia: response to antipsychotic medications. Front. Med. (Lausanne) 8, 757655.

Chen, L.W., Soh, S.E., Tint, M.T., Loy, S.L., Yap, F., et al., 2021b. Combined analysis of gestational diabetes and maternal weight status from pre-pregnancy through post-delivery in future development of type 2 diabetes. Sci. Rep. 11, 5021.

Chen, R., Wang, K., Yu, J., Howard, D., French, L., et al., 2021c. The spatial and cell-type distribution of SARS-CoV-2 receptor ACE2 in the human and mouse brains. Front. Neurol. 11, 573095.

Chen, S., Lai, S.W.T., Brown, C.E., Feng, M., 2021d. Harnessing and enhancing macrophage phagocytosis for cancer therapy. Front. Immunol. 12, 635173.

Chen, P., Wang, C., Ren, Y.N., Ye, Z.J., Jiang, C., Wu, Z.B., 2021e. Alterations in the gut microbiota and metabolite profiles in the context of neuropathic pain. Mol. Brain 14 (1), 50.

Chen, J., Zhang, Z.Z., Luo, B.L., Yang, Q.G., Ni, M.Z., et al., 2022a. Prenatal exposure to inflammation increases anxiety-like behaviors in F1 and F2 generations: possible links to decreased FABP7 in hippocampus. Front. Behav. Neurosci. 16, 973069.

Chen, S., Aruldass, A.R., Cardinal, R.N., 2022b. Mental health outcomes after SARS-CoV-2 vaccination in the United States: a national cross-sectional study. J. Affect. Disord. 298, 396–399.

Chen, W., Schilperoort, M., Cao, Y., Shi, J., Tabas, I., Tao, W., 2022c. Macrophage-targeted nanomedicine for the diagnosis and treatment of atherosclerosis. Nat. Rev. Cardiol. 19, 228–249.

Chen, J., Chen, J., Lei, Z., Zhang, F., Zeng, L.H., et al., 2023. Amyloid precursor protein facilitates SARS-CoV-2 virus entry into cells and enhances amyloid-β-associated pathology in APP/PS1 mouse model of Alzheimer's disease. Transl. Psychiatry 13, 396.

Chen, G., Han, Q., Li, W.X., Hai, R., Ding, S.W., 2024. Live-attenuated virus vaccine defective in RNAi suppression induces rapid protection in neonatal and adult mice lacking mature B and T cells. Proc. Natl. Acad. Sci. USA 121, e2321170121.

Cheng, C., Lau, H.P.B., Chan, M.P.S., 2014. Coping flexibility and psychological adjustment to stressful life changes: a meta-analytic review. Psychol. Bull. 140, 1582–1607.

Cheng, Y., Jope, R.S., Beurel, E., 2015. A pre-conditioning stress accelerates increases in mouse plasma inflammatory cytokines induced by stress. BMC Neurosci. 16, 31.

Cheng, Y., Tran Minh, N., Tran Minh, Q., Khandelwal, S., Clapham, H. E., 2022. Estimates of Japanese encephalitis mortality and morbidity: a systematic review and modeling analysis. PLoS Negl. Trop. Dis. 16, e0010361.

Cheng, X., Xie, Y., Wang, A., Zhu, C., Yan, F., et al., 2023. Correlation between elevated serum interleukin-1β, interleukin-16 levels and psychiatric symptoms in patients with schizophrenia at different stages. BMC Psychiatry 23, 396.

Cheng, X., Meng, X., Chen, R., Song, Z., Li, S., et al., 2024. The molecular subtypes of autoimmune diseases. Comput. Struct. Biotechnol. J. 23, 1348–1363.

Cheong, J.G., Ravishankar, A., Sharma, S., Parkhurst, C.N., Grassmann, S.A., et al., 2023. Epigenetic memory of coronavirus infection in innate immune cells and their progenitors. Cell 186, 3882–3902.e24.

Cherkin, D.C., Sherman, K.J., Balderson, B.H., Cook, A.J., Anderson, M. L., et al., 2016. Effect of mindfulness-based stress reduction vs cognitive behavioral therapy or usual care on back pain and functional limitations in adults with chronic low back pain: a randomized clinical trial. JAMA 315, 1240–1249.

Cherry, J.D., Olschowka, J.A., O'Banion, M.K., 2014. Neuroinflammation and M2 microglia: the good, the bad, and the inflamed. J. Neuroinflammation 11, 98.

Chesnokova, V., Pechnick, R.N., Wawrowsky, K., 2016. Chronic peripheral inflammation, hippocampal neurogenesis, and behavior. Brain Behav. Immun. 58, 1–8.

Chetty, R., Stepner, M., Abraham, S., Lin, S., Scuderi, B., et al., 2016. The association between income and life expectancy in the United States, 2001-2014. JAMA 315, 1750–1766.

Cheval, B., Darrous, L., Choi, K.W., Klimentidis, Y.C., Raichlen, D. A., et al., 2023. Genetic insights into the causal relationship between physical activity and cognitive functioning. Sci. Rep. 13, 5310.

Chi, H., Pepper, M., Thomas, P.G., 2024. Principles and therapeutic applications of adaptive immunity. Cell 187, 2052–2078.

Chida, Y., Steptoe, A., 2008. Positive psychological well-being and mortality: a quantitative review of prospective observational studies. Psychosom. Med. 70, 741–756.

Childs, E., Lutz, J.A., de Wit, H., 2017. Dose-related effects of delta-9-THC on emotional responses to acute psychosocial stress. Drug Alcohol Depend. 177, 136–144.

Chin-Chan, M., Navarro-Yepes, J., Quintanilla-Vega, B., 2015. Environmental pollutants as risk factors for neurodegenerative disorders: Alzheimer and Parkinson diseases. Front. Cell. Neurosci. 9, 124.

Chinna Meyyappan, A., Forth, E., Wallace, C.J.K., Milev, R., 2020. Effect of fecal microbiota transplant on symptoms of psychiatric disorders: a systematic review. BMC Psychiatry 20, 299.

Chita-Tegmark, M., 2016. Social attention in ASD: a review and meta-analysis of eye-tracking studies. Res. Dev. Disabil. 48, 79–93.

Chiu, W.C., Su, Y.P., Su, K.P., Chen, P.C., 2017. Recurrence of depressive disorders after interferon-induced depression. Transl. Psychiatry 7, e1026.

Choe, J.Y., Nair, M., Basha, R., Kim, B.J., Jones, H.P., 2019. Defining early life stress as a precursor for autoimmune disease. Crit. Rev. Immunol. 39, 329–342.

Choi, H.H., Cho, Y.S., 2016. Fecal microbiota transplantation: current applications, effectiveness, and future perspectives. Clin. Endosc. 49, 257–265.

Choi, S.H., Aid, S., Caracciolo, L., Minami, S.S., et al., 2013. Cyclooxygenase-1 inhibition reduces amyloid pathology and improves memory deficits in a mouse model of Alzheimer's disease. J. Neurochem. 124, 59–68.

Choi, G.E., Chae, C.W., Park, M.R., Yoon, J.H., Jung, Y.H., et al., 2022. Prenatal glucocorticoid exposure selectively impairs neuroligin 1-dependent neurogenesis by suppressing astrocytic FGF2-neuronal FGFR1 axis. Cell. Mol. Life Sci. 79, 294.

Choi, Y., Lichterman, J.N., Coughlin, L.A., Poulides, N., Li, W., et al., 2023a. Immune checkpoint blockade induces gut microbiota translocation that augments extraintestinal antitumor immunity. Sci. Immunol. 8, eabo2003.

Choi, I., Wang, M., Yoo, S., Xu, P., Seegobin, S.P., et al., 2023b. Autophagy enables microglia to engage amyloid plaques and prevents microglial senescence. Nat. Cell Biol. 25, 963–974.

Cholesterol Treatment Trialists' Collaboration, 2019. Efficacy and safety of statin therapy in older people: a meta-analysis of individual participant data from 28 randomised controlled trials. Lancet 393, 407–415.

Choudary, P.V., Molnar, M., Evans, S.J., Tomita, H., Li, J.Z., et al., 2005. Altered cortical glutamatergic and GABAergic signal transmission with glial involvement in depression. Proc. Natl. Acad. Sci. USA 102, 15653–15658.

Choudhury, A., Mukherjee, S., 2020. In silico studies on the comparative characterization of the interactions of SARS-CoV-2 spike glycoprotein with ACE-2 receptor homologs and human TLRs. J. Med. Virol. 92, 2105–2113.

Choudhury, M.E., Sugimoto, K., Kubo, M., Nagai, M., Nomoto, M., et al., 2011. A cytokine mixture of GM-CSF and IL-3 that induces a neuroprotective phenotype of microglia leading to amelioration of (6-OHDA)-induced Parkinsonism of rats. Brain Behav. 1, 26–43.

Choutka, J., Jansari, V., Hornig, M., Iwasaki, A., 2022. Unexplained post-acute infection syndromes. Nat. Med. 28, 911–923.

Chow, L.S., Gerszten, R.E., Taylor, J.M., Pedersen, B.K., van Praag, H., et al., 2022. Exerkines in health, resilience and disease. Nat. Rev. Endocrinol. 18, 273–289.

Chowdhury, S., Chowdhury, N.S., 2023. Novel anti-amyloid-beta (Aβ) monoclonal antibody lecanemab for Alzheimer's disease: a systematic review. Int. J. Immunopathol. Pharmacol. 37. 3946320231209839.

Chowdhury, R., Warnakula, S., Kunutsor, S., Crowe, F., et al., 2014. Association of dietary, circulating, and supplement fatty acids with coronary risk: a systematic review and meta-analysis. Ann. Intern. Med. 160, 398–406.

Christensen, J., Pedersen, H.S., Fenger-Grøn, M., Fann, J.R., Jones, N.C., Vestergaard, M., 2019. Selective serotonin reuptake inhibitors and risk of epilepsy after traumatic brain injury - a population based cohort study. PLoS One 14, e0219137.

Christovich, A., Luo, X.M., 2022. Gut microbiota, leaky gut, and autoimmune diseases. Front. Immunol. 13, 946248.

Chriswell, M.E., Lefferts, A.R., Clay, M.R., Hsu, A.R., Seifert, J., et al., 2022. Clonal IgA and IgG autoantibodies from individuals at risk for rheumatoid arthritis identify an arthritogenic strain of Subdoligranulum. Sci. Transl. Med. 14, eabn5166.

Chu, H., Mazmanian, S.K., 2013. Innate immune recognition of the microbiota promotes host-microbial symbiosis. Nat. Immunol. 14, 668–675.

Chu, A.L., Stochl, J., Lewis, G., Zammit, S., Jones, P.B., Khandaker, G.M., 2019a. Longitudinal association between inflammatory markers and specific symptoms of depression in a prospective birth cohort. Brain Behav. Immun. 76, 74–81.

Chu, C., Murdock, M.H., Jing, D., Won, T.H., Chung, H., et al., 2019b. The microbiota regulate neuronal function and fear extinction learning. Nature 574, 543–548.

Chua, M.C., Ben-Amor, K., Lay, C., Neo, A.G.E., Chiang, W.C., et al., 2017. Effect of synbiotic on the gut microbiota of cesarean delivered infants: a randomized, double-blind, multicenter study. J. Pediatr. Gastroenterol. Nutr. 65, 102–106.

Chuang, Y.H., Lill, C.M., Lee, P.C., Hansen, J., Lassen, C.F., Bertram, L., et al., 2016. Gene-environment interaction in Parkinson's disease: coffee, ADORA2A, and CYP1A2. Neuroepidemiology 47, 192–200.

Chung, W.S., Welsh, C.A., Barres, B.A., Stevens, B., 2015. Do glia drive synaptic and cognitive impairment in disease? Nat. Neurosci. 18, 1539–1545.

Chung, W.S., Verghese, P.B., Chakraborty, C., Joung, J., Hyman, B.T., et al., 2016. Novel allele-dependent role for APOE in controlling the rate of synapse pruning by astrocytes. Proc. Natl. Acad. Sci. USA 113, 10186–10191.

Chung, H.C., Keiller, D.R., Roberts, J.D., Gordon, D.A., 2021. Do exercise-associated genes explain phenotypic variance in the three components of fitness? A systematic review & meta-analysis. PLoS One 16, e0249501.

Cichon, J., Wasilczuk, A.Z., Looger, L.L., Contreras, D., Kelz, M.B., Proekt, A., 2023. Ketamine triggers a switch in excitatory neuronal activity across neocortex. Nat. Neurosci. 26, 39–52.

Cicognola, C., Chiasserini, D., Parnetti, L., 2015. Preanalytical confounding factors in the analysis of cerebrospinal fluid biomarkers for Alzheimer's disease: the issue of diurnal variation. Front. Neurol. 6, 143.

Cipriani, A., Zhou, X., Del Giovane, C., Hetrick, S.E., Qin, B., et al., 2016. Comparative efficacy and tolerability of antidepressants for major depressive disorder in children and adolescents: a network meta-analysis. Lancet 388, 881–890.

Cirillo, P., Gold, A.K., Nardi, A.E., Ornelas, A.C., Nierenberg, A.A., et al., 2019. Transcranial magnetic stimulation in anxiety and trauma-related disorders: a systematic review and meta-analysis. Brain Behav. 9, e01284.

Cirulli, F., Francia, N., Branchi, I., Antonucci, M.T., Aloe, L., et al., 2009. Changes in plasma levels of BDNF and NGF reveal a gender-selective vulnerability to early adversity in rhesus macaques. Psychoneuroendocrinology 34, 172–180.

Cisler, J.M., Esbensen, K., Sellnow, K., Ross, M., Weaver, S., et al., 2019. Differential roles of the salience network during prediction error encoding and facial emotion processing among female adolescent assault victims. Biol. Psychiatry Cogn. Neurosci. Neuroimaging 4, 371–380.

Civelek, M., Wu, Y., Pan, C., Raulerson, C.K., Ko, A., et al., 2017. Genetic regulation of adipose gene expression and cardio-metabolic traits. Am. J. Hum. Genet. 100, 428–443.

Civieri, G., Osborne, M., Abohashem, S., Grewal, S., Gharios, C., et al., 2023. Depression and anxiety accelerate the rate of gain of cardiovascular risk factors: mechanism leading to increased risk of cardiac events. Circulation 148, A12440.

Claassen, D.O., Josephs, K.A., Ahlskog, J.E., Silber, M.H., Tippmann-Peikert, M., et al., 2010. REM sleep behavior disorder preceding other aspects of synucleinopathies by up to half a century. Neurology 75, 494–499.

Clark, D.A., Beck, A.T., 2010. Cognitive theory and therapy of anxiety and depression: convergence with neurobiological findings. Trends Cogn. Sci. 14, 418–424.

Clark, A., Mach, N., 2017. The crosstalk between the gut microbiota and mitochondria during exercise. Front. Physiol. 8, 319.

Clark, S.M., Sand, J., Francis, T.C., Nagaraju, A., Michael, K.C., et al., 2014. Immune status influences fear and anxiety responses in mice after acute stress exposure. Brain Behav. Immun. 38, 192–201.

Clark, P.J., Amat, J., McConnell, S.O., Ghasem, P.R., Greenwood, B.N., et al., 2015. Running reduces uncontrollable stress-evoked serotonin and potentiates stress-evoked dopamine concentrations in the rat dorsal striatum. PLoS One 10, e0141898.

Clark, S.M., Pocivavsek, A., Nicholson, J.D., Notarangelo, F.M., Langenberg, P., et al., 2016. Reduced kynurenine pathway metabolism and cytokine expression in the prefrontal cortex of depressed individuals. J. Psychiatry Neurosci. 41, 386–394.

Clark, S.L., Hattab, M.W., Chan, R.F., Shabalin, A.A., Han, L.K.M., et al., 2020. A methylation study of long-term depression risk. Mol. Psychiatry 25, 1334–1343.

Clarke, G., Grenham, S., Scully, P., Fitzgerald, P., Moloney, R.D., et al., 2013. The microbiome-gut-brain axis during early life regulates the hippocampal serotonergic system in a sex-dependent manner. Mol. Psychiatry 18, 666–673.

Clark-Raymond, A., Halaris, A., 2013. VEGF and depression: a comprehensive assessment of clinical data. J. Psychiatr. Res. 47, 1080–1087.

Clausen, A.N., Francisco, A.J., Thelen, J., Bruce, J., Martin, L.E., et al., 2017. PTSD and cognitive symptoms relate to inhibition-related prefrontal activation and functional connectivity. Depress. Anxiety 24, 427–436.

Claussnitzer, M., Susztak, K., 2021. Gaining insight into metabolic diseases from human genetic discoveries. Trends Genet. 37, 1081–1094.

Claussnitzer, M., Hui, C.C., Kellis, M., 2016. FTO obesity variant and adipocyte browning in humans. N. Engl. J. Med. 374, 192–193.

Clavaguera, F., Akatsu, H., Fraser, G., Crowther, R.A., Frank, S., et al., 2013. Brain homogenates from human tauopathies induce tau inclusions in mouse brain. Proc. Natl. Acad. Sci. USA 110, 9535–9540.

Claypoole, L.D., Zimmerberg, B., Williamson, L.L., 2017. Neonatal lipopolysaccharide treatment alters hippocampal neuroinflammation, microglia morphology and anxiety-like behavior in rats selectively bred for an infantile trait. Brain Behav. Immun. 59, 135–146.

Clayton, A.H., Lasser, R., Parikh, S.V., Iosifescu, D.V., Jung, J., et al., 2023. Zuranolone for the treatment of adults with major depressive disorder: a randomized, placebo-controlled phase 3 trial. Am. J. Psychiatry 180, 676–684.

Clinard, C.T., Bader, L.R., Sullivan, M.A., Cooper, M.A., 2015. Activation of 5-HT2a receptors in the basolateral amygdala promotes defeat-induced anxiety and the acquisition of conditioned defeat in Syrian hamsters. Neuropharmacology 90, 102–112.

Cobia, D.J., Smith, M.J., Wang, L., Csernansky, J.G., 2012. Longitudinal progression of frontal and temporal lobe changes in schizophrenia. Schizophr. Res. 139, 1–6.

Cobo, M.M., Green, G., Andritsou, F., Baxter, L., Evans Fry, R., et al., 2022. Early life inflammation is associated with spinal cord excitability and nociceptive sensitivity in human infants. Nat. Commun. 13 (1), 3943.

Cochemé, H.M., Murphy, M.P., 2008. Complex I is the major site of mitochondrial superoxide production by paraquat. J. Biol. Chem. 283, 1786–1798.

Cohen, S., Frank, E., Doyle, W.J., Skoner, D.P., Rabin, B.S., et al., 1998. Types of stressors that increase susceptibility to the common cold in healthy adults. Health Psychol. 17, 214–223.

Cohen, M., Meir, T., Klein, E., Volpin, G., Assaf, M., et al., 2011. Cytokine levels as potential biomarkers for predicting the development of posttraumatic stress symptoms in casualties of accidents. Int. J. Psychiatry Med. 42, 117–131.

Cohen, S., Janicki-Deverts, D., Doyle, W.J., Miller, G.E., Frank, E., et al., 2012a. Chronic stress, glucocorticoid receptor resistance, inflammation, and disease risk. Proc. Natl. Acad. Sci. USA 109, 5995–5999.

Cohen, H., Liu, T., Kozlovsky, N., Kaplan, Z., Zohar, J., Mathé, A.A., 2012b. The neuropeptide Y (NPY)-ergic system is associated with behavioral resilience to stress exposure in an animal model of post-traumatic stress disorder. Neuropsychopharmacology 37, 350–363.

Cohen, S., Janicki-Deverts, D., Doyle, W.J., 2015a. Self-rated health in healthy adults and susceptibility to the common cold. Psychosom. Med. 77, 959–968.

Cohen, S., Janicki-Deverts, D., Turner, R.B., Doyle, W.J., 2015b. Does hugging provide stress-buffering social support? A study of susceptibility to upper respiratory infection and illness. Psychol. Sci. 26, 135–147.

Cohen, I.V., Makunts, T., Atayee, R., Abagyan, R., 2017a. Population scale data reveals the antidepressant effects of ketamine and other therapeutics approved for non-psychiatric indications. Sci. Rep. 7, 1450.

Cohen, L.J., Esterhazy, D., Kim, S.H., Lemetre, C., Aguilar, R.R., et al., 2017b. Commensal bacteria make GPCR ligands that mimic human signalling molecules. Nature 549, 48–53.

Çolak Sivri, R., Bilgiç, A., Kılınç, İ., 2018. Cytokine, chemokine and BDNF levels in medication-free pediatric patients with obsessive-compulsive disorder. Eur. Child Adolesc. Psychiatry 27, 977–984.

Colasanto, M., Madigan, S., Korczak, D.J., 2020. Depression and inflammation among children and adolescents: a meta-analysis. J. Affect. Disord. 277, 940–948.

Colleluori, G., Villareal, D.T., 2021. Aging, obesity, sarcopenia and the effect of diet and exercise intervention. Exp. Gerontol. 155, 111561.

Collins, S.M., Surette, M., Bercik, P., 2012. The interplay between the intestinal microbiota and the brain. Nat. Rev. Microbiol. 10, 735–742.

Colonna, M., Wang, Y., 2016. TREM2 variants: new keys to decipher Alzheimer disease pathogenesis. Nat. Rev. Neurosci. 17, 201–207.

Colpo, G.D., Leboyer, M., Dantzer, R., Trivedi, M.H., Teixeira, A.L., 2018. Immune-based strategies for mood disorders: facts and challenges. Expert. Rev. Neurother. 18, 139–152.

Colvonen, P.J., Glassman, L.H., Crocker, L.D., Buttner, M.M., Orff, H., et al., 2017. Pretreatment biomarkers predicting PTSD psychotherapy outcomes: a systematic review. Neurosci. Biobehav. Rev. 75, 140–156.

Comer, A.L., Carrier, M., Tremblay, M.È., Cruz-Martín, A., 2020. The inflamed brain in schizophrenia: the convergence of genetic and environmental risk factors that lead to uncontrolled neuroinflammation. Front. Cell. Neurosci. 14, 274.

Comim, C.M., Bussmann, R.M., Simão, S.R., Ventura, L., Freiberger, V., et al., 2016. Experimental neonatal sepsis causes long-term cognitive impairment. Mol. Neurobiol. 53, 5928–5934.

Committee for the Assessment of NIH Research on Autoimmune Diseases, Board on Population Health and Public Health Practice, Health and Medicine Division, National Academies of Sciences, Engineering, and Medicine, 2022. Enhancing NIH Research on Autoimmune Disease. National Academies Press (US), Washington, DC. PMID: 35593778. Copyright © 2022, National Academy of Sciences.

Compare, A., Mommersteeg, P.M., Faletra, F., Grossi, E., Pasotti, E., et al., 2014. Personality traits, cardiac risk factors, and their association with presence and severity of coronary artery plaque in people with no history of cardiovascular disease. J. Cardiovasc. Med. (Hagerstown) 15, 423–430.

Conen, S., Gregory, C.J., Hinz, R., Smallman, R., Corsi-Zuelli, F., et al., 2021. Neuroinflammation as measured by positron emission tomography in patients with recent onset and established schizophrenia: implications for immune pathogenesis. Mol. Psychiatry 26, 5398–5406.

Conneely, K., Powers, A., Duncan, E., Almli, L., Massa, N., et al., 2019. Genome-wide association study in two populations to determine genetic variants associated with Toxoplasma gondii infection and relationship to schizophrenia risk. Prog. Neuro-Psychopharmacol. Biol. Psychiatry 92, 133–147.

Contopoulos-Ioannidis, D.G., Gianniki, M., Ai-Nhi Truong, A., Montoya, J.G., 2022. Toxoplasmosis and schizophrenia: a systematic review and meta-analysis of prevalence and associations and future directions. Psychiatr. Res. Clin. Pract. 4, 48–60.

Cookson, M.R., 2015. LRRK2 pathways leading to neurodegeneration. Curr. Neurol. Neurosci. Rep. 15, 42.

Cookson, M.R., 2017. Mechanisms of mutant LRRK2 neurodegeneration. Adv. Neurobiol. 14, 227–239.

Coperchini, F., Chiovato, L., Rotondi, M., 2021. Interleukin-6, CXCL10 and infiltrating macrophages in COVID-19-related cytokine storm: not one for all but all for one! Front. Immunol. 12, 668507.

Coppola, G., Rurak, G.M., Simard, S., Salmaso, N., 2019. A further analysis and commentary on: profiling changes in cortical astroglial cells following chronic stress. J. Exp. Neurosci. 13. 1179069519870182.

Corallo, K.L., Lyle, S.M., Carlock, M.A., Ross, T.M., Ehrlich, K.B., 2022. Emotional distress, targeted rejection, and antibody production after influenza vaccination in adolescence. Psychosom. Med. 84, 429–436.

Corbett, D., Carmichael, S.T., Murphy, T.H., Jones, T.A., Schwab, M.E., et al., 2017. Enhancing the alignment of the preclinical and clinical stroke recovery research pipeline: consensus-based core recommendations from the stroke recovery and rehabilitation roundtable translational working group. Neurorehabil. Neural Repair 31, 699–707.

Cordella, F., Sanchini, C., Rosito, M., Ferrucci, L., Pediconi, N., et al., 2021. Antibiotics treatment modulates microglia-synapses interaction. Cells 10, 2648.

Cordero, M.I., Stenz, L., Moser, D.A., Rusconi Serpa, S., Paoloni-Giacobino, A., Schechter, D.S., 2022. The relationship of maternal and child methylation of the glucocorticoid receptor NR3C1 during early childhood and subsequent child psychopathology at school-age in the context of maternal interpersonal violence-related post-traumatic stress disorder. Front. Psychiatry 13, 919820.

Cordova, M.J., Riba, M.B., Spiegel, D., 2017. Post-traumatic stress disorder and cancer. Lancet Psychiatry 4, 330–338.

Corrigan, P.W., Rao, D., 2012. On the self-stigma of mental illness: stages, disclosure, and strategies for change. Can. J. Psychiatry 57, 464–469.

Corsi-Zuelli, F., Deakin, B., 2021. Impaired regulatory T cell control of astroglial overdrive and microglial pruning in schizophrenia. Neurosci. Biobehav. Rev. 125, 637–653.

Corsi-Zuelli, F., Fachim, H.A., Loureiro, C.M., Shuhama, R., Bertozi, G., et al., 2019. Prolonged periods of social isolation from weaning reduce the anti-inflammatory cytokine IL-10 in blood and brain. Front. Neurosci. 12, 1011.

Corsi-Zuelli, F., Schneider, A.H., Santos-Silva, T., Loureiro, C.M., Shuhama, R., et al., 2022. Increased blood neutrophil extracellular traps (NETs) associated with early life stress: translational findings in recent-onset schizophrenia and rodent model. Transl. Psychiatry 12, 526.

Cortes-Altamirano, J.L., Olmos-Hernandez, A., Jaime, H.B., Carrillo-Mora, P., Bandala, C., et al., 2018. Review: 5-HT1, 5-HT2, 5-HT3 and 5-HT7 receptors and their role in the modulation of pain response in the central nervous system. Curr. Neuropharmacol. 16 (2), 210–221.

Cortese, F., Scicchitano, P., Gesualdo, M., Filaninno, A., De Giorgi, E., et al., 2016. Early and late infections in newborns: where do we stand? A review. Pediatr. Neonatol. 57, 265–273.

Corvalan, C., Gray, B., Villalobos Prats, E., Sena, A., Hanna, F., Campbell-Lendrum, D., 2022. Mental health and the global climate crisis. Epidemiol. Psychiatr. Sci. 31, e86.

Costa, R.J.S., Snipe, R.M.J., Kitic, C.M., Gibson, P.R., 2017. Systematic review: exercise-induced gastrointestinal syndrome-implications for health and intestinal disease. Aliment. Pharmacol. Ther. 46, 246–265.

Costa, A.N., Ferguson, B.J., Hawkins, E., Coman, A., Schauer, J., et al., 2023. The relationship between maternal antibodies to fetal brain and prenatal stress exposure in autism spectrum disorder. Metabolites 13, 663.

Costello, H., Gould, R.L., Abrol, E., Howard, R., 2019. Systematic review and meta-analysis of the association between peripheral inflammatory cytokines and generalized anxiety disorder. BMJ Open 9, e027925.

Cotman, C.W., Berchtold, N.C., Christie, L.A., 2007. Exercise builds brain health: key roles of growth factor cascades and inflammation. Trends Neurosci. 30, 464–472.

Courchesne, E., Gazestani, V.H., Lewis, N.E., 2020. Prenatal origins of ASD: the when, what, and how of ASD development. Trends Neurosci. 43, 326–342.

Coureuil, M., Lécuyer, H., Bourdoulous, S., Nassif, X., 2017. A journey into the brain: insight into how bacterial pathogens cross blood-brain barriers. Nat. Rev. Microbiol. 15, 149–159.

Courtney, C.M., Goodman, S.M., Nagy, T.A., Levy, M., Bhusal, P., et al., 2017. Potentiating antibiotics in drug-resistant clinical isolates via stimuli-activated superoxide generation. Sci. Adv. 3, e1701776.

Coutinho, L.G., Grandgirard, D., Leib, S.L., Agnez-Lima, L.F., 2013. Cerebrospinal-fluid cytokine and chemokine profile in patients with pneumococcal and meningococcal meningitis. BMC Infect. Dis. 13, 326.

Coventry, P.A., Meader, N., Melton, H., Temple, M., Dale, H., et al., 2020. Psychological and pharmacological interventions for posttraumatic stress disorder and comorbid mental health problems following complex traumatic events: systematic review and component network meta-analysis. PLoS Med. 17, e1003262.

Cowardin, C.A., Buonomo, E.L., Saleh, M.M., 2016. The binary toxin CDT enhances Clostridium difficile virulence by suppressing protective colonic eosinophilia. Nat. Microbiol. 1, 16108.

Cowger, T.L., Wortham, J.M., Burton, D.C., 2019. Epidemiology of tuberculosis among children and adolescents in the USA, 2007-17: an analysis of national surveillance data. Lancet Public Health 4, e506–e516.

Cox, L.M., Blaser, M.J., 2015. Antibiotics in early life and obesity. Nat. Rev. Endocrinol. 11, 182–190.

Crabbe, J.C., Wahlsten, D., Dudek, B.C., 1999. Genetics of mouse behavior: interactions with laboratory environment. Science 284, 1670–1672.

Craft, T.K., DeVries, A.C., 2006. Role of IL-1 in poststroke depressive-like behavior in mice. Biol. Psychiatry 60, 812–818.

Crane, J.D., Palanivel, R., Mottillo, E.P., et al., 2015. Inhibiting peripheral serotonin synthesis reduces obesity and metabolic dysfunction by promoting brown adipose tissue thermogenesis. Nat. Med. 21, 166–172.

Craveiro, M., Cretenet, G., Mongellaz, C., Matias, M.I., Caron, O., et al., 2017. Resveratrol stimulates the metabolic reprogramming of human CD4+ T cells to enhance effector function. Sci. Signal. 10. pii: eaal3024.

Crawford, J.D., Wang, H., Trejo-Zambrano, D., Cimbro, R., Talbot Jr., C.C., et al., 2023. The XIST lncRNA is a sex-specific reservoir of TLR7 ligands in SLE. JCI Insight 8, e169344.

Cribb, L., Hodge, A.M., Yu, C., Li, S.X., English, D.R., et al., 2022. Inflammation and epigenetic aging are largely independent markers of biological aging and mortality. J. Gerontol. A Biol. Sci. Med. Sci. 77, 2378–2386.

Cristofori, F., Dargenio, V.N., Dargenio, C., Miniello, V.L., Barone, M., Francavilla, R., 2021. Anti-inflammatory and immunomodulatory effects of probiotics in gut inflammation: a door to the body. Front. Immunol. 12, 578386.

Crotty, S., 2015. A brief history of T cell help to B cells. Nat. Rev. Immunol. 15, 185–189.

Crouse, D.L., Pinault, L., Balram, A., Hystad, P., Peters, P.A., et al., 2017. Urban greenness and mortality in Canada's largest cities: a national cohort study. Lancet Planet. Heath 1, e289–e297.

Crowley, N.A., Bloodgood, D.W., Hardaway, J.A., Kendra, A.M., McCall, J.G., et al., 2016. Dynorphin controls the gain of an amygdalar anxiety circuit. Cell Rep. 14, 2774–2783.

Cruchaga, C., Del-Aguila, J.L., Saef, B., Black, K., Fernandez, M.V., et al., 2017. Polygenic risk score of sporadic late-onset Alzheimer's disease reveals a shared architecture with the familial and early-onset forms. Alzheimers Dement. pii: S1552-5260(17)33708-1.

Crumeyrolle-Arias, M., Jaglin, M., Bruneau, A., Vancassel, S., Cardona, A., et al., 2014. Absence of the gut microbiota enhances anxiety-like behavior and neuroendocrine response to acute stress in rats. Psychoneuroendocrinology 42, 207–217.

Crunfli, F., Carregari, V.C., Veras, F.P., Silva, L.S., Nogueira, M.H., et al., 2022. Morphological, cellular, and molecular basis of brain infection in COVID-19 patients. Proc. Natl. Acad. Sci. USA 119, e2200960119.

Cruwys, T., Haslam, C., Steffens, N.K., Haslam, S.A., Fong, P., Lam, B.C.P., 2019. Friendships that money can buy: financial security protects health in retirement by enabling social connectedness. BMC Geriatr. 19, 319.

Cruz, A.A., Cooper, P.J., Figueiredo, C.A., Alcantara-Neves, N.M., Rodrigues, L.C., Barreto, M.L., 2017. Global issues in allergy and immunology: parasitic infections and allergy. J. Allergy Clin. Immunol. 140, 1217–1228.

Cryan, J.F., Dinan, T.G., 2012. Mind-altering microorganisms: the impact of the gut microbiota on brain and behavior. Nat. Rev. Neurosci. 13, 701–712.

Cryan, J.F., O'Riordan, K.J., Cowan, C.S.M., Sandhu, K.V., Bastiaanssen, T.F.S., et al., 2019. The microbiota-gut-brain axis. Physiol. Rev. 99, 1877–2013.

Cuddapah, V.A., Zhang, S.L., Sehgal, A., 2019. regulation of the blood-brain barrier by circadian rhythms and sleep. Trends Neurosci. 42, 500–510.

Cuello, A.C., 2017. Early and late CNS inflammation in Alzheimer's disease: two extremes of a continuum? Trends Pharmacol. Sci. 38, 956–966.

Cui, M., Dai, W., Kong, J., Chen, H., 2021. Th17 cells in depression: are they crucial for the antidepressant effect of ketamine? Front. Pharmacol. 12, 649144.

Cui, A., Huang, T., Li, S., Ma, A., Pérez, J.L., et al., 2024. Dictionary of immune responses to cytokines at single-cell resolution. Nature 625 (7994), 377–384.

Cuitavi, J., Torres-Pérez, J.V., Lorente, J.D., Campos-Jurado, Y., Andrés-Herrera, P., et al., 2023. Crosstalk between Mu-Opioid receptors and neuroinflammation: consequences for drug addiction and pain. Neurosci. Biobehav. Rev. 145, 105011.

Cullen, J.M.A., Shahzad, S., Dhillon, J., 2023. A systematic review on the effects of exercise on gut microbial diversity, taxonomic composition, and microbial metabolites: identifying research gaps and future directions. Front. Physiol. 14, 1292673.

Cullen, N.C., Novak, P., Tosun, D., Kovacech, B., Hanes, J., et al., 2024. Efficacy assessment of an active tau immunotherapy in Alzheimer's disease patients with amyloid and tau pathology: a post hoc analysis of the "ADAMANT" randomised, placebo-controlled, double-blind, multi-centre, phase 2 clinical trial. EBioMedicine 99, 104923.

Culverhouse, R.C., Saccone, N.L., Horton, A.C., Ma, Y., Anstey, K.J., et al., 2018. Collaborative meta-analysis finds no evidence of a strong interaction between stress and 5-HTTLPR genotype contributing to the development of depression. Mol. Psychiatry 23, 133–142.

Cunsolo, A., Ellis, N.R., 2018. Ecological grief as a mental health response to climate change-related loss. Nat. Clim. Chang. 8, 275–281.

Curtin, N.M., Boyle, N.T., Mills, K.H., Connor, T.J., 2009a. Psychological stress suppresses innate IFN-gamma production via glucocorticoid receptor activation: reversal by the anxiolytic chlordiazepoxide. Brain Behav. Immun. 23, 535–547.

Curtin, N.M., Mills, K.H., Connor, T.J., 2009b. Psychological stress increases expression of IL-10 and its homolog IL-19 via beta-adrenoceptor activation: reversal by the anxiolytic chlordiazepoxide. Brain Behav. Immun. 23, 371–379.

Curtiss, J., Andrews, L., Davis, M., Smits, J., Hofmann, S.G., 2017. A meta-analysis of pharmacotherapy for social anxiety disorder: an examination of efficacy, moderators, and mediators. Expert. Opin. Pharmacother. 18, 243–251.

Cuthbert, B.N., Insel, T.R., 2013. Toward the future of psychiatric diagnosis: the seven pillars of RDoC. BMC Med. 11, 126.

Daffis, S., Samuel, M.A., Suthar, M.S., Gale Jr., M., Diamond, M.S., 2008. Toll-like receptor 3 has a protective role against West Nile virus infection. J. Virol. 82, 10349–10358.

Dahan, S., Bragazzi, N.L., Yogev, A., Bar-Gad, M., Barak, V., et al., 2018. The relationship between serum cytokine levels and degree of psychosis in patients with schizophrenia. Psychiatry Res. 268, 467–472.

Daher, J.P.L., Abdelmotilib, H.A., Hu, X., Volpicelli-Daley, L.A., Moehle, M.S., et al., 2015. Leucine-rich repeat kinase 2 (LRRK2) pharmacological inhibition abates α-Synuclein gene-induced neurodegeneration. J. Biol. Chem. 290, 19433–19444.

Dahl, W.J., Rivero Mendoza, D., Lambert, J.M., 2020. Diet, nutrients and the microbiome. Prog. Mol. Biol. Transl. Sci. 171, 237–263.

Dahlén, A.D., Miguet, M., Schiöth, H.B., Rukh, G., 2022. The influence of personality on the risk of myocardial infarction in UK Biobank cohort. Sci. Rep. 12, 6706.

Dahlgren, M.K., Laifer, L.M., VanElzakker, M.B., Offringa, R., Hughes, K.C., 2018. Diminished medial prefrontal cortex activation during the recollection of stressful events is an acquired characteristic of PTSD. Psychol. Med. 48, 1128–1138.

Dahmen, B., Puetz, V.B., Scharke, W., von Polier, G.G., Herpertz-Dahlmann, B., Konrad, K., 2018. Effects of early-life adversity on hippocampal structures and associated HPA axis functions. Dev. Neurosci. 40, 13–22.

Dalamaga, M., Christodoulatos, G.S., Karampela, I., Vallianou, N., Apovian, C.M., 2021. Understanding the co-epidemic of obesity and covid-19: Current evidence, comparison with previous epidemics, mechanisms, and preventive and therapeutic perspectives. Curr. Obes. Rep. 10, 214–243.

Dalile, B., Van Oudenhove, L., Vervliet, B., Verbeke, K., 2019. The role of short-chain fatty acids in microbiota–gut–brain communication. Nat. Rev. Gastroenterol. Hepatol. 16, 461–478.

Dallé, E., Mabandla, M.V., 2018. Early life stress, depression and Parkinson's disease: a new approach. Mol. Brain 11, 18.

Dallman, M.F., 2010. Stress-induced obesity and the emotional nervous system. Trends Endocrinol. Metab. 21, 159–165.

Daly, M., Sutin, A.R., Robinson, E., 2019. Perceived weight discrimination mediates the prospective association between obesity and physiological dysregulation: evidence from a population-based cohort. Psychol. Sci. 30, 1030–1039.

Dambrova, M., Latkovskis, G., Kuka, J., Strele, I., Konrade, I., et al., 2016. Diabetes is associated with higher trimethylamine N-oxide plasma levels. Exp. Clin. Endocrinol. Diabetes 124, 251–256.

Dan, R., Růžička, F., Bezdicek, O., Růžička, E., Roth, J., et al., 2017. Separate neural representations of depression, anxiety and apathy in Parkinson's disease. Sci. Rep. 7, 12164.

Dandekar, M.P., Saxena, A., Scaini, G., Shin, J.H., Migut, A., et al., 2019. Medial forebrain bundle deep brain stimulation reverses anhedonic-like behavior in a chronic model of depression: importance of BDNF and inflammatory cytokines. Mol. Neurobiol. 56, 4364–4380.

Danese, A., McEwen, B.S., 2012. Adverse childhood experiences, allostasis, allostatic load, and age-related disease. Physiol. Behav. 106, 29–39.

Danese, A., Lewis, J., S., 2017. Psychoneuroimmunology of early-life stress: the hidden wounds of childhood trauma? Neuropsychopharmacology 42, 99–114.

Danieli, Y., 1998. International Handbook of Multigenerational Legacies of Trauma. Plenum, New York.

Daniels, M.J., Rivers-Auty, J., Schilling, T., Spencer, N.G., Watremez, W., et al., 2016. Fenamate NSAIDs inhibit the NLRP3 inflammasome and protect against Alzheimer's disease in rodent models. Nat. Commun. 7, 12504.

Danielson, A.M., Matheson, K., Anisman, H., 2011. Cytokine levels at a single time point following a reminder stimulus among women in abusive dating relationships: relationship to emotional states. Psychoneuroendocrinology 36, 40–50.

Danneskiold-Samsøe, N.B., de Freitas, D., Queiroz, B.H., Santos, R., Bicas, J.L., et al., 2019. Interplay between food and gut microbiota in health and disease. Food Res. Int. 115, 23–31.

Dantzer, R., 2017. Role of the kynurenine metabolism pathway in inflammation-induced depression: preclinical approaches. Curr. Top. Behav. Neurosci. 31, 117–138.

Dantzer, R., 2019. From stress sensitization to microglial priming and vice versa: a new era of research in biological psychiatry. Biol. Psychiatry 85, 619–620.

Dantzer, R., O'Connor, J.C., Freund, G.G., Johnson, R.W., Kelley, K.W., 2008. From inflammation to sickness and depression: when the immune system subjugates the brain. Nat. Rev. Neurosci. 9, 46.

Daria, A., Colombo, A., Llovera, G., Hampel, H., Willem, M., et al., 2017. Young microglia restore amyloid plaque clearance of aged microglia. EMBO J. 36, 583–603.

Dashorst, P., Mooren, T.M., Kleber, R.J., de Jong, P.J., Huntjens, R.J.C., 2019. Intergenerational consequences of the Holocaust on offspring

mental health: a systematic review of associated factors and mechanisms. Eur. J. Psychotraumatol. 10, 1654065.

Daskalakis, N.P., Cohen, H., Nievergelt, C.M., Baker, D.G., Buxbaum, J. D., et al., 2016. New translational perspectives for blood-based biomarkers of PTSD: from glucocorticoid to immune mediators of stress susceptibility. Exp. Neurol. 284, 133–140.

Daskalakis, N.P., Xu, C., Bader, H.N., Chatzinakos, C., Weber, P., et al., 2021. Intergenerational trauma is associated with expression alterations in glucocorticoid-and immune-related genes. Neuropsychopharmacology 46, 763–773.

Dauer, W., Przedborski, S., 2003. Parkinson's disease: mechanisms and models. Neuron 39, 889–909.

Davalos, D., Grutzendler, J., Yang, G., Kim, J.V., Zuo, Y., et al., 2005. ATP mediates rapid microglial response to local brain injury in vivo. Nat. Neurosci. 8, 752–758.

Davar, D., Dzutsev, A.K., McCulloch, J.A., Rodrigues, R.R., Chauvin, J.-M., et al., 2021. Fecal microbiota transplant overcomes resistance to anti-PD-1 therapy in melanoma patients. Science 371, 595.

Davis, M., Walker, D.L., Miles, L., Grillon, C., 2010. Phasic vs sustained fear in rats and humans: role of the extended amygdala in fear vs anxiety. Neuropsychopharmacology 35 (1), 105–135.

Davis, K.D., Flor, H., Greely, H.T., Iannetti, G.D., Mackey, S., et al., 2017. Brain imaging tests for chronic pain: medical, legal and ethical issues and recommendations. Nat. Rev. Neurol. 13, 624–638.

Davis, M.T., Hillmer, A., Holmes, S.E., Pietrzak, R.H., DellaGioia, N., et al., 2019. In vivo evidence for dysregulation of mGluR5 as a biomarker of suicidal ideation. PNAS 116, 11490–11495.

Davis, A.K., Barrett, F.S., May, D.G., Cosimano, M.P., Sepeda, N.D., 2021. Effects of psilocybin-assisted therapy on major depressive disorder: a randomized clinical trial. JAMA Psychiatry 78, 481–489.

Davis, H.E., McCorkell, L., Vogel, J.M., Topol, E.J., 2023. Long COVID: major findings, mechanisms and recommendations. Nat. Rev. Microbiol. 21, 133–146.

Daviu, N., Bruchas, M.R., Moghaddam, B., Sandi, C., Beyeler, A., 2019. Neurobiological links between stress and anxiety. Neurobiol. Stress 11, 100191.

Dawidowski, B., Górniak, A., Podwalski, P., Lebiecka, Z., Misiak, B., Samochowiec, J., 2021. The role of cytokines in the pathogenesis of schizophrenia. J. Clin. Med. 10, 3849.

Daws, R.E., Timmermann, C., Giribaldi, B., Sexton, J.D., Wall, M.B., et al., 2022. Increased global integration in the brain after psilocybin therapy for depression. Nat. Med. 28, 844–851.

Dayananda, P., Wilcox, M.H., 2019. A review of mixed strain Clostridium difficile colonization and infection. Front. Microbiol. 10, 692.

de Bartolomeis, A., Barone, A., Vellucci, L., Mazza, B., Austin, M.C., et al., 2022. Linking inflammation, aberrant glutamate-dopamine interaction, and post-synaptic changes: translational relevance for schizophrenia and antipsychotic treatment: a systematic review. Mol. Neurobiol. 59, 6460–6501.

de Berker, A.O., Rutledge, R.B., Mathys, C., Marshall, L., Cross, G.F., et al., 2016. Computations of uncertainty mediate acute stress responses in humans. Nat. Commun. 7, 10996.

de Bilbao, F., Arsenijevic, D., Moll, T., Garcia-Gabay, I., Vallet, P., et al., 2009. In vivo over-expression of interleukin-10 increases resistance to focal brain ischemia in mice. J. Neurochem. 110, 12–22.

de Bree, L.C.J., Mourits, V.P., Koeken, V.A.C.M., Moorlag, S.J., Janssen, R., Folkman, L., et al., 2020. Circadian rhythm influences induction of trained immunity by BCG vaccination. J. Clin. Invest. 130, 5603–5617.

de Cossío, L.F., Fourrier, C., Sauvant, J., Everard, A., Capuron, L., et al., 2017. Impact of prebiotics on metabolic and behavioral alterations in a mouse model of metabolic syndrome. Brain Behav. Immun. 64, 33–49.

De Dreu, C.K., Kret, M.E., 2016. Oxytocin conditions intergroup relations through upregulated in-group empathy, cooperation, conformity, and defense. Biol. Psychiatry 79, 165–173.

de Groot, J., Boersma, W.J., Scholten, J.W., Koolhaas, J.M., 2002. Social stress in male mice impairs long-term antiviral immunity selectively in wounded subjects. Physiol. Behav. 75, 277–285.

de Haan, C.A., Te Lintelo, E., Li, Z., Raaben, M., Wurdinger, T., et al., 2006. Cooperative involvement of the S1 and S2 subunits of the murine coronavirus spike protein in receptor binding and extended host range. J. Virol. 80, 10909–10918.

de Jong, D., Jansen, R., Hoefnagels, W., Jellesma-Eggenkamp, M., Verbeek, M., et al., 2008. No effect of one-year treatment with indomethacin on Alzheimer's disease progression: a randomized controlled trial. PLoS One 3, e1475.

de Kluiver, H., Jansen, R., Milaneschi, Y., Penninx, B.W.J.H., 2019. Involvement of inflammatory gene expression pathways in depressed patients with hyperphagia. Transl. Psychiatry 9, 193.

de Koning, R.M., Kuzminskaite, E., Vinkers, C.H., Giltay, E.J., Penninx, B.W.J.H., 2022. Childhood trauma and LPS-stimulated inflammation in adulthood: results from the Netherlands Study of Depression and Anxiety. Brain Behav. Immun. 106, 21–29.

De La Garza, R. 2nd, Asnis, G.M., Pedrosa, E., Stearns, C., Migdal, A.L., et al., 2005. Recombinant human interferon-alpha does not alter reward behavior, or neuroimmune and neuroendocrine activation in rats. Prog. Neuropsychopharmacol. Biol. Psychiatry 29, 781–792.

de Mello, A.J., Moretti, M., Rodrigues, A.L.S., 2022. SARS-CoV-2 consequences for mental health: neuroinflammatory pathways linking COVID-19 to anxiety and depression. World J. Psychiatry 12, 874–883.

De Miguel, Z., Haditsch, U., Palmer, T.D., Azpiroz, A., Sapolsky, R.M., 2018. Adult-generated neurons born during chronic social stress are uniquely adapted to respond to subsequent chronic social stress. Mol. Psychiatry. https://doi.org/10.1038/s41380-017-0013-1.

de Miranda, A.S., Lacerda-Queiroz, N., de Carvalho Vilela, M., Rodrigues, D.H., Rachid, M.A., et al., 2011. Anxiety-like behavior and proinflammatory cytokine levels in the brain of C57BL/6 mice infected with Plasmodium berghei (strain ANKA). Neurosci. Lett. 491, 202–206.

de Nies, L., Kobras, C.M., Stracy, M., 2023. Antibiotic-induced collateral damage to the microbiota and associated infections. Nat. Rev. Microbiol. 12, 789–804.

De Oliveira, F.L., Salgaço, M.K., de Oliveira, M.T., Mesa, V., Sartoratto, A., et al., 2023. Exploring the potential of Lactobacillus helveticus R0052 and Bifidobacterium longum R0175 as promising psychobiotics using SHIME. Nutrients 15, 1521.

De Pablos, R.M., Herrera, A.J., Espinosa-Oliva, A.M., Sarmiento, M., Muñoz, M.F., et al., 2014. Chronic stress enhances microglia activation and exacerbates death of nigral dopaminergic neurons under conditions of inflammation. J. Neuroinflammation 11, 34.

De Padova, S., Grassi, L., Vagheggini, A., Belvederi Murri, M., Folesani, F., et al., 2021. Post-traumatic stress symptoms in long-term disease-free cancer survivors and their family caregivers. Cancer Med. 10, 3974–3985.

De Palma, G., Blennerhassett, P., Lu, J., Deng, Y., Park, A.J., et al., 2015. Microbiota and host determinants of behavioural phenotype in maternally separated mice. Nat. Commun. 6, 7735.

De Vita, M.J., Maisto, S.A., Gilmour, C.E., McGuire, L., Tarvin, E., Moskal, D., 2021. The effects of cannabidiol and analgesic expectancies on experimental pain reactivity in healthy adults: a balanced placebo design trial. Exp. Clin. Psychopharmacol. 30 (5), 536–546.

de Roos, C., van der Oord, S., Zijlstra, B., Lucassen, S., Perrin, S., et al., 2017. Comparison of eye movement desensitization and reprocessing therapy, cognitive behavioral writing therapy, and wait-list in pediatric posttraumatic stress disorder following single-incident trauma: a multicenter randomized clinical trial. J. Child Psychol. Psychiatry 58, 1219–1228.

de Sousa, G.M., de Oliveira Tavares, V.D., de Menezes Galvão, A.C., de Almeida, R.N., et al., 2022. Moderators of ayahuasca's biological antidepressant action. Front. Psychiatry 13, 1033816.

Dean, R.L., Hurducas, C., Hawton, K., Spyridi, S., Cowen, P.J., et al., 2021. Ketamine and other glutamate receptor modulators for depression in adults with unipolar major depressive disorder. Cochrane Database Syst. Rev. 9, CD011612.

Dębiec, J., Bush, D.E., LeDoux, J.E., 2011. Noradrenergic enhancement of reconsolidation in the amygdala impairs extinction of conditioned fear in rats—a possible mechanism for the persistence of traumatic memories in PTSD. Depress. Anxiety 28, 186–193.

Deckert, J., Weber, H., Villmann, C., Lonsdorf, T.B., Richter, J., et al., 2017. GLRB allelic variation associated with agoraphobic cognitions, increased startle response and fear network activation: a potential neurogenetic pathway to panic disorder. Mol. Psychiatry 22 (10), 1431–1439.

Decout, A., Katz, J.D., Venkatraman, S., Ablasser, A., 2021. The cGAS-STING pathway as a therapeutic target in inflammatory diseases. Nat. Rev. Immunol. 21, 548–569.

Dedek, A., Xu, J., Lorenzo, L.É., Godin, A.G., Kandegedara, C.M., et al., 2022. Sexual dimorphism in a neuronal mechanism of spinal hyperexcitability across rodent and human models of pathological pain. Brain 145 (3), 1124–1138.

Dedic, N., Chen, A., Deussing, J.M., 2018. The CRF family of neuropeptides and their receptors – mediators of the central stress response. Curr. Mol. Pharmacol. 11, 4–31.

Dedmon, L.E., 2020. The genetics of rheumatoid arthritis. Rheumatology (Oxford) 59, 2661–2670.

Dehghan, M., Mente, A., Zhang, X., Swaminathan, S., Li, W., et al., 2017. Associations of fats and carbohydrate intake with cardiovascular disease and mortality in 18 countries from five continents (PURE): a prospective cohort study. Lancet 390 (10107), 2050–2062.

Dejanovic, B., Wu, T., Tsai, M.C., Graykowski, D., Gandham, V.D., et al., 2022. Complement C1q-dependent excitatory and inhibitory synapse elimination by astrocytes and microglia in Alzheimer's disease mouse models. Nat. Aging 2, 837–850.

Del Casale, A., Sorice, S., Padovano, A., Simmaco, M., Ferracuti, S., et al., 2019. Psychopharmacological treatment of obsessive-compulsive disorder (OCD). Curr. Neuropharmacol. 17, 710–736.

Del Gobbo, L.C., Imamura, F., Aslibekyan, S., Marklund, M., Virtanen, J.K., et al., 2016. ω-3 polyunsaturated fatty acid biomarkers and coronary heart disease: pooling project of 19 cohort studies. JAMA Intern. Med. 176 (8), 1155–1166.

Del Grande da Silva, G., Wiener, C.D., Barbosa, L.P., Gonçalves Araujo, J.M., Molina, M.L., et al., 2016. Pro-inflammatory cytokines and psychotherapy in depression: results from a randomized clinical trial. J. Psychiatr. Res. 75, 57–64.

Del Tredici, K., Braak, H., 2012. Lewy pathology and neurodegeneration in premotor Parkinson's disease. Mov. Disord. 27 (5), 597607.

Deleidi, M., Gasser, T., 2013. The role of inflammation in sporadic and familial Parkinson's disease. Cell. Mol. Life Sci. 70 (22), 42594273.

Dell'Oste, V., Fantasia, S., Gravina, D., Palego, L., Betti, L., et al., 2023. Metabolic and inflammatory response in post-traumatic stress disorder (PTSD): a systematic review on peripheral neuroimmune biomarkers. Int. J. Environ. Res. Public Health 20, 2937.

Delpech, J.C., Madore, C., Nadjar, A., Joffre, C., Wohleb, E.S., et al., 2015. Microglia in neuronal plasticity: influence of stress. Neuropharmacology 96 (Pt A), 1928.

DeLuca, P.F., Buist, S., Johnston, N., 2012. The Code Red Project: engaging communities in health system change in Hamilton, Canada. Social Indic. Res. 108, 317–327.

Delva, N.C., Stanwood, G.D., 2021. Dysregulation of brain dopamine systems in major depressive disorder. Exp. Biol. Med. (Maywood) 246, 1084–1093.

Delvaux, E., Mastroeni, D., Nolz, J., Chow, N., Sabbagh, M., et al., 2017. Multivariate analyses of peripheral blood leukocyte transcripts distinguish Alzheimer's, Parkinson's, control, and those at risk for developing Alzheimer's. Neurobiol. Aging 58, 225–237.

Dema, B., Charles, N., 2016. Autoantibodies in SLE: Specificities, isotypes and receptors. Antibodies 5, 2.

Demoruelle, M.K., Deane, K.D., Holers, V.M., 2014. When and where does inflammation begin in rheumatoid arthritis? Curr. Opin. Rheumatol. 26 (1), 6471.

Deng, G., 2019. Integrative medicine therapies for pain management in cancer patients. Cancer J. 25 (5), 343–348.

Deng, Y., Lu, X., Wang, L., Li, T., Ding, Y., et al., 2014. Curcumin inhibits the AKT/NF-κB signaling via CpG demethylation of the promoter and restoration of NEP in the N2a cell line. AAPS J. 16 (4), 649–657.

Denk, F., McMahon, S.B., Tracey, I., 2014. Pain vulnerability: a neurobiological perspective. Nat. Neurosci. 17 (2), 192–200.

Dennis, J., Sealock, J., Levinson, R.T., Farber-Eger, E., Franco, J., et al., 2021. Genetic risk for major depressive disorder and loneliness in sex-specific associations with coronary artery disease. Mol. Psychiatry 26, 4254–4264.

Denollet, J., Gidron, Y., Vrints, C.J., Conraads, V.M., 2010. Anger, suppressed anger, and risk of adverse events in patients with coronary artery disease. Am. J. Cardiol. 105, 1555–1560.

Denollet, J., van Felius, R.A., Lodder, P., Mommersteeg, P.M., Goovaerts, I., 2018. Predictive value of Type D personality for impaired endothelial function in patients with coronary artery disease. Int. J. Cardiol. 259, 205–210.

Depino, A.M., 2015. Early prenatal exposure to LPS results in anxiety- and depression-related behaviors in adulthood. Neuroscience 299, 56–65.

Depino, A.M., 2017. Perinatal inflammation and adult psychopathology: from preclinical models to humans. Semin. Cell Dev. Biol. pii: S1084-9521(17)30309-9.

Derks, N.A., Krugers, H.J., Hoogenraad, C.C., Joëls, M., Sarabdjitsingh, R.A., 2016. Effects of early life stress on synaptic plasticity in the developing hippocampus of male and female rats. PLoS One 11, e0164551.

Derks, E.M., Thorp, J.G., Gerring, Z.F., 2022. Ten challenges for clinical translation in psychiatric genetics. Nat. Genet. 54, 1457–1465.

Derry, C.J., Derry, S., Moore, R.A., 2015. Single dose oral analgesics for acute postoperative pain in adults - an overview of Cochrane reviews. Cochrane Database Syst. Rev. CD008659.

Desbonnet, L., Clarke, G., Traplin, A., O'Sullivan, O., Crispie, F., et al., 2015. Gut microbiota depletion from early adolescence in mice: implications for brain and behaviour. Brain Behav. Immun. 48, 165–173.

Deslauriers, J., Powell, S., Risbrough, V.B., 2017. Immune signaling mechanisms of PTSD risk and symptom development: insights from animal models. Curr. Opin. Behav. Sci. 14, 123–132.

Destras, G., Bal, A., Escuret, V., Morfin, F., Lina, B., COVID-Diagnosis HCL Study Group, et al., 2020. Systematic SARS-CoV-2 screening in cerebrospinal fluid during the COVID-19 pandemic. Lancet Microbe 1, e149.

DeTure, M.A., Dickson, D.W., 2019. The neuropathological diagnosis of Alzheimer's disease. Mol. Neurodegener. 14, 32.

Devanand, D.P., 2018. Viral hypothesis and antiviral treatment in Alzheimer's disease. Curr. Neurol. Neurosci. Rep. 18, 55.

Devi, S., Alexandre, Y.O., Loi, J.K., Gillis, R., Ghazanfari, N., et al., 2021. Adrenergic regulation of the vasculature impairs leukocyte interstitial migration and suppresses immune responses. Immunity 54, 1219–1230.e7.

Devor, A., Andreassen, O.A., Wang, Y., Maki-Marttunen, T., Smeland, O.B., et al., 2017. Genetic evidence for role of integration of fast and slow neurotransmission in schizophrenia. Mol. Psychiatry 22, 792–801.

DeVries, A., Shambhu, S., Sloop, S., Overhage, J.M., 2023. One-year adverse outcomes among US adults with post-COVID-19 condition

vs those without COVID-19 in a large commercial insurance database. JAMA Health Forum 4, e230010.

DeWall, C.N., MacDonald, G., Webster, G.D., Masten, C.L., Baumeister, R.F., et al., 2010. Acetaminophen reduces social pain: behavioral and neural evidence. Psychol. Sci. 2, 931–937.

Dewey, F.E., Murray, M.F., Overton, J.D., Habegger, L., Leader, J.B., et al., 2016. Distribution and clinical impact of functional variants in 50,726 whole-exome sequences from the DiscovEHR study. Science 354, aaf6814.

Deyama, S., Bang, E., Kato, T., Li, X.Y., Duman, R.S., 2019. Neurotrophic and antidepressant actions of brain-derived neurotrophic factor require vascular endothelial growth factor. Biol. Psychiatry 86, 143–152.

Dhabhar, F.S., 2009. Enhancing versus suppressive effects of stress on immune function: implications for immunoprotection and immunopathology. Neuroimmunomodulation 16, 300–317.

Dhabhar, F.S., 2014. Effects of stress on immune function: the good, the bad, and the beautiful. Immunol. Res. 58, 193–210.

Dhabhar, F.S., McEwen, B.S., 1999. Enhancing versus suppressive effects of stress hormones on skin immune function. Proc. Natl. Acad. Sci. USA 96, 1059–1064.

Dhabhar, F.S., Viswanathan, K., 2005. Short-term stress experienced at time of immunization induces a long-lasting increase in immunologic memory. Am. J. Phys. Regul. Integr. Comp. Phys. 289, R738–R744.

Dhabhar, F.S., Saul, A.N., Daugherty, C., Holmes, T.H., Bouley, D.M., et al., 2010. Short-term stress enhances cellular immunity and increases early resistance to squamous cell carcinoma. Brain Behav. Immun. 24, 127–137.

Di Battista, A.M., Heinsinger, N.M., Rebeck, G.W., 2016. Alzheimer's disease genetic risk factor APOE-ε4 also affects normal brain function. Curr. Alzheimer Res. 13, 1200–1207.

di Filippo, L., Frara, S., Nannipieri, F., Cotellessa, A., Locatelli, M., et al., 2023. Low vitamin D levels are associated with Long COVID syndrome in COVID-19 survivors. J. Clin. Endocrinol. Metab. dgad207.

Dib, P., Zhang, Y., Ihnat, M.A., Gallucci, R.M., Standifer, K.M., 2021. TNF-alpha as an initiator of allodynia and anxiety-like behaviors in a preclinical model of PTSD and comorbid pain. Front. Psychiatry 12, 721999.

Dickerson, S., Kemeny, M., 2004. Acute stressors and cortisol responses: a theoretical integration and synthesis of laboratory research. Psychol. Bull. 130, 355–391.

Dickerson, D.D., Wolff, A.R., Bilkey, D.K., 2010. Abnormal long-range neural synchrony in a maternal immune activation animal model of schizophrenia. J. Neurosci. 30, 12424–12431.

Dickerson, F., Stallings, C., Origoni, A., Vaughan, C., Katsafanas, E., et al., 2013. A combined marker of inflammation in individuals with mania. PLoS One 8, e73520.

Dicks, L.M.T., Hurn, D., Hermanus, D., 2021. Gut bacteria and neuropsychiatric disorders. Microorganisms 9, 2583.

Dietz, D.M., Nestler, E.J., 2012. From father to offspring: paternal transmission of depressive-like behaviors. Neuropsychopharmacology 37, 311–312.

Dietz, D.M., Laplant, Q., Watts, E.L., Hodes, G.E., Russo, S.J., et al., 2011. Paternal transmission of stress-induced pathologies. Biol. Psychiatry 70, 408–414.

Dietz, A.G., Goldman, S.A., Nedergaard, M., 2020. Glial cells in schizophrenia: a unified hypothesis. Lancet Psychiatry 7, 272–281.

Dietz, B., Katz, P., Dall'Era, M., Murphy, L.B., Lanata, C., et al., 2021. Major depression and adverse patient-reported outcomes in systemic lupus erythematosus: results from a prospective longitudinal cohort. Arthritis Care Res. 73, 48–54.

Díez-Solinska, A., Lebeña, A., Garmendia, L., Labaka, A., Azkona, G., et al., 2022. Chronic social instability stress down-regulates IL-10 and up-regulates CX3CR1 in tumor-bearing and non-tumor-bearing female mice. Behav. Brain Res. 435, 114063.

Dimatelis, J.J., Pillay, N.S., Mutyaba, A.K., Russell, V.A., Daniels, W.M. U., et al., 2012. Early maternal separation leads to down-regulation of cytokine gene expression. Metab. Brain Dis. 27, 393–397.

Dimitriadis, M., van den Brink, R.H.S., Comijs, H.C., Oude Voshaar, R. C., 2019. Prognostic effect of serum BDNF levels in late-life depression: moderated by childhood trauma and SSRI usage? Psychoneuroendocrinology 103, 276–283.

Dimitrijević, M., Stanojević, S., 2013. The intriguing mission of neuropeptide Y in the immune system. Amino Acids 45, 41–53.

Dinan, T.G., Cryan, J.F., 2016. Microbes, immunity, and behavior: psychoneuroimmunology meets the microbiome. Neuropsychopharmacology 42, 178–192.

Dinan, T.G., Cryan, J.F., 2017. Gut-brain axis in 2016: brain-gut-microbiota axis–mood, metabolism and behaviour. Nat. Rev. Gastroenterol. Hepatol. 14, 69–70.

Dinan, K., Dinan, T., 2022. Antibiotics and mental health: the good, the bad and the ugly. J. Intern. Med. 292, 858–869.

Dinel, A.L., Joffre, C., Trifilieff, P., Aubert, A., Foury, A., et al., 2014. Inflammation early in life is a vulnerability factor for emotional behavior at adolescence and for lipopolysaccharide-induced spatial memory and neurogenesis alteration at adulthood. J. Neuroinflammation 11, 155.

Ding, H.T., Taur, Y., Walkup, J.T., 2017. Gut microbiota and autism: key concepts and findings. J. Autism Dev. Disord. 47 (2), 480–489.

Ding, D., Yong, H., You, N., Lu, W., Yang, X., et al., 2022. Prospective study reveals host microbial determinants of clinical response to fecal microbiota transplant therapy in type 2 diabetes patients. Front. Cell. Infect. Microbiol. 12, 820367.

Dintica, C.S., Bahorik, A., Xia, F., Kind, A., Yaffe, K., 2023. Dementia risk and disadvantaged neighborhoods. JAMA Neurol. 80, 903–909.

Dionisie, V., Filip, G.A., Manea, M.C., Manea, M., Riga, S., 2021. The anti-inflammatory role of SSRI and SNRI in the treatment of depression: a review of human and rodent research studies. Inflammopharmacology 29, 75–90.

DiSabato, D.J., Nemeth, D.P., Liu, X., Witcher, K.G., O'Neil, S.M., et al., 2021. Interleukin-1 receptor on hippocampal neurons drives social withdrawal and cognitive deficits after chronic social stress. Mol. Psychiatry 26, 4770–4782.

Disner, S.G., Beevers, C.G., Haigh, E.A., Beck, A.T., 2011. Neural mechanisms of the cognitive model of depression. Nat. Rev. Neurosci. 12, 467–477.

D'Mello, C., Ronaghan, N., Zaheer, R., Dicay, M., Le, T., et al., 2015. Probiotics improve inflammation-associated sickness behavior by altering communication between the peripheral immune system and the brain. J. Neurosci. 35, 10821–10830.

Doan, L., Manders, T., Wang, J., 2015. Neuroplasticity underlying the comorbidity of pain and depression. Neural Plast. 2015, 504691.

Dodd, G.T., Andrews, Z.B., Simonds, S.E., Michael, N.J., DeVeer, M., et al., 2017. A hypothalamic phosphatase switch coordinates energy expenditure with feeding. Cell Metab. 26.

Doenni, V.M., Song, C.M., Hill, M.N., Pittman, Q.J., 2017. Early-life inflammation with LPS delays fear extinction in adult rodents. Brain Behav. Immun. 63, 176–185.

Doenyas, C., 2018. Gut microbiota, inflammation, and probiotics on neural development in autism spectrum disorder. Neuroscience 374, 271286. pii: S0306-4522(18) 30099-X.

Dohnalová, L., Lundgren, P., Carty, J.R.E., Goldstein, N., Wenski, S.L., et al., 2022. A microbiome-dependent gut-brain pathway regulates motivation for exercise. Nature 612, 739–747.

Dolzani, S.D., Baratta, M.V., Amat, J., Agster, K.L., Saddoris, M.P., et al., 2016. Activation of a habenulo-raphe circuit is critical for the behavioral and neurochemical consequences of uncontrollable stress in the male rat. eNeuro 3. ENEURO.0229-16.2016.

Domènech, L., Willis, J., Alemany-Navarro, M., Morell, M., Real, E., et al., 2022. Changes in the stool and oropharyngeal microbiome in obsessive-compulsive disorder. Sci. Rep. 12, 1448.

Domenighetti, C., Sugier, P.E., Sreelatha, A.A.K., Schulte, C., et al., 2022. Mendelian randomisation study of smoking, alcohol, and coffee drinking in relation to Parkinson's disease. J. Parkinsons Dis. 12, 267–282.

Domingueti, C.P., Dusse, L.M., Carvalho, M.D., de Sousa, L.P., Gomes, K.B., et al., 2016. Diabetes mellitus: the linkage between oxidative stress, inflammation, hypercoagulaility and vascular complications. J. Diabetes Complicat. 30, 738–745.

Dominguez-Bello, M.G., De Jesus-Laboy, K.M., Shen, N., Cox, L.M., Amir, A., et al., 2016. Partial restoration of the microbiota of caesarean-born infants via vaginal microbial transfer. Nat. Med. 22, 250–253.

Dondo, T.B., Hall, M., West, R.M., Jernberg, T., Lindahl, B., et al., 2017. β-blockers and mortality after acute myocardial infarction in patients without heart failure or ventricular dysfunction. J. Am. Coll. Cardiol. 69, 2710–2720.

Doney, E., Cadoret, A., Dion-Albert, L., Lebel, M., Menard, C., 2022. Inflammation-driven brain and gut barrier dysfunction in stress and mood disorders. Eur. J. Neurosci. 55, 2851–2894.

Dong, M., Zhou, C., Ji, L., Pan, B., Zheng, L., 2017a. AG1296 enhances plaque stability via inhibiting inflammatory responses and decreasing MMP-2 and MMP-9 expression in ApoE-/- mice. Biochem. Biophys. Res. Commun. 489, 426–431.

Dong, S., Maniar, S., Manole, M.D., Sun, D., 2017b. Cerebral hypoperfusion and other shared brain pathologies in ischemic stroke and Alzheimer's disease. Transl. Stroke Res.

Dooley, J., Tian, L., Schonefeldt, S., Delghingaro-Augusto, V., Garcia-Perez, J.E., et al., 2016. Genetic predisposition for beta cell fragility underlies type 1 and type 2 diabetes. Nat. Genet. 48, 519–527.

Dorfschmidt, L., Bethlehem, R.A., Seidlitz, J., Váša, F., White, S.R., et al., 2022. Sexually divergent development of depression-related brain networks during healthy human adolescence. Sci. Adv. 8, eabm7825.

Dos Santos, R.G., Balthazar, F.M., Bouso, J.C., Hallak, J.E., et al., 2016. The current state of research on ayahuasca: a systematic review of human studies assessing psychiatric symptoms, neuropsychological functioning, and neuroimaging. J. Psychopharmacol. 30, 1230–1247.

Dos Santos, L.R., Pimassoni, L.H.S., Sena, G.G.S., Camporez, D., Belcavello, L., et al., 2017. Validating GWAS variants from microglial genes implicated in Alzheimer's disease. J. Mol. Neurosci. 62, 215–221.

Doshi, G., Bhatia, N., Ved, H., Pandya, A., Kulkarni, D., et al., 2023. Update on oxytocin, phosphodiesterase, neurokinin, glycine as a therapeutic approach in the treatment of schizophrenia. CNS Neurol. Disord. Drug Targets 22, 994–1007.

Doty, R.L., 2012. Olfactory dysfunction in Parkinson disease. Nat. Rev. Neurol. 8, 329–339.

Dou, H., Feher, A., Davila, A.C., Romero, M.J., Patel, V.S., et al., 2017. Role of adipose tissue endothelial ADAM17 in age-related coronary microvascular dysfunction. Arterioscler. Thromb. Vasc. Biol. 37, 1180–1193.

Dou, D.R., Zhao, Y., Belk, J.A., Zhao, Y., Casey, K.M., et al., 2023. XIST ribonucleoproteins promote female sex-biased autoimmunity. Cell 187, 733–749.

Douaud, G., Lee, S., Alfaro-Almagro, F., Arthofer, C., Wang, C., et al., 2022. SARS-CoV-2 is associated with changes in brain structure in UK Biobank. Nature 604, 697–707.

Dougherty, D.D., Rezai, A.R., Carpenter, L.L., Howland, R.H., Bhati, M. T., et al., 2015. A randomized sham-controlled trial of deep brain stimulation of the ventral capsule/ventral striatum for chronic treatment-resistant depression. Biol. Psychiatry 78, 240–248.

Douma, E.H., de Kloet, E.R., 2020. Stress-induced plasticity and functioning of ventral tegmental dopamine neurons. Neurosci. Biobehav. Rev. 108, 48–77.

Dourson, A.J., Fadaka, A.O., Warshak, A.M., Paranjpe, A., Weinhaus, B., et al., 2024. Macrophage memories of early-life injury drive neonatal nociceptive priming. Cell Rep. 43, 114–129.

Dowell, D., Haegerich, T.M., Chou, R., 2016. CDC guideline for prescribing opioids for chronic pain—United States, 2016. JAMA 315, 1624–1645.

Doyle, D.M., Molix, L., 2014. Perceived discrimination and well-being in gay men: the protective role of behavioural identification. Psychol. Sex. 5, 117–130.

Drevets, W.C., Bullmore, E.T., 2020. Major depressive disorder is associated with differential expression of innate immune and neutrophil-related gene networks in peripheral blood: a quantitative review of whole-genome transcriptional data from case-control studies. Biol. Psychiatry 88, 625–637.

Drevets, W.C., Wittenberg, G.M., Bullmore, E.T., Manji, H.K., 2022. Immune targets for therapeutic development in depression: towards precision medicine. Nat. Rev. Drug Discov. 21, 224–244.

Drury, J., Carter, H., Cocking, C., Ntontis, E., Tekin, G.S., Amlôt, R., 2019. Facilitating collective psychosocial resilience in the public in emergencies: twelve recommendations based on the social identity approach. Front. Public Health 7, 141.

Drysdale, A.T., Grosenick, L., Downar, J., Dunlop, K., Mansouri, F., et al., 2017. Resting-state connectivity biomarkers define neurophysiological subtypes of depression. Nat. Med. 23, 28–38.

Du, R., Han, R., Niu, K., Xu, J., Zhao, Z., et al., 2022. The multivariate effect of ketamine on PTSD: systematic review and meta-analysis. Front. Psychiatry 13, 813103.

Du, Y., Li, Y., Zhao, X., Yao, Y., Wang, B., et al., 2023. Psilocybin facilitates fear extinction in mice by promoting hippocampal neuroplasticity. Chin. Med. J. 136, 2983–2992.

Duan, F.F., Liu, J.H., March, J.C., 2015. Engineered commensal bacteria reprogram intestinal cells into glucose-responsive insulin-secreting cells for the treatment of diabetes. Diabetes 64, 1794–1803.

Duan, Y., Young, R., Schnabl, B., 2022a. Bacteriophages and their potential for treatment of gastrointestinal diseases. Nat. Rev. Gastroenterol. Hepatol. 19, 135–144.

Duan, N., Zhang, Y., Tan, S., Sun, J., Ye, M., et al., 2022b. Therapeutic targeting of STING-TBK1-IRF3 signalling ameliorates chronic stress induced depression-like behaviours by modulating neuroinflammation and microglia phagocytosis. Neurobiol. Dis. 169, 105739.

Duclot, F., Kabbaj, M., 2015. Epigenetic mechanisms underlying the role of brain-derived neurotrophic factor in depression and response to antidepressants. J. Exp. Biol. 218, 21–31.

Duda-Chodak, A., Tarko, T., Satora, P., Sroka, P., 2015. Interaction of dietary compounds, especially polyphenols, with the intestinal microbiota: a review. Eur. J. Nutr. 54, 325–341.

Dudek, K.A., Dion-Albert, L., Lebel, M., LeClair, K., Labrecque, S., et al., 2020. Molecular adaptations of the blood-brain barrier promote stress resilience vs. depression. Proc. Natl. Acad. Sci. USA 117, 3326–3336.

Dudek, K.A., Kaufmann, F.N., Lavoie, O., Menard, C., 2021. Central and peripheral stress-induced epigenetic mechanisms of resilience. Curr. Opin. Psychiatry 34, 1–9.

Duko, B., Bedaso, A., Ayano, G., 2020. The prevalence of depression among patients with tuberculosis: a systematic review and meta-analysis. Ann. General Psychiatry 19, 30.

Duman, R.S., Monteggia, L.M., 2006. A neurotrophic model for stress-related mood disorders. Biol. Psychiatry 9 (12), 1116–1127.

Duman, R.S., Sanacora, G., Krystal, J.H., 2019. Altered connectivity in depression: GABA and glutamate neurotransmitter deficits and reversal by novel treatments. Neuron 102, 75–90.

Duman, R.S., Deyama, S., Fogaça, M.V., 2021. Role of BDNF in the pathophysiology and treatment of depression: activity-dependent effects distinguish rapid-acting antidepressants. Eur. J. Neurosci. 53, 126–139.

Dumornay, N.M., Lebois, L.A.M., Ressler, K.J., Harnett, N.G., 2023. Racial disparities in adversity during childhood and the false appearance of race-related differences in brain structure. Am. J. Psychiatry 180, 127–138.

DuPont, H.L., Suescun, J., Jiang, Z.D., Brown, E.L., Essigmann, H.T., et al., 2023. Fecal microbiota transplantation in Parkinson's disease-a randomized repeat-dose, placebo-controlled clinical pilot study. Front. Neurol. 14, 1104759.

Duque-Wilckens, N., Teis, R., Sarno, E., Stoelting, F., Khalid, S., et al., 2022. Early life adversity drives sex-specific anhedonia and meningeal immune gene expression through mast cell activation. Brain Behav. Immun. 103, 73–84.

Duran, E., Duran, B., Heart, M.Y.H.B., Horse-Davis, S.Y., 1998. Healing the American Indian soul wound. In: International Handbook of Multigenerational Legacies of Trauma. Springer US, pp. 341–354.

Duranti, S., Ferrario, C., van Sinderen, D., Ventura, M., Turroni, F., 2017. Obesity and microbiota: an example of an intricate relationship. Genes Nutr. 12, 18.

Dursun, E., Gezen-Ak, D., Hanağası, H., Bilgiç, B., Lohmann, E., et al., 2015. The interleukin 1 alpha, interleukin 1 beta, interleukin 6 and alpha-2-macroglobulin serum levels in patients with early or late onset Alzheimer's disease, mild cognitive impairment or Parkinson's disease. J. Neuroimmunol. 283, 50–57.

Düsedau, H.P., Steffen, J., Figueiredo, C.A., Boehme, J.D., Schultz, K., et al., 2021. Influenza A virus (H1N1) infection induces microglial activation and temporal dysbalance in glutamatergic synaptic transmission. MBio 12, e0177621.

Dutcher, E.G., Pama, E.A.C., Lynall, M.E., Khan, S., Clatworthy, M.R., et al., 2020. Early-life stress and inflammation: a systematic review of a key experimental approach in rodents. Brain Neurosci. Adv. 4. 2398212820978049.

Duty, S., Jenner, P., 2011. Animal models of Parkinson's disease: a source of novel treatments and clues to the cause of the disease. Br. J. Pharmacol. 164, 1357–1391.

Dworkin, R.H., Anderson, B.T., Andrews, N., Edwards, R.R., Grob, C.S., Ross, S., et al., 2022. If the doors of perception were cleansed, would chronic pain be relieved? Evaluating the benefits and risks of psychedelics. J. Pain 23 (10), 1666–1679.

Dwyer, J.B., Bloch, M.H., 2019. Antidepressants for pediatric patients. Curr. Psychiatr. 18, 26F–42F.

Dwyer, Z., Chaiquin, M., Landrigan, J., Ayoub, K., Shail, P., et al., 2021. The impact of dextran sodium sulphate and probiotic pre-treatment in a murine model of Parkinson's disease. J. Neuroinflammation 18, 1–15.

Dwyer, D.B., Buciuman, M.O., Ruef, A., Kambeitz, J., Sen Dong, M., et al., 2022. Clinical, brain, and multilevel clustering in early psychosis and affective stages. JAMA Psychiatry 79, 677–689.

Dye, C., Lenz, K.M., Leuner, B., 2022. Immune system alterations and postpartum mental illness: evidence from basic and clinical research. Front. Glob. Womens Health 2, 758748.

Dzamko, N., Gysbers, A., Perera, G., Bahar, A., Shankar, A., et al., 2017. Toll-like receptor 2 is increased in neurons in Parkinson's disease brain and may contribute to alpha-synuclein pathology. Acta Neuropathol. 133, 303–319.

Dzierzewski, J.M., Song, Y., Fung, C.H., Rodriguez, J.C., Jouldjian, S., et al., 2015. Self-reported sleep duration mitigates the association between inflammation and cognitive functioning in hospitalized older men. Front. Psychol. 6, 1004.

Earnshaw, V.A., Elliott, M.N., Reisner, S.L., Mrug, S., Windle, M., et al., 2017. Peer victimization, depressive symptoms, and substance use: a longitudinal analysis. Pediatrics, e20163426.

Eaton, W.W., Rodriguez, K.M., Thomas, M.A., Johnson, J., Talor, M.V., et al., 2022. Immunologic profiling in schizophrenia and rheumatoid arthritis. Psychiatry Res. 317, 114812.

Eban-Rothschild, A., Rothschild, G., Giardino, W.J., Jones, J.R., de Lecea, L., 2016. VTA dopaminergic neurons regulate ethologically relevant sleep-wake behaviors. Nat. Neurosci. 19, 1356–1366.

Eckel, N., Li, Y., Kuxhaus, O., Stefan, N., Hu, F.B., Schulze, M.B., 2018. Transition from metabolic healthy to unhealthy phenotypes and association with cardiovascular disease risk across BMI categories in 90257 women (the Nurses' Health Study): 30 year follow-up from a prospective cohort study. Lancet Diabetes Endocrinol. 6, 714–724.

Edelmann, S., Wiegand, A., Hentrich, T., Pasche, S., Schulze-Hentrich, J. M., et al., 2023. Blood transcriptome analysis suggests an indirect molecular association of early life adversities and adult social anxiety disorder by immune-related signal transduction. Front. Psychiatry 14, 1125553.

Edén, A., Grahn, A., Bremell, D., Aghvanyan, A., Bathala, P., et al., 2022. Viral antigen and inflammatory biomarkers in cerebrospinal fluid in patients with COVID-19 infection and neurologic symptoms compared with control participants without infection or neurologic symptoms. JAMA Netw. Open 5, e2213253.

Edgar, R.S., Stangherlin, A., Nagy, A.D., Nicoll, M.P., Efstathiou, S., et al., 2016. Cell autonomous regulation of herpes and influenza virus infection by the circadian clock. Proc. Natl. Acad. Sci. USA 113, 10085–10090.

Edlow, A.G., Castro, V.M., Shook, L.L., Haneuse, S., Kaimal, A.J., Perlis, R.H., 2023. Sex-specific neurodevelopmental outcomes among offspring of mothers with SARS-CoV-2 infection during pregnancy. JAMA Netw. Open 6, e234415.

Edmiston, E., Ashwood, P., Van de Water, J., 2017. Autoimmunity, autoantibodies, and autism spectrum disorder. Biol. Psychiatry 81 (5), 383–390.

Efstathiou, C., Abidi, S.H., Harker, J., Stevenson, N.J., 2020. Revisiting respiratory syncytial virus's interaction with host immunity, towards novel therapeutics. Cell. Mol. Life Sci. 77, 5045–5058.

Egorov, A.I., Converse, R.R., Griffin, S.M., Styles, J.N., Sams, E., et al., 2021. Latent Toxoplasma gondii infections are associated with elevated biomarkers of inflammation and vascular injury. BMC Infect. Dis. 21, 188.

Ehrlich, A.T., Kieffer, B.L., Darcq, E., 2019. Current strategies toward safer mu opioid receptor drugs for pain management. Expert Opin. Ther. Targets 23, 315–326.

Eijsbouts, C., Zheng, T., Kennedy, N.A., Bonfiglio, F., Anderson, C.A., et al., 2021. Genome-wide analysis of 53,400 people with irritable bowel syndrome highlights shared genetic pathways with mood and anxiety disorders. Nat. Genet. 53, 1543–1552.

Eisele, Y.S., Bolmont, T., Heikenwalder, M., Langer, F., Jacobson, L.H., et al., 2009. Induction of cerebral beta-amyloidosis: intracerebral versus systemic Abeta inoculation. Proc. Natl. Acad. Sci. USA 106, 12926–12931.

Eisenberg, D.T., Borja, J.B., Hayes, M.G., Kuzawa, C.W., 2017. Early life infection, but not breastfeeding, predicts adult blood telomere lengths in the Philippines. Am. J. Hum. Biol. 29.

Eisenberger, N.I., 2013. An empirical review of the neural underpinnings of receiving and giving social support: implications for health. Psychosom. Med. 75, 545–556.

Eisenberger, N.I., Moieni, M., Inagaki, T.K., Muscatell, K.A., Irwin, M.R., 2017. In sickness and in health: the co-regulation of inflammation and social behavior. Neuropsychopharmacology 42, 242–253.

Eisenstein, M., 2016. Microbiome: bacterial broadband. Nature 533, S104–S106.

Eisenstein, S.A., Bischoff, A.N., Gredysa, D.M., Antenor-Dorsey, J.A., Koller, J.M., et al., 2015. Emotional eating phenotype is associated

with central dopamine d2 receptor binding independent of body mass index. Sci. Rep. 5, 11283.

Ejlerskov, P., Hultberg, J.G., Wang, J., Carlsson, R., Ambjorn, M., Kuss, M., et al., 2015. Lack of neuronal IFN-beta-IFNAR causes Lewy body- and Parkinson's disease-like dementia. Cell 163, 324–339.

Ekelund, U., Steene-Johannessen, J., Brown, W.J., Fagerland, M.W., Owen, N., et al., 2016. Does physical activity attenuate, or even eliminate, the detrimental association of sitting time with mortality? A harmonised meta-analysis of data from more than 1 million men and women. Lancet 388, 1302–1310.

El Ali, A., Rivest, S., 2016. Microglia in Alzheimer's disease: a multifaceted relationship. Brain Behav. Immun. 55, 138–150.

El Baou, C., Desai, R., Cooper, C., Marchant, N.L., Pilling, S., et al., 2023. Psychological therapies for depression and cardiovascular risk: evidence from national healthcare records in England. Eur. Heart J. 44, 1650–1662.

Elabbadi, A., Turpin, M., Gerotziafas, G.T., Teulier, M., Voiriot, G., Fartoukh, M., 2021. Bacterial coinfection in critically ill COVID-19 patients with severe pneumonia. Infection 49, 559–562.

Elesawy, B.H., Raafat, B.M., Muqbali, A.A., Abbas, A.M., Sakr, H.F., 2021. The impact of intermittent fasting on brain-derived neurotrophic factor, neurotrophin 3, and rat behavior in a rat model of type 2 diabetes mellitus. Brain Sci. 11, 242.

Elgueta, D., Murgas, P., Riquelme, E., Yang, G., Cancino, G.I., 2022. Consequences of viral infection and cytokine production during pregnancy on brain development in offspring. Front. Immunol. 13, 816619.

Ellenbogen, M.A., Linnen, A.M., Cardoso, C., Joober, R., 2013. Intranasal oxytocin impedes the ability to ignore task-irrelevant facial expressions of sadness in students with depressive symptoms. Psychoneuroendocrinology 38, 387–398.

Ellingsen, D.M., Isenburg, K., Jung, C., Lee, J., Gerber, J., et al., 2023. Brain-to-brain mechanisms underlying pain empathy and social modulation of pain in the patient-clinician interaction. Proc. Natl. Acad. Sci. USA 120, e2212910120.

Ellis, A., Grace, P.M., Wieseler, J., Favret, J., Springer, K., et al., 2016. Morphine amplifies mechanical allodynia via TLR4 in a rat model of spinal cord injury. Brain Behav. Immun. 58, 348–356.

Elman, I., Pustilnik, A., Borsook, D., 2022. Beating pain with psychedelics: matter over mind? Neurosci. Biobehav. Rev. 134, 104482.

Elmore, M.R.P., Najafi, A.R., Koike, M.A., Dagher, N.N., Spangenberg, E.E., et al., 2014. Colony-stimulating factor 1 receptor signaling is necessary for microglia viability, unmasking a microglia progenitor cell in the adult brain. Neuron 82, 380–397.

Elsayed, M., Banasr, M., Duric, V., Fournier, N.M., Licznerski, P., Duman, R.S., 2012. Antidepressant effects of fibroblast growth factor-2 in behavioral and cellular models of depression. Biol. Psychiatry 72, 258–265.

Emerging Risk Factors Collaboration, Kaptoge, S., Srk Seshasai, L., Sun, M.W., et al., 2023. Life expectancy associated with different ages at diagnosis of type 2 diabetes in high-income countries: 23 million person-years of observation. Lancet Diabetes Endocrinol. 11, 731–742.

Endesfelder, S., Weichelt, U., Strauß, E., Schlör, A., Sifringer, M., et al., 2017. Neuroprotection by caffeine in hyperoxia-induced neonatal brain injury. Int. J. Mol. Sci. 18.

Endres, D., Pollak, T.A., Bechter, K., Denzel, D., Pitsch, K., et al., 2022. Immunological causes of obsessive-compulsive disorder: is it time for the concept of an "autoimmune OCD" subtype? Transl. Psychiatry 12, 5.

Engelhart, M.J., Geerlings, M.I., Meijer, J., Kiliaan, A., Ruitenberg, A., et al., 2004. Inflammatory proteins in plasma and the risk of dementia: the rotterdam study. Arch. Neurol. 61, 668–672.

Engemann, K., Pedersen, C.B., Arge, L., Tsirogiannis, C., Mortensen, P. B., Svenning, J.C., 2019. Residential green space in childhood is associated with lower risk of psychiatric disorders from adolescence into adulthood. Proc. Natl. Acad. Sci. USA 116, 5188–5193.

Engin, A., 2017. The pathogenesis of obesity-associated adipose tissue inflammation. Adv. Exp. Med. Biol. 960, 221–245.

Engin, E., Benham, R.S., Rudolph, U., 2018. An emerging circuit pharmacology of GABAA receptors. Trends Pharmacol. Sci. 39, 710–732.

Engler, H., Bailey, M.T., Engler, A., Stiner-Jones, L.M., Quan, N., et al., 2008. Interleukin-1 receptor type 1-deficient mice fail to develop social stress-associated glucocorticoid resistance in the spleen. Psychoneuroendocrinology 33, 108–117.

Englund, J., Haikonen, J., Shteinikov, V., Amarilla, S.P., Atanasova, T., et al., 2021. Downregulation of kainate receptors regulating GABAergic transmission in amygdala after early life stress is associated with anxiety-like behavior in rodents. Transl. Psychiatry 11, 538.

Enomoto, S., Kato, T.A., 2021. Involvement of microglia in disturbed fear memory regulation: possible microglial contribution to the pathophysiology of posttraumatic stress disorder. Neurochem. Int. 142, 104921.

Entringer, S., Epel, E.S., Lin, J., Buss, C., Shahbaba, B., et al., 2013. Maternal psychosocial stress during pregnancy is associated with newborn leukocyte telomere length. Am. J. Obstet. Gynecol. 208, 134. e1–7.

Entringer, S., Buss, C., Wadhwa, P.D., 2015. Prenatal stress, development, health and disease risk: a psychobiological perspective-2015 Curt Richter Award Paper. Psychoneuroendocrinology 62, 366–375.

Eraly, S.A., Nievergelt, C.M., Maihofer, A.X., Barkauskas, D.A., Biswas, N., et al., 2014. Assessment of plasma C-reactive protein as a biomarker of posttraumatic stress disorder risk. JAMA Psychiatry 71, 423–431.

Erbescu, A., Papuc, S.M., Budisteanu, M., Arghir, A., Neagu, M., 2022. Re-emerging concepts of immune dysregulation in autism spectrum disorders. Front. Psychiatry 13, 1006612.

Eren-Yazicioglu, C.Y., Yigit, A., Dogruoz, R.E., Yapici-Eser, H., 2021. Can GLP-1 be a target for reward system related disorders? A qualitative synthesis and systematic review analysis of studies on palatable food, drugs of abuse, and alcohol. Front. Behav. Neurosci. 14, 614884.

Erickson, M.A., Banks, W.A., 2018. Neuroimmune axes of the blood-brain barriers and blood-brain interfaces: bases for physiological regulation, disease states, and pharmacological interventions. Pharmacol. Rev. 70, 278–314.

Erkkilä, J., Punkanen, M., Fachner, J., Ala-Ruona, E., Pöntiö, I., et al., 2011. Individual music therapy for depression: randomised controlled trial. Br. J. Psychiatry 199, 132–139.

Ermakov, E.A., Mednova, I.A., Boiko, A.S., Buneva, V.N., Ivanova, S.A., 2023. Chemokine dysregulation and neuroinflammation in schizophrenia: a systematic review. Int. J. Mol. Sci. 24, 2215.

Erny, D., Hrabě de Angelis, A.L., Jaitin, D., Wieghofer, P., Staszewski, O., et al., 2015. Host microbiota constantly control maturation and function of microglia in the CNS. Nat. Neurosci. 18, 965–977.

Esalatmanesh, S., Abrishami, Z., Zeinoddini, A., Rahiminejad, F., Sadeghi, M., et al., 2016. Minocycline combination therapy with fluvoxamine in moderate to severe obsessive-compulsive disorder: a placebo-controlled, double-blind, randomized trial. Psychiatry Clin. Neurosci. 70, 517–526.

Escarfulleri, S., Ellickson-Larew, S., Fein-Schaffer, D., Mitchell, K.S., Wolf, E.J., 2021. Emotion regulation and the association between PTSD, diet, and exercise: a longitudinal evaluation among US military veterans. Eur. J. Psychotraumatol. 12, 1895515.

Esser, N., Legrand-Poels, S., Piette, J., Scheen, A.J., Paquot, N., 2014. Inflammation as a link between obesity, metabolic syndrome and type 2 diabetes. Diabetes Res. Clin. Pract. 105, 141–150.

Estes, M.L., McAllister, A.K., 2015. Immune mediators in the brain and peripheral tissues in autism spectrum disorder. Nat. Rev. Neurosci. 16, 469–486.

Esteves, K.C., Jones, C.W., Wade, M., Callerame, K., Smith, A., et al., 2020. Adverse childhood experiences: implications for offspring telomere length and psychopathology. Am. J. Psychiatry 177, 47–57.

Estévez-López, F., Dall'Aglio, L., Rodriguez-Ayllon, M., Xu, B., You, Y., et al., 2023. Levels of physical activity at age 10 years and brain morphology changes from ages 10 to 14 years. JAMA Netw. Open 6, e2333157.

Evangelou, E., Warren, H.R., Mosen-Ansorena, D., Mifsud, B., Pazoki, R., et al., 2018. Genetic analysis of over 1 million people identifies 535 new loci associated with blood pressure traits. Nat. Genet. 50, 1412–1425.

Evans, S.J., Choudary, P.V., Neal, C.R., Li, J.Z., Vawter, M.P., et al., 2004. Dysregulation of the fibroblast growth factor system in major depression. Proc. Natl. Acad. Sci. USA 101, 15506–15511.

Evans, S.J., Bassis, C.M., Hein, R., Assari, S., Flowers, S.A., et al., 2017. The gut microbiome composition associates with bipolar disorder and illness severity. J. Psychiatr. Res. 87, 23–29.

Everett, B.M., Brooks, M.M., Vlachos, H.E., Chaitman, B.R., Frye, R.L., et al., 2015. Troponin and cardiac events in stable ischemic heart disease and diabetes. N. Engl. J. Med. 373, 610–620.

Eyre, H.A., Air, T., Proctor, S., Rositano, S., Baune, B.T., 2015. A critical review of the efficacy of non-steroidal anti-inflammatory drugs in depression. Prog. Neuropharmacol. Biol. Psychiatry 57, 11–16.

Fabbri, C., Kasper, S., Kautzky, A., Zohar, J., Souery, D., et al., 2020. A polygenic predictor of treatment-resistant depression using whole exome sequencing and genome-wide genotyping. Transl. Psychiatry 10, 50.

Fabricius, R.A., Sørensen, C.B., Skov, L., Debes, N.M., 2022. Cytokine profile of pediatric patients with obsessive-compulsive and/or movement disorder symptoms: a review. Front. Pediatr. 10, 893815.

Fagundes, C.P., Glaser, R., Kiecolt-Glaser, J.K., 2013. Stressful early life experiences and immune dysregulation across the lifespan. Brain Behav. Immun. 27, 8–12.

Fairhurst, A.M., Hwang, S.H., Wang, A., Tian, X.H., Boudreaux, C., et al., 2008. Yaa autoimmune phenotypes are conferred by overexpression of TLR7. Eur. J. Immunol. 38, 1971–1978.

Falcone, M., Fousteri, G., 2020. Role of the PD-1/PD-L1 dyad in the maintenance of pancreatic immune tolerance for prevention of type 1 diabetes. Front. Endocrinol. 11, 569.

Fali, T., Vallet, H., Sauce, D., 2018. Impact of stress on aged immune system compartments: overview from fundamental to clinical data. Exp. Gerontol. pii: S0531-5565 (17)30782-9.

Fan, R., Toubal, A., Goñi, S., Drareni, K., Huang, Z., et al., 2016. Loss of the co-repressor GPS2 sensitizes macrophage activation upon metabolic stress induced by obesity and type 2 diabets. Nat. Med. 22, 780–791.

Fan, Y., Gao, Y., Ma, Q., Yang, Z., Zhao, B., et al., 2022a. Multi-omics analysis reveals aberrant gut-metabolome-immune network in schizophrenia. Front. Immunol. 13, 812293.

Fan, Y., Gao, Y., Ma, Q., Zhao, B., He, X., et al., 2022b. Grey matter volume and its association with cognitive impairment and peripheral cytokines in excited individuals with schizophrenia. Brain Imaging Behav. 16, 2618–2626.

Fang, S., Suh, J.M., Reilly, S.M., Yu, E., Osborn, O., et al., 2015. Intestinal FXR agonism promotes adipose tissue browning and reduces obesity and insulin resistance. Nat. Med. 21, 159–165.

Fann, J.R., Bombardier, C.H., Temkin, N., Esselman, P., Warms, C., et al., 2017. Sertraline for major depression during the year following traumatic brain injury: a randomized controlled trial. J. Head Trauma Rehabil. 32, 332–342.

Faria, V., Ahs, F., Appel, L., Linnman, C., Bani, M., et al., 2014. Amygdala-frontal couplings characterizing SSRI and placebo response in social anxiety disorder. Int. J. Neuropsychopharmacol. 17, 1149–1157.

Farina, N., Rusted, J., Tabet, N., 2014. The effect of exercise interventions on cognitive outcome in Alzheimer's disease: a systematic review. Int. Psychogeriatr. 26, 9–18.

Farmen, K., Tofiño-Vian, M., Iovino, F., 2021. Neuronal damage and neuroinflammation, a bridge between bacterial meningitis and neurodegenerative diseases. Front. Cell. Neurosci. 15, 680858.

Farmer, K., Rudyk, C., Prowse, N.A., Hayley, S., 2015. Hematopoietic cytokines as therapeutic players in early stages Parkinson's disease. Front. Aging Neurosci. 7, 126.

Faron-Górecka, A., Latocha, K., Pabian, P., Kolasa, M., Sobczyk-Krupiarz, I., Dziedzicka-Wasylewska, M., 2023. The involvement of prolactin in stress-related disorders. Int. J. Environ. Res. Public Health 20, 3257.

Farooqi, A., Khunti, K., Abner, S., Gillies, C., Morriss, R., Seidu, S., 2019. Comorbid depression and risk of cardiac events and cardiac mortality in people with diabetes: a systematic review and meta-analysis. Diabetes Res. Clin. Pract. 156, 107816.

Farrell, S.W., Barlow, C.E., Willis, B.L., Leonard, D., Pavlovic, A., et al., 2020. Cardiorespiratory fitness, different measures of adiposity, and cardiovascular disease mortality risk in women. J. Women's Health (Larchmt) 29, 319–326.

Farrell, K.R., Karey, E., Xu, S., Gibbon, G., Gordon, T., Weitzman, M., 2021. E-cigarette use, systemic inflammation, and depression. Int. J. Environ. Res. Public Health 18, 10402.

Farrugia, F., Aquilina, A., Vassallo, J., Pace, N.P., 2021. Bisphenol A and type 2 diabetes mellitus: a review of epidemiologic, functional, and early life factors. Int. J. Environ. Res. Public Health 18 (2), 716.

Farsi, Z., Sheng, M., 2023. Molecular mechanisms of schizophrenia: insights from human genetics. Curr. Opin. Neurobiol. 81, 102731.

Fasano, S., Milone, A., Nicoletti, G.F., Isenberg, D.A., Ciccia, F., 2023. Precision medicine in systemic lupus erythematosus. Nat. Rev. Rheumatol. (Advance online publication).

Fatouros-Bergman, H., Cervenka, S., Flyckt, L., Edman, G., Farde, L., 2014. Meta-analysis of cognitive performance in drug-naïve patients with schizophrenia. Schizophr. Res. 158, 156–162.

Faux, P., Ding, L., Ramirez-Aristeguieta, L.M., Chacón-Duque, J.C., Comini, M., et al., 2023. Neanderthal introgression in SCN9A impacts mechanical pain sensitivity. Commun. Biol. 6, 958.

Fawcett, J., Rush, A.J., Vukelich, J., Diaz, S.H., Dunklee, L., et al., 2016. Clinical experience with high-dosage pramipexole in patients with treatment-resistant depressive episodes in unipolar and bipolar depression. Am. J. Psychiatry 173, 107–111.

Faye, C., Mcgowan, J.C., Denny, C.A., David, D.J., 2018. Neurobiological mechanisms of stress resilience and implications for the aged population. Curr. Neuropharmacol. 16, 234–270.

Fazel, P., Sedighian, H., Behzadi, E., Kachuei, R., Imani Fooladi, A.A., 2023. Interaction between SARS-CoV-2 and pathogenic bacteria. Curr. Microbiol. 80, 223.

Febbraio, M.A., Karin, M., 2021. "Sweet death": fructose as a metabolic toxin that targets the gut-liver axis. Cell Metab. 33, 2316–2328.

Feduccia, A.A., Mithoefer, M.C., 2018. MDMA-assisted psychotherapy for PTSD: are memory reconsolidation and fear extinction underlying mechanisms? Prog. Neuro-Psychopharmacol. Biol. Psychiatry 84, 221–228.

Feehily, C., O'Neill, I.J., Walsh, C.J., Moore, R.L., Killeen, S.L., et al., 2023. Detailed mapping of Bifidobacterium strain transmission from mother to infant via a dual culture-based and metagenomic approach. Nat. Commun. 14, 3015.

Feigenson, K.A., Kusnecov, A.W., Silverstein, S.M., 2014. Inflammation and the two-hit hypothesis of schizophrenia. Neurosci. Biobehav. Rev. 38, 72–93.

Feldman, C.H., Malspeis, S., Leatherwood, C., Kubzansky, L., Costenbader, K.H., Roberts, A.L., 2019. Association of childhood abuse with incident systemic lupus erythematosus in adulthood in a longitudinal cohort of women. J. Rheumatol. 46, 1589–1596.

Felger, J.C., 2018. Imaging the role of inflammation in mood and anxiety-related disorders. Curr. Neuropharmacol. 16, 533–558.

Felger, J.C., Li, Z., Haroon, E., Woolwine, B.J., Jung, M.Y., et al., 2016. Inflammation is associated with decreased functional connectivity within corticostriatal reward circuitry in depression. Mol. Psychiatry 21, 1358–1365.

Fellows, R., Denizot, J., Stellato, C., Cuomo, A., Jain, P., et al., 2018. Microbiota derived short chain fatty acids promote histone crotonylation in the colon through histone deacetylases. Nat. Commun. 9, 105.

Feng, C., Fang, M., Liu, X.Y., 2014. The neurobiological pathogenesis of poststroke depression. Sci. World J. 2014, 521349.

Feng, L., Lo, H., Hong, Z., Zheng, J., Yan, Y., et al., 2023. Microglial LRRK2-mediated NFATc1 attenuates α-synuclein immunotoxicity in association with CX3CR1-induced migration and the lysosome-initiated degradation. Glia 71, 2266–2284.

Fenn, A.M., Gensel, J.C., Huang, Y., Popovich, P.G., Lifshitz, J., et al., 2014. Immune activation promotes depression 1 month after diffuse brain injury: a role for primed microglia. Biol. Psychiatry 76, 575–584.

Fenster, R.J., Lebois, L.A.M., Ressler, K.J., Suh, J., 2018. Brain circuit dysfunction in post-traumatic stress disorder: from mouse to man. Nat. Rev. Neurosci. 19, 535–551.

Feranmi, F., 2022. Universal flu vaccine protects against influenza A and B. Lancet Microbe 3, e902.

Ferić Bojić, E., Kučukalić, S., Džubur Kulenović, A., Avdibegović, E., Babić, D., et al., 2019. Associations of gene variations in neuropeptide y and brain derived neurotrophic factor genes with posttraumatic stress disorder. Psychiatr. Danub. 31, 227–234.

Ferland, J.N., Ellis, R.J., Betts, G., Silveira, M.M., de Firmino, J.B., 2023. Long-term outcomes of adolescent THC exposure on translational cognitive measures in adulthood in an animal model and computational assessment of human data. JAMA Psychiatry 80, 66–76.

Fernandes, S.B., Patil, N.D., Meriaux, S., Theresine, M., Muller, C.P., et al., 2021. Unbiased screening identifies functional differences in NK cells after early life psychosocial stress. Front. Immunol. 12, 674532.

Fernandez, H., Cevallos, A., Jimbo Sotomayor, R., Naranjo-Saltos, F., Mera Orces, D., Basantes, E., 2019. Mental disorders in systemic lupus erythematosus: a cohort study. Rheumatol. Int. 39, 1689–1695.

Fernández-Fillol, C., Pitsiakou, C., Perez-Garcia, M., Teva, I., Hidalgo-Ruzzante, N., 2021. Complex PTSD in survivors of intimate partner violence: risk factors related to symptoms and diagnoses. Eur. J. Psychotraumatol. 12, 2003616.

Fernández-Rodríguez, R., Martínez-Vizcaíno, V., Mesas, A.E., Notario-Pacheco, B., Medrano, M., Heilbronn, L.K., 2022. Does intermittent fasting impact mental disorders? A systematic review with meta-analysis. Crit. Rev. Food Sci. Nutr. 1–16.

Ferrante, P., Saresella, M., Guerini, F.R., Marzorati, M., Musetti, M.C., Cazzullo, A.G., 2003. Significant association of HLA A2–DR11 with CD4 naive decrease in autistic children. Biomed. Pharmacother. 57, 372–374.

Ferrari, A., Whang, E., Xiao, X., Kennelly, J.P., Romartinez-Alonso, B., et al., 2023. Aster-dependent nonvesicular transport facilitates dietary cholesterol uptake. Science 382, eadf0966.

Ferreira, G.E., Abdel-Shaheed, C., Underwood, M., Finnerup, N.B., Day, R.O., et al., 2023. Efficacy, safety, and tolerability of antidepressants for pain in adults: overview of systematic reviews. BMJ 380, e072415.

Ferreiro, A.L., Choi, J., Ryou, J., Newcomer, E.P., Thompson, R., et al., 2023. Gut microbiome composition may be an indicator of preclinical Alzheimer's disease. Sci. Transl. Med. 15, eabo2984.

Ferrer, A., Labad, J., Salvat-Pujol, N., Barrachina, M., Costas, J., et al., 2019. BDNF genetic variants and methylation: effects on cognition in major depressive disorder. Transl. Psychiatry 9, 265.

Ferrucci, L., Fabbri, E., 2018. Inflammageing: chronic inflammation in ageing, cardiovascular disease, and frailty. Nat. Rev. Cardiol. 15, 505–522.

Fetit, R., Hillary, R.F., Price, D.J., Lawrie, S.M., 2021. The neuropathology of autism: a systematic review of post-mortem studies of autism and related disorders. Neurosci. Biobehav. Rev. 129, 35–62.

Ficek-Tani, B., Horien, C., Ju, S., Xu, W., Li, N., et al., 2023. Sex differences in default mode network connectivity in healthy aging adults. Cereb. Cortex 33, 6139–6151.

Figueiredo-Braga, M., Cornaby, C., Cortez, A., Bernardes, M., Terroso, G., et al., 2018. Depression and anxiety in systemic lupus erythematosus: the crosstalk between immunological, clinical, and psychosocial factors. Medicine (Baltimore) 97, e11376.

Figueroa, C.A., Mocking, R.J.T., van Wingen, G., Martens, S., Ruhé, H.G., Schene, A.H., 2017. Aberrant default-mode network-hippocampus connectivity after sad memory-recall in remitted-depression. Soc. Cogn. Affect. Neurosci. 12, 1803–1813.

Filgueira, L., Larionov, A., Lannes, N., 2021. The influence of virus infection on microglia and accelerated brain aging. Cells 10, 1836.

Filiano, A.J., Xu, Y., Tustison, N.J., Marsh, R.L., Baker, W., et al., 2016. Unexpected role of interferon-gamma in regulating neuronal connectivity and social behaviour. Nature 535 (7612), 425–429.

Filiano, A.J., Gadani, S.P., Kipnis, J., 2017. How and why do T cells and their derived cytokines affect the injured and healthy brain? Nat. Rev. Neurosci. 18, 375–384.

Fillman, S.G., Cloonan, N., Catts, V.S., Miller, L.C., Wong, J., et al., 2013. Increased inflammatory markers identified in the dorsolateral prefrontal cortex of individuals with schizophrenia. Mol. Psychiatry 18, 206–214.

Fineberg, N.A., Baldwin, D.S., Drummond, L.M., Wyatt, S., Hanson, J., et al., 2018. Optimal treatment for obsessive compulsive disorder: a randomized controlled feasibility study of the clinical-effectiveness and cost-effectiveness of cognitive-behavioural therapy, selective serotonin reuptake inhibitors and their combination in the management of obsessive compulsive disorder. Int. Clin. Psychopharmacol. 33, 334–348.

Finnell, J.E., Moffitt, C.M., Hesser, L.A., Harrington, E., Melson, M.N., et al., 2019. The contribution of the locus coeruleus-norepinephrine system in the emergence of defeat-induced inflammatory priming. Brain Behav. Immun. 79, 102–113.

Fioranelli, M., Bottaccioli, A.G., Bottaccioli, F., Bianchi, M., Rovesti, M., Roccia, M.G., 2018. Stress and inflammation in coronary artery disease: a review psychoneuroendocrineimmunology-based. Front. Immunol. 9, 2031.

Fiore, N.T., Debs, S.R., Hayes, J.P., Duffy, S.S., Moalem-Taylor, G., 2023. Pain-resolving immune mechanisms in neuropathic pain. Nat. Rev. Neurol. 19, 199–220.

Fischer Fumeaux, C.J., Morisod Harari, M., Weisskopf, E., Eap, C.B., Epiney, M., et al., 2019. Risk-benefit balance assessment of SSRI antidepressant use during pregnancy and lactation based on best available evidence - an update. Expert Opin. Drug Saf. 18, 949–963.

Fisher, L.B., Pedrelli, P., Iverson, G.L., Bergquist, T.F., Bombardier, C.H., et al., 2016. Prevalence of suicidal behaviour following traumatic brain injury: longitudinal follow-up data from the NIDRR Traumatic Brain Injury Model Systems. Brain Inj. 30, 1311–1318.

Fitton, R., Sweetman, J., Heseltine-Carp, W., van der Feltz-Cornelis, C., 2022. Anti-inflammatory medications for the treatment of mental disorders: a scoping review. Brain Behav. Immun. Health 26, 100518.

Fitzgerald, P.J., Giustino, T.F., Seemann, J.R., Maren, S., 2015. Noradrenergic blockade stabilizes prefrontal activity and enables fear extinction under stress. Proc. Natl. Acad. Sci. USA 112, E3729–E3737.

Fitzgerald, M.L., Kassir, S.A., Underwood, M.D., Bakalian, M.J., Mann, J.J., et al., 2016. Dysregulation of striatal dopamine receptor binding in suicide. Neuropsychopharmacology 42, 974–982.

Fitzpatrick, J.K., Downer, E.J., 2017. Toll-like receptor signalling as a cannabinoid target in Multiple Sclerosis. Neuropharmacology 3 (Pt B), 618–626.

Fiuza-Luces, C., Valenzuela, P.L., Gálvez, B.G., Ramírez, M., López-Soto, A., et al., 2023. The effect of physical exercise on anticancer immunity. Nat. Rev. Immunol. (Epub ahead of print).

Fjell, A.M., Idland, A.V., Sala-Llonch, R., Watne, L.O., Borza, T., et al., 2017. Neuroinflammation and Tau interact with amyloid in predicting sleep problems in aging independently of atrophy. Cereb. Cortex, 1–11.

Flack, J.M., Ference, B.A., Levy, P., 2014. Should African Americans with hypertension be treated differently than non-African Americans? Curr. Hypertens. Rep. 16, 409.

Flanagan, T.W., Nichols, C.D., 2022. Psychedelics and anti-inflammatory activity in animal models. Curr. Top. Behav. Neurosci. 56, 229–245.

Flanagan, J.C., Korte, K.J., Killeen, T.K., Back, S.E., 2016. Concurrent treatment of substance use and PTSD. Curr. Psychiatry Rep. 18, 70.

Flannick, J., Thorleifsson, G., Beer, N.L., Jacobs, S.B., Grarup, N., et al., 2014. Loss-of-function mutations in SLC30A8 protect against type 2 diabetes. Nat. Genet. 46, 357–363.

Flegal, K.M., Kit, B.K., Orpana, H., Graubard, B.I., 2013. Association of all-cause mortality with overweight and obesity using standard body mass index categories: a systematic review and meta-analysis. JAMA 309, 71–82.

Fleshner, M., 2013. Stress-evoked sterile inflammation, danger associated molecular patterns (DAMPs), microbial associated molecular patterns (MAMPs) and the inflammasome. Brain Behav. Immun. 27, 1–7.

Flint, M.S., Valosen, J.M., Johnson, E.A., Miller, D.B., Tinkle, S.S., 2001. Restraint stress applied prior to chemical sensitization modulates the development of allergic contact dermatitis differently than restraint prior to challenge. J. Neuroimmunol. 113, 72–80.

Flores, C., Stewart, J., 2000. Basic fibroblast growth factor as a mediator of the effects of glutamate in the development of long-lasting sensitization to stimulant drugs: studies in the rat. Psychopharmacology 151, 152–165.

Flores-Pajot, M.C., Ofner, M., Do, M.T., Lavigne, E., Villeneuve, P.J., 2016. Childhood autism spectrum disorders and exposure to nitrogen dioxide, and particulate matter air pollution: a review and meta-analysis. Environ. Res. 151, 763–776.

Florido, A., Velasco, E.R., Monari, S., Cano, M., Cardoner, N., et al., 2023. Glucocorticoid-based pharmacotherapies preventing PTSD. Neuropharmacology 224, 109344.

Florsheim, E.B., Bachtel, N.D., Cullen, J., Lima, B.G.C., Godazgar, M., et al., 2023. Immune sensing of food allergens promotes avoidance behaviour. Nature 620, 643–650.

Fodoulian, L., Tuberosa, J., Rossier, D., Boillat, M., Kan, C., et al., 2020. SARS-CoV-2 receptors and entry genes are expressed in the human olfactory neuroepithelium and brain. iScience 23, 101839.

Fogaça, M.V., Duman, R.S., 2019. Cortical GABAergic dysfunction in stress and depression: new insights for therapeutic interventions. Front. Cell. Neurosci. 13, 448587.

Foland-Ross, L.C., Cooney, R.E., Joormann, J., Henry, M.L., Gotlib, I.H., 2014. Recalling happy memories in remitted depression: a neuroimaging investigation of the repair of sad mood. Cogn. Affect. Behav. Neurosci. 14, 818–826.

Foley, P.B., 2009. Encephalitis lethargica and influenza. I. The role of the influenza virus in the influenza pandemic of 1918/1919. J. Neural Transm. (Vienna) 116, 143–150.

Foley, É.M., Parkinson, J.T., Kappelmann, N., Khandaker, G.M., 2021. Clinical phenotypes of depressed patients with evidence of inflammation and somatic symptoms. Compr. Psychoneuroendocrinol. 8, 100079.

Fond, G., Lançon, C., Auquier, P., Boyer, L., 2018. C-Reactive protein as a peripheral biomarker in schizophrenia. An updated systematic review. Front. Psychiatry 9, 392.

Fond, G., Godin, O., Boyer, L., Berna, F., Andrianarisoa, M., et al., 2019. Chronic low-grade peripheral inflammation is associated with ultra resistant schizophrenia. Results from the FACE-SZ cohort. Eur. Arch. Psychiatry Clin. Neurosci. 269, 985–992.

Fonken, L.K., Frank, M.G., Gaudet, A.D., D'Angelo, H.M., Daut, R.A., et al., 2018. Neuroinflammatory priming to stress is differentially regulated in male and female rats. Brain Behav. Immun. 70, 257–267.

Fonkoue, I.T., Marvar, P.J., Norrholm, S., Li, Y., Kankam, M.L., et al., 2020. Symptom severity impacts sympathetic dysregulation and inflammation in post-traumatic stress disorder (PTSD). Brain Behav. Immun., 83, 260–269.

Fontenelle, L.F., Barbosa, I.G., Luna, J.V., de Sousa, L.P., Abreu, M.N., et al., 2012. A cytokine study of adult patients with obsessive-compulsive disorder. Compr. Psychiatry 53, 797–804.

Ford, B.N., Savitz, J., 2022. Depression, aging, and immunity: implications for COVID-19 vaccine immunogenicity. Immun. Ageing 19 (1), 32.

Forni, D., Cagliani, R., Clerici, M., Sironi, M., 2022. Disease-causing human viruses: novelty and legacy. Trends Microbiol. 30, 1232–1242.

Forouhi, N.G., Imamura, F., Sharp, S.J., Koulman, A., Schulze, M.B., et al., 2016. Association of plasma phospholipid n-3 and n-6 polyunsaturated fatty acids with type 2 diabetes: the EPIC-InterAct Case-Cohort Study. PLoS Med. 13, e1002094.

Forslund, K., Hildebrand, F., Nielsen, T., Falony, G., Le Chatelier, E., et al., 2015. Disentangling type 2 diabetes and metformin treatment signatures in the human gut microbiota. Nature 528, 262–266.

Forstner, A.J., Rambau, S., Friedrich, N., Ludwig, K.U., Böhmer, A.C., et al., 2017. Further evidence for genetic variation at the serotonin transporter gene SLC6A4 contributing toward anxiety. Psychiatr. Genet. 27, 96–102.

Forsyth, K.S., Jiwrajka, N., Lovell, C.D., Toothacre, N.E., Anguera, M.C., 2024. The conneXion between sex and immune responses. Nat. Rev. Immunol. 24, 487–502.

Forsythe, M.L., Boileau, A.J., 2021. Use of cannabinoids for the treatment of patients with post-traumatic stress disorder. J. Basic Clin. Physiol. Pharmacol. 33, 121–132.

Fortea, A., van Eijndhoven, P., Calvet-Mirabent, A., Ilzarbe, D., Batalla, A., et al., 2023. Age-related change in cortical thickness in adolescents at clinical high risk for psychosis: a longitudinal study. Eur. Child Adolesc. Psychiatry 33, 1837–1846.

Foster, J.A., McVey Neufeld, K.A., 2013. Gut–brain axis: how the microbiome influences anxiety and depression. Trends Neurosci. 36, 305–312.

Fraguas, D., Díaz-Caneja, C.M., Ayora, M., Hernández-Álvarez, F., Rodríguez-Quiroga, A., et al., 2019. Oxidative stress and inflammation in first-episode psychosis: a systematic review and meta-analysis. Schizophr. Bull. 45, 742–751.

Frank, M.G., Hershman, S.A., Weber, M.D., Watkins, L.R., Maier, S.F., 2014. Chronic exposure to exogenous glucocorticoids primes microglia to pro-inflammatory stimuli and induces NLRP3 mRNA in the hippocampus. Psychoneuroendocrinology 40, 191–200.

Frank, M.G., Fonken, L.K., Dolzani, S.D., Annis, J.L., Siebler, P.H., et al., 2018. Immunization with Mycobacterium vaccae induces an anti-inflammatory milieu in the CNS: attenuation of stress-induced microglial priming, alarmins and anxiety-like behavior. Brain Behav. Immun. 73, 352–363.

Frank, M.G., Fonken, L.K., Watkins, L.R., Maier, S.F., 2019. Microglia: neuroimmune-sensors of stress. Semin. Cell Dev. Biol. 94, 176–185.

Frank, M.G., Fonken, L.K., Watkins, L.R., Maier, S.F., 2020. Acute stress induces chronic neuroinflammatory, microglial and behavioral priming: a role for potentiated NLRP3 inflammasome activation. Brain Behav. Immun. 89, 32–42.

Frank, M.G., Nguyen, K.H., Ball, J.B., Hopkins, S., Kelley, T., et al., 2022. SARS-CoV-2 spike S1 subunit induces neuroinflammatory, microglial and behavioral sickness responses: evidence of PAMP-like properties. Brain Behav. Immun. 100, 267–277.

Frankl, V.E., 1959. Man's Search for Meaning. Beacon Press, Boston, MA.

Franklin, T.B., Russig, H., Weiss, I.C., Gräff, J., Linder, N., et al., 2010. Epigenetic transmission of the impact of early stress across generations. Biol. Psychiatry 68, 408–415.

Franks, N.P., Wisden, W., 2021. The inescapable drive to sleep: overlapping mechanisms of sleep and sedation. Science 374, 556–559.

Fransen, F., van Beek, A.A., Borghuis, T., El Aidy, S., Hugenholtz, F., et al., 2017. Aged gut Microbiota contributes to systemical inflammaging after Transfer to germ-Free Mice. Front. Immunol. 8, 1385.

Fredrikson, M., Faria, V., 2014. Neuroimaging in anxiety disorders. In: Baldwin, D.S., Leonard, B.E. (Eds.), Anxiety Disorders. Mod Tends Pharmacopsychiatry. 29. Karger Publishers, Basel, pp. 47–66.

Freedberg, D.E., Salmasian, H., Cohen, B., Abrams, J.A., Larson, E.L., 2016. Receipt of antibiotics in hospitalized patients and risk for Clostridium difficile infection in subsequent patients who occupy the same bed. JAMA Intern. Med. 176, 1801–1808.

Friedman, M., Rosenman, R.H., 1971. Type A behavior pattern: its association with coronary heart disease. Ann. Clin. Res. 3, 300–312.

Friedman, M.J., The Traumatic Stress Brain Research Group, Huber, B. R., Brady, C.B., Ursano, R.J., et al., 2017. VA's national PTSD brain bank: a national resource for research. Curr. Psychiatry Rep. 19, 73.

Friend, S.F., Nachnani, R., Powell, S.B., Risbrough, V.B., 2022. C-Reactive Protein: marker of risk for post-traumatic stress disorder and its potential for a mechanistic role in trauma response and recovery. Eur. J. Neurosci. 55, 2297–2310.

Frimpong, A., Owusu, E.D.A., Amponsah, J.A., Obeng-Aboagye, E., van der Puije, W., et al., 2022. Cytokines as potential biomarkers for differential diagnosis of sepsis and other non-septic disease conditions. Front. Cell. Infect. Microbiol. 12, 901433.

Frisch, S.M., MacFawn, I.P., 2020. Type I interferons and related pathways in cell senescence. Aging Cell 19, e13234.

Frøbert, O., Götberg, M., Erlinge, D., Akhtar, Z., Christiansen, E.H., et al., 2021. Influenza vaccination after myocardial infarction: a randomized, double-blind, placebo-controlled, multicenter trial. Circulation 144, 1476–1484.

Frydecka, D., Krzystek-Korpacka, M., Lubeiro, A., Stramecki, F., Stańczykiewicz, B., et al., 2018. Profiling inflammatory signatures of schizophrenia: a cross-sectional and meta-analysis study. Brain Behav. Immun. 71, 28–36.

Fu, L., Li, Y., Hu, Y., Zheng, Y., Yu, B., et al., 2017. Norovirus P particle-based active Aβ immunotherapy elicits sufficient immunogenicity and improves cognitive capacity in a mouse model of Alzheimer's disease. Sci. Rep. 7, 41041.

Fujimura, K.E., Sitarik, A.R., Havstad, S., Lin, D.L., Levan, S., et al., 2016. Neonatal gut microbiota associates with childhood multisensitized atopy and T cell differentiation. Nat. Med. 22, 1187–1191.

Fujioka, H., Phelix, C.F., Friedland, R.P., Zhu, X., Perry, E.A., et al., 2013. Apolipoprotein E4 prevents growth of malaria at the intraerythrocyte stage: implications for differences in racial susceptibility to Alzheimer's disease. J. Health Care Poor Underserved 24, 70–78.

Fujita, M., Tamano, R., Yoneda, S., Omachi, S., Yogo, E., et al., 2018. Ibudilast produces anti-allodynic effects at the persistent phase of peripheral or central neuropathic pain in rats: different inhibitory mechanism on spinal microglia from minocycline and propentofylline. Eur. J. Pharmacol. 833, 263–274.

Fuller-Thomson, E., Ramzan, N., Baird, S.L., 2016. Arthritis and suicide attempts: findings from a large nationally representative Canadian survey. Rheumatol. Int. 36, 1237–1248.

Fülling, C., Dinan, T.G., Cryan, J.F., 2019. Gut microbe to brain signaling: what happens in vagus…. Neuron 101, 998–1002.

Fulton, S., Pissios, P., Manchon, R.P., Stiles, L., Frank, L., et al., 2006. Leptin regulation of the mesoaccumbens dopamine pathway. Neuron 51, 811–822.

Fulton, S., Décarie-Spain, L., Fioramonti, X., Guiard, B., Nakajima, S., 2022. The menace of obesity to depression and anxiety prevalence. Trends Endocrinol. Metab. 33, 18–35.

Fumagalli, M., Moltke, I., Grarup, N., Racimo, F., Bjerregaard, P., et al., 2015. Greenlandic Inuit show genetic signatures of diet and climate adaptation. Science 349, 1343–1347.

Fund, N., Ash, N., Porath, A., Shalev, V., Koren, G., 2019. Comparison of mortality and comorbidity rates between Holocaust survivors and individuals in the general population in Israel. JAMA Netw. Open 2, e186643.

Fung, T.C., Olson, C.A., Hsiao, E.Y., 2017. Interactions between the microbiota, immune and nervous systems in health and disease. Nat. Neurosci. 20, 145–155.

Furukawa, T.A., Cipriani, A., Cowen, P.J., Leucht, S., Egger, M., Salanti, G., 2019. Optimal dose of selective serotonin reuptake inhibitors, venlafaxine, and mirtazapine in major depression: a systematic review and dose-response meta-analysis. Lancet Psychiatry 6, 601–609.

Fyfe, I., 2020. Aspirin and ibuprofen could lower risk of LRRK2 Parkinson disease. Nat. Rev. Neurol. 16, 460.

Gabanyi, I., Lepousez, G., Wheeler, R., Vieites-Prado, A., Nissant, A., et al., 2022. Bacterial sensing via neuronal Nod2 regulates appetite and body temperature. Science 376, eabj3986.

Gabbert, C., König, I.R., Lüth, T., Kolms, B., Kasten, M., et al., 2022. Coffee, smoking and aspirin are associated with age at onset in idiopathic Parkinson's disease. J. Neurol. 269, 4195–4203.

Gabrys, R.L., Tabri, N., Anisman, H., Matheson, K., 2018. Cognitive control and flexibility in the context of stress and depressive symptoms: the cognitive control and flexibility questionnaire. Front. Psychol. 9, 2219.

Gacias, M., Gaspari, S., Santos, P.M., Tamburini, S., Andrade, M., et al., 2016. Microbiota-driven transcriptional changes in prefrontal cortex override genetic differences in social behavior. elife 5, e13442.

Gad, E.S., Zaitone, S.A., Moustafa, Y.M., 2016. Pioglitazone and exenatide enhance cognition and downregulate hippocampal beta amyloid oligomer and microglia expression in insulin-resistant rats. Can. J. Physiol. Pharmacol. 94, 819–828.

Gadani, S., Singh, S., Kim, S., Calabresi, P., Smith, M., Bhargava, P., 2023. Spatial transcriptomics of meningeal inflammation reveals variable penetrance of inflammatory gene signatures into adjacent brain parenchyma. bioRxiv. 2023-06.

Gagne, J.J., Power, M.C., 2010. Anti-inflammatory drugs and risk of Parkinson disease: a meta-analysis. Neurology 74, 995–1002.

Galanter, J.M., Gignoux, C.R., Oh, S.S., Torgerson, D., Pino-Yanes, M., et al., 2017. Differential methylation between ethnic sub-groups reflects the effect of genetic ancestry and environmental exposures. elife 6, e20532.

Gallagher, S., Kate, M.B., 2021. Caregiving and allostatic load predict future illness and disability: a population-based study. Brain Behav. Immun. Health 16, 100295.

Gallagher, S., Howard, S., Muldoon, O.T., Whittaker, A.C., 2022. Social cohesion and loneliness are associated with the antibody response to COVID-19 vaccination. Brain. Behav. Immun. 103, 79–185.

Galland, L., 2014. The gut microbiome and the brain. J. Med. Food 17, 1261–1272.

Gallegos, A.M., Lytle, M.C., Moynihan, J.A., Talbot, N.L., 2015. Mindfulness-based stress reduction to enhance psychological functioning and improve inflammatory biomarkers in trauma-exposed women: a pilot study. Psychol. Trauma 7, 525–532.

Galvão-Coelho, N.L., de Menezes Galvão, A.C., de Almeida, R.N., Palhano-Fontes, F., Campos Braga, I., et al., 2020. Changes in inflammatory biomarkers are related to the antidepressant effects of Ayahuasca. J. Psychopharmacol. 34, 1125–1133.

Galvez-Sánchez, C.M., Montoro, C.I., Moreno-Padilla, M., Reyes Del Paso, G.A., de la Coba, P., 2021. Effectiveness of acceptance and commitment therapy in central pain sensitization syndromes: a systematic review. J. Clin. Med. 10, 2706.

Gan, H., Hou, X., Wang, Y., Xu, G., Huang, Z., et al., 2023. Global burden of rabies in 204 countries and territories, from 1990 to 2019: results from the Global Burden of Disease Study 2019. Int. J. Infect. Dis. 126, 136–144.

Gananção, L., Oquendo, M.A., Tyrka, A.R., Cisneros-Trujillo, S., Mann, J. J., Sublette, M.E., 2016. The role of cytokines in the pathophysiology of suicidal behavior. Psychoneuroendocrinology 63, 296–310.

Ganaraja, V.H., Jamuna, R., Nagarathna, C., Saini, J., Netravathi, M., 2021. Long-term cognitive outcomes in tuberculous meningitis. Neurol. Clin. Pract. 11, e222–e231.

Gandal, M.J., Haney, J.R., Parikshak, N.N., Leppa, V., Ramaswami, G., et al., 2018. Shared molecular neuropathology across major psychiatric disorders parallels polygenic overlap. Science 359, 693–697.

Gandhi, R., Hayley, S., Gibb, J., Merali, Z., Anisman, H., 2007. Influence of poly I:C on sickness behaviors, plasma cytokines, corticosterone and central monoamine activity: moderation by social stressors. Brain Behav. Immun. 21, 477–489.

Ganea, D., Hooper, K.M., Kong, W., 2015. The neuropeptide vasoactive intestinal peptide: direct effects on immune cells and involvement in inflammatory and autoimmune diseases. Acta Physiol. 213, 442–452.

Ganguli, R., Brar, J.S., Chengappa, K.R., DeLeo, M., Yang, Z.W., Shurin, G., Rabin, B.S., 1995. Mitogen-stimulated interleukin-2 production in never-medicated, firstepisode schizophrenic patients. the influence of age at onset and negative symptoms. Arch. Gen. Psychiatry 52 (8), 668–672.

Ganusov, V.V., Auerbach, J., 2014. Mathematical modeling reveals kinetics of lymphocyte recirculation in the whole organism. PLoS Comput. Biol. 10, e1003586.

Gao, H.M., Hong, J.S., 2008. Why neurodegenerative diseases are progressive: uncontrolled inflammation drives disease progression. Trends Immunol. 29, 357–365.

Gao, J., Wang, H., Liu, Y., Li, Y.Y., Chen, C., et al., 2014. Glutamate and GABA imbalance promotes neuronal apoptosis in hippocampus after stress. Med. Sci. Monit. 20, 499–512.

Gao, F., Guo, R., Ma, Q., Li, Y., Wang, W., et al., 2022. Stressful events induce long-term gut microbiota dysbiosis and associated post-traumatic stress symptoms in healthcare workers fighting against COVID-19. J. Affect. Disord. 303, 187–195.

Gao, C., Jiang, J., Tan, Y., Chen, S., 2023a. Microglia in neurodegenerative diseases: mechanism and potential therapeutic targets. Signal Transduct. Target. Ther. 8, 359.

Gao, J., Feng, L., Li, Y., Lowe, S., Guo, Z., et al., 2023b. A systematic review and meta-analysis of the association between SARS-CoV-2 vaccination and myocarditis or pericarditis. Am. J. Prev. Med. 64, 275–284.

Gao, W., Sanna, M., Chen, Y.H., Tsai, M.K., Wen, C.P., 2024. Occupational sitting time, leisure physical activity, and all-cause and cardiovascular disease mortality. JAMA Netw. Open 7 (1), e2350680.

Gapp, K., Bohacek, J., Grossmann, J., Brunner, A.M., Manuella, F., et al., 2016. Potential of environmental enrichment to prevent transgenerational effects of paternal trauma. Neuropsychopharmacology 41, 2749–2758.

Gapstur, S.M., Bouvard, V., Nethan, S.T., Freudenheim, J.L., Abnet, C. C., et al., 2023. The IARC perspective on alcohol reduction or cessation and cancer risk. N. Engl. J. Med. 389, 2486–2494.

Garakani, A., Alexander, J.L., Sumner, C.R., Pine, J.H., Gross, L.S., et al., 2023. Psychedelics, with a focus on psilocybin: issues for the clinician. J. Psychiatr. Pract. 29, 345–353.

Garaschuk, O., Verkhratsky, A., 2019. GABAergic astrocytes in Alzheimer's disease. Aging (Albany NY) 11, 1602–1604.

Garay, J.A.R., Turpin, W., Lee, S.H., Smith, M.I., Goethel, A., et al., 2023. Gut microbiome composition is associated with future onset of Crohn's disease in healthy first-degree relatives. Gastroenterology 165, 670–681.

Garbarino, S., Lanteri, P., Bragazzi, N.L., Magnavita, N., Scoditti, E., 2021. Role of sleep deprivation in immune-related disease risk and outcomes. Commun. Biol. 4, 1304.

Garcia, R., 2017. Neurobiology of fear and specific phobias. Learn. Mem. 24, 462–471.

Garcia, M.A., Junglen, A., Ceroni, T., Johnson, D., Ciesla, J., Delahanty, D.L., 2020. The mediating impact of PTSD symptoms on cortisol awakening response in the context of intimate partner violence. Biol. Psychol. 152, 107873.

Garcia-Flores, V., Romero, R., Xu, Y., Theis, K.R., Arenas-Hernandez, M., 2022. Maternal-fetal immune responses in pregnant women infected with SARS-CoV-2. Nat. Commun. 13, 320.

García-Gutiérrez, M.S., Navarrete, F., Gasparyan, A., Austrich-Olivares, A., Sala, F., Manzanares, J., 2020. Cannabidiol: a potential new alternative for the treatment of anxiety, depression, and psychotic disorders. Biomolecules 10, 1575.

Garcia-Ruiz, B., Moreno, L., Muntané, G., Sánchez-Gistau, V., Gutiérrez-Zotes, A., et al., 2020. Leukocyte and brain DDR1 hypermethylation is altered in psychosis and is correlated with stress and inflammatory markers. Epigenomics 12, 251–265.

Garcia-Vallejo, J.J., van Kooyk, Y., 2015. DC-SIGN: the strange case of Dr. Jekyll and Mr. Hyde. Immunity 42, 983–985.

Garrett, M.E., Qin, X.J., Mehta, D., Dennis, M.F., Marx, C.E., et al., 2021. Gene expression analysis in three posttraumatic stress disorder cohorts implicates inflammation and innate immunity pathways and uncovers shared genetic risk with major depressive disorder. Front. Neurosci. 15, 678548.

Garvin, M.M., Bolton, J.L., 2022. Sex-specific behavioral outcomes of early-life adversity and emerging microglia-dependent mechanisms. Front. Behav. Neurosci. 16, 1013865.

Gaspar, R., Soares-Cunha, C., Domingues, A.V., Coimbra, B., Baptista, F.I., et al., 2021. Resilience to stress and sex-specific remodeling of microglia and neuronal morphology in a rat model of anxiety and anhedonia. Neurobiol. Stress 14, 100302.

Gasparotto, J., Senger, M.R., de Sá, T., Moreira, E., Brum, P.O., Carazza Kessler, F.G., et al., 2021. Neurological impairment caused by Schistosoma mansoni systemic infection exhibits early features of idiopathic neurodegenerative disease. J. Biol. Chem. 297, 100979.

Gasser, E., Moutos, C.P., Downes, M., Evans, R.M., 2017. FGF1 – a new weapon to control type 2 diabetes mellitus. Nat. Rev. Endocrinol. 13, 599–609.

Gata-Garcia, A., Diamond, B., 2019. Maternal antibody and ASD: clinical data and animal models. Front. Immunol. 10, 1129.

Gazda, L.S., Smith, T., Watkins, L.R., Maier, S.F., Fleshner, M., 2003. Stressor exposure produces long-term reductions in antigen-specific T and B cell responses. Stress 6, 259–267.

GBD 2019 Antimicrobial Resistance Collaborators, 2022. Global mortality associated with 33 bacterial pathogens in 2019: a systematic analysis for the Global Burden of Disease Study 2019. Lancet 400, 2221–2248.

Ge, L., Sadeghirad, B., Ball, G.D.C., da Costa, B.R., Hitchcock, C.L., et al., 2020. Comparison of dietary macronutrient patterns of 14 popular named dietary programmes for weight and cardiovascular risk factor reduction in adults: systematic review and network meta-analysis of randomised trials. BMJ 369, m696.

Gelernter, J., Sun, N., Polimanti, R., Pietrzak, R., Levey, D.F., et al., 2019. Genome-wide association study of post-traumatic stress disorder reexperiencing symptoms in 165,000 US veterans. Nat. Neurosci. 22, 1394–1401.

Geller, L.T., Barzily-Rokni, M., Danino, T., Jonas, O.H., Shental, N., et al., 2017. Potential role of intratumor bacteria in mediating tumor resistance to the chemotherapeutic drug gemcitabine. Science 357, 1156–1160.

Gendelman, H.E., Appel, S.H., 2011. Neuroprotective activities of regulatory T cells. Trends Mol. Med. 17, 687–688.

Generoso, M.B., Trevizol, A.P., Kasper, S., Cho, H.J., Cordeiro, Q., Shiozawa, P., 2017. Pregabalin for generalized anxiety disorder: an updated systematic review and meta-analysis. Int. Clin. Psychopharmacol. 32, 49–55.

Geng, X., Kang, X., Wong, P.C.M., 2020. Autism spectrum disorder risk prediction: a systematic review of behavioral and neural investigations. Prog. Mol. Biol. Transl. Sci. 173, 91–137.

Genovese, A., Butler, M.G., 2023. The autism spectrum: behavioral, psychiatric and genetic associations. Genes (Basel) 14, 677.

Gensollen, T., Iyer, S.S., Kasper, D.L., Blumberg, R.S., 2016. How colonization by microbiota in early life shapes the immune system. Science 352, 539–544.

Gentile, A., De Vito, F., Fresegna, D., Musella, A., Buttari, F., et al., 2015. Exploring the role of microglia in mood disorders associated with experimental multiple sclerosis. Front. Cell. Neurosci. 9, 243.

George, C., Gontier, G., Lacube, P., François, J.C., Holzenberger, M., et al., 2017. The Alzheimer's disease transcriptome mimics the neuroprotective signature of IGF-1 receptor-deficient neurons. Brain J. Neurol. 140, 2012–2027.

Geraghty, A.C., Muroy, S.E., Zhao, S., Bentley, G.E., Kriegsfeld, L.J., et al., 2015. Knockdown of hypothalamic RFRP3 prevents chronic stress-induced infertility and embryo resorption. elife 4, e04316.

Gerentes, M., Pelissolo, A., Rajagopal, K., Tamouza, R., Hamdani, N., 2019. Obsessive-compulsive disorder: autoimmunity and neuroinflammation. Curr. Psychiatry Rep. 21, 78.

Gerngroß, C., Schretter, J., Klingenspor, M., Schwaiger, M., Fromme, T., 2017. Active brown fat during 18FDG-PET/CT imageing defines a patient group with characteristic traits and an increased probability of brown fat redetection. J. Nucl. Med. 58, 1104–1110.

Ghaemmaghami, P., Dainese, S.M., La Marca, R., Zimmermann, R., Ehlert, U., 2014. The association between the acute psychobiological stress response in second trimester pregnant women, amniotic fluid glucocorticoids, and neonatal birth outcome. Dev. Psychobiol. 56, 734–747.

Ghahari, S., Mohammadi-Hasel, K., Malakouti, S.K., Roshanpajouh, M., 2020. Mindfulness-based cognitive therapy for generalised anxiety disorder: a systematic review and meta-analysis. East Asian Arch. Psychiatr. 30, 52–56.

Ghahremani, M., Smith, E.E., Chen, H.Y., Creese, B., Goodarzi, Z., Ismail, Z., 2023. Vitamin D supplementation and incident dementia: effects of sex, APOE, and baseline cognitive status. Alzheimers Dement. 15, e12404.

Gheorghe, C.E., Ritz, N.L., Martin, J.A., Wardill, H.R., Cryan, J.F., Clarke, G., 2021. Investigating causality with fecal microbiota transplantation in rodents: applications, recommendations and pitfalls. Gut Microbes 13, 1941711.

Ghosal, S., Hare, B., Duman, R.S., 2017. Prefrontal cortex GABAergic deficits and circuit dysfunction in the pathophysiology and treatment of chronic stress and depression. Curr. Opin. Behav. Sci. 14, 1–8.

Ghosh, A., Roy, A., Matras, J., Brahmachari, S., Gendelman, H.E., et al., 2009. Simvastatin inhibits the activation of p21ras and prevents the loss of dopaminergic neurons in a mouse model of Parkinson's disease. J. Neurosci. 29, 13543–13556.

Ghosh, S., Wu, M.D., Shaftel, S.S., Kyrkanides, S., LaFerla, F.M., et al., 2013. Sustained interleukin-1β overexpression exacerbates tau pathology despite reduced amyloid burden in an Alzheimer's mouse model. J. Neurosci. 33, 5053–5064.

Ghosh, T.S., Rampelli, S., Jeffery, I.B., et al., 2020. Mediterranean diet intervention alters the gut microbiome in older people reducing frailty and improving health status: the NU-AGE 1-year dietary intervention across five European countries. Gut 69, 1218–1228.

Ghosh, T.S., Shanahan, F., O'Toole, P.W., 2022. The gut microbiome as a modulator of healthy ageing. Nat. Rev. Gastroenterol. Hepatol. 19, 565–584.

Giacomin, P.R., Kraeuter, A.K., Albornoz, E.A., Jin, S., Bengtsson, M., et al., 2018. Chronic helminth infection perturbs the gut-brain axis, promotes neuropathology, and alters behavior. J. Infect. Dis. 218, 1511–1516.

Gianaros, P.J., Marsland, A.L., Kuan, D.C., Schirda, B.L., Jennings, J.R., et al., 2014. An inflammatory pathway links atherosclerotic cardiovascular disease risk to neural activity evoked by the cognitive regulation of emotion. Biol. Psychiatry 75, 738–745.

Giannitrapani, L., Mirarchi, L., Amodeo, S., Licata, A., Soresi, M., et al., 2023. Can baseline il-6 levels predict long COVID in subjects hospitalized for SARS-CoV-2 disease? Int. J. Mol. Sci. 24, 1731.

Gibb, J., Hayley, S., Gandhi, R., Poulter, M.O., Anisman, H., 2008. Synergistic and additive actions of a psychosocial stressor and endotoxin challenge: circulating and brain cytokines, plasma corticosterone and behavioral changes in mice. Brain Behav. Immun. 22, 573–589.

Gibb, J., Hayley, S., Poulter, M.O., Anisman, H., 2011. Effects of stressors and immune activating agents on peripheral and central cytokines in mouse strains that differ in stressor responsivity. Brain Behav. Immun. 25, 468–482.

Gibb, J., Al-Yawer, F., Anisman, H., 2013. Synergistic and antagonistic actions of acute or chronic social stressors and an endotoxin challenge vary over time following the challenge. Brain Behav. Immun. 28, 149–158.

Gibney, S.M., McGuinness, B., Prendergast, C., Harkin, A., Connor, T.J., 2013. Poly I: C-induced activation of the immune response is accompanied by depression and anxiety-like behaviours, kynurenine pathway activation and reduced BDNF expression. Brain Behav. Immun. 28, 170–181.

Gibson, M.K., Wang, B., Ahmadi, S., Burnham, C.A.D., Tar, P.I., et al., 2016. Developmental dynamics of the preterm infant gut microbiota and antibiotic resistome. Nat. Microbiol. 1, 16024.

Gilabert-Juan, J., Castillo-Gomez, E., Guirado, R., Moltó, M.D., Nacher, J., 2013. Chronic stress alters inhibitory networks in the medial prefrontal cortex of adult mice. Brain Struct. Funct. 218, 1591–1605.

Gilfarb, R.A., Leuner, B., 2022. GABA system modifications during periods of hormonal flux across the female lifespan. Front. Behav. Neurosci. 16, 802530.

Gillardon, F., Schmid, R., Draheim, H., 2012. Parkinson's disease-linked leucine-rich repeat kinase 2 (R1441G) mutation increases proinflammatory cytokine release from activated primary microglial cells and resultant neurotoxicity. Neuroscience 208, 41–48.

Gilman, S.E., Cherkerzian, S., Buka, S.L., Hahn, J., Hornig, M., et al., 2016. Prenatal immune programming of the sex-dependent risk for major depression. Transl. Psychiatry 6, e822.

Giloteaux, L., Goodrich, J.K., Walters, W.A., Levine, S.M., Ley, R.E., et al., 2016. Reduced diversity and altered composition of the gut microbiome in individuals with myalgic encephalomyelitis/chronic fatigue syndrome. Microbiome 4, 30.

Gilsbach, B.K., Kortholt, A., 2014. Structural biology of the LRRK2 GTPase and kinase domains: implications for regulation. Front. Mol. Neurosci. 7, 32.

Gilvarry, C.M., Sham, P.C., Jones, P.B., Cannon, M., Wright, P., et al., 1996. Family history of autoimmune diseases in psychosis. Schizophr. Res. 19 (1), 33–40.

Gimblet, C., Meisel, J.S., Loesche, M.A., Cole, S.D., Horwinski, J., et al., 2017. Cutaneous leishmaniasis induces a transmissible dysbiotic skin microbiota that promotes skin inflammation. Cell Host Microbe 22, 13–24.

Gimeno, D., Kivimäki, M., Brunner, E.J., Elovainio, M., De Vogli, R., et al., 2009. Associations of C-reactive protein and interleukin-6 with cognitive symptoms of depression: 12-year follow-up of the Whitehall II study. Psychol. Med. 39, 413–423.

Giorgi, S., Nikolaeva-Koleva, M., Alarcón-Alarcón, D., Butrón, L., González-Rodríguez, S., 2019. Is TRPA1 burning down TRPV1 as druggable target for the treatment of chronic pain? Int. J. Mol. Sci. 20, 2906.

Girard, D., Tardif, J.C., Boisclair Demarble, J., D'Antono, B., 2016. Trait hostility and acute inflammatory responses to stress in the laboratory. PLoS One 11, e0156329.

Girgenti, M.J., Ghosal, S., LoPresto, D., Taylor, J.R., Duman, R.S., 2017. Ketamine accelerates fear extinction via mTORC1 signaling. Neurobiol. Dis. 100, 1–8.

Giustino, T.F., Fitzgerald, P.J., Maren, S., 2016. Revisiting propranolol and PTSD: memory erasure or extinction enhancement? Neurobiol. Learn. Mem. 130, 26–33.

Giustino, T.F., Ramanathan, K.R., Totty, M.S., Miles, O.W., Maren, S., 2020. Locus coeruleus norepinephrine drives stress-induced increases in basolateral amygdala firing and impairs extinction learning. J. Neurosci. 40, 907–916.

Gładysz, D., Krzywdzińska, A., Hozyasz, K.K., 2018. Immune abnormalities in autism spectrum disorder-could they hold promise for causative treatment? Mol. Neurobiol. 55, 6387–6435.

Glaser, R., Kiecolt-Glaser, J.K., 2005. Stress-induced immune dysfunction: implications for health. Nat. Rev. Immunol. 5, 243–251.

Glaser, R., MacCallum, R.C., Laskowski, B.F., Malarkey, W.B., Sheridan, J.F., et al., 2001. Evidence for a shift in the Th-1 to Th-2 cytokine response associated with chronic stress and aging. J. Gerontol. Ser. A: Biol. Sci. Med. Sci. 56, M477–M482.

Glausier, J.R., Lewis, D.A., 2012. Dendritic spine pathology in schizophrenia. Neuroscience 251, 90–107.

Gloeckner, C.J., Schumacher, A., Boldt, K., Ueffing, M., 2009. The Parkinson disease-associated protein kinase LRRK2 exhibits MAPKKK activity and phosphorylates MKK3/6 and MKK4/7, in vitro. J. Neurochem. 109, 959–968.

Glover, V., 2011. Annual research review: prenatal stress and the origins of psychopathology: an evolutionary perspective. J. Child Psychol. Psychiatry 52, 356–367.

Gluckman, P.D., Hanson, M.A., Buklijas, T., 2010. A conceptual framework for the developmental origins of health and disease. J. Dev. Orig. Health Dis. 1, 6–18.

Gobbi, G., Atkin, T., Zytynski, T., Wang, S., Askari, S., et al., 2019. Association of cannabis use in adolescence and risk of depression, anxiety, and suicidality in young adulthood: a systematic review and meta-analysis. JAMA Psychiatry 76, 426–434.

Godfrey, K.E.M., Gardner, A.C., Kwon, S., Chea, W., Muthukumaraswamy, S.D., 2018. Differences in excitatory and inhibitory neurotransmitter levels between depressed patients and healthy controls: a systematic review and meta-analysis. J. Psychiatr. Res. 105, 33–44.

Godinho-Silva, C., Domingues, R.G., Rendas, M., Raposo, B., Ribeiro, H., et al., 2019. Light-entrained and brain-tuned circadian circuits regulate ILC3s and gut homeostasis. Nature 574, 254–258.

Goebel, M.U., Mills, P.J., Irwin, M.R., Ziegler, M.G., 2000. Interleukin-6 and tumor necrosis factor-alpha production after acute psychological stress, exercise, and infused isoproterenol: differential effects and pathways. Psychosom. Med. 62, 591–598.

Gogolla, N., 2021. The brain remembers where and how inflammation struck. Cell 184, 5851–5853.

Gogry, F.A., Siddiqui, M.T., Sultan, I., Haq, Q.M.R., 2021. Current update on intrinsic and acquired colistin resistance mechanisms in bacteria. Front. Med. (Lausanne) 8, 677720. https://doi.org/10.3389/fmed.2021.677720.

Goh, K.K., Lu, M.L., 2022. Relationship between the domains of theory of mind, social dysfunction, and oxytocin in schizophrenia. J. Psychiatr. Res. 155, 420–429.

Gola, H., Engler, H., Sommershof, A., Adenauer, H., Kolassa, S., et al., 2013. Posttraumatic stress disorder is associated with an enhanced spontaneous production of proinflammatory cytokines by peripheral blood mononuclear cells. BMC Psychiatry 13, 40.

Golaszewski, N.M., LaCroix, A.Z., Godino, J.G., Allison, M.A., Manson, J. E., et al., 2022. Evaluation of social isolation, loneliness, and cardiovascular disease among older women in the US. JAMA Netw. Open 5, e2146461.

Gold, S.M., Köhler-Forsberg, O., Moss-Morris, R., Mehnert, A., Miranda, J.J., et al., 2020. Comorbid depression in medical diseases. Nat. Rev. Dis. Primers 6, 69.

Golde, T.E., Streit, W.J., Chakrabarty, P., 2013. Alzheimer's disease risk alleles in TREM2 illuminate innate immunity in Alzheimer's disease. Alzheimers Res. Ther. 5, 24.

Goldin, P.R., Thurston, M., Allende, S., Moodie, C., Dixon, M.L., et al., 2021. Evaluation of cognitive behavioral therapy vs mindfulness meditation in brain changes during reappraisal and acceptance among patients with social anxiety disorder: a randomized clinical trial. JAMA Psychiatry 78, 1134–1142.

Goldman, N., Chen, M., Fujita, T., Xu, Q., Peng, W., et al., 2010. Adenosine A1 receptors mediate local anti-nociceptive effects of acupuncture. Nat. Neurosci. 13, 883–888.

Goldsmith, D.R., Rapaport, M.H., Miller, B.J., 2016. A meta-analysis of blood cytokine network alterations in psychiatric patients: comparisons between schizophrenia, bipolar disorder and depression. Mol. Psychiatry 21, 1696–1709.

Goldsmith, D.R., Haroon, E., Miller, A.H., Strauss, G.P., Buckley, P.F., Miller, B.J., 2018. TNF-α and IL-6 are associated with the deficit syndrome and negative symptoms in patients with chronic schizophrenia. Schizophr. Res. 199, 281–284.

Goldsmith, D.R., Bekhbat, M., Mehta, N.D., Felger, J.C., 2023. Inflammation-related functional and structural dysconnectivity as a pathway to psychopathology. Biol. Psychiatry 93, 405–418.

Goldstein, J.M., Hale, T., Foster, S.L., Tobet, S.A., Handa, R.J., 2019. Sex differences in major depression and comorbidity of cardiometabolic disorders: impact of prenatal stress and immune exposures. Neuropsychopharmacology 44, 59–70.

Golzari, S.E., Mahmoodpoor, A., 2014. Sepsis-associated encephalopathy versus sepsis-induced encephalopathy. Lancet Neurol. 13, 967–968. https://doi.org/10.1016/S1474-4422(14)70205-4.

Gomez, K., Stratton, H.J., Duran, P., Loya, S., Tang, C., et al., 2023. Identification and targeting of a unique NaV1.7 domain driving chronic pain. Proc. Natl. Acad. Sci. USA 120, e2217800120.

Gomez-Delgado, F., Torres-Peña, J.D., Gutierrez-Lara, G., Romero-Cabrera, J.L., Perez-Martinez, P., 2023. Artificial sweeteners and cardiovascular risk. Curr. Opin. Cardiol. 38, 344–351.

Gong, W., Guo, P., Li, Y., Liu, L., Yan, R., et al., 2023. Role of the gut-brain axis in the shared genetic etiology between gastrointestinal tract diseases and psychiatric disorders: a genome-wide pleiotropic analysis. JAMA Psychiatry 80, 360–370.

González Ramírez, C., Villavicencio Queijeiro, A., Jiménez Morales, S., Bárcenas López, D., Hidalgo Miranda, A., et al., 2020. The NR3C1 gene expression is a potential surrogate biomarker for risk and diagnosis of posttraumatic stress disorder. Psychiatry Res. 284, 112797.

Goodwin, G.M., Aaronson, S.T., Alvarez, O., Arden, P.C., Baker, A., et al., 2022. Single-dose psilocybin for a treatment-resistant episode of major depression. N. Engl. J. Med. 387, 1637–1648.

Gopalakrishnan, V., Spencer, C.N., Nezi, L., Reuben, A., Andrews, M.C., et al., 2018. Gut microbiome modulates response to anti-PD-1 immunotherapy in melanoma patients. Science 359, 97–103.

Gordon, L.B., Knopf, P.M., Cserr, H.F., 1992. Ovalbumin is more immunogenic when introduced into brain or cerebrospinal fluid than into extracerebral sites. J. Neuroimmunol. 40, 81–87.

Gorka, S.M., Fitzgerald, D.A., Labuschagne, I., Hosanagar, A., Wood, A. G., et al., 2015. Oxytocin modulation of amygdala functional connectivity to fearful faces in generalized social anxiety disorder. Neuropsychopharmacology 40, 278–286.

Gorski, D.H., Novella, S.P., 2014. Clinical trials of integrative medicine: testing whether magic works? Trends Mol. Med. 20, 473–476.

Goshen, I., Yirmiya, R., 2009. Interleukin-1 (IL-1): a central regulator of stress responses. Front. Neuroendocrinol. 30, 30–45.

Goshen, I., Kreisel, T., Ben-Menachem-Zidon, O., Licht, T., Weidenfeld, J., et al., 2008. Brain interleukin-1 mediates chronic stress-induced depression in mice via adrenocortical activation and hippocampal neurogenesis suppression. Mol. Psychiatry 13, 717–728.

Gosmann, N.P., Costa, M.A., Jaeger, M.B., Motta, L.S., Frozi, J., et al., 2021. Selective serotonin reuptake inhibitors, and serotonin and norepinephrine reuptake inhibitors for anxiety, obsessive-compulsive, and stress disorders: a 3-level network meta-analysis. PLoS Med. 18, e1003664.

Gottesman, I.I., Gould, T.D., 2003. The endophenotype concept in psychiatry: etymology and strategic intentions. Am. J. Psychiatry 160, 636–645.

Gottschalk, M.G., Domschke, K., 2017. Genetics of generalized anxiety disorder and related traits. Dialogues Clin. Neurosci. 19, 159–168.

Gould, F., Harvey, P.D., Hodgins, G., Jones, M.T., Michopoulos, V., 2021. Prior trauma-related experiences predict the development of posttraumatic stress disorder after a new traumatic event. Depress. Anxiety 38, 40–47.

Grabe, H.J., Wittfeld, K., Van der Auwera, S., Janowitz, D., Hegenscheid, K., et al., 2016. Effect of the interaction between childhood abuse and rs1360780 of the FKBP5 gene on gray matter volume in a general population sample. Hum. Brain Mapp. 37, 1602–1613.

Grabrucker, S., Marizzoni, M., Silajdžić, E., Lopizzo, N., Mombelli, E., et al., 2023. Microbiota from Alzheimer's patients induce deficits in cognition and hippocampal neurogenesis. Brain 146, 4916–4934.

Grace, P.M., Hutchinson, M.R., Maier, S.F., Watkins, L.R., 2014. Pathological pain and the neuroimmune interface. Nat. Rev. Immunol. 14, 217–231.

Grace, P.M., Strand, K.A., Galer, E.L., Urban, D.J., Wang, X., et al., 2016. Morphine paradoxically prolongs neuropathic pain in rats by amplifying spinal NLRP3 inflammasome activation. Proc. Natl. Acad. Sci. USA 113, E3441–E5340.

Graham, B.S., Ambrosino, D.M., 2015. History of passive antibody administration for prevention and treatment of infectious diseases. Curr. Opin. HIV AIDS 10, 129–134.

Graham, L.C., Harder, J.M., Soto, I., de Vries, W.N., John, S.W., et al., 2016. Chronic consumption of a western diet induces robust glial activation in ageing mice and in a mouse model of Alzheimer's disease. Sci. Rep. 6, 21568.

Graham, A.M., Doyle, O., Tilden, E.L., Sullivan, E.L., Gustafsson, H.C., et al., 2022. Effects of maternal psychological stress during pregnancy on offspring brain development: considering the role of inflammation and potential for preventive intervention. Biol. Psychiatry Cogn. Neurosci. Neuroimaging 7, 461–470.

Grandinetti, A., Morens, D.M., Reed, D., MacEachern, D., 1994. Prospective study of cigarette smoking and the risk of developing idiopathic Parkinson's disease. Am. J. Epidemiol. 139, 1129–1138.

Grasmann, J., Almenräder, F., Voracek, M., Tran, U.S., 2023. Only small effects of mindfulness-based interventions on biomarker levels of inflammation and stress: a preregistered systematic review and two three-level meta-analyses. Int. J. Mol. Sci. 24, 4445.

Grassi, G., Cecchelli, C., Vignozzi, L., Pacini, S., 2021. Investigational and experimental drugs to treat obsessive-compulsive disorder. J. Exp. Pharmacol. 12, 695–706.

Grasso, D.J., Drury, S., Briggs-Gowan, M., Johnson, A., Ford, J., et al., 2020. Adverse childhood experiences, posttraumatic stress, and FKBP5 methylation patterns in postpartum women and their newborn infants. Psychoneuroendocrinology 114, 104604.

Gratuze, M., Leyns, C.E.G., Holtzman, D.M., 2018. New insights into the role of TREM2 in Alzheimer's disease. Mol. Neurodegener. 13, 66.

Grau, A.S., Xie, H., Redfern, R.E., Moussa, M., Wang, X., Shih, C.H., 2022. Effects of acute pain medications on posttraumatic stress symptoms in early aftermath of trauma. Int. Clin. Psychopharmacol. 37, 201–205.

Graves, J.M., Dilley, J.A., Klein, T., Liebelt, E., 2023. Suspected suicidal cannabis exposures reported to US Poison Centers, 2009-2021. JAMA Netw. Open 6, e239044.

Grech, L.B., Butler, E., Stuckey, S., Hester, R., 2019. Neuroprotective benefits of antidepressants in multiple sclerosis: are we missing the mark? J. Neuropsychiatr. Clin. Neurosci. 31, 289–297.

Green, C., Shen, X., Stevenson, A.J., Conole, E.L.S., Harris, M.A., 2021. Structural brain correlates of serum and epigenetic markers of inflammation in major depressive disorder. Brain Behav. Immun. 92, 39–48.

Greenamyre, J.T., Cannon, J.R., Drolet, R., Mastroberardino, P.G., 2010. Lessons from the rotenone model of Parkinson's disease. Trends Pharmacol. Sci. 31, 141–142. author reply 142-143.

Greenberg, M.V., Glaser, J., Borsos, M., El Marjou, F., Walter, M., et al., 2017. Transient transcription in the early embryo sets an epigenetic state that programs postnatal growth. Nat. Genet. 49, 110–118.

Greenhill, C., 2017. Obesity: fermentable carbohydrates increase satiety signals. Nat. Rev. Endocrinol. 13, 3.

Greer, S.M., Goldstein, A.N., Walker, M.P., 2013. The impact of sleep deprivation on food desire in the human brain. Nat. Commun. 4, 2259.

Greggio, E., Civiero, L., Bisaglia, M., Bubacco, L., 2012. Parkinson's disease and immune system: is the culprit LRRKing in the periphery? J. Neuroinflammation 9, 1–7.

Griciuc, A., Serrano-Pozo, A., Parrado, A.R., Lesinski, A.N., Asselin, C. N., et al., 2013. Alzheimer's disease risk gene CD33 inhibits microglial uptake of amyloid beta. Neuron 78, 631–643.

Griciuc, A., Patel, S., Federico, A.N., Choi, S.H., Innes, B.J., et al., 2019. TREM2 acts downstream of CD33 in modulating microglial pathology in Alzheimer's disease. Neuron 103, 820–835.e7.

Griffiths, K., Smart, S.E., Barker, G.J., Deakin, B., Lawrie, S.M., et al., 2023. Treatment resistance NMDA receptor pathway polygenic score is associated with brain glutamate in schizophrenia. Schizophr. Res. 260, 152–159.

Grigorian-Shamagian, L., Liu, W., Fereydooni, S., Middleton, R.C., Valle, J., et al., 2017. Cardiac and systemic rejuvenation after cardiosphere-derived cell therapy in senescent rats. Eur. Heart J. 38, 2957–2967.

Grilli, M., Diodato, E., Lozza, G., Brusa, R., Casarini, M., et al., 2000. Presenilin-1 regulates the neuronal threshold to excitotoxicity both physiologically and pathologically. Proc. Natl. Acad. Sci. USA 97, 12822–12827.

Grollman, A.P., Marcus, D.M., 2016. Global hazards of herbal remedies: lessons from Aristolochia: the lesson from the health hazards of Aristolochia should lead to more research into the safety and efficacy of medicinal plants. EMBO Rep. 17, 619–625.

Grossman, N., Bono, D., Dedic, N., Kodandaramaiah, S.B., Rudenko, A., et al., 2017. Noninvasive deep brain stimulation via temporally interfering electric fields. Cell 169, 1029–1041.

Grundwald, N.J., Brunton, P.J., 2015. Prenatal stress programs neuroendocrine stress responses and affective behaviors in second generation rats in a sex-dependent manner. Psychoneuroendocrinology 62, 204–216.

Grzadzinski, R., Lord, C., Sanders, S.J., Werling, D., Bal, V.H., 2017. Children with autism spectrum disorder who improve with fever: Insights from the Simons Simplex Collection. Autism Res. 11, 175–184.

Gu, D., Dong, N., Zheng, Z., Lin, D., Huang, M., et al., 2018. A fatal outbreak of ST11 carbapenem-resistant hypervirulent Klebsiella pneumoniae in a Chinese hospital: a molecular epidemiological study. Lancet Infect. Dis. 18, 37–46.

Guarente, L., 2014. The many faces of sirtuins: sirtuins and the Warburg effect. Nat. Med. 20, 24–25.

Guerrin, C.G.J., Doorduin, J., Prasad, K., Vazquez-Matias, D.A., Barazzuol, L., de Vries, E.F.J., 2023. Social adversity during juvenile age but not adulthood increases susceptibility to an immune challenge later in life. Neurobiol. Stress 23, 100526.

Guida, F., Turco, F., Iannotta, M., De Gregorio, D., Palumbo, I., et al., 2017. Antibiotic-induced microbiota perturbation causes gut endocannabinoidome changes, hippocampal neuroglial reorganization and depression in mice. Brain Behav. Immun. 67, 230–245.

Guimarães, A.L., Richer Araujo Coelho, D., Scoriels, L., Mambrini, J., Ribeiro do Valle Antonelli, L., et al., 2022. Effects of Toxoplasma gondii infection on cognition, symptoms, and response to digital cognitive training in schizophrenia. Schizophrenia 8, 104.

Gulen, M.F., Samson, N., Keller, A., Schwabenland, M., Liu, C., et al., 2023. cGAS-STING drives ageing-related inflammation and neurodegeneration. Nature 620, 374–380.

Gumusoglu, S.B., Fine, R.S., Murray, S.J., Bittle, J.L., Stevens, H.E., 2017. The role of IL-6 in neurodevelopment after prenatal stress. Brain Behav. Immun. 65, 274–283.

Guo, J., Zhang, W., Zhang, L., Ding, H., Zhang, J., et al., 2016. Probable involvement of p11 with interferon alpha induced depression. Sci. Rep. 6, 17029.

Guo, Y., Kuang, Y.S., Li, S.H., Yuan, M.Y., He, J.R., et al., 2017. Connections between the gut microbiome and gestational diabetes mellitus. Am. J. Obstet. Gynecol. 216, S293–S294.

Guo, R., Chen, L.H., Xing, C., Liu, T., 2019. Pain regulation by gut microbiota: molecular mechanisms and therapeutic potential. Br. J. Anaesth. 123, 637–654.

Guo, P., Chen, S., Wang, H., Wang, Y., Wang, J., 2022a. A systematic analysis on the genes and their interaction underlying the comorbidity of Alzheimer's disease and major depressive disorder. Front. Aging Neurosci. 13, 789698.

Guo, Z., Ruan, Z., Zhang, D., Liu, X., Hou, L., Wang, Q., 2022b. Rotenone impairs learning and memory in mice through microglia-mediated blood brain barrier disruption and neuronal apoptosis. Chemosphere 291, 132982.

Guo, C., Che, X., Briese, T., Ranjan, A., Allicock, O., et al., 2023a. Deficient butyrate-producing capacity in the gut microbiome is associated with bacterial network disturbances and fatigue symptoms in ME/CFS. Cell Host Microbe 31 (2), 288–304.e8.

Guo, J., Marseglia, A., Shang, Y., Dove, A., Grande, G., et al., 2023b. Association between late-life weight change and dementia: a population-based cohort study. J. Gerontol. A Biol. Sci. Med. Sci. 78, 143–150.

Guo, W., He, Y., Zhang, W., Sun, Y., Wang, J., Liu, S., Ming, D., 2023c. A novel non-invasive brain stimulation technique: "temporally interfering electrical stimulation". Front. Neurosci. 17, 1092539.

Guo, Y.C., Cao, H.D., Lian, X.F., Wu, P.X., Zhang, F., et al., 2023d. Molecular mechanisms of noncoding RNA and epigenetic regulation in obesity with consequent diabetes mellitus development. World J. Diabetes 14, 1621–1631.

Gupta, S., Guleria, R.S., 2022. Involvement of nuclear factor-κb in inflammation and neuronal plasticity associated with post-traumatic stress disorder. Cells 11, 2034.

Gupta, M.A., Jarosz, P., Gupta, A.K., 2017. Posttraumatic stress disorder (PTSD) and the dermatology patient. Clin. Dermatol. 35, 260–266.

Gur, T.L., Bailey, M.T., 2016. Effects of stress on commensal microbes and immune system activity. Adv. Exp. Med. Biol. 874, 289–300.

Gur, T.L., Shay, L., Palkar, A.V., Fisher, S., Varaljay, V.A., et al., 2017. Prenatal stress affects placental cytokines and neurotrophins, commensal microbes, and anxiety-like behavior in adult female offspring. Brain Behav. Immun. 64, 50–58.

Gurfein, B.T., Hasdemir, B., Milush, J.M., Touma, C., Palme, R., et al., 2017. Enriched environment and stress exposure influence splenic B lymphocyte composition. PLoS One 12, e0180771.

Gustafsson, H., Nordström, A., Nordström, P., 2015. Depression and subsequent risk of Parkinson disease: a nationwide cohort study. Neurology 84, 2422–2429.

Guzzetta, K.E., Cryan, J.F., O'Leary, O.F., 2022. Microbiota-gut-brain axis regulation of adult hippocampal neurogenesis. Brain Plast. 8, 97–119.

Haaker, J., Yi, J., Petrovic, P., Olsson, A., 2017. Endogenous opioids regulate social threat learning in humans. Nat. Commun. 8, 15495.

Hablitz, L.M., Nedergaard, M., 2021. The glymphatic system: a novel component of fundamental neurobiology. J. Neurosci. 41, 7698–7711.

Hablitz, L.M., Pla, V., Giannetto, M., Vinitsky, H.S., Staeger, F.F., et al., 2020. Circadian control of brain glymphatic and lymphatic fluid flow. Nat. Commun. 11, 4411.

Hachmann, N.P., Miller, J., Collier, A.Y., Barouch, D.H., 2022a. Neutralization escape by SARS-CoV-2 omicron subvariant BA.4.6. N. Engl. J. Med. 387 (20), 1904–1906.

Hachmann, N.P., Miller, J., Collier, A.Y., Ventura, J.D., Yu, J., et al., 2022b. Neutralization Escape by SARS-CoV-2 omicron subvariants BA.2.12.1, BA.4, and BA.5. N. Engl. J. Med. 387, 86–88.

Hadad-Ophir, O., Ardi, Z., Brande-Eilat, N., Kehat, O., Anunu, R., et al., 2017. Exposure to prolonged controllable or uncontrollable stress affects GABAergic function in sub-regions of the hippocampus and the amygdala. Neurobiol. Learn. Mem. 138, 271–280.

Haddad, P.M., Correll, C.U., 2018. The acute efficacy of antipsychotics in schizophrenia: a review of recent meta-analyses. Ther. Adv. Psychopharmacol. 8, 303–318.

Haddad, F.L., Patel, S.V., Schmid, S., 2020. Maternal immune activation by poly I:C as a preclinical model for neurodevelopmental disorders: a focus on autism and schizophrenia. Neurosci. Biobehav. Rev. 113, 546–567.

Hains, L.E., Loram, L.C., Weiseler, J.L., Frank, M.G., Bloss, E.B., et al., 2010. Pain intensity and duration can be enhanced by prior challenge: initial evidence suggestive of a role of microglial priming. J. Pain 11, 1004–1014.

Haji, J., Hamilton, J.K., Ye, C., Swaminathan, B., Hanley, A.J., et al., 2014. Delivery by Caesarean section and infant cardiometabolic status at one year of age. J. Obstet. Gynaecol. Can. 36, 864–869.

Hakamata, Y., Suzuki, Y., Kobashikawa, H., Hori, H., 2022. Neurobiology of early life adversity: a systematic review of meta-analyses towards an integrative account of its neurobiological trajectories to mental disorders. Front. Neuroendocrinol. 65, 100994.

Halaris, A., 2017. Inflammation-associated co-morbidity between depression and cardiovascular disease. Curr. Top. Behav. Neurosci. 31, 45–70.

Hale, K.D., Weigent, D.A., Gauthier, D.K., Hiramoto, R.N., Ghanta, V.K., 2003. Cytokine and hormone profiles in mice subjected to handling combined with rectal temperature measurement stress and handling only stress. Life Sci. 72, 1495–1508.

Halfvarson, J., Brislawn, C.J., Lamendella, R., Vázquez-Baeza, Y., Walters, W.A., et al., 2017. Dynamics of the human gut microbiome in inflammatory bowel disease. Nat. Microbiol. 2, 17004.

Hall, H., 2013. Uncertainty in medicine. Skeptic Magazine 18, 4. Retrieved from http://www.skeptic.com/reading_room/uncertainty-in-medicine/.

Hall, A.B., Tolonen, A.C., Xavier, R.J., 2017. Human genetic variation and the gut microbiome in disease. Nat. Rev. Genet. 1, 690–699.

Hall, V., Foulkes, S., Insalata, F., Kirwan, P., Saei, A., et al., 2022. Protection against SARS-CoV-2 after Covid-19 vaccination and previous infection. N. Engl. J. Med. 386 (13), 1207–1220.

Hall, M.B., Willis, D.E., Rodriguez, E.L., Schwarz, J.M., 2023. Maternal immune activation as an epidemiological risk factor for neurodevelopmental disorders: considerations of timing, severity, individual differences, and sex in human and rodent studies. Front. Neurosci. 17, 1135559.

Halliday, G., Herrero, M.T., Murphy, K., McCann, H., Ros-Bernal, F., et al., 2009. No Lewy pathology in monkeys with over 10 years of severe MPTP Parkinsonism. Mov. Disord. 24, 1519–1523.

Halstead, S., Siskind, D., Amft, M., Wagner, E., Yakimov, V., et al., 2023. Alteration patterns of peripheral concentrations of cytokines and associated inflammatory proteins in acute and chronic stages of schizophrenia: a systematic review and network meta-analysis. Lancet Psychiatry 10, 260–271.

Hamers, L., 2016. Big biological datasets map life's networks: multiomics offers a new way of doing biology. Retrieved from https://www.sciencenews.org/article/big-biological-datasets-map-lifes-networks.

Hammerschlag, A.R., Stringer, S., de Leeuw, C.A., Sniekers, S., Taskesen, E., et al., 2017. Genome-wide association analysis of insomnia complaints identifies risk genes and genetic overlap with psychiatric and metabolic traits. Nat. Genet. 49, 1584–1592.

Hampel, H., Caraci, F., Cuello, A.C., Caruso, G., Nisticò, R., et al., 2020. A path toward precision medicine for neuroinflammatory mechanisms in Alzheimer's disease. Front. Immunol. 11, 456.

Hampel, H., Hu, Y., Hardy, J., Blennow, K., Chen, C., et al., 2023. The amyloid-β pathway in Alzheimer's disease: a plain language summary. Neurodegener. Dis. Manag. 13, 141–149.

Han, Y.W., Choi, J.Y., Uyangaa, E., Kim, S.B., Kim, J.H., et al., 2014. Distinct dictation of Japanese encephalitis virus-induced neuroinflammation and lethality via triggering TLR3 and TLR4 signal pathways. PLoS Pathog. 10, e1004319.

Han, F., Xiao, B., Wen, L., Shi, Y., 2015a. Effects of fluoxetine on the amygdala and the hippocampus after administration of a single prolonged stress to male Wistar rates: In vivo proton magnetic resonance spectroscopy findings. Psychiatry Res. 232, 154–161.

Han, K., Chapman, S.B., Krawczyk, D.C., 2015b. Altered amygdala connectivity in individuals with chronic traumatic brain injury and comorbid depressive symptoms. Front. Neurol. 6, 231.

Han, Y., Liu, D., Li, L., 2020. PD-1/PD-L1 pathway: current researches in cancer. Am. J. Cancer Res. 10, 727–742.

Han, Y., Xia, G., Srisai, D., Meng, F., He, Y., et al., 2021. Deciphering an AgRP-serotoninergic neural circuit in distinct control of energy metabolism from feeding. Nat. Commun. 12, 3525.

Han, K.M., Choi, K.W., Kim, A., Kang, W., Kang, Y., et al., 2022a. Association of DNA methylation of the NLRP3 gene with changes in cortical thickness in major depressive disorder. Int. J. Mol. Sci. 23, 5768.

Han, S., Xu, Y., Guo, H.R., Fang, K., Wei, Y., et al., 2022b. Two distinct subtypes of obsessive compulsive disorder revealed by heterogeneity through discriminative analysis. Hum. Brain Mapp. 43 (10), 3037–3046.

Hanagasi, H.A., Tufekcioglu, Z., Emre, M., 2017. Dementia in Parkinson's disease. J. Neurol. Sci. 374, 26–31.

Hanna, S.J., Codd, A.S., Gea-Mallorqui, E., Scourfield, D.O., Richter, F.C., et al., 2020. T cell phenotypes in COVID-19 - a living review. Oxf. Open Immunol. 2, iqaa007.

Hannestad, J., DellaGioia, N., Ortiz, N., Pittman, B., Bhagwagar, Z., 2011. Citalopram reduces endotoxin-induced fatigue. Brain Behav. Immun. 25, 256–259.

Hanscom, M., Loane, D.J., Shea-Donohue, T., 2021. Brain-gut axis dysfunction in the pathogenesis of traumatic brain injury. J. Clin. Invest. 131, e143777.

Hanson, P., Weickert, M.O., Barber, T.M., 2020. Obesity: novel and unusual predisposing factors. Ther. Adv. Endocrinol. Metab. 11. 2042018820922018.

Hantsoo, L., Jašarević, E., Criniti, S., McGeehan, B., Tanes, C., et al., 2019a. Childhood adversity impact on gut microbiota and inflammatory response to stress during pregnancy. Brain Behav. Immun. 75, 240–250.

Hantsoo, L., Kornfield, S., Anguera, M.C., Epperson, C.N., 2019b. Inflammation: a proposed intermediary between maternal stress and offspring neuropsychiatric risk. Biol. Psychiatry 85, 97–106.

Hao, Z., Zhang, X., Chen, P., 2022. Effects of different exercise therapies on balance function and functional walking ability in multiple sclerosis disease patients-a network meta-analysis of randomized controlled trials. Int. J. Environ. Res. Public Health 19, 7175.

Haq, S.U., Bhat, U.A., Kumar, A., 2021. Prenatal stress effects on off-spring brain and behavior: mediators, alterations and dysregulated epigenetic mechanisms. J. Biosci. 46, 34.

Harkness, K.L., Washburn, D., 2016. Stress generation. In: Stress: Concepts, Cognition, Emotion, and Behavior. Elsevier, Academic Press, pp. 331–338.

Harley, J.B., Chen, X., Pujato, M., Miller, D., Maddox, A., et al., 2018. Transcription factors operate across disease loci, with EBNA2 implicated in autoimmunity. Nat. Genet. 50, 699–707.

Harnett, N.G., Wood, K.H., Ference 3rd, E.W., Reid, M.A., Lahti, A.C., et al., 2017. Glutamate/glutamine concentrations in the dorsal anterior cingulate vary with post-traumatic stress disorder symptoms. J. Psychiatr. Res. 91, 169–176.

Haroon, E., Miller, A.H., 2016. Inflammation effects on brain glutamate in depression: mechanistic considerations and treatment implications. Curr. Top. Behav. Neurosci. 31, 173–198.

Haroon, F., Handel, U., Angenstein, F., Goldschmidt, J., Kreutzmann, P., et al., 2012. Toxoplasma gondii actively inhibits neuronal function in chronically infected mice. PLoS One 7, e35516.

Haroon, E., Welle, J.R., Woolwine, B.J., Goldsmith, D.R., Baer, W., et al., 2020. Associations among peripheral and central kynurenine pathway metabolites and inflammation in depression. Neuropsychopharmacology 45, 998–1007.

Harper, J.A., South, C., Trivedi, M.H., Toups, M.S., 2017. Pilot investigation into the sickness response to influenza vaccination in adults: effect of depression and anxiety. Gen. Hosp. Psychiatry 48, 56–61.

Harrell, C.S., Zainaldin, C., McFarlane, D., Hyer, M.M., Stein, D., et al., 2018. High-fructose diet during adolescent development increases neuroinflammation and depressive-like behavior without exacerbating outcomes after stroke. Brain Behav. Immun. 73, 340–351.

Harrington, W.E., Kanaan, S.B., Muehlenbachs, A., Morrison, R., Stevenson, P., et al., 2017. Maternal microchimerism predicts increased infection but decreased disease due to plasmodium falciparum during early childhood. J. Infect. Dis. 215, 1445–1451.

Harris, H.R., Willett, W.C., Vaidya, R.L., Michels, K.B., 2017a. An adolescent and early adulthood dietary pattern associated with inflammation and the incidence of breast cancer. Cancer Res. 77, 1179–1187.

Harris, M.L., Oldmeadow, C., Hure, A., Luu, J., Loxton, D., et al., 2017b. Stress increases the risk of type 2 diabetes onset in women: a 12-year longitudinal study using causal modelling. PLoS One 12, e0172126.

Harrison, N.A., Brydon, L., Walker, C., Gray, M.A., Steptoe, A., Critchley, H.D., 2009. Inflammation causes mood changes through alterations in subgenual cingulated activity and mesolimbic connectivity. Biol. Psychiatry 66, 407–414.

Hartmann, J., Wagner, K.V., Gaali, S., Kirschner, A., Kozany, C., et al., 2015. Pharmacological inhibition of the psychiatric risk factor FKBP51 has anxiolytic properties. J. Neurosci. 35, 9007–9016.

Hartmann, J., Dedic, N., Pöhlmann, M.L., Häusl, A., Karst, H., et al., 2017. Forebrain glutamatergic, but not GABAergic, neurons mediate anxiogenic effects of the glucocorticoid receptor. Mol. Psychiatry 22, 466–475.

Hartnett, K.B., Ferguson, B.J., Hecht, P.M., Schuster, L.E., Shenker, J.I., et al., 2023. Potential neuroprotective effects of dietary Omega-3 fatty acids on stress in Alzheimer's disease. Biomolecules 13, 1096.

Harvey, L., Boksa, P., 2012. Prenatal and postnatal animal models of immune activation: relevance to a range of neurodevelopmental disorders. Dev. Neurobiol. 72, 1335–1348.

Harvey, S.B., Øverland, S., Hatch, S.L., Wessely, S., Mykletun, A., et al., 2017. Exercise and the prevention of depression: results of the HUNT cohort study. Am. J. Psychiatry 175, 28–36.

Hasavci, D., Blank, T., 2022. Age-dependent effects of gut microbiota metabolites on brain resident macrophages. Front. Cell. Neurosci. 16, 944526.

Hashimoto, K., Yamawaki, Y., Yamaoka, K., Yoshida, T., Okada, K., et al., 2021. Spike firing attenuation of serotonin neurons in learned helplessness rats is reversed by ketamine. Brain Commun. 3, fcab285.

Hashioka, S., McGeer, E.G., Miyaoka, T., Wake, R., Horiguchi, J., McGeer, P.L., 2015. Interferon-γ-induced neurotoxicity of human astrocytes. CNS Neurol. Disord. Drug Targets 14, 251–256.

Haslam, C., Cruwys, T., Chang, M.X., Bentley, S.V., Haslam, S.A., 2019. GROUPS 4 HEALTH reduces loneliness and social anxiety in adults with psychological distress: findings from a randomized controlled trial. J. Consult. Clin. Psychol. 87, 787–801.

Haslam, S.A., Fransen, K., Boen, F., 2020. The New Psychology of Stress and Exercise. Sage, London.

Hatcher, J.M., Choi, H.G., Alessi, D.R., Gray, N.S., 2017. Small-molecule inhibitors of LRRK2. Adv. Neurobiol. 14, 241–264.

Havekes, R., Park, A.J., Tudor, J.C., Luczak, V.G., Hansen, R.T., et al., 2016. Sleep deprivation causes memory deficits by negatively impacting neuronal connectivity in hippocampal area CA1. elife 5, e13424.

Hawn, S.E., Neale, Z., Wolf, E.J., Zhao, X., Pierce, M., et al., 2022. Methylation of the AIM2 gene: an epigenetic mediator of PTSD-related inflammation and neuropathology plasma biomarkers. Depress. Anxiety 39, 323–333.

Hayes, J.P., Logue, M.W., Reagan, A., Salat, D., Wolf, E.J., et al., 2017. COMT Val158Met polymorphism moderates the association between PTSD symptom severity and hippocampal volume. J. Psychiatry Neurosci. 42, 95–102.

Hayley, S., Litteljohn, D., 2013. Neuroplasticity and the next wave of antidepressant strategies. Front. Cell. Neurosci. 7, 218.

Hayley, S., Sun, H., 2021. Neuroimmune multi-hit perspective of coronaviral infection. J. Neuroinflammation 18 (1), 231.

Hayley, S., Staines, W., Merali, Z., Anisman, H., 2001. Time-dependent sensitization of corticotropin-releasing hormone, arginine vasopressin and c-fos immunoreactivity within the mouse brain in response to tumor necrosis factor-alpha. Neuroscience 106, 137–148.

Hayley, S., Lacosta, S., Merali, Z., van Rooijen, N., Anisman, H., 2014. Central monoamine and plasma corticosterone changes induced by a bacterial endotoxin: sensitization and cross-sensitization effects. Eur. J. Neurosci. 13, 1155–1165.

Hayley, S., Du, L., Litteljohn, D., Palkovits, M., Faludi, G., et al., 2015. Gender and brain regions specific differences in brain derived neurotrophic factor protein levels of depressed individuals who died through suicide. Neurosci. Lett. 600, 12–16.

Hayward, A.C., Fragaszy, E.B., Bermingham, A., Wang, L., Copas, A., et al., 2014. Comparative community burden and severity of seasonal and pandemic influenza: results of the Flu Watch cohort study. Lancet Respir. Med. 2, 445–454.

Hazlett, H.C., Gu, H., Munsell, B.C., Kim, S.H., Styner, M., et al., 2017. Early brain development in infants at high risk for autism spectrum disorder. Nature 542, 348–351.

He, Y., Gao, H., Li, X., Zhao, Y., 2014. Psychological stress exerts effects on pathogenesis of hepatitis B via type-1/type-2 cytokines shift toward type-2 cytokine response. PLoS One 9, e105530.

He, Q., Zhang, P., Li, G., Dai, H., Shi, J., 2017. The association between insomnia symptoms and risk of cardio-cerebral vascular events: a meta-analysis of prospective cohort studies. Eur. J. Prev. Cardiol. 24, 1071–1082.

He, X., Ma, Q., Fan, Y., Zhao, B., Wang, W., et al., 2020. The role of cytokines in predicting the efficacy of acute stage treatment in patients with schizophrenia. Neuropsychiatr. Dis. Treat. 16, 191–199.

He, H., Yu, Y., Liew, Z., Gissler, M., László, K.D., et al., 2022. Association of maternal autoimmune diseases with risk of mental disorders in offspring in Denmark. JAMA Netw. Open 5, e227503.

He, J., Wei, Y., Li, J., Tang, Y., Liu, J., et al., 2023a. Sex differences in the association of treatment-resistant schizophrenia and serum interleukin-6 levels. BMC Psychiatry 23, 470.

He, J., Zhong, X., Cheng, C., Dong, D., Zhang, B., et al., 2023b. Characteristics of white matter structural connectivity in healthy adults with childhood maltreatment. Eur. J. Psychotraumatol. 14, 2179278.

Hees, J.T., Wanderoy, S., Lindner, J., Helms, M., Murali Mahadevan, H., Harbauer, A.B., 2024. Insulin signalling regulates Pink1 mRNA localization via modulation of AMPK activity to support PINK1 function in neurons. Nat. Metab. 6, 514–530.

Heijmans, B.T., Tobi, E.W., Stein, A.D., Putter, H., Blauw, G.J., et al., 2008. Persistent epigenetic differences associated with prenatal exposure to famine in humans. Proc. Natl. Acad. Sci. USA 105, 17046–17049.

Heim, C., Nemeroff, C.B., 2002. Neurobiology of early life stress: clinical studies. Semin. Clin. Neuropsychiatry 7, 147–159.

Heim, C., Newport, D.J., Mletzko, T., Miller, A.H., Nemeroff, C.B., 2008. The link between childhood trauma and depression: insights from HPA axis studies in humans. Psychoneuroendocrinology 33, 693–710.

Heimfarth, L., Carvalho, A.M.S., Quintans, J.S.S., Pereira, E.W.M., Lima, N.T., et al., 2020. Indole-3-guanylhydrazone hydrochloride mitigates long-term cognitive impairment in a neonatal sepsis model with involvement of MAPK and NFκB pathways. Neurochem. Int. 134, 104647.

Heithoff, D.M., Mahan, S.P., Barnes, L., Leyn, S.A., George, C.X., et al., 2023. A broad-spectrum synthetic antibiotic that does not evoke bacterial resistance. EBioMedicine 89.

Held, B.S., 2002. The tyranny of the positive attitude in America: observation and speculation. J. Clin. Psychol. 58, 965–991.

Hellmann-Regen, J., Clemens, V., Grözinger, M., Kornhuber, J., Reif, A., et al., 2022. Effect of minocycline on depressive symptoms in patients with treatment-resistant depression: a randomized clinical trial. JAMA Netw. Open 5, e2230367.

Helms, J., Kremer, S., Merdji, H., Clere-Jehl, R., Schenck, M., et al., 2020. Neurologic features in severe SARS-CoV-2 infection. N. Engl. J. Med. 382, 2268–2270.

Helmy, E., Elnakib, A., ElNakieb, Y., Khudri, M., Abdelrahim, M., et al., 2023. Role of artificial intelligence for autism diagnosis using DTI and fMRI: a survey. Biomedicines 11, 1858.

Helpman, L., Marin, M.F., Papini, S., Zhu, X., Sullivan, G.M., et al., 2016. Neural changes in extinction recall following prolonged exposure treatment for PTSD: a longitudinal fMRI study. Neuroimage Clin. 12, 715–723.

Hemmerle, A.M., Dickerson, J.W., Herman, J.P., Seroogy, K.B., 2014. Stress exacerbates experimental Parkinson's disease. Mol. Psychiatry 19, 638–640.

Hemmingsson, E., Johansson, K., Reynisdottir, S., 2014. Effects of childhood abuse on adult obesity: a systematic review and meta-analysis. Obes. Rev. 15, 882–893.

Henckens, M.J., Printz, Y., Shamgar, U., Dine, J., Lebow, M., et al., 2017. CRF receptor type 2 neurons in the posterior bed nucleus of the stria terminalis critically contribute to stress recovery. Mol. Psychiatry 22, 1691–1700.

Hendrickson, R.C., Raskind, M.A., Millard, S.P., Sikkema, C., Terry, G. E., et al., 2018. Evidence for altered brain reactivity to norepinephrine in Veterans with a history of traumatic stress. Neurobiol. Stress 8, 103–111.

Hendriksen, H., Bink, D.I., Daniels, E.G., Pandit, R., Piriou, C., et al., 2014. Re-exposure and environmental enrichment reveal NPY-Y1 as a possible target for post-traumatic stress disorder. Neuropharmacology 63, 733–742.

Heneka, M.T., Golenbock, D., Latz, E., Morgan, D., Brown, R., 2020. Immediate and long-term consequences of COVID-19 infections for the development of neurological disease. Alzheimers Res. Ther. 12, 69.

Hennessy, E., Gormley, S., Lopez-Rodriguez, A.B., Murray, C., Murray, C., et al., 2017. Systemic TNF-α produces acute cognitive dysfunction and exaggerated sickness behavior when superimposed upon progressive neurodegeneration. Brain Behav. Immun. 59, 233–244.

Hepgul, N., Cattaneo, A., Agarwal, K., Baraldi, S., Borsini, A., et al., 2016. Transcriptomics in interferon-α-treated patients identifies inflammation-, neuroplasticity-and oxidative stress-related signatures as predictors and correlates of depression. Neuropsychopharmacology 41, 2502–2511.

Heppner, F.L., Ransohoff, R.M., Becher, B., 2015. Immune attack: the role of inflammation in Alzheimer disease. Nat. Rev. Neurosci. 16, 358–372.

Herman, J.P., Tasker, J.G., 2016. Paraventricular hypothalamic mechanisms of chronic stress adaptation. Front. Endocrinol. 7, 137.

Herman, J.P., McKlveen, J.M., Ghosal, S., Kopp, B., Wulsin, A., et al., 2016. Regulation of the hypothalamic-pituitary-adrenocortical stress response. Compr. Physiol. 6, 603–621.

Hernán, M.A., Takkouche, B., Caamaño-Isorna, F., Gestal-Otero, J.J., 2002. A meta-analysis of coffee drinking, cigarette smoking, and the risk of Parkinson's disease. Ann. Neurol. 52, 276–284.

Hernández-Parra, H., Reyes-Hernández, O.D., Figueroa-González, G., González-Del Carmen, M., González-Torres, M., et al., 2023. Alteration of the blood-brain barrier by COVID-19 and its implication in the permeation of drugs into the brain. Front. Cell. Neurosci. 17, 1125109.

Herrick, M.K., Tansey, M.G., 2021. Is LRRK2 the missing link between inflammatory bowel disease and Parkinson's disease? NPJ Parkinsons Dis. 7, 26.

Herz, J., Filiano, A.J., Wiltbank, A.T., Yogev, N., Kipnis, J., 2017. Myeloid cells in the central nervous system. Immunity 46, 943–956.

Hessami, K., Norooznezhad, A.H., Monteiro, S., Barrozo, E.R., Abdolmaleki, A.S., et al., 2022. COVID-19 pandemic and infant neurodevelopmental impairment: a systematic review and meta-analysis. JAMA Netw. Open 5, e2238941.

Hickman, B., Pourkazemi, F., Pebdani, R.N., Hiller, C.E., Fong, Y.A., 2022. Dance for chronic pain conditions: a systematic review. Pain Med. 23, 2022–2041.

Hidalgo, J.L., Sotos, J.R., DEP-EXERCISE Group, 2021. Effectiveness of physical exercise in older adults with mild to moderate depression. Ann. Fam. Med. 19, 302–309.

Hidese, S., Hattori, K., Sasayama, D., Tsumagari, T., Miyakawa, T., et al., 2021. Cerebrospinal fluid inflammatory cytokine levels in patients with major psychiatric disorders: a multiplex immunoassay study. Front. Pharmacol. 11, 594394.

Higgins, G.A., Sellers, E.M., Fletcher, P.J., 2013. From obesity to substance abuse: therapeutic opportunities for 5-HT 2C receptor agonists. Trends Pharmacol. Sci. 34, 560–570.

Hill, L.K., Thayer, J.F., 2019. The autonomic nervous system and hypertension: ethnic differences and psychosocial factors. Curr. Cardiol. Rep. 21, 15.

Hill, J., Zawia, N.H., 2021. Fenamates as potential therapeutics for neurodegenerative disorders. Cells 10, 702.

Hill, M.N., Campolongo, P., Yehuda, R., Patel, S., 2018. Integrating endocannabinoid signaling and cannabinoids into the biology and treatment of posttraumatic stress disorder. Neuropsychopharmacology 43, 80–102.

Himmerich, H., Fischer, J., Bauer, K., Kirkby, K.C., Sack, U., Krügel, U., 2013. Stress-induced cytokine changes in rats. Eur. Cytokine Netw. 24, 97–103.

Hinkle, S.N., Mumford, S.L., Grantz, K.L., Mendola, P., Mills, J.L., et al., 2023. Gestational weight change in a diverse pregnancy cohort and mortality over 50 years: a prospective observational cohort study. Lancet 402, 1857–1865.

Hinojosa, C.A., George, G.C., Ben-Zion, Z., 2024. Neuroimaging of posttraumatic stress disorder in adults and youth: progress over the last decade on three leading questions of the field. Mol. Psychiatry.

Hirabayashi, N., Honda, T., Hata, J., Furuta, Y., Shibata, M., et al., 2023. Association between frequency of social contact and brain atrophy in community-dwelling older people without dementia: the JPSC-AD study. Neurology 101, e1108–e1117.

Hirsch, M., 2001. Surviving images: Holocaust photographs and the work of postmemory. Yale J. Criticism 14, 5–37.

Hirsch, L., Jette, N., Frolkis, A., Steeves, T., Pringsheim, T., 2016. The Incidence of Parkinson's disease: a systematic review and meta-analysis. Neuroepidemiology 46, 292–300.

Hiscox, L.V., Fairchild, G., Donald, K.A., Groenewold, N.A., Koen, N., et al., 2023. Antenatal maternal intimate partner violence exposure is associated with sex-specific alterations in brain structure among young infants: evidence from a South African birth cohort. Dev. Cogn. Neurosci. 60, 101210.

Hiser, J., Koenigs, M., 2018. The multifaceted role of the ventromedial prefrontal cortex in emotion, decision making, social cognition, and psychopathology. Biol. Psychiatry 83, 638–647.

Hjort, L., Rushiti, F., Wang, S.J., Fransquet, P., Krasniqi, S.P., et al., 2021. Intergenerational effects of maternal post-traumatic stress disorder on offspring epigenetic patterns and cortisol levels. Epigenomics 13, 967–980.

Hjorth, M.F., Roager, H.M., Larsen, T.M., Poulsen, S.K., Licht, T.R., et al., 2018. Pre-treatment microbial Prevotella-to-Bacteroides ratio, determines body fat loss success during a 6-month randomized controlled diet intervention. Int. J. Obes. 42, 580–583.

Ho, C.S., Lopez, J.A., Vuckovic, S., Pyke, C.M., Hockey, R.L., et al., 2001. Surgical and physical stress increases circulating blood dendritic cell counts independently of monocyte counts. Blood 98, 140–145.

Ho, P., Cahir-McFarland, E., Fontenot, J.D., Lodie, T., Nada, A., et al., 2024. Harnessing regulatory T cells to establish immune tolerance. Sci. Transl. Med. 16, eadm8859.

Hoban, A.E., Moloney, R.D., Golubeva, A.V., Neufeld, K.M., O'Sullivan, O., et al., 2016a. Behavioural and neurochemical consequences of chronic gut microbiota depletion during adulthood in the rat. Neuroscience 339, 463–477.

Hoban, A.E., Stilling, R.M., Ryan, F.J., Shanahan, F., Dinan, T.G., et al., 2016b. Regulation of prefrontal cortex myelination by the microbiota. Transl. Psychiatry 6, e774.

Hoban, A.E., Stilling, R.M., Moloney, M., G., Moloney, R.D., Shanahan, F., et al., 2017a. Microbial regulation of microRNA expression in the amygdala and prefrontal cortex. Microbiome 5, 102.

Hoban, A.E., Stilling, R.M., Moloney, G., Shanahan, F., Dinan, T.G., et al., 2017b. The microbiome regulates amygdala-dependent fear recall. Mol. Psychiatry 23, 1134–1144.

Hodes, G.E., Epperson, C.N., 2019. Sex differences in vulnerability and resilience to stress across the life span. Biol. Psychiatry 86, 421–432.

Hodes, G.E., Pfau, M.L., Leboeuf, M., Golden, S.A., Christoffel, D.J., et al., 2014. Individual differences in the peripheral immune system promote resilience versus susceptibility to social stress. Proc. Natl. Acad. Sci. USA 111, 16136–16141.

Hodes, G.E., Kana, V., Menard, C., Merad, M., Russo, S.J., 2015. Neuroimmune mechanisms of depression. Nat. Neurosci. 18, 1386–1393.

Hodgkin, P.D., Heath, W.R., Baxter, A.G., 2007. The clonal selection theory: 50 years since the revolution. Nat. Immunol. 8, 1019–1026.

Hodson, R., 2016. Precision medicine. Nature 537, S49.

Hoeksema, M.A., de Winther, M.P., 2016. Epigenetic regulation of monocyte and macrophage function. Antioxid. Redox Signal. 25, 758–774.

Hoffman, R.W., Gazitt, T., Foecking, M.F., Ortmann, R.A., Misfeldt, M., et al., 2004. U1 RNA induces innate immunity signaling. Arthritis Rheum. 50, 2891–2896.

Hoffman, K.M., Trawalter, S., Axt, J.R., Oliver, M.N., 2016. Racial bias in pain assessment and treatment recommendations, and false beliefs about biological differences between blacks and whites. Proc. Natl. Acad. Sci. USA 113, 4296–4301.

Hofmann, S.G., Sawyer, A.T., Korte, K.J., Smits, J.A., 2009. Is it beneficial to add pharmacotherapy to cognitive-behavioral therapy when treating anxiety disorders? A meta-analytic review. Int. J. Cogn. Ther. 2, 162–178.

Hoge, E.A., Bui, E., Marques, L., Metcalf, C.A., Morris, L.K., et al., 2013. Randomized controlled trial of mindfulness meditation for generalized anxiety disorder: effects on anxiety and stress reactivity. J. Clin. Psychiatry 74, 786–792.

Hoge, E.A., Bui, E., Palitz, S.A., Schwarz, N.R., Owens, M.E., et al., 2017. The effect of mindfulness meditation training on biological acute stress responses in generalized anxiety disorder. Psychiatry Res. pii: S0165-1781(16)30847-2.

Holingue, C., Budavari, A.C., Rodriguez, K.M., Zisman, C.R., Windheim, G., Fallin, M.D., 2020. Sex differences in the gut-brain axis: implications for mental health. Curr. Psychiatry Rep. 22, 83.

Hollands, G.J., French, D.P., Griffin, S.J., Prevost, A.T., Sutton, S., et al., 2016. The impact of communicating genetic risks of disease on risk-reducing health behaviour: systematic review with meta-analysis. BMJ 352, i1102.

Holley, M.M., Kielian, T., 2012. Th1 and Th17 cells regulate innate immune responses and bacterial clearance during central nervous system infection. J. Immunol. 188, 1360–1370.

Holm, C.K., Rahbek, S.H., Gad, H.H., Bak, R.O., Jakobsen, M.R., et al., 2016. Influenza A virus targets a cGAS-independent STING pathway that controls enveloped RNA viruses. Nat. Commun. 7, 10680.

Holmes, C., Boche, D., Wilkinson, D., Yadegarfar, G., Hopkins, V., 2008. Long-term effects of Abeta42 immunisation in Alzheimer's disease: follow-up of a randomised, placebo-controlled phase I trial. Lancet 372, 216–223.

Holmes, C., Cunningham, C., Zotova, E., Culliford, D., Perry, V.H., 2011. Proinflammatory cytokines, sickness behavior, and Alzheimer disease. Neurology 77, 212–218.

Holmes, S.E., Hinz, R., Drake, R.J., Gregory, C.J., Conen, S., et al., 2016. In vivo imaging of brain microglial activity in antipsychotic-free and medicated schizophrenia: a 11C-PK11195 positron emission tomography study. Mol. Psychiatry 21, 1672–1679.

Holmes, S.E., Girgenti, M.J., Davis, M.T., Pietrzak, R.H., DellaGioia, N., 2017. Altered metabotropic glutamate receptor 5 markers in PTSD: in vivo and postmortem evidence. Proc. Natl. Acad. Sci. USA 114, 8390–8395.

Holmes, S.E., Hinz, R., Conen, S., Gregory, C.J., Matthews, J.C., et al., 2018. Elevated translocator protein in anterior cingulate in major depression and a role for inflammation in suicidal thinking: a positron emission tomography study. Biol. Psychiatry 83, 61–69.

Holsboer, F., Ising, M., 2008. Central CRH system in depression and anxiety—evidence from clinical studies with CRH1 receptor antagonists. Eur. J. Pharmacol. 583, 350–357.

Holt-Lunstad, J., Smith, T.B., Baker, M., Harris, T., Stephenson, D., 2015. Loneliness and social isolation as risk factors for mortality: a meta-analytic review. Perspect. Psychol. Sci. 10, 7237.

Holtman, I.R., Raj, D.D., Miller, J.A., Schaafsma, W., Yin, Z., et al., 2015. Induction of a common microglia gene expression signature by aging and neurodegenerative conditions: a co-expression meta-analysis. Acta Neuropathol. Commun. 3, 31.

Holtzheimer, P.E., Husain, M.M., Lisanby, S.H., Taylor, S.F., Whitworth, L.A., 2017. Subcallosal cingulate deep brain stimulation for treatment-resistant depression: a multisite, randomised, sham-controlled trial. Lancet Psychiatry 4, 839–849.

Honarmand Tamizkar, K., Badrlou, E., Aslani, T., Brand, S., Arsang-Jang, S., et al., 2021. Dysregulation of NF-κB-associated LncRNAs in autism spectrum disorder. Front. Mol. Neurosci. 14, 747785.

Hones, V.I., Mizumori, S.J.Y., 2022. Response flexibility: the role of the lateral habenula. Front. Behav. Neurosci. 16, 852235.

Hong, S., Beja-Glasser, V.F., Nfonoyim, B.M., Frouin, A., Li, S., et al., 2016. Complement and microglia mediate early synapse loss in Alzheimer mouse models. Science 352, 712–716.

Hong, C.T., Chan, L., Bai, C.H., 2020. The Effect of caffeine on the risk and progression of Parkinson's disease: a meta-analysis. Nutrients 12, 1860.

Honigsbaum, M., 2020. Revisiting the 1957 and 1968 influenza pandemics. Lancet 395, 1824–1826.

Hood, S., Amir, S., 2017. Neurodegeneration and the circadian clock. Front. Ageing Neurosci. 9, 170.

Hood, L., Auffray, C., 2013. Participatory medicine: a driving force for revolutionizing healthcare. Genome Med. 5, 1–4.

Hooten, W.M., Townsend, C.O., Sletten, C.D., 2017. The triallelic serotonin transporter gene polymorphism is associated with depressive symptoms in adults with chronic pain. J. Pain Res. 10, 1071–1078.

Hori, H., Kim, Y., 2019. Inflammation and post-traumatic stress disorder. Psychiatry Clin. Neurosci. 73, 143–153.

Horn, S.R., Charney, D.S., Feder, A., 2016. Understanding resilience: new approaches for preventing and treating PTSD. Exp. Neurol. 284, 119–132.

Horsthemke, B.A., 2018. Critical view on transgenerational epigenetic inheritance in humans. Nat. Commun. 9, 2973.

Horvath, S., Raj, K., 2018. DNA methylation-based biomarkers and the epigenetic clock theory of ageing. Nat. Rev. Genet. 19, 371–384.

Horvath, I., Iashchishyn, I.A., Forsgren, L., Morozova-Roche, L.A., 2017. Immunochemical detection of α-synuclein autoantibodies in Parkinson's disease: correlation between plasma and cerebrospinal fluid levels. ACS Chem. Neurosci. 8, 1170–1176.

Hosang, G.M., Shiles, C., Tansey, K.E., McGuffin, P., Uher, R., 2014. Interaction between stress and the BDNF Val66Met polymorphism in depression: a systematic review and meta-analysis. BMC Med. 12, 7.

Hoskinson, C., Dai, D.L.Y., Del Bel, K.L., Becker, A.B., Moraes, T.J., et al., 2023. Delayed gut microbiota maturation in the first year of life is a hallmark of pediatric allergic disease. Nat. Commun. 14, 4785.

Hosseini, S., Wilk, E., Michaelsen-Preusse, K., Gerhauser, I., Baumgärtner, W., et al., 2018. Long-term neuroinflammation induced by influenza A virus infection and the impact on hippocampal neuron morphology and function. J. Neurosci. 38, 3060–3080.

Hostinar, C.E., Davidson, R.J., Graham, E.K., Mroczek, D.K., Lachman, M.E., et al., 2017. Frontal brain asymmetry, childhood maltreatment, and low-grade inflammation at midlife. Psychoneuroendocrinology 75, 152–163.

Hou, R., Garner, M., Holmes, C., Osmond, C., Teeling, J., et al., 2017. Peripheral inflammatory cytokines and immune balance in generalised anxiety disorder: case-controlled study. Brain Behav. Immun. 62, 212–218.

Hou, R., Ye, G., Liu, Y., Chen, X., Pan, M., et al., 2019. Effects of SSRIs on peripheral inflammatory cytokines in patients with generalized anxiety disorder. Brain Behav. Immun. 81, 105–110.

Houle, S., Kokiko-Cochran, O.N., 2022. A levee to the flood: pre-injury neuroinflammation and immune stress influence traumatic brain injury outcome. Front. Aging Neurosci. 13, 788055.

Howard, C.D., Li, H., Geddes, C.E., Jin, X., 2017. Dynamic nigrostriatal dopamine biases action selection. Neuron 93, 1436–1450.

Howard, D.M., Adams, M.J., Clarke, T.K., Hafferty, J.D., Gibson, J., et al., 2019. Genome-wide meta-analysis of depression identifies 102 independent variants and highlights the importance of the prefrontal brain regions. Nat. Neurosci. 22, 343–352.

Howes, O.D., McCutcheon, R., 2017. Inflammation and the neural diathesis-stress hypothesis of schizophrenia: a reconceptualization. Transl. Psychiatry 7, e1024.

Howes, O.D., Shatalina, E., 2022. Integrating the neurodevelopmental and dopamine hypotheses of schizophrenia and the role of cortical excitation-inhibition balance. Biol. Psychiatry 92, 501–513.

Howes, O.D., McCutcheon, R., Owen, M.J., Murray, R.M., 2017. The role of genes, stress, and dopamine in the development of schizophrenia. Biol. Psychiatry 81, 9–20.

Howlett, J.R., Nelson, L.D., Stein, M.B., 2022. Mental health consequences of traumatic brain injury. Biol. Psychiatry 91, 413–420.

Hryhorczuk, C., Florea, M., Rodaros, D., Poirier, I., Daneault, C., et al., 2015. Dampened mesolimbic dopamine function and signaling by saturated but not monounsaturated dietary lipids. Neuropsychopharmacology 41, 811–821.

Hsiao, C.L., Katsuumi, G., Suda, M., Shimizu, I., Yoshida, Y., et al., 2023. Abstract P3004. Vaccination targets senescence-associated glycoprotein ameliorates Alzheimer's pathology and cognitive behavior in mice. Circ. Res. 133, AP3004.

Hu, Y., Dolcos, S., 2017. Trait anxiety mediates the link between inferior frontal cortex volume and negative affective bias in healthy adults. Soc. Cogn. Affect. Neurosci. 12, 775–782.

Hu, X., Li, J., Fu, M., Zhao, X., Wang, W., 2021a. The JAK/STAT signaling pathway: from bench to clinic. Signal Transduct. Target. Ther. 6, 402.

Hu, X.Y., Wu, Y.L., Cheng, C.H., Liu, X.X., Zhou, L., 2021b. Association of brain-derived neurotrophic factor rs6265 G>A polymorphism and post-traumatic stress disorder susceptibility: a systematic review and meta-analysis. Brain Behav. 11, e02118.

Hu, P., Lu, Y., Pan, B.X., Zhang, W.H., 2022. New insights into the pivotal role of the amygdala in inflammation-related depression and anxiety disorder. Int. J. Mol. Sci. 23, 11076.

Hu, J., Xie, S., Zhang, H., Wang, X., Meng, B., et al., 2023. Microglial activation: Key players in sepsis-associated encephalopathy. Brain Sci. 13, 1453.

Hua, Y., Yang, Y., Sun, S., Iwanowycz, S., Westwater, C., et al., 2017. Gut homeostasis and regulatory T cell induction depend on molecular chaperone gp96 in CD11c+ cells. Sci. Rep. 7, 2171.

Huang, L.Z., Parameswaran, N., Bordia, T., Michael McIntosh, J., Quik, M., 2009. Nicotine is neuroprotective when administered before but not after nigrostriatal damage in rats and monkeys. J. Neurochem. 109, 826–837.

Huang, C.L., Chao, C.C., Lee, Y.C., Lu, M.K., Cheng, J.J., et al., 2016a. Paraquat induces cell death through impairing mitochondrial membrane permeability. Mol. Neurobiol. 53, 2169–2188.

Huang, Y., Shen, Z., Hu, L., Xia, F., Li, Y., et al., 2016b. Exposure of mother rats to chronic unpredictable stress before pregnancy alters the metabolism of gamma-aminobutyric acid and glutamate in the right hippocampus of offspring in early adolescence in a sexually dimorphic manner. Psychiatry Res. 246, 236–245.

Huang, R., Ning, H., Shen, M., Li, J., Zhang, J., Chen, X., 2017. Probiotics for the treatment of atopic dermatitis in children: a systematic review and meta-analysis of randomized controlled trials. Front. Cell. Infect. Microbiol. 7, 392.

Huang, J., Xu, F., Yang, L., Tuolihong, L., Wang, X., et al., 2023a. Involvement of the GABAergic system in PTSD and its therapeutic significance. Front. Mol. Neurosci. 16, 1052288.

Huang, M., Malovic, E., Ealy, A., Jin, H., Anantharam, V., et al., 2023b. Microglial immune regulation by epigenetic reprogramming through histone H3K27 acetylation in neuroinflammation. Front. Immunol. 14, 1052925.

Huang, Z., Jordan, J.D., Zhang, Q., 2023c. Early life adversity as a risk factor for cognitive impairment and Alzheimer's disease. Transl. Neurodegener. 12, 25.

Huat, T.J., Camats-Perna, J., Newcombe, E.A., Valmas, N., Kitazawa, M., Medeiros, R., 2019. Metal toxicity links to Alzheimer's disease and neuroinflammation. J. Mol. Biol. 431, 1843–1868.

Hughes, C.E., Nibbs, R.J.B., 2018. A guide to chemokines and their receptors. FEBS J. 285, 2944–2971.

Hughes, K., Bellis, M.A., Hardcastle, K.A., Sethi, D., Butchart, A., et al., 2017. The effect of multiple adverse childhood experiences on health: a systematic review and meta-analysis. Lancet Public Health 2, e356–e366.

Hughes, H.K., Mills-Ko, E., Yang, H., Lesh, T.A., Carter, C.S., Ashwood, P., 2021. Differential macrophage responses in affective versus nonaffective first-episode psychosis patients. Front. Cell. Neurosci. 15, 583351.

Hughes, H.K., Rowland, M.E., Onore, C.E., Rogers, S., Ciernia, A.V., Ashwood, P., 2022. Dysregulated gene expression associated with inflammatory and translation pathways in activated monocytes from children with autism spectrum disorder. Transl. Psychiatry 12, 39.

Hughes, H.K., Moreno, R.J., Ashwood, P., 2023. Innate immune dysfunction and neuroinflammation in autism spectrum disorder (ASD). Brain Behav. Immun. 108, 245–254.

Hunsberger, J.G., Rao, M., Kurtzberg, J., Bulte, J.W., Atala, A., et al., 2016. Accelerating stem cell trials for Alzheimer's disease. Lancet Neurol. 15, 219–230.

Hunt, C., Macedo, E., Cordeiro, T., Suchting, R., de Dios, C., et al., 2020. Effect of immune activation on the kynurenine pathway and depression symptoms - a systematic review and meta-analysis. Neurosci. Biobehav. Rev. 118, 514–523.

Hur, J., Stockbridge, M.D., Fox, A.S., Shackman, A.J., 2019. Dispositional negativity, cognition, and anxiety disorders: an integrative translational neuroscience framework. Prog. Brain Res. 247, 375–436.

Hurley, M.J., Menozzi, E., Koletsi, S., Bates, R., Gegg, M.E., et al., 2023. α-synuclein expression in response to bacterial ligands and metabolites in gut enteroendocrine cells. bioRxiv. 2023-04.

Hussey, S., Purves, J., Allcock, N., Fernandes, V.E., Monks, P.S., et al., 2017. Air pollution alters Staphylococcus aureus and Streptococcus pneumoniae biofilms, antibiotic tolerance and colonisation. Environ. Microbiol. 19, 1868–1880.

Hutchinson, M.R., Zhang, Y., Shridhar, M., Evans, J.H., Buchanan, M.M., et al., 2010. Evidence that opioids may have toll-like receptor 4 and MD-2 effects. Brain Behav. Immun. 24, 83–95.

Huypens, P., Sass, S., Wu, W., Dyckhoff, D., Tschöp, M., et al., 2016. Epigenetic germline inheritance of diet-induced obesity and insulin resistance. Nat. Genet. 48, 497–499.

Hwang, P.W., Braun, K.L., 2015. The effectiveness of dance interventions to improve older adults' health: a systematic literature review. Altern. Ther. Health Med. 21 (5), 64–70.

Hyde, C.L., Nagle, M.W., Tian, C., Chen, X., Paciga, S.A., et al., 2016. Identification of 15 genetic loci associated with risk of major depression in individuals of European descent. Nat. Genet. 48, 1031–1036.

Hyland, N.P., Cryan, J.F., 2016. Microbe-host interactions: influence of the gut microbiota on the enteric nervous system. Dev. Biol. 417, 182–187.

Iannucci, J., Sen, A., Grammas, P., 2021. Isoform-specific effects of apolipoprotein E on markers of inflammation and toxicity in brain glia and neuronal cells in vitro. Curr. Issues Mol. Biol. 43, 215–225.

Igarashi, K.M., 2023. Entorhinal cortex dysfunction in Alzheimer's disease. Trends Neurosci. 46, 124–136.

Iggman, D., Ärnlöv, J., Cederholm, T., Risérus, U., 2016. Association of adipose tissue fatty acids with cardiovascular and all-cause mortality in elderly men. JAMA Cardiol. 1, 745–753.

Ihara, F., Nishimura, M., Muroi, Y., Mahmoud, M.E., Yokoyama, N., et al., 2016. Toxoplasma gondii Infection in mice impairs long-term fear memory consolidation through dysfunction of the cortex and amygdala. Infect. Immun. 84, 2861–2870.

Iida, N., Dzutsev, A., Stewart, C.A., Smith, L., Bouladoux, N., et al., 2013. Commensal bacteria control cancer response to therapy by modulating the tumor microenvironment. Science 342, 967–970.

Ilchmann-Diounou, H., Menard, S., 2020. Psychological stress, intestinal barrier dysfunctions, and autoimmune disorders: an overview. Front. Immunol. 11, 1823.

Imperatore, F., Maurizio, J., Vargas Aguilar, S., Busch, C.J., Favret, J., et al., 2017. SIRT1 regulates macrophage self-renewal. EMBO J. 36, 2353–2372.

Inal-Emiroglu, F.N., Karabay, N., Resmi, H., Guleryuz, H., Baykara, B., et al., 2015. Correlations between amygdala volumes and serum levels of BDNF and NGF as a neurobiological markerin adolescents with bipolar disorder. J. Affect. Disord. 182, 50–56.

Inan-Eroglu, E., Ayaz, A., 2018. Is aluminum exposure a risk factor for neurological disorders? J. Res. Med. Sci. 23, 51.

Ince, L.M., Weber, J., Scheiermann, C., 2019. Control of leukocyte trafficking by stress-associated hormones. Front. Immunol. 9, 3143.

Infante-Garcia, C., Ramos-Rodriguez, J.J., Galindo-Gonzalez, L., Garcia-Alloza, M., 2016. Long-term central pathology and cognitive impairment are exacerbated in a mixed model of Alzheimer's disease and type 2 diabetes. Psychoneuroendocrinology 65, 15–25.

Ingram, K.M., Betz, N.E., Mindes, E.J., Schmitt, M.M., Smith, N.G., 2001. Unsupportive responses from others concerning a stressful life event: development of the unsupportive social interactions inventory. J. Soc. Clin. Psychol. 20, 173–207.

Innominato, P.F., Roche, V.P., Palesh, O.G., Ulusakarya, A., Spiegel, D., et al., 2014. The circadian timing system in clinical oncology. Ann. Med. 46, 191–207.

Insel, T.R., 2014. The NIMH Research Domain Criteria (RDoC) Project: precision medicine for psychiatry. Am. J. Psychiatry 171, 395–397.

Insel, T.R., Hulihan, T.J., 1995. A gender-specific mechanism for pair bonding: oxytocin and partner preference formation in monogamous voles. Behav. Neurosci. 109, 782–789.

Insel, T., Cuthbert, B., Garvey, M., Heinssen, R., Pine, D.S., et al., 2010. Research domain criteria (RDoC): toward a new classification framework for research on mental disorders. Am. J. Psychiatry 167, 748–751.

Inserra, A., Choo, J.M., Lewis, M.D., Rogers, G.B., Wong, M.L., Licinio, J., 2019. Mice lacking Casp1, Ifngr and Nos2 genes exhibit altered depressive- and anxiety-like behaviour, and gut microbiome composition. Sci. Rep. 9, 6456.

Inslicht, S.S., Niles, A.N., Metzler, T.J., Lipshitz, S.L., Otte, C., et al., 2022. Randomized controlled experimental study of hydrocortisone and D-cycloserine effects on fear extinction in PTSD. Neuropsychopharmacology 47, 1945–1952.

Ioannidis, J.P.A., 2013. Implausible results in human nutrition research. BMJ 347, f6698.

Ip, C.K., Rezitis, J., Qi, Y., Bajaj, N., Koller, J., et al., 2023. Critical role of lateral habenula circuits in the control of stress-induced palatable food consumption. Neuron 111, 2583–2600.

Iravanpour, F., Farrokhi, M.R., Jafarinia, M., Oliaee, R.T., 2024. The effect of SARS-CoV-2 on the development of Parkinson's disease: the role of α-synuclein. Human Cell 37, 1–8.

Irwin, M.R., 2015. Why sleep is important for health: a psychoneuroimmunology perspective. Annu. Rev. Psychol. 66, 143–172.

Irwin, M.R., 2019. Sleep and inflammation: partners in sickness and in health. Nat. Rev. Immunol. 19, 702–715.

Irwin, M.R., Opp, M.R., 2017. Sleep health: reciprocal regulation of sleep and innate immunity. Neuropsychopharmacology 42, 129–155.

Irwin, M.R., Olmstead, R., Carrillo, C., Sadeghi, N., Breen, E.C., et al., 2014. Cognitive behavioral therapy versus tai chi for late life insomnia and inflammation: a randomized controlled comparative efficacy trial. Sleep 37, 1543–1552.

Irwin, M.R., Olmstead, R., Carroll, J.E., 2016. Sleep disturbance, sleep duration, and inflammation: a systematic review and meta-analysis of cohort studies and experimental sleep deprivation. Biol. Psychiatry 80, 40–52.

Irwin, M.R., Cole, S., Olmstead, R., Breen, E.C., Cho, J.J., Moieni, M., Eisenberger, N.I., 2019. Moderators for depressed mood and systemic and transcriptional inflammatory responses: a randomized controlled trial of endotoxin. Neuropsychopharmacology 44, 635–641.

Isacsson, G., Rich, C.L., 2014. Antidepressant drugs and the risk of suicide in children and adolescents. Pediatr. Drugs 16, 115–122.

Isaevska, E., Moccia, C., Asta, F., Cibella, F., Gagliardi, L., et al., 2021. Exposure to ambient air pollution in the first 1000 days of life and alterations in the DNA methylome and telomere length in children: a systematic review. Environ. Res. 193, 110504.

Ishikawa, Y., Kitaoka, S., Kawano, Y., Ishii, S., Suzuki, T., et al., 2021. Repeated social defeat stress induces neutrophil mobilization in mice: maintenance after cessation of stress and strain-dependent difference in response. Br. J. Pharmacol. 178, 827–844.

Isingrini, E., Perret, L., Rainer, Q., Amilhon, B., Guma, E., et al., 2016. Resilience to chronic stress is mediated by noradrenergic regulation of dopamine neurons. Nat. Neurosci. 19, 560–563.

Islam, H., Chamberlain, T.C., Mui, A.L., Little, J.P., 2021. Elevated Interleukin-10 levels in COVID-19: potentiation of pro-inflammatory responses or impaired anti-inflammatory action? Front. Immunol. 12, 677008.

Italiani, P., Puxeddu, I., Napoletano, S., Scala, E., Melillo, D., et al., 2018. Circulating levels of IL-1 family cytokines and receptors in Alzheimer's disease: new markers of disease progression? J. Neuroinflammation 15, 342.

Itani, O., Jike, M., Watanabe, N., Kaneita, Y., 2017. Short sleep duration and health outcomes: a systematic review, meta-analysis, and meta-regression. Sleep Med. 32, 246–256.

Itzhaki, R.F., 2017. Herpes simplex virus type 1 and Alzheimer's disease: possible mechanisms and signposts. FASEB J. 31, 3216–3226.

Itzhaki, R.F., Lathe, R., Balin, B.J., Ball, M.J., Bearer, E.L., et al., 2016. Microbes and Alzheimer's disease. J. Alzheimers Dis. 51, 979–984.

Ives, A., Le Roy, D., Théroude, C., Bernhagen, J., Roger, T., Calandra, T., 2021. Macrophage migration inhibitory factor promotes the migration of dendritic cells through CD74 and the activation of the Src/PI3K/myosin II pathway. FASEB J. 35, e21418.

Iwasaki, A., Putrino, D., 2023. Why we need a deeper understanding of the pathophysiology of long COVID. Lancet Infect. Dis. 23, 393–395.

Iwata, M., Ota, K.T., Duman, R.S., 2013. The inflammasome: pathways linking psychological stress, depression, and systemic illnesses. Brain Behav. Immun. 31, 105–114.

Iyadurai, L., Blackwell, S.E., Meiser-Stedman, R., Watson, P.C., Bonsall, M.B., et al., 2016. Preventing intrusive memories after trauma via a brief intervention involving Tetris computer game play in the emergency department: a proof-of-concept randomized controlled trial. Mol. Psychiatry. https://doi.org/10.1038/mp.2017.23.

Izvolskaia, M., Sharova, V., Zakharova, L., 2020. Perinatal inflammation reprograms neuroendocrine, immune, and reproductive functions: Profile of cytokine biomarkers. Inflammation 43, 1175–1183.

Jacka, F.N., 2017. Nutritional psychiatry: where to next? EBioMedicine 17, 24–29.

Jacka, F.N., O'Neil, A., Opie, R., Itsiopoulos, C., Cotton, S., et al., 2017. A randomised controlled trial of dietary improvement for adults with major depression (the 'SMILES' trial). BMC Med. 15, 23.

Jackson, S.E., Kirschbaum, C., Steptoe, A., 2017. Hair cortisol and adiposity in a population based sample of 2,527 men and women aged 54 to 87 years. Obesity 25, 539–544.

Jackson-Lewis, V., Przedborski, S., 2007. Protocol for the MPTP mouse model of Parkinson's disease. Nat. Protoc. 2, 141–151.

Jacob, J.A., 2016. As opioid prescribing guidelines tighten, mindfulness meditation holds promise for pain relief. JAMA 315, 2385–2387.

Jacobs, R.H., Jenkins, L.M., Gabriel, L.B., Barba, A., Ryan, K.A., et al., 2014. Increased coupling of intrinsic networks in remitted depressed youth predicts rumination and cognitive control. PLoS One 9, e104366.

Jacobson, M.L., Browne, C.A., Lucki, I., 2020. Kappa opioid receptor antagonists as potential therapeutics for stress-related disorders. Annu. Rev. Pharmacol. Toxicol. 60, 615–636.

Jacobson, M.H., Stein, C.R., Liu, M., Ackerman, M.G., Blakemore, J.K., et al., 2021. Prenatal exposure to bisphenols and phthalates and postpartum depression: the role of neurosteroid hormone disruption. J. Clin. Endocrinol. Metab. 106, 1887–1899.

Jahankhani, K., Ahangari, F., Adcock, I.M., Mortaz, E., 2023. Possible cancer-causing capacity of COVID-19: Is SARS-CoV-2 an oncogenic agent? Biochimie 213, 130–138.

Jahrami, H.A., Alhaj, O.A., Humood, A.M., Alenezi, A.F., Fekih-Romdhane, F., et al., 2022. Sleep disturbances during the COVID-19 pandemic: a systematic review, meta-analysis, and meta-regression. Sleep Med. Rev. 62, 101591.

Jain, S., Marotta, F., Haghshenas, L., Yadav, H., 2023. Treating leaky syndrome in the over 65s: progress and challenges. Clin. Interv. Aging 18, 1447–1451.

Jakubovski, E., Johnson, J.A., Nasir, M., Müller-Vahl, K., Bloch, M.H., 2019. Systematic review and meta-analysis: dose-response curve of SSRIs and SNRIs in anxiety disorders. Depress. Anxiety 36, 198–212.

James, S.N., Chiou, Y.J., Fatih, N., Needham, L.P., Schott, J.M., Richards, M., 2023. Timing of physical activity across adulthood on later-life cognition: 30 years follow-up in the 1946 British birth cohort. J. Neurol. Neurosurg. Psychiatry 94, 349–356.

Janakiraman, U., Manivasagam, T., Thenmozhi, A.J., Essa, M.M., Barathidasan, R., et al., 2016. Influences of chronic mild stress exposure on motor, non-motor impairments and neurochemical variables in specific brain areas of MPTP/probenecid induced neurotoxicity in mice. PLoS One 11, e0146671.

Janelidze, S., Suchankova, P., Ekman, A., Erhardt, S., Sellgren, C., et al., 2015. Low IL-8 is associated with anxiety in suicidal patients: genetic variation and decreased protein levels. Acta Psychiatr. Scand. 131, 269–278.

Jang, H., Boltz, D., Sturm-Ramirez, K., Shepherd, K.R., Jiao, Y., et al., 2009. Highly pathogenic H5N1 influenza virus can enter the central nervous system and induce neuroinflammation and neurodegeneration. Proc. Natl. Acad. Sci. USA 106, 14063–14068.

Jang, W., Kim, H.J., Li, H., Jo, K.D., Lee, M.K., Yang, H.O., 2016. The neuroprotective effect of erythropoietin on rotenone-induced neurotoxicity in SH-SY5Y cells through the induction of autophagy. Mol. Neurobiol. 53 (6), 3812–3821.

Jangi, S., Gandhi, R., Cox, L.M., Li, N., Von Glehn, F., et al., 2016. Alterations of the human gut microbiome in multiple sclerosis. Nat. Commun. 7, 12015.

Janicki Deverts, D., Cohen, S., Doyle, W.J., 2017. Dispositional affect moderates the stress-buffering effect of social support on risk for developing the common cold. J. Pers. 85, 675–686.

Janova, H., Böttcher, C., Holtman, I.R., Regen, T., van Rossum, D., et al., 2016. CD14 is a key organizer of microglial responses to CNS infection and injury. Glia 64, 635–649.

Jansen, R., Penninx, B.W., Madar, V., Xia, K., Milaneschi, Y., et al., 2016. Gene expression in major depressive disorder. Mol. Psychiatry 21, 339–347.

Jaroudi, W., Garami, J., Garrido, S., Hornberger, M., Keri, S., et al., 2017. Factors underlying cognitive decline in old age and Alzheimer's disease: the role of the hippocampus. Rev. Neurosci. 28, 705–714.

Jastreboff, A.M., Kaplan, L.M., Frías, J.P., Wu, Q., Du, Y., et al., 2023. Triple-hormone-receptor agonist Retatrutide for obesity - a phase 2 trial. N. Engl. J. Med. 10, 514–526.

Jaunmuktane, Z., Mead, S., Ellis, M., Wadsworth, J.D.F., Nicoll, A.J., et al., 2015. Evidence for human transmission of amyloid-β pathology and cerebral amyloid angiopathy. Nature 525, 247–250.

Jay, T.R., Miller, C.M., Cheng, P.J., Graham, L.C., Bemiller, S., et al., 2015. TREM2 deficiency eliminates TREM2+ inflammatory macrophages and ameliorates pathology in Alzheimer's disease mouse models. J. Exp. Med. 212, 287–295.

Jeffers, A.M., Glantz, S., Byers, A.L., Keyhani, S., 2024. Association of cannabis use with cardiovascular outcomes among US adults. J. Am. Heart Assoc. e030178.

Jeffery, I.B., O'Toole, P.W., Öhman, L., Claesson, M.J., Deane, J., et al., 2012. An irritable bowel syndrome subtype defined by species-specific alterations in faecal microbiota. Gut 61, 997–1006.

Jeličić, L., Veselinović, A., Ćirović, M., Jakovljević, V., Raičević, S., Subotić, M., 2022. Maternal distress during pregnancy and the postpartum period: underlying mechanisms and child's developmental outcomes—a narrative review. Int. J. Mol. Sci. 23, 13932.

Jembrek, M.J., Auteri, M., Serio, R., Vlainić, J., 2017. GABAergic system in action: connection to gastrointestinal stress-related disorders. Curr. Pharm. Des. 23, 4003–4011.

Jensen, M.P., Day, M.A., Miró, J., 2014. Neuromodulatory treatments for chronic pain: efficacy and mechanisms. Nat. Rev. Neurol. 10 (3), 167–178.

Jensen, M.P., Jamieson, G.A., Lutz, A., Mazzoni, G., McGeown, W.J., et al., 2017a. New directions in hypnosis research: strategies for advancing the cognitive and clinical neuroscience of hypnosis. Neurosci. Conscious. 3 (1), nix004.

Jensen, P.S., Zhu, Z., van Opijnen, T., 2017b. Antibiotics disrupt coordination between transcriptional and phenotypic stress responses in pathogenic bacteria. Cell Rep. 20, 1705–1716.

Jeon, O.H., Mehdipour, M., Gil, T.H., Kang, M., Aguirre, N.W., et al., 2022. Systemic induction of senescence in young mice after single heterochronic blood exchange. Nat. Metab. 4, 995–1006.

Jeppesen, R., Benros, M.E., 2019. Autoimmune diseases and psychotic disorders. Front. Psychiatry 10, 131.

Jergović, M., Bendelja, K., Vidović, A., Savić, A., Vojvoda, V., et al., 2014. Patients with posttraumatic stress disorder exhibit an altered phenotype of regulatory T cells. Allergy, Asthma Clin. Immunol. 10, 43.

Jha, H.C., Mehta, D., Lu, J., El-Naccache, D., Shukla, S.K., et al., 2015. Gammaherpesvirus infection of human neuronal cells. mBio 6, e01844-15.

Jiang, T., Yu, J.T., Hu, N., Tan, M.S., Zhu, X.C., et al., 2014. CD33 in Alzheimer's disease. Mol. Neurobiol. 49, 529–535.

Jiang, H.Y., Xu, L.L., Shao, L., Xia, R.M., Yu, Z.H., Ling, Z.X., et al., 2016. Maternal infection during pregnancy and risk of autism spectrum disorders: a systematic review and meta-analysis. Brain Behav. Immun. 58, 165–172.

Jiang, H., Wang, J., Rogers, J., Xie, J., 2017a. Brain iron metabolism dysfunction in Parkinson's disease. Mol. Neurobiol. 54, 3078–3101.

Jiang, S., Gao, H., Luo, Q., Wang, P., Yang, X., 2017b. The correlation of lymphocyte subsets, natural killer cell, and Parkinson's disease: a meta-analysis. Neurol. Sci. 38, 1373–1380.

Jiang, C., Ma, X., Qi, S., Han, G., Li, Y., Liu, Y., Liu, L., 2018a. Association between TNF-α-238G/A gene polymorphism and OCD susceptibility: a meta-analysis. Medicine (Baltimore) 97 (5), e9769.

Jiang, H.Y., Zhang, X., Yu, Z.H., Zhang, Z., Deng, M., Zhao, J.H., Ruan, B., 2018b. Altered gut microbiota profile in patients with generalized anxiety disorder. J. Psychiatr. Res. 104, 130–136.

Jiang, T., Farkas, D.K., Ahern, T.P., Lash, T.L., Sørensen, H.T., Gradus, J. L., 2019. Posttraumatic stress disorder and incident infections: a nationwide cohort study. Epidemiology 30, 911–917.

Jiang, W., Han, T., Duan, W., Dong, Q., Hou, W., et al., 2020. Prenatal famine exposure and estimated glomerular filtration rate across consecutive generations: association and epigenetic mediation in a population-based cohort study in Suihua China. Aging 12, 12206–12221.

Jiang, S., Huang, H., Zhou, J., Li, H., Duan, M., et al., 2023. Progressive trajectories of schizophrenia across symptoms, genes, and the brain. BMC Med. 21, 237.

Jiao, S., Cao, T., Cai, H., 2022. Peripheral biomarkers of treatment-resistant schizophrenia: genetic, inflammation and stress perspectives. Front. Pharmacol. 13, 1005702.

Jimenez, J.C., Su, K., Goldberg, A.R., Luna, V.M., Biane, J.S., et al., 2018. Anxiety cells in a hippocampal-hypothalamic circuit. Neuron 97, 670–683.

Jiménez-Castellanos, J.C., Wan Nur Ismah, W.A.K., Takebayashi, Y., Findlay, J., Schneiders, T., et al., 2018. Envelope proteome changes driven by RamA overproduction in Klebsiella pneumoniae that enhance acquired β-lactam resistance. J. Antimicrob. Chemother. 73, 88–94.

Jiménez-Sánchez, L., Castañé, A., Pérez-Caballero, L., Grifoll-Escoda, M., López-Gil, X., et al., 2016. Activation of AMPA receptors mediates the antidepressant action of deep brain stimulation of the infralimbic prefrontal cortex. Cereb. Cortex 26, 2778–2789.

Jin Yoon, E., Ismail, Z., Kathol, I., Kibreab, M., Hammer, T., et al., 2021. Patterns of brain activity during a set-shifting task linked to mild behavioral impairment in Parkinson's disease. Neuroimage Clin. 30, 102590.

Jin, M.J., Jeon, H., Hyun, M.H., Lee, S.H., 2019a. Influence of childhood trauma and brain-derived neurotrophic factor Val66Met polymorphism on posttraumatic stress symptoms and cortical thickness. Sci. Rep. 9, 6028.

Jin, Y., Sun, L.H., Yang, W., Cui, R.J., Xu, S.B., 2019b. The role of BDNF in the neuroimmune axis regulation of mood disorders. Front. Neurol. 10, 515.

Jin, H., Xie, J., Mao, L., Zhao, M., Bai, X., et al., 2020. Bisphenol analogue concentrations in human breast milk and their associations with postnatal infant growth. Environ. Pollut. 259, 113779.

Johansen, A., Armand, S., Plavén-Sigray, P., Nasser, A., Ozenne, B., et al., 2023a. Effects of escitalopram on synaptic density in the healthy human brain: a randomized controlled trial. Mol. Psychiatry 28, 4272–4279.

Johansen, J., Atarashi, K., Arai, Y., Hirose, N., Sørensen, S.J., et al., 2023b. Centenarians have a diverse gut virome with the potential to modulate metabolism and promote healthy lifespan. Nat. Microbiol. 8, 1064–1078.

John-Henderson, N.A., Henderson-Matthews, B., Ollinger, S.R., Racine, J., Gordon, M.R., et al., 2020. Adverse childhood experiences and immune system inflammation in adults residing on the blackfeet reservation: the moderating role of sense of belonging to the community. Ann. Behav. Med. 54, 87–93.

Johnson, F.K., Kaffman, A., 2018. Early life stress perturbs the function of microglia in the developing rodent brain: new insights and future challenges. Brain Behav. Immun. 69, 18–27.

Johnson, J.D., O'Connor, K.A., Deak, T., Stark, M., Watkins, L.R., et al., 2002. Prior stressor exposure sensitizes LPS-induced cytokine production. Brain Behav. Immun. 16, 461–476.

Johnson, J.D., O'Connor, K.A., Hansen, M.K., Watkins, L.R., Maier, S.F., 2003. Effects of prior stress on LPS-induced cytokine and sickness responses. Am. J. Phys. Regul. Integr. Comp. Phys. 284, R422–R432.

Johnson, J.D., Barnard, D.F., Kulp, A.C., Mehta, D.M., 2019. Neuroendocrine regulation of brain cytokines after psychological stress. J. Endocr. Soc. 3, 1302–1320.

Johnson, A.J., Buchanan, T., Laffitte Nodarse, C., Valdes Hernandez, P. A., Huo, Z., et al., 2022. Cross-sectional brain-predicted age differences in community-dwelling middle-aged and older adults with high impact knee pain. J. Pain Res. 15, 3575–3587.

Johnson, R.J., Lanaspa, M.A., Sanchez-Lozada, L.G., Tolan, D., Nakagawa, T., et al., 2023. The fructose survival hypothesis for obesity. Philos. Trans. R. Soc. Lond. Ser. B Biol. Sci. 378, 20220230.

Johnson, K.A., Okun, M.S., Scangos, K.W., Mayberg, H.S., de Hemptinne, C., et al., 2024. Deep brain stimulation for refractory major depressive disorder: a comprehensive review. Mol. Psychiatry.

Johnston, B.C., Kanters, S., Bandayrel, K., Wu, P., Naji, F., et al., 2014. Comparison of weight loss among named diet programs in overweight and obese adults: a meta-analysis. JAMA 312, 923–933.

Jokela, M., Laakasuo, M., 2023. Obesity as a causal risk factor for depression: systematic review and meta-analysis of Mendelian Randomization studies and implications for population mental health. J. Psychiatr. Res. 163, 86–92.

Jollant, F., Colle, R., Nguyen, T.M.L., Corruble, E., Gardier, A.M., et al., 2023. Ketamine and esketamine in suicidal thoughts and behaviors: a systematic review. Ther. Adv. Psychopharmacol. 13. 20451253231151327.

Jonas, W., Woodside, B., 2016. Physiological mechanisms and behavioral and psychological factors influencing the transfer of milk from mothers to their young. Horm. Behav. 77, 167–181.

Jones, K.L., Van de Water, J., 2019. Maternal autoantibody related autism: mechanisms and pathways. Mol. Psychiatry 24, 252–265.

Jones, M.E., Lebonville, C.L., Paniccia, J.E., Balentine, M.E., Reissner, K. J., et al., 2018. Hippocampal interleukin-1 mediates stress-enhanced fear learning: a potential role for astrocyte-derived interleukin-1β. Brain Behav. Immun. 67, 355–363.

Jones, H.F., Han, V.X., Patel, S., Gloss, B.S., Soler, N., et al., 2021. Maternal autoimmunity and inflammation are associated with childhood tics and obsessive-compulsive disorder: transcriptomic data show common enriched innate immune pathways. Brain Behav. Immun. 94, 308–317.

Jones, C.M., Day, R.O., Koes, B.W., Latimer, J., Maher, C.G., et al., 2023a. Opioid analgesia for acute low back pain and neck pain (the OPAL trial): a randomised placebo-controlled trial. Lancet 402, 304–312.

Jones, J., Correll, D.J., Lechner, S.M., Jazic, I., Miao, X., et al., 2023b. Selective Inhibition of NaV1.8 with VX-548 for acute pain. N. Engl. J. Med. 389 (5), 393–405.

Jong, C.J., Sandal, P., Schaffer, S.W., 2021. The role of taurine in mitochondria health: more than just an antioxidant. Molecules 26, 4913.

Jonker, I., Rosmalen, J.G., Schoevers, R.A., 2017. Childhood life events, immune activation and the development of mood and anxiety disorders: the TRAILS study. Transl. Psychiatry 7, e1112.

Jøntvedt Jørgensen, M., Holter, J.C., Christensen, E.E., Schjalm, C., Tonby, K., et al., 2020. Increased interleukin-6 and macrophage chemoattractant protein-1 are associated with respiratory failure in COVID-19. Sci. Rep. 10, 21697.

Jorcano, A., Lubczyńska, M.J., Pierotti, L., Altug, H., Ballester, F., et al., 2019. Prenatal and postnatal exposure to air pollution and emotional and aggressive symptoms in children from 8 European birth cohorts. Environ. Int. 131, 104927.

Jorden, S., Matheson, K., Anisman, H., 2009. Supportive and unsupportive social interactions in relation to cultural adaptation and psychological distress among Somali refugees exposed to collective or personal traumas. J. Cross-Cult. Psychol. 40, 853–874.

Jorfi, M., Maaser-Hecker, A., Tanzi, R.E., 2023. The neuroimmune axis of Alzheimer's disease. Genome Med. 15, 6.

Jorgačevski, J., Potokar, M., 2023. Immune functions of astrocytes in viral neuroinfections. Int. J. Mol. Sci. 24 (4), 3514.

Jorge, R.E., Acion, L., Burin, D.I., Robinson, R.G., 2016. Sertraline for preventing mood disorders following traumatic brain injury. JAMA Psychiatry 73, 1041–1047.

Joseph, J., 2015. The Trouble with Twin Studies: A Reassessment of Twin Research in the Social and Behavioral Sciences. Routledge, New York.

Josephs, R.A., Cobb, A.R., Lancaster, C.L., Lee, H.J., Telch, M.J., 2017. Dual-hormone stress reactivity predicts downstream war-zone stress-evoked PTSD. Psychoneuroendocrinology 78, 76–84.

Ju, Y.E.S., Ooms, S.J., Sutphen, C., Macauley, S.L., Zangrilli, M.A., et al., 2017. Slow wave sleep disruption increases cerebrospinal fluid amyloid-β levels. Brain J. Neurol. 140, 2104–2111.

Ju, Y., Ou, W., Su, J., Averill, C.L., Liu, J., Wang, M., et al., 2020. White matter microstructural alterations in posttraumatic stress disorder: an ROI and whole-brain based meta-analysis. J. Affect. Disord. 266, 655–670.

Juhasz, G., Gonda, X., Hullam, G., Eszlari, N., Kovacs, D., et al., 2015. Variability in the effect of 5-HTTLPR on depression in a large European population: the role of age, symptom profile, type and intensity of life stressors. PLoS One 10, 1–15.

Jung, S., Bae, H., Song, W.S., Jang, C., 2022. Dietary fructose and fructose-induced pathologies. Annu. Rev. Nutr. 42, 45–66.

Jung, S.Y., Chun, S., Cho, E.B., Han, K., Yoo, J., et al., 2023. Changes in smoking, alcohol consumption, and the risk of Parkinson's disease. Front. Aging Neurosci. 15, 1223310.

Jurgens, H.A., Amancherla, K., Johnson, R.W., 2012. Influenza infection induces neuroinflammation, alters hippocampal neuron morphology, and impairs cognition in adult mice. J. Neurosci. 32, 3958–3968.

Jyonouchi, H., 2009. Food allergy and autism spectrum disorders: is there a link? Curr Allergy Asthma Rep 9, 194–201.

Jyonouchi, H., Geng, L., Streck, D.L., Toruner, G.A., 2011. Children with autism spectrum disorders (ASD) who exhibit chronic gastrointestinal (GI) symptoms and marked fluctuation of behavioral symptoms exhibit distinct innate immune abnormalities and transcriptional profiles of peripheral blood (PB) monocytes. J. Neuroimmunol. 238 (12), 7380.

Kabat-Zinn, J., 1990. Full Catastrophe Living: Using the Wisdom of Your Body and Mind to Face Stress, Pain, and Illness. Delacourt, New York, NY.

Kabat-Zinn, J., 2019. Foreword: seeds of a necessary global renaissance in the making: the refining of psychology's understanding of the nature of mind, self, and embodiment through the lens of mindfulness and its origins at a key inflection point for the species. Curr. Opin. Psychol. 28, xi–xvii.

Kadam, R.U., Juraszek, J., Brandenburg, B., Buyck, C., Schepens, W.B.G., 2017. Potent peptidic fusion inhibitors of influenza virus. Science 358, 496–502.

Kadokura, A., Yamazaki, T., Lemere, C.A., Takatama, M., Okamoto, K., 2009. Regional distribution of TDP-43 inclusions in Alzheimer disease (AD) brains: their relation to AD common pathology. Neuropathology 29, 566–573.

Kagan, J.C., 2023. Infection infidelities drive innate immunity. Science 379, 333–335.

Kahlenberg, J.M., Kaplan, M.J., 2013. Little peptide, big effects: the role of LL-37 in inflammation and autoimmune disease. J. Immunol. 191, 4895–4901.

Kahn, M.S., Kranjac, D., Alonzo, C.A., Haase, J.H., Cedillos, R.O., et al., 2012. Prolonged elevation in hippocampal Aβ and cognitive deficits following repeated endotoxin exposure in the mouse. Behav. Brain Res. 229, 176–184.

Kahneman, D., 2011. Thinking, Fast and Slow. Farrar, Straus and Giroux, New York.

Kahneman, D., Tversky, A., 1996. On the reality of cognitive illusions. Psychol. Rev. 103, 582–591.

Kaim, A., Saban, M., 2022. Are we suffering from the Peltzman effect? Risk perception among recovered and vaccinated people during the COVID-19 pandemic in Israel. Public Health 209, 19–22.

Kajitani, N., Hisaoka-Nakashima, K., Okada-Tsuchioka, M., Hosoi, M., Yokoe, T., et al., 2015. Fibroblast growth factor 2 mRNA expression evoked by amitriptyline involves extracellular signal-regulated kinase-dependent early growth response 1 production in rat primary cultured astrocytes. J. Neurochem. 135, 27–37.

Kalia, L.V., Lang, A.E., 2015. Parkinson's disease. Lancet 386, 896–912.

Kalin, N.H., Fox, A.S., Kovner, R., Riedel, M.K., Fekete, E.M., et al., 2016. Overexpressing corticotropin-releasing factor in the primate amygdala increases anxious temperament and alters its neural circuit. Biol. Psychiatry 80, 345–355.

Kalmakis, K.A., Chandler, G.E., 2015. Health consequences of adverse childhood experiences: a systematic review. J. Am. Assoc. Nurse Pract. 27, 457–465.

Kam, T.I., Hinkle, J.T., Dawson, T.M., Dawson, V.L., 2020. Microglia and astrocyte dysfunction in parkinson's disease. Neurobiol. Dis. 144, 105028.

Kamagata, K., Andica, C., Takabayashi, K., Saito, Y., Taoka, T., et al., 2022. Association of MRI indices of glymphatic system with amyloid deposition and cognition in mild cognitive impairment and Alzheimer disease. Neurology 99, e2648–e2660.

Kamal, H., Tan, G.C., Ibrahim, S.F., Shaikh, M.F., Mohamed, I.N., et al., 2020. Alcohol use disorder, neurodegeneration, Alzheimer's and Parkinson's disease: interplay between oxidative stress, neuroimmune response and excitotoxicity. Front. Cell. Neurosci. 14, 282.

Kami, K., Tajima, F., Senba, E., 2017. Exercise-induced hypoalgesia: potential mechanisms in animal models of neuropathic pain. Anat. Sci. Int. 92, 79–90.

Kammel, A., Saussenthaler, A., Jähnert, M., Jonas, W., Stirm, L., et al., 2016. Early hypermethylation of hepatic Igfbp2 results in its reduced expression preceding fatty liver in mice. Hum. Mol. Genet. 25, 2588–2599.

Kanarek, N., Petrova, B., Sabatini, D.M., 2020. Dietary modifications for enhanced cancer therapy. Nature 579, 507–517.

Kanchanatawan, B., Sirivichayakul, S., Carvalho, A.F., Anderson, G., Galecki, P., et al., 2018. Depressive, anxiety and hypomanic symptoms in schizophrenia may be driven by tryptophan catabolite (TRYCAT) patterning of IgA and IgM responses directed to TRYCATs. Prog. Neuro-Psychopharmacol. Biol. Psychiatry 80, 205–216.

Kandel, E.R., Dudai, Y., Mayford, M.R., 2014. The molecular and systems biology of memory. Cell 157, 163–186.

Kang, V., Wagner, G.C., Ming, X., 2014. Gastrointestinal dysfunction in children with autism spectrum disorders. Autism Res. 7 (4), 501–506.

Kang, H.J., Kim, J.M., Bae, K.Y., Kim, S.W., Shin, I.S., et al., 2015. Longitudinal associations between BDNF promoter methylation and late-life depression. Neurobiol. Aging 36, 1764.e1–1764.e7.

Kang, D.W., Adams, J.B., Gregory, A.C., Borody, T., Chittick, L., et al., 2017. Microbiota transfer therapy alters gut ecosystem and improves gastrointestinal and autism symptoms: an open-label study. Microbiome 5, 10.

Kang, D.W., Adams, J.B., Coleman, D.M., Pollard, E.L., Maldonado, J., et al., 2019. Long-term benefit of Microbiota Transfer Therapy on autism symptoms and gut microbiota. Sci. Rep. 9, 5821.

Kang, S.J., Kim, J.H., Kim, D.I., Roberts, B.Z., Han, S., 2024. A pontomesencephalic PACAPergic pathway underlying panic-like behavioral and somatic symptoms in mice. Nat. Neurosci. 27, 90–101.

Kanner, L., 1946. Irrelevant and metaphorical language in early infantile autism. Am. J. Psychiatry 103 (2), 242–246.

Kanner, J., Pollack, A.Z., Ranasinghe, S., Stevens, D.R., Nobles, C., et al., 2021. Chronic exposure to air pollution and risk of mental health disorders complicating pregnancy. Environ. Res. 196, 110937.

Kao, C.Y., Stalla, G., Stalla, J., Wotjak, C.T., Anderzhanova, E., 2015. Norepinephrine and corticosterone in the medial prefrontal cortex and hippocampus predict PTSD-like symptoms in mice. Eur. J. Neurosci. 41, 1139–1148.

Kaplan, G.B., Leite-Morris, K.A., Wang, L., Rumbika, K.K., Heinrichs, S. C., et al., 2018. Pathophysiological bases of comorbidity: traumatic brain injury and post-traumatic stress disorder. J. Neurotrauma 35, 210–225.

Kappelmann, N., Lewis, G., Dantzer, R., Jones, P.B., Khandaker, G.M., 2016. Antidepressant activity of anti-cytokine treatment: a systematic review and meta-analysis of clinical trials of chronic inflammatory conditions. Mol. Psychiatry 23, 335–343.

Karagiannis, T.T., Dowrey, T.W., Villacorta-Martin, C., Montano, M., Reed, E., et al., 2023. Multi-modal profiling of peripheral blood cells across the human lifespan reveals distinct immune cell signatures of aging and longevity. EBioMedicine 90, 104514.

Karagkouni, A., Alevizos, M., Theoharides, T.C., 2013. Effect of stress on brain inflammation and multiple sclerosis. Autoimmun. Rev. 12, 947–953.

Karagüzel, E.Ö., Arslan, F.C., Uysal, E.K., Demir, S., Aykut, D.S., et al., 2019. Blood levels of interleukin-1 beta, interleukin-6 and tumor necrosis factor-alpha and cognitive functions in patients with obsessive compulsive disorder. Compr. Psychiatry 89, 61–66.

Karg, K., Burmeister, M., Shedden, K., Sen, S., 2011. The serotonin transporter promoter variant (5-HTTLPR), stress, and depression metaanalysis revisited: evidence of genetic moderation. Arch. Gen. Psychiatry 68, 444–454.

Karikari, A.A., McFleder, R.L., Ribechini, E., Blum, R., Bruttel, V., et al., 2022. Neurodegeneration by α-synuclein-specific T cells in AAV-A53T-α-synuclein Parkinson's disease mice. Brain Behav. Immun. 101, 194–210.

Karkhanis, A.N., Rose, J.H., Weiner, J.L., Jones, S.R., 2016. Early-life social isolation stress increases kappa opioid receptor responsiveness and downregulates the dopamine system. Neuropsychopharmacology 41, 2263–2274.

Karlsson, H.K., Tuominen, L., Tuulari, J.J., Hirvonen, J., Parkkola, R., et al., 2015. Obesity is associated with decreased μ-opioid but unaltered dopamine D2 receptor availability in the brain. J. Neurosci. 35, 3959–3965.

Karlsson, H., Sjöqvist, H., Brynge, M., Gardner, R., Dalman, C., 2022. Childhood infections and autism spectrum disorders and/or intellectual disability: a register-based cohort study. J. Neurodev. Disord. 14, 12.

Karshikoff, B., Jensen, K.B., Kosek, E., Kalpouzos, G., Soop, A., et al., 2016. Why sickness hurts: a central mechanism for pain induced by peripheral inflammation. Brain Behav. Immun. 57, 38–46.

Karuppagounder, S.S., Xiong, Y., Lee, Y., Lawless, M.C., Kim, D., et al., 2016. LRRK2 G2019S transgenic mice display increased susceptibility to 1-methyl-4-phenyl-1, 2, 3, 6-tetrahydropyridine (MPTP)-mediated neurotoxicity. J. Chem. Neuroanat. 76, 90–97.

Kasai, K., Yamasue, H., Gilbertson, M.W., Shenton, M.E., Rauch, S.L., Pitman, R.K., 2008. Evidence for acquired pregenual anterior cingulate gray matter loss from a twin study of combat-related posttraumatic stress disorder. Biol. Psychiatry 63, 550–556.

Kastanenka, K.V., Hou, S.S., Shakerdge, N., Logan, R., Feng, D., et al., 2017. Optogenetic restoration of disrupted slow oscillations halts amyloid deposition and restores calcium homeostasis in an animal model of Alzheimer's disease. PLoS One 12, e0170275.

Kato, H., Kimura, H., Kushima, I., Takahashi, N., Aleksic, B., Ozaki, N., 2023. The genetic architecture of schizophrenia: review of large-scale genetic studies. J. Hum. Genet. 68, 175–182.

Katrinli, S., Maihofer, A.X., Wani, A.H., Pfeiffer, J.R., Ketema, E., et al., 2022a. Epigenome-wide meta-analysis of PTSD symptom severity in three military cohorts implicates DNA methylation changes in genes involved in immune system and oxidative stress. Mol. Psychiatry 27, 1720–1728.

Katrinli, S., Oliveira, N.C.S., Felger, J.C., Michopoulos, V., Smith, A.K., 2022b. The role of the immune system in posttraumatic stress disorder. Transl. Psychiatry 12, 313.

Katsi, V., Didagelos, M., Skevofilax, S., Armenis, I., Kartalis, A., et al., 2019. GUT microbiome - GUT dysbiosis-arterial hypertension: new horizons. Curr. Hypertens. Rev. 15, 40–46.

Kaufmann, T., Elvsåshagen, T., Alnæs, D., Zak, N., Pedersen, P.Ø., et al., 2016. The brain functional connectome is robustly altered by lack of sleep. NeuroImage 127, 324–332.

Kaur, K., Fayad, R., Saxena, A., Frizzell, N., Chanda, A., et al., 2016. Fluoroquinolone-related neuropsychiatric and mitochondrial toxicity: a collaborative investigation by scientists and members of a social network. J Community Support Oncol. 14, 54–65.

Kautz, M., Charney, D.S., Murrough, J.W., 2017. Neuropeptide Y, resilience, and PTSD therapeutics. Neurosci. Lett. 649, 164–169.

Kawaguchi, Y., Okada, T., Konishi, H., Fujino, M., Asai, J., et al., 1997. Reduction of the DTH response is related to morphological changes of Langerhans cells in mice exposed to acute immobilization stress. Clin. Exp. Immunol. 109, 397–401.

Kawahata, I., Finkelstein, D.I., Fukunaga, K., 2022. Pathogenic impact of α-Synuclein phosphorylation and its kinases in α-Synucleinopathies. Int. J. Mol. Sci. 23, 6216.

Kawai, T., Ikegawa, M., Ori, D., Akira, S., 2024. Decoding Toll-like receptors: recent insights and perspectives in innate immunity. Immunity 57, 649–673.

Ke, S., Wang, X.W., Ratanatharathorn, A., Huang, T., Roberts, A.L., et al., 2023. Association of probable post-traumatic stress disorder with dietary pattern and gut microbiome in a cohort of women. Nat. Ment. Health 1, 900–913.

Keijser, R., Åslund, C., Nilsson, K.W., Olofsdotter, S., 2021. Gene-environment interaction: oxytocin receptor (OXTR) polymorphisms and parenting style as potential predictors for depressive symptoms. Psychiatry Res. 303, 114057.

Keil, A., Daniels, J.L., Forssen, U., Hultman, C., Cnattingius, S., et al., 2010. Parental autoimmune diseases associated with autism spectrum disorders in offspring. Epidemiology 21, 805–808.

Kekow, J., Moots, R., Khandker, R., Melin, J., Freundlich, B., Singh, A., 2011. Improvements in patient-reported outcomes, symptoms of depression and anxiety, and their association with clinical remission among patients with moderate-to-severe active early rheumatoid arthritis. Rheumatology 50, 401–409.

Kelada, S.N.P., Checkoway, H., Kardia, S.L.R., Carlson, C.S., Costa-Mallen, P., et al., 2006. 5′ and 3′ region variability in the dopamine transporter gene (SLC6A3), pesticide exposure and Parkinson's disease risk: a hypothesis-generating study. Hum. Mol. Genet. 15, 3055–3062.

Kelleher, R.J., Shen, J., 2017. Presenilin-1 mutations and Alzheimer's disease. Proc. Natl. Acad. Sci. USA 114, 629–631.

Kellum, C.E., Kemp, K.M., Mrug, S., Pollock, J.S., Seifert, M.E., Feig, D.I., 2023. Adverse childhood experiences are associated with vascular changes in adolescents that are risk factors for future cardiovascular disease. Pediatr. Nephrol. 38, 2155–2163.

Kelly, J.R., Kennedy, P.J., Cryan, J.F., Dinan, T.G., Clarke, G., et al., 2015. Breaking down the barriers: the gut microbiome, intestinal permeability and stress-related psychiatric disorders. Front. Cell. Neurosci. 9, 392.

Kelly, J.R., Clarke, G., Cryan, J.F., Dinan, T.G., 2016. Brain-gut-microbiota axis: challenges for translation in psychiatry. Ann. Epidemiol. 26, 366–372.

Kelly, J.R., Borre, Y., O′ Brien, C., Patterson, E., El Aidy, S., et al., 2016a. Transferring the blues: depression-associated gut microbiota induces neurobehavioural changes in the rat. J. Psychiatr. Res., 109–118.

Kelly, J.R., Minuto, C., Cryan, J.F., Clarke, G., Dinan, T.G., 2017. Cross talk: the microbiota and neurodevelopmental disorders. Front. Neurosci. 11, 490.

Kelly, E., Meng, F., Fujita, H., Morgado, F., Kazemi, Y., et al., 2020. Regulation of autism-relevant behaviors by cerebellar-prefrontal cortical circuits. Nat. Neurosci. 23, 1102–1110.

Kelly, K.M., Smith, J.A., Mezuk, B., 2021. Depression and interleukin-6 signaling: a Mendelian randomization study. Brain Behav. Immun. 95, 106–114.

Kemp, G.M., Altimimi, H.F., Nho, Y., Heir, R., Klyczek, A., Stellwagen, D., 2022. Sustained TNF signaling is required for the synaptic and anxiety-like behavioral response to acute stress. Mol. Psychiatry 27, 4474–4484.

Kempuraj, D., Selvakumar, G.P., Ahmed, M.E., Raikwar, S.P., Thangavel, R., et al., 2020. COVID-19, mast cells, cytokine storm, psychological stress, and neuroinflammation. Neuroscientist 26, 402–414.

Kennedy, E., Niedzwiedz, C.L., 2021. The association of anxiety and stress-related disorders with C-reactive protein (CRP) within UK Biobank. Brain Behav. Immun. Health 19, 100410.

Kennedy, P.J., Cryan, J.F., Dinan, T.G., Clarke, G., 2017. Kynurenine pathway metabolism and the microbiota-gut-brain axis. Neuropharmacology 112 (Pt B), 399–412.

Kennis, M., van Rooij, S.J., Reijnen, A., Geuze, E., 2017. The predictive value of dorsal cingulate activity and fractional anisotropy on long-term PTSD symptom severity. Depress. Anxiety 34, 410–418.

Kępińska, A.P., Iyegbe, C.O., Vernon, A.C., Yolken, R., Murray, R.M., Pollak, T.A., 2020. Schizophrenia and Influenza at the Centenary of the 1918-1919 Spanish Influenza pandemic: mechanisms of psychosis risk. Front. Psychiatry 11, 72.

Kerner, G., Neehus, A.L., Philippot, Q., Bohlen, J., Rinchai, D., et al., 2023. Genetic adaptation to pathogens and increased risk of inflammatory disorders in post-Neolithic Europe. Cell Genom. 3, 100248.

Kerr, D.M., McDonald, J., Minnis, H., 2021. The association of child maltreatment and systemic inflammation in adulthood: a systematic review. PLoS One 16, e0243685.

Kertes, D.A., Bhatt, S.S., Kamin, H.S., Hughes, D.A., Rodney, N.C., et al., 2017. BNDF methylation in mothers and newborns is associated with maternal exposure to war trauma. Clin. Epigenetics 9, 68.

Kessler, R.C., Aguilar-Gaxiola, S., Alonso, J., Bromet, E.J., Gureje, O., et al., 2018. The associations of earlier trauma exposures and history of mental disorders with PTSD after subsequent traumas. Mol. Psychiatry 23, 1892–1899.

Kessner, S., Forkmann, K., Ritter, C., Wiech, K., Ploner, M., et al., 2014. The effect of treatment history on therapeutic outcome: psychological and neurobiological underpinnings. PLoS One 9, e109014.

Keymeulen, B., De Groot, K., Jacobs-Tulleneers-Thevissen, D., Thompson, D.M., Bellin, M.D., et al., 2023. Encapsulated stem cell-derived β cells exert glucose control in patients with type 1 diabetes. Nat. Biotechnol. 41, 1477–1486.

Khachadourian, V., Mahjani, B., Sandin, S., Kolevzon, A., Buxbaum, J. D., Reichenberg, A., Janecka, M., 2023. Comorbidities in autism spectrum disorder and their etiologies. Transl. Psychiatry 13, 71.

Khalaf, F.K., Connolly, J., Khatib-Shahidi, B., Albehadili, A., Tassavvor, I., et al., 2023. Paraoxonases at the heart of neurological disorders. Int. J. Mol. Sci. 24, 6881.

Khan, A.M., Dar, S., Ahmed, R., Bachu, R., Adnan, M., Kotapati, V.P., 2018. Cognitive behavioral therapy versus eye movement desensitization and reprocessing in patients with post-traumatic stress disorder: systematic review and meta-analysis of randomized clinical trials. Cureus 10, e3250.

Khan, A.J., Bradley, E., O′Donovan, A., Woolley, J., 2022. Psilocybin for trauma-related disorders. Curr. Top. Behav. Neurosci. 56, 319–332.

Khandaker, G.M., Zuber, V., Rees, J.M.B., Carvalho, L., Mason, A.M., et al., 2020. Shared mechanisms between coronary heart disease and depression: findings from a large UK general population-based cohort. Mol. Psychiatry 25, 1477–1486.

Khandaker, G.M., Stochl, J., Zammit, S., Lewis, G., Dantzer, R., Jones, P. B., 2021. Association between circulating levels of C-reactive protein and positive and negative symptoms of psychosis in adolescents in a general population birth cohort. J. Psychiatr. Res. 143, 534–542.

Khavari, B., Cairns, M.J., 2020. Epigenomic dysregulation in schizophrenia: in search of disease etiology and biomarkers. Cells 9, 1837.

Khera, A.V., Kathiresan, S., 2017. Genetics of coronary artery disease: discovery, biology and clinical translation. Nat. Rev. Genet. 18, 331–344.

Khera, A.V., Emdin, C.A., Drake, I., Natarajan, P., Bick, A.G., et al., 2016. Genetic risk, adherence to a healthy life-style, and coronary disease. N. Engl. J. Med. 375, 2349–2358.

Khoury, M.J., Galea, S., 2016. Will precision medicine improve population health? JAMA 316, 1357–1358.

Khoury, R., Nasrallah, H.A., 2018. Inflammatory biomarkers in individuals at clinical high risk for psychosis (CHR-P): state or trait? Schizophr. Res. 199, 31–38.

Khoury, M.J., Iademarco, M.F., Riley, W.T., 2016. Precision public health for the era of precision medicine. Am. J. Prev. Med. 50, 398–401.

Kidron, C.A., Kotliar, D.M., Kirmayer, L.J., 2019. Transmitted trauma as badge of honor: phenomenological accounts of Holocaust descendant resilient vulnerability. Soc. Sci. Med. 239, 112524.

Kiecolt-Glaser, J.K., Preacher, K.J., MacCallum, R.C., Atkinson, C., Malarkey, W.B., Glaser, R., 2003. Chronic stress and age-related increases in the proinflammatory cytokine IL-6. Proc. Natl. Acad. Sci. USA 100, 9090–9095.

Kiecolt-Glaser, J.K., Derry, H.M., Fagundes, C.P., 2015. Inflammation: depression fans the flames and feasts on the heat. Am. J. Psychiatry 172, 1075–1091.

Kierkegaard, M., Lundberg, I.E., Olsson, T., Johansson, S., Ygberg, S., et al., 2016. High-intensity resistance training in multiple sclerosis - an exploratory study of effects on immune markers in blood and cerebrospinal fluid, and on mood, fatigue, health-related quality of life, muscle strength, walking and cognition. J. Neurol. Sci. 362, 251–257.

Kikuchi, K., Galera-Laporta, L., Weatherwax, C., Lam, J.Y., Moon, E.C., et al., 2022. Electrochemical potential enables dormant spores to integrate environmental signals. Science 378, 43–49.

Kilian, L.S., Frank, D., Rangrez, A.Y., 2021. RhoA signaling in immune cell response and cardiac disease. Cells 10, 1681.

Kim, H.D., Cao, Y., Kong, F.K., Van Kampen, K.R., Lewis, T.L., et al., 2005. Induction of a Th2 immune response by co-administration of recombinant adenovirus vectors encoding amyloid beta-protein and GM-CSF. Vaccine 23, 2977–2986.

Kim, N.K., Choi, B.H., Huang, X., Snyder, B.J., Bukhari, S., et al., 2009. Granulocyte-macrophage colony-stimulating factor promotes survival of dopaminergic neurons in the 1-methyl-4-phenyl-1,2,3,6-tetrahydropyridine-induced murine Parkinson's disease model. Eur. J. Neurosci. 29, 891–900.

Kim, H.S., Sherman, D.K., Mojaverian, T., Sasaki, J.Y., Park, J., et al., 2011. Gene-culture interaction: oxytocin receptor polymorphism (OXTR) and emotion regulation. Soc. Psychol. Personal. Sci. 2, 665–672.

Kim, D., Kubzansky, L.D., Baccarelli, A., Sparrow, D., Spiro, A., et al., 2016a. Psychological factors and DNA methylation of genes related to immune/inflammatory system markers: the VA Normative Aging Study. BMJ Open 6, e009790.

Kim, K.S., Hong, S.W., Han, D., Yi, J., Jung, J., et al., 2016b. Dietary antigens limit mucosal immunity by inducing regulatory T cells in the small intestine. Science 51, 858–863.

Kim, S.K., Hayashi, H., Ishikawa, T., Shibata, K., Shigetomi, E., et al., 2016c. Cortical astrocytes rewire somatosensory cortical circuits for peripheral neuropathic pain. J. Clin. Invest. 126, 1983–1997.

Kim, J.H., Choi, S.H., Jang, J.H., Lee, D.H., Lee, K.J., et al., 2017a. Impaired insula functional connectivity associated with persistent pain perception in patients with complex regional pain syndrome. PLoS One 12 (7), e0180479.

Kim, K.H., Kim, Y.H., Son, J.E., Lee, J.H., Kim, S., et al., 2017b. Intermittent fasting promotes adipose thermogenesis and metabolic homeostasis via VEGF-mediated alternative activation of macrophage. Cell Res. 27, 1309–1326.

Kim, M.J., Shin, J., Taylor, J.M., Mattek, A.M., Chavez, S.J., et al., 2017c. Intolerance of uncertainty predicts increased striatal volume. Emotion 17, 895–899.

Kim, T.Y., Kim, S.J., Chung, H.G., Choi, J.H., Kim, S.H., Kang, J.I., 2017d. Epigenetic alterations of the BDNF gene in combat-related post-traumatic stress disorder. Acta Psychiatr. Scand. 135, 170–179.

Kim, Y.J., Seo, S.W., Park, S.B., Yang, J.J., Lee, J.S., et al., 2017e. Protective effects of APOE e2 against disease progression in subcortical vascular mild cognitive impairment patients: a three-year longitudinal study. Sci. Rep. 7, 1910.

Kim, C.K., Choi, Y.M., Bae, E., Jue, M.S., So, H.S., Hwang, E.S., 2018. Reduced NK cell IFN-γ secretion and psychological stress are independently associated with herpes zoster. PLoS One 13, e0193299.

Kim, K., Jang, E.H., Kim, A.Y., Fava, M., Mischoulon, D., et al., 2019. Pretreatment peripheral biomarkers associated with treatment response in panic symptoms in patients with major depressive disorder and panic disorder: a 12-week follow-up study. Compr. Psychiatry 95, 152140.

Kim, Y., Kim, H., Suh, S.Y., Park, H., Lee, H., 2022. Association between inflammatory cytokines and caregiving distress in family caregivers of cancer patients. Support Care Cancer 30, 1715–1722.

Kim, I.B., Park, S.C., Kim, Y.K., 2023a. Microbiota-gut-brain axis in major depression: a new therapeutic approach. Adv. Exp. Med. Biol. 1411, 209–224.

Kim, J., Song, J., Kambari, Y., Plitman, E., Shah, P., et al., 2023b. Cortical thinning in relation to impaired insight into illness in patients with treatment resistant schizophrenia. Schizophrenia (Heidelb) 9, 27.

Kim, K., Joyce, B.T., Nannini, D.R., Zheng, Y., Gordon-Larsen, P., et al., 2023c. Inequalities in urban greenness and epigenetic aging: different associations by race and neighborhood socioeconomic status. Sci. Adv. 9, eadf8140.

Kim, S.Y., An, S.J., Han, J.H., Kang, Y., Bae, E.B., et al., 2023d. Childhood abuse and cortical gray matter volume in patients with major depressive disorder. Psychiatry Res. 319, 114990.

Kimmel, M., Clive, M., Gispen, F., Guintivano, J., Brown, T., et al., 2016. Oxytocin receptor DNA methylation in postpartum depression. Psychoneuroendocrinology 69, 150–160.

King, S.E., Skinner, M.K., 2020. Epigenetic transgenerational inheritance of obesity susceptibility. Trends Endocrinol. Metab. 31, 478–494.

King, S., Holleran, L., Mothersill, D., Patlola, S., Rokita, K., et al., 2021. Early life adversity, functional connectivity and cognitive performance in schizophrenia: the mediating role of IL-6. Brain Behav. Immun. 98, 388–396.

Kingsbury, M.A., Bilbo, S.D., 2019. The inflammatory event of birth: how oxytocin signaling may guide the development of the brain and gastrointestinal system. Front. Neuroendocrinol. 55, 100794.

Kinnally, E.L., Capitanio, J.P., Leibel, R., Deng, L., LeDuc, C., et al., 2010. Epigenetic regulation of serotonin transporter expression and behavior in infant rhesus macaques. Genes Brain Behav. 9, 575–582.

Kinney, J.W., Bemiller, S.M., Murtishaw, A.S., Leisgang, A.M., Salazar, A.M., Lamb, B.T., 2018. Inflammation as a central mechanism in Alzheimer's disease. Alzheimers Dement. (N Y) 4, 575–590.

Kiraly, D.D., Walker, D.M., Calipari, E.S., Labonte, B., Issler, O., et al., 2016. Alterations of the host microbiome affect behavioral responses to cocaine. Sci. Rep. 6, 35455.

Kirk, J.A., Gebhart, D., Buckley, A.M., Lok, S., Scholl, D., et al., 2017. New class of precision antimicrobials redefines role of clostridium difficile S-layer in virulence and viability. Sci. Transl. Med. 9, eaah6813.

Kirmayer, L.J., Crafa, D., 2014. What kind of science for psychiatry? Front. Hum. Neurosci. 8, 435.

Kirsch, I., 2014. The emperor's new drugs: medication and placebo in the treatment of depression. Handb. Exp. Pharmacol. 225, 291–303.

Kishi, T., Yoshimura, R., Ikuta, T., Iwata, N., 2018. Brain-derived neurotrophic factor and major depressive disorder: evidence from meta-analyses. Front. Psychiatry 8, 308.

Kitazawa, M., Cheng, D., Tsukamoto, M.R., Koike, M.A., Wes, P.D., et al., 2011. Blocking IL-1 signaling rescues cognition, attenuates tau pathology, and restores neuronal β-catenin pathway function in an Alzheimer's disease model. J. Immunol. 187, 6539–6549.

Kitzbichler, M.G., Aruldass, A.R., Barker, G.J., Wood, T.C., Dowell, N.G., et al., 2021. Peripheral inflammation is associated with microstructural and functional connectivity changes in depression-related brain networks. Mol. Psychiatry 26, 7346–7354.

Kiyota, T., Machhi, J., Lu, Y., Dyavarshetty, B., Nemati, M., et al., 2018. Granulocyte-macrophage colony-stimulating factor neuroprotective activities in Alzheimer's disease mice. J. Neuroimmunol. 319, 80–92.

Kleeman, E.A., Gubert, C., Hannan, A.J., 2022. Transgenerational epigenetic impacts of parental infection on offspring health and disease susceptibility. Trends Genet. 38, 662–675.

Kleih, T.S., Entringer, S., Scholaske, L., Kathmann, N., DePunder, K., et al., 2022. Exposure to childhood maltreatment and systemic inflammation across pregnancy: the moderating role of depressive symptomatology. Brain Behav. Immun. 101, 397–409.

Klein, J., Wood, J., Jaycox, J., Dhodapkar, R.M., Lu, P., et al., 2023. Distinguishing features of Long COVID identified through immune profiling. Nature 23, 139–148.

Kleinberger, G., Yamanishi, Y., Suárez-Calvet, M., Czirr, E., Lohmann, E., et al., 2014. TREM2 mutations implicated in neurodegeneration impair cell surface transport and phagocytosis. Sci. Transl. Med. 6, 243ra86.

Kleyn, C.E., Schneider, L., Saraceno, R., Mantovani, C., Richards, H.L., et al., 2008. The effects of acute social stress on epidermal Langerhans' cell frequency and expression of cutaneous neuropeptides. J. Invest. Dermatol. 128, 1273–1279.

Klopack, E.T., Crimmins, E.M., Cole, S.W., Seeman, T.E., Carroll, J.E., 2022. Social stressors associated with age-related T lymphocyte percentages in older US adults: evidence from the US Health and Retirement Study. Proc. Natl. Acad. Sci. USA 119, e2202780119.

Klumpers, F., Kroes, M.C.W., Baas, J.M.P., Fernández, G., 2017. How human amygdala and bed nucleus of the stria terminalis may drive distinct defensive responses. J. Neurosci. 37, 9645–9656.

Klumpp, H., Fitzgerald, D.A., Cook, E., Shankman, S.A., Angstadt, M., et al., 2014. Serotonin transporter gene alters insula activity to threat in social anxiety disorder. Neuroreport 25, 926–931.

Klumpp, H., Roberts, J., Kapella, M.C., Kennedy, A.E., Kumar, A., et al., 2017. Subjective and objective sleep quality modulate emotion regulatory brain function in anxiety and depression. Depress. Anxiety 34, 651–660.

Knezevic, E., Nenic, K., Milanovic, V., Knezevic, N.N., 2023. The role of cortisol in chronic stress, neurodegenerative diseases, and psychological disorders. Cells 12, 2726.

Knight, L.K., Depue, B.E., 2019. New frontiers in anxiety research: the translational potential of the bed nucleus of the stria terminalis. Front. Psychiatry 10, 510.

Knight, L.K., Naaz, F., Stoica, T., Depue, B.E., Alzheimer's Disease Neuroimaging Initiative, 2017. Lifetime PTSD and geriatric depression symptomatology relate to altered dorsomedial frontal and amygdala morphometry. Psychiatry Res. 267, 59–68.

Knowles, K.A., Cox, R.C., Armstrong, T., Olatunji, B.O., 2019. Cognitive mechanisms of disgust in the development and maintenance of psychopathology: a qualitative review and synthesis. Clin. Psychol. Rev. 69, 30–50.

Koban, L., Kross, E., Woo, C.W., Ruzic, L., Wager, T.D., 2017. Frontal-brainstem pathways mediating placebo effects on social rejection. J. Neurosci. 37, 3621–3631.

Kobayashi, R., Ogawa, Y., Hashizume-Takizawa, T., Kurita-Ochiai, T., 2020. Oral bacteria affect the gut microbiome and intestinal immunity. Pathog. Dis. 78, ftaa024.

Koc, D., Tiemeier, H., Stricker, B.H., Muetzel, R.L., Hillegers, M., El Marroun, H., 2023. Prenatal antidepressant exposure and offspring brain morphologic trajectory. JAMA Psychiatry 80, 1208–1217.

Kodila, A., Franko, N., Sollner Dolenc, M., 2023. A review on immunomodulatory effects of BPA analogues. Arch. Toxicol. 97, 1831–1846.

Koenen, K.C., Ratanatharathorn, A., Ng, L., McLaughlin, K.A., Bromet, E.J., et al., 2017. Posttraumatic stress disorder in the World Mental Health Surveys. Psychol. Med. 47, 2260–2274.

Köhler, O., Benros, M.E., Nordentoft, M., Farkouh, M.E., Iyengar, R.L., et al., 2014. Effect of anti-inflammatory treatment on depression, depressive symptoms, and adverse effects: a systematic review and meta-analysis of randomized clinical trials. JAMA Psychiatry 71, 1381–1391.

Köhler, O., Krogh, J., Mors, O., Benros, M.E., 2016a. Inflammation in depression and the potential for anti-inflammatory treatment. Curr. Neuropharmacol. 14, 732–742.

Köhler, O., Petersen, L., Benros, M.E., Mors, O., Gasse, C., 2016b. Concomitant NSAID use during antipsychotic treatment and risk of 2-year relapse—a population-based study of 16,253 incident patients with schizophrenia. Expert Opin. Pharmacother. 17, 1055–1062.

Kohli, K., Pillarisetty, V.G., Kim, T.S., 2022. Key chemokines direct migration of immune cells in solid tumors. Cancer Gene Ther. 29, 10–21.

Kohman, R., Kusnecov, A.W., 2009. Stress, immunity and dendritic cells in cancer. In: Salter, R., Shurin, M. (Eds.), Dendritic Cells in Cancer. Springer, New York, NY.

Kohut, M.L., Cooper, M.M., Nickolaus, M.S., Russell, D.R., Cunnick, J.E., 2002. Exercise and psychosocial factors modulate immunity to influenza vaccine in elderly individuals. J. Gerontol. A Biol. Sci. Med. Sci. 57, M557–M562.

Kojima, Y., Volkmer, J.P., McKenna, K., Civelek, M., Lusis, A.J., 2016. CD47-blocking antibodies restore phagocytosis and prevent antherosclerosis. Nature 536, 86–90.

Kok, E.H., Paetau, A., Martiskainen, M., Lyytikäinen, L.P., Lehtimäki, T., et al., 2024. Accumulation of Lewy-related pathology starts in middle age: the Tampere sudden death study. Ann. Neurol.

Kolodziejczyk, A.A., Zheng, D., Elinav, E., 2019. Diet–microbiota interactions and personalized nutrition. Nat. Rev. Microbiol. 17, 742–753.

Komuro, J., Kaneko, H., Suzuki, Y., Okada, A., Fujiu, K., et al., 2024. Sex differences in the relationship between schizophrenia and the development of cardiovascular disease. J. Am. Heart Assoc. 13, e032625.

Konings, B., Villatoro, L., Van den Eynde, J., Barahona, G., Burns, R., et al., 2023. Gastrointestinal syndromes preceding a diagnosis of Parkinson's disease: testing Braak's hypothesis using a nationwide database for comparison with Alzheimer's disease and cerebrovascular diseases. Gut 72, 2103–2111.

Koo, J.W., Duman, R.S., 2008. IL-1beta is an essential mediator of the antineurogenic and anhedonic effects of stress. Proc. Natl. Acad. Sci. USA 105, 751–756.

Koo, J.W., Labonté, B., Engmann, O., Calipari, E.S., Juarez, B., et al., 2016. Essential role of mesolimbic brain-derived neurotrophic factor in chronic social stress–induced depressive behaviors. Biol. Psychiatry 80, 469–478.

Kootte, R.S., Levin, E., Salojärvi, J., Smits, L.P., Hartstra, A.V., et al., 2017. Improvement of insulin sensitivity after lean donor feces in metabolic syndrome is driven by baseline intestinal microbiota composition. Cell Metab. 26, 611–619.

Kop, W.J., Mommersteeg, P.M.C., 2014. Psychoneuroimmunological processes in coronary artery disease and heart failure. In: Kusnecov, A., Anisman, H. (Eds.), The Wiley-Blackwell Handbook of Psychoneuroimmunology. Chichester, UK, John Wiley & Sons Ltd.

Kordower, J.H., Chu, Y., Hauser, R.A., Freeman, T.B., Olanow, C.W., 2008. Lewy body-like pathology in long-term embryonic nigral transplants in Parkinson's disease. Nat. Med. 14, 504–506.

Koren, T., Rolls, A., 2022. Immunoception: defining brain-regulated immunity. Neuron 110 (21), 3425–3428.

Koren, T., Amer, M., Krot, M., Boshnak, N., Ben-Shaanan, T.L., et al., 2021a. Insular cortex neurons encode and retrieve specific immune responses. Cell 184, 5902–5915.

Koren, T., Yifa, R., Amer, M., Krot, M., Boshnak, N., et al., 2021b. Insular cortex neurons encode and retrieve specific immune responses. Cell 184, 5902–5915.e17.

Körner, R.W., Majjouti, M., Alcazar, M.A.A., Mahabir, E., 2020. Of mice and men: the coronavirus MHV and mouse models as a translational approach to understand SARS-CoV-2. Viruses 12, 880.

Korpela, K., Helve, O., Kolho, K.L., Saisto, T., Skogberg, K., et al., 2020. Maternal fecal microbiota transplantation in cesarean-born infants rapidly restores normal gut microbial development: a proof-of-concept study. Cell 183, 324–334.

Kosloski, L.M., Kosmacek, E.A., Olson, K.E., Mosley, R.L., Gendelman, H.E., 2013. GM-CSF induces neuroprotective and anti-inflammatory responses in 1-methyl-4-phenyl-1,2,3,6-tetrahydropyridine intoxicated mice. J. Neuroimmunol. 265, 1–10.

Kostuk, E.W., Cai, J., Iacovitti, L., 2019. Subregional differences in astrocytes underlie selective neurodegeneration or protection in Parkinson's disease models in culture. Glia 67, 1542–1557.

Kotsiri, I., Resta, P., Spyrantis, A., Panotopoulos, C., Chaniotis, D., et al., 2023. Viral infections and schizophrenia: a comprehensive review. Viruses 15, 1345.

Koutsouleris, N., Dwyer, D.B., Degenhardt, F., Maj, C., Urquijo-Castro, M.F., et al., 2021a. Multimodal machine learning workflows for prediction of psychosis in patients with clinical high-risk syndromes and recent-onset depression. JAMA Psychiatry 78, 95–209.

Koutsouleris, N., Worthington, M., Dwyer, D.B., Kambeitz-Ilankovic, L., Sanfelici, R., et al., 2021b. Toward generalizable and transdiagnostic tools for psychosis prediction: an independent validation and improvement of the NAPLS-2 risk calculator in the multisite PRONIA cohort. Biol. Psychiatry 90, 632–642.

Kovács, P., Pánczél, G., Balatoni, T., Liszkay, G., Gonda, X., et al., 2015. Social support decreases depressogenic effect of low-dose interferon alpha treatment in melanoma patients. J. Psychosom. Res. 78, 579–584.

Kovacs, D., Eszlari, N., Petschner, P., Pap, D., Vas, S., et al., 2016. Effects of IL1B single nucleotide polymorphisms on depressive and anxiety symptoms are determined by severity and type of life stress. Brain Behav. Immun. 56, 96–104.

Kovács, T., Mikó, E., Ujlaki, G., Sári, Z., Bai, P., 2020. The microbiome as a component of the tumor microenvironment. Adv. Exp. Med. Biol. 1225, 137–153.

Koyama, Y., Nawa, N., Yamaoka, Y., Nishimura, H., Sonoda, S., et al., 2021. Interplay between social isolation and loneliness and chronic systemic inflammation during the COVID-19 pandemic in Japan: results from U-CORONA study. Brain Behav. Immun. 94, 51–59.

Kozina, E., Byrne, M., Smeyne, R.J., 2022. Mutant LRRK2 in lymphocytes regulates neurodegeneration via IL-6 in an inflammatory model of Parkinson's disease. NPJ Parkinson's Dis. 8, 24.

Koziol, M.J., Bradshaw, C.R., Allen, G.E., Costa, A.S.H., Frezza, C., et al., 2016. Identification of methylated deoxyadenosines in vertebrates reveals diversity in DNA modifications. Nat. Struct. Mol. Biol. 23, 24–30.

Kozlovsky, N., Zohar, J., Kaplan, Z., Cohen, H., 2012. Microinfusion of a corticotrophin-releasing hormone receptor 1 antisense oligodeoxynucleotide into the dorsal hippocampus attenuates stress responses at specific times after stress exposure. J. Neuroendocrinol. 24, 489–503.

Kozłowska, E., Agier, J., Wysokiński, A., Łucka, A., Sobierajska, K., Brzezińska-Błaszczyk, E., 2019. The expression of toll-like receptors in peripheral blood mononuclear cells is altered in schizophrenia. Psychiatry Res. 272, 540–550.

Krajewska, M., Witkowska-Sędek, E., Rumińska, M., Stelmaszczyk-Emmel, A., Sobol, M., et al., 2022. Vitamin D effects on selected anti-inflammatory and pro-inflammatory markers of obesity-related chronic inflammation. Front. Endocrinol. (Lausanne) 13, 920340.

Krakowiak, P., Goines, P.E., Tancredi, D.J., Ashwood, P., Hansen, R.L., Hertz-Picciotto, I., Van de Water, J., 2017a. Neonatal cytokine profiles associated with autism spectrum disorder. Biol. Psychiatry 81 (5), 442–451.

Krakowiak, P., Goines, P.E., Tancredi, D.J., Ashwood, P., Hansen, R.L., Hertz-Picciotto, I., et al., 2017b. Neonatal cytokine profiles associated with autism spectrum disorder. Biol. Psychiatry 81, 442–451.

Krantz, D.S., Shank, L.M., Goodie, J.L., 2022. Post-traumatic stress disorder (PTSD) as a systemic disorder: pathways to cardiovascular disease. Health Psychol. 41, 651–662.

Krasemann, S., Madore, C., Cialic, R., Baufeld, C., Calcagno, N., et al., 2017. The TREM2-APOE pathway drives the transcriptional phenotype of dysfunctional microglia in neurodegenerative diseases. Immunity 47, 566–581.e9.

Kraus, M.R., Al-Taie, O., Schäfer, A., Pfersdorff, M., Lesch, K.P., et al., 2007. Serotonin-1A receptor gene HTR1A variation predicts interferon-induced depression in chronic hepatitis C. Gastroenterology 132, 1279–1286.

Krause, A.J., Simon, E.B., Mander, B.A., Greer, S.M., Saletin, J.M., Goldstein-Piekarski, A.N., Walker, M.P., 2017. The sleep-deprived human brain. Nat. Rev. Neurosci. 18, 404–418.

Kraynak, T.E., Marsland, A.L., Gianaros, P.J., 2018. Neural mechanisms linking emotion with cardiovascular disease. Curr. Cardiol. Rep. 20, 128.

Kredlow, A., Fenster, R.J., Laurent, E.S., Ressler, K.J., Phelps, E.A., 2022. Prefrontal cortex, amygdala, and threat processing: implications for PTSD. Neuropsychopharmacology 47, 247–259.

Krishnan, V., Nestler, E.J., 2011. Animal models of depression: molecular perspectives. Curr. Top. Behav. Neurosci. 7, 121–147.

Kristof, Z., Gal, Z., Torok, D., Eszlari, N., Sutori, S., et al., 2023. Variation along P2RX7 interacts with early traumas on severity of anxiety suggesting a role for neuroinflammation. Sci. Rep. 13, 7757.

Kroenke, K., Alford, D.P., Argoff, C., et al., 2019. Challenges with implementing the Centers for Disease Control and Prevention opioid guideline: a Consensus Panel report. Pain Med. 20, 724–735.

Krontira, A.C., Cruceanu, C., Binder, E.B., 2020. Glucocorticoids as mediators of adverse outcomes of prenatal stress. Trends Neurosci. 43, 394–405.

Kross, E., Berman, M.G., Mischel, W., Smith, E.E., Wager, T.D., 2011. Social rejection shares somatosensory representations with physical pain. Proc. Natl. Acad. Sci. USA 108, 6270–6275.

Krueger, J.M., Opp, M.R., 2016. Sleep and microbes. Int. Rev. Neurobiol. 131, 207–225.

Krystal, J.H., Kaye, A.P., Jefferson, S., Girgenti, M.J., Wilkinson, S.T., et al., 2023. Ketamine and the neurobiology of depression: toward next-generation rapid-acting antidepressant treatments. Proc. Natl. Acad. Sci. USA 120, e2305772120.

Kuang, Z., Wang, Y., Li, Y., Ye, C., Ruhn, K.A., et al., 2019. The intestinal microbiota programs diurnal rhythms in host metabolism through histone deacetylase 3. Science 365, 1428–1434.

Kübler, R., Ormel, P.R., Sommer, I.E.C., Kahn, R.S., de Witte, L.D., 2023. Gene expression profiling of monocytes in recent-onset schizophrenia. Brain Behav. Immun. 111, 334–342.

Kucukkarapinar, M., Yay-Pence, A., Yildiz, Y., Buyukkoruk, M., Yaz-Aydin, G., et al., 2022. Psychological outcomes of COVID-19 survivors at sixth months after diagnose: the role of kynurenine pathway metabolites in depression, anxiety, and stress. J. Neural Transm. 129, 1077–1089.

Kudielka, B.M., Kirschbaum, C., 2005. Sex differences in HPA axis responses to stress: a review. Biol. Psychol. 69, 113–132.

Kuebler, U., Zuccarella-Hackl, C., Arpagaus, A., Wolf, J.M., Farahmand, F., et al., 2015. Stress-induced modulation of NF-κB activation, inflammation-associated gene expression, and cytokine levels in blood of healthy men. Brain Behav. Immun. 46, 87–95.

Kulcsarova, K., Bang, C., Berg, D., Schaeffer, E., 2023. Pesticides and the microbiome-gut-brain axis: convergent pathways in the pathogenesis of Parkinson's disease. J. Parkinsons Dis. 13, 1079–1106.

Kumar, D.K., Choi, S.H., Washicosky, K.J., Eimer, W.A., Tucker, S., 2016. Amyloid-β peptide protects against microbial infection in mouse and worm models of Alzheimer's disease. Sci. Transl. Med. 8, 340ra72.

Kumari, P., Rothan, H.A., Natekar, J.P., Stone, S., Pathak, H., et al., 2021. Neuroinvasion and encephalitis following intranasal inoculation of SARS-CoV-2 in K18-hACE2 mice. Viruses 13 (1), 132.

Kummer, K.K., Zeidler, M., Kalpachidou, T., Kress, M., 2021. Role of IL-6 in the regulation of neuronal development, survival and function. Cytokine 144, 155582.

Kundakovic, M., Jaric, I., 2017. The epigenetic link between prenatal adverse environments and neurodevelopmental disorders. Genes (Basel) 8, 104.

Kundakovic, M., Gudsnuk, K., Herbstman, J.B., Tang, D., Perera, F.P., et al., 2015. DNA methylation of BDNF as a biomarker of early-life adversity. Proc. Natl. Acad. Sci. USA 112, 6807–6813.

Kuner, R., Kuner, T., 2021. Cellular circuits in the brain and their modulation in acute and chronic pain. Physiol. Rev. 101, 213–258.

Kunze, R., Fischer, S., Marti, H.H., Preissner, K.T., 2023. Brain alarm by self-extracellular nucleic acids: from neuroinflammation to neurodegeneration. J. Biomed. Sci. 30, 64.

Kupper, N., Denollet, J., 2018. Type D personality as a risk factor in coronary heart disease: a review of current evidence. Curr. Cardiol. Rep. 20, 104.

Kusakabe, T., Lin, W.Y., Cheong, J.G., Singh, G., Ravishankar, A., et al., 2023. Fungal microbiota sustains lasting immune activation of neutrophils and their progenitors in severe COVID-19. Nat. Immunol. 24, 1879–1889.

Kusnecov, A.W., Rabin, B.S., 1993. Inescapable footshock exposure differentially alters antigen- and mitogen-stimulated spleen cell proliferation in rats. J. Neuroimmunol. 44, 33–42.

Kusnecov, A.W., Rossi-George, A., 2001. Potentiation of interleukin-1beta adjuvant effects on the humoral immune response to antigen in adrenalectomized mice. Neuroimmunomodulation 9, 109–118.

Kusnecov, A.W., Rossi-George, A., 2002. Stressor-induced modulation of immune function: a review of acute, chronic effects in animals. Acta Neuropsychiatr. 14, 279–291.

Kwon, D., 2023. Your brain could be controlling how sick you get - and how you recover. Nature 614, 613–615.

Kwon, Y.C., Chun, S., Kim, K., Mak, A., 2019. Update on the genetics of systemic lupus erythematosus: genome-wide association studies and beyond. Cells 8, 1180.

Kyriakoulis, P., Kyrios, M., 2023. Biological and cognitive theories explaining panic disorder: a narrative review. Front. Psychiatry 14, 957515.

Kyrozis, A., Ghika, A., Stathopoulos, P., Vassilopoulos, D., Trichopoulos, D., et al., 2013. Dietary and life-style variables in relation to incidence of Parkinson's disease in Greece. Eur. J. Epidemiol. 28, 67–77.

Labaka, A., Gómez-Lázaro, E., Vegas, O., Pérez-Tejada, J., Arregi, A., Garmendia, L., 2017. Reduced hippocampal IL-10 expression, altered monoaminergic activity and anxiety and depressive-like behavior in female mice subjected to chronic social instability stress. Behav. Brain Res. 335, 8–18.

Labonte, B., Azoulay, N., Yerko, V., Turecki, G., Brunet, A., 2014. Epigenetic modulation of glucocorticoid receptors in posttraumatic stress disorder. Transl. Psychiatry 4, e368.

Labonté, B., Engmann, O., Purushothaman, I., Menard, C., Wang, J., et al., 2017. Sex-specific transcriptional signatures in human depression. Nat. Med. 23, 1102–1111.

Labonté, B., Suderman, M., Maussion, G., Lopez, J.P., Navarro-Sánchez, L., et al., 2013. Genome-wide methylation changes in the brains of suicide completers. Am. J. Psychiatry 170, 511–520.

Labouesse, M.A., Dong, E., Grayson, D.R., Guidotti, A., Meyer, U., 2015. Maternal immune activation induces GAD1 and GAD2 promoter remodeling in the offspring prefrontal cortex. Epigenetics 10, 1143–1155.

Labrakakis, C., 2023. The role of the insular cortex in pain. Int. J. Mol. Sci. 24, 5736.

Labrecque, N., Whitfield, L.S., Obst, R., Waltzinger, C., Benoist, C., et al., 2001. How much TCR does a T cell need? Immunity 15, 71–82.

Labuschagne, I., Phan, K.L., Wood, A., Angstadt, M., Chua, P., et al., 2010. Oxytocin attenuates amygdala reactivity to fear in generalized social anxiety disorder. Neuropsychopharmacology 35, 2403–2413.

Lacagnina, M.J., Heijnen, C.J., Watkins, L.R., Grace, P.M., 2021. Autoimmune regulation of chronic pain. Pain Rep. 6, e905.

Lacagnina, M.J., Watkins, L.R., Grace, P.M., 2018. Toll-like receptors and their role in persistent pain. Pharmacol. Ther. 184, 145–158.

Lach, G., Schellekens, H., Dinan, T.G., Cryan, J.F., 2018. Anxiety, depression, and the microbiome: a role for gut peptides. Neurotherapeutics 15, 36–59.

Lafortune, S., Laplante, D.P., Elgbeili, G., Li, X., Lebel, S., Dagenais, C., King, S., 2021. Effect of natural disaster-related prenatal maternal stress on child development and health: a meta-analytic review. Int. J. Environ. Res. Public Health 18, 8332.

Lafuse, W.P., Gearinger, R., Fisher, S., Nealer, C., Mackos, A.R., et al., 2017. Exposure to a social stressor induces translocation of commensal lactobacilli to the spleen and priming of the innate immune system. J. Immunol. 198, 2383–2393.

Lafuse, W.P., Wu, Q., Kumar, N., Saljoughian, N., Sunkum, S., et al., 2022. Psychological stress creates an immune suppressive environment in the lung that increases susceptibility of aged mice to Mycobacterium tuberculosis infection. Front. Cell. Infect. Microbiol. 12, 990402.

Lagunas-Rangel, F.A., 2020. Neutrophil-to-lymphocyte ratio and lymphocyte-to-C-reactive protein ratio in patients with severe coronavirus disease 2019 (COVID-19): a meta-analysis. J. Med. Virol. 92, 1733–1734.

Laham, B.J., Gould, E., 2022. How stress influences the dynamic plasticity of the brain's extracellular matrix. Front. Cell. Neurosci. 15, 814287.

Lai, N.S., Yu, H.C., Huang Tseng, H.Y., Hsu, C.W., et al., 2021. Increased serum levels of brain-derived neurotrophic factor contribute to inflammatory responses in patients with rheumatoid arthritis. Int. J. Mol. Sci. 22, 1841.

Laine, M.A., Sokolowska, E., Dudek, M., Callan, S.A., Hyytia, P., et al., 2017. Brain activation induced by chronic psychosocial stress in mice. Sci. Rep. 7, 15061.

Laing, R.D., 1960. The Divided Self: An Existential Study in Sanity and Madness. Penguin, Harmondsworth.

Lalonde, C.S., Mekawi, Y., Ethun, K.F., Beurel, E., Gould, F., et al., 2021. Sex differences in peritraumatic inflammatory cytokines and steroid hormones contribute to prospective risk for nonremitting posttraumatic stress disorder. Chronic Stress 5. 24705470211032208.

Lamers, F., Milaneschi, Y., Smit, J.H., Schoevers, R.A., Wittenberg, G., Penninx, B.W.J.H., 2019. Longitudinal association between depression and inflammatory markers: results from the netherlands study of depression and anxiety. Biol. Psychiatry 85, 829–837.

Lamers, F., Milaneschi, Y., Vinkers, C.H., Schoevers, R.A., Giltay, E.J., Penninx, B.W.J.H., 2020. Depression profilers and immunometabolic dysregulation: longitudinal results from the NESDA study. Brain Behav. Immun. 88, 174–183.

Lancet, 2017. Life, death, and disability in 2016. Lancet 390, 1083.

Landolfo, E., Cutuli, D., Decandia, D., Balsamo, F., Petrosini, L., Gelfo, F., 2023. Environmental enrichment protects against neurotoxic effects of lipopolysaccharide: a comprehensive overview. Int. J. Mol. Sci. 24, 5404.

Lane, M.M., Gamage, E., Du, S., Ashtree, D.N., McGuinness, A.J., et al., 2024. Ultra-processed food exposure and adverse health outcomes: umbrella review of epidemiological meta-analyses. BMJ 384, e077310.

Lane, M.M., Lotfaliany, M., Forbes, M., Loughman, A., Rocks, T., et al., 2022. Higher ultra-processed food consumption is associated with greater high-sensitivity c-reactive protein concentration in adults: cross-sectional results from the melbourne collaborative cohort study. Nutrients 14, 3309.

Lane, M.M., Lotfaliany, M., Hodge, A.M., O'Neil, A., Travica, N., et al., 2023. High ultra-processed food consumption is associated with elevated psychological distress as an indicator of depression in adults from the Melbourne Collaborative Cohort Study. J. Affect. Disord. 335, 57–66.

Lang, A.E., Lozano, A.M., 1998. Parkinson's disease: first of two parts. N. Engl. J. Med. 339, 1044–1053.

Langham, J., Gurol-Urganci, I., Muller, P., Webster, K., Tassie, E., et al., 2023. Obstetric and neonatal outcomes in pregnant women with and without a history of specialist mental health care: a national population-based cohort study using linked routinely collected data in England. Lancet Psychiatry 10, 748–759.

Langston, P.K., Sun, Y., Ryback, B.A., Mueller, A.L., Spiegelman, B.M., et al., 2023. Regulatory T cells shield muscle mitochondria from interferon-γ-mediated damage to promote the beneficial effects of exercise. Sci. Immunol. 8, eadi5377.

Langston, J.W., Forno, L.S., Tetrud, J., Reeves, A.G., Kaplan, J.A., et al., 1999. Evidence of active nerve cell degeneration in the substantia nigra of humans years after 1-methyl-4-phenyl-1,2,3,6-tetrahydropyridine exposure. Ann. Neurol. 46, 598–605.

Lanoiselée, H.-M., Nicolas, G., Wallon, D., Rovelet-Lecrux, A., Lacour, M., et al., 2017. APP, PSEN1, and PSEN2 mutations in early-onset Alzheimer disease: a genetic screening study of familial and sporadic cases. PLoS Med. 14, e1002270.

Lanz, T.A., Reinhart, V., Sheehan, M.J., Rizzo, S.J.S., Bove, S.E., et al., 2019. Postmortem transcriptional profiling reveals widespread increase in inflammation in schizophrenia: a comparison of prefrontal cortex, striatum, and hippocampus among matched tetrads of controls with subjects diagnosed with schizophrenia, bipolar or major depressive disorder. Transl. Psychiatry 9, 151.

Lanz, T.V., Brewer, R.C., Ho, P.P., Moon, J.S., Jude, K.M., et al., 2022. Clonally expanded B cells in multiple sclerosis bind EBV EBNA1 and GlialCAM. Nature 603, 321–327.

Lao, X.Q., Liu, X., Deng, H.B., Chan, T.C., Ho, K.F., et al., 2018. Sleep quality, sleep duration, and the risk of coronary heart disease: a prospective cohort study with 60,586 adults. J. Clin. Sleep Med. 14, 109–117.

Larvin, H., Kang, J., Aggarwal, V.R., Pavitt, S., Wu, J., 2021. Risk of incident cardiovascular disease in people with periodontal disease: a systematic review and meta-analysis. Clin. Exp. Dent. Res. 7, 109–122.

Lasselin, J., Elsenbruch, S., Lekander, M., Axelsson, J., Karshikoff, B., et al., 2016. Mood disturbance during experimental endotoxemia: predictors of state anxiety as a psychological component of sickness behavior. Brain Behav. Immun. 57, 30–37.

Lateef, A., Petri, M., 2017. Systemic lupus erythematosus and pregnancy. Rheum. Dis. Clin. N. Am. 43, 215–226.

Lau, R.I., Su, Q., Lau, I.S., Ching, J.Y., Wong, M.C., et al., 2023. A synbiotic preparation (SIM01) for post-acute COVID-19 syndrome in Hong Kong (RECOVERY): a randomised, double-blind, placebo-controlled trial. Lancet Infect. Dis. Click or tap here to enter text.

Laudani, S., Torrisi, S.A., Alboni, S., Bastiaanssen, T.F.S., Benatti, C., et al., 2023. Gut microbiota alterations promote traumatic stress susceptibility associated with p-cresol-induced dopaminergic dysfunctions. Brain Behav. Immun. 107, 385–396.

Lauden, A., Geishin, A., Merzon, E., Korobeinikov, A., Green, I., et al., 2021. Higher rates of allergies, autoimmune diseases and low-grade inflammation markers in treatment-resistant major depression. Brain Behav. Immun. Health 16, 100313.

Laukova, M., Alaluf, L.G., Serova, L.I., Arango, V., Sabban, E.L., 2014. Early intervention with intranasal NPY prevents single prolonged stress-triggered impairments in hypothalamus and ventral hippocampus in male rats. Endocrinology 155, 3920–3933.

Laurikainen, H., Vuorela, A., Toivonen, A., Reinert-Hartwall, L., Trontti, K., et al., 2020. Elevated serum chemokine CCL22 levels in first-episode psychosis: associations with symptoms, peripheral immune state and in vivo brain glial cell function. Transl. Psychiatry 10, 94.

Lawrence, J.A., Kawachi, I., White, K., Bassett, M.T., Priest, N., et al., 2022. A systematic review and meta-analysis of the Everyday Discrimination Scale and biomarker outcomes. Psychoneuroendocrinology 142, 105772.

Lawrence, J.M., Schardien, K., Wigdahl, B., Nonnemacher, M.R., 2023. Roles of neuropathology-associated reactive astrocytes: a systematic review. Acta Neuropathol. Commun. 11, 42.

Lazaridis, I., Tzortzi, O., Weglage, M., Märtin, A., Xuan, Y., et al., 2019. A hypothalamus-habenula circuit controls aversion. Mol. Psychiatry 24, 1351–1368.

Lazarou, M., 2015. Keeping the immune system in check: a role for mitophagy. Immunol. Cell Biol. 93, 3–10.

Lazarus, R.S., Folkman, S., 1984. Stress, Appraisal, and Coping. Springer, New York.

Le Bastard, Q., Al-Ghalith, G.A., Grégoire, M., Chapalet, G., Javaudin, F., et al., 2018. Systematic review: human gut dysbiosis induced by non-antibiotic prescription medications. Aliment. Pharmacol. Ther. 47 (3), 332–345.

Lechner, L., Opitz, R., Silver, M.J., Krabusch, P.M., Prentice, A.M., et al., 2023. Early-set POMC methylation variability is accompanied by increased risk for obesity and is addressable by MC4R agonist treatment. Sci. Transl. Med. 15, eadg1659.

Leclercq, S., Forsythe, P., Bienenstock, J., 2016. Posttraumatic stress disorder: does the gut microbiome hold the key? Can. J. Psychiatry 61, 204–213.

LeDoux, J.E., 2012. Evolution of human emotion: a view through fear. Prog. Brain Res. 195, 431–442.

LeDoux, J.E., 2014. Coming to terms with fear. Proc. Natl. Acad. Sci. USA 111, 2871–2878.

LeDoux, J.E., Pine, D.S., 2016. Using neuroscience to help understand fear and anxiety: a two-system framework. Am. J. Psychiatry 173, 1083–1093.

LeDoux, J.E., Brown, R., 2017. A higher-order theory of emotional consciousness. Proc. Natl. Acad. Sci. USA 114, E2016–E2025.

Lee, B., Sur, B., Yeom, M., Shim, I., Lee, H., et al., 2016a. Effects of systemic administration of ibuprofen on stress response in a rat model of post-traumatic stress disorder. Korean J. Physiol. Pharmacol. 20, 357–366.

Lee, B.K., Magnusson, C., Gardner, R.M., Blomström, Å., Newschaffer, C.J., et al., 2015. Maternal hospitalization with infection during pregnancy and risk of autism spectrum disorders. Brain Behav. Immun. 44, 100–105.

Lee, D.H., Lee, J.Y., Hong, D.Y., Lee, E.C., Park, S.W., et al., 2022a. Neuroinflammation in post-traumatic stress disorder. Biomedicines 10, 953.

Lee, D.J., Schnitzlein, C.W., Wolf, J.P., Vythilingam, M., Rasmusson, A. M., et al., 2016b. Psychotherapy versus pharmacotherapy for post-traumatic stress disorder: systemic review and meta-analyses to determine first-line treatments. Depress. Anxiety 33, 792–806.

Lee, D.Y., Im, S.C., Kang, N.Y., Kim, K., 2023a. Analysis of effect of intensity of aerobic exercise on cognitive and motor functions and neurotrophic factor expression patterns in an Alzheimer's disease rat model. J. Pers. Med. 13, 1622.

Lee, H., James, W.S., Cowley, S.A., 2017a. LRRK2 in peripheral and central nervous system innate immunity: Its link to Parkinson's disease. Biochem. Soc. Trans. 45, 131–139.

Lee, H.S., Min, D., Baik, S.Y., Kwon, A., Jin, M.J., Lee, S.H., 2022b. Association between dissociative symptoms and morning cortisol levels in patients with post-traumatic stress disorder. Clin. Psychopharmacol. Neurosci. 20, 292–299.

Lee, J.J., Piras, E., Tamburini, S., Bu, K., Wallach, D.S., et al., 2023b. Gut and oral microbiome modulate molecular and clinical markers of schizophrenia-related symptoms: a transdiagnostic, multilevel pilot study. Psychiatry Res. 326, 115279.

Lee, J.L.C., Nader, K., Schiller, D., 2017b. An update on memory reconsolidation updating. Trends Cogn. Sci. 21, 531–545.

Lee, K.M., Hunger, J.M., Tomiyama, A.J., 2021a. Weight stigma and health behaviors: evidence from the Eating in America Study. Int. J. Obes. 45, 1499–1509.

Lee, M., Krishnamurthy, J., Susi, A., Sullivan, C., Gorman, G.H., Hisle-Gorman, E., Nylund, C.M., 2017c. Association of autism spectrum disorders and inflammatory bowel disease. J. Autism Dev. Disord. 48 (5), 1523–1529.

Lee, M.B., Hill C.M., Bitto A., Kaeberlein M., 2021c. Antiaging diets: separating fact from fiction. Science, 374, eabe7365.

Lee, M.R., Abshire, K.M., Farokhnia, M., Akhlaghi, F., Leggio, L., 2021b. Effect of oral alcohol administration on plasma cytokine concentrations in heavy drinking individuals. Drug Alcohol Depend. 225, 108771.

Lee, P.C., Raaschou-Nielsen, O., Lill, C.M., Bertram, L., Sinsheimer, J.S., et al., 2016c. Gene-environment interactions linking air pollution and inflammation in Parkinson's disease. Environ. Res. 151, 713–720.

Lee, S., Yoon, S., Namgung, E., Kim, T.D., Hong, H., et al., 2023c. Distinctively different human neurobiological responses after trauma exposure and implications for posttraumatic stress disorder subtyping. Mol. Psychiatry 28, 2964–2974.

Lee, S.E., Shim, S.R., Youn, J.H., Han, H.W., 2023d. COVID-19 vaccination is not associated with psychiatric adverse events: a meta-analysis. Vaccines (Basel) 11, 194.

Lee, Y.H., Song, G.G., 2022. Mendelian randomization research on the relationship between rheumatoid arthritis and systemic lupus erythematosus and the risk of autistic spectrum disorder. J. Rheum. Dis. 29, 46–51.

Lehouritis, P., Cummins, J., Stanton, M., et al., 2015. Local bacteria affect the efficacy of chemotherapeutic drugs. Sci. Rep. 5, 14554.

Lehrner, A., Yehuda, R., 2018. Trauma across generations and paths to adaptation and resilience. Psychol. Trauma 10, 22–29.

Lei, D., Li, L., Li, L., Suo, X., Huang, X., et al., 2015. Microstructural abnormalities in children with post-traumatic stress disorder: a diffusion tensor imaging study at 3.0T. Sci. Rep. 5, 8933.

Lei, Y., Chen, C.J., Yan, X.X., Li, Z., Deng, X.H., 2017. Early-life lipopolysaccharide exposure potentiates forebrain expression of NLRP3 inflammasome proteins and anxiety-like behavior in adolescent rats. Brain Res. pii: S0006-8993(17)30262-7.

Lei, Y.M., Chen, L., Wang, Y., Stefka, A.T., Molinero, L.L., et al., 2016. The composition of the microbiota modulates allograft rejection. J. Clin. Invest. 126, 2736–2744.

Leigh, S.J., Uhlig, F., Wilmes, L., Sanchez-Diaz, P., Gheorghe, C.E., et al., 2023. The impact of acute and chronic stress on gastrointestinal physiology and function: a microbiota-gut-brain axis perspective. J. Physiol. 601, 4491–4538.

Leinenga, G., Koh, W.K., Götz, J., 2021. A comparative study of the effects of aducanumab and scanning ultrasound on amyloid plaques and behavior in the APP23 mouse model of Alzheimer disease. Alzheimers Res. Ther. 13, 76.

Leite Dantas, R., Freff, J., Ambrée, O., Beins, E.C., Forstner, A.J., et al., 2021. Dendritic cells: neglected modulators of peripheral immune responses and neuroinflammation in mood disorders? Cells 10, 941.

Leiter, O., Brici, D., Fletcher, S.J., Yong, X.L.H., Widagdo, J., et al., 2023. Platelet-derived exerkine CXCL4/platelet factor 4 rejuvenates hippocampal neurogenesis and restores cognitive function in aged mice. Nat. Commun. 14, 4375.

Lemos, J.C., Wanat, M.J., Smith, J.S., Reyes, B.A., Hollon, N.G., et al., 2012. Severe stress switches CRF action in the nucleus accumbens from appetitive to aversive. Nature 490, 402–406.

Lempka, S.F., Malone, D.A., Hu, B., Baker, K.B., Wyant, A., et al., 2017. Randomized clinical trial of deep brain stimulation for poststroke pain. Ann. Neurol. 81, 653–663.

Leng, Y., Hu, Q., Ling, Q., Yao, X., Liu, M., et al., 2023. Periodontal disease is associated with the risk of cardiovascular disease independent of sex: a meta-analysis. Front. Cardiovasc. Med. 10, 1114927.

Lenharo, M., 2024. Obesity drugs have another superpower: taming inflammation. Nature 626, 246.

Lennox, B.R., Palmer-Cooper, E.C., Pollak, T., Hainsworth, J., Marks, J., et al., 2017. Prevalence and clinical characteristics of serum neuronal cell surface antibodies in first-episode psychosis: a case-control study. Lancet Psychiatry 4, 42–48.

Lerner, M.J., Montada, L., 1998. An overview: advances in belief in a just world theory and methods. In: Montada, L., Lerner, M.J. (Eds.), Responses to Victimizations and Belief in a Just World. Plenum Press, New York, pp. 1–7.

Leschik, J., Gentile, A., Cicek, C., Péron, S., Tevosian, M., et al., 2022. Brain-derived neurotrophic factor expression in serotonergic neurons improves stress resilience and promotes adult hippocampal neurogenesis. Prog. Neurobiol. 217, 102333.

Leslie, D.L., Kobre, R.A., Richmand, B.J., Aktan Guloksuz, S., Leckman, J.F., 2017. Temporal association of certain neuropsychiatric disorders following vaccination of children and adolescents: a pilot case-control study. Front. Psychiatry 8, 3.

Lesuis, S.L., Lucassen, P.J., Krugers, H.J., 2019. Early life stress impairs fear memory and synaptic plasticity; a potential role for GluN2B. Neuropharmacology 149, 195–203.

Letchumanan, G., Abdullah, N., Marlini, M., Baharom, N., Lawley, B., et al., 2022. Gut microbiota composition in prediabetes and newly diagnosed type 2 diabetes: a systematic review of observational studies. Front. Cell. Infect. Microbiol. 12, 943427.

Leucht, S., Leucht, C., Huhn, M., Chaimani, A., Mavridis, D., et al., 2017. Sixty years of placebo-controlled antipsychotic drug trials in acute schizophrenia: systematic review, bayesian meta-analysis, and meta-regression of efficacy predictors. Am. J. Psychiatry 174, 927–942.

Levada, O.A., Troyan, A.S., 2017. Insulin-like growth factor-1: a possible marker for emotional and cognitive disturbances, and treatment effectiveness in major depressive disorder. Ann. General Psychiatry 16, 38.

Levert-Levitt, E., Shapira, G., Sragovich, S., Shomron, N., Lam, J.C.K., et al., 2022. Oral microbiota signatures in post-traumatic stress disorder (PTSD) veterans. Mol. Psychiatry 27, 4590–4598.

Levine, K.S., Leonard, H.L., Blauwendraat, C., Iwaki, H., Johnson, N., et al., 2023. Virus exposure and neurodegenerative disease risk across national biobanks. Neuron 111, 1086–1093.

Levine, S.Z., Kodesh, A., Viktorin, A., Smith, L., Uher, R., et al., 2018. Association of maternal use of folic acid and multivitamin supplements in the periods before and during pregnancy with the risk of autism spectrum disorder in offspring. JAMA Psychiatry 75, 176–184.

Levine, S.Z., Levav, I., Goldberg, Y., Pugachova, I., Becher, Y., Yoffe, R., 2016. Exposure to genocide and the risk of schizophrenia: a population-based study. Psychol. Med. 246, 855–863.

Levis, S.C., Birnie, M.T., Bolton, J.L., Perrone, C.R., Montesinos, J.S., et al., 2022. Enduring disruption of reward and stress circuit activities by early-life adversity in male rats. Transl. Psychiatry 12, 251.

Levkovitz, Y., Fenchel, D., Kaplan, Z., Zohar, J., Cohen, H., 2015. Early post-stressor intervention with minocycline, a second-generation tetracycline, attenuates post-traumatic stress response in an animal model of PTSD. Eur. Neuropsychopharmacol. 25), 124–132.

Lewis, C., Roberts, N.P., Andrew, M., Starling, E., Bisson, J.I., 2020. Psychological therapies for post-traumatic stress disorder in adults: systematic review and meta-analysis. Eur. J. Psychotraumatol. 11, 1729633.

Lewis, C.R., Tafur, J., Spencer, S., Green, J.M., Harrison, C., et al., 2023a. Pilot study suggests DNA methylation of the glucocorticoid receptor gene (NR3C1) is associated with MDMA-assisted therapy treatment response for severe PTSD. Front. Psychiatry 14, 959590.

Lewis, J.R., 2020. What is driving the decline in people's willingness to take the COVID-19 vaccine in the United States? JAMA Health Forum 1, e201393.

Lewis, P.A., Manzoni, C., 2012. LRRK2 and human disease: a complicated question or a question of complexes? Sci. Signal. 5, pe2.

Lewis, V., Bonniwell, E.M., Lanham, J.K., Ghaffari, A., Sheshbaradaran, H., et al., 2023b. A non-hallucinogenic LSD analog with therapeutic potential for mood disorders. Cell Rep. 42, 112203.

Ley, R.E., Bäckhed, F., Turnbaugh, P., Lozupone, C.A., Knight, R.D., et al., 2005. Obesity alters gut microbial ecology. Proc. Natl. Acad. Sci. USA, 11070–11075.

Leyns, C.E.G., Ulrich, J.D., Finn, M.B., Stewart, F.R., Koscal, L.J., et al., 2017. TREM2 deficiency attenuates neuroinflammation and protects against neurodegeneration in a mouse model of tauopathy. Proc. Natl. Acad. Sci. USA 114, 11524–11529.

Li, D., Li, Y., Yang, S., Lu, J., Jin, X., Wu, M., 2022a. Diet-gut microbiota-epigenetics in metabolic diseases: from mechanisms to therapeutics. Biomed. Pharmacother. 153, 113290.

Li, H., Xia, N., Hasselwander, S., Daiber, A., 2019a. Resveratrol and vascular function. Int. J. Mol. Sci. 20, 2155.

Li Hi Shing, S., Chipika, R.H., Finegan, E., Murray, D., Hardiman, O., Bede, P., 2019. Post-polio syndrome: more than just a lower motor neuron disease. Front. Neurol. 10, 773.

Li, J.M., Yu, R., Zhang, L.P., Wen, S.Y., Wang, S.J., et al., 2019b. Dietary fructose-induced gut dysbiosis promotes mouse hippocampal neuroinflammation: a benefit of short-chain fatty acids. Microbiome 7, 98.

Li, C., Liu, Y., Yin, S., Lu, C., Liu, D., et al., 2015a. Long-term effects of early adolescent stress: dysregulation of hypothalamic-pituitary-adrenal axis and central corticotropin releasing factor receptor 1 expression in adult male rats. Behav. Brain Res. 288, 39–49.

Li, D., Zhong, J., Zhang, Q., Zhang, J., 2023a. Effects of anti-inflammatory therapies on glycemic control in type 2 diabetes mellitus. Front. Immunol. 14, 1125116.

Li, F., Ke, H., Wang, S., Mao, W., Fu, C., et al., 2023b. Leaky gut plays a critical role in the pathophysiology of autism in mice by activating the lipopolysaccharide-mediated toll-like receptor 4-myeloid differentiation factor 88-nuclear factor kappa b signaling pathway. Neurosci. Bull. 39, 911–928.

Li, H., Achour, I., Bastarache, L., Berghout, J., Gardeux, V., et al., 2016. Integrative genomics analyses unveil downstream biological effectors of disease-specific polymorphisms buried in intergenic regions. NPJ Genom. Med. 1, 16006.

Li, H., Namburi, P., Olson, J.M., Borio, M., Lemieux, M.E., et al., 2022b. Neurotensin orchestrates valence assignment in the amygdala. Nature 608, 586–592.

Li, H.Y., Zhou, D.D., Gan, R.Y., Huang, S.Y., Zhao, C.N., et al., 2021a. Effects and mechanisms of probiotics, prebiotics, synbiotics, and postbiotics on metabolic diseases targeting gut microbiota: a narrative review. Nutrients 13, 3211.

Li, J., Liu, S., Li, S., Feng, R., Na, L., et al., 2017a. Prenatal exposure to famine and the development of hyperglycemia and type 2 diabetes in adulthood across consecutive generations: a population-based cohort study of families in Suihua, China. Am. J. Clin. Nutr. 105, 221–227.

Li, J., Olsen, J., Vestergaard, M., Kristensen, J.K., Olsen, J., 2012. Prenatal exposure to bereavement and type-2 diabetes: a Danish longitudinal population based study. PLoS One 7, e43508.

Li, J., Ruggiero-Ruff, R.E., He, Y., Qiu, X., Lainez, N., et al., 2023c. Sexual dimorphism in obesity is governed by RELMα regulation of adipose macrophages and eosinophils. elife 12, e86001.

Li, J., Tong, L., Schock, B.C., Ji, L.L., 2023d. Post-traumatic stress disorder: focus on neuroinflammation. Mol. Neurobiol. 60, 3963–3978.

Li, L., Ni, L., Heary, R.F., Elkabes, S., 2020a. Astroglial TLR9 antagonism promotes chemotaxis and alternative activation of macrophages via modulation of astrocyte-derived signals: implications for spinal cord injury. J. Neuroinflammation 17 (1), 73.

Li, M., Peng, H., Duan, G., Wang, J., Yu, Z., et al., 2022c. Older age and depressive state are risk factors for re-positivity with SARS-CoV-2 Omicron variant. Front. Public Health 10, 1014470.

Li, M.X., Zheng, H.L., Luo, Y., He, J.G., Wang, W., et al., 2018. Gene deficiency and pharmacological inhibition of caspase-1 confers resilience to chronic social defeat stress via regulating the stability of surface AMPARs. Mol. Psychiatry 23, 556–568.

Li, P., Bandyopadhyay, G., Lagakos, W.S., Lagakos, W.S., Talukdar, S., et al., 2015b. LTB4 promotes insulin resistance in obese mice by acting on macrophages, hepatocytes and myocytes. Nat. Med. 21, 239–247.

Li, Q., Park, K., Xia, Y., Matsumoto, M., Qi, W., et al., 2017b. Regulation of macrophage apoptosis and atherosclerosis by lipid-induced PKCδ isoform activation. Circ. Res. 121, 1153–1167.

Li, S., Fang, Y., Zhang, Y., Song, M., Zhang, X., Ding, X., et al., 2022d. Microglial NLRP3 inflammasome activates neurotoxic astrocytes in depression-like mice. Cell Rep. 41, 111532.

Li, T., He, X., Thomas, J.M., Yang, D., Zhong, S., et al., 2015c. A novel GTP-binding inhibitor, FX2149, attenuates LRRK2 toxicity in Parkinson's disease models. PLoS One 10, e0122461.

Li, W., Yi, G., Chen, Z., Dai, X., Wu, J., et al., 2021b. Is job strain associated with a higher risk of type 2 diabetes mellitus? A systematic review and meta-analysis of prospective cohort studies. Scand. J. Work Environ. Health 47, 249–257.

Li, X., Chauhan, A., Sheikh, A.M., Patil, S., Chauhan, V., et al., 2009. Elevated immune response in the brain of autistic patients. J. Neuroimmunol. 207, 111–116.

Li, X., Li, M., Tian, L., Chen, J., Liu, R., Ning, B., 2020b. Reactive astrogliosis: implications in spinal cord injury progression and therapy. Oxidative Med. Cell. Longev. 2020, 9494352.

Li, Y., Ji, M., Yang, J., 2022e. Current understanding of long-term cognitive impairment after sepsis. Front. Immunol. 13, 855006.

Li, Y., Jinxiang, T., Shu, Y., Yadong, P., Ying, L., et al., 2022f. Childhood trauma and the plasma levels of IL-6, TNF-α are risk factors for major depressive disorder and schizophrenia in adolescents: a cross-sectional and case-control study. J. Affect. Disord. 305, 227–232.

Li, Y., Shen, R., Wen, G., Ding, R., Du, A., et al., 2017c. Effects of ketamine on levels of inflammatory cytokines IL-6, IL-1β, and TNF-α in the hippocampus of mice following acute or chronic administration. Front. Pharmacol. 8, 139.

Li, S., Hafeez, A., Noorulla, F., Geng, X., Shao, G., et al., 2017d. Preconditioning in neuroprotection: from hypoxia to ischemia. Prog. Neurobiol. 157, 79–91.

Li, Y., Yue, Y., Chen, S., Jiang, W., Xu, Z., et al., 2022g. Combined serum IL-6, C-reactive protein, and cortisol may distinguish patients with anhedonia in major depressive disorder. Front. Mol. Neurosci. 15, 935031.

Li, Z.A., Cai, Y., Taylor, R.L., Eisenstein, S.A., Barch, D.M., et al., 2023e. Associations between socioeconomic status, obesity, cognition, and white matter microstructure in children. JAMA Netw. Open 6, e2320276.

Li, Z., Ma, L., Kulesskaya, N., Võikar, V., Tian, L., 2014. Microglia are polarized to M1 type in high-anxiety inbred mice in response to lipopolysaccharide challenge. Brain Behav. Immun. 38, 237–248.

Li, Z., Yan, H., Zhang, X., Shah, S., Yang, G., 2021c. Air pollution interacts with genetic risk to influence cortical networks implicated in depression. Proc. Natl. Acad. Sci. USA 118, e2109310118.

Lian, H., Litvinchuk, A., Chiang, A.C.A., Aithmitti, N., Jankowsky, J.L., et al., 2016. Astrocyte-microglia cross talk through complement activation modulates amyloid pathology in mouse models of Alzheimer's disease. J. Neurosci. 36, 577–589.

Liang, S., Wang, T., Hu, X., Luo, J., Li, W., et al., 2015. Administration of *Lactobacillus helveticus* NS8 improves behavioral, cognitive, and biochemical aberrations caused by chronic restraint stress. Neuroscience 310, 561–577.

Liang, Y.Y., Feng, H., Chen, Y., Jin, X., Xue, H., et al., 2023. Joint association of physical activity and sleep duration with risk of all-cause and cause-specific mortality: a population-based cohort study using accelerometry. Eur. J. Prev. Cardiol. 30, 832–843.

Liao, S., von der Weid, P.Y., 2015. Lymphatic system: an active pathway for immune protection. Semin. Cell Dev. Biol. 38, 83–89.

Liao, X., Liu, Y., Fu, X., Li, Y., 2020. Postmortem studies of neuroinflammation in autism spectrum disorder: a systematic review. Mol. Neurobiol. 57, 3424–3438.

Liberzon, I., Abelson, J.L., 2016. Context processing and the neurobiology of post-traumatic stress disorder. Neuron 92, 14–30.

Licznerski, P., Duric, V., Banasr, M., Alavian, K.N., Ota, K.T., et al., 2015. Decreased SGK1 expression and function contributes to behavioral deficits induced by traumatic stress. PLoS Biol. 13, e1002282.

Liddelow, S.A., Guttenplan, K.A., Clarke, L.E., Bennett, F.C., Bohlen, C. J., et al., 2017. Neurotoxic reactive astrocytes are induced by activated microglia. Nature 541, 481–487.

Lieberman, M.D., Eisenberger, N.I., 2015. The dorsal anterior cingulate cortex is selective for pain: results from large-scale reverse inference. Proc. Natl. Acad. Sci. USA 112, 15250–15255.

Liguori, C., Chiaravalloti, A., Nuccetelli, M., Izzi, F., Sancesario, G., et al., 2017. Hypothalamic dysfunction is related to sleep impairment and CSF biomarkers in Alzheimer's disease. J. Neurol. 264, 2215–2223.

Liguori, C., Romigi, A., Nuccetelli, M., Zannino, S., Sancesario, G., et al., 2014. Orexinergic system dysregulation, sleep impairment, and cognitive decline in Alzheimer disease. JAMA Neurol. 71, 1498–1505.

Lii, T.R., Smith, A.E., Flohr, J.R., Okada, R.L., Nyongesa, C.A., et al., 2023. Randomized trial of ketamine masked by surgical anesthesia in patients with depression. Nat. Mental Health 1, 876–886.

Likhtik, E., Paz, R., 2015. Amygdala-prefrontal interactions in (mal)adaptive learning. Trends Neurosci. 38, 158–166.

Lim, J., Altman, M.D., Baker, J., Brubaker, J.D., Chen, H., et al., 2015. Identification of N-(1H-pyrazol-4-yl)carboxamide inhibitors of interleukin-1 receptor associated kinase 4: bicyclic core modifications. Bioorg. Med. Chem. Lett. 25, 5384–5388.

Lima, L.V., Abner, T.S.S., Sluka, K.A., 2017. Does exercise increase or decrease pain? Central mechanisms underlying these two phenomena. J. Physiol. 595, 4141–4150.

Limoges, A., Yarur, H.E., Tejeda, H.A., 2022. Dynorphin/kappa opioid receptor system regulation on amygdaloid circuitry: implications for neuropsychiatric disorders. Front. Syst. Neurosci. 16, 963691.

Lin, C., Chen, K., Yu, J., Feng, W., Fu, W., et al., 2021. Relationship between TNF-α levels and psychiatric symptoms in first-episode drug-naïve patients with schizophrenia before and after risperidone treatment and in chronic patients. BMC Psychiatry 21, 561.

Lin, C.H., Lin, H.I., Chen, M.L., Lai, T.T., Cao, L.P., et al., 2016. Lovastatin protects neurite degeneration in LRRK2-G2019S parkinsonism through activating the Akt/Nrf pathway and inhibiting GSK3β activity. Hum. Mol. Genet. 25, 1965–1978.

Lin, C.H., Lin, H.Y., Ho, E.P., Ke, Y.C., Cheng, M.F., et al., 2022. Mild chronic colitis triggers parkinsonism in LRRK2 mutant mice through activating TNF-α pathway. Mov. Disord. 37, 745–757.

Lin, C.H., Shao, L., Zhang, Y.M., Tu, Y.J., Zhang, Y., et al., 2020. An evaluation of liraglutide including its efficacy and safety for the treatment of obesity. Expert. Opin. Pharmacother. 21, 275–285.

Lin, J., Epel, E., 2022. Stress and telomere shortening: insights from cellular mechanisms. Ageing Res. Rev. 73, 101507.

Lin, J., Sun, J., Wang, S., Milush, J.M., Baker, C.A., et al., 2018. In vitro proinflammatory gene expression predicts in vivo telomere shortening: a preliminary study. Psychoneuroendocrinology 96, 179–187.

Lin, L., Jung, K.M., Lee, H.L., Le, J., Colleluori, G., et al., 2023. Adolescent exposure to low-dose THC disrupts energy balance and adipose organ homeostasis in adulthood. Cell Metab. S1550-413100179-1.

Lin, P.Y., Chang, C.H., Chong, M.F., Chen, H., Su, K.P., 2017. Polyunsaturated fatty acids in perinatal depression: a systematic review and meta-analysis. Biol. Psychiatry 82, 560–569.

Lindqvist, D., Dhabhar, F.S., Mellon, S.H., Yehuda, R., Grenon, S.M., et al., 2017. Increased proinflammatory milieu in combat related PTSD - A new cohort replication study. Brain Behav. Immun. 59, 260–264.

Lindvall, O., Björklund, A., 1989. Transplantation strategies in the treatment of Parkinson's disease: experimental basis and clinical trials. Acta Neurol. Scand. 126, 197–210.

Lindvall, O., Kokaia, Z., 2015. Neurogenesis following stroke affecting the adult brain. Cold Spring Harb. Perspect. Biol. 7, 1–19.

Ling, C., Rönn, T., 2019. Epigenetics in human obesity and type 2 diabetes. Cell Metab. 29, 1028–1044.

Ling, C., Bacos, K., Rönn, T., 2022. Epigenetics of type 2 diabetes mellitus and weight change—a tool for precision medicine? Nat. Rev. Endocrinol. 18, 433–448.

Linnstaedt, S.D., Bortsov, A.V., Soward, A.C., Swor, R., Peak, D.A., et al., 2016. CRHBP polymorphisms predict chronic pain development following motor vehicle collision. Pain 157, 273–279.

Lionnet, A., Leclair-Visonneau, L., Neunlist, M., Murayama, S., Takao, M., et al., 2017. Does Parkinson's disease start in the gut? Acta Neuropathol. 135, 1–12.

Lipoldová, M., Demant, P., 2021. Gene-specific sex effects on susceptibility to infectious diseases. Front. Immunol. 12, 712688.

Lippi, G., Mattiuzzi, C., Sanchis-Gomar, F., 2020. Updated overview on interplay between physical exercise, neurotrophins, and cognitive function in humans. J. Sport Health Sci. 9 (1), 74–81.

Lisboa, S.F., Gomes, F.V., Guimaraes, F.S., Campos, A.C., 2016. Microglial cells as a link between cannabinoids and the immune hypothesis of psychiatric disorders. Front. Neurol. 7, 5.

Lisnevskaia, L., Murphy, G., Isenberg, D., 2014. Systemic lupus erythematosus. Lancet 384, 1878–1888.

Litteljohn, D., Mangano, E.N., Hayley, S., 2008. Cyclooxygenase-2 deficiency modifies the neurochemical effects, motor impairment and comorbid anxiety provoked by paraquat administration in mice. Eur. J. Neurosci. 28, 707–716.

Litteljohn, D., Nelson, E., Hayley, S., 2014. IFN-γ differentially modulates memory-related processes under basal and chronic stressor conditions. Front. Cell. Neurosci. 8, 391.

Litteljohn, D., Rudyk, C., Dwyer, Z., Farmer, K., Fortin, T., et al., 2017. The impact of murine LRRK2 G2019S transgene overexpression on acute responses to inflammatory challenge. Brain Behav. Immun. 67, 246–256.

Littlefield, A.M., Setti, S.E., Prister, C., Kohman, R.A., 2015. Voluntary exercise attenuates LPS-induced reductions in neurogenesis and increases microglia expression of proneurogenic phenotype in aged mice. J. Neuroinflammation 12, 138.

Liu, C.C., Liu, C.C., Kanekiyo, T., Xu, H., Bu, G., 2013. Apolipoprotein E and Alzheimer disease: risk, mechanisms and therapy. Nat. Rev. Neurol. 9, 106–118.

Liu, C.H., Hua, N., Yang, H.Y., 2021a. Alterations in peripheral c-reactive protein and inflammatory cytokine levels in patients with panic disorder: a systematic review and meta-analysis. Neuropsychiatr. Dis. Treat. 17, 3539–3558.

Liu, C.H., Zhang, G.Z., Li, B., Li, M., Woelfer, M., Walter, M., Wang, L., 2019. Role of inflammation in depression relapse. J. Neuroinflammation 16, 90.

Liu, F.C., Huang, W.Y., Lin, T.Y., Shen, C.H., Chou, Y.C., et al., 2015a. Inverse association of parkinson disease with systemic lupus erythematosus: a nationwide population-based study. Medicine (Baltimore) 94, e2097.

Liu, G., Zhang, L., Fan, Y., Ji, W., 2023a. The pathogenesis in Alzheimer's disease: TREM2 as a potential target. J. Integr. Neurosci. 22, 150.

Liu, L., Cui, Y., Chang, Y.Z., Yu, P., 2023b. Ferroptosis-related factors in the substantia nigra are associated with Parkinson's disease. Sci. Rep. 13, 15365.

Liu, L., Wang, H., Chen, X., Zhang, Y., Zhang, H., Xie, P., 2023c. Gut microbiota and its metabolites in depression: from pathogenesis to treatment. EBioMedicine 90, 104527.

Liu, L.Y., Coe, C.L., Swenson, C.A., Kelly, E.A., Kita, H., et al., 2002. School examinations enhance airway inflammation to antigen challenge. Am. J. Respir. Crit. Care Med. 165, 1062–1067.

Liu, R.S., Aiello, A.E., Mensah, F.K., Gasser, C.E., Rueb, K., et al., 2017a. Socioeconomic status in childhood and C reactive protein in adulthood: a systematic review and meta-analysis. J. Epidemiol. Community Health 71, 817–826.

Liu, S.H., Du, Y., Chen, L., Cheng, Y., 2022a. Glial cell abnormalities in major psychiatric diseases: a systematic review of postmortem brain studies. Mol. Neurobiol. 59, 1665–1692.

Liu, T., Gu, X., Li, L.X., Li, M., Li, B., et al., 2020. Microbial and metabolomic profiles in correlation with depression and anxiety comorbidities in diarrhoea-predominant IBS patients. BMC Microbiol. 20, 168.

Liu, X., Agerbo, E., Ingstrup, K.G., Musliner, K., Meltzer-Brody, S., et al., 2017b. Antidepressant use during pregnancy and psychiatric disorders in offspring: Danish nationwide register based cohort study. BMJ 358, j3668.

Liu, X., Yang, W., Zhu, C., Sun, S., Wu, S., et al., 2022b. Toll-like receptors and their role in neuropathic pain and migraine. Mol. Brain 15 (1), 73.

Liu, Y., Dai, M., 2020. Trimethylamine N-oxide generated by the gut microbiota is associated with vascular inflammation: new insights into atherosclerosis. Mediat. Inflamm. 2020, 4634172.

Liu, Y., Li, H.T., Zhou, S.J., Zhou, H.H., Xiong, Y., et al., 2023d. Effects of vaginal seeding on gut microbiota, body mass index, and allergy risks in infants born through cesarean delivery: a randomized clinical trial. Am. J. Obstet. Gynecol. MFM 5, 100793.

Liu, Y., Lin, W., Xu, P., Zhang, D., Luo, Y., 2015b. Neural basis of disgust perception in racial prejudice. Hum. Brain Mapp. 36 (12), 5275–5286.

Liu, Z., Mao, X., Dan, Z., Pei, Y., Xu, R., et al., 2021b. Gene variations in autism spectrum disorder are associated with alteration of gut microbiota, metabolites and cytokines. Gut Microbes 13, 1–16.

Liu, Z., Zhu, F., Wang, G., Xiao, Z., Tang, J., et al., 2007. Association study of corticotropin-releasing hormone receptor1 gene polymorphisms and antidepressant response in major depressive disorders. Neurosci. Lett. 414, 155–158.

Livanos, A.E., Greiner, T.U., Vangay, P., Pathmasiri, W., Stewart, D., et al., 2016. Antibiotic-mediated gut microbiome perturbation accelerates development of type 1 diabetes in mice. Nat. Microbiol. 1, 16140.

Livneh, Y., Sugden, A.U., Madara, J.C., Essner, R.A., Flores, V.I., et al., 2020. Estimation of current and future physiological states in insular cortex. Neuron 105, 1094–1111.e10.

Lizano, P., Lutz, O., Xu, Y., Rubin, L.H., Paskowitz, L., et al., 2021. Multivariate relationships between peripheral inflammatory marker subtypes and cognitive and brain structural measures in psychosis. Mol. Psychiatry 26, 3430–3443.

Llewelyn, M.J., Fitzpatrick, J.M., Darwin, E., Tonkin-Crine, S., Gorton, C., et al., 2017. The antibiotic course has had its day. BMJ 358, j3418.

Lloyd, C.E., Dyer, P.H., Lancashire, R.J., Harris, T., Daniels, J.E., et al., 1999. Association between stress and glycemic control in adults with type 1 (insulin-dependent) diabetes. Diabetes Care 22, 1278–1283.

Loane, D.J., Kumar, A., 2016. Microglia in the TBI brain: the good, the bad, and the dysregulated. Exp. Neurol. 275, 316–327.

Lobo, J.J., Ayoub, L.J., Moayedi, M., Linnstaedt, S.D., 2022. Hippocampal volume, FKBP5 genetic risk alleles, and childhood trauma interact to increase vulnerability to chronic multisite musculoskeletal pain. Sci. Rep. 12, 6511.

Lodhi, I.J., Dean, J.M., He, A., Park, H., Tan, M., et al., 2017. PexRAP inhibits PRDM16-mediated thermogenic gene expression. Cell Rep. 20, 2766–2774.

Logue, M.W., Miller, M.W., Wolf, E.J., Huber, B.R., Morrison, F.G., et al., 2020. An epigenome-wide association study of posttraumatic stress disorder in US veterans implicates several new DNA methylation loci. Clin. Epigenetics 12, 46.

Loh, M.K., Stickling, C., Schrank, S., Hanshaw, M., Ritger, A.C., et al., 2023. Liposaccharide-induced sustained mild inflammation fragments social behavior and alters basolateral amygdala activity. Psychopharmacology 240, 647–671.

Lombardi, A.L., Manfredi, L., Conversi, D., 2022. How does IL-6 change after combined treatment in MDD patients? A systematic review. Brain Behav. Immun. Health 27, 100579.

Lombardo, G., Nettis, M.A., Hastings, C., Zajkowska, Z., Mariani, N., et al., 2022. Sex differences in a double-blind randomized clinical trial with minocycline in treatment-resistant depressed patients: CRP and IL-6 as sex-specific predictors of treatment response. Brain Behav. Immun. Health 26, 100561.

Lombardo, M.V., Moon, H.M., Su, J., Palmer, T.D., Courchesne, E., Pramparo, T., 2018. Maternal immune activation dysregulation of the fetal brain transcriptome and relevance to the pathophysiology of autism spectrum disorder. Mol. Psychiatry 23, 1001–1013.

Long, J.E., Drayson, M.T., Taylor, A.E., Toellner, K.M., Lord, J.M., Phillips, A.C., 2016. Morning vaccination enhances antibody response over afternoon vaccination: a cluster-randomised trial. Vaccine 34, 2679–2685.

Long, J.Y., Wang, X.J., Li, X.Y., Kong, X.H., Yang, G., et al., 2022. Spinal microglia and astrocytes: two key players in chronic visceral pain pathogenesis. Neurochem. Res. 47, 545–551.

Longhi, G., van Sinderen, D., Ventura, M., Turroni, F., 2020. Microbiota and cancer: the emerging beneficial role of bifidobacteria in cancer immunotherapy. Front. Microbiol. 11, 575072.

Longo, V.D., Di Tano, M., Mattson, M.P., Guidi, N., 2021. Intermittent and periodic fasting, longevity and disease. Nat. Aging 1, 47–59.

Loos, R.J.F., Yeo, G.S.H., 2022. The genetics of obesity: from discovery to biology. Nat. Rev. Genet. 23, 120–133.

Lopatkin, J.A., Meredith, H.R., Srimani, J.K., Pfeiffer, C., Durrett, R., et al., 2017. Persistence and reversal of plasmid-mediated antibiotic resistance. Nat. Commun. 8, 1689.

López-Otín, C., Blasco, M.A., Partridge, L., Serrano, M., Kroemer, G., 2023. Hallmarks of aging: an expanding universe. Cell 186, 243–278.

Lopez-Tello, J., Yong, H.E., Sandovici, I., Dowsett, G.K., Christoforou, E. R., et al., 2023. Fetal manipulation of maternal metabolism is a critical function of the imprinted Igf2 gene. Cell Metab. 35, 1195–1208.

Lorkiewicz, P., Waszkiewicz, N., 2021. Biomarkers of post-COVID depression. J. Clin. Med. 10, 4142.

Losenkov, I.S., Mulder, N.J.V., Levchuk, L.A., Vyalova, N.M., Loonen, A.J.M., et al., 2020. Association between BDNF gene variant Rs6265 and the severity of depression in antidepressant treatment-free depressed patients. Front. Psychiatry 11, 38.

Losin, E.A.R., Anderson, S.R., Wager, T.D., 2017. Feelings of clinician-patient similarity and trust influence pain: evidence from simulated clinical interactions. J. Pain 18, 787–799.

Lou, Q.Y., Li, Z., Teng, Y., Xie, Q.M., Zhang, M., et al., 2021. Associations of FKBP4 and FKBP5 gene polymorphisms with disease susceptibility, glucocorticoid efficacy, anxiety, depression, and health-related quality of life in systemic lupus erythematosus patients. Clin. Rheumatol. 40, 167–179.

Lourbopoulos, A., Ertürk, A., Hellal, F., 2015. Microglia in action: how aging and injury can change the brain's guardians. Front. Cell. Neurosci. 9, 54.

Louveau, A., Plog, B.A., Antila, S., Alitalo, K., Nedergaard, M., et al., 2017. Understanding the functions and relationships of the glymphatic system and meningeal lymphatics. J. Clin. Invest. 127, 3210–3219.

Lovell, B., Wetherell, M.A., 2011. The cost of caregiving: endocrine and immune implications in elderly and non-elderly caregivers. Neurosci. Biobehav. Rev. 35, 1342–1352.

Lowry, C.A., Smith, D.G., Siebler, P.H., Schmidt, D., Stamper, C.E., et al., 2016. The microbiota, immunoregulation, and mental health: implications for public health. Curr. Environ. Health Rep. 3, 270–286.

Lu, C.Y., Penfold, R.B., Wallace, J., Lupton, C., Libby, A.M., Soumerai, S. B., 2020. Increases in suicide deaths among adolescents and young adults following US food and drug administration antidepressant boxed warnings and declines in depression care. Psychiatr. Res. Clin. Pract. 2, 43–52.

Lu, C., Zhang, Y., Li, B., Zhao, Z., Huang, C., et al., 2022a. Interaction effect of prenatal and postnatal exposure to ambient air pollution and temperature on childhood asthma. Environ. Int. 167, 107456.

Lu, J.Y.D., Su, P., Barber, J.E.M., Nash, J.E., Le, A.D., et al., 2017. The neuroprotective effect of nicotine in Parkinson's disease models is associated with inhibiting PARP-1 and caspase-3 cleavage. PeerJ 5, e3933.

Lu, S.S.M., Mohammed, Z., Häggström, C., Myte, R., Lindquist, E., et al., 2022b. Antibiotics use and subsequent risk of colorectal cancer: a swedish nationwide population-based study. J. Natl. Cancer Inst. 114, 38–46.

Lu, Y., Jin, X., Feng, L.W., Tang, C., Neo, M., Ho, R.C., 2022c. Effects of illness perception on negative emotions and fatigue in chronic rheumatic diseases: rumination as a possible mediator. World J. Clin. Cases 10 (34), 12515–12531.

Luby, J.L., Baram, T.Z., Rogers, C.E., Barch, D.M., 2020. Neurodevelopmental optimization after early-life adversity: cross-species studies to elucidate sensitive periods and brain mechanisms to inform early intervention. Trends Neurosci. 43, 744–751.

Luby, J.L., Belden, A., Harms, M.P., Tillman, R., Barch, D.M., 2016. Preschool is a sensitive period for the influence of maternal support on the trajectory of hippocampal development. Proc. Natl. Acad. Sci. USA 113, 5742–5747.

Lucas, K., Morris, G., Anderson, G., Maes, M., 2015. The Toll-like receptor radical cycle pathway: a new drug target in immune-related chronic fatigue. CNS Neurol. Disord. Drug Targets 14, 838–854.

Lucidi, L., Pettorruso, M., Vellante, F., Di Carlo, F., Ceci, F., et al., 2021. Gut microbiota and bipolar disorder: an overview on a novel biomarker for diagnosis and treatment. Int. J. Mol. Sci. 22, 3723.

Lucido, M.J., Bekhbat, M., Goldsmith, D.R., Treadway, M.T., Haroon, E., et al., 2021. Aiding and abetting anhedonia: impact of inflammation on the brain and pharmacological implications. Pharmacol. Rev. 73, 1084–1117.

Luck, H., Tsai, S., Chung, J., Clemente-Casares, X., Ghazarian, M., et al., 2015. Regulation of obesity related insulin resistance with gut anti-inflammatory agents. Cell Metab. 21, 527–542.

Luczynski, P., McVey Neufeld, K.A., Oriach, C.S., Clarke, G., Dinan, T.G., Cryan, J.F., 2016a. Growing up in a bubble: using germ-free animals to assess the influence of the gut microbiota on brain and behavior. Int. J. Neuropsychopharmacol. 19, pyw020.

Luczynski, P., Tramullas, M., Viola, M., Shanahan, F., Clarke, G., et al., 2017. Microbiota regulates visceral pain in the mouse. elife. pii: e25887.

Luczynski, P., Whelan, S.O., O'Sullivan, C., Clarke, G., Shanahan, F., et al., 2016b. Adult microbiota-deficient mice have distinct dendritic morphology changes: differential effects in the amygdala and hippocampus. Eur. J. Neurosci. 44, 2654–2666.

Lue, L.F., Schmitz, C., Walker, D.G., 2015. What happens to microglial TREM2 in Alzheimer's disease: immunoregulatory turned into immunopathogenic? Neuroscience 302, 138–150.

Lueken, U., Zierhut, K.C., Hahn, T., Straube, B., Kircher, T., et al., 2016. Neurobiological markers predicting treatment response in anxiety disorders: a systematic review and implications for clinical application. Neurosci. Biobehav. Rev. 66, 143–162.

Łukasik, J., Patro-Gołąb, B., Horvath, A., Baron, R., Szajewska, H., et al., 2019. Early life exposure to antibiotics and autism spectrum disorders: a systematic review. J. Autism Dev. Disord. 49, 3866–3876.

Lukiw, W.J., Pogue, A., Hill, J.M., 2022. SARS-CoV-2 infectivity and neurological targets in the brain. Cell. Mol. Neurobiol. 42 (1), 217–224.

Lukkes, J.L., Meda, S., Thompson, B.S., Freund, N., Andersen, S.L., 2017. Early life stress and later peer distress on depressive behavior in adolescent female rats: effects of a novel intervention on GABA and D2 receptors. Behav. Brain Res. 330, 37–45.

Lund-Sørensen, H., Benros, M.E., Madsen, T., Sørensen, H.J., Eaton, W.W., 2016. A nationwide cohort study of the association between hospitalization with infection and risk of death by suicide. JAMA Psychiatry 73, 912–919.

Lund, F.E., 2008. Cytokine-producing B lymphocytes-key regulators of immunity. Curr. Opin. Immunol. 20, 332–338.

Luning Prak, E.T., Brooks, T., Makhoul, W., Beer, J.C., Zhao, L., et al., 2022. No increase in inflammation in late-life major depression screened to exclude physical illness. Transl. Psychiatry 12, 118.

Luo, Y., Kusay, A.S., Jiang, T., Chebib, M., Balle, T., 2021. Delta-containing GABAA receptors in pain management: promising targets for novel analgesics. Neuropharmacology 195, 108675.

Luo, Y., Zeng, B., Zeng, L., Du, X., Li, B., et al., 2018. Gut microbiota regulates mouse behaviors through glucocorticoid receptor pathway genes in the hippocampus. Transl. Psychiatry 8, 187.

Luo, Z., Liu, L.F., Jiang, Y.N., Tang, L.P., Li, W., et al., 2020. Novel insights into stress-induced susceptibility to influenza: corticosterone impacts interferon-β responses by Mfn2-mediated ubiquitin degradation of MAVS. Signal Transduct. Target. Ther. 5, 202.

Luongo, L., Maione, S., Di Marzo, V., 2014. Endocannabinoids and neuropathic pain: focus on neuron-glia and endocannabinoid-neurotrophin interactions. Eur. J. Neurosci. 39, 401–408.

Luoni, A., Berry, A., Raggi, C., Bellisario, V., Cirulli, F., Riva, M.A., 2016. Sex-specific effects of prenatal stress on Bdnf expression in response to an acute challenge in rats: a role for Gadd45β. Mol. Neurobiol. 53, 7037–7047.

Lupien, S.J., Ouellet-Morin, I., Herba, C.M., Juster, R., McEwen, B.S., 2016. From vulnerability to neurotoxicity: a developmental approach to the effects of stress on the brain and behavior. Epigenetics Neuroendocrinol. 1, 3–48.

Lupinsky, D., Moquin, L., Gratton, A., 2017. Interhemispheric regulation of the rat medial prefrontal cortical glutamate stress response: role of local GABA- and dopamine-sensitive mechanisms. Psychopharmacology 234, 353–363.

Lupo, G.F.D., Rocchetti, G., Lucini, L., Lorusso, L., Manara, E., et al., 2021. Potential role of microbiome in chronic fatigue syndrome/myalgic encephalomyelits (CFS/ME). Sci. Rep. 11, 7043.

Lurie, I., Yang, Y.X., Haynes, K., Mamtani, R., Boursi, B., 2015. Antibiotic exposure and the risk for depression, anxiety, or psychosis: a nested case-control study. J. Clin. Psychiatry 76, 1522–1528.

Luscher, B., Maguire, J.L., Rudolph, U., Sibille, E., 2023. GABAA receptors as targets for treating affective and cognitive symptoms of depression. Trends Pharmacol. Sci. 44, 586–600.

Lutz, P.E., Tanti, A., Gasecka, A., Barnett-Burns, S., Kim, J.J., et al., 2017. Association of a history of child abuse with impaired myelination in the anterior cingulate cortex: convergent epigenetic, transcriptional, and morphological evidence. Am. J. Psychiatry 174, 1185–1194.

Lwin, M.N., Serhal, L., Holroyd, C., Edwards, C.J., 2020. Rheumatoid arthritis: the impact of mental health on disease: a narrative review. Rheumatol. Ther. 7, 457–471.

Ly, M., Yu, G.Z., Mian, A., Cramer, A., Meysami, S., et al., 2023. Neuroinflammation: a modifiable pathway linking obesity, Alzheimer's disease, and depression. Am. J. Geriatr. Psychiatry 31, 853–866.

Lyman, M., Forsberg, K., Sexton, D.J., Chow, N.A., Lockhart, S.R., et al., 2023. Worsening spread of Candida auris in the United States, 2019 to 2021. Ann. Intern. Med. 176, 489–495.

Lynall, M.E., Kigar, S.L., Lehmann, M.L., DePuyt, A.E., Tuong, Z.K., et al., 2021. B-cells are abnormal in psychosocial stress and regulate meningeal myeloid cell activation. Brain Behav. Immun. 97, 226–238.

Lynch, C.M., Cowan, C.S., Bastiaanssen, T.F., Moloney, G.M., Theune, N., et al., 2023. Critical windows of early-life microbiota disruption on behaviour, neuroimmune function, and neurodevelopment. Brain Behav. Immun. 108, 309–327.

Lynch, J.F., Winiecki, P., Gilman, T.L., Adkins, J.M., Jasnow, A.M., 2017. Hippocampal GABAB(1a) receptors constrain generalized contextual fear. Neuropsychopharmacology 42, 914–924.

Lynch, M.A., 2015. Neuroinflammatory changes negatively impact on LTP: a focus on IL-1β. Brain Res. 1621, 197–204.

Lynch, S.V., Pedersen, O., 2016. The human intestinal microbiome in health and disease. N. Engl. J. Med. 375, 2369–2379.

Lynn, D.J., Pulendran, B., 2017. The potential of the microbiota to influence vaccine responses. J. Leukoc. Biol. pii: jlb.5MR0617-216R.

Lyra e Silva, N.M., Gonçalves, R.A., Pascoal, T.A., Lima-Filho, R.A.S., Resende, E.P.F., et al., 2021. Pro-inflammatory interleukin-6 signaling links cognitive impairments and peripheral metabolic alterations in Alzheimer's disease. Transl. Psychiatry 11, 251.

Løfblad, L., Hov, G.G., Åsberg, A., Videm, V., 2021. Inflammatory markers and risk of cardiovascular mortality in relation to diabetes status in the HUNT study. Sci. Rep. 11, 15644.

Løkhammer, S., Stavrum, A.K., Polushina, T., Aas, M., Ottesen, A.A., et al., 2022. An epigenetic association analysis of childhood trauma in psychosis reveals possible overlap with methylation changes associated with PTSD. Transl. Psychiatry 12, 177.

Ma, K., Xu, A., Cui, S., Sun, M.R., Xue, Y.C., et al., 2016. Impaired GABA synthesis, uptake and release are associated with depression-like behaviors induced by chronic mild stress. Transl. Psychiatry 6, e910.

Ma, P., Mo, R., Liao, H., Qiu, C., Wu, G., et al., 2022. Gut microbiota depletion by antibiotics ameliorates somatic neuropathic pain induced by nerve injury, chemotherapy, and diabetes in mice. J. Neuroinflammation 19, 169.

Ma, Q., 2022. A functional subdivision within the somatosensory system and its implications for pain research. Neuron 110, 749–769.

Ma, Y., Sun, L., Mu, Z., 2024. Network meta-analysis of three different forms of intermittent energy restrictions for overweight or obese adults. Int. J. Obes. 48, 55–64.

Mac Giollabhui, N., Ng, T.H., Ellman, L.M., Alloy, L.B., 2021. The longitudinal associations of inflammatory biomarkers and depression revisited: systematic review, meta-analysis, and meta-regression. Mol. Psychiatry 26, 3302–3314.

Machado-Filho, J.A., Correia, A.O., Montenegro, A.B.A., Nobre, M.E.P., Cerqueira, G.S., et al., 2014. Caffeine neuroprotective effects on 6-OHDA-lesioned rats are mediated by several factors, including pro-inflammatory cytokines and histone deacetylase inhibitions. Behav. Brain Res. 264, 116–125.

Machado, M.O., Oriolo, G., Bortolato, B., Köhler, C.A., Maes, M., et al., 2016. Biological mechanisms of depression following treatment with interferon for chronic hepatitis C: a critical systematic review. J. Affect. Disord. 209, 235–245.

MacIntyre, C.R., Mahimbo, A., Moa, A.M., Barnes, M., 2016. Influenza vaccine as a coronary intervention for prevention of myocardial infarction. Heart 102, 1953–1955.

Mackensen, A., Müller, F., Mougiakakos, D., Böltz, S., Wilhelm, A., et al., 2022. Anti-CD19 CAR T cell therapy for refractory systemic lupus erythematosus. Nat. Med. 28, 2124–2132.

Mackenzie, J.S., Jeggo, M., 2019. The One Health approach—why is it so important? Trop. Med. Infect. Dis. 4, 88.

Maclean, C.J., Baker, H.F., Ridley, R.M., Mori, H., 2000. Naturally occurring and experimentally induced beta-amyloid deposits in the brains of marmosets (Callithrix jacchus). J. Neural Transm. (Vienna) 107, 799–814.

Macpherson, A.J., de Agüero, M.G., Ganal-Vonarburg, S.C., 2017. How nutrition and the maternal microbiota shape the neonatal immune system. Nat. Rev. Immunol. 17, 508–517.

Maddox, S.A., Schafe, G.E., Ressler, K.J., 2015. Exploring epigenetic regulation of fear memory and biomarkers associated with post-traumatic stress disorder. Front. Psychiatry 4, 62.

Madigan, S., Wade, M., Plamondon, A., Maguire, J.L., Jenkins, J.M., 2017. Maternal adverse childhood experience and infant health: biomedical and psychosocial risks as intermediary mechanisms. J. Pediatr. 187, 282–289.e1.

Madore, M.R., Kozel, F.A., Williams, L.M., Green, L.C., George, M.S., et al., 2022. Prefrontal transcranial magnetic stimulation for depression in US military veterans - a naturalistic cohort study in the veterans health administration. J. Affect. Disord. 297, 671–678.

Maekawa, M., Watanabe, A., Iwayama, Y., Kimura, T., Hamazaki, K., et al., 2017. Polyunsaturated fatty acid deficiency during neurodevelopment in mice models the prodromal state of schizophrenia through epigenetic changes in nuclear receptor genes. Transl. Psychiatry 7, e1229.

Maenner, M.J., Warren, Z., Williams, A.R., Amoakohene, E., Bakian, A. V., et al., 2023. Prevalence and characteristics of autism spectrum disorder among children aged 8 years - autism and developmental disabilities monitoring network, 11 sites, United States, 2020. MMWR Surveill. Summ. 72, 1–14.

Maes, M., 1995. Evidence for an immune response in major depression: a review and hypothesis. Prog. Neuro-Psychopharmacol. Biol. Psychiatry 19, 11–38.

Maes, M., Song, C., Lin, A., De Jongh, R., Van Gastel, A., et al., 1998. The effects of psychological stress on humans: increased production of pro-inflammatory cytokines and a Th1-like response in stress-induced anxiety. Cytokine 10, 313–318.

Maes, M., Christophe, A., Bosmans, E., Lin, A., Neels, H., 2000. In humans, serum polyunsaturated fatty acid levels predict the response of proinflammatory cytokines to psychologic stress. Biol. Psychiatry 47, 910–920.

Maes, M., Galecki, P., Chang, Y.S., Berk, M., 2011. A review on the oxidative and nitrosative stress (O&NS) pathways in major depression and their possible contribution to the (neuro)degenerative processes in that illness. Prog. Neuro-Psychopharmacol. Biol. Psychiatry 35, 676–692.

Maetzler, W., Apel, A., Langkamp, M., Deuschle, C., Dilger, S.S., et al., 2014. Comparable autoantibody serum levels against amyloid- and inflammation-associated proteins in Parkinson's disease patients and controls. PLoS One 9, e88604.

Magalhães, J., Candeias, E., Melo-Marques, I., Silva, D.F., Esteves, A.R., et al., 2023. Intestinal infection triggers mitochondria-mediated α-synuclein pathology: relevance to Parkinson's disease. Cell. Mol. Life Sci. 80, 166.

Magalhães, J., Tresse, E., Ejlerskov, P., Hu, E., Liu, Y., Marin, A., et al., 2021. PIAS2-mediated blockade of IFN-beta signaling: a basis for sporadic Parkinson disease dementia. Mol. Psychiatry 26, 6083–6099.

Magnus, P., Gunnes, N., Tveito, K., Bakken, I.J., Ghaderi, S., et al., 2015. Chronic fatigue syndrome/myalgic encephalomyelitis (CFS/ME) is associated with pandemic influenza infection, but not with an adjuvanted pandemic influenza vaccine. Vaccine 33, 6173–6177.

Magri, G., Comerma, L., Pybus, M., Sintes, J., Lligé, D., et al., 2017. Human secretory IgM emerges from plasma cells clonally related to gut memory B cells and targets highly diverse commensals. Immunity 47, 118–134.

Maheshwari, P., Eslick, G.D., 2015. Bacterial infection and Alzheimer's disease: a meta-analysis. J. Alzheimers Dis. 43, 957–966.

Mahajan, G.J., Vallender, E.J., Garrett, M.R., Challagundla, L., Overholser, J.C., et al., 2018. Altered neuro-inflammatory gene expression in hippocampus in major depressive disorder. Prog. Neuro-Psychopharmacol. Biol. Psychiatry 82, 177–186.

Mahalingam, G., Samtani, S., Lam, B.C.P., Lipnicki, D.M., Lima-Costa, M.F., et al., 2023. Social connections and risk of incident mild cognitive impairment, dementia, and mortality in 13 longitudinal cohort studies of ageing. Alzheimers Dement. 19, 5114–5128.

Maheu, M., Lopez, J.P., Crapper, L., Davoli, M.A., Turecki, G., Mechawar, N., 2015. MicroRNA regulation of central glial cell line-derived neurotrophic factor (GDNF) signaling in depression. Transl. Psychiatry 5, e511.

Mahmoud, R., Wainwright, S.R., Galea, L.A., 2016. Sex hormones and adult hippocampal neurogenesis: regulation, implications, and potential mechanisms. Front. Neuroendocrinol. 41, 129–152.

Mahmoudvand, H., Ziaali, N., Aghaei, I., Sheibani, V., Shojaee, S., et al., 2015. The possible association between Toxoplasma gondii infection and risk of anxiety and cognitive disorders in BALB/c mice. Pathog. Glob. Health 109, 369–376.

Mahrer, N.E., Guardino, C.M., Hobel, C., Dunkel Schetter, C., 2021. Maternal stress before conception is associated with shorter gestation. Ann. Behav. Med. 55, 242–252.

Maier, S.F., Seligman, M.E., 2016. Learned helplessness at fifty: insights from neuroscience. Psychol. Rev. 123, 349–367.

Maiese, K., 2023. Cellular metabolism: a fundamental component of degeneration in the nervous system. Biomolecules 13, 816.

Maiese, K., Chong, Z.Z., Shang, Y.C., 2008. Raves and risks for erythropoietin. Cytokine Growth Factor Rev. 19, 145–155.

Maihofer, A.X., Choi, K.W., Coleman, J.R.I., Daskalakis, N.P., Denckla, C. A., et al., 2022. Enhancing discovery of genetic variants for posttraumatic stress disorder through integration of quantitative phenotypes and trauma exposure information. Biol. Psychiatry 91, 626–636.

Mailing, L.J., Allen, J.M., Buford, T.W., Fields, C.J., Woods, J.A., 2019. Exercise and the gut microbiome: a review of the evidence, potential mechanisms, and implications for human health. Exerc. Sport Sci. Rev. 47, 75–85.

Majerczyk, D., Ayad, E.G., Brewton, K.L., Saing, P., Hart, P.C., 2022. Systemic maternal inflammation promotes ASD via IL-6 and IFN-γ. Biosci. Rep. 42. BSR20220713.

Majidi, J., Kosari-Nasab, M., Salari, A.A., 2016. Developmental minocycline treatment reverses the effects of neonatal immune activation on anxiety- and depression-like behaviors, hippocampal inflammation, and HPA axis activity in adult mice. Brain Res. Bull. 120, 1–13.

Majoros, A., Platanitis, E., Kernbauer-Hölzl, E., Rosebrock, F., Müller, M., et al., 2017. Canonical and non-canonical aspects of JAK-STAT signaling: lessons from interferons for cytokine responses. Front. Immunol. 8, 29.

Makovac, E., Meeten, F., Watson, D.R., Herman, A., Garfinkel, S.N., et al., 2016. Alterations in amygdala-prefrontal functional connectivity account for excessive worry and autonomic dysregulation in generalized anxiety disorder. Biol. Psychiatry 80, 786–795.

Makrgeorgou, A., Leonardi-Bee, J., Bath-Hextall, F.J., Murrell, D.F., Tang, M.L., et al., 2018. Probiotics for treating eczema. Cochrane Database Syst. Rev. 11, CD006135.

Malan-Muller, S., Valles-Colomer, M., Foxx, C.L., Vieira-Silva, S., van den Heuvel, L.L., et al., 2022. Exploring the relationship between the gut microbiome and mental health outcomes in a posttraumatic stress disorder cohort relative to trauma-exposed controls. Eur. Neuropsychopharmacol. 56, 24–38.

Malan-Müller, S., Valles-Colomer, M., Palomo, T., Leza, J.C., 2023. The gut-microbiota-brain axis in a Spanish population in the aftermath of the COVID-19 pandemic: microbiota composition linked to anxiety, trauma, and depression profiles. Gut Microbes 15, 2162306.

Malberg, J.E., Duman, R.S., 2003. Cell proliferation in adult hippocampus is decreased by inescapable stress: reversal by fluoxetine treatment. Neuropsychopharmacology 28, 1562–1671.

Male, D.K., Brostoff, J., Roth, D., Roitt, I., 2006. Immunology, seventh ed. Mosby.

Male, V., 2022. SARS-CoV-2 infection and COVID-19 vaccination in pregnancy. Nat. Rev. Immunol. 22, 277–282.

Maletic, V., Eramo, A., Gwin, K., Offord, S.J., Duffy, R.A., 2017. The role of norepinephrine and its α-Adrenergic receptors in the pathophysiology and treatment of major depressive disorder and schizophrenia: a systematic review. Front. Psychiatry 8, 42.

Malhotra, R., Tareque, M.I., Saito, Y., Ma, S., Chiu, C.T., Chan, A., 2021. Loneliness and health expectancy among older adults: a longitudinal population-based study. J. Am. Geriatr. Soc. 69, 3092–3102.

Malik, S., Alnaji, O., Malik, M., Gambale, T., Rathbone, M.P., 2022. Correlation between mild traumatic brain injury-induced inflammatory cytokines and emotional symptom traits: a systematic review. Brain Sci. 12, 102.

Malkus, K.A., Tsika, E., Ischiropoulos, H., 2009. Oxidative modifications, mitochondrial dysfunction, and impaired protein degradation in Parkinson's disease: how neurons are lost in the Bermuda triangle. Mol. Neurodegener. 4, 24.

Mangalam, A., Shahi, S.K., Luckey, D., Karau, M., Marietta, E., et al., 2017. Human gut-derived commensal bacteria suppress CNS inflammatory and demyelinating disease. Cell Rep. 20, 1269–1277.

Mangale, V., Syage, A.R., Ekiz, H.A., Skinner, D.D., Cheng, Y., et al., 2020. Microglia influence host defense, disease, and repair following murine coronavirus infection of the central nervous system. Glia 68, 2345–2360.

Mangano, E.N., Hayley, S., 2009. Inflammatory priming of the substantia nigra influences the impact of later paraquat exposure: neuroimmune sensitization of neurodegeneration. Neurobiol. Aging 30, 1361–1378.

Mangano, E.N., Litteljohn, D., So, R., Nelson, E., Peters, S., et al., 2012. Interferon-γ plays a role in paraquat-induced neurodegeneration involving oxidative and proinflammatory pathways. Neurobiol. Aging 33, 1411–1426.

Mangano, E.N., Peters, S., Litteljohn, D., So, R., Bethune, C., et al., 2011. Granulocyte macrophage-colony stimulating factor protects against substantia nigra dopaminergic cell loss in an environmental toxin model of Parkinson's disease. Neurobiol. Dis. 43, 99–112.

Mangino, M., Roederer, M., Beddall, M.H., Nestle, F.O., Spector, T.D., 2017. Innate and adaptive immune traits are differentially affected by genetic and environmental factors. Nat. Commun. 8, 13850.

Mani, B.K., Zigman, J.M., 2017. Ghrelin as a survival hormone. Trends Endocrinol. Metab. 28, 843–854.

Manousaki, D., Paternoster, L., Standl, M., Moffatt, M.F., Farrall, M., et al., 2017. Vitamin D levels and susceptibility to asthma, elevated immunoglobulin E levels, and atopic dermatitis: a Mendelian randomization study. PLoS Med. 14, e1002294.

Mansfeld, J., Urban, N., Priebe, S., Groth, M., Frahm, C., et al., 2015. Branched-chain amino acid catabolism is a conserved regulator of physiological ageing. Nat. Commun. 6, 10043.

Manthey, A., Sierk, A., Brakemeier, E.L., Walter, H., Daniels, J.K., 2021. Does trauma-focused psychotherapy change the brain? A systematic review of neural correlates of therapeutic gains in PTSD. Eur. J. Psychotraumatol. 12, 1929025.

Mao, L., Jin, H., Wang, M., Hu, Y., Chen, S., et al., 2020. Neurologic manifestations of hospitalized patients with coronavirus disease 2019 in Wuhan, China. JAMA Neurol. 77, 683–690.

Marazziti, D., Palermo, S., Arone, A., Massa, L., Parra, E., et al., 2023. Obsessive-compulsive disorder, PANDAS, and tourette syndrome: immuno-inflammatory disorders. Adv. Exp. Med. Biol. 1411, 275–300.

Marche, T.A., Briere, J.L., von Baeyer, C.L., 2016. Children's forgetting of pain-related memories. J. Pediatr. Psychol. 41, 220–231.

Marcinkiewcz, C.A., Mazzone, C.M., D'Agostino, G., Halladay, L.R., Hardaway, J.A., et al., 2016. Serotonin engages an anxiety and fear-promoting circuit in the extended amygdala. Nature 537, 97–101.

Marcinowicz, P., Więdłocha, M., Zborowska, N., Dębowska, W., Podwalski, P., et al., 2021. A meta-analysis of the influence of antipsychotics on cytokines levels in first episode psychosis. J. Clin. Med. 10, 2488.

Marcocci, M.E., Napoletani, G., Protto, V., Kolesova, O., Piacentini, R., et al., 2020. Herpes Simplex Virus-1 in the brain: the dark side of a sneaky infection. Trends Microbiol. 28, 808–820.

Margolis, A.E., Herbstman, J.B., Davis, K.S., Thomas, V.K., Tang, D., et al., 2016. Longitudinal effects of prenatal exposure to air pollutants on self-regulatory capacities and social competence. J. Child Psychol. Psychiatry 57, 851–860.

Marietta, E.V., Murray, J.A., Luckey, D.H., Jeraldo, P.R., Lamba, A., et al., 2016. Suppression of inflammatory arthritis by human gut-derived *Prevotella histicola* in humanized mice. Arthritis Rheumatol. 68, 2878–2888.

Marin, T.J., Chen, E., Munch, J.A., Miller, G.E., 2009. Double-exposure to acute stress and chronic family stress is associated with immune changes in children with asthma. Psychosom. Med. 71, 378–384.

Marinova-Mutafchieva, L., Sadeghian, M., Broom, L., Davis, J.B., Medhurst, A.D., et al., 2009. Relationship between microglial activation and dopaminergic neuronal loss in the substantia nigra: a time course study in a 6-hydroxydopamine model of Parkinson's disease. J. Neurochem. 110, 966–975.

Marinova, Z., Chuang, D.M., Fineberg, N., 2017. Glutamate-modulating drugs as a potential therapeutic strategy in obsessive-compulsive disorder. Curr. Neuropharmacol. 15, 977–995.

Marinovic, D.A., Hunter, R.L., 2022. Examining the interrelationships between mindfulness-based interventions, depression, inflammation, and cancer survival. CA Cancer J. Clin. 72, 490–502.

Marizzoni, M., Cattaneo, A., Mirabelli, P., Festari, C., Lopizzo, N., et al., 2020. Short-chain fatty acids and lipopolysaccharide as mediators between gut dysbiosis and amyloid pathology in Alzheimer's disease. J. Alzheimer's Dis. 78, 683–697.

Markle, J.G., Frank, D.N., Mortin-Toth, S., Robertson, C.E., Feazel, L.M., et al., 2013. Sex differences in the gut microbiome drive hormone-dependent regulation of autoimmunity. Science 339, 1084–1088.

Marks, E., Hickman, C., Pihkala, P., Clayton, S., Lewandowski, E.R., et al., 2021. Young people's voices on climate anxiety, government betrayal and moral injury: a global phenomenon. Lancet Planet. Heath 5, 863–873.

Marmot, M.G., Rose, G., Shipley, M., Hamilton, P.J., 1978. Employment grade and coronary heart disease in British civil servants. J. Epidemiol. Community Health 32, 244–249.

Marotta, R., Risoleo, M.C., Messina, G., Parisi, L., Carotenuto, M., Vetri, L., Roccella, M., 2020. The neurochemistry of autism. Brain Sci. 10, 163.

Marques, T.R., Ashok, A.H., Pillinger, T., Veronese, M., Turkheimer, F. E., et al., 2019. Neuroinflammation in schizophrenia: meta-analysis of in vivo microglial imaging studies. Psychol. Med. 49, 2186–2196.

Marra, A.R., Perencevich, E.N., Nelson, R.E., Samore, M., Khader, K., et al., 2020. Incidence and outcomes associated with Clostridium difficile infections: a systematic review and meta-analysis. JAMA Netw. Open 3, e1917597.

Marras, C., Alcalay, R.N., Caspell-Garcia, C., Coffey, C., Chan, P., et al., 2016. Motor and nonmotor heterogeneity of LRRK2-related and idiopathic Parkinson's disease. Mov. Disord. 31, 1192–1202.

Marrie, R.A., Walld, R., Bolton, J.M., Sareen, J., Walker, J.R., et al., 2017. Increased incidence of psychiatric disorders in immune-mediated inflammatory disease. J. Psychosom. Res. 101, 17–23.

Marrie, R.A., Walld, R., Bolton, J.M., Sareen, J., Walker, J.R., et al., 2019. Rising incidence of psychiatric disorders before diagnosis of immune-mediated inflammatory disease. Epidemiol. Psychiatr. Sci. 28, 333–342.

Marrocco, J., Reynaert, M.L., Gatta, E., Gabriel, C., Mocaër, E., et al., 2014. The effects of antidepressant treatment in prenatally stressed rats support the glutamatergic hypothesis of stress-related disorders. J. Neurosci. 34, 2015–2024.

Marsland, A.L., Walsh, C., Lockwood, K., John-Henderson, N.A., 2017. The effects of acute psychological stress on circulating and stimulated inflammatory markers: a systematic review and meta-analysis. Brain Behav. Immun. 64, 208–219.

Martel, J., Ojcius, D.M., Chang, C.J., Lin, C.S., Lu, C.C., et al., 2016. Anti-obesogenic and antidiabetic effects of plants and mushrooms. Nat. Rev. Endocrinol. 13, 149–160.

Martikainen, I.K., Hagelberg, N., Jääskeläinen, S.K., Hietala, J., Pertovaara, A., 2018. Dopaminergic and serotonergic mechanisms in the modulation of pain: In vivo studies in human brain. Eur. J. Pharmacol. 834, 337–345.

Martin, J.B., 2002. The integration of neurology, psychiatry, and neuroscience in the 21st century. Am. J. Psychiatry 159, 695–704.

Martin, P., Kurth, E.A., Budean, D., Momplaisir, N., Qu, E., et al., 2024. Biophysical characterization of the CXC chemokine receptor 2 ligands. PLoS One 19, e0298418.

Martineau, A.R., Jolliffe, D.A., Hooper, R.L., Greenberg, L., Aloia, J.F., et al., 2017. Vitamin D supplementation to prevent acute respiratory tract infections: systematic review and meta-analysis of individual participant data. BMJ 356, i6583.

Martinello, K.A., Meehan, C., Avdic-Belltheus, A., Lingam, I., Ragab, S., et al., 2019. Acute LPS sensitization and continuous infusion exacerbates hypoxic brain injury in a piglet model of neonatal encephalopathy. Sci. Rep. 9, 10184.

Martínez-Albert, E., Lutz, N.D., Hübener, R., Dimitrov, S., Lange, T., et al., 2024. Sleep promotes T-cell migration towards CCL19 via growth hormone and prolactin signaling in humans. Brain Behav. Immun. 118, 69–77.

Martínez-Cerdeño, V., Camacho, J., Fox, E., Miller, E., Ariza, J., et al., 2016. Prenatal exposure to autism-specific maternal autoantibodies alters proliferation of cortical neural precursor cells, enlarges brain, and increases neuronal size in adult animals. Cereb. Cortex 26, 374–383.

Martínez-Mármol, R., Giordano-Santini, R., Kaulich, E., Cho, A.N., Przybyla, M., et al., 2023. SARS-CoV-2 infection and viral fusogens cause neuronal and glial fusion that compromises neuronal activity. Sci. Adv. 9, eadg2248.

Martinez-Muniz, G.A., Wood, S.K., 2020. Sex differences in the inflammatory consequences of stress: implications for pharmacotherapy. J. Pharmacol. Exp. Ther. 375, 161–174.

Martinez, C.A., Marteinsdottir, I., Josefsson, A., Sydsjö, G., Theodorsson, E., Rodriguez-Martinez, H., 2022. Prenatal stress, anxiety and depression alter transcripts, proteins and pathways associated with immune responses at the maternal-fetal interface. Biol. Reprod. 106, 449–462.

Martínez, M., Postolache, T.T., García-Bueno, B., Leza, J.C., Figuero, E., et al., 2022. The role of the oral microbiota related to periodontal diseases in anxiety, mood and trauma- and stress-related disorders. Front. Psychiatry 12, 814177.

Martins de Carvalho, L., Chen, W.Y., Lasek, A.W., 2021. Epigenetic mechanisms underlying stress-induced depression. Int. Rev. Neurobiol. 156, 87–126.

Marusak, H.A., Furman, D.J., Kuruvadi, N., Shattuck, D.W., Joshi, S.H., et al., 2015. Amygdala responses to salient social cues vary with oxytocin receptor genotype in youth. Neuropsychologia 79 (Pt A), 1–9.

Marx, W., Lane, M., Hockey, M., Aslam, H., Berk, M., et al., 2021. Diet and depression: exploring the biological mechanisms of action. Mol. Psychiatry 26, 134–150.

Masanetz, R.K., Winkler, J., Winner, B., Günther, C., Süß, P., et al., 2022. The gut-immune-brain axis: an important route for neuropsychiatric morbidity in inflammatory bowel disease. Int. J. Mol. Sci. 23, 11111.

Masi, A., Glozier, N., Dale, R., Guastella, A.J., 2017. The immune system, cytokines, and biomarkers in autism spectrum disorder. Neurosci. Bull. 33, 194–204.

Maslanik, T., Bernstein-Hanley, I., Helwig, B., Fleshner, M., 2012. The impact of acute-stressor exposure on splenic innate immunity: a gene expression analysis. Brain Behav. Immun. 26, 142–149.

Mason, B.L., Li, Q., Minhajuddin, A., Czysz, A.H., Coughlin, L.A., et al., 2020. Reduced anti-inflammatory gut microbiota are associated with depression and anhedonia. J. Affect. Disord. 266, 394–401.

Mason, N.L., Szabo, A., Kuypers, K.P.C., Mallaroni, P.A., de la Torre Fornell, R., et al., 2023. Psilocybin induces acute and persisting alterations in immune status in healthy volunteers: an experimental, placebo-controlled study. Brain Behav. Immun. 114, 299–310.

Masri, S., Kinouchi, K., Sassone-Corsi, P., 2015. Circadian clocks, epigenetics, and cancer. Curr. Opin. Oncol. 27, 50–56.

Massey, S.H., Backes, K.A., Schuette, S.A., 2016. Plasma oxytocin concentration and depressive symptoms: a review of current evidence and directions for future research. Depress. Anxiety 33, 316–322.

Mataix-Cols, D., Fernández de la Cruz, L., Monzani, B., Rosenfield, D., Andersson, E., et al., 2017. D-Cycloserine augmentation of exposure-based cognitive behavior therapy for anxiety, obsessive-compulsive, and posttraumatic stress disorders: a systematic review and meta-analysis of individual participant data. JAMA Psychiatry 74, 501–510.

Mataix-Cols, D., Frans, E., Pérez-Vigil, A., Kuja-Halkola, R., Gromark, C., et al., 2018. A total-population multigenerational family clustering study of autoimmune diseases in obsessive–compulsive disorder and Tourette's/chronic tic disorders. Mol. Psychiatry 23, 1652–1658.

Matejuk, A., Ransohoff, R.M., 2020. Crosstalk between astrocytes and microglia: an overview. Front. Immunol. 11, 1416.

Mateo, D., Marquès, M., Torrente, M., 2023. Metals linked with the most prevalent primary neurodegenerative dementias in the elderly: a narrative review. Environ. Res. 236, 116722.

Matheoud, D., Sugiura, A., Bellemare-Pelletier, A., Laplante, A., Rondeau, C., et al., 2016. Parkinson's disease-related proteins PINK1 and parkin repress mitochondrial antigen presentation. Cell 166, 314–327.

Matheson, K., Anisman, H., 2012. Biological and psychosocial responses to discrimination. In: Jetten, J., Haslam, C., Haslam, S.A. (Eds.), The Social Cure. Psychology Press, New York, pp. 133–154.

Matheson, K., Anisman, H., 2003. Systems of coping associated with dysphoria, anxiety and depressive illness: a multivariate profile perspective. Stress 6, 223–234.

Matheson, K., Asokumar, A., Anisman, H., 2020. Resilience: safety in the aftermath of traumatic stressor experiences. Front. Behav. Neurosci. 14, 596919.

Matheson, K., Bombay, A., Dixon, K., Anisman, H., 2018a. Intergenerational communication regarding Indian Residential Schools: implications for cultural identity, perceived discrimination, and depressive symptoms. Transcult. Psychiatry. in press.

Matheson, K., Bombay, A., Anisman, H., 2018b. Culture as an ingredient of personalized medicine. J. Psychiatry Neurosci. 43, 3–6.

Matheson, K., Pierre, A., Foster, M.D., Kent, M., Anisman, H., 2021. Untangling racism: stress reactions in response to variations of racism against Black Canadians. Humanities Social Sci. Commun. 8, 1–12.

Matheson, K., Skomorovsky, A., Fiocco, A., Anisman, H., 2007. The limits of 'adaptive' coping: well-being and affective reactions to stressors among women in abusive dating relationships. Stress 10, 75–92.

Matrisciano, F., 2023. Epigenetic regulation of metabotropic glutamate 2/3 receptors: potential role for ultra-resistant schizophrenia? Pharmacol. Biochem. Behav. 229, 173589.

Matschke, J., Lütgehetmann, M., Hagel, C., Sperhake, J.P., Schröder, A. S., et al., 2020. Neuropathology of patients with COVID-19 in Germany: a post-mortem case series. Lancet Neurol. 19, 919–929.

Matson, V., Fessler, J., Bao, R., Chongsuwat, T., Zha, Y., et al., 2018. The commensal microbiome is associated with anti-PD-1 efficacy in metastatic melanoma patients. Science 359, 104–108.

Matsumoto, R., Gray, J., Rybkina, K., Oppenheimer, H., Levy, L., et al., 2023. Induction of bronchus-associated lymphoid tissue is an early life adaptation for promoting human B cell immunity. Nat. Immunol. 24.

Matt, S.M., Allen, J.M., Lawson, M.A., Mailing, L.J., Woods, J.A., Johnson, R.W., 2018. Butyrate and dietary soluble fiber improve neuroinflammation associated with aging in mice. Front. Immunol. 14, 1832.

Matthews, C.E., Moore, S.C., Arem, H., Cook, M.B., Trabert, B., et al., 2020. Amount and intensity of leisure-time physical activity and lower cancer risk. J. Clin. Oncol. 38, 686–697.

Mattson, M.P., Longo, V.D., Harvie, M., 2017. Impact of intermittent fasting on health and disease processes. Ageing Res. Rev. 39, 46–58.

Maul, S., Giegling, I., Fabbri, C., Corponi, F., Serretti, A., Rujescu, D., 2020. Genetics of resilience: implications from genome-wide association studies and candidate genes of the stress response system in posttraumatic stress disorder and depression. Am. J. Med. Genet. B Neuropsychiatr. Genet. 183, 77–94.

Maurer, M.H., Schäbitz, W.R., Schneider, A., 2008. Old friends in new constellations—the hematopoetic growth factors G-CSF, GM-CSF, and EPO for the treatment of neurological diseases. Curr. Med. Chem. 15, 1407–1411.

Mausbach, B.T., von Känel, R., Roepke, S.K., Moore, R., Patterson, T.L., et al., 2011. Self-efficacy buffers the relationship between dementia caregiving stress and circulating concentrations of the proinflammatory cytokine interleukin-6. Am. J. Geriatr. Psychiatry 19, 64–71.

Mavromatis, L.A., Rosoff, D.B., Bell, A.S., Jung, J., Wagner, J., Lohoff, F. W., 2023. Multi-omic underpinnings of epigenetic aging and human longevity. Nat. Commun. 14, 2236.

Mawson, E.R., Morris, B.J., 2023. A consideration of the increased risk of schizophrenia due to prenatal maternal stress, and the possible role of microglia. Prog. Neuro-Psychopharmacol. Biol. Psychiatry 125, 110773.

Mayberg, H.S., Lozano, A.M., Voon, V., McNeely, H.E., Seminowicz, D., et al., 2005. Deep brain stimulation for depression. Neuron 45, 651–660.

Maydych, V., Claus, M., Dychus, N., Ebel, M., Damaschke, J., et al., 2017. Impact of chronic and acute academic stress on lymphocyte subsets and monocyte function. PLoS One 12, e0188108.

Maymon, N., Mizrachi Zer-Aviv, T., Sabban, E.L., Akirav, I., 2020. Neuropeptide Y and cannabinoids interaction in the amygdala after exposure to shock and reminders model of PTSD. Neuropharmacology 162, 107804.

Mayo-Wilson, E., Dias, S., Mavranezouli, I., Kew, K., Clark, D.M., et al., 2014. Psychological and pharmacological interventions for social anxiety disorder in adults: a systematic review and network meta-analysis. Lancet Psychiatry 1, 368–376.

Mayorova, M.A., Butoma, B.G., Churilov, L.P., Gilburd, B., Petrova, N. N., Shoenfeld, Y., 2021. Autoimmune concept of schizophrenia: historical roots and current facets. Psychiatr. Danub. 33, 3–17.

Mazier, W., Le Corf, K., Martinez, C., Tudela, H., Kissi, D., et al., 2021. A new strain of Christensenella minuta as a potential biotherapy for obesity and associated metabolic diseases. Cells 10, 823.

Mazza, M.G., De Lorenzo, R., Conte, C., Poletti, S., Vai, B., et al., 2020. Anxiety and depression in COVID-19 survivors: role of inflammatory and clinical predictors. Brain Behav. Immun. 89, 594–600.

Mazza, M.G., Palladini, M., Villa, G., Agnoletto, E., Harrington, Y., et al., 2023. Prevalence of depression in SARS-CoV-2 infected patients: an umbrella review of meta-analyses. Gen. Hosp. Psychiatry 80, 17–25.

Mazzone, L., Dooling, S.W., Volpe, E., Uljarević, M., Waters, J.L., et al., 2023. Precision microbial intervention improves social behavior but not autism severity: a pilot double-blind randomized placebo-controlled trial. Cell Host Microbe 32, 106–116.

McAuley, P.A., Artero, E.G., Sui, X., Lee, D., Church, T.S., et al., 2012. The obesity paradox, cardiorespiratory fitness, and coronary heart disease. Mayo Clin. Proc. 87, 443–451.

McCarty, R.J., Downing, S.T., Daley, M.L., McNamara, J.P.H., Guastello, A.D., 2023. Relationships between stress appraisals and intolerance of uncertainty with psychological health during early COVID-19 in the USA. Anxiety Stress Coping 36, 97–109.

McCrory, E.J., De Brito, S.A., Sebastian, C.L., Mechelli, A., Bird, G., et al., 2011. Heightened neural reactivity to threat in child victims of family violence. Curr. Biol. 21, R947–R948.

McCutcheon, R.A., Krystal, J.H., Howes, O.D., 2020. Dopamine and glutamate in schizophrenia: biology, symptoms and treatment. World Psychiatry 19, 15–33.

McDade, T.W., Ryan, C., Jones, M.J., MacIsaac, J.L., Morin, A.M., et al., 2017. Social and physical environments early in development predict DNA methylation of inflammatory genes in young adulthood. Proc. Natl. Acad. Sci. USA 114, 7611–7616.

McDaid, A.F., Joshi, P.K., Porcu, E., Komljenovic, A., Li, H., et al., 2017. Bayesian association scan reveals loci associated with human lifespan and linked biomarkers. Nat. Commun. 8, 15842.

McDonald, D., Ackermann, G., Khailova, L., Baird, C., Heyland, D., et al., 2016. Extreme dysbiosis of the microbiome in critical illness. mSphere 1, e00199–e00216.

McDougle, D.R., Watson, J.E., Adbeen, A.A., Abdeen, A.A., Adili, R., et al., 2017. Anti-inflammatory ω-3 endocannabinoid epoxides. Proc. Natl. Acad. Sci. 114, E6034–E6043.

McEwen, B.S., 2000. Allostasis and allostatic load: implications for neuropsychopharmacology. Neuropsychopharmacology 22, 108–124.

McEwen, B.S., Gianaros, P.J., 2011. Stress- and allostasis-induced brain plasticity. Annu. Rev. Med. 62, 431–445.

McEwen, B.S., Karatsoreos, I.N., 2015. Sleep deprivation and circadian disruption: stress, allostasis, and allostatic load. Sleep Med. Clin. 10, 1–10.

McEwen, B.S., Wingfield, J.C., 2003. The concept of allostasis in biology and biomedicine. Horm. Behav. 43, 2–15.

McEwen, B.S., Bowles, N.P., Gray, J.D., Hill, M.N., Hunter, R.G., et al., 2015. Mechanisms of stress in the brain. Nat. Neurosci. 18, 1353–1363.

McEwen, C.A., McEwen, B.S., 2017. Social structure, adversity, toxic stress, and intergenerational poverty: an early childhood model. Annu. Rev. Sociol. 43, 445–472.

McGaunn, J., Baur, J.A., 2023. Taurine linked with healthy aging. Science 380, 1010–1011.

McGeer, P.L., McGeer, E.G., 2004. Inflammation and the degenerative diseases of aging. Ann. N. Y. Acad. Sci. 1035, 104–116.

McGovern, N., Shin, A., Low, G., Low, D., Duan, K., et al., 2017. Human fetal dendritic cells promote prenatal T-cell immune suppression through arginase-2. Nature 546, 662–666.

McGowan, P.O., Szyf, M., 2010. The epigenetics of social adversity in early life: implications for mental health outcomes. Neurobiol. Dis. 39, 66–72.

McGowan, P.O., Sasaki, A., D'Alessio, A.C., Dymov, S., Labonté, B., et al., 2009. Epigenetic regulation of the glucocorticoid receptor in human brain associates with childhood abuse. Nat. Neurosci. 12, 342–348.

McGrath, J.J., 2017. Vitamin D and mental health–the scrutiny of science delivers a sober message. Acta Psychiatr. Scand. 135, 183–184.

McGrath, J.J., Al-Hamzawi, A., Alonso, J., Altwaijri, Y., Andrade, L.H., et al., 2023. Age of onset and cumulative risk of mental disorders: a cross-national analysis of population surveys from 29 countries. Lancet Psychiatry 10, 668–681.

McGuinness, A.J., Davis, J.A., Dawson, S.L., Loughman, A., Collier, F., et al., 2022. A systematic review of gut microbiota composition in observational studies of major depressive disorder, bipolar disorder and schizophrenia. Mol. Psychiatry 27, 1920–1935.

McGuire, A., Steele, R.G., Singh, M.N., 2021a. Systematic review on the application of trauma-focused cognitive behavioral therapy (TF-CBT) for preschool-aged children. Clin. Child. Fam. Psychol. Rev. 24, 20–37.

McGuire, D.K., Shih, W.J., Cosentino, F., Charbonnel, B., Cherney, D.Z.I., et al., 2021b. Association of SGLT2 inhibitors with cardiovascular and kidney outcomes in patients with type 2 diabetes: a metaanalysis. JAMA Cardiol. 6, 148–158.

McIntosh, J., Anisman, H., Merali, Z., 1999. Short-and long-periods of neonatal maternal separation differentially affect anxiety and feeding in adult rats: gender-dependent effects. Dev. Brain Res. 113, 97–106.

McKenna, B.G., Mekawi, Y., Katrinli, S., Carter, S., Stevens, J.S., et al., 2021. When anger remains unspoken: anger and accelerated epigenetic aging among stress-exposed black Americans. Psychosom. Med. 83, 949–958.

McKenzie, J.A., Spielman, L.J., Pointer, C.B., Lowry, J.R., Bajwa, E., et al., 2017. Neuroinflammation as a common mechanism associated with the modifiable risk factors for Alzheimer's and Parkinson's diseases. Curr. Aging Sci. 10, 158–176.

McKim, D.B., Weber, M.D., Niraula, A., Sawicki, C.M., Liu, X., et al., 2017. Microglial recruitment of IL-1β-producing monocytes to brain endothelium causes stress-induced anxiety. Mol. Psychiatry 7, 64.

McKim, D.B., Weber, M.D., Niraula, A., Sawicki, C.M., Liu, X., et al., 2018. Microglial recruitment of IL-1β-producing monocytes to brain endothelium causes stress-induced anxiety. Mol. Psychiatry 23, 1421–1431.

McKlveen, J.M., Morano, R.L., Fitzgerald, M., Zoubovsky, S., Cassella, S. N., et al., 2016. Chronic stress increases prefrontal inhibition: a mechanism for stress-induced prefrontal dysfunction. Biol. Psychiatry 80, 754–764.

McLaughlin, A.P., Nikkheslat, N., Hastings, C., Nettis, M.A., Kose, M., et al., 2022. The influence of comorbid depression and overweight status on peripheral inflammation and cortisol levels. Psychol. Med. 52, 3289–3296.

McLeod, M., Hong, M., Mukhida, K., Sadi, D., Ulalia, R., Mendez, I., 2006. Erythropoietin and GDNF enhance ventral mesencephalic fiber outgrowth and capillary proliferation following neural transplantation in a rodent model of Parkinson's disease. Eur. J. Neurosci. 24 (2), 361370.

McMurray, K.M., Ramaker, M.J., Barkley-Levenson, A.M., Sidhu, P.S., Elkin, P.K., et al., 2017. Identification of a novel, fast-acting GABAergic antidepressant. Mol. Psychiatry 23, 384–391.

McMurray, K.M.J., Sah, R., 2022. Neuroimmune mechanisms in fear and panic pathophysiology. Front. Psychiatry 13, 1015349.

McNerney, M.W., Sheng, T., Nechvatal, J.M., Lee, A.G., Lyons, D.M., et al., 2018. Integration of neural and epigenetic contributions to posttraumatic stress symptoms: the role of hippocampal volume and glucocorticoid receptor gene methylation. PLoS One 13, e0192222.

McNicol, E.D., Ferguson, M.C., Haroutounian, S., Carr, D.B., Schumann, R., et al., 2016. Single dose intravenous paracetamol or intravenous propacetamol for postoperative pain. Cochrane Database Syst. Rev., CD007126.

McNulty, C.J., Fallon, I.P., Amat, J., Sanchez, R.J., Leslie, N.R., et al., 2023. Elevated prefrontal dopamine interferes with the stressbuffering properties of behavioral control in female rats. Neuropsychopharmacology 48, 498–507.

McQuaid, R.J., 2021. Transdiagnostic biomarker approaches to mental health disorders: consideration of symptom complexity, comorbidity and context. Brain Behav. Immun. Health 16, 100303.

McQuaid, R.J., McInnis, O.A., Abizaid, A., Anisman, H., 2014. Making room for oxytocin in understanding depression. Neurosci. Biobehav. Rev. 45, 305–322.

McQuaid, R.J., McInnis, O.A., Matheson, K., Anisman, H., 2015. Distress of ostracism: oxytocin receptor gene polymorphism confers sensitivity to social exclusion. Soc. Cogn. Affect. Neurosci. 10, 1153–1159.

McQuaid, R.J., McInnis, O.A., Paric, A., Al-Yawer, F., Matheson, K., et al., 2016. Relations between plasma oxytocin and cortisol: the stress buffering role of social support. Neurobiol. Stress 3, 52–60.

McWhirt, J., Sathyanesan, M., Sampath, D., Newton, S.S., 2019. Effects of restraint stress on the regulation of hippocampal glutamate receptor and inflammation genes in female C57BL/6 and BALB/c mice. Neurobiol. Stress 10, 100169.

Mead, J., Ashwood, P., 2015. Evidence supporting an altered immune response in ASD. Immunol. Lett. 163, 49–55.

Meagher, M.W., Johnson, R.R., Young, E.E., Vichaya, E.G., Lunt, S., et al., 2007. Interleukin-6 as a mechanism for the adverse effects of social stress on acute Theiler's virus infection. Brain Behav. Immun. 21, 1083–1095.

Medawar, E., Witte, A.V., 2022. Impact of obesity and diet on brain structure and function: a gut-brain-body crosstalk. Proc. Nutr. Soc. 81, 306–316.

Medrihan, L., Sagi, Y., Inde, Z., Krupa, O., Daniels, C., et al., 2017. Initiation of behavioral response to antidepressants by cholecystokinin neurons of the dentate gyrus. Neuron 95, 564–576.

Medzhitov, R., 2009. Approaching the asymptote: 20 years later. Immunity 30, 766–775.

Meftahi, G.H., Jangravi, Z., Sahraei, H., Bahari, Z., 2020. The possible pathophysiology mechanism of cytokine storm in elderly adults with COVID-19 infection: the contribution of "inflame-aging". Inflamm. Res. 69 (9), 825–839.

Megli, C.J., Coyne, C.B., 2022. Infections at the maternal-fetal interface: an overview of pathogenesis and defence. Nat. Rev. Microbiol. 20, 67–82.

Mehta, D., Bruenig, D., Pierce, J., Sathyanarayanan, A., Stringfellow, R., et al., 2022. Recalibrating the epigenetic clock after exposure to trauma: the role of risk and protective psychosocial factors. J. Psychiatr. Res. 149, 374–381.

Mehta, D., Grewen, K., Pearson, B., Wani, S., Wallace, L., et al., 2021. Genome-wide gene expression changes in postpartum depression point towards an altered immune landscape. Transl. Psychiatry 11, 155.

Mehta, D., Miller, O., Bruenig, D., David, G., Shakespeare-Finch, J., 2020. A systematic review of DNA methylation and gene expression studies in posttraumatic stress disorder, posttraumatic growth, and resilience. J. Trauma. Stress. 33 (2), 171–180.

Mehta, D., Voisey, J., Bruenig, D., Harvey, W., Morris, C.P., et al., 2018a. Transcriptome analysis reveals novel genes and immune networks dysregulated in veterans with PTSD. Brain Behav. Immun. 74, 133–142.

Mehta, N.D., Haroon, E., Xu, X., Woolwine, B.J., Li, Z., Felger, J.C., 2018b. Inflammation negatively correlates with amygdala-ventromedial prefrontal functional connectivity in association with anxiety in patients with depression: preliminary results. Brain Behav. Immun. 73, 725–730.

Mehterov, N., Minchev, D., Gevezova, M., Sarafian, V., Maes, M., 2022. Interactions among brain-derived neurotrophic factor and neuroimmune pathways are key components of the major psychiatric disorders. Mol. Neurobiol. 59, 4926–4952.

Mei, Y.Y., Wu, D.C., Zhou, N., 2018. Astrocytic regulation of glutamate transmission in schizophrenia. Front. Psychiatry 9, 544.

Meinhardt, J., Radke, J., Dittmayer, C., Franz, J., Thomas, C., et al., 2021. Olfactory transmucosal SARS-CoV-2 invasion as a port of central nervous system entry in individuals with COVID-19. Nat. Neurosci. 24, 168–175.

Meinlschmidt, G., Guemghar, S., Roemmel, N., Battegay, E., Hunziker, S., Schaefert, R., 2022. Depressive symptoms, but not anxiety, predict subsequent diagnosis of Coronavirus disease 19: a national cohort study. Epidemiol. Psychiatr. Sci. 31, e16.

Meir Drexler, S., Merz, C.J., Hamacher-Dang, T.C., Wolf, O.T., 2016. Cortisol effects on fear memory reconsolidation in women. Psychopharmacology 233, 2687–2697.

Meirlaen, L., Levy, E.I., Vandenplas, Y., 2021. Prevention and management with pro-, pre and synbiotics in children with asthma and allergic rhinitis: a narrative review. Nutrients 13, 934.

Mekli, K., Lophatananon, A., Maharani, A., Nazroo, J.Y., Muir, K.R., 2023. Association between an inflammatory biomarker score and future dementia diagnosis in the population-based UK Biobank cohort of 500,000 people. PLoS One 18, e0288045.

Melnikov, M., Lopatina, A., 2022. Th17-cells in depression: implication in multiple sclerosis. Front. Immunol. 13, 1010304.

Meltzer, A., Van de Water, J., 2017. The role of the immune system in autism spectrum disorder. Neuropsychopharmacology 42 (1), 284–298.

Melzack, R., Casey, K.L., 1968. Sensory, motivational, and central control determinants of pain. In: Kenshalo, D.R. (Ed.), The Skin Senses. Charles C. Thomas, Springfield, IL, pp. 423–439.

Melzack, R., Wall, P.D., 1965. Pain mechanisms: a new theory. Science 150, 971–979.

Menard, C., Pfau, M.L., Hodes, G.E., Kana, V., Wang, V.X., et al., 2017. Social stress induces neurovascular pathology promoting depression. Nat. Neurosci. 20, 1752–1760.

Mendelsohn, A.R., Larrick, J.W., 2013. Sleep facilitates clearance of metabolites from the brain: glymphatic function in ageing and neurodegenerative diseases. Rejuvenation Res. 16, 518–523.

Meng, Q., Lin, M.S., Tzeng, I.S., 2020. Relationship between exercise and Alzheimer's disease: a narrative literature review. Front. Neurosci. 14, 131.

Méquinion, M., Foldi, C.J., Andrews, Z.B., 2020. The ghrelin-AgRP neuron nexus in anorexia nervosa: Implications for metabolic and behavioral adaptations. Front. Nutr. 6, 190.

Merali, Z., Anisman, H., 2016. Deconstructing the mental health crisis: 5 uneasy pieces. J. Psychiatry Neurosci. 41, 219–221.

Merali, Z., Du, L., Hrdina, P., Palkovits, M., Faludi, G., et al., 2004. Dysregulation in the suicide brain: mRNA expression of corticotropin releasing hormone receptors and $GABA_A$ receptor subunits in frontal cortical brain region. J. Neurosci. 24, 1478–1485.

Merali, Z., Brennan, K., Brau, P., Anisman, H., 2003. Dissociating anorexia and anhedonia elicited by interleukin-1beta: antidepressant and gender effects on responding for "free chow" and "earned" sucrose intake. Psychopharmacology 165, 413–418.

Merali, Z., Graitson, S., Mackay, J.C., Kent, P., 2013. Stress and eating: a dual role for bombesin like peptides. Front. Neurosci. 7, 193.

Merali, Z., Kent, P., Du, L., Hrdina, P., Palkovits, M., et al., 2006. Corticotropin-releasing hormone, arginine vasopressin, gastrin-releasing peptide, and neuromedin B alterations in stress-relevant brain regions of suicides and control subjects. Biol. Psychiatry 59, 594–602.

Merali, Z., McIntosh, J., Kent, P., Michaud, D., Anisman, H., 1998. Aversive and appetitive events evoke the release of corticotropin-releasing hormone and bombesin-like peptides at the central nucleus of the amygdala. J. Neurosci. 18, 4758–4766.

Merino, J.J., Muñetón-Gómez, V., Alvárez, M.I., Toledano-Díaz, A., 2016. Effects of CX3CR1 and fractalkine chemokines in amyloid beta clearance and p-Tau accumulation in Alzheimer's disease (AD) rodent models: is fractalkine a systemic biomarker for AD? Curr. Alzheimer Res. 13, 403–412.

Merkl, A., Aust, S., Schneider, G.H., Visser-Vandewalle, V., Horn, A., et al., 2018. Deep brain stimulation of the subcallosal cingulate gyrus in patients with treatment-resistant depression: a double-blinded randomized controlled study and long-term follow-up in eight patients. J. Affect. Disord. 227, 521–529.

Merlot, E., Moze, E., Dantzer, R., Neveu, P.J., 2004. Cytokine production by spleen cells after social defeat in mice: activation of T cells and reduced inhibition by glucocorticoids. Stress 7, 55–61.

Merritt, K., McCutcheon, R.A., Aleman, A., Ashley, S., Beck, K., et al., 2023. Variability and magnitude of brain glutamate levels in schizophrenia: a meta and mega-analysis. Mol. Psychiatry (Epub ahead of print).

Messina, A., Concerto, C., Rodolico, A., Petralia, A., Caraci, F., Signorelli, M.S., 2023. Is It time for a paradigm shift in the treatment of schizophrenia? The use of inflammation-reducing and neuroprotective drugs-A review. Brain Sci. 13, 957.

Mestre, H., Mori, Y., Nedergaard, M., 2020. The brain's glymphatic system: current controversies. Trends Neurosci. 43, 458–466.

Metrik, J., Stevens, A.K., Gunn, R.L., Borsari, B., Jackson, K.M., 2022. Cannabis use and posttraumatic stress disorder: prospective evidence from a longitudinal study of veterans. Psychol. Med. 52, 446–456.

Meyer, U., 2019. Neurodevelopmental resilience and susceptibility to maternal immune activation. Trends Neurosci. 42, 793–806.

Meyer, U., Feldon, J., 2009. Prenatal exposure to infection: a primary mechanism for abnormal dopaminergic development in schizophrenia. Psychopharmacology 206, 587–602.

Meyer-Luehmann, M., Coomaraswamy, J., Bolmont, T., Kaeser, S., Schaefer, C., et al., 2006. Exogenous induction of cerebral beta-amyloidogenesis is governed by agent and host. Science 313, 1781–1784.

Meyer, J., 2021. Inflammation, obsessive-compulsive disorder, and related disorders. Curr. Top. Behav. Neurosci. 49, 31–53.

Meyer, J.H., McMain, S., Kennedy, S.H., Korman, L., Brown, G.M., et al., 2003. Dysfunctional attitudes and 5-HT2 receptors during depression and self-harm. Am. J. Psychiatry 160, 90–99.

Meyer, U., Feldon, J., 2012. To poly(I:C) or not to poly(I:C): advancing preclinical schizophrenia research through the use of prenatal immune activation models. Neuropharmacology 62, 1308–1321.

Mezhlumyan, A.G., Tallerova, A.V., Povarnina, P.Y., Tarasiuk, A.V., Sazonova, N.M., et al., 2022. Antidepressant-like effects of BDNF and NGF individual loop dipeptide mimetics depend on the signal transmission patterns associated with trk. Pharmaceuticals 15, 284.

Miao, J., Ma, H., Yang, Y., Liao, Y., Lin, C., et al., 2023. Microglia in Alzheimer's disease: pathogenesis, mechanisms, and therapeutic potentials. Front. Aging Neurosci. 15, 1201982.

Michael, B.D., Dunai, C., Needham, E.J., Tharmaratnam, K., Williams, R., et al., 2023. Para-infectious brain injury in COVID-19 persists at follow-up despite attenuated cytokine and autoantibody responses. Nat. Commun. 14, 8487.

Michaud, J.P., Hallé, M., Lampron, A., Thériault, P., Préfontaine, P., et al., 2013. Toll-like receptor 4 stimulation with the detoxified ligand monophosphoryl lipid A improves Alzheimer's disease-related pathology. Proc. Natl. Acad. Sci. USA 110, 1941–1946.

Michaud, J.P., Richard, K.L., Rivest, S., 2012. Hematopoietic MyD88-adaptor protein acts as a natural defense mechanism for cognitive deficits in Alzheimer's disease. Stem Cell Rev. 8, 898–904.

Michaud, K., Matheson, K., Kelly, O., Anisman, H., 2008. Impact of stressors in a natural context on release of cortisol in healthy adult humans: a meta-analysis. Stress 11, 177–197.

Michel, T.M., Frangou, S., Camara, S., Thiemeyer, D., Jecel, J., et al., 2008. Altered glial cell line-derived neurotrophic factor (GDNF) concentrations in the brain of patients with depressive disorder: a comparative post-mortem study. Eur. Psychiatry 23, 413–420.

Michopoulos, V., Beurel, E., Gould, F., Dhabhar, F.S., Schultebraucks, K., et al., 2020. Association of prospective risk for chronic PTSD symptoms with low TNFα and IFNγ concentrations in the immediate aftermath of trauma exposure. Am. J. Psychiatry 177, 58–65.

Michopoulos, V., Powers, A., Gillespie, C.F., Ressler, K.J., Jovanovic, T., 2017. Inflammation in fear- and anxiety-based disorders: PTSD, GAD, and beyond. Neuropsychopharmacology 42, 254–270.

Michopoulos, V., Rothbaum, A.O., Jovanovic, T., Almli, L.M., Bradley, B., et al., 2015. Association of CRP genetic variation and CRP level with elevated PTSD symptoms and physiological responses in a civilian population with high levels of trauma. Am. J. Psychiatry 172, 353–362.

Michopoulos, V., Vester, A., Neigh, G., 2016. Posttraumatic stress disorder: a metabolic disorder in disguise? Exp. Neurol. 284, 220–229.

Mieling, M., Meier, H., Bunzeck, N., 2023. Structural degeneration of the nucleus basalis of Meynert in mild cognitive impairment and Alzheimer's disease - evidence from an MRI-based meta-analysis. Neurosci. Biobehav. Rev. 154, 105393.

Mifflin, K.A., Kerr, B.J., 2017. Pain in autoimmune disorders. J. Neurosci. Res. 95, 1282–1294.

Migheli, R., Del Giudice, M.G., Spissu, Y., Sanna, G., Xiong, Y., et al., 2013. LRRK2 affects vesicle trafficking, neurotransmitter extracellular level and membrane receptor localization. PLoS One 8, e77198.

Miguel-Álvarez, M., Santos-Lozano, A., Sanchis-Gomar, F., Fiuza-Luces, C., Pareja-Galeano, H., et al., 2015. Non-steroidal anti-inflammatory drugs as a treatment for Alzheimer's disease: a systematic review and meta-analysis of treatment effect. Drugs Aging 32, 139–147.

Mika, A., Fleshner, M., 2016. Early-life exercise may promote lasting brain and metabolic health through gut bacterial metabolites. Immunol. Cell Biol. 94, 151–157.

Mika, A., Day, H.E., Martinez, A., Rumian, N.L., Greenwood, B.N., et al., 2017. Early life diets with prebiotics and bioactive milk fractions attenuate the impact of stress on learned helplessness behaviours and alter gene expression within neural circuits important for stress resistance. Eur. J. Neurosci. 45, 342–357.

Mika, A., Rumian, N., Loughridge, A.B., Fleshner, M., 2016. Exercise and prebiotics produce stress resistance: Converging impacts on stress-protective and butyrate-producing gut bacteria. Int. Rev. Neurobiol. 131, 165–191.

Mikkelsen, K.H., Knop, F.K., Frost, M., Hallas, J., Pottega°rd, A., 2015. Use of antibiotics and risk of type 2 diabetes: a population-based case-control study. J. Clin. Endocrinol. Metab. 100, 36333640.

Milad, M.R., Rauch, S.L., 2011. Obsessive-compulsive disorder: Beyond segregated corticostriatal pathways. Trends Cogn. Sci. 16, 43–51.

Milaneschi, Y., Kappelmann, N., Ye, Z., Lamers, F., Moser, S., et al., 2021. Association of inflammation with depression and anxiety: evidence for symptom-specificity and potential causality from UK Biobank and NESDA cohorts. Mol. Psychiatry 26, 7393–7402.

Milaneschi, Y., Lamers, F., Bot, M., Drent, M.L., Penninx, B.W., 2017. Leptin dysregulation is specifically associated with major depression with atypical features: Evidence for a mechanism connecting obesity and depression. Biol. Psychiatry 81, 807–814.

Milbocker, K.A., Campbell, T.S., Collins, N., Kim, S., Smith, I.F., Roth, T.L., Klintsova, A.Y., 2021. Glia-driven brain circuit refinement is altered by early-life adversity: behavioral outcomes. Front. Behav. Neurosci. 15, 786234.

Mildenberger, J., Johansson, I., Sergin, I., Kjøbli, E., Damås, J.K., et al., 2017. N-3 PUFAs induce inflammatory tolerance by formation of KEAP1-containing SQSTM1/p62-bodies and activation of NFE2L2. Autophagy 13, 1664–1678.

Miljevic, C., Munjiza-Jovanovic, A., Jovanovic, T., 2023. Impact of childhood adversity, as early life distress, on cytokine alterations in schizophrenia. Neuropsychiatr. Dis. Treat. 19, 579–586.

Miller, A.H., Raison, C.L., 2016. The role of inflammation in depression: from evolutionary imperative to modern treatment target. Nat. Rev. Immunol. 16, 22–34.

Miller, B.J., Goldsmith, D.R., 2017. Towards an immunophenotype of schizophrenia: progress, potential mechanisms, and future directions. Neuropsychopharmacology 42, 299–317.

Miller, G.E., Cohen, S., Pressman, S., Barkin, A., Rabin, B.S., et al., 2004. Psychological stress and antibody response to influenza vaccination: when is the critical period for stress, and how does it get inside the body? Psychosom. Med. 66, 215–223.

Miller, G.E., Lachman, M.E., Chen, E., Gruenewald, T.L., Karlamangla, A.S., et al., 2011. Pathways to resilience: maternal nurturance as a buffer against the effects of childhood poverty on metabolic syndrome at midlife. Psychol. Sci. 22, 1591–1599.

Miller, L., Bansal, R., Wickramaratne, P., Hao, X., Tenke, C.E., et al., 2014. Neuroanatomical correlates of religiosity and spirituality: a study in adults at high and low familial risk for depression. JAMA Psychiatry 71, 128–135.

Miller, M.W., Maniates, H., Wolf, E.J., Logue, M.W., Schichman, S.A., et al., 2018. CRP polymorphisms and DNA methylation of the AIM2 gene influence associations between trauma exposure, PTSD, and C-reactive protein. Brain Behav. Immun. 67, 194–202.

Miller, G.E., Cohen, S., Ritchey, A.K., 2022. Chronic psychological stress and the regulation of pro-inflammatory cytokines: a glucocorticoid-resistance model. Health Psychol. 21, 531–541.

Milton, R., Gillespie, D., Dyer, C., Taiyari, K., Carvalho, M.J., et al., 2022. Neonatal sepsis and mortality in low-income and middle-income countries from a facility-based birth cohort: an international multisite prospective observational study. Lancet Glob. Health 10, e661–e672.

Mina, M.J., Metcalf, C.J., de Swart, R.L., Osterhaus, A.D., Grenfell, B.T., 2015. Vaccines: long-term measles-induced immunomodulation increases overall childhood infectious disease mortality. Science 348, 694–699.

Mina, Y., Enose-Akahata, Y., Hammoud, D.A., Videckis, A.J., Narpala, S.R., et al., 2023. Deep phenotyping of neurologic postacute sequelae of SARS-CoV-2 infection. Neurol. Neuroimmunol. Neuroinflamm. 10, e200097.

Minelli, A., Maffioletti, E., Cloninger, C.R., Magri, C., Sartori, R., et al., 2013. Role of allelic variants of FK506-binding protein 51 (FKBP5) gene in the development of anxiety disorders. Depress. Anxiety 30, 1170–1176.

Minett, T., Classey, J., Matthews, F.E., Fahrenhold, M., Taga, M., et al., 2016. Microglial immunophenotype in dementia with Alzheimer's pathology. J. Neuroinflammation 13, 135.

Mischkowski, D., Crocker, J., Way, B.M., 2016. From painkiller to empathy killer: acetaminophen (paracetamol) reduces empathy for pain. Soc. Cogn. Affect. Neurosci. 11, 1345–1353.

Mischkowski, D., Crocker, J., Way, B.M., 2019. A social analgesic? Acetaminophen (Paracetamol) reduces positive empathy. Front. Psychol. 10, 538.

Mishra, P., Silva, A., Sharma, J., Nguyen, J., Pizzo, D.P., et al., 2023. Rescue of Alzheimer's disease phenotype in a mouse model by transplantation of wild-type hematopoietic stem and progenitor cells. Cell Rep. 42, 112956.

Mishra, S.K., Hoon, M.A., 2013. The cells and circuitry for itch responses in mice. Science, 968–971.

Misiak, B., Bartoli, F., Carrà, G., Stańczykiewicz, B., Gładka, A., et al., 2021. Immune-inflammatory markers and psychosis risk: a systematic review and meta-analysis. Psychoneuroendocrinology 127, 105200.

Mitchell, C.A., Verovskaya, E.V., Calero-Nieto, F.J., Olson, O.C., Swann, J.W., et al., 2023. Stromal niche inflammation mediated by IL-1 signalling is a targetable driver of haematopoietic ageing. Nat. Cell Biol. 25, 30–41.

Mitchell, J.M., Bogenschutz, M., Lilienstein, A., Harrison, C., Kleiman, S., et al., 2021. MDMA-assisted therapy for severe PTSD: a randomized, double-blind, placebo-controlled phase 3 study. Nat. Med. 27 (6), 1025–1033.

Mitra, A., Raichle, M.E., Geoly, A.D., Kratter, I.H., Williams, N.R., 2023. Targeted neurostimulation reverses a spatiotemporal biomarker of treatment-resistant depression. Proc. Natl. Acad. Sci. USA 120, e2218958120.

Mitsi, V., Terzi, D., Purushothaman, I., Manouras, L., Gaspari, S., et al., 2015. RGS9-2-controlled adaptations in the striatum determine the onset of action and efficacy of antidepressants in neuropathic pain states. Proc. Natl. Acad. Sci. USA 112, E5088–E5097.

Mittli, D., 2023. Inflammatory processes in the prefrontal cortex induced by systemic immune challenge: focusing on neurons. Brain Behav. Immun. Health 34, 100703.

Miyake, S., Kim, S., Suda, W., Oshima, K., Nakamura, M., et al., 2015. Dysbiosis in the gut microbiota of patients with multiple sclerosis, with a striking depletion of species belonging to clostridia XIVa and IV clusters. PLoS One 10, e0137429.

Miyasaka, M., Tanaka, T., 2004. Lymphocyte trafficking across high endothelial venules: dogmas and enigmas. Nat. Rev. Immunol. 4, 360–370.

Mizuno, T., Zhang, G., Takeuchi, H., Kawanokuchi, J., Wang, J., et al., 2008. Interferon-gamma directly induces neurotoxicity through a neuron specific, calcium-permeable complex of IFN-gamma receptor and AMPA GluR1 receptor. FASEB J. 22, 1797–1806.

Moayedi, M., Davis, K.D., 2013. Theories of pain: from specificity to gate control. J. Neurophysiol. 109, 5–12.

Mochcovitch, M.D., da Rocha Freire, R.C., Garcia, R.F., Nardi, A.E., 2014. A systematic review of fMRI studies in generalized anxiety disorder: evaluating its neural and cognitive basis. J. Affect. Disord. 167, 336–342.

Modarresi, C.A., Masoumi, S.A., Bigdeloo, M., Arsad, H., Lim, V., 2022. The effect of exercise on patients with rheumatoid arthritis on the modulation of inflammation. Clin. Exp. Rheumatol. 40, 1420–1431.

Modin, D., Claggett, B., Jørgensen, M.E., Køber, L., Benfield, T., et al., 2022. Flu vaccine and mortality in hypertension: a nationwide cohort study. J. Am. Heart Assoc. 11, e021715.

Moehle, M.S., Daher, J.P., Hull, T.D., Boddu, R., Abdelmotilib, H.A., Mobley, J., West, A.B., 2015. The G2019S LRRK2 mutation increases myeloid cell chemotactic responses and enhances LRRK2 binding to actin-regulatory proteins. Hum. Mol. Genet. 24 (15), 4250–4267.

Moehle, M.S., Webber, P.J., Tse, T., Sukar, N., Standaert, D.G., et al., 2012. LRRK2 inhibition attenuates microglial inflammatory responses. J. Neurosci. 32, 1602–1611.

Mogil, J.S., 2016. Perspective: equality need not be painful. Nature 535, S7.

Mohr, D.C., Goodkin, D.E., Nelson, S., Cox, D., Weiner, M., 2002. Moderating effects of coping on the relationship between stress and the development of new brain lesions in multiple sclerosis. Psychosom. Med. 64, 803–809.

Mohr, D.C., Hart, S.L., Julian, L., Cox, D., Pelletier, D., 2004. Association between stressful life events and exacerbation in multiple sclerosis: a meta-analysis. BMJ 328, 731.

Moieni, M., Eisenberger, N.I., 2018. Effects of inflammation on social processes and implications for health. Ann. N. Y. Acad. Sci. 1428, 5–13.

Moir, R.D., Lathe, R., Tanzi, R.E., 2018. The antimicrobial protection hypothesis of Alzheimer's disease. Alzheimers Dement. 14, 1602–1614.

Moisan, A., Lee, Y.K., Zhang, J.D., Hudak, C.S., Meyer, C.A., et al., 2015. White-to-brown metabolic conversion of human adipocytes by JAK inhibition. Nat. Cell Biol. 17, 57–67.

Moisset, X., de Andrade, D.C., Bouhassira, D., 2016. From pulses to pain relief: an update on the mechanisms of rTMS-induced analgesic effects. Eur. J. Pain 20, 689–700.

Mojtabavi, H., Saghazadeh, A., van den Heuvel, L., Bucker, J., Rezaei, N., 2020. Peripheral blood levels of brain-derived neurotrophic factor in patients with post-traumatic stress disorder (PTSD): a systematic review and meta-analysis. PLoS One 15, e0241928.

Mokhtari, R., Lachman, H.M., 2016. The major histocompatibility complex (MHC) in schizophrenia: a review. J. Clin. Cell. Immunol. 7.

Mold, M., Linhart, C., Gómez-Ramírez, J., Villegas-Lanau, A., Exley, C., 2020. Aluminum and amyloid-β in familial Alzheimer's disease. J. Alzheimers Dis. 73, 1627–1635.

Moldin, S.O., Rubenstein, J.L., Hyman, S.E., 2006. Can autism speak to neuroscience? J. Neurosci. 26, 6893–6896.

Molendijk, M.L., Spinhoven, P., Polak, M., Bus, B.A., Penninx, B.W., et al., 2014. Serum bdnf concentrations as peripheral manifestations of depression: evidence from a systematic review and meta-analyses on 179 associations (n = 9484). Mol. Psychiatry 19, 791–800.

Molina, E., Gould, N., Lee, K., Krimins, R., Hardenbergh, D., Timlin, H., 2022. Stress, mindfulness, and systemic lupus erythematosus: an overview and directions for future research. Lupus 31, 1549–1562.

Moliner, R., Girych, M., Brunello, C.A., Kovaleva, V., Biojone, C., et al., 2023. Psychedelics promote plasticity by directly binding to BDNF receptor TrkB. Nat. Neurosci. 26, 1032–1041.

Molla, H., Lee, R., Lyubomirsky, S., de Wit, H., 2023. Drug-induced social connection: both MDMA and methamphetamine increase feelings of connectedness during controlled dyadic conversations. Sci. Rep. 13, 15846.

Moloney, R.D., Sajjad, J., Foley, T., Felice, V.D., Dinan, T.G., et al., 2016. Estrous cycle influences excitatory amino acid transport and visceral pain sensitivity in the rat: effects of early-life stress. Biol. Sex Differ. 7, 33.

Molot, J., Sears, M., Anisman, H., 2023. Multiple chemical sensitivity: It's time to catch up to the science. Neurosci. Biobehav. Rev. 151, 105227.

Molteni, M., Gemma, S., Rossetti, C., 2016a. The role of Toll-Like Receptor 4 in infectious and noninfectious inflammation. Mediat. Inflamm. 2016, 6978936.

Molteni, R., Rossetti, A.C., Savino, E., Racagni, G., Calabrese, F., 2016b. Chronic mild stress modulates activity-dependent transcription of BDNF in rat hippocampal slices. Neural Plast. 2016, 2592319.

Monari, S., Guillot de Suduiraut, I., Grosse, J., Zanoletti, O., Walker, S.E., et al., 2023. Blunted glucocorticoid responsiveness to stress causes behavioral and biological alterations that lead to posttraumatic stress disorder vulnerability. Biol. Psychiatry 95, 762–773.

Mondal, A.C., Fatima, M., 2019. Direct and indirect evidences of BDNF and NGF as key modulators in depression: role of antidepressants treatment. Int. J. Neurosci. 129, 283–296.

Mondelli, V., Vernon, A.C., Turkheimer, F., Dazzan, P., Pariante, C.M., 2017. Brain microglia in psychiatric disorders. Lancet Psychiatry 4, 563–572.

Montalvo-Ortiz, J.L., Gelernter, J., Cheng, Z., Girgenti, M.J., Xu, K., et al., 2022. Epigenome-wide association study of posttraumatic stress disorder identifies novel loci in U.S. military veterans. Transl. Psychiatry 12, 65.

Montero-Cosme, T.G., Pascual-Mathey, L.I., Hernández-Aguilar, M.E., Herrera-Covarrubias, D., Rojas-Durán, F., Aranda-Abreu, G.E., 2023. Potential drugs for the treatment of Alzheimer's disease. Pharmacol. Rep. 75, 544–559.

Montoy, J.C.C., Ford, J., Yu, H., Gottlieb, M., Morse, D., et al., 2023. Prevalence of symptoms ≤12 months after acute illness, by COVID-19 testing status among adults - United States, December 2020-March 2023. MMWR Morb. Mortal Wkly. Rep. 72, 859–865.

Montoya, J.G., Holmes, T.H., Anderson, J.N., Maecker, H.T., Rosenberg-Hasson, Y., et al., 2017. Cytokine signature associated with disease severity in chronic fatigue syndrome patients. Proc. Natl. Acad. Sci. USA 114, E7150–E7158.

Moody, L., Chen, H., Pan, Y.X., 2017. Postnatal diet remodels hepatic DNA methylation in metabolic pathways established by a maternal high-fat diet. Epigenomics 9, 1387–1402.

Moon, M.L., Joesting, J.J., Blevins, N.A., Lawson, M.A., Gainey, S.J., et al., 2015. IL-4 knock out mice display anxiety-like behavior. Behav. Genet. 45, 451–460.

Moon, S., Yu, S.H., Lee, C.B., Park, Y.J., Yoo, H.J., Kim, D.S., 2021. Effects of bisphenol A on cardiovascular disease: an epidemiological study using National Health and Nutrition Examination Survey 2003-2016 and meta-analysis. Sci. Total Environ. 763, 142941.

Moons, W.G., Shields, G.S., 2015. Anxiety, not anger, induces inflammatory activity: an avoidance/approach model of immune system activation. Emotion 15, 463–476.

Moore, B.B., Kunkel, S.L., 2019. Attracting attention: discovery of IL-8/CXCL8 and the birth of the chemokine field. J. Immunol. 202, 3–4.

Moore, K.W., de Waal Malefyt, R., Coffman, R.L., O'Garra, A., 2001. Interleukin-10 and the interleukin-10 receptor. Annu. Rev. Immunol. 19, 683–765.

Moore, A.R., Rosenberg, S.C., McCormick, F., Malek, S., 2020. RAS-targeted therapies: is the undruggable drugged? Nat. Rev. Drug Discov. 19, 533–552.

Morales, I., 2022. Brain regulation of hunger and motivation: the case for integrating homeostatic and hedonic concepts and its implications for obesity and addiction. Appetite 177, 106146.

Morales, I., Berridge, K.C., 2020. 'Liking' and 'wanting' in eating and food reward: Brain mechanisms and clinical implications. Physiol. Behav. 227, 113152.

Morales-Soto, W., Gonzales, J., Jackson, W.F., Gulbransen, B.D., 2023. Enteric glia promote visceral hypersensitivity during inflammation through intercellular signaling with gut nociceptors. Sci. Signal. 16, eadg1668.

Morassi, M., Bagatto, D., Cobelli, M., D'Agostini, S., Gigli, G.L., et al., 2020. Stroke in patients with SARS-CoV-2 infection: case series. J. Neurol. 267, 2185–2192.

Morath, J., Gola, H., Sommershof, A., Hamuni, G., Kolassa, S., et al., 2014. The effect of trauma-focused therapy on the altered T cell distribution in individuals with PTSD: evidence from a randomized controlled trial. J. Psychiatr. Res. 54, 1–10.

Mordelt, A., de Witte, L.D., 2023. Microglia-mediated synaptic pruning as a key deficit in neurodevelopmental disorders: Hype or hope? Curr. Opin. Neurobiol. 79, 102674.

Morel, L., Domingues, O., Zimmer, J., Michel, T., 2020. Revisiting the role of neurotrophic factors in inflammation. Cells 9, 865.

Morena, M., Patel, S., Bains, J.S., Hill, M.N., 2016. Neurobiological interactions between stress and the endocannabinoid system. Neuropsychopharmacology 41, 80–102.

Morens, D.M., Folkers, G.K., Fauci, A.S., 2022. The concept of classical herd immunity may not apply to COVID-19. J. Infect. Dis. 226, 195–198.

Mores, K.L., Cummins, B.R., Cassell, R.J., van Rijn, R.M., 2019. A review of the therapeutic potential of recently developed G protein-biased kappa agonists. Front. Pharmacol. 10, 407.

Morgan, B.P., 2015. The role of complement in neurological and neuropsychiatric diseases. Expert Rev. Clin. Immunol. 11, 1109–1119.

Morgan 3rd, C.A., Wang, S., Southwick, S.M., Rasmusson, A., Hazlett, G., et al., 2000. Plasma neuropeptide-Y concentrations in humans exposed to military survival training. Biol. Psychiatry 47, 902–909.

Morgan, C.P., Bale, T.L., 2017. Sex differences in microRNA-mRNA networks: examination of novel epigenetic programming mechanisms in the sexually dimorphic neonatal hypothalamus. Biol. Sex Differ. 8, 27.

Morimoto, Y., Izumi, H., Kuroda, E., 2014. Significance of persistent inflammation in respiratory disorders induced by nanoparticles. J Immunol Res 2014, 962871.

Morreall, J., Kim, A., Liu, Y., Degtyareva, N., Weiss, B., et al., 2015. Evidence for retromutagenesis as a mechanism for adaptive mutation in Escherichia coli. PLoS Genet. 11, e1005477.

Morris, G., Berk, M., Carvalho, A., Caso, J.R., Sanz, Y., et al., 2016a. The role of the microbial metabolites including tryptophan catabolites and short chain fatty acids in the pathophysiology of immune-inflammatory and neuroimmune disease. Mol. Neurobiol. 54, 4432–4451.

Morris, G., Berk, M., Galecki, P., Maes, M., 2014. The emerging role of autoimmunity in myalgic encephalomyelitis/chronic fatigue syndrome (ME/cfs). Mol. Neurobiol. 49, 741–756.

Morris, G., Reiche, E.M.V., Murru, A., Carvalho, A.F., Maes, M., et al., 2018. Multiple immune-inflammatory and oxidative and nitrosative stress pathways explain the frequent presence of depression in multiple sclerosis. Mol. Neurobiol. 55, 6282–6306.

Morris, M.J., Beilharz, J.E., Maniam, J., Reichelt, A.C., Westbrook, R.F., et al., 2016b. Why is obesity such a problem in the 21st century? The intersection of palatable food, cues and reward pathways, stress, and cognition. Neurosci. Biobehav. Rev. 58, 36–45.

Morrison, F.G., Miller, M.W., Logue, M.W., Assef, M., Wolf, E.J., 2019. DNA methylation correlates of PTSD: recent findings and technical challenges. Prog. Neuro-Psychopharmacol. Biol. Psychiatry 90, 223–234.

Morrow, C.B., Leoutsakos, J., Yan, H., Onyike, C., Kamath, V., 2023. Weight change and neuropsychiatric symptoms in Alzheimer's disease and frontotemporal dementia: associations with cognitive decline. J. Alzheimers Dis. Rep. 7, 767–774.

Morton, J.T., Jin, D.M., Mills, R.H., Shao, Y., Rahman, G., et al., 2023. Multi-level analysis of the gut–brain axis shows autism spectrum disorder-associated molecular and microbial profiles. Nat. Neurosci. 26, 1208–1217.

Moser, S., Martins, J., Czamara, D., Lange, J., Müller-Myhsok, B., Erhardt, A., 2022. DNA-methylation dynamics across short-term, exposure-containing CBT in patients with panic disorder. Transl. Psychiatry 12, 46.

Moshtael, R., Lynch, M.E., Duncan, G.E., Beam, C.R., 2024. A genetically informed study of the association between perceived stress and loneliness. Behav. Genet. https://doi.org/10.1007/s10519-024-10144-9.

Mostafa, G.A., Shehab, A.A., 2010. The link of C4B null allele to autism and to a family history of autoimmunity in Egyptian autistic children. J. Neuroimmunol. 223 (1-2), 115–119.

Mostafalou, S., Abdollahi, M., 2017. Pesticides: an update of human exposure and toxicity. Arch. Toxicol. 91, 549–599.

Moulton, V.R., 2018. Sex hormones in acquired immunity and autoimmune disease. Front. Immunol. 9, 2279.

Mount, M.P., Lira, A., Grimes, D., Smith, P.D., Faucher, S., et al., 2007. Involvement of interferon-gamma in microglial-mediated loss of dopaminergic neurons. J. Neurosci. 27, 3328–3337.

Moussaoui, N., Jacobs, J.P., Larauche, M., Biraud, M., Million, M., et al., 2017. Chronic early-life stress in rat pups alters basal corticosterone, intestinal permeability, and fecal microbiota at weaning: influence of sex. J. Neurogastroenterol. Motil. 23, 135–143.

Moya-Pérez, A., Perez-Villalba, A., Benítez-Páez, A., Campillo, I., Sanz, Y., et al., 2017. Bifidobacterium CECT 7765 modulates early stress-induced immune, neuroendocrine and behavioral alterations in mice. Brain Behav. Immun. 65, 43–56.

Muhie, S., Gautam, A., Chakraborty, N., Hoke, A., Meyerhoff, J., et al., 2017. Molecular indicators of stress-induced neuroinflammation in a mouse model simulating features of post-traumatic stress disorder. Transl. Psychiatry 7, e1135.

Muller, M.B., Zimmermann, S., Sillaber, I., Hagemeyer, T.P., Deussing, J.M., et al., 2003. Limbic corticotropin-releasing hormone receptor 1 mediates anxiety-related behavior and hormonal adaptation to stress. Nat. Neurosci. 6, 1100–1107.

Munawar, N., Ahsan, K., Muhammad, K., Ahmad, A., Anwar, M.A., et al., 2021. Hidden role of gut microbiome dysbiosis in schizophrenia: antipsychotics or psychobiotics as therapeutics? Int. J. Mol. Sci. 22, 7671.

Mundula, T., Baldi, S., Gerace, E., Amedei, A., 2023. Role of the intestinal microbiota in the genesis of major depression and the response to antidepressant drug therapy: a narrative review. Biomedicines 11, 550.

Muñoz Aguilera, E., Leira, Y., Miró Catalina, Q., Orlandi, M., Czesnikiewicz-Guzik, M., et al., 2021. Is systemic inflammation a missing link between periodontitis and hypertension? Results from two large population-based surveys. J. Intern. Med. 289, 532–546.

Muñoz-Delgado, L., Macías-García, D., Periñán, M.T., Jesús, S., Adarmes-Gómez, A.D., et al., 2023. Peripheral inflammatory immune response differs among sporadic and familial Parkinson's disease. npj Parkinson's Dis. 9, 12.

Munshi, S., Loh, M.K., Ferrara, N., DeJoseph, M.R., Ritger, A., et al., 2020. Repeated stress induces a pro-inflammatory state, increases amygdala neuronal and microglial activation, and causes anxiety in adult male rats. Brain Behav. Immun. 84, 180–199.

Murakami, M., Tognini, P., 2019. The circadian clock as an essential molecular link between host physiology and microorganisms. Front. Cell. Infect. Microbiol. 9, 469.

Murdock, K.W., Stowe, R.P., Peek, M.K., Lawrence, S.L., Fagundes, C.P., 2017. An elevation of perceived health risk and depressive symptoms prior to a disaster in predicting post-disaster inflammation. Psychosom. Med. 80, 49–54.

Murkar, A., Kendzerska, T., Shlik, J., Quilty, L., Saad, M., Robillard, R., 2022. Increased cannabis intake during the COVID-19 pandemic is associated with worsening of depression symptoms in people with PTSD. BMC Psychiatry 22, 554.

Murnane, K.S., 2019. Serotonin 2A receptors are a stress response system: implications for post-traumatic stress disorder. Behav. Pharmacol. 30, 151–162.

Murphy, T.M., O'Donovan, A., Mullins, N., O'Farrelly, C., McCann, A., et al., 2015. Anxiety is associated with higher levels of global DNA methylation and altered expression of epigenetic and interleukin-6 genes. Psychiatr. Genet. 25, 71–78.

Murray, E., Sharma, R., Smith, K.B., Mar, K.D., Barve, R., et al., 2019. Probiotic consumption during puberty mitigates LPS-induced immune responses and protects against stress-induced depression- and anxiety-like behaviors in adulthood in a sex-specific manner. Brain Behav. Immun. 81, 198–212.

Murray, E., Smith, K.B., Stoby, K.S., Thomas, B.J., Swenson, M.J., et al., 2020. Pubertal probiotic blocks LPS-induced anxiety and the associated neurochemical and microbial outcomes, in a sex-dependent manner. Psychoneuroendocrinology 112, 104481.

Murrough, J.W., Soleimani, L., DeWilde, K.E., Collins, K.A., Lapidus, K.A., et al., 2015. Ketamine for rapid reduction of suicidal ideation: a randomized controlled trial. Psychol. Med. 45, 3571–3580.

Musanabaganwa, C., Wani, A.H., Donglasan, J., Fatumo, S., Jansen, S., et al., 2022. Leukocyte methylomic imprints of exposure to the genocide against the Tutsi in Rwanda: a pilot epigenome-wide analysis. Epigenomics 14, 11–25.

Musardo, S., Marcello, E., 2017. Synaptic dysfunction in Alzheimer's disease: from the role of amyloid β-peptide to the α-secretase ADAM10. Eur. J. Pharmacol. 817, 30–37.

Musazzi, L., Tornese, P., Sala, N., Popoli, M., 2014. Acute stress is not acute: sustained enhancement of glutamate release after acute stress involves readily releasable pool size and synapsin I activation. Mol. Psychiatry 22, 1226–1227.

Musazzi, L., Treccani, G., Mallei, A., Popoli, M., 2013. The action of antidepressants on the glutamate system: regulation of glutamate release and glutamate receptors. Biol. Psychiatry 73, 1180–1188.

Mustač, F., Pašić, H., Medić, F., Bjedov, B., Vujević, L., et al., 2021. Anxiety and depression as comorbidities of multiple sclerosis. Psychiatr. Danub. 33, 480–485.

Musunuru, K., Kathiresan, S., 2017. Cardiovascular endocrinology: is ANGPTL3 the next PCSK9? Nat. Rev. Endocrinol. 13, 503–504.

Muto, V., Jaspar, M., Meyer, C., Kussé, C., Chellappa, S.L., et al., 2016. Local modulation of human brain responses by circadian rhythmicity and sleep debt. Science 353, 687–690.

Mutz, J., Vipulananthan, V., Carter, B., Hurlemann, R., Fu, C.H.Y., Young, A.H., 2019. Comparative efficacy and acceptability of non-surgical brain stimulation for the acute treatment of major depressive episodes in adults: systematic review and network meta-analysis. BMJ 364, l1079.

Myasoedova, E., Davis, J., Matteson, E.L., Crowson, C.S., 2020. Is the epidemiology of rheumatoid arthritis changing? Results from a population-based incidence study, 1985-2014. Ann. Rheum. Dis. 79, 440–444.

Mychasiuk, R., Schmold, N., Ilnytskyy, S., Kovalchuk, O., Kolb, B., et al., 2011. Prenatal bystander stress alters brain, behavior, and the epigenome of developing rat offspring. Dev. Neurosci. 33, 159–169.

Mörkl, S., Butler, M.I., Holl, A., Cryan, J.F., Dinan, T.G., 2020. Probiotics and the microbiota-gut-brain axis: focus on psychiatry. Curr. Nutr. Rep. 9, 171–182.

Månsson, K.N., Salami, A., Carlbring, P., Boraxbekk, C.J., Andersson, G., et al., 2017. Structural but not functional neuroplasticity one year after effective cognitive behaviour therapy for social anxiety disorder. Behav. Brain Res. 318, 45–51.

Maarouf, M., Maarouf, C.L., Yosipovitch, G., Shi, V.Y., 2019. The impact of stress on epidermal barrier function: an evidence-based review. Br. J. Dermatol. 181, 1129–1137.

Nacasch, N., Avni, C., Toren, P., 2023. Medical cannabis for treatment-resistant combat PTSD. Front. Psychiatry 13, 1014630.

Nadeem, A., Ahmad, S.F., Attia, S.M., Bakheet, S.A., AlHarbi, N.O., Al-Ayadhi, L.Y., 2018. Activation of IL-17 receptor leads to increased oxidative inflammation in peripheral monocytes of autistic children. Brain Behav. Immun. 67, 335–344.

Nader, K., 2015. Reconsolidation and the dynamic nature of memory. Cold Spring Harb. Perspect. Biol. 7, a021782.

Naegeli, C., Zeffiro, T., Piccirelli, M., Jaillard, A., Weilenmann, A., et al., 2017. Locus coeruleus activity mediates hyper-responsiveness in posttraumatic stress disorder. Biol. Psychiatry 83, 254–262.

Nagarajan, N., Capecchi, M.R., 2023. Optogenetic stimulation of mouse Hoxb8 microglia in specific regions of the brain induces anxiety, grooming, or both. Mol. Psychiatry 27, 90–101.

Nagarajan, R., Krishnamoorthy, Y., Basavarachar, V., Dakshinamoorthy, R., 2022. Prevalence of post-traumatic stress disorder among survivors of severe COVID-19 infections: a systematic review and meta-analysis. J. Affect. Disord. 299, 52–59.

Nagasaka, M., Sexton, R., Alhasan, R., Rahman, S., Azmi, A.S., Sukari, A., 2020. Gut microbiome and response to checkpoint inhibitors in non-small cell lung cancer-a review. Crit. Rev. Oncol. Hematol. 145, 102841. https://doi.org/10.1016/j.critrevonc.2019.102841.

Nagele, E.P., Han, M., Acharya, N.K., DeMarshall, C., Kosciuk, M.C., et al., 2013. Natural IgG autoantibodies are abundant and ubiquitous in human sera, and their number is influenced by age, gender, and disease. PLoS One 8, e60726.

Nahvi, R.J., Sabban, E.L., 2020. Sex differences in the neuropeptide Y system and implications for stress related disorders. Biomolecules 10, 1248.

Nair, S., Jao Keehn, R.J., Berkebile, M.M., Maximo, J.O., Witkowska, N., et al., 2018. Local resting state functional connectivity in autism: site and cohort variability and the effect of eye status. Brain Imaging Behav. 12, 168–179.

Najjar, S., Najjar, A., Chong, D.J., Pramanik, B.K., Kirsch, C., et al., 2020. Central nervous system complications associated with SARS-CoV-2 infection: integrative concepts of pathophysiology and case reports. J. Neuroinflammation 17, 231.

Nakhaee, H., Zangiabadian, M., Bayati, R., Rahmanian, M., Ghaffari Jolfayi, A., Rakhshanderou, S., 2022. The effect of antidepressants on the severity of COVID-19 in hospitalized patients: a systematic review and meta-analysis. PLoS One 17, e0267423.

Nallu, A., Sharma, S., Ramezani, A., Muralidharan, J., Raj, D., 2017. Gut microbiome in chronic kidney disease: challenges and opportunities. Transl. Res. 179, 24–37.

Nam, S., Jeon, S., Lee, S.J., Ash, G., Nelson, L.E., Granger, D.A., 2022. Real-time racial discrimination, affective states, salivary cortisol and alpha-amylase in Black adults. PLoS One 17, e0273081.

Naninck, E.F., Oosterink, J.E., Yam, K.Y., de Vries, L.P., et al., 2017. Early micronutrient supplementation protects against early stress–induced cognitive impairments. FASEB J. 31, 505–518.

Napso, T., Yong, H.E., Lopez-Tello, J., Sferruzzi-Perri, A.N., 2018. The role of placental hormones in mediating maternal adaptations to support pregnancy and lactation. Front. Physiol. 9, 1091.

Nara, H., Watanabe, R., 2021. Anti-inflammatory effect of muscle-derived interleukin-6 and its involvement in lipid metabolism. Int. J. Mol. Sci. 22, 9889.

Narasimhan, J., Letinski, S., Jung, S.P., Gerasyuto, A., Wang, J., et al., 2022. Ribonucleotide reductase, a novel drug target for gonorrhea. elife 11, e67447.

Narayan, S., Sinsheimer, J.S., Paul, K.C., Liew, Z., Cockburn, M., et al., 2015. Genetic variability in ABCB1, occupational pesticide exposure, and Parkinson's disease. Environ. Res. 143, 98–106.

Nasca, C., Zelli, D., Bigio, B., Piccinin, S., Scaccianoce, S., et al., 2015. Stress dynamically regulates behavior and glutamatergic gene expression in hippocampus by opening a window of epigenetic plasticity. Proc. Natl. Acad. Sci. USA 112, 14960–14965.

Natarajan, P., Kathiresan, S., 2016. PCSK9. Cell 165, 1037.

Natarajan, R., Forrester, L., Chiaia, N.L., Yamamoto, B.K., 2017. Chronic-stress-induced behavioral changes associated with subregion-selective serotonin cell death in the dorsal raphe. J. Neurosci. 37, 6214–6223.

Nathan, D.E., Bellgowan, J.A.F., French, L.M., Wolf, J., Oakes, T.R., et al., 2017. Assessing the impact of post-traumatic stress symptoms on the resting-state default mode network in a military chronic mild traumatic brain injury sample. Brain Connect. 7, 236–249.

National Institutes of Health, 2023. Dietary Supplements in the Time of COVID-19. https://ods.od.nih.gov/factsheets/COVID19-HealthProfessional/. (Accessed May 2023).

Natrajan, R., Sailem, H., Mardakheh, F.K., Arias Garcia, M., Tape, C.J., et al., 2016. Microenvironmental heterogeneity parallels breast cancer progression: a histology-genomic integration analysis. PLoS Med. 13, e1001961.

Navaneethabalakrishnan, S., Smith, H.L., Arenaz, C.M., Goodlett, B.L., McDermott, J.G., Mitchell, B.M., 2022. Update on immune mechanisms in hypertension. Am. J. Hypertens. 35, 842–851.

Navarro, V., Sanchez-Mejias, E., Jimenez, S., Muñoz-Castro, C., Sanchez-Varo, R., et al., 2018. Microglia in Alzheimer's disease: activated, dysfunctional or degenerative. Front. Aging Neurosci. 10, 140.

Naveed, Z., Velásquez García, H.A., Wong, S., Wilton, J., McKee, G., et al., 2023. Association of COVID-19 infection with incident diabetes. JAMA Netw. Open 6, e238866.

Needham, B.D., Tang, W., Wu, W.L., 2018. Searching for the gut microbial contributing factors to social behavior in rodent models of autism spectrum disorder. Dev. Neurobiol. 78 (5), 474–499.

Neher, J.J., Cunningham, C., 2019. Priming microglia for innate immune memory in the brain. Trends Immunol. 40, 358–374.

Neigh, G.N., Ali, F.F., 2016. Co-morbidity of PTSD and immune system dysfunction: opportunities for treatment. Curr. Opin. Pharmacol. 29, 104–110.

Neklyudova, A., Smirnov, K., Rebreikina, A., Martynova, O., Sysoeva, O., 2022. Electrophysiological and behavioral evidence for hyper- and hyposensitivity in rare genetic syndromes associated with autism. Genes (Basel) 13, 671.

Nelson 3rd, C.A., Gabard-Durnam, L.J., 2020. Early adversity and critical periods: neurodevelopmental consequences of violating the expectable environment. Trends Neurosci. 43, 133–143.

Nelson, M.D., Tumpap, A.M., 2016. Posttraumatic stress disorder symptom severity is associated with left hippocampal volume reduction: a meta-analytic study. CNS Spectr. 22, 363–372.

Nemirovsky, A., Ilan, K., Lerner, L., Cohen-Lavi, L., Schwartz, D., et al., 2021. Brain-immune axis regulation is responsive to cognitive behavioral therapy and mindfulness intervention: observations from a randomized controlled trial in patients with Crohn's disease. Brain Behav. Immun. Health 19, 100407.

Nerurkar, L., Siebert, S., McInnes, I.B., Cavanagh, J., 2019. Rheumatoid arthritis and depression: an inflammatory perspective. Lancet Psychiatry 6, 164–173.

Nestler, E.J., Waxman, S.G., 2020. Resilience to stress and resilience to pain: lessons from molecular neurobiology and genetics. Trends Mol. Med. 26, 924–935.

Netea, M.G., Latz, E., Mills, K.H., O'Neill, L.A., 2015. Innate immune memory: a paradigm shift in understanding host defense. Nat. Immunol. 16, 675–679.

Nettis, M.A., Pariante, C.M., Mondelli, V., 2020. Early-life adversity, systemic inflammation and comorbid physical and psychiatric illnesses of adult life. Curr. Top. Behav. Neurosci. 44, 207–225.

Nettis, M.A., Lombardo, G., Hastings, C., Zajkowska, Z., Mariani, N., et al., 2021. Augmentation therapy with minocycline in treatment-resistant depression patients with low-grade peripheral inflammation: results from a double-blind randomised clinical trial. Neuropsychopharmacology 46, 939–948.

Neufeld, K.M., Kang, N., Bienenstock, J., Foster, J.A., 2011. Reduced anxiety-like behavior and central neurochemical change in germ-free mice. Neurogastroenterol. Motil. 23, 255–264.

Neumann, I.D., Slattery, D.A., 2016. Oxytocin in general anxiety and social fear: a translational approach. Biol. Psychiatry 79, 213–221.

Nevriana, A., Pierce, M., Abel, K.M., Rossides, M., Wicks, S., et al., 2022. Association between parental mental illness and autoimmune diseases in the offspring - A nationwide register-based cohort study in Sweden. J. Psychiatr. Res. 151, 122–130.

Newman, T.M., Vitolins, M.Z., Cook, K.L., 2019. From the table to the tumor: the role of mediterranean and western dietary patterns in shifting microbial-mediated signaling to impact breast cancer risk. Nutrients 11, 2565.

Newton, T.L., Fernandez-Botran, R., Miller, J.J., Burns, V.E., 2014. Interleukin-6 and soluble interleukin-6 receptor levels in posttraumatic stress disorder: associations with lifetime diagnostic status and psychological context. Biol. Psychol. 99, 150–159.

Neyazi, A., Theilmann, W., Brandt, C., Rantamäki, T., Matsui, N., et al., 2018. P11 promoter methylation predicts the antidepressant effect of electroconvulsive therapy. Transl. Psychiatry 8, 25.

Nezgovorova, V., Ferretti, C.J., Pallanti, S., Hollander, E., 2022. Modulating neuroinflammation in COVID-19 patients with obsessive-compulsive disorder. J. Psychiatr. Res. 149, 367–373.

Ng, Q.X., Venkatanarayanan, N., Ho, C.Y., 2017. Clinical use of *Hypericum perforatum* (St John's wort) in depression: a meta-analysis. J. Affect. Disord. 210, 211–221.

Ng, A., Tam, W.W., Zhang, M.W., Ho, C.S., Husain, S.F., et al., 2018. IL-1β, IL-6, TNF-α and CRP in elderly patients with depression or Alzheimer's disease: Systematic review and meta-analysis. Sci. Rep. 8, 12050.

Ni, L., Wu, J., Long, Y., Tao, J., Xu, J., et al., 2019. Mortality of site-specific cancer in patients with schizophrenia: a systematic review and meta-analysis. BMC Psychiatry 19, 323.

Ni, Y.X., Ma, L., Li, J.P., 2021. Effects of mindfulness-based intervention on glycemic control and psychological outcomes in people with diabetes: a systematic review and meta-analysis. J. Diabetes Investig. 12, 1092–1103.

Ni, J.J., Xu, Q., Yan, S.S., Han, B.X., Zhang, H., et al., 2022. Gut microbiota and psychiatric disorders: a two-sample mendelian randomization study. Front. Microbiol. 12, 737197.

Nicholson, E.L., Bryant, R.A., Felmingham, K.L., 2014. Interaction of noradrenaline and cortisol predicts negative intrusive memories in posttraumatic stress disorder. Neurobiol. Learn. Mem. 112, 204–211.

Nicholson, A.A., Harricharan, S., Densmore, M., Neufeld, R.W.J., Ros, T., et al., 2020. Classifying heterogeneous presentations of PTSD via the default mode, central executive, and salience networks with machine learning. Neuroimage Clin. 27, 102262.

Nicoll, J.A.R., Wilkinson, D., Holmes, C., Steart, P., Markham, H., et al., 2003. Neuropathology of human Alzheimer disease after immunization with amyloid-beta peptide: a case report. Nat. Med. 9, 448–452.

Nicolucci, A.C., Hume, M.P., Martínez, I., Mayengbam, S., Walter, J., et al., 2017. Prebiotic reduces body fat and alters intestinal microbiota in children with overweight or obesity. Gastroenterology 153, 711–722.

Niculescu, A.B., Le-Niculescu, H., Levey, D.F., Phalen, P.L., Dainton, H.L., et al., 2017. Precision medicine for suicidality: from universality to subtypes and personalization. Mol. Psychiatry 22, 1250–1273.

Niedzwiecki, M.M., Rosa, M.J., Solano-González, M., Kloog, I., Just, A.C., et al., 2020. Particulate air pollution exposure during pregnancy and postpartum depression symptoms in women in Mexico City. Environ. Int. 134, 105325.

Nievergelt, C.M., Maihofer, A.X., Klengel, T., Atkinson, E.G., Chen, C.Y., et al., 2019. International meta-analysis of PTSD genome-wide association studies identifies sex- and ancestry-specific genetic risk loci. Nat. Commun. 10, 4558.

Nievergelt, C.M., Maihofer, A.X., Atkinson, E.G., Chen, C.Y., Choi, K.W., et al., 2024. Genome-wide association analyses identify 95 risk loci and provide insights into the neurobiology of post-traumatic stress disorder. Nat. Genet. 56, 792–808.

Nikolova, V.L., Smith, M.R.B., Hall, L.J., Cleare, A.J., Stone, J.M., Young, A.H., 2021. Perturbations in gut microbiota composition in psychiatric disorders: a review and meta-analysis. JAMA Psychiatry 78, 1343–1354.

Nikolova, V.L., Cleare, A.J., Young, A.H., Stone, J.M., 2023. Acceptability, tolerability, and estimates of putative treatment effects of probiotics as adjunctive treatment in patients with depression: a randomized clinical trial. JAMA Psychiatry, e231817.

Nilsson, E.E., McBirney, M., De Santos, S., King, S.E., Beck, D., et al., 2023. Multiple generation distinct toxicant exposures induce epigenetic transgenerational inheritance of enhanced pathology and obesity. Environ. Epigenet. 9, dvad006.

Niraula, A., Witcher, K.G., Sheridan, J.F., Godbout, J.P., 2019. Interleukin-6 induced by social stress promotes a unique transcriptional signature in the monocytes that facilitate anxiety. Biol. Psychiatry 85, 679–689.

Nisar, S., Bhat, A.A., Hashem, S., Syed, N., Yadav, S.K., et al., 2020. Genetic and neuroimaging approaches to understanding post-traumatic stress disorder. Int. J. Mol. Sci. 21, 4503.

Nishi, M., 2020. Effects of early-life stress on the brain and behaviors: implications of early maternal separation in rodents. Int. J. Mol. Sci. 21, 7212.

Nishimi, K., Neylan, T.C., Bertenthal, D., Seal, K.H., O'Donovan, A., 2022. Association of psychiatric disorders with incidence of SARS-CoV-2 breakthrough infection among vaccinated adults. JAMA Netw. Open 5, e227287.

Niu, Y., Hua, Y.Q., Wang, L.L., 2013. Imbalance between Th17 and Treg cells may play an important role in the development of chronic unpredictable mild stress-induced depression in mice. Neuroimmunomodulation 20, 39–50.

Niu, M., Kasai, A., Tanuma, M., Seiriki, K., Igarashi, H., et al., 2022. Claustrum mediates bidirectional and reversible control of stress-induced anxiety responses. Sci. Adv. 8, eabi6375.

Njau, S., Joshi, S.H., Espinoza, R., Leaver, A.M., Vasavada, M., et al., 2017. Neurochemical correlates of rapid treatment response to electroconvulsive therapy in patients with major depression. J. Psychiatry Neurosci. 42, 6–16.

Njotto, L.L., Simin, J., Fornes, R., Odsbu, I., Mussche, I., et al., 2023. Maternal and early-life exposure to antibiotics and the risk of autism and attention-deficit hyperactivity disorder in childhood: a Swedish population-based cohort study. Drug Saf. 46, 467–478.

Nobs, S.P., Tuganbaev, T., Elinav, E., 2019. Microbiome diurnal rhythmicity and its impact on host physiology and disease risk. EMBO Rep. 20, e47129.

Nocera, A., Nasrallah, H.A., 2022. The association of the gut microbiota with clinical features in schizophrenia. Behav. Sci. (Basel) 12, 89.

Nohesara, S., Abdolmaleky, H.M., Thiagalingam, S., 2023. Epigenetic aberrations in major psychiatric diseases related to diet and gut microbiome alterations. Genes 14, 1506.

Nolan, M., Roman, E., Nasa, A., Levins, K.J., O'Hanlon, E., et al., 2020. Hippocampal and amygdalar volume changes in major depressive disorder: a targeted review and focus on stress. Chronic Stress (Thousand Oaks) 4. 2470547020944553.

Nolen-Hoeksema, S., 1998. Ruminative coping with depression. In: Heckhausen, J., Dweck, C.S. (Eds.), Motivation and Self-Regulation across the Life Span. Cambridge University Press, Cambridge, UK, pp. 237–256.

Nolvi, S., Merz, E.C., Kataja, E.L., Parsons, C.E., 2022. Prenatal stress and the developing brain: postnatal environments promoting resilience. Biol. Psychiatry. S0006-3223(22)01853-4.

Nomura, Y., Rompala, G., Pritchett, L., Aushev, V., Chen, J., Hurd, Y.L., 2021. Natural disaster stress during pregnancy is linked to reprogramming of the placenta transcriptome in relation to anxiety and stress hormones in young offspring. Mol. Psychiatry 26, 6520–6530.

Nordmann, T.M., Dror, E., Schulze, F., Traub, S., Berishvili, E., et al., 2017. The role of inflammation in β-cell dedifferentiation. Sci. Rep. 7, 6285.

Northoff, G., Sibille, E., 2014. Why are cortical GABA neurons relevant to internal focus in depression? A cross-level model linking cellular, biochemical and neural network findings. Mol. Psychiatry 19, 966–977.

Northoff, G., Wainio-Theberge, S., Evers, K., 2020. Is temporo-spatial dynamics the "common currency" of brain and mind? In Quest of "Spatiotemporal Neuroscience". Phys. Life Rev. 33, 34–54.

Noshadi, N., Heidari, M., Naemi Kermanshahi, M., Zarezadeh, M., Sanaie, S., Ebrahimi-Mameghani, M., 2022. Effects of probiotics supplementation on CRP, IL-6, and length of ICU stay in traumatic brain injuries and multiple trauma patients: a systematic review and meta-analysis of randomized controlled trials. Evid. Based Complement. Alternat. Med., 4674000.

Notzon, S., Domschke, K., Holitschke, K., Ziegler, C., Arolt, V., et al., 2016. Attachment style and oxytocin receptor gene variation interact in influencing social anxiety. World J. Biol. Psychiatry 17, 76–83.

Noubiap, J.J., Nansseu, J.R., Nyaga, U.F., Sime, P.S., Francis, I., Bigna, J. J., 2019. Global prevalence of resistant hypertension: a meta-analysis of data from 3.2 million patients. Heart 105, 98–105.

Novais, F., Capela, J., Machado, S., Murillo-Rodriguez, E., Telles-Correia, D., 2021. Does dysbiosis increase the risk of developing schizophrenia? - a comprehensive narrative review. Curr. Top. Med. Chem. 21, 976–984.

Novak, P., Schmidt, R., Kontsekova, E., Zilka, N., Kovacech, B., et al., 2017. Safety and immunogenicity of the tau vaccine AADvac1 in patients with Alzheimer's disease: a randomised, double-blind, placebo-controlled, phase 1 trial. Lancet Neurol. 16, 123–134.

Novak, P., Schmidt, R., Kontsekova, E., Kovacech, B., Smolek, T., et al., 2018. FUNDAMANT: an interventional 72-week phase 1 follow-up study of AADvac1, an active immunotherapy against tau protein pathology in Alzheimer's disease. Alzheimers Res. Ther. 10, 108.

Novak, P., Kovacech, B., Katina, S., Schmidt, R., Scheltens, P., et al., 2021. ADAMANT: a placebo-controlled randomized phase 2 study of AADvac1, an active immunotherapy against pathological tau in Alzheimer's disease. Nat. Aging 1, 521–534.

Novoa, C., Salazar, P., Cisternas, P., Gherardelli, C., Vera-Salazar, R., et al., 2022. Inflammation context in Alzheimer's disease, a relationship intricate to define. Biol. Res. 55, 39.

Nowacka-Chmielewska, M.M., Paul-Samojedny, M., Bielecka-Wajdman, A.M., Barski, J.J., Obuchowicz, E., 2017. Alterations in VEGF expression induced by antidepressant drugs in female rats under chronic social stress. Exp. Ther. Med. 13, 723–730.

Nowak, G.J., Sheedy, K., Bursey, K., Smith, T.M., Basket, M., 2015. Promoting influenza vaccination: insights from a qualitative meta-analysis of 14 years of influenza-related communications research

by US Centers for Disease Control and Prevention (CDC). Vaccine 33, 2741–2756.

Noya, S.B., Colameo, D., Brüning, F., Spinnler, A., Mircsof, D., et al., 2019. The forebrain synaptic transcriptome is organized by clocks but its proteome is driven by sleep. Science 366, eaav2642.

Noyan, H., Erdağ, E., Tüzün, E., Yaylım, İ., Küçükhüseyin, Ö., et al., 2021. Association of the kynurenine pathway metabolites with clinical, cognitive features and IL-1β levels in patients with schizophrenia spectrum disorder and their siblings. Schizophr. Res. 229, 27–37.

Nummenmaa, L., Tuominen, L., 2018. Opioid system and human emotions. Br. J. Pharmacol. 175, 2737–2749.

Nuncio-Mora, L., Lanzagorta, N., Nicolini, H., Sarmiento, E., Ortiz, G., et al., 2023. The role of the microbiome in first episode of psychosis. Biomedicines 11, 1770.

Núñez-Rios, D.L., Martínez-Magaña, J.J., Nagamatsu, S.T., Andrade-Brito, D.E., Forero, D.A., et al., 2022. Central and peripheral immune dysregulation in posttraumatic stress disorder: convergent multi-omics evidence. Biomedicines 10, 1107.

Nuovo, G.J., Magro, C., Shaffer, T., Awad, H., Suster, D., et al., 2021. Endothelial cell damage is the central part of COVID-19 and a mouse model induced by injection of the S1 subunit of the spike protein. Ann. Diagn. Pathol. 51, 151682.

Nuss, P., 2015. Anxiety disorders and GABA neurotransmission: a disturbance of modulation. Neuropsychiatr. Dis. Treat. 11, 165–175.

Nutt, D., Erritzoe, D., Carhart-Harris, R., 2020. Psychedelic psychiatry's brave new world. Cell 181, 24–28.

Nwanaji-Enwerem, J.C., Van Der Laan, L., Kogut, K., Eskenazi, B., Holland, N., et al., 2021. Maternal adverse childhood experiences before pregnancy are associated with epigenetic aging changes in their children. Aging 13, 25653.

Nygren, M., Carstensen, J., Koch, F., Ludvigsson, J., Frostell, A., 2015. Experience of a serious life event increases the risk for childhood type 1 diabetes: the ABIS population-based prospective cohort study. Diabetologia 58, 1188–1197.

O'Connell, C.P., Goldstein-Piekarski, A.N., Nemeroff, C.B., Schatzberg, A.F., Debattista, C., et al., 2018. Antidepressant outcomes predicted by genetic variation in corticotropin-releasing hormone binding protein. Am. J. Psychiatry 175, 251–261.

O'Mahony, S.M., Felice, V.D., Nally, K., Savignac, H.M., Claesson, M.J., Scully, P., et al., 2014. Disturbance of the gut microbiota in early-life selectively affects visceral pain in adulthood without impacting cognitive or anxiety-related behaviors. Neuroscience 277, 885–901.

O'Mahony, S.M., Dinan, T.G., Cryan, J.F., 2017. The gut microbiota as a key regulator of visceral pain. Pain 158, S19–S28.

Obradovich, N., Migliorini, R., Paulus, M.P., Rahwan, I., 2018. Empirical evidence of mental health risks posed by climate change. Proc. Natl. Acad. Sci. USA 115, 10953–10958.

O'Bryant, S.E., Zhang, F., Johnson, L.A., Hall, J., Edwards, M., et al., 2018. A precision medicine model for targeted NSAID therapy in Alzheimer's disease. J. Alzheimers Dis. 66, 97–104.

O'Connor, T.G., Willoughby, M.T., Moynihan, J.A., Messing, S., Vallejo Sefair, A., et al., 2020. Early childhood risk exposures and inflammation in early adolescence. Brain Behav. Immun. 86, 22–29. https://doi.org/10.1016/j.bbi.2019.05.001.

O'Connor, D.B., Thayer, J.F., Vedhara, K., 2021. Stress and health: a review of psychobiological processes. Annu. Rev. Psychol. 72, 663–688.

O'Doherty, D.C., Chitty, K.M., Saddiqui, S., Bennett, M.R., Lagopoulos, J., 2015. A systematic review and meta-analysis of magnetic resonance imaging measurement of structural volumes in posttraumatic stress disorder. Psychiatry Res. 232, 1–33.

O'Doherty, D.C.M., Ryder, W., Paquola, C., Tickell, A., Chan, C., et al., 2018. White matter integrity alterations in post-traumatic stress disorder. Hum. Brain Mapp. 39, 1327–1338.

Oertel, W.H., Müller, H.H., Unger, M.M., Schade-Brittinger, C., Balthasar, K., et al., 2023. Transdermal nicotine treatment and progression of early Parkinson's disease. NEJM Evidence 2. EVIDoa2200311.

Oftedal, B.E., Hellesen, A., Erichsen, M.M., Bratland, E., Vardi, A., et al., 2015. Dominant mutations in the autoimmune regulator AIRE are associated with common organ-specific autoimmune diseases. Immunity 42, 1185–1196.

Oh, H.S.H., Rutledge, J., Nachun, D., Pálovics, R., Abiose, O., et al., 2023. Organ aging signatures in the plasma proteome track health and disease. Nature 624, 164–172.

Ohara, T., Doi, Y., Ninomiya, T., Hirakawa, Y., Hata, J., Iwaki, T., et al., 2011. Glucose tolerance status and risk of dementia in the community: the Hisayama Study. Neurology 77, 1126–1134.

Ohja, K., Gozal, E., Fahnestock, M., Cai, L., Cai, J., et al., 2018. Neuroimmunologic and neurotrophic interactions in autism spectrum disorders: Relationship to neuroinflammation. NeuroMolecular Med. 20, 161–173.

Ohka, S., Hao Tan, S., Kaneda, S., Fujii, T., Schiavo, G., 2022. Retrograde axonal transport of poliovirus and EV71 in motor neurons. Biochem. Biophys. Res. Commun. 626, 72–78.

Ojalehto, E., Zhan, Y., Jylhävä, J., Reynolds, C.A., Aslan, A.K.D., Karlsson, I.K., 2023. Genetically and environmentally predicted obesity in relation to cardiovascular disease: a nationwide cohort study. eClinicalMedicine 58, 101943.

Okada, N.J., Liu, J., Tsang, T., Nosco, E., McDonald, N.M., et al., 2022. Atypical cerebellar functional connectivity at 9 months of age predicts delayed socio-communicative profiles in infants at high and low risk for autism. J. Child Psychol. Psychiatry 63, 1002–1016.

Okamura, H., Yasugaki, S., Suzuki-Abe, H., Arai, Y., Sakurai, K., et al., 2022. Long-term effects of repeated social defeat stress on brain activity during social interaction in BALB/c mice. eNeuro 9. ENEURO.0068-22.2022.

O'Keane, V., Lightman, S., Patrick, K., Marsh, M., Papadopoulos, A.S., et al., 2013. Changes in the maternal hypothalamic-pituitary-adrenal axis during the early puerperium may be related to the postpartum 'blues'. J. Neuroendocrinol. 23, 1149–1155.

Oksanen, M., Petersen, A.J., Naumenko, N., Puttonen, K., Lehtonen, Š., et al., 2017. PSEN1 mutant iPSC-derived model reveals severe astrocyte pathology in Alzheimer's disease. Stem Cell Rep. 9, 1885–1897.

Okun, E., Mattson, M.P., Arumugam, T.V., 2010. Involvement of Fc receptors in disorders of the central nervous system. NeuroMolecular Med. 12, 164–178.

Okuyama, K., Ohwada, K., Sakurada, S., Sato, N., Sora, I., et al., 2007. The distinctive effects of acute and chronic psychological stress on airway inflammation in a murine model of allergic asthma. Allergol. Int. 56, 29–35.

Olajide, O.A., Iwuanyanwu, V.U., Adegbola, O.D., Al-Hindawi, A.A., 2022. SARS-CoV-2 spike glycoprotein S1 induces neuroinflammation in BV-2 microglia. Mol. Neurobiol. 59, 445–458.

Oliva, A., Torre, S., Taranto, P., Delvecchio, G., Brambilla, P., 2021. Neural correlates of emotional processing in panic disorder: a mini review of functional magnetic resonance imaging studies. J. Affect. Disord. 282, 906–914.

Olney, J.J., Warlow, S.M., Naffziger, E.E., Berridge, K.C., 2018. Current perspectives on incentive salience and applications to clinical disorders. Curr. Opin. Behav. Sci. 22, 59–69.

Olsen, A.L., Clemens, S.G., Feany, M.B., 2023. Nicotine-mediated rescue of α-synuclein toxicity requires synaptic vesicle glycoprotein 2 in Drosophila. Mov. Disord. 38, 244–255.

Olson, V.G., Rockett, H.R., Reh, R.K., Redila, V.A., Tran, P.M., et al., 2011. The role of norepinephrine in differential response to stress in an animal model of posttraumatic stress disorder. Biol. Psychiatry 70, 441–448.

Olson, K.E., Kosloski-Bilek, L.M., Anderson, K.M., Diggs, B.J., Clark, B.E., et al., 2015. Selective VIP receptor agonists facilitate immune transformation for dopaminergic neuroprotection in MPTP-intoxicated mice. J. Neurosci. 35, 16463–16478.

Olson, K.E., Namminga, K.L., Lu, Y., Schwab, A.D., Thurston, M.J., et al., 2021a. Safety, tolerability, and immune-biomarker profiling for year-long sargramostim treatment of Parkinson's disease. EBioMedicine 67, 103380.

Olson, K.E., Namminga, K.L., Lu, Y., Thurston, M.J., Schwab, A.D., et al., 2021b. Granulocyte-macrophage colony-stimulating factor mRNA and Neuroprotective Immunity in Parkinson's disease. Biomaterials 272, 120786.

Olson, K.E., Abdelmoaty, M.M., Namminga, K.L., Lu, Y., Obaro, H., et al., 2023. An open-label multiyear study of sargramostim-treated Parkinson's disease patients examining drug safety, tolerability, and immune biomarkers from limited case numbers. Transl. Neurodegener. 12, 26.

Olsson, A., Kross, E., Nordberg, S.S., Weinberg, A., Weber, J., et al., 2015. Neural and genetic markers of vulnerability to post-traumatic stress symptoms among survivors of the World Trade Center attacks. Soc. Cogn. Affect. Neurosci. 10, 863–868.

Olvera Alvarez, H.A., Kubzansky, L.D., Campen, M.J., Slavich, G.M., 2018. Early life stress, air pollution, inflammation, and disease: an integrative review and immunologic model of social-environmental adversity and lifespan health. Neurosci. Biobehav. Rev. 92, 226–242.

O'Mahony, S.M., Clarke, G., Dinan, T.G., Cryan, J.F., 2017a. Early-life adversity and brain development: is the microbiome a missing piece of the puzzle? Neuroscience 342, 37–54.

O'Mahony, S.M., Clarke, G., Dinan, T.G., Cryan, J.F., 2017b. Irritable bowel syndrome and stress-related psychiatric co-morbidities: focus on early life stress. Handb. Exp. Pharmacol. 239, 219–246.

Oosterhof, C.A., El Mansari, M., Merali, Z., Blier, P., 2016. Altered monoamine system activities after prenatal and adult stress: a role for stress resilience? Brain Res. 1642, 409–418.

Opel, N., Cearns, M., Clark, S., Toben, C., Grotegerd, D., et al., 2019. Large-scale evidence for an association between low-grade peripheral inflammation and brain structural alterations in major depression in the BiDirect study. J. Psychiatry Neurosci. 44, 423–431.

Orenstein, S.J., Kuo, S.H., Tasset, I., Arias, E., Koga, H., et al., 2013. Interplay of LRRK2 with chaperone-mediated autophagy. Nat. Neurosci. 16, 394–406.

Orr, C.F., Rowe, D.B., Mizuno, Y., Mori, H., Halliday, G.M., 2005. A possible role for humoral immunity in the pathogenesis of Parkinson's disease. Brain J. Neurol. 128, 2665–2674.

Orsolini, L., Sarchione, F., Vellante, F., Fornaro, M., Matarazzo, I., et al., 2018. Protein-C reactive as biomarker predictor of schizophrenia phases of illness? A systematic review. Curr. Neuropharmacol. 16, 583–606.

Orsolini, L., Chiappini, S., Volpe, U., Berardis, D., Latini, R., et al., 2019. Use of medicinal cannabis and synthetic cannabinoids in post-traumatic stress disorder (PTSD): a systematic review. Medicina (Kaunas) 55, 525.

Orsolini, L., Pompili, S., Volpe, U., 2023. C-Reactive Protein (CRP): a potent inflammation biomarker in psychiatric disorders. Adv. Exp. Med. Biol. 1411, 135–160.

Ortega, M.A., Álvarez-Mon, M.A., García-Montero, C., Fraile-Martínez, Ó., Monserrat, J., et al., 2023. Microbiota-gut-brain axis mechanisms in the complex network of bipolar disorders: potential clinical implications and translational opportunities. Mol. Psychiatry 28, 2645–2673.

Osborn, M., Rustom, N., Clarke, M., Litteljohn, D., Rudyk, C., et al., 2013. Antidepressant-like effects of erythropoietin: a focus on behavioural and hippocampal processes. PLoS One 8, e72813.

Osborne, L.M., Gilden, J., Kamperman, A.M., Hoogendijk, W.J.G., Spicer, J., Drexhage, H.A., et al., 2020a. T-cell defects and postpartum depression. Brain Behav. Immun. 87, 7.

Osborne, M.T., Shin, L.M., Mehta, N.N., Pitman, R.K., Fayad, Z.A., Tawakol, A., 2020b. Disentangling the links between psychosocial stress and cardiovascular disease. Circ. Cardiovasc. Imaging 13, e010931.

Osimo, E.F., Pillinger, T., Rodriguez, I.M., Khandaker, G.M., Pariante, C. M., Howes, O.D., 2020. Inflammatory markers in depression: a meta-analysis of mean differences and variability in 5,166 patients and 5,083 controls. Brain Behav. Immun. 87, 901–909.

Osorio, C., Sfera, A., Anton, J.J., Thomas, K.G., Andronescu, C.V., 2022. Virus-induced membrane fusion in neurodegenerative disorders. Front. Cell. Infect. Microbiol. 12, 845580.

Ostuzzi, G., Matcham, F., Dauchy, S., Barbui, C., Hotopf, M., 2018. Anti-depressants for the treatment of depression in people with cancer. Cochrane Database Syst. Rev. 4, CD011006.

Otsuka, T., Hori, H., Yoshida, F., Itoh, M., Lin, M., et al., 2021. Association of CRP genetic variation with symptomatology, cognitive function, and circulating proinflammatory markers in civilian women with PTSD. J. Affect. Disord. 279, 640–649.

Ott, B., Skurk, T., Hastreiter, L., Lagkouvardos, I., Fischer, S., 2017. Effect of caloric restriction on gut permeability, inflammation markers, and fecal microbiota in obese women. Sci. Rep. 7, 11955.

Ousdal, O.T., Milde, A.M., Craven, A.R., Ersland, L., Endestad, T., et al., 2019. Prefrontal glutamate levels predict altered amygdala–prefrontal connectivity in traumatized youths. Psychol. Med. 49, 1822–1830.

Ouyang, L., Li, D., Li, Z., Ma, X., Yuan, L., et al., 2022. IL-17 and TNF-β: Predictive biomarkers for transition to psychosis in ultra-high risk individuals. Front. Psychiatry 13, 1072380.

Owen, M.J., Legge, S.E., Rees, E., Walters, J.T., O'Donovan, M.C., 2023. Genomic findings in schizophrenia and their implications. Mol. Psychiatry 28, 3638–3647.

Oyola, M.G., Handa, R.J., 2017. Hypothalamic-pituitary-adrenal and hypothalamic-pituitary-gonadal axes: sex differences in regulation of stress responsivity. Stress 20, 476–494.

Ozsoy, S., Olguner Eker, O., Abdulrezzak, U., 2016. The effects of anti-depressants on neuropeptide Y in patients with depression and anxiety. Pharmacopsychiatry 49, 26–31.

Pabalan, N., Singian, E., Tabangay, L., Jarjanazi, H., Boivin, M.J., Ezeamama, A.E., 2018. Soil-transmitted helminth infection, loss of education and cognitive impairment in school-aged children: a systematic review and meta-analysis. PLoS Negl. Trop. Dis. 12, e0005523.

Pabst, O., Mowat, A.M., 2012. Oral tolerance to food protein. Mucosal Immunol. 5, 232–239.

Pace, T.W., Wingenfeld, K., Schmidt, I., Meinlschmidt, G., Hellhammer, D.H., et al., 2012. Increased peripheral NF-κB pathway activity in women with childhood abuse-related posttraumatic stress disorder. Brain Behav. Immun. 26, 13–17.

Padget, R.L., Zeitz, M.J., Blair, G.A., Wu, X., North, M.D., et al., 2024. Acute adenoviral infection elicits an arrhythmogenic substrate prior to myocarditis. Circ. Res. 134, 892–912.

Padrick, S.B., Doolittle, L.K., Brautigam, C.A., King, D.S., Rosen, M.K., 2011. Arp2/3 complex is bound and activated by two WASP proteins. Proc. Natl. Acad. Sci. USA 108, E472–E479.

Padro, C.J., Sanders, V.M., 2014. Neuroendocrine regulation of inflammation. Semin. Immunol. 26, 357–368.

Paik, D., Yao, L., Zhang, Y., Bae, S., D'Agostino, G.D., et al., 2022. Human gut bacteria produce TꞪ17-modulating bile acid metabolites. Nature 603, 907–912.

Painold, A., Mörkl, S., Kashofer, K., Halwachs, B., Dalkner, N., et al., 2019. A step ahead: exploring the gut microbiota in inpatients with bipolar disorder during a depressive episode. Bipolar Disord. 21, 40–49.

Paliwal, V.K., Garg, R.K., Gupta, A., Tejan, N., 2020. Neuromuscular presentations in patients with COVID-19. Neurol. Sci. 41, 3039–3056.

Palma-Gudiel, H., Prather, A.A., Lin, J., Oxendine, J.D., Guintivano, J., et al., 2021. HPA axis regulation and epigenetic programming of immune-related genes in chronically stressed and non-stressed mid-life women. Brain Behav. Immun. 92, 49–56.

Palmqvist, S., Schöll, M., Strandberg, O., Mattsson, N., Stomrud, E., et al., 2017. Earliest accumulation of β-amyloid occurs within the default-mode network and concurrently affects brain connectivity. Nat. Commun. 8, 1214.

Pan, X., Kaminga, A.C., Wen, S.W., Liu, A., 2018. Catecholamines in post-traumatic stress disorder: a systematic review and meta-analysis. Front. Mol. Neurosci. 11, 450.

Pandey, G.N., Rizavi, H.S., Ren, X., Bhaumik, R., Dwivedi, Y., 2014. Toll-like receptors in the depressed and suicide brain. J. Psychiatr. Res. 53, 62–68.

Pandey, G.N., Ren, X., Rizavi, H.S., Zhang, H., 2015. Abnormal gene expression of proinflammatory cytokines and their receptors in the lymphocytes of patients with bipolar disorder. Bipolar Disord. 17, 636–644.

Pandey, G.N., Rizavi, H.S., Zhang, H., Ren, X., 2018. Abnormal gene and protein expression of inflammatory cytokines in the postmortem brain of schizophrenia patients. Schizophr. Res. 192, 247–254.

Pandey, G.N., Rizavi, H.S., Bhaumik, R., Ren, X., 2019. Innate immunity in the postmortem brain of depressed and suicide subjects: role of Toll-like receptors. Brain Behav. Immun. 75, 101–111.

Pandey, G.N., Zhang, H., Sharma, A., Ren, X., 2021. Innate immunity receptors in depression and suicide: upregulated NOD-like receptors containing pyrin (NLRPs) and hyperactive inflammasomes in the postmortem brains of people who were depressed and died by suicide. J. Psychiatry Neurosci. 46, E538–E547.

Papachroni, K.K., Ninkina, N., Papapanagiotou, A., Hadjigeorgiou, G. M., Xiromerisiou, G., et al., 2007. Autoantibodies to alpha-synuclein in inherited Parkinson's disease. J. Neurochem. 101, 749–756.

Papadea, D., Dalla, C., Tata, D.A., 2023. Exploring a possible interplay between schizophrenia, oxytocin, and estrogens: a narrative review. Brain Sci. 13, 461.

Pape, J.C., Carrillo-Roa, T., Rothbaum, B.O., Nemeroff, C.B., Czamara, D., et al., 2018. DNA methylation levels are associated with CRF1 receptor antagonist treatment outcome in women with post-traumatic stress disorder. Clin. Epigenetics 10, 1–11.

Papuć, E., Kurys-Denis, E., Krupski, W., Rejdak, K., 2015. Humoral response against small heat shock proteins in Parkinson's disease. PLoS One 10, e0115480.

Parade, S.H., Novick, A.M., Parent, J., Seifer, R., Klaver, S.J., et al., 2017. Stress exposure and psychopathology alter methylation of the serotonin receptor 2A (HTR2A) gene in preschoolers. Dev. Psychopathol. 29, 1619–1626.

Paradis, J., Boureau, P., Moyon, T., Nicklaus, S., Parnet, P., et al., 2017. Perinatal western diet consumption leads to profound plasticity and GABAergic phenotype changes within hypothalamus and reward pathway from birth to sexual maturity in rat. Front. Endocrinol. 8, 216.

Parbo, P., Ismail, R., Hansen, K.V., Amidi, A., Mårup, F.H., et al., 2017. Brain inflammation accompanies amyloid in the majority of mild cognitive impairment cases due to Alzheimer's disease. Brain J. Neurol. 140, 2002–2011.

Pardos-Gascón, E.M., Narambuena, L., Leal-Costa, C., van-der Hofstadt-Román, C.J., 2021. Differential efficacy between cognitive-behavioral therapy and mindfulness-based therapies for chronic pain: Systematic review. Int. J. Clin. Health Psychol. 21, 100197.

Pardridge, W.M., 2020. Treatment of Alzheimer's disease and blood-brain barrier drug delivery. Pharmaceuticals (Basel) 13, 394.

Paredes, S., Cantillo, S., Candido, K.D., Knezevic, N.N., 2019. An association of serotonin with pain disorders and its modulation by estrogens. Int. J. Mol. Sci. 20, 5729.

Parel, S.T., Bennett, S.N., Cheng, C.J., Timmermans, O.C., Fiori, L.M., et al., 2023. Transcriptional signatures of early-life stress and antidepressant treatment efficacy. Proc. Natl. Acad. Sci. USA 120, e2305776120.

Parellada, E., Gassó, P., 2021. Glutamate and microglia activation as a driver of dendritic apoptosis: a core pathophysiological mechanism to understand schizophrenia. Transl. Psychiatry 11, 271.

Parellada, M., Andreu-Bernabeu, Á., Burdeus, M., San José Cáceres, A., Urbiola, E., et al., 2023. In search of biomarkers to guide interventions in autism spectrum disorder: a systematic review. Am. J. Psychiatry 180, 23–40.

Parisien, M., Lima, L.V., Dagostino, C., El-Hachem, N., Drury, G.L., et al., 2022. Acute inflammatory response via neutrophil activation protects against the development of chronic pain. Sci. Transl. Med. 14, eabj9954.

Park, C.L., 2016. Meaning making in the context of disasters. J. Clin. Psychol. 72, 1234–1246.

Park, C.H., Ishinaka, M., Takada, A., Kida, H., Kimura, T., et al., 2002. The invasion routes of neurovirulent A/Hong Kong/483/97 (H5N1) influenza virus into the central nervous system after respiratory infection in mice. Arch. Virol. 147, 1425–1436.

Park, C., Rosenblat, J.D., Brietzke, E., Pan, Z., Lee, Y., et al., 2019. Stress, epigenetics and depression: a systematic review. Neurosci. Biobehav. Rev. 102, 139–152.

Park, K., Park, S., Nagappan, A., Ray, N., Kim, J., et al., 2021. Probiotic Escherichia coli ameliorates antibiotic-associated anxiety responses in mice. Nutrients 13, 811.

Park, K.W., Ryu, H.S., Shin, E., Park, Y., Jeon, S.R., et al., 2023a. Ethnicity- and sex-specific genome wide association study on Parkinson's disease. NPJ Parkinsons Dis. 9, 141.

Park, T.Y., Jeon, J., Lee, N., Kim, J., Song, B., et al., 2023b. Co-transplantation of autologous Treg cells in a cell therapy for Parkinson's disease. Nature 619, 606–615.

Parker, A., Fonseca, S., Carding, S.R., 2020. Gut microbes and metabolites as modulators of blood-brain barrier integrity and brain health. Gut Microbes 11, 135–157.

Partiot, E., Hirschler, A., Colomb, S., Lutz, W., Claeys, T., et al., 2024. Brain exposure to SARS-CoV-2 virions perturbs synaptic homeostasis. Nat. Microbiol. 9, 1189–1206.

Partridge, J.G., Forcelli, P.A., Luo, R., Cashdan, J.M., Schulkin, J., et al., 2016. Stress increases GABAergic neurotransmission in CRF neurons of the central amygdala and bed nucleus stria terminalis. Neuropharmacology 107, 239–250.

Pase, M.P., Himali, J.J., Beiser, A.S., Aparicio, H.J., Satizabal, C.L., et al., 2017. Sugar- and artificially sweetened beverages and the risks of incident stroke and dementia: a prospective cohort study. Stroke 48, 1139–1146.

Passos, I.C., Vasconcelos-Moreno, M.P., Costa, L.G., Kunz, M., Brietzke, E., et al., 2015. Inflammatory markers in post-traumatic stress disorder: a systematic review, meta-analysis, and meta-regression. Lancet Psychiatry 2, 1002–1012.

Patas, K., Penninx, B.W., Bus, B.A., Vogelzangs, N., Molendijk, M.L., et al., 2014. Association between serum brain-derived neurotrophic factor and plasma interleukin-6 in major depressive disorder with melancholic features. Brain Behav. Immun. 36, 71–79.

Patel, K., Allen, S., Haque, M.N., Angelescu, I., Baumeister, D., et al., 2016a. Bupropion: a systematic review and meta-analysis of effectiveness as an antidepressant. Ther. Adv. Psychopharmacol. 6, 99–144.

Patel, N., Crider, A., Pandya, C.D., Ahmed, A.O., Pillai, A., 2016b. Altered mRNA levels of glucocorticoid receptor, mineralocorticoid receptor, and co-chaperones (FKBP5 and PTGES3) in the middle frontal gyrus of autism spectrum disorder subjects. Mol. Neurobiol. 53, 2090–2099.

Patel, S., Hill, M.N., Cheer, J.F., Wotjak, C.T., Holmes, A., 2017. The endocannabinoid system as a target for novel anxiolytic drugs. Neurosci. Biobehav. Rev. 76, 56–66.

Patel, D.R., Feucht, C., Brown, K., Ramsay, J., 2018a. Pharmacological treatment of anxiety disorders in children and adolescents: a review for practitioners. Transl. Pediatr. 7 (1), 23–35.

Patel, S., Masi, A., Dale, R.C., Whitehouse, A.J.O., Pokorski, I., et al., 2018b. Social impairments in autism spectrum disorder are related to maternal immune history profile. Mol. Psychiatry 23, 1794–1797.

Patel, V., Ts, J., Kamble, N., Yadav, R., Thennarassu, K., Pal, P.K., Reddy Yc, J., 2023. Prevalence and correlates of psychiatric comorbidity and multimorbidity in Parkinson's Disease and atypical parkinsonian syndromes. J. Geriatr. Psychiatry Neurol. 36, 155–163.

Pateras, I.S., Williams, C., Gianniou, D.D., Margetis, A.T., Avgeris, M., et al., 2023. Short term starvation potentiates the efficacy of chemotherapy in triple negative breast cancer via metabolic reprogramming. J. Transl. Med. 21, 169.

Pati, S., Irfan, W., Jameel, A., Ahmed, S., Shahid, R.K., 2023. Obesity and cancer: a current overview of epidemiology, pathogenesis, outcomes, and management. Cancers (Basel) 15, 485.

Patlola, S.R., Donohoe, G., McKernan, D.P., 2023a. Anti-inflammatory effects of 2nd generation antipsychotics in patients with schizophrenia: a systematic review and meta-analysis. J. Psychiatr. Res. 160, 126–136.

Patlola, S.R., Donohoe, G., McKernan, D.P., 2023b. Counting the toll of inflammation on schizophrenia-a potential role for toll-like receptors. Biomolecules 13, 1188.

Patrick, K.L., Bell, S.L., Weindel, C.G., Watson, R.O., 2019. Exploring the "Multiple-Hit Hypothesis" of neurodegenerative disease: bacterial infection comes up to bat. Front. Cell. Infect. Microbiol. 9, 138.

Patten, D.K., Schultz, B.G., Berlau, D.J., 2018. The safety and efficacy of low-dose naltrexone in the management of chronic pain and inflammation in multiple sclerosis, fibromyalgia, crohn's disease, and other chronic pain disorders. Pharmacotherapy 38, 382–389.

Patterson, Z.R., Abizaid, A., 2013. Stress induced obesity: lessons from rodent models of stress. Front. Neurosci. 7, 130.

Patterson, Z.R., Khazall, R., Mackay, H., Anisman, H., Abizaid, A., 2013. Central ghrelin signaling mediates the metabolic response of C57BL/6 male mice to chronic social defeat stress. Endocrinology 154, 1080–1091.

Patterson, S., Trupin, L., Hartogensis, W., DeQuattro, K., Lanata, C., et al., 2022. Perceived stress and prediction of worse disease activity and symptoms in a multiracial, multiethnic systemic lupus erythematosus cohort. Arthritis Care Res. https://doi.org/10.1002/acr.25076.

Paul, B.D., Snyder, S.H., Bohr, V.A., 2021. Signaling by cGAS-STING in neurodegeneration, neuroinflammation, and aging. Trends Neurosci. 44, 83–96.

Paul, K.C., Kusters, C., Furlong, M., Zhang, K., Yu, Y., et al., 2022a. Immune system disruptions implicated in whole blood epigenome-wide association study of depression among Parkinson's disease patients. Brain Behav. Immun. Health 26, 100530.

Paul, P., Kaul, R., Harfouche, M., Arabi, M., Al-Najjar, Y., et al., 2022b. The effect of microbiome-modulating probiotics, prebiotics and synbiotics on glucose homeostasis in type 2 diabetes: a systematic review, meta-analysis, and meta-regression of clinical trials. Pharmacol. Res. 185, 106520.

Paul, K.C., Krolewski, R.C., Lucumi Moreno, E., Blank, J., Holton, K.M., et al., 2023. A pesticide and iPSC dopaminergic neuron screen identifies and classifies Parkinson-relevant pesticides. Nat. Commun. 14, 2803.

Paulose, J.K., Cassone, C.V., Graniczkowska, K.B., Cassone, V.M., 2019. Entrainment of the circadian clock of the enteric Bacterium Klebsiella aerogenes by temperature cycles. iScience 19, 1202–1213.

Paxman, E.J., Boora, N.S., Kiss, D., Laplante, D.P., King, S., et al., 2018. Prenatal maternal stress from a natural disaster alters urinary metabolomic profiles in Project Ice Storm participants. Sci. Rep. 8, 12932.

Payant, M.A., Chee, M.J., 2021. Neural mechanisms underlying the role of fructose in overeating. Neurosci. Biobehav. Rev. 128, 346–357.

Paz Levy, D., Sheiner, E., Wainstock, T., Sergienko, R., Landau, D., et al., 2017. Evidence that children born at early term (37-38 6/7 weeks) are at increased risk for diabetes and obesity-related disorders. Am. J. Obstet. Gynecol. 217, 588.e1–588.e11.

Pedersen, A.F., Zachariae, R., Bovbjerg, D.H., 2009. Psychological stress and antibody response to influenza vaccination: a meta-analysis. Brain Behav. Immun. 23, 427–433.

Pedersen, L., Idorn, M., Olofsson, G.H., Lauenborg, B., Nookaew, I., et al., 2016. Voluntary running suppresses tumor growth through epinephrine-and IL-6-dependent NK cell mobilization and redistribution. Cell Metab. 23, 554–562.

Pedraz-Petrozzi, B., Elyamany, O., Rummel, C., Mulert, C., 2020. Effects of inflammation on the kynurenine pathway in schizophrenia – a systematic review. J. Neuroinflammation 17, 56.

Peh, A., O'Donnell, J.A., Broughton, B.R.S., Marques, F.Z., 2022. Gut microbiota and their metabolites in stroke: a double-edged sword. Stroke 53, 1788–1801.

Pellegrini, C., Antonioli, L., Calderone, V., Colucci, R., Fornai, M., Blandizzi, C., 2020. Microbiota-gut-brain axis in health and disease: is NLRP3 inflammasome at the crossroads of microbiota-gut-brain communications? Prog. Neurobiol. 191, 101806.

Peña, C.J., Kronman, H.G., Walker, D.M., Cates, H.M., Bagot, R.C., et al., 2017. Early life stress confers lifelong stress susceptibility in mice via ventral tegmental area OTX2. Science 356, 1158–1188.

Pendlebury, S.T., Rothwell, P.M., 2009. Prevalence, incidence, and factors associated with pre-stroke and post-stroke dementia: a systematic review and meta-analysis. Lancet Neurol. 8, 1006–1018.

Penedo, M.A., Rivera-Baltanás, T., Pérez-Rodríguez, D., Allen, J., Borrajo, A., et al., 2021. The role of dopamine receptors in lymphocytes and their changes in schizophrenia. Brain Behav. Immun. Health 12, 100199.

Peng, J., Peng, L., Stevenson, F.F., Doctrow, S.R., Andersen, J.K., 2007. Iron and paraquat as synergistic environmental risk factors in sporadic Parkinson's disease accelerate age-related neurodegeneration. J. Neurosci. 27, 6914–6922.

Peng, W., Xing, Z., Yang, J., Wang, Y., Wang, W., et al., 2014. The efficacy of erythropoietin in treating experimental traumatic brain injury: a systematic review of controlled trials in animal models. J. Neurosurg. 121, 653–664.

Peng, J., Gu, N., Zhou, L., Eyo, U.B., Murugan, M., et al., 2016. Microglia and monocytes synergistically promote the transition from acute to chronic pain after nerve injury. Nat. Commun. 7, 12029.

Peng, Z., Peng, S., Lin, K., Zhao, B., Wei, L., et al., 2022. Chronic stress-induced depression requires the recruitment of peripheral Th17 cells into the brain. J. Neuroinflammation 19, 186.

Peng, K., Li, X., Yang, D., Chan, S.C.W., Zhou, J., et al., 2023. Risk of autoimmune diseases following COVID-19 and the potential protective effect from vaccination: a population-based cohort study. EClinicalMedicine 63, 102154.

Penninx, B.W., 2017. Depression and cardiovascular disease: epidemiological evidence on their linking mechanisms. Neurosci. Biobehav. Rev. 74, 277–286.

Penteado, S.H., Teodorov, E., Kirsten, T.B., Eluf, B.P., Reis-Silva, T.M., et al., 2014. Prenatal lipopolysaccharide disrupts maternal behavior, reduces nest odor preference in pups, and induces anxiety: Studies of F1 and F2 generations. Eur. J. Pharmacol. 738, 342–351.

Peralta-Marzal, L.N., Prince, N., Bajic, D., Roussin, L., Naudon, L., et al., 2021. The impact of gut microbiota-derived metabolites in autism spectrum disorders. Int. J. Mol. Sci. 22, 10052.

Perego, C., Fumagalli, S., De Simoni, M.G., 2013. Three-dimensional confocal analysis of microglia/macrophage markers of polarization in experimental brain injury. J. Vis. Exp. 79.

Pereira, O.C., Bernardi, M.M., Gerardin, D.C., 2006. Could neonatal testosterone replacement prevent alterations induced by prenatal stress in male rats? Life Sci. 78, 2767–2771.

Perez-Rando, M., Carceller, H., Castillo-Gomez, E., Bueno-Fernandez, C., García-Mompó, C., et al., 2022. Impact of stress on inhibitory neuronal circuits, our tribute to Bruce McEwen. Neurobiol. Stress 19, 100460.

Perkins, D.O., Jeffries, C.D., Addington, J., Bearden, C.E., Cadenhead, K. S., et al., 2015. Towards a psychosis risk blood diagnostic for persons experiencing high-risk symptoms: preliminary results from the NAPLS project. Schizophr. Bull. 41, 419–428.

Perl, O., Duek, O., Kulkarni, K.R., Gordon, C., Krystal, J.H., et al., 2023. Neural patterns differentiate traumatic from sad autobiographical memories in PTSD. Nat. Neurosci. 26, 2226–2236.

Perron, H., Lang, A., 2010. The human endogenous retrovirus link between genes and environment in multiple sclerosis and in multifactorial diseases associating neuroinflammation. Clin. Rev. Allergy Immunol. 39, 51–61.

Perron, H., Hamdani, N., Faucard, R., Lajnef, M., Jamain, S., et al., 2012. Molecular characteristics of Human Endogenous Retrovirus type-W in schizophrenia and bipolar disorder. Transl. Psychiatry 2, e201.

Perroud, N., Rutembesa, E., Paoloni-Giacobino, A., Mutabaruka, J., Mutesa, L., et al., 2014. The Tutsi genocide and transgenerational transmission of maternal stress: epigenetics and biology of the HPA axis. World J. Biol. Psychiatry 15, 334–345.

Perry, V.H., Nicoll, J.A.R., Holmes, C., 2010. Microglia in neurodegenerative disease. Nat. Rev. Neurol. 6, 193–201.

Persaud, N.S., Cates, H.M., 2022. The epigenetics of anxiety pathophysiology: a DNA methylation and histone modification focused review. eNeuro. ENEURO.0109-21.2021.

Peruzzolo, T.L., Pinto, J.V., Roza, T.H., Shintani, A.O., Anzolin, A.P., et al., 2022. Inflammatory and oxidative stress markers in posttraumatic stress disorder: a systematic review and meta-analysis. Mol. Psychiatry 27, 3150–3163.

Peters, A., 2023. Ambient air pollution and Alzheimer's disease: the role of the composition of fine particles. Proc. Natl. Acad. Sci. USA 120, e2220028120.

Peters, L.A., Perrigoue, J., Mortha, A., Iuga, A., Song, W., et al., 2017. A functional genomics predictive network model identifies regulators of inflammatory bowel disease. Nat. Genet. 49, 1437–1449.

Peters, K., McDonald, T., Muhammad, F., Walsh, M., Drenen, K., et al., 2023. A2Ar-dependent PD-1+ and TIGIT+ Treg cells have distinct homing requirements to suppress autoimmune uveitis in mice. Mucosal Immunol. 16, 422–431.

Petersen, C.L., Chen, J.Q., Salas, L.A., Christensen, B.C., 2020. Altered immune phenotype and DNA methylation in panic disorder. Clin. Epigenetics 12, 177.

Peterson, A., Thome, J., Frewen, P., Lanius, R.A., 2014. Resting-state neuroimaging studies: a new way of identifying differences and similarities among the anxiety disorders? Can. J. Psychiatr. 59, 294–300.

Petralia, M.C., Mazzon, E., Fagone, P., Basile, M.S., Lenzo, V., et al., 2020. Pathogenic contribution of the Macrophage migration inhibitory factor family to major depressive disorder and emerging tailored therapeutic approaches. J. Affect. Disord. 263, 15–24.

Pflug, N., Kluth, S., Vehreschild, J.J., Bahlo, J., Tacke, D., et al., 2016. Efficacy of antineoplastic treatment is associated with the use of antibiotics that modulate intestinal microbiota. Onco Targets Ther 5, e1150399.

Philbert, J., Beeské, S., Belzung, C., Griebel, G., 2015. The CRF receptor antagonist SSR125543 prevents stress-induced long-lasting sleep disturbances in a mouse model of PTSD: comparison with paroxetine and d-cycloserine. Behav. Brain Res. 279, 41–46.

Philip, N.S., Barredo, J., Aiken, E., Larson, V., Jones, R.N., et al., 2019. Theta-burst transcranial magnetic stimulation for posttraumatic stress disorder. Am. J. Psychiatry 176, 939–948.

Philip, N.S., Ramanathan, D., Gamboa, B., Brennan, M.C., Kozel, F.A., et al., 2023. Repetitive transcranial magnetic stimulation for depression and posttraumatic stress disorder in veterans with mild traumatic brain injury. Neuromodulation 26, 878–884.

Philippens, I., Böszörményi, K.P., Wubben, J.A.M., Fagrouch, Z.C., van Driel, N., et al., 2022. Brain inflammation and intracellular α-Synuclein aggregates in macaques after SARS-CoV-2 infection. Viruses 14, 776.

Phillips, C., Fahimi, A., 2018. Immune and neuroprotective effects of physical activity on the brain in depression. Front. Neurosci. 12, 498.

Phillips, J.L., Batten, L.A., Tremblay, P., Aldosary, F., Du, L., et al., 2015. Impact of monoamine-related gene polymorphisms on hippocampal volume in treatment-resistant depression. Acta Neuropsychiatr. 27, 353–361.

Phillips, C.M., Shivappa, N., Hébert, J.R., Perry, I.J., 2017. Dietary inflammatory index and mental health: a cross-sectional analysis of the relationship with depressive symptoms, anxiety and well-being in adults. Clin. Nutr. pii: S0261-5614(17)30312-6.

Phillips, J.L., Norris, S., Talbot, J., Hatchard, T., Ortiz, A., et al., 2020. Single and repeated ketamine infusions for reduction of suicidal ideation in treatment-resistant depression. Neuropsychopharmacology 45, 606–612.

Phillips, M.C., LaRocque, R.C., Thompson, G.R., 2024. Infectious diseases in a changing climate. JAMA 331, 1318–1319.

Piantadosi, S.C., French, B.J., Poe, M.M., Timić, T., Marković, B.D., et al., 2016. Sex-dependent anti-stress effect of an α5 subunit containing GABAA receptor positive allosteric modulator. Front. Pharmacol. 7, 446.

Piaserico, S., Papadavid, E., Cecere, A., Orlando, G., Theodoropoulos, K., et al., 2023. Coronary microvascular dysfunction in asymptomatic patients with severe psoriasis. J. Invest. Dermatol. 143, 1929–1936.e2.

Piber, D., Cho, J.H., Lee, O., Lamkin, D.M., Olmstead, R., Irwin, M.R., 2022. Sleep disturbance and activation of cellular and transcriptional mechanisms of inflammation in older adults. Brain Behav. Immun. 106, 67–75.

Picard, M., McEwen, B.S., 2018. Psychological stress and mitochondria: a systematic review. Psychosom. Med. 80, 141–153.

Picard, K., Bisht, K., Poggini, S., Garofalo, S., Golia, M.T., et al., 2021. Microglial-glucocorticoid receptor depletion alters the response of hippocampal microglia and neurons in a chronic unpredictable mild stress paradigm in female mice. Brain Behav. Immun. 97, 423–439.

Pierce, Z.P., Black, J.M., 2022. Stress and susceptibility: a systematic review of prenatal epigenetic risks for developing post-traumatic stress disorder. Trauma Violence Abuse 24, 2648–2660.

Pierce, Z.P., Johnson, E.R., Kim, I.A., Lear, B.E., Mast, A.M., Black, J.M., 2023. Therapeutic interventions impact brain function and promote post-traumatic growth in adults living with post-traumatic stress disorder: a systematic review and meta-analysis of functional magnetic resonance imaging studies. Front. Psychol. 14, 1074972.

Pierrehumbert, B., Torrisi, R., Laufer, D., Halfon, O., Ansermet, F., et al., 2010. Oxytocin response to an experimental psychosocial challenge in adults exposed to traumatic experiences during childhood or adolescence. Neuroscience 166, 168–177.

Pigeon, S., Lonergan, M., Rotondo, O., Pitman, R.K., Brunet, A., 2022. Impairing memory reconsolidation with propranolol in healthy and clinical samples: a meta-analysis. J. Psychiatry Neurosci. 47, 109–122.

Pigneur, B., Sokol, H., 2016. Fecal microbiota transplantation in inflammatory bowel disease: the quest for the holy grail. Mucosal Immunol. 9, 1360–1365.

Piirainen, S., Youssef, A., Song, C., Kalueff, A.V., Landreth, G.E., et al., 2017. Psychosocial stress on neuroinflammation and cognitive dysfunctions in Alzheimer's disease: the emerging role for microglia? Neurosci. Biobehav. Rev. 77, 148–164.

Pilkington, P.D., Bishop, A., Younan, R., 2021. Adverse childhood experiences and early maladaptive schemas in adulthood: a systematic review and meta-analysis. Clin. Psychol. Psychother. 28, 569–584.

Pilsner, J.R., Parker, M., Sergeyev, O., Suvorov, A., 2017. Spermatogenesis disruption by dioxins: epigenetic reprograming and windows of susceptibility. Reprod. Toxicol. 69, 221–229.

Pinciotti, C.M., Riemann, B.C., Abramowitz, J.S., 2021. Intolerance of uncertainty and obsessive-compulsive disorder dimensions. J. Anxiety Disord. 81, 102417.

Pinho-Ribeiro, F.A., Deng, L., Neel, D.V., Erdogan, O., Basu, H., et al., 2023. Bacteria hijack a meningeal neuroimmune axis to facilitate brain invasion. Nature 615, 472–481.

Pinto-Sanchez, M.I., Hall, G.B., Ghajar, K., Nardelli, A., Bolino, C., et al., 2017. Probiotic bifidobacterium longum NCC3001 reduces depression scores and alters brain activity: a pilot study in patients with irritable bowel syndrome. Gastroenterology 153, 448–459.

Pisanu, A., Lecca, D., Mulas, G., Wardas, J., Simbula, G., et al., 2014. Dynamic changes in pro- and anti-inflammatory cytokines in microglia after PPAR-γ agonist neuroprotective treatment in the MPTPp mouse model of progressive Parkinson's disease. Neurobiol. Dis. 71, 280–291.

Pitake, S., Debrecht, J., Mishra, S.K., 2017. Brain natriuretic peptide (BNP) expressing sensory neurons are not involved in acute, inflammatory or neuropathic pain. Mol. Pain 13. 1744806917736993.

Pitman, R.K., Gilbertson, M.W., Gurvits, T.V., May, F.S., Lasko, N.B., Orr, S.P., 2006. Harvard/VA PTSD twin study investigators. Ann. N. Y. Acad. Sci. 1071, 242–254.

Pitman, R.K., Rasmusson, A.M., Koenen, K.C., Shin, L.M., Orr, S.P., 2012. Biological studies of post-traumatic stress disorder. Nat. Rev. Neurosci. 13, 769–787.

Pittenger, C., 2021. Pharmacotherapeutic strategies and new targets in OCD. Curr. Top. Behav. Neurosci. 49, 331–384.

Pivac, N., Vuic, B., Sagud, M., Nedic Erjavec, G., Nikolac Perkovic, M., et al., 2023. PTSD, immune system, and inflammation. Adv. Exp. Med. Biol. 1411, 225–262.

Pizarro-Cerdá, J., Chorev, D.S., Geiger, B., Cossart, P., 2017. The diverse family of Arp2/3 complexes. Trends Cell Biol. 27, 93–100.

Plantinga, L., Bremner, J.D., Miller, A.H., Jones, D.P., Veledar, E., et al., 2013. Association between posttraumatic stress disorder and inflammation: a twin study. Brain Behav. Immun. 30, 125–132.

Playfair, J.H.L., Chain, B.M., 2013. Immunology at a Glance, tenth ed. Wiley Blackwell, Oxford, UK.

Plog, B.A., Nedergaard, M., 2017. The glymphatic system in central nervous system health and disease: past, present, and future. Annu. Rev. Pathol. 13, 379–394.

Plovier, H., Everard, A., Druart, C., Depommier, C., Van Hul, M., et al., 2017. A purified membrane protein from akkermansia muciniphila or the pasteurized bacterium improves metabolism in obese and diabetic mice. Nat. Med. 23, 107–113.

Poganik, J.R., Zhang, B., Baht, G.S., Tyshkovskiy, A., Deik, A., et al., 2023. Biological age is increased by stress and restored upon recovery. Cell Metab. 35 (5), 807–820.e5.

Pohl, C.S., Medland, J.E., Moeser, A.J., 2015. Early-life stress origins of gastrointestinal disease: animal models, intestinal pathophysiology, and translational implications. Am. J. Physiol. Gastrointest. Liver Physiol. 309, G927–G941.

Polansky, H., Javaherian, A., 2015. The latent cytomegalovirus decreases telomere length by microcompetition. Open Med. (Wars) 10, 294–296.

Polheber, J.P., Matchock, R.L., 2014. The presence of a dog attenuates cortisol and heart rate in the Trier Social Stress Test compared to human friends. J. Behav. Med. 37, 860–867.

Poli, A., Pozza, A., Orrù, G., Conversano, C., Ciacchini, R., et al., 2022. Neurobiological outcomes of cognitive behavioral therapy for obsessive-compulsive disorder: a systematic review. Front. Psychiatry 13, 1063116.

Policicchio, S., Ahmad, A.N., Powell, J.F., Proitsi, P., 2017. Rheumatoid arthritis and risk for Alzheimer's disease: a systematic review and meta-analysis and a Mendelian Randomization study. Sci. Rep. 7, 12861.

Poller, W.C., Downey, J., Mooslechner, A.A., Khan, N., Li, L., et al., 2022. Brain motor and fear circuits regulate leukocytes during acute stress. Nature 607, 578–584.

Poloni, T.E., Medici, V., Moretti, M., Visonà, S.D., Cirrincione, A., et al., 2021. COVID-19-related neuropathology and microglial activation in elderly with and without dementia. Brain Pathol. 31, e12997.

Pompoli, A., Furukawa, T.A., Efthimiou, O., Imai, H., Tajika, A., Salanti, G., 2018. Dismantling cognitive-behaviour therapy for panic disorder: a systematic review and component network meta-analysis. Psychol. Med. 48, 1945–1953.

Pope, B.S., Wood, S.K., 2020. Advances in understanding mechanisms and therapeutic targets to treat comorbid depression and cardiovascular disease. Neurosci. Biobehav. Rev. 116, 337–349.

Popkes, M., Valenzano, D.R., 2020. Microbiota-host interactions shape ageing dynamics. Philos. Trans. R. Soc. Lond. B Biol. Sci. 375 (1808), 20190596.

Portugal, B., Artaud, F., Degaey, I., Roze, E., Fournier, A., et al., 2023. Association of physical activity and Parkinson disease in women: long-term follow-up of the E3N cohort study. Neurology 101, e386–e398.

Posillico, C.K., Schwarz, J.M., 2016. An investigation into the effects of antenatal stressors on the postpartum neuroimmune profile and depressive-like behaviors. Behav. Brain Res. 298, 218–228.

Potier, F., Degryse, J.M., de Saint-Hubert, M., 2018. Impact of caregiving for older people and pro-inflammatory biomarkers among caregivers: a systematic review. Aging Clin. Exp. Res. 30, 119–132.

Pouget, J.G., Han, B., Wu, Y., Mignot, E., et al., 2019. Cross-disorder analysis of schizophrenia and 19 immune-mediated diseases identifies shared genetic risk. Hum. Mol. Genet. 28, 3498–3513.

Poulter, M.O., Du, L., Weaver, I.C., Palkovits, M., Faludi, G., et al., 2008. GABAA receptor promoter hypermethylation in suicide brain: implications for the involvement of epigenetic processes. Biol. Psychiatry 64, 645–652.

Poulter, M.O., Du, L., Zhurov, V., Merali, Z., Anisman, H., 2010. Plasticity of the GABA(A) receptor subunit cassette in response to stressors in reactive versus resilient mice. Neuroscience 165, 1039–1051.

Pournajafi-Nazarloo, H., Kenkel, W., Mohsenpour, S.R., Sanzenbacher, L., Saadat, H., et al., 2013. Exposure to chronic isolation modulates receptors mRNAs for oxytocin and vasopressin in the hypothalamus and heart. Peptides 43, 20–26.

Powell, N.D., Bailey, M.T., Mays, J.W., Stiner-Jones, L.M., Hanke, M.L., et al., 2009. Repeated social defeat activates dendritic cells and enhances Toll-like receptor dependent cytokine secretion. Brain Behav. Immun. 23, 225–231.

Powell, N., Walker, M.M., Talley, N.J., 2017. The mucosal immune system: master regulator of bidirectional gut-brain communications. Nat. Rev. Gastroenterol. Hepatol. 14, 143–159.

Powers, K.M., Kay, D.M., Factor, S.A., Zabetian, C.P., Higgins, D.S., et al., 2008. Combined effects of smoking, coffee, and NSAIDs on Parkinson's disease risk. Mov. Disord. 23, 88–95.

Pratchett, L.C., Yehuda, R., 2011. Foundations of posttraumatic stress disorder: does early life trauma lead to adult posttraumatic stress disorder?. Dev. Psychopathol. 23, 477–491.

Prather, A.A., Pressman, S.D., Miller, G.E., Cohen, S., 2021. Temporal links between self-reported sleep and antibody responses to the influenza vaccine. Int. J. Behav. Med. 28, 151–158.

Prem, S., Millonig, J.H., DiCicco-Bloom, E., 2020. Dysregulation of neurite outgrowth and cell migration in autism and other neurodevelopmental disorders. Adv. Neurobiol. 25, 109–153.

Pressman, S.D., Cohen, S., Miller, G.E., Barkin, A., Rabin, B.S., et al., 2005. Loneliness, social network size, and immune response to influenza vaccination in college freshmen. Health Psychol. 24, 297–306.

Pressman, S.D., Jenkins, B.N., Moskowitz, J.T., 2019. Positive affect and health: What do we know and where next should we go? Annu. Rev. Psychol. 70, 627–650.

Prieto, G.A., Snigdha, S., Baglietto-Vargas, D., Smith, E.D., Berchtold, N.C., et al., 2015. Synapse-specific IL-1 receptor subunit reconfiguration augments vulnerability to IL-1β in the aged hippocampus. Proc. Natl. Acad. Sci. USA 112, E5078–E5087.

Priya, P.K., Rajappa, M., Kattimani, S., Mohanraj, P.S., Revathy, G., 2016. Association of neurotrophins, inflammation and stress with suicide risk in young adults. Clin. Chim. Acta 457, 41–45.

Proal, A.D., VanElzakker, M.B., Aleman, S., Bach, K., Boribong, B.P., et al., 2023. SARS-CoV-2 reservoir in post-acute sequelae of COVID-19 (PASC). Nat. Immunol. 24, 1616–1627.

Prowse, N., Hayley, S., 2021. Microglia and BDNF at the crossroads of stressor related disorders: Towards a unique trophic phenotype. Neurosci. Biobehav. Rev. 131, 135–163.

Punnonen, J., Miller, J.L., Collier, T.J., Spencer, J.R., 2015. Agonists of the tissue-protective erythropoietin receptor in the treatment of Parkinson's disease. Curr. Top. Med. Chem. 15, 955–969.

Pusceddu, M.M., Kelly, P., Ariffin, N., Cryan, J.F., Clarke, G., et al., 2015. n-3 PUFAs have beneficial effects on anxiety and cognition in female rats: effects of early life stress. Psychoneuroendocrinology 58, 79–90.

Pyrkov, T.V., Avchaciov, K., Tarkhov, A.E., Menshikov, L.I., Gudkov, A.V., Fedichev, P.O., 2021. Longitudinal analysis of blood markers reveals progressive loss of resilience and predicts human lifespan limit. Nat. Commun. 12, 2765.

Qi, K., Yu, Y., Guan, J., Zhang, J., Lu, W., Wei, Y., 2022. The inflammatory effect of epigenetic factors and modifications in depressive disorder: a review. Cent. Nerv. Syst. Agents Med. Chem. 22, 15–30.

Qian, L., He, X., Gao, F., Fan, Y., Zhao, B., et al., 2022. Estimation of the bidirectional relationship between schizophrenia and inflammatory bowel disease using the mendelian randomization approach. Schizophrenia (Heidelb) 8, 31.

Qian, F., Tintle, N., Jensen, P.N., Lemaitre, R.N., Imamura, F., et al., 2023. Omega-3 fatty acid biomarkers and incident atrial fibrillation. J. Am. Coll. Cardiol. 82, 336–349.

Qiao, Y., Wu, M., Feng, Y., Zhou, Z., Chen, L., Chen, F., 2018. Alterations of oral microbiota distinguish children with autism spectrum disorders from healthy controls. Sci. Rep. 8, 1597.

Qin, Y., Liu, Y., Hao, W., Decker, Y., Tomic, I., et al., 2016. Stimulation of TLR4 attenuates Alzheimer's disease-related symptoms and pathology in Tau-transgenic mice. J. Immunol. 197, 3281–3292.

Qin, Q., Zhi, L.T., Li, X.T., Yue, Z.Y., Li, G.Z., et al., 2017. Effects of LRRK2 inhibitors on nigrostriatal dopaminergic neurotransmission. CNS Neurosci. Ther. 23, 162–173.

Qin, Q., Teng, Z., Liu, C., Li, Q., Yin, Y., Tang, Y., 2021. TREM2, microglia, and Alzheimer's disease. Mech. Ageing Dev. 195, 111438.

Qing, H., Desrouleaux, R., Israni-Winger, K., Mineur, Y.S., Fogelman, N., et al., 2020. Origin and function of stress-induced IL-6 in murine models. Cell 182, 372–387.e14.

Quadros, A., Patel, N., Crescentini, R., Crawford, F., Paris, D., et al., 2003. Increased TNFalpha production and Cox-2 activity in organotypic brain slice cultures from APPsw transgenic mice. Neurosci. Lett. 353, 66–68.

Quagliato, L.A., Nardi, A.E., 2018. Cytokine alterations in panic disorder: a systematic review. J. Affect. Disord. 228, 91–96.

Quagliato, L.A., Nardi, A.E., 2022. Cytokine profile in drug-naïve panic disorder patients. Transl. Psychiatry 12, 75.

Quagliato, L.A., de Matos, U., Nardi, A.E., 2021. Maternal immune activation generates anxiety in offspring: a translational meta-analysis. Transl. Psychiatry 11, 245.

Quah, S.K.L., McIver, L., Roberts, A.C., Santangelo, A.M., 2020. Trait anxiety mediated by amygdala serotonin transporter in the common marmoset. J. Neurosci. 40, 4739–4749.

Quan, L., Xu, X., Cui, Y., Han, H., Hendren, R.L., Zhao, L., You, X., 2022. A systematic review and meta-analysis of the benefits of a gluten-free diet and/or casein-free diet for children with autism spectrum disorder. Nutr. Rev. 80, 1237–1246.

Quattrone, A., Barbagallo, G., Cerasa, A., Stoessl, A.J., 2018. Neurobiology of placebo effect in Parkinson's disease: what we have learned and where we are going. Mov. Disord. 33, 1213–1227.

Quddos, F., Hubshman, Z., Tegge, A., Sane, D., Marti, E., et al., 2023. Semaglutide and Tirzepatide reduce alcohol consumption in individuals with obesity. Sci. Rep. 13, 20998.

Querdasi, F.R., Vogel, S.C., Thomason, M.E., Callaghan, B.L., Brito, N. H., 2023. A comparison of the infant gut microbiome before versus after the start of the COVID-19 pandemic. Sci. Rep. 13, 13289.

Quidé, Y., Bortolasci, C.C., Spolding, B., Kidnapillai, S., Watkeys, O.J., et al., 2019. Association between childhood trauma exposure and pro-inflammatory cytokines in schizophrenia and bipolar-I disorder. Psychol. Med. 49, 2736–2744.

Quidé, Y., Bortolasci, C.C., Spolding, B., Kidnapillai, S., Watkeys, O.J., et al., 2021. Systemic inflammation and grey matter volume in schizophrenia and bipolar disorder: moderation by childhood trauma severity. Prog. Neuro-Psychopharmacol. Biol. Psychiatry 105, 110013.

Quik, M., Chen, L., Parameswaran, N., Xie, X., Langston, J.W., McCallum, S.E., 2006. Chronic oral nicotine normalizes dopaminergic function and synaptic plasticity in 1-methyl-4-phenyl-1,2,3,6-tetrahydropyridinelesioned primates. J. Neurosci. 26, 4681–4689.

Qureshi, F., Adams, J., Hanagan, K., Kang, D.W., Krajmalnik-Brown, R., Hahn, J., 2020. Multivariate analysis of fecal metabolites from children with autism spectrum disorder and gastrointestinal symptoms before and after microbiota transfer therapy. J. Pers. Med. 10, 152.

Rabl, U., Meyer, B.M., Diers, K., Bartova, L., Berger, A., et al., 2014. Additive gene–environment effects on hippocampal structure in healthy humans. J. Neurosci. 34, 9917–9926.

Racette, B.A., Gross, A., Vouri, S.M., Camacho-Soto, A., Willis, A.W., Searles Nielsen, S., 2018. Immunosuppressants and risk of Parkinson disease. Ann. Clin. Transl. Neurol. 5, 870–875.

Racicot, K., Mor, G., 2017. Risks associated with viral infections during pregnancy. J. Clin. Invest. 127, 1591–1599.

Radad, K., Moldzio, R., Al-Shraim, M., Kranner, B., Krewenka, C., et al., 2017. Recent advances on the role of neurogenesis in the adult brain: therapeutic potential in Parkinson's and Alzheimer's diseases. CNS Neurol. Disord. Drug Targets 16, 740–748.

Radke, J., Meinhardt, J., Aschman, T., Chua, R.L., Farztdinov, V., et al., 2024. Proteomic and transcriptomic profiling of brainstem, cerebellum and olfactory tissues in early- and late-phase COVID-19. Nat. Neurosci. 27, 409–420.

Radulovic, J., Jovasevic, V., Meyer, M.A., 2017. Neurobiological mechanisms of state-dependent learning. Curr. Opin. Neurobiol. 45, 92–98.

Raha, A.A., Henderson, J.W., Stott, S.R., Vuono, R., Foscarin, S., et al., 2017. Neuroprotective effect of TREM-2 in aging and Alzheimer's disease model. J. Alzheimer's Dis. 55, 199–217.

Rahman, T., Purves-Tyson, T., Geddes, A.E., Huang, X.F., Newell, K.A., Weickert, C.S., 2022. N-Methyl-d-Aspartate receptor and inflammation in dorsolateral prefrontal cortex in schizophrenia. Schizophr. Res. 240, 61–70.

Rai, A., Narisawa, M., Li, P., Piao, L., Li, Y., Yang, G., et al., 2020. Adaptive immune disorders in hypertension and heart failure: focusing on

T-cell subset activation and clinical implications. J. Hypertens. 38, 1878–1889.

Raison, C.L., Miller, A.H., 2003. When not enough is too much: the role of insufficient glucocorticoid signaling in the pathophysiology of stress-related disorders. Am. J. Psychiatry 160, 1554–1565.

Raison, C.L., Miller, A.H., 2013. The evolutionary significance of depression in Pathogen Host Defense (PATHOS-D). Mol. Psychiatry 18, 15–37.

Raison, C.L., Dantzer, R., Kelley, K.W., Lawson, M.A., Woolwine, B.J., et al., 2010. CSF concentrations of brain tryptophan and kynurenines during immune stimulation with IFN-alpha: relationship to CNS immune responses and depression. Mol. Psychiatry 15, 393–403.

Raison, C.L., Rutherford, R.E., Woolwine, B.J., Shuo, C., Schettler, P., et al., 2013. A randomized controlled trial of the tumor necrosis factor antagonist infliximab for treatment-resistant depression: the role of baseline inflammatory biomarkers. JAMA Psychiatry 70, 31–41.

Raison, C.L., Sanacora, G., Woolley, J., Heinzerling, K., Dunlop, B.W., et al., 2023. Single-dose psilocybin treatment for major depressive disorder: a randomized clinical trial. JAMA 330, 843–853.

Rajbhandari, A.K., Baldo, B.A., Bakshi, V.P., 2015. Predator stress-induced CRF release causes enduring sensitization of basolateral amygdala norepinephrine systems that promote PTSD-like startle abnormalities. J. Neurosci. 35, 14270–14285.

Raji, C.A., Meysami, S., Hashemi, S., Garg, S., Akbari, N., et al., 2023. Exercise-related physical activity relates to brain volumes in 10,125 individuals. J. Alzheimers Dis. 97, 829–839.

Rakers, F., Rupprecht, S., Dreiling, M., Bergmeier, C., Witte, O.W., Schwab, M., 2020. Transfer of maternal psychosocial stress to the fetus. Neurosci. Biobehav. Rev. 117, 185–197.

Rakoff-Nahoum, S., Foster, K.R., Comstock, L.E., 2016. The evolution of cooperation within the gut microbiota. Nature 533, 255–259.

Ramaekers, J.G., Hutten, N., Mason, N.L., Dolder, P., Theunissen, E.L., et al., 2021. A low dose of lysergic acid diethylamide decreases pain perception in healthy volunteers. J. Psychopharmacol. 35, 398–405.

Ramakrishnan, A., Piehl, N., Simonton, B., Parikh, M., Zhang, Z., Teregulova, V., van Olst, L., Gate, D., 2024. Epigenetic dysregulation in Alzheimer's disease peripheral immunity. Neuron 112, 1135–1248.

Ramasubbu, R., Clark, D.L., Golding, S., Dobson, K.S., Mackie, A., et al., 2020. Long versus short pulse width subcallosal cingulate stimulation for treatment-resistant depression: a randomised, double-blind, crossover trial. Lancet Psychiatry 7, 29–40.

Ramirez, K., Sheridan, J.F., 2016. Antidepressant imipramine diminishes stress-induced inflammation in the periphery and central nervous system and related anxiety-and depressive-like behaviors. Brain Behav. Immun. 57, 293–303.

Ramirez, K., Niraula, A., Sheridan, J.F., 2016. GABAergic modulation with classical benzodiazepines prevent stress-induced neuroimmune dysregulation and behavioral alterations. Brain Behav. Immun. 51, 154–168.

Ramirez, A.I., de Hoz, R., Salobrar-Garcia, E., Salazar, J.J., Rojas, B., et al., 2017. The role of microglia in retinal neurodegeneration: Alzheimer's disease, Parkinson, and Glaucoma. Front. Aging Neurosci. 9, 214.

Ramos-Lopez, O., Milagro, F.I., Riezu-Boj, J.I., Martinez, J.A., 2021. Epigenetic signatures underlying inflammation: an interplay of nutrition, physical activity, metabolic diseases, and environmental factors for personalized nutrition. Inflamm. Res. 70, 29–49.

Ramot, Y., Nyska, A., Maronpot, R.R., Shaltiel-Karyo, R., Tsarfati, Y., et al., 2017. Ninety-day local tolerability and toxicity study of ND0612, a novel formulation of levodopa/carbidopa, administered by subcutaneous continuous infusion in minipigs. Toxicol. Pathol. 45, 764–773.

Ramsaran, A.I., Wang, Y., Golbabaei, A., Aleshin, S., de Snoo, M.L., et al., 2023. A shift in the mechanisms controlling hippocampal engram formation during brain maturation. Science 380, 543–551.

Ransohoff, R.M., 2016. A polarizing question: do M1 and M2 microglia exist? Nat. Neurosci. 19, 987–991.

Rantala, M.J., Luoto, S., Borráz-León, J.I., Krams, I., 2022. Schizophrenia: the new etiological synthesis. Neurosci. Biobehav. Rev. 142, 104894.

Rao, M., Gershon, M.D., 2016. The bowel and beyond: the enteric nervous system in neurological disorders. Nat. Rev. Gastroenterol. Hepatol. 13, 517–528.

Rao, M., Gershon, M.D., 2017. Neurogastroenterology: the dynamic cycle of life in the enteric nervous system. Nat. Rev. Gastroenterol. Hepatol. 14, 453–454.

Raphael, I., Nalawade, S., Eagar, T.N., Forsthuber, T.G., 2015. T cell subsets and their signature cytokines in autoimmune and inflammatory diseases. Cytokine 74, 5–17.

Rapinesi, C., Kotzalidis, G.D., Ferracuti, S., Sani, G., Girardi, P., Del Casale, A., 2019. Brain stimulation in obsessive-compulsive disorder (OCD): a systematic review. Curr. Neuropharmacol. 17, 787–807.

Rappeneau, V., Blaker, A., Petro, J.R., Yamamoto, B.K., Shimamoto, A., 2016. Disruption of the glutamate-glutamine cycle involving astrocytes in an animal model of depression for males and females. Front. Behav. Neurosci. 10, 231.

Rappold, P.M., Cui, M., Chesser, A.S., Tibbett, J., Grima, J.C., et al., 2011. Paraquat neurotoxicity is mediated by the dopamine transporter and organic cation transporter-3. Proc. Natl. Acad. Sci. USA 108, 20766–20771.

Rasmusson, A.M., Hauger, R.L., Morgan, C.A., Bremner, J.D., Charney, D.S., et al., 2000. Low baseline and yohimbine-stimulated plasma neuropeptide Y (NPY) levels in combat-related PTSD. Biol. Psychiatry 47, 526–539.

Rasmusson, A.M., Marx, C.E., Pineles, S.L., Locci, A., Scioli-Salter, E.R., et al., 2017. Neuroactive steroids and PTSD treatment. Neurosci. Lett. 649, 156–163.

Rasooly, D., Peloso, G.M., Pereira, A.C., Dashti, H., Giambartolomei, C., et al., 2023. Genome-wide association analysis and Mendelian randomization proteomics identify drug targets for heart failure. Nat. Commun. 14, 3826.

Raspopow, K., Matheson, K., Abizaid, A., Anisman, H., 2013. Unsupportive social interactions influence emotional eating behaviors. The role of coping styles as mediators. Appetite 62, 143–149.

Raspopow, K., Abizaid, A., Matheson, K., Anisman, H., 2014. Anticipation of a psychosocial stressor differentially influences ghrelin, cortisol and food intake among emotional and non-emotional eaters. Appetite 74, 35–43.

Ratiner, K., Ciocan, D., Abdeen, S.K., Elinav, E., 2024. Utilization of the microbiome in personalized medicine. Nat. Rev. Microbiol. 22, 291–308.

Ratsika, A., Cruz Pereira, J.S., Lynch, C.M.K., Clarke, G., Cryan, J.F., 2023. Microbiota-immune-brain interactions: a lifespan perspective. Curr. Opin. Neurobiol. 78, 102652.

Rauch, S.A.M., Kim, H.M., Powell, C., Tuerk, P.W., Simon, N.M., et al., 2019. Efficacy of prolonged exposure therapy, sertraline hydrochloride, and their combination among combat veterans with posttraumatic stress disorder: a randomized clinical trial. JAMA Psychiatry 76, 117–126.

Ravenholt, R.T., Foege, W.H., 1982. 1918 influenza, encephalitis lethargica, parkinsonism. Lancet 2, 860–864.

Ravichandran, K.A., Heneka, M.T., 2024. Inflammasomes in neurological disorders—mechanisms and therapeutic potential. Nat. Rev. Neurol. 20, 67–83.

Ravindran, A.V., Balneaves, L.G., Faulkner, G., Ortiz, A., McIntosh, D., 2016. Canadian Network for Mood and Anxiety Treatments (CANMAT) 2016 clinical guidelines for the management of adults with major depressive disorder: section 5. complementary and alternative medicine treatments. Can. J. Psychiatry 61, 576–587.

Raymann, S., Schalbetter, S.M., Schaer, R., Bernhardt, A.C., Mueller, F. S., et al., 2023. Late prenatal immune activation in mice induces transgenerational effects via the maternal and paternal lineages. Cereb. Cortex 33, 2273–2286.

Raymundi, A.M., da Silva, T.R., Sohn, J.M.B., Bertoglio, L.J., Stern, C.A., 2020. Effects of Δ9-tetrahydrocannabinol on aversive memories and anxiety: a review from human studies. BMC Psychiatry 20, 420.

Rebecchini, L., 2021. Music, mental health, and immunity. Brain Behav. Immun. Health 18, 100374.

Reber, S.O., Siebler, P.H., Donner, N.C., Morton, J.T., Smith, D.G., et al., 2016. Immunization with a heat-killed preparation of the environmental bacterium Mycobacterium vaccae promotes stress resilience in mice. Proc. Natl. Acad. Sci. USA 113, E3130–E3139.

Rebo, J., Mehdipour, M., Gathwala, R., Causey, K., Liu, Y., et al., 2016. A single heterochronic blood exchange reveals rapid inhibition of multiple tissues by old blood. Nat. Commun. 7, 13363.

Recalde, M., Pistillo, A., Davila-Batista, V., Leitzmann, M., Romieu, I., et al., 2023. Longitudinal body mass index and cancer risk: a cohort study of 2.6 million Catalan adults. Nat. Commun. 14, 3816.

Recchia, F., Leung, C.K., Chin, E.C., Fong, D.Y., Montero, D., et al., 2022. Comparative effectiveness of exercise, antidepressants and their combination in treating non-severe depression: a systematic review and network meta-analysis of randomised controlled trials. Br. J. Sports Med. 56, 1375–1380.

Recker, M., Laabei, M., Toleman, M.S., Reuter, S., Saunderson, R.B., et al., 2017. Clonal differences in Staphylococcus aureus bacteraemia-associated mortality. Nat. Microbiol. 2, 1381–1388.

Reddy-Thootkur, M., Kraguljac, N.V., Lahti, A.C., 2022. The role of glutamate and GABA in cognitive dysfunction in schizophrenia and mood disorders - a systematic review of magnetic resonance spectroscopy studies. Schizophr. Res. 249, 74–84.

Redwine, L., Hauger, R.L., Gillin, J.C., Irwin, M., 2000. Effects of sleep and sleep deprivation on interleukin-6, growth hormone, cortisol, and melatonin levels in humans. J. Clin. Endocrinol. Metab. 85, 3597–3603.

Reed, R.G., Presnell, S.R., Al-Attar, A., Lutz, C.T., Segerstrom, S.C., 2019. Perceived stress, cytomegalovirus titers, and late-differentiated T and NK cells: Between-, within-person associations in a longitudinal study of older adults. Brain Behav. Immun. 80, 266–274.

Reemst, K., Kracht, L., Kotah, J.M., Rahimian, R., van Irsen, A.A.S., et al., 2022a. Early-life stress lastingly impacts microglial transcriptome and function under basal and immune-challenged conditions. Transl. Psychiatry 12, 507.

Reemst, K., Kracht, L., Kotah, J.M., Rahimian, R., van Irsen, A.A.S., et al., 2022b. Early-life stress lastingly impacts microglial transcriptome and function under basal and immune-challenged conditions. Transl. Psychiatry 12, 507.

Rees, E., Creeth, H.D.J., Hwu, H.G., Chen, W.J., Tsuang, M., et al., 2021. Schizophrenia, autism spectrum disorders and developmental disorders share specific disruptive coding mutations. Nat. Commun. 12, 5353.

Reeve, S.M., Scocchera, E.W., G-Dayanandan, N., Keshipeddy, S., Krucinska, J., et al., 2016. MRSA isolates from United States hospitals carry dfrG and dfrK resistance genes and succumb to propargyl-linked antifolates. Cell Chem. Biol. 23, 1458–1467.

Rehman, Y., Saini, A., Huang, S., Sood, E., Gill, R., Yanikomeroglu, S., 2021. Cannabis in the management of PTSD: a systematic review. AIMS Neurosci. 8, 414–434.

Reichenberg, A., Yirmiya, R., Schuld, A., Kraus, T., Haack, M., et al., 2001. Cytokine-associated emotional and cognitive disturbances in humans. Arch. Gen. Psychiatry 58, 445–452.

Reichetzeder, C., 2021. Overweight and obesity in pregnancy: their impact on epigenetics. Eur. J. Clin. Nutr. 75, 1710–1722.

Reid, B.M., Doom, J.R., Argote, R.B., Correa-Burrows, P., Lozoff, B., et al., 2020. Pathways to inflammation in adolescence through early adversity, childhood depressive symptoms, and body mass index: a prospective longitudinal study of Chilean infants. Brain Behav. Immun. 86, 4–13.

Reid, J.E., Laws, K.R., Drummond, L., Vismara, M., Grancini, B., et al., 2021. Cognitive behavioural therapy with exposure and response prevention in the treatment of obsessive-compulsive disorder: a systematic review and meta-analysis of randomised controlled trials. Compr. Psychiatry 106, 152223.

Reif, A., Bitter, I., Buyze, J., Cebulla, K., Frey, R., et al., 2023. Esketamine nasal spray versus quetiapine for treatment-resistant depression. N. Engl. J. Med. 389, 1298–1309.

Reinhart, K., Daniels, R., Kissoon, N., Machado, F.R., Schachter, R.D., Finfer, S., 2017. Recognizing sepsis as a global health priority—a WHO resolution. N. Engl. J. Med. 377, 414–417.

Reinhorn, I.M., Bernstein, C.N., Graff, L.A., Patten, S.B., Sareen, J., 2020. Social phobia in immune-mediated inflammatory diseases. J. Psychosom. Res. 128, 109890.

Reiss, A.B., Glass, D.S., Lam, E., Glass, A.D., De Leon, J., Kasselman, L.J., 2019. Oxytocin: potential to mitigate cardiovascular risk. Peptides 117, 170089.

Remely, M., Hippe, B., Zanner, J., Aumueller, E., Brath, H., et al., 2016. Gut microbiota of obese, type 2 diabetic individuals is enriched in Faecalibacterium prausnitzii, Akkermansia muciniphila and Peptostreptococcus anaerobius after weight loss. Endocr Metab Immune Disord Drug Targets 16.

Rempelos, L., Wang, J., Barański, M., Watson, A., Volakakis, N., et al., 2021. Diet and food type affect urinary pesticide residue excretion profiles in healthy individuals: results of a randomized controlled dietary intervention trial. Am. J. Clin. Nutr. 115, 364–377.

Ren, Z., Yuan, J., Luo, Y., Wang, J., Li, Y., 2023. Association of air pollution and fine particulate matter (PM2.5) exposure with gestational diabetes: a systematic review and meta-analysis. Ann. Transl. Med. 11, 23.

Renard, J., Rushlow, W.J., Laviolette, S.R., 2018. Effects of adolescent THC exposure on the prefrontal GABAergic system: implications for schizophrenia-related psychopathology. Front. Psychiatry 9, 281.

Rengasamy, M., Brundin, L., Griffo, A., Panny, B., Capan, C., et al., 2022. Cytokine and reward circuitry relationships in treatment-resistant depression. Biol. Psychiatry Glob. Open Sci. 2, 45–53.

Renner, V., Joraschky, P., Kirschbaum, C., Schellong, J., Petrowski, K., 2022. Pro- and anti-inflammatory cytokines Interleukin-6 and Interleukin-10 predict therapy outcome of female patients with post-traumatic stress disorder. Transl. Psychiatry 12, 472.

Ressler, K.J., Berretta, S., Bolshakov, V.Y., Rosso, I.M., Meloni, E.G., et al., 2022. Post-traumatic stress disorder: clinical and translational neuroscience from cells to circuits. Nat. Rev. Neurol. 18, 273–288.

Reul, J.M., Holsboer, F., 2022. On the role of corticotropin-releasing hormone receptors in anxiety and depression. Dialogues Clin. Neurosci. 4, 31–46.

Reuveni, I., Nugent, A.C., Gill, J., Vythilingam, M., Carlson, P.J., et al., 2018. Altered cerebral benzodiazepine receptor binding in post-traumatic stress disorder. Transl. Psychiatry 8, 206.

Revitt-Mills, S.A., Robinson, A., 2020. Antibiotic-induced mutagenesis: under the microscope. Front. Microbiol. 11, 585175.

Rey, F., Ottolenghi, S., Giallongo, T., Balsari, A., Martinelli, C., et al., 2021. Mitochondrial metabolism as target of the neuroprotective role of erythropoietin in parkinson's disease. Antioxidants 10, 121.

Reyes-Castillo, Z., Valdés-Miramontes, E., Llamas-Covarrubias, M., Muñoz-Valle, J.F., 2021. Troublesome friends within us: the role of gut microbiota on rheumatoid arthritis etiopathogenesis and its clinical and therapeutic relevance. Clin. Exp. Med. 21, 1–13.

Reynolds, A.D., Stone, D.K., Hutter, J.A., Benner, E.J., Mosley, R.L., Gendelman, H.E., 2010. Regulatory T cells attenuate Th17 cell-mediated nigrostriatal dopaminergic neurodegeneration in a model of Parkinson's disease. J. Immunol. 184, 2261–2271.

Reynolds, A., Mann, J., Cummings, J., Winter, N., Mete, E., Te Morenga, L., 2019. Carbohydrate quality and human health: a series of systematic reviews and meta-analyses. Lancet 393, 434–445.

Rezai, A.R., D'Haese, P.F., Finomore, V., Carpenter, J., Ranjan, M., et al., 2024. Ultrasound blood–brain barrier opening and Aducanumab in Alzheimer's disease. N. Engl. J. Med. 390, 55–62.

Reza-Zaldívar, E.E., Hernández-Sapiéns, M.A., Minjarez, B., Gómez-Pinedo, U., Márquez-Aguirre, A.L., et al., 2021. Infection mechanism of SARS-COV-2 and its implication on the nervous system. Front. Immunol. 11, 621735.

Rhie, S.J., Jung, E.Y., Shim, I., 2020. The role of neuroinflammation on pathogenesis of affective disorders. J. Exerc. Rehabil. 16, 2–9.

Ricci, S., Businaro, R., Ippoliti, F., Lo Vasco, V.R., Massoni, F., et al., 2013. Altered cytokine and BDNF levels in autism spectrum disorder. Neurotox. Res. 24, 491–501.

Rice, D., Nijs, J., Kosek, E., Wideman, T., Hasenbring, M.I., et al., 2019. Exercise-induced hypoalgesia in pain-free and chronic pain populations: state of the art and future directions. J. Pain 20, 1249–1266.

Richards, D.A., Ekers, D., McMillan, D., Taylor, R.S., Byford, S., et al., 2016. Cost and outcome of behavioural activation versus cognitive behavioural therapy for depression (COBRA): a randomised, controlled, non-inferiority trial. Lancet 388, 871–880.

Richardson Jr., T.L., Halvorson, A.E., Hackstadt, A.J., Hung, A.M., Greevy, R., et al., 2023. Primary occurrence of cardiovascular events after adding sodium-glucose cotransporter-2 inhibitors or glucagon-like peptide-1 receptor agonists compared with dipeptidyl peptidase-4 inhibitors: a cohort study in veterans with diabetes. Ann. Intern. Med. 176, 751–760.

Richardson, J.R., Quan, Y., Sherer, T.B., Greenamyre, J.T., Miller, G.W., 2005. Paraquat neurotoxicity is distinct from that of MPTP and rotenone. Toxicol. Sci. 88, 193–201.

Richetto, J., Calabrese, F., Riva, M.A., Meyer, U., 2014. Prenatal immune activation induces maturation-dependent alterations in the prefrontal GABAergic transcriptome. Schizophr. Bull. 40, 351–361.

Richter, C., Woods, I.G., Schier, A.F., 2014. Neuropeptidergic control of sleep and wakefulness. Annu. Rev. Neurosci. 37, 503–531.

Richter-Levin, G., Stork, O., Schmidt, M.V., 2019. Animal models of PTSD: a challenge to be met. Mol. Psychiatry 24, 1135–1156.

Ridaura, V.K., Faith, J.J., Rey, F.E., Cheng, J., Duncan, A.E., et al., 2013. Gut microbiota from twins discordant for obesity modulate metabolism in mice. Science 341, 1241214.

Ridker, P.M., Lonn, E., Paynter, N.P., Glynn, R., Yusuf, S., 2017. Primary prevention with statin therapy in the elderly: new meta-analyses from the contemporary JUPITER and HOPE-3 randomized trials. Circulation 135, 1979–1981.

Riederer, P., Sian-Hülsmann, J., 2012. The significance of neuronal lateralisation in Parkinson's disease. J. Neural Transm. 119, 953–962.

Riera Romo, M., Perez-Martinez, D., Castillo Ferrer, C., 2016. Innate immunity in vertebrates: an overview. Immunology 148, 125–139.

Rietdijk, C.D., Perez-Pardo, P., Garssen, J., van Wezel, R.J.A., Kraneveld, A.D., 2017. Exploring Braak's hypothesis of Parkinson's disease. Front. Neurol. 8, 37.

Rigney, N., de Vries, G.J., Petrulis, A., 2023. Modulation of social behavior by distinct vasopressin sources. Front. Endocrinol. 14, 1127792.

Rijlaarsdam, J., Cecil, C.A., Walton, E., Mesirow, M.S., Relton, C.L., et al., 2017. Prenatal unhealthy diet, insulin-like growth factor 2 gene (IGF2) methylation, and attention deficit hyperactivity disorder symptoms in youth with early-onset conduct problems. J. Child Psychol. Psychiatry 58, 19–27.

Rijo-Ferreira, F., Takahashi, J.S., 2019. Genomics of circadian rhythms in health and disease. Genome Med. 11, 82.

Ringman, J.M., Frautschy, S.A., Teng, E., Begum, A.N., Bardens, J., et al., 2012. Oral curcumin for Alzheimer's disease: tolerability and efficacy in a 24-week randomized, double blind, placebo-controlled study. Alzheimers Res. Ther. 4, 43.

Rinninella, E., Raoul, P., Cintoni, M., Franceschi, F., Miggiano, G., et al., 2019. What is the healthy gut microbiota composition? A changing ecosystem across age, environment, diet, and diseases. Microorganisms 7, 14.

Rinninella, E., Mele, M.C., Cintoni, M., et al., 2020. The facts about food after cancer diagnosis: a systematic review of prospective cohort studies. Nutrients 12, E2345.

Ritvo, A.D., Foster, D.E., Huff, C., Finlayson, A.J.R., Silvernail, B., Martin, P.R., 2023. Long-term consequences of benzodiazepine-induced neurological dysfunction: a survey. PLoS One 18, e0285584.

Ritz, B.R., Paul, K.C., Bronstein, J.M., 2016. Of pesticides and men: a California story of genes and environment in Parkinson's disease. Curr. Environ. Health Rep. 3, 40–52.

Ritz, N.L., Brocka, M., Butler, M.I., Cowan, C.S.M., Barrera-Bugueño, C., Turkington, C.J.R., et al., 2024. Social anxiety disorder-associated gut microbiota increases social fear. Proc. Natl. Acad. Sci. USA 121, e2308706120.

Riva, P., Romero Lauro, L.J., Vergallito, A., DeWall, C.N., Bushman, B.J., 2015. Electrified emotions: modulatory effects of transcranial direct stimulation on negative emotional reactions to social exclusion. Soc. Neurosci. 10, 46–54.

Rivas-Larrauri, F., Yamazaki-Nakashimada, M.A., 2016. Systemic lupus erythematosus: is it one disease? Reumatol. Clin. 12, 274–281.

Robakis, T.K., Roth, M.C., King, L.S., Humphreys, K.L., Ho, M., et al., 2022. Maternal attachment insecurity, maltreatment history, and depressive symptoms are associated with broad DNA methylation signatures in infants. Mol. Psychiatry 27, 3306–3315.

Robbins, T.W., Banca, P., Belin, D., 2024. From compulsivity to compulsion: the neural basis of compulsive disorders. Nat. Rev. Neurosci. 25, 313–333.

Robciuc, M.R., Kivelä, R., Williams, I.M., de Boer, J.F., van Dijk, T.H., et al., 2016. VEGFB/VEGFR1-Induced expansion of adipose vasculature counteracts obesity and related metabolic complications. Cell Metab. 23, 712–724.

Roberts, R.M., Ezashi, T., Schulz, L.C., Sugimoto, J., Schust, D.J., Khan, T., Zhou, J., 2021. Syncytins expressed in human placental trophoblast. Placenta 113, 8–14.

Robinson, R.G., Jorge, R.E., 2016. Post-stroke depression: a review. Am. J. Psychiatry 173, 221–231.

Robinson, M., Carter, K.W., Pennell, C.E., Jacoby, P., Moore, H.C., et al., 2021. Maternal prenatal stress exposure and sex-specific risk of severe infection in offspring. PLoS One 16, e0245747.

Rodgers, K.R., Lin, Y., Langan, T.J., Iwakura, Y., Chou, R.C., 2020. Innate immune functions of astrocytes are dependent upon tumor necrosis factor-alpha. Sci. Rep. 10, 7047.

Rodgman, C., Verrico, C.D., Holst, M., Thompson-Lake, D., Haile, C.N., et al., 2016. Doxazosin XL reduces symptoms of posttraumatic stress disorder in veterans with PTSD: a pilot clinical trial. J. Clin. Psychiatry 77, e561–e565.

Rodríguez Murúa, S., Farez, M.F., Quintana, F.J., 2022. The immune response in multiple sclerosis. Annu. Rev. Pathol. 17, 121–139.

Rodríguez, N., Morer, A., González-Navarro, E.A., Serra-Pages, C., Boloc, D., et al., 2019. Altered frequencies of Th17 and Treg cells in children and adolescents with obsessive-compulsive disorder. Brain Behav. Immun. 81, 608–616.

Rodriguez, N., Tun, H.M., Field, C.J., Mandhane, P.J., Scott, J.A., Kozyrskyj, A.L., 2021. Prenatal depression, breastfeeding, and infant gut microbiota. Front. Microbiol. 12, 664257.

Rodríguez, N., Lázaro, L., Ortiz, A.E., Morer, A., Martínez-Pinteño, A., et al., 2022. Gene expression study in monocytes: evidence of inflammatory dysregulation in early-onset obsessive-compulsive disorder. Transl. Psychiatry 12, 134.

Rodriguez-Sanchez, F., Rodriguez-Blazquez, C., Bielza, C., Larrañaga, P., Weintraub, D., et al., 2021. Identifying Parkinson's disease subtypes with motor and non-motor symptoms via model-based multi-partition clustering. Sci. Rep. 11, 23645.

Rojas, J.M., Avia, M., Martin, V., Sevilla, N., 2017. IL-10: a multifunctional cytokine in viral infections. J Immunol Res 2017, 6104054.

Rolls, E.T., Cheng, W., Feng, J., 2020. The orbitofrontal cortex: reward, emotion and depression. Brain Commun. 2, fcaa196.

Romero Núñez, E., Blanco Ayala, T., Vázquez Cervantes, G.I., Roldán-Roldán, G., González Esquivel, D.F., et al., 2022. Pregestational exposure to T. gondii produces maternal antibodies that recognize fetal brain mimotopes and induces neurochemical and behavioral dysfunction in the offspring. Cells 11, 3819.

Romero, C., Werme, J., Jansen, P.R., Gelernter, J., Stein, M.B., et al., 2022. Exploring the genetic overlap between twelve psychiatric disorders. Nat. Genet. 54, 1795–1802.

Romero-Figueroa, M.D.S., Ramírez-Durán, N., Montiel-Jarquín, A.J., Horta-Baas, G., 2023. Gut-joint axis: gut dysbiosis can contribute to the onset of rheumatoid arthritis via multiple pathways. Front. Cell. Infect. Microbiol. 13, 1092118.

Rompala, G., Nomura, Y., Hurd, Y.L., 2021. Maternal cannabis use is associated with suppression of immune gene networks in placenta and increased anxiety phenotypes in offspring. Proc. Natl. Acad. Sci. USA 118, e2106115118.

Ronaldson, A., Gazali, A.M., Zalli, A., Kaiser, F., Thompson, S.J., et al., 2016. Increased percentages of regulatory T cells are associated with inflammatory and neuroendocrine responses to acute psychological stress and poorer health status in older men and women. Psychopharmacology 233, 1661–1668.

Rong, Z., Chen, H., Zhang, Z., Zhang, Y., Ge, L., et al., 2022. Identification of cardiomyopathy-related core genes through human metabolic networks and expression data. BMC Genomics 23, 47.

Rong, Z., Mai, H., Kapoor, S., Puelles, V., Czogalla, J., et al., 2023. SARS-CoV-2 spike protein accumulation in the skull-meninges-brain axis: potential implications for long-term neurological complications in post-COVID-19. bioRxiv. 2023-04.

Ronin, I., Katsowich, N., Rosenshine, I., Balaban, N.Q., 2017. A long-term epigenetic memory switch controls bacterial virulence bimodality. elife 6, e19599.

Ronovsky, M., Berger, S., Zambon, A., Reisinger, S.N., Horvath, O., et al., 2017. Maternal immune activation transgenerationally modulates maternal care and offspring depression-like behavior. Brain Behav. Immun. 63, 127–136.

Roodveldt, C., Labrador-Garrido, A., Gonzalez-Rey, E., Lachaud, C.C., Guilliams, T., et al., 2013. Preconditioning of microglia by α-synuclein strongly affects the response induced by toll-like receptor (TLR) stimulation. PLoS One 8, e79160.

Rook, G.A.W., 2023. The old friends hypothesis: evolution, immunoregulation and essential microbial inputs. Front. Allergy 4, 1220481.

Rook, G.A., Lowry, C.A., Raison, C.L., 2015. Hygiene and other early childhood influences on the subsequent function of the immune system. Brain Res. 1617, 47–62.

Roomruangwong, C., Anderson, G., Berk, M., Stoyanov, D., Carvalho, A.F., et al., 2018. A neuro-immune, neuro-oxidative and neuro-nitrosative model of prenatal and postpartum depression. Prog. Neuro-Psychopharmacol. Biol. Psychiatry 81, 262–274.

Roomruangwong, C., Noto, C., Kanchanatawan, B., Anderson, G., Kubera, M., et al., 2020. The role of aberrations in the immune-inflammatory response system (IRS) and the compensatory immune-regulatory reflex system (CIRS) in different phenotypes of schizophrenia: the IRS-CIRS theory of schizophrenia. Mol. Neurobiol. 57, 778–797.

Roosen, D.A., Cookson, M.R., 2016. LRRK2 at the interface of autophagosomes, endosomes and lysosomes. Mol. Neurodegener. 11, 73.

Rosa, J.M., Formolo, D.A., Yu, J., Lee, T.H., Yau, S.Y., 2022. The role of microRNA and microbiota in depression and anxiety. Front. Behav. Neurosci. 16, 828258.

Rosa-Gonçalves, P., Ribeiro-Gomes, F.L., Daniel-Ribeiro, C.T., 2022. Malaria related neurocognitive deficits and behavioral alterations. Front. Cell. Infect. Microbiol. 12, 829413.

Rose, D.R., Careaga, M., Van de Water, J., McAllister, K., Bauman, M.D., Ashwood, P., 2017. Long-term altered immune responses following fetal priming in a non-human primate model of maternal immune activation. Brain Behav. Immun. 63, 60–70.

Roseboom, T.J., Painter, R.C., van Abeelen, A.F., Veenendaal, M.V., de Rooij, S.R., 2011. Hungry in the womb: What are the consequences? Lessons from the Dutch famine. Maturitas 70, 141–145.

Rosén, A., Yi, J., Kirsch, I., Kaptchuk, T.J., Ingvar, M., et al., 2017. Effects of subtle cognitive manipulations on placebo analgesia - an implicit priming study. Eur. J. Pain 21, 594–604.

Ross, A.G., McManus, D.P., Farrar, J., Hunstman, R.J., Gray, D.J., Li, Y. S., 2012. Neuroschistosomiasis. J. Neurol. 259, 22–32.

Ross, R.E., VanDerwerker, C.J., Saladin, M.E., Gregory, C.M., 2023. The role of exercise in the treatment of depression: biological underpinnings and clinical outcomes. Mol. Psychiatry 28, 298–328.

Rosshart, S.P., Vassallo, B.G., Angeletti, D., Hutchinson, D.S., Morgan, A.P., et al., 2017. Wild mouse gut microbiota promotes host fitness and improves disease resistance. Cell 171, 1015–1028.

Rossi, S., Studer, V., Motta, C., Polidoro, S., Perugini, J., et al., 2017. Neuroinflammation drives anxiety and depression in relapsing-remitting multiple sclerosis. Neurology 89, 1338–1347.

Rossi-George, A., LeBlanc, F., Kaneta, T., Urbach, D., Kusnecov, A.W., 2004. Effects of bacterial superantigens on behavior of mice in the elevated plus maze and light–dark box. Brain Behav. Immun. 18, 46–54.

Rosso, I.M., Crowley, D.J., Silveri, M.M., Rauch, S.L., Jensen, J.E., et al., 2017. Hippocampus glutamate and N-acetyl aspartate markers of excitotoxic neuronal compromise in posttraumatic stress disorder. Neuropsychopharmacology 42, 1698–1705.

Rosso, I.M., Silveri, M.M., Olson, E.A., Eric Jensen, J., Ren, B., 2021. Regional specificity and clinical correlates of cortical GABA alterations in posttraumatic stress disorder. Neuropsychopharmacology 47, 1055–1062.

Roth, K., Petriello, M.C., 2022. Exposure to per- and polyfluoroalkyl substances (PFAS) and type 2 diabetes risk. Front. Endocrinol. (Lausanne) 13, 965384.

Roth, T.L., Lubin, F.D., Funk, A.J., Sweatt, J.D., 2009. Lasting epigenetic influence of early- life adversity on the BDNF gene. Biol. Psychiatry 65, 760–769.

Roth, D.L., Sheehan, O.C., Haley, W.E., Jenny, N.S., Cushman, M., Walston, J.D., 2019. Is family caregiving associated with inflammation or compromised immunity? A meta-analysis. Gerontologist 59, e521–e534.

Rothaug, M., Becker-Pauly, C., Rose-John, S., 2016. The role of interleukin-6 signaling in nervous tissue. Biochim. Biophys. Acta 1863, 1218–1227.

Rothstein, T.L., 2023. Cortical Grey matter volume depletion links to neurological sequelae in post COVID-19 "long haulers". BMC Neurol. 23, 22.

Roullet, P., Vaiva, G., Véry, E., Bourcier, A., Yrondi, A., et al., 2021. Traumatic memory reactivation with or without propranolol for PTSD and comorbid MD symptoms: a randomised clinical trial. Neuropsychopharmacology 46, 1643–1649.

Rousseau, P.F., El Khoury-Malhame, M., Reynaud, E., Boukezzi, S., Cancel, A., et al., 2019. Fear extinction learning improvement in PTSD after EMDR therapy: an fMRI study. Eur. J. Psychotraumatol. 10, 1568132.

Routy, B., Gopalakrishnan, V., Daillère, R., Zitvogel, L., Wargo, J.A., Kroemer, G., 2018a. The gut microbiota influences anticancer immunosurveillance and general health. Nat. Rev. Clin. Oncol. 15, 382–396.

Routy, B., Le Chatelier, E., Derosa, L., Duong, C.P.M., Alou, M.T., et al., 2018b. Gut microbiome influences efficacy of PD-1-based immunotherapy against epithelial tumors. Science 359, 91–97.

Rouxel, O., Da Silva, J., Beaudoin, L., Nel, I., Tard, C., et al., 2017. Cytotoxic and regulatory roles of mucosal-associated invariant T cells in type 1 diabetes. Nat. Immunol. 18, 1321–1331.

Roy, B., Dunbar, M., Shelton, R.C., Dwivedi, Y., 2016. Identification of microRNA-124-3p as a putative epigenetic signature of major depressive disorder. Neuropsychopharmacology 42, 864–875.

Roy, D.S., Park, Y.G., Kim, M.E., Zhang, Y., Ogawa, S.K., et al., 2022. Brain-wide mapping reveals that engrams for a single memory are distributed across multiple brain regions. Nat. Commun. 13, 1799.

Rozeske, R.R., Evans, A.K., Frank, M.G., Watkins, L.R., Lowry, C.A., et al., 2011. Uncontrollable, but not controllable, stress desensitizes 5-HT1A receptors in the dorsal raphe nucleus. J. Neurosci. 31, 14107–14115.

Rozing, M.P., Veerhuis, R., Westendorp, R.G.J., Eikelenboom, P., Stek, M., et al., 2019. Inflammation in older subjects with early- and late-onset depression in the NESDO study: a cross-sectional and longitudinal case-only design. Psychoneuroendocrinology 99, 20–27.

Ruan, H.B., Dietrich, M.O., Liu, Z.W., Zimmer, M.R., Li, M.D., et al., 2014. O-GlcNAc transferase enables AgRP neurons to suppress browning of white fat. Cell 159, 306–317.

Ruben, M.D., Smith, D.F., FitzGerald, G.A., Hogenesch, J.B., 2019. Dosing time matters. Science 365, 547–549.

Rubin, M., Shvil, E., Papini, S., Chhetry, B.T., Helpman, L., et al., 2016. Greater hippocampal volume is associated with PTSD treatment response. Psychiatry Res. 252, 36–39.

Rubio-Casillas, A., Fernández-Guasti, A., 2016. The dose makes the poison: from glutamate-mediated neurogenesis to neuronal atrophy and depression. Rev. Neurosci. 27 (6), 599–622.

Rubio-Casillas, A., Redwan, E.M., Uversky, V.N., 2022. SARS-CoV-2: a master of immune evasion. Biomedicines 10, 1339.

Rudd, K.E., Johnson, S.C., Agesa, K.M., Shackelford, K.A., Tsoi, D., et al., 2020. Global, regional, and national sepsis incidence and mortality, 1990-2017: analysis for the Global Burden of Disease Study. Lancet 395, 200–211.

Rudenko, I.N., Chia, R., Cookson, M.R., 2012. Is inhibition of kinase activity the only therapeutic strategy for LRRK2-associated Parkinson's disease? BMC Med. 10, 20.

Rudyk, C., Litteljohn, D., Syed, S., Dwyer, Z., Hayley, S., 2015. Paraquat and psychological stressor interactions as pertains to Parkinsonian co-morbidity. Neurobiol. Stress 2, 85–93.

Rui, Q., Ni, H., Li, D., Gao, R., Chen, G., 2018. The role of LRRK2 in neurodegeneration of Parkinson disease. Curr. Neuropharmacol. 16, 1348–1357.

Ruigrok, S.R., Abbink, M.R., Geertsema, J., Kuindersma, J.E., Stöberl, N., et al., 2021. Effects of early-life stress, postnatal diet modulation and long-term Western-style diet on peripheral and central inflammatory markers. Nutrients 13, 288.

Ruiz-Grosso, P., Cachay, R., de la Flor, A., Schwalb, A., Ugarte-Gil, C., 2020. Association between tuberculosis and depression on negative outcomes of tuberculosis treatment: a systematic review and meta-analysis. PLoS One 15, e0227472.

Rumgay, H., Shield, K., Charvat, H., Ferrari, P., Sornpaisarn, B., et al., 2021. Global burden of cancer in 2020 attributable to alcohol consumption: a population-based study. Lancet Oncol. 22, 1071–1080.

Rundqvist, H., Velica, P., Barbieri, L., Gameiro, P.A., Bargiela, D., et al., 2020. Cytotoxic T-cells mediate exercise-induced reductions in tumor growth. elife 9, e59996.

Runge, K., Fiebich, B.L., Kuzior, H., Saliba, S.W., Yousif, N.M., et al., 2021. An observational study investigating cytokine levels in the cerebrospinal fluid of patients with schizophrenia spectrum disorders. Schizophr. Res. 231, 205–213.

Rusiecki, J.A., Byrne, C., Galdzicki, Z., Srikantan, V., Chen, L., et al., 2013. PTSD and DNA methylation in select immune function gene promoter regions: a repeated measures case-control study of U.S. military service members. Front. Psychiatry 4, 56.

Russo, M.V., McGavern, D.B., 2016. Inflammatory neuroprotection following traumatic brain injury. Science 353, 783–785.

Russo, E., Citraro, R., Constanti, A., De Sarro, G., 2012a. The mTOR signaling pathway in the brain: focus on epilepsy and epileptogenesis. Mol. Neurobiol. 46, 662–681.

Russo, S.J., Murrough, J.W., Han, M.H., Charney, D.S., Nestler, E.J., 2012b. Neurobiology of resilience. Nat. Neurosci. 15, 1475–1484.

Russo, R., Cristiano, C., Avagliano, C., De Caro, C., La Rana, G., et al., 2017. Gut-brain axis: role of lipids in the regulation of inflammation, pain and CNS diseases. Curr. Med. Chem. 25, 3930–3952.

Rutayisire, E., Huang, K., Liu, Y., Tao, F., 2016. The mode of delivery affects the diversity and colonization pattern of the gut microbiota during the first year of infants' life: a systematic review. BMC Gastroenterol. 16, 86.

Rutishauser, U., Mamelak, A.N., Adolphs, R., 2015. The primate amygdala in social perception—insights from electrophysiological recordings and stimulation. Trends Neurosci. 38, 295–306.

Rutsch, A., Kantsjö, J.B., Ronchi, F., 2020. The gut-brain axis: how microbiota and host inflammasome influence brain physiology and pathology. Front. Immunol. 11, 604179.

Ryan, T.J., Frankland, P.W., 2022. Forgetting as a form of adaptive engram cell plasticity. Nat. Rev. Neurosci. 23, 173–186.

Saavedra, K., Salazar, L.A., 2021. Epigenetics: a missing link between early life stress and depression. Adv. Exp. Med. Biol. 1305, 117–128.

Sabban, E.L., Alaluf, L.G., Serova, L.I., 2016. Potential of neuropeptide Y for preventing or treating post-traumatic stress disorder. Neuropeptides 56, 19–24.

Saben, J.L., Boudoures, A.L., Asghar, Z., Thompson, A., Drury, A., et al., 2016. Maternal metabolic syndrome programs mitochondrial dysfunction via germline changes across three generations. Cell Rep. 16, 1–8.

Sabia, S., Fayosse, A., Dumurgier, J., van Hees, V.T., Paquet, C., et al., 2021. Association of sleep duration in middle and old age with incidence of dementia. Nat. Commun. 12, 2289.

Sacks, F.M., Lichtenstein, A.H., Wu, J.H., Appel, L.J., Creager, M.A., et al., 2017. Dietary fats and cardiovascular disease: a presidential advisory from the American Heart Association. Circulation 136, e1–e23.

Sadabadi, F., Darroudi, S., Esmaily, H., Asadi, Z., Ferns, G.A., et al., 2023. The importance of sleep patterns in the incidence of coronary heart disease: a 6-year prospective study in Mashhad, Iran. Sci. Rep. 13, 2903.

Sada-Fuente, E., Aranda, S., Papiol, S., Heilbronner, U., Moltó, M.D., et al., 2023. Common genetic variants contribute to heritability of age at onset of schizophrenia. Transl. Psychiatry 13, 201.

Saeed, A., Lopez, O., Cohen, A., Reis, S.E., 2023. Cardiovascular disease and Alzheimer's disease: the heart-brain axis. J. Am. Heart Assoc. 12, e030780.

Safaiyan, S., Kannaiyan, N., Snaidero, N., Brioschi, S., Biber, K., et al., 2016. Age-related myelin degradation burdens the clearance function of microglia during aging. Nat. Neurosci. 19, 995–998.

Sah, R., Ekhator, N.N., Jefferson-Wilson, L., Horn, P.S., Geracioti, T.D., 2014. Cerebrospinal fluid neuropeptide Y in combat veterans with and without posttraumatic stress disorder. Psychoneuroendocrinology 40, 277–283.

Sahoo, S., Padhy, S.K., Padhee, B., Singla, N., Sarkar, S., 2018. Role of personality in cardiovascular diseases: an issue that needs to be focused too! Indian Heart J. 70, S471–S477.

Saidman, J., Rubin, S., Swaminath, A., 2023. Inflammatory bowel disease and cannabis: key counseling strategies. Curr. Opin. Gastroenterol. 39, 301–307.

Sainsbury, B., Bloxham, J., Pour, M.H., Padilla, M., Enciso, R., 2021. Efficacy of cannabis-based medications compared to placebo for the treatment of chronic neuropathic pain: a systematic review with meta-analysis. J. Dent. Anesth. Pain Med. 21, 479–506.

Saint-Mezard, P., Chavagnac, C., Bosset, S., Ionescu, M., Peyron, E., et al., 2003. Psychological stress exerts an adjuvant effect on skin dendritic cell functions in vivo. J. Immunol. 171, 4073–4080.

Sait, A., Angeli, C., Doig, A.J., Day, P.J.R., 2021. Viral involvement in Alzheimer's disease. ACS Chem. Neurosci. 12, 1049–1060.

Sakamoto, S., Zhu, X., Hasegawa, Y., Karma, S., Obayashi, M., et al., 2021. Inflamed brain: targeting immune changes and inflammation for treatment of depression. Psychiatry Clin. Neurosci. 75, 304–311.

Sakayori, N., Kikkawa, T., Tokuda, H., Kiryu, E., Yoshizaki, K., et al., 2016. Maternal dietary imbalance between omega-6 and omega-3 polyunsaturated fatty acids impairs neocortical development via epoxy metabolites. Stem Cells 34, 470–482.

Salam, A.P., Borsini, A., Zunszain, P.A., 2018. Trained innate immunity: a salient factor in the pathogenesis of neuroimmune psychiatric disorders. Mol. Psychiatry 23, 170–176.

Salari, A.A., Amani, M., 2017. Neonatal blockade of GABA-A receptors alters behavioral and physiological phenotypes in adult mice. Int. J. Dev. Neurosci. 57, 62–71.

Salberg, S., Yamakawa, G.R., Griep, Y., Bain, J., Beveridge, J.K., et al., 2021. Pain in the developing brain: early life factors alter nociception and neurobiological function in adolescent rats. Cereb. Cortex Comm. 2, tgab014.

Sallenave, J.M., Guillot, L., 2020. Innate immune signaling and proteolytic pathways in the resolution or exacerbation of SARS-CoV-2 in Covid-19: key therapeutic targets? Front. Immunol. 11, 1229.

Salter, M.W., Stevens, B., 2017. Microglia emerge as central players in brain disease. Nat. Med. 23, 1018–1027.

Salvemini, D., Jacobson, K.A., 2017. Highly selective A3 adenosine receptor agonists relieve chronic neuropathic pain. Expert Opin. Ther. Pat. 27, 967.

Samara, A., Li, Z., Rutlin, J., Raji, C.A., Sun, P., et al., 2021. Nucleus accumbens microstructure mediates the relationship between obesity and eating behavior in adults. Obesity (Silver Spring) 29, 1328–1337.

Samii, A., Etminan, M., Wiens, M.O., Jafari, S., 2009. NSAID use and the risk of Parkinson's disease: systematic review and meta-analysis of observational studies. Drugs Aging 26, 769–779.

Sampaio, V.S., Pinto, A.L.C.B., Barboza, L.L., Mouta, G.S., Silva, E.L., et al., 2022. Impact of Plasmodium vivax malaria on executive and cognitive functions in elderlies in the Brazilian Amazon. Sci. Rep. 12, 10361.

Sampath, P., Rajamanickam, A., Thiruvengadam, K., Natarajan, A.P., Hissar, S., et al., 2023. Cytokine upsurge among drug-resistant tuberculosis endorse the signatures of hyper inflammation and disease severity. Sci. Rep. 13, 785.

Sampson, T.R., Debelius, J.W., Thron, T., Janssen, S., Shastri, G.G., et al., 2016. Gut microbiota regulate motor deficits and neuroinflammation in a model of Parkinson's disease. Cell 167, 1469–1480.e12.

Samsam, M., Ahangari, R., Naser, S.A., 2014. Pathophysiology of autism spectrum disorders: revisiting gastrointestinal involvement and immune imbalance. World J. Gastroenterol. 20, 9942–9951.

Samuels, B.A., Anacker, C., Hu, A., Levinstein, M.R., Pickenhagen, A., et al., 2015. 5-HT1A receptors on mature dentate gyrus granule cells are critical for the antidepressant response. Nat. Neurosci. 18, 1606–1616.

Samuthpongtorn, C., Nguyen, L.H., Okereke, O.I., Wang, D.D., Song, M., et al., 2023. Consumption of ultraprocessed food and risk of depression. JAMA Netw. Open 6, e2334770.

Sánchez, K.E., Rosenberg, G.A., 2022. Shared inflammatory pathology of stroke and COVID-19. Int. J. Mol. Sci. 23, 5150.

Sánchez-Santed, F., Colomina, M.T., Hernández, E.H., 2016. Organophosphate pesticide exposure and neurodegeneration. Cortex 74, 417–426.

Sánchez-Vidaña, D.I., Chan, N.J., Chan, A.H., Hui, K.K., Lee, S., et al., 2016. Repeated treatment with oxytocin promotes hippocampal cell proliferation, dendritic maturation and affects socio-emotional behavior. Neuroscience 333, 65–77.

Sandhu, A., Chopra, T., 2021. Fecal microbiota transplantation for recurrent Clostridioides difficile, safety, and pitfalls. Ther. Adv. Gastroenterol. 14. 17562848211053105.

Sangah, A.B., Jabeen, S., Hunde, M.Z., Devi, S., Mumtaz, H., Shaikh, S. S., 2023. Maternal and fetal outcomes of SLE in pregnancy: a literature review. J. Obstet. Gynaecol. 43, 2205513.

Sangaralingham, S.J., Kuhn, M., Cannone, V., Chen, H.H., Burnett, J.C., 2023. Natriuretic peptide pathways in heart failure: further therapeutic possibilities. Cardiovasc. Res. 118, 3416–3433.

Sankaranarayanan, K., Ozga, A.T., Warinner, C., Tito, R.Y., Obregon-Tito, A.J., et al., 2015. Gut microbiome diversity among Cheyenne and Arapaho individuals from Western Oklahoma. Curr. Biol. 25, 3161–3169.

Sanson, A., Bosch, O.J., 2022. Dysfunctions of brain oxytocin signaling: Implications for poor mothering. Neuropharmacology 211, 109049.

Santarelli, S., Zimmermann, C., Kalideris, G., Lesuis, S.L., Arloth, J., et al., 2017. An adverse early life environment can enhance stress resilience in adulthood. Psychoneuroendocrinology 78, 213–221.

Santhanam, P., Teslovich, T., Wilson, S.H., Yeh, P.H., Oakes, T.R., Weaver, L.K., 2019. Decreases in white matter integrity of ventro-limbic pathway linked to post-traumatic stress disorder in mild traumatic brain injury. J. Neurotrauma 36, 1093–1098.

Santoft, F., Hedman-Lagerlöf, E., Salomonsson, S., Lindsäter, E., Ljótsson, B., et al., 2020. Inflammatory cytokines in patients with common mental disorders treated with cognitive behavior therapy. Brain Behav. Immun. Health 3, 100045.

Santos Jr., H.P., Nephew, B.C., Bhattacharya, A., Tan, X., Smith, L., et al., 2018. Discrimination exposure and DNA methylation of stress-related genes in Latina mothers. Psychoneuroendocrinology 98, 131–138.

Santosa, A., Rosengren, A., Ramasundarahettige, C., Rangarajan, S., Gulec, S., et al., 2021. Psychosocial risk factors and cardiovascular disease and death in a population-based cohort from 21 low-, middle-, and high-income countries. JAMA Netw. Open 4, e2138920.

Sapienza, J., Spangaro, M., Guillemin, G.J., Comai, S., Bosia, M., 2023. Importance of the dysregulation of the kynurenine pathway on cognition in schizophrenia: a systematic review of clinical studies. Eur. Arch. Psychiatry Clin. Neurosci. 273, 1317–1328.

Sapolsky, R.M., 2023. Determined: A Science of Life Without Free Will. Penguin Press, New York.

Sapolsky, R.M., Romero, L.M., Munck, A.U., 2000. How do glucocorticoids influence stress responses? Integrating permissive, suppressive, stimulatory, and preparative actions. Endocr. Rev. 21, 55–89.

Sarailoo, M., Afshari, S., Asghariazar, V., Safarzadeh, E., Dadkhah, M., 2022. Cognitive impairment and neurodegenerative diseases development associated with organophosphate pesticides exposure: a review study. Neurotox. Res. 40, 1624–1643.

Saresella, M., Marventano, I., Guerini, F.R., Mancuso, R., Ceresa, L., et al., 2009. An autistic endophenotype results in complex immune dysfunction in healthy siblings of autistic children. Biol. Psychiatry 66, 978–984.

Sarkar, A., Lehto, S.M., Harty, S., Dinan, T.G., Cryan, J.F., Burnet, P.W.J., 2016. Psychobiotics and the manipulation of bacteria-gut-brain signals. Trends Neurosci. 39, 763–781.

Sarkar, A., Harty, S., Johnson, K.V., Moeller, A.H., Carmody, R.N., et al., 2020. The role of the microbiome in the neurobiology of social behaviour. Biol. Rev. Camb. Philos. Soc. 95, 1131–1166.

Sarris, J., Murphy, J., Mischoulon, D., Papakostas, G.I., Fava, M., et al., 2016. Adjunctive nutraceuticals for depression: a systematic review and meta-analyses. Am. J. Psychiatry 173, 575–587.

Sarris, J., Sinclair, J., Karamacoska, D., Davidson, M., Firth, J., 2020. Medicinal cannabis for psychiatric disorders: a clinically-focused systematic review. BMC Psychiatry 20, 24.

Sarubbo, F., Cavallucci, V., Pani, G., 2022. The influence of gut microbiota on neurogenesis: evidence and hopes. Cells 11, 382.

Sasaki, A., 2017. Microglia and brain macrophages: an update. Neuropathology 37, 452–464.

Sasaki, M., Schwab, C., Ramirez Garcia, A., Li, Q., Ferstl, R., et al., 2022. The abundance of Ruminococcus bromii is associated with faecal butyrate levels and atopic dermatitis in infancy. Allergy 77, 3629–3640.

Sassone-Corsi, M., Raffatellu, M., 2015. No vacancy: how beneficial microbes cooperate with immunity to provide colonization resistance to pathogens. J. Immunol. 194, 4081–4087.

Sathyanesan, M., Haiar, J.M., Watt, M.J., Newton, S.S., 2017. Restraint stress differentially regulates inflammation and glutamate receptor gene expression in the hippocampus of C57BL/6 and BALB/c mice. Stress 20, 197–204.

Sato, H., Zhang, L.S., Martinez, K., Chang, E.B., Yang, Q., et al., 2016. Antibiotics suppress activation of intestinal mucosal mast cells and reduce dietary lipid absorption in Sprague-Dawley rats. Gastroenterology 151, 923–932.

Sato, W., Uono, S., Kochiyama, T., 2020. Neurocognitive mechanisms underlying social atypicalities in autism: weak amygdala's emotional modulation hypothesis. Front. Psychiatry 11, 864.

Sato, Y., Atarashi, K., Plichta, D.R., Arai, Y., Sasajima, S., et al., 2021. Novel bile acid biosynthetic pathways are enriched in the microbiome of centenarians. Nature 599, 458–464.

Sato, M., Nakai, N., Fujima, S., Choe, K.Y., Takumi, T., 2023. Social circuits and their dysfunction in autism spectrum disorder. Mol. Psychiatry 28, 3194–3206.

Satterlee, J.S., Chadwick, L.H., Tyson, F.L., McAllister, K., Beaver, J., et al., 2019. The NIH Common Fund/Roadmap Epigenomics Program: successes of a comprehensive consortium. Sci. Adv. 5, eaaw6507.

Saul, A.N., Oberyszyn, T.M., Daugherty, C., Kusewitt, D., Jones, S., et al., 2005. Chronic stress and susceptibility to skin cancer. J. Natl. Cancer Inst. 97, 1760–1767.

Saunders, G.R.B., Wang, X., Chen, F., Jang, S.K., Liu, M., et al., 2022. Genetic diversity fuels gene discovery for tobacco and alcohol use. Nature 612, 720–724.

Savcisens, G., Eliassi-Rad, T., Hansen, L.K., Mortensen, L.H., Lilleholt, L., et al., 2024. Using sequences of life-events to predict human lives. Nat. Comput. Sci. 4, 43–56.

Savignac, H.M., Couch, Y., Stratford, M., Bannerman, D.M., Tzortzis, G., et al., 2016. Prebiotic administration normalizes lipopolysaccharide (LPS)-induced anxiety and cortical 5-HT2A receptor and IL1-β levels in male mice. Brain Behav. Immun. 52, 120–131.

Savino, R., Carotenuto, M., Polito, A.N., Di Noia, S., Albenzio, M., et al., 2020. Analyzing the potential biological determinants of autism spectrum disorder: from neuroinflammation to the kynurenine pathway. Brain Sci. 10, 631.

Savitz, J., 2020. The kynurenine pathway: a finger in every pie. Mol. Psychiatry 25, 131–147.

Savoy, C., Ferro, M.A., Schmidt, L.A., Saigal, S., Van Lieshout, R.J., 2016. Prenatal betamethasone exposure and psychopathology risk in extremely low birth weight survivors in the third and fourth decades of life. Psychoneuroendocrinology 74, 278–285.

Sawada, Y., Nishio, Y., Suzuki, K., Hirayama, K., Takeda, A., et al., 2012. Attentional set-shifting deficit in Parkinson's disease is associated with prefrontal dysfunction: an FDG-PET study. PLoS One 7, e38498.

Sawyer, K.M., 2021. The role of inflammation in the pathogenesis of perinatal depression and offspring outcomes. Brain Behav. Immun. Health 18, 100390.

Sayed, N., Huang, Y., Nguyen, K., Krejciova-Rajaniemi, Z., Grawe, A.P., et al., 2021. An inflammatory aging clock (iAge) based on deep learning tracks multimorbidity, immunosenescence, frailty and cardiovascular aging. Nat. Aging 1, 598–615.

Saylor, D., Dickens, A.M., Sacktor, N., Haughey, N., Slusher, B., et al., 2016. HIV-associated neurocognitive disorder—pathogenesis and prospects for treatment. Nat. Rev. Neurol. 12, 234–248.

Scarante, F.F., Vila-Verde, C., Ferreira-Junior, N.C., Detoni, V.L., Guimaraes, F., et al., 2017. Cannabinoid modulation of the stressed hippocampus. Front. Mol. Neurosci. 10, 411.

Scardua-Silva, L., Amorim da Costa, B., Karmann Aventurato, Í., Batista Joao, R., Machado de Campos, B., et al., 2024. Microstructural brain abnormalities, fatigue, and cognitive dysfunction after mild COVID-19. Sci. Rep. 14, 1758.

Schafer, M.J., White, T.A., Evans, G., Tonne, J.M., Verzosa, G.C., et al., 2016. Exercise prevents diet-induced cellular senescence in adipose tissue. Diabetes 65, 1606–1615.

Scharko, A.M., 2011. The infection hypothesis of schizophrenia: a systematic review. Behav. Brain Sci. 1, 47–56.

Schechter, D.S., Moser, D.A., Paoloni-Giacobino, A., Stenz, L., Gex-Fabry, M., et al., 2015. Methylation of NR3C1 is related to maternal PTSD, parenting stress and maternal medial prefrontal cortical activity in response to child separation among mothers with histories of violence exposure. Front. Psychol. 6, 690.

Scheier, M.F., Carver, C.S., 2018. Dispositional optimism and physical health: a long look back, a quick look forward. Am. Psychol. 73, 1082–1094.

Schenk, D., Barbour, R., Dunn, W., Gordon, G., Grajeda, H., 1999. Immunization with amyloid-beta attenuates Alzheimer-disease-like pathology in the PDAPP mouse. Nature 400, 173–177.

Schepanski, S., Buss, C., Hanganu-Opatz, I.L., Arck, P.C., 2018. Prenatal immune and endocrine modulators of offspring's brain development and cognitive functions later in life. Front. Immunol. 9, 2186.

Scheyer, A., Yasmin, F., Naskar, S., Patel, S., 2023. Endocannabinoids at the synapse and beyond: implications for neuropsychiatric disease pathophysiology and treatment. Neuropsychopharmacology 48, 37–53.

Schiess, N., Cataldi, R., Okun, M.S., Fothergill-Misbah, N., Dorsey, E.R., 2022. Six action steps to address global disparities in Parkinson disease: a World Health Organization priority. JAMA Neurol. 79, 929–936.

Schiff, N.D., Giacino, J.T., Butson, C.R., Choi, E.Y., Baker, J.L., et al., 2023. Thalamic deep brain stimulation in traumatic brain injury: a phase 1, randomized feasibility study. Nat. Med. 29, 3162–3174.

Schiffman, S.S., Scholl, E.H., Furey, T.S., Nagle, H.T., 2023. Toxicological and pharmacokinetic properties of sucralose-6-acetate and its parent sucralose: in vitro screening assays. J. Toxicol. Environ. Health B Crit. Rev., 1–35.

Schilling, S., Chausse, B., Dikmen, H.O., Almouhanna, F., Hollnagel, J.O., et al., 2021. TLR2- and TLR3-activated microglia induce different levels of neuronal network dysfunction in a context-dependent manner. Brain Behav. Immun. 96, 80–91.

Schlossmacher, M.G., Tomlinson, J.J., Santos, G., Shutinoski, B., Brown, E.G., et al., 2017. Modelling idiopathic Parkinson disease as a complex illness can inform incidence rate in healthy adults: the PR EDIGT score. Eur. J. Neurosci. 45, 175–191.

Schmeltzer, S.N., Herman, J.P., Sah, R., 2016. Neuropeptide Y (NPY) and posttraumatic stress disorder (PTSD): a translational update. Exp. Neurol. 284, 196–210.

Schmidt, N.B., Keough, M.E., 2010. Treatment of panic. Annu. Rev. Clin. Psychol. 6 (1), 241–256.

Schmidt, E.D., Janszen, A.W., Wouterlood, F.G., Tilders, F.J., 1995. Interleukin-1-induced long-lasting changes in hypothalamic corticotropin-releasing hormone (CRH)—neurons and hyperresponsiveness of the hypothalamus-pituitary-adrenal axis. J. Neurosci. 15, 7417–7426.

Schmidt, K., Cowen, P.J., Harmer, C.J., Tzortzis, G., Errington, S., Burnet, P.W., 2015. Prebiotic intake reduces the waking cortisol response and alters emotional bias in healthy volunteers. Psychopharmacology 232, 1793–1801.

Schmidt, S.A.J., Sørensen, H.T., Langan, S.M., Vestergaard, M., 2021. Perceived psychological stress and risk of herpes zoster: a nationwide population-based cohort study. Br. J. Dermatol. 185, 130–138.

Schmidt-Reinwald, A., Pruessner, J.C., Hellhammer, D.H., Federenko, I., Rohleder, N., et al., 1999. The cortisol response to awakening in relation to different challenge tests and a 12-hour cortisol rhythm. Life Sci. 64, 1653–1660.

Schneider, K.M., Blank, N., Alvarez, Y., Thum, K., Lundgren, P., et al., 2023. The enteric nervous system relays psychological stress to intestinal inflammation. Cell 186, 2823–2838.

Schoenfeld, J.D., Ioannidis, J.P., 2013. Is everything we eat associated with cancer? A systematic cookbook review. Am. J. Clin. Nutr. 97, 127–134.

Schoenfeld, T.J., Rada, P., Pieruzzini, P.R., Hsueh, B., Gould, E., 2013. Physical exercise prevents stress-induced activation of granule neurons and enhances local inhibitory mechanisms in the dentate gyrus. J. Neurosci. 33, 7770–7777.

Schonhoff, A.M., Figge, D.A., Williams, G.P., Jurkuvenaite, A., Gallups, N.J., et al., 2023. Border-associated macrophages mediate the neuroinflammatory response in an alpha-synuclein model of Parkinson disease. Nat. Commun. 14, 3754.

Schormair, B., Kemlink, D., Mollenhauer, B., Fiala, O., Machetanz, G., et al., 2017. Diagnostic exome sequencing in early-onset Parkinson's disease confirms VPS13C as a rare cause of autosomal-recessive Parkinson's disease. Clin. Genet. https://doi.org/10.1111/cge.13124.

Schou, T.M., Joca, S., Wegener, G., Bay-Richter, C., 2021. Psychiatric and neuropsychiatric sequelae of COVID-19 - a systematic review. Brain Behav. Immun. 97, 328–348.

Schreiber, S., Schreiber, F., Lockhart, S.N., Horng, A., Bejanin, A., et al., 2017. Alzheimer disease signature neurodegeneration and APOE genotype in mild cognitive impairment with suspected non-Alzheimer disease pathophysiology. JAMA Neurol. 74, 650–659.

Schrock, J.M., McDade, T.W., Carrico, A.W., D'Aquila, R.T., Mustanski, B., 2021. Traumatic events and mental health: the amplifying effects of pre-trauma systemic inflammation. Brain Behav. Immun. 98, 173–184.

Schroder, K., Hertzog, P.J., Ravasi, T., Hume, D.A., 2004. Interferon-γ: an overview of signals, mechanisms and functions. J. Leucocyte Biol. 75, 63–189.

Schroeder, B.O., Bäckhed, F., 2016. Signals from the gut microbiota to distant organs in physiology and disease. Nat. Med. 22, 1079–1089.

Schroer, A.B., Ventura, P.B., Sucharov, J., Misra, R., Chui, M.K.K., et al., 2023. Platelet factors attenuate inflammation and rescue cognition in ageing. Nature 620, 1071–1079.

Schultebraucks, K., Qian, M., Abu-Amara, D., Dean, K., Laska, E., et al., 2021. Pre-deployment risk factors for PTSD in active-duty personnel deployed to Afghanistan: a machine-learning approach for analyzing multivariate predictors. Mol. Psychiatry 26, 5011–5022.

Schulz, K.M., Sisk, C.L., 2016. The organizing actions of adolescent gonadal steroid hormones on brain and behavioral development. Neurosci. Biobehav. Rev. 70, 148–158.

Schür, R.R., Boks, M.P., Geuze, E., Prinsen, H.C., Verhoeven-Duif, N.M., et al., 2016. Development of psychopathology in deployed armed forces in relation to plasma GABA levels. Psychoneuroendocrinology 73, 263–270.

Schwabenland, M., Salié, H., Tanevski, J., Killmer, S., Lago, M.S., et al., 2021. Deep spatial profiling of human COVID-19 brains reveals neuroinflammation with distinct microanatomical microglia-T-cell interactions. Immunity 54, 1594–1610.e11.

Schwingshackl, L., Schwedhelm, C., Hoffmann, G., Lampousi, A.M., Knuppel, S., et al., 2017. Food groups and risk of all-cause mortality:

a systematic review and meta-analysis of prospective studies. Am. J. Clin. Nutr. 105, 1462–1473.

Scioli-Salter, E.R., Forman, D.E., Otis, J.D., Gregor, K., Valovski, I., Rasmusson, A.M., 2015. The shared neuroanatomy and neurobiology of comorbid chronic pain and PTSD: therapeutic implications. Clin. J. Pain 31, 363–374.

Scott, K.A., Ida, M., Peterson, V.L., Prenderville, J.A., Moloney, G.M., et al., 2017. Revisiting Metchnikoff: age-related alterations in microbiota-gut-brain axis in the mouse. Brain Behav. Immun. 65, 20–32.

Scott, G.A., Terstege, D.J., Vu, A.P., Law, S., Evans, A., Epp, J.R., 2020. Disrupted neurogenesis in germ-free mice: effects of age and sex. Front. Cell Dev. Biol. 8, 407.

Scott, H.R., Stevelink, S.A.M., Gafoor, R., Lamb, D., Carr, E., et al., 2023. Prevalence of post-traumatic stress disorder and common mental disorders in health-care workers in England during the COVID-19 pandemic: a two-phase cross-sectional study. Lancet Psychiatry 10, 40–49.

Scotti, M.A., Carlton, E.D., Demas, G.E., Grippo, A.J., 2015. Social isolation disrupts innate immune responses in both male and female prairie voles and enhances agonistic behavior in female prairie voles (Microtus ochrogaster). Horm. Behav. 70, 7–13.

Sebastiani, P., Gurinovich, A., Bae, H., Andersen, S., Malovini, A., et al., 2017. Four genome-wide association studies identify new extreme longevity variants. J. Gerontol. A Biol. Sci. Med. Sci. 72, 1453–1464.

Secinti, E., Thompson, E.J., Richards, M., Gaysina, D., 2017. Research review: childhood chronic physical illness and adult emotional health - a systematic review and meta-analysis. J. Child Psychol. Psychiatry 58, 753–769.

Seet, D., Allameen, N.A., Tay, S.H., Cho, J., Mak, A., 2021. Cognitive dysfunction in systemic lupus erythematosus: immunopathology, clinical manifestations, neuroimaging and management. Rheumatol. Ther. 8, 651–679.

Segal, Z.V., Walsh, K.M., 2016. Mindfulness-based cognitive therapy for residual depressive symptoms and relapse prophylaxis. Curr. Opin. Psychiatry 29, 7–12.

Segal, Z.V., Williams, J.M.G., Teasdale, J., 2002. Mindfulness-Based Cognitive Therapy for Depression: A New Approach to Preventing Relapse. Guilford, New York.

Segerstrom, S.C., Miller, G.E., 2004. Psychological stress and the human immune system: a meta-analytic study of 30 years of inquiry. Psychol. Bull. 130, 601–630.

Segerstrom, S.C., Schipper, L.J., Greenberg, R.N., 2008. Caregiving, repetitive thought, and immune response to vaccination in older adults. Brain Behav. Immun. 22, 744–752.

Seiler, A., Fagundes, C.P., Christian, L.M., 2020. The impact of everyday stressors on the immune system and health. In: Stress Challenges and Immunity in Space: From Mechanisms to Monitoring and Preventive Strategies. Springer, pp. 71–92.

Sekar, A., Bialas, A.R., de Rivera, H., Davis, A., Hammond, T.R., et al., 2016. Schizophrenia risk from complex variation of complement component 4. Nature 530, 177–183.

Sela, U., Euler, C.W., Correa da Rosa, J., Fischetti, V.A., 2018. Strains of bacterial species induce a greatly varied acute adaptive immune response: the contribution of the accessory genome. PLoS Pathog. 14, e1006726.

Seligman, M.E., Csikszentmihalyi, M., 2000. Positive psychology: an introduction. Am. Psychol. 55, 5–14.

Sellgren, C.M., Gracias, J., Watmuff, B., Biag, J.D., Thanos, J.M., et al., 2019. Increased synapse elimination by microglia in schizophrenia patient-derived models of synaptic pruning. Nat. Neurosci. 22, 374–385.

Senese, V.P., Shinohara, K., Venuti, P., Bornstein, M.H., Rosanio, V., et al., 2022. The interaction effect of parental rejection and oxytocin receptor gene polymorphism on depression: a cross-cultural study in non-clinical samples. Int. J. Environ. Res. Public Health 19, 5566.

Seng, J., Low, L., Sperlich, M., Ronis, D., Liberzon, I., 2011. Post-traumatic stress disorder, child abuse history, birthweight and gestational age: a prospective cohort study. BJOG 118, 1329–1339.

Sensi, S.L., Granzotto, A., Siotto, M., Squitti, R., 2018. Copper and zinc dysregulation in Alzheimer's disease. Trends Pharmacol. Sci. 39, 1049–1063.

Sepehrinezhad, A., Shahbazi, A., Negah, S.S., 2020. COVID-19 virus may have neuroinvasive potential and cause neurological complications: a perspective review. J. Neurovirol. 26, 324–329.

Sepich-Poore, G.D., Zitvogel, L., Straussman, R., Hasty, J., Wargo, J.A., Knight, R., 2021. The microbiome and human cancer. Science 371, eabc4552.

Sequeira, A., Mamdani, F., Ernst, C., Vawter, M.P., Bunney, W.E., et al., 2009. Global brain gene expression analysis links glutamatergic and GABAergic alterations to suicide and major depression. PLoS One 4, e6585.

Serchuk, M.D., Corrigan, P.W., Reed, S., Ohan, J.L., 2021. Vicarious stigma and self-stigma experienced by parents of children with mental health and/or neurodevelopmental disorders. Community Ment. Health J. 57, 1537–1546.

Serin, Y., Acar Tek, N., 2019. Effect of circadian rhythm on metabolic processes and the regulation of energy balance. Ann. Nutr. Metab. 74 (4), 322–330.

Serova, L., Mulhall, H., Sabban, E., 2017. NPY1 receptor agonist modulates development of depressive-like behavior and gene expression in hypothalamus in SPS rodent PTSD model. Front. Neurosci. 11, 203.

Serpa, M., Doshi, J., Joaquim, H.P.G., Vieira, E.L.M., Erus, G., et al., 2023. Inflammatory cytokines and white matter microstructure in the acute phase of first-episode psychosis: a longitudinal study. Schizophr. Res. 257, 5–18.

Serrats, J., Grigoleit, J.S., Alvarez-Salas, E., Sawchenko, P.E., 2017. Pro-inflammatory immune-to-brain signaling is involved in neuroendocrine responses to acute emotional stress. Brain Behav. Immun. 62, 53–63.

Seth, P., Rudd, R.A., Noonan, R.K., Haegerich, T.M., 2018. Quantifying the epidemic of prescription opioid overdose deaths. Am. J. Public Health 108, 500–502.

Sevelsted, A., Stokholm, J., Bonnelykke, K., Bisgaard, H., 2015. Caesarean section and chronic immune disorders. Pediatrics 135, e92–e98.

Severance, E.G., Yolken, R.H., 2020. Deciphering microbiome and neuroactive immune gene interactions in schizophrenia. Neurobiol. Dis. 135, 104331.

Severance, E.G., Gressitt, K.L., Stallings, C.R., Katsafanas, E., Schweinfurth, L.A., et al., 2016. Candida albicans exposures, sex specificity and cognitive deficits in schizophrenia and bipolar disorder. NPJ Schizophr. 2, 16018.

Severance, E.G., Gressitt, K.L., Stallings, C.R., Katsafanas, E., Schweinfurth, L.A., et al., 2017. Probiotic normalization of Candida albicans in schizophrenia: a randomized, placebo-controlled, longitudinal pilot study. Brain Behav. Immun. 62, 41–45.

Severinsen, M.C.K., Pedersen, B.K., 2020. Muscle-organ crosstalk: the emerging roles of myokines. Endocr. Rev. 41, 594–609.

Sewell, E., Roberts, J., Mukhopadhyay, S., 2021. Association of infection in neonates and long-term neurodevelopmental outcome. Clin. Perinatol. 48, 251–261.

Sha, Z., Versace, A., Edmiston, E.K., Fournier, J., Graur, S., et al., 2020. Functional disruption in prefrontal-striatal network in obsessive-compulsive disorder. Psychiatry Res. Neuroimaging 300, 111081.

Shackman, A.J., Stockbridge, M.D., Tillman, R.M., Kaplan, C.M., Tromp, D.P., et al., 2016. The neurobiology of dispositional negativity and attentional biases to threat: Implications for understanding anxiety disorders in adults and youth. J. Exp. Psychopathol. 7, 311–342.

Shafiee-Kandjani, A.R., Nezhadettehad, N., Farhang, S., Bruggeman, R., Shanebandi, D., et al., 2023. MicroRNAs and pro-inflammatory cytokines as candidate biomarkers for recent-onset psychosis. BMC Psychiatry 23, 631.

Shah, P., Plitman, E., Iwata, Y., Kim, J., Nakajima, S., et al., 2020. Glutamatergic neurometabolites and cortical thickness in treatment-resistant schizophrenia: implications for glutamate-mediated excitotoxicity. J. Psychiatr. Res. 124, 151–158.

Shah, A., Kishore, U., Shastri, A., 2021. Complement system in Alzheimer's disease. Int. J. Mol. Sci. 22, 13647.

Shahbazi, S., Kaur, J., Kuanar, A., Kar, D., Singh, S., et al., 2017. Risk of late-onset Alzheimer's disease by plasma cholesterol: rational in silico drug investigation of pyrrole-based HMG-CoA reductase inhibitors. Assay Drug Dev. Technol. 15, 342–351.

Shaikh, S.R., MacIver, N.J., Beck, M.A., 2022. Obesity dysregulates the immune response to influenza infection and vaccination through metabolic and inflammatory mechanisms. Annu. Rev. Nutr. 42, 67–89.

Shalbafan, M., Mohammadinejad, P., Shariat, S.V., Alavi, K., Zeinoddini, A., et al., 2015. Celecoxib as an adjuvant to fluvoxamine in moderate to severe obsessive-compulsive disorder: a double-blind, placebo-controlled, randomized trial. Pharmacopsychiatry 48, 136–140.

Shalev, A.Y., Ankri, Y., Gilad, M., Israeli-Shalev, Y., Adessky, R., et al., 2016. Long-term outcome of early interventions to prevent posttraumatic stress disorder. J. Clin. Psychiatry 77, e580–e587.

Shalev, A., Liberzon, I., Marmar, C., 2017. Post-traumatic stress disorder. N. Engl. J. Med. 376, 2459–2469.

Sham, H.P., Yu, E.Y., Gulen, M.F., Bhinder, G., Stahl, M., et al., 2013. SIGIRR, a negative regulator of TLR/IL-1R signalling promotes microbiota dependent resistance to colonization by enteric bacterial pathogens. PLoS Pathog. 9, e1003539.

Shamir, D.B., Deng, Y., Wu, Q., Modak, S., Congdon, E.E., Sigurdsson, E.M., 2020. Dynamics of internalization and intracellular interaction of tau antibodies and human pathological tau protein in a human neuron-like model. Front. Neurol. 11, 602292.

Shang, D.S., Yang, Y.M., Zhang, H., Tian, L., Jiang, J.S., et al., 2016. Intracerebral GM-CSF contributes to transendothelial monocyte migration in APP/PS1 Alzheimer's disease mice. J. Cereb. Blood Flow Metab. 36, 1978–1991.

Shanks, N., Kusnecov, A.W., 1998. Differential immune reactivity to stress in BALB/cByJ and C57BL/6J mice: in vivo dependence on macrophages. Physiol. Behav. 65, 95–103.

Shanks, N., Renton, C., Zalcman, S., Anisman, H., 1994. Influence of change from grouped to individual housing on a T-cell-dependent immune response in mice: antagonism by diazepam. Pharmacol. Biochem. Behav. 47, 497–502.

Shanks, N., Kusnecov, A., Pezzone, M., Berkun, J., Rabin, B.S., 1997. Lactation alters the effects of conditioned stress on immune function. Am. J. Phys. 272, R16–R25.

Shanks, N., Windle, R.J., Perks, P.A., Harbuz, M.S., Jessop, D.S., et al., 2000. Early-life exposure to endotoxin alters hypothalamic-pituitary-adrenal function and predisposition to inflammation. Proc. Natl. Acad. Sci. USA 97, 5645–5650.

Shanley, M.R., Miura, Y., Guevara, C.A., Onoichenco, A., Kore, R., et al., 2023. Estrous cycle mediates midbrain neuron excitability altering social behavior upon stress. J. Neurosci. 43, 736–748.

Shao, Y., Forster, S.C., Tsaliki, E., Vervier, K., Strang, A., et al., 2019. Stunted microbiota and opportunistic pathogen colonization in caesarean-section birth. Nature 574, 117–121.

Shao, L.X., Liao, C., Gregg, I., Davoudian, P.A., Savalia, N.K., et al., 2021. Psilocybin induces rapid and persistent growth of dendritic spines in frontal cortex in vivo. Neuron 109, 2535–2544.e4.

Sharifi-Sanjani, M., Oyster, N.M., Tichy, E.D., Bedi Jr., K.C., Harel, O., et al., 2017. Cardiomyocyte-specific telomere shortening is a distinct signature of heart failure in humans. J. Am. Heart Assoc. 6, e005086.

Sharkey, K.A., Wiley, J.W., 2016. The role of the endocannabinoid system in the gut-brain axis. Gastroenterology 151, 252–266.

Sharma, A., 2016. Systems genomics support for immune and inflammation hypothesis of depression. Curr. Neuropharmacol. 14, 749–758.

Sharma, N., Sharma, S., Nehru, B., 2017. Curcumin protects dopaminergic neurons against inflammation-mediated damage and improves motor dysfunction induced by single intranigral lipopolysaccharide injection. Inflammopharmacology 25, 351–368.

Sharma, A., Ren, X., Zhang, H., Pandey, G.N., 2022. Effect of depression and suicidal behavior on neuropeptide Y (NPY) and its receptors in the adult human brain: a postmortem study. Prog. Neuro-Psychopharmacol. Biol. Psychiatry 112, 110428.

Sharoar, M.G., Hu, X., Ma, X.M., Zhu, X., Yan, R., 2019. Sequential formation of different layers of dystrophic neurites in Alzheimer's brains. Mol. Psychiatry 24, 1369–1382.

Sharpe, M., Goldsmith, K.A., Johnson, A.L., Chalder, T., Walker, J., et al., 2015. Rehabilitative treatments for chronic fatigue syndrome: long-term follow-up from the PACE trial. Lancet Psychiatry 2, 1067–1074.

Sharpe, L., Sinclair, J., Kramer, A., de Manincor, M., Sarris, J., 2020. Cannabis, a cause for anxiety? A critical appraisal of the anxiogenic and anxiolytic properties. J. Transl. Med. 18, 374.

Sheerin, C.M., Lind, M.J., Bountress, K.E., Marraccini, M.E., Amstadter, A.B., Bacanu, S.-A., et al., 2020. Meta-analysis of associations between hypothalamic-pituitary-adrenal axis genes and risk of post-traumatic stress disorder. J. Trauma. Stress. 33, 688–698.

Sheffler, J.L., Piazza, J.R., Quinn, J.M., Sachs-Ericsson, N.J., Stanley, I.H., 2019. Adverse childhood experiences and coping strategies: identifying pathways to resiliency in adulthood. Anxiety Stress Coping 32, 594–609.

Shelton, R.C., Sanders-Bush, E., Manier, D.H., Lewis, D.A., 2009. Elevated 5-HT 2A receptors in postmortem prefrontal cortex in major depression is associated with reduced activity of protein kinase A. Neuroscience 158, 1406–1415.

Shen, L., Li, C., Wang, Z., Zhang, R., Shen, Y., et al., 2019. Early-life exposure to severe famine is associated with higher methylation level in the IGF2 gene and higher total cholesterol in late adulthood: the Genomic Research of the Chinese Famine (GRECF) study. Clin. Epigenetics 11, 88.

Shen, S., Gong, C., Jin, K., Zhou, L., Xiao, Y., Ma, L., 2022a. Omega-3 fatty acid supplementation and coronary heart disease risks: a meta-analysis of randomized controlled clinical trials. Front. Nutr. 9, 809311.

Shen, Z., Xiang, M., Chen, C., Ding, F., Wang, Y., et al., 2022b. Glutamate excitotoxicity: Potential therapeutic target for ischemic stroke. Biomed. Pharmacother. 151, 113125.

Shenk, C.E., O'Donnell, K.J., Pokhvisneva, I., Kobor, M.S., Meaney, M.J., et al., 2022. Epigenetic age acceleration and risk for posttraumatic stress disorder following exposure to substantiated child maltreatment. J. Clin. Child Adolesc. Psychol. 51, 651–661.

Sherer, M.L., Posillico, C.K., Schwarz, J.M., 2017. An examination of changes in maternal neuroimmune function during pregnancy and the postpartum period. Brain Behav. Immun. 66, 201–209.

Sherer, M.L., Voegtline, K.M., Park, H.S., Miller, K.N., Shuffrey, L.C., et al., 2022. The immune phenotype of perinatal anxiety. Brain Behav. Immun. 106, 280–288.

Sherman, M.P., Sherman, J., Arcinue, R., Niklas, V., 2016. Randomized control trial of human recombinant lactoferrin: a substudy reveals effects on the fecal microbiome of very low birth weight infants. J. Pediatr. 173, S37–S42.

Sherwin, E., Bordenstein, S.R., Quinn, J.L., Dinan, T.G., Cryan, J.F., 2019. Microbiota and the social brain. Science 366, eaar2016.

Sherwood, P.R., Price, T.J., Weimer, J., Ren, D., Donovan, H.S., et al., 2016. Neuro-oncology family caregivers are at risk for systemic inflammation. J. Neuro-Oncol. 128, 109–118.

Sheth, C., Prescot, A.P., Legarreta, M., Renshaw, P.F., McGlade, E., Yurgelun-Todd, D., 2019. Reduced gamma-amino butyric acid (GABA) and glutamine in the anterior cingulate cortex (ACC) of veterans exposed to trauma. J. Affect. Disord. 248, 166–174.

Sheynin, J., Lokshina, Y., Ahrari, S., Nickelsen, T., Duval, E.R., et al., 2023. Greater early posttrauma activation in the right inferior frontal gyrus predicts recovery from posttraumatic stress disorder symptoms. Biol. Psychiatry Cogn. Neurosci. Neuroimaging 9, 91–100.

Shi, Q., Colodner, K.J., Matousek, S.B., Merry, K., Hong, S., et al., 2015. Complement C3-deficient mice fail to display age-related hippocampal decline. J. Neurosci. 35, 13029–13042.

Shi, Q., Chowdhury, S., Ma, R., Le, K.X., Hong, S., et al., 2017. Complement C3 deficiency protects against neurodegeneration in aged plaque-rich APP/PS1 mice. Sci. Transl. Med. 9.

Shi, Y., Luan, D., Song, R., Zhang, Z., 2020. Value of peripheral neurotrophin levels for the diagnosis of depression and response to treatment: a systematic review and meta-analysis. Eur. Neuropsychopharmacol. 41, 40–51.

Shi, Y., Zhang, H., Huang, S., Yin, L., Wang, F., et al., 2022. Epigenetic regulation in cardiovascular disease: mechanisms and advances in clinical trials. Signal Transduct. Target Ther. 7, 200.

Shields, G.S., Slavich, G.M., 2017. Lifetime stress exposure and health: a review of contemporary assessment methods and biological mechanisms. Soc. Personal. Psychol. Compass 11, e12335.

Shields, G.S., Kuchenbecker, S.Y., Pressman, S.D., Sumida, K.D., Slavich, G.M., 2016. Better cognitive control of emotional information is associated with reduced pro-inflammatory cytokine reactivity to emotional stress. Stress 19, 63–68.

Shields, G.S., Spahr, C.M., Slavich, G.M., 2020. Psychosocial interventions and immune system function: a systematic review and meta-analysis of randomized clinical trials. JAMA Psychiatry 77, 1031–1043.

Shih, I.F., Liew, Z., Krause, N., Ritz, B., 2016. Lifetime occupational and leisure time physical activity and risk of Parkinson's disease. Parkinsonism Relat. Disord. 28, 112–117.

Shimanoe, C., Hara, M., Nishida, Y., Nanri, H., Otsuka, Y., et al., 2018. Coping strategy and social support modify the association between perceived stress and C-reactive protein: a longitudinal study of healthy men and women. Stress 21, 237–246.

Shimo, Y., Cathomas, F., Lin, H.Y., Chan, K.L., Parise, L.F., et al., 2023. Social stress induces autoimmune responses against the brain. Proc. Natl. Acad. Sci. USA 120, e2305778120.

Shin, S.H., Kim, Y.K., 2023. Early life stress, neuroinflammation, and psychiatric illness of adulthood. Adv. Exp. Med. Biol. 1411, 105–134.

Shirafkan, F., Hensel, L., Rattay, K., 2024. Immune tolerance and the prevention of autoimmune diseases essentially depend on thymic tissue homeostasis. Front. Immunol. 15, 1339714.

Shirahige, L., Melo, L., Nogueira, F., Rocha, S., Monte-Silva, K., 2016. Efficacy of noninvasive brain stimulation on pain control in migraine patients: a systematic review and meta-analysis. Headache 56, 1565–1596.

Shirato, K., Kizaki, T., 2021. SARS-CoV-2 spike protein S1 subunit induces pro-inflammatory responses via toll-like receptor 4 signaling in murine and human macrophages. Heliyon 7, e06187.

Shirvalkar, P., Prosky, J., Chin, G., Ahmadipour, P., Sani, O.G., et al., 2023. First-in-human prediction of chronic pain state using intracranial neural biomarkers. Nat. Neurosci. 26, 1090–1099.

Shively, C.A., Register, T.C., Appt, S.E., Clarkson, T.B., Uberseder, B., et al., 2018. Consumption of mediterranean versus western diet leads to distinct mammary gland microbiome populations. Cell Rep. 25, 47–56.e3.

Shobeiri, P., Kalantari, A., Teixeira, A.L., Rezaei, N., 2022. Shedding light on biological sex differences and microbiota-gut-brain axis: a comprehensive review of its roles in neuropsychiatric disorders. Biol. Sex Differ. 13, 12.

Shomron, N., Lam, J.C.K., Li, V.O.K., Heimesaat, M.M., Bereswill, S., et al., 2022. Oral microbiota signatures in post-traumatic stress disorder (PTSD) veterans. Mol. Psychiatry 27, 4590–4598.

Shonkoff, J.P., Boyce, W.T., McEwen, B.S., 2009. Neuroscience, molecular biology, and the childhood roots of health disparities: building a new framework for health promotion and disease prevention. JAMA 301, 2252–2259.

Shono, Y., van den Brink, M.R.M., 2018. Gut microbiota injury in allogeneic haematopoietic stem cell transplantation. Nat. Rev. Cancer 18, 283–295.

Shook, L.L., Sullivan, E.L., Lo, J.O., Perlis, R.H., Edlow, A.G., 2022. COVID-19 in pregnancy: implications for fetal brain development. Trends Mol. Med. 28, 319–330.

Short, S.J., Lubach, G.R., Karasin, A.I., Olsen, C.W., Styner, M., et al., 2010. Maternal influenza infection during pregnancy impacts postnatal brain development in the rhesus monkey. Biol. Psychiatry 67, 965–973.

Short, B., Fong, J., Galvez, V., Shelker, W., Loo, C.K., 2018. Side-effects associated with ketamine use in depression: a systematic review. Lancet Psychiatry 5, 65–78.

Shrestha, S., Hirvonen, J., Hines, C.S., Henter, I.D., Svenningsson, P., et al., 2011. Serotonin-1A receptors in major depression quantified using PET: controversies, confounds, and recommendations. NeuroImage 59, 3243–3251.

Shukla, R., Lavore, F., Maity, S., Derks, M.G.N., Jones, C.R., et al., 2022. Teixobactin kills bacteria by a two-pronged attack on the cell envelope. Nature 608, 390–396.

Shultz, B., Farkash, A., Collins, B., Mohammadmirzaei, N., Knox, D., 2022. Fear learning-induced changes in AMPAR and NMDAR expression in the fear circuit. Learn. Mem. 29, 83–92.

Shushtari, N., Abtahi Froushani, S.M., 2017. Caffeine augments the instruction of anti-inflammatory macrophages by the conditioned medium of mesenchymal stem cells. Cell J. 19, 415–424.

Shuwen, H., Kefeng, D., 2022. Intestinal phages interact with bacteria and are involved in human diseases. Gut Microbes 14, 2113717.

Sibille, E., Arango, V., Galfalvy, H.C., Pavlidis, P., Erraji-Benchekroun, L., et al., 2004. Gene expression profiling of depression and suicide in human prefrontal cortex. Neuropsychopharmacology 29, 351–361.

Siddiqi, S.H., Kandala, S., Hacker, C.D., Trapp, N.T., Leuthardt, E.C., et al., 2023. Individualized precision targeting of dorsal attention and default mode networks with rTMS in traumatic brain injury-associated depression. Sci. Rep. 13, 4052.

Siddiqui, S.S., Springer, S.A., Verhagen, A., Sundaramurthy, V., Alisson-Silva, F., et al., 2017. The Alzheimer's disease-protective CD33 splice variant mediates adaptive loss of function via diversion to an intracellular pool. J. Biol. Chem. 292, 15312–15320.

Siedlik, J.A., Benedict, S.H., Landes, E.J., Weir, J.P., Vardiman, J.P., et al., 2016. Acute bouts of exercise induce a suppressive effect on lymphocyte proliferation in human subjects: a meta-analysis. Brain Behav. Immun. 56, 343–351.

Siegel, G., Gerber, H., Koch, P., Bruestle, O., Fraering, P.C., Rajendran, L., 2017. The Alzheimer's disease γ-secretase generates higher 42:40 ratios for β-amyloid than for p3 peptides. Cell Rep. 19, 1967–1976.

Sierra, A., Paolicelli, R.C., Kettenmann, H., 2019. Cien Años de Microglía: milestones in a century of microglial research. Trends Neurosci. 42, 778–792.

Silberstein, S., Liberman, A.C., Dos Santos Claro, P.A., Ugo, M.B., Deussing, J.M., Arzt, E., 2021. Stress-related brain neuroinflammation impact in depression: role of the corticotropin-releasing hormone system and P2X7 receptor. Neuroimmunomodulation 28, 52–60.

Silva, L.C., de Araújo, A.L., Fernandes, J.R., Matias, M.S., Silva, P.R., et al., 2016. Moderate and intense exercise lifestyles attenuate the effects of aging on telomere length and the survival and composition of T cell subpopulations. Age 38, 24.

Silveira, P.P., Gaudreau, H., Atkinson, L., Fleming, A.S., Sokolowski, M. B., 2016. Genetic differential susceptibility to socioeconomic status and childhood obesogenic behavior: why targeted prevention may be the best societal investment. JAMA Pediatr. 170, 359–364.

Silverberg, J.I., 2016. Atopic disease and cardiovascular risk factors in US children. J. Allergy Clin. Immunol. 137, 938–940.

Silverberg, N.D., Panenka, W.J., 2019. Antidepressants for depression after concussion and traumatic brain injury are still best practice. BMC Psychiatry 19, 100.

Silverberg, N.D., Iaccarino, M.A., Panenka, W.J., Iverson, G.L., McCulloch, L., et al., 2020. Management of concussion and mild traumatic brain injury: a synthesis of practice guidelines. Arch. Phys. Med. Rehabil. 101, 382–393.

Silverman, J.D., Washburne, A.D., Mukherjee, S., David, L.A., 2017a. A phylogenetic transform enhances analysis of compositional microbiota data. elife 6, e21887.

Silverman, M., Kua, L., Tanca, A., Pala, M., Palomba, A., et al., 2017b. Protective major histocompatibility complex allele prevents type 1 diabetes by shaping the intestinal microbiota early in ontogeny. Proc. Natl. Acad. Sci. USA 114, 9671–9676.

Silverstein, A.M., 2003. Cellular versus humoral immunology: a century-long dispute. Nat. Immunol. 4, 425–428.

Simmons, W.K., Burrows, K., Avery, J.A., Kerr, K.L., Bodurka, J., et al., 2016. Depression-related increases and decreases in appetite: dissociable patterns of aberrant activity in reward and interoceptive neurocircuitry. Am. J. Psychiatry 173, 418–428.

Simmons, W.K., Burrows, K., Avery, J.A., Kerr, K.L., Taylor, A., et al., 2020. Appetite changes reveal depression subgroups with distinct endocrine, metabolic, and immune states. Mol. Psychiatry 25, 1457–1468.

Simon, P.Y., Rousseau, P.F., 2017. Treatment of post-traumatic stress disorders with the Alpha-1 Adrenergic antagonist Prazosin. Can. J. Psychiatr. 62, 186–198.

Simon, D.W., McGeachy, M.J., Bayır, H., Clark, R.S., Loane, D.J., Kochanek, P.M., 2017. The far-reaching scope of neuroinflammation after traumatic brain injury. Nat. Rev. Neurol. 13, 171–191.

Simon, M.S., Arteaga-Henríquez, G., Fouad Algendy, A., Siepmann, T., Illigens, B.M.W., 2023. Anti-inflammatory treatment efficacy in major depressive disorder: a systematic review of meta-analyses. Neuropsychiatr. Dis. Treat. 19, 1–25.

Simons, R.L., Lei, M.K., Klopack, E., Zhang, Y., Gibbons, F.X., Beach, S.R.H., 2021. Racial discrimination, inflammation, and chronic illness among African American women at midlife: support for the weathering perspective. J. Racial Ethn. Health Disparities 8, 339–349.

Simpson, R.J., Kunz, H., Agha, N., Graff, R., 2015. Exercise and the regulation of immune functions. Prog. Mol. Biol. Transl. Sci. 135, 355–380.

Sims, M., Glover, L.S.M., Gebreab, S.Y., Spruill, T.M., 2020. Cumulative psychosocial factors are associated with cardiovascular disease risk factors and management among African Americans in the Jackson Heart Study. BMC Public Health 20, 556.

Singh, I., Sagare, A.P., Coma, M., Perlmutter, D., Gelein, R., et al., 2013. Low levels of copper disrupt brain amyloid-β homeostasis by altering its production and clearance. Proc. Natl. Acad. Sci. USA 110, 14771–14776.

Singh, V., Roth, S., Llovera, G., Sadler, R., Garzetti, D., et al., 2016. Microbiota dysbiosis controls the neuroinflammatory response after stroke. J. Neurosci. 36, 7428–7440.

Singh, A.K., Mishra, G., Maurya, A., Awasthi, R., Kumari, K., et al., 2019. Role of TREM2 in Alzheimer's disease and its consequences on β-amyloid, tau and neurofibrillary tangles. Curr. Alzheimer Res. 16, 1216–1229.

Singh, D., Guest, P.C., Dobrowolny, H., Vasilevska, V., Meyer-Lotz, G., et al., 2022. Changes in leukocytes and CRP in different stages of major depression. J. Neuroinflammation 19, 74.

Singh, P., Gollapalli, K., Mangiola, S., Schranner, D., Yusuf, M.A., et al., 2023. Taurine deficiency as a driver of aging. Science 380, eabn9257.

Sinha, P., Matthay, M.A., Calfee, C.S., 2020. Is a "cytokine storm" relevant to COVID-19? JAMA Intern. Med. 180 (9), 1152–1154.

Siopi, E., Chevalier, G., Katsimpardi, L., Saha, S., Bigot, M., et al., 2020. Changes in gut microbiota by chronic stress impair the efficacy of fluoxetine. Cell Rep. 30, 3682–3690.e6.

Skilbeck, K.J., Johnston, G.A., Hinton, T., 2010. Stress and GABA receptors. J. Neurochem. 112, 1115–1130.

Slater, P.G., Yarur, H.E., Gysling, K., 2016. Corticotropin-releasing factor receptors and their interacting proteins: functional consequences. Mol. Pharmacol. 90, 627–632.

Slavich, G.M., Irwin, M.R., 2014. From stress to inflammation and major depressive disorder: a social signal transduction theory of depression. Psychol. Bull. 140, 774–815.

Slavich, G.M., Shields, G.S., Deal, B.D., Gregory, A., Toussaint, L.L., 2019. Alleviating social pain: a double-blind, randomized, placebo-controlled trial of forgiveness and acetaminophen. Ann. Behav. Med. 53, 1045–1054.

Sloan, E.K., Capitanio, J.P., Cole, S.W., 2008. Stress-induced remodeling of lymphoid innervation. Brain Behav. Immun. 22, 15–21.

Slokar, G., Hasler, G., 2015. Human endogenous retroviruses as pathogenic factors in the development of schizophrenia. Front. Psychiatry 6, 183.

Slopen, N., Loucks, E.B., Appleton, A.A., Kawachi, I., Kubzansky, L.D., et al., 2015. Early origins of inflammation: an examination of prenatal and childhood social adversity in a prospective cohort study. Psychoneuroendocrinology 51, 403–413.

Sluiter, F., Incollingo Rodriguez, A.C., Nephew, B.C., Cali, R., Murgatroyd, C., Santos Jr., H.P., 2020. Pregnancy associated epigenetic markers of inflammation predict depression and anxiety symptoms in response to discrimination. Neurobiol. Stress 13, 100273.

Slyepchenko, A., Maes, M., Köhler, C.A., Anderson, G., Quevedo, J., et al., 2016. T helper 17 cells may drive neuroprogression in major depressive disorder: proposal of an integrative model. Neurosci. Biobehav. Rev. 64, 83–100.

Smearman, E.L., Almli, L.M., Conneely, K.N., Brody, G.H., Sales, J.M., et al., 2016. Oxytocin receptor genetic and epigenetic variations: association with child abuse and adult psychiatric symptoms. Child Dev. 87, 122–134.

Smeland, O.B., Frei, O., Dale, A.M., Andreassen, O.A., 2020. The polygenic architecture of schizophrenia—rethinking pathogenesis and nosology. Nat. Rev. Neurol. 16, 366–379.

Smeyne, M., Jiao, Y., Shepherd, K.R., Smeyne, R.J., 2005. Glia cell number modulates sensitivity to MPTP in mice. Glia 52, 144–152.

Smigielski, L., Jagannath, V., Rössler, W., Walitza, S., Grünblatt, E., 2020. Epigenetic mechanisms in schizophrenia and other psychotic disorders: a systematic review of empirical human findings. Mol. Psychiatry 25, 1718–1748.

Smith, M.J., White Jr., K.L., 2010. Establishment and comparison of delayed-type hypersensitivity models in the B(6)C(3)F(1) mouse. J. Immunotoxicol. 7, 308–317.

Smith, M.L., Hostetler, C.M., Heinricher, M.M., Ryabinin, A.E., 2016. Social transfer of pain in mice. Sci. Adv. 2, e1600855.

Smith, K.L., Kassem, M.S., Clarke, D.J., Kuligowski, M.P., Bedoya-Pérez, M.A., et al., 2019. Microglial cell hyper-ramification and neuronal dendritic spine loss in the hippocampus and medial prefrontal cortex in a mouse model of PTSD. Brain Behav. Immun. 80, 889–899.

Smith, A.K., Ratanatharathorn, A., Maihofer, A.X., Naviaux, R.K., Aiello, A.E., et al., 2020. Epigenome-wide meta-analysis of PTSD across 10 military and civilian cohorts identifies methylation changes in AHRR. Nat. Commun. 11, 5965.

Smith, A., Woodside, B., Abizaid, A., 2022. Ghrelin and the control of energy balance in females. Front. Endocrinol. (Lausanne) 13, 904754.

Smyth, A., O'Donnell, M., Hankey, G.J., Rangarajan, S., Lopez-Jaramillo, P., et al., 2022. Anger or emotional upset and heavy physical exertion as triggers of stroke: the INTERSTROKE study. Eur. Heart J. 43, 202–209.

Snijders, A.M., Langley, S.A., Kim, Y.M., Brislawn, C.J., Noecker, C., et al., 2016. Influence of early life exposure, host genetics and diet on the mouse gut microbiome and metabolome. Nat. Microbiol. 2, 16221.

Snijders, C., Maihofer, A.X., Ratanatharathorn, A., Baker, D.G., Boks, M.P., et al., 2020. Longitudinal epigenome-wide association studies of three male military cohorts reveal multiple CpG sites associated with post-traumatic stress disorder. Clin. Epigenetics 12, 11.

Soares, S., Rocha, V., Kelly-Irving, M., Stringhini, S., Fraga, S., 2021. Adverse childhood events and health biomarkers: a systematic review. Front. Public Health 9, 649825.

Sochocka, M., Zwolińska, K., Leszek, J., 2017. The infectious etiology of Alzheimer's disease. Curr. Neuropharmacol. 15, 996–1009.

Soga, T., Teo, C.H., Parhar, I., 2021. Genetic and epigenetic consequence of early-life social stress on depression: role of serotonin-associated genes. Front. Genet. 11, 601868.

Soheili-Nezhad, S., Ibáñez-Solé, O., Izeta, A., Hoeijmakers, J.H., Stoeger, T., 2024. Time is ticking faster for long genes in aging. Trends Genet. 40, 299–312.

Solati, K., Mousavi, M., Kheiri, S., Hasanpour-Dehkordi, A., 2017. The effectiveness of mindfulness-based cognitive therapy on psychological symptoms and quality of life in systemic lupus erythematosus patients: a randomized controlled trial. Oman Med. J. 32, 378–385.

Soleimani, L., Oquendo, M.A., Sullivan, G.M., Mathé, A.A., Mann, J.J., 2014. Cerebrospinal fluid neuropeptide Y levels in major depression and reported childhood trauma. Int. J. Neuropsychopharmacol. 18, pyu023.

Sołtys, A., Bidzan, M., Tyburski, E., 2021. The moderating effects of personal resources on caregiver burden in carers of Alzheimer's patients. Front. Psychiatry 12, 772050.

Sommer, I.E., de Witte, L., Begemann, M., Kahn, R.S., 2012. Nonsteroidal anti-inflammatory drugs in schizophrenia: ready for practice or a good start? a meta-analysis. J. Clin. Psychiatry 73, 414–419.

Sompayrac, L., 2016. How the Immune System Works, fifth ed. Wiley-Blackwell.

Somvanshi, P.R., Mellon, S.H., Yehuda, R., Flory, J.D., Makotkine, I., et al., 2020. Role of enhanced glucocorticoid receptor sensitivity in inflammation in PTSD: insights from computational model for circadian-neuroendocrine-immune interactions. Am. J. Physiol. Endocrinol. Metab. 319, E48–E66.

Son, Y.J., Song, E.K., 2012. The impact of type D personality and high-sensitivity C-reactive protein on health-related quality of life in patients with atrial fibrillation. Eur. J. Cardiovasc. Nurs. 11, 304–312.

Song, M., Hu, F.B., Wu, K., Must, A., Chan, A.T., et al., 2016. Trajectory of body shape in early and middle life and all cause and cause specific mortality: results from two prospective US cohort studies. BMJ 353, i2195.

Song, H., Fang, F., Tomasson, G., Arnberg, F.K., Mataix-Cols, D., et al., 2018. Association of stress-related disorders with subsequent autoimmune disease. JAMA 319, 2388–2400.

Song, H., Fall, K., Fang, F., Erlendsdóttir, H., Lu, D., et al., 2019a. Stress related disorders and subsequent risk of life threatening infections: population based sibling controlled cohort study. BMJ 367, l5784.

Song, P., Zhang, Y., Yu, J., Zha, M., Zhu, Y., et al., 2019b. Global prevalence of hypertension in children: a systematic review and meta-analysis. JAMA Pediatr. 173, 1154–1163.

Song, P., Li, W., Xie, J., Hou, Y., You, C., 2020. Cytokine storm induced by SARS-CoV-2. Clin. Chim. Acta 509, 280–287.

Song, S.J., Wang, J., Martino, C., Jiang, L., Wesley, K., et al., 2021. Naturalization of the microbiota developmental trajectory of Cesarean-born neonates after vaginal seeding. Med 13, 951–964.e1.

Soni, S., Jiang, Y., Tesfaigzi, Y., Hornick, J.L., Çataltepe, S., 2021. Comparative analysis of ACE2 protein expression in rodent, non-human primate, and human respiratory tract at baseline and after injury: a conundrum for COVID-19 pathogenesis. PLoS One 16, e0247510.

Sonnenburg, J.L., Bäckhed, F., 2016. Diet-microbiota interactions as moderators of human metabolism. Nature 535, 56–64.

Sonnenburg, E.D., Smits, S.A., Tikhonov, M., Higginbottom, S.K., Wingreen, N.S., et al., 2016. Diet-induced extinctions in the gut microbiota compound over generations. Nature 529, 212–215.

Sonnenschein, S.F., Mayeli, A., Yushmanov, V.E., Blazer, A., Calabro, F.J., et al., 2022. A longitudinal investigation of GABA, glutamate, and glutamine across the insula during antipsychotic treatment of first-episode schizophrenia. Schizophr. Res. 248, 98–106.

Sorboni, S.G., Moghaddam, H.S., Jafarzadeh-Esfehani, R., Soleimanpour, S., 2022. A comprehensive review on the role of the gut microbiome in human neurological disorders. Clin. Microbiol. Rev. 35, e0033820.

Sorg, R.A., Lin, L., van Doorn, G.S., Sorg, M., Olson, J., et al., 2016. Collective resistance in microbial communities by intracellular antibiotic deactivation. PLoS Biol. 14, e2000631.

Sorge, R.E., Totsch, S.K., 2017. Sex differences in pain. J. Neurosci. Res. 95, 1271–1281.

Soto, J.S., Jami-Alahmadi, Y., Chacon, J., Moye, S.L., Diaz-Castro, B., et al., 2023. Astrocyte-neuron subproteomes and obsessive-compulsive disorder mechanisms. Nature 616, 764–773.

Sousa, V.C., Mantas, I., Stroth, N., Hager, T., Pereira, M., et al., 2021. P11 deficiency increases stress reactivity along with HPA axis and autonomic hyperresponsiveness. Mol. Psychiatry 26, 3253–3265.

Southwick, S.M., Charney, D.D., 2018. Resilience: The Science of Mastering Life's Greatest Challenges. Cambridge University Press, Cambridge, UK.

Souza, L.C., Peny, J.A., Fortuna, J.T.S., Furigo, I.C., Hashiguchi, D., et al., 2021. Pro-inflammatory interleukin-6 signaling links cognitive impairments and peripheral metabolic alterations in Alzheimer's disease. Transl. Psychiatry 11, 251.

Soyman, E., Bruls, R., Ioumpa, K., Müller-Pinzler, L., Gallo, S., et al., 2022. Intracranial human recordings reveal association between neural activity and perceived intensity for the pain of others in the insula. elife 11, e75197.

Spann, M.N., Timonen-Soivio, L., Suominen, A., Cheslack-Postava, K., McKeague, I.W., et al., 2019. Proband and familial autoimmune diseases are associated with proband diagnosis of autism spectrum disorders. J. Am. Acad. Child Adolesc. Psychiatry 58, 496–505.

Spellberg, B., 2016. The new antibiotic mantra - "shorter is better". JAMA Intern. Med. 176, 1254–1255.

Spencer, S.P., Belkaid, Y., 2012. Dietary and commensal derived nutrients: shaping mucosal and systemic immunity. Curr. Opin. Immunol. 24, 379–384.

Spencer, S.J., Emmerzaal, T.L., Kozicz, T., Andrews, Z.B., 2015. Ghrelin's role in the hypothalamic-pituitary-adrenal axis stress response: implications for mood disorders. Biol. Psychiatry 78, 19–27.

Spencer, L., Olawuni, B., Singh, P., 2022. Gut virome: role and distribution in health and gastrointestinal diseases. Front. Cell. Infect. Microbiol. 12, 836706.

Sperdin, H.F., Coito, A., Kojovic, N., Rihs, T.A., Jan, R.K., et al., 2018. Early alterations of social brain networks in young children with autism. elife 7, e31670.

Spiegel, K., Rey, A.E.R., Cheylus, A., Ayling, K., Benedict, C., et al., 2023. A meta-analysis of the associations between insufficient sleep duration and antibody response to vaccination. Curr. Biol. 33, 998–1005.

Spinler, J.K., Ross, C.L., Savidge, T.C., 2016. Probiotics as adjunctive therapy for preventing Clostridium difficile infection - what are we waiting for? Anaerobe 41, 51–57.

Spohn, S.N., Mawe, G.M., 2017. Non-conventional features of peripheral serotonin signaling – the gut and beyond. Nat. Rev. Gastroenterol. Hepatol. 14, 412–420.

Spor, A., Koren, O., Ley, R., 2011. Unravelling the effects of the environment and host genotype on the gut microbiome. Nat. Rev. Microbiol. 9, 279–290.

Spragge, F., Bakkeren, E., Jahn, M.T., Araujo, E., Pearson, C.F., et al., 2023. Microbiome diversity protects against pathogens by nutrient blocking. Science 382, eadj3502.

Spreng, R.N., Dimas, E., Mwilambwe-Tshilobo, L., Dagher, A., Koellinger, P., et al., 2020. The default network of the human brain is associated with perceived social isolation. Nat. Commun. 11, 6393.

Srikantha, P., Mohajeri, M.H., 2019. The possible role of the microbiota-gut-brain-axis in autism spectrum disorder. Int. J. Mol. Sci. 20, 2115.

Srivastava, A., Chaudhary, Z., Qian, J., Al Chalabi, N., Burhan, A.M., et al., 2022. Genome-wide methylation analysis of early-onset schizophrenia. Psychiatr. Genet. 32, 214–220.

Stafoggia, M., Oftedal, B., Chen, J., Rodopoulou, S., Renzi, M., et al., 2022. Long-term exposure to low ambient air pollution concentrations and mortality among 28 million people: results from seven large European cohorts within the ELAPSE project. Lancet Planet. Heath 6, 9–18.

Stalnaker, T.A., Cooch, N.K., Schoenbaum, G., 2015. What the orbitofrontal cortex does not do. Nat. Neurosci. 18, 620–627.

Stampanoni Bassi, M., Garofalo, S., Marfia, G.A., Gilio, L., Simonelli, I., et al., 2017. Amyloid-β homeostasis bridges inflammation, synaptic plasticity deficits and cognitive dysfunction in multiple sclerosis. Front. Mol. Neurosci. 10, 390.

Stanley, D., Moore, R.J., Wong, C.H.Y., 2018. An insight into intestinal mucosal microbiota disruption after stroke. Sci. Rep. 8, 568.

Starcevic, V., 2014. The reappraisal of benzodiazepines in the treatment of anxiety and related disorders. Expert. Rev. Neurother. 14, 1275–1286.

Stark, J.L., Avitsur, R., Hunzeker, J., Padgett, D.A., Sheridan, J.F., 2002. Interleukin-6 and the development of social disruption-induced glucocorticoid resistance. J. Neuroimmunol. 124, 9–15.

Steardo Jr., L., Carbone, E.A., Menculini, G., Moretti, P., Steardo, L., Tortorella, A., 2021. Endocannabinoid system as therapeutic target of PTSD: a systematic review. Life (Basel) 11, 214.

Steenen, S.A., van Wijk, A.J., van der Heijden, G.J., van Westrhenen, R., de Lange, J., et al., 2016. Propranolol for the treatment of anxiety disorders: Systematic review and meta-analysis. J. Psychopharmacol. 30, 128–139.

Stegemann, A., Liu, S., Retana Romero, O.A., Oswald, M.J., Han, Y., et al., 2023. Prefrontal engrams of long-term fear memory perpetuate pain perception. Nat. Neurosci. 26, 820–829.

Steiger, V.R., Brühl, A.B., Weidt, S., Delsignore, A., Rufer, M., et al., 2017. Pattern of structural brain changes in social anxiety disorder after cognitive behavioral group therapy: a longitudinal multimodal MRI study. Mol. Psychiatry 22, 1164–1171.

Stein, M.B., Chen, C.Y., Ursano, R.J., Cai, T., Gelernter, J., et al., 2016. Genome-wide association studies of posttraumatic stress disorder in 2 cohorts of US Army soldiers. JAMA Psychiatry 73, 695–704.

Stein, M.B., Jain, S., Giacino, J.T., Levin, H., Dikmen, S., et al., 2019. Risk of posttraumatic stress disorder and major depression in civilian patients after mild traumatic brain injury: a TRACK-TBI Study. JAMA Psychiatry 76, 249–258.

Stein, M.B., Levey, D.F., Cheng, Z., Wendt, F.R., Harrington, K., et al., 2021a. Genome-wide association analyses of post-traumatic stress disorder and its symptom subdomains in the Million Veteran Program. Nat. Genet. 53, 174–184.

Stein, M.B., Yuh, E., Jain, S., Okonkwo, D.O., Mac Donald, C.L., et al., 2021b. Smaller regional brain volumes predict posttraumatic stress disorder at 3 months after mild traumatic brain injury. Biol. Psychiatry Cogn. Neurosci. Neuroimaging 6, 352–359.

Stein, M.B., Jain, S., Parodi, L., Choi, K.W., Maihofer, A.X., et al., 2023. Polygenic risk for mental disorders as predictors of posttraumatic stress disorder after mild traumatic brain injury. Transl. Psychiatry 13, 24.

Steiner, J., Walter, M., Glanz, W., Sarnyai, Z., Bernstein, H.G., et al., 2013. Increased prevalence of diverse N-methyl-D-aspartate glutamate receptor antibodies in patients with an initial diagnosis of schizophrenia: specific relevance of IgG NR1a antibodies for distinction from N-methyl-D-aspartate glutamate receptor encephalitis. JAMA Psychiatry 70, 271–278.

Steinman, M.Q., Duque-Wilckens, N., Greenberg, G.D., Hao, R., Campi, K.L., et al., 2016. Sex-specific effects of stress on oxytocin neurons correspond with responses to intranasal oxytocin. Biol. Psychiatry 80, 406–414.

Stelmashook, E.V., Isaev, N.K., Genrikhs, E.E., Amelkina, G.A., Khaspekov, L.G., et al., 2014. Role of zinc and copper ions in the pathogenetic mechanisms of Alzheimer's and Parkinson's diseases. Biochemistry 79, 391–396.

Stenvers, D.J., Scheer, F.A., Schrauwen, P., la Fleur, S.E., Kalsbeek, A., 2019. Circadian clocks and insulin resistance. Nat. Rev. Endocrinol. 15, 75–89.

Stephan, A.H., Barres, B.A., Stevens, B., 2012. The complement system: an unexpected role in synaptic pruning during development and disease. Annu. Rev. Neurosci. 35, 369–389.

Stephan, A.H., Madison, D.V., Mateos, J.M., Fraser, D.A., Lovelett, E.A., et al., 2013. A dramatic increase of C1q protein in the CNS during normal aging. J. Neurosci. 33, 13460–13474.

Steptoe, A., Hamer, M., Chida, Y., 2007. The effects of acute psychological stress on circulating inflammatory factors in humans: a review and meta-analysis. Brain Behav. Immun. 21, 901–912.

Steptoe, A., Shamaei-Tousi, A., Gylfe, Å., Henderson, B., Bergström, S., et al., 2008. Socioeconomic status, pathogen burden and cardiovascular disease risk. Heart 93, 1567–1570.

Stern, Y., 2012. Cognitive reserve in ageing and Alzheimer's disease. Lancet Neurol. 11, 1006–1012.

Stern, C., Schwarz, S., Moser, G., Cvitic, S., Jantscher-Krenn, E., et al., 2021. Placental endocrine activity: adaptation and disruption of maternal glucose metabolism in pregnancy and the influence of fetal sex. Int. J. Mol. Sci. 22, 12722.

Stetler, C., Chen, E., Miller, G.E., 2006. Written disclosure of experiences with racial discrimination and antibody response to an influenza vaccine. Int. J. Behav. Med. 13, 60–68.

Stevens, J.S., Kim, Y.J., Galatzer-Levy, I.R., Reddy, R., Ely, T.D., et al., 2017a. Amygdala reactivity and anterior cingulate habituation predict PTSD symptom maintenance after acute civilian trauma. Biol. Psychiatry 81, 1023–1029.

Stevens, J.S., Kim, Y.J., Galatzer-Levy, I.R., Reddy, R., Ely, T.D., et al., 2017b. Amygdala reactivity and anterior cingulate habituation predict posttraumatic stress disorder symptom maintenance after acute civilian trauma. Biol. Psychiatry 81, 1023–1029.

Stilling, R.M., Dinan, T.G., Cryan, J.F., 2014. Microbial genes, brain & behaviour–epigenetic regulation of the gut–brain axis. Genes Brain Behav. 13, 69–86.

Stilling, R.M., Ryan, F.J., Hoban, A.E., Shanahan, F., Clarke, G., et al., 2015. Microbes & neurodevelopment–absence of microbiota during early life increases activity-related transcriptional pathways in the amygdala. Brain Behav. Immun. 50, 209–220.

Stilling, R.M., van de Wouw, M., Clarke, G., Stanton, C., Dinan, T.G., et al., 2016. The neuropharmacology of butyrate: the bread and butter of the microbiota-gut-brain axis? Neurochem. Int. 99, 110–132.

Stock, S.J., Carruthers, J., Calvert, C., Denny, C., Donaghy, J., et al., 2022. SARS-CoV-2 infection and COVID-19 vaccination rates in pregnant women in Scotland. Nat. Med. 28, 504–512.

Stokes, A., Preston, S.H., 2016. Revealing the burden of obesity using weight histories. Proc. Natl. Acad. Sci. USA 113, 572–577.

Stokes, J.M., MacNair, C.R., Ilyas, B., French, S., Côté, J.P., et al., 2017a. Pentamidine sensitizes Gram-negative pathogens to antibiotics and overcomes acquired colistin resistance. Nat. Microbiol. 2, 17028.

Stokes, L., Layhadi, J.A., Bibic, L., Dhuna, K., Fountain, S.J., 2017b. P2X4 receptor function in the nervous system and current breakthroughs in pharmacology. Front. Pharmacol. 23, 291.

Stokes, J.M., Yang, K., Swanson, K., Jin, W., Cubillos-Ruiz, A., et al., 2020. A deep learning approach to antibiotic discovery. Cell 180, 688–702.

Stokowska, A., Aswendt, M., Zucha, D., Lohmann, S., Wieters, F., et al., 2023. Complement C3a treatment accelerates recovery after stroke via modulation of astrocyte reactivity and cortical connectivity. J. Clin. Invest. 30.

Stopschinski, B.E., Weideman, R.A., McMahan, D., Jacob, D.A., Little, B. B., et al., 2023. Microglia as a cellular target of diclofenac therapy in Alzheimer's disease. Ther. Adv. Neurol. Disord. 16. 17562864231156674.

Stouffer, M.A., Woods, C.A., Patel, J.C., Lee, C.R., Witkovsky, P., et al., 2015. Insulin enhances striatal dopamine release by activating cholinergic interneurons and thereby signals reward. Nat. Commun. 6, 8543.

Strachan, D.P., 2000. Family size, infection and atopy: the first decade of the 'hygiene hypothesis'. Thorax 55, S2.

Strati, F., Cavalieri, D., Albanese, D., De Felice, C., Donati, C., et al., 2017. New evidences on the altered gut microbiota in autism spectrum disorders. Microbiome 5, 24.

Straub, R.H., 2014. Rheumatoid arthritis—a neuroendocrine immune disorder: glucocorticoid resistance, relative glucocorticoid deficiency, low-dose glucocorticoid therapy, and insulin resistance. Arthritis Res. Ther. 16 (Suppl 2), I1.

Strawn, J.R., Vollmer, L.L., McMurray, K.M.J., Mills, J.A., Mossman, S. A., et al., 2018. Acid-sensing T cell death associated gene-8 receptor expression in panic disorder. Brain Behav. Immun. 67, 36–41.

Strehl, C., Ehlers, L., Gaber, T., Buttgereit, F., 2019. Glucocorticoids-all-rounders tackling the versatile players of the immune system. Front. Immunol. 10, 1744.

Stubbs, T.M., Bonder, M.J., Stark, A.K., Krueger, F., Ageing Clock Team, B.I., et al., 2017. Multi-tissue DNA methylation age predictor in mouse. Genome Biol. 18, 68.

Sturm, R., Xanthopoulos, L., Heftrig, D., Oppermann, E., Vrdoljak, T., et al., 2020. Regulatory T cells modulate CD4 proliferation after severe trauma via IL-10. J. Clin. Med. 9, 1052.

Su, S., Chen, R., Zhang, S., Shu, H., Luo, J., 2022a. Immune system changes in those with hypertension when infected with SARS-CoV-2. Cell. Immunol. 378, 104562.

Su, Y., Lian, J., Hodgson, J., Zhang, W., Deng, C., 2022b. Prenatal Poly I: C challenge affects behaviors and neurotransmission via elevated neuroinflammation responses in female juvenile rats. Int. J. Neuropsychopharmacol. 25, 160–171.

Suárez-Zamorano, N., Fabbiano, S., Chevalier, C., Stojanović, O., Colin, D.J., et al., 2015. Microbiota depletion promotes browning of white adipose tissue and reduces obesity. Nat. Med. 21, 1497–1501.

Subbanna, M., Shivakumar, V., Jose, D., Venkataswamy, M., Debnath, M., et al., 2021. Reduced T cell immunity in unmedicated, comorbidity-free obsessive-compulsive disorder: an immunophenotyping study. J. Psychiatr. Res. 137, 521–524.

Subramaniam, S.R., Federoff, H.J., 2017. Targeting microglial activation states as a therapeutic avenue in Parkinson's disease. Front. Aging Neurosci. 9, 176.

Suda, K., Matsuda, K., 2022. How microbes affect depression: underlying mechanisms via the gut-brain axis and the modulating role of probiotics. Int. J. Mol. Sci. 23, 1172.

Suderman, M., Borghol, N., Pappas, J.J., Pinto Pereira, S.M., Pembrey, M., et al., 2014. Childhood abuse is associated with methylation of multiple loci in adult DNA. BMC Med. Genet. 7, 13.

Suez, J., Zmora, N., Zilberman-Schapira, G., Mor, U., Dori-Bachash, M., et al., 2018. Post-antibiotic gut mucosal microbiome reconstitution is impaired by probiotics and improved by autologous FMT. Cell 174, 1406–1423.e16.

Sugama, S., Kakinuma, Y., 2016. Loss of dopaminergic neurons occurs in the ventral tegmental area and hypothalamus of rats following chronic stress: possible pathogenetic loci for depression involved in Parkinson's disease. Neurosci. Res. 111, 48–55.

Sugama, S., Sekiyama, K., Kodama, T., Takamatsu, Y., Takenouchi, T., et al., 2016. Chronic restraint stress triggers dopaminergic and noradrenergic neurodegeneration: possible role of chronic stress in the onset of Parkinson's disease. Brain Behav. Immun. 51, 39–46.

Sulakhiya, K., Keshavlal, G.P., Bezbaruah, B.B., Dwivedi, S., Gurjar, S.S., et al., 2016. Lipopolysaccharide induced anxiety-and depressive-like behaviour in mice are prevented by chronic pre-treatment of esculetin. Neurosci. Lett. 611, 106–111.

Sultan, R.S., Zhang, A.W., Olfson, M., Kwizera, M.H., Levin, F.R., 2023. Nondisordered cannabis use among US adolescents. JAMA Netw. Open 6, e2311294.

Summers, P.M., Hartmann, D.A., Hui, E.S., Nie, X., Deardorff, R.L., et al., 2017. Functional deficits induced by cortical microinfarcts. J. Cereb. Blood Flow Metab. 37, 3599–3614.

Sumner, J.A., Chen, Q., Roberts, A.L., Winning, A., Rimm, E.B., et al., 2018. Posttraumatic stress disorder onset and inflammatory and endothelial function biomarkers in women. Brain Behav. Immun. 69, 203–209.

Sun, J., Wang, F., Hong, G., Pang, M., Xu, H., et al., 2016a. Antidepressant-like effects of sodium butyrate and its possible mechanisms of action in mice exposed to chronic unpredictable mild stress. Neurosci. Lett. 618, 159–166.

Sun, J., Ma, H., Yu, C., Lv, J., Guo, Y., et al., 2016b. Association of major depressive episodes with stroke risk in a prospective study of 0.5 million Chinese adults. Stroke 47, 2203–2208.

Sun, L., Li, Y., Jia, X., Wang, Q., Li, Y., et al., 2017a. Neuroprotection by IFN-gamma via astrocyte-secreted IL-6 in acute neuroinflammation. Oncotarget 8, 40065–40078.

Sun, Q., Xie, N., Tang, B., Li, R., Shen, Y., 2017b. Alzheimer's disease: From genetic variants to the distinct pathological mechanisms. Front. Mol. Neurosci. 10, 319.

Sun, M.F., Zhu, Y.L., Zhou, Z.L., Jia, X.B., Xu, Y.D., et al., 2018. Neuroprotective effects of fecal microbiota transplantation on MPTP-induced Parkinson's disease mice: gut microbiota, glial reaction and TLR4/TNF-α signaling pathway. Brain Behav. Immun. 70, 48–60.

Sun, Y., Li, L., Xie, R., Wang, B., Jiang, K., Cao, H., 2019. Stress triggers flare of inflammatory bowel disease in children and adults. Front. Pediatr. 7, 432.

Sun, Y., Qu, Y., Zhu, J., 2021. The relationship between inflammation and post-traumatic stress disorder. Front. Psychiatry 12, 707543.

Sun, H., Liu, M., Liu, J., 2023a. Association of influenza vaccination and dementia risk: a meta-analysis of cohort studies. J. Alzheimers Dis. 92, 667–678.

Sun, J., Qiu, L., Zhang, H., Zhou, Z., Ju, L., Yang, J., 2023b. CRHR1 antagonist alleviates LPS-induced depression-like behaviour in mice. BMC Psychiatry 23 (1), 17.

Sun, W., Advani, M., Spruston, N., Saxe, A., Fitzgerald, J.E., 2023c. Organizing memories for generalization in complementary learning systems. Nat. Neurosci. 26, 1438–1448.

Sun, X., Sun, J., Lu, X., Dong, Q., Zhang, L., et al., 2023d. Mapping neurophysiological subtypes of major depressive disorder using normative models of the functional connectome. Biol. Psychiatry 94, 936–947.

Sun, Y., Wu, Y., Fan, S., Dal Santo, T., Li, L., et al., 2023e. Comparison of mental health symptoms before and during the covid-19 pandemic: evidence from a systematic review and meta-analysis of 134 cohorts. BMJ 380, e074224.

Sun, Y.Y., Wang, Z., Huang, H.C., 2023f. Roles of ApoE4 on the pathogenesis in Alzheimer's disease and the potential therapeutic approaches. Cell. Mol. Neurobiol. 43, 3115–3136.

Sundaram, J.R., Poore, C.P., Sulaimee, N.H.B., Pareek, T., Cheong, W.F., et al., 2017. Curcumin ameliorates neuroinflammation, neurodegeneration, and memory deficits in p25 transgenic mouse model that bears hallmarks of Alzheimer's disease. J. Alzheimers Dis. 60, 1429–1442.

Sundararajan, T., Manzardo, A.M., Butler, M.G., 2018. Functional analysis of schizophrenia genes using GeneAnalytics program and integrated databases. Gene 641, 25–34.

Sundquist, J., Palmér, K., Johansson, L.M., Sundquist, K., 2017. The effect of mindfulness group therapy on a broad range of psychiatric symptoms: a randomised controlled trial in primary health care. Eur. Psychiatry 43, 19–27.

Sundquist, K., Palmér, K., Memon, A.A., Sundquist, J., Wang, X., 2020. Macrophage migration inhibitory factor as a predictor for long-term improvements after mindfulness-based group therapy or treatment as usual for depression, anxiety or stress and adjustment disorders. Mindfulness 11, 1370–1377.

Sutin, A.R., Stephan, Y., Terracciano, A., 2015. Weight discrimination and risk of mortality. Psychol. Sci. 26, 1803–1811.

Sutton, L.P., Ostrovskaya, O., Dao, M., Xie, K., Orlandi, C., et al., 2016. Regulator of G-protein signaling 7 regulates reward behavior by controlling opioid signaling in the striatum. Biol. Psychiatry 80, 235–245.

Suzuki, H., Ohgidani, M., Kuwano, N., Chrétien, F., Lorin de la Grandmaison, G., et al., 2019. Suicide and microglia: recent findings and future perspectives based on human studies. Front. Cell. Neurosci. 13, 31.

Svenningsson, P., Kim, Y., Warner-Schmidt, J., Oh, Y.S., Greengard, P., 2013. p11 and its role in depression and therapeutic responses to antidepressants. Nat. Rev. Neurosci. 14, 673–680.

Svenningsson, P., Berg, L., Matthews, D., Ionescu, D.F., Richards, E.M., et al., 2014. Preliminary evidence that early reduction in p11 levels in natural killer cells and monocytes predicts the likelihood of antidepressant response to chronic citalopram. Mol. Psychiatry 19, 962–964.

Svensson, T., Kitlinski, M., Engström, G., Melander, O., 2017. A genetic risk score for CAD, psychological stress, and their interaction as predictors of CAD, fatal MI, non-fatal MI and cardiovascular death. PLoS One 12, e0176029.

Swartz, J.R., Prather, A.A., Di Iorio, C.R., Bogdan, R., Hariri, A.R., 2016. A functional interleukin-18 haplotype predicts depression and anxiety through increased threat-related amygdala reactivity in women but not men. Neuropsychopharmacology 42, 419–426.

Swartz, J.R., Prather, A.A., Hariri, A.R., 2017. Threat-related amygdala activity is associated with peripheral CRP concentrations in men but not women. Psychoneuroendocrinology 78, 93–96.

Swinford, C.G., Risacher, S.L., Wu, Y.C., Apostolova, L.G., Gao, S., et al., 2023. Altered cerebral blood flow in older adults with Alzheimer's disease: a systematic review. Brain Imaging Behav. 17, 223–256.

Sydnor, V.J., Bouix, S., Pasternak, O., Hartl, E., Levin-Gleba, L., et al., 2020. Mild traumatic brain injury impacts associations between limbic system microstructure and post-traumatic stress disorder symptomatology. Neuroimage Clin. 26, 102190.

Sydnor, V.J., Larsen, B., Bassett, D.S., Alexander-Bloch, A., Fair, D.A., et al., 2021. Neurodevelopment of the association cortices: Patterns, mechanisms, and implications for psychopathology. Neuron 109, 2820–2846.

Sylvester, S.V., Rusu, R., Chan, B., Bellows, M., O'Keefe, C., Nicholson, S., 2022. Sex differences in sequelae from COVID-19 infection and in long COVID syndrome: a review. Curr. Med. Res. Opin. 38, 1391–1399.

Szabo, Y.Z., Slavish, D.C., Graham-Engeland, J.E., 2020. The effect of acute stress on salivary markers of inflammation: a systematic review and meta-analysis. Brain Behav. Immun. 88, 887–900.

Szabo, Y.Z., Burns, C.M., Lantrip, C., 2022. Understanding associations between rumination and inflammation: a scoping review. Neurosci. Biobehav. Rev. 135, 104523.

Szaruga, M., Munteanu, B., Lismont, S., Veugelen, S., Horré, K., et al., 2017. Alzheimer's-causing mutations shift Aβ length by destabilizing γ-secretase-Aβn interactions. Cell 170, 443–456.e14.

Szasz, T., 1973. The Second Sin. Doubleday, New York, p. 113.

Szasz, T., 1974. The Myth of Mental Illness: Foundations of a Theory of Personal Conduct. Harper & Row, New York.

Szentirmai, E., Millican, N.S., Massie, A.R., Kapas, L., 2019. Butyrate, a metabolite of intestinal bacteria, enhances sleep. Sci. Rep. 9, 7035.

Szeszko, P.R., Lehrner, A., Yehuda, R., 2018. Glucocorticoids and hippocampal structure and function in PTSD. Harv. Rev. Psychiatry 26, 142–157.

Szyf, M., 2015. Epigenetics, a key for unlocking complex CNS disorders? Therapeutic implications. Eur. Neuropsychopharmacol. 25, 682–702.

Szyf, M., 2019. The epigenetics of perinatal stress. Dialogues Clin. Neurosci. 21, 369–378.

Szyf, M., Bick, J., 2013. DNA methylation: a mechanism for embedding early life experiences in the genome. Child Dev. 84, 49–57.

Szyf, M., Tang, Y.Y., Hill, K.G., Musci, R., 2016. The dynamic epigenome and its implications for behavioral interventions: a role for epigenetics to inform disorder prevention and health promotion. Transl. Behav. Med. 6, 55–62.

Tabak, B.A., Meyer, M.L., Castle, E., Dutcher, J.M., Irwin, M.R., et al., 2015. Vasopressin, but not oxytocin, increases empathic concern among individuals who received higher levels of paternal warmth: a randomized controlled trial. Psychoneuroendocrinology 51, 253–261.

Tabak, B.A., Meyer, M.L., Dutcher, J.M., Castle, E., Irwin, M.R., et al., 2016a. Oxytocin, but not vasopressin, impairs social cognitive ability among individuals with higher levels of social anxiety: a randomized controlled trial. Soc. Cogn. Affect. Neurosci. 11, 1272–1279.

Tabak, B.A., Vrshek-Schallhorn, S., Zinbarg, R.E., Prenoveau, J.M., Mineka, S., et al., 2016b. Interaction of CD38 variant and chronic interpersonal stress prospectively predicts social anxiety and depression symptoms over six years. Clin. Psychol. Sci. 4, 17–27.

Tabibzadeh, S., 2021. Signaling pathways and effectors of aging. Front. Biosci. (Landmark Ed) 26, 50–96.

Tabung, F.K., Steck, S.E., Liese, A.D., Zhang, J., Ma, Y., et al., 2016. Association between dietary inflammatory potential and breast cancer incidence and death: results from the women's health initiative. Br. J. Cancer 114, 1277–1285.

Taché, Y., Larauche, M., Yuan, P.Q., Million, M., 2017. Brain and gut CRF signaling: biological actions and role in the gastrointestinal tract. Curr. Mol. Pharmacol. 11, 51–71.

Taha, S.A., Matheson, K., Anisman, H., 2013. The 2009 H1N1 influenza pandemic: the role of threat, coping, and media trust on vaccination intentions in Canada. J. Health Commun. 18, 278–290.

Taha, S.A., Matheson, K., Anisman, H., 2014. H1N1 was not all that scary: uncertainty and stressor appraisals predict anxiety related to a coming viral threat. Stress. Health 30, 149–157.

Tak, C.R., Job, K.M., Schoen-Gentry, K., Campbell, S.C., Carroll, P., et al., 2017. The impact of exposure to antidepressant medications during pregnancy on neonatal outcomes: a review of retrospective database cohort studies. Eur. J. Clin. Pharmacol. 73, 1055–1069.

Takao, Y., Okuno, Y., Mori, Y., Asada, H., Yamanishi, K., Iso, H., 2018. Associations of perceived mental stress, sense of purpose in life, and negative life events with the risk of incident herpes zoster and post-herpetic neuralgia: the SHEZ study. Am. J. Epidemiol. 187, 251–259.

Takemori, Y., Sasayama, D., Toida, Y., Kotagiri, M., Sugiyama, N., et al., 2021. Possible utilization of salivary IFN-γ/IL-4 ratio as a marker of chronic stress in healthy individuals. Neuropsychopharmacol. Rep. 41, 65–72.

Takeuchi, T., Kameyama, K., Miyauchi, E., Nakanishi, Y., Kanaya, T., et al., 2023a. Fatty acid overproduction by gut commensal microbiota exacerbates obesity. Cell Metab. 35, 361–375.e9.

Takeuchi, T., Kubota, T., Nakanishi, Y., Tsugawa, H., Suda, W., et al., 2023b. Gut microbial carbohydrate metabolism contributes to insulin resistance. Nature 621, 389–395.

Tam, V., Patel, N., Turcotte, M., Bossé, Y., Paré, G., Meyre, D., 2019. Benefits and limitations of genome-wide association studies. Nat. Rev. Genet. 20, 467–484.

Tamouza, R., Meyer, U., Foiselle, M., Richard, J.R., Wu, C.L., et al., 2021. Identification of inflammatory subgroups of schizophrenia and bipolar disorder patients with HERV-W ENV antigenemia by unsupervised cluster analysis. Transl. Psychiatry 11, 37.

Tan, L.L., Pelzer, P., Heinl, C., Tang, W., Gangadharan, V., et al., 2017. A pathway from midcingulate cortex to posterior insula gates nociceptive hypersensitivity. Nat. Neurosci. 20, 1591–1601.

Tanelian, A., Nankova, B., Miari, M., Nahvi, R.J., Sabban, E.L., 2022. Resilience or susceptibility to traumatic stress: potential influence of the microbiome. Neurobiol. Stress 19, 100461.

Tang, W.W., Hazen, S.L., 2017. Microbiome, trimethylamine N-oxide, and cardiometabolic disease. Transl. Res. 179, 108–115.

Tang, C., Ma, Y., Lei, X., Ding, Y., Yang, S., He, D., 2023a. Hypertension linked to Alzheimer's disease via stroke: Mendelian randomization. Sci. Rep. 13, 21606.

Tang, C.F., Wang, C.Y., Wang, J.H., Wang, Q.N., Li, S.J., et al., 2022a. Short-chain fatty acids ameliorate depressive-like behaviors of high fructose-fed mice by rescuing hippocampal neurogenesis decline and blood-brain barrier damage. Nutrients 14, 1882.

Tang, H., Liang, J., Chai, K., Gu, H., Ye, W., Cao, P., Chen, S., Shen, D., 2023b. Artificial intelligence and bioinformatics analyze markers of children's transcriptional genome to predict autism spectrum disorder. Front. Neurol. 14, 1203375.

Tang, J., Huang, M., He, S., Zeng, J., Zhu, H., 2022b. Uncovering the extensive trade-off between adaptive evolution and disease susceptibility. Cell Rep. 40 (11), 111351.

Tang, L., Cai, N., Zhou, Y., Liu, Y., Hu, J., et al., 2022c. Acute stress induces an inflammation dominated by innate immunity represented by neutrophils in mice. Front. Immunol. 13, 1014296.

Tang, M.M., Lin, W.J., Pan, Y.Q., Li, Y.C., 2018a. Fibroblast growth factor 2 modulates hippocampal microglia activation in a neuroinflammation induced model of depression. Front. Cell. Neurosci. 12, 255.

Tang, W.W., Wang, Z., Li, X.S., Fan, Y., Li, D.S., et al., 2017. Increased trimethylamine N-oxide portends high mortality risk independent of glycemic control in patients with type 2 diabetes mellitus. Clin. Chem. 63, 297–306.

Tang, Y., Le, W., 2016. Differential roles of M1 and M2 microglia in neurodegenerative diseases. Mol. Neurobiol. 53, 1181–1194.

Tang, Z., Ye, G., Chen, X., Pan, M., Fu, J., et al., 2018b. Peripheral proinflammatory cytokines in Chinese patients with generalised anxiety disorder. J. Affect. Disord. 225, 593.

Taniya, M.A., Chung, H.J., Al Mamun, A., Alam, S., Aziz, M.A., et al., 2022. Role of gut microbiome in autism spectrum disorder and its therapeutic regulation. Front. Cell. Infect. Microbiol. 12, 915701.

Tanner, C.M., Kamel, F., Ross, G.W., Hoppin, J.A., Goldman, S.M., et al., 2011. Rotenone, paraquat, and Parkinson's disease. Environ. Health Perspect. 119, 866–872.

Tansey, M.G., Wallings, R.L., Houser, M.C., Herrick, M.K., Keating, C. E., Joers, V., 2022. Inflammation and immune dysfunction in Parkinson disease. Nat. Rev. Immunol. 22, 657–673.

Tansley, S., Gu, N., Guzmán, A.U., Cai, W., Wong, C., et al., 2022a. Microglia-mediated degradation of perineuronal nets promotes pain. Science 377 (6601), 80–86.

Tansley, S., Uttam, S., Ureña Guzmán, A., Yaqubi, M., Pacis, A., et al., 2022b. Single-cell RNA sequencing reveals time- and sex-specific responses of mouse spinal cord microglia to peripheral nerve injury and links ApoE to chronic pain. Nat. Commun. 13, 843.

Tao, S., Chen, H., Li, N., Liang, W., 2022. The application of the CRISPR-Cas system in antibiotic resistance. Infect. Drug Resist. 15, 4155–4168.

Tao, Z.Y., Wang, P.X., Wei, S.Q., Traub, R.J., Li, J.F., Cao, D.Y., 2019. The role of descending pain modulation in chronic primary pain: potential application of drugs targeting serotonergic system. Neural Plast. 2019, 1389296.

Tapia-Rojas, C., Aranguiz, F., Varela-Nallar, L., Inestrosa, N.C., 2016. Voluntary running attenuates memory loss, decreases neuropathological changes and induces neurogenesis in a mouse model of Alzheimer's disease. Brain Pathol. 26, 62–74.

Taquet, M., Skorniewska, Z., Hampshire, A., Chalmers, J.D., Ho, L.P., et al., 2023. Acute blood biomarker profiles predict cognitive deficits 6 and 12 months after COVID-19 hospitalization. Nat. Med. 29, 2498–2508.

Taraskina, A., Ignatyeva, O., Lisovaya, D., Ivanov, M., Ivanova, L., et al., 2022. Effects of traumatic brain injury on the gut microbiota composition and serum amino acid profile in rats. Cells 11, 1409.

Tarr, A.J., Galley, J.D., Fisher, S.E., Chichlowski, M., Berg, B.M., et al., 2015. The prebiotics 3'Sialyllactose and 6'Sialyllactose diminish stressor-induced anxiety-like behavior and colonic microbiota alterations: evidence for effects on the gut-brain axis. Brain Behav. Immun. 50, 166–177.

Taskinen, M.R., Packard, C.J., Borén, J., 2019. Dietary fructose and the metabolic syndrome. Nutrients 11, 1987.

Tate, W., Walker, M., Sweetman, E., Helliwell, A., Peppercorn, K., et al., 2022. Molecular mechanisms of neuroinflammation in ME/CFS and long COVID to sustain disease and promote relapses. Front. Neurol. 13, 877772.

Tateishi, H., Mizoguchi, Y., Monji, A., 2022. Is the therapeutic mechanism of repetitive transcranial magnetic stimulation in cognitive dysfunctions of depression related to the neuroinflammatory processes in depression? Front. Psychiatry 13, 834425.

Tavakoli, P., Vollmer-Conna, U., Hadzi-Pavlovic, D., Grimm, M.C., 2021. A review of inflammatory bowel disease: a model of microbial, immune and neuropsychological integration. Public Health Rev. 42, 1603990.

Taylor-Robinson, D.C., Maayan, N., Soares-Weiser, K., Donegan, S., Garner, P., 2015. Deworming drugs for soil-transmitted intestinal worms in children: effects on nutritional indicators, haemoglobin, and school performance. Cochrane Database Syst. Rev. 7, CD000371.

Taylor, J., de Vries, Y.A., van Loo, H.M., Kendler, K.S., 2023. Clinical characteristics indexing genetic differences in schizophrenia: a systematic review. Mol. Psychiatry 28, 883–890.

Taylor, M.P., Enquist, L.W., 2015. Axonal spread of neuroinvasive viral infections. Trends Microbiol. 23, 283–288.

Taylor, S.E., Klein, L.C., Lewis, B.P., Gruenewald, T.L., Gurung, R.A., et al., 2000. Biobehavioral responses to stress in females: tend-and-befriend, not fight-or-flight. Psychol. Rev. 107, 411.

Taylor, S.S., Noor, N., Urits, I., Paladini, A., Sadhu, M.S., et al., 2021. Complex regional pain syndrome: a comprehensive review. Pain Ther. 10, 875–892.

Teche, S.P., Rovaris, D.L., Aguiar, B.W., Hauck, S., Vitola, E.S., et al., 2017. Resilience to traumatic events related to urban violence and increased IL10 serum levels. Psychiatry Res. 250, 136–140.

Tellez, L.A., Han, W., Zhang, X., Ferreira, T.L., Perez, I.O., et al., 2016. Separate circuitries encode the hedonic and nutritional values of sugar. Nat. Neurosci. 19, 465–470.

Tercan, H., Riksen, N.P., Joosten, L.A.B., Netea, M.G., Bekkering, S., 2021. Trained immunity: long-term adaptation in innate immune responses. Arterioscler. Thromb. Vasc. Biol. 41, 55–61.

Terracciano, A., Luchetti, M., Karakose, S., Stephan, Y., Sutin, A.R., 2023. Loneliness and risk of Parkinson disease. JAMA Neurol. 80, 1138–1144.

Tétreault, P., Mansour, A., Vachon-Presseau, E., Schnitzer, T.J., Apkarian, A.V., et al., 2016. Brain connectivity predicts placebo response across chronic pain clinical trials (2016). PLoS Biol. 14, e1002570.

Thaiss, C.A., Levy, M., Korem, T., Dohnalová, L., Shapiro, H., et al., 2016a. Microbiota diurnal rhythmicity programs host transcriptome oscillations. Cell 167, 1495–1510.

Thaiss, C.A., Zmora, N., Levy, M., Elinav, E., 2016b. The microbiome and innate immunity. Nature 535, 65–74.

Thakur, A., Bogati, S., Pandey, S., 2023. Attempts to develop vaccines against Alzheimer's disease: a systematic review of ongoing and completed vaccination trials in humans. Cureus 15, e40138.

Thakur, P., Breger, L.S., Lundblad, M., Wan, O.W., Mattsson, B., et al., 2017. Modeling Parkinson's disease pathology by combination of fibril seeds and α-synuclein overexpression in the rat brain. Proc. Natl. Acad. Sci. USA 114, E8284–E8293.

Thaler, R., Sunstein, C.R., 2008. Nudge: Improving Decisions About Health, Wealth, and Happiness. Yale University Press, New Haven, CT.

The Global BMI Mortality Collaboration, Di Angelantonio, E., Bhupathiraju, S.N., Wormser, D., Gao, P., et al., 2016. Body-mass index and all-cause mortality: individual-participant-data meta-analysis of 239 prospective studies in four continents. Lancet 388, 776–786.

The Lancet, 2017. The health inequalities and ill-health of children in the UK. Lancet 389, 477.

The Lancet, 2023. Loneliness as a health issue. Lancet 402, 79.

Themelis, K., Gillett, J.L., Karadag, P., Cheatle, M.D., Giordano, N.A., et al., 2023. Mental defeat and suicidality in chronic pain: a prospective analysis. J. Pain 24, 2079–2092.

Theoharides, T.C., Tsilioni, I., Patel, A.B., Doyle, R., 2016. Atopic diseases and inflammation of the brain in the pathogenesis of autism spectrum disorders. Transl. Psychiatry 6, e844.

Thevaranjan, N., Puchta, A., Schulz, C., Naidoo, A., Szamosi, J.C., et al., 2017. Age-associated microbial dysbiosis promotes intestinal permeability, systemic inflammation, and macrophage dysfunction. Cell Host Microbe 21, 455–466.

Thomaes, K., Engelhard, I.M., Sijbrandij, M., Cath, D.C., Van den Heuvel, O.A., 2016. Degrading traumatic memories with eye movements: a pilot functional MRI study in PTSD. Eur. J. Psychotraumatol. 7, 31371.

Thomas, A., Ryan, C.P., Caspi, A., Liu, Z., Moffitt, T.E., et al., 2024. Diet, pace of biological aging, and risk of dementia in the Framingham Heart Study. Ann. Neurol.

Thomas, C.S., Mohammadkhani, A., Rana, M., Qiao, M., Baimel, C., Borgland, S.L., 2022a. Optogenetic stimulation of lateral hypothalamic orexin/dynorphin inputs in the ventral tegmental area potentiates mesolimbic dopamine neurotransmission and promotes reward-seeking behaviours. Neuropsychopharmacology 47, 728–740.

Thomas, E.G., Rhodius-Meester, H., Exalto, L., Peters, S.A.E., van Bloemendaal, L., et al., 2022b. Sex-specific associations of diabetes with brain structure and function in a geriatric population. Front. Aging Neurosci. 14, 885787.

Thomas, P.A., Carter, G.T., Bombardier, C.H., 2022c. A scoping review on the effect of cannabis on pain intensity in people with spinal cord injury. J. Spinal Cord Med. 45 (5), 656–667.

Thomas, R., Connolly, K.J., Brekk, O.R., Hinrich, A.J., Hastings, M.L., et al., 2023. Viral-like TLR3 induction of cytokine networks and α-synuclein are reduced by complement C3 blockade in mouse brain. Sci. Rep. 13, 15164.

Thompson, B., Waterhouse, M., English, D.R., McLeod, D.S., Armstrong, B.K., et al., 2023. Vitamin D supplementation and major cardiovascular events: D-Health randomised controlled trial. BMJ 381, e075230.

Thompson, K.I., Chau, M., Lorenzetti, M.S., Hill, L.D., Fins, A.I., Tartar, J.L., 2022. Acute sleep deprivation disrupts emotion, cognition, inflammation, and cortisol in young healthy adults. Front. Behav. Neurosci. 16, 945661.

Thompson, R.S., Roller, R., Mika, A., Greenwood, B.N., Knight, R., et al., 2017. Dietary prebiotics and bioactive milk fractions improve NREM sleep, enhance REM sleep rebound and attenuate the stress-induced decrease in diurnal temperature and gut microbial alpha diversity. Front. Behav. Neurosci. 10, 240.

Thompson, T., Terhune, D.B., Oram, C., Sharangparni, J., Rouf, R., et al., 2019. The effectiveness of hypnosis for pain relief: a systematic review and meta-analysis of 85 controlled experimental trials. Neurosci. Biobehav. Rev. 99, 298–310.

Thul, T.A., Corwin, E.J., Carlson, N.S., Brennan, P.A., Young, L.J., 2020. Oxytocin and postpartum depression: a systematic review. Psychoneuroendocrinology 120, 104793.

Tian, Y.E., Cropley, V., Maier, A.B., Lautenschlager, N.T., Breakspear, M., Zalesky, A., 2023. Heterogeneous aging across multiple organ systems and prediction of chronic disease and mortality. Nat. Med. 29, 1221–1231.

Tillisch, K., Labus, J., Kilpatrick, L., Jiang, Z., Stains, J., et al., 2013. Consumption of fermented milk product with probiotic modulates brain activity. Gastroenterology 144, 1394–1401.

Tindle, R., Hemi, A., Moustafa, A.A., 2022. Social support, psychological flexibility and coping mediate the association between COVID-19 related stress exposure and psychological distress. Sci. Rep. 12, 8688.

Tioleco, N., Silberman, A.E., Stratigos, K., Banerjee-Basu, S., Spann, M.N., et al., 2021. Prenatal maternal infection and risk for autism in offspring: a meta-analysis. Autism Res. 14, 1296–1316.

Tiwari, V., Guan, Y., Raja, S.N., 2014. Modulating the delicate glial-neuronal interactions in neuropathic pain: promises and potential caveats. Neurosci. Biobehav. Rev. 45, 19–27.

Tobi, E.W., Goeman, J.J., Monajemi, R., Gu, H., Putter, H., 2014. DNA methylation signatures link prenatal famine exposure to growth and metabolism. Nat. Commun. 5, 5592.

Toenders, Y.J., Schmaal, L., Nawijn, L., Han, L.K.M., Binnewies, J., et al., 2022. The association between clinical and biological characteristics of depression and structural brain alterations. J. Affect. Disord. 312, 268–274.

Tomasetti, C., Vogelstein, B., 2015. Cancer etiology. Variation in cancer risk among tissues can be explained by the number of stem cell divisions. Science 347, 78–81.

Tomiyama, H., Nakao, T., Murayama, K., Nemoto, K., Ikari, K., et al., 2019. Dysfunction between dorsal caudate and salience network associated with impaired cognitive flexibility in obsessive-compulsive disorder: a resting-state fMRI study. Neuroimage Clin. 24, 102004.

Tomova, L., Tye, K., Saxe, R., 2021. The neuroscience of unmet social needs. Soc. Neurosci. 16, 221–231.

Tomova, L., Wang, K.L., Thompson, T., Matthews, G.A., Takahashi, A., et al., 2020. Acute social isolation evokes midbrain craving responses similar to hunger. Nat. Neurosci. 23, 1597–1605.

Tonelli, L.H., Stiller, J., Rujescu, D., Giegling, I., Schneider, B., et al., 2008. Elevated cytokine expression in the orbitofrontal cortex of victims of suicide. Acta Psychiatr. Scand. 117, 198–206.

Tong, T., Duan, W., Xu, Y., Hong, H., Xu, J., et al., 2022. Paraquat exposure induces Parkinsonism by altering lipid profile and evoking neuroinflammation in the midbrain. Environ. Int. 169, 107512.

Tong, X., Wu, J., Sun, R., Li, H., Hong, Y., et al., 2024. Elevated dorsal medial prefrontal cortex to lateral habenula pathway activity mediates chronic stress-induced depressive and anxiety-like behaviors. Neuropsychopharmacology 49, 1402–1411.

Torkamani, A., Andersen, K.G., Steinhubl, S.R., Topol, E.J., 2017. High-definition medicine. Cell 170, 828–843.

Torres-Acosta, N., O'Keefe, J.H., O'Keefe, E.L., Isaacson, R., Small, G., 2020. Therapeutic potential of TNF-α inhibition for Alzheimer's disease prevention. J. Alzheimers Dis. 78, 619–626.

Torres-Berrío, A., Issler, O., Parise, E.M., Nestler, E.J., 2019. Unraveling the epigenetic landscape of depression: focus on early life stress. Dialogues Clin. Neurosci. 21, 341–357.

Torres-Fuentes, C., Schellekens, H., Dinan, T.G., Cryan, J.F., 2017. The microbiota–gut–brain axis in obesity. Lancet Gastroenterol. Hepatol. 2, 747–756.

Torres-Ruiz, J., Lomelín-Gascón, J., Lira Luna, J., Vargas-Castro, A.S., Pérez-Fragoso, A., et al., 2023. Novel clinical and immunological features associated with persistent post-acute sequelae of COVID-19 after six months of follow-up: a pilot study. Infect. Dis. 55, 243–254.

Torres-Sánchez, E.D., Ortiz, C.G., Reyes-Uribe, E., Torres-Jasso, J.H., Salazar-Flores, J., 2023. Effect of pesticides on phosphorylation of tau protein, and its influence on Alzheimer's disease. World J. Clin. Cases 11, 5628–5642.

Torres, A.R., Sweeten, T.L., Johnson, R.C., Odell, D., Westover, J.B., et al., 2016. Common genetic variants found in HLA and KIR immune genes in autism spectrum disorder. Front. Neurosci. 10, 463.

Toth, M., Flandreau, E.I., Deslauriers, J., Geyer, M.A., Mansuy, I.M., et al., 2016. Overexpression of forebrain CRH during early life increases trauma susceptibility in adulthood. Neuropsychopharmacology 41, 1681–1690.

Tournier, J.N., Mathieu, J., Mailfert, Y., Multon, E., Drouet, C., et al., 2001. Chronic restraint stress induces severe disruption of the T-cell specific response to tetanus toxin vaccine. Immunology 102, 87–93.

Treaster, S., Deelen, J., Daane, J.M., Murabito, J., Karasik, D., Harris, M.P., 2023. Convergent genomics of longevity in rockfishes highlights the genetics of human life span variation. Sci. Adv. 9, eadd2743.

Treble-Barna, A., Heinsberg, L.W., Stec, Z., Breazeale, S., Davis, T.S., et al., 2023. Brain-derived neurotrophic factor (BDNF) epigenomic modifications and brain-related phenotypes in humans: a systematic review. Neurosci. Biobehav. Rev. 147, 105078.

Tremlett, H., Fadrosh, D.W., Faruqi, A.A., Hart, J., Roalstad, S., et al., 2016. Gut microbiota composition and relapse risk in pediatric MS: a pilot study. J. Neurol. Sci. 363, 153–157.

Trépanier, M.O., Hopperton, K.E., Mizrahi, R., Mechawar, N., Bazinet, R.P., 2016. Postmortem evidence of cerebral inflammation in schizophrenia: a systematic review. Mol. Psychiatry 21, 1009–1026.

Tresse, E., Marturia-Navarro, J., Sew, W.Q.G., Cisquella-Serra, M., Jaberi, E., et al., 2023. Mitochondrial DNA damage triggers spread of Parkinson's disease-like pathology. Mol. Psychiatry (Advance online publication).

Treutlein, J., Strohmaier, J., Frank, J., Witt, S.H., Rietschel, L., et al., 2017. Association between neuropeptide Y receptor Y2 promoter variant rs6857715 and major depressive disorder. Psychiatr. Genet. 27, 34–37.

Trifonova, E.A., Klimenko, A.I., Mustafin, Z.S., Lashin, S.A., Kochetov, A.V., 2021. Do autism spectrum and autoimmune disorders share predisposition gene signature due to mTOR signaling pathway controlling expression? Int. J. Mol. Sci. 22, 5248.

Trikojat, K., Luksch, H., Rösen-Wolff, A., Plessow, F., Schmitt, J., et al., 2017. "Allergic mood" – depressive and anxiety symptoms in patients with seasonal allergic rhinitis (SAR) and their association to inflammatory, endocrine, and allergic markers. Brain Behav. Immun. 65, 202–209.

Tripp, A., Oh, H., Guilloux, J.P., Martinowich, K., Lewis, D.A., Sibille, E., 2012. Brain-derived neurotrophic factor signaling and subgenual anterior cingulate cortex dysfunction in major depressive disorder. Am. J. Psychiatry 169, 1194–1202.

Troubat, R., Leman, S., Pinchaud, K., Surget, A., Barone, P., et al., 2021. Brain immune cells characterization in UCMS exposed P2X7 knockout mouse. Brain Behav. Immun. 94, 159–174.

Troyer, E.A., Kohn, J.N., Ecklu-Mensah, G., Aleti, G., Rosenberg, D.R., Hong, S., 2021. Searching for host immune-microbiome mechanisms in obsessive-compulsive disorder: a narrative literature review and future directions. Neurosci. Biobehav. Rev. 125, 517–534.

Trubetskoy, V., Pardiñas, A.F., Qi, T., Panagiotaropoulou, G., Awasthi, S., Bigdeli, T.B., et al., 2022. Mapping genomic loci implicates genes and synaptic biology in schizophrenia. Nature 604, 502–508.

Tränkner, D., Boulet, A., Peden, E., Focht, R., Van Deren, D., Capecchi, M., 2019. A microglia sublineage protects from sex-linked anxiety symptoms and obsessive compulsion. Cell Rep. 29, 791–799.e3.

Tsai, S.Y., Yang, T.Y., Chen, H.J., Chen, C.S., Lin, W.M., et al., 2014. Increased risk of chronic fatigue syndrome following herpes zoster: a population-based study. Eur. J. Clin. Microbiol. Infect. Dis. 33, 1653–1659.

Tsai, T.Y., Tseng, H.H., Chi, M.H., Chang, H.H., Wu, C.K., et al., 2019. The interaction of oxytocin and social support, loneliness, and cortisol level in major depression. Clin. Psychopharmacol. Neurosci. 17, 487–494.

Tsampasian, V., Elghazaly, H., Chattopadhyay, R., Debski, M., Naing, T.K.P., et al., 2023. Risk factors associated with post − COVID-19 condition: a systematic review and meta-analysis. JAMA Intern. Med. 183, 566–580.

Tsen, C.M., Liu, J.H., Yang, D.P., Chao, H.R., Chen, J.L., et al., 2021. Study on the correlation of bisphenol A exposure, pro-inflammatory gene expression, and C-reactive protein with potential cardiovascular disease symptoms in young adults. Environ. Sci. Pollut. Res. Int.

Tsuda, M., 2017. P2 receptors, microglial cytokines and chemokines, and neuropathic pain. J. Neurosci. Res. 95, 1319–1329.

Tu, H.Y., Yuan, B.S., Hou, X.O., Zhang, X.J., Pei, C.S., et al., 2021. α-synuclein suppresses microglial autophagy and promotes neurodegeneration in a mouse model of Parkinson's disease. Aging Cell 20, e13522.

Tu, M.T., Lupien, S.J., Walker, C.D., 2005. Measuring stress responses in postpartum mothers: Perspectives from studies in human and animal populations. Stress 8, 19–34.

Tubbs, J.D., Ding, J., Baum, L., Sham, P.C., 2020. Immune dysregulation in depression: evidence from genome-wide association. Brain Behav. Immun. Health 7, 100108.

Tuganbaev, T., Yoshida, K., Honda, K., 2022. The effects of oral microbiota on health. Science 376, 934–936.

Tuikhar, N., Keisam, S., Labala, R.K., Imrat, R.P., Arunkumar, M.C., et al., 2019. Comparative analysis of the gut microbiota in centenarians and young adults shows a common signature across genotypically non-related populations. Mech. Ageing Dev. 179, 23–35.

Tully, P.J., Ang, S.Y., Lee, E.J., Bendig, E., Bauereiß, N., et al., 2021. Psychological and pharmacological interventions for depression in patients with coronary artery disease. Cochrane Database Syst. Rev. 12, CD008012.

Turecki, G., Meaney, M.J., 2016. Effects of the social environment and stress on glucocorticoid receptor gene methylation: a systematic review. Biol. Psychiatry 79, 87–96.

Turek, J., Gąsior, Ł., 2023. Estrogen fluctuations during the menopausal transition are a risk factor for depressive disorders. Pharmacol. Rep. 75, 32–43.

Turk Wensveen, T., Gašparini, D., Rahelić, D., Wensveen, F.M., 2021. Type 2 diabetes and viral infection; cause and effect of disease. Diabetes Res. Clin. Pract. 172, 108637.

Turna, J., Grosman Kaplan, K., Anglin, R., Patterson, B., Soreni, N., et al., 2020. The gut microbiome and inflammation in obsessive-compulsive disorder patients compared to age- and sex-matched controls: a pilot study. Acta Psychiatr. Scand. 142, 337–347.

Turnbaugh, P.J., Ley, R.E., Mahowald, M.A., Magrini, V., Mardis, E.R., et al., 2006. An obesity-associated gut microbiome with increased capacity for energy harvest. Nature 444, 1027–1031.

Turner, C.A., Khalil, H., Murphy-Weinberg, V., Hagenauer, M.H., Gates, L., et al., 2023. The impact of COVID-19 on a college freshman sample reveals genetic and nongenetic forms of susceptibility and resilience to stress. Proc. Natl. Acad. Sci. USA 120, e2305779120.

Turner, J.E., 2016. Is immunosenescence influenced by our lifetime "dose" of exercise? Biogerontology 17, 581–602.

Turner, N.A., Sharma-Kuinkel, B.K., Maskarinec, S.A., Eichenberger, E. M., Shah, P.P., et al., 2019. Methicillin-resistant Staphylococcus aureus: an overview of basic and clinical research. Nat. Rev. Microbiol. 17, 203–218.

Tylee, D.S., Chandler, S.D., Nievergelt, C.M3, Liu, X., Pazol, J., et al., 2015. Blood-based gene-expression biomarkers of post-traumatic stress disorder among deployed marines: a pilot study. Psychoneuroendocrinology 51, 472–494.

Tylee, D.S., Sun, J., Hess, J.L., Tahir, M.A., Sharma, E., et al., 2018. Genetic correlations among psychiatric and immune-related phenotypes based on genome-wide association data. Am. J. Med. Genet. B Neuropsychiatr. Genet. 177, 641–657.

Tyml, K., Swarbreck, S., Pape, C., Secor, D., Koropatnick, J., et al., 2017. Voluntary running exercise protects against sepsis-induced early inflammatory and pro-coagulant responses in aged mice. Crit. Care 21, 210.

Uddin, L.Q., 2015. Salience processing and insular cortical function and dysfunction. Nat. Rev. Neurosci. 16, 55–61.

Uddin, M., Koenen, K.C., Aiello, A.E., Wildman, D.E., de los Santos, R., Galea, S., 2011. Epigenetic and inflammatory marker profiles associated with depression in a community-based epidemiologic sample. Psychol. Med. 41, 997–1007.

Uddin, M., Ratanatharathorn, A., Armstrong, D., Kuan, P.-F., Aiello, A. E., et al., 2018. Epigenetic meta-analysis across three civilian cohorts identifies NRG1 and HGS as blood-based biomarkers for post-traumatic stress disorder. Epigenomics 10, 1585–1601.

Uddin, M.S., Lim, L.W., 2022. Glial cells in Alzheimer's disease: From neuropathological changes to therapeutic implications. Ageing Res. Rev. 78, 101622.

Udina, M., Moreno-España, J., Navinés, R., Giménez, D., Langohr, K., et al., 2013. Serotonin and interleukin-6: the role of genetic polymorphisms in IFN-induced neuropsychiatric symptoms. Psychoneuroendocrinology 38, 1803–1813.

Udina, M., Navinés, R., Egmond, E., Oriolo, G., Langohr, K., et al., 2016. Glucocorticoid receptors, brain-derived neurotrophic factor, serotonin and dopamine neurotransmission are associated with interferon-induced depression. Int. J. Neuropsychopharmacol. 19, 135.

Uhlhaas, P.J., Davey, C.G., Mehta, U.M., Shah, J., Torous, J., et al., 2023. Towards a youth mental health paradigm: a perspective and roadmap. Mol. Psychiatry 28, 3171–3181.

Ulland, T.K., Song, W.M., Huang, S.C.C., Ulrich, J.D., Sergushichev, A., et al., 2017. TREM2 maintains microglial metabolic fitness in Alzheimer's disease. Cell 170, 649–663.

Ullrich, C., Humpel, C., 2009. Rotenone induces cell death of cholinergic neurons in an organotypic co-culture brain slice model. Neurochem. Res. 34, 2147–2153.

Umaña, K.L., Biarnes, M.C., Canet-Avilés, R.M., Jack Jr., C.R., Breton, Y. A., et al., 2021. Longitudinal CSF proteomics identifies NPTX2 as a prognostic biomarker of Alzheimer's disease. Alzheimers Dement. 17, 1976–1987.

Umehara, H., Numata, S., Watanabe, S.Y., Hatakeyama, Y., Kinoshita, M., et al., 2017. Altered KYN/TRP, Gln/Glu, and Met/methionine sulfoxide ratios in the blood plasma of medication-free patients with major depressive disorder. Sci. Rep. 7, 4855.

Unemo, M., Bradshaw, C.S., Hocking, J.S., de Vries, H.J.C., Fancis, S.C., et al., 2017. Sexually transmitted infections: challenges ahead. Lancet Infect. Dis. 17, e235–e279.

Unternaehrer, E., Bolten, M., Nast, I., Staehli, S., Meyer, A.H., et al., 2016. Maternal adversities during pregnancy and cord blood oxytocin receptor (OXTR) DNA methylation. Soc. Cogn. Affect. Neurosci. 11 (9), 1460–1470.

Upthegrove, R., Khandaker, G.M., 2020. Cytokines, oxidative stress and cellular markers of inflammation in schizophrenia. Curr. Top. Behav. Neurosci. 44, 49–66.

Uranga, A., España, P.P., Bilbao, A., Quintana, J.M., Arriaga, I., et al., 2016. Duration of antibiotic treatment in community-acquired pneumonia: a multicenter randomized clinical trial. JAMA Intern. Med. 176, 1257–1265.

Uschold-Schmidt, N., Nyuyki, K.D., Füchsl, A.M., Neumann, I.D., Reber, S.O., 2012. Chronic psychosocial stress results in sensitization of the HPA axis to acute heterotypic stressors despite a reduction of adrenal in vitro ACTH responsiveness. Psychoneuroendocrinology 37, 1676–1687.

Usmani, A., Shavarebi, F., Hiniker, A., 2021. The cell biology of LRRK2 in Parkinson's disease. Mol. Cell. Biol. 41, e00660-20.

Usui, N., Kobayashi, H., Shimada, S., 2023. Neuroinflammation and oxidative stress in the pathogenesis of autism spectrum disorder. Int. J. Mol. Sci. 24, 5487.

Uversky, V.N., Elrashdy, F., Aljadawi, A., Ali, S.M., Khan, R.H., Redwan, E.M., 2021. Severe acute respiratory syndrome coronavirus 2 infection reaches the human nervous system: how? J. Neurosci. Res. 99, 750–777.

Uversky, V.N., Li, J., Bower, K., Fink, A.L., 2002. Synergistic effects of pesticides and metals on the fibrillation of alpha-synuclein: implications for Parkinson's disease. Neurotoxicology 23, 527–536.

Uzun, N., Akça, Ö.F., Kılınç, İ., Balcı, T., 2022. Oxytocin and vasopressin levels and related factors in adolescents with social phobia and other anxiety disorders. Clin. Psychopharmacol. Neurosci. 20, 330–342.

Vachon-Presseau, E., Berger, S.E., Abdullah, T.B., Huang, L., Cecchi, G. A., et al., 2018. Brain and psychological determinants of placebo pill response in chronic pain patients. Nat. Commun. 9, 3397.

Vaduganathan, M., Mensah, G.A., Turco, J.V., Fuster, V., Roth, G.A., 2022. The global burden of cardiovascular diseases and risk: a compass for future health. J. Am. Coll. Cardiol. 80, 2361–2371.

Vahid-Ansari, F., Zhang, M., Zahrai, A., Albert, P.R., 2019. Overcoming resistance to selective serotonin reuptake inhibitors: targeting serotonin, serotonin-1A receptors and adult neuroplasticity. Front. Neurosci. 13, 404.

Valastro, B., Ghribi, O., Poirier, J., Krzywkowski, P., Massicotte, G., 2001. AMPA receptor regulation and LTP in the hippocampus of young and aged apolipoprotein E-deficient mice. Neurobiol. Aging 22, 9–15.

Valdearcos, M., Douglass, J.D., Robblee, M.M., Dorfman, M.D., Stifler, D.R., et al., 2017. Microglial inflammatory signaling orchestrates the hypothalamic immune response to dietary excess and mediates obesity susceptibility. Cell Metab. 26, 185–197.

Valdes, A.M., Walter, J., Segal, E., Spector, T.D., 2018. Role of the gut microbiota in nutrition and health. BMJ 361, k2179.

Valentine, T.R., Kuzu, D., Kratz, A.L., 2023. Coping as a moderator of associations between symptoms and functional and affective outcomes in the daily lives of individuals with multiple sclerosis. Ann. Behav. Med. 57, 249–259.

Valenzano, D.R., Benayoun, B.A., Singh, P.P., Zhang, E., Etter, P.D., et al., 2015. The African Turquoise Killifish genome provides insights into evolution and genetic architecture of lifespan. Cell 163, 15.

Valenzuela, P.L., Santos-Lozano, A., Barrán, A.T., Fernández-Navarro, P., Castillo-García, A., et al., 2022. Joint association of physical activity and body mass index with cardiovascular risk: a nationwide population-based cross-sectional study. Eur. J. Prev. Cardiol. 29, e50–e52.

Valero, J., Mastrella, G., Neiva, I., Sánchez, S., Malva, J.O., 2014. Long-term effects of an acute and systemic administration of LPS on adult neurogenesis and spatial memory. Front. Neurosci. 8, 83.

Valiuliene, G., Valiulis, V., Dapsys, K., Vitkeviciene, A., Gerulskis, G., et al., 2021. Brain stimulation effects on serum BDNF, VEGF, and TNFα in treatment-resistant psychiatric disorders. Eur. J. Neurosci. 53, 3791–3802.

Vallejo-Vaz, A.J., Robertson, M., Catapano, A.L., Watts, G.F., Kastelein, J.J., et al., 2017. Low-density lipoprotein cholesterol lowering for the primary prevention of cardiovascular disease among men with primary elevations of low-density lipoprotein cholesterol levels of 190 mg/dL or above: analyses from the WOSCOPS (West of Scotland Coronary Prevention Study) 5-year randomized trial and 20-Year observational follow-up. Circulation 136, 1878–1891.

Valtorta, N.K., Kanaan, M., Gilbody, S., Ronzi, S., Hanratty, B., 2016. Loneliness and social isolation as risk factors for coronary heart disease and stroke: systematic review and meta-analysis of longitudinal observational studies. Heart 102, 1009–1016.

van Bloemendaal, L., Ten Kulve, J.S., la Fleur, S.E., Ijzerman, R.G., Diamant, M., 2014. Effects of glucagon-like peptide 1 on appetite and body weight: focus on the CNS. J. Endocrinol. 221, T1–16.

van de Giessen, E.M., de Win, M.M.L., Tanck, M.W.T., van den Brink, W., Baas, F., et al., 2009. Striatal dopamine transporter availability associated with polymorphisms in the dopamine transporter gene SLC6A3. J. Nucl. Med. 50, 45–52.

van de Wouw, M., Boehme, M., Lyte, J.M., Wiley, N., Strain, C., et al., 2018. Short-chain fatty acids: microbial metabolites that alleviate stress-induced brain-gut axis alterations. J. Physiol. 596, 4923–4944.

van den Berk-Clark, C., Secrest, S., Walls, J., Hallberg, E., Lustman, P.J., et al., 2018. Association between posttraumatic stress disorder and lack of exercise, poor diet, obesity, and co-occurring smoking: a systematic review and meta-analysis. Health Psychol. 37, 407–416.

van den Elsen, L.W.J., Garssen, J., Burcelin, R., Verhasselt, V., 2019. Shaping the gut microbiota by breastfeeding: the gateway to allergy prevention? Front. Pediatr. 7, 47.

van der Klaauw, A.A., Horner, E.C., Pereyra-Gerber, P., Agrawal, U., Foster, W.S., et al., 2023. Accelerated waning of the humoral response to COVID-19 vaccines in obesity. Nat. Med. 29, 1146–1156.

van der Meer, D., Hartman, C.A., Pruim, R.H., Mennes, M., Heslenfeld, D., et al., 2017. The interaction between 5-HTTLPR and stress exposure influences connectivity of the executive control and default mode brain networks. Brain Imaging Behav. 11, 1486–1496.

van der Vlist, M., Raoof, R., Willemen, H.L.D.M., Prado, J., Versteeg, S., et al., 2022. Macrophages transfer mitochondria to sensory neurons to resolve inflammatory pain. Neuron 110, 613–626.e9.

Van Dyck, C.H., Swanson, C.J., Aisen, P., Bateman, R.J., Chen, C., et al., 2023. Lecanemab in early Alzheimer's disease. N. Engl. J. Med. 388, 9–21.

Van Dyke, N., Yenugadhati, N., Birkett, N.J., Lindsay, J., Turner, M.C., et al., 2021. Association between aluminum in drinking water and incident Alzheimer's disease in the Canadian Study of Health and Aging cohort. Neurotoxicology 83, 157–165.

van Gelderen, M.J., Nijdam, M.J., de Vries, F., Meijer, O.C., Vermetten, E., 2020. Exposure-related cortisol predicts outcome of psychotherapy in veterans with treatment-resistant posttraumatic stress disorder. J. Psychiatr. Res. 130, 387–393.

van Kesteren, C.F., Gremmels, H., de Witte, L.D., Hol, E.M., Van Gool, A.R., et al., 2017. Immune involvement in the pathogenesis of schizophrenia: a meta-analysis on postmortem brain studies. Transl. Psychiatry 7, e1075.

van Leeuwen, W.M., Lehto, M., Karisola, P., Lindholm, H., Luukkonen, R., et al., 2009. Sleep restriction increases the risk of developing cardiovascular diseases by augmenting proinflammatory responses through IL-17 and CRP. PLoS One 4, e4589.

Van Laar, A.D., Webb, K.R., Keeney, M.T., Van Laar, V.S., Zharikov, A., et al., 2023. Transient exposure to rotenone causes degeneration and progressive parkinsonian motor deficits, neuroinflammation, and synucleinopathy. npj Parkinson's Dis. 9, 121.

Van Minnen, A., Voorendonk, E.M., Rozendaal, L., de Jongh, A., 2020. Sequence matters: combining prolonged exposure and EMDR therapy for PTSD. Psychiatry Res. 290, 113032.

van Nuenen, B.F.L., Helmich, R.C., Ferraye, M., Thaler, A., Hendler, T., et al., 2012. Cerebral pathological and compensatory mechanisms in the premotor phase of leucine-rich repeat kinase 2 parkinsonism. Brain J. Neurol. 135, 3687–3698.

van Roessel, P.J., Grassi, G., Aboujaoude, E.N., Menchón, J.M., Van Ameringen, M., Rodríguez, C.I., 2023. Treatment-resistant OCD: Pharmacotherapies in adults. Compr. Psychiatry 120, 152352.

van Rooij, S.J., Stevens, J.S., Ely, T.D., Fani, N., Smith, A.K., et al., 2016. Childhood trauma and COMT genotype interact to increase hippocampal activation in resilient individuals. Front. Psychiatry 7, 156.

Van Someren, E.J.W., 2021. Brain mechanisms of insomnia: new perspectives on causes and consequences. Physiol. Rev. 101, 995–1046.

Van Wijngaarden-Cremers, P.J., van Eeten, E., Groen, W.B., Van Deurzen, P.A., Oosterling, I.J., Van der Gaag, R.J., 2014. Gender and age differences in the core triad of impairments in autism spectrum disorders: a systematic review and meta-analysis. J. Autism Dev. Disord. 44, 627–635.

Van't Wout, M., Longo, S.M., Reddy, M.K., Philip, N.S., Bowker, M.T., et al., 2017. Transcranial direct current stimulation may modulate extinction memory in posttraumatic stress disorder. Brain Behav. 7, e00681.

Vanhaecke, T., Aubert, P., Grohard, P.A., Durand, T., Hulin, P., et al., 2017. L. fermentum CECT 5716 prevents stress-induced intestinal barrier dysfunction in newborn rats. Neurogastroenterol. Motil. 29.

Varga, Z., Csabai, D., Miseta, A., Wiborg, O., Czéh, B., 2017. Chronic stress affects the number of GABAergic neurons in the orbitofrontal cortex of rats. Behav. Brain Res. 316, 104–114.

Vargas, A.S., Luis, A., Barroso, M., Gallardo, E., Pereira, L., 2020. Psilocybin as a new approach to treat depression and anxiety in the context of life-threatening diseases: a systematic review and meta-analysis of clinical trials. Biomedicines 8, 331.

Vargas, D.L., Nascimbene, C., Krishnan, C., Zimmerman, A.W., Pardo, C.A., 2005. Neuroglial activation and neuroinflammation in the brain of patients with autism. Ann. Neurol. 57, 67–81.

Vargas, M.V., Dunlap, L.E., Dong, C., Carter, S.J., Tombari, R.J., et al., 2023. Psychedelics promote neuroplasticity through the activation of intracellular 5-HT2A receptors. Science 379, 700–706.

Varian, B.J., Poutahidis, T., DiBenedictis, B.T., Levkovich, T., Ibrahim, Y., et al., 2017. Microbial lysate upregulates host oxytocin. Brain Behav. Immun. 61, 36–49.

Varrassi, G., Pergolizzi, J.V., Dowling, P., Paladini, A., 2020. Ibuprofen safety at the golden anniversary: are all NSAIDs the same? A narrative review. Adv. Ther. 37, 61–82.

Vasileva, S.S., Tucker, J., Siskind, D., Eyles, D., 2022. Does the gut microbiome mediate antipsychotic-induced metabolic side effects in schizophrenia? Expert Opin. Drug Saf. 21, 625–639.

Vedantam, A., Brennan, J., Levin, H.S., McCarthy, J.J., Dash, P.K., et al., 2021. Early versus late profiles of inflammatory cytokines after mild traumatic brain injury and their association with neuropsychological outcomes. J. Neurotrauma 38, 53–62.

Veer, I.M., Oei, N.Y., van Buchem, M.A., Spinhoven, P., Elzinga, B.M., et al., 2015. Evidence for smaller right amygdala volumes in posttraumatic stress disorder following childhood trauma. Psychiatry Res. 233, 436–442.

Vegeto, E., Villa, A., Della Torre, S., Crippa, V., Rusmini, P., et al., 2020. The role of sex and sex hormones in neurodegenerative diseases. Endocr. Rev. 41, 273–319.

Veldman, E.R., Mamula, D., Jiang, H., Tiger, M., Ekman, C.J., et al., 2021. P11 (S100A10) as a potential predictor of ketamine response in patients with SSRI-resistant depression. J. Affect. Disord. 290, 240–244.

Venegas, C., Heneka, M.T., 2017. Danger-associated molecular patterns in Alzheimer's disease. J. Leukoc. Biol. 101, 87–98.

Verhoeven, J.E., Han, L.K.M., Lever-van Milligen, B.A., Hu, M.X., Révész, D., et al., 2023. Antidepressants or running therapy: comparing effects on mental and physical health in patients with depression and anxiety disorders. J. Affect. Disord. 329, 19–29.

Verner, G., Epel, E., Lahti-Pulkkinen, M., Kajantie, E., Buss, C., et al., 2021. Maternal psychological resilience during pregnancy and newborn telomere length: a prospective study. Am. J. Psychiatry 178, 183–192.

Veronese, N., Demurtas, J., Smith, L., Michel, J.P., Barbagallo, M., et al., 2022. Influenza vaccination reduces dementia risk: a systematic review and meta-analysis. Ageing Res. Rev. 73, 101534.

Veru, F., Laplante, D.P., Luheshi, G., King, S., 2014. Prenatal maternal stress exposure and immune function in the offspring. Stress 17, 133–148.

Vétizou, M., Pitt, J.M., Daillère, R., Lepage, P., Waldschmitt, N., et al., 2015. Anticancer immunotherapy by CTLA-4 blockade relies on the gut microbiota. Science 350, 1079–1084.

Vidarsson, G., Dekkers, G., Rispens, T., 2014. IgG subclasses and allotypes: from structure to effector functions. Front. Immunol. 5, 520.

Vieira, M.M., Ferreira, T.B., Pacheco, P.A., Barros, P.O., Almeida, C.R., et al., 2010. Enhanced Th17 phenotype in individuals with generalized anxiety disorder. J. Neuroimmunol. 229, 212–218.

Villafane, G., Thiriez, C., Audureau, E., Straczek, C., Kerschen, P., et al., 2017. High-dose transdermal nicotine in Parkinson's disease patients: a randomized, open-label, blinded-endpoint evaluation phase 2 study. Eur. J. Neurol. 25, 120–127.

Villeda, S.A., Plambeck, K.E., Middeldorp, J., Castellano, J.M., Mosher, K.I., et al., 2014. Young blood reverses age-related impairments in cognitive function and synaptic plasticity in mice. Nat. Med. 20, 659–663.

Vinet, E., Pineau, C.A., Clarke, A.E., Scott, S., Fombonne, E., et al., 2015. Increased risk of autism spectrum disorders in children born to women with systemic lupus erythematosus: results from a large population-based cohort. Arthritis Rheumatol. 67, 3201–3208.

Vingeliene, S., Hiyoshi, A., Lentjes, M., Fall, K., Montgomery, S., 2019. Longitudinal analysis of loneliness and inflammation at older ages: English longitudinal study of ageing. Psychoneuroendocrinology 110, 104421.

Visser, K., Koggel, M., Blaauw, J., van der Horn, H.J., Jacobs, B., van der Naalt, J., 2022. Blood-based biomarkers of inflammation in mild traumatic brain injury: a systematic review. Neurosci. Biobehav. Rev. 132, 154–168.

Viswanathan, K., Daugherty, C., Dhabhar, F.S., 2005. Stress as an endogenous adjuvant: augmentation of the immunization phase of cell-mediated immunity. Int. Immunol. 17, 1059–1069.

Vlasova, R.M., Iosif, A.M., Ryan, A.M., Funk, L.H., Murai, T., et al., 2021. Maternal immune activation during pregnancy alters postnatal brain growth and cognitive development in nonhuman primate offspring. J. Neurosci. 41, 9971–9987.

Vogelzangs, N., Beekman, A.T.F., De Jonge, P., Penninx, B.W.J.H., 2013. Anxiety disorders and inflammation in a large adult cohort. Transl. Psychiatry 3 (4), e249.

Vogelzangs, N., de Jonge, P., Smit, J.H., Bahn, S., Penninx, B.W., 2016. Cytokine production capacity in depression and anxiety. Transl. Psychiatry 6, e825.

Vogl, T., Kalka, I.N., Klompus, S., Leviatan, S., Weinberger, A., Segal, E., 2022. Systemic antibody responses against human microbiota flagellins are overrepresented in chronic fatigue syndrome patients. Sci. Adv. 8, eabq2422.

Vogt, A.S., Jennings, G.T., Mohsen, M.O., Vogel, M., Bachmann, M.F., 2023. Alzheimer's disease: a brief history of immunotherapies targeting amyloid β. Int. J. Mol. Sci. 24, 3895.

Vohra, K., Vodonos, A., Schwartz, J., Marais, E.A., Sulprizio, M.P., Mickley, L.J., 2021. Global mortality from outdoor fine particle pollution generated by fossil fuel combustion: results from GEOS-Chem. Environ. Res. 195, 11075.

Voigt, R.M., Zalta, A.K., Raeisi, S., Zhang, L., Brown, J.M., et al., 2022. Abnormal intestinal milieu in posttraumatic stress disorder is not impacted by treatment that improves symptoms. Am. J. Physiol. Gastrointest. Liver Physiol. 323, G61–G70.

Volkow, N., Benveniste, H., McLellan, A.T., 2017. Use and misuse of opioids in chronic pain. Annu. Rev. Med. 69, 451–465.

Volkow, N.D., Baler, R.D., Compton, W.M., Weiss, S.R., 2014. Adverse health effects of marijuana use. N. Engl. J. Med. 370, 2219–2227.

Volkow, N.D., Wang, G.J., Fowler, J.S., Tomasi, D., Telang, F., et al., 2010. Addiction: decreased reward sensitivity and increased expectation sensitivity conspire to overwhelm the brain's control circuit. BioEssays 32, 748–755.

Vollmer, L.L., Schmeltzer, S., Schurdak, J., Ahlbrand, R., Rush, J., et al., 2016. Neuropeptide Y impairs retrieval of extinguished fear and modulates excitability of neurons in the infralimbic prefrontal cortex. J. Neurosci. 36, 1306–1315.

Volmar, C.H., Ait-Ghezala, G., Frieling, J., Paris, D., Mullan, M.J., 2008. The granulocyte macrophage colony stimulating factor (GM-CSF) regulates amyloid beta (Abeta) production. Cytokine 42, 336–344.

Voorhees, J.L., Tarr, A.J., Wohleb, E.S., Godbout, J.P., Mo, X., et al., 2013. Prolonged restraint stress increases IL-6, reduces IL-10, and causes persistent depressive-like behavior that is reversed by recombinant IL-10. PLoS One 8, e58488.

Vorstman, J.A.S., Parr, J.R., Moreno-De-Luca, D., Anney, R.J.L., Nurnberger Jr., J.I., et al., 2017. Autism genetics: opportunities and challenges for clinical translation. Nat. Rev. Genet. 18, 362–376.

Voss, M.W., Weng, T.B., Burzynska, A.Z., Wong, C.N., Cooke, G.E., et al., 2015. Fitness, but not physical activity, is related to functional integrity of brain networks associated with ageing. NeuroImage 131, 113–125.

Vyas, C.M., Manson, J.E., Sesso, H.D., Cook, N.R., Rist, P.M., et al., 2024. Effect of multivitamin-mineral supplementation versus placebo on cognitive function: results from the clinic subcohort of the COcoa Supplement and Multivitamin Outcomes Study (COSMOS) randomized clinical trial and meta-analysis of 3 cognitive studies within COSMOS. Am. J. Clin. Nutr. 119, 692–701.

Wada, M., Noda, Y., Iwata, Y., Tsugawa, S., Yoshida, K., et al., 2022. Dopaminergic dysfunction and excitatory/inhibitory imbalance in treatment-resistant schizophrenia and novel neuromodulatory treatment. Mol. Psychiatry 27, 2950–2967.

Wadji, D.L., Tandon, T., Ketcha Wanda, G.J.M., Wicky, C., Dentz, A., et al., 2021. Child maltreatment and NR3C1 exon 1F methylation, link with deregulated hypothalamus-pituitary-adrenal axis and psychopathology: a systematic review. Child Abuse Negl. 122, 105304.

Wager, T.D., Atlas, L.Y., Lindquist, M.A., Roy, M., Woo, C.W., Kross, E., 2013. An fMRI-based neurologic signature of physical pain. N. Engl. J. Med. 368, 1388–1397.

Wager, T.D., Atlas, L.Y., Botvinick, M.M., 2016. Pain in the ACC? Proc. Natl. Acad. Sci. USA 113, E2474–E2475.

Wahbeh, M.H., Avramopoulos, D., 2021. Gene-environment interactions in schizophrenia: a literature review. Genes (Basel) 12, 1850.

Walitt, B., Singh, K., LaMunion, S.R., Hallett, M., Jacobson, S., et al., 2024. Deep phenotyping of post-infectious myalgic encephalomyelitis/chronic fatigue syndrome. Nat. Commun. 15, 907.

Walker, C.D., 2010. Maternal touch and feed as critical regulators of behavioral and stress responses in the offspring. Dev. Psychobiol. 52, 638–650.

Walker, C.D., Bath, K.G., Joels, M., Korosi, A., Larauche, M., et al., 2017a. Chronic early life stress induced by limited bedding and nesting (LBN) material in rodents: critical considerations of methodology, outcomes and translational potential. Stress 20, 421–448.

Walker, K.A., Hoogeveen, R.C., Folsom, A.R., Ballantyne, C.M., Knopman, D.S., et al., 2017b. Midlife systemic inflammatory markers are associated with late-life brain volume: the ARIC study. Neurology 89, 2262–2270.

Walker, E., Ploubidis, G., Fancourt, D., 2019. Social engagement and loneliness are differentially associated with neuro-immune markers in older age: time-varying associations from the English Longitudinal Study of Ageing. Brain Behav. Immun. 82, 224–229.

Walker, A.J., Mohebbi, M., Maes, M., Berk, M., Walder, K., et al., 2022. Adjunctive minocycline for major depressive disorder: a sub-study exploring peripheral immune-inflammatory markers and associated treatment response. Brain Behav. Immun. Health 27, 100581.

Walker, K.A., Chen, J., Shi, L., Yang, Y., Fornage, M., et al., 2023. Proteomics analysis of plasma from middle-aged adults identifies protein markers of dementia risk in later life. Sci. Transl. Med. 15, eadf5681.

Wallace, S., Nazroo, J., Bécares, L., 2016. Cumulative effect of racial discrimination on the mental health of ethnic minorities in the United Kingdom. Am. J. Public Health 106, 1294–1300.

Wallace-Wells, D., 2019. The Uninhabitable Earth. Tim Guggan Books, New York.

Wallin, C., Sholts, S.B., Österlund, N., Luo, J., Jarvet, J., et al., 2017. Alzheimer's disease and cigarette smoke components: effects of nicotine, PAHs, and Cd(II), Cr(III), Pb(II), Pb(IV) ions on amyloid-β peptide aggregation. Sci. Rep. 7, 14423.

Wallings, R.L., Tansey, M.G., 2019. LRRK2 regulation of immune-pathways and inflammatory disease. Biochem. Soc. Trans. 47, 1581–1595.

Walsh, N.P., Oliver, S.J., 2016. Exercise, immune function and respiratory infection: an update on the influence of training and environmental stress. Immunol. Cell Biol. 94, 132–139.

Walsh, C.P., Bovbjerg, D.H., Marsland, A.L., 2021. Glucocorticoid resistance and β2-adrenergic receptor signaling pathways promote peripheral pro-inflammatory conditions associated with chronic psychological stress: a systematic review across species. Neurosci. Biobehav. Rev. 128, 117–135.

Walters, E.T., 2014. Neuroinflammatory contributions to pain after SCI: roles for central glial mechanisms and nociceptor-mediated host defense. Exp. Neurol. 258, 48–61.

Waltl, I., Kalinke, U., 2022. Beneficial and detrimental functions of microglia during viral encephalitis. Trends Neurosci. 45 (2), 158–170.

Wang, A.K., Miller, B.J., 2018. Meta-analysis of cerebrospinal fluid cytokine and tryptophan catabolite alterations in psychiatric patients: comparisons between schizophrenia, bipolar disorder, and depression. Schizophr. Bull. 44, 75–83.

Wang, T., Town, T., Alexopoulou, L., Anderson, J.F., Fikrig, E., Flavell, R.A., 2004. Toll-like receptor 3 mediates West Nile virus entry into the brain causing lethal encephalitis. Nat. Med. 10, 1366–1373.

Wang, Z., Klipfell, E., Bennett, B.J., Koeth, R., Levison, B.S., et al., 2011. Gut flora metabolism of phosphatidylcholine promotes cardiovascular disease. Nature 472, 57–63.

Wang, X., Piñol, R.A., Byrne, P., Mendelowitz, D., 2014. Optogenetic stimulation of locus ceruleus neurons augments inhibitory transmission to parasympathetic cardiac vagal neurons via activation of brainstem α1 and β1 receptors. J. Neurosci. 34, 6182–6189.

Wang, X., Lopez, O.L., Sweet, R.A., Becker, J.T., DeKosky, S.T., et al., 2015. Genetic determinants of disease progression in Alzheimer's disease. J. Alzheimers Dis. 43, 649–655.

Wang, D.D., Li, Y., Stampfer, M.J., Stampfer, M.J., Manson, J.E., et al., 2016a. Association of specific dietary fats with total and cause-specific mortality. JAMA Intern. Med. 176, 1134–1145.

Wang, Q., Ao, Y., Yang, K., Tang, H., Chen, D., 2016b. Circadian clock gene Per2 plays an important role in cell proliferation, apoptosis and cell cycle progression in human oral squamous cell carcinoma. Oncol. Rep. 35, 3387–3394.

Wang, Y., Zhao, Z., Rege, S.V., Wang, M., Si, G., et al., 2016c. 3K3A-activated protein C stimulates postischemic neuronal repair by human neural stem cells in mice. Nat. Med. 22, 1050–1055.

Wang, Z., Wei, X., Yang, J., Suo, J., Chen, J., et al., 2016d. Chronic exposure to aluminum and risk of Alzheimer's disease: a meta-analysis. Neurosci. Lett. 610, 200–206.

Wang, H.T., Huang, F.L., Hu, Z.L., Zhang, W.J., Qiao, X.Q., et al., 2017a. Early-life social isolation-induced depressive-like behavior in rats results in microglial activation and neuronal histone methylation that are mitigated by minocycline. Neurotox. Res. 31, 505–520.

Wang, J., Hossain, M., Thanabalasuriar, A., Gunzer, M., Meininger, C., et al., 2017b. Visualizing the function and fate of neutrophils in sterile injury and repair. Science 358, 111–116.

Wang, M., Yan, W., Liu, Y., Hu, H., Sun, Q., et al., 2017c. Erythropoietin ameliorates diabetes-associated cognitive dysfunction in vitro and in vivo. Sci. Rep. 7, 2801.

Wang, Q., Dong, X., Wang, Y., Liu, M., Sun, A., et al., 2017d. Adolescent escitalopram prevents the effects of maternal separation on depression- and anxiety-like behaviours and regulates the levels of inflammatory cytokines in adult male mice. Int. J. Dev. Neurosci. 62, 37–45.

Wang, W.Y., Liu, Y., Wang, H.F., Tan, L., Sun, F.R., et al., 2017e. Impacts of CD33 genetic variations on the atrophy rates of hippocampus and parahippocampal gyrus in normal aging and mild cognitive impairment. Mol. Neurobiol. 54, 1111–1118.

Wang, J., Hodes, G.E., Zhang, H., Zhang, S., Zhao, W., et al., 2018a. Epigenetic modulation of inflammation and synaptic plasticity promotes resilience against stress in mice. Nat. Commun. 9, 477.

Wang, J., Jia, Y., Li, G., Wang, B., Zhou, T., et al., 2018b. The dopamine receptor D3 regulates lipopolysaccharide-induced depressive-like behavior in mice. Int. J. Neuropsychopharmacol. 21 (5), 448–460.

Wang, L., Li, X.X., Chen, X., Qin, X.Y., Kardami, E., Cheng, Y., 2018c. Antidepressant-like effects of low- and high-molecular weight FGF-2 on chronic unpredictable mild stress mice. Front. Mol. Neurosci. 11, 377.

Wang, W., Wang, L., Xu, H., Cao, C., Liu, P., et al., 2019. Characteristics of pro- and anti-inflammatory cytokines alteration in PTSD patients exposed to a deadly earthquake. J. Affect. Disord. 248, 52–58.

Wang, R., Yang, B., Liu, P., Zhang, J., Liu, Y., et al., 2020. The longitudinal relationship between exposure to air pollution and depression in older adults. Int. J. Geriatr. Psychiatry 35, 610–616.

Wang, H., Wang, G., Banerjee, N., Liang, Y., Du, X., et al., 2021a. Aberrant gut microbiome contributes to intestinal oxidative stress, barrier dysfunction, inflammation and systemic autoimmune responses in mrl/lpr mice. Front. Immunol. 12, 651191.

Wang, H.Y., Xu, G.Q., Ni, M.F., Zhang, C.H., Li, X.L., Chang, Y., Sun, X. P., Zhang, B.W., 2021b. Neural basis of implicit cognitive reappraisal in panic disorder: an event-related fMRI study. J. Transl. Med. 19 (1), 304.

Wang, L., Hong, P.J., May, C., Rehman, Y., Oparin, Y., et al., 2021c. Medical cannabis or cannabinoids for chronic non-cancer and cancer related pain: a systematic review and meta-analysis of randomised clinical trials. BMJ 374, n1034.

Wang, D., Wang, Y., Chen, Y., Yu, L., Wu, Z., et al., 2022a. Differences in inflammatory marker profiles and cognitive functioning between deficit and nondeficit schizophrenia. Front. Immunol. 13, 958972.

Wang, H., He, Y., Sun, Z., Ren, S., Liu, M., et al., 2022b. Microglia in depression: an overview of microglia in the pathogenesis and treatment of depression. J. Neuroinflammation 19, 132.

Wang, M., Pan, W., Xu, Y., Zhang, J., Wan, J., Jiang, H., 2022c. Microglia-mediated neuroinflammation: a potential target for the treatment of cardiovascular diseases. J. Inflamm. Res. 15, 3083–3094.

Wang, S., Quan, L., Chavarro, J.E., Slopen, N., Kubzansky, L.D., et al., 2022d. Associations of depression, anxiety, worry, perceived stress, and prior to infection with risk of post-covid-19 conditions. JAMA Psychiatry 79 (11), 1081–1091.

Wang, W.C., Sayedahmed, E.E., Sambhara, S., Mittal, S.K., 2022e. Progress towards the Development of a Universal Influenza Vaccine. Viruses 14, 1684.

Wang, Z., Emmerich, A., Pillon, N.J., Moore, T., Hemerich, D., et al., 2022f. Genome-wide association analyses of physical activity and sedentary behavior provide insights into underlying mechanisms and roles in disease prevention. Nat. Genet. 54, 1332–1344.

Wang, B., Liang, J., Lu, C., Lu, A., Wang, C., 2023a. Exercise regulates myokines on aging-related diseases through muscle-brain crosstalk. Gerontology 24.

Wang, J., Zhou, T., Liu, F., Huang, Y., Xiao, Z., et al., 2023b. Influence of gut microbiota on resilience and its possible mechanisms. Int. J. Biol. Sci. 19, 2588–2598.

Wang, M., Zhang, H., Liang, J., Huang, J., Chen, N., 2023c. Exercise suppresses neuroinflammation for alleviating Alzheimer's disease. J. Neuroinflammation 20, 76.

Wang, W., Li, Y., Ma, F., Sheng, X., Chen, K., et al., 2023d. Microglial repopulation reverses cognitive and synaptic deficits in an Alzheimer's disease model by restoring BDNF signaling. Brain Behav. Immun. 113, 275–288.

Wardhana, M., 2016. Role of psychological stress on plasma cortisol and interleukin-4 in atopic dermatitis. Eur. J. Biomed. Pharm. Sci. 3, 25–29.

Ware, J.L., Li, R., Chen, A., Nelson, J.M., Kmet, J.M., et al., 2023. Associations between breastfeeding and post-perinatal infant deaths in the US. Am. J. Prev. Med. 65, 763–774.

Warren, H.R., Evangelou, E., Cabrera, C.P., Gao, H., Ren, M., et al., 2017. Genome-wide association analysis identifies novel blood pressure loci and offers biological insights into cardiovascular risk. Nat. Genet. 49, 403–415.

Warrener, C.D., Valentin, E.M., Gallin, C., Richey, L., Ross, D.B., et al., 2021. The role of oxytocin signaling in depression and suicidality in returning war veterans. Psychoneuroendocrinology 126, 105085.

Wassmann, C.S., Højrup, P., Klitgaard, J.K., 2020. Cannabidiol is an effective helper compound in combination with bacitracin to kill Gram-positive bacteria. Sci. Rep. 10, 4112.

Waszak, S.M., Delaneau, O., Gschwind, A.R., Kilpinen, H., Raghav, S.K., et al., 2015. Population variation and genetic control of modular chromatin architecture in humans. Cell 162, 1039–1050.

Watanabe, Y., Nunokawa, A., Kaneko, N., Muratake, T., Arinami, T., Ujike, H., Someya, T., 2009. Two-stage case-control association study of polymorphisms in rheumatoid arthritis susceptibility genes with schizophrenia. J. Hum. Genet. 54, 62–65.

Watkins, C.C., Andrews, S.R., 2016. Clinical studies of neuroinflammatory mechanisms in schizophrenia. Schizophr. Res. 176, 14–22.

Watkins, C., Stanton, C., Ryan, C.A., Ross, R.P., 2017. Microbial therapeutics designed for infant health. Front. Nutr. 4, 48.

Watts, N., Amann, M., Arnell, N., Ayeb-Karlsson, S., Beagley, J., 2021. The 2020 report of the Lancet Countdown on health and climate change: responding to converging crises. Lancet 397, 129–170.

Wazir, U., Mokbel, K., 2019. Bisphenol A: a concise review of literature and a discussion of health and regulatory implications. In Vivo 33, 1421–1423.

Weaver, I.C., Champagne, F.A., Brown, S.E., Dymov, S., Sharma, S., et al., 2005. Reversal of maternal programming of stress responses in adult offspring through methyl supplementation: altering epigenetic marking later in life. J. Neurosci. 25, 11045–11054.

Webb, E.K., Harnett, N.G., 2024. The biological embedding of structural inequities: new insight from neuroscience. Neuropsychopharmacology 49, 337–338.

Weber, M.D., Frank, M.G., Sobesky, J.L., Watkins, L.R., Maier, S.F., 2013. Blocking toll-like receptor 2 and 4 signaling during a stressor prevents stress-induced priming of neuroinflammatory responses to a subsequent immune challenge. Brain Behav. Immun. 32, 112–121.

Weber, M.D., Frank, M.G., Tracey, K.J., Watkins, L.R., Maier, S.F., 2015. Stress induces the danger-associated molecular pattern HMGB-1 in the hippocampus of male Sprague Dawley rats: a priming stimulus of microglia and the NLRP3 inflammasome. J. Neurosci. 35, 316–324.

Weber, M.D., Godbout, J.P., Sheridan, J.F., 2017. Repeated social defeat, neuroinflammation, and behavior: monocytes carry the signal. Neuropsychopharmacology 42, 46–61.

Weber-Stadlbauer, U., 2017. Epigenetic and transgenerational mechanisms in infection-mediated neurodevelopmental disorders. Transl. Psychiatry 7, e1113.

Webster, M.J., 2023. Infections, inflammation, and psychiatric illness: review of postmortem evidence. Curr. Top. Behav. Neurosci. 61, 35–48.

Weed, D.L., 2021. Does paraquat cause Parkinson's disease? A review of reviews. Neurotoxicology 86, 180–184.

Wegman, M.P., Guo, M., Bennion, D.M., Shankar, M.N., Chrzanowski, S.M., et al., 2015. Practicality of intermittent fasting in humans and its effect on oxidative stress and genes related to ageing and metabolism. Rejuvenation Res. 18, 162–172.

Wei, R.Q., Guo, W.L., Wu, Y.T., Alarcòn Rodrìguez, R., Requena Mullor, M.D.M., et al., 2022. Bioinformatics analysis of genomic and immune infiltration patterns in autism spectrum disorder. Ann. Transl. Med. 10, 1013.

Wei, Y., Yazdi, M.D., Ma, T., Castro, E., Liu, C.S., et al., 2023. Additive effects of 10-year exposures to PM2.5 and NO2 and primary cancer incidence in American older adults. Environ. Epidemiol. 7, e265.

Weichhart, T., 2018. mTOR as regulator of lifespan, aging, and cellular senescence: a mini-review. Gerontology 64, 127–134.

Weigert Muñoz, A., Hoyer, E., Schumacher, K., Grognot, M., Taute, K.M., et al., 2022. Eukaryotic catecholamine hormones influence the chemotactic control of vibrio campbellii by binding to the coupling protein CheW. Proc. Natl. Acad. Sci. USA 119, e2118227119.

Weintraub, M.K., Kranjac, D., Eimerbrink, M.J., Pearson, S.J., Vinson, B.T., et al., 2014. Peripheral administration of poly I:C leads to increased hippocampal amyloid-beta and cognitive deficits in a non-transgenic mouse. Behav. Brain Res. 266, 183–187.

Weir, R.K., Forghany, R., Smith, S.E., Patterson, P.H., McAllister, A.K., et al., 2015. Preliminary evidence of neuropathology in nonhuman primates prenatally exposed to maternal immune activation. Brain Behav. Immun. 48, 139–146.

Weiss, J.T., Donlea, J.M., 2022. Roles for sleep in neural and behavioral plasticity: reviewing variation in the consequences of sleep loss. Front. Behav. Neurosci. 15, 777799.

Weissman, M.M., Wickramaratne, P., Gameroff, M.J., Warner, V., Pilowsky, D., et al., 2016. Offspring of depressed parents: 30 years later. Am. J. Psychiatry 173, 1024–1032.

Weisz, J.R., Kuppens, S., Ng, M.Y., Eckshtain, D., Ugueto, A.M., et al., 2017. What five decades of research tells us about the effects of youth psychological therapy: a multilevel meta-analysis and implications for science and practice. Am. Psychol. 72, 79–117.

Welc, S.S., Clanton, T.L., 2013. The regulation of interleukin-6 implicates skeletal muscle as an integrative stress sensor and endocrine organ. Exp. Physiol. 98, 359–371.

Welch, M.D., Way, M., 2013. Arp2/3-mediated actin-based motility: a tail of pathogen abuse. Cell Host Microbe 14, 242–255.

Welly, R.J., Liu, T.W., Zidon, T.M., Rowles 3rd, J.L., Park, Y.M., et al., 2016. Comparison of diet versus exercise on metabolic function and gut microbiota in obese rats. Med. Sci. Sports Exerc. 48, 1688–1698.

Welser-Alves, J.V., Milner, R., 2013. Microglia are the major source of TNF-α and TGF-β1 in postnatal glial cultures; regulation by cytokines, lipopolysaccharide, and vitronectin. Neurochem. Int. 63, 47–53.

Wen, D.J., Poh, J.S., Ni, S.N., Chong, Y.S., Chen, H., et al., 2017. Influences of prenatal and postnatal maternal depression on amygdala volume and microstructure in young children. Transl. Psychiatry 7, e1103.

West, A.B., 2017. Achieving neuroprotection with LRRK2 kinase inhibitors in Parkinson disease. Exp. Neurol. 298, 236–245.

Westermann, J., Lange, T., Textor, J., Born, J., 2015. System consolidation during sleep - a common principle underlying psychological and immunological memory formation. Trends Neurosci. 38, 585–597.

Westfall, S., Caracci, F., Zhao, D., Wu, Q.L., Frolinger, T., et al., 2021. Microbiota metabolites modulate the T helper 17 to regulatory T cell (Th17/Treg) imbalance promoting resilience to stress-induced anxiety- and depressive-like behaviors. Brain Behav. Immun. 91, 350–368.

Weston, C.S.E., 2019. Four social brain regions, their dysfunctions, and sequelae, extensively explain autism spectrum disorder symptomatology. Brain Sci. 9, 130.

Westwell-Roper, C., Best, J.R., Naqqash, Z., Au, A., Lin, B., et al., 2022. Severe symptoms predict salivary interleukin-6, interleukin-1β, and tumor necrosis factor-α levels in children and youth with obsessive-compulsive disorder. J. Psychosom. Res. 155, 110743.

Wheaton, M.G., Galfalvy, H., Steinman, S.A., Wall, M.M., Foa, E.B., et al., 2016. Patient adherence and treatment outcome with exposure and response prevention for OCD: which components of adherence matter and who becomes well? Behav. Res. Ther. 85, 6–12.

Wheeler, D.L., Sariol, A., Meyerholz, D.K., Perlman, S., 2018. Microglia are required for protection against lethal coronavirus encephalitis in mice. J. Clin. Invest. 128, 931–943.

White, J.D., Kaffman, A., 2019. The moderating effects of sex on consequences of childhood maltreatment: from clinical studies to animal models. Front. Neurosci. 13, 1082.

White, S., Acierno, R., Ruggiero, K.J., Koenen, K.C., Kilpatrick, D.G., et al., 2013. Association of CRHR1 variants and posttraumatic stress symptoms in hurricane exposed adults. J. Anxiety Disord. 27, 678–683.

White, J., Kivimäki, M., Jokela, M., Batty, G.D., 2017. Association of inflammation with specific symptoms of depression in a general population of older people: the English Longitudinal Study of Ageing. Brain Behav. Immun. 61, 27–30.

White, C.A., Uttl, B., Holder, M.D., 2019. Meta-analyses of positive psychology interventions: the effects are much smaller than previously reported. PLoS One 14, e0216588.

Whittaker, A.C., Gallagher, S., 2019. Caregiving alters immunity and stress hormones: a review of recent research. Curr. Opin. Behav. Sci. 28, 93–97.

WHO, 2023a. Malaria. [Online]. Available: https://www.who.int/news-room/fact-sheets/detail/malaria#:~:text=Globally%20in%202022%2C%20there%20were,580%20000)%20of%20malaria%20deaths. (Accessed January 2024).

WHO, 2023b. Tuberculosis. [Online]. Available: https://www.who.int/news-room/fact-sheets/detail/tuberculosis. (Accessed January 2024).

Wiebe, N., Lloyd, A., Crumley, E.T., Tonelli, M., 2023. Associations between body mass index and all-cause mortality: a systematic review and meta-analysis. Obes. Rev., e13588.

Wijenayake, S., Rahman, M.F., Lum, C.M.W., De Vega, W.C., Sasaki, A., McGowan, P.O., 2020. Maternal high-fat diet induces sex-specific changes to glucocorticoid and inflammatory signaling in response to corticosterone and lipopolysaccharide challenge in adult rat offspring. J. Neuroinflammation 17, 116.

Wilding, J.P.H., Batterham, R.L., Davies, M., Van Gaal, L.F., Kandler, K., et al., 2022. Weight regain and cardiometabolic effects after withdrawal of semaglutide: the STEP 1 trial extension. Diabetes Obes. Metab. 24, 1553–1564.

Wiles, N.J., Thomas, L., Turner, N., Garfield, K., Kounali, D., et al., 2016. Long-term effectiveness and cost-effectiveness of cognitive behavioural therapy as an adjunct to pharmacotherapy for treatment-resistant depression in primary care: follow-up of the CoBalT randomised controlled trial. Lancet Psychiatry 3, 137–144.

Wilker, S., Vukojevic, V., Schneider, A., Pfeiffer, A., Inerle, S., et al., 2023. Epigenetics of traumatic stress: the association of NR3C1 methylation and posttraumatic stress disorder symptom changes in response to narrative exposure therapy. Transl. Psychiatry 13.

Willemen, H.L.D.M., Santos Ribeiro, P.S., Broeks, M., Meijer, N., Versteeg, S., Tiggeler, A., et al., 2023. Inflammation-induced mitochondrial and metabolic disturbances in sensory neurons control the switch from acute to chronic pain. Cell Rep. Med. 4, 101265.

Williams, L.M., Debattista, C., Duchemin, A.M., Schatzberg, A.F., Nemeroff, C.B., 2016a. Childhood trauma predicts antidepressant response in adults with major depression: data from the randomized international study to predict optimized treatment for depression. Transl. Psychiatry 6, e799.

Williams, S., Chen, L., Savignac, H.M., Tzortzis, G., Anthony, D.C., et al., 2016b. Neonatal prebiotic (BGOS) supplementation increases the levels of synaptophysin, GluN2A-subunits and BDNF proteins in the adult rat hippocampus. Synapse 70, 121–124.

Williams, J.A., Burgess, S., Suckling, J., Lalousis, P.A., Batool, F., et al., 2022a. Inflammation and brain structure in schizophrenia and other neuropsychiatric disorders: a mendelian randomization study. JAMA Psychiatry 79, 498–507.

Williams, T., Phillips, N.J., Stein, D.J., Ipser, J.C., 2022b. Pharmacotherapy for post traumatic stress disorder (PTSD). Cochrane Database Syst. Rev. 3, CD002795.

Williams, Z.A., Lang, L., Nicolas, S., Clarke, G., Cryan, J., et al., 2024. Do microbes play a role in Alzheimer's disease? Microb. Biotechnol. 17, e14462.

Willing, B.P., Vacharaksa, A., Croxen, M., Thanachayanont, T., Finlay, B. B., 2011. Altering host resistance to infections through microbial transplantation. PLoS One 6, e26988.

Willner, P., 2016. Reliability of the chronic mild stress model of depression: a user survey. Neurobiol. Stress 6, 68–77.

Wills, S., Cabanlit, M., Bennett, J., Ashwood, P., Amaral, D.G., Van de Water, J., 2009. Detection of autoantibodies to neural cells of the cerebellum in the plasma of subjects with autism spectrum disorders. Brain Behav. Immun. 23, 67–74.

Willyard, C., 2017. An epigenetics gold rush: new controls for gene expression. Nature 542, 406–408.

Wilmanski, T., Diener, C., Rappaport, N., Patwardhan, S., Wiedrick, J., et al., 2021. Gut microbiome pattern reflects healthy ageing and predicts survival in humans. Nat. Metab. 3, 274–286.

Wilson, S.J., Padin, A.C., Bailey, B.E., Laskowski, B., Andridge, R., et al., 2020. Spousal bereavement after dementia caregiving: a turning point for immune health. Psychoneuroendocrinology 118, 104717.

Wingfield, J.C., Sapolsky, R.M., 2003. Reproduction and resistance to stress: when and how. J. Neuroendocrinol. 15, 711–724.

Winglee, K., Howard, A.G., Sha, W., Gharaibeh, R.Z., Liu, J., et al., 2017. Recent urbanization in China is correlated with a westernized microbiome encoding increased virulence and antibiotic resistance genes. Microbiome 5, 121.

Wingo, A.P., Liu, Y., Gerasimov, E.S., Vattathil, S.M., Liu, J., et al., 2023. Sex differences in brain protein expression and disease. Nat. Med. 29, 2224–2232.

Winter, C., Djodari-Irani, A., Sohr, R., Morgenstern, R., Feldon, J., et al., 2009. Prenatal immune activation leads to multiple changes in basal neurotransmitter levels in the adult brain: implications for brain disorders of neurodevelopmental origin such as schizophrenia. Int. J. Neuropsychopharmacol. 12, 513–524.

Winters, B.L., Lau, B.K., Vaughan, C.W., 2022. Cannabinoids and opioids differentially target extrinsic and intrinsic GABAergic inputs onto the periaqueductal grey descending pathway. J. Neurosci. 42, 7744–7756.

Winterton, A., Bettella, F., Beck, D., Gurholt, T.P., Steen, N.E., et al., 2022. The oxytocin signalling gene pathway contributes to the association between loneliness and cardiometabolic health. Psychoneuroendocrinology 144, 105875.

Wischhof, L., Irrsack, E., Osorio, C., Koch, M., 2015. Prenatal LPS-exposure–a neurodevelopmental rat model of schizophrenia–differentially affects cognitive functions, myelination and parvalbumin expression in male and female offspring. Prog. Neuro-Psychopharmacol. Biol. Psychiatry 57, 17–30.

Wiseman, H., Barber, P., Raz, A., Yam, I., Foltz, C., et al., 2002. Parental communication of Holocaust experiences and interpersonal patterns in offspring of Holocaust survivors. Int. J. Behav. Dev. 26, 371–381.

Wissler, A., DeWitte, S.N., 2023. Frailty and survival in the 1918 influenza pandemic. Proc. Natl. Acad. Sci. USA 120, e2304545120.

Witcher, K.G., Eiferman, D.S., Godbout, J.P., 2015. Priming the inflammatory pump of the CNS after traumatic brain injury. Trends Neurosci. 38, 609–620.

Witkowski, M., Nemet, I., Alamri, H., Wilcox, J., Gupta, N., et al., 2023. The artificial sweetener erythritol and cardiovascular event risk. Nat. Med. 29 (3), 710–718.

Wittenberg, G.M., Stylianou, A., Zhang, Y., Sun, Y., Gupta, A., et al., 2020. Effects of immunomodulatory drugs on depressive symptoms: a mega-analysis of randomized, placebo-controlled clinical trials in inflammatory disorders. Mol. Psychiatry 25, 1275–1285.

Witthauer, C., Gloster, A.T., Meyer, A.H., Goodwin, R.D., Lieb, R., 2014. Comorbidity of infectious diseases and anxiety disorders in adults and its association with quality of life: a community study. Front. Public Health 2, 80.

Wohleb, E.S., Patterson, J.M., Sharma, V., Quan, N., Godbout, J.P., Sheridan, J.F., 2014. knockdown of interleukin-1 receptor type-1 on endothelial cells attenuated stress-induced neuroinflammation and prevented anxiety-like behavior. J. Neurosci. 34, 2583–2591.

Wohleb, E.S., McKim, D.B., Sheridan, J.F., Godbout, J.P., 2015. Monocyte trafficking to the brain with stress and inflammation: a novel axis of immune-to-brain communication that influences mood and behavior. Front. Neurosci. 8, 447.

Wohleb, E.S., Franklin, T., Iwata, M., Duman, R.S., 2016. Integrating neuroimmune systems in the neurobiology of depression. Nat. Rev. Neurosci. 17, 497–511.

Wohlfert, E.A., Blader, I.J., Wilson, E.H., 2017. Brains and brawn: toxoplasma infections of the central nervous system and skeletal muscle. Trends Parasitol. 33, 519–531.

Wolf, J.M., Miller, G.E., Chen, E., 2008. Parent psychological states predict changes in inflammatory markers in children with asthma and healthy children. Brain Behav. Immun. 22, 433–441.

Wolf, E.J., Mitchell, K.S., Logue, M.W., Baldwin, C.T., Reardon, A.F., et al., 2013. Corticotropin releasing hormone receptor 2 (CRHR-2) gene is associated with decreased risk and severity of posttraumatic stress disorder in women. Depress. Anxiety 30, 1161–1169.

Wolf, E.J., Zhao, X., Hawn, S.E., Morrison, F.G., Zhou, Z., et al., 2021. Gene expression correlates of advanced epigenetic age and psychopathology in postmortem cortical tissue. Neurobiol. Stress 15, 100371.

Wolkowitz, O.M., Wolf, J., Shelly, W., Rosser, R., Burke, H.M., et al., 2011. Serum BDNF levels before treatment predict SSRI response in depression. Prog. Neuropsychopharmacol. Biol. Psychiatry 35, 1623–1630.

Wolters, M., Ahrens, J., Romaní-Pérez, M., Watkins, C., Sanz, Y., et al., 2019. Dietary fat, the gut microbiota, and metabolic health - a systematic review conducted within the MyNewGut project. Clin. Nutr. 38, 2504–2520.

Wong, C., 2024. Measles outbreaks cause alarm: what the data say. Nature, Online ahead of print.

Wong, H., Hoeffer, C., 2018. Maternal IL-17A in autism. Exp. Neurol. 299, 228–240.

Wong, P.T.P., Roy, S., 2017. Critique of positive psychology and positive interventions. In: Brown, N.J.L., Lomas, T., Eiroa-Orosa, F.J. (Eds.), The Routledge International Handbook of Critical Positive Psychology. Routledge, London, pp. 142–160.

Wong, P., Sze, Y., Gray, L.J., Chang, C.C., Cai, S., Zhang, X., 2015. Early life environmental and pharmacological stressors result in persistent dysregulations of the serotonergic system. Front. Behav. Neurosci. 9, 94.

Wong, M.L., Inserra, A., Lewis, M.D., Mastronardi, C.A., Leong, L., et al., 2016. Inflammasome signaling affects anxiety- and depressive-like behavior and gut microbiome composition. Mol. Psychiatry 21, 797–805.

Wong, R.T., Cafferky, B.M., Alejandro, J.P., 2022. Chronic disease and elder mistreatment: a meta-analysis. Int. J. Geriatr. Psychiatry 37.

Wong, A.C., Devason, A.S., Umana, I.C., Cox, T.O., Dohnalová, L., et al., 2023a. Serotonin reduction in post-acute sequelae of viral infection. Cell 186, 4851–4867.

Wong, F., Zheng, E.J., Valeri, J.A., Donghia, N.M., Anahtar, M.N., et al., 2023b. Discovery of a structural class of antibiotics with explainable deep learning. Nature 626, 177–185.

Wood, H., 2017. Parkinson disease: caffeine and nicotine do not provide symptomatic relief in Parkinson disease. Nat. Rev. Neurol. 13, 707.

Woodburn, S.C., Bollinger, J.L., Wohleb, E.S., 2021. The semantics of microglia activation: neuroinflammation, homeostasis, and stress. J. Neuroinflammation 18, 258.

Woodhams, S.G., Sagar, D.R., Burston, J.J., Chapman, V., 2015. The role of the endocannabinoid system in pain. Handb. Exp. Pharmacol. 227, 119–143.

Woodrow, R.E., Winzeck, S., Luppi, A.I., Kelleher-Unger, I.R., Spindler, L.R.B., et al., 2023. Acute thalamic connectivity precedes chronic post-concussive symptoms in mild traumatic brain injury. Brain 146, 3484–3499.

Woods, R.M., Lorusso, J.M., Potter, H.G., Neill, J.C., Glazier, J.D., Hager, R., 2021. Maternal immune activation in rodent models: a systematic review of neurodevelopmental changes in gene expression and epigenetic modulation in the offspring brain. Neurosci. Biobehav. Rev. 129, 389–421.

Woodward, E., Rangel-Barajas, C., Ringland, A., Logrip, M.L., Coutellier, L., 2023. Sex-specific timelines for adaptations of prefrontal parvalbumin neurons in response to stress and changes in anxiety- and depressive-like behaviors. eNeuro 10. ENEURO.0300-22.2023.

Woody, E.Z., Hoffman, K.L., Szechtman, H., 2019. Obsessive compulsive disorder (OCD): Current treatments and a framework for neurotherapeutic research. Adv. Pharmacol. 86, 237–271.

World Health Organization, 2014. Global Tuberculosis Report 2014. Retrieved October 2016, from www.who.int/tb/publications/global_report/en/.

World Health Organization, 2016a. Investing in Treatment for Depression and Anxiety Leads to Fourfold Return. Retrieved from https://www.who.int/news/item/13-04-2016-investing-in-treatment-for-depression-and-anxiety-leads-to-fourfold-return. Accessed Jan, 2024.

World Health Organization, 2016b. Tuberculosis: Fact Sheet. Retrieved May 2015, from http://www.who.int/mediacentre/factsheets/fs104/en/.

Worley, N.B., Hill, M.N., Christianson, J.P., 2018. Prefrontal endocannabinoids, stress controllability and resilience: a hypothesis. Prog. Neuro-Psychopharmacol. Biol. Psychiatry 85, 180–188.

Worthen, R.J., Beurel, E., 2022. Inflammatory and neurodegenerative pathophysiology implicated in postpartum depression. Neurobiol. Dis. 165, 105646.

Wray, N.R., Ripke, S., Mattheisen, M., Trzaskowski, M., Byrne, E.M., et al., 2018. Genome-wide association analyses identify 44 risk variants and refine the genetic architecture of major depression. Nat. Genet. 50, 668–681.

Wright, N.D., Bahrami, B., Johnson, E., Di Malta, G., Rees, G., et al., 2012. Testosterone disrupts human collaboration by increasing egocentric choices. Proc. Biol. Sci. 279, 2275–2280.

Wrocklage, K.M., Averill, L.A., Cobb Scott, J., Averill, C.L., Schweinsburg, B., et al., 2017. Cortical thickness reduction in combat exposed U.S. veterans with and without PTSD. Eur. Neuropsychopharmacol. 27, 515–525.

Wu, S., Ding, Y., Wu, F., Li, R., Xie, G., et al., 2015. Family history of autoimmune diseases is associated with an increased risk of autism in children: a systematic review and meta-analysis. Neurosci. Biobehav. Rev. 55, 322–332.

Wu, Z.M., Zheng, C.H., Zhu, Z.H., Wu, F.T., et al., 2016. SiRNA-mediated serotonin transporter knockdown in the dorsal raphe nucleus rescues single prolonged stress-induced hippocampal autophagy in rats. J. Neurol. Sci. 360, 133–140.

Wu, H., Esteve, E., Tremaroli, V., Khan, M.T., Caesar, R., et al., 2017a. Metformin alters the gut microbiome of individuals with treatment-naive type 2 diabetes, contributing to the therapeutic effects of the drug. Nat. Med. 23, 850–858.

Wu, S.C., Cao, Z.S., Chang, K.M., Juang, J.L., 2017b. Intestinal microbial dysbiosis aggravates the progression of Alzheimer's disease in Drosophila. Nat. Commun. 8, 24.

Wu, D., Lv, P., Li, F., Zhang, W., Fu, G., et al., 2019a. Association of peripheral cytokine levels with cerebral structural abnormalities in schizophrenia. Brain Res. 1724, 146463.

Wu, L., Zeng, T., Zinellu, A., Rubino, S., Kelvin, D.J., Carru, C., 2019b. A cross-sectional study of compositional and functional profiles of gut microbiota in sardinian centenarians. mSystems 4, e00325-19.

Wu, Y., Lu, Y.C., Jacobs, M., Pradhan, S., Kapse, K., et al., 2020. Association of prenatal maternal psychological distress with fetal brain growth, metabolism, and cortical maturation. JAMA Netw. Open 3, e1919940.

Wu, C., Zhou, Z., Ni, L., Cao, J., Tan, M., et al., 2021. Correlation between anxiety-depression symptoms and immune characteristics in inpatients with 2019 novel coronavirus in Wuhan, China. J. Psychiatr. Res. 141, 378–384.

Wu, J., Li, X., Zhang, X., Wang, W., You, X., 2022a. What role of the cGAS-STING pathway plays in chronic pain? Front. Mol. Neurosci. 15, 963206.

Wu, X., Yang, H., He, S., Xia, T., Chen, D., et al., 2022b. Adult vaccination as a protective factor for dementia: a meta-analysis and systematic review of population-based observational studies. Front. Immunol. 13, 872542.

Wuchty, S., Myers, A.J., Ramirez-Restrepo, M., Huentelman, M., Richolt, R., et al., 2021. Integration of peripheral transcriptomics, genomics, and interactomics following trauma identifies causal genes for symptoms of post-traumatic stress and major depression. Mol. Psychiatry 26, 3077–3092.

Xia, C.Y., Zhang, S., Gao, Y., Wang, Z.Z., Chen, N.H., 2015. Selective modulation of microglia polarization to M2 phenotype for stroke treatment. Int. Immunopharmacol. 25, 377–382.

Xia, C.Y., Guo, Y.X., Lian, W.W., Yan, Y., Ma, B.Z., et al., 2023. The NLRP3 inflammasome in depression: potential mechanisms and therapies. Pharmacol. Res. 187, 106625.

Xia, W., Veeragandham, P., Cao, Y., Xu, Y., Rhyne, T.E., et al., 2024. Obesity causes mitochondrial fragmentation and dysfunction in white adipocytes due to RalA activation. Nat. Metab. 6, 273–289.

Xiao, L., Priest, M.F., Nasenbeny, J., Lu, T., Kozorovitskiy, Y., 2017. Biased oxytocinergic modulation of midbrain dopamine systems. Neuron 95, 368–384.

Xiao, L., Zheng, H., Li, J., Wang, Q., Sun, H., 2020. Neuroinflammation mediated by NLRP3 inflammasome after intracerebral hemorrhage and potential therapeutic targets. Mol. Neurobiol. 57, 5130–5149.

Xie, T., Stathopoulou, M., de Andrés, F., et al., 2017. VEGF-related polymorphisms identified by GWAS and risk for major depression. Transl. Psychiatry 7, e1055.

Xie, Z., Jiang, W., Deng, M., Wang, W., Xie, X., et al., 2021. Alterations of oral microbiota in patients with panic disorder. Bioengineered 12, 9103–9112.

Xie, Y., Xu, E., Bowe, B., Al-Aly, Z., 2022. Long-term cardiovascular outcomes of COVID-19. Nat. Med. 28, 583–590.

Xie, H., Wang, H., Wu, Z., Li, W., Liu, Y., Wang, N., 2023a. The association of dietary inflammatory potential with skeletal muscle strength, mass, and sarcopenia: a meta-analysis. Front. Nutr. 10, 1100918.

Xie, M., Cai, J., Liu, Y., Wei, W., Zhao, Z., et al., 2023b. Association between childhood trauma and white matter deficits in first-episode schizophrenia. Psychiatry Res. 323, 115111.

Xie, Y., Choi, T., Al-Aly, Z., 2023c. Long-term outcomes following hospital admission for COVID-19 versus seasonal influenza: a cohort study. Lancet Infect. Dis. 14.

Xiong, Z., Zhang, K., Ishima, T., Ren, Q., Chang, L., Chen, J., Hashimoto, K., 2018. Comparison of rapid and long-lasting antidepressant effects of negative modulators of α5-containing GABAA receptors and (R)-ketamine in a chronic social defeat stress model. Pharmacol. Biochem. Behav. 175, 139–145.

Xu, J., Lazartigues, E., 2022. Expression of ACE2 in human neurons supports the neuro-invasive potential of COVID-19 virus. Cell. Mol. Neurobiol. 42, 305–309.

Xu, N., Li, X., Zhong, Y., 2015. Inflammatory cytokines: potential biomarkers of immunologic dysfunction in autism spectrum disorders. Mediat. Inflamm. 2015, 531518.

Xu, J., Huang, G., Guo, T.L., 2016. Developmental bisphenol A exposure modulates immune-related diseases. Toxics 4, 23.

Xu, J., Marshall, J.J., Fernandes, H.B., Nomura, T., Copits, B.A., et al., 2017a. Complete disruption of the kainate receptor gene family results in corticostriatal dysfunction in mice. Cell Rep. 18, 1848–1857.

Xu, Y., Zhou, H., Zhu, Q., 2017b. The impact of microbiota-gut-brain axis on diabetic cognition impairment. Front. Aging Neurosci. 9, 1–18.

Xu, H.B., Xu, Y.H., He, Y., Xue, F., Wei, J., et al., 2018a. Decreased serum brain-derived neurotrophic factor may indicate the development of poststroke depression in patients with acute ischemic stroke: a meta-analysis. J. Stroke Cerebrovasc. Dis. 27, 709–715.

Xu, T., Magnusson Hanson, L.L., Lange, T., Starkopf, L., Westerlund, H., et al., 2018b. Workplace bullying and violence as risk factors for type 2 diabetes: a multicohort study and meta-analysis. Diabetologia 61, 75–83.

Xu, E., Xie, Y., Al-Aly, Z., 2022. Long-term neurologic outcomes of COVID-19. Nat. Med. 28, 2406–2415.

Xu, K., Wu, Y., Tian, Z., Xu, Y., Wu, C., Wang, Z., 2023a. Microglial cannabinoid CB2 receptors in pain modulation. Int. J. Mol. Sci. 24 (3), 2348.

Xu, Y., Kusuyama, J., Osana, S., Matsuhashi, S., Li, L., et al., 2023b. Lactate promotes neuronal differentiation of SH-SY5Y cells by lactate-responsive gene sets through NDRG3-dependent and-independent manners. J. Biol. Chem. 299, 104802.

Xue, Y., Xu, X., Zhang, X.Q., Farokhzad, O.C., Langer, R., 2016. Preventing diet-induced obesity in mice by adipose tissue transformation and angiogenesis using targeted nanoparticles. Proc. Natl. Acad. Sci. USA 113, 5552–5557.

Xue, A., Wu, Y., Zhu, Z., Zhang, F., Kemper, K.E., et al., 2018. Genome-wide association analyses identify 143 risk variants and putative regulatory mechanisms for type 2 diabetes. Nat. Commun. 9, 2941.

Xue, L.J., Yang, X.Z., Tong, Q., Shen, P., Ma, S.J., et al., 2020. Fecal microbiota transplantation therapy for Parkinson's disease: a preliminary study. Medicine (Baltimore) 99, e22035.

Xue, Y., Zhou, J., Wang, P., Lan, J.H., Lian, W.Q., et al., 2022. Burden of tuberculosis and its association with socio-economic development status in 204 countries and territories, 1990-2019. Front. Med. 9, 905245.

Xue, J.R., Mackay-Smith, A., Mouri, K., Garcia, M.F., Dong, M.X., et al., 2023. The functional and evolutionary impacts of human-specific deletions in conserved elements. Science 380, eabn2253.

Yamanashi, T., Iwata, M., Kamiya, N., Tsunetomi, K., Kajitani, N., et al., 2017. Beta-hydroxybutyrate, an endogenic NLRP3 inflammasome inhibitor, attenuates stress-induced behavioral and inflammatory responses. Sci. Rep. 7, 7677.

Yamazaki, Y., Zhao, N., Caulfield, T.R., Liu, C.C., Bu, G., 2019. Apolipoprotein E and Alzheimer disease: pathobiology and targeting strategies. Nat. Rev. Neurol. 15, 501–518.

Yan, J.Q., Sun, J.C., Zhai, M.M., Cheng, L.N., Bai, X.L., et al., 2015. Lovastatin induces neuroprotection by inhibiting inflammatory cytokines in 6-hydroxydopamine treated microglia cells. Int. J. Clin. Exp. Med. 8, 9030–9037.

Yan, D., Zhang, Y., Liu, L., Yan, H., 2016a. Pesticide exposure and risk of Alzheimer's disease: a systematic review and meta-analysis. Sci. Rep. 6, 32222.

Yan, M., Audet-Walsh, É., Manteghi, S., Dufour, C.R., Walker, B., et al., 2016b. Chronic AMPK activation via loss of FLCN induces functional beige adipose tissue through PGC-1α/ERRα. Genes Dev. 30, 1034–1046.

Yan, S., Zheng, C., Paranjpe, M.D., Li, Y., Li, W., et al., 2021. Sex modifies APOE ε4 dose effect on brain tau deposition in cognitively impaired individuals. Brain 144, 3201–3211.

Yan, Q., Wu, X., Zhou, P., Zhou, Y., Li, X., et al., 2022a. HERV-W envelope triggers abnormal dopaminergic neuron process through DRD2/PP2A/AKT1/GSK3 for schizophrenia risk. Viruses 14, 145.

Yan, X., Yang, K., Xiao, Q., Hou, R., Pan, X., Zhu, X., 2022b. Central role of microglia in sepsis-associated encephalopathy: from mechanism to therapy. Front. Immunol. 13, 929316.

Yang, G.R., Murray, J.D., Wang, X.J., 2016. A dendritic disinhibitory circuit mechanism for pathway-specific gating. Nat. Commun. 7, 12815.

Yang, Y., Tian, J., Yang, B., 2018. Targeting gut microbiome: a novel and potential therapy for autism. Life Sci. 194, 111–119.

Yang, B., Wei, J., Ju, P., Chen, J., 2019. Effects of regulating intestinal microbiota on anxiety symptoms: a systematic review. Gen. Psychiatr. 32, e100056.

Yang, A.C., Kern, F., Losada, P.M., Agam, M.R., Maat, C.A., et al., 2021a. Dysregulation of brain and choroid plexus cell types in severe COVID-19. Nature 595 (7868), 565–571.

Yang, A.J.T., Bagit, A., MacPherson, R.E.K., 2021b. Resveratrol, metabolic dysregulation, and Alzheimer's disease: considerations for neurogenerative disease. Int. J. Mol. Sci. 22, 4628.

Yang, L., Xie, X., Tu, Z., Fu, J., Xu, D., Zhou, Y., 2021c. The signal pathways and treatment of cytokine storm in COVID-19. Signal Transduct. Target Ther. 6, 255.

Yang, R., Wu, G.W.Y., Verhoeven, J.E., Gautam, A., Reus, V.I., et al., 2021d. A DNA methylation clock associated with age-related illnesses and mortality is accelerated in men with combat PTSD. Mol. Psychiatry 26, 4999–5009.

Yang, R., Xu, C., Bierer, L.M., Flory, J.D., Gautam, A., et al., 2021e. Longitudinal genome-wide methylation study of PTSD treatment using prolonged exposure and hydrocortisone. Transl. Psychiatry 11, 398.

Yang, J., Wei, H., Zhou, Y., Szeto, C.H., Li, C., et al., 2022a. High-fat diet promotes colorectal tumorigenesis through modulating gut microbiota and metabolites. Gastroenterology 162, 135–149.e2.

Yang, J., Zou, M., Chu, X., Floess, S., Li, Y., et al., 2022b. Inflammatory perturbations in early life long-lastingly shape the transcriptome and TCR repertoire of the first wave of regulatory T cells. Front. Immunol. 13, 991671.

Yang, J.H., Petty, C.A., Dixon-McDougall, T., Lopez, M.V., Tyshkovskiy, A., et al., 2023a. Chemically induced reprogramming to reverse cellular aging. Aging 15.

Yang, W., Jiao, H., Xue, Y., Wang, L., Zhang, Y., et al., 2023b. A meta-analysis of the influence on inflammatory factors in type 2 diabetes among middle-aged and elderly patients by various exercise modalities. Int. J. Environ. Res. Public Health 20, 1783.

Yang, L., Kim, T.W., Han, Y., Nair, M.S., Harschnitz, O., et al., 2024a. SARS-CoV-2 infection causes dopaminergic neuron senescence. Cell Stem Cell 31, 196–211.e6.

Yang, T.T., Liu, P.J., Sun, Q.Y., Wang, Z.Y., Yuan, G.B., et al., 2024b. CD4(+)CD25(+) regulatory T cells ex vivo generated from autologous naïve CD4(+) T cells suppress EAE progression. Sci. Rep. 14, 6262.

Yasmin, F., Saxena, K., McEwen, B.S., Chattarji, S., 2016. The delayed strengthening of synaptic connectivity in the amygdala depends on NMDA receptor activation during acute stress. Phys. Rep. 4, e13002.

Yavorsky, V.A., Rozumna, N.M., Lukyanetz, E.A., 2023. Influence of amyloid beta on impulse spiking of isolated hippocampal neurons. Front. Cell. Neurosci. 17, 1132092.

Yavropoulou, M.P., Tsokos, G.C., Chrousos, G.P., Sfikakis, P.P., 2022. Protracted stress-induced hypocortisolemia may account for the clinical and immune manifestations of Long COVID. Clin. Immunol. 245, 109133.

Ye, J., Wen, Y., Chu, X., Li, P., Cheng, B., et al., 2020. Association between herpes simplex virus 1 exposure and the risk of depression in UK Biobank. Clin. Transl. Med. 10, e108.

Ye, Z., Kappelmann, N., Moser, S., Davey Smith, G., Burgess, S., et al., 2021. Role of inflammation in depression and anxiety: tests for disorder specificity, linearity and potential causality of association in the UK Biobank. EClinicalMedicine 38, 100992.

Yehuda, R., 2002. Current status of cortisol findings in post-traumatic stress disorder. Psychiatr. Clin. North Am. 25, 341–368.

Yehuda, R., Bierer, L.M., 2009. The relevance of epigenetics to PTSD: implications for the DSM-V. J. Trauma. Stress. 22, 427–434.

Yehuda, R., Seckl, J., 2011. Minireview: stress-related psychiatric disorders with low cortisol levels: a metabolic hypothesis. Endocrinology 152, 4496–4503.

Yehuda, R., Brand, S., Yang, R.K., 2006. Plasma neuropeptide Y concentrations in combat exposed veterans: relationship to trauma exposure, recovery from PTSD, and coping. Biol. Psychiatry 59, 660–663.

Yehuda, R., Daskalakis, N.P., Desarnaud, F., Makotkine, I., Lehrner, A., et al., 2013. Epigenetic biomarkers as predictors and correlates of symptom improvement following psychotherapy in combat veterans with PTSD. Front. Psychiatry 4, 118.

Yehuda, R., Daskalakis, N.P., Lehrner, A., Desarnaud, F., Bader, H.N., et al., 2014. Influences of maternal and paternal PTSD on epigenetic regulation of the glucocorticoid receptor gene in Holocaust survivor offspring. Am. J. Psychiatry 171, 872–880.

Yehuda, R., Bierer, L.M., Pratchett, L.C., Lehrner, A., Koch, E.C., et al., 2015. Cortisol augmentation of a psychological treatment for warfighters with posttraumatic stress disorder: randomized trial showing improved treatment retention and outcome. Psychoneuroendocrinology 51, 589–597.

Yehuda, R., Daskalakis, N.P., Bierer, L.M., Bader, H.N., Klengel, T., et al., 2016. Holocaust exposure induced intergenerational effects on FKBP5 methylation. Biol. Psychiatry 80, 372–380.

Yelin, I., Flett, K.B., Merakou, C., Mehrotra, P., Stam, J., et al., 2019. Genomic and epidemiological evidence of bacterial transmission from probiotic capsule to blood in ICU patients. Nat. Med. 25, 1728–1732.

Yeoh, Y.K., Zuo, T., Lui, G.C., Zhang, F., Liu, Q., et al., 2021. Gut microbiota composition reflects disease severity and dysfunctional immune responses in patients with COVID-19. Gut 70, 698–706.

Yeramilli, V., Cheddadi, R., Shah, J., Brawner, K., Martin, C., 2023. A review of the impact of maternal prenatal stress on offspring microbiota and metabolites. Metabolites 13, 535.

Yeung, L.K., Alschuler, D.M., Wall, M., Luttmann-Gibson, H., Copeland, T., et al., 2023. Multivitamin supplementation improves memory in older adults: a randomized clinical trial. Am. J. Clin. Nutr. 118, 273–282.

Yi, M., Jiao, D., Qin, S., Chu, Q., Li, A., Wu, K., 2019. Manipulating gut microbiota composition to enhance the therapeutic effect of cancer immunotherapy. Integr. Cancer Ther. 18. 1534735419876351.

Yi, M., Zheng, X., Niu, M., Zhu, S., Ge, H., Wu, K., 2022a. Combination strategies with PD-1/PD-L1 blockade: current advances and future directions. Mol. Cancer 21, 28.

Yi, S., Wang, L., Wang, H., Ho, M.S., Zhang, S., 2022b. Pathogenesis of α-synuclein in Parkinson's disease: from a neuron-glia crosstalk perspective. Int. J. Mol. Sci. 23, 14753.

Yin, J., Jin, X., Shan, Z., Li, S., Huang, H., et al., 2017. Relationship of sleep duration with all-cause mortality and cardiovascular events: a systematic review and dose-response meta-analysis of prospective cohort studies. J. Am. Heart Assoc. 6, e005947.

Yin, L., Xu, X., Chen, G., Mehta, N.D., Haroon, E., Miller, A.H., et al., 2019. Inflammation and decreased functional connectivity in a widely-distributed network in depression: centralized effects in the ventral medial prefrontal cortex. Brain Behav. Immun. 80, 657–666.

Yin, K., Peluso, M.J., Luo, X., Thomas, R., Shin, M.G., et al., 2024. Long COVID manifests with T cell dysregulation, inflammation and an uncoordinated adaptive immune response to SARS-CoV-2. Nat. Immunol. 25, 218–225.

Yip, A.Y.G., King, O.G., Omelchenko, O., Kurkimat, S., Horrocks, V., et al., 2023. Antibiotics promote intestinal growth of carbapenem-resistant Enterobacteriaceae by enriching nutrients and depleting microbial metabolites. Nat. Commun. 14, 5094.

Yirmiya, R., 2000. Depression in medical illness: the role of the immune system. West. J. Med. 173, 333–336.

Yirmiya, R., Pollak, Y., Morag, M., Reichenberg, A., Barak, O., 2000. Illness, cytokines, and depression. Ann. N. Y. Acad. Sci. 917, 478–487.

Yirmiya, R., Rimmerman, N., Reshef, R., 2015. Depression as a microglial disease. Trends Neurosci. 38, 637–658.

Yohn, N.L., Blendy, J.A., 2017. Adolescent chronic unpredictable stress exposure is a sensitive window for long-term changes in adult behavior in mice. Neuropsychopharmacology 42, 1670–1678.

Yohn, S.E., Arif, Y., Haley, A., Tripodi, G.1., Baqi, Y., et al., 2016. Effort-related motivational effects of the pro-inflammatory cytokine interleukin-6: pharmacological and neurochemical characterization. Psychopharmacology 233, 3575–3586.

Yohn, C.N., Gergues, M.M., Samuels, B.A., 2017. The role of 5-HT receptors in depression. Mol. Brain 10, 28.

Yolken, R., Adamos, M., Katsafanas, E., Khushalani, S., Origoni, A., et al., 2016. Individuals hospitalized with acute mania have increased exposure to antimicrobial medications. Bipolar Disord. 18, 404–409.

Yoon, S., Cho, H., Kim, J., Lee, D.W., Kim, G.H., et al., 2017. Brain changes in overweight/obese and normal-weight adults with type 2 diabetes mellitus. Diabetologia 60, 1207–1217.

Yoon, H.J., Seo, E.H., Kim, J.J., Choo, I.H., 2019. Neural correlates of self-referential processing and their clinical implications in social anxiety disorder. Clin. Psychopharmacol. Neurosci. 17, 12–24.

Yoon, S.Y., Heo, S.J., Kim, Y.W., Lee, S.C., Shin, J., Lee, J.W., 2023. Depressive symptoms and the subsequent risk of Parkinson's disease: a nationwide cohort study. Am. J. Geriatr. Psychiatry 32, 339–348.

You, Z., Luo, C., Zhang, W., Chen, Y., He, J., et al., 2011. Pro- and anti-inflammatory cytokines expression in rat's brain and spleen exposed to chronic mild stress: involvement in depression. Behav. Brain Res. 225, 135–141.

You, I.J., Bae, Y., Beck, A.R., Shin, S., 2023. Lateral hypothalamic proenkephalin neurons drive threat-induced overeating associated with a negative emotional state. Nat. Commun. 14, 6875.

Youm, Y.H., Nguyen, K.Y., Grant, R.W., Goldberg, E.L., Bodogai, M., et al., 2015. The ketone metabolite β-hydroxybutyrate blocks NLRP3 inflammasome-mediated inflammatory disease. Nat. Med. 21, 263–269.

Yousef Yengej, F.A., van Royen-Kerkhof, A., Derksen, R., Fritsch-Stork, R.D.E., 2017. The development of offspring from mothers with systemic lupus erythematosus. A systematic review. Autoimmun. Rev. 16, 701–711.

Yshii, L., Pasciuto, E., Bielefeld, P., Mascali, L., Lemaitre, P., et al., 2022. Astrocyte-targeted gene delivery of interleukin 2 specifically increases brain-resident regulatory T cell numbers and protects against pathological neuroinflammation. Nat. Immunol. 23, 878–889.

Ysseldyk, R., Matheson, K., Anisman, H., 2010. Religiosity as identity: toward an understanding of religion from a social identity perspective. Personal. Soc. Psychol. Rev. 14, 60–71.

Yu, A., Dang, W., 2017. Regulation of stem cell aging by SIRT1—linking metabolic signaling to epigenetic modifications. Mol. Cell. Endocrinol. 455, 75–82.

Yu, X., Li, W., Ma, Y., Tossell, K., Harris, J.J., et al., 2019. GABA and glutamate neurons in the VTA regulate sleep and wakefulness. Nat. Neurosci. 22, 106–119.

Yu, Y., Huang, J., Chen, X., Fu, J., Wang, X., et al., 2022. Efficacy and safety of diet therapies in children with autism spectrum disorder: a systematic literature review and meta-analysis. Front. Neurol. 13, 844117.

Yu, X., Mostafijur Rahman, M., Carter, S.A., Lin, J.C., Zhuang, Z., et al., 2023. Prenatal air pollution, maternal immune activation, and autism spectrum disorder. Environ. Int. 179, 108148.

Yuan, L., Liu, S., Bai, X., Gao, Y., Liu, G., et al., 2016. Oxytocin inhibits lipopolysaccharide-induced inflammation in microglial cells and attenuates microglial activation in lipopolysaccharide-treated mice. J. Neuroinflammation 13, 77.

Yuan, X., Kang, Y., Zhuo, C., Huang, X.F., Song, X., 2019. The gut microbiota promotes the pathogenesis of schizophrenia via multiple pathways. Biochem. Biophys. Res. Commun. 512, 373–380.

Yuan, B., Li, W., Liu, H., Cai, X., Song, S., et al., 2020. Correlation between immune response and self-reported depression during convalescence from COVID-19. Brain Behav. Immun. 88, 39–43.

Yuen, E.Y., Wei, J., Yan, Z., 2016. Estrogen in prefrontal cortex blocks stress-induced cognitive impairments in female rats. J. Steroid Biochem. Mol. Biol. 160, 221–226.

Yulug, B., Hanoglu, L., Kilic, E., 2017. Does sleep disturbance affect the amyloid clearance mechanisms in Alzheimer's disease? Psychiatry Clin. Neurosci. 71, 673–677.

Zada, D., Bronshtein, I., Lerer-Goldshtein, T., Garini, Y., Appelbaum, L., 2019. Sleep increases chromosome dynamics to enable reduction of accumulating DNA damage in single neurons. Nat. Commun. 10, 895.

Zafiriou, E., Daponte, A.I., Siokas, V., Tsigalou, C., Dardiotis, E., Bogdanos, D.P., 2021. Depression and obesity in patients with psoriasis and psoriatic arthritis: Is IL-17-mediated immune dysregulation the connecting link? Front. Immunol. 12, 699848.

Zahid, U., Onwordi, E.C., Hedges, E.P., Wall, M.B., Modinos, G., et al., 2023. Neurofunctional correlates of glutamate and GABA imbalance in psychosis: a systematic review. Neurosci. Biobehav. Rev. 144, 105010.

Zai, G., Robbins, T.W., Sahakian, B.J., Kennedy, J.L., 2017. A review of molecular genetic studies of neurocognitive deficits in schizophrenia. Neurosci. Biobehav. Rev. 72, 50–67.

Zalcman, S., Anisman, H., 1993. Acute and chronic stressor effects on the antibody response to sheep red blood cells. Pharmacol. Biochem. Behav. 46, 445–452.

Zalli, A., Carvalho, L.A., Lin, J., Hamer, M., Erusalimsky, J.D., et al., 2014. Shorter telomeres with high telomerase activity are associated with raised allostatic load and impoverished psychosocial resources. Proc. Natl. Acad. Sci. USA 111, 4519–4524.

Zamanian, J.L., Xu, L., Foo, L.C., Nouri, N., Zhou, L., et al., 2012. Genomic analysis of reactive astrogliosis. J. Neurosci. 32, 6391–6410.

Zamanpoor, M., Ghaedi, H., Omrani, M.D., 2020. The genetic basis for the inverse relationship between rheumatoid arthritis and schizophrenia. Mol. Genet. Genomic Med. 8, e1483.

Zani, F., Blagih, J., Gruber, T., Buck, M.D., Jones, N., et al., 2023. The dietary sweetener sucralose is a negative modulator of T cell-mediated responses. Nature 615, 705–711.

Zannas, A.S., Jia, M., Hafner, K., Baumert, J., Wiechmann, T., et al., 2019. Epigenetic upregulation of FKBP5 by aging and stress contributes to NF-κB-driven inflammation and cardiovascular risk. Proc. Natl. Acad. Sci. USA 116, 11370–11379.

Zannas, A.S., Linnstaedt, S.D., An, X., Stevens, J.S., Harnett, N.G., et al., 2023. Epigenetic aging and PTSD outcomes in the immediate aftermath of trauma. Psychol. Med. 53, 7170–7179.

Zanoni, P., Khetarpal, S.A., Larach, D.B., Hancock-Cerutti, W.F., Millar, J.S., et al., 2016. Rare variant in scavenger receptor BI raises HDL cholesterol and increases risk of coronary heart disease. Science 351, 1166–1171.

Zanos, P., Gould, T.D., 2018. Mechanisms of ketamine action as an antidepressant. Mol. Psychiatry 23, 801–811.

Zareie, M., Johnson-Henry, K., Jury, J., Yang, P.C., Ngan, B.Y., et al., 2006. Probiotics prevent bacterial translocation and improve intestinal barrier function in rats following chronic psychological stress. Gut 55, 1553–1560.

Zawadzka, A., Cieślik, M., Adamczyk, A., 2021. The role of maternal immune activation in the pathogenesis of autism: a review of the evidence, proposed mechanisms and implications for treatment. Int. J. Mol. Sci. 22, 11516.

Zeevi, D., Korem, T., Zmora, N., Israeli, D., Rothschild, D., et al., 2015. Personalized nutrition by prediction of glycemic responses. Cell 163, 1079–1094.

Zeidan, F., Emerson, N.M., Farris, S.R., Ray, J.N., Jung, Y., et al., 2015. Mindfulness meditation-based pain relief employs different neural mechanisms than placebo and sham mindfulness meditation-induced analgesia. J. Neurosci. 35, 15307–15325.

Zeidan, J., Fombonne, E., Scorah, J., Ibrahim, A., Durkin, M.S., Saxena, S., Yusuf, A., Shih, A., Elsabbagh, M., 2022. Global prevalence of autism: a systematic review update. Autism Res. 15, 778–790.

Zeineh, M.M., Kang, J., Atlas, S.W., Raman, M.M., Reiss, A.L., et al., 2014. Right arcuate fasciculus abnormality in chronic fatigue syndrome. Radiology 274, 517–526.

Zelada, M.I., Garrido, V., Liberona, A., Jones, N., Zúñiga, K., et al., 2023. Brain-derived neurotrophic factor (BDNF) as a predictor of treatment response in major depressive disorder (MDD): a systematic review. Int. J. Mol. Sci. 24, 14810.

Zeng, Y., Suo, C., Yao, S., Lu, D., Larsson, H., et al., 2023. Genetic associations between stress-related disorders and autoimmune disease. Am. J. Psychiatry 180, 294–304.

Zeng, Y., Wu, Y., Zhang, Q., Xiao, X., 2024. Crosstalk between glucagon-like peptide 1 and gut microbiota in metabolic diseases. MBio 15, e0203223.

Zengeler, K.E., Lukens, J.R., 2021. Innate immunity at the crossroads of healthy brain maturation and neurodevelopmental disorders. Nat. Rev. Immunol. 21, 454–468.

Zenk, F., Loeser, E., Schiavo, R., Kilpert, F., Bogdanović, O., et al., 2017. Germ line–inherited H3K27me3 restricts enhancer function during maternal-to-zygotic transition. Science 357, 212–216.

Zerbo, O., Qian, Y., Yoshida, C., Fireman, B.H., Klein, N.P., Croen, L.A., 2017. Association between influenza infection and vaccination during pregnancy and risk of autism spectrum disorder. JAMA Pediatr. 171, e163609.

Zhang, B., Gems, D., 2021. Gross ways to live long: parasitic worms as an anti-inflammaging therapy? elife 10, e65180.

Zhang, Y., Liu, C., 2022. Evaluating the challenges and reproducibility of studies investigating DNA methylation signatures of psychological stress. Epigenomics 14, 405–421.

Zhang, C., Qin, G., 2023. Irregular sleep and cardiometabolic risk: clinical evidence and mechanisms. Front. Cardiovasc. Med. 10, 1059257.

Zhang, W., Rosenkranz, J.A., 2016. Effects of repeated stress on age-dependent GABAergic regulation of the lateral nucleus of the amygdala. Neuropsychopharmacology 41, 2309–2323.

Zhang, D., Kishihara, K., Wang, B., Mizobe, K., Kubo, C., et al., 1998. Restraint stress-induced immunosuppression by inhibiting leukocyte migration and Th1 cytokine expression during the intraperitoneal infection of Listeria monocytogenes. J. Neuroimmunol. 92, 139–151.

Zhang, L., Hu, X.Z., Li, X., Li, H., Smerin, S., et al., 2014. Telomere length - a cellular aging marker for depression and post-traumatic stress disorder. Med. Hypotheses 83, 182–185.

Zhang, D., Wang, X., Wang, B., Garza, J.C., Fang, X., et al., 2016. Adiponectin regulates contextual fear extinction and intrinsic excitability of dentate gyrus granule neurons through AdipoR2 receptors. Mol. Psychiatry 22, 1044–1055.

Zhang, J.C., Yao, W., Dong, C., Yang, C., Ren, Q., et al., 2017a. Blockade of interleukin-6 receptor in the periphery promotes rapid and sustained antidepressant actions: a possible role of gut-microbiota-brain axis. Transl. Psychiatry 7, e1138.

Zhang, Y., Shao, F., Wang, Q., Xie, X., Wang, W., 2017b. Neuroplastic correlates in the mPFC underlying the impairment of stress-coping ability and cognitive flexibility in adult rats exposed to chronic mild stress during adolescence. Neural Plast., 9382797.

Zhang, J., Luo, W., Huang, P., Peng, L., Huang, Q., 2018a. Maternal C-reactive protein and cytokine levels during pregnancy and the risk of selected neuropsychiatric disorders in offspring: a systematic review and meta-analysis. J. Psychiatr. Res. 105, 86–94.

Zhang, L., Zheng, H., Wu, R., Zhu, F., Kosten, T., et al., 2018b. Minocycline adjunctive treatment to risperidone for negative symptoms in schizophrenia: association with pro-inflammatory cytokine levels. Prog. Neuro-Psychopharmacol. Biol. Psychiatry 85, 69–76.

Zhang, B.Z., Chu, H., Han, S., Shuai, H., Deng, J., et al., 2020a. SARS-CoV-2 infects human neural progenitor cells and brain organoids. Cell Res. 30, 928–931.

Zhang, Q., Sun, Y., He, Z., Xu, Y., Li, X., et al., 2020b. Kynurenine regulates NLRP2 inflammasome in astrocytes and its implications in depression. Brain Behav. Immun. 88, 471–481.

Zhang, Y., Wang, Y., Ke, B., Du, J., 2021. TMAO: how gut microbiota contributes to heart failure. Transl. Res. 228, 109–125.

Zhang, M.M., Geng, A.Q., Chen, K., Wang, J., Wang, P., et al., 2022a. Glutamatergic synapses from the insular cortex to the basolateral amygdala encode observational pain. Neuron 110 (12), 1993–2008.e6.

Zhang, Q., Cheng, X., Wu, W., Yang, S., You, H., et al., 2022b. Age-related LRRK2 G2019S mutation impacts microglial dopaminergic fiber refinement and synaptic pruning involved in abnormal behaviors. J. Mol. Neurosci. 72, 527–543.

Zhang, Y., Dong, Y., Zhu, Y., Sun, D., Wang, S., et al., 2022c. Microglia-specific transcriptional repression of interferon-regulated genes after prolonged stress in mice. Neurobiol. Stress 21, 100495.

Zhang, H.C., Du, Y., Chen, L., Yuan, Z.Q., Cheng, Y., 2023a. MicroRNA schizophrenia: etiology, biomarkers and therapeutic targets. Neurosci. Biobehav. Rev. 146, 105064.

Zhang, N., Qi, X., Chang, H., Li, C., Qin, X., et al., 2023b. Combined effects of inflammation and coronavirus disease 2019 (COVID-19) on the risks of anxiety and depression: a cross-sectional study based on UK Biobank. J. Med. Virol. 95 (4), e28726.

Zhang, X., Zhou, J., Chen, Y., Guo, L., Yang, Z., et al., 2023c. Pathological networking of gray matter dendritic density with classic brain morphometries in OCD. JAMA Netw. Open 6 (11), e2343208.

Zhang, Z., Jackson, S.L., Gillespie, C., Merritt, R., Yang, Q., 2023d. Depressive symptoms and mortality among US adults. JAMA Netw. Open 6, e2337011.

Zhang, N., Yan, Z., Xin, H., Shao, S., Xue, S., et al., 2024a. Relationship among α-synuclein, aging and inflammation in Parkinson's disease. Exp. Ther. Med. 27, 23.

Zhang, R., Zhang, B., Shen, C., Sahakian, B.J., Li, Z., et al., 2024b. Associations of dietary patterns with brain health from behavioral, neuroimaging, biochemical and genetic analyses. Nat. Mental Health 2, 535–552.

Zhang, Y., Guo, Z., Liu, Y., Zhou, Y., Jing, L., 2024c. Is dancing an effective intervention for fat loss? A systematic review and meta-analysis of dance interventions on body composition. PLoS One 19, e0296089.

Zhao, L., 2019. CD33 in Alzheimer's disease - biology, pathogenesis, and therapeutics: a mini-review. Gerontology 65, 323–331.

Zhao, S.C., Ma, L.S., Chu, Z.H., Xu, H., Wu, W.Q., et al., 2017a. Regulation of microglial activation in stroke. Acta Pharmacol. Sin. 38, 445–458.

Zhao, W., Rasheed, A., Tikkanen, E., Lee, J.J., Butterworth, A.S., et al., 2017b. Identification of new susceptibility loci for type 2 diabetes and shared etiological pathways with coronary heart disease. Nat. Genet. 49, 1450–1457.

Zhao, Y., Jaber, V., Lukiw, W.J., 2017c. Secretory products of the human GI tract microbiome and their potential impact on Alzheimer's disease (AD): detection of lipopolysaccharide (LPS) in AD hippocampus. Front. Cell. Infect. Microbiol. 7, 318.

Zhao, M., Yang, J., Wang, W., Ma, J., Zhang, J., et al., 2017d. Meta-analysis of the interaction between serotonin transporter promoter variant, stress, and posttraumatic stress disorder. Sci. Rep. 7, 16532.

Zhao, M., Chen, L., Yang, J., Han, D., Fang, D., et al., 2018. BDNF Val66Met polymorphism, life stress and depression: a meta-analysis of gene-environment interaction. J. Affect. Disord. 227, 226–235.

Zhao, J., Bi, W., Xiao, S., Lan, X., Cheng, X., et al., 2019a. Neuroinflammation induced by lipopolysaccharide causes cognitive impairment in mice. Sci. Rep. 9 (1), 5790.

Zhao, X., Li, Y., Tian, Q., Zhu, B., Zhao, Z., 2019b. Repetitive transcranial magnetic stimulation increases serum brain-derived neurotrophic factor and decreases interleukin-1β and tumor necrosis factor-α in elderly patients with refractory depression. J. Int. Med. Res. 47, 1848–1855.

Zhao, R., Bu, W., Chen, Y., Chen, X., 2020a. The dose-response associations of sedentary time with chronic diseases and the risk for all-cause mortality affected by different health status: a systematic review and meta-analysis. J. Nutr. Health Aging 24, 63–70.

Zhao, Y., Keshiya, S., Perera, G., Schramko, L., Halliday, G.M., Dzamko, N., 2020b. LRRK2 kinase inhibitors reduce alpha-synuclein in human neuronal cell lines with the G2019S mutation. Neurobiol. Dis. 144, 105049.

Zhao, M., Wang, W., Jiang, Z., Zhu, Z., Liu, D., Pan, F., 2020c. Long-term effect of post-traumatic stress in adolescence on dendrite development and H3K9me2/BDNF expression in male rat hippocampus and prefrontal cortex. Front. Cell Dev. Biol. 8, 682.

Zhao, H., He, Z., Yun, H., Wang, R., Liu, C., 2022. A meta-analysis of the effects of different exercise modes on inflammatory response in the elderly. Int. J. Environ. Res. Public Health 19, 10451.

Zhao, J., Zeng, L., Liang, G., Dou, Y., Zhou, G., et al., 2023a. Higher systemic immune-inflammation index is associated with sarcopenia in individuals aged 18-59 years: a population-based study. Sci. Rep. 13, 22156.

Zhao, Y., Yang, L., Sahakian, B.J., Langley, C., Zhang, W., et al., 2023b. The brain structure, immunometabolic and genetic mechanisms underlying the association between lifestyle and depression. Nat. Mental Health 1, 736–750.

Zhao, J., Andreev, I., Silva, H.M., 2024a. Resident tissue macrophages: key coordinators of tissue homeostasis beyond immunity. Sci. Immunol. 9, eadd1967.

Zhao, L., Liu, J., Zhao, W., Chen, J., Fan, J., et al., 2024b. Morphological and genetic decoding shows heterogeneous patterns of brain aging in chronic musculoskeletal pain. Nat. Ment. Health 2, 435–449.

Zheng, Y., Manson, J.E., Yuan, C., Liang, M.H., Grodstein, F., et al., 2017. Associations of weight gain from early to middle adulthood with major health outcomes later in life. JAMA 318, 255–269.

Zheng, P., Zeng, B., Liu, M., Chen, J., Pan, J., et al., 2019a. The gut microbiome from patients with schizophrenia modulates the glutamate-glutamine-GABA cycle and schizophrenia-relevant behaviors in mice. Sci. Adv. 5, eaau8317.

Zheng, Y., Wang, T., Tu, X., Huang, Y., Zhang, H., Tan, D., et al., 2019b. Gut microbiome affects the response to anti-PD-1 immunotherapy in patients with hepatocellular carcinoma. J. Immunother. Cancer 7, 193.

Zhernakova, A., Kurilshikov, A., Bonder, M.J., Tigchelaar, E.F., Schirmer, M., et al., 2016. Population-based metagenomics analysis reveals markers for gut microbiome composition and diversity. Science 352, 565–569.

Zhou, R., Chen, F., Feng, X., Zhou, L., Li, Y., et al., 2015a. Perinatal exposure to low-dose of bisphenol A causes anxiety-like alteration in adrenal axis regulation and behaviors of rat offspring: a potential role for metabotropic glutamate 2/3 receptors. J. Psychiatr. Res. 64, 121–129.

Zhou, W., Lv, H., Li, M.X., Su, H., Huang, L.G., et al., 2015b. Protective effects of bifidobacteria on intestines in newborn rats with necrotizing enterocolitis and its regulation on TLR2 and TLR4. Genet. Mol. Res. 14, 11505–11514.

Zhou, J., Nagarkatti, P., Zhong, Y., Ginsberg, J.P., Singh, N.P., et al., 2016. Dysregulation in microRNA expression is associated with alterations in immune functions in combat veterans with post-traumatic stress disorder. PLoS One 9, e94075.

Zhou, C., Zhong, J., Zou, B., Fang, L., Chen, J., et al., 2017a. Meta-analyses of comparative efficacy of antidepressant medications on peripheral BDNF concentration in patients with depression. PLoS One 12, e0172270.

Zhou, T., Huang, Z., Sun, X., Zhu, X., Zhou, L., et al., 2017b. Microglia polarization with M1/M2 phenotype changes in rd1 mouse model of retinal degeneration. Front. Neuroanat. 11, 77.

Zhou, Y.Q., Liu, D.Q., Chen, S.P., Sun, J., Wang, X.M., et al., 2018. Minocycline as a promising therapeutic strategy for chronic pain. Pharmacol. Res. 134, 305–310.

Zhou, D.D., Zhou, X.X., Lv, Z., Chen, X.R., Wang, W., et al., 2019. Comparative efficacy and tolerability of antipsychotics as augmentations in adults with treatment-resistant obsessive-compulsive disorder: a network meta-analysis. J. Psychiatr. Res. 111, 51–58.

Zhou, S., Liu, S., Liu, X., Zhuang, W., 2020. Selective serotonin reuptake inhibitors for functional independence and depression prevention in early stage of post-stroke: a meta-analysis. Medicine (Baltimore) 99, e19062.

Zhou, J., Wang, J., Zhang, X., Tang, Q., 2021a. New insights into cancer chronotherapies. Front. Pharmacol. 12, 741295.

Zhou, W., Xie, Z., Li, C., Xing, Z., Xie, S., et al., 2021b. Driving effect of BDNF in the spinal dorsal horn on neuropathic pain. Neurosci. Lett. 756, 135965.

Zhou, X., Tian, B., Han, H.B., 2021c. Serum interleukin-6 in schizophrenia: a system review and meta-analysis. Cytokine 141, 155441.

Zhou, Z., Sun, B., Yu, D., Zhu, C., 2022. Gut microbiota: an important player in type 2 diabetes mellitus. Front. Cell. Infect. Microbiol. 12, 834485.

Zhou, A., Ancelin, M.L., Ritchie, K., Ryan, J., 2023a. Childhood adverse events and BDNF promoter methylation in later-life. Front. Psychiatry 14, 1108485.

Zhou, L., Qiu, W., Wang, J., Zhao, A., Zhou, C., et al., 2023b. Effects of vaginal microbiota transfer on the neurodevelopment and microbiome of cesarean-born infants: a blinded randomized controlled trial. Cell Host Microbe 31, 1232–1247.e5.

Zhou, M., Wu, J., Chang, H., Fang, Y., Zhang, D., Guo, Y., 2023c. Adenosine signaling mediate pain transmission in the central nervous system. Purinergic Signal 19 (1), 245–254.

Zhou, Z., Shuai, X., Lin, Z., Yu, X., Ba, X., et al., 2023d. Association between particulate matter (PM) 2·5 air pollution and clinical antibiotic resistance: a global analysis. Lancet Planet. Heath 7, e649–e659.

Zhu, F., Ju, Y., Wang, W., Wang, Q., Guo, R., et al., 2020. Metagenome-wide association of gut microbiome features for schizophrenia. Nat. Commun. 11, 1612.

Zhu, J., Jin, J., Tang, J., 2022a. Inflammatory pathophysiological mechanisms implicated in postpartum depression. Front. Pharmacol. 13, 955672.

Zhu, L., Wang, F., Huang, J., Wang, H., Wang, G., et al., 2022b. Inflammatory aging clock: a cancer clock to characterize the patients' subtypes and predict the overall survival in glioblastoma. Front. Genet. 13, 925469.

Zhu, B., Wei, X., Narasimhan, H., Qian, W., Zhang, R., et al., 2023a. Inhibition of the mitochondrial pyruvate carrier simultaneously mitigates hyperinflammation and hyperglycemia in COVID-19. Sci. Immunol. 8, eadf0348.

Zhu, X., Li, R., Zhu, Y., Zhou, J., Huang, J., et al., 2023b. Changes in inflammatory biomarkers in patients with schizophrenia: a 3-year retrospective study. Neuropsychiatr. Dis. Treat. 19, 1597–1604.

Zhu, X., Sakamoto, S., Ishii, C., Smith, M.D., Ito, K., et al., 2023c. Dectin-1 signaling on colonic γδ T cells promotes psychosocial stress responses. Nat. Immunol. 24, 625–636.

Zhu, Y., Guan, Y., Xiao, X., Jiao, B., Liao, X., et al., 2023d. Mendelian randomization analyses of smoking and Alzheimer's disease in Chinese and Japanese populations. Front. Aging Neurosci. 15, 1157051.

Zhuo, C., Tian, H., Song, X., Jiang, D., Chen, G., Cai, Z., Ping, J., Cheng, L., Zhou, C., Chen, C., 2023a. Microglia and cognitive impairment in schizophrenia: translating scientific progress into novel therapeutic interventions. Schizophrenia 9, 42.

Zhuo, Y., Li, X., He, Z., Lu, M., 2023b. Pathological mechanisms of neuroimmune response and multitarget disease-modifying therapies of mesenchymal stem cells in Parkinson's disease. Stem Cell Res Ther 14, 80.

Zia, F.Z., Baumann, M.H., Belouin, S.J., Dworkin, R.H., Ghauri, M.H., Hendricks, P.S., et al., 2023. Are psychedelic medicines the reset for chronic pain? Preliminary findings and research needs. Neuropharmacology 233, 109528.

Zięba, A., Stępnicki, P., Matosiuk, D., Kaczor, A.A., 2021. Overcoming depression with 5-HT2A receptor ligands. Int. J. Mol. Sci. 23, 10.

Ziegler, T.E., Crockford, C., 2017. Neuroendocrine control in social relationships in non-human primates: field based evidence. Horm. Behav. 91, 107–121.

Ziegler, C., Dannlowski, U., Bräuer, D., Stevens, S., Laeger, I., et al., 2015. Oxytocin receptor gene methylation: converging multilevel evidence for a role in social anxiety. Neuropsychopharmacology 40, 1528–1538.

Ziegler, C.G.K., Allon, S.J., Nyquist, S.K., Mbano, I.M., Miao, V.N., et al., 2020. SARS-CoV-2 receptor ACE2 is an interferon-stimulated gene in

human airway epithelial cells and is detected in specific cell subsets across tissues. Cell 181 (5), 1016–1035.e19.

Ziemichod, W., Kotlinska, J., Gibula-Tarlowska, E., Karkoszka, N., Kedzierska, E., 2022. Cebranopadol as a novel promising agent for the treatment of pain. Molecules 27, 3987.

Zimmerman, A.W., Jyonouchi, H., Comi, A.M., Connors, S.L., Milstien, S., et al., 2005. Cerebrospinal fluid and serum markers of inflammation in autism. Pediatr. Neurol. 33, 195–201.

Zimmermann, M., Zimmermann-Kogadeeva, M., Wegmann, R., Goodman, A.L., 2019. Mapping human microbiome drug metabolism by gut bacteria and their genes. Nature 570, 462–467.

Zimmet, P., Shi, Z., El-Osta, A., Ji, L., 2018. Epidemic T2DM, early development and epigenetics: Implications of the Chinese Famine. Nat. Rev. Endocrinol. 14, 738–746.

Zimprich, A., Biskup, S., Leitner, P., Lichtner, P., Farrer, M., et al., 2004. Mutations in LRRK2 cause autosomal-dominant parkinsonism with pleomorphic pathology. Neuron 44, 601–607.

Zincir, S., Öztürk, P., Bilgen, A.E., İzci, F., Yükselir, C., 2016. Levels of serum immunomodulators and alterations with electroconvulsive therapy in treatment-resistant major depression. Neuropsychiatr. Dis. Treat. 12, 1389–1396.

Zipursky, R.B., Odejayi, G., Agid, O., Remington, G., 2020. You say "schizophrenia" and I say "psychosis": just tell me when I can come off this medication. Schizophr. Res. 225, 39–46.

Zitvogel, L., Daillere, R., Roberti, M.P., Routy, B., Kroemer, G., 2017. Anticancer effects of the microbiome and its products. Nat. Rev. Microbiol. 15, 465–478.

Zitzmann, M., 2020. Testosterone, mood, behaviour and quality of life. Andrology 8, 1598–1605.

Zivadinov, R., Jakimovski, D., Gandhi, S., Ahmed, R., Dwyer, M.G., et al., 2016. Clinical relevance of brain atrophy assessment in multiple sclerosis. Implications for its use in a clinical routine. Expert. Rev. Neurother. 16, 777–793.

Zmora, N., Zilberman-Schapira, G., Suez, J., Mor, U., Dori-Bachash, M., et al., 2018. Personalized gut mucosal colonization resistance to empiric probiotics is associated with unique host and microbiome features. Cell 174, 1388–1405.

Zmora, N., Suez, J., Elinav, E., 2019. You are what you eat: diet, health and the gut microbiota. Nat. Rev. Gastroenterol. Hepatol. 16, 35–56.

Zobel, A.W., Nickel, T., Sonntag, A., Uhr, M., Holsboer, F., et al., 2001. Cortisol response in the combined dexamethasone/CRH test as predictor of relapse in patients with remitted depression: a prospective study. J. Psychiatr. Res. 35, 83–94.

Zoccola, P.M., Figueroa, W.S., Rabideau, E.M., Woody, A., Benencia, F., 2014. Differential effects of poststressor rumination and distraction on cortisol and C-reactive protein. Health Psychol. 33, 1606–1609.

Zohar, J., Fostick, L., Juven-Wetzler, A., Kaplan, Z., Shalev, H., et al., 2017. Secondary prevention of chronic PTSD by early and short-term administration of escitalopram: a prospective randomized, placebo-controlled, double-blind trial. J. Clin. Psychiatry. pii: 16m10730.

Zou, Z., Zhou, B., Huang, Y., Wang, J., Min, W., Li, T., 2020. Differences in cytokines between patients with generalised anxiety disorder and panic disorder. J. Psychosom. Res. 133, 109975.

Zou, J., Shang, W., Yang, L., Liu, T., Wang, L., et al., 2022. Microglia activation in the mPFC mediates anxiety-like behaviors caused by Staphylococcus aureus strain USA300. Brain Behav. 12, e2715.

Zou, B., Li, J., Ma, R.X., Cheng, X.Y., Ma, R.Y., et al., 2023a. Gut microbiota is an impact factor based on the brain-gut axis to alzheimer's disease: a systematic review. Aging Dis. 14, 964–1678.

Zou, Y., Guo, Q., Chang, Y., Zhong, Y., Cheng, L., Wei, W., 2023b. Effects of maternal high-fructose diet on long non-coding RNAs and anxiety-like behaviors in offspring. Int. J. Mol. Sci. 24, 4460.

Zovkic, I.B., Sweatt, J.D., 2013. Epigenetic mechanisms in learned fear: implications for PTSD. Neuropsychopharmacology 38, 77–93.

Zucker, R., Kovalerchik, M., Linial, M., 2023. Gene-based association study reveals a distinct female genetic signal in primary hypertension. Hum. Genet. 142, 863–878.

Zunhammer, M., Spisák, T., Wager, T.D., Bingel, U., 2021. Meta-analysis of neural systems underlying placebo analgesia from individual participant fMRI data. Nat. Commun. 12, 1391.

Zuo, C., Cao, H., Feng, F., Li, G., Huang, Y., et al., 2022. Repetitive transcranial magnetic stimulation exerts anti-inflammatory effects via modulating glial activation in mice with chronic unpredictable mild stress-induced depression. Int. Immunopharmacol. 109, 108788.

Index

Note: Page numbers followed by *f* indicate figures, *t* indicate tables, and *b* indicate boxes.

A

Acceptance and commitment therapy (ACT), 232–233, 341–343
Acquired/adaptive immunity, 29, 36–39, 39–40b, 165–167, 176, 185–186
  classes of, 36–37
  cognitive phase, 38–39
  diversity, 37–38
  effector phase, 38–39
  features of, 37–38
  memory, 38
  Parkinson's disease (PD), 406–408
  phases, 38–39
  self-limitation, 38
  self-/nonself-discrimination, 38
  specificity, 37
Acute stress disorder, 294
Adiposity, coronary artery disease (CAD), 454b
Adrenocorticotropic hormone (ACTH), 150t
Adverse childhood experiences (ACEs), 153, 207–208
Aerobic exercise, 117–119. *See also* Exercise
Afferent lymph, 36
Aging process
  Alzheimer's disease (AD), 423–424
  longevity, 449–450
  mental illness, 10–15, 10b
  Parkinson's disease (PD), 398–401
  posttraumatic stress disorder (PTSD), 315–316b
  psychological challenges, 121b
Agouti-related peptide (AGRP), 108t
Akkermansia muciniphila, 69–70
All-cause mortality, 101–102, 110–111, 120
Allodynia, 324–325, 331, 340
Allostatic overload, 141
Alpha-synuclein, 394–395, 402–403, 405–407, 409–411
Alzheimer's disease (AD), 103, 419–423, 419b
  aging process, 423–424
  β-amyloid plaques, 420f, 422–423, 423b, 432–433b
  antiinflammatory treatments, 436–437
  cluster of differentiation 33 (CD33), 429
  combined therapy approach for, 441b
  cytokines, 437–438
  default mode network (DMN), 422
  environmental factors, 430–433
  gastrointestinal problems, 425
  genetics of, 426–430, 435f

heavy metals of, 430
  hippocampal processes and, 421–422
  histopathological feature of, 422–423
  immunomodulatory treatments for, 436–441
  inflammatory mechanisms of, 433–436
  insulin growth factor-1 (IGF-1), 424
  lactic acid bacteria and bifidobacteria, 425
  lifestyle factors and, 423–426
  long-term potentiation (LTP), 421–422
  microglia role in, 433–436
  mild cognitive impairment (MCI), 420–421
  Mini-Mental State Examination, 424–425
  neurofibrillary tangles (NFTs), 422–423
  pathogens, 431–433
  peripheral inflammation, 433
  personalized treatment of, 435f
  positron emission tomography (PET), 426
  senescence-associated glycoprotein (SAGP), 440
  TREM2, 428–429, 428f, 435
  trophic factors, 437–438
  type 2 diabetes, 424
  vaccine therapy, 438–441
γ-Aminobutyric Acid, 145–146
β-Amyloid plaques, 420f, 422–423, 423b, 432–433b
Amyloid precursor protein (APP), 355, 422
Anaerobic exercise, 117–118. *See also* Exercise
Angiotensin-converting enzyme 2 (ACE2), 89–90
Animal studies, stressor and immunity, 166–179
  antibody production, 174–176
  assessment, 167
  B cell function, 174–176
  cell-mediated immune responses, 170–174, 172–173f
  characteristics, 167–169
  dendritic cells, 177–178
  effects, 169–170
  infection, 178–179
  mononuclear phagocytic cells, 176–177
Antibiotic resistance
  bacteria, 58–61, 60b, 71b
  in farm animals, 59
  "out-of-the-box" approaches, 61
Antibody
  animal studies, stressor and immunity, 174–176
  B lymphocytes and, 40–43

class switching, 44–45
  distribution of, 42–43
  function of, 44
  molecule, 43–45
  specificity, 43–44
  structure of, 43, 43f
Antibody-forming cell (AFC), 42
Anticipatory stressors, 134
Antidepressants
  exercise as, 260
  for pain, 339–340
  selective serotonin reuptake inhibitors (SSRIs)
    effectiveness, 238–239, 239b
    fast acting, 241–242
    during pregnancy, 239b
    treatments, 242b
Antigen, 28–29
Antiinflammatory treatments
  Alzheimer's disease (AD), 436–437
  depression, 259–260
  Parkinson's disease (PD), 415–416
  posttraumatic stress disorder (PTSD), 320
Antimicrobial resistance, 58–59
Anti-NMDA receptor antibodies, 386
Antioxidants
  broccoli, 112–113b
  foods, 112
Anxiety/anxiety disorders, 11–12, 72, 269b, 274f
  bidirectional communication mechanisms, 286, 288f
  cannabinoids, 273–275
  Caspase-1 antagonist, 288–289
  corticotrophin-releasing hormone (CRH), 271–272
  and depression, 128, 164–165
  fear and, 269–270
  gamma-aminobutyric acid (GABA), 272–273
  generalized, 275–276
    benzodiazepines, 275
    inflammatory processes, 283–284
    treatments, 275–276
  germ-free mice, 286–290
  glutamate, 272–273
  inflammatory processes, 283–286
    in animal, 280–282
    with chronic illness, 282–283
    genetic correlates, 283
    in human, 282–283
    immune and, 283–286

580 Index

Anxiety/anxiety disorders (Continued)
  microbiota, 286–290
  neurobiology of, 269–275
  neuroinflammation, 286, 287f
  norepinephrine, 272
  obsessive-compulsive disorder (OCD)
    (see Obsessive-compulsive disorder
    (OCD))
  panic disorder, 276–277, 284
  phobias, 279
  posttraumatic stress disorder (PTSD),
    311–312
  serotonin, 272
  social anxiety, 279–280, 285–286
  stress-induced periphery inflammation,
    286, 287f
  types, 270t
Apolipoprotein (APOE), 326b
Appraisals
  early life adverse events, 209–210
  and misappraisals, 134
Arginine vasopressin (AVP), 150t, 243
Artificial sweeteners, 102–103
Assorted stressors, 142. See also Stress/stressor
Asthma, vulnerability to, 218–220
Atypical depression, 230t, 231.
    See also Depression
Autism spectrum disorder (ASD), 221, 345b,
    363–364b
  autoimmune disease and, 351–353
  behavioral deficits, 347–349
  brain development and, 347–349
  cytokines and, 353–357
  effector immune function and, 357–359
  epidemic, 346
  etiology of, 346, 349–364
  ex vivo measures, 356–357
  features of, 345–347
  food allergies, 362–363
  gastrointestinal immunity and, 359–361
  gut-bacterial dependent metabolic
    pathways, 359, 360f
  human leukocyte antigen (HLA) system,
    350–351
  immune dysregulation in, 353
  immunogenetics, 350–351
  kynurenine pathway, 353–354
  major histocompatibility complex (MHC),
    350–351
  maternal immune activation (MIA),
    350–351, 356
  microbiota, 361–363
  motor behaviors, 348
  neural eccentricity, 348
  perinatal immune activation, 363–364
  prevalence of, 345–346
  regressive, 360
  social behavior, 348
  Swedish Hospital Discharge Register, 352–353
Autoimmune disease, 54, 191–193, 444–448
  autism spectrum disorder (ASD), 351–353
    development, 351
    Swedish Hospital Discharge Register,
      352–353
  multiple sclerosis (MS), 193–194
  Parkinson's disease (PD) (see Parkinson's
    disease (PD))

rheumatoid arthritis (RA), 195
schizophrenia, 383–386
systemic lupus erythematosus (SLE), 194
tolerance and, 54–55
Autonomic nervous system (ANS), 10–11,
  164–165

B

Bacteria
  affect hormonal and CNS processes, 66–79
    gut bacteria and obesity, 66–70, 68f
    gut microbiota dysbiosis, 70–71, 74–75
    microbiota, 66, 67f, 69–70b
    well-being, implications for, 70–71
  antibiotic resistance, 58–61, 60b, 71b
  antibiotics, 58
  challenges, 57–61
Bacterial infections
  central nervous system (CNS), 83–84
  postpartum, 57b
Bacterial meningitis, 83
Basal ganglia, 393–394
Basic fibroblast growth factor (bFGF),
  14t, 148t
B-cells
  activation, 41–42
  animal studies, stressor and immunity,
    174–176
  cognate T cell with, 41–42
  cross-linked activation, 41
  development of, 40–41
  effector, response, 42
  induction, 41
  memory, 42
Bed nucleus of the stria terminalis (BNST),
  269–270
Behavioral activation, 232–233
Behavioral deficits, autism spectrum disorder
  (ASD), 347–349
Behaviors, 72
  autism spectrum disorder (ASD) and,
    346–347
  blaming victim, 100–101b
  in decision-making situations, 134
  risky, 99–101
  sickness, 74–75b
Bermuda triangle, 403, 404f
Beta-endorphin, 148t
Beta2-microglobulin, 47
Bifidobacterium species, 115, 264
Bipolar disorder, 267b
Bisphenol A (BPA), 221b
Bleuler, Eugene, 345b
Blood–brain barrier (BBB), 14, 164
B lymphocytes, 31–32, 40–43
Bombesin, 108t
Brain-derived neurotrophic factor (BDNF),
  14t, 148t
  depression, 245–246, 246f
  posttraumatic stress disorder (PTSD), 308
  schizophrenia, 374, 379
Brain development, and autism spectrum
  disorder (ASD), 347–349
Brain microvascular endothelial cells
  (BMVECs), 86
Breastfeeding, 212–213b
Bush, H.W., 112–113b

C

Caenorhabditis elegans, 69–70b
Cancer therapy, 75–78
Candida auris, 60b
Cannabinoids, 155
  anxiety disorders, 273–275
  pain, 328–329
  posttraumatic stress disorder (PTSD), 320
Cardinal motor symptoms, 393
Cardiovascular disease (CVD), 451–461
Caregiving stress, 180
Catechol-O-methyltransferase (COMT), 297
CC-chemokine ligand 2 (CCL2), 171
Cell-mediated immune responses, 170–174
Cell trafficking/migration, 34–35
Cellular immunity, 36–37
Central nervous system (CNS), 165–166
  bacterial infections, 83–84
  direct infection, 88–89
  glial cells and infection, 87–88
  mental illness, 2, 11–12
  neuropsychiatric illnesses, 82, 94–95
  parasitic infection, 84–85, 85b
  pathogens and functioning, 82–85
  proinflammatory cytokines, 164
  viral-induced cell fusion, 87
  viral infections, 82–83, 85–89
Chemokines, 53–54
Chlamydophila pneumoniae, 431–432
Chloramphenicol acetyltransferase (CAT),
  59–60
Chronic fatigue syndrome (CFS), 184, 336–337b
Chronic obstructive pulmonary disease
  (COPD), 450
Chronic pain, 323
  acceptance and commitment therapy (ACT),
    341–343
  behavioral and cognitive approaches,
    341–343
  cognitive behavior therapy (CBT), 341–343
  complementary and alternative medicine
    (CAM), 343–344
  contextual factors, 325
  depression, 325
  gate control theory, 325–326
  neurological pain signature, 325
  posttraumatic stress disorder (PTSD), 325
  psychological factors, 325
  risk factors, 324f
  sensory component of, 325
Chronic stressor, 141, 142f
  challenges, 144–145, 168, 179–180
  coronary artery disease (CAD), 455
  discrimination as, 153–154b
  impact of, 151, 181
Circadian cycles, 122–124, 126–127b
Climate change, affects physical/mental
  health, 127–128, 129f
Clone, 37
Clostridium difficile, 60b, 71b
Clostridium histolyticum, 360
Cluster of differentiation 33 (CD33), 429
Cognitive behavioral therapy (CBT)
  depression, 232–233
  pain, 341–343
  posttraumatic stress disorder (PTSD), 317
Cognitive model, depression, 231

cognitive behavioral therapy (CBT), 232–233
framing hopelessness, 232
"hopelessness" hypothesis, 231–232
Cohen, Leonard, 81b
Collective trauma, 225–226, 225–226b
Colony-stimulating factors (CSF), 148t
Combination antiretroviral therapy (CART), 83
Communicable diseases, 81b
Comorbidity illness, 396, 443–444, 443b, 444f
  autoimmune disorders, 444–448
  diabetes, 463–464
  drug repurposing, 448–449
  heart disease, 451–461
    adiposity, 454b
    cytokines, 454b
    danger-associated molecular patterns
      (DAMPs), 455–456, 457f
    depression, 456–458
    elevated CRP, 456
    genetic factors in, 454–455
    gum disease, 460b
    inflammatory processes in, 455–456, 456f
    loneliness, 458b
    microbiota, 459–460
    personality factors, 458–459
    prevention, 460–461
    risk factors for, 454–459
    sex differences in, 459
    sleep disturbances, 459
    stressor effects, 455
  hypertension, 451–453, 452f
  inflammatory factors in, 444–451
  Parkinson's disease (PD), 395–396
  posttraumatic stress disorder (PTSD),
    321–322
  predicting and increasing longevity, 449–451
  psychiatric disorders, 451–461
  stroke, 461–467
    depression and, 461–462
    diabetes and, 462–463
    synaptic plasticity, advantage, 462b
Complementary and alternative medicine
  (CAM), 116, 343–344
Complex regional pain syndrome (CRPS), 324
Coping methods
  early life adverse events, 209–210
  posttraumatic stress disorder (PTSD),
    295, 308
  with stressor, 136–138, 137t
Coronary artery disease (CAD), 453–454
  adiposity, 454b
  cardiomyopathy, 455–456, 457f
  cytokines, 454b
  danger-associated molecular patterns
    (DAMPs), 455–456, 457f
  depression, 456–458
  elevated CRP, 456
  flavin-containing monooxygenase 3
    (FMO3), 459
  genetic factors in, 454–455
  gum disease, 460b
  inflammatory processes in, 455–456, 456f
  JUPITER trials, 460
  loneliness, 458b
  microbiota, 459–460
  omega-3 fatty acid supplements, 460
  personality factors, 458–459

prevention, 460–461
  risk factors for, 454–459
  sex differences in, 459
  sleep disturbances, 459
  stressor effects, 455
  trimethylamine N-oxide (TMAO), 459–460
Corticoid responses, stressor, 149–152
Corticotropin-releasing hormone (CRH),
    145, 150t
  anxiety disorders, 271–272
  biological effect, 159t
  depression, 242–243
  energy balances, 154–155
  posttraumatic stress disorder (PTSD), 304–305
  stress, 154–155, 160
Cortisol, 148t
  biological effect, 159t
  posttraumatic stress disorder (PTSD),
    152–153, 305–306, 314, 321
  stress and energy balances, 160
COVID-19, 81b, 89–95, 100
  brain and, 91b, 94–95
  long, 96–97
  neuroimmunity
    ACE2 and CD209 in, 92–94, 93f
    olfactory and vascular/lymphatic
      crossroads, 95
  posttraumatic stress disorder (PTSD), 295b,
    316, 320
  stress and, 183–185
  systemic clinical manifestation, 90, 90f
  vaccine hesitancy, 92b
C-reactive protein (CRP), 111, 153, 374–376
Crohn's disease, 447–448b
Culture, and personalized medicine, 24b
Cytokines, 50–54, 262b
  Alzheimer's disease (AD), 437–438
  and autism, 353–357
    brain measures, 353–354
    ex vivo measures of immune function,
      356–357
    immune dysregulation in, 358f
    periphery measures, 355–356
  chemokines, 53–54
  coronary artery disease (CAD), 454b
  network, 50–52, 51f
  in neural-immune investigations, 52–53t
  in Parkinson's disease (PD), 411, 416–418
  schizophrenia, 374–376
  storm, 90
  stress and, 185–190
  trophic (see Trophic cytokines)
Cytomegalovirus (CMV), 178–179
Cytotoxic T-lymphocyte-associated antigen-4
  (CTLA-4), 77

D

Damage-associated molecular pattern
  (DAMP), 74
    coronary artery disease (CAD), 455–456, 457f
    Parkinson's disease (PD), 408–409
    schizophrenia, 373
    sterile inflammation, 34
Decision-making process, 134
Deep brain stimulation (DBS), 340b
Default mode network (DMN), 422
Dehydroepiandrosterone (DHEA), 157t

Delayed hypersensitivity (DTH), 170–171
Dementia, 423–424. See also Alzheimer's
  disease (AD)
Dendritic cells (DCs), stressor effects on,
  177–178
Depression, 5, 10, 13, 19, 23, 229–231
  antiinflammatory treatments, 259–260
  and anxiety, 128, 164–165
  arginine vasopressin (AVP), 243
  Bifidobacterium longum, 264
  brain-derived neurotrophic factor (BDNF),
    245–246, 246f
  burden of, 229b
  chronic pain, 325
  cognitive model, 231
    cognitive behavioral therapy (CBT),
      232–233
    framing hopelessness, 232
    "hopelessness" hypothesis, 231–232
  coronary artery disease (CAD), 456–458
  corticotropin-releasing hormone (CRH),
    242–243
  dopamine in, 235
  epigenetic processes and, 247–249
  gamma-aminobutyric acid (GABA),
    239–244, 240b
  genetic links to, 236–238
  glial cell line-derived neurotrophic factor
    (GDNF), 247
  glutamate process, 240
  gut bacteria and immune variations,
    264–267, 265f
  and heart disease, 443, 445f
  immunotherapy, 260–262
  inflammation, 249–263
    case–control studies, 250–251
    imaging and postmortem analyses,
      257–258
    infection and, 256–257
    Mendelian randomization analysis, 251
    recurrence of, 253
    rodents, 249–250
    stress, 255–256, 261f
    and suicidality, 256b
  interferon-α (IFN-α), 260–263
  learned helplessness, 233
  microbiome-immune interactions, 263–267
  microglia, 254–255b
  mindfulness, 233b
  monoamine processes, 234
  nerve growth factor (NGF), 247
  neurochemical effects, 233–239, 253–255
  neuronal stimulation to attenuate, 249b
  neuropeptide Y (NPY), 244
  neurotrophins, 245–249
  norepinephrine in, 234–235
  obesity, 250b
  oxytocin, 243–244
  in Parkinson's disease (PD), 396–397
  perinatal, 253b
  during pregnancy, 200–201
  prenatal stress, 200–201
  repetitive transcranial magnetic stimulation
    (rTMS), 246–247
  serotonin in, 235–236
  single nucleotide polymorphism (SNP),
    242–243

Depression (*Continued*)
and stroke, 461–462
subtypes of, 230*t*
symptoms, 136–138
theoretical perspectives, 231–233
traumatic brain injury (TBI), 258*b*
treatments affect, 258–259
typical *vs.* atypical, 231
Determinants of health, dietary factors, 101–117
Developmental disorders, 8
Diabetes. *See also Specific diabetes*
gut bacteria and, 466–467
and stroke, 462–463
Dietary factors, 101–117
Dietary fructose, 103–104
Diets
adaptations, 101–103
fasting, 107–108*b*
and immunity, 112–115
inflammatory index, 76
prenatal stress, 201*b*
Direct central nervous system (CNS) infection, 88–89
Discrimination, as chronic stressor, 153–154*b*
Disturbed sleep, 123, 123–124*b*
Diurnal variations, sleep, 127
Dizygotic twins, 6
Dopamine
in depression, 235
Parkinson's disease (PD), 401*b*
schizophrenia, 368–369, 368–369*b*, 371*f*
stressor, 143–144
Drug repurposing, 448–449
Dysbiosis, 70–71, 73–75
Dysthymia, 230*t*

E
Early development, 197–198
Early life adverse events, 207–215, 208*f*
adolescence and early adulthood, 214–215*b*
allergies, 218–220
brain neurochemical variations, 214–215
developmental conditions, 218–220
experiences, 110
growth factors and, 213–214
gut and immunity, 217–218
hypothalamic–pituitary–adrenal (HPA) activation, 166*b*
intergenerational and transgenerational effects, 222
microbial factors, 215–220
poor appraisal and coping methods, 209–210
poverty during, 210
stressors, 210–213
toxic challenges, 208
vulnerability to asthma, 218–220
Eating process
hormonal regulation of, 106–109, 108*t*
stress and, 109–110
Eco-anxiety, 128
Effector function, 28–29
autism spectrum disorder (ASD), 357–359
B cells response, 42
T-cell activation, 49–50
Eicosapentaenoic acid (EPA), 424
Electroconvulsive therapy (ECT), 237, 241

Emotion
affects pain, 332–333
coping strategies, 136, 137*t*
Endogenous retroviruses, 383
Endophenotypic approach, 19–24, 20*f*
Endotoxin, 34
Energy balances
hormones related to, 159*t*
stress and, 159–160
corticotropin-releasing hormone (CRH), 160
cortisol, 160
leptin and ghrelin, 159–160
English Longitudinal Study of Aging (ELSA), 111
Enteric nervous system, 16–19
Environmental factors
Alzheimer's disease (AD), 430–433
Parkinson's disease (PD), 403–412
Epidermal growth factor (EGF), 148*t*
Epigenetic process
collective/historical trauma, 225–226, 225–226*b*
and depression, 247–249
environmental toxicants, 221–222
and intergenerational actions, 220–226
mental illness, 8–10, 9*f*
posttraumatic stress disorder (PTSD), 314–317
aging, 315–316*b*
glucocorticoids, 314
inflammation, 314–316
microbiota, 316–317
prenatal and early postnatal factors, 221–222, 223*f*
schizophrenia, 373
stressor-related effects, 220
Epigenome, 3*t*
Epinephrine, 148*t*
Epstein–Barr Virus (EBV), 178
Erythropoietin (EPO), 148*t*, 260, 417–418
ESKAPE organisms, 60
Estrogens, 157*t*
estrogen receptor alpha (ERα), 220
stressor, 155–156
Exercise, 117–122, 119*f*, 121*b*
as antidepressant, 260
benefits of, 117, 119
cognitive and affective benefits, 119–120
dance as, 118*b*
diet and, 118
forms of, 117–118
and microbiota, 121–122
and positive health, 120
Experimental autoimmune encephalitis (EAE), 73–74
Exposure therapy, 317–318
Eye movement desensitization and reprocessing (EMDR), 318

F
Fasting diets, 107–108*b*
Fecal transplants, *Clostridium difficile* through, 71*b*
Fibroblast growth factor (FGF)-21, 103–104
Flavin-containing monooxygenase 3 (FMO3), 459

Fleming, Alexander, 58
Follicle-stimulating hormone (FSH), 157*t*
Food allergies, 362–363
Freudian psychoanalytic thinking, 365

G
Gamma-aminobutyric acid (GABA), 239–244
anxiety disorders, 272–273
depression, 239–244, 240*b*
posttraumatic stress disorder (PTSD), 306–307
schizophrenia, 369–370
Gastrin-releasing peptide (GRP), 159*t*
Gastrointestinal problems
Alzheimer's disease (AD), 425
autism spectrum disorder (ASD), 359–361
Gate control theory, 325–326
Gene polymorphisms, 7–8
Generalized anxiety disorders (GAD), 275–276
benzodiazepines, 275
inflammatory processes, 283–284
microbiota, 289
treatment of, 275–276
Genetics
of Alzheimer's disease (AD), 426–430, 435*f*
constituency, 4
coronary artery disease (CAD), 454–455
mental illness, 5–10
Parkinson's disease (PD), 412–414
posttraumatic stress disorder (PTSD), 312–314
schizophrenia, 367–368, 370
type 2 diabetes, 464–465, 464*f*
Gene transcription, 197
Genome, 3*t*
Genome-wide association studies (GWAS)
of Alzheimer's disease (AD), 429
autoimmune disorders, 444
coronary artery disease (CAD), 454–455
posttraumatic stress disorder (PTSD), 297, 305, 312–313
schizophrenia, 367–368
twin research, 6
type 2 diabetes, 464–465
Germ theory, 57*b*
Gestational diabetes, 462–463
Ghrelin, 106–108, 106*f*
behavioral outcome, 108*t*
biological effect, 108*t*, 159*t*
change in eating and energy regulation, 106–108, 106*f*
orexin cells, 106
stress and energy balances, 159–160
Glial cell line-derived neurotrophic factor (GDNF), 148*t*, 247
Glial cells
central nervous system (CNS), 87–88
cerebrospinal fluid (CSF), 12–13
mental illness, 12–13
Glial fibrillary acidic protein (GFAP), 353–354, 430
Glucagon-like peptide-1 (GLP-1), 105, 115–116
behavioral outcome, 108*t*
biological effect, 108*t*, 159*t*

Glucocorticoids
  posttraumatic stress disorder (PTSD),
      305–306
    epigenetic changes, 314
    manipulations, 320–321
    prenatal stress, 199, 200*f*
Glutamate
  anxiety disorders, 272–273
  depression, 240
  hypofunction hypothesis, schizophrenia,
      384–385
  posttraumatic stress disorder (PTSD), 307,
      319–320
  schizophrenia, 369–370, 371*f*, 384–386
  stressor, 146–147
G-protein pathway suppressor 2 (GPS2), 465
Granulocyte-macrophage colony-stimulating
      factor (GM-CSF), 354, 416–417
Greedy genes, 198*b*
Greenland Inuit, 101
Growth hormone, 159*t*
Gum disease, 460*b*
Gut bacteria/microbiota, 66, 67*f*, 69–70*b*, 72–78
  affect hormonal and CNS processes, 66, 67*f*,
      69–70*b*
  anxiety disorders through, 286–290
  autism spectrum disorder (ASD), 361–363
  and brain, 66, 67*f*
  coronary artery disease (CAD), 459–460
  diet, and cancer, 75–78
  diurnal variations, 127
  dysbiosis, 70–71, 73–75
  and genes, 71
  for health benefits, 78–79
  and health outcomes, 79
  and host adipose tissue, 104*f*
  and immune variations, 264–267, 265*f*
  mental illness, 16–19, 17–18*f*
  and obesity, 66–70, 68*f*
  and Parkinson's disease (PD), 405, 405*b*
  and peptides, 68*f*
  physical illness, 73–78
  posttraumatic stress disorder (PTSD),
      316–317
  psychological functioning associated with,
      72
  and schizophrenia, 387–389, 390*f*
  stressor and, 161
  stroke, 462
  and well-being, implications for, 70–71

H
Head injury, posttraumatic stress disorder
      (PTSD), 311–312
Heart disease, 451–461
  adiposity, 454*b*
  cytokines, 454*b*
  danger-associated molecular patterns
      (DAMPs), 455–456, 457*f*
  depression, 456–458
  elevated CRP, 456
  genetic factors in, 454–455
  gum disease, 460*b*
  inflammatory processes in, 455–456, 456*f*
  loneliness, 458*b*
  microbiota, 459–460
  personality factors, 458–459

prevention, 460–461
  risk factors for, 454–459
  sex differences in, 459
  sleep disturbances, 459
  stressor effects, 455
Helminths, 85
Hematopoietic stem cell (HSC), 29, 30*f*, 31–32
Herpes simplex virus type 1/2 (HSV-1/2), 86
  acquired immune response, 37
  schizophrenia, 379
Heuristics, role, 134
High-definition medicine approach, 24
High-fructose corn syrup, 103–104
"Hijack" cognitive control circuits, 100–101*b*
Hippocampal processes, Alzheimer's disease
      (AD), 421–422
HIV-associated neurocognitive disorder
      (HAND), 83
H1N1 Flu pandemic, 65
Hopelessness, depression
  framing, 232
  hypothesis, 231–232
Hormonal changes
  early life stressors and, 210–213
  of eating processes, 106–109, 108*t*
  mental illness, 15
Hostile environments, mutations in, 7*b*
Human leukocyte antigen (HLA)
      system, 350–351
Humoral immunity, 174–176
6-Hydroxydopamine (6-OHDA), 401–403
Hygiene hypothesis, 219–220*b*
Hypericum perforatum (St John's wort), 116
Hypertension, 451–453, 452*f*, 461
Hypothalamic–pituitary–adrenal (HPA)
      activation, 166*b*, 211–212*b*

I
Identical twins, 6
Ileocolonic lymphoid nodular hyperplasia
      (LNH), 361
Illness, 123–124*b*
  characteristics, 194
  obesity and, 110–112
  risky, 99–101
  in women, 138
Immunity, 27*b*, 28–29
  acquired/adaptive, 36–39
  autism spectrum disorder (ASD), 353,
      363–364*b*
  B lymphocytes and antibody, 40–43
  cells and organs of, 32–34
  cell senescence, 91
  cells formation, 29, 30*f*
  diurnal variations, 127
  exercise and, 118–119, 119*f*, 121*b*
  hematopoietic stem cell (HSC), 29, 30*f*, 32
  infectious microorganisms, 32
  information-processing and mobile
      surveillance, 28–29
  innate, 32–33
  lymph nodes, 35*f*, 36
  lymphoid system, 34–36, 35*f*
  mental illness, 15–16
  and microbial factors, 215–220
  microbiota and, 72–78
  sterile inflammation, 33–34

stress and (*see* Stress and immunity)
  types of, 31–32
  white blood cells, 29–31
Immunoglobulin, 43–45
  class switching, 44–45
  specificity, 43–44
  structure of, 43, 43*f*
Immunomodulatory treatments
  for Alzheimer's disease (AD), 436–441
  parkinson's disease (PD), 414–418
Immunotherapy
  Alzheimer's disease (AD), 436–441
  depression, 260–262
  Parkinson's disease (PD), 414–418
Indoleamine-2,3-dioxygenase (IDO), 262–263
Infections. *See also Specific infections*
  animal studies, stressor and immunity,
      178–179
  Parkinson's disease (PD), 404–405
  postnatal, 381–383
  puerperal, 57*b*
  respiratory, 82
  stress and, 178–179
Inflammation, 2, 13–15, 24–25, 83, 96, 105, 107,
      112–113, 118–119, 122, 247, 255, 267
  of Alzheimer's disease (AD), 433–436
  anxiety disorders, 283–286
    in animal, 280–282
    with chronic illness, 282–283
    genetic correlates, 283
    in human, 282–283
    immune and, 283–286
  chronic, 97
  in comorbidity illness, 444–451
  coronary artery disease (CAD), 455–456, 456*f*
  cytokines, 312, 313*f*
  depression, 249–263
    case–control studies, 250–251
    imaging and postmortem analyses,
        257–258
    infection and, 256–257
    Mendelian randomization analysis, 251
    recurrence of, 253
    rodents, 249–250
    stress, 255–256, 261*f*
    and suicidality, 256*b*
  pain, 329–332, 340
  Parkinson's disease (PD), 406
  posttraumatic stress disorder (PTSD),
      308–314
  schizophrenia, 384–386
  type 2 diabetes, 465–466
Inflammatory bowel disease (IBD), 192–193,
      394*f*, 447–448*b*
Influenza viruses, 63, 81*b*, 82–83
Information-processing system, 28–29
Innate immunity, 32–33, 165–167, 173, 176
Insulin
  behavioral outcome, 108*t*
  biological effect, 108*t*, 159*t*
Insulin-like growth factor 1/2 (IGF-1/2), 14*t*,
      148*t*, 198*b*, 424
Interfering RNAs (RNAi), 63–64
Interferons (IFNs)
  interferon-α, 260–263
  interferon-γ, 48, 186
  Parkinson's disease (PD), 411–412

Intergenerational actions, epigenetics and, 220–226
Interleukin (IL), 50–52
  autism spectrum disorder (ASD), 353–357
  IL-1, 189–190
  IL-2, 186–187
  IL-4, 187
  IL-5, 188
  IL-6, 188–189
  IL-10, 187–188
  Parkinson's disease (PD), 398
  schizophrenia, 374–376
Intermittent fasting, 107–108b
Intracellular processes, pain, 329–330
Ischemic stroke, 461
Isotype switching, 43–44

J
Japanese encephalitis virus (JEV), 83

K
Ketamine, 241–242
Keyhole limpet hemocyanin (KLH), 170–171
Killer-cell immunoglobulin-like receptors (KIRs), 350–351
Kynurenic acid (KYNA), 385–386
Kynurenine, 260, 262–263

L
Lack of sleep, 123–124, 123–124b
Lactobacillus reuteri, 287–288, 362
Lancet Countdown, 128
Langerhans cells (LCs), 177
Large granular lymphocytes (LGLs), 179–180
Leaky gut hypothesis, 359
Learned helplessness, 135, 233
Leptin
  behavioral outcome, 108t
  biological effect, 108t, 159t
  change in eating and energy regulation, 106–108, 106f
  stress and energy balances, 159–160
Leukocytes, 29–32
Lifestyle factors
  Alzheimer's disease (AD), 423–426
  Parkinson's disease (PD), 398–401
Lipopolysaccharide (LPS), 34, 83, 87–88
Listeria monocytogenes, 176
Loneliness, coronary artery disease (CAD), 458b
Longevity, 449–451
Long-term potentiation (LTP), 421–422
Lorcaserin (Belviq), 115
Low socioeconomic status, 197b
LRRK2, Parkinson's disease (PD), 413
  CNS-immune system interaction, 413–414
  inhibition and recovery, 416
Luteinizing hormone (LH), 157t
Lymphatic vessel, afferent *vs.* efferent, 36
Lymph nodes, immunity, 35f, 36
Lymphocytes
  B lymphocytes, 31–32, 40–43
  T lymphocytes, 31–32, 45–50
  types of, 36
Lymphoid system, 34–36, 35f
Lymphokines, 50–52

M
Macrophages, 48, 176–177
Major histocompatibility complex (MHC), 46–48, 350–351
Malaria, 85
Mammalian target of rapamycin (mTOR) complex, 241
Maternal immune activation (MIA), 350–351, 356, 380, 380f
Maternal obesity, 111
Matrix metalloproteinase-8 (MMP-8), 164
Measles, mumps, and rubella (MMR), 99–100
Mechanistic target of rapamycin (mTOR), 107–108b
Mediterranean diets, 76
Melancholic depression, 230t, 250–251
Melanocyte-stimulating hormone (MSH), 108t, 159t
Memory
  acquired immune response, 38
  B cells, 42
  in innate immunity, 33
  posttraumatic stress disorder (PTSD), 299–300
Mendelian randomization analysis, 251
Mental disorders, 81b
  age of onset of, 229–231, 230f
  central nervous system (CNS)
    bacterial infections, 83–84
    direct infection, 88–89
    glial cells and infection, 87–88
    neuropsychiatric illnesses, 82, 94–95
    parasitic infection, 84–85, 85b
    pathogens and functioning, 82–85
    viral-induced cell fusion, 87
    viral infections, 82–83, 85–89
  COVID-19, 81b, 89–95
    affect brain, 91b, 94–95
    long, 96–97
    neuroimmunity, 92–95, 93f
    systemic clinical manifestation, 90, 90f
    vaccine hesitancy, 92b
Mental illness
  autonomic nervous system (ANS), 10–11
  biological process, 10–15, 10b
  brain and, 13–14
  central nervous system (CNS), 11–12
  culture and personalized medicine, 24b
  enteric nervous system, 16–19
  epigenetics, 8–10, 9f
  evaluating biological substrates of, 5
  gene polymorphisms, 7–8
  genetics, 5–10
  glial cells, 12–13
  gut microbiota and health, 16–19, 18f
  hormones, 15
  immune processes, 15–16
  impact of, 1b
  microbiota, 16–19, 17f
  molecular genetic approaches, 6–7
  multisystem coordination, 3–5
  mutations in hostile environments, 7b
  neuroimmune links, 15–16
  neurotransmission, 11–12
  neurotrophins, 14, 14t
  personalized treatment strategies, 19–24
    endophenotypic analyses, 19–24, 20f
    RDoC matrix, 22t

perspective of, 2–3
sequential/concurrent influences on, 3–4
social and environmental moderation, 7
twin research, 6
Mesenteric lymph nodes (MLN), 359
Messenger RNA (mRNA), 6
Metabolic syndrome, 463, 466–467
Metabolome, 3t
Methicillin-resistant *Staphylococcus aureus* (MRSA), 60b
1-Methyl-4-phenyl-1,2,3,6-tetrahydropyridine (MPTP), 401–403
Microbe-associated molecular patterns (MAMPs), 74
Microbiome-immune interactions, 263–267
Microbiota dysbiosis, 289
Microbiota, gut. *See* Gut bacteria/microbiota
Microbiota–gut–immune–brain axis, 189, 192f, 447–448b, 447f
Microglia
  Alzheimer's disease (AD), 433–436
  depression, 254–255b
  and Parkinson's disease (PD), 408–411, 410f
  and schizophrenia, 386–387, 388f
  sensitization effects, 173–174, 174f
MicroRNAs (miRNAs), 8–9
Middle East Respiratory Syndrome (MERS), 63b, 64–65
Migration inhibitory factor (MIF), 176–177
Mild cognitive impairment (MCI), 420–421
Mindful meditation, pain, 341–343
Mineralocorticoids, 148t
Minor depression, 230t
Misappraisals, 134
Mitochondrial antigen presentation (MITAP), 407–409, 408f
Mitochondrial-derived vesicles (MDVs), 407–408
Molecular genetic approaches, 6–7
Monoamine, depression, 234
Monokines, 50–52
Mononuclear phagocytic cells, 176–177
Monozygotic twins, 6
Mood disorders, 443b
Motor behaviors, 348
Multidrug-resistant tuberculosis (MDR-TB), 83–84
Multihit hypothesis, 4
  Parkinson's disease (PD), 393b, 399
Multiple sclerosis (MS), 164, 193–194, 444–446
Multisystem coordination, mental illness, 3–5
Multivitamin, 114b
Myalgic encephalomyelitis (ME), 184, 336–337b
Mycobacterium tuberculosis, 83–84, 179

N
National Institutes of Mental Health (NIMH), 2–3
Natural killer (NK) cells, 32
Natural products, 116–117, 117b
Nerve growth factor (NGF), 14t, 148t, 247
Neural-immune interactions, 28, 52–53t, 53
Neuraminidase inhibitor, 62–63
Neurobiological process
  prenatal stress, 202, 203f
  stressor, 140–141

Neurochemical processes, 5, 11
Neurodegenerative diseases, 82
Neurodevelopmental disorders (NDDs), 380, 380f
Neurofibrillary tangles (NFTs), 422–423
Neuroimmunity
    ACE2 and CD209 in, 92–94, 93f
    mental illness, 15–16
    olfactory and vascular/lymphatic crossroads, 95
Neuroinflammation, 103–105
    anxiety disorders, 286, 287f
    schizophrenia, 384–385, 385f
Neurological illnesses
    Alzheimer's disease (*see* Alzheimer's disease (AD))
    evaluating biological substrates of, 5
    mental and (*see* Mental disorders)
    pain signature, 325
Neuromedin B (NMB), 159t
Neuronal pentraxin 2 (NPTX2), 430
Neuropathic pain, 323–324
Neuropeptide Y (NPY), 158–159, 159t
    biological effect and behavioral outcome, 108t
    depression, 244
    posttraumatic stress disorder (PTSD), 307–308
    stressor, 158–159
Neuroplasticity, 411–412, 462b
Neuropsychiatric illnesses, 82, 94–95
Neurotransmission, mental illness, 11–12
Neurotrophins
    depression, 245–249
    measured in blood, 247
    mental illness, 14, 14t
    neurotrophin-3/4, 14t, 148t
    prenatal stress, 203–204
    stressor, 147–149
Neutralize antigen, 44
Neutrophil extracellular traps (NETs), 378
Nocebo, 334
Nociceptive pain, 323–324
Nocturnal sleep, 124
Noncommunicable diseases, 101
Nonsleep, sleep and, 123–124
Norepinephrine (NE)
    anxiety disorders, 272
    in depression, 234–235
    norepinephrine (NE) receptor antagonists, 318–319
    Parkinson's disease (PD), 396
    posttraumatic stress disorder (PTSD), 302–304
    stressor, 143–144
Normal sleep, 124. *See also* Sleep

O

Obesity
    brown fat in, 104–105
    depression, 250b
    and gestational diabetes, 198
    gut bacteria and, 66–70, 68f
    and illness, 110–112, 112b
    microbiome and, 67–68
    and mortality, 111
    and physical illnesses, 111
    and type 2 diabetes, 466

Obsessive-compulsive disorder (OCD), 277–279
    inflammatory processes, 284–285
    microbiota, 290
    treatment of, 278–279
Omega-3 polyunsaturated fatty acids (PUFAs), 114–115
Opioids, pain, 327–328, 338–339
Opsonization process, 44
Orexin (hypocretin), 108t
Ostracism, social rejection and, 140
Oxytocin, 148t
    depression, 243–244
    social anxiety, 290
    stressor, 158

P

Pacific Yew Tree, 116
Pain, 323–324, 323b
    acceptance and commitment therapy (ACT), 341–343
    acute experiences, 337b
    antidepressants for, 339–340
    cannabinoids, 328–329
    classification systems, 323–324
    cognitive behavior therapy (CBT), 341–343
    deep brain stimulation (DBS), 340b
    drug treatments, 337
    emotional processing affects, 332–333
    glial functioning, 329–332
    inflammatory process, 329–332, 340
    inhibition of proinflammatory signaling, 340
    intracellular processes, 329–330
    itch and scratch, 338–339b
    management methods, 337–344
    memories of, 343b
    microbiota and, 330f, 331, 332b
    mindful meditation, 341–343
    neurophysiological/psychological process, 325–337
    opioids, 327–328, 338–339
    placebo effects, 333–334
        neurobiological correlates, 334–337
        variations in brain responses, 335f
    and posttraumatic stress disorder (PTSD), 322b
    psychedelics in, 339–340b
    repetitive transcranial magnetic stimulation (rTMS), 340b
    sex differences in, 326b
    spinal cord, 329, 330f
Panic disorder, 276–277
    inflammatory processes, 284
    treatment of, 276–277
Parasitic infection, central nervous system (CNS), 84–85
Pareto principle, 63
Parkinson's disease (PD), 393–398, 393b, 408b
    adaptive immune processes, 406–408
    antiinflammatory treatments, 415–416
    Bermuda triangle, 403, 404f
    characteristic of, 394
    comorbidity illness, 395–396
    cytokines in, 411
    depression in, 396–397
    dopamine, 401b
    environmental stresses, 403–412

experimental animal models of, 401–403
future immunomodulatory treatments, 414–418
genetic vulnerability, 412–414
gut and, 405, 405b
6-hydroxydopamine (6-OHDA), 401–403
incidence of, 399
infectious diseases, 404–405
inflammatory mechanisms, 406
interferons (IFNs), 411–412
Korean longitudinal cohort study, 396–397
lifestyle and aging, 398–401
LRRK2, 413–414, 416
1-methyl-4-phenyl-1,2,3,6-tetrahydropyridine (MPTP), 401–403
microglia and, 408–411, 410f
mitochondrial antigen induced autoimmunity and, 407, 408f
mitochondrial-derived vesicles (MDVs), 407–408
multihit hypothesis, 393b, 399
norepinephrine (NE), 396
pathogenesis of, 394f
pathology, 395f
pattern-recognition receptors (PRRs), 405
peripheral immune processes, 395b
pesticides and, 403, 404f
PINK1 and Parkin mutations, 407–408
positron emission tomography (PET), 402–403
risk factors, 400–401, 401f
stress, 395–398
toxicant-based models, 401–403
trophic cytokines, 416–418
Pathogen-associated molecular pattern (PAMP), 29, 74
    Alzheimer's disease (AD), 431–433
    coronary artery disease (CAD), 455–456
    innate immunity, 33
    Parkinson's disease (PD), 408–410
    sterile inflammation, 34
Pattern-recognition receptors (PRRs), 74, 86, 405
Pediatric Autoimmune Neuropsychiatric Disorders Associated with Streptococcal Infections (PANDAS), 284–285
Peltzman effect, 100
Penetrance, 5
Perinatal diet, 201b
Perinatal infection, and schizophrenia, 379–383, 380f
Perineuronal nets, 329–330
Peripheral inflammation
    Alzheimer's disease (AD), 433
    anxiety disorders, 286, 287f
Peripheral nervous system (PNS), 86
Personality factor
    coronary artery disease (CAD), 458–459
    twin research, 6
Personality traits, 138
Personalized medicine, 2, 19–24
    culture, 24b
    endophenotypic analyses, 19–24, 20f
    mental illness, 24b
    RDoC matrix, 22t

586 Index

Pesticides, and Parkinson's disease (PD), 403, 404f
Peyer's patches, 34–35
Phagocytes, 32–33
Pharmacological therapy, 318
Phenome, 3t
Phobias, 279
Physical illness
  obesity and, 111
  and posttraumatic stress disorder (PTSD), 295b
Placebo-controlled trials, 424
Placebo effects, pain, 333–334
  neurobiological correlates, 334–337
  variations in brain responses, 335f
Plaque-forming cells (PFC), 42
Plasma cells, 28–29, 42
Plasmodium falciparum, 85
Plasticity, synaptic, 462b
Platelet factor 4 (PF4), 119
Pluripotent hematopoietic stem cell (HSC), 32
Pokeweed Mitogen (PWM), 356
Poliomyelitis, 86
Polymorphism, 6
Polyunsaturated fats (PUFAs), 76, 102
Ponce de León, Juan, 449
Porphyromonadaceae, 69–70b
Postacute infection syndromes (PAISs), 87
Postnatal infections, 381–383
Postpartum bacterial infections, 57b
Postpartum depression, 230t
Postpolio syndrome (PPS), 86–87
Posttraumatic stress disorder (PTSD), 11, 167–168, 293–294, 293b
  antiinflammatory agents, 320
  anxiety, 311–312
  attention and salience, 301
  biochemical correlates, 301–308
  brain-derived neurotrophic factor (BDNF), 308
  cannabinoids, 320
  chronic pain, 325
  cognitive behavioral therapy (CBT), 317
  comorbid illnesses, 321–322
  corticotropin-releasing hormone (CRH), 304–305
  cortisol in relation to, 152–153
  disturbance of memory processes, 299–300
  dynamic neurobiological changes, 319b
  epigenetic changes, 314–317
    aging, 315–316b
    glucocorticoids, 314
    inflammation, 314–316
    microbiota, 316–317
  exposure therapy, 317–318
  eye movement desensitization and reprocessing (EMDR), 318
  failure of recovery systems, 299
  gamma-aminobutyric acid (GABA), 306–307
  genetic influences, 312–314
  glucocorticoids, 305–306, 320–321
  glutamate, 307, 319–320
  inflammatory processes, 308–314, 313f
  memory reconsolidation, 300
  neurochemical changes with, 303f
  neuropeptide Y (NPY), 307–308
  norepinephrine, 302–304

norepinephrine receptor antagonists, 318–319
  pain and, 322b
  pharmacological therapies, 318
  physical illness and, 295b
  pretrauma experiences, 295
  related to head injury, 311–312
  resilience, 294–295
  selective serotonin reuptake inhibitors, 318
  sensitized responses, 301
  serotonin, 304
  structural brain alteration, 295–298
  theoretical perspectives on, 298–301
  threat detection, 299
  transcranial magnetic stimulation (TMS), 321
  traumatic brain injury (TBI), 311–312
  treatment, 317–321
  vulnerability, 294–295, 298b, 298f
Preclinical studies, schizophrenia, 376–379
Prenatal development, pregnancy
  impact of, 153–159, 210
  resources, 198
  stress, 198–201
    challenges, 198–207, 201f
    depression, 200–201
    diet, 201b
    glucocorticoid, 199, 200f, 202
    in humans, 202
    inflammatory factors, 204
    neurobiological processes, 203f
    neurotrophins, 203–204
    schizophrenia, 378–379
    sex-dependent effects, 202
    timing of, 202
  viral infection, 204–206, 205f
    impact of, 204–206
    schizophrenia, 206–207
Pretrauma experiences, 295
Prevotella histicola, 445
Probiotics, 162f
  off-the-shelf, 79
  usefulness of, 266b
Problem-solving coping methods, 136, 137t
Progesterone, 157t
Progressive motor disability, 86–87
Proinflammatory cytokines, 84, 104, 164, 172f, 174f, 182f
  central nervous system (CNS), 164
  depression, 253–254
  Parkinson's disease (PD), 409
  stressors and illness, 173–174, 182f
Prolactin, 157–158, 157t
Proteome, 3t
Prototypical brain-gut axis disorder, 447–448b
Psilocybin, 242b
Psychedelics, in pain, 339–340b
Psychiatric disorders, 367, 451–461
Psychogenic stressors, 167
Psychological illnesses, 140–141, 140–141b, 181, 398
Psychopathology, animal models of, 143b
Psychosis, 366–367, 369–370
Puerperal infections. *See* Postpartum bacterial infections

Q
Quinolones, 60

R
Rabies, 83
Racial discrimination, on hormonal processes, 181
Rapid eye movement (REM), 124
Reactive oxygen species (ROS), 112
Reappraisals, 132–141
Recuperation, 123–124b
Recurrent brief depression, 230t
Regressive autism spectrum disorder (ASD), 360
Relapsing–remitting multiple sclerosis, 118–119
Repetitive transcranial magnetic stimulation (rTMS), 246–247, 249b, 340b
Reproduction, stressor and, 156–157
Repurposing of drugs, 448–449
Research Domain Criteria (RDoC) approach, 2–3, 21, 22t
Resilience
  posttraumatic stress disorder (PTSD), 294–295
  stress-related, 132, 133f
  vulnerability *vs*., 131–132
Respiratory infection, 82
Resveratrol, 112–113b
Retatrutide, 115
Retinoic acid, 361
Retroviruses, 383
Rheumatoid arthritis (RA), 195, 384, 444–445
Risk compensation theory, 100

S
Salience, posttraumatic stress disorder (PTSD), 301
Salmonella enterica, 69–70
SARS-CoV-1/2, 60b, 63–64, 81b, 82–84, 89–90, 380
  brain viral infection, 94–95
  cognitive disturbances, 94–95
  encephalopathy, 94
  infection, 183–184
  in mouse models, 93–94b
  vaccination, 92b, 125–127
Schizophrenia, 84, 164–165, 365–367, 365b, 443b
  anti-NMDA receptor antibodies, 386
  autoimmunity, 383–386
  bidirectional communication, 386
  biochemical process, 368
  chlorpromazine, 368–369
  dopamine hypothesis, 368–369, 368–369b, 371f
  endogenous retroviruses, 383
  epigenetic changes, 373
  gamma-aminobutyric acid (GABA), 369–370
  genetic contributions, 367–368
  genome-wide association studies (GWAS), 367–368
  glutamate hypothesis, 369–370, 371f, 384–386
  immunological theory of, 372
  inflammation, 384–386
  kynurenic acid (KYNA), 385–386
  maternal immune activation (MIA), 380, 380f
  microbiota and, 387–389, 390f
  microglial activity and, 386–387, 388f

neurodevelopmental disorders (NDDs), 380, 380f
neuroinflammatory mechanism, 384–385, 385f
pathophysiology of, 367–371
perinatal infection and, 379–383, 380f
positive and negative symptoms, 366
postnatal infections, 381–383
prenatal viral infection, 206–207
psychopharmacological treatment, 384–385
rheumatoid arthritis (RA) and, 384
toxoplasmosis, 379, 382–383
two-hit hypothesis, 370–379
   C-reactive protein, 374–376
   cytokines, 374–376
   preclinical studies, 376–379
   prenatal stressors, 378–379
   viral infection, 377–378
Seasonal affective disorder (SAD), 230t
Secondary progressive multiple sclerosis (MS), 193
Selective serotonin reuptake inhibitors (SSRIs)
antidepressant, 238–239, 239b
anxiety disorders, 272
in depression, 235–236
effectiveness, 238–239, 239b
posttraumatic stress disorder (PTSD), 318
resistance to effects of, 263
Semaglutide (Ozempic), 115
Senescence-associated glycoprotein (SAGP), 440
Sepsis, 84
Serotonin
anxiety disorders, 272
in depression, 235–236
posttraumatic stress disorder (PTSD), 304
stressor, 143–144
Sex-dependent effects, 202
Sex differences
coronary artery disease (CAD), 459
in pain, 326b
Sex hormones, 157t
Sexual dimorphism, 193, 326b
Sexually transmitted diseases (STDs), 100
Short-chain fatty acids (SCFAs), 165
Sickness behaviors, 74–75b, 82
Single nucleotide polymorphism (SNP), 7–8, 242–243
Single-stranded RNA (ssRNA), 88
Slebsiella pneumoniae, 69–70
Sleep, 122
biological systems during, 123–124b
disorders, 324
disturbance, 124, 126f, 127, 459
diurnal variations, 127
emotional regulation, 124
habitual short, 124–125, 125f
impact of, 124–127
mental health, 124
neurobiological factors associated with, 122–124
and nonsleep, 123–124
Small for gestational age, 352
Smoking, 100–101b, 430
Social anxiety, 279–280
inflammatory processes, 285–286
microbiota diversity, 290

oxytocin, 290
treatment of, 280
Social interaction, autism spectrum disorder (ASD), 346–347
Social rejection, 140
Social stressors, 181, 187
Social support
coping methods, 137t, 138–139
unsupportive relations, 139–140
Soil-transmitted helminths (STH), 85
Spanish Flu, 81b
Specificity
acquired immune response, 37
antibody, 43–44
in innate immunity, 33
Staphylococcus aureus (S. aureus), 60b, 83
Sterile inflammation, 33–34
Stigma, 100–101b
Stress and immunity, 163b, 166b
animal studies, 166–179
   antibody production, 174–176
   assessment, 167
   B cell function, 174–176
   cell-mediated immune responses, 170–174, 172–173f
   characteristics, 167–169
   dendritic cells, 177–178
   effects, 169–170
   infection, 178–179
   mononuclear phagocytic cells, 176–177
assessment, 167
characteristics, 167–169
dynamic interplay between, 165–166
in humans, 179–185
implications of, 163–165
inflammatory signals to brain, 172f
proinflammatory cytokines, 172f, 174f, 182f
Stress/stressor
acute, 142f, 171
appraisals, 132–141
assorted, 142
attributes of, 132
cannabinoids, 155
caregiving, 180
characteristics, impact, 134–136
chronic (*see* Chronic stressor)
coping with, 136–138
coronary artery disease (CAD) and, 455
corticoid responses, 149–151
corticotropin-releasing hormone (CRH), 154–155
cortisol changes, 151–152
and COVID-19 pandemic, 183–185
cross-sensitization effects, 145
cumulative experiences, 140–141b
depression, 255–256, 261f
dopamine, 143–144
early life adverse events, 210–213
and eating, 109–110
elimination/attenuation of, 168
and energy balances, 159–160
   corticotropin-releasing hormone (CRH), 160
   cortisol, 160
   leptin and ghrelin, 159–160
environmental, 403–412
estrogen, 155–156
experiences, 145

gender differences in, 138
glutamate, 146–147
growth factors and functions, 148t
homeostatic state in, 173–174, 174f
hormonal changes, 149–153, 150t
and immune alterations, 151f, 160–161
and infection, 178–179
inflammation, 161, 255–256, 261f
and microbiota, 161
neurobiological changes, 140–141
neuropeptide Y (NPY), 158–159
neurotransmitter alterations provoked, 141–149
neurotrophins, 147–149
norepinephrine, 143–144
oxytocin, 158
Parkinson's disease (PD), 395–398
prenatal development, pregnancy, 198–201
   challenges, 198–207, 201f
   depression, 200–201
   diet, 201b
   glucocorticoid, 199, 200f, 202
   in humans, 202
   inflammatory factors, 204
   neurobiological processes, 203f
   neurotrophins, 203–204
   sex-dependent effects, 202
   timing of, 202
prolactin, 157–158
psychogenic, 167
psychological factors, 140–141
and reproduction, 156–157
self-reported, 138
sensitization, 145
serotonin, 143–144
testosterone, 156
type 2 diabetes and, 465
uncontrollable, 142–143
Striatum, 393–394
Stroke, 461–467
depression and, 461–462
diabetes and, 462–463
gestational diabetes, 462–463
synaptic plasticity, advantage, 462b
type 1 diabetes, 463
type 2 diabetes, 463
   development of, 464–466
   environmental toxicants, 466b
   genetic contributions, 464–465, 464f
   impact of stressors, 465
   inflammatory immune factors, 465–466
   obesity and, 466
Suicidality, depression and, 256b
Superspreaders, 63
Synaptic plasticity, 462b
Synaptic-vesicle glycoprotein 2C (SV2C), 399–400
Syncytium, 87
Systemic lupus erythematosus (SLE), 194, 352

T

Taurine supplementation, 450
T-cell receptor (TCR), 38, 45–46, 48, 83
acquired immune response, 37
functional complex, 48
induction, 45–46

T cells
- activation, 49–50, 49f
- antigen presentation to, 48–49
- development, 45
- effector function, 49–50

Teixobactin, 61

Telomeres, 140–141b

Tend-and-befriend characteristics, 158

Testosterone, 156, 157t

Theiler's murine encephalomyelitis virus (TMEV) infection, 193

T helper (Th) cytokine responses, 185–190
- interferon-γ, 186
- interleukin-1, 189–190
- interleukin-2, 186–187
- interleukin-4, 187
- interleukin-5, 188
- interleukin-6, 188–189
- interleukin-10, 187–188
- tumor necrosis factor, 189–190

T lymphocytes, 31–32, 45–50, 83

Tolerance, and autoimmune disease, 54–55

Toll-like receptors (TLRs), 86–88
- coronary artery disease (CAD), 455–456
- Parkinson's disease (PD), 405
- schizophrenia, 373

Toxicants, 4
- Parkinson's disease (PD), 401–403
- type 2 diabetes, 466b

Toxoplasma Gondii infection, 379, 382–383

Traditional Chinese Medicines (TCMs), 117b

Trained innate immunity, 196

Transcranial direct-current stimulation (tDCS), 249b

Transcranial magnetic stimulation (TMS), 321

Transcriptome, 3t

Transforming growth factor (TGF)-β, 148t

Transient ischemic attack (TIA), 461

Translation process, twin research, 6

Translocator protein (TSPO), 184

Trauma
- collective, 225–226, 225–226b
- psychological and physical sequelae of, 222–225

Traumatic brain injury (TBI)
- depression, 258b
- posttraumatic stress disorder (PTSD), 311–312

Treatment-resistant depression, 230t

TREM2, Alzheimer's disease (AD), 428–429, 428f, 435

Triad of impairments, 346

Trier Social Stress Test (TSST), 152, 186

Trimethylamine N-oxide (TMAO), 459–460

Trojan horse strategy, 82

Trophic cytokines
- Alzheimer's disease (AD), 437–438
- Parkinson's disease (PD), 416–418

Tryptophan (TRP)-kynurenine (KYN) pathway, 184

Tumor necrosis factor (TNF), 189–190
- autism spectrum disorder (ASD), 353–354
- tumor necrosis factor alpha (TNF-α), 374–376
- tumor necrosis factor receptors (TNFR), 353

Turquoise killifish, 69–70b

Twin research, 6

Two-hit hypothesis, schizophrenia, 370–379
- C-reactive protein, 374–376
- cytokines, 374–376
- genetic risk, 370
- preclinical studies, 376–379
- prenatal stressors, 378–379
- viral infection, 377–378

Type 2 allostatic overload, 141

Type 1/2 diabetes, 3–4, 463
- Alzheimer's disease (AD), 424
- development of, 464–466
- environmental toxicants, 466b
- genetic contributions, 464–465, 464f
- impact of stressors, 465
- inflammatory immune factors, 465–466
- metabolic syndrome, 463, 466–467
- metformin, 449
- mitochondrial pyruvate carrier (MPC), 464
- obesity and, 466
- semaglutide (Ozempic), 448

Typhoid Mary, 63b

Typical depression, 230t, 231

U

Ulcerative colitis, 447–448b

Ultra-processed foods, 101–102

Uncontrollable stressors, 135

Unsaturated fats, 76

V

Vaccine hesitancy, 92b

Vaccines/vaccination
- Alzheimer's disease (AD), 438–441
- immune system responds to, 63, 64f
- response to, 182–183
- viruses, 63–65

Vagal nerve, 72

Varicella-zoster virus, 86

Vascular endothelial growth factors (VEGF), 14t, 148t

Vasoactive intestinal peptide (VIP), 165

Vector-borne diseases, 62, 127–128

Viral encephalitis (VE), 88, 88f

Viral illnesses, 61–65

Viral-induced cell fusion, 87

Viral infection
- central nervous system (CNS), 82–83, 85–89
- influence of, 82–83
- prenatal, 204–206, 205f
- impact of, 204–206
- schizophrenia from, 206–207

Viruses
- affect hormonal and CNS processes, 66–79
  - gut bacteria and obesity, 66–70, 68f
  - gut microbiota dysbiosis, 70–71, 74–75
  - microbiota, 66, 67f, 69–70b
  - well-being, implications for, 70–71
- immune system responds to, 64f
- primary targets, 62, 62f
- tracking deadly and beneficial, 82
- vaccines, 63–65

Vitamins, 114b

Vulnerability, vs. resilience, 131–132

W

Water-borne diseases, 127–128

Well-being
- microbiota and, 70–71
- positivity and, 139b
- threat/risk, 133f

Western diet, 103

West Nile virus (WNV), 85–86

White blood cells, 29–31

White fat, in obesity, 104–105